Forest Entomology and Pathology

Jeremy D. Allison · Timothy D. Paine ·
Bernard Slippers · Michael J. Wingfield
Editors

Forest Entomology and Pathology

Volume 1: Entomology

 Springer

Editors
Jeremy D. Allison
Natural Resources Canada
Sault Ste. Marie, ON, Canada

Bernard Slippers
Forestry and Agricultural Biotechnology
Institute
University of Pretoria
Hatfield, South Africa

Timothy D. Paine
Department of Entomology
University of California
Riverside, CA, USA

Michael J. Wingfield
Forestry and Agricultural Biotechnology
Institute
University of Pretoria
Hatfield, South Africa

Funded by the Canadian Forest Service and the Forestry and Agricultural Biotechnology Institute, University of Pretoria, South Africa

ISBN 978-3-031-11555-4   ISBN 978-3-031-11553-0 (eBook)
https://doi.org/10.1007/978-3-031-11553-0

This Springer imprint is published by the registered company Springer Nature Switzerland AG
The registered company address is: Gewerbestrasse 11, 6330 Cham, Switzerland

Preface

The idea for this book emerged from conversations between the editors, in particular during visits by JDA to South Africa to work with BS on plantation pests and the ecology of invasive forest insects. In the process of these discussions, we realized that the field of forest entomology and pathology had changed dramatically in recent years. This was primarily due to altered distributions and patterns of interactions among insects, fungal pathogens, and trees, the emergence of new technologies and increased emphasis on multidisciplinary solutions to problems. In light of these changes, we felt it was time for an update.

The scope of this book is intentionally broad, introducing the audience to the diversity of insects and the roles that they play in forest ecosystems. Although much of the impetus to study insects in forest ecosystems comes from the premise that an understanding of their ecology would facilitate management of pest species, this volume covers the beneficial and negative impacts insects have on forest health. There are several excellent books and reviews that provide more in-depth treatment of many of the topics covered in this book. This book is intended, however, to be a comprehensive introduction to the discipline of forest entomology.

Recognition of the value of forest and woodland ecosystem services continues to increase. For example, it has been estimated that globally ca. 1 in 6 people rely on forests for food and many more rely on non-food forest ecosystem services (e.g. carbon storage, wood, and wood products resources). The significance of forest and woodland ecosystem services is expected to increase as human population levels increase globally. Coincident with the increase in demand for forest ecosystem services has been an increase in the frequency and severity of disturbances experienced by forest ecosystems. Altered distributions and patterns of interactions among forest insects, fungal pathogens, and trees, as a consequence of these disturbances, have contributed to these dramatic increases. Considerable research has been conducted to understand the drivers of these disturbances, their impacts, and how to prevent them and mitigate their impact. There are numerous new and emerging technologies that have increased our understanding of the importance and impacts of insects in forest ecosystems, the factors that influence their distribution and abundance and how to mitigate the impacts of pest species. This book provides an introduction to this literature.

This book is organized into four sections with different learning objectives.

- The first section is a series of eight chapters designed to introduce the reader to the discipline of forest entomology. First, the reader is introduced to insects and their importance (Chapter 1), followed by a general discussion of direct relationship between insect morphology and how they function (Chapter 2) and the diversity of arthropods (with an emphasis on insects) in forest ecosystems (Chapter 3). The reader is then introduced to the topics of insect ecology (Chapter 4), population dynamics (Chapter 5), insect–natural enemy interactions (Chapter 6), and insect–plant interactions (Chapter 7), all with an emphasis on forest insects. Section one ends with a comprehensive treatment of insect and forest succession (Chapter 8).
- The next section introduces the reader to the principal insect feeding groups: foliage feeders (Chapter 9), bark beetles (Chapter 10), ambrosia beetles (Chapter 11), woodborers (Chapter 12), sapsuckers (Chapter 13), gall formers (Chapter 14), tip, shoot, root, and regeneration pests (Chapter 15), and insects of reproductive structures (Chapter 16). This goal of this section is to provide a general introduction to the primary insect species that impact forest ecosystems.
- The third section consists of four chapters that introduce the reader to the management of forest insects. Topics covered include the application of Integrated Pest Management (IPM) to forest ecosystems (Chapter 17), the spatial dynamics of forest insects (Chapter 18), monitoring and surveillance of forest insects (Chapter 19), and the growth and management of trees, silviculture (Chapter 20). The goal of this section is to introduce the reader to the spatial dynamics of forest insects and its impact on approaches to insect monitoring in the context of IPM and silviculture.
- The last section focuses on significant issues and concepts likely to increase in importance. Specific topics covered in this section include forest health (Chapter 21), impact of climate change on forest insects and their impacts (Chapter 22), and forest biosecurity (Chapter 23).

All of the editors of this book have taught aspects of forest health and discussions with graduate and undergraduate students have contributed to our understanding of the topics covered. The content has also been shaped by feedback from a number of colleagues including Juan Corley, Brett Hurley, Maartje Klapwijk, Brian Strom, James Meeker, Kevin Dodds, Ring Cardé, Jolanda Roux as well as discussions with many members of IUFRO Division 7 (Tree Health) over long periods of time.

Sault Ste. Marie, Canada   Jeremy D. Allison
Riverside, USA   Timothy D. Paine
Hatfield, South Africa   Bernard Slippers
Hatfield, South Africa   Michael J. Wingfield
December 2022

Contents

Contributors

Jeremy D. Allison Natural Resources Canada, Canadian Forest Service, Great Lakes Forestry Centre, Sault Ste. Marie, ON, Canada;
Department of Zoology and Entomology, Forestry and Agricultural Biotechnology Institute, University of Pretoria, Pretoria, Gauteng, South Africa

Matthew P. Ayres Department of Biological Sciences, Dartmouth College, Hanover, NH, USA

Andrea Battisti University of Padova, DAFNAE, Legnaro, Padova, Italy

Jörg Bohlmann Michael Smith Laboratories, University of British Columbia, Vancouver, BC, Canada

Manuela Branco Forest Research Center, School of Agriculture, University of Lisbon, Lisbon, Portugal

Eckehard G. Brockerhoff Swiss Federal Research Institute WSL, Birmensdorf, Switzerland

Ethan Bucholz Northern Arizona University, Flagstaff, AZ, USA;
Colorado State Forest Service, Fort Collins, CO, USA

Jean-Noel Candau Natural Resources Canada, Canadian Forest Service, Great Lakes Forestry Centre, Sault Ste. Marie, ON, Canada

Angus J. Carnegie Forest Science, Department of Primary Industries, Sydney, NSW, Australia

Allan L. Carroll Faculty of Forestry, Department of Forest & Conservation Sciences, University of British Columbia, Vancouver, BC, Canada

Anthony I. Cognato Michigan State University, East Lansing, MI, USA

Juan C. Corley IFAB (INTA EEA Bariloche-CONICET) and Departamento de Ecología, Universidad Nacional del Comahue, Bariloche, Argentina

David R. Coyle Department of Forestry and Environmental Conservation, Clemson University, Clemson, SC, USA

Gudrun Dittrich-Schröder Department of Zoology and Entomology, Forestry and Agricultural Biotechnology Institute (FABI), University of Pretoria, Pretoria, South Africa

Kevin J. Dodds U.S. Forest Service, Eastern Region, State, Private, and Tribal Forestry, Durham, NH, USA

Daniel Doucet Canadian Forest Service, Great Lakes Forestry Centre, Sault Ste. Marie, ON, Canada

Joseph Elkinton University of Massachusetts, Amherst, MA, USA

Christopher J. Fettig U.S. Forest Service, Pacific Southwest Research Station, Davis, CA, USA

José Carlos Franco Forest Research Center, School of Agriculture, University of Lisbon, Lisbon, Portugal

Jeff R. Garnas Natural Resources and the Environment, University of New Hampshire, Durham, NH, USA;
Department of Zoology and Entomology, Forestry and Agricultural Biotechnology Institute (FABI), University of Pretoria, Pretoria, South Africa;
Forestry and Agricultural Biotechnology Institute (FABI), University of Pretoria, Pretoria, South Africa

Caitlin R. Gevers Department of Zoology and Entomology, Forestry and Agricultural Biotechnology Institute (FABI), University of Pretoria, Pretoria, South Africa

Demian F. Gomez Texas A&M Forest Service, Austin, TX, USA

Juli R. Gould USDA APHIS PPQ CPHST Laboratory, Buzzards Bay, MA, USA

Jean-Claude Grégoire Université Libre de Bruxelles, Bruxelles, Belgium

Laurel J. Haavik USDA Forest Service, Washington, DC, USA

Kyle J. Haynes Department of Environmental Sciences, University of Virginia, Boyce, VA, USA

Jiri Hulcr School of Forest, Fisheries, and Geomatics Sciences, University of Florida, Gainesville, FL, USA

Brett P. Hurley Department of Zoology and Entomology, Forestry and Agricultural Biotechnology Institute (FABI), University of Pretoria, Pretoria, South Africa

Christian Hébert Canadian Forest Service, Laurentian Forestry Centre, Québec City, QC, Canada

Hervé Jactel INRAE, University of Bordeaux, BIOGECO, Cestas, Bordeaux, France

Paal Krokene Division of Biotechnology and Plant Health, Norwegian Institute of Bioeconomy Research, Ås, Norway

Stig Larsson Swedish University of Agricultural Sciences, Uppsala, Sweden

Andrew M. Liebhold US Forest Service Northern Research Station, Morgantown, WV, USA

Maria J. Lombardero Unidade de Xestion Ambiental E Forestal Sostible, Universidade de Santiago de Compostela, Lugo, Spain

Alex C. Mangini Southern Region, Forest Health Protection, USDA Forest Service, Pineville, LA, USA

Deborah G. McCullough Department of Entomology and Department of Forestry, Michigan State University, East Lansing, MI, USA

Zvi Mendel Department of Entomology, Agriculture Research Organization, The Volcani Center, Bet Dagan, Israel

Daniel R. Miller USDA Forest Service, Southern Research Station, Athens, GA, USA

Timothy D. Paine Department of Entomology, University of California-Riverside, Riverside, CA, USA

Robert J. Rabaglia USDA Forest Service, State and Private Forestry, Forest Health Protection, Washington, DC, USA

Richard Redak University of California-Riverside, Riverside, CA, USA

John J. Riggins Mississippi State University, Starkville, MS, USA

Artemis Roehrig University of Massachusetts, Amherst, MA, USA

James Skelton Biology Department, William and Mary, Williamsburg, VA, USA

Fred M. Stephen University of Arkansas, Fayetteville, AR, USA

Ward B. Strong British Columbia Ministry of Forests, Lands, Natural Resource Operations, and Rural Development (retired), Vernon, BC, Canada

Jon Sweeney Natural Resources Canada, Canadian Forest Service, Atlantic Forestry Centre, Fredericton, NB, Canada

Sean C. Thomas Institute of Forestry and Conservation, University of Toronto, Toronto, ON, Canada

Patrick C. Tobin School of Environmental and Forest Sciences, University of Washington, Seattle, WA, USA

Kristen M. Waring Northern Arizona University, Flagstaff, AZ, USA

Justin G. A. Whitehill College of Natural Resources, NC State University, Raleigh, NC, USA

Chapter 1
Introduction to and Importance of Insects

Richard Redak

1.1 Introduction

Insects and closely related arthropods are the dominant and most diverse forms of terrestrial and aquatic (non-marine) animal life on the planet. Other than marine systems, insects occupy every conceivable environment and habitat on the Earth. Crustaceans and Annelids (worms) are the dominant and most diverse groups of animals in marine systems. The dominance of insects is true in terms of diversity (number of species), numbers of individuals, and total biomass within a given area. As of this writing, there are approximately one million known species of insects (species that have been scientifically described—they have been provided a scientific name and their evolutionary relationship to other species is relatively well established). The known number of insect species is only a fraction of the estimated total number of species. The total number of insect species has been estimated to be between five and ten million species; most of which have yet to be discovered and scientifically described.

When all of the described species on the planet are considered, the number of insect species accounts for more than 50% of the total (Fig. 1.1; Purvis and Hector 2000; Stork 2018). As more species are discovered, the proportion of insects is likely to increase as our knowledge of the biodiversity of other plant and animal species is much more complete. One caveat to the claim of "dominance" is that there are other very poorly described groups with a large number of undescribed species: the prokaryotes (Bacteria and Archaea), many groups of protozoa, as well as the fungi and nematodes. Perhaps, with a better understanding of life's overall diversity, other groups will rival the apparent dominance on Earth by the insects.

R. Redak (✉)
University of California-Riverside, Riverside, CA, USA
e-mail: richard.redak@ucr.edu

© The Author(s) 2023
J. D. Allison et al. (eds.), *Forest Entomology and Pathology*,
https://doi.org/10.1007/978-3-031-11553-0_1

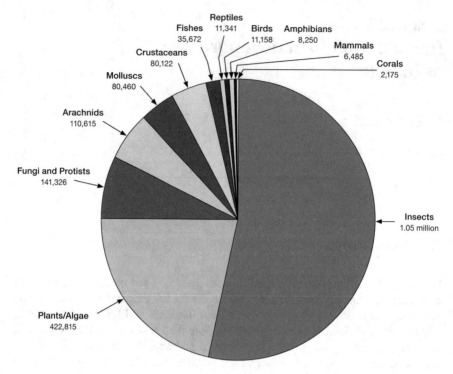

Fig. 1.1 The relative proportions of described species on the planet. © Matt Leatherman

The overwhelming majority of insect species are not harmful to human health, commerce, or agriculture, and in some form or fashion are actually beneficial. Worldwide, less than 1% of all known insect species are pests either destroying/damaging food and fiber resources, stored products, structures, or transmitting diseases. Within forest ecosystems, several insect species are serious pests that threaten our use of these natural resources and the ecosystem services they provide. In short, insects are our greatest competitors for the resources required to sustain our lives. Those few species that are pests result in tremendous efforts to manage their populations and limit the damage they cause. As the impacts of forest pests (especially undergoing outbreak infestations as shown in Fig. 1.8) are exceptionally visible, the value and benefit of most insect species is largely unknown to the general public.

1.2   What Is an Insect?

To fully understand the importance of insects to Earth's ecosystems, one must first understand where insects are placed in the phylogenetic tree of life. All species are placed within one of three large domains: Eukarya, Archaea, and Bacteria. Archaea and Bacteria are referred to as prokaryotic species. They are small (0.5–5 μm) unicellular organisms that lack true nuclei and other membrane covered organelles. Prokaryotic cells typically possess cell walls comprised of predominantly peptidoglycan (Bacteria) or a mix of other polysaccharides and proteins (Archaea). Prokaryotic DNA is not found within a membrane-bound nucleus. The diversity of prokaryotic species is exceedingly large and likely, when fully understood, much greater than any other life form on the planet, including insects. The domain Eukarya contains all other species on the planet and is characterized by possessing numerous membrane-bound organelles including a nucleus containing the cell's DNA. Eukaryotic cells are large (10–100 μm) compared to prokaryotic species and may lack cell walls. If cell walls are present, they are typically made up of primarily of cellulose (e.g. plants) or chitin (e.g. fungi) often mixed with other polysaccharides and glycoproteins. Each Domain is divided into smaller and smaller evolutionarily related groups (=clades) of related species. Each group is defined by shared traits (e.g. All insects possess a head, thorax, abdomen, three pair of legs and usually a pair of wings. All Coleoptera [beetles] possess these same traits and are typically convex in shape with the first set of wings being shell-like and protective; the second set of wings are membranous and held folded under the first set, Fig. 1.2).

This categorization of species is somewhat analogous to a set of Russian nesting dolls; below the level of Domain, in increasing specificity are the categories of Kingdom, Phyla, Class, Order, Family, Genus, and Species. Each Domain encompasses several Kingdoms, each Kingdom encompasses several Phyla, and so forth down to the level of a single species. As species are discovered, they are placed within this classification framework known as the Linnaean system of classification, named for Carolus Linnaeus who first proposed the system in the eighteenth century. Individual species are provided a two-word descriptor: the genus and the specific epithet, both of which are italicized when written (e.g. *Choristoneura fumiferana* (Tortricidae) for the eastern spruce budworm, Fig. 1.3).

Within this classification system, Insects (Class Insecta) are found within the Animal Kingdom and within that, the Phylum Arthropoda. Arthropoda includes not only insects, but also other Classes including Arachnida (spiders, scorpions, ticks,

Fig. 1.2 A beetle known as a firefly, *Photinus pyralis* (Coleoptera: Lampyridae) showing the two sets of wings. The forewings known as elytra are tough and protective, while the hindwings are membranous. © Alex Wild, used by permission

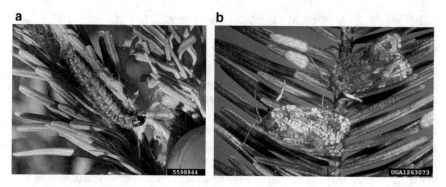

Fig. 1.3 The moth, *Choristoneura fumiferana* (Lepidoptera: Tortricidae). **a** Larval stage. © Neil Thompson, University of Maine at Fort Kent. Bugwood.org; **b** Adults © K. B. Jamieson, Canadian Forest Service, Bugwood.org

mites, and others), Crustacea (crabs, lobsters, shrimp, isopods, copepods, and others), Diplopoda (millipedes), Chilopoda (centipedes), and several other groups. Within the Arthropoda, insects are found in the Subphylum Hexapoda and the Class Insecta. Table 1.1 provides a general classification system for the Arthropoda. It is important to note that any classification system is subject to continual modification as new species are described and a better understanding of evolutionary relationships within and between groups is acquired.

Arthropods are generally described as **bilaterally symmetrical** and **segmented** creatures possessing an **exoskeleton.** The exoskeleton lines the entirety of the outside of the body and almost all of the internal portions of the digestive, excretory, respiratory and reproductive systems. The exoskeleton provides structural support for the animal as well as providing internal and external protection against predators, parasites, physical shock, and desiccation. The segments of the body may have undergone **tagmosis** (fusion) to form distinct body sections or **tagma** (e.g. a head). Internally, the circulatory system of arthropods is an **open system** lacking true veins and arteries—the blood is simply pumped around inside the body cavity by a structure called the **dorsal vessel**. There is no spinal cord; however, there is a **ventral nerve cord** comprised of a pair of ganglia located approximately in each body segment. Ganglia are connected in a chain-like manner by nerves. The foremost ganglion is multi-lobed and is referred to as the brain. The appendages of arthropods are referred to as jointed; "Arthropod" literally means "jointed foot" in Greek. The various classes of animals found within the Arthropoda are variations of the above characteristics. Within the Class Insecta, there are **29** orders of insects (Table 1.2), most of which can be found within forest ecosystems.

Insects are characterized by possessing **three body tagmata** (head, thorax, and abdomen, each of which is the result of tagmosis of multiple segments), **three pair of legs, two pair of wings as adults**, and **one pair of antennae** (Fig. 1.4). Within the class Insecta, there is a tremendous variety in appendages (e.g. antennae, legs and

Table 1.1 General classification system for the extant major groups within the Phylum Arthropoda. The listing below does not include extinct groups

Phylum Arthropoda
Subphylum Chilicerata
Class Arachnida: Spiders, Scorpions, Wind Scorpions, Sun spiders Ticks, Mites
Class Xiphosura: Horseshoe crabs
Class Pycnogonida: Sea Spiders
Subphylum Diplopoda: Millipedes
Subphylum Chilopoda: Centipedes
Subphylum Pauropoda: Pauropods
Subphylum Symphyla: Symphylans
Subphylum Crustacea*: Lobster, Crab, Shrimp, Copepods, Brachiopods, Barnacles, Sea lice
Subphylum Hexapod*
Class Collembola: Springtails
Class Protura: Proturans or Coneheads
Class Diplura: Diplurans
Class Insecta: The Insects

* Currently, many systematists group the Crustaceans and Hexapods into a single group known as the Pancrustacea. The combination of molecular and morphological evidence for doing so is strong. The resulting classification of the these subphyla (Oakley et al. 2013; Rota-Stabelli et al. 2013), as well as the Diplopoda and the Chilopoda is complicated and beyond the scope of this chapter. At this level the reader is urged to simply understand the characteristics that define the class Insecta

wings). These have been modified through evolutionary time for specific functions (Fig. 1.5).

Natatorial legs are oar-like in shape and used for swimming (e.g. water boatmen). Raptorial legs are used for grasping prey (e.g. mantids). Saltatorial legs have evolved for jumping (e.g. grasshoppers). Cursorial legs are used for running (e.g. carpenter ants), and fossorial legs are specialized for digging/burrowing in the soil (e.g. mole crickets). Not all insects may have wings. Juvenile insects lack wings. Almost all adult insect possess wings; however, some species, through the process of evolution, have entirely lost the need for and the ability to develop wings (e.g. fleas, adult worker ants). Insect mouthparts (Fig. 1.6) also show great variation. Mouthparts may be modified for chewing (e.g. beetles among many), sucking plant fluids (e.g. aphids, whiteflies, leafhoppers), sucking blood (mosquitoes), lapping up liquids (e.g. carrion flies), and combinations of the aforementioned (bees).

The possession of an exoskeleton presents several challenges. The exoskeleton cannot grow in the traditional sense; it does not and cannot stretch. During growth in the immature phases, the insect must shed its old exoskeleton and produce a new larger one. This process is called **molting** or **ecdysis** (see Chapter 2). After the old

Table 1.2 List of extant orders of the Class Insecta

Classs Insecta
Order Archaeognatha: Jumping Bristletails
Order Zygentoma: Silverfish and Firebrats
Order Ephemeroptera: Mayflies
Order Odonata: Dragonflies and Damselflies
Order Orthoptera: Grasshoppers, Crickets, Katydids
Order Phasmatodea: Walkingsticks and Leaf insects
Order Embioptera: Webspinners
Order Notoptera: Ice Crawlers, Gladiators
Order Dermaptera: Earwigs
Order Plecoptera: Stoneflies
Order Zoraptera: Angel Insects
Order Mantodea: Mantids
Order Blattodea: Roaches and Termites
Order Psocoptera: Booklice
Order Phthiraptera: Lice
Order Thysanoptera: Thrips
Order Hemiptera: True bugs, Leafhoppers, Aphids, Whiteflies, Psyllids, Scales
Order Coleoptera: Beetles
Order Raphidioptera: Snakeflies
Order Neuroptera: Lacewings and Antlions
Order Megaloptera: Alderflies and Dobsonflies
Order Strepsiptera: Twisted-wing Parasites
Order Trichoptera: Caddisflies
Order Lepidoptera: Butterflies and Moths
Order Siphonaptera: Fleas
Order Mecoptera: Scorpionflies
Order Diptera: Flies
Order Hymenoptera: Bees, Wasps, Ants, and Sawflies

exoskeleton is shed and prior to the hardening of the new exoskeleton, the animal will expand the volume of its body, thus providing new internal space for growth. Ultimately, the new exoskeleton will harden and external growth will cease until the next molt. During this process, with the expansion of the exoskeleton, space is also made available for internal growth of organs. Before the exoskeleton has hardened into a protective shell, the insect is at its most vulnerable to predation, parasistism, disease, and physical shock. The molting process is under tight control by the endocrine system of the insect and only occurs within the juvenile stages of the animal (Fig. 1.7).

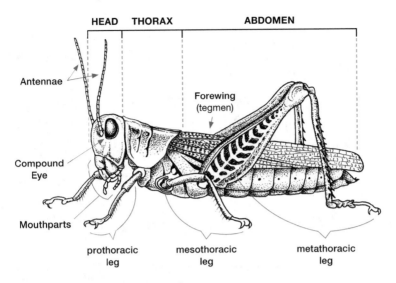

HEAD   THORAX   ABDOMEN

Antennae

Forewing
(tegmen)

Compound
Eye

Mouthparts

prothoracic
leg

mesothoracic
leg

metathoracic
leg

Fig. 1.4 Basic insect body plan. ©MattLeatherman

Once the adult stage is reached, molting and growth cease. Each juvenile stage in the life cycle of an insect is referred to as an instar. The time it takes to develop from one instar to the next is known as a stadium. The number of instars and the length of stadia vary tremendously among the insects. For example, depending on the species, insects may have one, two, or many generations per year (univoltine, bivoltine, and multivoltine, respectively). Conversely, many species will require many years to develop (e.g. several groups of aquatic insects and some wood boring beetles).

As insects are ectotherms ("cold-blooded") the process of growth is also dependent on environmental temperature. Typically, within limits, the warmer the environment is above a species-specific developmental threshold temperature, up to a maximum optimal temperature, the faster growth will occur. Below or above this optimum, growth will be slower. Below the threshold temperature, growth will cease. Temperatures more than a few degrees above the optimum are often fatal. For many species, this relationship between development time and temperature has been accurately quantified. With this information, one can predict the emergence of insect populations—a very useful tool in managing pestiferous species.

1.3   The Importance of Insects

Given the diversity and abundance of insects (and their near relatives), it is not surprising that they play essential and critical roles in the functioning of terrestrial and freshwater ecosystems and provide what are known as ecosystem services (Noriega et al. 2018).

Fig. 1.5 Insect legs. **a** A water boatman showing natatorial legs for swimming. © Kansas Department of Agriculture, Bugwood.org. **b** A mantidfly showing raptorial front legs for grasping prey. © Jon Yuschock, Bugwood.org. **c** Saltatorial rear legs for jumping in a grasshopper. © Whitney Cranshaw, Colorado State Univerisity, Bugwood.org. **d** a tiger beetle showing cursorial legs for running. © Susan Ellis, Bugwood.org. **e** Fossorial legs for digging in a mole cricket. ©Fir0002/Fflagstaffotos under GFDL

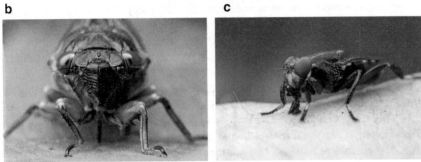

Fig. 1.6 Most common insect mouthparts. **a** Chewing (grasshopper. ©Alex Wild, used by permission), **b** Piercing-Sucking (periodical cicada, ©Alex Wild, used by permission), **c** Sponging-lapping (fly, Ropalomeridae: Diptera. ©Alex Wild, used by permission)

Fig. 1.7 A periodical cicada, *Magicicada* (Hemiptera: Cicadidae) undergoing a molt. © Alex Wild, used by permission

1.3.1 Decomposition, Nutrient Recycling, and Soil Formation

As organisms release waste products or die, they ultimately leave behind an abundance of organic material that can be used by other organisms as both energy and nutrient resources. Species that specialize on feeding on dead organisms and waste products are known as decomposers and/or detritivores (detritus is simply dead organic material). Detritivores are critical in physically and chemically breaking down and recycling organic material such that it can be, in turn, used by other organisms or returned to the abiotic environment. Indeed, much of the inorganic nutrients required by plants are derived by the decomposition of dead organisms. Similarly, all species occupying trophic levels which are dependent on plants for energy, nutrients and habitats, are indirectly supported by the recycling activities of detritivores. The processes of decomposition are complex. In short, it is a step-wise series of processes, by which dead organic material (a dead animal body or a fallen tree) is sequentially broken down into smaller and simpler particles, which are utilized by a succession of species, each specializing on a particular particle size with a particular nutrient value for that species.

In forest ecosystems, decomposition is critical in breaking down and recycling the complex macromolecules (cellulose, lignins, hemicellulose, etc.) found in plant cell walls. This is especially critical for the woody portions of the plant. As this material is broken down, it forms smaller and smaller particles of organic matter which are then utilized as sources of nutrition by additional species of decomposers. With respect to the role of insects, the process of plant decomposition starts with herbivorous insects feeding on the live structures of plants. During the process of ingestion and digestion, the plant material is physically and chemically broken down. The herbivore will absorb necessary nutrients from this material, metabolize organic compounds as sources of energy, and then will expel as waste undigestible/unused material. This expelled material (referred to as *frass*) is often still rich in organic and inorganic nutrients which, in turn, are utilized by additional organisms (including, ultimately, plants). CO_2 generated through metabolism is released during respiration into the atmosphere. As a result of the damage inflicted by herbivores, the plant is subject to additional attack by other organisms including herbivores that may accelerate both plant death and decomposition (other insects, fungi, prokaryotes). Often attack by one species leads to subsequent attack by others.

Decomposition need not start with a live plant and the breakdown of woody material often begins with plant death. Here, the material is initially attacked by a variety of species (e.g. termites, beetles) that have evolved symbiotic associations with microorganisms which allow them to digest cellulose. While feeding these wood-feeding (xylophagous) species tunnel into their food resource, opening it up for further feeding and decomposition by additional species.

Insects also play a role in the decomposition of animals. Once dead, vertebrate animals are subject to being fed upon by both other vertebrates (carrion feeders) as well as insects. Like plants, decomposition in animals follows a series of overlapping stages. Typically, the first insects that infest an animal corpse are evolutionarily specialized fly species (Order Diptera). Adult female flies responding to volatile cues will lay eggs on an animal corpse. The eggs hatch, and the developing larvae burrow into and feed on the dead animal tissues. This feeding opens the corpse up to inoculation and subsequent decomposition by a variety of microorganisms and prokaryotes. Subsequent to the initial feeding by flies, the physical and chemical properties of the carrion render it susceptible to colonization by a variety of beetle species (Coleoptera). As with plant decomposition, although the stages of animal decomposition are not distinct, each species of insect has its preferred type and quality of tissue on which to feed and develop.

The decomposition of both plants and animals attracts suites of predatory and parasitic insects and other arthropods that specialize on the insects engaged in decomposition. In both cases, there is a unique environment supporting insects that feed upon dead tissues while being fed upon by insect predators and parasites. Ultimately, with the decomposition of both plants and animals, microorganisms and prokaryotes break the remaining biological material down to simple organic and inorganic chemical constituents.

Linked to the process of decomposition is that of soil formation. Soil structure, texture, nutrient content, and water holding capacity are all emergent properties of a variety of factors including climate, parent material of the underlying bedrock, topography, the organisms associated with the soil, and time. Decomposition, and the roles that insects play in that ecological process, are responsible for much of the organic matter found in soil. The importance of fine-grained organic matter as a component of soil is critical to all of the aforementioned properties of soil that are necessary for supporting the diversity of terrestrial life on the planet (Jackson et al. 2017; Lehmann and Kleber 2015; Obalum et al. 2017).

1.3.2 Ecological Roles and Interactions

Like the diversity of the Class Insecta, the ecological roles that insects play in the Earth's ecosystems and interactions that insects are involved in are similarly diverse. Insects occupy virtually every ecological role in the planet's terrestrial ecosystems with the exception of being photosynthetic producers. As life on the planet relies on energy provided by the sun, from an ecological perspective, photosynthetic producers form the critical link between this ultimate source of energy and the rest of the organisms on the planet via a complex network of food webs (Schlesinger and Bernhardt 2020), exceptions being unique isolated systems that are reliant on deep ocean thermal vents for energy. Photosynthesis captures the energy from the Sun using carbon dioxide from the atmosphere and water to form a variety of energy rich compounds [e.g. carbohydrates, lipids, and proteins (using nitrogen sources extracted from the soil)] that form the fundamental building block for plants and are stored within plant cells. Thus, producers form the base of any food web. Above the level of producers are the herbivores—animals that feed on plants. Herbivores, consequently, are the first step in redistributing "captured energy" and nutrients to the rest of the living portions of any ecosystem. As a variety of animals consume herbivores and are in turn consumed themselves, captured solar energy and plant-derived resources are ultimately distributed through an ecosystem via complex food webs. Insects are the dominant set of herbivores (~25% of insect species are herbivores) and thus form the critical energy and nutrient linkages between plants and the rest of animal life.

As a group, insect herbivores feed on all parts and structures of plants (roots, stems, reproductive structures, etc.). Indeed, all parts of every terrestrial and freshwater aquatic plant are likely to be fed upon by at least one insect herbivore. There are a wide variety of types of insect herbivores and most can be categorized by the plant tissues on which they feed. Folivores are adapted to feed on the leafy components of the plant. Frugivores specialize upon fruit, while granivores feed upon seeds. Plant fluid-feeding insects specialize on extracting the fluid components of either or both xylem and phloem; the fluid conducting vessels of the plant. This latter category of insects is also important as they may transmit many pathogenic microorganisms that cause viral, fungal, and bacterial diseases reducing plant health and, in some cases, causing plant mortality.

Within each of these categories of insect herbivory there is a tremendous level of variation in the degree of host plant specialization. Many insect herbivores may feed on a specific tissue associated with only a few species of plants. Others may feed on many species of plants. Through herbivory, insects are important in the overall regulation of plant communities by thinning overly dense plant populations and removing stressed and diseased individuals. Under poor forest management, where plant densities are allowed to become extreme, or when extensive drought conditions persist for years reducing plant defensive capabilities, populations of insect herbivores can rapidly increase in density leading to massive forest die offs (Fig. 1.8).

The importance of insects in food webs goes much further than just herbivory. Insects, not just herbivores, are a vital source of food (energy and nutrients) to a tremendous variety of vertebrates including fish, amphibians, reptiles, birds, and mammals (Capinera 2011).

In addition to partially regulating plant communities through herbivory, insects also provide a variety of beneficial services. One of the more important of these services is biological control. When presented with optimum growing conditions coupled with limited or absent predation and parasitism, herbivorous insect populations can explode in density and geographical expanse while significantly reducing the capacity of forest ecosystems to provide services and/or triggering large-scale ecosystem changes. This is especially the case with successful insect invasions in which existing natural predators and parasites are absent (Brockerhoff and Liebhold 2017). Importation and release of specifically adapted insect predators and parasites from the areas of origin of the invasive species may restore the ecological balance such that the densities of the invasive species are held below damaging levels. In normal functioning forests without invasive species, insect predators and parasites play key roles in naturally managing herbivorous insects below damaging densities (Kidd and Jervis 1997, but see Rosenheim 1998) (Fig. 1.9).

Fig. 1.8 Mountain pine beetle infestation. © Dezene Huber, University of Northern British Columbia, used by permission

Granivorous insects, through the process of harvesting seed on which to feed, often inadvertently disperse viable, undamaged seed. Although some seed will be consumed, some will be dispersed to new unoccupied habitats; thus providing a benefit to the plant.

Although wind-pollination is critical for coniferous forests and grasslands, non-grass flowering Angiosperms rely on mutualistic pollination by animals, the majority of which are insects. Indeed, the success of flowering plants is partially the result of tens of millions of years of coevolution between insects and plants. Most pollinating insects can be found within the Orders Hymenoptera (ants, bees, stinging wasps), Diptera (flies), Lepidoptera (moths and butterflies), and Coleoptera (beetles) (Fig. 1.10); although, any insect feeding on flowers, pollen, or nectar, has the potential to provide pollination services.

Many insect-plant pollination associations are mutualistic in nature by which plants require and benefit from insect-transfer of pollen, and insects receive flower nectar and/or pollen as a food resource. Additionally, insect predators that forage on insect pollinators while pollinating may inadvertently assist in the movement of pollen between flowers or individual plants.

In addition to affecting plants by their feeding activities, insects (largely herbivores) also play an extremely important role as vectors of a wide variety of plant diseases caused by viruses, bacteria, fungi, and nematodes. Such diseases may be

Fig. 1.9 a Mantid (a predatory insect with raptorial front legs) capturing and feeding on a cricket. © Ian Wright, used by permission. **b** *Aphidius ervi* (Hymenoptera: Braconidae) (a parasitic wasp) attacking an aphid. The wasp deposits and egg into the body of the aphid, and the developing wasp larvae feeds upon and ultimately kills the aphid. ©Alex Wild, used by permission. **c** Harvester ant collecting seed. © Alex Wild, used by permission

fatal and wide-spread resulting in complete loss of tree species (e.g. the Dutch Elm disease fungi transmitted by bark beetles). Others may only result in poor tree growth and branch die back (e.g. *Xylella* bacterial diseases in many species of Eastern North American hardwood forests). Many plant viruses that affect trees and understory plant species are transmitted by a host of aphids. Several species of bark beetles not only directly damage their hosts through feeding, but rely on mutualistic associations with fungal species to overcome host tree defenses. Cerambycid beetles are the primary insect vector of pine wood nematode which causes pine wilt disease. It should also not be overlooked that a number of insect species (e.g. mosquitoes and other biting flies) found in forests may transmit diseases that affect humans, domestic livestock, and wildlife (various species of *Plasmodium* [(i.e. malaria], West Nile and other arboviruses, plague).

1.3.3   Insect Decline

There has been growing concern and evidence for global declines in insect biodiversity (Wagner 2020). Loss of insect biodiversity has been detected on every continent

Fig. 1.10 Representative pollinators. **a** A native leafcutter bee (Hymenoptera). © Alex Wild, used by permission. **b** A syrphid fly (Diptera). © Ansel Oommen, Bugwood.org. **c** A skipper butterfly (Lepidoptera). © Ansel Oommen, Bugwood.org. **d** A flower-feeding blister beetle (Coleoptera). © Alex Wild, used by permission

where it has been examined. These losses have been documented for multiple types of terrestrial and aquatic insect communities. Potential causes of these declines are many and it is likely that no one factor is responsible for declines everywhere. The causes are not unique to the loss of insect species and are generally attributable to human activities: habitat loss and fragmentation due to agriculture, urbanization, recreation, pollution; climate change; and increase in transport and establishment of invasive species. The implications of these findings are serious as the degradation or loss of many of the ecosystem services that insects provide, including linking the earth's food webs (which include humans), would be catastrophic. Hopefully, with additional research and monitoring, both global and local factors responsible for these declines can be clearly identified and mitigated.

1.4 Summary

Insects are found in almost all ecological niches within the forests of the world and forest animal life, similar to other terrestrial ecosystems, is dominated by insects. Collectively, insects perform critical ecosystem services that maintain the health of the planet. The overwhelming majority of forest insect species are beneficial or are neutral in their impact on humans. Indeed without insects, most terrestrial ecosystems would likely collapse. Nonetheless, there are a relative very few, but very important, forest insect pests that either directly damage trees and understory plants or transmit damaging plant pathogens via their feeding behaviors (e.g. spruce budworms, various defoliators and sucking insects, several species of bark beetles and other wood boring insects, and newly arrived invasive species). It is not unusual to see outbreak populations of these pest species in overgrown, unnaturally dense, poorly managed forests and/or in forests subjected to long periods of drought. Often, these few pestiferous species must be managed in order to protect natural resources, and the management often involves the application of insecticides which have their own set of broad spectrum deleterious impacts. Failure to successfully manage pest species often leads to additional forest decline and further threatens overall species diversity ecosystem health. In 1987, the famed biologist E. O. Wilson published a paper entitled "The Little things that run the world" in which he emphasized the global importance of insects to the health of the planet (Wilson 1987). That statement is even more true today than it was 35 years ago.

References

Brockerhoff EG, Liebhold AM (2017) Ecology of forest insect invasions. Biol Invasions 19:3141–3159

Capinera J (2011) Insects and wildlife: arthropods and the relationships with wild vertebrate animals. John Wiley & Sons, Hoboken

Jackson RB, Lajtha K, Crow SE, Hugelius G, Kramer MG, Pineiro G (2017) The ecology of soil carbon: pools, vulnerabilities, and biotic and abiotic controls. Annu Rev Ecol Syst 48:419–445

Kidd NAC, Jervis MA (1997) The impact of parasitoid and predators on forest insect populations. In: Watt AD, Stork NE, Hunter MD (eds) Forests and insects. Chapman and Hall, London, pp 49–65

Lehmann J, Kleber M (2015) The contentious nature of soil organic matter. Nature 528:60–68

Noriega JA, Horta J, Azcarate FM, Berg MP, Bonada N, Briones MJL, Del Toro I, Goulson D, Ibanez S, Landis D, Moretti M, Potts SG, Slade EM, Stout JC, Ulyshen MD, Wackers FL, Woodcock BA, Santos AMC (2018) Research trends in ecosystem services provided by insects. Basic Applied Ecol 26:8–23

Oakley TH, Wolfe JM, Lindgren AR, Zaharoff AK (2013) Phylotranscriptomics to bring the understudied into the fold: monophyletic ostracoda fossil placement, and pancrustaccean phylogeny. Mol Biol Evol 30:215–233

Obalum SE, Chibuke GU, Peth S, Ouyang Y (2017) Soil organic matter as sole indicator of soil degradation. Environ Monit Assess 189:176–195

Purvis A, Hector A (2000) Getting the measure of biodiversity. Nature 405:212–219

Rosenheim JA (1998) Higher-order predator and the regulation of insect herbivore populations. Annu Rev Entomol 43:421–447

Rota-Stabelli O, Lartillot N, Philippe H, Pisani D (2013) Serine codon-usage bias in deep phylogenomics: pancrustacean relationships as case study. Syst Biol 62:121–133

Schlesinger WH, Bernhardt ES (2020) The carbon cycle of terrestrial ecosystems. In: Schlesinger WH, Bernhardt ES (eds) Biogeochemistry: an analysis of global change, 4th edn. Academic Press, Cambridge, pp 141–182

Stork NE (2018) How many species of insects and other terrestrial arthropods are there on Earth? Annu Rev Entomol 63:31–45

Wagner DL (2020) Insect declines in the Anthropocene. Annu Rev Entomol 65:457–480

Wilson EO (1987) The little things that run the world (the importance and conservation of invertebrates). Conserv Biol 1:244–346

Chapter 2
Form and Function

Daniel Doucet and Timothy D. Paine

Insect form and function is a vast field of knowledge that covers the relationship between the structure and physiology of insects and how they interact with their biotic and abiotic environment to survive and reproduce. Foundational textbooks in this area have been written by Snodgrass (1935) and Wigglesworth (1950) and comprehensive updates have been written by Klowden (2013) or incorporated within contemporary general entomology textbooks, such as by Gillott (2005). In the last thirty years, insect physiology has increasingly been studied through the lens of molecular biology, genetics and genomics (Hoy 2018). It is obviously too large a body of knowledge to address in depth in a single chapter concerning (or in a book on) forest entomology and the reader is invited to consult these works for a deeper understanding of the topic.

Advances in insect physiology have been dependant on the study of a few "model" insect species that provide plenty of biological material at a reasonable cost, usually reared under controlled laboratory settings. As very few insects relevant to forestry fit these requirements, the physiology of insects is largely known from agriculturally- and medically-important species. Nevertheless, the major organs and mechanisms by which specific insect physiological systems operate are conserved across species, such that generalizations can be made.

D. Doucet (✉)
Canadian Forest Service, Great Lakes Forestry Centre, Sault Ste. Marie, ON, Canada
e-mail: daniel.doucet@nrcan-rncan.gc.ca

T. D. Paine
Department of Entomology, University of California-Riverside, Riverside, CA, USA
e-mail: timothy.paine@ucr.edu

J. D. Allison et al. (eds.), *Forest Entomology and Pathology*,
https://doi.org/10.1007/978-3-031-11553-0_2

2.1 Insect Development

Developmental trajectories in insects and related hexapod taxa can be classified along three main types, depending on the degree of morphological difference between immature and adult stages, and the presence or absence of metamorphosis. In ametabolous hexapods, larvae reach the adult stage through a series of molts without exhibiting significant morphological changes, except for the presence of genitalia in adults. Metamorphosis is absent in these insects and molting can continue to occur in adults. This type of developmental program is restricted to the basal hexapod orders Protura, Diplura and Collembola, all soil-dwelling organisms.

Hemimetabolous insects also grow through a series of molts in which the nymphs (immatures) are morphologically similar to the adults. However, adults gain their wings and genitalia after a single molt. Hemimetabolous insects are represented by eleven recognized orders (Song et al. 2016), including some of the largest such as Hemiptera (true bugs, scale insects, aphids) and Orthoptera (crickets and grasshoppers).

Holometabolous insects display radically different morphologies between immatures and adults. The transformation between immature and adult takes place in what is called the pupal stage, a generally inactive stage where extensive organ and tissue remodeling occurs. Complete metamorphosis has been deemed an "evolutionary innovation" among insects that enabled the occupation of different ecological niches by adult and immature forms, and explains some of the evolutionary success of the major modern insect orders i.e. Coleoptera, Lepidoptera, Hymenoptra and Diptera (Ureña et al. 2016).

2.1.1 Eggs

In most insects, progeny are deposited in the environment in the form of eggs, a reproductive pattern known as oviparity. Insect eggs consist of a developing embryo accompanied by yolk, a maternally-secreted substance rich in proteins that fuels growth until hatching. The embryo and the yolk are enclosed in protective layers originating from two sources: a maternally-derived eggshell synthesized before fertilisation and two epithelia, the amnion and the serosa, produced by the embryo. The eggshell is comprised of an innermost vitelline membrane on top of which sits the chorion, a multilayered, proteinaceous cover. In some species a wax layer is present between the vitelline membrane and the chorion to prevent desiccation (Klowden 2013). Egg chorions vary in the number of layers and internal architecture between species, but the presence of meshwork of airspaces between the inner and outer chorionic layers is common. These pockets of air facilitate gas exchange for the developing embryo and are connected to the outside world via openings called aeropyles. The eggshell also presents a micropyle, an opening that enables access of the sperm to the egg (Zeh et al. 1989). The embryo-derived amnion envelops the ventral side of the embryo, while the serosa surrounds both the yolk and the embryo, just underneath the vitelline envelope (Panfilio 2008). The amnion and the serosa provide additional protection against desiccation, act as a barrier against microbial pathogens and in many insects one or both layers can synthesize a chitinous cuticle to enhance mechanical rigidity (Jacobs et al. 2013; Rezende et al. 2008.).

Eggs often require further maternally-derived products to ensure their survival. Glue-like secretions from the ovipositor can be added by the female to attach the egg to a substrate, such as the abaxial surface of leaves. Some Lepidoptera also cover their eggs with hairs or scales to deter potential predators (Floater 1998). The forest tent caterpillar (*Malacososma disstria*) attaches eggs in masses around branches of host trees, using a foamy substance called spumaline that also protects the eggs against parasitism (Williams and Langor 2011).

2.1.2 Viviparity

In some insects, embryonic development proceeds inside the female and free-living nymphs or larvae are laid instead of eggs. This reproductive pattern is known as viviparity and is classified in four types depending on the amount of yolk, the body cavity where the embryo is incubated, and the manner in which supplementary nutrients are acquired if yolk is absent or reduced. Ovoviviparity and pseudoplacental viviparity are by far the most common types in insects of relevance to forestry. Ovoviviparous insects produce embryos covered with a thin and elastic eggshell that also encloses yolk. Eggs hatch in the uterus after a period of incubation. Species from various lineages employ this reproductive pattern, including thrips (Thysanoptera), Lepidoptera, Coleoptera, Hymenoptera, Homoptera and several families of Diptera such as parasitic tachinid flies (Hagan 1948). In pseudoplacental viviparity embryos also develop in the reproductive tract but a significant amount of nutrients is acquired through placenta-like structures of maternal or embryonic origin, in the latter from the amnion or serosa. All aphids (Aphididae) and some other Hemiptera reproduce in this fashion, along with a few species of barklice (Psocoptera). The other two types of viviparity are hemoceolic and adenotrophic viviparity. In hemocelic viviparity, embryos develop in the mother's hemocoel and absorb nutrients from the hemolymph. This viviparity type is found among the parasitic Strepsiptera and in some gall midges (Cecidomyiidae). In adenotrophic viviparity, hatched larvae feed on nutritive secretions produced by maternal glands, and this form of viviparity is found only in some families of dipteran parasites of mammals (e.g. Tsetse flies, Hagan 1948).

2.1.3 Post-embryonic Development and Larval Morphology

Insect post-embryonic growth proceeds in discrete steps marked by the shedding of the exoskeleton, an event known as molting. The form assumed by immatures between two molts is called a larval or nymphal instar (or simply instar) preceded by a number identifying its order of appearance after egg hatching (e.g. 1st larval instar or 1st instar) (Chapman and Chapman 1998). The term "stage" generally refers to the major ontogenetic divisions of the life cycle (larval, pupal and adult stages), but here again the numbering system can be applied for nymphs and larvae (e.g. 1st larval stage). The term "stadium" is applied strictly in reference to the duration of an instar (Carlson 1983).

As mentioned previously, hemimetabolous nymphs molt progressively into adults, however in some taxa there are deviations from this basic growth pattern. In the Plecoptera, Odonata and Ephemeroptera, the aquatic nymphs (also called naiads) have gills that are lost at metamorphosis. In the thrips (Thysanoptera), whiteflies (Alyrodidae) and male scale insects (Coccoidea), the transition from nymph to adults is interrupted by immobile non-feeding stages which functionally resemble holometabolous pupae. There can be up to three such stages in thrips of the suborder Tubulifera, named propupa and pupa I and pupa II (Moritz 1997).

The larvae of holometabolous insects display a variety of morphologies but convergence of form is observed for many distinct taxa that feed on the same food type. In general, cryptic feeders such as leaf miners, skeletonizers, wood borers, gall-forming insects and endoparasitoids show the simplest overall shape, are most often apodous with greatly reduced sensory appendages and cephalic structures. Extreme minimalism is observed in Dipteran larvae, such as the Agromyzidae (leaf miners) and Tachinidae (endoparasitoids) where the only distinguishing feature of the maggots is the sclerotized cephalopharyngeal skeleton (mouth hooks) (Feener and Brown 1997; Teskey 1981). In parasitic Hymenoptera (e.g. Ichneumonidae), young instars are also apodous but the terminal segment can extend as a tail in which case larvae are called "caudate" or in addition show developed mouthparts, in which case they are termed "caudate-mandibulate". The first instar of some species, particularly in the Cynipidae and Figitidae are termed "eucoiliform" for the long flexible processes that they present on their ventral side, whose function is unknown. Eventually these hymenopteran larvae transition to featureless, grub-like "hymenopteriform" shapes in later instars (Gordh et al. 1999).

The immatures of woodboring insect taxa are also generally larviform, but harbor more robust cephalic structures to consume hard woody substrates. In the Cerambycidae, larvae are cylindrical and slightly dorsoventrally compressed with a sclerotized head capsule retracted within the prothorax, with mouthparts oriented forward (prognathous, e.g. *Eutrypanus dorsalis*, Casari and Teixeira 2014). Buprestid immatures are typically elongate and the dorsoventral flattening is more pronounced (e.g. emerald ash borer *Agrilus planipennis*, Chamorro et al. 2012) while Curculionids are compressed along the antero-posterior axis (Chamorro 2019). The thoracic legs of woodborer larvae are usually small or absent, but in the latter case locomotion can be aided by protuberances present on the abdomen or the thorax, called ambulatory ampullae.

Soil dwelling insects that feed on roots or rotten wood adopt two distinctive larval shapes, the scarabeiform or elateriform-type. Scarabeiform larvae are comma-shaped, with highly sclerotized head capsules and developed thoracic legs, taking its name from immatures of the scarab beetle family. Elateriform larvae are slender, heavily sclerotized with powerful legs and mouthparts, adaptations which allow them to move rapidly in the soil and cope with abrasion stress.

Larvae from holometabolous insects feeding on aerial plant parts have the familiar "caterpillar" shape, also known as eruciform (latin *eruca-*: caterpillar). Lepidoptera, Trichoptera, some species of the basal Hymneopteran suborder Symphyta (sawflies), and Chrysomelidae adopt this form. Eruciform larvae are characterized by elongate and cylindrical bodies, three pairs of segmented thoracic legs and a variable number (2–5) of pairs of unsegmented abdominal prolegs, adaptations that allow them to move rapidly between patches of food (Kou and Hua 2016). They also have a head capsule and highly sclerotized mandibles to crush foliage or other plant structures (e.g. seeds, buds, cones). The mouthparts can be prognathous (oriented forward) or hypognathous (oriented ventrally).

The larvae of highly mobile insects often display a form called campodeiform, characterized by an overall flattened shape and well-developed legs and antennae (Krafka 1923). This shape is more often associated with the obligate or facultative predatory lifestyle of certain families in diverse orders (e.g. Carabidae and Staphylinidae in the Coleoptera, Chrysopidae in the Neuroptera, Winterton et al. 2018), but is also encountered among filter-feeding insects, such as in the Trichoptera.

2.1.4 Molting and Metamorphosis

Insects benefit from the presence of a chitinous exoskeleton, which acts primarily as a barrier against external biotic and abiotic insults. However, this barrier is incompatible with continuous growth. Insects, as well as all the arthropods, have solved the growth-protection conundrum by introducing molting as an elaborate mechanism to ensure that the exoskeleton is replaced rapidly. The cellular and molecular aspects of molting have been studied in representative species of several insect orders, particularly the Diptera, Coleoptera and Lepidoptera and the topic has been extensively reviewed elsewhere (Truman and Riddiford 2002; Belles 2011). Molting is a chain of events that culminates in the synthesis and tanning of a new cuticle. Two hormones orchestrate this process: the steroid 20-hydroxyecdysone (20E) and the sesquiterpenoid juvenile hormone (JH). 20E is normally present at low levels throughout the immature stages, but its titers increase and decrease rapidly as a "pulse" before each molt. Therefore, 20E determines the timing of the molt. JH for its part is present at high levels throughout the larval stages, but disappears as the larva reaches the species-specific critical weight necessary for metamorphosis (Nijhout and Callier 2015). Under high JH, a pulse of 20E directs the larva to molt into another larva, but in the absence of JH, a pulse of 20E will direct the larva to molt into a pupa and a pupa to molt into an adult.

The increase in 20E titers originates from the activation of neurosecretory cells in the brain that release the prothoracicotropic hormone (PTTH). PTTH activates the production of the 20E precursor ecdysone in the prothoracic glands and, upon reaching peripheral tissues (e.g. the epidermis), ecdysone is converted into 20E, the active version of the hormone. The physiological conditions that trigger the molting cascade vary between insects and can involve multiple stimuli announcing the need to "change suit". Some hemipterans use the stretching of the abdomen that occurs after feeding as a cue to molt. In the hornworm *Manduca sexta*, the sensing of oxygen limitation to growing tissues is also a trigger, since the chitin-lined tracheal system becomes unable to adequately facilitate gas exchange as the larva grows in volume (Callier and Nijhout 2013).

Metamorphosis involves a much more extensive remodeling of the body plan unfolding over two consecutive molts (larval-pupal and pupal-adult). In *Manduca*, two pulses of 20E occur in the last larval instar. The first one, called the "commitment peak", is a brief low amplitude elevation of 20E titers that irreversibly changes the gene expression program of epidermal cells from larval to pupal. The second peak, much larger, directs epidermal cells to synthesize pupal cuticle (Riddiford 1976). The morphogenesis of the adult appendages and internal organs during pupation varies substantially between endopterygote insects. In the Diptera, Hymenoptera and Lepidoptera, most larval tissues are completely dissolved (histolysed) while adult appendages such as the wings, legs, antennae and eyes, arise from the rapid growth and differentiation from clusters of cells of embryonic origin called imaginal discs. Likewise, many of the adult's internal organs in these orders are formed from undifferentiated cells, the histoblasts. In the beetles (Coleoptera), metamorphosis is more progressive and is reminiscent of the changes observed in exopterygotes, with the most notable change being in the development of the adult flight mechanism (Gillott 2005).

2.2   Sensory Perception

Sensory perception involves the detection of electromagnetic radiation (vision), diverse chemicals (olfaction, gustation), temperature (thermoreception) and changes in mechanical pressure on or distortion of (touch, proprioception and hearing) the cuticle. Some insects, particularly eusocial species, can also use the earth's magnetic field as a sensory input during foraging (Wajnberg et al. 2010; Fleischmann et al. 2018). Signals are detected by specialized sensory neurons that convert the stimulus into an electrical response (signal transduction) carried from the peripheral to the central nervous system (CNS) (Torre et al. 1995). The CNS integrates all these diverse sensory modalities to drive physiological and or behavioral responses.

The physiological and molecular features of sensory perception in forest insects have received a great deal of attention, particularly olfaction and vision in moths and beetles. In many insects, mating partners must be located over long distances (relative to insect body size) and, in herbivores, acceptable host plants must be found among a diversity of non-host species. Much of the impetus for the study of insect sensory physiology emerges from practical considerations. Senses that can detect stimuli over long distances, such as vision and olfaction, can be exploited for the survey of insect spatial and temporal abundance via visually attractive and/or semiochemical-baited traps (Grant 1991; Brockerhoff et al. 2006; Cavaletto et al. 2020; Thistle and Strom 2006). Additionally, pest management tactics that directly suppress insect populations based on these stimuli also exist (e.g. pheromone-based mating disruption).

In adult insects, olfaction is mediated primarily by antennae. Antennae are composed of three sections: a basal segment, the scape, anchoring the rest of the antenna to the cranium. The next is the pedicel, which acts as like a hinge joint between the scape and the last section, the flagellum. The flagellum is constituted of units called antennomeres (Minelli 2017). Beyond this basic segmental arrangement, antennal morphology is extremely varied among taxa, the result of natural and sexual selection (Elgar et al. 2018).

Insect antennae are populated by microscopic protruding structures, called sensillae, which serve to detect odor. A typical olfactory sensilla consists of a fine and porous hair-like extension (seta) rising from the antennal cuticle. The pores serve to let volatile odors in, where they will be dissolved in an aqueous fluid (the sensillar lymph) before reaching the plasma membrane of sensory receptor neurons located inside. As many volatiles are hydrophobic, their transport within the sensillar lymph is mediated by specialized odor-binding proteins (odorant-binding proteins, OBPs and chemosensory proteins, CSPs, Pelosi et al. 2018). Sensillae have been classified depending on their external appearance and internal morphology, such as the number of pores and seta shape. Sensilla type, location on antennae and differential abundance between the sexes are important pieces of information in order to understand the chemical ecology of a given insect species.

The ultrastructure of the antennae has been characterized using electron microscopy techniques in numerous forest pest insects of economic importance. Examples include the *Dendroctonus* bark beetle species complex [*D. valens* (Chen et al. 2010), *D. frontalis* (Dickens and Payne 1978) and *D. ponderosae* (Whitehead 1981)], the emerald ash borer *Agrilus planipennis* (Crook et al. 2008), the eucalyptus borer *Phoracantha semipunctata* (Lopes et al. 2002) and the brown spruce longhorned beetle, *Tetropium fuscum* (Mackay et al. 2014). Likewise, similar attention has been given to antennal sensillar structures in lepidopteran forest pest species. They include the teak skeletonizer *Eutectona machaeralis* (Lan et al. 2020), the spruce budworm *Choristoneura fumiferana* (Albert and Seabrook 1973) and the Chinese pine caterpillar *Dendrolimus tabulaeformis* (Zhang et al. 2013). A comparative analysis of the antennae of Trichoptera, and basal and derived Lepidoptera species, revealed that a relationship exists between the proportions of certain sensilla types, and the type of sex pheromone used (Yuvaraj et al. 2018).

Gustation is used to perceive surface chemicals that can mediate acceptance or rejection of food sources. In herbivore species, many types of sugars act as phagostimulants while plant secondary metabolites can act as deterrents. Gustation is mediated by structures analogous to the olfactory system, i.e. via sensilla housing gustatory receptor neurons. Most gustatory sensillae are located on dedicated sensory appendages around the mouth (labial and maxillary palps) and inside the mouth itself, but can be found in other locations including the tarsi and ovipositor (Seada et al. 2018). Gustatory sensillae have been studied in larvae of the spruce budworm, *Choristoneura fumiferana*. The L1 sensilla and the lateral styloconic sensilla (LST) located respectively at the tips of the maxillary palp and on the galea, enable the detection of sugars. While the L1 sensillum detects furanose sugars, LST detects pyranose-type sugars (Hock et al. 2007). Interestingly furanoses, either as monosaccharides (fructose) or as subunits of disaccharides (e.g. sucrose) are indicators of plant stress. Thus L1 sensilla may assist in the identification of vulnerable host trees (Albert 2003).

Insect vision is accomplished by two types of ocular structures: simple eyes and compound eyes. Two types of simple eyes are further recognized: the ocelli and the stemmata. Ocelli are located dorsally on the head and are present in many insect orders in both adults and larvae, although they are absent from holometabolous larvae (Stehr 2009). In *Drosophila*, ocelli are composed of a corneal lens located above a thin layer of corneagenous cells (which secretes the lens), itself located above a group of 80 photoreceptor cells (Sabat et al. 2016). In general, ocelli cannot form images at high resolution and serve mostly to perceive rapid and slow (e.g. day/night cycles) changes in light intensity.

Stemmata are simple eyes located on the lateral sides of the head of holometabolous insect larvae. Like ocelli, they are composed of a corneal lens and a layer of photoreceptor cells, but also present a transparent crystalline cone as an intermediate layer. The structural organization of stemmata is reminiscent of the ommatidia of compound eyes, and indeed good molecular evidence suggests that stemmata are derived from a compound eye ancestor existing before the split of the holometabola and hemimetabola lineages (Buschbeck 2014). In insects with rudimentary stemmata, these simple eyes fulfill a light intensity detection function similar to the ocelli, but on the horizontal plane rather than above the head. However, in some predatory insects such as tiger beetle larvae, stemmata are sophisticated enough that they can be used to locate and capture prey (Buschbeck 2014).

Compound eyes are present in adult insect species and in immature hemimetabolous species. They occupy the lateral portion of the head, and in some good fliers (e.g. Tabanid flies, species in the order Odonata) they can extend to meet on the dorsal section of the head. The basic functional unit of compound eyes is the ommatidium and its architecture is well conserved among insects (Friedrich et al. 2006). The external facet of the ommatidium, also called the corneal lens, is made of transparent cuticle. Situated directly underneath is another transparent structure called the crystalline lens, flanked by four secretory cells, the Semper's cells. Both lenses form the dioptric apparatus that refracts incident light toward a layer of eight photoreceptor cells occupying the basal section of the ommatidium. The dioptric apparatus is sheathed by primary pigment cells, and secondary and tertiary pigment cells can occur in some species to surround the photoreceptors. Light sensation is concentrated in an area around the central axis of the ommatidium where the cell membrane of each photoreceptor come in close proximity, called the rhabdom, each cell contributing a "rhabdomere". Rhabdomeres display dense microvilli and are enriched in opsins, the visual pigments responsible for the conversion of light into an electric signal. Insects active during the day and the night show important structural differences in their compound eyes. Daytime active insects have apposition compound eyes, where photoreceptors only receive light penetrating through the lens directly above them. In these insects the ommatidial pigment cells are optically opaque to light coming from neighboring ommatidia. In contrast, nighttime active insects have superposition compound eyes, where the walls of each ommatidium are made of transparent pigment cells. Superposition eyes enable the collection of a much larger number of photons by the photoreceptor cells (Warrant 2017).

Insects show enormous species-to-species differences in their ability to perceive the various properties of light, such as light intensity, spectral composition and polarization. Herbivorous insects present a variety of adaptations of their visual system, ranging from eye structure, compound eye facet arrangement and opsin gene content that match the requirements to select acceptable hosts. For instance, scolytine beetles display reduced numbers of compound eye facets, indicative of a secondary dependence on visual cues compared to olfactory ones (Chapman 1972). In contrast, buprestid beetles show extremely good visual abilities, mediated in part by opsin gene sequence diversity and expression patterns in photoreceptors (Lord et al. 2016). Several reviews on the anatomical and molecular adaptations of insect visual systems have been published elsewhere (Briscoe and Chittka 2001; Egelhaaf and Kern 2002; Warrant and Dacke 2011; Cheng and Frye 2019) and provide comprehensive treatments of this most complex sensory organ.

2.3 Food Acquisition, Consumption and Utilization

Food intake, processing and utilization take place along the subdivisions of the insect alimentary canal. Physical breakdown of the food into smaller particles is helped by the crushing action of the mandibles in leaf- and wood-feeding species, while obviously little such modification is necessary in sap-feeders. The extraction of nutriments then proceeds along the three main regions of the gut: the foregut, midgut and hindgut. An example of these broad alimentary canal divisions in the larva of a xylophagous insect, the Brown Spruce longhorned beetle (*Tetropium fuscum*), is provided in Fig. 2.1. Organic polymers such as proteins and cellulose are broken down into their respective amino acids and sugar units through the action of digestive enzymes. In some xylophagous insect species, cellulose degradation requires a supply of enzymes secreted by microbial symbionts, harbored in specific regions of the gut (Martin 1991).

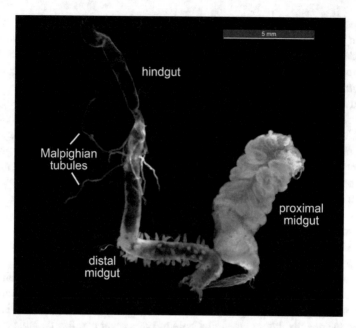

Fig. 2.1 Morphology of the alimentary canal of the brown spruce longhorn beetle (*T. fuscum*) larva. Two major divisions of the midgut, proximal and distal, form the majority of the digestive system in this species. The Malpighian tubule openings define the boundary between the midgut and the hindgut. Photo credit: Susan Bowman, Natural Resources Canada

In insects, five mouthpart regions are involved in food manipulation: (1) the labrum, (2) the hypopharynx, (3) the mandibles, (4) the maxillae, and (5) the labium. Borrowing from vertebrate anatomy, the labrum and labium can be thought as analogous to the upper and lower lips respectively, while the hypopharynx would be closest to the tongue. The pair of mandibles and maxillae are positioned caudal to the labrum and their task is food crushing (primarily the mandibles) and manipulation. The morphology of insect mouthparts, which is classically illustrated by using chewing insects as examples (e.g. orthopterans, Coleoptera) is actually remarkably varied (Labandeira 1997). Insects that feed on a strict diet of liquids show a range of mouthpart shapes adapted to facilitate liquid uptake. Nectar-feeding hymenoptera such as orchid bees have "lapping" mouthparts that have evolved by greatly extending the palps of the labium and the lobes (galea) of the maxillary palps. Joined together, they form the proboscis, which encloses a similarly extended lobe of the labium called the glossa (Düster et al. 2018). Adult cytclorraphan Diptera have a highly specialized "sponging" structure (the labellum) formed by the fusion of labial palps. In adult Lepidoptera, the coiled maxillary galea are joined in a proboscis.

The foregut and the hindgut of immature insects originate from the embryonic ectoderm and for this reason are able to secrete a cuticle, just like the epidermis (Reuter 1994). The midgut for its part develops from the endoderm (Stainier 2002). The foregut is subdivided into four anatomical regions: the pharynx, the oesophagus, the crop and the proventriculus. Collectively these regions serve to further breakdown food particles and regulate the flow of food before its entry in the midgut. The pharynx is populated with sensory neurons and assists in the decision of rejecting or accepting food, while the oesophagus is primarily involved in pushing food down the alimentary canal. The foregut crop is generally present as an enlargement where food is stored and can be further processed before entering the midgut. The proventriculus acts like a valve to regulate food entry into the midgut. Cuticular hair-like projections present on the crop side of the proventriculus help separate food based on particle size (Serrão 2005). In some species of ants, the proventriculus is also instrumental in preventing bacteria from entering the midgut (Lanan et al. 2016).

The midgut is in most insects the primary site for enzymatic breakdown of ingested food and the absorption of nutriments. The midgut epithelium is constituted of two types of differentiated cells, the columnar cells and endocrine cells, whose renewal is ensured by undifferentiated stem cells (Caccia et al. 2019). The side of the columnar cells facing the lumen of the gut (called the apical side) is extensively folded into microvilli to increase the effective surface of secretion and absorption functions. As their name implies, endocrine cells assist to regulate gut function via the secretion of peptide hormones. These hormones can affect nearby cells (paracrine signalling) or other organs in the insect. An important structure present in the midgut, but absent from both the foregut and the hindgut is the peritrophic matrix. The peritrophic matrix is composed of a thin layer of proteins attached to chitin fibers (Tellam 1996). The structure of the peritrophic matrix is highly variable among insects, and in some groups such as many sap sucking insects, it is totally absent. The peritrophic matrix has several protective functions on the midgut epithelium, mainly from physical damage by food particles, but also from digestive enzymes and it also prevents the entry of pathogenic microbes such as viruses (Hegedus et al. 2009).

The insect hindgut is generally divided in three sections called the pylorus, ileum and rectum. The pylorus typically regulates the flow of contents exiting the midgut by way of a muscular pyloric sphincter (Dallai 1976). It is also the section where in most insects the proximal ends of the Malpighian tubules connect with the digestive system. In several insect species the ileum section of the foregut is involved in ion and water transport. In termites, the entire hindgut, including the ileum, has been extensively modified to handle the digestion of cellulose. Five distinct proctodeal segments are present (P1 to P5), with the ileum being the first one (Rocha and Constantini 2015). The third proctodeal segment (P3) harbors most of the microbial symbionts responsible for the enzymatic breakdown of cellulose and the hydrolysis of xylan (e.g. *Nasutitermes*, Warnecke et al. 2007).

2.4 Nervous System

The nervous system and the endocrine system function as coordination centers of the insect. The insect sensory system receives stimuli from both the internal and external environments, integrates this information within the central nervous system, and processes the information to make the appropriate responses to the stimuli. The insect nervous system is comprised of nerve cells or neurons that are very similar in structure and physiology to vertebrate neurons. They are elongate cells that propagate a nerve impulse (a wave of electrical depolarization of the cell membrane) from one end of the elongate cell (the dendrite end) to the other end (the axon end). Information is coded in the frequency and temporal pattern of nerve impulses and by which neuron the nerve impulses originate (e.g. the same pattern of nerve impulses coming from the optic nerve are interpreted by the brain differently than the same pattern of nerve impulses coming from the auditory nerve). Neurons are arranged sequentially (end to end) so that nerve impulses can be transmitted from one neuron to the next. When the nerve impulse reaches the axon end of the neuron, it stops; however, it causes the release of a chemical neurotransmitter from the cell membrane which then diffuses across the very tiny gap called a synapse between the two neurons. When the neurotransmitter reaches the dendrites on the other neuron across the synapse, it stimulates an electrical nerve impulse to travel down the length of the next neuron.

There are several different types of neurons. Sensory neurons are afferent neurons that conduct nerve impulses initiated at sensory organs towards the central nervous system (CNS). Motor neurons are efferent neurons that carry nerve impulses from the CNS and transmit them to the effector organs (e.g. muscles or glands) to stimulate the effector organs to respond (e.g. muscles contract or glands secrete). Internuncial neurons are located entirely within the CNS and interconnect neurons with each other.

The insect nervous system is generally organized as a connected series of ganglia, a nerve cord, and peripheral nerves. Ganglia are groups of neuron cell bodies. The segmental ganglia ("segmental brains") are the functional units of the central nervous system and individual segments often have their own ganglion. This reflects the ancestral trait organization inherited from the proto-annelid ancestors of arthropods. However, there is fusion of adjacent segmental ganglia into larger compound ganglia in many advanced insect taxa illustrating a higher degree of centralization (Niven et al. 2008). Synapses, the connections between neurons, occur only in ganglia; consequently, all communication between neurons takes place in the ganglia. The ganglia are where sensory information is received and processed and where the appropriate motor responses are initiated and coordinated. Segmental ganglia function much like "segmental brains". The ventral nerve cord is a paired structure that connects ganglia with one another and allows ganglia to communicate and coordinate with each other. Peripheral nerves enter and leave the ganglia. These neurons innervate the various parts of the body (sensory organs, muscles, glands, etc.). They bring sensory information from the body into ganglia and transmit motor signals from ganglia out to parts of the body.

The insect central nervous system has a number of specialized ganglia. The brain is a fusion of ganglia of the anterior-most segments of the body. The brain controls movement of the antennae and labrum, receives some of the most important sensory information from the eyes, ocelli, antennae, and labrum, and is the most important ganglion for processing that sensory information and initiating the appropriate behavioral or physiological responses. The brain has considerable influence over other ganglia and plays a major role in coordinating the ganglia so that the insect functions as a unit rather that a collection of individual segments. The sub-esophageal ganglion is a fusion of mandibular, maxillary and labial segmental ganglia. It receives sensory information from these mouthparts and coordinates and initiates their movement. Thoracic ganglia receive sensory information from legs, wings, and other structures on the thorax, and coordinate and initiate their movement. The ancestral condition for the thoracic ganglia is one per segment but they are sometimes fused together as a compound thoracic ganglion in more advanced groups. Abdominal ganglia receive sensory information from abdominal structures, and coordinate and initiate their operation.

2.5   Epidermis and Cuticle

The integument is the outer body covering of arthropods and functions as an exoskeleton. It is one of the major reasons why arthropods are the most diverse and successful Phylum. The exoskeleton gives arthropods an enormous advantage over other invertebrates due to the efficient locomotion that a skeleton can provide. The lever-like mechanics of the exoskeleton allows a small muscle contraction to cause a large movement of an appendage. The arthropod integument serves as rigid skeleton, provides tough protective covering (armor), gives protection against water loss (an absolutely critical necessity for terrestrial organisms), and facilitates perception of the environment through sensilla embedded in the cuticle.

The insect integument has a characteristic layered microstructure. There are two major divisions: (1) the interior epidermis, the living, cellular part, and (2) the exterior cuticle, the non-living part secreted by the epidermis. The epidermis is a single layer of cells beneath the cuticle comprised of several cell types. The epidermal cells secrete the non-living cuticle and the dermal glands secrete defensive fluids, pheromones, etc. to the exterior surface of the body. There are also specialized epidermal cells that form at least part of many sensory organs located on the integument. The cuticle is non-living and comprises the bulk of the integument (Moussian 2010). The cuticle is initially secreted as procuticle that differentiates into the endocuticle and exocuticle layers. Endocuticle is located immediately above the epidermis and is tough and flexible. The major chemical constituent is chitin, a complex carbohydrate similar to cellulose that forms fibrils. The chitin fibrils of the procuticle are laid down in distinct layers with the fibrils within each layer oriented in the same direction, but the orientation of each layer is at a slightly different direction. This provides a lot of structural strength and is similar in principle to why plywood is so strong: the grain of each layer in plywood is oriented in a different direction. The exocuticle layer is above the endocuticle and is tough and rigid. In addition to flexible chitin, the other major chemical component is sclerotin (a protein that binds with the chitin fibers). The sclerotin bound to the chitin fibrils becomes cross-linked to each other by quinones; this prevents the chitin fibrils from moving relative to each other and thus the cuticle is no longer flexible, it becomes hard and rigid. The cross-linking process is referred to as hardening or as sclerotization. The third layer, the epicuticle, is the exterior covering of the cuticle. In insects, this layer is very thin, but very complex. It is composed of four separate layers including a waterproofing wax layer (a critical adaptation to terrestrial life) and a cuticulin layer that provides a critical protective barrier during molting.

The macrostructure of the integument is organized into sclerites and membranes, which join adjacent sclerites and enable the articulated movement of a rigid exoskeleton. Sclerites are hardened plates comprised of heavily sclerotized exocuticle. Membranes are flexible areas composed mostly of flexible endocuticle. The sclerites provide protection while the membranes provide flexibility of joints.

As mentioned earlier in this chapter, there are a number of advantages and disadvantages of having an exoskeleton relative to an internal skeleton. An exoskeleton is lightweight and strong; the tubular structure provides maximal strength with minimal skeletal material. The exoskeleton also provides a protective armor covering. However, the exoskeleton must be periodically shed (molted) in order to permit growth; even unsclerotized cuticle is relatively "unstretchable". The necessity to molt, limits the maximum size that can be attained. Immediately after a molt, the integument is not yet hardened (because it must expand to stretch bigger than the previous integument); consequently, if the arthropod is too big and heavy, the integument will not be able to support the body weight right after a molt. It will bend, buckle and distort, and then eventually harden in a malformed shape. This limitation is more restrictive for terrestrial arthropods than for aquatic or marine arthropods.

The process of molting is critical in the life history of insects. At the start of molting, the epidermis separates from the endocuticle, a process referred to as *apolysis*, and epidermal cells divide to increase their number in order to accommodate the upcoming larger body size. The epidermal cells expand in size so that the new epidermis is now larger than it was before. However, in order to fit within the confines of the old cuticle, the newly expanded epidermis is folded and pleated beneath the old cuticle. Next, the epidermis secretes the cuticulin layer of the new epicuticle. The epidermal cells then secrete molting fluid that passes through pores in the cuticulin into the gap between the epidermis and the old cuticle; molting fluid contains enzymes that digest the old endocuticle. As the old endocuticle is being digested by molting fluid, the new procuticle is synthesized and laid down by the epidermal cells. Much of the breakdown products from enzymatic digestion of the old cuticle are absorbed by the epidermal cells and recycled for the synthesis of the new cuticle; this conserves a lot of the building blocks from the old cuticle, but it is not enough to complete the new cuticle, so additional building blocks are also synthesized by the epidermis.

The cuticulin layer plays a critical role as a barrier between the molting fluid and the newly forming cuticle; without the cuticulin layer, the new cuticle would also be digested by the molting fluid. Only the endocuticle of the old cuticle is digested by the molting fluid. The exocuticle resists digestion (probably due to the sclerotin) and remains as a thin shell enclosing the insect, which must be shed. Following the deposition of the new procuticle, the insect then expands its body, mostly by muscle contraction. This expansion causes the old exocuticle to split open, allowing the insect to escape the confines of its old cuticle.

The process of shedding the old cuticle is called ecdysis, and old cuticle which is shed is called an exuvium. The old cuticle splits along pre-formed lines of weakness, called ecdysial sutures, generally occurring longitudinally along the dorsal midline. Ecdysial sutures are simply a line along the cuticle where there is a sharp break in the exocuticle and the endocuticle fills this break; as long as the endocuticle is intact, this is not a weak spot, but when the endocuticle is digested during molting, this becomes a line of weakness that splits apart when the insect begins to expand its body. The insect then continues to expand its body by muscle contraction, pumping blood into its extremities (i.e. legs, wings, antennae) and swallowing air (or water for aquatic insects) until the folds and pleats in the new cuticle are all stretched out and the insect has reached its new, larger size. The procuticle is still completely flexible and non-rigid, so it cannot function as an exoskeleton; consequently, the insect is very vulnerable at this point. Shortly after the insect expands its new cuticle, quinones are secreted by the epidermis which crosslink the sclerotin-chitin complex in the exocuticle, making the exocuticle hard and rigid with all the functions of an exoskeleton and the insect can efficiently move around.

2.6 Neuroendocrine System

Insects have a complex endocrine system that regulates physiology and behavior. Hormones are chemical messengers secreted into the insect body (Gilbert 2011). They are released by cells in endocrine glands and cause a physiological response in target tissues. Many, but not all, of the endocrine glands are associated with neurosecretory tissues. They travel from endocrine glands to the target tissues by circulating in the blood within the body cavity. Hormones regulate many physiological and behavioral functions in insects. In addition to metamorphosis and molting described previously, other functions under hormonal control include heart rate, pigmentation, blood sugar level, egg development, water excretion, cuticle sclerotization, and diapause (insect version of hibernation).

2.7 Circulation and Immunity

The insect circulatory system almost never is involved in transport of respiratory gasses (O_2, CO_2) and there is nothing equivalent to our red blood cells in insects. The insect circulatory system is an open system. The body is a contiguous open cavity, called a hemocoel ("blood cavity"), from head to abdomen (Hillyer 2015). Blood is pumped from the posterior to anterior end of the body in the dorsal vessel. In the return trip (anterior to posterior), blood is not confined to vessels; it percolates through the hemocoel, bathing all the tissues and organs in blood. The dorsal vessel forms a long, narrow tube located along dorsal midline. The heart is the posterior part and is contractile, providing the pumping force to propel the blood forward. The aorta is the anterior part, and serves as an artery for blood to travel from the heart to the anterior end of the insect. When the heart fills with blood during the diastole phase of a heartbeat, alary muscles attaching the heart to the lateral body walls contract causing the heart to dilate. Blood flows from the hemocoel into the heart through a series of openings, called ostia, located on each side of the heart in the abdominal region. Ostia are one-way valves; they allow blood to enter the heart from hemocoel, but do not permit blood flow in the opposite direction from the heart out to the hemocoel. During the systole phase of a heartbeat, the muscles that form the wall of the heart contract, forcing blood out of the heart. Blood cannot be forced out of the ostia (one-way valves), so it must be forced out through one of the two ends of the heart (almost always the anterior end). Blood is pumped from the posterior to the anterior end of the heart by peristaltic waves of systole followed by diastole traveling the length of the heart from posterior to anterior end. This creates an area of high pressure in the head and an area of low pressure in the abdomen. After leaving the aorta at the anterior end of the body, blood simply flows through the hemocoel down a pressure gradient from head to abdomen, bathing the organs and tissues in a current of blood.

The appendages (legs, antennae, wings, etc.) are not in the main current of blood through the hemocoel so an additional mechanism is needed to circulate blood through the appendages. In the wings, blood circulates through some of the wing veins. Septa divide most other appendages longitudinally into two channels but these two channels connect at the apex of the appendage. When muscles in the appendage contract, they put more pressure on one side of the septum than on the other side, depending which side of the septum the contracting muscle is located. This creates a pressure gradient between the channels on either side of the septum, and blood will flow from the channel under high pressure to the apex of the appendage and into the channel on the other side of the septum. Some appendages have accessory pulsatile organs at their base to assist circulation (Pass 2000). The accessory pulsatile organs are contractile organs that pump blood into one of the channels along the septum causing a circulation of blood down one channel and back out the other; this makes circulation through the appendage very efficient.

Insect blood (hemolymph) consists of plasma and blood cells (Mullins 1985). Plasma functions to transport nutrients, hormones, and waste throughout the body. The blood cells (hemocytes) function in clotting (wound healing), phagocytosis of histolysing tissue (clean up broken-down tissues), immunity to microorganisms (fight infections), and encapsulation of parasitoid eggs (a very important defense mechanism against parasitoids) (Hillyer 2016). In encapsulation, blood cells aggregate around the parasitoid egg, cutting it off from nutrients in the blood and from access to oxygen. The phagocytosis function is especially active during molting and metamorphosis when some tissues and organs from the previous developmental stage that do not occur in the next developmental stage are being broken down.

2.8 Respiration and Gas Exchange

Insects breathe using a tracheal system comprised of a branching network of cuticle-lined tubes, called trachea, that reach every cell in the body (Locke 1957). Spiracles are the openings of the tracheal system to the atmosphere. The maximum number is 10 pair; each pair is laterally positioned (left and right) on the meso- and metathorax and on abdominal segments 1–8. The openings have valves that can be closed to reduce respiratory water loss. The valves are also used in controlling ventilation or the movement of air through the body. Trachea are the main air tubes extending from the spiracles to all parts of the body. These tubes branch extensively and as they branch, their diameter generally gets smaller and smaller. Trachea are reinforced with taenidia, which are thickened cuticular rings in the tracheal walls that strengthen the trachea and prevent their collapse. However, trachea do not function in gas exchange with respiring tissues. Gas exchange with respiring cells occurs via the tracheoles, which are very narrow diameter tubular terminal ends of the tracheal system. Tracheoles are not lined with cuticle and are anatomically, functionally, and physiologically different from trachea.

Ventilation of the insect body occurs through the longitudinal tracheal trunks. These are wide diameter trachea that run longitudinally along the body, connecting the trachea that originate at the spiracles in each spiracle-bearing segment, and also extend into the head where there are no spiracles. Generally, there are three pair of longitudinal tracheal trunks: lateral, dorsal, and ventral. The tracheal trunks are interconnected with each other via additional trachea. Tracheal air sacs are parts of tracheal trunks that are very wide diameter, enclosing a relatively large volume of air. These play an important role in ventilation of the tracheal system. When muscles in surrounding parts of the body contract, the air sacs get compressed, forcing air out of the sacks; when these muscles relax, the air sacs return to their normal, wide-diameter shape, drawing air into the sacs. The abdomen can compress and decompress by contraction and relaxation of dorsal–ventral muscles and dorsal longitudinal muscles; thus compressing and decompressing the tracheal air sacs in the abdomen. Thoracic tracheal air sacs are often in close contact with flight muscles and compress and decompress as the adjacent flight muscles contract and relax.

Movement of respiratory gasses (O_2, CO_2) through the tracheal system is a combination of diffusion and active ventilation (Buck 1962). Transport of O_2 and CO_2 between spiracles and tracheoles is very dependent on diffusion. However, diffusion is a relatively slow process and is efficient only over short distances. Active ventilation (mechanical air movement) of the tracheal system can reduce the reliance on diffusion to move O_2 and CO_2 between spiracles and tracheoles. Compression and decompression of air sacs can move air at least the distance from the spiracles to the air sacs. The rest of the route for gas movement (air sacs to tracheoles) still relies on diffusion. If air sacs move air in and out of the same spiracles, there is not a 100% air exchange due to the residual volume of air in the sacs and trachea (i.e. the sacs and trachea cannot compress down to zero volume). However, timing the opening and closing of spiracular valves to coordinate with the compression and decompression of air sacs can result in nearly 100% air exchange throughout the tracheal trunks. The trachea from all spiracles are interconnected via tracheal trunks. During a ventilation cycle, the anterior spiracular valves close and posterior valves open, and abdominal muscles relax causing the abdominal air sacs to decompress (expand) causing air to be drawn into the tracheal system through the posterior spiracles. When abdominal muscles contract causing the abdominal air sacs to compress, the anterior spiracular valves open and posterior valves close forcing air out of the tracheal system through the anterior spiracles. By this rhythmic coordination of muscle contractions with the opening and closing of the spiracles, a steady stream of fresh air flows through the tracheal trunks without leaving any residual volume of "old air". The rest of the route for O_2 and CO_2 (tracheal trunks to tracheoles) still relies on diffusion.

2.9 Locomotion

The structures associated with insect locomotion are located on the thorax. In the adult insect, there are three pairs of legs, one pair associated with each body segment. The legs have five components; each element is unsegmented except the most distal. Beginning at the body, they are the coxa, trochanter, femur, tibia, and tarsus. The tarsus can be made up of 3–5 segments called tarsomeres. While insect legs have the same structural organization, the legs have been modified in form through natural selection to adapt to a wide range of life history strategies. For example, long and slender cursorial legs are adapted for running, natatorial legs with expanded and flattened oar-like femur or tibia are adapted for swimming, raptorial legs are adapted for grasping prey, saltatorial legs have enlarged femur with powerful muscles for jumping, and fossorial have shovel-like shapes for digging.

Insects are the only group of invertebrate animals that have evolved the capability of powered flight. The evolution of wings gives insects a significant advantage in exploiting their environment. Wings, when present on the adult insect, are found on the meso- and meta-thoracic segments, but never on the prothoracic segment. They are composed of two layers of integument (exoskeleton) with heavier veins in the wings providing stability and rigidity (Wootton 1992). Veins contain nerves, trachaea and haemolymph. Some orders of wingless insects, the Apterygota, evolved before the advent of wings, and within the winged orders, the Pterygota, some orders have lost their wings through natural selection (i.e. Siphonaptera). Like the legs, the wings have been subject to intense natural selection for adaptation to specific life histories. Consequently, there have been significant modifications. The winged Diptera have a pair of mesothoracic wings, but the metathoracic wings have been modified into club-like halteres that have numerous sensillae that respond to body position in flight. The mesothoracic wings (the elytra) of many species of Coleoptera are hard and rigid, protecting the underlying metathoracic wings and abdomen from physical damage and enabling the insects to use a wide range of habitats or niches. The leathery mesothoracic wings (the tegmina) of many species of Orthoptera and related groups have a similar function. The wings of the Lepidoptera are covered with scales that are often colored and can provide crypsis or advertise their presence, while the wings of the Thysanoptera are narrow and covered with long hairs that provide surface area for lift. The tiny parasitic Hymenoptera have greatly reduced wing venation.

2.10   Excretion and Osmoregulation Systems

In most insects, the excretory and osmoregulation systems involve Malpighian tubules working in concert with the hindgut. However, other organs such as salivary glands play a role in excretion and/or osmoregulation in some insects. The Malpighian tubules are hollow, blind ended tubes extending from the digestive system near the midgut/hindgut junction. The walls are 1 cell thick the number of tubules can vary from 0 to 250. Malpighian tubules generally float freely in hemolymph where they filter out wastes from the blood (analogous to vertebrate kidneys). They remove nitrogenous waste (usually uric acid), salts, and water from the hemolymph and transport them (the primary filtrate) into the hollow lumen of the tubule (Beyenbach et al. 2010). The contents of the tubule lumen flow to the base of the Malpighian tubule and empty into the gut near the hindgut/midgut junction. The Maligian tubules also function in reabsorption of vital salts; in order to maintain proper osmolarity of the blood, water and salts are selectively resorbed from the primary filtrate. Reabsorption takes place in the hindgut, and in some insects, it also takes place in all or part of the Malpighian tubules. As water is resorbed, uric acid precipitates out as a solid because it is not very water soluble; the precipitated uric acid mixes with the contents of the hindgut and is passed out the anus with the feces.

The insect fat body can also be important for excretion and osmoregulation. It is a very diffuse, amorphous organ located throughout the hemocoel, primarily in the abdomen. It generally appears as a mass of whitish or yellowish globules that float in the hemocoel and is continuously bathed in hemolymph. The fat body serves many different physiological functions including storage of fat, glycogen, and protein. In some insects, specialized fat body cells store nitrogenous waste such as uric acid. It also serves as a metabolic center controlling intermediate metabolism (e.g. amino acid conversions, glycogen synthesis and breakdown, fat metabolism, etc.). The fat body can also provide functions analogous to the vertebrate liver by detoxifying poisons and metabolizing hormones (Li et al. 2019).

2.11   Reproduction

The female reproductive system is located in the abdomen. It includes a pair of ovaries each made up of one, a few, or many ovarioles. The ovarioles are elongate tubes that are the functional unit of the ovaries and produce the eggs (Hodin 2009). As eggs develop, they travel down the length of the ovariole from the distal end (the germarium) where meiosis and egg cell formation occurs to the proximal end (the vitillarium) where eggs grow, accumulate yolk, and mature before they leave the ovariole and enter the lateral oviduct. A specialized storage organ, the spermatheca also opens into the oviduct (Pascini and Martins 2017). The spermatheca stores sperm for days to years depending on the species of insect; fertilization does not necessarily occur shortly after copulation. It has a valve to let sperm in during copulation and to regulate the release of sperm when eggs are ready to be fertilized. This is a critical fitness advantage for haplo-diploid insects (see below) like some social and parasitic Hymenoptera that can control the sex of their offspring. Fertilization is regulated by females that can withhold or release sperm when an egg is present. In addition to the spermatheca, accessory glands also release a variety of secretions associated with oviposition into the oviduct. Secretions from the accessory gland can produce egg cases which enclose and protect a clutch of eggs produced by some insects from desiccation, predators, and disease. The accessory gland can produce adhesive for eggs to glue the eggs to a substrate and produce venom for bee and wasp sting. In these cases, the sting is a modified ovipositor.

The male reproductive system is composed of a pair of testes that produce sperm, the vas deferens, which are tubes to transport sperm from testes to the ejaculatory duct, seminal vesicles that store mature sperm, an ejaculatory duct through which sperm leave the body, and accessory glands. The accessory glands in the male reproductive system produce a variety of secretions associated with copulation including seminal fluid, which is a liquid medium for sperm motility, but may also provide nourishment for sperm. Accessory glands also produce spermatophores which are enclosed packets of sperm and are thought to be an early adaptation for fertilization by terrestrial arthropods. Spermatophores do not occur in all insects. In many early terrestrial arthropods and insects, males deposit a spermatophore on the substrate and then the female picks it up off the substrate with her genitals. In these species there is no copulation associated with sperm transfer. In more advanced groups, fertilization became more efficient by the male directly placing the spermatophore into the female's genitals. More derived insects have lost the spermatophore altogether, and the male has a penis to deposit the sperm in a non-encapsulated form directly into the female's genital opening. In a few insect groups, accessory glands can produce "mating plugs". These are gel plugs that seal the female's genital opening after copulation to prevent other males from copulating with her, thus providing a mechanism for ensuring paternity.

There are many examples of insect groups that reproduce through sexual reproduction or through parthenogenesis (reproduction without fertilization). Sexual reproduction is the ancestral means of reproduction and it is still the most common strategy. Nonetheless, parthenogenesis has evolved independently in many different groups of insects, in some cases multiple times within groups. Hymenoptera (bees, wasps, ants), whiteflies, scale insects, thrips, and a few others have a rather unusual reproduction process; female offspring are produced by sexual reproduction and male offspring are primarily produced parthenogenically. In these groups, females develop from fertilized eggs (fertilized by standard sexual reproduction) and are diploid (2n chromosomes) while unfertilized eggs develop into males which are haploid (1n chromosomes), known as haplo-diploid sex determination. As a consequence, in many species with haplo-diploid sex determination, the mother can choose the sex of her offspring according to current needs. For example, female insects generally are bigger and require more food to reach maturity than male insects; consequently, many Hymenoptera parasitoids deposit male eggs (unfertilized) in small hosts and female eggs (fertilized) in large hosts. In many social Hymenoptera, workers are all female. The queen produces male eggs only right before the mating season. The rest of the year, she produces only female eggs.

2.12 Conclusions

Insects have successfully adapted to virtually every environment on the planet. Their basic body plan and physiology has been modified through evolutionary selection to allow them to exploit a wide variety of habitats and the ecological niches within those habitats. Many of the insect groups have highly specialized feeding ecologies, while many others are extreme generalists in their requirements. Most importantly, they have demonstrated exceptional capacity to adapt to environmental change. This has served the insects, as a taxonomic group, very well in evolutionary history and suggests that they have the capacity to adapt to the current pattern of global change.

References

Albert PJ, Seabrook WD (1973) Morphology and histology of the antenna of the male eastern spruce budworm, Choristoneura fumiferana (Clem.) (Lepidoptera: Tortricidae). Can J Zool 51(4):443–448

Albert PJ (2003) Electrophysiological responses to sucrose from a gustatory sensillum on the larval maxillary palp of the spruce budworm, Choristoneura fumiferana (Clem.) (Lepidoptera: Tortricidae). J Insect Physiol 49(8):733–738

Belles X (2011) Origin and evolution of insect metamorphosis. In: Encyclopedia of life sciences (eLS). Wiley, Chichester

Beyenbach KW, Skaer H, Dow JA (2010) The developmental, molecular, and transport biology of Malpighian tubules. Annu Rev Entomol 55:351–374

Brockerhoff EG, Jones DC, Kimberley MO, Suckling DM, Donaldson T (2006) Nationwide survey for invasive wood-boring and bark beetles (Coleoptera) using traps baited with pheromones and kairomones. For Ecol Manage 228(1–3):234–240

Briscoe AD, Chittka L (2001) The evolution of color vision in insects. Annu Rev Entomol 46(1):471–510

Buck J (1962) Some physical aspects of insect respiration. Annu Rev Entomol 7(1):27–56

Buschbeck EK (2014) Escaping compound eye ancestry: the evolution of single-chamber eyes in holometabolous larvae. J Exp Biol 217(16):2818–2824

Caccia S, Casartelli M, Tettamanti G (2019) The amazing complexity of insect midgut cells: types, peculiarities, and functions. Cell Tissue Res 377(3):505–525

Callier V, Nijhout HF (2013) Body size determination in insects: a review and synthesis of size-and brain-dependent and independent mechanisms. Biol Rev 88(4):944–954

Carlson RW (1983) Instar, stadium, and stage: definitions to fit usage. Ann Entomol Soc Am 76(3):319–319

Casari SA, Teixeira ÉP (2014) Immatures of Acanthocinini (Coleoptera, Cerambycidae, Lamiinae). Rev Bras Entomol 58(2):107–128

Cavaletto G, Faccoli M, Marini L, Spaethe J, Giannone F, Moino S, Rassati D (2020) Exploiting trap color to improve surveys of longhorn beetles. J Pest Sci 94:871–883

Chamorro ML (2019) An illustrated synoptic key and comparative morphology of the larvae of Dryophthorinae (Coleoptera, Curculionidae) genera with emphasis on the mouthparts. Diversity 11(1):4

Chamorro ML, Volkovitsh MG, Poland TM, Haack RA, Lingafelter SW (2012) Preimaginal stages of the emerald ash borer, *Agrilus planipennis* Fairmaire (Coleoptera: Buprestidae): an invasive pest on ash trees (*Fraxinus*). PLoS ONE 7(3):e33185

Chapman JA (1972) Ommatidia numbers and eyes in scolytid beetles. Ann Entomol Soc Am 65(3):550–553

Chapman RF, Chapman RF (1998) The insects: structure and function. Cambridge University Press, Cambridge

Chen HB, Zhang Z, Wang HB, Kong XB (2010) Antennal morphology and sensilla ultrastructure of Dendroctonus valens LeConte (Coleoptera: Curculionidae, Seolytinae), an invasive forest pest in China. Micron 41(7):735–741

Cheng KY, Frye MA (2019) Neuromodulation of insect motion vision. J Comp Physiol A 206:125–137

Crook DJ, Kerr LM, Mastro VC (2008) Distribution and fine structure of antennal sensilla in emerald ash borer (Coleoptera: Buprestidae). Ann Entomol Soc Am 101(6):1103–1111

Dallai R (1976) Fine structure of the pyloric region and Malpighian papillae of Protura (Insecta Apterygota). J Morphol 150(3):727–761

Dickens JC, Payne TL (1978) Structure and function of the sensilla on the antennal club of the southern pine beetle, Dendroctonus frontalis (Zimmerman) (Coleoptera: Scolytidae). Int J Insect Morphol Embryol 7(3):251–265

Düster JV, Gruber MH, Karolyi F, Plant JD, Krenn HW (2018) Drinking with a very long proboscis: functional morphology of orchid bee mouthparts (Euglossini, Apidae, Hymenoptera). Arthropod Struct Dev 47(1):25–35

Egelhaaf M, Kern R (2002) Vision in flying insects. Curr Opin Neurobiol 12(6):699–706

Elgar MA, Zhang D, Wang Q, Wittwer B, Pham HT, Johnson TL, Freelance CB, Coquilleau M (2018) Focus: ecology and evolution: insect antennal morphology: the evolution of diverse solutions to odorant perception. Yale J Biol Med 91(4):457

Feener DH Jr, Brown BV (1997) Diptera as parasitoids. Annu Rev Entomol 42(1):73–97

Fleischmann PN, Grob R, Müller VL, Wehner R, Rössler W (2018) The geomagnetic field is a compass cue in Cataglyphis ant navigation. Curr Biol 28(9):1440–1444

Floater GJ (1998) Tuft scales and egg protection in *Ochrogaster lunifer* Herrich-Schäffer (Lepidoptera: Thaumetopoeidae). Aust J Entomol 37(1):34–39

Friedrich M, Dong Y, Jackowska M (2006) Insect interordinal relationships: evidence from the visual system. Arthropod Syst Phylogeny 64(2):133–148

Gilbert LI (ed) (2011) Insect endocrinology. Academic Press, Amsterdam

Gillott C (2005) Entomology. Springer, Dordrecht

Gordh G, Legner EF, Caltagirone LE (1999) Biology of parasitic Hymenoptera. In: Fisher TW, Bellows TS, Caltagirone LE, Dahlsten DL, Huffaker CB, Gordh G (eds) Handbook of biological control: principles and applications of biological control. Elsevier, San Diego, pp 355–381

Grant GG (1991) Development and use of pheromones for monitoring lepidopteran forest defoliators in North America. For Ecol Manage 39:153–162

Hagan HR (1948) A brief analysis of viviparity in insects. J N Y Entomol Soc 56(1):63–68

Hegedus D, Erlandson M, Gillott C, Toprak U (2009) New insights into peritrophic matrix synthesis, architecture, and function. Annu Rev Entomol 54:285–302

Hillyer JF (2015) Integrated immune and cardiovascular function in Pancrustacea: lessons from the insects. Integr Comp Biol 55(5):843–855

Hillyer JF (2016) Insect immunology and hematopoiesis. Dev Comp Immunol 58:102–118

Hock V, Albert PJ, Sandoval M (2007) Physiological differences between two sugar-sensitive neurons in the galea and the maxillary palp of the spruce budworm larva Choristoneura fumiferana (Clem.) (Lepidoptera: Tortricidae). J Insect Physiol 53(1):59–66

Hodin J (2009) She shapes events as they come: plasticity in female insect reproduction. In: Phenotypic plasticity of insects: mechanisms and consequences. Science Publishers, Enfield, pp 423–521

Hoy MA (2018) Insect molecular genetics: an introduction to principles and applications, 4th edn. Academic Press, London

Jacobs CG, Rezende GL, Lamers GE, van der Zee M (2013) The extraembryonic serosa protects the insect egg against desiccation. Proc R Soc B: Biol Sci 280(1764):20131082

Klowden MJ (2013) Physiological systems in insects. Academic Press, London

Kou LX, Hua BZ (2016) Comparative embryogenesis of Mecoptera and Lepidoptera with special reference to the abdominal prolegs. J Morphol 277(5):585–593

Krafka J (1923) Morphology of the head of trichopterous larvae as a basis for the revision of the family relationships. J N Y Entomol Soc 31(1):31–52

Labandeira CC (1997) Insect mouthparts: ascertaining the paleobiology of insect feeding strategies. Annu Rev Ecol Syst 28(1):153–193

Lan L, Wang S, Hu K, Ma T, Wen X (2020) Ultrastructure of antennal morphology and sensilla of teak skeletonizer, *Eutectona machaeralis* Walker (Lepidoptera: Crambidae). Microsc Microanal 26(6):1274–1282

Lanan MC, Rodrigues PAP, Agellon A, Jansma P, Wheeler DE (2016) A bacterial filter protects and structures the gut microbiome of an insect. ISME J 10(8):1866–1876

Li S, Yu X, Feng Q (2019) Fat body biology in the last decade. Annu Rev Entomol 64:315–333

Locke M (1957) The structure of insect tracheae. J Cell Sci 3(44):487–492

Lopes O, Barata EN, Mustaparta H, Araújo J (2002) Fine structure of antennal sensilla basiconica and their detection of plant volatiles in the eucalyptus woodborer, Phoracantha semipunctata Fabricius (Coleoptera: Cerambycidae). Arthropod Struct Dev 31(1):1–13

Lord NP, Plimpton RL, Sharkey CR, Suvorov A, Lelito JP, Willardson BM, Bybee SM (2016) A cure for the blues: opsin duplication and subfunctionalization for short-wavelength sensitivity in jewel beetles (Coleoptera: Buprestidae). BMC Evol Biol 16(1):1–17

MacKay CA, Sweeney JD, Hillier NK (2014) Morphology of antennal sensilla of the brown spruce longhorn beetle, Tetropium fuscum (Fabr.) (Coleoptera: Cerambycidae). Arthropod Struct Dev 43(5):469–475

Martin MM (1991) The evolution of cellulose digestion in insects. Philos Trans R Soc London Ser B: Biol Sci 333(1267):281–288

Minelli A (2017) The insect antenna: segmentation, patterning and positional homology. J Entomol Acarol Res 49(1). https://doi.org/10.4081/jear.2017.6680

Moritz G (1997) Structure, growth and development. In: Lewis T (ed) Thrips as crop pests. CAB International, Wallingford, pp 15–63

Moussian B (2010) Recent advances in understanding mechanisms of insect cuticle differentiation. Insect Biochem Mol Biol 40(5):363–375

Mullins DE (1985) Chemistry and physiology of the hemolymph. Compr Insect Physiol Biochem Pharmacol 3:355–400

Nijhout HF, Callier V (2015) Developmental mechanisms of body size and wing-body scaling in insects. Annu Rev Entomol 60:141–156

Niven JE, Graham CM, Burrows M (2008) Diversity and evolution of the insect ventral nerve cord. Annu Rev Entomol 53:253–271

Panfilio KA (2008) Extraembryonic development in insects and the acrobatics of blastokinesis. Dev Biol 313(2):471–491

Pascini TV, Martins GF (2017) The insect spermatheca: an overview. Zoology 121:56–71

Pass G (2000) Accessory pulsatile organs: evolutionary innovations in insects. Annu Rev Entomol 45(1):495–518

Pelosi P, Iovinella I, Zhu J, Wang G, Dani FR (2018) Beyond chemoreception: diverse tasks of soluble olfactory proteins in insects. Biol Rev 93(1):184–200

Reuter R (1994) The gene serpent has homeotic properties and specifies endoderm versus ectoderm within the Drosophila gut. Development 120(5):1123–1135

Rezende GL, Martins AJ, Gentile C, Farnesi LC, Pelajo-Machado M, Peixoto AA, Valle D (2008) Embryonic desiccation resistance in Aedes aegypti: presumptive role of the chitinized serosal cuticle. BMC Dev Biol 8(1):82

Riddiford LM (1976) Hormonal control of insect epidermal cell commitment in vitro. Nature 259(5539):115–117

Rocha M, Constantini JP (2015) Internal ornamentation of the first proctodeal segment of the digestive tube of Syntermitinae (Isoptera, Termitidae). Deutsche Entomologische Zeitschrift 62:29

Sabat D, Priyadarsini S, Mishra M (2016) Understanding the structural and developmental aspect of simple eye of Drosophila: the ocelli. J Cell Signal 1(109):2

Seada MA, Ignell R, Assiuty A, Naieem A, Anderson P (2018) Functional characterization of the gustatory sensilla of tarsi of the female polyphagous moth Spodoptera littoralis. Front Physiol 9:1606

Serrão JE (2005) Proventricular structure in solitary bees (Hymenoptera: Apoidea). Org Divers Evol 5(2):125–133

Snodgrass RE (1935) Principles of insect morphology. McGraw-Hill, New York

Song N, Li H, Song F, Cai W (2016) Molecular phylogeny of Coleoptera (Insecta) inferred from expanded mitogenomic data. Sci Rep 6(1):1–10

Stainier DY (2002) A glimpse into the molecular entrails of endoderm formation. Genes Dev 16(8):893–907

Stehr FW (2009) Ocelli and stemmata. In: Encyclopedia of insects. Academic Press, San Diego, p 721

Tellam RL (1996) The peritrophic matrix. In: Biology of the insect midgut. Springer, Dordrecht, pp 86–114

Teskey HJ (1981) Morphology and terminology-larvae. In: McAlpine JF (ed) Manual of Nearctic Diptera, vol. 1. Agriculture Canada, Ottawa, ON, pp 65–88

Thistle HW, Strom BL (2006) Optical cues in forest insect host homing: an overview. In: 2006 ASAE Annual Meeting. American Society of Agricultural and Biological Engineers, p 1

Torre V, Ashmore JF, Lamb TD, Menini A (1995) Transduction and adaptation in sensory receptor cells. J Neurosci 15(12):7757–7768

Truman JW, Riddiford LM (2002) Endocrine insights into the evolution of metamorphosis in insects. Annu Rev Entomol 47(1):467–500

Ureña E, Chafino S, Manjón C, Franch-Marro X, Martín D (2016) The occurrence of the holometabolous pupal stage requires the interaction between E93, Krüppel-homolog 1 and Broad-complex. PLoS Genet 12(5):e1006020

Wajnberg E, Acosta-Avalos D, Alves OC, de Oliveira JF, Srygley RB, Esquivel DM (2010) Magnetoreception in eusocial insects: an updatefried. J R Soc Interface 7(Suppl. 2):S207–S225

Warnecke F, Luginbühl P, Ivanova N, Ghassemian M, Richardson TH, Stege JT, Cayouette M, McHardy AC, Djordjevic G, Aboushadi N, Sorek R, Tringe SG, Podar M, Martin HG, Kunin V, Dalevi D, Madejska J, Kirton E, Platt D, Szeto E, Salamov A, Barry K, Mikhailova N, Kyrpides NC, Matson EG, Ottesen EA, Zhang X, Hernández M, Murillo C, Acosta LG, Rigoutsos I, Tamayo G, Green BD, Chang C, Rubin EM, Mathur EJ, Robertson DE, Hugenholtz P, Leadbetter JR (2007) Metagenomic and functional analysis of hindgut microbiota of a wood-feeding higher termite. Nature 450(7169):560–565

Warrant EJ (2017) The remarkable visual capacities of nocturnal insects: vision at the limits with small eyes and tiny brains. Philos Trans R Soc B: Biol Sci 372(1717):20160063

Warrant EJ, Dacke M (2011) Vision and visual navigation in nocturnal insects. Annu Rev Entomol 56:239–254

Whitehead AT (1981) Ultrastructure of sensilla of the female mountain pine beetle, Dendroctonus ponderosae Hopkins (Coleoptera: Scolytidae). Int J Insect Morphol Embryol 10(1):19–28

Wigglesworth VB (1950) The principles of insect physiology, 4th edn. Springer, Dordrecht

Williams DJ, Langor DW (2011) Distribution, species composition, and incidence of egg parasitoids of the forest tent caterpillar (Lepidoptera: Lasiocampidae), during a widespread outbreak in the Canadian prairies. Can Entomol 143(3):272–278

Winterton SL, Lemmon AR, Gillung JP, Garzon IJ, Badano D, Bakkes DK, Breitkreuz LCV, Engel MS, Moriarty Lemmon E, Liu X, Machado RJP, Skevington JH, Oswald JD (2018) Evolution of lacewings and allied orders using anchored phylogenomics (Neuroptera, Megaloptera, Raphidioptera). Syst Entomol 43(2):330–354

Wootton RJ (1992) Functional morphology of insect wings. Annu Rev Entomol 37(1):113–140

Yuvaraj JK, Andersson MN, Corcoran JA, Anderbrant O, Löfstedt C (2018) Functional characterization of odorant receptors from Lampronia capitella suggests a non-ditrysian origin of the lepidopteran pheromone receptor clade. Insect Biochem Mol Biol 100:39–47

Zhang S, Zhang Z, Kong X, Wang H (2013) Sexual dimorphism in antennal morphology and sensilla ultrastructure of Dendrolimus tabulaeformis Tsai et Liu (Lepidoptera: Lasiocampidae). Microsc Res Tech 76(1):50–57

Zeh DW, Zeh JA, Smith RL (1989) Ovipositors, amnions and eggshell architecture in the diversification of terrestrial arthropods. Q Rev Biol 64(2):147–168

Chapter 3
Forest Arthropod Diversity

Christian Hébert

3.1 Introduction

Insects are the most diverse group of organisms on Earth with 952,794 described species (Roskov et al. 2021). They account for 85% of arthropod species, 67% of animal species and 47% of all species currently known on the planet (Roskov et al. 2021) (Fig. 3.1). However, this is an underestimate as the number of species of insects and other arthropods living on Earth is still unknown. More than 30 years ago, Robert May published a paper entitled *"How many species are there on Earth"* and concluded that the number of species living on Earth was not even known within an order of magnitude (May 1988). The most recent estimates of richness suggest that there are approximately 5.5 and 7 million species of terrestrial insects and arthropods, respectively (Stork 2018). This suggests that over 80% of species remain to be found and described. Although knowledge of the diversity of species present is fundamental information for managing natural ecosystems, determining the number of insect or arthropod species existing on Earth, in a biome or in any forest habitat is a great challenge for scientists. It is concerning to realize that forest ecosystems are managed without accurate knowledge of the diversity involved in the ecological processes critical to healthy forest ecosystems.

3.1.1 Plant–Insect Coevolution as a Driver for Diversification

Arthropods have existed on Earth for at least 400 million years and they are among the earliest animals known to have colonised terrestrial habitats, where they have co-evolved with plants (Grimaldi and Engel 2005). Insects arose before the Devonian,

C. Hébert (✉)
Canadian Forest Service, Laurentian Forestry Centre, Québec City, QC, Canada
e-mail: christian.hebert@nrcan-rncan.gc.ca

© The Author(s) 2023
J. D. Allison et al. (eds.), *Forest Entomology and Pathology*,
https://doi.org/10.1007/978-3-031-11553-0_3

Fig. 3.1 Relative
importance of major groups
of organisms, based on a
total of 2,026,387 described
species, as of 10 June 2021.
Data from the Catalog of
Life website

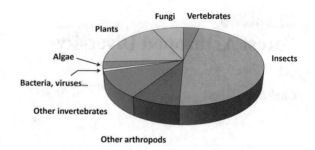

from a Silurian aquatic arthropod (Gaunt and Miles 2002), after the first fossils of
terrestrial plants were found from the Ordovician (Knoll and Nowak 2017) (Fig. 3.2).
Fossil records showed that the first trees were recorded in the late Devonian and
they diversified during the Carboniferous, which was followed by diversification of
insects in the Late Carboniferous (Retallack 1997). Signs of attack by phytophagous
arthropods have been recorded on fossil roots, leaves, wood and seeds, and the first
wood boring Coleoptera were reported from the early Permian (Labandeira 2006).
All phytophagous groups were present by the mid-Triassic; at this time, the dominant
taxon was Coleoptera (Labandeira 2006). The type of leaf feeding revealed by fossils
showed increasing complexity of interactions between arthropods and plants. For
instance, the earliest leaves showed only marginal feeding while non-marginal leaf
feeding, which requires specialized mouthparts, came later in the mid-Cretaceous
after the arrival of angiosperms (Scott et al. 1992). Leaf-mining and gall production
also coincided with plant diversification during the Cretaceous and Tertiary.

Today, plants (18.5%) and phytophagous insects (21.4%) represent about 40%
of known terrestrial species. Also, it is estimated that at least one predacious or
saprophagous insect species exists for every phytophagous insect species (Strong
et al. 1984). Thus, globally, nearly 2 terrestrial species out of 3 depend on plants. This
supports Ehrlich and Raven (1964) conclusion that *"the plant–herbivore interface
may be the major zone of interaction responsible for generating terrestrial organic
diversity"*. They suggested that the evolution of plant chemical defense in response
to insect phytophagy resulted in a co-evolutionary arms race that generated high
biodiversity in these two groups of organisms. However, it has been suggested that
this coevolutionary arms race has been overemphasized and that deterrent effects
of plant secondary chemicals for some phytophagous insects may have arisen from
the need to avoid plants on which they were easily found by predators (Bernays and
Graham 1988). In fact, coevolution is extremely difficult to demonstrate as it involves
reciprocal adaptive changes in interacting species and this change must result from
selection exerted by the other species (Thompson 1994). Nevertheless, the concept
of coevolution between plants and phytophagous insects has been generally accepted
as the basis of arthropod diversity (Janz 2011). Plant diversity was also shown to be
a powerful predictor of the richness of other arthropod guilds (Basset et al. 2012).

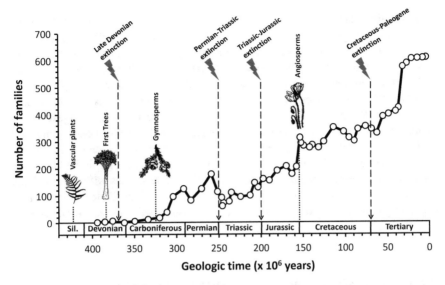

Fig. 3.2 Insect diversity, expressed as the number of insect families along Geological Time. Four major extinction events and the onsets of major groups of plants are highlighted. The geologic time events are from Figure 1 in Condamine et al. (2016) (reprinted with permission from Springer). The curve comes from Labandeira and Sepkoski (1993) (reprinted with permission from AAAS). Plant drawings were done by Jean-Michel Béland from the Canadian Forest Service (reprinted with permission from Jean-Michel Béland)

3.1.2 Wood as a Distinctive Forest Attribute and a Powerful Driver for Diversification

The most distinctive feature of trees comes from their vertical structure, which result from woody tissues that provide the mechanical support to permit their vertical growth. This allows trees to outcompete herbaceous plants and shrubs for light and produce the greatest amount of biomass among vascular plants. The resource abundance hypothesis suggests that plants offering greater amounts of resources should support more species and higher abundance of arthropod herbivores (De Alckmin Marques et al. 2000). The great aboveground biomass produced in forests may thus explain why these biotopes support so many species. Trees also tend to house more pest species than shrubs, which in turn have more than herbs (Strong and Levin 1979). Morphologically complex hosts provide diverse ecological niches and larger hosts are easier to find by arthropods. The greater size and morphological complexity of trees compared with shrubs and herbs likely explains the higher number of pest species on trees.

The structural heterogeneity of forests is both vertical, and horizontal, particularly in primary forests where closed areas alternate with clearings, which occur when trees die (Kuuluvainen 1994). Although forest ecosystems are often perceived as homogeneous at large scales, at smaller scales, forests show important horizontal heterogeneity. Closed-canopy areas alternate with forest gaps resulting from tree death. In forest gaps, local abiotic conditions differ from those in closed canopy areas (Ritter et al. 2005). These gaps influence forest dynamics and provide a succession of microhabitats that promote biodiversity. Abiotic variables interact with biotic variables such as tree species, tree size and bark thickness to provide ecological niches to arthropods. For instance, vertical segregation of bark beetles has been reported in *Pinus taeda* (Paine et al. 1981) and *Pinus strobus* (Price 1984). The largest species (genus *Dendroctonus*) are found at the tree base while the smallest ones (genus *Ips*) are found higher on the bole and even in the canopy where species such as *Pityogenes hopkinsi* feed on small branches (Price 1984). Beetles compete for limited resources (phloem) and their interactions result in partitioning resources within trees (Paine et al. 1981). This might be driven by bark thickness, as this attribute is important for explaining community composition of early-arriving beetles in recently dead Scots pine (Foit 2010).

3.1.3   Latitudinal Gradient of Arthropod Diversity

There is no complete inventory of arthropods in any biome or in any type of forest ecosystem. However, the latitudinal gradient theory predicts decreasing species richness with increasing latitude (Pianka 1966; Hillebrand 2004). Latitude is a surrogate for environmental gradients (e.g. temperature, insolation and precipitation) (Willig et al. 2003), which also vary with elevation. Tropical regions receive more solar energy and precipitation, so they should be more productive than temperate regions (Pianka 1966; Willig et al. 2003). Habitats showing greater plant species richness usually exhibit greater arthropod richness (Speight et al. 2008). In addition, glaciation events have had negative impacts on biodiversity in temperate regions, but they have not had similar effects in tropical regions (Willig et al. 2003). Also, the warmer climate and higher moisture levels in tropical regions are not only more favorable for the growth and survival of most plant species, but also for groups such as fungi on which arthropods feed. Similarly, the importance of temperature for biodiversity diversification has been highlighted along a 3.7 km elevation gradient at Mt. Kilimanjaro, Tanzania (Peters et al. 2016). Species richness of single taxa vary in complex distribution patterns along elevation, according to their tolerance limits to environmental gradients (Peters et al. 2016). Similarly, the Ichneumonidae (Hymenoptera), a family of parasitoid wasps, do not follow the usual latitudinal gradient of biodiversity, their diversity peaking at mid-latitudes (Janzen 1981; Skillen et al. 2000).

3.2 Feeding Guilds of Arthropods Living in Forests

Traditionally, arthropods are described taxonomically but they can also be described on the basis of their diet and functional role. Those that feed on living plants are generally called phytophagous (Bernays 2009), while those that feed on living animals are called zoophagous. The term saprophagous is used for organisms that feed on decaying plants or animals, but can also include feeding on fungi since they are often interlinked with decaying organic matter (Natural Resources Canada 2015). Combining functional roles and niches allows grouping arthropods among guilds, which are groups of species that exploit the same type of resources in a similar way (Root 1967). Guilds help to structure ecological communities (Simberloff and Dayan 1991) and will be used to describe arthropods living in forests.

3.2.1 Phytophagous Arthropods

Phytophagous arthropods can be grouped into guilds according to their feeding mode, the plant part they exploit and whether they feed internally or externally on the plant (Novotny et al. 2010). To illustrate the concept and give an overview of the taxonomic composition of various guilds, the 116 most damaging phytophagous arthropods attacking trees in Quebec, Canada, were classified according to their feeding behavior on different parts of trees (Hébert et al. 2017) (Table 3.1). Arthropods that feed on tree foliage are called phyllophagous and they mainly belong to a few higher orders of insects, which have been able to overcome the defenses of higher plants (Strong et al. 1984). This is one of the largest group of arthropods found on trees and the largest single group damaging trees (Ciesla 2011). Overall, 65% of the most important arthropods attacking trees in Quebec are foliage feeders and 70% of them are larvae from the orders Lepidoptera and Hymenoptera (mainly sawflies). Among Lepidoptera, the Tortricidae and Geometridae are the most important phyllophagous families, with 13 and 7 species out of 40 pest species of trees in Quebec, while among the Hymenoptera, the Diprionidae and Tenthredinidade account for 8 and 7 species. Other phyllophagous taxa belong to several families of Lepidoptera as well as to several families of Coleoptera and Hemiptera, which include leaf-miners, leaf-suckers and gall-makers. Hemiptera often feed on tree sap by inserting their piercing-sucking mouthparts into most tree tissues. They are most often hidden (Table 3.1) under scales or galls caused by a mechanical disruption of vascular tissues or a physiological reaction from the tree to insect saliva (Barbosa and Wagner 1989). Phyllophagous arthropods will be discussed in detail in Chapter 9 and other groups briefly presented here are treated in Chapters 13–16.

Phytophagous arthropods also include those that feed on woody tissues, which are dominated by Coleoptera (Table 3.1). Those feeding on nutrient-rich subcortical tissues (phloem and cambium) are called phloeophagous and most belong to Curculionidae/Scolytinae (see Chapter 10), which are highly host-specific, at least

Table 3.1 Number of phytophagous arthropod taxa of various orders and feeding exposed or hidden on different parts of trees in Quebec, Canada

Order	Leaves or needles		Shoots or twigs		Woody tissues		Total
	Exposed feeder	Hidden feeder	Exposed feeder	Hidden feeder	Exposed feeder	Hidden feeder	
Coleoptera	5[a]	2[a]		3[g]	2[g]	16[b, c, d]	28
Diptera		3[f]					3
Hemiptera	1[e]	5[e, f]	2[e]	5[e, f]		6[e, f]	19
Hymenoptera	13[a]	3[a]				1[d]	17
Lepidoptera	17[a]	23[a]		4[g, h]		1[d]	45
Prostigmata	1[e]	2[f]					3
Thysanoptera	1[a]						1
Total	**38**	**38**	**2**	**12**	**2**	**21**	**116**

[a–h]See Chapters 9–16
Source of data: Hébert et al. (2017)

at the family level (Novotny et al. 2010). Buprestidae and Cerambycidae also bore galleries under bark or at the wood surface. Some Scolytinae and Platypodinae, another subfamily of Curculionidae, belong to a group called ambrosia beetles, which bore into the sapwood and feed on introduced symbiotic fungi (see Chapter 11). Insects that bore into the sapwood and even into the xylem are called xylophagous (see Chapter 12). They have strong mandibles and the most distinctive species belong to Cerambycidae and Siricidae (Hymenoptera), which are much larger than Scolytinae. Among wood boring insects, carpenterworms (Lepidoptera: Cossidae) are exceptions as they belong to an order of predominately phyllophagous insects.

3.2.2 Zoophagous Arthropods

Three types of zoophagous arthropods exist:

1. Predators, adults or larvae (but not necessarily both stages for a species) hunt, attack, kill and feed directly on prey. Predators are generally not host specific and they are larger than their prey or attack them in large numbers (e.g. ants).
2. Parasites, feed on a host without killing it. Parasites are generally smaller than their host and they can live at the expense of both, invertebrates or vertebrates. They can feed externally (often occasionally, such as mosquitoes) or internally on the host. Many parasites have claws or hooks to grasp their host, and often have piercing-sucking mouthparts.
3. Parasitoids, free-living adults locate a host, deposit their eggs on or in it, and larvae feed on and kill the host at the end of their development. Generally, parasitoids are smaller than their host and are selective, attacking specific life stages of one or closely related species.

Zoophagous arthropods from several orders and families feed on phytophagous and saprophagous arthropods that live in different microhabitats (e.g. canopy, trunks, litter, etc.). For instance, ladybird beetles, syrphid flies and lacewings prey on aphids and other insects in tree canopies while many carabid beetles and spiders are voracious predators of invertebrates on the forest floor. Ants can prey on various arthropods in the tree canopy or at the ground level. Most predators use an active hunting strategy but web-spinning spiders use a sit and wait hunting strategy in various vegetation strata (Michalko et al. 2019). In dead and dying trees, predators of phloeophagous and xylophagous insects belong to several beetle families, Cleridae and Monotomidae being the most well-known.

Parasitoids are a diverse group of insects with most species belonging to the Diptera and Hymenoptera. The Hymenoptera have received more attention than the Diptera and they exhibit sophisticated host selection behaviors which involve olfactory responses by adult parasitoids to specific semiochemicals emitted by hosts or by damaged plants (Godfray 1994; Stireman 2002). Host selection is less-well known in Diptera, but some search visually, responding to host movement, while their response to plant odors is generally weak (Stireman 2002). Most families of Hymenoptera use a parasitic mode of life and parasitic Hymenoptera could represent up to 20% of all insect species (Gaston 1991). However, at least 75% of the parasitic Hymenoptera had not yet been described in the early 1990's (Lasalle 1993). Recent estimates suggest that Hymenoptera may have 2.5–3.2 times more species than Coleoptera, and thus, could be the most speciose animal order (Forbes et al. 2018). The Ichneumonidae and Braconidae are probably the most diversified families of parasitoids but many poorly known families of micro-Hymenoptera are also important in regulating arthropod populations. The full spectrum of host specificity can be found in the Ichneumonidae, with species that attack a single host known for species of *Megarhyssa* (Pook et al. 2016) to the highly polyphagous *Itoplectis conquisitor*, which attacks hundreds of Lepidoptera species (Townes and Townes 1960). Natural enemies will be further discussed in Chapter 6.

3.2.3 *Saprophagous Arthropods*

Saprophagous arthropods which feed on rapidly decaying vegetation such as dead leaves are called detritivorous while those feeding on slowly decaying vegetation such as woody debris are called saproxylophagous. With the notable exception of the ambrosia beetles, arthropods that feed on fungi are traditionally included in the saprophagous group as they often feed on a mixture of mycelium and dead leaves or wood. More technically, species feeding on the aerial and visible parts of fungi are called fungivorous while those which feed on non-visible parts of fungi are either mycetophagous if they feed on fungal mycelium in the soil/litter or mycophagous if they feed on molds (Natural Resources Canada 2015). Arthropods feeding on dead animals are called necrophagous with those feeding more specifically on feces being called coprophagous or scatophagous.

3.2.3.1 Soil and Litter Feeders

The soil fauna is usually split among three groups according to their size: microfauna, which include invertebrates of less than 0.2 mm (mainly nematodes) that live in the water present between soil particles, mesofauna (0.2–2 mm) to which belong Enchytraeidae (not arthropods), Collembola and Acari (both arthropods), and macrofauna (>2 mm diameter), which include large oligochaetes (earthworms), most insects and large arthropods such as Diplopoda and Chilopoda (Brussaard et al. 1997; Lavelle 1997). A well-illustrated synthesis on soil organisms and associated food webs is provided by Zanella et al. (2018).

Generally, mesofauna dominates northern coniferous forests while macrofauna dominates temperate deciduous and tropical forests (Shaw et al. 1991; Lal 1988). Densities of 1 million arthropods/m^2 have been reported in black spruce forest soils (Behan et al. 1978), with 200,000 arthropods/m^2 being common in Canadian soils (Marshall et al. 1982). Most Acari living in the soil belong to the suborder Cryptostigmata (formerly called oribatid mites), and can account for up to 90% of estimated biomass in coniferous soils (Shaw et al. 1991). Collembola living in humus are called endogenous while those living in the litter are called epigeous. Endogenous species measure less than 1 mm, have an elongate form, very small appendices and non-pigmented eyes while epigeous species are larger, often of globular form and have well-developed appendices and eyes (Dajoz 1998). Collembola and Acari are wingless but mobility is not a major issue for species feeding on predictable and abundant resources. Dipterous larvae are also abundant and diverse in forest soils, the most prevalent families being Sciaridae, Cecidomyidae, Phoridae and Mycetophilidae (Hibbert 2010). Earthworms (Oligochaetes) account for the highest biomass among groups forming the macrofauna and are dominant in Mull humus of temperate deciduous forests with 5,300 mg/m^2, their biomass falling to 200 mg/m^2 in Mor humus (Shaw et al. 1991). In the latter forests, Diplopoda and Chilopoda are prevalent (Shaw et al. 1991), while in tropical forests, termites and ants play important roles, where they are dominant in arid and semi-arid regions while earthworms are mainly important in humid and subhumid regions (Lal 1988).

3.2.3.2 Dead Wood Feeders

Dead wood is the habitat of numerous saproxylic species, which are defined as *"species that are dependent, during some part of their life cycle, upon the dead or dying wood of moribund or dead trees (standing or fallen), wood-inhabiting fungi, or the presence of other saproxylic organisms"* (Speight 1989). There is overlap between arthropods feeding on woody tissues (Sect. 3.2.1) and dead wood, particularly among phloeophagous species (Stokland 2012). Most bark beetles (Scolytinae) and many longhorn beetles (Cerambycidae) feed on phloem of moribund trees, which are technically still alive. These beetles are early colonizers of dead wood and the resource remains suitable for them until the phloem dries up and the bark gradually comes off the wood. Xylophagous species include insects of several orders: Coleoptera (mainly Cerambycidae), Hymenoptera (Siricidae), Lepidoptera (Cossidae, Hepialidae, Sesiidae) and Diptera (Tipulidae and Chironomidae). Many of these species mainly feed on fungal mycelium involved in wood decay. Species of several families of Coleoptera (e.g. Ciidae, Anobiidae, Tenebrionidae, Tetratomidae) also feed and reproduce in bracket fungi, which develop on dead trees, and usually with much higher levels of host specificity than those feeding on mushrooms (Jonsell and Nordlander 2004). Numerous species of various orders also live in tree hollows (Ferro 2018), which highlights the diversity of microhabitats associated with dead wood.

3.2.3.3 Dung and Carrion Feeders

Animals return organic matter to the ecosystem throughout their lives by the dung or feces they produce and also when they die through their carcasses. Small detritivorous arthropods (e.g. collembola and acari) feed on dead organic matter, which is often mixed with soil, fungi and bacteria, particularly in advanced stages of decomposition. Woodlice (Crustacea: Isopods), feed preferentially on feces produced by *Operophthera fagata* caterpillars, a Geometrid that feeds on beech (*Fagus sylvatica*), rather than on the beech litter itself (Zimmer and Topp 2002).

A specialised fauna composed of larger arthropods develop in vertebrate dung and carrion, with Scarabaeinae, a subfamily of Scarabaeidae, being the most prevalent group of coprophagous beetles. They are commonly called dung beetles and are widely distributed, although they are most diverse in tropical forests where their burying behavior has been widely studied (Braack 1987). Dung beetle larvae feed on the microorganism-rich liquid component of dung, mainly of mammals but also from other vertebrates or from rotting fruits, fungus and carrion (Nichols et al. 2008).

Another type of organic matter provided by vertebrate animals is carrion. Blowflies (Diptera: Calliphoridae) are usually the first to colonize new carcasses (Paula et al. 2016) but over 20 families of flies feed in vertebrate carcasses (Payne 1965). Flies have good flight ability, and have developed efficient host selection behavior primarily based on olfactory and visual stimuli associated with carcasses. Conversely, ants and beetles are typically generalists that exploit carrion opportunistically, have a more limited dispersal capacity and often use habitat features for orientation (Barton and Evans 2017).

The most common beetles feeding on carrion belong to the family Silphidae. In addition to feeding on carrion they also prey on other species exploiting carrion. There are two subfamilies of Silphidae with different biologies. More is known about the Nicrophorinae, or burying beetles, than about the Silphinae because of their unusual behavior. One of the most striking behaviors of burying beetles is their reproductive cooperation and the extended adult biparental care of their progeny (Scott 1998), which is not observed in Silphinae. Nicrophorinae breed and feed in small carcasses such as mice and birds (<300 g), while Silphinae breed and feed in large carcasses where they compete with blow flies (Dekeirsschieter et al. 2011). Adult Nicrophorinae use olfactory stimuli to locate carrion (Scott 1998). Vertebrate carcasses are rare and unpredictable spatially and temporally. Progeny care in burying beetles may be an adaptation to maximize fitness in these habitats (Scott and Gladstein 1993). When a carcass is located, males and females work together to move it to a suitable environment and dig beneath it to bury the carcass and prepare it as food for their progeny (Scott 1998). Because of the low number of available carcasses, several adult pairs may converge on fresh carcasses. If a carcass is large enough to support reproduction by several adult pairs, burying beetles work cooperatively to bury the carcass. If the carcass is too small, intraspecific fights occur and only the winners will reproduce. The burial chamber varies from a simple depression under leaf litter up to 60 cm underground (Scott 1998). Burying the carcass protects it from fly colonisation (Suzuki 2000) and reduces detection by other competitors (Shubeck 1985; Trumbo 1994). Beetles remove feathers or hair, shape the carcass as a ball and take care of it through regular cleaning and depositing anal and oral anti-microbial secretions, which suppress fungal and bacterial growth (Suzuki 2001) and reduce rates of decomposition (Hoback et al. 2004). Eggs are laid nearby and the newly hatched larvae require parental care for feeding (Scott 1998). About 75 species belong to the genus *Nicrophorus*, which is only present in the northern hemisphere (Scott 1998).

3.3   Functional Roles and Ecosystem Services

Arthropods are involved in nearly all ecological processes that drive ecosystem functioning (Jones et al. 1994). However, they represent less than 0.2% of the total biomass on the planet, dwarfed by plants, microbes and fungi (Bar-On et al. 2018). Uncertainty exists whether they are important drivers of ecological processes or whether they play only minor roles (Schmitz et al. 2014; Yang and Gratton 2014). The functional importance of arthropods in ecological processes has primarily been assumed and not based on experimental work quantifying the value of these functions. The few studies that do exist have primarily been conducted in agroecosystems (Noriega et al. 2018).

Arthropods are primary (herbivores) and secondary (carnivores) consumers in the food chain and thus, they depend on the production of primary producers, mainly trees in forest ecosystems. Thus, biomass transformation of living and dead plants and animals appears to be the most important functional role of arthropods in forest ecosystems (Yang and Gratton 2014) and as a result, they are involved in nutrient cycling and energy fluxes. These important ecosystem services are critical to ensure forest productivity but they are often overlooked. Apart from this central role in ecosystem functioning, arthropods are also involved in promoting plant reproduction through pollination and seed dispersal. Combined with insects that kill trees over wide areas, which strongly modify environmental conditions, these phytophagous insects influence forest succession. Secondary consumers (predators and parasitoids) account for a large part of forest arthropod diversity (Strong et al. 1984) and they are instrumental in regulating food webs.

Noriega et al. (2018) defined ecosystem services as *"the beneficial functions and goods that humans obtain from ecosystems, that support directly or indirectly their quality of life"*. Arthropods provide ecosystem services in all categories recognized by the Common International Classification of Ecosystem Services (CICES), i.e. regulation and maintenance (pollination, biological control, recycling organic matter), provisioning and cultural services (Ameixa et al. 2018). Pollination, biological control, recycling organic matter, and food provisioning have been the most studied ecosystem services but arthropods also provide cultural services, whether they be religious, artistic or recreational (Noriega et al. 2018). The latter includes hunting, fishing and wildlife observation, activities in which insects, as food sources, are estimated to account annually for $2.7 billion in USA alone (Losey and Vaughan 2006). Insects are also used in arts and crafts, as cultural icons or religious symbols, and are often associated with tourist destinations (e.g. the Monarch Butterfly Reserves in Mexico) (Schowalter et al. 2018).

Although arthropods play key roles in the regulation and maintenance of several ecosystem services, these roles are usually assumed and their value has rarely been quantified experimentally (Noriega et al. 2018). Losey and Vaughan (2006) were the first to estimate the economic value of ecosystem services provided by insects to be at least $57 billion annually in the United States and this only considers four ecosystem services provided by "wild" and native insects, for which data were available: pollination, pest control, wildlife nutrition and dung burial.

3.3.1 Regulating Primary Production

Phyllophagous arthropods feed on highly nutritious tissues, which are the basis of tree photosynthesis, namely leaves (Vergutz et al. 2012) and needles (Moreau et al. 2003). A low rate of herbivory stimulates primary production in natural forests while a high rate suppresses it (Mattson and Addy 1975). Defoliation of mature trees increases sunlight penetration to understory trees and saplings, which typically increase their growth as competition for light from overstory trees decreases (Mattson and Addy 1975). Moderate defoliation ($\leq 50\%$) from *Orgyia pseudotsugata* stimulates Douglas-fir growth (Alfaro and Shepherd 1991), compensating for losses of severely defoliated trees. Herbivory appears to reduce variation in primary production and helps maintain it at intermediate levels (Schowalter 2012). Indeed, phytophagous insects have been presented as "regulators" of forest primary production (Mattson and Addy 1975; Belovsky and Slade 2000; Schowalter 2012), but this should be considered over long time intervals (see Sect. 3.4.2).

3.3.2 Decomposition and Nutrient Cycling

3.3.2.1 Insect Feces and Cadavers

Phyllophagous insects contribute to the cycling of rich organic matter produced by trees. Their feces provide high quality but ephemeral nitrogen pulses to soils, which are rapidly recycled by soil biota and assimilated into the foliage, often within the same season (Belovsky and Slade 2000; Frost and Hunter 2007). Zimmer and Topp (2002) recognized a "fast nutrient cycle" for feces of phytophagous animals (sensu McNaughton et al. (1988) who studied vertebrate herbivory in African grasslands) and a "slow cycle" for leaf litter and wood decomposition (plant material). Indeed, microbial degradation of *Operophthera fagata* feces took approximately half the time of beech leaf litter in microcosms (Zimmer and Topp 2002). Moreover, the addition of woodlice (isopods) tripled the rate of mass loss for both feces and litter. In fact, meso and macrofauna often reingest their faecal pellets a few days after deposition (Hassall and Rushton 1982). They then absorb organic compounds that have been released by microbial activity (Lavelle 1997). This is considered as a type of mutualism and referred to as external rumen digestion (Swift et al. 1979). Internal rumen digestion also exists in earthworms, termites and, to a lesser extent ants, as they interact internally with micro-organisms to produce various organo-mineral structures (Lavelle 1997).

Phyllophagous insects also return nutrients to the detritus pool when they die (Gessner et al. 2010). For instance, during outbreaks, insect cadavers are a major pulse of resources for detritivorous communities. However, models of ecological processes rarely consider this resource. Indeed, many predators are in fact omnivores and predation rates are often inflated in food-web research while scavenging is largely underestimated (Wilson and Wolkovitch 2011). For instance, ants are active scavengers of entomopathogenic nematode-killed insects (Baur et al. 1998) and it has been estimated that they account for 52% bait removal in tropical rain forests (Griffiths et al. 2018). This is particularly important as no other scavenger group compensated when ants were excluded, indicating a low functional redundancy of this important ecological role. Ants are estimated to make up 25% of animal biomass in tropical forests (Hölldobler and Wilson 1990) and are recognised as ecosystem engineers (Folgarait 1998). In North American temperate forests, the 17-year periodical emergence of cicadas (*Magicicada* spp.) provides a massive addition of insect cadavers and this increases bacterial and fungal abundance by 12 and 28% respectively (Yang 2004). The herbaceous plant *Campanulastrum americanum* then produce 9% larger seeds, highlighting the reciprocal links between above and belowground components of the ecosystem. Cicadas have patchy distributions and these resource pulses generate spatial and temporal heterogeneity in ecosystems (Yang 2004).

3.3.2.2   Leaf Litter

Less than 10% of the foliage produced by trees is consumed by phytophagous arthropods, over 90% entering the detritus pool as leaf litter (Gessner et al. 2010). Standardized litter types used in 336 sites across 9 biomes, showed that litter quality explained 65% of the variability in the early stages of decomposition, climate only having a significant effect when data were aggregated at the biome scale (Djukic et al. 2018). A meta-analysis also carried out at the biome scale showed that abiotic conditions controlled decomposition in cold and dry (harsh conditions) biomes while soil fauna had an important role in warm and wet (mild conditions) biomes (García Palacios et al. 2013). In tropical ecosystems, climate is likely less important than soil macrofauna (González and Seastedt 2001).

Soils are usually classified according to their physical and chemical properties, but they are regulated by complex interactions among the soil biota (Brussaard et al. 1997; Barrios 2007; Schmitz et al. 2014). Decomposition of dead organic matter involves the physical fragmentation of dead organic matter through feeding by arthropods and other invertebrates. This increases the surface/volume ratio of dead organic matter, which enhances fungal and microbial activity, releasing nutrients and making them readily available to plants (Barrios 2007). The decomposition of dead organic matter by living organisms and the progressive incorporation of released nutrients into the pool available in soils is at the basis of forest primary production and thus central to forest ecosystem functioning (Swift et al. 1979). Nevertheless, dead organic matter is rarely considered in ecological models and when included, it is usually treated as a single resource (detritus) that does not vary. However, nitrogen content varies widely among different types of dead organic matter (Fig. 3.3) and decay rates increase with nitrogen content. For instance, dead wood takes decades to decompose (Harmon et al. 1986) compared to months or years for leaf litter and days or weeks for animal dung and carrion (Wilson and Wolkovitch 2011).

3.3.2.3   Dead Wood

Severe and repeated defoliation by phyllophagous insects often results in tree death over wide areas. Before dying, trees progressively weaken and become vulnerable to wood feeding insects called secondary insects as they usually colonize trees physiologically stressed by another agent. This has been observed during and after spruce budworm (Belyea 1952a, b; Régnier 2020) and hemlock looper outbreaks (Béland et al. 2019). Stressed trees may emit volatiles that are attractive to secondary insects (Faiola and Taipale 2020) and thus, the functional role of secondary insects in forest ecosystems is to accelerate death of weakened trees and initiate the process of wood decomposition. Secondary insects generate openings in forest stands and thus increase ecosystem heterogeneity and promote plant succession. As herbaceous plants, shrubs and tree seedlings compete for light and nutrients, it results in complex successional dynamics that characterize different forest types. In natural

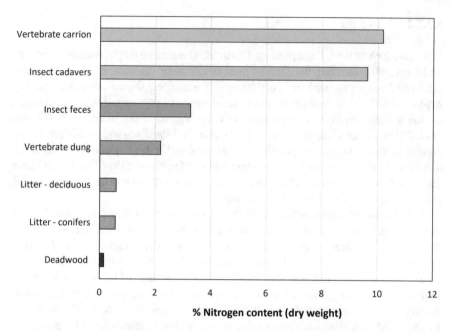

Fig. 3.3 Nitrogen content (% of dry weight) of various types of dead organic matter, both from animal and plant origin. Data from Parmenter and MacMahon (2009)—vertebrate carrion; Rafes (1971)—insect cadavers and insect feces; Holter and Scholtz (2007)—vertebrate dung; Taylor et al. (1989)—litter, deciduous and conifers; Piaszczyk et al. (2020)—deadwood

forest ecosystems, such dynamics also ensure continuity in dead wood stocks, which is important for maintaining diversity of saproxylic arthropods (Grove 2002).

The greatest amount of forest biomass is stored in woody tissues (Dajoz 1998) and thus wood decomposition after tree death is an important ecological process in forests (Harmon et al. 1986). Bark is a major physical barrier to the establishment of fungi, among which basidiomycetes are instrumental for decomposing the various structural components of wood (Strid et al. 2014). The first insects to colonize dying or recently dead trees are phloem feeders (Ulyshen 2016), and they bore holes through the bark to breed and feed on the nutritious phloem beneath the bark. Many woodboring insects transport fungi beneath the bark and their boring also provides access for fungi. Obligate insect-fungus mutualism increases the probability that fungi reach a suitable substrate (Birkemoe et al. 2018). Insects that have developed obligate-mutualisms with fungi, such as ambrosia beetles, are known to farm fungi within their galleries. These fungi possess wood-degrading enzymes which make essential nutrients from the wood available for insects. Similarly, the symbiotic fungi of wood wasps (Siricidae) serve as an "external rumen" for insects (Birkemoe et al. 2018). They produce enzymes that digest lignocellulosic compounds in the wood, which are then ingested by growing larvae (Thompson et al. 2014; Kukor and Martin 1983). According to Filipiak and Weiner (2014), wood-feeding insects are in fact fungivorous species or at least xylomycetophagous as their wood diet is supplemented with fungi found in decaying wood. Without the essential nutritional elements provided by fungi, they estimated that the cerambycid *Stictoleptura rubra* would need between 40 (males) to 85 (females) years to develop into an adult.

Bark beetles also defecate under the bark, thus providing rich organic matter which contributes to fungal and microbial growth (Birkemoe et al. 2018). By feeding on the protein-rich subcortical tissues at the phloem/cambium interface and inoculating fungi, early colonizing bark beetles accelerate bark loss of dead trees (Ulyshen 2016), which is a type of insect-mediated ecosystem engineering (Birkemoe et al. 2018). Tunneling by wood-boring insects provides access into the xylem for fungi and improves aeration, which increases rates of decomposition (Dighton 2003). In temperate deciduous forests, bacterial and fungal densities increase with decay stages and reach their maximum during the "invertebrate channelization" stage. This stage occurs when logs are colonized by termites, carpenter ants and Passalid beetles which, as a community, can regulate the process of wood decomposition (Ausmus 1977).

A recent experimental study on the contribution of insects to forest deadwood decomposition, carried out in 55 sites on six continents, estimated that insects account for 29% of the carbon flux from deadwood, highlighting their functional importance in the process of wood decomposition (Seibold et al. 2021). Direct and indirect effects of insects accelerate decomposition in tropical forests but have weak positive or negative effects in temperate and boreal forests (Seibold et al. 2021). Termites and fungi are the most important determinant of wood decay in tropical regions while in temperate and boreal forests it appears to be moisture (González et al. 2008).

3.3.2.4   Vertebrate Dung and Carrion

By dispersing and incorporating vertebrate dung into the soil, dung beetles are involved in nutrient cycling, soil aeration, seed burial and parasite suppression. Several experimental studies have linked dung beetle effects on soil structure and nutrient content to increases in plant height and above-ground biomass (Nichols et al. 2008). Their activity increases soil porosity and soil water retention, which alleviates water stress on plants, even during a severe drought (Johnson et al. 2016). The effects of dung beetles on nutrient availability and ultimately plant growth may rival chemical fertilizers in agriculture. Further research is thus needed, particularly in tropical forests, where dung beetles can transfer mammal feces into the soil within a few hours (Slade et al. 2007).

Vertebrate carcasses do not provide major pulses but a rather low and steady supply of resources as it represents less than 1% of the overall nutrient budget of ecosystems (Hoback et al. 2020). However, locally, they significantly improve soil conditions. Carrion has a higher nutritional value than dung as the latter is composed of metabolic waste products and undigested remains of the original food (Frank et al. 2017). Vertebrate carrion decomposes faster than plant material as carrion N content (6–12%) is much higher than for plant litter (typically 1–2%) (Parmenter and MacMahon 2009). In tropical regions, blow flies can eat all soft-tissues of a carcass within four days during warm weather (Braack 1987). The decomposition of a carcass produces an island of soil fertility, which increases stand heterogeneity. Soil nitrogen increases significantly under carcasses and large ones modify soil temperature, moisture and physical structure. Roots of neighboring plants that reach the modified soil area are influenced by these new micro-environmental conditions, which produce a "halo" effect (Parmenter and MacMahon 2009). Nutrients can be dispersed by insects, mainly ants and burying beetles, while bacteria and fungi may only increase nutrients in the soil under the carcass (Barton et al. 2013).

3.3.3 Seed Dispersal

Myrmecochory, or ant-mediated seed dispersal, is a widespread mutualistic interaction between ants and plants (Wenny 2001; Ness and Bressmer 2005). Seeds of myrmecochorous plants have lipid-rich appendages called elaiosomes, which are highly nutritious and attractive to ants (Ness and Bressmer 2005). Ant workers harvest seeds of these plants and bring them back to their nests. Unlike vertebrate frugivores which eat fruit pulp before dispersing seeds randomly, often far from the parent tree, ants typically disperse seeds over shorter distances but in more predictable and rich habitats, i.e. their nests (Wilson and Traverset 2000). The rich elaiosomes are then provided to the developing progeny and seeds are simply abandoned in the nest or discarded in middens outside the nest (Wenny 2001). This produces rich micro-environments where nutrient concentration is higher than in the surrounding soil, often resulting in higher rates of seed germination and seedling growth (Wenny 2001). It has also been suggested that ants could be responsible for seed arrival in rich and humid substrates favorable to seed germination and seedling growth, such as pits and rotting logs (Wenny 2001). By harvesting seeds, ants make them unavailable to vertebrates, lower the density of seeds beneath trees and, ultimately, increase seed germination rates and reduce competition among seedlings. Seed dispersal by ants is an important mechanism for increasing tree reproduction, particularly in tropical and temperate forests (Wilson and Traverset 2000).

3.3.4 Pollination

It has been estimated that 87.5% of angiosperms are pollinated by animals, ranging from 78% in temperate-zone communities to 94% in tropical ones (Ollerton et al. 2011). Most plants in tropical forests are pollinated by insects, with bees being the most important group of pollinators (Bawa 1990). Medium to large-sized bees are important in the forest canopy while small bees are prevalent in the subcanopy and understory (Bawa et al. 1985; Bawa 1990). Moths are the second most important pollinators in tropical forests, with sphinx moths being particularly active in the subcanopy (Bawa et al. 1985). Surprisingly, little is known about fly pollination in tropical forests (Bawa 1990). Bees dominate in tropical forests but flies outnumber bees in both diversity and abundance as pollinators in cold regions (IPBES 2016). Although Diptera are known as the second most important order of insect pollinators, their role in pollination has been unappreciated (Larson et al. 2001; Orford et al. 2015). In recent years, concern has been expressed about the conservation of wild pollinators in North American forests and literature reviews have revealed significant knowledge gaps on forest pollinators (Hanula et al. 2016; Rivers et al. 2018).

3.3.5 Top-Down Regulation of Phytophagous Arthropods

Phytophagous arthropods experience strong selective pressures from the trees on which they feed (bottom-up pressure) and from organisms that feed on them (top-down pressure), including numerous invertebrate predators and insect parasitoids. A meta-analysis of the population ecology of phytophagous arthropods suggests that top-down forces have stronger effects than bottom-up forces, for chewing, sucking or gall-making arthropods (Vidal and Murphy 2018). Natural enemy communities can be complex and often overlap among arthropods. For example, the spruce budworm, *Choristoneura fumiferana* (Clemens) (Lepidoptera: Tortricidae) is an important pest of conifers in North America and part of a complex food web in which most parasitoids have at least two generations per year and need alternate hosts to complete their life cycle (Eveleigh et al. 2007). Requiring an alternate host limits the regulating potential of a parasitoid (Maltais et al. 1989), but parasitism by *M. trachynotus* was reported to increase up to 50% near the end of most outbreaks (McGugan and Blais 1959; Blais 1960). This may result from a slower development of the spruce budworm near the end of outbreak (Wilson 1973), which widens the window of availability of budworm larvae to parasitoids (Hébert 1989). Budworm larvae develop slower when they are affected by sublethal doses of a microsporidian (Bauer and Nordin 1989) or when they feed on needles with higher fiber content (Bauce and Hardy 1988), both of which become more common as outbreaks progress.

Egg parasitoids can be efficient natural enemies of phyllophagous insects but their importance for regulating pests has been overlooked in the past because of our poor knowledge of their biology and systematics (Anderson 1976). Some of the most efficient egg parasitoids of forest defoliating Lepidoptera belong to the genus *Telenomus* (Hymenoptera: Platygastridae) (Anderson 1976; Bin and Johnson 1982; Hirose 1986; Orr 1988), which have contributed to the collapse of outbreaks of several lepidopteran pests (Hébert et al. 2001). These ecosystem services were overlooked for decades for the hemlock looper, *Lambdina fiscellaria* (Guenée) (Lepidoptera: Geometridae). A systematic study showed that previous identifications (*T. dalmani*) were incorrect and that three species were attacking the looper, one of these being new to science (Pelletier and Piché 2003). Moreover, most attacks were recorded in spring (50–100% parasitism), rather than fall ($\leq$3%) (Hébert et al. 2001). However, egg parasitism was estimated from fall eggs, when overwintering hemlock looper populations were sampled as part of control programs (Otvos and Bryant 1972; Otvos

1973; Hartling et al. 1991). Fall estimates only provided a partial estimate of egg parasitism.

Arthropod predators are also involved in the natural regulation of phytophagous arthropods, but they have been much less studied than parasitoids. However, Holling (1961) provided an excellent conceptual framework for the response of predators to prey species. The type II functional response, in which predators respond strongly to increasing prey density to a saturation level, is most common in predatory insects and parasitoids. The regulatory potential of predators then depends on the searching capacity and the attack rate but also involves handling and ingestion times. The behavior of a predator is thus important when evaluating its potential as a natural control agent. For instance, the carabid *Calosoma frigidum* Kirby kills more larvae of the spongy moth, *Lymantria dispar* L. (Lepidoptera: Erebidae) that it eats (Hébert 1983). Similarly, in Europe, an adult *Calosoma sycophanta* can annually kill up to 280 pine processionary moth larvae (Kanat and Mol 2008) or 336 larvae or pupae of the spongy moth (Dajoz 1998). This killing/feeding behavior is often observed in carabids when prey populations are abundant, and with their mobility, this make them efficient predators of pest insects (Allen 1973).

Ants are omnivores but in some instances they were shown to be important primary predators of insect pests. They contributed up to 80% predation of prepupae of the spruce budmoth, *Zeiraphera canadensis* Mutuura and Freeman (Lepidoptera: Tortricidae), in young white spruce plantations in Quebec (Pilon 1965; Hébert 1990). They were opportunists as prepupae fall to the ground during about one hour each day in late afternoon for about one week. Once on the ground, prepupae rapidly wander through the litter to find a hidden location for pupation, 50% being no longer visible after 90s (Hébert 1990). Ants are social insects that use pheromone trails to rapidly locate food sources. This behavior explains their success in taking advantage of suddenly available resources.

Spiders are probably the most abundant and diverse group of generalist predators in terrestrial ecosystems and there is growing evidence that their communities play key roles in limiting arthropod populations (Riechert and Lockley 1984; Michalko et al. 2019). They consume up to 800 million metric tons of prey annually and they are also prey for other animals, attesting to their important functional role in food webs and ecological processes (Oxbrough and Ziesche 2013; Nyffeler and Birkhofer 2017). Spiders limit population growth of soil invertebrates and stabilize their populations (Clarke and Grant 1968) and through complex interactions with microarthropods, litter and fungi, they can slow down or speed up litter decomposition by preventing overgrazing of fungal populations (Lawrence and Wise 2004).

3.3.6 Food Provisioning and Medicines

Over 50% of bird food requirements are fulfilled by insects (Ollerton et al. 2011), but food provisioning to humans is another ecosystem service provided by forest insects for which interest is rapidly increasing. Historically, most insects consumed by humans were harvested from trees or wood (Schabel 2010), attesting for the importance of forest conservation. Insects are 5 times more efficient than beef cattle at converting vegetation into tissues that can be consumed by others and as result, they could help reduce the human environmental footprint (Durst and Shono 2010). In addition to the nutritional value of insects, entomophagy could secure food supply for rural populations, reduce poverty and generate income (Schabel 2010). Entomophagy may need reduced pesticide use, reduced logging and thus favor biodiversity conservation of natural forests (DeFoliart 2005). For example, in the 1980's, the Native Paiute community succeeded in stopping US governmental agencies from spraying insecticides against Pandora moth caterpillars (*Coloradia pandora*), a Saturniid defoliator of pines, which is also a traditional food for this community (DeFoliart 1991). Edible Saturniid caterpillars are also of great value to indigenous cultures in Zambian forests, where the activity of harvesting caterpillars is ritually regulated (Mbata et al. 2002), attesting for the importance of this provisioning ecosystem service. In Thailand, forest insects are a preferred food source of local people, not just a cheap, nutritious and environmentally-responsible food source (Durst and Shono 2010).

Arthropods also have medicinal properties (Meyer-Rochow 2017). Recently, some novel antimicrobial anionic cecropins were found in the spruce budworm and could provide templates for the development of new anticancer drugs (Maaroufi et al. 2021). It has been suggested that systematic screening of forest insects would undoubtedly yield more species for entomophagy and medicine similar to bioprospecting in fungi and plants which has resulted in the identification of numerous new medicinal compounds.

3.4 Effects of Natural Disturbances on Forest Arthropods

Natural forests are dynamic ecosystems that always change as a result of tree growth and death and arthropods respond rapidly to these changes. Tree death is probably the most important mechanism for maintaining biodiversity in old forests as it produces gaps which increase light penetration to the forest floor and initiates succession (Watkins et al. 2017). Gap dynamics have been documented for both tropical and boreal forests and in both cases over 65% of the gaps were smaller than 100 m^2 (Brokaw 1982; Pham et al. 2004). Vegetation gradually recovers in these gaps and because gaps of varying sizes are added each year, they generate high levels of heterogeneity, especially in old-growth stands in which dead tree recruitment is continuous. In old-growth boreal forest, the richness of ground-dwelling beetles is best predicted by the composition component (i.e. number of tree species) of heterogeneity at the stand scale while richness of flying beetles is rather linked to the combined influence of structural (i.e. number of tree diameter classes) and compositional heterogeneity at both the stand and landscape scales (Janssen et al. 2009).

Forest ecosystems are also driven by stand-replacing natural disturbances which kill trees over large areas. They are caused by abiotic or biotic factors, which alter

Table 3.2 Comparison of conditions generated by abiotic and biotic disturbances for arthropods

	Abiotic disturbances	Biotic disturbances
Spatial and temporal predictability	• Can be forecasted few hours or days before the event	• Can be forecasted several weeks or months before with efficient monitoring
Time duration	• Last few hours or days; kill most trees in a short period of time	• Last months or years; kill trees progressively over a long period of time
Selectivity	• Affect all tree species to varying degrees	• Affect only host tree species
	• Kill all types of trees, including healthy ones	• Kill weak trees first, healthy ones dying only later
Soil disturbance	• Physically disturb soils	• Do not physically disturb soils

environmental conditions and forest attributes in specific ways (Table 3.2) to which arthropod communities respond differently.

3.4.1 Abiotic Disturbances

Forest fires are probably the stand-replacing natural disturbance that has been most studied by forest ecologists and entomologists and contrary to popular belief, burned forests are not biodiversity deserts. For instance, the number of beetle species caught in recently burned boreal forests is more than twice that in unburned forests (Saint-Germain et al. 2004; Johansson et al. 2011). Certain insects have developed the ability to exploit recently burned trees, a resource that becomes available in large amounts after wildfire. Insects may be attracted to burned trees using cues coming from them (e.g. smoke). For example, the buprestid *Melanophila acuminata* uses paired pit sensory organs located on its mesothorax to detect infrared radiation (Evans 1964). This beetle may use these organs to locate burns from as far away as 5 km (Evans 1966). Moreover, this buprestid has antennal receptors that respond to methoxylated phenols released during the incomplete combustion of lignin (Schütz et al. 1999). Pyrophilic habits have been reported in several other insect orders: Hemiptera with *Aradus* (Aradidae) flat bugs (Wikars 1997a), Diptera with *Microsania* (Platypezida) smoke flies (Komarek 1969) and Lepidoptera with the Black Army Cutworm, *Actebia fennica* (Noctuidae) (Everaerts et al. 2000). Thus, several insects can take advantage of recently burned forests, making them unique habitats for specialised insect communities.

Not all beetles found in recently burned forest are "burned forest specialists" (Muona and Rutanen 1994) or "pyrophilic" species (Wikars 1997a, 2002; Saint-Germain et al. 2004). Some species found in recent burns are simply opportunists that take advantage of an abundant resource. For example, *Monochamus scutellatus*, which is abundant in burned trees, is also commonly found in trees stressed by insect outbreaks (Régnier 2020), windthrows (Murillas Gómez 2013) and after logging (Bloin 2021). However, other species are closely associated with burned forests and are rarely found elsewhere. For instance, the small predator *Sphaeriestes virescens* LeConte (Coleoptera: Salpingidae) peaks one year postfire and then declines in abundance as time elapses (Jeffrey 2013). The Cerambycidae *Gnathacmaeops pratensis* (Laicharting), which is red-listed in Europe (as *Acmaeops pratensis*; Moretti et al. 2010) and rarely found in unburned boreal forests of eastern Canada, is also closely associated with wildfire. Unlike *S. virescens*, the strength of the relationship between *G. pratensis* and burned forests increases as time elapses (Boucher et al. 2012). Such species might become an interesting indicator of sustainable management in burned boreal forests (Boucher et al. 2016). The current hypothesis used to explain these pulses in insect populations is that forest fires generate optimal conditions for species associated with burns, and that these population increases could be important to maintain low insect populations in unburned forests until the next fire event. This suggests that species associated with burned forests have strong dispersal capacity as wildfires are stochastic unpredictable events (Wikars 1997b).

The bark provides efficient insulation against heat and phloem tissue often remains nutritious for many insects, particularly for trees with thick bark, or when burn severity is low to moderate (Cadorette-Breton et al. 2016). Indeed, burn severity is a determinant variable for predicting successful beetle colonization (Azeria et al. 2012; Boucher et al. 2012, 2016, 2020; Boulanger et al. 2010, 2013). Colonization by large numbers of phloeophagous and xylophagous insects is the first step in insect succession after a wildfire and it promotes secondary succession and wood decomposition (Boulanger et al. 2011). In the boreal forest, post-fire ant colonization of burned woody debris is positively related with woodborer boring activity and it influences decomposition as indicated by lower C:N ratios compared to uncolonized woody debris (Boucher et al. 2015).

Windthrow, another important abiotic disturbance, is less prevalent than fire in boreal forests but it is the most important driver of European temperate forest dynamics (Wermelinger et al. 2017). Climate change will likely favour more frequent and severe windstorms and as a result, windthrows will increase the amount of dead wood in forest landscapes. In addition to making dead wood available for arthropods, windthrows generate gaps which stimulate vegetation growth and promote the growth of flowering herbaceous plants that many saproxylic arthropods feed on to mature their eggs. Species assemblages differ between gaps and non-gap areas (Bouget and Duelli 2004), and twice as many species were found in windthrows than in undisturbed forests (Wermelinger et al. 2017). In Switzerland, during the first 10 years after a windthrow event, longhorn and buprestid beetles were 30–500 times more abundant and species richness was 2–4 times higher than in non-affected portions of the forest (Wermelinger et al. 2002). Overall arthropod richness increased by 17% and original species composition did not show any sign of recovery 10 years after the storm event (Duelli et al. 2002), indicating that windthrows initiate new successions that may have long lasting effects on biodiversity. Sun-exposed snags and large woody debris observed in windthrow gaps are rarely found in managed stands. In Sweden, where forests are managed intensively and dead wood has rarefied, 59%

of the 542 red-listed saproxylic invertebrates prefer sun-exposed sites (Jonsell et al. 1998). Windthrows also provide important habitat for wildlife including nesting sites for Megachiliid bees (Warren and Key 1991) and shelter for many overwintering invertebrates (Alexander 1995). As windthrow provides abundant resources to saproxylics, it facilitates population growth and rare species can become more apparent (Wermelinger et al. 2002). Extensive windthrows have positive effects on the abundance of 20% of Swedish red-listed beetles and negative effects on only 4% (Berg et al. 1994).

Trees may be weakened by several agents among which drought is one of the most widely known. In many parts of the world, drought has become more frequent in recent decades, as a consequence of ongoing climate change (Moore and Allard 2011). By reducing root water uptake, drought induces stress for trees, mainly those which have shallow rooting systems. As a result, tree seedlings and saplings are much more vulnerable to drought than mature trees which have deeper rooting systems. Forest stands growing on shallow soils are also more susceptible to water deficits (Moore and Allard 2011). However, these general patterns may vary according to tree species. For example, after an extreme 4-yr drought in California, native bark beetles were instrumental in killing trees but important differences were noticed between tree species. Bark beetles killed mature pines regardless of their level of decline while the most affected firs were killed regardless of their age (Stephenson et al. 2018). Other extreme weather events linked with climate change will likely stress and weaken trees, making them more susceptible to secondary insects, including flooding and excessive rainfall that saturates poorly drained soils, leading roots to suffocate. Physical damage to roots or disturbances that interfere with water and nutrient uptake may result in tree dieback, thus increasing susceptibility and vulnerability to secondary insects.

3.4.2 Biotic Disturbances

Insect outbreaks are the most common biotic disturbance but their ecological impacts have received much less attention than their control. Like other types of disturbances, by killing trees, insect outbreaks influence forest structure and composition. Their impacts vary according to their severity, which in turn vary with forest composition (De Grandpré et al. 2018). These reciprocal interactions between forests and insect pests result, at the landscape scale, in forest mosaics with variable levels of heterogeneity. This is true in the boreal forest with the spruce budworm and the Mountain Pine Beetle, *Dendroctonus ponderosa* Hopkins (Coleoptera: Curculionidae) but also in temperate and Mediterranean forests where outbreaks from other species also occur but over smaller areas because of greater fragmentation of forest matrices. Tropical forests, previously thought to be free of outbreaks (Elton 1958), are also affected by insect outbreaks, but they are likely less frequent and extensive, as high tree diversity reduces risk (Dyer et al. 2012).

Few studies have documented the effects of biotic disturbances on insect communities. Those who did used flight interception traps 3 to 15 years after the outbreak.

Only weak responses have been reported from saproxylic beetles (Barnouin 2005; Vindstad et al. 2014). The impact of the two most damaging insect defoliators in Canada, the spruce budworm and the hemlock looper, differ. Both insects affect balsam fir but they produce different temporal patterns of tree mortality (Fig. 3.4). The spruce budworm feeds mainly on current-year foliage and tree mortality begins only after 4–5 years of heavy defoliation, following a progressive weakening of trees (MacLean 1980). Once tree mortality begins in a stand, it continues for up to a decade and even more (Taylor and MacLean 2009). On the other hand, hemlock looper larvae feed on needles of all age classes without eating them completely (Hébert and Jobin 2001; Iqbal and MacLean 2010). Affected needles then dry and fall in late summer-early fall. If trees are heavily defoliated, they may even die after a single year of defoliation (Fig. 3.5). As tree mortality is spread over a longer period during spruce budworm outbreak, the window of availability of suitable trees for secondary insects is much longer than during hemlock looper outbreaks. Greater diversity in arthropod communities is expected from disturbances that generate greater heterogeneity. For instance, the striped ambrosia beetle, *Trypodendron lineatum* (Olivier), was the only species to respond to balsam fir affected by the hemlock looper (Béland et al. 2019) while it was secondary to a melandryid, a sirex and another bark beetle in firs affected by the spruce budworm (Belyea 1952a, b; Régnier 2020).

The Cerambycid *Monochamus scutellatus* is also found in trees killed by the spruce budworm and the hemlock looper, but in much lower abundance than in trees killed by fire. The dominance of Cerambycidae, including *M. scutellatus*, in burned

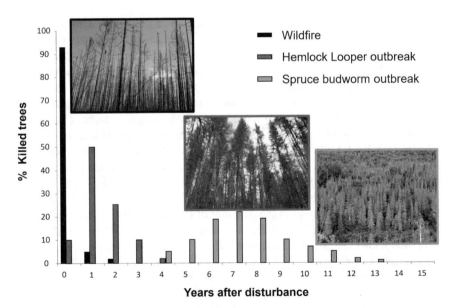

Fig. 3.4 Temporal patterns of tree mortality for three different natural disturbances in eastern Canada. Photos of insect outbreaks from C. Hébert and of wildfire from S. Bélanger (reprinted with permission of C. Hébert and S. Bélanger)

Fig. 3.5 Young stand of white spruce regenerating 20 years after a severe outbreak of the hemlock looper in an old-growth balsam fir stand on Anticosti Island. Photo from C. Hébert (reprinted with permission of Le Naturaliste Canadien and C Hébert)

trees, may be explained by the Jarman-Bell principle, a concept in herbivore nutritional ecology which states that the body size of herbivores is negatively correlated with diet quality (Steuer et al. 2014). After severe wildfire in boreal forests, large Cerambycidae dominate the habitat and small bark beetles are much less abundant. After insect outbreaks the reverse is true, suggesting that subcortical food quality might be poor after fire. Wood water content is a useful proxy for assessing food quality of subcortical tissues and it decreases with increasing fire severity (Cadorette-Breton et al. 2016). In trees recently killed by fire, water content is always below 30% (Jeffrey 2013; Cadorette-Breton et al. 2016) while it remains well-above 50% in trees defoliated by the spruce budworm (unpublished data). Even trees affected by non-lethal fires show phloem/cambium necrosis and misshapen xylem vessels, which lead to hydrolic dysfunction (Bar et al. 2019). On the other hand, defoliation induces a 20% reduction in the diameter of phloem channels, likely impacting sap transportation capability of trees and increasing the risk of vascular dysfunction (Hillabrand et al. 2019). Although defoliation reduces subcortical tissue quality, fire reduces it more extensively and more rapidly. Obviously, this influences the successional dynamics of saproxylic insect communities.

Bark beetles are also important pests of coniferous forests in many regions of the world (Morris et al. 2017). In western North America, the Mountain Pine Beetle has affected >27 M ha of mature forest stands and has had major impacts on forest ecosystem dynamics, biodiversity (Bunnell et al. 2011; Saab et al. 2014) and ecosystem services (Dhar et al. 2016; Audley et al. 2020). The outbreak has increased diversity of understory plants and this certainly has affected arthropod communities, but this has not been documented. However, higher diversity of alpine bees was linked with the increased availability of floral resources in post-outbreak stands affected by the spruce beetle, *Dendroctonus rufipennis* Kirby (Davis et al. 2020), a similar species.

3.5 Effects of Forest Logging on Arthropods

Remote-sensing assessments showed that only 22% of the world's forest landscape was classified as intact in 2000 and had decreased to 20.4% between 2000 and 2013 (Potapov et al. 2017). Expansion of agriculture and pasture in tropical regions were responsible for 60% of this reduction. Old-growth forests have virtually disappeared from Europe (Wirth et al. 2009) and they have become rare in many parts of North America, mainly due to timber harvesting (Potapov et al. 2017; Schowalter 2017). Where they still exist, old-growth forests are often limited to small remnant areas which might not be representative of the original forest matrix.

3.5.1 Clear-Cuts

The first reported impacts of logging on biodiversity were associated with the widespread use of clearcutting which resulted in the loss and fragmentation of old-growth forests. From the perspective of biodiversity conservation, clear-cuts are inappropriate for maintaining some forest species (Spence 2001), particularly those which are closely associated with old-growth forests (Spence et al. 1996; Niemelä 1997; Siitonen and Saaristo 2000; Buddle et al. 2006; Pohl et al. 2007). Clear-cuts initiate forest succession and homogenize stand structure and composition for several decades, often over large areas, and thus rarify old-growth forest attributes, which are important drivers of arthropod diversity (Janssen et al. 2009). Moreover, intensive forestry has used short rotations in order to optimize wood production and avoid reaching the senescent forest stage in which a certain amount of trees die, i.e. when forests recover certain attributes characterizing old-growth forests.

In Scandinavia, where boreal forests have been managed intensively, the amount of dead wood has decreased to extremely low levels, severely impacting numerous saproxylic species (Kaila et al. 1997; Grove 2002; Stenbacka et al. 2010). In the late 1990s, nearly 70% of red-listed forest invertebrates were saproxylics (Jonsell et al. 1998). In Canada, mature balsam fir stands are usually harvested at 50 years of age as they are then highly vulnerable to the spruce budworm and also because they are considered to have reached their silvicultural maturity. Dead trees are rare in 50 year old balsam fir stands and short rotations could thus lead to a rarefaction of dead wood (Norvez et al. 2013). This is a first step towards breakage of forest continuity, a concept which refers to the continuous availability of a certain amount of micro-habitats (e.g. dead wood) or appropriate conditions (e.g. close-canopy cover) to ensure survival of living organisms (Jonsell and Nordlander 2002). Populations of several saproxylic beetles are still less abundant in 50-yrs post-harvest balsam fir stands than in older stands regulated by spruce budworm outbreaks (Bouchard 2000), suggesting that forest continuity in dead wood may be broken by short rotations. This may result in a subtle erosion of saproxylic insect diversity characterizing naturally disturbed forests (Norvez et al. 2013).

Arthropods with poor dispersal ability are particularly vulnerable to the loss of old-growth forests, to habitat fragmentation and to a breakage in forest continuity (Koivula 2002). Several carabid beetles, common in old-growth forests, persist temporarily in recent clear-cuts but they were scarce or had disappeared from stands by 27 years post-harvest in Alberta, Canada (Spence et al. 1996). In addition to drastically modified environmental conditions, old-growth specialists face competition from open-habitat species that heavily colonize clear-cut patches. The surrounding landscape is important as the impact of clear-cuts is lower in a matrix of old-growth boreal forest stands, highlighting the importance of source habitats for recolonizing harvested stands (Le Borgne et al. 2018). In heterogeneous landscapes, beetle community assembly is mainly driven by interspecific interactions rather than by habitat attributes (Le Borgne et al. 2018).

3.5.2   Salvage Logging

For economic and phytosanitary reasons, salvage logging after natural disturbances has become increasingly prevalent all over the world (Lindenmayer et al. 2008). A meta-analysis revealed that salvage logging significantly decreases species richness of saproxylic beetles, which is not surprising as habitat is removed (Thorn et al. 2018). It has been estimated that to maintain 90% of saproxylic beetle richness, 85% of these disturbed forests would need to be retained (Thorn et al. 2018). Richness of springtails also decreases after salvage logging, these micro-arthropods being very sensitive to the drying out of the soil following canopy and tree removal. However, richness of ground-dwelling spiders and carabids increase, many species of these groups being typically associated with open habitats (Thorn et al. 2018). Indeed carabid recovery is typically rapid with the retention of almost any disturbed patches in postfire forests (Koivula and Spence 2006).

As in clear-cutting, the impact of salvage logging lasts decades. For instance, habitat attributes still differ between unsalvaged and salvaged balsam fir stands, 20 years after the end of a spruce budworm outbreak (Norvez et al. 2013). As in most natural disturbances, insect outbreaks rarely kill all trees, survivors being important legacies in forest dynamics as they contribute to the maintenance of ecological continuity in dead wood recruitment. This legacy is illustrated through the larger amount and greater diversity of coarse woody debris in unsalvaged stands compared with salvaged ones (Norvez et al. 2013) (Fig. 3.6).

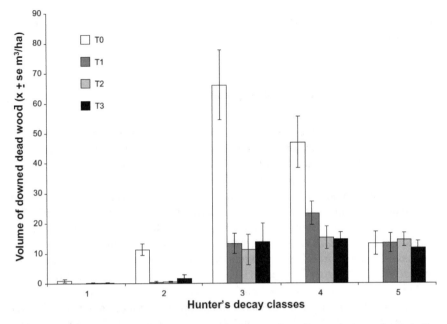

Fig. 3.6 Distribution of downed dead wood volumes according to Hunter's decay classes in four types of stands, 20 years after the end of a spruce budworm outbreak in balsam fir forest. T_0: unsalvaged stands, T_1: salvage logging only, T_2: salvage logging followed by a pre-commercial thinning, T_3: salvage logging followed by scarification and black spruce plantation and mechanical release. From Norvez et al. (2013) (reprinted with permission of Elsevier)

3.5.3 Partial Cuts

In recent years, partial cutting has been used as a more socially and environmentally acceptable silvicultural treatment than clear-cuts (Franklin et al. 1997; Harvey et al. 2002). Partial cutting removes only a portion of the trees (usually up to 45%) and thus, it maintains a forest cover useful for biodiversity. It also ensures a continuous recruitment of dead trees necessary to maintain unique elements of forest biodiversity, namely saproxylic organisms that only live in this habitat. These dead trees are also important components for generating heterogeneity in forest ecosystems and providing habitats for non-saproxylic organisms. By maintaining structural and compositional attributes of mature forests (Harvey et al. 2002), partial cuts limit landscape fragmentation (Warkentin and Bradshaw 2012) and maintain ecological functions of forest ecosystems. Partial cuts generate new niches that are absent in closed canopy forests, maintain similar amounts of snags and coarse woody debris as in closed canopy forests and as a result, beetle communities in partial cuts remain similar to those found in old-growth boreal forest (Légaré et al. 2011). Similarly, in Finland, the carabid assemblages of thinned (10–30% tree removal) and mature stands were similar (Koivula 2002). Partial cuts help to reduce the impact of logging on forest ecosystems and have become a useful tool for implementing ecosystem-based forest management (see Sect. 3.6.2).

3.6 Conservation and Management

The concept of sustainable development defined by Brundtland (1987) has led to changes in forest management. Henceforth, development must meet the needs of the present without compromising the ability of future generations to meet their own needs. Applying this concept requires a better balance between economic, environmental and social issues. The importance of biodiversity conservation to achieve the environmental goal of sustainable development was recognized with the agreement of the Convention on Biodiversity signed after the Earth Summit in 1992 (United Nations 1992). The implementation of sustainable forest management then required different approaches to integrate the objective of maintaining biodiversity while continuing logging (Thorpe and Thomas 2007). Although in some rare cases single species are protected by regulations, conservation of arthropod diversity is usually approached globally. Protecting single species through regulation forbids their capture and trade, and sometimes protects their habitat. This approach has rarely been successful (Samways 2018). Conservation strategies now look at larger spatial scales. Developing resilient forest landscapes is an emerging field of interest in conservation biology, but preserving forest arthropods remains a challenge for scientists and policy-makers.

3.6.1 Protected Areas

The establishment of protected areas is the first measure proposed by conservation biologists to protect biodiversity as it maintains habitats (Jenkins and Joppa 2009; Samways 2007). It requires protecting large areas of primary forests, which are rapidly declining in tropical areas, and which are already small or strongly modified in most north-temperate areas (Samways et al. 2020). Also, as biodiversity is never fully inventoried, delimiting protected areas is usually based on surrogates (plant endemism, vegetation classification), which assume that this approach is efficient for protecting non-sampled and poorly known biodiversity, which include arthropods (Rodrigues and Brooks 2007). The use of vascular plant endemism to delineate hotspots of biodiversity is controversial (Marchese 2015), but it was shown to be efficient in protecting bush crickets (Orthoptera: Tettigoniidea) in South Africa (Bazelet et al. 2016). Also, an analysis of the efficacy of protected areas in Italy showed that 91% of the 150 red-listed saproxylic beetles were present (D'Amen et al. 2013). Foresters could consider it as a success while it could be viewed as a failure for conservationists as 9% of red-listed species are still absent. However, as the full extent of a species geographic range was captured for only 7% of these red-listed species, the reserve network was considered inadequate to protect Italian saproxylic insect diversity (D'Amen et al. 2013). It is possible that protected areas designed to be representative of large-scale vegetation regions are less effective than a small-scale inventory of endemic plants. In addition, protected areas are often established on the basis of aesthetic criteria (e.g. spectacular landscapes), or for political reasons, which rarely meet biodiversity conservation objectives. Protected areas may also complicate political decisions related with land use planning, particularly when, without appropriate pest management, insect outbreaks may kill trees. Decisions regarding whether and how to manage protected areas to preserve biodiversity and sustain ecosystem services, usually involve public debate. In Germany, numerous trees died during a large-scale outbreak of *Ips typographus* but red-listed species populations increased as well as the overall biodiversity, providing support for the policy of allowing the natural course of natural disturbances in protected areas and promoting recovery processes that characterize post-natural disturbance successional stages (Beudert et al. 2015).

The ability of current networks of protected areas to protect biodiversity and ecological processes will undoubtedly be affected by changing climate. These areas are spatially fixed and may not host the same species in the future as climate change will cause range shifts or reductions for many species (Hannah et al. 2007). The functional connectivity between protected areas should be improved to enable species range expansion in response to climate change (Samways et al. 2020).

3.6.2 Ecosystem-Based Forest Management

The concept of ecosystem-based forest management aims to maintain forest ecosystems within their natural range of variability, using natural disturbance regimes as references, with the underlying idea that species should not experience conditions they never faced before (Hunter 1990). Natural processes that regulate forest ecosystem dynamics should be preserved, thus ensuring progress toward sustainable forest management (Attiwill 1994; Angelstam 1998; Bergeron et al. 1999; Gauthier et al. 2008).

Unlike natural disturbances, forest logging typically (in particular clear-cuts) reduces heterogeneity and the amount of dead wood for decades. Thus, adapting silvicultural practices so that managed forests more closely replicate natural forests has been, and continues to be, a major challenge for forest managers. Ecosystem-based forest management is primarily implemented by mimicking the spatial arrangement produced by natural disturbances in terms of size and distribution of logging patches. At the stand level, it attempts to maintain key structural elements produced by natural disturbances such as snags and coarse woody debris (Niemelä 1997; Harvey et al. 2002; Bauhus et al. 2009). It is difficult to mimic the conditions generated by natural disturbances with logging as tree harvesting reduces the future amount of dead wood in the forest while natural disturbances do the opposite. Dead wood still present after logging is almost entirely under the form of woody debris on the ground, where it decays rapidly (Grove 2002). Thus, saproxylic organisms using snags lose their habitat and those that use dead wood on the ground will lose it soon after logging as decomposition progresses. It is possible to increase dead wood stocks by girdling trees (Dufour-Pelletier et al. 2020) or by leaving a certain amount of high stumps when stands are harvested (Jonsell et al. 2004). Supplementing them with logs of various tree species on the ground is recommended (Andersson et al. 2015). However, in western North America and in other dry regions of the world where wildfire is already a major issue, and likely to worsen with climate change, this approach should not be used as it would increase fuel loading. In these regions, silvicultural practices aimed to reduce fire risk by managing fuel loads is a critical forest management objective.

In areas where dead wood has rarefied, short-term measures aimed to increase the amount of dead wood must be accompanied by medium and long-term measures to avoid critical gaps in the continuity of deadwood (Grove 2002). Therefore, it is crucial to determine the minimum amount of dead wood, under various forms, necessary to maintain biodiversity along all post-harvest successional stages. This requires leaving enough live trees to ensure continuous recruitment of dead wood to avoid breakage of forest continuity. The retention of patches of varying sizes is now used in the context of ecosystem-based boreal forest management. In the short-term, 2.5 ha patches are efficient to maintain beetle communities in boreal forest but negative effects could increase with time (Bouchard and Hébert 2016), highlighting the importance of long-term studies.

3.6.3 Restoration

Adapting forestry practices to maintain biodiversity associated with old-growth forests or at least with natural mature forests is a major challenge of contemporary forestry (Niemelä 1997). In regions where old-growth forests no longer exist, the challenge is two-fold: first, it is recommended to lengthen rotations to reach the senescent stage in which certain old-growth forest attributes are recovered and second implementation of restoration programs.

Tree planting after clear-cutting or salvage logging is the most widely used restoration method. It is usually applied when natural regeneration will result in seedling density too low to maintain stand productivity. Although tree planting provides habitat to maintain certain forest arthropods, planted forests have been reported to have lower abundance and species richness of beetles, compared to old-growth forests, and up to a 40% difference in species composition (Albert et al. 2021). Tree planting appears particularly unsuited to tropical forests where it does not reproduce the complex and diverse microhabitats and biotic interactions of old-growth tropical forests (Gibson et al. 2011). Monospecific plantations of exotic trees have strong negative effects on beetle communities and they should be restricted to areas where old-growth forests are scarce and highly fragmented, and where planting native trees is not an option (Albert et al. 2021). Negative effects of forest plantations are less significant in other biomes, particularly when native tree species are planted (Albert et al. 2021). Thus, when used in combination with ecosystem-based forest management and protected areas, native tree plantations could be helpful to help restoring a portion of arthropod diversity in temperate and boreal forests. However, stand conversion toward another native tree species can also have detrimental effects. For example, converting mature balsam fir stands severely affected by the spruce budworm to black spruce plantations after salvage logging moves beetle communities farther away from the original stands than salvage logging alone or salvage logging followed by pre-commercial thinning (Norvez et al. 2013).

Another approach that can be used in forest restoration and which is consistent with ecosystem-based forest management is the reintroduction of natural processes through direct intervention. For example, prescribed burning is used to manage forest fuel and reduce fire risks (Fernandes and Botelho 2003), but it is also used to regenerate certain pine species, such as the eastern white pine, *Pinus strobus* L. in Canadian National Parks (Hébert et al. 2019). Eastern white pine was much more prevalent in pre-settlement forests of eastern North America than it is today (Doyon and Bouffard 2009). It is well adapted to low-severity surface fires as it has thick bark that efficiently insulates subcortical tissues (Hengst and Dawson 1994), but also because this tall tree has deep roots and a branch-free lower trunk (Farrar 1995). By reducing competition from saplings of other shade-tolerant tree species, and increasing light penetration, prescribed burning improves seedbed quality and helps white pine seedlings to sprout and grow (Hébert et al. 2019). Prescribed burning was shown to be efficient to increase the richness of both saproxylic and non-saproxylic beetles, suggesting that burning treatments do not only increase the amount of dead wood but also favour other attributes found in post-fire environments (Domaine 2009). Moreover, prescribed burning significantly increased the number of rare beetles, attesting for the usefulness of this restoration practice for biodiversity conservation.

3.7 New Challenges

The efficacy of management and conservation measures presented in the previous sections are challenged by arthropod declines highlighted in recent reports (Hallmann et al. 2017; Kunin 2019; Seibold et al. 2019; Wagner 2020). In 2017, the publication of a paper reporting a 76% drop in insect biomass in protected areas of Germany (Hallmann et al. 2017), received attention in the media and raised awareness of the general population. Reports of insect decline have existed for decades, perhaps best documented with light traps within the Rothamsted Insect Survey network which has sampled moths in Great Britain since 1968 (Conrad et al. 2006). The magnitude and geographic extent of arthropod decline remains largely unknown and vigorously debated (Wagner 2020).

Most data showing arthropod decline mainly come from open habitats, but arthropod decline has been also reported in forest habitats, although the effects do not appear as strong (Kunin 2019; Seibold et al. 2019). Drivers of arthropod decline in forests is unclear but in grasslands, it is associated with the importance of agriculture in the landscape (Seibold et al. 2019). Apart from agricultural intensification (including pesticide use), factors suggested as possible causes of arthropod decline include habitat destruction (including deforestation), climate change, invasive species, atmospheric nitrification from burning fossil fuels and drought (Wagner 2020). Arthropod decline raises important ecological and economic issues as it will generate unpredictable cascading effects on ecosystems linked with the expected losses of ecological services provided by arthropods (Hallmann et al. 2017). Monitoring biodiversity and climate appear more important than ever as the impacts of these ecological crises intensify (O'Connor et al. 2020). This highlights the importance of long-term data using standardized methods and appropriate tools to manage and analyse these data, and ensure their long-term storage (Kunin 2019).

Climate change and biodiversity issues are closely linked and both the Intergovernmental Panel on Climate Change (IPCC) and the Intergovernmental Science-Policy Platform on Biodiversity and Ecosystem Services (IPBES) have called for urgent action to reduce the human ecological footprint (IPBES 2019; IPCC 2018). Without a doubt, the crises of climate change and biodiversity will be at the heart of the ecological agenda for the next decade.

References

Albert G, Gallegos SC, Greig KE, Hanisch M, de la Fuente DL, Fost S, Maier SD, Sarathchandra C, Phillips HRP, Kambach S (2021) The conservation value of forests and tree plantations for beetle (Coleoptera) communities: a global meta-analysis. For Ecol Manage 491:119201

Alexander K (1995) The use of freshly downed timber by insects following the 1987 storm. In: Buckley G (ed) Ecological responses to the 1987 great storm in the woods of Southeast England. English Nature Science 23:134–150

Alfaro RI, Shepherd RF (1991) Tree-ring growth of interior Douglas-fir after one year's defoliation by Douglas-fir tussock moth. For Sci 37:959–964

Allen DC (1973) Review of some parasite/predator-prey interactions in the forest soil microcommunity. Proceedings of the first soil microcommunities conference. Syracuse, New York, 18–20 Oct 1971. State University, New York, pp 218–226

Ameixa OMCC, Soares AO, Soares AMVM, Lillebø AI (2018). Ecosystem services provided by the little things that run the World. In: Şen B, Grillo O (eds) Selected studies in biodiversity. IntechOpen, Chapter 13, pp 267–302. https://doi.org/10.5772/intechopen.74847

Anderson JF (1976) Egg parasitoids of forest defoliating Lepidoptera. In: Anderson JF, Kaya HK (eds) Perspectives in forest entomology. Academic, New York, pp 233–249

Andersson J, Hjältén J, Dynesius M (2015) Wood-inhabiting beetles in low stumps, high stumps and logs on boreal clear-cuts: implications for dead wood management. PLoS ONE 10(3):e0118896. https://doi.org/10.1371/journal.pone.0118896

Angelstam PK (1998) Maintaining and restoring biodiversity in European boreal forests by developing natural disturbance regimes. J Veg Sci 9:593–602

Attiwill PM (1994) The disturbance of forest ecosystems: the ecological basis for conservative management. For Ecol Manage 63:247–300

Audley JP, Fettig CJ, Munson AS, Runyon JB, Mortenson LA, Steed BE, Gibson KE, Jorgensen CL, McKelvey SR, McMillin JD, Negron JF (2020) Impacts of mountain pine beetle outbreaks on lodgepole pine forests in the Intermountain West, U.S., 2004–2019. For Ecol Manag 475:118403

Ausmus BS (1977) Regulation of wood decomposition rates by arthropod and annelid populations. Ecological Bulletins No 25, Soil Organisms as Components of Ecosystems, pp 180–192

Azeria ET, Ibarzabal J, Hébert C (2012) Effects of habitat characteristics and interspecific interactions on co-occurrence patterns of saproxylic beetles breeding in tree boles after forest fire: null model analyses. Oecologia 168:1123–1135

Bar A, Michaletz ST, Mayr S (2019) Fire effects on tree physiology. New Phytol 223:1728–1741

Bar-On YM, Phillips R, Miloa R (2018) The biomass distribution on Earth. PNAS 115:6506–6511

Barbosa P, Wagner MR (1989) Introduction to forest and shade tree insects. Academic Press, Inc., San Diego, CA

Barnouin T (2005) Évaluation de l'importance des forêts affectées par la tordeuse des bourgeons de l'épinette. *Choristoneura fumiferana* (Clem.), dans le maintien de la diversité des coléoptères saproxyliques. MSc Thesis, Université Laval, 59 p

Barrios E (2007) Soil biota, ecosystem services and land productivity. Ecol Econ 64:269–285

Barton PS, Cunningham SA, Lindenmayer DB, Manning AD (2013) The role of carrion in maintaining biodiversity and ecological processes in terrestrial ecosystems. Oecologia 171:761–772

Barton PS, Evans JM (2017) Insect biodiversity meets ecosystem function: differential effects of habitat and insects on carrion decomposition. Ecol Entomol 42:364–374

Basset Y, Cizek L, Cuénoud P, Didham RK, Guilhaumon F, Missa O, Novotny V, Ødegaard F, Roslin T, Schmidl J, Tishechkin AK, Winchester NN, Roubik DW, Aberlenc H-P, Bail J, Barrios H, Bridle JR, Castaño-Meneses G, Corbara B, Curletti G, Duarte da Rocha W, De Bakker D, Delabie JHC, Dejean A, Fagan LL, Floren A, Kitching RL, Medianero E, Miller SE, de Oliveira EG, Orivel J, Pollet M, Rapp M, Ribeiro SP, Roisin Y, Schmidt JB, Sørensen L, Leponce M (2012) Arthropod diversity in a tropical forest. Science 338:1481–1484

Bauce E, Hardy Y (1988) Effects of drainage and severe defoliation on the raw fiber content of balsam fir needles and growth of the spruce budworm (Lepidoptera: Torfricidae). Environ Entomol 17:671–674

Bauer LS, Nordin GL (1989) Effect of *Nosema fumifennae* (Microsporida) on fecundity, fertility, and progeny performance of *Choristoneura fumiferana* (Lepidoptera: Tortricidae). Environ Entomol 18:261–265

Bauhus J, Puettmann K, Messier C (2009) Silviculture for old-growth attributes. For Ecol Manage 258:525–537

Baur ME, Kaya HK, Strong D (1998) Foraging ants as scavengers on entomopathogenic nematode-killed insects. Biol Control 12:231–236

Bawa KS (1990) Plant-pollinator interactions in tropical rain forests. Annu Rev Ecol Syst 21:399–422

Bawa KS, Bullock SH, Perry DR, Coville RE, Grayum MH (1985) Reproductive biology of tropical lowland rain forest trees. II. Pollination systems. Am J Bot 72:346–356

Bazelet CS, Thompson AC, Naskrecki P (2016) Testing the efficacy of global biodiversity hotspots for insect conservation: the case of South African Katydids. PLoS ONE 11(9):e0160630

Behan VM, Hill SB, Kevan DKMcE (1978) Effects of nitrogen fertilizers, as urea, on Acarina and other arthropods in Quebec black spruce humus. Pedobiologia 18:249–263

Béland J-M, Bauce É, Cloutier C, Berthiaume R, Hébert C (2019) Early responses of bark and wood boring beetles during an outbreak of the hemlock looper *Lambdina fiscellaria* (Guenée) (Lepidoptera: Geometridae) in a boreal balsam fir forest of North America. Agric For Entomol 21:407–416. https://doi.org/10.1111/afe.12347

Belovsky GE, Slade JB (2000) Insect herbivory accelerates nutrient cycling and increases plant production. PNAS 97:14412–14417

Belyea RM (1952a) Death and deterioration of balsam fir weakened by spruce budworm defoliation in Ontario. Can Entomol 89:325–335

Belyea RM (1952b) Death and deterioration of balsam fir weakened by spruce budworm defoliation in Ontario. Part II. An assessment of the role of associated insect species in the death of severely weakened trees. J Forest 50:729–738

Berg A, Ehnstrom B, Gustafsson L, Hallingback T, Jonsell M, Weslien J (1994) Threatened plant, animal, and fungus species in Swedish forests: distribution and habitat associations. Conserv Biol 8:718–731

Bergeron Y, Harvey B, Leduc A, Gauthier S (1999) Stratégies d'aménagement forestier qui s'inspirent de la dynamique des perturbations naturelles: considérations à l'échelle du peuplement et de la forêt. For Chron 75:55–61

Bernays E (2009) Phytophagous insects. In: Resh VH, Cardé RT (eds) Encyclopedia of insects. Academic Press, Burlington, MA, Chapter 201, pp 798–800

Bernays E, Graham M (1988) On the evolution of host specificity in phytophagous arthropods. Ecology 69:886–892

Beudert B, Baussler C, Thorn S, Noss R, Schroder B, Dieffenbach-Fries H, Foullois N, Muller J (2015) Bark beetles increase biodiversity while maintaining drinking water quality. Conserv Lett 8:272–281

Bin F, Johnson NF (1982) Potential of Telenominae in biocontrol with egg parasitoids (Hym., Scelionidae). In: INRA Publlic (ed) Les Trichogrammes, Antibes (France), 20–23 Avril 1982. Coll. INRA 9, pp 275–287

Birkemoe T, Jacobsen RM, Sverdrup-Thygeson A, Biedermann PHW (2018) Insect-fungus interactions in dead wood systems. In: Ulyshen MD (ed) Saproxylic insects—diversity, ecology and conservation. Springer, Chapter 12

Blais JR (1960) Spruce budworm parasite investigations in the Lower St. Lawrence and Gaspé region of Quebec. Can Entomol 92:384–396

Bloin P (2021) Réponse à court terme des insectes saproxyliques à la disponibilité du bois mort. MSc Thesis, Université Laval, 79 p

Bouchard M (2000) Effets de la coupe à blanc et de la structure de la forêt sur les communautés de léiodides (Coleoptera: Leiodidae) dans des sapinières boréales. MSc Thesis, Université Laval, Québec, Canada, 33 p

Bouchard M, Hébert C (2016) Beetle community response to residual forest patch size in managed boreal forest landscapes: feeding habits matter. For Ecol Manage 368:63–70. https://doi.org/10.1016/j.foreco.2016.02.029

Boucher J, Azeria ET, Ibarzabal J, Hébert C (2012) Saproxylic beetles in disturbed boreal forest: temporal dynamics, habitat associations and community structure. Ecoscience 19:328–343

Boucher J, Hébert C, Ibarzabal J, Bauce E (2016) High conservation value forests for burn-associated saproxylic beetles: an approach for developing sustainable post-fire salvage logging in boreal forest. Insect Conserv Divers 9:402–415

Boucher J, Bauce E, Hébert C (2020) A flexible approach for predicting and mapping postfire wood borer attacks in black spruce and jack pine forests using the differenced normalized burn ratio (dNBR). Can J For Res 50:880–889

Boucher P, Hébert C, Francoeur A, Sirois L (2015) Postfire succession of ants (Hymenoptera: Formicidae) nesting in dead wood of northern boreal forest. Environ Entomol 44:1316–1327

Bouget C, Duelli P (2004) The effects of windthrow on forest insect communities: a literature review. Biol Cons 118:281–299

Boulanger Y, Sirois L, Hébert C (2010) Distribution of saproxylic beetles in a recently burnt landscape of the northern boreal forest of Québec. For Ecol Manage 260:1114–1123

Boulanger Y, Sirois L, Hébert C (2011) Fire severity as a determinant factor of the decomposition rate of fire-killed black spruce in the northern boreal forest. Can J For Res 41:370–379

Boulanger Y, Sirois L, Hébert C (2013) Distribution patterns of three long-horned beetles (Coleoptera: Cerambycidae) shortly after fire in boreal forest: adults colonizing stands versus progeny emerging from trees. Environ Entomol 42:17–28

Braack LEO (1987) Community dynamics of carrion-attendant arthropods in tropical african woodland. Oecologia 72:402–409

Brokaw NVL (1982) The definition of treefall gap and its effect on measures of forest dynamics. Biotropica 14:158–160

Brundtland GH (1987) Our common future. The world commission on environment and development. Oxford University Press, New York

Brussaard L, Behan-Pelletier VM, Bignell DE, Brown VK, Didden W, Folgarait P, Fragoso C, Wall Freckman D, Gupta VVSR, Hattori T, Hawksworth DL, Klopatek C, Lavelle P, Malloch DW, Rusek J, Soderstrom B, Tiedje JM, Virginia RA (1997) Biodiversity and ecosystem functioning in soil. Ambio 26:563–570

Buddle CM, Langor DW, Pohl GR, Spence JR (2006) Arthropod responses to harvesting and wildfire: implications for emulation of natural disturbance in forest management. Biol Cons 128:346–357

Bunnell FL, Kremsater LL, Houde I (2011) Mountain pine beetle: a synthesis of the ecological consequences of large-scale disturbances on sustainable forest management, with emphasis on biodiversity. Natural Resources Canada, Canadian Forest Service, Pacific Forestry Centre, Victoria, BC. Information Report BC-X-426, 99 p

Cadorette-Breton Y, Hébert C, Ibarzabal J, Berthiaume R, Bauce E (2016) Vertical distribution of three longhorned beetle species (Coleoptera: Cerambycidae) in burned trees of the boreal forest. Can J For Res 46:564–571

Ciesla WM (2011) Forest entomology: a global perspective. Wiley-Blackwell, Hoboken NJ

Clarke RD, Grant PR (1968) An experimental study of the role of spiders as predators in a forest litter community. Part 1. Ecology 49:1152–1154

Condamine FL, Clapham ME, Kergoat GK (2016) Global patterns of insect diversification: towards a reconciliation of fossil and molecular evidence? Scientific Reports 6:19208

Conrad KF, Warren MS, Fox R, Parsons MS, Woiwod IP (2006) Rapid declines of common, widespread British moths provide evidence of an insect biodiversity crisis. Biol Cons 132:279–291. https://doi.org/10.1016/j.biocon.2006.04.020

D'Amen M, Bombi P, Campanaro A, Zapponi L, Bologna MA, Mason F (2013) Protected areas and insect conservation: questioning the effectiveness of Natura 2000 network for saproxylic beetles in Italy. Anim Conserv 16:370–378

Dajoz R (1998) Les insectes et la forêt : rôle et diversité des insectes dans le milieu forestier. Editions Tec & Doc Lavoisier Londres, Paris, New York, 594 p

Davis TS, Rhoades PR, Mann AJ, Griswold T (2020) Bark beetle outbreak enhances biodiversity and foraging habitat of native bees in alpine landscapes of the southern Rocky Mountains. Sci Rep 10:16400

De Alckmin Marques ES, Price PW, Cobb NS (2000) Resource abundance and insect herbivore diversity on woody Fabaceous desert plants. Environ Entomol 29:696–703

De Grandpré L, Waldron K, Bouchard M, Gauthier S, Beaudet M, Ruel J-C, Hébert C, Kneeshaw D (2018) Incorporating insect and wind disturbances in a natural disturbance-based management framework for the boreal forest. Forests 9(8):471. https://doi.org/10.3390/f9080471

DeFoliart G (1991) Forest management for the protection of edible caterpillars in Africa. The Food Insects Newsletter 4(2):1–2

DeFoliart G (2005) Overview of role of edible insects in preserving biodiversity. In: Paoletti MG (ed) Ecological implications of mini livestock. Science Publisher, Enfield, NH, pp 123–140

Dekeirsschieter J, Verheggen F, Lognay G, Haubruge E (2011) Large carrion beetles (Coleoptera, Silphidae) in Western Europe: a review. Biotechnol Agron Soc Environ 15:435–447

Dhar A, Parrott L, Heckbert S (2016) Consequences of mountain pine beetle outbreak on forest ecosystem services in western Canada. Can J For Res 46:987–999

Dighton J (2003) Fungi in ecosystem processes, 2nd edn. CRC Press, New York

Djukic I, Kepfer-Rojas S, Kappel Schmidt I, Steenberg Larsen K, Beier C, Berg B, Verheyen K, Composition T (2018) Early stage litter decomposition across biomes. Sci Total Environ 628–629:1369–1394

Domaine E (2009) Effets des brûlages dirigés sur la régénération du pin blanc et la diversité des coléoptères du Parc national du Canada de la Mauricie. MSc Thesis, Université Laval, 116 p

Doyon F, Bouffard D (2009). Enjeux écologiques de la forêt feuillue tempérée québécoise, Québec. Pour le ministère des Ressources naturelles et de la Faune, Direction de l'environnement et de la protection des forêts. 63 p. Available from: http://www.mrnf.gouv.qc.ca/forets/amenagement/amenagement-ecosystemique.jsp

Duelli P, Obrist MK, Wermelinger B (2002) Windthrow-induced changes in faunistic biodiversity in alpine spruce forests. For Snow Landsc Res 77:117–131

Dufour-Pelletier S, Tremblay JA, Hébert C, Lachat T, Ibarzabal J (2020) Testing the effect of snag and cavity supply on deadwood-associated species in a managed boreal forest. Forests 11:424. https://doi.org/10.3390/f11040424

Durst PB, Shono K (2010) Edible forest insects: exploring new horizons and traditional practices. In: Durst PB, Johnson DV, Leslie RN, Shono K (eds) Forest insects as food: humans bite back. Proceedings of a workshop on Asia-Pacific resources and their potential for development. 19–21 Feb 2008, Chiang Mai, Thailand, pp 1–4

Dyer LA, Carson WP, Leigh Jr EG (2012) Insect outbreaks in tropical forests: patterns, mechanisms, and consequences. In: Barbosa P, Letourneau DK, Agrawal AA (eds) Insect outbreaks revisited. Wiley-Blackwell, Hoboken, NJ, Chapter 11

Ehrlich PR, Raven PH (1964) Butterflies and plants: a study in coevolution. Evolution 18:586–608

Elton CS (1958) The ecology of invasions by animals and plants. Chapman & Hall, London

Everaerts C, Cusson M, McNeil JN (2000) The influence of smoke volatiles on sexual maturation and juvenile hormone biosynthesis in the black army cutworm, *Actebia fennica* (Lepidoptera: Noctuidae). Insect Biochem Mol Biol 30:855–862

Evans WG (1964) Infrared receptors in *Melanophila acuminata* De Geer. Nature 202:211

Evans WG (1966) Perception of infrared radiation from forest fires by *Melanophila acuminata* De Geer (Buprestidae, Coleoptera). Ecology 47:1061–1065

Eveleigh ES, McCann KS, McCarthy PC, Pollock SJ, Lucarotti CJ, Morin B, McDougall GA, Strongman DB, Huber JT, Umbanhowar J, Faria LDB (2007) Fluctuations in density of an outbreak species drive diversity cascades in food webs. PNAS 104:16976–16981

Faiola C, Taipale D (2020) Impact of insect herbivory on plant stress volatile emissions from trees: a synthesis of quantitative measurements and recommendations for future research. Atmos Environ 5:100060

Farrar JL (1995) Les Arbres du Canada. Service canadien des forêts, Ressources Naturelles Canada et Fides, Saint-Laurent, Canada

Fernandes PM, Botelho HS (2003) A review of prescribed burning effectiveness in fire hazard reduction. Int J Wildland Fire 12:117–128

Ferro ML (2018) It's the end of the wood as we know it: insects in veteris (highly decomposed) wood. In: Ulyshen MD (ed) Saproxylic insects, diversity, ecology and conservation. Zoological Monographs 1, Springer

Filipiak M, Weiner J (2014) How to make a beetle out of wood: multi-elemental stoichiometry of wood decay, xylophagy and fungivory. PLoS ONE 9(12):e115104

Foit J (2010) Distribution of early-arriving saproxylic beetles on standing dead Scots pine trees. Agric For Entomol 12:133–141

Folgarait PJ (1998) Ant biodiversity and its relationship to ecosystem functioning: a review. Biodivers Conserv 7:1221–1244

Forbes AA, Bagley RK, Beer MA, Hippee AC, Widmayer HA (2018) Quantifying the unquantifiable: why Hymenoptera, not Coleoptera, is the most speciose animal order. BMC Ecol 18:21. https://doi.org/10.1186/s12898-018-0176-x

Frank K, Brückner A, Hilpert A, Heethoff M, Blüthgen N (2017) Nutrient quality of vertebrate dung as a diet for dung beetles. Sci Rep 7:12141

Franklin JF, Berg DR, Thornburgh DA, Tappeiner JC (1997) Alternative silvicultural approaches to timber harvesting: variable retention harvest systems. In: Kohm KA, Franklin JF (eds) Creating a forestry for the 21st century: the science of ecosystem management. Island Press, Washington, DC, pp 111–139

Frost CJ, Hunter MD (2007) Recycling of nitrogen in herbivore feces: plant recovery, herbivore assimilation, soil retention, and leaching losses. Oecologia 151:42–53

García Palacios P, Maestre FT, Kattge J, Wall DH (2013) Climate and litter quality differently modulate the effects of soil fauna on litter decomposition across biomes. Ecol Lett 16:1045–1053

Gaston KJ (1991) The magnitude of global insect species richness. Conserv Biol 5:283–296

Gaunt MW, Miles MA (2002) An insect molecular clock dates the origin of the insects and accords with palaeontological and biogeographic landmarks. Mol Biol Evol 19:748–761

Gauthier S, Vaillancourt M-A, Leduc A, De Granpré L, Kneeshaw D, Morin H, Drapeau P, Bergeron Y (2008) Aménagement écosystémique en forêt boréale. Presses de l'Université du Québec, Québec, 568 p

Gessner MO, Swan CM, Dang CK, McKie BG, Bardgett RD, Wall DH, Hattenschwiler S (2010) Diversity meets decomposition. Trends Ecol Evol 25:372–380

Gibson L, Lee TM, Koh LP, Brook BW, Gardner TA, Barlow J, Peres CA, Bradshaw CJA, Laurance WF, Lovejoy TE, Sodhi NS (2011) Primary forests are irreplaceable for sustaining tropical biodiversity. Nature 478(7369):378–381

Godfray HCJ (1994) Parasitoids: behavioral and evolutionary ecology. Princeton University Press, Princeton, NJ

González G, Seastedt TR (2001) Soil fauna and plant litter decomposition in tropical and subalpine forests. Ecology 82:955–964

González G, Gould WA, Hudak AT, Hollingsworth TN (2008) Decay of Aspen (*Populus tremuloides* Michx.) wood in moist and dry boreal, temperate, and tropical forest fragments. Ambio 37:588–597

Griffiths HM, Ashton LA, Walker AE, Hasan F, Evans TA, Eggleton P, Parr CL (2018) Ants are the major agents of resource removal from tropical rainforests. J Anim Ecol 87:293–300

Grimaldi D, Engel M (2005) Evolution of the insects. Cambridge University Press, New York and Cambridge

Grove SJ (2002) Saproxylic insect ecology and the sustainable management of forests. Annu Rev Ecol Syst 33:1–23

Hallmann CA, Sorg M, Jongejans E, Siepel H, Hofland N, Schwan H, Stenmans W, Müller A, Sumser H, Hörren T, Goulson D, de Kroon H (2017) More than 75 percent decline over 27 years in total flying insect biomass in protected areas. PLoS ONE 12:e0185809

Hannah L, Midgley G, Andelman S, Araújo M, Hughes G, Martinez-Meyer E, Pearson R, Williams P (2007) Protected area needs in a changing climate. Front Ecol Environ 5:131–138

Hanula JL, Ulyshen MD, Horn S (2016) Conserving pollinators in North American forests: a review. Nat Areas J 36:427–439

Harmon ME, Franklin JF, Swanson FJ, Sollins P, Gregory SV, Lattin JD, Anderson NH, Cline SP, Aumen NG, Sedell JR, Lienkaemper GW, Cromack K Jr, Cummins KW (1986) Ecology of coarse woody debris in temperate ecosystems. Adv Ecol Res 15:133–302

Hartling L, MacNutt P, Carter N (1991) Hemlock looper in New Brunswick: notes on biology and survey methods. Department of Natural Resources and Energy. Timber Manage. Branch For. Pest Manage. Sec. Rep. 1991

Harvey BD, Leduc A, Gauthier S, Bergeron Y (2002) Stand-landscape integration in natural disturbance-based management of the southern boreal forest. For Ecol Manage 155:369–385

Hassall M, Rushton SP (1982) The role of coprophagy in the feeding strategies of terrestrial Isopods. Oecologia 53:374–381

Hébert C (1983) Bio-écologie et contrôle naturel d'une population de *Lymantria dispar* L. (Lepidoptera: Lymantriidae) de la Mauricie, Québec. Mémoire de Maîtrise, Université du Québec à Trois-Rivières, 231 p

Hébert C (1989) Bio-écologie de la mouche *Winthemia fumiferanae* et comparaison de sa relation parasitique avec la tordeuse des bourgeons de l'épinette à celle de l'hyménoptère braconide Meteorus trachynotus. Thèse de Doctorat, Université Laval, 138 p

Hébert C (1990) Biologie et contrôle naturel de *Zeiraphera canadensis* Mut. & Free. (Lepidoptera: Tortricidae) dans la région de Matapédia, Québec. Rapport présenté au Ministère de l'énergie et des ressources du Québec, Service de la protection contre les insectes et les maladies, 61 p

Hébert C, Jobin L (2001) The Hemlock looper. Ressour. nat. Can., Serv. can. for., Cent. for. Laurentides, Sainte-Foy, Qc. Inf. Leaf. LFC 4E

Hébert C, Berthiaume R, Dupont A, Auger M (2001) Population collapses in a forecasted outbreak of *Lambdina fiscellaria* (Lepidoptera: Geometridae) caused by spring egg parasitism by Telenomus spp. (Hymenoptera: Scelionidae). Environ Entomol 30:37–43

Hébert C, Comtois B, Morneau L (2017) Insectes des arbres du Québec. Les Publications du Québec, Québec, 299 p

Hébert C, Domaine É, Bélanger L (2019) Prescribed burning used to restore white pine forests in La Mauricie National Park of Canada. Book Chapter In: National Parks, IntechOpen. https://doi.org/10.5772/intechopen.86224

Hengst GE, Dawson JO (1994) Bark properties and fire resistance of selected tree species from the central hardwood region of North America. Can J For Res 24:688–696

Hibbert A (2010) Importance of fallen coarse woody debris to the diversity of saproxylic Diptera in the boreal mixedwood forests of Eastern North America. MSc Thesis, Université du Québec à Montréal, 87 p

Hillabrand RM, Hacke UG, Lieffers VJ (2019) Defoliation constrains xylem and phloem functionality. Tree Physiol 39:1099–1108

Hillebrand H (2004) On the generality of the latitudinal diversity gradient. Am Nat 163:192–211

Hirose Y (1986) Biological and ecological comparison of *Trichogramma* and *Telenomus* as control agents of lepidopterous pests. J Appl Entomol 101:39–47

Hoback WW, Bishop AA, Kroemer J, Scalzitti J, Shaffer JJ (2004) Differences among antimicrobial properties of carrion beetle secretions reflect phylogeny and ecology. J Chem Ecol 30:719–729

Hoback WW, Freeman L, Payton M, Peterson BC (2020) Burying beetle (Coleoptera: Silphidae: *Nicrophorus* Fabricius) brooding improves soil fertility. Coleopt Bull 74:427–433

Hölldobler B, Wilson EO (1990) The ants. Belknap Press of Harvard. University Press, Cambridge, MA

Holling CS (1961) Principles of insect predation. Annu Rev Entomol 6:163–182

Holter P, Scholtz CH (2007) What do dung beetles eat? Ecol Entomol 32:690–697

Hunter ML (1990) Wildlife, forests and forestry. Principles of managing forests for biological diversity. Prentice-Hall, Englewood Cliffs, NJ, 370 p

IPBES (2016) The assessment report of the intergovernmental science-policy platform on biodiversity and ecosystem services on pollinators, pollination and food production. In: Potts SG, Imperatriz-Fonseca VL, Ngo HT, (eds) Secretariat of the intergovernmental science-policy platform on biodiversity and ecosystem services. Bonn, Germany, 552 p

IPBES (2019) Global assessment report on biodiversity and ecosystem services of the intergovernmental science-policy platform on biodiversity and ecosystem services. In: Brondizio ES, Settele J, Díaz S, Ngo HT (eds) IPBES secretariat, Bonn, Germany, 1148 p. https://doi.org/10.5281/zenodo.3831673

IPCC (2018) Summary for policymakers. In: Masson-Delmotte V, Zhai P, Pörtner H-O, Roberts D, Skea J, Shukla PR, Pirani A, Moufouma-Okia W, Péan C, Pidcock R, Connors S, Matthews JBR, Chen Y, Zhou X, Gomis MI, Lonnoy E, Maycock T, Tignor M, Waterfield T (eds) Global warming of 1.5°C. An IPCC special report on the impacts of global warming of 1.5°C above pre-industrial levels and related global greenhouse gas emission pathways, in the context of strengthening the global response to the threat of climate change, sustainable development, and efforts to eradicate poverty. World Meteorological Organization, Geneva, Switzerland, 32 pp

Iqbal J, MacLean DA (2010) Estimating cumulative defoliation of balsam fir from hemlock looper and balsam fir sawfly using aerial defoliation survey in western Newfoundland, Canada. For Ecol Manage 259:591–597

Janssen P, Fortin D, Hébert C (2009) Beetle diversity in a matrix of old-growth boreal forest: influence of habitat heterogeneity at multiple scales. Ecography 32:423–432

Janz N (2011) Ehrlich and Raven revisited: mechanisms underlying codiversification of plants and enemies. Annu Rev Ecol Evol Syst 42:71–89

Janzen DH (1981) The peak in North American Ichneumonid species richness lies between 38° and 42° North. Ecology 62:532–537

Jeffrey O (2013) Effets des coupes de récupération sur les successions naturelles de coléoptères saproxyliques le long d'une chronoséquence de 15 ans après feu en forêt boréale commerciale. MSc Thesis, UQAC, 96 p

Jenkins CN, Joppa L (2009) Expansion of the global terrestrial protected area system. Biol Cons 142:2166–2174

Jones CG, Lawton JH, Shachak M (1994) Organisms as ecosystem engineers. Oikos 69:373–386

Jonsell M, Nordlander G (2002) Insects in polypore fungi as indicator species: a comparison between forest sites differing in amounts and continuity of dead wood. For Ecol Manage 157:101–118

Jonsell M, Nordlander G (2004) Host selection patterns in insects breeding in bracket fungi. Ecol Entomol 29:697–705

Jonsell M, Weslien J, Ehnstrom B (1998) Substrate requirements of red-listed saproxylic invertebrates in Sweden. Biodivers Conserv 7:749–764

Jonsell M, Nittérus K, Stighäll K (2004) Saproxylic beetles in natural and man-made deciduous high stumps retained for conservation. Biol Cons 118:163–173

Johansson T, Andersson J, Hjalten J, Dynesisus M, Ecke F (2011) Short-term responses of beetle assemblages to wildfire in a region with more than 100 years of fire suppression. Insect Conserv Diver 4:142–151

Johnson SN, Lopaticki G, Barnett K, Facey SL, Powell JR, Hartley SE (2016) An insect ecosystem engineer alleviates drought stress in plants without increasing plant susceptibility to an aboveground herbivore. Funct Ecol 30:894–902

Kaila L, Martikainen P, Punttila P (1997) Dead trees left in clear-cuts benefit saproxylic Coleoptera adapted to natural disturbances in boreal forest. Biodivers Conserv 6:1–18

Kanat M, Mol T (2008) The effect of Calosoma sycophanta L. (Coleoptera: Carabidae) feeding on the pine processionary moth, Thaumetopoea pityocampa (Denis & Schiffermüller) (Lepidoptera: Thaumetopoeidae), in the laboratory. Turk J Zool 32:367–372

Knoll AH, Nowak MA (2017) The timetable of evolution. Science. Advances 3:e1603076

Koivula M (2002) Boreal carabid-beetle (Coleoptera, Carabidae) assemblages in thinned uneven-aged and clear-cut spruce stands. Annales Zoology Fennici 39:131–149

Koivula M, Spence JR (2006) Effects of post-fire salvage logging on boreal mixed-wood ground beetle assemblages (Coleoptera, Carabidae). For Ecol Manage 236:102–112

Komarek EV (1969) Fire and animal behavior. In: Komarek EV (conference chairman). Proceedings tall timbers fire ecology conference: No 9. Tall Timbers Research Station, Tallahassee, FL, pp 160–207

Kukor JJ, Martin MM (1983) Acquisition of digestive enzymes by siricid woodwasps from their fungal symbiont. Science 220(4602):1161–1163

Kunin WE (2019) Robust evidence of insect declines. Nature 574:641–642

Kuuluvainen T (1994) Gap disturbance, ground microtopography, and the regeneration dynamics of boreal coniferous forests in Finland: a review. Ann Zool Fennici 31:35–51

Labandeira CC (2006) The four phases of plant-arthropod associations in deep time. Geol Acta 4:409–438

Labandeira CC, Sepkoski JJ Jr (1993) Insect diversity in the fossil record. Science 261(5119):310–315

Lal R (1988) Effects of macrofauna on soil properties in tropical ecosystems. Agr Ecosyst Environ 24:101–116

Larson BMH, Kevan PG, Inouye DW (2001) Flies and flowers: taxonomic diversity of anthophiles and pollinators. Can Entomol 133:439–465

Lasalle J (1993) Parasitic hymenoptera, biological control and biodiversity. In: Lasalle J, Gauld ID (eds) Hymenoptera and biodiversity. CAB international, pp 197–215

Lavelle P (1997) Faunal activities and soil processes: adaptive strategies that determine ecosystem function. Adv Ecol Res 27:93–132

Lawrence KL, Wise DH (2004) Unexpected indirect effect of spiders on the rate of litter disappearance in a deciduous forest. Pedobiologia 48:149–157

Le Borgne H, Hébert C, Dupuch A, Bichet O, Pinaud D, Fortin D (2018) Temporal dynamics in animal community assembly during post-logging succession in boreal forest. PLoS ONE 13(9):e0204445. https://doi.org/10.1371/journal.pone.0204445

Légaré J-P, Hébert C, Ruel J-C (2011) Alternative silvicultural practices in irregular boreal forests: response of beetle assemblages. Silva Fenn 45:937–956

Lindenmayer DB, Burton PJ, Franklin J (2008) Salvage logging and its ecological consequences. Island Press, Washington, DC

Losey JE, Vaughan M (2006) The economic value of ecological services provided by insects. Bioscience 56:311–323

Maaroufi H, Potvin M, Cusson M, Levesque RC (2021) Novel antimicrobial anionic cecropins from the spruce budworm feature a poly-L-aspartic acid C-terminus. Proteins 89:1205–1215

MacLean DA (1980) Vulnerability of fir-spruce stands during uncontrolled spruce budworm outbreaks: a review and discussion. For Chron 56:213–221

Maltais J, Régnière J, Cloutier C, Hébert C, Perry DF (1989) Seasonal biology of Meteorus trachynotus Vier. (Hymenoptera: Braconidae) and of its overwintering host Choristoneura rosaceana (Harr.) (Lepidoptera: Tortricidae). Can Entomol 121:745–756

Marchese C (2015) Biodiversity hotspots: a shortcut for a more complicated concept. Glob Ecol Conserv 3:297–309

Marshall VG, Kevan DKM, Matthews JV, Tomlin AD (1982) Status and research needs of Canadian soil arthropods. Bulletin of the Entomological Society of Canada, Supplement 14: 5 p

Mattson WJ, Addy ND (1975) Phytophagous insects as regulators of forest primary production. Science 190:515–522

May RM (1988) How many species are there on Earth? Science 241:1441–1449

Mbata KJ, Chidumayo EN, Lwatula CM (2002) Traditional regulation of edible caterpillar exploitation in the Kopa area of Mpika district in northern Zambia. J Insect Conserv 6:115–130

McGugan BM, Blais JR (1959) Spruce budworm parasite studies in Northwestern Ontario. Can Entomol 91:758–783

McNaughton SJ, Ruess RW, Seagle SW (1988) Large mammals and process dynamics in African ecosystems. Bioscience 38:794–800

Meyer-Rochow VB (2017) Therapeutic arthropods and other, largely terrestrial, folk-medicinally important invertebrates: a comparative survey and review. J Ethnobiol Ethnomed 13:31 p. https://doi.org/10.1186/s13002-017-0136-0

Michalko R, Pekár S, Entling MH (2019) An updated perspective on spiders as generalist predators in biological control. Oecologia 189:21–36

Moore BA, Allard G (2011) Abiotic disturbances and their influence on forest health. A review. Forests Health & Biosecurity Working Paper FBS/35E, Rome, Italy

Moreau G, Quiring DT, Eveleigh ES, Bauce E (2003) Advantages of a mixed diet: feeding on several foliar age classes increases the performance of a specialist insect herbivore. Oecologia 135:391–399

Moretti M, De Caceres M, Pradella C, Obrist MK, Wermelinger B, Legendre P, Duelli P (2010) Fire-induced taxonomic and functional changes in saproxylic beetle communities in fire sensitive regions. Ecography 33:760–777

Morris JL, Cottrell S, Fettig CJ, Hansen WD, Sherriff RL, Carter VA, Clear JL, Clement J, DeRose RJ, Hicke JA, Higuera PE, Mattor KM, Seddon AWR, Seppa HT, Stednick JD, Seybold SJ (2017) Managing bark beetle impacts on ecosystems and society: priority questions to motivate future research. J Appl Ecol 54:750–760

Muona J, Rutanen I (1994) The short-term impact of fire on the beetle fauna in boreal coniferous forest. Ann Zool Fenn 31:109–121

Murillas Gómez M (2013) Impact du longicorne noir, Monochamus scutellatus scutellatus, sur l'épinette noire et le sapin baumier à la suite de chablis en forêt boréale irrégulière. MSc Thesis, Université Laval, 46 p

Natural Resources Canada (2015) Trees, insects and diseases of Canada's forests. https://tidcf.nrcan.gc.ca/en/insects/diet

Ness JH, Bressmer K (2005) Abiotic influences on the behaviour of rodents, ants, and plants affect an ant-seed mutualism. Ecoscience 12:76–81

Nichols E, Spector S, Louzada J, Larsen T, Amezquita S, Favila ME, The Scarabaeinae Research Network (2008) Ecological functions and ecosystem services provided by Scarabaeinae dung beetles. Biol Cons 141:1461–1474

Niemelä J (1997) Invertebrates and boreal forest management. Conserv Biol 11:601–610

Noriega JA, Hortala J, Azcárate FM, Berg MP, Bonada N, Briones MJI, Del Toro I, Goulson D, Ibanez S, Landis DA, Moretti M, Potts SG, Slade EM, Stout JC, Ulyshen MD, Wackers FL, Woodcock BA, Santos AMC (2018) Research trends in ecosystem services provided by insects. Basic Appl Ecol 26:8–23

Norvez O, Bélanger L, Hébert C (2013) Impact of salvage logging on stand structure and beetle diversity in boreal balsam fir forest, 20 years after a spruce budworm outbreak. For Ecol Manage 302:122–132

Novotny V, Miller SE, Baje L, Balagawi S, Basset Y, Cizek L, Craft KJ, Dem F, Drew RAI, Hulcr J, Leps J, Lewis OT, Pokon R, Stewart AJA, Samuelson GA, Weiblen GD (2010) Guild-specific patterns of species richness and host specialization in plant–herbivore food webs from a tropical forest. J Anim Ecol 79:1193–1203

Nyffeler M, Birkhofer K (2017) An estimated 400–800 million tons of prey are annually killed by the global spider community. Sci Nat 104:30

O'Connor B, Bojinski S, Roosli C, Schaepman ME (2020) Monitoring global changes in biodiversity and climate essential as ecological crisis intensifies. Eco Inform 55:101033

Ollerton J, Winfree R, Tarrant S (2011) How many flowering plants are pollinated by animals? Oikos 120:321–326

Orford KA, Vaughan IP, Memmott J (2015) The forgotten flies: the importance of non-syrphid Diptera as pollinators. Proc R Soc B 282:20142934

Orr DB (1988) Scelionid wasps as biological control agents: a review. Fla Entomol 71:506–528

Otvos IS (1973) Biological control agents and their role in the population fluctuation of the eastern hemlock looper in Newfoundland. Environ. Can. Can. For. Serv. Newfoundland For. Res. Cent. St. John's, NF. N-X-102

Otvos IS, Bryant DG (1972) An extraction method for rapid sampling of eastern hemlock looper eggs, *Lambdina fiscellaria fiscellaria* (Lepidoptera: Geometridae). Can Entomol 104:1511–1514

Oxbrough A, Ziesche TM (2013) Spiders in forest ecosystems. In: Kraus D, Krumm F (eds) Integrative approaches as an opportunity for the conservation of forest biodiversity. European Forest Institute, Chapter 3.6, pp 186–193

Paine TD, Birch MC, Svihra P (1981) Niche breadth and resource partitioning by four sympatric species of bark beetles (Coleoptera: Scolytidae). Oecologia 48:1–6

Parmenter RR, MacMahon JA (2009) Carrion decomposition and nutrient cycling in a semiarid shrub-steppe ecosystem. Ecol Monogr 79:637–661

Paula MC, Morishita GM, Cavarson CH, Gonçalves CR, Tavares PRA, Mendonça A, Suarez YR, Antonialli WF Jr (2016) Action of ants on vertebrate carcasses and blow flies (Calliphoridae). J Med Entomol 53:1283–1291

Payne JA (1965) A summer carrion study of the baby pig *Sus scrofa* Linnaeus. Ecology 46:592–602

Pelletier G, Piché C (2003) Species of *Telenomus* (Hymenoptera: Scelionidae) associated with the hemlock looper (Lepidoptera: Geometridae) in Canada. Can Entomol 135:23–39

Peters MK, Hemp A, Appelhans T, Behler C, Classen A, Detsch F, Ensslin A, Ferger SW, Frederiksen SB, Gebert F, Haas M, Helbig-Bonitz M, Hemp C, Kindeketa WJ, Mwangomo E, Ngereza C, Otte I, Röder J, Rutten G, Schellenberger Costa D, Tardanico J, Zancolli G, Deckert J, Eardley CD, Peters RS, Rödel M-O, Schleuning M, Ssymank A, Kakengi V, Zhang J, Böhning-Gaese K, Brandl R, Kalko EKV, Kleyer M, Nauss T, Tschapka M, Fischer M, Steffan-Dewenter I (2016) Predictors of elevational biodiversity gradients change from single taxa to the multi-taxa community level. Nat Commun 7:13736. https://doi.org/10.1038/ncomms13736

Pham AT, De Grandpré L, Gauthier S, Bergeron Y (2004) Gap dynamics and replacement patterns in gaps of the northeastern boreal forest of Quebec. Can J For Res 34:353–364

Pianka ER (1966) Latitudinal gradients in species diversity: a review of concepts. Am Nat 100:33–46

Piaszczyk W, Lasota J, Blonska E (2020) Effect of organic matter released from deadwood at different decomposition stages on physical properties of forest soils. Forests 11:24

Pilon J-G (1965) Bionomics of the spruce budmoth, *Zeiraphera ratzeburgiana* (Ratz.) (Lepidoptera: Olethreutidae). Phytoprotection 46:5–13

Pohl GR, Langor DW, Spence JR (2007) Rove beetles and ground beetles (Coleoptera: Staphylinidae, Carabidae) as indicators of harvest and regeneration practices in western Canadian foothills forests. Biol Cons 137:294–307

Pook VG, Sharkey MJ, Wahl DB (2016) Key to the Species of *Megarhyssa* (Hymenoptera, Ichneumonidae, Rhyssinae) in America, North of Mexico. Deutsche Entomologische Zeitschrift 63:137–148

Potapov P, Hansen MC, Laestadius L, Turubanova S, Yaroshenko A, Thies C, Smith W, Zhuravleva I, Komarova A, Minnemeyer S, Esipova E (2017) The last frontiers of wilderness: tracking loss of intact forest landscapes from 2000 to 2013. Sci Adv 3:e1600821

Price PW (1984) Insect ecology, 2nd edn. Wiley, New York, NY

Rafes P-M (1971) Pests and the damage which they cause to forests. In: Duvigneaud P (ed) Productivité des écosystèmes forestiers. Unesco, Paris, pp 357–367

Régnier M (2020) Colonisation, par les insectes xylophages, du sapin baumier défolié par la tordeuse des bourgeons de l'épinette. MSc Thesis, Université Laval, 72 p

Retallack GJ (1997) Early forest soils and their role in Devonian global change. Science 276:583–585

Riechert SE, Lockley T (1984) Spiders as biological control agents. Annu Rev Entomol 29:299–320

Ritter E, Dalsgaard L, Einhorn KS (2005) Light, temperature and soil moisture regimes following gap formation in a semi-natural beech-dominated forest in Denmark. For Ecol Manage 206:15–33

Rivers JW, Galbraith SM, Cane JH, Schultz CB, Ulyshen MD, Kormann UG (2018) A review of research needs for pollinators in managed conifer forests. J Forest 116:563–572

Rodrigues ASL, Brooks TM (2007) Shortcuts for biodiversity conservation planning: the effectiveness of surrogates. Annu Rev Ecol Evol Syst 38:713–737

Root RB (1967) The niche exploitation pattern of the Blue-Gray Gnatcatcher. Ecol Monogr 37:317–350

Roskov Y, Ower G, Orrell T, Nicolson D, Bailly N, Kirk PM, Bourgoin T, DeWalt RE, Decock W, van Nieukerken E, Zarucchi J, Penev L (eds) (2021) Species 2000 & ITIS catalogue of life, 2021-06-10. Digital resource at www.catalogueoflife.org. Species 2000: Naturalis, Leiden, the Netherlands. ISSN 2405-8858

Saab VA, Latif QS, Rowland MM, Johnson TN, Chalfoun AD, Buskirk SW, Heyward JE, Dresser MA (2014) Ecological consequences of Mountain Pine Beetle outbreaks for wildlife in Western North American forests. For Sci 60:539–559

Saint-Germain M, Drapeau P, Hébert C (2004) Comparison of Coleoptera assemblages from a recently burned and unburned black spruce forests of northeastern North America. Biol Cons 118:583–592

Samways MJ (2007) Insect conservation: a synthetic management approach. Annu Rev Entomol 52:465–487

Samways MJ (2018) Insect conservation for the twenty-first century. In: Insect science—diversity, conservation and nutrition. Intech Open, Chapter 2

Samways MJ, Barton PS, Birkhofer K, Chichorro F, Deacon C, Fartmann T, Fukushima CS, Gaigher R, Habel JC, Hallmann CA, Hill MJ, Hochkirch A, Kaila L, Kwak ML, Maes D, Mammola S, Noriega JA, Orfinger AB, Pedraza F, Pryke JS, Roque FO, Settele J, Simaika JP, Stork NE, Suhling F, Vorster C, Cardoso P (2020) Solutions for humanity on how to conserve insects. Biol Cons 242:108427

Schabel H (2010) Forest insects as food: a global review. In: Durst PB, Johnson DV, Leslie RN, Shono K (eds) Forest insects as food: humans bite back. Proceedings of a workshop on Asia-Pacific resources and their potential for development. 19–21 Feb 2008, Chiang Mai, Thailand, pp 37–64

Schmitz OJ, Raymond PA, Estes JA, Kurz WA, Holtgrieve GW, Ritchie ME, Schindler DE, Spivak AC, Wilson RW, Bradford MA, Christensen V, Deegan L, Smetacek V, Vanni MJ, Wilmer CC (2014) Animating the Carbon cycle. Ecosystems 17:344–359

Schowalter TD (2012) Insect herbivore effects on forest ecosystem services. J Sustain For 31:518–536

Schowalter TD (2017) Arthropod diversity and functional importance in old-growth forests of North America. Forests 8:97. https://doi.org/10.3390/f8040097

Schowalter TD, Noriega JA, Tscharntke T (2018) Insect effects on ecosystem services – Introduction. Basic Appl Ecol 26:1–7

Schütz S, Weissbecker B, Hummel HE, Apel K-H, Schmitz H, Bleckmann H (1999) Insect antenna as a smoke detector. Nature 398:298

Scott MP (1998) The ecology and behavior of burying beetles. Annu Rev Entomol 43:595–618

Scott MP, Gladstein D (1993) Calculating males? An empirical and theoretical examination of the duration of paternal care in burying beetles. Evol Ecol 7:362–378

Scott AC, Stephenson J, Chaloner WG (1992) Interaction and coevolution of plants and arthropods during the Palaeozoic and Mesozoic. Philos Trans R Soc Lond B 335:129–165

Seibold S, Gossner MM, Simons NK, Blüthgen N, Müller J, Ambar D, Ammer C, Bauhus J, Fischer M, Habel JC, Linsenmair KE, Nauss T, Penone C, Prati D, Schall P, Schulze E-D, Vogt J, Wöllauer S, Weisser WW (2019) Arthropod decline in grasslands and forests is associated with landscape-level drivers. Nature 574:671–674

Seibold S, Rammer W, Hothorn T, Seidl R, Ulyshen MD, Lorz J, Cadotte MW, Lindenmayer DB, Adhikari YP, Aragón R, Bae S, Baldrian P, Barimani Varandi H, Barlow J, Bässler C, Beauchêne J, Berenguer E, Bergamin RS, Birkemoe T, Boros G, Brandl R, Brustel H, Burton PJ, Cakpo-Tossou YT, Castro J, Cateau E, Cobb TP, Farwig N, Fernández RD, Firn J, Gan KS, González G, Gossner MM, Habel JC, Hébert C, Heibl C, Heikkala O, Hemp A, Hemp C, Hjältén J, Hotes S, Kouki J, Lachat T, Yu Liu JL, Luo Y-H, Macandog DM, Martina PE, Mukul SA, Nachin B, Nisbet K, O'Halloran J, Oxbrough A, Pandey JN, Pavlíček T Pawson SM, Rakotondranary JS, Ramanamanjato J-B, Rossi L, Schmidl J, Schulze M, Seaton S, Stone MJ, Stork NE, Suran B, Sverdrup-Thygeson A, Thorn S, Thyagarajan G, Wardlaw TJ, Weisser WW, Yoon S, Zhang N, Müller J 2021 The contribution of insects to global forest deadwood decomposition Seibold S, Rammer W, Hothorn T, Seidl R, Ulyshen MD, Lorz J, Cadotte MW, Lindenmayer DB, Adhikari YP, Aragón R, Bae S, Baldrian P, Barimani Varandi H, Barlow J, Bässler C, Beauchêne J, Berenguer E, Bergamin RS, Birkemoe T, Boros G, Brandl R, Brustel H, Burton PJ, Cakpo-Tossou YT, Castro J, Cateau E, Cobb TP, Farwig N, Fernández RD, Firn J, Gan KS, González G, Gossner MM, Habel JC, Hébert C, Heibl C, Heikkala O, Hemp A, Hemp C, Hjältén J, Hotes S, Kouki J, Lachat T, Yu Liu JL, Luo Y-H, Macandog DM, Martina PE, Mukul SA, Nachin B, Nisbet K, O'Halloran J, Oxbrough A, Pandey JN, Pavlíček T Pawson SM, Rakotondranary JS, Ramanamanjato J-B, Rossi L, Schmidl J, Schulze M, Seaton S, Stone MJ, Stork NE, Suran B, Sverdrup-Thygeson A, Thorn S, Thyagarajan G, Wardlaw TJ, Weisser WW, Yoon S, Zhang N, Müller J (2021) The contribution of insects to global forest deadwood decomposition. Nature 597:77–81. https://doi.org/10.1038/s41586-021-03740-8

Shaw CH, Lundkvist H, Moldenke A, Boyle JR (1991) The relationships of soil fauna to long-term forest productivity in temperate and boreal ecosystems: processes and research strategies. In Dyck WJ, Mess CA (eds) Long-term field trials to assess environmental impacts of harvesting. Proceedings, IEA/BE T6/A6 Workshop, Florida, February 1990. IEA/BE T6/A6 Report No 5. Forest Research Institute, Rotorua, New Zealand, FRI Bulletin No 161, pp 39–77

Shubeck PP (1985) Orientation of carrion beetles to carrion buried under shallow layers of sand (Coleoptera: Silphidae). Entomol News 96:163–166

Siitonen J, Saaristo L (2000) Habitat requirements and conservation of *Pytho kolwensis*, a beetle species of old-growth boreal forest. Biol Cons 94:211–220

Simberloff D, Dayan T (1991) The guild concept and the structure of ecological communities. Annu Rev Ecol Syst 22:115–143

Skillen EL, Pickering J, Sharkey MJ (2000) Species richness of the Campopleginae and Ichneumoninae (Hymenoptera: Ichneumonidae) along a latitudinal gradient in eastern North America old-growth forests. Environ Entomol 29:460–466

Slade EM, Mann DJ, Villanueva JF, Lewis OT (2007) Experimental evidence for the effects of dung beetle functional group richness and composition on ecosystem function in a tropical forest. J Anim Ecol 76:1094–1104

Speight MC (1989) Saproxylic invertebrates and their conservation. Nature and Environment Series, No 42. Council of Europe, Strasbourg. 79 p

Speight MR, Hunter MD, Watt AD (2008) Ecology of insects: concepts and applications, 2nd edn. Wiley-Blackwell, Oxford, UK

Spence JR (2001) The new boreal forestry: adjusting timber management to accommodate biodiversity. Trends Ecol Evol 16:591–593

Spence JR, Langor DW, Niemelä JK, Cárcamo HA, Currie CR (1996) Northern forestry and carabids: the case for concern about old-growth species. Ann Zool Fenn 33:173–184

Stenbacka F, Hjalten J, Hilszcazanski J, Ball JP, Gibb H, Johansson T, Pettersson RB, Danell K (2010) Saproxylic parasitoid (Hymenoptera, Ichneumonidea) communities in managed boreal forest landscapes. Insect Conserv Divers 3:114–123

Stephenson NL, Das AJ, Ampersee NJ, Bulaon BM, Yee JL (2018) Which trees die during drought? The key role of insect host tree selection. J Ecol 107:2383–2401

Steuer P, Sudekum K-H, Tutken T, Muller DWF, Kaandorp J, Bucher M, Clauss M, Hummel J (2014) Does body mass convey a digestive advantage for large herbivores? Funct Ecol 28:1127–1134

Stireman JO III (2002) Host location and selection cues in a generalist tachinid parasitoid. Entomol Exp Appl 103:23–34

Stokland JN (2012) The saproxylic food web. In: Stokland JN, Siitonen J Jonsson BG (eds) Biodiversity in dead wood. Cambridge University Press, Cambridge.

Stork NE (2018) How many species of insects and other terrestrial arthropods are there on Earth? Annu Rev Entomol 63:31–45

Strid Y, Schroeder M, Lindahl B, Ihrmark K, Stenlid J (2014) Bark beetles have a decisive impact on fungal communities in Norway spruce stem sections. Fungal Ecol 7:47–58

Strong DR, Levin DA (1979) Species richness of plant parasites and growth form of their hosts. Am Nat 114:1–22

Strong DR, Lawton JH, Southwood R (1984) Insects on plants: community patterns and mechanisms. Blackwell Scientific Publications, Harvard University Press, Cambridge, MA

Suzuki S (2000) Carrion burial by *Nicrophorus vespilloides* (Coleoptera: Silphidae) prevents fly infestation. Entomol Sci 3:269–272

Suzuki S (2001) Suppression of fungal development on carcasses by the burying beetle *Nicrophorus quadripunctatus* (Coleoptera: Silphidae). Entomol Sci 4:403–405

Swift MJ, Heal OW, Anderson JM (1979) Decomposition in terrestrial ecosystems. University of California Press, Berkeley and Los Angeles

Taylor BR, Parkinson D, Parsons WFJ (1989) Nitrogen and lignin content as predictors of litter decay rates: a microcosm test. Ecology 70:97–104

Taylor SL, MacLean DA (2009) Legacy of insect defoliators: increased wind-related mortality two decades after a spruce budworm outbreak. For Sci 55:256–267

Thompson BM, Bodart J, McEwen C, Gruner DS (2014) Adaptations for symbiont-mediated external digestion in *Sirex noctilio* (Hymenoptera: Siricidae). Ann Entomol Soc Am 107:453–460

Thompson JN (1994) The coevolutionary process. University of Chicago Press, Chicago, IL

Thorn S, Bässler C, Brandl R, Burton PJ, Cahall R, Campbell JL, Castro J, Choi C-Y, Cobb T, Donato DC, Durska E, Fontaine JB, Gauthier S, Hébert C, Hothorn T, Hutto RL, Lee E-J, Leverkus AB, Lindenmayer DB, Obrist MK, Rost J, Seibold S, Seidl R, Thorn D, Waldron K, Wermelinger

B, Winter MB, Zmihorski M, Müller J (2018) Impacts of salvage logging on biodiversity: a meta-analysis. J Appl Ecol 55:279–289. https://doi.org/10.1111/1365-2664.12945

Thorpe HC, Thomas SC (2007) Partial harvesting in the Canadian boreal: success will depend on stand dynamic responses. For Chron 83:319–325

Townes H, Townes, M (1960) Ichneumon-flies of America North of Mexico: subfamilies Ephialtinae, Xoridinae, Acaenitinae. United States National Museum Bulletin 216, Part 2. Smithsonian Institution Press, Washington, 676 p

Trumbo ST (1994) Interspecific competition, brood parasitism, and the evolution of biparental cooperation in burying beetles. Oikos 69:241–249

Ulyshen MD (2016) Wood decomposition as influenced by invertebrates. Biol Rev 91:70–85

United Nations (1992) Convention on biological diversity. 28 p. https://www.cbd.int/convention/text/

Vergutz L, Manzoni S, Porporato A, Ferreira Novas R, Jackson RB (2012) Global resorption efficiencies and concentrations of carbon and nutrients in leaves of terrestrial plants. Ecol Monogr 82:205–220

Vidal MC, Murphy SM (2018) Bottom-up vs. top-down effects on terrestrial insect herbivores: a meta-analysis. Ecol Lett 21:138–150

Vindstad OPL, Schultze S, Jepsen JU, Biuw M, Kapari L, Sverdrup-Thygeson A, Ims RA (2014) Numerical responses of saproxylic beetles to rapid increases in dead wood availability following Geometrid moth outbreaks in sub-arctic mountain birch forest. PLoS ONE 9:e99624

Wagner DL (2020) Insect declines in the Anthropocene. Annu Rev Entomol 65:457–480

Warkentin IG, Bradshaw CJA (2012) A tropical perspective on conserving the boreal 'lung of the planet.' Biol Cons 151:50–52

Warren MS, Key RS (1991) Woodlands: past, present and potential for insects. In: Collins NM, Thomas JA (eds) The conservation of insects and their habitats. Academic Press, London, pp 155–212

Watkins E, Kitching RL, Nakamura A, Stork NE (2017) Beetle assemblages in rainforest gaps along a subtropical to tropical latitudinal gradient. Biodivers Conserv 26:1689–1703

Wenny DG (2001) Advantages of seed dispersal: a re-evaluation of directed dispersal. Evol Ecol Res 3:51–74

Wermelinger B, Duelli P, Obrist MK (2002) Dynamics of saproxylic beetles (Coleoptera) in windthrow areas in alpine spruce forests. For Snow Landsc Res 77:133–148

Wermelinger B, Moretti M, Duelli P, Lachat T, Pezzatti GB, Obrist MK (2017) Impact of windthrow and salvage-logging on taxonomic and functional diversity of forest arthropods. For Ecol Manage 391:9–18

Wikars L-O (1997a) Pyrophilous insects in orsa Finnark, central Sweden: biology, distribution and conservation. Entomol Tidskr 118:155–169

Wikars L-O (1997b) Effects of forest fire and the ecology of fire-adapted insects. Acta Univ. Ups. Comprehensive Summaries of Uppsala Dissertations fiom the Faculty of Science and Technology 272.35 pp. Uppsala. ISBN 91 -554-3954-3

Wikars L-O (2002) Dependence on fire in wood-living insects: an experiment with burned and unburned spruce and birch logs. J Insect Conserv 6:1–12

Willig MR, Kaufman DM, Stevens RD (2003) Latitudinal gradients of biodiversity: pattern, process, scale and synthesis. Annu Rev Ecol Evol Syst 34:273–309

Wilson GG (1973) Incidence of microsporidia in a field population of spruce budworm. Environ. Can. For. Serv. Bi-Monthly Research Notes 29:35–36

Wilson EE, Wolkovitch EM (2011) Scavenging: how carnivores and carrion structure communities. Trends Ecol Evol 26:129–135

Wilson MF, Traverset A (2000) The ecology of seed dispersal. In: Fenner M (ed) Seeds: the ecology of regeneration in plant communities, 2nd edn, Chapter 4, pp 85–110

Wirth C, Gleixner G, Heimann M (2009) Old-growth forests: function, fate and value—an overview. In: Wirth C, et al (eds) Old-growth forests. Ecological Studies 207. Springer-Verlag, Berlin Heidelberg, Chapter 1

Yang LH (2004) Periodical Cicadas as resource pulses in North American forests. Science 306:1565–1567
Yang LH, Gratton C (2014) Insects as drivers of ecosystem processes. Curr Opin Insect Sci 2:26–32
Zanella A, Ponge J-F, Briones MJI (2018) Humusica 1, article 8: terrestrial humus systems and forms – Biological activity and soil aggregates, space-time dynamics. Appl Soil Ecol 122:103–137
Zimmer M, Topp W (2002) The role of coprophagy in nutrient release from feces of phytophagous insects. Soil Biol Biochem 34:1093–1099

Chapter 4
Insect Ecology

Laurel J. Haavik and Fred M. Stephen

4.1 Introduction

Insect ecology is the study of how insects interact with the environment. The environment consists of both physical characteristics (abiotic) and other organisms (biotic). Insects are natural components of forests and perform a variety of essential functions that help maintain forests as ecosystems. As consumers of forest products, people sometimes compete with insects for forest resources. Most research and management efforts in forest entomology have focused on insects that damage or kill large numbers of ecologically or economically important trees. In this chapter, we consider the various environmental challenges that confront forest insects, and the adaptations they have evolved to be successful in forest ecosystems.

4.2 Insects Assume Many Roles in Forests

Insects are ubiquitous in forests because of many remarkable adaptations that allow them to survive and reproduce. They perform a wide variety of functions that influence and maintain ecosystem services. These functional roles fit into a hierarchy of trophic levels, characterized by who eats who (Price et al. 2011; Speight et al. 2008). Plants are primary producers because they convert electromagnetic energy (light) into chemical energy through photosynthesis. Herbivorous insects (and other animals) that eat seeds, flowers, leaves, stems, roots, or other plant parts are primary consumers. Insects (and other animals) that are predators and parasitoids that prey

L. J. Haavik (✉)
USDA Forest Service, Washington, DC, USA
e-mail: laurel.haavik@usda.gov

F. M. Stephen
University of Arkansas, Fayetteville, AR, USA

© The Author(s) 2023
J. D. Allison et al. (eds.), *Forest Entomology and Pathology*,
https://doi.org/10.1007/978-3-031-11553-0_4

on herbivores are secondary consumers. A hyperparasitoid, a parasitoid that attacks another parasitoid, is an example of a tertiary consumer (Hajek 2004). Insects serve as prey for many other tertiary and quaternary consumers (mammals, reptiles, and birds). Some insects are detritivores (also called saprophages or decomposers) that consume and break apart organic matter (dead plants, animals, and fungi). The organic matter is subsequently recycled into its nutrient components by microbes (bacteria and fungi) and primary producers ultimately use the nutrient components. Feeding guilds are composed of consumers at the same trophic level, which in the case of forest insects, may be even further specialized. For example, seed and cone insects feed on reproductive tissue of trees, woodborers feed in woody tissue of trees, and sap feeders extract liquid from inside leaves or bark. Producers and consumers in a forest community form complex networks, or webs, rather than a simple food chain, because consumers often feed on more than one species of prey (and trophic level). Fundamentally, this web explains nutrient and energy flow, and cycles within forest ecosystems.

Forest insects can affect the balance in nutrient and energy flow from primary producers through all levels of consumers to decomposers. A natural component of forest ecosystems, insect populations (a group of individuals of the same species that inhabit an area) that increase to outbreak levels and cause landscape-scale tree mortality are agents of disturbance that can selectively kill certain tree species. Insect outbreaks can alter the structure, age class diversity, and composition of forest patches on the landscape, and in extreme cases this may re-set forest succession to an earlier stage (Coulson and Stephen 2006). Succession is the natural and predictable process of change in the forest community over time, from the earliest colonizers (e.g. fast-growing trees and other plants that are poor competitors and thrive in sunny conditions), to the latest (climax) colonizers (e.g. slow-growing trees and other plants that are good competitors and tolerant of shade).

The balance in nutrient and energy flow may be altered if a non-native species is introduced to a forest (Gandhi and Herms 2010), or if a native species expands its range into a forest it has not previously occupied. Some non-native species are more disruptive to nutrient and energy flow than others. For instance, the hemlock woolly adelgid, *Adelges tsugae* (Annand), threatens to kill nearly all eastern hemlock in North America (Ellison et al. 2005). In hemlock forests, wildlife and ecosystem processes (nutrient cycling, hydrology) depend heavily on eastern hemlock as a foundation species. Hemlock mediates soil moisture levels, stabilizes stream flow, and decreases daily variation in stream temperatures, which results in a community of freshwater invertebrates and other animals that cannot survive in a forest without hemlock (Ellison et al. 2005). In contrast, some invasive species seem to pose little threat to critical ecosystem functions and behave much like naturalized residents. For instance, the European woodwasp, *Sirex noctilio* F., mainly kills non-competitive and otherwise stressed pines in northeastern North America, and seems to coexist with a suite of other subcortical pine insects, essentially functioning as a forest thinning agent (Dodds et al. 2010; Foelker 2016; Haavik et al. 2018). When considering an insect a forest pest, it is important to keep in mind its natural functional role in the food web. If its presence or activity has altered the function/s of other members of the web, such as with hemlock woolly adelgid, then the ecological balance of the system will be disrupted.

4.3   Species Interact in Many Ways

Forest insects have a variety of different relationships with other species. These symbiotic relationships often facilitate acquisition of resources for one or more members of the association. They are best considered on a continuum, in which the relationship between two species ranges from positive, to neutral, to negative (Price et al. 2011) (Fig. 4.1). These relationships are fluid, and sometimes difficult to delineate. They can change throughout insect life cycles, in different environments, and throughout evolutionary time in response to varying selection pressures.

Interactions between herbivorous insects and trees are of particular concern to forest scientists and managers. These interactions might have negative outcomes for the tree and the forest products it provides. Insects gain nutrition from feeding on a tree organ (e.g. cones or acorns, leaves, bark, phloem, wood, or roots); as a result, the tree can lose essential reproductive, photosynthetic, vascular, structural, or nutrient-acquiring tissue. Alternatively, the interaction could be positive. Pollination is a classic example of mutualism: pollinating insects (e.g. beetles, butterflies, moths, bees, ants) gain nutrition from the plant by feeding on nectar or pollen, and the plant gains a method of dispersal for its genes, as the insect carries pollen grains from one plant to fertilize another. In mutualistic partnerships, insects often serve as dispersal

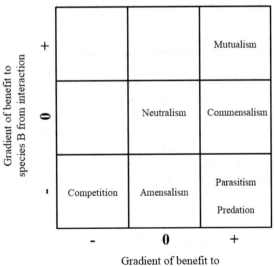

Fig. 4.1 Possible relationships between two species, A and B. Adapted from and see Price et al. (2011) page 234 for details

agents for organisms that lack or have limited mobility, such as fungi, mites, plants, or nematodes.

More complex, multi-species symbiotic relationships among forest insects and other organisms involve bark beetles, several other insects, mites, and fungi (Hofstetter et al. 2015). A well-known example is the southern pine beetle, *Dendroctonus frontalis* Zimmerman, system. Southern pine beetles often carry three different species of fungi on their bodies or in specialized structures termed mycangia. All three fungi gain dispersal from the beetles and grow in phloem or xylem tissues. Two fungi compete with one another for phloem and are clear mutualists for the beetle, creating a more nutritious substrate than pine phloem for beetle larvae to eat. The role of the third fungal species is under debate, and seems to be important for initial colonization of the tree (Klepzig and Hofstetter 2011). Southern pine beetles also carry mites, in a potentially commensal manner, because the mites benefit from transportation by the beetles, though a strong positive or negative effect on the beetles has not yet been observed (Klepzig and Hofstetter 2011). In addition, by colonizing and killing pines, southern pine beetles facilitate a suitable food resource (dying trees) for over 100 other insects, including *Ips* bark beetles and wood borers (primarily *Monochamus* species), some of which will ultimately compete with one another for phloem (Dixon and Payne 1979).

Competitive interactions beneath the bark of southern pines have been difficult to specify, partly because the habitat is cryptic and partly because the resource is ephemeral. The southern pine sawyer, *Monochamus titillator* (F.), is attracted to and develops in stressed, damaged and weakened pine trees. Adult *Monochamus* are attracted to host trees by a combination of host volatiles, including α-pinene and ethanol. Ipsenol, the aggregation pheromone of *Ips* bark beetles, is also a powerful attractant to many *Monochamus* species (Allison et al. 2001; Miller et al. 2013). *Monochamus* beetles are much larger than bark beetles, and consequently consume much more phloem. Laboratory and field studies found that *Monochamus* larvae will feed on bark beetle larvae in the phloem, termed facultative intraguild predation (Dodds et al. 2001; Schoeller et al. 2012). *Monochamus* larvae can thus outcompete bark beetle larvae for phloem indirectly and feed on them directly. *Monochamus titillator* may be an important facultative natural enemy of bark beetles, potentially contributing to the collapse of southern pine beetle infestations (Stephen 2011). In addition, there are numerous parasitoids, as well as predators, that comprise the natural enemy complex that preys upon southern pine beetle and its phloem-consuming associates (Stephen et al. 1993). Many other complex relationships among species in forest ecosystems remain undescribed.

4.4 Life Histories Vary

In order to survive from egg to adult, and to reproduce successfully, insects must escape or endure environmental extremes, avoid predation, avoid or endure parasitism, acquire the nutrients necessary to grow, and find mates. Forest insects have evolved a myriad of adaptations to cope with these environmental challenges. Many theoretical categories have been developed to group species with similar adaptations, life history traits, and the trade-offs that accompany them. Below, we consider some of these important ideas as they relate to forest insects.

4.4.1 K- and r-Selection: Forces in the Environment Dictate Reproductive Adaptations

MacArthur and Wilson (1967) introduced the idea of natural selection operating to favor high reproductive ability (r) for individuals occurring in uncrowded populations, and to favor competitive ability (K) in crowded populations (see Fig. 4). This idea has been modified, adapted and criticized by numerous authors since its inception. Interpreted as a general framework, it can be a useful tool to evaluate the relative importance of challenges posed to insects by biotic versus abiotic components of the environment (Table 4.1).

Species that reside in harsh habitats, where climatic conditions may be extreme or unpredictable, may share some common life history characteristics (Table 4.1). Species that are small in size, short lived, have high dispersal abilities, and a high population growth rate are said to be r-selected. They are likely to have high fecundity, reproduce early in their life, and reproduce only once. They often exist in early stage successional environments, at population levels well below the carrying capacity of the environment, and the mortality they incur is often from density-independent factors (see Chapter 5).

r-Selected species can be contrasted with species living in habitats that are environmentally stable. For these K-selected species, body size tends to be larger, and individuals live longer, disperse less, and have a lower population growth rate. They may reproduce later in life, and more than once. They produce fewer eggs but invest more energy in each one. They are effective competitors, and their population densities are often nearer to the carrying capacity of the environment (see Sect. 4.6.1). The mortality factors that affect their populations are normally from biotic, density-dependent agents (see Chapter 5).

Some bark beetles could be considered r-selected species, because they are extremely small (only a few mm in length), reproduce in great numbers (100 or so eggs per female) and develop in a nutrient-poor, ephemeral environment (phloem of dying trees). In comparison, their parasitoid natural enemies could be considered K-selected, because they reproduce in fewer numbers (10s of eggs per female), often have lower population growth rates, and develop in a nutrient-rich, relatively stable environment (often feeding within the bodies of developing bark beetles). It is important to remember that the concepts of r and K strategies in relation to life history traits are meaningful only in a relative sense. A given organism is more or less an r strategist only in comparison with another organism, for example.

	Characteristic	r-selected	K-selected
Table 4.1 Generalized characteristics of insects, populations, processes and environments in relation to r- and K-selection	Body size	Small	Large
	Colonization ability	Opportunistic	Non opportunistic
	Dispersal ability	High	Low
	Development rate	Fast	Slow
	Egg size	Small	Large
	Fecundity	High	Low
	Parental investment in offspring	Small	Large
	Longevity	Short	Long
	Age of reproduction	Early	Late
	Frequency of reproduction	Once (few)	Repeated (many)
	Intrinsic rate of increase	High	Low
	Population density level	Fluctuating	Stable
	Intraspecific competition	Scramble	Contest
	Sex ratio	Female biased	Neutral
	Ecological succession	Early seres	Late (climax)
	Density in relation to carrying capacity	Well below	At or near
	Importance of density-dependent processes	Less important	Very important

Adapted from MacArthur and Wilson (1967), Price et al. (2011), and Speight et al. (2008)

4.4.2 Some Insects Specialize by Feeding on Trees in a Particular Condition

Some herbivorous forest insects have been categorized by the condition of the host tree that they typically colonize. This helps forest managers predict which trees and whether a large number of them are likely to be damaged or killed. Bark and woodboring beetles have been grouped as follows: (1) primary bark beetles and wood borers are capable of colonizing healthy trees; (2) secondary bark beetles and wood borers colonize trees that have been stressed or weakened by some other biotic agent or abiotic factor; and (3) saprophytic bark beetles and wood borers colonize dead, or extremely moribund, trees (Hanks 1999; Lindgren and Raffa 2013; Raffa et al. 2015). These groupings can be further subdivided or even considered fluid for some species, especially those that typically attack stressed trees, though are able to colonize healthy trees during outbreaks.

There are far fewer species of primary bark beetles than secondary bark beetles. It has been hypothesized that this may be because primary species have highly specific adaptations to overcome tree resistance mechanisms and to establish associations with symbionts (Lindgren and Raffa 2013). These adaptations likely evolved from intense competition with other species for ephemeral host resources, i.e. dying trees (Lindgren and Raffa 2013). Primary bark beetles such as the southern and mountain pine beetles, *D. frontalis* and *D. ponderosae* (Hopkins), are of significant concern to forest managers because when their densities reach outbreak levels, they have the capacity to kill large numbers of trees very quickly, and outbreaks continue until all suitable trees have been killed. Secondary bark beetles, such as engravers, e.g. *Ips pini* (Say), *Ips grandicollis* (Eichhoff), and *Ips confusus* (LeConte), can also be a threat, especially following environmental disturbance, such as wildfire or drought, although their populations return to low levels once environmental stress has abated.

Much less is known regarding the primary vs. secondary nature of wood borers. Most economically important wood borers are secondary mortality agents that can become aggressive during periods of environmental stress, especially in their native habitats on their co-evolved hosts. Although some, such as the emerald ash borer, *Agrilus planipennis* Fairmaire, Asian longhorned beetle, *Anoplophora glabripennis* (Motschulsky), and the European woodwasp, *Sirex noctilio* F., have aggressively killed apparently healthy trees outside of their native geographical and host ranges. Wood borers that colonize healthy trees usually inhabit branches or twigs and rarely outbreak or kill trees (Solomon 1995). When outbreaks of wood borers do occur, the impact is usually far less severe than an outbreak of primary bark beetles.

Foliage feeders have been termed primary insects (Jactel et al. 2012; Manion 1991), because they also feed on healthy trees and can have landscape-scale impacts. It is not clear whether defoliators consistently prefer trees of a particular condition (Jactel et al. 2012), though outbreaks are often linked to weather (Haynes et al. 2014; Myers 1998), and weakened trees are usually the first and most likely individuals to die from defoliation (Davidson et al. 1999). Major defoliators in North America, such as eastern and western spruce budworms, *Choristoneura fumiferana* (Clemens) and *C. freemani* (syn *occidentalis*) (Freeman), respectively; the non-native European spongy moth, *Lymantria dispar dispar* (L.); Douglas-fir tussock moth, *Orgyia pseudotsugata* (McDunnough); and forest tent caterpillar, *Malacocoma disstria* Hübner, can consume entire forest canopies during outbreaks. Many of these species repeatedly defoliate the same trees for several successive years, which leads to branch dieback and top kill, and sometimes mortality. Most importantly, though, repeated defoliation weakens trees and makes them more susceptible to secondary insects and diseases (Manion 1991). For instance, even though the polyphagous European spongy moth will consume foliage from trees of all susceptible species in a stand, stressed or suppressed trees, especially oaks, will die first (Davidson et al. 1999). Stressed oaks are then usually killed by the secondary mortality agents, twolined chestnut borer, *Agrilus bilineatus* (Weber), and *Armillaria* spp. root disease (Wargo 1977). Healthy oaks that are completely defoliated in spring can draw on stored carbon reserves to re-foliate by summer (Davidson et al. 1999). However, repeated severe defoliation weakens healthy oaks enough that they can also be killed by twolined chestnut borer or *Armillaria* spp. or both (Wargo 1977).

All herbivorous insects face many environmental challenges that affect their ability to colonize and gain sufficient nutrition from their hosts including weather, natural enemies, and plant resistance (defense) (Cornell and Hawkins 1995; Herms and Mattson 1992); which of those is most influential depends on the insect species, environmental conditions, and how environmental conditions impact the host plant. There is strong selection for various adaptations to avoid, tolerate, overcome, or detoxify physical and chemical mechanisms of plant resistance (see Chapter 7). It is useful to think of primary, secondary, and saprophytic forest insects on a continuum, where a species may tend towards one or the other extreme depending on environmental factors and how those factors affect the insects and the condition of their host trees.

4.5   Abiotic Conditions Alter Insect Growth and Survival

In addition to biotic elements, insects are challenged by the physical or environmental characteristics of forest habitats. Important abiotic factors that affect insects are temperature and moisture (precipitation), which we will consider directly in relation to forest insects and indirectly through the trees that they eat. Environmental conditions that influence insects as individuals or at the population level include extreme weather and regional climate or weather patterns. If abiotic conditions trigger an increase or decrease in the size of a forest insect population, the amount of damage to trees is also likely to change.

4.5.1   Temperature Affects Behavior and Development

Insects are poikilothermic animals, meaning they do not regulate their own body temperature. Consequently, ambient temperatures dictate aspects of insect behavior and development. Many insects in temperate climates possess behavioral and physiological adaptations to tolerate or avoid extreme cold (Danks 1978). Insects may undergo a dormant period during winter. This escape of harsh environmental conditions in time is categorized as a diapause that is genetically programmed and is either obligate, occurring at a specific time during development, or facultative, dictated by adverse environmental conditions. Alternatively, some insects only undergo a quiescence, initiated by unfavorable conditions, after which development resumes. During the dormant period, a series of energetically expensive biochemical changes occur that involve synthesis of glycerol and other cryoprotectants that act through several mechanisms as solutes to slow the formation of ice within cells (Danks 1978). The supercooling point, the temperature at which insect body fluids begin to freeze, and the lower lethal temperature, the point at which mortality occurs, both vary seasonally within and among insect species and populations. For instance, cold tolerance of larch casebearer, *Coleophora laricella* Hübner, was greatest in mid-winter and reduced in spring and autumn (Ward et al. 2019). Investigating insect cold tolerance can be especially useful for predicting suitable geographic range of introduced insects or even for native species undergoing range expansion.

Migration represents the most extreme behavioral adaptation to avoid cold in space. Monarch butterflies, *Danaus plexippus* (L.), are an excellent example of insects that overwinter in forest canopies in southern California, Florida, or Mexico, and travel northward throughout the growing season, with some populations reaching Canada (Solensky and Oberhauser 2004). On a more local scale, the microhabitat in which forest insects overwinter can provide some protection from the cold. Many bark- or wood-boring insects overwinter as adults or larvae under the bark, which can be several degrees warmer than the surrounding air temperature (Vermunt et al. 2012). Some foliage feeding insects overwinter at the base of trees in the soil or leaf litter, which also offers insulation from the cold. Others overwinter in the egg stage in bark crevices or other protected places.

Adult insects are often more active during warm, favorable weather. Females may seek warm, bright locations to lay eggs. For instance, emerald ash borer and bronze birch borer, *Agrilus anxius* Gory, females prefer to oviposit on the sunny, southern exposures of tree trunks (Akers and Nielsen 1990; Timms et al. 2006).

Temperature also regulates the rate at which insects develop. Typically, there is a lower temperature threshold below which development does not occur, an optimum temperature at which development is most rapid, and above the optimum, development rate slows until the upper temperature threshold is reached and mortality occurs (Fig. 4.2). For instance, larvae of the Pales weevil, *Hylobius pales* (Herbst), require about 220 days to complete development at 9 °C, 27 days at 30 °C, and 37 days at 32 °C (Salom et al. 1987). Small changes in weather and climate can therefore translate to large changes in generation time. For example, larval development in the six-spined engraver, *Ips calligraphus* (Germar), can range from 18 to 224 days, depending on temperature (Wagner et al. 1987). Generation time is somewhat plastic for many subcortical forest insects, and variations are usually related to regional weather or climate, which are predictable by latitude and elevation. The southern pine beetle is an extreme example, and completes between one and nine generation(s) per year, developing from egg to adult in 26–54 days, dependent upon geographic location in its range (Hain et al. 2011; Thatcher 1960, 1967), which is directly related to growing season length and ultimately temperature.

The tight link between temperature and insect development rate can facilitate accurate predictions of seasonal timing of different insect life stages, particularly adult flight. Degree-day models rely on this premise. The number of heat units (°F or °C) that accumulate above a certain minimum threshold temperature at which insect development proceeds—or conversely, is halted below that threshold (Higley et al. 1986)—is typically calculated during a growing season and can be used to predict when certain insect life stages are present in the forest.

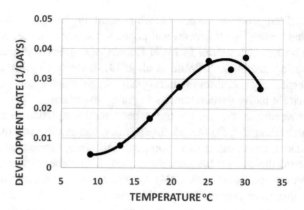

Fig. 4.2 Development rate curve for pales weevil, *Hylobius pales*, larvae at constant temperatures. Dots represent the observed median development rates and the solid line the predicted values over the range of observed development. Lower development threshold (or base temperature) is ca. 9 °C, optimum is ca. 27 °C and upper threshold is ca. 32 °C. Data and adapted figure from Salom et al. (1987)

4.5.2 Precipitation Indirectly Affects Insects by Its Impacts on Trees

Some forest insect outbreaks are linked to drought (Jactel et al. 2012; Mattson and Haack 1987a). Outbreaks of several *Dendroctonus*, *Ips*, *Scolytus*, and *Agrilus* species are often preceded by warm and dry environmental conditions (Mattson and Haack 1987a). It is thought that changes in tree physiology induced by periods of insufficient soil moisture improve suitability of trees for insect growth and development, which in turn results in greater insect survival and reproduction, and increased population growth (see Sect. 4.6). These physiological changes are either related to compromised resistance (defense) or enhanced nutritional value of drought-stressed trees (Mattson and Haack 1987b; Rhoades 1979, 1985; White 1984).

Trees have various mechanisms to tolerate or minimize the effects of drought (Bréda et al. 2006; Pallardy 2008). One mechanism is to adjust solute content in cells, which prevents water loss, by break-down and mobilization of sugars and proteins (i.e. osmotic adjustment). This process presumably makes these essential nutrients more readily available for insect consumption (White 1984). There has been indirect, observational evidence in several feeding guilds that supports this claim and the general theory that plant stress results in improved insect performance (White 2015), but experimental support is lacking. Relationships among environmental stress, insects, and their host trees that result in altered nutrition for insects are likely complex and variable in time.

Trees can invest vast amounts of energy into resisting or tolerating insect attack. For instance, the outcome of bark beetle attack and subsequent colonization of conifers is completely dependent on tree resistance, usually through two systems, constitutive and induced, related to resin production (Berryman 1972; Raffa and

Berryman 1983) (see Chapter 7). If one or both of those systems is compromised as soil moisture becomes limiting for trees, then bark beetles may be more likely to successfully colonize trees or produce more progeny or both. For example, experimental drought predisposed piñon pine, *Pinus edulis*, to bark beetle attack and subsequent mortality (Gaylord et al. 2013), although reduced resin flow was only partially responsible (Gaylord et al. 2015). Resistance in hardwoods has been less studied, but appears to be related to overall tree health and vigor. For instance, landscape-scale oak mortality may result from repeated drought that eventually causes imbalance in carbon storage and use, which reduces tolerance to colonization by secondary wood borers (Haavik et al. 2015).

The degree to which enhanced nutritional value or reduced resistance capacity, or both, contribute to improved conditions for insects when trees are stressed by drought is not completely understood, and is likely variable spatially and temporally. Ultimately, trees must balance energy investment among growth, maintenance, defense, and reproduction (Herms and Mattson 1992). Inadequate precipitation can cause the re-distribution or depletion of energy supplies and stores or both, and some insects may take advantage of this situation. Not all forest insects benefit from drought. The relationship between drought and insect damage to trees is complex and seemingly related to whether the insect species is primary or secondary in nature, its feeding guild, and the severity of drought (Huberty and Denno 2004; Jactel et al. 2012).

4.5.3   Extreme Weather Can Have Indirect Effects Through Trees

Weather events of shorter duration than prolonged stressors like drought can also weaken trees and render them more suitable hosts for insects. For example, frosts that occur late in spring, once hardwoods have already leafed out, temporarily alter normal physiological functions, and may deplete energy reserves needed to defend against insects and diseases. Late-spring frosts have contributed to outbreaks or greater abundance of several forest insects, including twolined chestnut borer (Nichols 1968; Staley 1965), oak splendor beetle, *A. biguttatus* (Fabricius) (Hartmann and Blank 1992; Thomas et al. 2002), and sugar maple borer, *Glycobius speciosus* (Say) (Horsley et al. 2002). Ice storms, tornadoes, or other blowdowns that cause breakage of tree limbs and branches also facilitate increased survival and population growth of insects by providing abundant, yet ephemeral host material for reproduction and development. Some forest insects specialize on broken branches and stems. Pine engravers, *Ips* spp.; Douglas-fir beetle, *Dendroctonus pseudotsugae* Hopkins; the European spruce bark beetle, *Ips typographus* (L.); and the European pine shoot beetle, *Tomicus piniperda* (L.) readily colonize windthrown and freshly cut trees (Gothlin et al. 2000; Rudinsky 1966; Schlyter and Lofqvist 1990).

4.5.4 Climate and Weather Patterns Affect Population Density of Insects Regionally

Fluctuations in insect population density often occur at the same time over large geographic areas, in regions of forest not immediately adjacent to one another. This synchronous timing of landscape-scale insect outbreaks may be driven by large-scale weather and climate patterns. This phenomenon is termed the Moran effect (Moran 1953) and emphasizes the influence of factors acting at larger scales than within a single stand or forest on outbreaks of forest insects. The Moran effect seems to be important in the population dynamics of several different defoliator species, and weather plays at least a partial role. For example, spatially synchronous outbreaks of European spongy moth were determined to be most likely driven by patterns of rainfall (Haynes et al. 2013). Jack pine budworm, *Choristoneura pinus pinus* Freeman, outbreaks are often synchronous across the landscape as well, which could signify Moran processes, but other factors such as budworm dispersal have not been ruled out (McCullough 2000). Similarly, climatic variation was correlated with regional outbreaks of eastern spruce budworm, though spatial variation in outbreaks was more closely linked to forest landscape structure and management history (Robert et al. 2018). Cool springs, which were also associated with a certain point in the sunspot cycle, coincided with outbreaks of several different forest caterpillars on three continents (Myers 1998).

Weather patterns could influence insect development directly or indirectly through the effects on hosts or natural enemies. Spring temperatures are especially important, as insects exit diapause or quiescence, and resume development (see Sect. 4.5.1). Cool springs could directly cause insects to become active later in the season, whereas warm springs may initiate insect activity earlier. Indirectly, timing of budburst is important for foliage feeders that eat new buds or flowers. To avoid reduced fitness (survival or fecundity), these insects, such as western and eastern spruce budworms and winter moth, *Operophtera brumata* (L.), need to be phenologically synchronized with their hosts (van Asch and Visser 2007). A departure from normal spring temperatures can cause a phenological mismatch with budburst for some of these defoliators, which could affect outbreak frequency or intensity (Pureswaran et al. 2015; Visser and Holleman 2018). Prevalence or spread of natural enemies can also be affected by regional or local weather patterns. Cool, wet springs are favorable for spread of *Entomophaga miamaiga*, a fungal pathogen of European spongy moth (Hajek and Tobin 2011), which may influence the collapse of spongy moth outbreaks (Hajek et al. 2015). Thus, in a variety of ways, regional weather patterns can have landscape-scale influence over forest insect activity.

4.6 Insect Population Growth Is a Function of Births, Deaths, and Movement

Abundance of herbivorous insects in a forest usually indicates whether they are likely to be a problem. In other words, are numbers high enough that they will damage or kill an economically or ecologically significant number of trees? Insect abundance per unit area, or population density, and what factors drive changes in population density over time are of critical importance to understanding and predicting the status of pest insects. The number of insects in a population at a given time (N_{t+1}) is determined by the number at a previous time (N_t, when the population was last measured) plus the number of new young ($B = births$), minus the number that perished ($D = deaths$), plus the number that migrated into the population ($I = immigration$), and minus the number that migrated out of the population ($E = emigration$).

$$N_{t+1} = N_t + B - D + I - E \qquad (4.1)$$

A population can grow exponentially. If every female is replaced by two females in the subsequent time period (or generation), then the population is growing by a factor of two (i.e. the growth multiplier or finite rate of increase, λ, is two). This is often expressed as the natural logarithm of λ, and defined as the intrinsic rate of increase (r), which is "the rate of increase per [individual] under specified physical conditions, in an unlimited environment where the effects of increasing density do not need to be considered." (Birch 1948). Exponential population growth rate is described by

$$\frac{dN}{dt} = rN \qquad (4.2)$$

where the change in number of insects at any given time $\left(\frac{dN}{dt}\right)$, is determined by the number in the population (N) multiplied by r per individual (Fig. 4.3a). A critical component of population growth rate, r is sometimes called the exponential rate of increase, the intrinsic rate of increase, the instantaneous rate of increase, or the Malthusian parameter (Price et al. 2011). The intrinsic rate of increase is defined as the number of females produced per female per unit of time (e.g. females/female/year).

4.6.1 The Environment Can Support a Finite Number of Insects

Insect populations in nature do not undergo exponential growth indefinitely. Rapid growth is curtailed as the population density approaches the carrying capacity, K, the theoretical limit for population numbers given the resources of a particular habitat. One resource is often scarcer than all others; typically, this resource is food. The

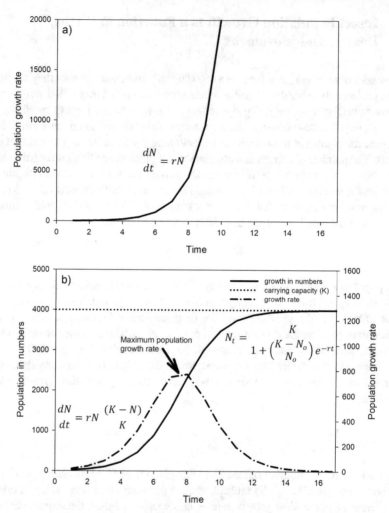

Fig. 4.3 Example of Exponential Population Growth and Population Growth Rate in an (**a**) Unlimited and a (**b**) Limited Environment According to Eqs. 4.2–4.4

combined influence of limited resources will dictate the theoretical numerical value of K, which is determined by both abiotic and biotic variables including weather, natural enemies, and presence of disease. The following equation represents the continuous rate of population growth in a limited environment

$$\frac{dN}{dt} = rN\frac{(K-N)}{K} \tag{4.3}$$

and the discrete numbers of individuals in the population at a particular time is given by the logistic equation:

$$N_t = \frac{K}{1 + \left(\frac{K - N_o}{N_o}\right)e^{-rt}} \tag{4.4}$$

As illustrated in Fig. 4.3a, population growth rate increases exponentially without limits. When limits to growth are considered, theoretical population growth rate (Eq. 4.3) and population numbers (Eq. 4.4) can be illustrated by the curves shown in Fig. 4.3b. In Fig. 4.3b, when N is low with respect to K, then rN is not affected, and population growth and growth rate appear exponential (both lines, left side of the figure). As N grows larger and approaches K, the population growth rate (rN) begins to slow (dashed line, middle of the figure), eventually approaching zero as the population approaches K (both lines, right side of the figure).

Below, we briefly examine factors that influence B, D, I, and E, how those factors affect population growth, and describe a useful way to measure them.

4.6.2 Births

Forest insect populations can grow exponentially in a short period of time (months or years), partly because insect generation times are short (relative to trees), and partly because females are highly fecund. Fecundity, measured by number of eggs, can be described as potential or realized. It is potential in regard to the total number of eggs a female can produce in her lifetime, and realized in regard to the number that she actually lays. Fecundity varies among and within species.

4.6.3 Deaths

There are a multitude of sources of insect mortality. Some important sources include natural enemies, intra- and interspecific competition, and failure to acquire necessary nutrients. Natural enemies (predators, parasitoids, and pathogens) kill herbivorous insects, and through negative feedback can regulate their population density. Biological control programs rely on the theory that natural enemies can effectively regulate populations of herbivorous insects by killing enough individuals to lower the population below a threshold that economically damages or kills plants. For example, in classical biological control it is thought that importation of natural enemies (which can include predators, parasitoids, and pathogens) from the native range of an exotic invasive pest will mitigate plant damage levels within the environment invaded by the pest (Hajek 2004). In practice, there are multiple factors that can limit the success of biological control efforts, yet many have been successful (Kenis et al. 2017). For example, winter moth, native to Europe, has twice been effectively controlled in North America by the introduction of two insect parasitoids, *Cyzenis albicans*

(Fallén) and *Agrypon flaveolatum* (Gravenhorst) that are effective mortality agents during specific stages of winter moth development (Roland and Embree 1995).

Competition for resources occurs among species (interspecific), but also within species (intraspecific), especially when population density approaches *K*. As resources become scarce, competition among individuals for limited resources can result in mortality and exponential declines in population size (see Chapter 5). Scarcity of food for the existing population density is often the primary cause of death among herbivorous forest insects. For example, there may be a limited number of trees in space and time that are nutritionally adequate or have depleted defenses against herbivory to support survival and development of enough insects for the population to grow.

4.6.4 Movement

Immigration to and emigration of individuals from a population can change its size. Natural dispersal is the movement of an individual (or group of individuals) away from the natal population to another location where it (they) will reproduce (Schowalter 2006). Individuals in a population may disperse for many reasons. For example, if their habitats are very patchy or ephemeral in nature, insects may disperse if quantity or quality of resources (e.g. suitable trees for feeding or oviposition) becomes scarce and/or no mates can be found. Crowding or other stimuli may also be important. Dispersal may also occur randomly. Some foliage feeding caterpillars, such as spongy moth, climb to the top of the tree canopy and produce a small thread of silk that "balloons" them to a new tree on wind currents (McManus 1973). Strong winds or weather fronts could carry insects hundreds of miles, and authors have hypothesized this for several species (Frank et al. 2013; Furniss and Furniss 1972; Sturtevant et al. 2013), though it is difficult to determine for sure. Displacement of insect populations across vast areas with no suitable hosts or other methods of transport implies weather could have carried them. Insects may also be transported long distances by humans, as larvae in firewood (e.g. emerald ash borer), or as eggs or other life stages on nursery trees, lumber, household goods, or vehicles (e.g. European spongy moth).

4.6.5 A Tool to Measure Population Growth and Regulation

An effective way to determine how fast insect populations grow and what factors inhibit or allow their growth is to follow a cohort (a cohort is a group of individuals, usually of the same species, born within a defined period of time) or many cohorts, throughout development and document the sources of mortality. Life tables accomplish this task; they determine the identity, timing, and relative importance of mortality factors. For example, parasitism (*Trichogramma* spp. and unidentified parasitoids), tree resistance (resinosis), and other (unknown) factors contributed to mortality during a generation of Nantucket pine tip moth, *Rhyacionia frustrana* (Comstock), in Georgia (Table 4.2). *Trichogramma* spp., parasitoids of tip moth eggs, were by far the most common mortality factor, responsible for 48.0% of all tip moth mortality during a generation. Though pupal mortality was high (49.7%) in relation to the numbers of pupae measured, it was low relative to the number of individuals present at the beginning of the cohort, a statistic termed real mortality. Real pupal mortality (10.7%) was much less than real egg mortality (48.0%). Generation mortality is the sum of all real mortality that occurred from the egg to adult stages, and indicates whether the population will be larger or smaller in the following generation (larger by 10.8% in this case).

A survivorship curve shows the number of individuals entering each successive life stage (Fig. 4.4), and is a simple way to examine mortality occurring through the different insect life stages. For the Nantucket pine tip moth, the precipitous drop in number of individuals between the egg stage and 1st larval instar shows that more

Table 4.2 Life Table of the 2nd 1979 Generation of the Nantucket Pine Tip Moth, *Rhyacionia frustrana*, in Oglethorpe County, Georgia, USA

Life stage (x)	No. entering life stage (l_x)	Mortality factor (d_xF)	No. dying during life stage (d_x)	Apparent mortality ($100q_x$)
Eggs	23,425	*Trichogramma* spp.	11,236	48.0
Instar 1	12,189	Resinosis	881	7.2
		Other	845	7.0
		Total	1,726	14.2
Instar 2	10,463	Resinosis	121	1.2
		Other	1,257	12.0
		Total	1,378	13.2
Instar 3	9,085	Resinosis	64	0.7
		Other	912	10.0
		Total	976	10.7
Instar 4	8,109	Unknown	881	10.9
Instar 5	7,228	Parasites	1,153	15.9
		Other	1,024	14.2
		Total	2,177	30.1
Pupae	5,051	Parasites	569	11.3
		Other	1,942	38.4
		Total	2,511	49.7
Moths	2,540			
Generation				89.2

Adapted from Gargiullo and Berisford (1983)

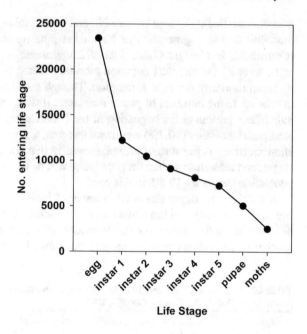

Fig. 4.4 Survivorship Curve Adapted from Life Table of the 2nd 1979 Generation of the Nantucket Pine Tip Moth, *Rhyacionia frustrana*, in Oglethorpe County, Georgia, USA. From Gargiullo and Berisford (1983)

mortality occurred between these two life stages than any other successive stages (Fig. 4.4).

4.7 How Global Change Affects Insects in Forest Ecosystems

Any change in abiotic conditions (temperature, precipitation, extreme weather) will result in a change in insect growth and survival, because insects are poikilothermic organisms and because their food sources may also be affected by such changes. Forest insects have the capacity to adapt more quickly (short generation time coupled with high fecundity, mobility, and genetic plasticity) to environmental change than trees (long generation time and limited capacity for dispersal). As a result, with a warming climate, herbivorous insects may expand beyond their historical ranges to greater latitudes and altitudes where they will encounter new tree populations and species, potentially causing extensive tree mortality. All of this may disrupt the balance of energy and nutrient flow within those forest ecosystems.

Some of these changes have already begun. Warmer temperatures and longer growing seasons have allowed the mountain pine beetle to expand its range to higher elevations, as well as northward and eastward, where it has encountered populations of whitebark, lodgepole, ponderosa, and jack pine that lack a co-evolutionary history with the beetle (Logan et al. 2010; Safranyik et al. 2010). Similarly, the southern pine beetle has moved north beyond its historical range in the southeastern

US, encountering several pines (red, eastern white, jack, and scots pine) in New England that lack historical exposure to the beetle (Dodds et al. 2018). As the range of eastern spruce budworm expands northward, it will encounter greater prevalence of black spruce than it has in the past. Black spruce is considered a host for spruce budworm, yet historically budburst occurred too late in the season to support sufficient budworm survival (Pureswaran et al. 2015). Range expansion, combined with changes in phenology, and a low diversity of natural enemies in much of the boreal forest, will result in different impacts and outbreak dynamics of spruce budworm in decades to come (Pureswaran et al. 2015). Also, engraver beetles (*Ips* spp.), which have mostly been minor pests historically, may become serious pests in the future, because they can reach outbreak populations in hot and dry conditions (Negrón et al. 2009) and after windstorms (Gothlin et al. 2000).

Forest pests are transported to new forest ecosystems as people move forest products from one place to another. These introductions can have significant negative impacts on tree and forest health if the non-native pest is an aggressive one or there is no biotic resistance in the new habitat, from the trees or the community of natural enemies and competitors. Some invasive species are so aggressive that they threaten to completely or functionally eliminate a tree species, or genus, from a continent (e.g. emerald ash borer, hemlock woolly adelgid). Whether the threatened trees are foundation species for the ecosystem or not, the pest invasion will alter forest communities and functional relationships, e.g. Gandhi and Herms (2010). There are numerous examples of non-native forest insect invasions worldwide (see Chapter 23). In the US, each introduced species that has become a major forest pest has engendered a massive research, regulatory, and management effort to understand its biology, eradicate it, slow its spread, or elsewise mitigate its impact. As people continue to mobilize and engage in inter-continental trade, the problem of forest insect invasions is unlikely to recede.

Acknowledgements Authors thank the editors and Steve Katovich for helpful comments on an earlier version of this chapter.

References

Akers RC, Nielsen DG (1990) Spatial emergence pattern of bronze birch borer (Coleoptera: Buprestidae) from European white birch. J Entomol Sci 25(1):150–157

Allison JD, Borden JH, McIntosh RL, de Groot P, Gries R (2001) Kairomonal responses by four *Monochamus* species (Coleoptera: Cerambycidae) to bark beetle pheromones. J Chem Ecol 27(4):633–646

Berryman AA (1972) Resistance of conifers to invasion by bark beetle-fungus associations. BioSci 22:598–602

Birch LC (1948) The intrinsic rate of natural increase of an insect population. J Anim Ecol 17:15–26

Bréda N, Huc R, Granier A, Dreyer E (2006) Temperate forest trees and stands under severe drought: a review of ecophysiological responses, adaptation processes and long-term consequences. Ann For Sci 63:625–644

Cornell HV, Hawkins BA (1995) Survival patterns and mortality sources of herbivorous insects: some demographic trends. Am Nat 145:563–593

Coulson RN, Stephen FM (2006) Impacts of insects in forest landscapes: implications for forest health management. In: Paine TD (ed) Invasive forest insects, introduced forest trees, and altered ecosystems. Springer, New York, pp 101–125

Danks HV (1978) Modes of seasonal adaptation in the insects. Can Entomol 110:1167–1205

Davidson CB, Gottschalk KW, Johnson JE (1999) Tree mortality following defoliation by the European gypsy moth (*Lymantria dispar* L.) in the United States: a review. For Sci 45:74–84

Dixon WN, Payne TL (1979) Sequence of arrival and spatial distribution of entomophagous and associate insects on southern pine beetle-infested trees. Texas Agricultural Experiment Station, Texas A&M University System, College Station, Texas, p 27

Dodds KJ, Graber C, Stephen FM (2001) Facultative intraguild predation by larval Cerambycidae (Coleoptera) on bark beetle larvae (Coleoptera: Scolytidae). Environ Entomol 30:17–22

Dodds KJ, de Groot P, Orwig D (2010) The impact of *Sirex noctilio* in *Pinus resinosa* and *Pinus sylvestris* stands in New York and Ontario. Can J For Res 40:212–223

Dodds KJ, Aoki CK, Arango-Velez A, Cancelliere J, D'Amato AW, DiGirolomo MF, Rabaglia RJ (2018) Expansion of the southern pine beetle into northeastern forests: management and impact of a primary bark beetle in a new region. J Forest 116:178–191

Ellison AM, Bank MS, Clinton BD, Colburn EA, Elliott K, Ford CR, Foster DR, Kloeppel BD, Knoepp JD, Lovett GD, Mohan J, Orwig DA, Rodenhouse NL, Sobczak WV, Stinson KA, Swan CM, Thompson J, Von Holle B, Webster JR (2005) Loss of foundation species: consequences for the structure and dynamics of forested ecosystems. Front Ecol Environ 3(9):479–486

Foelker CJ (2016) Beneath the bark: associations among *Sirex noctilio* development, bluestain fungi, and pine host species in North America. Ecol Entomol 41:676–684

Frank KL, Tobin PC, Thistle HWJ, Kalkstein LS (2013) Interpretation of gypsy moth frontal advance using meteorology in a conditional algorithm. Int J Biometeorol 57:459–473

Furniss MM, Furniss RL (1972) Scolytids (Coleoptera) on snowfields above timberline in Oregon and Washington. Can Entomol 104:1471–1478

Gandhi KJK, Herms DA (2010) Direct and indirect effects of alien insect herbivores on ecological processes and interactions in forests of eastern North America. Biol Invasions 12:389–405

Gargiullo PM, Berisford CW (1983) Life tables for the nantucket pine tip moth, *Rhyacionia frustrana* (Comstock), and the pitch pine tip moth, *Rhyacionia rigidana* (Fernald) (Lepidoptera: Tortricidae). Environ Entomol 12:1391–1402

Gaylord ML, Kolb TE, Pockman WT, Plaut JA, Yepez EA, Macalady AK, Pangle RE, McDowell NG (2013) Drought predisposes pinon-juniper woodlands to insect attacks and mortality. New Phytol 198:567–578

Gaylord ML, Kolb TE, McDowell NG (2015) Mechanisms of pinon pine mortality after severe drought: a retrospective study of mature trees. Tree Physiol 35:806–816

Gothlin E, Schroeder LM, Lindelow A (2000) Attacks by *Ips typographus* and *Pityogenes chalcographus* on windthrown spruces (*Picea abies*) during the two years following a storm felling. Scand J For Res 15:542–549

Haavik LJ, Billings SA, Guldin JM, Stephen FM (2015) Emergent insects, pathogens and drought shape changing patterns in oak decline in North America and Europe. For Ecol Manage 354:190–205

Haavik LJ, Dodds KJ, Allison JD (2018) *Sirex noctilio* (Hymenoptera: Siricidae) in Ontario (Canada) pine forests: observations over five years. Can Entomol 150:347–360

Hain FP, Duehl AJ, Gardner MJ, Payne TL (2011) Natural history of the southern pine beetle. In: Coulson RN, Klepzig KD (eds) Southern Pine Beetle II. USDA Forest Service, pp 13–24

Hajek AE (2004) Natural enemies: an introduction to biological control. Cambridge University Press, Cambridge

Hajek AE, Tobin PC (2011) Introduced pathogens follow the invasion front of a spreading alien host. J Anim Ecol 80:1217–1226

Hajek AE, Tobin PC, Haynes KJ (2015) Replacement of a dominant viral pathogen by a fungal pathogen does not alter the collapse of a regional forest insect outbreak. Oecologia 177:785–797

Hanks LM (1999) Influence of the larval host plant on reproductive strategies of cerambycid beetles. Annu Rev Entomol 44:483–505

Hartmann G, Blank R (1992) Winter frost, insect defoliation and *Agrilus biguttatus* attack as causal factors of oak decline in northern Germany. Forst Und Holz 15:443–4452

Haynes KJ, Bjornstad ON, Allstadt AJ, Liebhold AM (2013) Geographical variation in the spatial synchrony of a forest-defoliating insect: isolation of environmental and spatial drivers. Proc R Soc B 280:2012–2373

Haynes KJ, Allstadt AJ, Klimetzek D (2014) Forest defoliator outbreaks under climate change: effects on the frequency and severity of outbreaks of five pine insect pests. Glob Change Biol 20:2004–2018

Herms DA, Mattson WJ (1992) The dilemma of plants: to grow or defend. Q R Biol 67(3):283–335

Higley LG, Pedigo LP, Ostlie KR (1986) Degday: a program for calculating degree-days, and assumptions behind the degree-day approach. Environ Entomol 15:999–1016

Hofstetter RW, Dinkins-Bookwalter J, Davis TS, Klepzig KD (2015) Symbiotic associations of bark beetles. In: Vega FE, Hofstetter RW (eds) Bark beetles: biology and ecology of native and invasive species. Elsevier, New York, pp 209–246

Horsley SB, Long RP, Bailey SW, Hallett RA, Wargo PM (2002) Health of eastern North American sugar maple forests and factors affecting decline. North J Appl For 19:34–44

Huberty AF, Denno RF (2004) Plant water stress and its consequences for herbivorous insects: a new synthesis. Ecology 85(5):1383–1398

Jactel H, Petit J, Desprez-Loustau ML, Delzon S, Piou D, Battisti A, Koricheva J (2012) Drought effects on damage by forest insects and pathogens: a meta-analysis. Glob Change Biol 18:267–276

Kenis M, Hurley BP, Hajek AE, Cock MJW (2017) Classical biological control of insect pests of trees: facts and figures. Biol Invasions 19:3401–3417

Klepzig KD, Hofstetter RW (2011) From attack to emergence: interactions between Southern Pine Beetle, Mites, Microbes, and Trees. In: Coulson RN, Klepzig KD (eds) Southern Pine Beetle II. USDA Forest Service, Asheville, NC, pp 141–152

Lindgren BS, Raffa KF (2013) Evolution of tree killing in bark beetles (Coleoptera: Curculionidae): trade-offs between the maddening crowds and a sticky situation. Can Entomol 145:471–495

Logan JA, Macfarlane WW, Willcox L (2010) Whitebark pine vulnerability to climate-driven mountain pine beetle disturbance in the Greater Yellowstone Ecosystem. Ecol Appl 20(4):895–902

MacArthur RH, Wilson EO (1967) The theory of island biogeography. Princeton University Press, Princeton

Manion P (1991) Tree disease concepts. Prentice Hall Career & Technology, Upper Saddle River, NJ

Mattson WJ, Haack RA (1987a) The role of drought stress in provoking outbreaks of phytophagous insects. In: Barbosa P, Schultz JC (eds) Insect outbreaks. Academic Press, New York, pp 365–407

Mattson WJ, Haack RA (1987b) The role of drought in outbreaks of plant-eating insects. BioSci 37(2):110–118

McCullough DG (2000) A review of factors affecting the population dynamics of jack pine budworm (*Choristoneura pinus pinus* Freeman). Popul Ecol 42:243–256

McManus ML (1973) The role of behavior in the dispersal of newly hatched gypsy moth larvae. USDA Forest Service, Northeastern Experiment Station, p 10

Miller DR, Dodds KJ, Eglitis A, Fettig CJ, Hofstetter RW, Langor DW, Mayfield AE III, Munson AS, Poland TM, Raffa KF (2013) Trap lure blend of pine volatiles and bark beetle pheromones for *Monochamus* spp. (Coleoptera: Cerambycidae) in pine forests of Canada and the United States. J Econ Entomol 106(4):1684–1692

Moran PAP (1953) The statistical analysis of the Canadian lynx cycle. II. Synchronization and meteorology. Aust J Zool 1:291–298

Myers JH (1998) Synchrony in outbreaks of forest Lepidoptera: a possible example of the Moran effect. Ecology 79:1111–1117

Negrón JF, McMillin JD, Anhold JA, Coulson D (2009) Bark beetle-caused mortality in a drought-affected ponderosa pine landscape in Arizona, USA. For Ecol Manage 257:1353–1362

Nichols JO (1968) Oak mortality in Pennsylvania: a ten-year study. J Forest 66:681–694

Pallardy SG (2008) Physiology of woody plants. Academic Press, Burlington

Price PW, Denno RF, Eubanks MD, Finke DL, Kaplan I (2011) Insect ecology. Cambridge University Press, New York

Pureswaran DS, De Grandpré L, Paré D, Taylor A, Barrette M, Morin H, Régnière J, Kneeshaw DD (2015) Climate-induced changes in host tree-insect phenology may drive ecological state-shift in boreal forests. Ecology 96:1480–1491

Raffa KF, Berryman AA (1983) The role of host plant resistance in the colonization behavior and ecology of bark beetles (Coleoptera: Scolytidae). Ecol Monogr 53(1):27–49

Raffa KF, Grégoire JC, Lindgren BS (2015) Natural history and ecology of bark beetles. In: Vega FE, Hofstetter RW (eds) Bark beetles. Elsevier, New York, pp 1–40

Rhoades DF (1979) Evolution of plant chemical defenses against herbivores. In: Rosenthal GA, Janzen DH (eds) Herbivores: their interaction with secondary plant metabolites. Academic Press, New York, pp 3–54

Rhoades DF (1985) Offensive-defensive interactions between herbivores and plants: their relevance in herbivore population dynamics and ecological theory. Am Nat 125(2):205–238

Robert LE, Sturtevant BR, Cooke BJ, James PMA, Fortin MJ, Townsend PA, Wolter PT, Kneeshaw D (2018) Landscape host abundance and configuration regulate periodic outbreak behavior in spruce budworm *Choristoneura fumiferana*. Ecography 40:1–16

Roland J, Embree DG (1995) Biological control of the winter moth. Annu Rev Entomol 40:475–492

Rudinsky JA (1966) Host selection and invasion by the Douglas-fir beetle, *Dendroctonus pseudotsugae* Hopkins, in coastal Douglas-fir forests. Can Entomol 98(1):98–111

Safranyik L, Carroll AL, Régnière J, Langor DW, Riel WG, Shore TL, Peter B, Cooke BJ, Nealis VG, Taylor SW (2010) Potential for range expansion of mountain pine beetle into the boreal forest of North America. Can Entomol 142:415–442

Salom SM, Stephen FM, Thompson LC (1987) Development rates and a temperature-dependent model of pales weevil, *Hylobius pales* (Herbst), development. Environ Entomol 16:956–962

Schlyter F, Lofqvist J (1990) Colonization pattern in the pine shoot beetle, *Tomicus piniperda*: effects of host declination, structure and presence of conspecifics. Entomol Exp Appl 54:163–172

Schoeller EN, Husseneder C, Allison JD (2012) Molecular evidence of facultative intraguild predation by *Monochamus titillator* larvae (Coleoptera: Cerambycidae) on members of the southern pine beetle guild. Naturwissenschaften 99:913–924

Schowalter TD (2006) Insect ecology: an ecosystem approach. Academic Press, Cambridge

Solensky MJ, Oberhauser KS (2004) The monarch butterfly: biology and conservation. Cornell University Press, Ithaca

Solomon JD (1995) Guide to insect borers in North American broadleaf trees and shrubs. USDA Forest Service Agriculture Handbook, AH-706.

Speight MR, Hunter MD, Watt AD (2008) Ecology of insects: concepts and applications. Wiley-Blackwell, Hoboken

Staley JM (1965) Decline and mortality of red and scarlet oaks. For Sci 11(1):2–17

Stephen FM (2011) Southern pine beetle competitors. In: Coulson RN, Klepzig KD (eds) The southern pine beetle II. USDA Forest Service, pp 183–198

Stephen FM, Berisford CW, Dahlsten DL, Fenn P, Moser JC (1993) Invertebrate and microbial associates. In: Schowalter TD, Filip GM (eds) Beetle-pathogen interactions in conifer forests. Academic Press, London, pp 129–153

Sturtevant BR, Achtemeier GL, Charney JJ, Anderson DP, Cooke BJ, Townsend PA (2013) Long-distance dispersal of spruce budworm (*Choristoneura fumiferana* Clemens) in Minnesota (USA) and Ontario (Canada) via the atmospheric pathway. Agric For Meteorol 168:186–200

Thatcher RC (1960) Bark beetles affecting southern pines: a review of current knowledge. USDA Forest Service, Southern Forest Experiment Station

Thatcher RC (1967) Winter brood development of the southern pine beetle in southeast Texas. J Econ Entomol 60:599–600

Thomas FM, Blank R, Hartman G (2002) Abiotic and biotic factors and their interactions as causes of oak decline in Central Europe. Forest Pathol 32:277–307

Timms LL, Smith SM, de Groot P (2006) Patterns in the within-tree distribution of the emerald ash borer *Agrilus planipennis* (Fairmaire) in young, green-ash plantations of south-western Ontario, Canada. Agric For Entomol 8:313–321

van Asch M, Visser ME (2007) Phenology of forest caterpillars and their host trees: importance of synchrony. Annu Rev Entomol 52:37–55

Vermunt B, Cuddingham K, Sobek-Swant S, Crosthwaite JC, Lyons DB, Sinclair BJ (2012) Temperatures experienced by wood-boring beetles in the under-bark microclimate. For Ecol Manage 269:149–157

Visser ME, Holleman LJM (2018) Warmer springs disrupt the synchrony of oak and winter moth phenology. Proc R Soc B 268:289–294

Wagner TL, Flamm RO, Wu FI, Fargo WS, Coulson RN (1987) Temperature-dependent model of life cycle development of *Ips calligraphis* (Coleoptera: Scolytidae). Environ Entomol 16:497–502

Ward SF, Venette RC, Aukema BH (2019) Cold tolerance of the invasive larch casebearer and implications for invasion success. Agric For Entomol 21:88–98

Wargo PM (1977) *Armillariella mellea* and *Agrilus bilineatus* and mortality of defoliated oak trees. For Sci 23(4):485–492

White TCR (1984) The abundance of invertebrate herbivores in relation to the availability of nitrogen in stressed food plants. Oecologia 63:90–105

White TCR (2015) Senescence-feeders: a new trophic sub-guild of insect herbivores. J Appl Entomol 139:11–22

Chapter 5
Forest Insect Population Dynamics

Jeff R. Garnas, Matthew P. Ayres, and Maria J. Lombardero

5.1   Introduction

To the casual observer, the arthropod fauna of temperate forests may appear to be dominated by mosquitoes or other biting insects. Closer inspection of the leaf litter, the moss at the base of a tree, or leaf surfaces (or reading this book, in particular this chapter), quickly reveals that insect diversity in many forested landscapes can be considerable. Still, the degree to which insects interact with trees, stands and landscapes to drive forest community and ecosystem dynamics is rarely obvious without intensive study. In fact, most species of insects are rare most of the time.

Occasionally, insect populations increase to levels that are difficult or impossible to ignore. Such events, often referred to as "outbreaks," are characterized by explosive increases in abundance (Berryman 1987) which are often episodic (Myers 1988; Williams et al. 2000) and where population growth is largely unconstrained by the ecological forces that had held it in check at lower densities. By virtue of the sheer number of individuals they comprise, outbreaking populations can cause significant damage to forests, crops, and other ecosystems and can disrupt ecosystem services. In the most dramatic examples, outbreaking populations can reach abundances in the

J. R. Garnas (✉)
Natural Resources and the Environment, University of New Hampshire, Durham, NH, USA
e-mail: Jeff.Garnas@unh.edu

Department of Zoology and Entomology, University of Pretoria, Pretoria, South Africa

Forestry and Agricultural Biotechnology Institute (FABI), University of Pretoria, Pretoria, South Africa

M. P. Ayres
Department of Biological Sciences, Dartmouth College, Hanover, NH, USA

M. J. Lombardero
Unidade de Xestion Ambiental E Forestal Sostible, Universidade de Santiago de Compostela, Lugo, Spain

© The Author(s) 2023
J. D. Allison et al. (eds.), *Forest Entomology and Pathology*,
https://doi.org/10.1007/978-3-031-11553-0_5

tens of billions, capable of transforming whole landscapes in ways that can even be seen from space or that warrant multiple mentions in the Bible, as with the infamous plagues of desert locusts which continue to this day (Behmer 2009).

Outbreaks are also common in forest systems. Recently, an unprecedented outbreak of the Mountain pine beetle in the western United States and Canada produced tree mortality over 374,000 km^2 from 2000–2020; the ensuing fires, decay and growth losses are estimated to have released 270 megatons (Mt) of carbon, contributing measurably to global carbon dioxide pools (Aukema et al. 2006; Kurz et al. 2008; Reed et al. 2014). Some species experience cyclical dynamics with peaks and troughs in abundance that occur at strikingly regular intervals ranging from a few years to multiple decades (Baltensweiler and Fischlin 1988; Tenow et al. 2013; Pureswaran et al. 2016). Others experience yearly fluctuations that can appear random or chaotic and are much more difficult to predict. In this chapter we offer an exploration of the factors that influence population cycles and that lead to outbreaks along with some of some of the principal approaches to modeling such dynamics.

The field of population dynamics has deep roots in entomology. Studies of fluctuations in insect abundance—particularly of forest insects—represent some of the core empirical work in the discipline and have informed key theory in the field (Royama 1977, 1992; Speight et al. 1999; Liebhold and Kamata 2000; Abbott and Dwyer 2008; Price 2011; Isaev et al. 2017). This is due in part to the relative ease by which insects can be monitored (either directly via trapping or by measuring defoliation, for example). Long time series of population abundance spanning at least a few decades and/or detailed life tables (tallies of abundance across life stages) are required to effectively examine hypotheses relating to patterns of abundance over time. Contemporary abundance estimates of sufficient length exist for numerous insect species, particularly for pests of economic importance (Turchin 2003). Dendrochronological (tree ring) studies that cross-reference patterns of growth or xylem damage across living and dead trees (including naturally preserved wood or structural timber) allow researchers to reconstruct abundance time series over centuries (Esper et al. 2007), though interpretation of these data can be challenging (Trotter et al. 2002). Finally, paleoecological reconstruction of insect abundance (e.g. using insect head capsules, wing scales, frass, or damaged plants preserved in bogs or sediments) can even span millennia (Sonia et al. 2011; Montoro Girona et al. 2018; Navarro et al. 2018).

5.1.1 Forest Insects on Plantation Trees and on Evolutionarily Naïve Hosts

One increasingly common situation where herbivorous forest insects can become serious economic and/or ecological threats corresponds to the relatively small subset of species that respond to a super-abundant and often minimally defended resource. This occurs primarily (a) in plantation forestry where trees are typically grown in high-density, low-diversity monocultures, and (b) as a consequence of biological invasion in natural forests where native tree hosts are exposed to insects with which they have no evolutionary history and against which they have little capacity for defense. In the first case, any of the often globally distributed insects colonizing pine or *Eucalyptus* plantations [e.g. the Eurasian woodwasp (*Sirex noctilio*) or the Red gum lerp psyllid (*Glycaspis brimblecombei*)] could clearly be labeled pests as they reduce yields and negatively impact forest plantation profitability (Garnas et al. 2012; Hurley et al. 2016). Here, host trees are nearly always available as new compartments of even-aged cohorts are continuously being planted. As such, the plantation environment comprises a mosaic of different ages. This results in a relatively stable and renewable resource from the perspective of insects (see Box 5.1 for a detailed example). It is worthwhile to note that such sustained, elevated pest densities can also occur when both trees and insects are native, such as is the case with root weevils in North American pine plantations (Rieske and Raffa 1990), chrysomelid beetles on *Eucalyptus* in Australia (Strauss 2001), or pine shoot beetles in Europe (Schroeder 1987) among others.

The second case arises in large part as an unintended consequence of global trade whereby exotic organisms establish in forests or plantations worldwide. Where affected trees lack a co-evolutionary history with newly arrived insects, resistance to herbivory can be low or even absent. This is largely the situation with American ash (*Fraxinus* spp.) which lacks resistance to the Emerald ash borer (*Agrilus planipennis*) in the United States and Europe (Herms and McCullough 2014) or pine (*Pinus* spp.) and the Red turpentine beetle (*Dendroctonus valens*) in China (Wingfield et al. 2016). In such examples, insect populations can reach extremely high abundances that often result in widespread mortality of host trees. Consequently, novel insect pests often devastate the local tree resource after which their own populations crash due to the lack of available host material. While it's tempting to imagine that pest populations may go extinct once they have eaten all available trees, in practice, populations often persist on low-density "escape" trees (those that were missed by the initial wave of attack) or on the small tree cohort that survived as seeds or seedlings but become susceptible as they age. In this case, the "outbreak," while dramatic and devastating, is likely to be short-lived as it moves toward some new equilibrium density on the landscape.

5.1.2 Outbreak Dynamics as an Emergent Property of Insect-Host-Natural Enemy Interactions

While some insects emerge as pests principally as a consequence of specific ecological conditions (e.g. high host densities/low diversity of host and/or a lack of co-evolved responses as discussed in the previous section), an important subset of damaging insects includes a suite of species that are naturally prone to volatile population dynamics. This volatility, characterized by wide though often remarkably regular fluctuations in abundance, arises as a consequence of particular aspects of their biology, ecology, or community interactions. These so-called "outbreak species" are a relatively small, highly non-random subset of insects that may be either native or introduced. Species characterized by outbreak dynamics account for a highly dispro-portionate share of management budgets and have been the focus of intense study relative to non-outbreaking species. Examining the combinations of environmental conditions, life history traits and community interactions that give rise to outbreak dynamics, or lack thereof, has practical value for management and contributes to basic understanding of biological populations. Understanding the features of populations that promote outbreak behavior also helps us to understand why most populations do not display outbreak dynamics and instead are relatively rare and stable. Numerous books and journal articles have been written on the topic, which we broadly synthesize in this chapter. Much of this theory is rooted in classical population dynamics.

5.1.3 Introduction to Population Dynamics

Many textbooks address the dynamics of populations in great depth and from many different perspectives. The field is active with sustained, ongoing discovery and theoretical development (Nicholson 1954; Royama 1992; Berryman 1999; Turchin 2003; Gotelli 2008; Vandermeer and Goldberg 2013; Isaev et al. 2017). Much of the conceptual basis of our current understanding of how (self-regulated) populations behave is rooted in the simple equation:

$$N_t = N_0 e^{R_t} \tag{5.1}$$

where t is a discrete number of generations and N_t is the population abundance t generations from an arbitrary starting point ($t = 0$). Following this logic, N_0 is the "starting" abundance at time zero. In the final term, e^{Rt}, e is Euler's number (~2.178) and R_t is defined as the per capita population growth rate, measured as the number of individuals in the next generation for each individual in the current generation. The relationship between N and R is at the core of why such an apparently simple model can produce a wide range of ecologically plausible dynamics with minimal modification to its parameters. Both terms carry the subscript t which means that they vary in time, and as it turns out, they also vary as a function of one another. For N this relationship is transparent: abundance is clearly a function of the growth rate of populations (Eq. 5.1; left [blue] arrow in Fig. 5.1). Interestingly (and crucially for the dynamics of populations), R is also a function of N (Fig. 5.1). In other words, the per capita growth rate (individuals per individual per unit time) is dependent on the number (or density) of individuals in that population. This feedback between density and growth rate is at the very core of our understanding of population dynamics. Special cases within this feedback system produce outbreak dynamics in a subset of forest insects.

Why does R vary with population density? One major reason is simply compe-tition for resources. When populations have few individuals, resources (i.e. food,

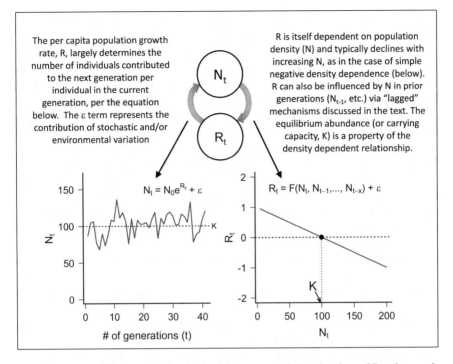

The per capita population growth rate, R, largely determines the number of individuals contributed to the next generation per individual in the current generation, per the equation below. The ε term represents the contribution of stochastic and/or environmental variation

N_t

R_t

R is itself dependent on population density (N) and typically declines with increasing N, as in the case of simple negative density dependence (below). R can also be influenced by N in prior generations (N_{t-1}, etc.) via "lagged" mechanisms discussed in the text. The equilibrium abundance (or carrying capacity, K) is a property of the density dependent relationship.

$$N_t = N_0 e^{R_t} + \varepsilon$$

$$R_t = F(N_t, N_{t-1},..., N_{t-x}) + \varepsilon$$

# of generations (t)

N_t

Fig. 5.1 Conceptual diagram showing feedback between population abundance (N) and per capita population growth rate (R). The simulated time series on the bottom left depicts population fluctuation under simple density dependence with the inclusion of a stochastic component (ε) that approximates the exogenous (e.g. climate or other abiotic effects, impact of generalist predators) contribution to interannual fluctuations in abundance. The graph in the bottom right shows negative density dependence while accommodating the potential for time-delayed feedbacks (lags) via the equation $R_t = F(N_t, N_{t-1},...,N_{t-x}) + \varepsilon$

oviposition sites, nutrients, etc.) are abundant. Thus, each individual is more likely to contribute maximally to population growth, either via increased birth rates, reduced death rates or both. At the other extreme, when *N* is high, resources become limiting and the average contribution of each individual to the next generation is reduced. Population regulation via competition for resources is dubbed "**bottom-up**" because the resource pool (often plants, as in the case of herbivorous insects) is usually depicted as below the consumer pool in visualizations of trophic (food) pyramids, webs or chains. There can also be "**top-down**" pressure from natural enemies (i.e. predators, parasitoids or pathogens) that sit "above" the consumer pool and respond to and sometimes suppress prey density. Bottom-up effects can also occur via the induction of plant defenses that limit resource *quality* or *availability* of plant tissues to herbivores. These defenses make plants more challenging or less profitable to eat. Top-down control by natural enemies as well as bottom-up control via inducible defenses can introduce a time lag (i.e. as predator populations respond to changes in prey density or as plants respond to herbivore attack). Such time lags turn out to be very important as they can result in predictable, cyclical fluctuations in abundance, which will be discussed in more detail below.

In many populations, the relationship between N and R is roughly linear and negative (Fig. 5.1, bottom right). In such cases it is referred to as **simple density dependence**. There are a few important things to recognize about the simple density dependent relationship, some of which require that we define a few new terms. First, note that R_t can be either positive, negative or zero (Fig. 5.1). It is intuitive that at high density, population growth becomes negative. Otherwise, populations would tend to grow forever and become infinitely abundant. Population growth rate must likewise be positive at low or intermediate density—species for which this is not the case would have gone extinct long ago. Where the density dependent line crosses the $R = 0$ line (dashed line in Fig. 5.1, right) is a stable equilibrium point; in the case of simple density dependence, this point has a special name: the **equilibrium abundance**, or K. The word "stable" when applied to an equilibrium point is another way of saying it is an **attractor**. An attractor in this context is an abundance toward which populations tend, as the term suggests. Looking again at Fig. 5.1, this is easy to visualize—when density is below K ($N < K$), R is positive and populations grow; when $N > K$, R is negative and populations shrink. In the absence of any stochastic variation, populations exactly at K ($N = K$) would neither grow nor shrink, though this rarely if ever occurs in nature over successive generations. In fact, anywhere the R function crosses the $R = 0$ line is an equilibrium point.

With simple (negative) density dependence, there is one additional parameter that emerges from the R function. Despite the potential to be confusing, this parameter uses the same letter as the per capita population growth rate, but in the lowercase: r. "Little r," as it is sometimes called, is the **intrinsic growth rate** of the population. Little r can be thought of as the maximum per capita growth rate when that growth rate is unaffected by any of the limitations imposed by density. In other words, r is the value of R for the special case when $N = 0$ (never mind that populations with zero individuals are technically extinct). Thus, r can be easily read as the Y intercept of the R by N function.

Figure 5.2 shows some of the possible relationships between r and K. Many of these concepts will have relevance in subsequent sections and so are worth examining here. In all cases, there are three primary aspects we are concerned with the: (1) intrinsic growth rate (r); (2) equilibrium abundance (K) of the population; and (3) the strength of the density dependent relationship, which can be understood as the slope of the line, and calculated as—r / K. In Fig. 5.2a, halving r from 3.0 to 1.5 while keeping the slope constant has the effect of shifting K to the left, from 100 to 50. In Fig. 5.2b, similar changes in r while holding K constant results in a significantly shallower slope (weaker density dependence). Finally, changing K from 100 to 50 while maintaining r at 3.0 leads to a doubling of the slope and the strength of density dependence (Fig. 5.2c). Of course, there are many examples where r and K are not tightly coupled, but it is useful to understand how each parameter influences model predictions independently.

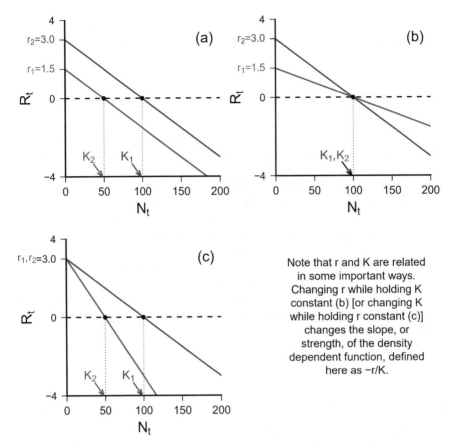

Fig. 5.2 Three graphical examples depicting the relationship between the per capita population growth rate (*R*) and population abundance (*N*) under simple (negative) density dependence using the Ricker model: $N_{t+1} = N_t e^{r\left(1-\frac{N_t}{K}\right)}$. In subfigure a, shifting from K_1 to K_2 while preserving the slope, or "strength," of the density dependent relationship has the consequence of reducing the intrinsic growth rate (*r*). In b and c, changes in either *r* or *K* while preserving the other results in changes in the density dependent slope, with consequences for population behavior or volatility

5.2 Drivers of Population Volatility

How do the models discussed above help us to understand or predict how real populations behave? In large part, the population dynamics of forest insects (and other organisms) can be understood with three relatively simple modifications of the parameters of Eq. 5.1 or to the nature or shape of endogenous feedback that defines the relationship between *N* and *R*. Together, the inclusion of (1) variation in intrinsic growth rates; (2) time-lagged endogenous feedbacks (between *N* and *R*); and (3) scramble competition (intraspecific competition defined by all-or-nothing survival or reproduction leading to decelerating non-linearity in the $R \sim N$ function) can produce dynamics that approximate those seen in forest insects.

5.2.1   Variation in the Intrinsic Growth Rate of Populations

Up to this point, we have dealt only with simple (i.e. linear) negative density dependence, which is a useful starting place but is not always a good match with natural populations (Turchin 2003). By changing the strength of density dependence (the slope of the density line, as in Fig. 5.2) we can produce a range of dynamics that approximates the range of dynamics seen in nature (May, 1976). Specifically, increasing the intrinsic growth rate (which as we saw, increases the steepness of the negative density dependent function) moves the dynamic feedback system in the direction of more volatile, complex dynamics. This shift is important from a management perspective, as increases in volatility/complexity inevitably result in lower predictability of populations (Berryman 1987).

Here we will use the mathematical formalizations of density dependent population growth known as the "Ricker model," originally developed for predicting fisheries stock (Ricker 1954):

$$N_{t+1} = N_t e^{r\left(1 - \frac{N_t}{K}\right)} \tag{5.2}$$

where N_{t+1} is the abundance in the next timestep, N_t is the current abundance, K is the equilibrium abundance (or carrying capacity) and r is the intrinsic growth rate of the population. Any model (such as this one) that considers changes in population abundance at regular time intervals (i.e. t, $t + 1$) is referred to as a *discrete time model*. The interval is arbitrary but usually takes a value with some biological meaning for the population in question, often one year for insects that reproduce annually. Semivoltine (those that take 2 years to develop) or multivoltine species (those with multiple generations per year) can be tracked annually or by using a longer or shorter time step as appropriate. The only requirement is that the tracking interval itself does not change over time. Most discrete time models have *continuous time* equivalents that employ calculus to model population abundance effectively "continuously," which is to say over infinitesimally small timesteps. Discrete time models are typically roughly (or precisely) equivalent to their continuous time counterparts, and for simplicity, this chapter presents only discrete time models.

Figure 5.3 shows five distinct outcomes that arise simply as a consequence of varying r, ranging from simple convergence (to the equilibrium abundance, or K) through damped oscillations, simple and complex cycles, to chaos. In this context, simple cycles refer to the situation where populations cycle between two abundances, one on each side of K, while in complex cycles there are four or more abundance values (for example, two high and two low) that repeat for as long as the models are run. The most volatile fluctuations are characterized as chaotic dynamics. All the models discussed are entirely deterministic with no stochastic, or random, elements. Here chaos does not refer to randomness. Rather, it refers to the fact that fluctuations in abundance are highly dependent on initial conditions where even slight differences (i.e. of a few individuals) predict vastly different abundances even a few time steps in the future. Thus, for chaotic systems accurate forecasting is nearly impossible (Hastings 1993).

Although intrinsic growth rates are of clear importance to population dynamics and species with higher intrinsic r values have a greater propensity toward rapid and dramatic changes in abundance, there is little support for the idea that population cycles are mainly a product of high r. To generate population cycles other mechanisms are needed—in particular, trophic dynamics.

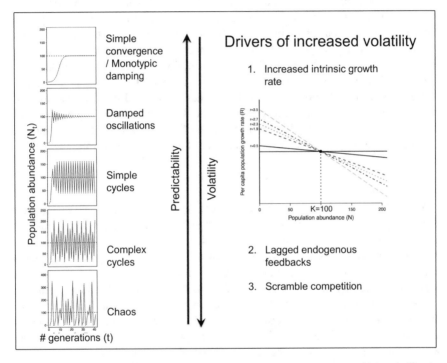

Fig. 5.3 Depiction of five distinct model behaviors (left) ranging from low to high volatility (or high to low predictability) using the Ricker model: $N_{t+1} = N_t e^{r\left(1-\frac{N_t}{K}\right)}$. Corresponding density dependent relationships are shown in the rightmost subfigure. Note that the only difference among the models is the value of little r (which drives the strength of density dependence [negative slope] at constant K, as in Fig. 5.2). Delayed feedbacks and scramble competition are likewise major contributors to population volatility—see text

5.2.2 *Lagged Endogenous Feedbacks*

Feedbacks between N and R are termed "endogenous" because not only does the per capita population growth rate (R) largely drive population abundance (N) in the next generation (an obvious and intuitive statement), but N also strongly influences R (Royama 1992). This feedback is the defining feature of endogenous population dynamics. The effects of population abundance on birth and death rates are not always instantaneous, particularly when the ecological mechanisms that drive such feedbacks involve additional species or trophic levels (Hunter and Price 1998). For example, some natural enemy populations (especially enemies that are relatively specialized on the focal prey species) respond to the high abundance of prey with increasing abundance (thereby reducing R for the prey). Changes in predator density in response to prey availability (referred to as a **numerical response**) are typically characterized by a delay, both in the initiation of population growth and decline as a consequence of prey surplus and scarcity, respectively. Predators may also respond **functionally** whereby the **rate** of prey consumption per predator individual (but not necessarily predator abundance) changes in response to changes in prey abundance. This can happen via prey switching or changes in handling efficiency and also involves a delayed, or **lagged,** response. Bottom-up effects can be lagged too, such as in the case of plant inducible defenses that take time to produce and accumulate. This delay can be generalized by the inclusion of a lagged term as follows:

$$R = F(N_{t-x}) + \epsilon \tag{5.3}$$

where R is a function of density (as before), but now the abundance that matters is not the present abundance, but rather the abundance x time steps (or generations) ago. In principle, lags can take any integer value, but in practice, lags of more than 2–3 timesteps in the past seem to be rare in nature (Turchin and Taylor 1992; Hunter and Price 1998). Lags in dynamic feedbacks have the consequence of elevating population volatility and can cause populations to cycle. Indeed, delayed impacts of specialist natural enemies and plant defenses have been regularly implicated as drivers of population cycles in forest insects, especially among defoliators (Liebhold and Kamata 2000). The increasingly volatile dynamics with increasing r values seen in Fig. 5.3 can result in simple or even complex cycles. However, these "first-order" cycles (those that derive from instantaneous feedbacks) have a period (distance between abundance peaks) that is too short to accurately describe oscillations in observed abundance in natural populations, which typically occur on the order of 8–12 years (Liebhold and Kamata 2000). In contrast, second-order feedbacks (those deriving from time delays in the relationship between N and R) can easily produce cyclical dynamics of much longer, biologically realistic time scales.

5.2.3 Scramble Competition

To this point, we have assumed for the sake of simplicity that the relationship between R and N is linear. This need not be the case. Intraspecific competition for resources is an important form of endogenous population regulation that can be modeled effectively under some conditions using the simple, first-order (non-lagged) models presented above. The linear R by N function assumes that organisms begin to compete even when densities are very low and that the effect of incremental increases in density are the same at high densities as they were at low. Neither assumption is unreasonable as a generality, but we know that linear density dependence is not universal. Instead, some populations display **scramble competition** (Royama 1992; Brännström and Sumpter 2005). Scramble competition refers to the phenomenon where at low to intermediate population densities, available food resources are sufficient for all individuals and should thus correspond to a weakly negative density dependent slope at low abundance values. At high densities, food quickly becomes insufficient for all individuals simultaneously and reduces individual survival and fecundity dramatically. This differs from contest competition where the strongest competitors, or those first to arrive and start feeding, gain sufficient resources while weaker or later-arriving individuals suffer. Scramble competition was famously described by Nicholson (1954) during his studies of sheep blowflies (*Lucilia cuprina*). Nicholson found stable population cycles when food (sheep brains) was supplied at a constant rate. He determined that blowflies had ample food resources and exhibited high survival and reproduction for a broad range of abundances from near 0 to near K. However, as densities approached and exceeded K, suddenly very few of the fly larvae had adequate food to complete development and therefore many died and few eggs were produced for the next generation. In short, high density populations tended to drop precipitously in abundance, or crash. This resulted in R vs. N being strongly decelerating in the region of K. Equation 5.4 allows for scramble competition of variable strength. Figure 5.4 uses this equation to show effects of varying the strength of nonlinearity in R vs. N.

$$N_{t+1} = N_t e^{r\left(1-\left(\frac{N_t}{K}\right)^b\right)} \tag{5.4}$$

When $b = 1$, the equation is equivalent to the Ricker model. As b increases, the R by N function becomes increasingly non-linear (Fig. 5.4a), and with increasingly nonlinear feedbacks comes greater population volatility (Fig. 5.4b).

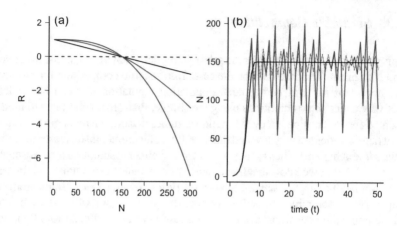

Fig. 5.4 Examples of nonlinear negative density dependence capturing the phenomenon of scramble competition (Nicholson 1954; May and McLean 2007). Both the density dependent relationship (between R and N) (a) and the resulting time series (b) were modeled using Eq. 5.4 $N_{t+1} = N_t e^{\left(r\left(1-\left(\frac{N_t}{K}\right)^b\right)\right)}$, and the following parameters (r = 1, K = 150, N_1 = 1, and either b = 1 [black], b = 2 [red], or b = 3 [blue lines]). Higher values of b correspond to stronger scramble competition (steeper nonlinearities in the R ~ N function in [a])

5.3   Broad Patterns and Real-World Examples

5.3.1   Cyclical Dynamics

Many populations from diverse animal groups display cyclical tendencies, including some small mammals and many forest insects. Often this phenomenon has been attributed to predator–prey dynamics, as with lynx and hare in the Arctic (Stenseth et al. 1999, but see Bryant et al. 1983; Elton and Nicholson 2007), moose on Isle Royale (Post et al. 2002), and lemmings in Scandinavia (Stenseth 1999; Forchhammer et al. 2008). Delayed density dependence arising from top-down pressure from specialist (and sometimes generalist) natural enemies at least partly explains this phenomenon for many forest insects.

Among forest insects, cyclical or outbreak dynamics are disproportionately common among defoliators, especially the Lepidoptera (moths and butterflies), though sawflies and some aphids/adelgids also exhibit similar densities and periodicities (Liebhold and Kamata 2000). Native lepidopterans such as the larch budmoth (*Zeiraphera diniana*), the autumnal moth (*Epirrita autumnata*), the winter moth (*Operophtera brumata*) in Europe, the eastern spruce budworm (*Choristoneura fumiferana*) and forest tent caterpillar (*Malacosoma disstria*) in North America have been extensively studied for their cyclic dynamics and their propensity to cause widespread defoliation during outbreak years (Varley et al. 1974; Ginzburg and Taneyhill 1994; Myers and Cory 2013; Pureswaran et al. 2016). Exotic species such as the spongy moth (*Lymantria dispar*) or the winter moth in North America (where both have been introduced) have also received considerable attention from population ecologists (Liebhold and Kamata 2000; Roland 2007). At least for spongy moth, cyclical outbreaks are evident across certain years (i.e. 1943–1965 and ca. 1978–1996) interspersed with periods of non-cyclical dynamics (Allstadt et al. 2013). It is important to recognize that population cycles, by virtue of their theoretical interest and practical importance, are likely more ubiquitous in the population dynamics literature than they are in nature. It is only a minority of leaf-eating insects that reach sufficient densities to completely defoliate trees, but nearly half (5 of 11, or ~ 45%) of the foliage-feeding forest insects included in a recent analysis displayed cyclical dynamics (Kendall et al. 1998; Liebhold and Kamata 2000). Cyclical dynamics are especially prevalent in Lepidopteran folivores. The proportion of tree-eating pests with cyclical dynamics dropped to 17% when all feeding guilds were considered (Kendall et al. 1998). Many well studied examples of cyclicity in population dynamics (including the autumnal moth, larch budmoth, and spruce budworm) are cyclical in the northern (poleward) part of their range in the Northern Hemisphere, but not in the southern parts (Ruohomäki et al. 2000). Likewise, historical patterns can be disrupted by changes in climate, host tree abundance or human activities or interventions. In fact, the larch budmoth cycles in parts of the insect's range (specifically the Tatra Mountains in southern Poland) ceased in 1981, despite tree ring records showing regular outbreaks every 8, 9 or 10 years over the last 12 centuries—a phenomenon that appears to reflect a phase shift driven by increasing temperatures (Iyengar et al. 2016). Understanding the context dependency of cyclicity and the relationship between cyclical dynamics and specific life history traits remains a central challenge for forest entomologists and population ecologists alike.

5.3.2   The Larch Budmoth in the European Alps

The larch budmoth (*Zeiraphera diniana*) (hereafter LBM) exhibits highly regular cycles of 8–10 years in the Swiss Alps (Fig. 5.5) and has been the subject of sustained study. Swiss researchers kept meticulous records over decades (Baltensweiler et al. 1977; Baltensweiler and Fischlin 1988), not only on caterpillar population densities, but also on tree responses to defoliation, as well as parasitism by a suite of over 100 species of parasitoids. Initial hypotheses emphasized parasitoids (especially the suite of eulophid and ichneumon wasps) and infection by a granulosis virus as a mechanism for observed population cycles, but later analyses indicated that fluctuations in parasitism or infection rates were more likely a consequence than a cause of moth density fluctuations (Baltensweiler and Fischlin 1988). Now, it appears that the cycles arise from density-dependent feedbacks involving both host plant quality and parasitoids.

Fig. 5.5 Population fluctuations in larch budmoth densities, measured as the number of caterpillars per kilogram of larch branch, from 1951–1992. Data are based on Baltensweiler and Fischlin (1988) as reported and analyzed by Turchin (2003)

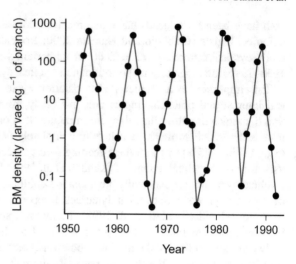

Box 5.1 When *K* is high. Case study *Sirex noctilio* in Southern Hemisphere pulp stands and of EAB on native ash in North America and Europe

The word "outbreak" has a specific meaning to population ecologists and to many forest entomologists, particularly those working with species exhibiting cyclical or chaotic dynamics (e.g. some bark beetles and defoliators). However, there is some ambiguity in the application of this term. There are many examples of forest insects that are apparently benign (or even difficult or impossible to find) in their native range that have become major pests when introduced into non-native managed or unmanaged landscapes. However, unlike SPB where high volatility complicates management, exotic pest populations are often relatively stable across years. Such stability simplifies management decisions, though stable populations may still cause considerable damage.

While it is tempting to see large numbers of insects and call it an outbreak, high population densities in pest insects may often be a predictable consequence of an abundance of susceptible host material (Örlander et al. 1997; Stenberg et al. 2010; Wainhouse et al. 2014; Krivak-Tetley et al. 2021). When coupled with a loss of natural enemies (as is the case with many introduced species), populations with large resource bases can become enormous, often as a predictable consequence of planting of susceptible species or genotypes under conditions that favor forest insect growth and survival (i.e. low diversity, high density hosts historically selected for growth and yield, often at the expense of defense). Discerning the effects of increased *K* stemming from massive increases in habitat or food availability from fluctuations in abundance arising from high r, lagged dynamics, or scramble competition (which can lead to outbreak dynamics; see "Drivers of population volatility" section above) can be critical to predicting population responsiveness to management, including biological control.

For insects that specifically utilize stressed or dying trees as part of their life history, elevated K at a plantation or landscape scale is often a direct consequence of planting practices. For example, in the Southern Hemisphere, North American pines are widely planted, often with extremely high initial planting densities of up to 1,600 stems per hectare. A few years after canopy closure, these trees begin to compete for light and below-ground resources and many experience elevated levels of stress. In the absence of *S. noctilio*, most trees are able to survive long enough to be harvested and processed. However, once *S. noctilio* is established (as has happened almost everywhere in the Southern Hemisphere where pine is grown commercially) wasp populations can increase dramatically in high-density pulp stands that contain many trees that are susceptible to wasp attack. Further, because compartments are continually being planted at the same densities, new compartments are regularly becoming stressed and vulnerable to attack. Thus, even as trees are killed, there is no negative feedback to bring populations down. Under a competing, more complex model, there may be an escape threshold, as with SPB, above which *S. noctilio* can attack and kill larger, healthier trees (Slippers et al. 2014). In the eastern US, where *S. noctilio* was discovered in 2005, populations grew quickly as wasps effectively attacked overstocked stands of Scots pine (Ayres et al. 2014). Now that this resource has been largely depleted, wasps have become rare and hard to find (Krivak-Tetley et al. 2021), though suppressive effects of native natural enemies and/or competitors may also play a role. Likewise, in Southern Hemisphere timber stands that are regularly thinned to reduce tree competition, wasps are rarely problematic. Similar effects of elevated carrying capacity on the dynamics of populations are evident across numerous managed forest landscapes (Örlander et al. 1997; Stenberg et al. 2010; Wainhouse et al. 2014).

Similar to *S. noctilio's* rise and fall in the eastern US, which appears to have tracked the abundance of overstocked pine, other invasive pests also show characteristic boom and bust dynamics that largely track resource availability. Emerald ash borer (EAB) was first detected in North America in Michigan in 2002 (Herms and McCullough 2014). Despite massive quarantine efforts, this insect has now spread to at least 35 states (as of 2021) and has been estimated to have killed over 1 billion ash trees (*Fraxinus* spp.). In this case it appears that there is very little natural resistance to EAB in the ash trees that are native to North America (Cipollini et al. 2011). Upon arriving in an area, EAB infests virtually all available ash trees, except for those with very small stems, which do not have sufficient phloem area to support gallery formation. As such, EAB populations reach incredibly high densities and then crash once they have killed all the available trees. It remains unknown whether populations will persist on the few escape trees and on smaller stems as they grow and become available to attack or whether EAB will go locally extinct once most of the ash trees are killed. To a large extent, the fate of ash on the continent depends on the long-term, endemic equilibria that establish in the aftermath of invasive spread and may also be influenced by the suite of native and introduced natural enemies that have established. Such is the case with many invasive insects for which high abundance post-arrival is more reflective of transient dynamics, namely a "feeding frenzy" on highly susceptible trees or genotypes on the way to a lower, stable, long-term equilibrium.

In the LBM system, host plant quality appears to change as a function of previous caterpillar density, making it a delayed feedback. Larch trees are deciduous conifers. Trees defoliated in a given year produce leaves in subsequent years that are shorter, less digestible, and contain less protein. Larch foliage becomes less nutritious for LBM populations for 1–4 years post-defoliation. This has consequences for larval survival and adult fecundity, which determine R in the years after defoliation. This feedback is crucial to the moth's ecology as it introduces 2nd-order (lagged) dynamics that can largely explain population oscillations. In this case, the length of lag associated with each feedback mechanism was also important to the dynamical behavior of LBM; induced effects on food quality persist for up to four years, while parasitism rates principally lag LBM densities by two years.

Interestingly, despite being a classic example of regular outbreak cycles, LBM population behavior abruptly and inexplicably changed around the 1980's such that these outbreak cycles have disappeared in recent years. Modeling efforts using population estimates from the past 1,200 years (Esper et al. 2007) clearly shows how outbreak epicenters regularly shift up and downslope in response to changes in temperature (Johnson et al. 2010). Recent warming has shifted optimal conditions for LBM population growth to the very edge of the range of host trees, dampening abundance fluctuations and disrupting ecological interactions (i.e. with natural enemies and competitors). In fact, this is among the strongest known examples of a climate change-driven collapse in population behavior (Esper et al. 2007; Johnson et al. 2010).

5.3.3   Tree-Killing Bark Beetles

Numerous species of tree-killing bark beetles also display outbreak dynamics, but the mechanisms appear to be different than for cyclical lepidoptera (Kausrud et al. 2011; Koricheva et al. 2012; Weed et al. 2015). The southern pine beetle (*Dendroctonus frontalis*; herein SPB) is a classic example of an insect that exhibits wide fluctuations in abundance (Fig. 5.6a). SPB is particularly useful to explore since many aspects of the biology and ecology of this insect have been studied in great detail, in large part because it is a major pest of highly productive pine forests in the southeastern United States (Coulson and Klepzig 2011). In fact, there are numerous species of bark beetles (Subfamily Scolytinae, within the weevil family, Curculionidae) that are important in different regions throughout the world, though the outbreak species are a small minority of the total scolytine fauna (see Chapters 10 and 11). We note that our perception of "importance," whether ecological or economic, is strongly linked with the propensity of a species to outbreak. Insects with populations that increase to outbreak status are particularly relevant to management since their impacts are often very difficult to predict in both space and time and can be locally or regionally devastating to a resource. Figure 5.6a shows the abundance of SPB infestations from 1958 to 2015. Though this behavior is not unique among the bark beetles, SPB is famous for its ability to rapidly aggregate on pine trees in huge numbers, which allows them to exhaust resin defenses and kill healthy, vigorously growing trees.

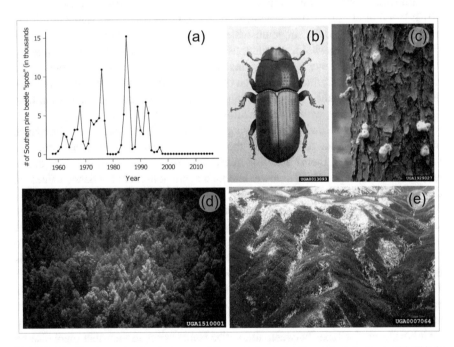

Fig. 5.6 The Southern pine beetle is one of the most damaging forest pests in the world. This is due in large part to its potential for outbreak where huge numbers of beetles mass-attack otherwise healthy trees, overcoming resin defenses and killing them, typically within a few weeks. Subfigures depict interannual fluctuations in the abundance of SPB "spots" (aggregations of beetle-killed trees) in Texas from 1958–2016 (a); an SPB adult (actual length = 2–4 mm; b); "pitch tubes," or resin defenses produced by trees in response to attack (c); aerial photo of an active SPB spot (d); widespread SPB damage that can result when outbreaks are left unmanaged (e). Photo credits (courtesy of forestry-images.com): (5.6b) UGA0013093: USDA Forest Service, USDA Forest Service, Bugwood.org; (5.6c) UGA1929027: Tim Tigner, Virginia Department of Forestry, Bugwood.org; (5.6d) UGA1510001: USDA Forest Service - Region 8 - Southern, USDA Forest Service, Bugwood.org; (5.6e) UGA0007064: Richard Spriggs, USDA Forest Service, Bugwood.org

Local outbreaks of SPB can be observed from the air due to the characteristic formation of beetle "spots," which are local aggregations of tens to hundreds of dead or dying pine trees that appear red against a sea of green trees/needles (Billings and Ward 1984). Why is it that in some forests in some years there are thousands of SPB spots, while in most forests in most years there are zero? It appears that the answer lies in some interesting population dynamical behavior whereby SPB populations can be regulated around two different equilibria and switch between them at unpredictable intervals (Martinson et al. 2012). More specifically, populations can be regulated at low, "endemic" levels where instead of attacking and killing healthy trees, they utilize primarily lightning-struck or other stressed trees that are at low density on the landscape. Eventually, via chance exogenous effects they exceed a numerical escape threshold (an unstable equilibrium) beyond which their deterministic tendency is to increase to an upper "epidemic" equilibrium. Figure 5.7a depicts this **alternative stable states** model as it is understood for SPB (Martinson et al. 2012; Weed et al. 2017). The graphical model represents the two stable equilibria as solid black dots and the single unstable equilibrium as an open circle (Fig. 5.7a). Below the escape threshold, populations tend to remain near the lower, endemic equilibrium, while above it, populations tend to "escape" the lower attractor and rise to epidemic equilibrium. The action of these two attractors results in a bi-modal distribution in abundance whereby low and high densities are more common than intermediate densities, which are transitional and rare (Fig. 5.7b).

This dynamical behavior is satisfying as it approximates observed abundance distributions. But what forces create these two equilibria and what accounts for the switches between them? The first question is equivalent to asking what drives negative density dependence at lower and then again at higher abundance values. In the case of SPB, it appears that the lower equilibrium is generated by predation by the clerid beetle, *Thanasimus dubius*, and competition from other bark beetle species (Martinson et al. 2012). The region of positive feedback (corresponding to a positive slope in R vs. N) generates an unstable equilibrium. The equilibrium is referred to as unstable since rather than acting as an attractor in itself, populations below this density tend to be drawn toward the lower attractor and above it to the higher attractor. This abundance value can also be thought of as an "escape threshold." Above this value there is a range of abundances for which SPB reproductive success continues to improve as there are more and more individuals available to join in mass attacks of their host trees. Switches between alternative stable states require that there also be important exogenous (density-independent) effects on abundance. In the case of SPB, this could come, for example, from changes in the abundance of a bluestain fungus (*Ophiostoma minus*), which is a powerful antagonist of SPB and whose abundance within trees seems largely independent of SPB abundance (Hofstetter et al. 2006; Weed et al. 2017).

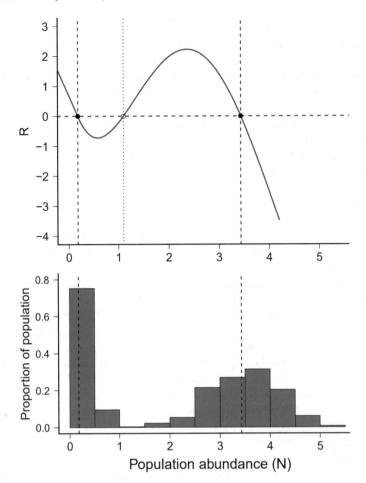

Fig. 5.7 Hypothesized dual equilibrium or "alternate attractors" model proposed for the Southern pine beetle in Martinson et al. (2012). Subfigure (a) shows the *r* by *N* function where two stable equilibria (solid points) represent attractors and predict two distinct abundances around which populations are predicted to fluctuate. An unstable equilibrium (open circle) exists between them and acts as a repellor. A frequency histogram (b) reveals two distinct peaks in expected abundances which correspond conceptually to observed beetle population behavior which tend to fluctuate between either low (endemic) or high (epidemic) abundances

5.3.4 Insect Population Dynamics in Managed Systems

In an increasingly globalized world where (a) high-density and high-yield production systems using a handful of tree species are relied upon to meet growing local, regional and global demand for fiber and fuel; (b) non-native pest insects are accumulating in natural and plantation forests; and (c) climate is changing, leading to shifting geographic ranges and altered dynamics, it is highly likely that managing damaging insects (and pathogens) will be of increasing importance in years to come. While outcomes of b and c above are generally difficult to predict, shifts toward monoculture plantations yield general predictions for short- and long-term impacts on insect populations. Perhaps most salient is the fact that conversion of ecosystems into monospecific production forests tends to increase the K for potential pests of the tree species that is being propagated (Box 5.1). If the K for an insect species exceeds economic damage thresholds (one definition of a pest species), then there may be need for active suppression. Since the natural tendency of populations is to grow toward K when populations are below it, it should be expected that control efforts will need to be sustained indefinitely. At the same time, homogenization of plant species and landscapes in such highly managed forests also tends to *decrease* K for pollinators, endangered species, generalist natural enemies and other elements of biodiversity. This could lead to an elevated extinction risk, especially where populations exhibit a tendency toward extinction when abundance falls below a minimum threshold. The existence of this extinction threshold, or more specifically the behavior of small populations to tend toward zero, is called an "Allee" effect.

Allee effects refer to the tendency of some populations to exhibit a positive corre-lation between abundance (N) and per capita growth rates at low population densities (Allee 1932). This region of positive density dependence (where the slope is posi-tive in the R ~ N function; Fig. 5.8) can arise via a suite of ecological mechanisms including cooperative behavior (e.g. herd vigilance, co-operative hunting, or mass attack on host trees), mate finding, or escape from the negative effects of inbreeding, all of which are particularly relevant when populations are small (Liebhold and Tobin 2008). In each case, higher population densities lead to increased per capita contri-butions to the next generation. In the case of insects, aposematically colored indi-viduals (brightly or conspicuously marked) experience lower predation rates when there are enough individuals for predators to effectively learn the warning signal (Sword 1999). Mate finding can likewise be important and may in part explain the over-representation of parthenogenetic, female-only species or races among inva-sive populations (Kanarek et al. 2015) which very often experience small population sizes at the time of introduction, or shortly thereafter. In fact, the successful "Slow the Spread" program targeting the spongy moth specifically takes advantage of Allee effects, exploiting the difficulty of individuals to locate mates in small, satellite popu-lations along the advancing front of the regional infestation. Intensive pheromone trap monitoring in these areas can detect incipient populations; aerial or ground-based spraying can then be used to reduce population size to near or below the Allee threshold (open circle; Fig. 5.8), below which the natural tendency of each local population is to go extinct (Liebhold and Tobin 2008, 2010).

In addition to changes in the equilibrium abundance, population behavior is predicted to respond to changes in habitat or community. For example, decreases in the abundance of generalist natural enemies can sometimes promote pest problems, not simply via the loss of their suppressive effects, but by altering the feedback system to produce population cycles. Decreases in immediate negative feedbacks (from generalist enemies) could increase the relative importance of delayed negative feed-back (from specialist enemies), which may cause increased population volatility and

Fig. 5.8 Density dependent population growth function showing a region of positive density dependence at low density, or an Allee effect. The lower, unstable equilibrium (open circle) represents the Allee, or extinction threshold. Populations below this threshold trend toward zero abundance. Populations exceeding the Allee threshold are regulated in this case by simple (negative) density dependence at the carrying capacity (K; solid circle)

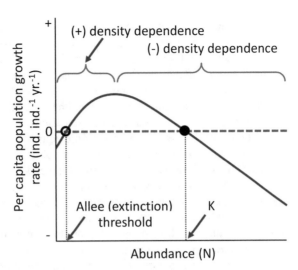

could induce cyclical or outbreak dynamics (Ruohomäki et al. 2000; Klemola et al. 2009). Interestingly, the intentional addition of specialist natural enemies for biological control could, in principle, have similar effects, increasing population volatility. Clear empirical examples or experimental demonstrations of this phenomenon are lacking, however (Myers 2018).

Finally, there is an unusually strong argument for considering active suppression when pest populations have alternative stable states (low abundance and high abundance separated by an unstable equilibrium) such as explained above for SPB. In this case, monitoring of abundance coupled with occasional suppression when populations first approach the escape threshold can hold potential pests at endemic levels (where they are regulated by natural forces) for sustained periods of time (Billings 2011). In contrast, active suppression of populations with naturally cyclical dynamics can theoretically have the undesirable effect of prolonging the outbreak phase by interrupting natural processes (i.e. top-down pressure from natural enemies) that would have led to declines without human intervention.

5.4 Conclusion

Forest insects represent some of the most well-studied organisms in the field of population ecology, due at least in part to their economic and ecological importance and amenability to monitoring and/or historical reconstruction of abundance. The availability of time series spanning decades or even millennia, together with comprehensive mechanistic studies particularly in outbreaking lepidopteran species, form a strong basis for forecasting from which key principles have been derived and tested. In this chapter we have reviewed some of the basic models of simple density dependent regulation, expanding on these ideas to include greater ecological complexity by incorporating lagged and nonlinear feedbacks. We demonstrate how to conceptualize and integrate stochastic variation into these models and discuss a suite of plausible model behaviors that approximate real-world fluctuations in abundance. Through case studies and examples, we explore the dominant ecological drivers of population dynamics in forest insects including interactions with host plants and especially specialist natural enemies that largely drive cyclical dynamics in many forest lepidopteran species. We consider multiple equilibria models or "alternative state" models that effectively approximate Southern pine beetle dynamics, and explore the role and functional form of positive density dependence when populations are small (Allee effects). Finally, we consider how population regulation can be conceptualized in highly managed systems such as high-yield, high-density monoculture plantation settings as well as in "naïve" ecosystems, where insects and trees interact under novel conditions with little co-evolutionary history, most often as a consequence of biological invasion. While this overview reflects many of the basic tenets of a field that has matured considerably, accurate forecasting of insects across time and space still represents a major challenge for forest managers and population ecologists alike, especially given complex, variable and changing environments.

References

Abbott KC, Dwyer G (2008) Using mechanistic models to understand synchrony in forest insect populations: the North American gypsy moth as a case study. Am Nat 172(5):613–624. https://doi.org/10.1086/591679

Allee W (1932) Animal aggregations: a study in general sociology. University of Chicago Press, Chicago

Allstadt AJ, Haynes KJ, Liebhold AM, Johnson DM (2013) Long-term shifts in the cyclicity of outbreaks of a forest-defoliating insect. Oecologia 172(1):141–151. https://doi.org/10.1007/s00 442-012-2474-x

Aukema B, Carroll A, Zhu J, Raffa K, Sickley T, Taylor S (2006) Landscape level analysis of mountain pine beetle in British Columbia, Canada: spatiotemporal development and spatial synchrony within the present outbreak. Ecography 29(3):427–441

Ayres MP, Pena R, Lombardo JA, Lombardero MJ (2014) Host use patterns by the European woodwasp, *Sirex noctilio*, in its native and invaded range. PLoS One 9(3):e90321. https://doi.org/10.1371/journal.pone.0090321

Baltensweiler W, Benz G, Bovey P, Delucchi V (1977) Dynamics of larch bud moth populations. Annu Rev Entomol 22:79–100. https://doi.org/10.1146/annurev.en.22.010177.000455

Baltensweiler W, Fischlin A (1988) The larch budmoth in the Alps. In: Berryman AA (ed) Dynamics of forest insect populations: patterns, causes, implications. Plenum Press, New York, pp 331–351

Behmer ST (2009) Insect herbivore nutrient regulation. Annu Rev Entomol 54:165–187. https://doi.org/10.1146/annurev.ento.54.110807.090537

Berryman A (1987) The theory and classification of outbreaks. In: Barbosa P, Schultz JC (eds.), Insect outbreaks. Academic Press, San Diego, California, USA

Berryman AA (1999). Principles of population dynamics and their application. Stanley Thornes, Cheltenham, U.K

Billings RF (2011). Mechanical control of Southern pine beetle infestations. In: Coulson R, Klepzig KD (eds.), Southern Pine Beetle II, General Technical Report SRS-140 pp 399–413, Asheville, NC: U.S. Forest Service Southern Research Station

Billings RF, Ward JD, (1984) How to conduct a southern pine beetle aerial detection survey. Circular 267, Texas Forest Service (USA).

Brännström Å, Sumpter DJ (2005) The role of competition and clustering in population dynamics. Proc Biol Sci 272(1576):2065–2072. https://doi.org/10.1098/rspb.2005.3185

Bryant J, Chapin F, Klein D (1983) Carbon nutrient balance of boreal plants in relation to vertebrate herbivory. Oikos 40(3):357–368

Cipollini D, Wang Q, Whitehill JGA, Powell JR, Bonello P, Herms DA (2011) Distinguishing defensive characteristics in the phloem of ash species resistant and susceptible to Emerald ash borer. J Chem Ecol 37(5):450–459. https://doi.org/10.1007/s10886-011-9954-z

Coulson RN, Klepzig K (2011) Southern Pine Beetle II, General Technical Report SRS-140. Asheville, NC: U.S. Department of Agriculture Forest Service, Southern Research Station, p. 512 Retrieved from https://www.fs.usda.gov/treesearch/pubs/39017

Elton, C., & Nicholson, M. (2007). The ten-year cycle in numbers of the lynx in Canada. 1–30.

Esper J, Büntgen U, Frank D, Nievergelt D, Liebhold A (2007) 1200 years of regular outbreaks in alpine insects. Proceedings of the Royal Society B: Biol Sci 274(1610):671

Forchhammer MC, Schmidt NM, Hoye TT, Berg TB, Hendrichsen DK, Post E (2008) Population dynamical responses to climate change. In: Advances in ecological research, Vol 40. Elsevier, pp 391–419

Garnas JR, Hurley BP, Slippers B, Wingfield MJ (2012) Biological control of forest plantation pests in an interconnected world requires greater international focus. Int J Pest Manag 58(3):211–223. https://doi.org/10.1080/09670874.2012.698764

Ginzburg L, Taneyhill D (1994) Population cycles of forest Lepidoptera—a maternal effect hypothesis. J Anim Ecol 63(1):79–92

Gotelli NJ (2008) A primer of ecology (4th ed.). Sinauer Associates, Sunderland, Mass

Hastings A, Hom CL, Ellner S, Turchin P, Godfray HCJ (1993) Chaos in ecology—is mother-nature a strange attractor. Annu Rev Ecol Syst 24:1–33. https://doi.org/10.1146/annurev.es.24.110193.000245

Herms DA, McCullough DG (2014) Emerald ash borer invasion of North America: History, biology, ecology, impacts, and management. Annu Rev Entomol 59:13–30. https://doi.org/10.1146/annurev-ento-011613-162051

Hofstetter RW, Cronin JT, Klepzig KD, Moser JC, Ayres MP (2006) Antagonisms, mutualisms and commensalisms affect outbreak dynamics of the Southern pine beetle. Oecologia 147(4):679–691. https://doi.org/10.1007/s00442-005-0312-0

Hunter MD, Price PW (1998) Cycles in insect populations: Delayed density dependence or exogenous driving variables? Ecol Entomol 23:216–222

Hurley BP, Garnas J, Wingfield MJ, Branco M, Richardson DM, Slippers B (2016) Increasing numbers and intercontinental spread of invasive insects on eucalypts. Biol Invasions 18(4):921–933. https://doi.org/10.1007/s10530-016-1081-x

Isaev AS, Soukhovolsky VG, Tarasova OV, Palnikova EN, Kovalev AV (2017) Forest insect population dynamics, outbreaks, and global warming effects. Scrivener Publishing, Beverly, MA

Iyengar SV, Balakrishnan J, Kurths J (2016) Impact of climate change on larch budmoth cyclic outbreaks. Sci Rep 6:1–8. https://doi.org/10.1038/srep27845

Johnson DM, Büntgen U, Frank DC, Kausrud K, Haynes KJ, Liebhold AM, Esper J, Stenseth NC (2010) Climatic warming disrupts recurrent alpine insect outbreaks. Proceedings of the national academy of sciences USA 107(47):20576–20581. https://doi.org/10.1073/pnas.1010270107

Kanarek AR, Webb CT, Barfield M, Holt RD (2015) Overcoming Allee effects through evolutionary, genetic, and demographic rescue. J Biol Dyn 9:15–33. https://doi.org/10.1080/17513758.2014.978399

Kausrud KL, Grégoire JC, Skarpaas O, Erbilgin N, Gilbert M, Økland B, Stenseth NC (2011) Trees wanted—dead or alive! Host selection and population dynamics in tree-killing bark beetles. PLoS One 6(5):e18274. https://doi.org/10.1371/journal.pone.0018274.t001

Kendall B, Prendergast J, Bjornstad ON (1998) The macroecology of population dynamics: taxonomic and biogeographic patterns in population cycles. Ecol Lett 1(3):160–164

Klemola N, Heisswolf A, Ammunet T, Ruohomaki K, Klemola T (2009) Reversed impacts by specialist parasitoids and generalist predators may explain a phase lag in moth cycles: a novel hypothesis and preliminary field tests. Ann Zool Fenn 46(5):380–393

Koricheva, J., Klapwijk, M. J., & Björkman, C. (2012). Life history traits and host plant use in defoliators and bark beetles: implications for population dynamics. In insect outbreaks revisited, pp 175–196 John Wiley & Sons, Ltd

Krivak-Tetley FE, Lantschner MV, Lombardero GJR, Hurley BP, Villacide JM, Slippers B, Corley JC, Liebhold AM, Ayres MP (2021) Aggressive tree killer or natural thinning agent? Assessing the impacts of a globally important forest insect. For Ecol Manage 483:118728

Kurz W, Dymond C, Stinson G, Rampley G, Neilson E, Carroll A, Ebata T, Safranyik L (2008) Mountain pine beetle and forest carbon feedback to climate change. Nature 452(7190):987–990

Liebhold A, Kamata N (2000) Are population cycles and spatial synchrony a universal characteristic of forest insect populations? Popul Ecol 42(3):205–209

Liebhold A, Tobin P (2008) Population ecology of insect invasions and their management. Annu Rev Entomol 53:387–408

Liebhold A, Tobin P (2010) Exploiting the achilles heels of pest invasions: Allee effects, stratified dispersal and management of forest insect establishment and spread. NZ J Forest Sci 40:S25–S33

Martinson SJ, Ylioja T, Sullivan BT, Billings RF, Ayres MP (2012) Alternate attractors in the population dynamics of a tree-killing bark beetle. Popul Ecol 55(1):95–106. https://doi.org/10.1007/s10144-012-0357-y

May R (1976) Simple mathematical models with very complicated dynamics. Nature 261(5560):459–467

May R, McLean AR (2007) Theoretical ecology: principles and applications. OUP Oxford, Oxford, UK

Montoro Girona M, Navarro L, Morin H (2018) A secret hidden in the sediments: Lepidoptera scales. Front Ecol Evol 6. https://doi.org/10.3389/fevo.2018.00002

Myers JH (1988) Can a general hypothesis explain population cycles of forest Lepidoptera? Adv Ecol Res 18:179–242. https://doi.org/10.1016/s0065-2504(08)60181-6

Myers JH (2018) Population cycles: Generalities, exceptions and remaining mysteries. Proc Biol Sci, 285(1875). https://doi.org/10.1098/rspb.2017.2841

Myers JH, Cory JS (2013) Population cycles in forest Lepidoptera revisited. Annu Rev Ecol Evol Syst 44:565–592. https://doi.org/10.1146/annurev-ecolsys-110512-135858

Navarro L, Harvey AE, Ali A, Bergeron Y, Morin H (2018) A Holocene landscape dynamic multi-proxy reconstruction: how do interactions between fire and insect outbreaks shape an ecosystem over long time scales? PLoS One 13(10):e0204316. https://doi.org/10.1371/journal.pone.0204316

Nicholson AJ (1954) An outline of the dynamics of animal populations. Aust J Zool 2(1):9–65. https://doi.org/10.1071/Zo9540009

Örlander G, Nilsson U, Nordlander G (1997) Pine weevil abundance on clear-cuttings of different ages: A 6-year study using pitfall traps. Scand J for Res 12(3):225–240. https://doi.org/10.1080/02827589709355405

Post E, Stenseth NC, Peterson RO, Vucetich JA, Ellis AM (2002) Phase dependence and population cycles in a large-mammal predator-prey system. Ecology 83(11):2997–3002

Price PW (2011) Insect ecology: Behavior, populations and communities. Cambridge University Press, Cambridge

Pureswaran DS, Johns R, Heard SB, Quiring D (2016) Paradigms in Eastern spruce budworm (Lepidoptera: Tortricidae) population ecology: a century of debate. Environ Entomol 45(6):1–10. https://doi.org/10.1093/ee/nvw103

Reed DE, Ewers BE, Pendall E (2014) Impact of mountain pine beetle induced mortality on forest carbon and water fluxes. Environ Res Lett, 9(10). doi:https://doi.org/10.1088/1748-9326/9/10/105004

Ricker WE (1954) Stock and recruitment. J Fish Res Board Can 11(5):559–623. https://doi.org/10.1139/f54-039

Rieske LK, Raffa KF (1990) Dispersal patterns and mark-and-recapture estimates of two pine root weevil species, *Hylobius pales* and *Pachylobius picivorus* (Coleoptera: Curculionidae). Christmas Tree Plantations. Environ Entomol 19(6):1829–1836. https://doi.org/10.1093/ee/19.6.1829

Roland J (2007) After the decline: what maintains low winter moth density after successful biological control? J Anim Ecol 63:392–398

Royama T (1977) Population persistence and density dependence. Ecol Monogr 47(1):1–35

Royama T (1992) Analytical population dynamics (1st. ed.). Chapman & Hall, London, New York

Ruohomäki K, Tanhuanpää M, Ayres MP, Kaitaniemi P, Tammaru T, Haukioja E (2000) Causes of cyclicity of *Epirrita autumnata* (Lepidoptera, Geometridae): grandiose theory and tedious practice. Popul Ecol 42:211–223

Schroeder LM (1987) Attraction of the bark beetle Tomicus piniperda to Scots pine trees in relation to tree vigor and attack density. Entomology Experimentalis Et Applicata. 44:53–58

Slippers B, Hurley BP, Wingfield MJ (2014) Sirex woodwasp: a model for evolving management paradigms of invasive forest pests. Annu Rev Entomol. https://doi.org/10.1146/annurev-ento-010814-021118

Sonia S, Morin H, Krause C (2011) Long-term spruce budworm outbreak dynamics reconstructed from subfossil trees. J Quat Sci 26(7):734–738. https://doi.org/10.1002/jqs.1492

Speight MR, Hunter MD, Watt AD, Southwood SR (1999) Ecology of insects: concepts and applications: Wiley-Blackwell.

Stenberg JA, Lehrman A, Björkman C (2010) Uncoupling direct and indirect plant defences: novel opportunities for improving crop security in willow plantations. Agr Ecosyst Environ 139(4):528–533. https://doi.org/10.1016/j.agee.2010.09.013

Stenseth NC (1999) Population cycles in voles and lemmings: density dependence and phase dependence in a stochastic world. Oikos 87(3):427–461

Stenseth NC, Chan KS, Tong H, Boonstra R, Boutin S, Krebs CJ, Post E, O'Donoghue M, Yoccoz NG, Forchhammer MC, Hurrell JW (1999) Common dynamic structure of Canada lynx populations within three climatic regions. Science 285(5430):1071–1073

Strauss S (2001) Benefits and risks of biotic exchange between Eucalyptus plantations and native Australian forests. Austral Ecol 26(5):447–457

Sword GA (1999) Density-dependent warning coloration. Nature 397(6716):217–217. https://doi.org/10.1038/16609

Tenow O, Nilssen AC, Bylund H, Pettersson R, Battisti A, Bohn U, Caroulle F, Ciornei C, Csoka G, Delb H, De Prins W, Glavendekic M, Gninenko YI, Hrasovec B, Matosevic D, Meshkova V, Moraal L, Netoiu C, Pajares J, Rubtsov V, Tomescu R, Utkina I (2013) Geometrid outbreak waves travel across Europe. J Anim Ecol 82(1):84–95. https://doi.org/10.1111/j.1365-2656.2012.02023.x

Trotter RT, Cobb NS, Whitham TG (2002) Herbivory, plant resistance, and climate in the tree ring record: interactions distort climatic reconstructions. Proc Natl Acad Sci 99(15):10197–10202. https://doi.org/10.1073/pnas.152030399

Turchin P (2003) Complex population dynamics: a theoretical/empirical synthesis. Princeton University Press, Princeton, N.J

Turchin P, Taylor A (1992) Complex dynamics in ecological time-series. Ecology 73(1):289–305

Vandermeer JH, Goldberg DE (2013) Population ecology: first principles (Second, edition. Princeton University Press, Princeton

Varley G, Gradwell G, Hassell M (1974) Insect population ecology: an analytical approach

Wainhouse D, Inward DJG, Morgan G (2014) Modelling geographical variation in voltinism of Hylobius abietis under climate change and implications for management. Agric for Entomol 16(2):136–146. https://doi.org/10.1111/afe.12043

Weed AS, Ayres MP, Bentz BJ (2015) Chapter 4 - Population dynamics of bark beetles. In: Hofstetter FEVW (ed) Bark Beetles. Academic Press, San Diego, pp 157–176

Weed AS, Ayres MP, Liebhold AM, Billings RF (2017) Spatio-temporal dynamics of a tree-killing beetle and its predator. Ecography 40(1):221–234. https://doi.org/10.1111/ecog.02046

Williams DW, Long RP, Wargo PM, Liebhold A (2000) Effects of climate change on forest insect and disease outbreaks. Responses of Northern U.S. Forests to Environmental Change. Ecol Stud 139:455–494

Wingfield MJ, Garnas JR, Hajek A, Hurley BP, De Beer ZW, Taerum SJ (2016) Novel and co-evolved associations between insects and microorganisms as drivers of forest pestilence. Biol Invasions 18(4):1045–1056. https://doi.org/10.1007/s10530-016-1084-7

Chapter 6
Forest Insect–Natural Enemy Interactions

Jean-Claude Grégoire and Juli R. Gould

6.1 Introduction

As illustrated in several other chapters of this book, "forest insects", including those linked to woody plants growing outside the forest environment *stricto sensu* (cities, field margins, hedgerows, river banks, roads, railway tracks, etc.), play various ecological and economic roles (pests, biocontrol agents, pollinators, recyclers of nutrients, key components of trophic webs, etc.). Often, the role of natural enemies in intricate food webs can be extremely complex and may change according to the presence and prevalence of other food web components. For example, the interactions of two prey species occupying the same niche and facing a common predator could result in a competitive advantage for one of the two prey species, if it suffers less damage from the predator (see Sect. 6.3).

Although the forest environment provides very specific habitats for natural enemies and their prey (see Sect. 6.4), in many respects natural enemies of forest insects are not different from species attacking prey or hosts in other habitats. Accordingly, ecological processes and behavioural traits such as specificity, prey/host location and exploitation, intra- and interspecific competition, multitrophic interactions, coevolutionary dynamics, can be found in any natural enemy in any habitat. Consequently, when relevant examples of these processes in forest natural enemies are not available, examples illustrating particular features of the complex relationships between insects and their natural enemies will sometimes be drawn from non-forest ecosystems.

J.-C. Grégoire (✉)
Université Libre de Bruxelles, Bruxelles, Belgium
e-mail: jean-claude.gregoire@ulb.be

J. R. Gould
USDA APHIS PPQ CPHST Laboratory, Buzzards Bay, MA, USA

© The Author(s) 2023
J. D. Allison et al. (eds.), *Forest Entomology and Pathology*,
https://doi.org/10.1007/978-3-031-11553-0_6

6.2   Natural Enemies

Any organism feeding on another species or group of species during at least one developmental stage can be described as a "natural enemy", a category to which predators, parasitoids and pathogens attacking forest insects obviously belong. To extend the label more widely, it could be argued that herbivores are natural enemies of the plant species they feed upon (see Sect. 6.3. Food webs). The categories: *predators* (mostly small mammals, birds, arthropods) and *parasitoids* (insects), *nematodes*, and *pathogens* (bacteria, fungi and viruses) are briefly discussed below. For comprehensive syntheses regarding natural enemies of insects in general, see Hajek and Eilenberg (2018) and Jervis (2012).

6.2.1   Predators

Predators kill, and feed on, live prey. Each individual consumes several prey during its development. Some species are predatory only at a given life stage. The adults of the common green lacewings (*Chrysoperla carnea*: Neuroptera, Chrysopidae) feed on pollen but their larvae consume a wide range of prey (aphids, scale insects, moth or butterfly eggs or larvae) (Huang and Enkegaard 2010), as well as extrafloral nectar (Limburg and Rosenheim 2001). Conversely, all life stages of the Monotomid beetle, *Rhizophagus grandis*, feed on the immature stages of the bark beetle *Dendroctonus micans* (Grégoire 1988).

Many predator species are *polyphagous* (attacking several families) or *oligophagous* (attacking several genera within one family). Small mammals and birds are notoriously polyphagous, shifting diet according to circumstances, even alternating between predation and herbivory. The white-footed mouse, *Peromyscus leucopus*, an important predator of the spongy moth, *Lymantria dispar*, is known to feed primarily on acorns and to expand its diet to include spongy moth pupae when they become locally available (Elkinton et al. 1996). The Clerid beetle, *Thanasimus formicarius*, is a good example of an oligophagous predator. It is restricted to Scolytinae but attacks at least 27 species within this this sub-family (Warzée et al. 2006). Some predators are *monophagous* (feeding on a few, or even one species within one genus). For example, *R. grandis* is known to attack only one species: *Dendroctonus micans*, but there are only very few such cases (Dohet and Grégoire 2017).

A wide variety of organisms exhibit a predatory life style. Wegensteiner et al. (2015) listed 218 species recorded as predators of bark- and ambrosia beetles in Europe and North America, including 168 insect species belonging to 4 orders and 21 families, 40 mites and ten woodpecker species. Among the insects, predators belong to many families, including the Carabidae, Cleridae, Cucujidae, Histeridae, Monotomidae, Nitidulidae, Staphylinidae, Tenebrionidae, Trogossitidae, and Zopheridae.

Dipteran predators most commonly belong to the families Asilidae, Dolichopodidae, Empididae, and Lonchaeidae.

A review of forest pest control by vertebrate predators is provided by Buckner (1966). Small mammals have been observed to exert strong predatory impacts, in particular on ground dwelling life stages (sawfly and moth pre-pupae and pupae). Two shrews, *Sorex cinereus cinereus* and *Blarina brevicauda talpoides* and a deer mouse, *Peromyscus maniculatus bairdii* are important predators of the European pine sawfly, *Neodiprion sertifer* in Canada (Holling 1959a). *Peromyscus leucopus* is recognised as the major mortality factor regulating low-density populations of the Spongy moth in the eastern US (Elkinton and Liebhold 1990; Liebhold et al. 2005). Various species of birds exert strong pressure on Lepidoptera (Seifert et al. 2015) and scolytine beetles (Karp et al. 2013). Woodpeckers (Picidae) play an important role in the population dynamics of the Emerald Ash Borer, *Agrilus planipennis* (Coleoptera, Buprestidae) in North America (Jennings et al. 2016).

6.2.2 *Parasitoids*

Parasitoids differ from true parasites (e.g. flatworms, *Tenia* spp.) in that they eventually kill their hosts at the end of their own development. Even though a host may be infested by a developing parasitoid, the hosts survive and can sometimes produce progeny before they are killed. There are internal (endo-), and external (ecto-) parasitoids.

Each parasitoid larva consumes one single host during its development, but, in *gregarious* parasitoid species, several parasitoid larvae can share the same host. Adults may also exert an impact on their hosts via *host-feeding*, during which they puncture the host cuticle and feed on its haemolymph. As described above for predators, parasitoids can be monophagous, oligophagous or polyphagous. An example of a monophagous parasitoid is *Avetianella longoi*, an Encyrtid parasitoid of the Eucalyptus longhorned borer in California, USA. This parasitoid was successful in controlling *Phoracantha semipunctata* (Paine et al. 1993). But when *Phoracantha recurva* (in the same genus) was introduced into California, *A. longoi* was not effective in attacking or controlling the new pest.

Parasitoids are generally classified as either *idiobiont* or *koinobiont*. The *idiobionts* attack mostly hidden hosts (e.g. xylophagous larvae feeding on the sapwood within trees or branches), which are first paralysed, after which one or several eggs are laid on or near (but not within) the host (Fig. 6.1). The *koinobionts* are generally endoparasitoids. The host is often immature and continues to develop, which allows the host to grow and provide a larger food supply to the parasitoid larvae. To take advantage of the increased resource from a larger host, some parasitoids delay development until the host pupates, even if oviposition occurred in the host egg. However, the koinobiont strategy also imparts some important constraints. Because the host is still active (as opposed to the paralysed hosts of the idiobionts), it has the opportunity to defend itself by encapsulating the eggs with melanocytes (see also Sect. 6.5.2).

Fig. 6.1 *Coeloides bostrichorum* (Hymenoptera, Braconidae). a. female ovipositing through the bark; b. egg (arrow), next to a paralysed *Ips typographus* larva; c. mature parasitoid larva ready to spin a cocoon; the remnants of the host are not visible. Photos: Courtesy of Evelyne Hougardy

Also, when the host is more active it is susceptible to predation, which would kill the parasitoid larvae as well as the host.

Females of some gregarious species lay several eggs in each host while others lay one single, *polyembryonic* egg which, after many divisions, will produce up to several hundred clonal larvae. In some species with polyembryonic eggs there is larval caste differentiation: short-lived "soldier" larvae hatching first and roaming the host in search of competitors to destroy, and reproductive larvae that hatch later and become reproductive adults (Cruz 1981; Giron et al. 2004). Some species are *pro-ovigenic*: the females emerge with a complete egg load that will not increase. Others are *synovigenic* and have only a limited set of eggs upon emergence and need to feed (e.g. nectar, pollen, host-feeding) in order to develop additional eggs. For example, females of *Scambus buoliana* (Hymenoptera, Ichneumonidae), a parasitoid of the European pine shoot moth, *Rhyacionia buoliana* (Lepidoptera, Tortricidae) must host-feed or feed on pollen to increase longevity and fecundity (Leius 1961; 1963). Similar results have been reported for hymenopteran parasitoids of bark beetles (Mathews and Stephen 1997; Hougardy and Grégoire 2000). When food or hosts are scarce, synovigenic parasitoids can resorb their eggs in order to redirect resources to other physiological functions and resume oviposition when resources are available again.

Many families of Hymenoptera are primarily or exclusively parasitoids, including the Ichneumonidae, Braconidae, Torymidae, Chalcididae, Eurytomidae, Pteromalidae, Encyrtidae, Eulophidae, Trichogrammatidae and Aphelinidae. There are also parasitoids among the Diptera (e.g. the Bombylidae and the Tachinidae) and the Coleoptera (e.g. some Staphylinidae, Bothrideridae, Carabidae and Meloidae). A comprehensive review of the biology and ecology of parasitoids is provided by Godfray (1994).

There are multiple forms of parasitism (see Box 6.1).

BOX 6.1—Forms of parasitism by insect parasitoids

Primary parasitoids. Species that develop on non-parasitoids.

Hyperparasitoids. (secondary, tertiary parasitoids). Develop on other parasitoids. There may be more than one level of hyperparasitism in a system. Some hyperparasitoids oviposit directly in or on a primary parasitoid, others oviposit on or in the host, and their larva search for larval primary parasitoid hosts.

Multiparasitism. Two or more species of primary parasitoids which concurrently attack the same host. This phenomenon creates a high level of interspecific competition. Sometimes, multiparasitism is obligatory (see cleptoparasitoids).

Superparasitism. Several parasitoids of the same species can oviposit in or on the same host.

Autoparasitsm. Some species lay female eggs in unparasitized Sternorrhynchan hosts but lay male eggs in the immature parasitoids (of the same or another species) already present inside of the host.

Cleptoparasitoids. "Host stealers". These species, unable to paralyze a host themselves, are obligatory multiparasitoids. They only select hosts already parasitised by another species.

6.2.3   Nematodes and Pathogens

Nematodes and entomopathogenic viruses, bacteria, fungi and microsporidia are widely present and active in the forest and, similarly to insect parasitoids and predators, some of them are mass-produced and released as biological control agents. Reviews on the use of pathogens against insects have been published by Lacey and Kaya (2007), Lacey et al. (2015), Lacey (2016) and Hajek and van Frankenhuyzen (2017). Nematodes were reviewed by Poinar (1975, 1991) and Kaya and Gaugler (1993).

Nematodes (roundworms) are long and thin worms, living in moist environments, including the soil or the body of plants or animals. The adults of some *Gordius* species (horsehair worms, Mermithidae) that parasitise locusts, crickets or roaches measure 30–120 cm. Other nematodes are microscopic.

There are many known cases of nematodes infesting forest insect pests. For example, *Deladenus siridicicola* (Neotylenchidae), has been introduced to many parts of the Southern Hemisphere to control the Eurasian and North-African woodwasp, *Sirex noctilio*. This nematode can sterilize female woodwasps (see also

Sect. 6.3). Also of particular interest here are the so-called "entomophilic" or "entomopathogenic" nematodes (Steinernematidae and Heterorhabditidae), which are entomopathogenic because of their association with mutualistic bacteria in the genus *Xenorhabdus*. The bacteria are introduced by the nematodes into the body of a living insect, kill the host and feed and multiply on its dead body. The nematodes feed on the bacteria which also produce antibiotics that inhibit the growth of competing bacteria. The bacteria can also attack other nematodes that compete with their own associates. For example, *Xenorhabdus bovienii*, a symbiont of *Steinernema affine* can directly attack its competitor, *S. feltiae* and thus reduce competition by this latter nematode species (Murfin et al. 2019). *Steinernema* spp. infest the soil-inhabiting life stages of various beetles, moths and sawflies. *Heterorhabditis* spp. attack the soil-dwelling larvae of various scarabeids and weevils. Some *Steinernema* species are "ambushers", waiting for an insect to cross their path. Other nematodes (e.g. *Heterorhabditis* spp.) are "cruisers". They move actively in the soil, using semiochemicals and vibrations to locate prey.

Bacteria are unicellular organisms 0.5–5.0 μm long, protected by a membrane and a cell wall, with a single, naked circular DNA chromosome. The bacteria reproduce by fission, but they can also produce spores. They occur in many shapes (spherical, linear, spiral-shaped), and they are extraordinarily abundant everywhere in the world. Some are saprophytes (feed on decaying plant matter), some are symbiotic, and others are pathogens of plants and animals. A common bacterial entomopathogen is *Bacillus thuringiensis*, with distinct subspecies infecting different insect orders. The bacteria produce *sporangia*, containing a spore and a *crystal*. When swallowed by an insect, the crystal is dissolved in the alkaline conditions of the gut, and the *protoxin* within, activated by the gut's enzymes, attaches to the gut wall, creating pores through which the bacteria invade the host's body.

Fungi Many species of fungi infect insects, in particular among the orders Entomophthorales (e.g. *Entomophthora* spp.) and Hypocreales (e.g. *Beauveria* spp. and *Metarhizium* spp.). Pathogenic fungi start colonizing a new host via a spore attaching itself to the cuticle. The spores germinate and produce hyphae that enter the host through the cuticle, often at a thinner location (ventral surface, spiracle, sensilla, or joints between appendices or segments). In some species of fungal pathogens, the hyphae start covering the host's body before penetration occurs. Penetration is facilitated by enzymatic processes and mechanical pressure. Once inside the host, the fungus most often kills the host and colonizes its entire body. In many cases, the host's behavior is manipulated by the fungus, so that it dies in an exposed position, from which the fungal spores will have improved opportunities to reach a new host. There are various forms of fungal spores, some short-lived that allow direct contamination of another insect, others more resistant to climate and long-lived.

There are numerous examples of fungi attacking forest insects, e.g. *Beauveria bassiana* colonizing bark beetles, *B. brongniartii* attacking cockchafers, and *Entomophaga maimaiga,* found since 1989 to cause important epizootics among North American populations of *Lymantria dispar.* A comprehensive review of the parasitic fungi has been provided by Boddy (2016).

Microsporidia are unicellular organisms previously classified among the Protozoa, but which now belong to their own phylum, the Microspora. They live as obligate parasites within the cells of a large array of animal hosts, primarily arthropods, including insects (e.g. bees, locusts, bark beetles, Lepidoptera), but also other organisms such as nematodes and man. They can alter the behavior of their hosts, seriously impair, or kill them. They produce spores which are ingested by a new host and, once inside its digestive tract, extrude a long *polar tube* to inject themselves directly into a host cell. Examples of microsporidia infecting forest insects include *Nosema* species infecting bark beetles and the *Nosema*, *Vairimorpha* and *Endoreticulatus* spp. infecting forest Lepidoptera. These organisms affect not only their hosts, but also the endoparasitoids infesting these hosts. It has consequently been proposed that they can exert an important influence on the population dynamics and life cycle of these different insects.

Viruses are very small particles (*virions*), ca 10–150 nm long, which replicate inside the living cells of other organisms including bacteria, fungi, animals or plants. They consist of genetic material (RNA or DNA), surrounded by a protein shell, the *capsid*, itself sometimes encased in a lipid layer. They reach a new host via contaminated food or water or are spread by vectors (e.g. insects). Among the entomopathogenic viruses, the most common are the *baculoviruses* (Baculoviridae), which have double-stranded DNA. Some baculovirus species infect the larvae of moths (e.g. *Lymantria dispar*; *L. monacha*) and sawflies (e.g. *Gilpinia hercyniae*; *Neodiprion sertifer*). Baculoviruses may be protected before they enter the host body by a protein inclusion body, resistant to desiccation, light etc. Among the Baculoviridae, the *polyhedrosis viruses* are protected by polyhedric inclusion bodies that may contain many virions. There are *nuclear polyhedrosis viruses* (NPVs), replicating in the cells' nuclei, and *cytoplasmic polyhedrosis viruses* (CPVs), replicating in the cells' cytoplasma. The virions of the *granuloviruses* (GVs) are each protected by a rounded, smaller inclusion body.

A review of the use of pathogens as biopesticides has been recently published by Senthil-Nathan (2015).

6.3 Food Webs

Each host plant, herbivore, parasitoid, predator and pathogen is part of an often very complex *food web*. Each natural enemy can feed on several target species, and is itself attacked by other organisms, which are often prey for other species. The successive trophic levels that constitute a food web start at the primary producer (host-plant) level, the organisms in each additional level feeding on those of the one below (Price et al. 1980), with top predators occupying the highest level (Rosenheim 1998). This structure is further complicated by horizontal competitive or aggressive relationships between species sharing any given trophic level. For example, there is evidence that the larvae of the pine sawyers *Monochamus carolinensis* and *M. titillator* (Coleoptera, Cerambycidae) exert *intraguild predation* (predation on other

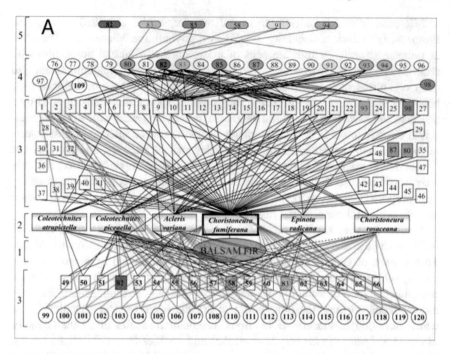

Fig. 6.2 A global food web established after ten years (1983–1993) of sampling in three balsam fir (*Abies balsamea*) stands infested by the spruce budworm, *Choristoneura fumiferana* in New Brunswick. Primary parasitoids are represented by squares, secondary parasitoids by ovals, tertiary parasitoids by octagons, and entomopathogens are represented by circles connected by red lines to hosts. The brackets and numbers on the far left identify trophic level. Vertebrate and invertebrate predators were not included in the study and therefore are missing in the figure. From Eveleigh et al. (2007)[1]

species sharing the same ecological niche) on bark-beetle species also feeding on phloem and sapwood (Dodds et al. 2001; Schoeller et al. 2012).

An existing food web can exert an adverse impact on exotic natural enemies introduced for classical biological control (see Sect. 6.6.5). When the ichneumonid parasitoid *Olesicampe benefactor* was introduced into Canada for the biological control of the European larch sawfly, *Pristiphora erichsoni*, it became a prey item of a local hyperparasitoid, the ichneumonid *Mesochorus dimidiatus*, which greatly reduced its impact (Ives and Muldrew 1984).

Figure 6.2 shows a complex global food web, suggesting the various feedback loops, negative and positive, that can arise from the interactions between organisms at the various levels. The primary hyperparasitoids ("secondary parasitoids" in the figure's legend) alleviate the burden of the parasitoids ("primary parasitoids") on the central species, *C. fumiferana*, and the secondary hyperparasitoids restore some of the impact of the parasitoids.

---

[1] Permission requests: http://www.pnas.org/page/about/rights-permissions.

Figure 6.2 also helps to understand the concept of *apparent competition*, which occurs when, at a given trophic level, several species share natural enemies. The species that produce more shared enemies are predicted to have a higher competitive impact on the other species at the same trophic level. This has been experimentally observed, e.g. by the artificial removal of herbivore species. Morris et al. (2004), studying a community of leaf-miners in a moist tropical forest in Central America, found that it was inhabited by 93 insect species from various orders, attacked by 84 species of hymenopteran parasitoid. After removing two of these leaf-mining species (by uprooting their specific host plants), the remaining species experienced significantly less parasitism by the parasitoids that they shared with the removed species. Conversely, increased but temporary, availability of an alternate food source can result in a larger reservoir of natural enemies, and increased predation on an insect pest after the alternate food has decreased in availability. Over a ten-year period in Massachusetts, Elkinton et al. (1996) recorded a negative correlation between spongy moth (*Lymantria dispar*) abundance and the abundance of an important predator, the white-footed mouse, *Peromyscus leucopus*, and a positive correlation between acorn crops (a basic food for *P. leucopus*) and population density changes in the mouse. So, *L. dispar* outbreaks, synchronised over large areas, appear lag correlated with periodical oak masting patterns (Liebhold et al. 2000).

Natural enemies in a food web can make complementary prey choices. Singer et al. (2017) censused the lepidopteran larvae in the canopies of eight deciduous tree species in northeastern USA after they excluded birds and reduced ant density. They found that birds selectively chose large generalist caterpillars while ants preferred the smaller host specialists, and that the combined impacts of the two types of predators were additive. Non-native species can alter food web dynamics and reduce the impact of biological control agents. For example, the bark beetle *Ips grandicollis* has invaded Australia and when it colonizes dying *Pinus* spp. it introduces the fungus *Ophiostoma ips* to this resource. The nematode *Deladenus siricidicola* is mycetophagous for part of its life cycle, feeding on *Amylostereum areolatum* which is the fungal symbiont of *Sirex noctilio*. The rest of its life cycle, *D. siricidicola* is parasitic on *S. noctilio* and this nematode is an important component of management programs for *S. noctilio* throughout the southern hemisphere. The presence of *O. ips* in dead pine reduces the availability of *A. areolatum* which in turn reduces the performance of *D. siricidicola* as a biocontrol agent for *S. noctilio* (Yousuf et al. 2018).

A striking example of a cascade of changes in the tritrophic interactions in a food web is provided by a study by Palmer et al. (2008) on the ant-acacias *Acacia drepanolobium* in a Kenyan savannah. The trees have extrafloral glands that produce nectar which attracts several ant species, and the tree provides *domatia,* small chambers that some of these ant species use as shelters. The ants protect the trees against large herbivores and woodboring insects. After ten years of exclusion of these herbivores, the trees had reduced their investment in nectar production, which had led to a shift in the dominant ant species towards a species nesting in Cerambycid galleries instead of domatia. This in turn resulted in higher colonization by woodborers and higher tree mortality.

6.4   The Forest Environment and Natural Enemies

The forest environment is generally favourable to many animal species because it is
(i) *long-lived*, (ii) *diversified* and (iii) it often *extends over large or very large areas.*

(i)   *Long-lived*—in strong contrast to insect life cycles which typically range from
      a few weeks to two or three years, most forest types remain in place for decades
      or centuries, with even "permanent" coverage in the case of unmanaged forests
      or of stands managed by selection cutting and natural regeneration. Even short
      rotation coppices (stands of willows or poplars harvested every 2–5 years for
      biomass) provide a longer lived (more stable?) habitat than agricultural land.
(ii) *Diversified*—one hectare of rainforest may contain several hundred tree and
      higher plant species vertically distributed in multiple layers. Even monospe-
      cific, even-aged plantations show a surprising level of complexity (Brockerhoff
      et al. 2008). For example, a survey of five 60–80 year old spruce plantations
      in Belgium identified 53 species of herbaceous plants belonging to more than
      20 families, sometimes in large numbers in clearings and gaps. These plants
      provide nectar and pollen to local synovigenic parasitoids that need to feed as
      adults in order to produce more eggs or to keep their existing load (Hougardy
      and Grégoire 2000). This high diversity of plants favours a high diversity of
      natural enemies feeding on multiple hosts, prey or other sources, and provides
      a large choice of habitats.
(iii) *Extends over large or very large areas*—forests cover *ca.* 4 billion ha in the
      world, i.e. 31% of the total land area (Keenan et al., 2015), with some extremely
      large, continuous coverage, and also with very small plots. The small forests
      are often located side by side, forming larger blocs with, from an insect's
      standpoint, no or little distinguishable boundaries between the individual units.
      The largest forest plantation in Europe (one million ha), the pine *Forêt des
      Landes* close to Bordeaux in south-western France, belongs largely (92%) to
      58,500 private owners, half of which own less than 1 ha (Pottier 2012), and yet
      pests and natural enemies roam the whole massif freely.

Several other forest attributes are important to natural enemies:

– *Forest fragmentation* (the extent and grain of the mosaic of cleared and forested
  land) has been shown to influence the parasitism rate of the forest tent caterpillar
  (*Malacosoma disstria*) by four dipteran parasitoids in Alberta, Canada. According
  to their relative body sizes (correlated to their dispersal capacity), the four species
  performed better at different levels of fragmentation because larger flies could
  fly further (Roland and Taylor 1997). Cronin et al. (2000) showed with mark-
  release-recapture experiments that the clerid predator *Thanasimus dubius* has a
  higher mobility than its prey, the bark beetle *Dendroctonus frontalis*. The radius
  containing 95% of the recaptured insects was 5.1 km for the predators, and 2.3 km
  for the prey, allowing the predators to forage in distant patches when experiencing
  patches of local prey extinction. Using examples taken from the host-parasitoid
  literature, Cronin and Reeve (2005) further argue that, because of local extinction

of either parasitoid and host or predator or prey, their interactions need to be studied at a scale sufficiently large to include the metapopulation level. From a review of theoretical work regarding the impact of habitat loss and fragmentation on predator–prey relationships, Ryall and Fahrig (2006) list a series of criteria that should be considered in further studies: prey and habitat specificity, extinction rates of prey-only and predator–prey patches, prey emigration rates from prey only *vs.* predator–prey patches.

– *Tree species composition* also has an impact on natural enemies. Because it needs the thick bark of pine for pupation and can less easily pupate in the thinner bark of spruce (Fig. 6.3), the oligophagous predator of Scolytinae, *Thanasimus formicarius*, was significantly more abundant in stands where spruce was mixed with pine in North-Eastern France than in pure spruce stands, and this higher frequency was associated with lower populations of the bark beetle *Ips typographus* (Warzée et al. 2006).

– *Forest type* also influences the abundance and impact of natural enemies. For example, Liebhold et al. (2005) observed that the abundance of *Peromyscus* sp. mice in Wisconsin and the level of control they exerted on the gyspy moth were higher in mesic sites than in urban and xeric forest types.

Forest can serve as *reservoirs of natural enemies,* spilling out towards cultivated areas, especially when, as measured by Cronin et al. (2000), natural enemies have a higher mobility than their prey. From a systematic literature review encompassing

Fig. 6.3 *Thanasimus formicarius.* a. An adult roaming the bark surface, either for oviposition below a bark scale, or hunting for adult *Ips typographus.* b. A pupa in its niche inside the bark. If the bark is thinner than 6 mm, pupation cannot occur, and the mature larvae leave the tree. Figure 6.3a: Courtesy of Nathalie Warzée; Fig. 6.3b: Jean-Claude Grégoire

158 studies, Boesing et al. (2017) concluded that in tropical areas, at least, avian predators that exert significant control on agricultural pests depend on native forests. For example, Karp et al. (2013) observed an increased abundance of avian consumers of the coffee berry borer beetle (the bark beetle *Hypothenemus hampei*), as well as lower infestation levels in Costa Rican coffee plantations established in more forested landscapes. *Natural enemy spillover*, however, can occur *in the other direction*, from cultivated landscapes to natural forests. Frost et al. (2015) used interception traps to quantify spillover of generalist predatory wasps (*Vespula* spp., Vespidae) and of 106 species of more specialized hymenopteran parasitoids of lepidopteran caterpillars between native forest, dominated by Nothofagaceae, and exotic *Pinus radiata* plantation forest in New Zealand. They found that spillover of both generalist and specialist predators was directed from plantation to native forest, with a greater trend among generalists. They interpreted this as the result of a higher productivity of caterpillars in the plantation forest. This hypothesis was verified for the *Vespula* spp. but not for the specialist parasitoid wasps, by selectively suppressing the caterpillars in the plantation forest plot by spraying a formulation of *Bacillus thuringiensis* var. *kurstaki*, which affects Lepidoptera but no other insect orders.

6.5 Predator–Prey Relationships

In this section and unless specified otherwise, predators, parasitoids and pathogens are all referred to as predators and prey and all host species as prey.

All predators need to locate, overcome and consume their food, and optimally exploit those species that are currently available. They rely for prey location on various stimuli: including visual cues, semiochemicals, sound, vibration, and heat. In many cases, the prey's host-plant is also involved in attracting or maintaining predators: they emit semiochemicals, provide alternate food (e.g. from extra floral nectaries), or offer shelters (the domatia of ant-acacias, see Sect. 6.3). Finding a prey, however, is only the beginning of a whole sequence of events. For example, parasitoids that oviposit in living hosts need to increase the survival chances of their eggs, specialised predators feeding on rare prey need to optimise their consumption, pathogens need to colonise their host and to propagate to other hosts. At a higher level, the population dynamics of predator–prey systems (the reciprocal influences of predator and prey population changes) is also important to understand natural population balances as well as the successes or failures of biocontrol introductions (see Chapter 5, Forest Insect Population Dynamics).

6.5.1 Prey Finding

Visual cues (shape, size, movement, colour -at least for birds) are often used by vertebrate predators. Visual stimuli alone suffice in some cases, as illustrated by bird predation studies relying only on artificial caterpillars made of plasticine (Seifert et al. 2015). *Public information* conveyed by the sight of other individuals in the act of feeding is another important visual stimulus, described in particular for birds (Danchin et al. 2004). Some birds also use olfactory cues when foraging. The great tit, *Parus major*, has been shown experimentally to orient to apple trees infested by the winter moth, *Operophtera brumata*, following semiochemicals released by the attacked plants but not by uninfested trees (Amo et al. 2013). Olfactory cues are used by a large range of other natural enemies, from cruising nematodes to insect predators. Small mammals detect insect cocoons in the ground by their odour (Holling 1959a). The checkered beetle *Thanasimus formicarius* has an adult life protracted over several months and thus needs to feed on several successive prey species with shorter life cycles. It has antennal receptor cells keyed to a vast number of bark beetle pheromones and host volatiles (Tømmerås 1985), and responds to the pheromones of 27 bark-beetle species, attacking either conifers or broadleaves (Warzée et al. 2006). Once on the trees under attack, it feeds on the landing bark beetles, oviposits on the bark, and its larvae enter the bark-beetle galleries where they feed on any insect inside, including conspecifics. Conversely, the monospecific *Rhizophagus grandis* locates its only prey, *D. micans,* with amazing accuracy, using a very attractive and discriminatory mixture of tree-produced monoterpenes and oxygenated monoterpenes produced by the prey (Grégoire et al. 1992). In Belgium, *D. micans* is very sparsely distributed in most spruce stands (1–5 brood systems per ha), but 90% of these broods are eventually colonised by the predators. This accurate and specific capacity to locate the host is certainly one of the major reasons explaining the high success of classical biological control (see Sect. 6.6.5) of *D. micans* using this predator (Grégoire 1988; Kenis et al. 2004). Parasitoids respond to a variety of olfactory cues, depending on the life stage they parasitise. Parasitoids of adult bark beetles, such as the Pteromalid wasp *Tomicobia* spp., respond to pheromones and oviposit in the landing hosts. Egg parasitoids use a variety of cues: sex-, anti-aphrodisiac- or aggregation pheromones, or volatiles emitted by host plants and triggered by herbivore oviposition (reviews by Fatouros et al. 2008; Hilker and Fatouros 2015). Some species among the Braconidae and the Trichogrammatidae even use phoresy on fertilised host females to make sure they are present when the eggs are laid (Fatouros et al. 2005). It has long been debated how parasitoids attacking bark-beetle late larval instars locate their hosts. Mills et al. (1991), studying *Coeloides bostrychorum* parasitising *Ips typographus*, developed an elegant series of experiments involving an infra-red scanner, thermistor probes, cellulose or wax barriers, and freezing infested logs before their exposure to parasitoids, and concluded that chemical cues and not sounds, vibrations or heat mediate host location by *C. bostrychorum*. A review of semiochemical-assisted prey location in tritrophic systems has been published by Vet and Dicke (1992).

6.5.2   Prey Exploitation and the Components of Predation

Once a prey has been located, important choices must be made. The females of
haplodiploid parasitoids can select the sex of each egg by deciding whether to
fertilise it (opening their *spermatheca*, resulting in a diploid female) or to lay it
unfertilised (resulting in a haploid male). The choice often depends on the host's
size, a larger host producing a larger parasitoid. As dispersal capacity, longevity
and fecundity are often positively correlated with body size, in many cases, large
hosts are devoted to the female offspring, which will have to disperse further and
live longer than males and produce eggs themselves. Insect prey are not passive
participants in parasitoid-prey interactions and attempt to defend themselves (e.g.
by encapsulating eggs with melanocytes). Some parasitoid species inject a venom,
and/or polydnaviruses together with their eggs, which inhibit the host's defenses
(Strand and Burke 2013). Remarkably it appears that herbivores that have been
injected with viruses and venom by their parasitoids, elicit different volatiles from
their host plants than uninfested herbivores and hyperparasitoids appear to be able
to exploit this information for host location (Zhu et al. 2018). Other information
used by natural enemies include oviposition stimuli or inhibitors, i.e. chemicals that
indicate the availability of prey for the predator's offspring or, on the contrary, the
local abundance of conspecific predators and hence a risk of intraspecific competition. Once in the brood chamber of its prey, *Rhizophagus grandis* uses chemical
information from its prey to assess the size of the local prey population and adjust
oviposition accordingly (more prey produce more semiochemicals and induce higher
oviposition). Conversely, the presence of conspecific predators leads to reduced egg
laying (Dohet and Grégoire 2017).

The mechanisms described above explain individual predator success in prey location and exploitation. Together with other interactions with the biotic (e.g. competition, hyperparasitism, host plant resistance) and abiotic (e.g. temperature, humidity,
thermoperiod, photoperiod) environment, they constitute the basic components of
the complex interactions that occur at the population level. These interactions grow
in complexity when several successive predator and prey generations are considered. Spatial constraints lead to additional levels of complexity, for example when
populations are constituted by smaller units (metapopulations) more or less loosely
connected together in fragmented habitats. The quantitative population changes
across space and time resulting from this whole set of interactions generally exerts
a profound influence on the population dynamics of forest insects as a whole (see
Chapter 5).

It is striking that much of the early pioneering work on these predator–prey relationships has been based on forest insects. Tinbergen (1960), studying the predation
behaviour of great tits (*Parus major*) on forest insects in Dutch pine forests, quantified
how the frequency and size of the various available prey influence predation rates.
He introduced the concept of *searching image*: vertebrate predators learn from experience and, with time, improve their efficiency at finding the most abundant or most
preferred prey. This concept has since influenced the behavioural sciences (Davies

et al. 2012). The seminal work by Holling (1959a, 1959b, 1961) used field and laboratory studies to quantify the prey consumption of several small predatory mammals in response to various cocoon densities of the European pine sawfly *Neodiprion sertifer*. This work also measured their responses when more or less palatable alternate food resources (respectively sunflower seeds or dog biscuits) were mixed with the cocoons. Holling (1959b) also used experiments with a blinded human subject asked to collect small sandpaper discs deployed at various densities on a table, to develop what has been since named the *Holling's disc equation*. This work described three possible quantitative responses of predators to increasing prey density (*functional responses*): (i) a theoretical, linear one (*type I functional response*) with a constant predation rate irrespective of prey density, (ii) a second type of response, with decreasing predation rates levelling off at a certain prey density (*type II functional response*), described by *Holling's disc equation*, and distinguishing between *searching time* which would decrease with increasing prey density, and a fixed *handling time* needed for either prey consumption or oviposition; and (iii) a sigmoid *type III functional response* that has been observed among vertebrate predators that learn (e.g. develop a searching image). In addition to these individual functional responses, predator populations also show *numerical responses* to prey density. They tend to aggregate and/or reproduce more abundantly in sites of higher prey density (respectively *aggregative-* and *reproductive numerical responses*). These two types of behaviours (functional and numerical responses) are further influenced by predator interactions that increase in frequency as predator density increases. These intraspecific interactions between predators can have adverse effects on individual predation and lead to different predator-dependent functional response models (see discussion in Skalski and Gilliam 2001). When multiple generations are considered, it becomes also possible to detect delayed impacts which would not occur immediately but at the following prey generation. For example, Turchin et al. (1999) suggested that the population cycles of the Southern pine beetle, *Dendroctonus frontalis*, are driven by a delayed density-dependent impact of natural enemies, in particular of the predatory checkered beetle, *Thanasimus dubius*.

Insects have developed many resistance mechanisms. *Immunity mechanisms* include phagocytosis or encapsulation by hemocytes (also valid for larger bodies such as parasitoid eggs), enzymatic proteolysis, and the synthesis by the fat body or the hematocytes of antimicrobial peptides that protect insects against viruses (Sparks et al. 2008), bacteria and fungi (Gillespie et al. 1997), and nematodes (Castillo et al., 2011). *Chemical defense* is common in insects and includes compounds sequestered from larval and adult diet or produced *de novo* (Pasteels et al. 1983). For example, larvae of Diprionid sawflies regurgitate monoterpene droplets collected from the host tree (Eisner et al., 1974), the nature of which can vary according to host tree species (Codella and Raffa 1995). Leaf beetle adults and larvae secrete defensive chemicals, often sequestered from their host plant (Laurent et al. 2005). Some caterpillars such as those of the processionary moths (*Thaumetopoea* spp.) or of the Siberian moth (*Dendrolimus sibiricus*) can project in the air hundreds of thousand minute (0.1 mm) hollow hairs containing allergenic proteins that can seriously harm vertebrate predators, although some insectivorous birds or parasitic wasps or flies do not seem sensitive. Some species escape because they are *cryptic*, difficult to distinguish from their environment, others are *mimetic*. *Batesian mimicry* corresponds to defenseless species resembling a defended insect. Clearwings moths (Sesiidae) look like wasps, with their transparent wings and transversely striped black and yellow abdomen. *Müllerian mimics* are species that are all chemically defended but have a similar appearance, thus sharing the cost of predator learning.

A clear, ancient but still very relevant, introduction to predator–prey relationships in a forest population ecology context is provided by Varley et al. (1973).

6.6 Biological Control

6.6.1 Definition

Eilenberg et al. (2001) proposed an operative and widely followed definition of biological control (or biocontrol): *"The use of living organisms to suppress the population density or impact of a specific pest organism, making it less abundant or less damaging than it would otherwise be"*. However, Heimpel and Mills (2017) remark that this definition excludes natural control, *"the use of"* referring to manipulative control, and that, taken literally, *"living organisms"* excludes viruses as biocontrol agents.

Biocontrol can involve native or exotic natural enemies, against native or exotic prey. The introductions may occur in one point in time, followed by long-term establishment, or may need to be repeated periodically.

Four types of biological control have been identified: inoculative-, inundative-, conservation- and classical biocontrol. The first two strategies rely on the long-term mass-production of beneficials which, in most cases, can only be justified economically if there is a stable demand. They thus fit well with the needs of agriculture and the greenhouse industry but presently are usually of lesser general relevance for forest insects.

Comprehensive reviews of biological control have been published by Van Driesche and Bellows (1996), Heimpel and Mills (2017), and Hajek and Eilenberg (2018).

6.6.2 Inoculative Biological Control

This approach consists of the periodical introduction of natural enemies that establish, multiply and spread. This strategy is widely used in glasshouses where several crops are cultivated each year and the pests reappear regularly after a new crop has been started. It is sometimes used against pests with populations that fluctuate dramatically in density. Nucleopolyhedroviruses (NPVs) are regularly used against the spongy moth, *Lymantria dispar*, the Douglas-fir tussock moth, *Orgyia pseudotsugata* and various sawflies (*Neodiprion sertifer*, *N. lecontei*) (van Frankenhuyzen et al. 2007). Entomopathogenic nematodes are used to kill the immature stages of the pine weevil, *Hylobius abietis* in the tree stumps where they develop (Dillon et al. 2006). Natural enemies and pests can be either exotic or native.

6.6.3 Inundative Biological Control

This type of biological control is based on the release of large numbers of natural enemies that should exert control immediately. No establishment or only limited reproduction is expected. For example, mass-releases of *Trichogramma* wasps (egg parasitoids) are made in maize fields at the time of oviposition of the maize corn borer, *Ostrinia nubilalis* (Razinger et al. 2016). There might be one or two wasp generations produced during this period if moth oviposition is protracted but the natural enemies disappear afterward. In the forest environment, inundative releases of *Bacillus thuringiensis* subsp. *kurstaki* (Btk) have been successful against the spruce budworm, *Choristoneura fumiferana* in Canada and the USA (van Frankenhuyzen et al. 2007). Target mortality is not caused by the bacteria but by the toxins liberated by the crystals in the released sporangia, and there is no evidence that the bacteria reproduce (Garczynski and Siegel 2007). Therefore, it could be argued that Bt is a biopesticide rather than a biocontrol agent. Natural enemies and pests can be either exotic or native.

6.6.4 Conservation Biological Control

This strategy includes habitat manipulation in order to maintain or increase the abundance of native natural enemies. The provision of alternate hosts or prey on alternate host plants, alternate food sources (e.g. pollen- or nectar-producing plants to sustain adult parasitoids; acorns for polyphagous mammals—see Sect. 6.2.1), pupation sites for insects (see Sect. 6.4), nesting sites for birds, or overwintering shelters, have all been used as components of conservation biocontrol. Improving inter-patch connectivity by creating vegetation corridors can also be a component of conservation biological control. Brockerhoff et al. (2008) remarked that, although plantation forests are poorer habitats than natural forests, they still provide suitable habitats to many species. Jactel and Brockerhoff (2007) showed in a meta-analysis of 119 studies that herbivory by oligophagous insects is significantly reduced in mixed forest as compared to monospecific stands, but the respective roles of host-tree dilution and natural enemy enhancement are unclear. Conservation biocontrol is thus very relevant for the control of forest pests but at the moment, we are still lacking most of the knowledge and mastery of ecosystem functioning necessary for a full use of this strategy.

6.6.5   Classical Biological Control

This approach usually targets exotic pests, often pests of woody plants, and involves the introduction of natural enemies collected in the area of origin of the target pests. There are cases, however, where exotic natural enemies were successfully introduced against native pests, or against exotic species with which they are not associated in their area of origin. These latter cases belong to a subcategory, "new association classical biological control". Once established, the biocontrol agent usually remains permanently present and does not need to be reintroduced.

Since the first successful introduction of two exotic natural enemies from Australia (the coccinellid beetle, *Rodolia cardinalis*, and the tachinid fly, *Cryptochetum iceryae*: Caltagirone and Doutt (1989)) into California in 1898 against an exotic pest of citrus, the cottony cushion scale *Icerya purchasi*, 6,158 introductions involving 2,384 exotic insect natural enemy species have been attempted against 588 exotic insect pests between 1886 and 2010 (Cock et al. 2016). Kenis et al. (2017) calculated that ca. 55% of these introductions targeted pests of woody plants, with an establishment rate of 37% *vs.* 30% with other pests, and a 34% success rate (i.e. efficient pest control) *vs.* 24% with other pests.

A comprehensive worldwide catalogue of the introductions of nematodes and pathogens against insects and mites exists (Hajek et al. 2016). Among 131 programmes using exotic pathogens and nematodes against 76 insect species and 3 mites, 75 programmes (57%) targeted woody plant pests (Hajek et al. 2007), with an establishment rate above 60% *vs.* to 40% for all other habitats. The basis for the higher rates of establishment and control on woody plants is hypothesized to be the favourable environment provided by forests (see Sect. 6.4), as well as the technical and regulatory obstacles to apply control methods widely used in agriculture, such as insecticide treatments, often prohibited and anyway often useless in the forest environment, or mating disruption which often needs to be applied over vast areas in order to prevent mated females from the neighbourhood to recolonise the treated zone.

The rationale behind this successful approach is that exotic species become pests in new areas because the coevolved natural enemies that control them in their area of origin were not also introduced. Therefore, the first step of any classical biological programme is to *identify the origin of the pest*, using literature records, museum collections, molecular phylogeography, etc.

Then, *foreign exploration* can start, in order to find natural enemies that could be taken to the area newly colonised by the pest. Because the pest is sometimes very tightly controlled by natural enemies in its original range, simply finding the pest can be difficult, not to mention collecting sufficiently high numbers. One approach to circumvent this difficulty is to rear large numbers of the host/pest in the laboratory and deploy them in the field in the area of origin in order to induce attacks from local natural enemies that could be reared out of the exposed insects. This approach was followed by Mills and Nealis (1992) who, searching for natural enemies to introduce in Canada for the biological control of *Lymantria dispar*, reared out a Tachinid parasitoid fly, *Aphantorhaphopsis samarensis* from spongy moth larvae exposed in European sites where the moth populations densities were very low. Another example concerns the Asian longhorned beetle, *Anoplophora glabripennis*, an introduced pest in North America and Europe, which is common in parts of China but rare in others. Adult beetles were collected in the field in China, allowed to oviposit in willow logs in the laboratory, and the logs hung from trees in areas of low density to attract natural enemies. Twelve species, many new to science, were recovered using this method (Li et al. 2020).

The more individuals are collected from as many origins as possible, the better, because this increases the diversity of the released biocontrol agents and their capacity to adapt to their new habitats. However, the successful introduction of *R. cardinalis* consisted of only 129 individuals, which successfully established (Caltagirone and Doutt 1989). Sometimes, individuals of different strains differ in their relationships to the prey. For example, an English strain of the ichneumonid parasitoid *Mesoleius tenthredinis* introduced in Canada to control *Pristiphora erichsoni*, proved suscep-tible to egg encapsulation by its host (Muldrew 1953), while a Bavarian strain released later was not encapsulated (Ives and Muldrew 1984). Similarly, while most strains of the nematode *Deladenus siridicicola* sterilize female *Sirex noctilio* (see Sect. 6.3), a strain unintentionally introduced in northeastern North America does not fully sterilize its hosts, resulting in less efficient biocontrol (Kroll et al. 2013).

After collection, natural enemies must then either be cultured locally or sent to the country of destination, to be reproduced, further identified (if necessary using molecular methods) and tested for non-target effects. Most countries will not allow the release of generalist natural enemies that will attack non-target organisms native to the country of release. All these steps are usually placed under strict administrative control in both the countries of origin and of destination. The candidate for release must then be kept in a quarantine facility (a high security laboratory with rigorous procedures accounting for all movements in and out) and reared for several generation in order to make sure that they are free of diseases or hyperparasites (hyperparasites: see Box 6.1). They must also be tested for their impact not only on the target species, but also on non-target organisms. Finally, if release is authorised, they must be mass-produced and submitted to quality control tests.

Usually, the higher the numbers released, the higher the chances of establishment. Impact assessments, including assuring that non-target attack is not occurring, must theoretically be performed at a later stage, but funding does not always allow for this last step.

One important prerequisite to biocontrol is to assess in advance potential environ-mental risks connected to the release of an exotic organism in a new environment. Past experience has repeatedly demonstrated that, once established, poorly selected biocontrol agents can become pests on their own. In Massachusetts, Boettner et al. (2000) found that a generalist parasitoid tachinid fly, *Compsilura concinnata*, intro-duced to North America in 1906 against different targets (including the spongy moth and the brown tailed moth), were heavily parasitizing three species of native saturnid moths, and they suggest that *C. concinnata* could be responsible for the observed local decline of silk moths. Another example is the recent expansion worldwide of the Asian lady beetle, *Harmonia axyridis*, mass-released in many countries in the world and which now exerts intense intraguild predation upon other aphidophagous species (Roy et al. 2012). A review of non-target impacts of classical biocontrol has been published by Myers and Cory (2017).

The many cases of successful or partly successful classical biocontrol of forest pests have been reviewed by Kenis et al. (2017), Hajek et al. (2007) and Hajek and van Frankenhuyzen (2017) and some examples are developed in more details by Van Driesche and Bellows (1996), Hajek and Eilenberg (2018), Van Driesche et al. (2010), Van Driesche and Reardon (2014) and MacQuarrie et al. (2016). A recent example is the introduction in Italy of the parasitoid wasp *Torymus sinensis* imported from Japan against the introduced Asian chestnut gall wasp, *Dryocosmus kuriphilus*, and which resulted in excellent control after 7–8 years (Ferracini et al. 2018). On the basis of this success *T. sinensis* has also been introduced in Croatia, France, Hungary, Portugal, Slovenia, Spain, and Turkey. Another recent example of promising classical biocontrol is the introduction in North America of exotic parasitoids against the emerald ash borer, *Agrilus planipennis* (Box 6.2).

BOX 6.2—Classical biological control of the emerald ash borer in North America

In response to the invasion of the United States by the destructive emerald ash borer (EAB), *Agrilus planipennis*, scientists from the U.S. and China, Korea, and Russia collaborated to discover promising natural enemies that could be released in a classical biological control program. Potential biocontrol agents were imported into quarantine and host specificity testing was conducted. In 2007, permits were issued for the release of three of the agents: two larval parasitoids, *Spathius agrili* (Hymenoptera: Braconidae) and *Tetrastichus planipennisi* (Hymenoptera: Eulophidae), and the egg parasitoid *Oobius agrili* (Hymenoptera: Encyrtidae). Releases began in the Midwest and have expanded to 30 states as the EAB population has spread throughout the country. Follow-up monitoring shows that *T. planipennisi* is establishing well in 18 mostly northern states, and although *O. agrili* is small and difficult to recover, it too seems to be establishing in 15 states. *Spathius agrili* populations have been recovered for a year or two after release, but populations do not persist in the north. Research on the phenology of EAB and its parasitoids (Jones et al. 2020) showed that *S. agrili* is better synchronized with EAB populations that have a one-year lifecycle (like what is found in the southern United States) and *T. planipennisi* does better where EAB has a two-year lifecycle (as in the northern United States). Gould et al. (2020) developed a model of EAB development based on summer temperatures that predicts the likelihood of parasitoid establishment throughout the country based on the availability of EAB larvae in the spring. A large parasitoid like *S. agrili* is needed in the northern United States, however, because *T. planipennisi* has a short ovipositor and can only parasitize EAB in branches less than 11 cm in diameter (Abell et al. 2012).

Scientists have discovered a new EAB parasitoid in the genus *Spathius, S. galinae*, from Russia. Climate matching indicates a better fit for the northern U.S. and early results indicate that this parasitoid is establishing well (Duan et al. 2019). The ultimate goal of releasing biocontrol agents is not just to get them established, but for the parasitoids to reduce EAB population density and ultimately improve the health of ash trees. Recent studies of the next generation of ash growing in sites where *T. planipennisi* has established indicate that this parasitoid, combined with predation by native woodpeckers, has the potential to maintain EAB at a low density following an outbreak (Duan et al. 2017). Work is underway to discover which parasitoids are best suited for the variety of climate conditions in the United States, to quantify the role that *O. agrili* is playing where it has established, and how to integrate the use of insecticides and biological control to save mature trees in urban and natural forests.

6.7 Synthesis and Perspectives

The rich and rather stable conditions generally provided by forest ecosystems and woody plants in general favor complex food webs where assemblages of herbivorous insects coexist with predators, parasitoids, nematodes and pathogens. In most cases, particularly when the forest itself is diversified in tree species and ages, the herbivore populations remain at low levels, with little economic or environmental impact. This balance can be upset when the status of some of the components of these communities change, for example when climatic factors (e.g. droughts, heat waves, storms) weaken the trees, or when anthropogenic actions (e.g. clear cuts, plantations, fire control) modify tree composition and resistance, or when an introduced hyperparasitoid modifies the impact of a natural enemy. Changes in tree resistance or tolerance to herbivores, or relief from natural enemy pressure, can allow herbivores to build up larger populations and acquire pest status, temporarily or permanently. The introduction of exotic herbivores constitutes another type of perturbation. Kept in check by natural enemies or host resistance in their areas of origin, some invasive species can severely harm the newly colonized forests, sometimes even threatening the survival of whole tree taxa. As illustrated above (see Sect. 6.6.5), classical biological control has often provided long-term and sustainable solutions against exotic pests.

The intricate relationships between the various species interacting in the forest environment provide rich ground for basic and applied biological and ecological studies, and for their application to forest management. However, our understanding of these systems is still extremely incomplete and there will even be levels of complexity that we shall never grasp fully, even though research regularly brings forward new and exciting results.

References

Abell KJ, Duan JJ, Bauer L, Lelito JP, Van Driesche RG (2012) The effect of bark thickness on host partitioning between *Tetrastichus planipennisi* (Hymen: Eulophidae) and *Atanycolus* spp. (Hymen: Braconidae), two parasitoids of emerald ash borer (Coleop: Buprestidae). Biol Control 63:320–325

Amo L, Jansen JJ, Van Dam NM, Dicke M, Visser ME (2013) Birds exploit herbivore-induced plant volatiles to locate herbivorous prey. Ecol Lett 16(11):1348–1355. https://doi.org/10.1111/ele.12177

Boddy L (2016) Interactions with Humans and Other Animals. In: Watkinson SC, Boddy L, Money NP (eds) The Fungi (Third Edition). Academic Press, pp 293–336, 466 pp

Boesing AL, Nichols E, Metzger JP (2017) Effects of landscape structure on avian-mediated insect pest control services: a review. Landscape Ecol 32(5):1–14. https://doi.org/10.1007/s10980-017-0503-1

Boettner GH, Elkinton JS, Boettner CJ (2000) Effects of biological control introduction on three nontarget native species of saturniid moths. Conserv Biol 14:1798–1806

Brockerhoff EG, Jactel H, Parrotta JA, Quine CP, Sayer J (2008) Plantation forests and biodiversity: oxymoron or opportunity? Biodivers Conserv 17(5):925–951

Buckner CH (1966) The role of vertebrate predators in the biological control of forest insects. Annu Rev Entomol 11(1):449–470

Caltagirone LE, Doutt RL (1989) The history of the vedalia beetle importation to California and its impact on the development of biological control. Annu Rev Entomol 34:1–16

Castillo JC, Reynolds SE, Eleftherianos I (2011) Insect immune responses to nematode parasites. Trends Parasitol 27(12):537–547

Cock MJ, Murphy ST, Kairo MT, Thompson E, Murphy RJ, Francis AW (2016) Trends in the classical biological control of insect pests by insects: an update of the BIOCAT database. Biocontrol 61(4):349–363

Codella SG, Raffa KF (1995) Host plant influence on chemical defense in conifer sawflies (Hymenoptera: Diprionidae). Oecol 104(1):1–11

Cronin JT, Reeve JD, Wilkens R, Turchin P (2000) The pattern and range of movement of a checkered beetle predator relative to its bark beetle prey. Oikos 90(1):127–138

Cronin JT, Reeve JD (2005) Host-parasitoid spatial ecology: a plea for a landscape-level synthesis. Proc R Soc Lond B Biol Sci 272:2225–2235

Cruz YP (1981) A sterile defender morph in a polyembryonic hymenopterous parasite. Nature 294(5840):446

Danchin É, Giraldeau LA, Valone TJ, Wagner RH (2004) Public information: from nosy neighbors to cultural evolution. Science 305(5683):487–491

Davies NB, Krebs JR, West SA (2012) An introduction to behavioural ecology. John Wiley & Sons

Dillon AB, Ward D, Downes MJ, Griffin CT (2006) Suppression of the large pine weevil *Hylobius abietis* (L.) (Coleoptera: Curculionidae) in pine stumps by entomopathogenic nematodes with different foraging strategies. Biol Control 38(2):217–226

Dodds KJ, Graber C, Stephen FM (2001) Facultative intraguild predation by larval Cerambycidae (Coleoptera) on bark beetle larvae (Coleoptera: Scolytidae). Environ Entomol 30(1):17–22

Dohet L, Grégoire J-C (2017) Is prey specificity constrained by geography? Semiochemically mediated oviposition in *Rhizophagus grandis* (Coleoptera: Monotomidae) with its specific prey, *Dendroctonus micans* (Coleoptera: Curculionidae: Scolytinae), and with exotic *Dendroctonus* species. J Chem Ecol 43(8):778–793. https://doi.org/10.1007/s10886-017-0869-1

Duan JJ, Van Driesche RG, Crandall RS, Schmude JM, Rutledge CE, Slager BH, Gould JR, Elkinton JS (2019) Establishment and early impact of *Spathius galinae* (Hymenoptera: Braconidae) on emerald ash borer (Coleoptera: Buprestidae) in the northeastern United States. J Econ Entomol 112:2121–2130

Duan JJ, Bauer LS, Van Driesche RG (2017) Emerald ash borer biocontrol in ash saplings: the potential for early stage recovery of North American ash trees. For Ecol Manage 394:64–72

Eilenberg J, Hajek A, Lomer C (2001) Suggestions for unifying the terminology in biological control. Biocontrol 46:387–400

Eisner T, Johnessee JS, Carrel J, Hendry LB, Meinwald J (1974) Defensive use by an insect of a plant resin. Science 184:996–999

Elkinton JS, Healy WM, Buonaccorsi JP, Boettner GH, Hazzard AM, Smith HR, Liebhold AM (1996) Interactions among gypsy moths, white-footed mice, and acorns. Ecol 77(8):2332–2342

Elkinton JS, Liebhold AM (1990) Population dynamics of gypsy moth in North America. Annu Rev Entomol 35(1):571–596

Eveleigh ES, McCann KS, McCarthy PC, Pollock SJ, Lucarotti CJ, Morin B et al (2007) Fluctuations in density of an outbreak species drive diversity cascades in food webs. Proc Natl Acad Sci USA 104:16976–16981

Fatouros NE, Dicke M, Mumm R, Meiners T, Hilker M (2008) Foraging behavior of egg parasitoids exploiting chemical information. Behav Ecol 19(3):677–689

Fatouros NE, Huigens ME, van Loon JJ, Dicke M, Hilker M (2005) Chemical communication: butterfly anti-aphrodisiac lures parasitic wasps. Nature 433(7027):704

Ferracini C, Ferrari E, Pontini M, Saladini MA, Alma A (2018) Effectiveness of *Torymus sinensis*: a successful long-term control of the Asian chestnut gall wasp in Italy. J Pest Sci 1–7

Frost C, Didham R, Rand T, Peralta G, Tylianakis J (2015) Community-level net spillover of natural enemies from managed to natural forest. Ecol 96(1):193–202

Garczynski SF, Siegel JP (2007) Bacteria. In: Lacey LA, Kaya HK (eds) Field manual of techniques in invertebrate pathology: application and evaluation of pathogens for control of insects and other invertebrate pests. Springer Science & Business Media, pp 175–197

Gillespie JP, Kanost MR, Trenczek T (1997) Biological mediators of insect immunity. Annu Rev Entomol 42(1):611–643

Giron D, Dunn DW, Hardy IC, Strand MR (2004) Aggression by polyembryonic wasp soldiers correlates with kinship but not resource competition. Nature 430(7000):676

Godfray HCJ (1994) Parasitoids: behavioral and evolutionary ecology. Princeton University Press, pp 473

Gould JR, Warden ML, Slager BH, Murphy TC (2020) Host overwintering phenology and climate change influence the establishment of *Tetrastichus planipennisi*, a larval parasitoid introduced for biocontrol of the emerald ash borer. J Econ Entomol. https://doi.org/10.1093/jee/toaa217

Grégoire JC (1988) The Greater European Spruce Beetle, *Dendroctonus micans* (Kug.). In: A.A. Berryman (ed) Population dynamics of forest insects, Plenum, New York, pp 455–478, 604 pp

Grégoire JC, Couillien D, Krebber R, König WA, Meyer H, Francke W (1992) Orientation of *Rhizophagus grandis* (Coleoptera: Rhizophagidae) to oxygenated monoterpenes in a species-specific predator-prey relationship. Chemoecology 3:14–18

Hajek AE, Gardescu S, Delalibera I Jr (2016) Classical biological control of insects and mites: a worldwide catalogue of pathogen and nematode introductions. USDA Forest Service. FHTET-2016-06.

Hajek AE, McManus ML, Delalibera I Jr (2007) A review of introductions of pathogens and nematodes for classical biological control of insects and mites. Biol Control 41:1–13

Hajek AE, Eilenberg J (2018) Natural enemies: an introduction to biological control. Cambridge University Press, 246 pp

Hajek AE, van Frankenhuyzen K (2017) Use of entomopathogens against forest pests. In microbial control of insect and mite pests, pp 313–330

Heimpel GE, Mills NJ (2017) Biological control: ecology and applications. Cambridge University Press, Cambridge, pp xii + 380

Hilker M, Fatouros NE (2015) Plant responses to insect egg deposition. Annu Rev Entomol 60:493–515

Holling CS (1959a) The components of predation as revealed by a study of small-mammal predation of the European pine sawfly. Can Entomol 91(5):293–320

Holling CS (1959b) Some characteristics of simple types of predation and parasitism. Can Entomol 91(7):385–398

Holling CS (1961) Principles of insect predation. Annu Rev Entomol 6(1):163–182

Hougardy E, Grégoire J-C (2000) Spruce stands provide natural food sources to adult hymenopteran parasitoids of bark beetles. Entomol Exp Appl 96:253–263

Huang N, Enkegaard A (2010) Predation capacity and prey preference of *Chrysoperla carnea* on *Pieris brassicae*. Biocontrol 55(3):379–385

Ives WGH, Muldrew JA (1984) *Pristiphora erichsonii* (Hartig), larch sawfly (Hymenoptera: Tenthredinidae). In: Kelleher JS, Hulme MA (eds) Biological control programmes against insects and weeds in Canada: 1969–1980. Commonwealth Agricultural Bureaux, Famham Royal, United Kingdom, pp 369–380

Jactel H, Brockerhoff EG (2007) Tree diversity reduces herbivory by forest insects. Ecol Lett 10(9):835–848

Jennings DE, Duan JJ, Bauer LS, Schmude JM, Wetherington MT, Shrewsbury PM (2016) Temporal dynamics of woodpecker predation on emerald ash borer (*Agrilus planipennis*) in the northeastern U.S.A. Agric For Entomol 18(2):174–181. https://doi.org/10.1111/afe.12142

Jervis MA (ed) (2012) Insect natural enemies: practical approaches to their study and evaluation. Springer Science & Business Media, p 491

Jones MI, Gould JR, Mahon HJ, Fierke MK (2020) Phenology of emerald ash borer (Coleoptera: Buprestidae) and its introduced larval parasitoids in the northeastern United States. J Econ Entomol 113:622–632

Karp DS, Mendenhall CD, Sandí RF, Chaumont N, Ehrlich PR, Hadly EA, Daily GC (2013) Forest bolsters bird abundance, pest control and coffee yield. Ecol Lett 16(11):1339–1347. https://doi.org/10.1111/ele.12173

Kaya HK, Gaugler R (1993) Entomopathogenic Nematodes. Annu Rev Entomol 38(1):181–206

Keenan RJ, Reams GA, Achard F, de Freitas JV, Grainger A, Lindquist E (2015) Dynamics of global forest area: results from the FAO Global Forest Resources Assessment 2015. For Ecol Manage 352:9–20

Kenis M, Hurley BP, Hajek AE, Cock MJW (2017) Classical biological control of insect pests of trees: facts and figures. Biol Invasions 19(11):3401–3417. https://doi.org/10.1007/s10530-017-1414-4

Kenis M, Wermelinger B, Grégoire JC (2004) Research on parasitoids and predators of Scolytidae in living trees in Europe—a review. In: Lieutier F, Day K, Battisti A, Grégoire JC, Evans H (eds) Bark and wood boring insects in living trees in Europe, a Synthesis. Kluwer, Dordrecht, pp 237–290, 583 pp

Kroll SA, Hajek AE, Morris EE, Long SJ (2013) Parasitism of *Sirex noctilio* by non-sterilizing *Deladenus siricidicola* in northeastern North America. Biol Control 67(2):203–211

Lacey LA (ed) (2016) Microbial control of insect and mite pests: from theory to practice. Academic Press.

Lacey LA, Grzywacz D, Shapiro-Ilan DI, Frutos R, Brownbridge M, Goettel MS (2015) Insect pathogens as biological control agents: back to the future. J Invertebr Pathol 132:1–41

Lacey LA, Kaya HK (eds) (2007) Field manual of techniques in invertebrate pathology: application and evaluation of pathogens for control of insects and other invertebrate pests. Springer Science & Business Media

Laurent P, Braekman JC, Daloze D (2005) Insect chemical defense. In: Schulz S. (eds) The chemistry of pheromones and other semiochemicals II. Topics in current chemistry, vol 240. Springer, Berlin, Heidelberg

Leius K (1963) Effects of pollen on fecundity and longevity of adult *Scambus buolianae* (Htg.) (Hymenoptera: Ichneumonidae). Can Entomol 95:202–207

Leius K (1961) Influence of various foods on fecundity and longevity of adults *Scambus buolianae* (Htg.) (Hymenoptera: Ichneumonidae). Can Entomol 93:1079–1084

Li F, Zhang YL, Wang XY, Cao LM, Yang ZQ, Gould JR, Duan JJ (2020) Discovery of Parasitoids of *Anoplophora glabripennis* (Coleoptera: Cerambycidae) and their seasonal abundance in China using sentinel Host Eggs and Larvae. J Econ Entomol 113(4):1656–1665. https://doi.org/10.1093/jee/toaa068

Liebhold AM, Raffa KF, Diss AL (2005) Forest type affects predation on gypsy moth pupae. Agric for Entomol 7(3):179–185

Liebhold AM, Elkinton J, Williams D, Muzika RM (2000) What causes outbreaks of the gypsy moth in North America? Popul Ecol 42(3):257–266

Limburg DD, Rosenheim JA (2001) Extrafloral nectar consumption and its influence on survival and development of an omnivorous predator, larval *Chrysoperla plorabunda* (Neuroptera: Chrysopidae). Environ Entomol 30(3):595–604

MacQuarrie C, Lyons D, Lukas Seehausen M, Smith S (2016) A history of biological control in Canadian forests, 1882–2014. Can Entomol 148(S1):S239–S269

Mathews PL, Stephen FM (1997) Effect of artificial diet on longevity of adult parasitoids of *Dendroctonus frontalis* (Coleoptera: Scolytidae). Environ Entomol 26(4):961–965

Mills NJ, Krüger K, Schlup J (1991) Short-range host location mechanisms of bark beetle parasitoids. J Appl Entomol 111(1–5):33–43

Mills NJ, Nealis VG (1992) European field collections and Canadian releases of *Ceranthia samarensis* (Dipt.: Tachinidae), a parasitoid of the gypsy moth. Entomophaga 37:181–191

Morris RJ, Lewis OT, Godfray HCJ (2004) Experimental evidence for apparent competition in a tropical forest food web. Nature 428(6980):310

Muldrew JA (1953) The natural immunity of the larch sawfly (*Pristiphora erichsonii* (Htg.)) to the introduced parasite *Mesoleius tenthredinis* Morley, in Manitoba and Saskatchewan. Can J Zool, 31(4):313–332.

Murfin KE, Ginete DR, Bashey F, Goodrich-Blair H (2019) Symbiont-mediated competition: *Xenorhabdus bovienii* confer an advantage to their nematode host *Steinernema affine* by killing competitor *Steinernema feltiae*. Environ Microbiol 21(9):3229–3243. https://doi.org/10.1111/1462-2920.14278

Myers JH, Cory JS (2017) Biological control agents: invasive species or valuable solutions? In: Vilà M, Hulme P (eds) Impact of biological invasions on ecosystem services. Invading Nature—Springer Series in Invasion Ecology, vol 12. Springer, Cham, pp 191–202

Paine TD, Millar JG, Bellows TS, Hanks LM, Gould JR (1993) Integrating classical biological control with plant health in the urban forest. J Arboric 19:125–130

Palmer TM, Stanton ML, Young TP, Goheen JR, Pringle RM, Karban R (2008) Breakdown of an ant-plant mutualism follows the loss of large herbivores from an African savanna. Science 319(5860):192–195

Pasteels JM, Grégoire JC, Rowell-Rahier M (1983) The chemical ecology of defense in arthropods. Annu Rev Entomol 28(1):263–289

Poinar GO Jr (1991) Nematoda and Nematomorpha. In: Thorp JH, Covich AP (eds) Ecology and classification of North American freshwater invertebrates. Academic Press, San Diego, CA, pp 249–283

Poinar GO Jr (1975) Entomogenous nematodes: a manual and host list of insect-nematode associations. E.J. Brill, Leiden, pp 317

Pottier A (2012) La forêt des Landes de Gascogne comme patrimoine naturel? Échelles, enjeux, valeurs. PhD dissertation, Université de Pau et des Pays de l'Adour, pp 483

Price PW, Bouton CE, Gross P, McPheron BA, Thompson JN, Weis AE (1980) Interactions among three trophic levels: influence of plants on interactions between insect herbivores and natural enemies. Annu Rev Ecol Syst 11(1):41–65

Razinger J, Vasileiadis VP, Giraud M, van Dijk W, Modic Š, Sattin M, Urek G (2016) On-farm evaluation of inundative biological control of *Ostrinia nubilalis* (Lepidoptera: Crambidae) by *Trichogramma brassicae* (Hymenoptera: Trichogrammatidae) in three European maize-producing regions. Pest Manag Sci 72:246–254. https://doi.org/10.1002/ps.4054

Roland J, Taylor PD (1997) Insect parasitoid species respond to forest structure at different spatial scales. Nature 386:710–713

Rosenheim JA (1998) Higher-order predators and the regulation of insect herbivore populations. Annu Rev Entomol 43(1):421–447

Roy HE, Adriaens T, Isaac NJB, Kenis M, Onkelinx T, San Martin G, Brown PMJ, Hautier L, Poland R, Roy DB, Comont R, Eschen R, Frost R, Zindel R, Van Vlaenderen J, Nedvěd O, Ravn HP, Grégoire JC, de Biseau JC, Maes D (2012) Invasive alien predator causes rapid declines of native European ladybirds. Divers Distrib 18(7):717–725. https://doi.org/10.1111/j.1472-4642. 2012.00883.x

Ryall KL, Fahrig L (2006) Response of predators to loss and fragmentation of prey habitat: a review of theory. Ecol 87(5):1086–1093

Schoeller EN, Husseneder C, Allison JD (2012) Molecular evidence of facultative intraguild predation by *Monochamus titillator* larvae (Coleoptera: Cerambycidae) on members of the southern pine beetle guild. Naturwissenschaften 99(11):913–924

Seifert CL, Lehner L, Adams M, Fiedler K (2015) Predation on artificial caterpillars is higher in countryside than near-natural forest habitat in lowland south-western Costa Rica. J Trop Ecol 31(3):281–284. https://doi.org/10.1017/S0266467415000012

Senthil-Nathan S (2015) A review of biopesticides and their mode of action against insect pests. In Environmental sustainability. Springer, New Delhi, pp 49–63

Singer MS, Clark RE, Lichter-Marck IH, Johnson ER, Mooney KA, Rodriguez-Cabal M (2017) Predatory birds and ants partition caterpillar prey by body size and diet breadth. J Anim Ecol 86(6):1363–1371. https://doi.org/10.1111/1365-2656.12727

Skalski GT, Gilliam JF (2001) Functional responses with predator interference: viable alternatives to the Holling type II model. Ecology 82(11):3083–3092

Sparks WO, Bartholomay LC, Bonning BC (2008) Insect immunity to viruses. In: Beckage NE (ed) Insect Immunology. Academic Press, pp 209–242, 360 p. https://doi.org/10.1016/b978-012 373976-6.50011-2

Strand MR, Burke GR (2013) Polydnavirus-wasp associations: evolution, genome organization, and function. Curr Opin Virol 3(5):587–594

Tinbergen L (1960) The natural control of insects in pinewoods. I. Factors influencing the intensity of predation by song birds. Arch. Neerl ZooL 13:265–343

Tømmerås BA (1985) Specialization of the olfactory receptor cells in the bark beetle *Ips typographus* and its predator *Thanasimus formicarius* to bark beetle pheromones and host tree volatiles. J Comp Physiol A 15:335–341

Turchin P, Taylor AD, Reeve JD (1999) Dynamical role of predators in population cycles of a forest insect: an experimental test. Science 285(5430):1068–1071

Van Driesche RG, Reardon R (eds) (2014) The use of classical biological control to preserve forests in North America. United States Department of Agriculture, Forest Service, Morgantown, WV, FHTET-2013–2. https://www.fs.fed.us/foresthealth/technology/pdfs/FHTET-2013-2.pdf

Van Driesche RG, Bellows TS Jr (1996) Biological Control. Chapman and Hall, New York, p 539

Van Driesche RG, Carruthers RI, Center T, Hoddle MS, Hough-Goldstein J, Morin L, Smith L, Wagner DL et al (2010) Classical biological control for the protection of natural ecosystems. Biol Control Suppl 1:S2–S33

van Frankenhuyzen K, Reardon RC, Dubois NR (2007) Forest defoliators. In: Lacey LA, Kaya HK, (eds) Field manual of techniques in invertebrate pathology: application and evaluation of pathogens for control of insects and other invertebrate pests. Springer Science & Business Media, pp 481–504

Varley GC, Gradwell GR, Hassell MP (1973). Insect population ecology: an Analytical Approach. Blackwell Scientific, UK

Vet LE, Dicke M (1992) Ecology of infochemical use by natural enemies in a tritrophic context. Annu Rev Entomol 37(1):141–172

Warzée N, Gilbert M, Grégoire JC (2006) Predator/prey ratios: a measure of bark-beetle population status influenced by stand composition in different French stands after the 1999 storms. Ann for Sci 63(3):301–308

Wegensteiner R, Wermelinger B, Herrmann M (2015) Natural enemies of bark beetles: predators, parasitoids, pathogens, and nematodes. In: Vega FE, Hofstetter RW (eds) Bark Beetles. Biology and Ecology of Native and Invasive Species. Elsevier/Academic Press, London, UK, pp 247–304

Yousuf F, Carnegie AJ, Bashford R, Nicol HI, Gurr GM (2018) The fungal matrices of *Ophiostoma ips* hinder movement of the biocontrol nematode agent, *Deladenus siricidicola*, disrupting management of the woodwasp, *Sirex noctilio*. Biocontrol 63:739–749. https://doi.org/10.1007/s10526-018-9897-1

Zhu F, Cusumano A, Bloem J, Weldegergis BT, Villela A, Fatouros NE, van Loon JJA, Dicke M, Harvey JA, Vogel H, Poelman EH (2018) Symbiotic polydnavirus and venom reveal parasitoid to its hyperparasitoids. Proc Natl Acad Sci 115(20):5205–5210

Chapter 7
Forest Insect—Plant Interactions

Justin G. A. Whitehill, Jörg Bohlmann, and Paal Krokene

7.1 The Ecology of Insect—Plant Interactions in Forests

Insects and plants dominate terrestrial ecosystems in terms of both species numbers and biomass. Ecological relationships between insects and plants are ubiquitous and insect-plant interactions are important for ecosystem structuring and functioning. Insects probably contain more species than any other group of organisms with an estimated 5.5 million species (Stork et al. 2015). They can affect plants positively, for example as pollinators, or negatively, as consumers of plant tissues and vectors of disease. Herbivorous species that consume living plant tissues make up nearly half of all known insect species. In this chapter, we describe the negative effects herbivorous insects can have on plant fitness and the mechanisms plants use to counter these effects.

Forests cover about 31% of the Earth's land surface (FAO UNEP 2020). A great diversity of forest types, with over 60,000 tree species, support ~ 80% of the planet's biodiversity and provide many vital ecosystem services (Bliss 2011; Beech et al. 2017). Trees exhibit considerable morphological diversity but typically have elongated stems, secondary (woody) growth, and long life spans. Healthy forests deliver global ecosystem services such as carbon storage, biodiversity, and natural climate regulation, while providing humans with building and other industrial materials, energy, and food (Raffa et al. 2009; Trumbore et al. 2015). Healthy forests are adapted

J. G. A. Whitehill (✉)
College of Natural Resources, NC State University, Raleigh, NC, USA
e-mail: jwhiteh2@ncsu.edu

J. Bohlmann
Michael Smith Laboratories, University of British Columbia, Vancouver, BC, Canada

P. Krokene
Division of Biotechnology and Plant Health, Norwegian Institute of Bioeconomy Research, Ås, Norway

© The Author(s) 2023
J. D. Allison et al. (eds.), *Forest Entomology and Pathology*,
https://doi.org/10.1007/978-3-031-11553-0_7

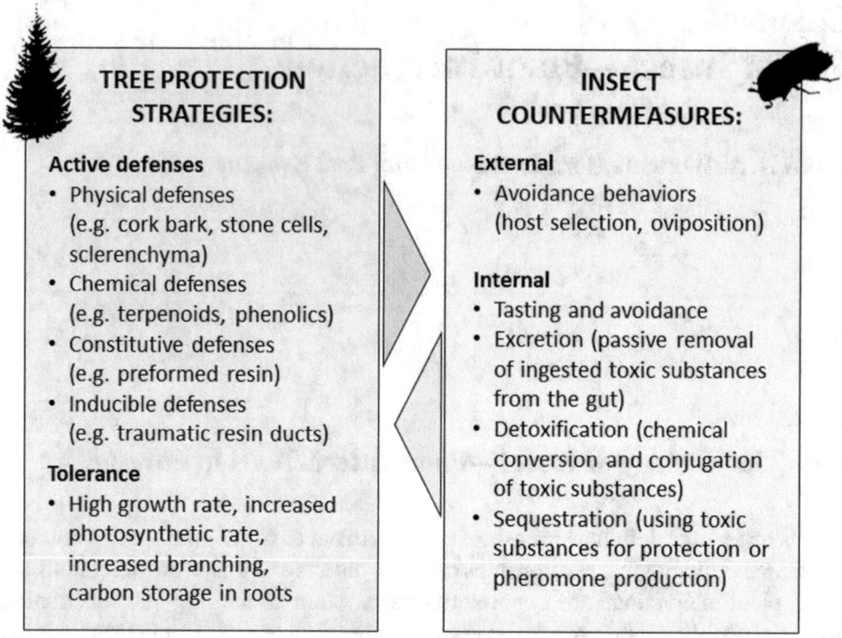

TREE PROTECTION STRATEGIES:	INSECT COUNTERMEASURES:
Active defenses • Physical defenses (e.g. cork bark, stone cells, sclerenchyma) • Chemical defenses (e.g. terpenoids, phenolics) • Constitutive defenses (e.g. preformed resin) • Inducible defenses (e.g. traumatic resin ducts) **Tolerance** • High growth rate, increased photosynthetic rate, increased branching, carbon storage in roots	**External** • Avoidance behaviors (host selection, oviposition) **Internal** • Tasting and avoidance • Excretion (passive removal of ingested toxic substances from the gut) • Detoxification (chemical conversion and conjugation of toxic substances) • Sequestration (using toxic substances for protection or pheromone production)

Fig. 7.1 Overview of tree protection strategies to minimize consumption by insects and insect countermeasures to avoid or tolerate tree defenses. The different strategies and countermeasures are presented in depth in the subchapters '7.2. The plant side—tree defenses against insects' and '7.3. The insect side—how insects cope with tree defenses'. © Justin Whitehill and Paal Krokene

to tolerate some level of stress caused by pests, pathogens and climate. A major challenge to forest health now and in the future is global climate change and mitigating the effects of climate change will be essential to maintaining healthy, resilient forests for future generations.

In this chapter, we describe insect-tree interactions from the perspective of both insects and trees (Fig. 7.1). We focus on interactions where insects use living trees as a food source and have to overcome or tolerate tree defenses. We first describe tree defense adaptations that minimize consumption by insects, including anatomical, mechanical, biochemical and molecular defenses. Then we explore how insects may counteract these defenses by different mechanisms that detoxify or provide tolerance against tree defenses, using examples of insects that feed internally and externally on both conifers and deciduous trees.

7.2   The Plant Side—Tree Defenses Against Insects

Co-evolution between plants and insects has driven the evolution of specialized plant defense mechanisms as well as insect counter-adaptations (Fraenkel 1959;

Janzen 1966; Walling 2000). Insect herbivory has thus been a major selection force behind species diversification in both plants and insects (Ehrlich and Raven 1964). Plant defenses can reduce the growth, survival, and fertility of attacking insects by disrupting insect feeding and oviposition preferences (Harborne 1993; Walling 2000). Plant defense traits are sometimes discussed without precise knowledge of specific traits, their ecological function, or the mechanisms through which they provide resistance against a pest. However, from an ecological perspective, a defense mechanism can be defined by how specific defense traits interact with and impact specific insect pests. Plants are said to be resistant to a specific insect species when plant defenses inhibit the insect's ability to utilize plant tissues for growth and survival.

Several previous reviews comprehensively explore the various aspects and intricacies of plant defenses against insects in both herbaceous and woody plant systems (Walling 2000; Franceschi et al. 2005; Howe and Jander 2008; Krokene 2015). We discuss the various terminologies associated with tree-insect interaction studies, while providing a conceptual framework to organize how different tree defense traits interact with insect herbivores (Fig. 7.2). This classification framework could be applied to describe any plant defense trait under investigation.

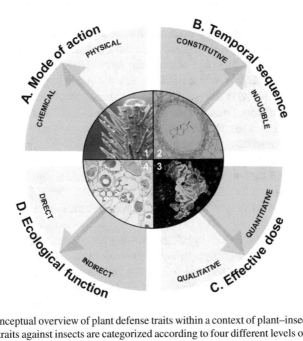

Fig. 7.2 A conceptual overview of plant defense traits within a context of plant–insect interactions. Plant defense traits against insects are categorized according to four different levels of organization. These include: (A) mode of action (plant-side); (B) temporal sequence (interaction between plant and insect); (C) effective dose (insect-side); and (D) ecological function (tritrophic interactions). Individual defense traits can be classified according to any of these categories and examples include: (1) oleoresin, (2) resin ducts and associated metabolites; (3) stone cells and other sclerified cell types; and (4) polyphenolic cells and associated metabolites. © Justin Whitehill and Paal Krokene

Tree defenses against insect pests are highly varied and combine chemical, physical, and molecular traits to resist attack (Franceschi et al. 2005; Krokene 2015; Whitehill et al. 2019). Tree defenses can be categorized in various ways, based on the compounds or structures by which they interfere with insects (physical versus chemical defenses), their effective doses (qualitative versus quantitative defenses), their ecological function (direct versus indirect defenses), or when they become active relative to insect attack (constitutive versus induced defenses) (Fig. 7.2). For example, categorizing tree defense traits by their mode of action contrasts structural and morphological traits that add toughness to tissues (physical defenses), and specialized (or secondary) metabolites that interfere with insect feeding and oviposition through toxic effects (chemical defenses).

7.2.1 Plant Defense Hypotheses

The theoretical framework of plant defense theory encompasses several independent but partially overlapping hypotheses. These include the *Optimal Defense (OD)* hypothesis, the *Carbon:Nutrient Balance (CNB)* hypothesis, the *Growth Rate (GR)* hypothesis, and the *Growth-Differentiation Balance (GDB)* hypothesis (Stamp 2003). The expanded *Growth-Differentiation Balance* hypothesis (Loomis 1932; Herms and Mattson 1992) may represent the most mature plant defense hypothesis, as it incorporates all the other hypotheses into its conceptual framework.

The GDB hypothesis provides a framework for predicting how plants balance resource allocation between differentiation-related and growth-related processes over a range of environments. Growth refers to the production of roots, stems and leaves, while differentiation is the process by which cells and tissues take on different functions. These functions can be transport of water and photosynthates or production of specialized metabolites and physical structures involved in defense against herbivory. The production of carbohydrates through photosynthesis represents the inflection point between growth and differentiation/defense. The GDB hypothesis predicts a trade-off in allocation to growth and defense that depends on resource availability (Stamp 2003).

Rigorously testing the GDB hypothesis in trees has proven difficult because trees have long lifespans and engage in complex ecological interactions. The diverse responses observed in tree chemical defenses to various nutrient levels in field studies suggest there is a need for comprehensive, multi-faceted experiments to test the GDB hypothesis. Such experiments should incorporate molecular, biochemical and ecological approaches to fully understand the subtle complexities of interactions that occur between herbivores and trees (Glynn et al. 2007; Kleczewski et al. 2010). Additionally, induced plant defenses play a critical role in many plant–insect interactions, but induced defenses have yet to be adequately incorporated into plant defense theories.

7.2.2 Defense, Resistance, Tolerance

Forestry and ecology are broad fields of study that each overlap with other disciplines. Each field approaches research questions from many angles and as a result can develop similar terminologies with very different meanings. The exact meaning of a term can vary based on the questions being explored, the lens through which the researcher studying these traits is viewing them, and the level of biological organization at which an interaction is being studied. For example, ecologists refer to quantitative and qualitative defense traits from the perspective of a trait and its dose-dependent direct impact on an insect, such as the effective lethal dose of a chemical required for mortality. Conversely, forest geneticists refer to quantitative and qualitative defense traits from the perspective of tree genetics. A quantitative defense 'trait' from the perspective of a geneticist refers to a phenotypic trait controlled through multiple genetic loci or nucleotides. We attempt to provide context to the area of tree defense traits and the intersection of terminologies across the major disciplines that study tree-insect interactions.

In this chapter, we distinguish between tree defense and tree resistance, although these terms are often used vaguely interchangeably. 'Defense' generally refers to the ways in which a tree defends itself from for example an insect attack. But just because defenses are present when an insect attacks, they may not be effective at protecting the tree. The absence of an effect may be due to insect counter-adaptations shaped through a shared co-evolutionary history with the tree. 'Resistance' is an observable phenotype that results from the interaction between the tree and an insect pest. Tree resistance occurs when one or several defense traits, working alone or together, provide complete or nearly complete protection from insect attack. For example, the resistance phenotype of Sitka spruce (*Picea sitchensis*) against spruce weevil (*Pissodes strobi*) is a result of multiple physical and chemical defense traits working together to provide resistance (Whitehill et al. 2019). However, while most Sitka spruce trees have chemical defense traits resembling those of resistant trees, the absence or reduction in a single physical defense trait may lead to susceptibility to insect attack (Whitehill et al. 2019). When multiple defense traits work together to provide resistance against an insect pest, the synergism between the traits is defined as a defense syndrome (Agrawal and Fishbein 2006; Raffa et al. 2017; Whitehill et al. 2019).

There is often no clear-cut line that separates resistance and susceptibility. Rather, complete resistance and complete susceptibility represent extremes along a continuum of tree phenotypes. To describe phenotypes that are neither completely resistant nor completely susceptible, the term tolerance is sometimes used. However, such intermediate phenotypes are usually categorized as partially resistant. Partial (or quantitative) resistance would describe a phenotype where a plant may not succumb completely to insect attack, but suffers a significant reduction in biomass compared to resistant genotypes. This type of resistance is typically due to many genes with small individual effects and appears to be the norm in insect-plant interactions (Kliebenstein 2014; French et al. 2016). For instance, induced terpene accumulation in Norway spruce (*Picea abies*) trees showed a negative relationship with attack success by the Eurasian spruce bark beetle (*Ips typographus*) (Zhao et al. 2011). Trees with high induced terpene levels had fewer and less successful beetle attacks than trees with low terpene levels. This example highlights the dose-dependent nature of plant defense traits against insects, because the level of resistance in individual spruce trees depended on the concentration of defensive terpenes in the attacked tissues.

The term tolerance is usually reserved for a clearly defined plant phenotype with compensatory responses to insect attack. Tolerance is achieved through mechanisms that modulate the plants' primary metabolism and is thus a distinct plant protection strategy that differs from the active defense strategies described above. Plants that are tolerant to herbivory are characterized by having: (1) high relative growth rates; (2) increased net photosynthetic rate after damage; (3) increased branching or tillering after release of apical dominance; (4) pre-existing high levels of carbon storage in roots available for allocation to above-ground reproduction; and (5) the ability to shunt carbon stores from roots to shoots after damage (Strauss and Agrawal 1999). Tolerance mechanisms thus involve changes in primary metabolism that mitigate negative effects of herbivore attack. We will not discuss tolerance further in this chapter, but rather focus on defense traits that actively protect trees against herbivory.

7.2.3 Mode of Action: Chemical and Physical Defenses in Trees

Plant defense traits can be distinguished by their mode of action of interfering with insects. Modes of action include chemical traits that have, for example, toxic effects and physical traits that provide a mechanical barrier, as well as traits that combine both modes (Fig. 7.2). Chemical and physical defense traits are considered the major components of a plant's defense system (Painter 1951; Gatehouse 2002).

Defensive plant chemicals may be species-specific and expressed in certain tissues or cell types (Walling, 2000). Chemical defense traits have received much attention since Gottfried Fraenkel's seminal 1959 paper 'The *raison d'etre* of secondary substances'. Fraenkel (1959) documented the defense chemistry of several common plant families and how these chemicals interact with known herbivore pests. He correctly highlighted that while many scientists had studied phytochemicals for their own purposes and applications, no one had accurately stated their intrinsic biological function and reason for existing—their *raison d'etre*. Several reviews have been published on the topic of chemical diversity, ecological function, and mechanisms of chemical defense in forest trees, including poplar (Phillippe and Bohlmann 2007), ash (Kostova and Iossifova 2007), oak (Salminen and Karonen 2011), eucalyptus (Naidoo et al. 2014), pine (Gijzen et al. 1993), and spruce (Keeling and Bohlmann 2006; Celedon and Bohlmann 2019). Well-studied defense chemicals in trees include terpenoids and phenolics.

Terpenoids make up the largest group of plant chemicals with tens of thousands of known compounds (Celedon and Bohlmann 2019; Fig. 7.3). They are structurally diverse, metabolically costly to produce, may occur in large quantities or as minor compounds, and can be toxic or inhibitory to a variety of insects and microorganisms (Raffa et al. 1985; Gershenzon 1994; Celedon and Bohlmann 2019). Terpenoids play important defensive roles in many conifers (Keeling and Bohlmann 2006). They are biosynthesized from five-carbon building blocks to produce monoterpenes (10 carbons), sesquiterpenes (15 carbons), diterpenes (20 carbons), and higher-order terpenes. Conifer resin typically consists predominantly of monoterpenes and diterpenes, and often-smaller amounts of sesquiterpenes and other compounds (Keeling and Bohlmann 2006). Different conifer species produce diverse resin mixtures containing dozens of individual terpenes (Schiebe et al. 2012). These compounds are produced by terpene synthases and cytochrome P450s that often make multiple products. A single terpene synthase (γ-humulene synthase) in grand fir (*Abies grandis*) can for example make 52 different sesquiterpene products (Steele et al. 1998). Such multiproduct enzymes contribute to the high biochemical diversity of conifer resin, and maintaining this chemical diversity seems to be an important part of the defense strategy of conifers (Ro et al. 2005; Keeling and Bohlmann 2006).

Phenolics in plants total several thousand compounds, including many with toxic or repellent effects towards insects and microorganisms (Lindroth and Hwang 1996; Zeneli et al. 2006; Fig. 7.3). Phenylalanine is a common precursor for the formation of

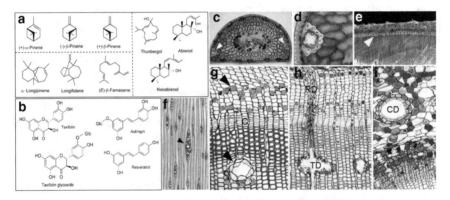

Fig. 7.3 Examples of chemical defenses in trees. (a) Monoterpenes (top left), sesquiterpenes (bottom left) and diterpenes (right) are the main constituents of conifer resin. (b) Soluble phenolics like flavonoids (left) and stilbenes (right) are important chemical defenses in many tree species. (c) Cross-section of a mountain pine (*Pinus mugo*) needle showing two resin ducts, with a close-up of one duct (d). (e) Cross-section of Norway spruce (*Picea abies*) stem showing a ring of traumatic resin ducts formed in response to external stress. (f) Tangential section of Scots pine (*Pinus sylvestris*) latewood showing a large radial ray with a resin canal in the center. (g) Cross-section of Scots pine stem showing an axial resin duct in the young sapwood and phenol-containing parenchyma cells in the young phloem (C: vascular cambium). (h) Cross-section of Norway spruce stem showing axially oriented traumatic resin ducts (TD) in the sapwood, interconnected with a radial resin duct (RD). (i) Cross-section of a balsam fir (*Abies balsamea*) stem showing a large cortical resin duct (CD) in the phloem surrounded by dark phenol-rich cells. © Justin Whitehill and Paal Krokene

phenolics, including flavonoids, stilbenes, condensed tannins and other polypheno-
lics, as well as the structural polymer lignin (Dixon et al. 2001). Beyond lignin, which
is the major phenolic in all trees, some tree species invest considerable resources
into phenolic defenses. For example, 35% of leaf dry weight in poplar may consist
of condensed tannins and other phenolics (Lindroth and Hwang 1996). So-called
soluble phenolics, which include stilbenes and flavonoids (Fig. 7.3), are abundant in
conifer bark and have been studied extensively. Stilbene production is inducible but
stilbene levels do not appear to increase following bark beetle attack or fungal infection (Zeneli et al. 2006; Schiebe et al. 2012), probably because the fungi metabolize
stilbenes at a faster rate than the tree can produce them (Hammerbacher et al. 2013).

Defense traits such as terpenoid resins, latexes and gums play well-documented
chemical roles in tree-insect interactions, but these traits can also be considered
physical defenses. The mechanical properties of these toxic substances can physically
trap or expel insects that attempt to bore into a tree. As an example, terpenoid resin is
stored under pressure in specialized resin ducts in many conifers. Tunneling insects
that rupture these ducts may be flushed out by the resin flow and trapped in the sticky,
toxic substance (Christiansen et al. 1987; Franceschi et al. 2005).

Some of the classical literature on plant–insect interactions emphasized physical
defenses, noting that 'repellent factors […] are very frequently physical in nature' and
that these factors influence feeding patterns of insects and other herbivores (Dethier
1941). Trees have several cell types and anatomical structures that reduce insect
feeding by providing physical toughness or thickness to tissues. Physical defense
traits may reside inside tissues and cells or they may be structures exposed on the
plant surface. They include spines, thorns, trichomes on leaf surfaces, bark texture,
leaf toughness, granular minerals incorporated into tissues, and increased quantities
of specialized sclerenchyma cells (Wainhouse et al. 1990; Franceschi et al. 2005;
Ferrenberg and Mitton 2014; Whitehill et al. 2016a, 2016b, 2019). The mode of action
of these traits is to disrupt feeding and tunneling of adult insects and larvae by wearing
down their mouthparts or interfering with digestion (Raupp, 1985; Wainhouse et al.
1990; Whitehill et al. 2016b). A number of studies have highlighted that physical
plant defenses play similarly important roles as chemical traits, depending on the
species under investigation (Massey and Hartley 2006; Hanley et al. 2007; Carmona
et al. 2011; Ferrenburg and Mitton 2014; Lopresti and Karban 2016).

The periderm, the tough outer surface of the bark, is the first line of physical
and chemical protection against insects and also protects trees against desiccation
and fire (Krokene 2015). The outermost part of the periderm is the cork, the dry
bark layer that is paper thin in young trees, but may be more than 30 cm thick in
older conifer trees. The cork consists of mostly dead cells reinforced with lignin and
lipophilic suberin polymers (Franceschi et al. 2005). The texture of the outer bark
surface may also serve as a physical defense. Trees with smooth, slippery bark have
been observed to have fewer bark beetle attacks compared to trees with rough bark.
Lower brood production under smooth outer bark that is more difficult for beetles
to grip suggests reduced oviposition on such slippery bark surfaces (Ferrenburg and
Mitton 2014).

Inside the periderm there are other more localized physical defenses, such as stone cells, fiber cells and calcium oxalate crystals. Stone cells are tough, highly lignified cells that function as a dose-dependent physical defense against insects (Wainhouse et al. 1990; Whitehill et al. 2016a, 2016b, 2019; Fig. 7.2). Fiber cells are lignified sclerenchyma cells that form densely spaced concentric sheets in the inner bark of many conifers. These sheets appear to be an effective barrier to bark beetles and other insects that attempt to penetrate the bark (Franceschi et al. 2005). Granular minerals such as calcium oxalate crystals are also interspersed throughout the bark of both angiosperm and conifer trees. These crystals are tough, pointed physical structures found inside and outside the cell walls in different plant tissues (Franceschi et al. 2005; Massey et al. 2007). The crystals are thought to provide protection from chewing insects.

While insects may adapt to chemical defenses, for example through mechanisms of secretion or detoxification (Despres et al. 2007), resistance based on anatomical defenses may be more difficult for insects to overcome (Whitehill et al. 2019). In conifers for example, stone cells have been recognized as a substantial determinant of resistance in different spruce species against several destructive forest pests, such as bark beetles and weevils (Wainhouse et al. 1990; Whitehill et al. 2016a, 2019; Whitehill & Bohlmann, 2019). Stone cells can provide resistance against phloem feeding weevils through at least three mechanisms: (i) they form a physical barrier that prevents establishment and movement of neonate larvae, (ii) they physically displace more nutritious host tissue and thereby reduce larval development, and (iii) they cause mandible damage to young larvae which affects feeding. By acting as a physical barrier that slows larval development, stone cells also increase larval exposure to other defenses such as resin (Whitehill et al. 2019). Such synergism between stone cells and resin-based defenses constitutes a robust defense syndrome that is difficult for insects to overcome (Whitehill and Bohlmann 2019).

7.2.4 Temporal Sequence: Constitutive, Induced and Primed Defenses in Trees

The distinguishing feature of constitutive and induced defenses is the time when they are deployed. Constitutive defenses are always present, even in the absence of insect attack. They can be viewed as an insurance against the attacks that almost inevitably will come during the long life of a tree (Franceschi et al. 2005). Examples of constitutive defenses in conifers are polyphenolic cells in the phloem that store phenolic metabolites, which are released upon insect feeding (Franceschi et al. 1998; Nagy et al. 2014), (ii) resin ducts filled with terpene-rich oleoresin (Celedon and Bohlmann 2019), and (iii) stone cells functioning as physical barriers (Whitehill et al. 2016a, 2019; Whitehill and Bohlmann 2019). In contrast, induced defenses are mobilized in response to an attack (Eyles et al. 2010). Examples of induced defenses are the formation of traumatic resin ducts in conifer wood and the hypersensitive response in foliage. Constitutive and induced plant defenses can be both physical and chemical in nature.

The concepts of constitutive and induced defenses play central roles in plant defense theory. Plant survival and competitive success require that plants optimize how they allocate the resources they have available. Resource allocation is primarily dependent upon carbon availability (i.e. photosynthate), which is used for two major purposes: growth or defense (Herms and Mattson 1992; Stamp 2003). Defense theory predicts that plant defense responses to insect attack are largely determined by the resources the plant has access to and how those resources are allocated within the plant. When resources are allocated to physical and chemical defenses, less are available to grow new leaves and other vegetative structures. This trade-off concept is crucial to understanding both the nature of present-day plant defenses and the evolutionary history of plant defense mechanisms. Inducible defenses are thought to have evolved as a means to reduce the overall costs associated with defense, since inducible defenses only are activated when they are needed, i.e. after an attack has occurred (Steppuhn and Baldwin 2007). Induction of plant defenses reduces the amount of resources diverted to specialized metabolism and facilitates a return to growth-dominated activities once a threat from an invading pest has been removed. Additionally, induced defenses can be targeted to the site of an ongoing attack and thereby further reduce resource allocation to defense, since the plant does not invest in defending tissues that are not being attacked.

In some cases, trees can trigger systemic defense responses in unattacked tissues following insect attack (Philippe and Bohlmann 2007; Eyles et al. 2010; Krokene 2015). Systemic induction of defense prepares plants for insect attack through signaling cascades involving the octadecanoid pathway, the plant hormone ethylene, or small peptides that induce defenses throughout the plant (Philippe and Bohlmann 2007; Eyles et al. 2010). Trees can also activate a form of delayed induced defense known as defense priming. Delayed or long-term defenses in trees are based on two, non-mutually exclusive mechanisms of induced defenses: prolonged upregulation of induced defenses and defense priming (Wilkinson et al. 2019). Prolonged up-regulation of induced defenses simply means that defenses induced by insect attack or fungal infection remain up-regulated for weeks or months and thus provide resistance to subsequent attacks. Because resources are diverted away from growth to defenses for a long time, prolonged up-regulation of induced defenses may be a costly defense strategy. A more cost-efficient mechanism of long-term induced defense is defense priming. When a plant is primed, induced defenses are sensitized in a way that provides faster and/or stronger activation of induced defenses in response to future attacks (Conrath et al. 2015). Following a priming stimulus, defenses are maintained at constitutive or weakly induced levels, but are then rapidly activated upon subsequent attack (Pastor et al. 2013). The priming stimuli may be wounding, colonization by insects, pathogens or beneficial organisms, or treatment with chemical compounds (Mauch-Mani et al. 2017).

Defense priming can provide very effective protection of forest trees. For example, Norway spruce trees in an area with epidemic bark beetle populations became almost completely resistant to attack when they had been treated with the wound hormone methyl jasmonate as a priming stimulus (Mageroy et al. 2020a). The molecular mechanisms responsible for defense priming in Norway spruce and other trees are still unclear, but many defense-related gene transcripts in spruce bark showed a primed response after methyl jasmonate treatment, including transcripts for Pathogenesis-Related (PR) proteins and epigenetic regulators (Mageroy et al. 2020b).

7.2.5   *Effective Dose: Qualitative and Quantitative Defenses in Trees*

Historically, the terms qualitative and quantitative defenses have been used mostly for chemical traits and refer to the dosage required for specific compounds to negatively affect a feeding insect. Toxic compounds that are effective in low amounts are said to be qualitative and compounds that must be ingested in high amounts to have an effect are considered quantitative. The terms were established and popularized by Feeny (1976) and Rhoades & Cates (1976) to explain the evolution of plant defenses based on plant apparency, i.e. how likely a plant is to be found by an herbivore. Large and long-lived plants that are easily found by herbivores are 'apparent', and small or ephemeral plants that are less likely to be found are 'unapparent'. Qualitative defense traits were predicted to be dominant in unapparent plants while quantitative defense traits were predicted to be dominant in apparent plants.

Qualitative chemical defenses are potent toxins that are effective at very small doses against most insect species, i.e. against generalist pests without co-evolved countermeasures. Examples of qualitative plant chemicals are small toxic molecules such as certain alkaloids and cyanogenic compounds. Insects that have co-evolved with their host plant may have adapted countermeasures to such qualitative defenses. Such specialist insects may for example sequester qualitative defense metabolites and use them for their own protection against predators and parasites (Rhoades and Cates 1976; Agrawal and Kurashige 2003). Strong selection pressures and short generation times may allow insect pests to rapidly evolve counter-adaptations and overcome tree defenses through specialization (Despres et al. 2007). Therefore, qualitative defenses in forest trees typically do not provide robust resistance against adapted insect pests, and the application of qualitative defenses for long-term pest management is not a viable strategy.

Quantitative chemical defenses, on the other hand, involve specialized metabolites such as tannins, with a dose-dependent effect and are generally effective against an herbivore only in high amounts. Due to the basic mechanisms by which quantitative defenses interfere with the physiology of an insect, it is difficult for insect pests to evolve countermeasures against these traits. Quantitative defenses thus tend to be effective against both specialist and generalist species. However, quantitative defenses may come at a high cost: because they are most effective in high concentrations they are energetically costly to produce and maintain.

Interestingly, in contrast to chemical defenses, physical defense traits have received less attention in plant defense hypotheses dealing with quantitative versus qualitative defense. Hay (2016) points out limitations of the plant apparency model and makes the case that 'plants are rarely defended by one compound or even by chemistry alone'. We propose that existing plant defense hypotheses incorporate physical defenses as an integral part of a synergistic plant defense system. As an example of a synergy between chemical and physical defense in trees, stone cells are a constitutive, quantitative and physical defense in Sitka spruce against the spruce weevil (Whitehill and Bohlmann 2019). Stone cells provide a robust resistance that synergizes the effect of a physical defense with terpenoid chemical defenses, which are both constitutive and induced and may be either quantitative or qualitative (Whitehill and Bohlmann 2019).

7.2.6 Ecological Function: Direct and Indirect Defenses in Trees and Tri-Trophic Interactions

Tree defenses that directly affect the physiology or behavior of an insect, and thus impair its growth, survival or reproduction, are defined as direct defenses (Fig. 7.2). However, a tree can also attract species in its environment to protect it against attackers. Such indirect defenses can involve the release of volatile metabolites, which may attract predators and parasitoids of plant-feeding insects. Such volatiles may be induced locally or systemically by activity of the insect and are then called herbivore-induced plant volatiles (HIPVs) (Turlings and Erb 2018; Wilkinson et al. 2019). When plants attract natural enemies of plant-feeding insects they engage in tri-trophic interactions, i.e. interactions with reciprocal ecological impacts between three trophic levels: a primary producer, a herbivore, and the herbivore's natural enemy. By engaging in tri-trophic interactions, plants can benefit from the vulnerability of plant-feeding insects to natural enemies. This is the premise for the tri-trophic niche concept, which states that certain plants may be an enemy-sparse or enemy-dense space for herbivores (Singer and Stireman 2005). Plants can increase or reduce the predation risk of an herbivore by releasing HIPVs or providing toxic plant metabolites that the herbivore can sequester and use in their anti-predator defense.

Tri-trophic interactions involving HIPV signaling have been mostly studied in herbaceous angiosperms but are also known from both angiosperm and gymnosperm trees (Turlings and Erb 2018). When Scots pine (*Pinus sylvestris*) needles are attacked by ovipositing sawflies, their foliage emits the sesquiterpene (E)-β-farnesene. This HIPV attracts a specialized egg parasitoid, which oviposits inside the sawfly eggs, thereby reducing the growth and ultimately survival of the sawfly larvae (Hilker et al. 2002). Similarly, black poplar (*Populus nigra*) responds to feeding by spongy moth (*Lymantria dispar*) larvae by releasing HIPVs that attract the spongy moth parasitoid *Glyptapanteles liparidis* (Clavijo-McCormick et al. 2014). Tri-trophic interactions have also been demonstrated belowground, at least in angiosperm systems. When insect larvae are feeding upon maize (*Zea mays*) roots, they emit a sesquiterpene that attracts nematodes, which then infect the larvae (Rasmann et al. 2005). Because indirect defenses involving tri-trophic interactions are found across the plant kingdom this is probably an ancient plant defense strategy that emerged early in the evolution of land plants (Mumm and Dicke 2010).

7.3 The Insect Side—How Insects Cope with Tree Defenses

Insect and host tree populations usually exist in some sort of equilibrium, where insect attacks are countered by tree defenses. Most insect herbivores subsist at low levels where they are rarely noticed, whereas others go through boom and bust cycles as part of their normal 'outbreak' behavior. Outbreak species are often referred to as 'pests', particularly if they damage economically important tree species. The delicate balances that regulate insect populations around an equilibrium are sometimes disrupted, for example if trees are suffering due to anthropogenic factors such as movement of species and climate change. Insect populations that are out of balance—because they are introduced into new environments or are favored by changing climates—often become pests.

Interactions between herbivorous insects and trees are highly variable. This diversity is a product of the enormous number of insect species that feed on trees and the many different ways that trees can be exploited by insects. Because of their large dimensions, long life cycles, and complex architecture, trees provide numerous niches that can be exploited by insects with many different lifestyles. Much of a tree consists of lignified organs and tissues, both above ground (main stems, branches, twigs) and below ground (roots in many different diameter classes). Wood may contain living cells, like the water-conducting sapwood, or consist mostly of dead cells, such as the heartwood. Bark, needles and leaves also offer a large and apparent array of living tissues that support many different insects. Tree-feeding insects subsist on their hosts by utilizing various feeding strategies and can be grouped into so-called feeding guilds. Feeding guild largely dictates the mechanisms by which different insects may cope with tree defenses. The oldest known fossil record of insects feeding on plants dates back approximately 400 million years and consists of fossilized insect guts or feces and feeding damage on fossilized plants (Labandeira 1998). Insect herbivory presumably originated as generalist feeding on foliage and diversified into specialized feeding guilds. The earliest fossils of insects feeding on living woody or wood-like tissues are 350 million years old. Wood-boring is considered a primitive life habit for beetles and their immediate ancestors have evolved into some of the most destructive present-day forest pests (Vega and Hofstetter 2015).

7.3.1 A Note on Generalist and Specialist Insect Herbivores

The mechanisms trees use to defend themselves are usually effective against most herbivorous insect species in the trees' natural environment. However, some insects have co-evolved with their host tree to overcome tree defenses. Such co-evolved species can successfully colonize unique niches that are not readily available to non-adapted competitors (Despres et al. 2007). Based on their host relationships herbivorous insects are often categorized as either specialist or generalist species. These terms are usually used within the context of chemical defenses as opposed to physical defenses. Specialist insects have evolved mechanisms that allow them to feed on a select set of plant species with a high concentration of a particular type of chemical defense, while these plants would not be suitable hosts for most other insects. In extreme cases, while increasing the insect's fitness on its preferred host(s), this specialization may have reduced its fitness on other plants. Generalist insects are species that have a much wider host range than specialists and are able to deal with more diverse chemical defenses, at least at low to moderate concentrations. The terms 'generalist' and 'specialist' are widely used in the literature but there are no defined set of criteria that clearly differentiates generalists from specialists (Ali and Agrawal 2012).

7.3.2 Insect Feeding Guilds and Their Interaction with Tree Defenses

Herbivore feeding or trophic guilds are groups of species that exploit the same kinds of plant resources in comparable ways. The major feeding guilds of insects that live on trees include foliage feeders (Chapter 9), bark beetles (Chapter 10), woodborers (Chapter 12), sucking insects (Chapter 13), and insects feeding on reproductive structures (Chapter 16). Here we briefly address how the major feeding guilds interact with tree defenses.

Insect-tree interactions are largely constrained by the physical and chemical properties of the tissues the insects feed upon and the physical closeness of the insect-tree association. Many insects live and feed inside trees and may thus remain in close contact with tree defenses for long periods. This is true for woodborers (Chapter 12) and bark beetles (Chapter 10) that feed and oviposit in tunnels in the bark or sapwood, and for ambrosia beetles (Chapter 11) that tunnel in the sapwood. Weevils feeding on tips, shoots, roots and reproductive organs (Chapter 15) and insects feeding on cones and seeds (Chapter 16) also spend most of their lives inside their host. The same is true for some foliage feeders, such as gall insects (Chapter 14) and leaf miners (Chapter 9). Most other foliage feeders feed externally in the canopy, such as sucking insects (Chapter 13) and some weevils feeding on tips, shoots and young plants. These external feeders have a looser physical association with their host tree and are exposed to tree defenses largely through the tissues they ingest. The herbivorous insects that tend to be the least exposed to tree defenses are sucking or piercing insects that ingest sap or xylem fluids.

7.3.3 Insect Strategies to Cope with Tree Defenses

Insects can overcome plant defenses through counter-adaptations that are genetically determined or due to behavioral plasticity (Fox et al. 2004). Insect strategies to cope with plant defenses can be classified as external or internal (Despres et al. 2007), depending on whether they operate before or after ingestion of plant tissues, respectively. Some insects, such as galling insects, actively suppress tree defenses prior to ingestion by manipulating host tissues externally (Samsone et al. 2012). Once an insect has ingested host tissues, it can excrete, sequester or detoxify chemical defenses internally. Such internal, post-ingestive counter-adaptations are well studied, especially against chemical defenses, and may involve the action of enzymes in the insect midgut, such as cytochromes P450 and glutathione S-transferases (Enyati et al. 2005; Feyereisen 2006; Despres et al. 2007; Che-Mendoza et al. 2009; Chiu et al. 2019). Insect counter-adaptations to physical defense traits, on the other hand, are not well studied.

7.3.3.1 External Strategies of Insects to Cope with Tree Defenses

Prior to feeding, an insect can respond to plant defense traits through behavioral avoidance mechanisms. These behaviors can reduce or completely bypass negative impacts of tree defenses. Insects actively evade defenses through avoidance in time (phenology) or by feeding on tissues that are less well defended. For example, many moths and butterflies that feed on leaves closely synchronize larval emergence with bud burst because emerging young leaves are less well defended chemically and physically than older leaves (Feeny 1970). Also, some leaf feeding insects cut through a primary leaf vein to reduce turgor pressure before they start to feed. This trenching behavior has been observed in insects feeding on plants that store highly toxic latex within specialized defense structures called laticifers (Doussard and Eisner 1987). Plant latexes and resins represent both chemical and physical defenses, as these fluids often contain toxic metabolites that are also highly viscous and sticky.

Insects use visual, olfactory or tactile cues from plant defense traits, volatile emissions or nutritional quality to avoid feeding or laying eggs on toxic plant tissues. Young larvae usually feed on the tissues where oviposition occurred and brood survival will thus be higher if optimal substrates are selected for oviposition. This is the premise for the "mother knows best" hypothesis which predicts that insects oviposit on hosts where their progeny will perform optimally (Bernays and Graham 1988). The use of chemical cues to avoid chemical defenses is often intertwined with the use of visual cues. For instance, woodboring beetles tend to rely first on visual cues to select potential host trees, before switching to tactile and olfactory cues when they land on the host. Ambrosia beetles also integrate visual and olfactory cues to differentiate host species from non-host species (Campbell and Borden 2009). The role of olfactory cues in host selection behaviors of bark beetles have been particularly well studied, since tree-killing bark beetles are important forest pests. Bark beetles utilize tree chemistry to identify suitable hosts for oviposition and brood development. Specifically, these beetles have evolved complex mechanisms to modify terpenes in the trees' chemical defenses for use in their own pheromone biosynthesis (Chiu et al. 2017). Instead of attempting to summarize the vast literature on this topic in a short paragraph, we refer the reader to some of the relevant literature that explores these well-documented interactions (Wood 1982; Raffa 2001; Zhang and Schlyter 2004; Blomquist et al. 2010).

Although we are not aware of studies that demonstrated active avoidance behaviors in forest pests to physical defense traits, observations of the spruce weevil have suggested that adult maturation feeding on spruce shoots prior to oviposition may improve brood fitness (Whitehill and Bohlmann 2019). Adult maturation feeding drains resin canals on the apical shoot and is hypothesized to reduce exposure of eggs and larvae to the toxic effects of oleoresin. This probably improves survival of young weevil larvae, although further experimental evidence is required to support this hypothesis. This behavior resembles the trenching behavior of insects that feed on herbaceous plants with toxic latex.

7.3.3.2 Internal Strategies of Insects to Cope with Tree Defenses

Insects have various internal mechanisms to circumvent the toxic effects of specialized plant metabolites. These mechanisms include tasting (gustation) and subsequent avoidance of toxic food, as well as excretion, detoxification, and sequestration of toxic plant metabolites after ingestion. In herbaceous systems, gustatory cues can deter continued insect feeding on plant tissues. For instance, cyanogenic glycosides deter further feeding by the alfalfa weevil (*Hypera brunneipennis*) even when glycoside levels are below the threshold of toxicity (Bernays and Cornelius 1992). In poplar (genus *Populus*), deterrents of insect feeding such as phenolic glycosides and salicinoids (glycosides of salicylic acid) are important for defense (Hwang and Lindroth 1997). Tasting and avoidance strategies are challenging behaviors to unravel, as they require careful observation combined with targeted bioassays of individual plant metabolites. The avoidance responses of insects to toxic metabolites are very simple: move on and feed on a different plant or tissue. Since avoidance is conceptually straightforward, we focus here instead on the more complex internal metabolic mechanisms insects use to cope with toxic plant compounds. Using forest insects as examples, we present the three non-behavioral metabolic coping mechanisms: excretion, detoxification, and sequestration of toxic plant metabolites.

Excretion—In the context of insect-plant interactions, excretion refers to the simple removal of ingested toxic plant metabolites from the insect gut with the feces

(Zagrobelny et al. 2004). Insects that are adapted to feed on plants with diverse chemical defenses tend to rely on excretion as their main mechanism to avoid potentially toxic metabolites. For instance, case moth (*Hyalarcta huebneri*) larvae that feed on chemically well-defended eucalyptus leaves excrete most of the toxic metabolites they ingest unchanged (Cooper 2001). Some ingested plant toxins are stopped by the peritrophic matrix in the insect midgut, acting as a barrier that prevents toxins from reaching the gut epithelium. The polarity of ingested compounds and the pH of the midgut can also influence the toxicity of certain plant metabolites. For instance, many lipophilic compounds do not interact readily with the insect midgut and therefore are passively excreted following ingestion (Barbehenn 1999). Conversely, hydrophilic compounds must be modified enzymatically in the midgut to reduce their toxicity and ease their removal from the digestive tract.

Detoxification—Detoxification involves biochemical processes to remove toxic compounds that have been ingested. Insect detoxification of plant defense compounds may involve variations and combinations of compounds being oxidized, hydrolyzed, or reduced, as well as conjugated to molecules that can be readily cleared from the insect body (Despres et al. 2007). Detoxification of plant metabolites by herbivorous insects has been described to involve a variety of different enzymes such as cytochrome P450 monooxygenases (CYP450s), glutathione-S-transferases, and carboxylesterases. Of these, CYP450s are perhaps the best studied and appear to play a key role in many plant–insect interactions (Feyereisen 2005). CYP450s are a diverse group of enzymes that are found throughout the animal and plant kingdoms (Li et al. 2007). In insects, CYP450s are essential to the function of certain organs such as antennae, where they clear old odorant molecules from the odorant receptors (Maïbèche-Coisne et al. 2005). CYP450s are also critical to insect metabolism and tolerance of anthropogenic chemicals such as insecticides (Petersen et al. 2001; Wondji et al. 2007). The important functions CYP450s have in detoxification are reflected in the large diversity and number of CYP450s in insect genomes. Glutathione-S-transferases are involved in detoxification of glucosinolates by making them more soluble and thus more easily excreted (Enayati et al. 2005). Insect carboxylesterases detoxify chemical insecticides and are therefore also thought to be involved in detoxification of other toxic substances, such as plant specialized metabolites (Yang et al. 2005).

Sequestration—Sequestration in insects is the process of utilizing plant metabolites for protection against predators or as precursors for pheromone production. Sequestration of plant metabolites is a highly specialized counter-adaptation to plant chemical defenses. The process may appear complex but only requires a few modifications of conserved molecular processes. Insect sequestration requires a selective import system that targets potentially harmful compounds, a safe transport mechanism through the body so the toxic metabolites do not harm the insect, and a site for safe, long-term storage (Kuhn et al. 2004). Sequestration processes are best documented in leaf beetles (Chrysomelidae) where the juvenile stages use sequestered plant compounds to defend themselves against predation (Meinwald et al. 1977; Pasteels et al. 1990; Gillespie et al. 2003). In trees, the poplar leaf beetle (*Chrysomela populii*) sequesters salicin in specialized defensive glands and excretes the toxin for its own protection (Strauss et al. 2013). Similarly, sawfly larvae feeding on pine foliage sequester diterpenes from the needles as a defense against predators (Eisner et al. 1974).

Sequestration versus detoxification: a closer look at the mountain pine beetle - The mountain pine beetle (*Dendroctonus ponderosae*) is a devastating forest pest with unique mechanisms to cope with the terpene-rich resin defenses of its host trees. Females initiate mass attacks on trees by releasing the aggregation pheromone *trans*-verbenol as they enter the bark. *Trans*-verbenol is formed by the hydroxylation of α-pinene, an abundant monoterpene in pine resin. This hydroxylation is catalyzed by a specific CYP450 in the beetle (Chiu et al. 2019). For attacking females it is essential to rapidly initiate mass attacks in order to overcome tree defenses and successfully colonize trees. Earlier, it was believed that females hydroxylated α-pinene into *trans*-verbenol immediately upon entering the bark. However, Chiu et al. (2019) found that the beetles lay the foundation for rapid pheromone production much earlier in life. As the larvae develop in the bark, they detoxify α-pinene and store it as monoterpenyl esters inside their body. These pheromone precursors are most abundant in female larvae around the time of pupation and are retained through to adult emergence and host finding. Detoxification of α-pinene and sequestration of pheromone precursors thus appears to provide a reservoir for the rapid female-specific release of *trans*-verbenol upon tree attack (Chiu et al. 2018). The mountain pine beetle example shows that sequestration and detoxification are not necessarily mutually exclusive mechanisms, but can be context dependent and open to interpretation; α-pinene is first detoxified, then sequestrated as monoterpenyl ester pheromone precursors, and finally converted to the aggregation pheromone *trans*-verbenol.

7.3.4 The Role of Symbiotic Microorganisms in Insect-Tree Interactions

Many herbivorous insects benefit from microorganisms in obtaining resources from well-defended and nutrient-poor tree tissues. It would therefore be oversimplified to consider insect-plant interactions as two-species interactions, as in reality they are likely complex insect-plant-microbiome interactions (Geib et al. 2008; Berasategui and Salem 2020; Frago et al. 2020). The insect microbiome includes the endo-microbiome (organisms living inside the insect, including in the gut) and the exo-microbiome (organisms living on the external surface of the insect). Bacteria and fungi in insect microbiomes may play essential roles in the breakdown of food (Scully et al. 2014; Lee et al. 2015; Berasategui and Salem 2020), defense against pathogens (Cardoza et al. 2006), and protection against plant defenses (Ceja-Navarro et al. 2015; Howe and Herde 2015; Frago et al. 2020). In the context of insect-tree interactions, the microbiome may significantly increase insect fitness by detoxifying tree defense metabolites and otherwise make plant tissues more suitable for feeding and reproduction. Large-scale mapping of insect microbiomes can be achieved by targeted sequencing of DNA barcoding regions of major microbial groups, such as bacteria, archaea, and fungi (Caporaso et al. 2012). Here, we present two examples that illustrate the intricate ways microbial symbionts may influence insect-tree interactions. First, we describe how fungal and bacterial symbionts may help bark beetles to colonize well-defended conifer trees, and secondly, how endosymbiotic bacteria are involved in a highly specialized nutritional mutualism with aphids.

7.3.4.1 Bark Beetles, Bluestain Fungi and Bacteria

A century-old paradigm in bark beetle ecology holds that fungi vectored by tree-killing bark beetles are critical for overwhelming host tree defenses and ultimately killing the tree (Six and Wingfield 2011; Krokene 2015). As early as 1928, F.C. Craighead suggested that ascomycete bluestain fungi carried by the beetles were important in tree killing (Craighead 1928), and historically most research on microorganisms involved in overwhelming tree defenses has focused on these fungi (Kirisits 2004). It has proved difficult to demonstrate experimentally that bluestain fungi are crucial for tree-killing, partly because it is difficult to separate the contribution of the fungi from that of the beetle itself. Even though it is hard to prove conclusively that microbionts are essential for tree-killing, fungi and bacteria have been shown to metabolize tree secondary metabolites and thus help detoxify tree defenses. In some North American bark beetle species, bacteria in the endo-microbiome have been demonstrated to help digest plant tissues and break down plant defenses (Adams et al. 2009, 2013; Boone et al. 2013). Also, bluestain fungi associated with the Eurasian spruce bark beetle rapidly break down phenolics in spruce bark and make the phloem more attractive to tunneling beetles (Hammerbacher et al. 2013; Kandasamy et al. 2019; Zhao et al. 2019a). Bluestain fungi may also produce components of bark beetle aggregation pheromones, suggesting that these fungi have a long co-evolutionary history with the beetle (Zhao et al. 2019b).

7.3.4.2 Aphids and Endosymbiotic Bacteria

Aphids are sap-sucking insects that feed externally on trees and other plants. Sap provides a very unbalanced diet consisting mostly of carbohydrates. It contains little nitrogen, and is a poor source of specific amino acids such as methionine and leucine (Sandström and Moran 1999). To overcome the nutritional deficiency of their diet, aphids harbor different species of endosymbiotic bacteria inside their cells. One species that is carried by almost all aphids is the endosymbiotic bacterium *Buchnera aphidicola*. This obligate intracellular endosymbiont provides essential amino acids that allow the aphids to survive on their carbohydrate-rich but nutrient-poor diet. In return, the bacterium receives all its other essential nutrients from its aphid host. The bacterium lives inside large specialized cells known as bacteriocytes and is vertically transmitted from mother to offspring with the egg. Since the bacterium cannot survive outside the cells of its aphid host, it essentially functions like an organelle. The aphid-Buchnera symbiosis is ancient and dates back at least 180 million years (Moran et al. 2008). Due to its obligatory endosymbiotic lifestyle the bacterium has lost many key genes for metabolic pathways and extracellular structures present in free-living bacteria. Because of this gene loss, the genome size of *Buchnera aphidicola* is only 15% of that of its close free-living relative *Escherichia coli* (Shigenobu et al. 2000). In addition to Buchnera, aphids harbor other bacteria such as *Hamiltonella defensa*, which may improve aphid fitness by providing protection against parasitic wasps and other natural enemies (Dion et al. 2011).

7.4 Case Studies: Major Forest Pest Issues Worldwide

Here we present examples of some major forest pest challenges. The selected insect-tree interactions highlight many of the tree defense mechanisms and insect adaptations described above. We present insect species with varied lifestyles and belonging to different feeding guilds, including species that feed internally or externally in conifer and broadleaved trees (Fig. 7.4). Also, since co-evolution between insect herbivores and trees is important in shaping insect-tree interactions, we present examples of both native and invasive forest pests.

7.4.1 Native Pests Living on Co-Evolved Host Trees

Interactions between native insects and their co-evolved host trees tend to be much more stable and predictable than interactions between invasive insects and evolutionary naïve tree species. Still, native insects such as sawflies and bark beetles may be opportunistic pests that go through boom-and-bust cycles and can have large-scale and long-lasting outbreaks.

7.4.1.1 The European Pine Sawfly: An Eruptive Defoliator with a Co-Evolved Tri-Trophic Niche

The European pine sawfly (*Neodiprion sertifer*) is native to Eurasia where it feeds on the needles of Scots pine and other two-needle pines. It is an early-season defoliator that occasionally undergoes short-lived outbreaks that may cover tens of thousands of hectares (Chorbadjian et al. 2019). The larvae feed on pine needles, starting with 1-year-old and older needles and only feeding on current-year needles if they run out of older needles. Larval development is completed relatively early in the summer and the mature larvae move down the stem and pupate in the forest litter. Adults emerge in the autumn and females lay eggs on current-year needles. Since the larvae rarely defoliate trees completely, tree mortality is low, but heavy attacks may cause

Native forest pests   **Invasive forest pests**

Fig. 7.4 Examples of native and invasive forest pests worldwide. Native insects living on co-evolved host trees: (a) the European pine sawfly (*Neodiprion sertifer*) is a native defoliator of pines in Europe and Asia; (b) a mountain pine beetle (*Dendroctonus ponderosae*) female is swimming through resin to enter and colonize a pine host in its native range in western North America; (c) the spruce weevil (*Pissodes strobi*) is a native regeneration pest across North America, ovipositing in the apical shoot of different spruce and pine species. Invasive insects attacking evolutionary naïve host trees: (d) the bark and wood boring emerald ash borer (*Agrilus planipennis*) is native to Asia but has invaded eastern North America where it is killing native ash trees; (e) the red turpentine beetle (*Dendroctonus valens*) is native to North America and has been introduced into China where it is killing millions of native pine trees; (f) the balsam woolly adelgid (*Adelges piceae*) is a small sap sucking insect of European origin that has been introduced into North America where it is killing native fir species. Photo credits: a © Erling Fløistad, Norwegian Institute of Bioeconomy Research; b © Christine Chiu, Natural Resources Canada; c and d © Justin Whitehill; e © (inset) Erich G. Vallery, USDA Forest Service—SRS-4552, https://doi.org/Bugwood.org and (damage) Bob Oakes, UGA1241449, USDA Forest Service, https://doi.org/Bugwood.org; f © Brad Edwards, North Carolina Cooperative Extension

significant growth losses. A complex relationship exists between the trees' chemical defenses, survival of sawfly larvae, and predation risk. As they feed, the larvae ingest diterpene resin acids stored in resin canals in the needles (Niemelä et al. 1982; Fig. 7.3). High concentrations of resin acids in the diet reduce larval growth, but resin acids may also improve larval survival. Larvae protect themselves against predators by sequestering ingested resin acids and storing them in specialized pouches in the foregut (Eisner et al. 1974). When challenged by birds or other predators, the larvae startle the attackers by synchronously waving their bodies and discharging a bubble of resin acid through their mouth. Ingestion of diterpene resin acids thus represents a trade-off for the larvae: in the absence of predation diterpenes negatively affect larval growth and survival, but diterpenes may increase larval survival when predators are present. This complex relationship between pine defenses and sawfly survival illustrates the tri-trophic niche concept and the intricate relationships that may exist between plants, herbivores and predators. As described above ('Plant side') the tri-trophic niche concept states that toxic specialized compounds and other plant characteristics may increase or decrease a herbivore's vulnerability to natural enemies by making the plant an enemy-sparse or enemy-dense space for the herbivore (Singer and Stireman 2005).

7.4.1.2   The Mountain Pine Beetle: Rapid Range Expansion by a Native Tree-Killing Bark Beetle

The mountain pine beetle is native to western North America, colonizing lodgepole pine (*Pinus contorta*) and other pine species throughout its large geographical range (Six and Bracewell 2015). The mountain pine beetle epitomizes the devastating effects tree-killing bark beetles can have on forest ecosystems, having killed 55% of all merchantable lodgepole pine over a 25 million hectare area since the 1990's (Meddens et al. 2012). Most of the time beetle population levels are low and oviposition occurs in the stem bark of weakened and dying trees. Following disturbances and favorable climatic conditions, beetle populations build up and massive outbreaks can occur, with an explosive increase in abundance over a short period of time. Beetle outbreaks may last several years, and during outbreaks the beetles are able to overwhelm the resistance of even healthy trees through mass-attacks coordinated by aggregation pheromones (Raffa et al. 2008; Boone et al. 2011). The last 20 years, climate change has been driving range expansions of this pest into higher altitudes and eastwards across the Rocky Mountains in Canada (Cudmore et al. 2010; Buotte et al. 2016). Warming temperatures have also favored beetle population growth and outbreak development by reducing winter mortality and causing drought stress that lowers tree defenses. The beetles vector a pathogenic fungal symbiont, the bluestain fungus *Grosmannia clavigera*, that colonizes the phloem and sapwood of attacked trees following beetle colonization. The combined effect of beetle mass-attacks and fungal infection ultimately overwhelms tree defenses and kills the trees. In an effort to mitigate the impacts of beetle outbreaks researchers are dissecting the complex three-way interactions between beetles, fungal symbionts and trees. This work has been facilitated by the development of genomic resources for both the fungal pathogen (DiGuistini et al. 2011) and the beetle (Keeling et al. 2013).

7.4.1.3 The Spruce Weevil: A Shoot-Feeding Reforestation Pest of North American Conifers

Pissodes strobi is a 'snout beetle' (family Curculionidae) colonizing various spruce and pine species across its wide range in North America. In western forests it attacks various spruce species and is known as the spruce weevil (Ebata 1991), whereas in the east it attacks primarily eastern white pine (*Pinus strobus*) and is referred to as white pine weevil. The beetles cause damage when females oviposit near the top of the apical shoot of young trees. The developing larvae tunnel downwards in the phloem, destroying the shoot in the process. Because of its abundance, wide geographical range, and ability to disrupt the height growth of young trees, the spruce weevil is considered the most important threat to reforestation of commercial spruce forests in western North America. Sitka spruce is particularly susceptible and very little reforestation has historically been attempted with this species, despite its intrinsically high economic value (King and Alfaro 2009). However, extensive research has identified weevil-resistant spruce genotypes that are now used actively in forest regeneration programs (Kiss and Yanchuk 1991; King and Alfaro 2009; King et al. 2011). Weevil-resistance in Sitka spruce results from a complex defense syndrome with synergism between chemical and physical defense traits that are both constitutively present and induced following insect attack. Specifically, resistant trees have more stone cells in the upper part of the shoot where the young larvae start their development. The stone cells slow down larval growth and increase larval exposure to the chemical toxicity and physical aspects of oleoresin (Whitehill et al. 2019). Resistant spruce genotypes have co-evolved with the insect in areas with high weevil densities (King et al. 2011). In contrast, a highly susceptible genotype was found on the remote Haida Gwaii Islands that have historically been free from weevils (King et al. 2011). Plants propagated from resistant and susceptible spruce genotypes have been used for detailed mechanistic studies of tree resistance (Robert and Bohlmann 2010; Robert et al. 2010; Hall et al. 2011; Whitehill et al. 2016a, 2016b, 2019). This research has generated important tools and resources, including one of the first sequenced conifer genomes that has been the basis for several genomic and gene sequence-based mechanistic studies (Birol et al. 2013; Celedon et al. 2017; Whitehill et al. 2019).

7.4.2 Invasive Pests Attacking Evolutionary Naïve Host Trees

Some of the most devastating insect-tree interactions involve insects that have been accidentally introduced into new areas where they interact with local tree species that lack effective defenses (Gandhi and Herms 2010). International trade with live plants, such as plants for planting, and the use of infested wood packaging materials are the main sources for the introduction of invasive tree pests to new areas (Aukema et al. 2010). Novel insect-tree associations may result in unpredictable and surprising outcomes due to the lack or reciprocal adaptations between insects and trees (Ploetz et al. 2013).

7.4.2.1 Emerald Ash Borer: An Invasive Stem Borer Ravaging Non-Adapted American Ash Species

The emerald ash borer (*Agrilus planipennis*) is an invasive bark- and wood-boring insect causing widespread mortality of ash (genus *Fraxinus*) in eastern North American forests. The beetle originates from East Asia and was accidentally introduced into North America in the 1990s (Herms and McCullough 2014). In its invasive range the beetle colonizes healthy ash trees and kills them within 2–3 years. The damage is done by the larvae as they feed on the inner bark and sapwood of the main stem, ultimately killing the trees by disrupting the flow of water and nutrients (McCullough and Katovich 2004). All North American ash species are susceptible to attack (Cappaert, et al. 2005; Poland and McCullough 2006). Detailed studies of the interaction between ash defenses and tunneling beetle larvae have shown that North American ash species are unable to confine and kill the young larvae. Thus, the evolutionary naïve ashes of North America lack effective defenses against this invasive pest. In contrast, Manchurian ash (*F. mandshurica*) native to Asia is resistant to attack, likely because it has targeted defenses developed over its co-evolutionary history with the insect (Bryant, et al. 1994; Rebek et al. 2008). Manchurian ash is less preferred for adult feeding and oviposition than susceptible ash species (Rebek et al. 2008), is more resistant to larval feeding (Chakraborty et al. 2014), and has higher constitutive concentrations of specialized metabolites and defensive proteins in the bark (Eyles et al. 2007; Whitehill et al. 2011, 2012, 2014; Hill et al. 2012). Interestingly, normally susceptible North American ash species can be made resistant to attack following external application of the wound hormone methyl jasmonate on the stem bark (Whitehill et al. 2014). Methyl jasmonate application increased the activity of trypsin inhibitors and concentrations of phenolics and lignin in the bark and decreased larval survival. This shows that even susceptible ash species have the defense machinery to prevent beetle infection, but they apparently are unable to induce these defenses under natural conditions, perhaps because they fail to recognize the feeding larvae or respond quickly enough to attack.

7.4.2.2 Red Turpentine Beetle: Novel Insect-Fungus Partnerships Are Invading Chinese Forests

Like the emerald ash borer, the red turpentine beetle (*Dendroctonus valens*) is mostly a secondary colonizer of weakened trees in its native range, but is a serious tree-killer in its invasive range. The red turpentine beetle is the largest and most widespread bark beetle in North America. It can breed in more than 40 conifer species in North America but is most common in different pine species. Although it normally colonizes weakened trees or trees attacked by other bark beetles, it may occasionally attack and kill apparently healthy trees in its native range (Sun et al. 2013). Unlike most other bark beetles, the larvae of the red turpentine beetle feed gregariously in groups of up to 100 larvae that excavate a large cave-like gallery in the bark of the lower stem. The beetle was accidentally introduced into China in the early 1980s, probably through import of unprocessed conifer logs from the western United States, and has killed millions of pine trees in China since its first outbreak in 1999 (Yan et al. 2005; Sun et al. 2013). The beetle's success in China appears to be due to a combination of naïve host trees, few natural enemies, and an ability to partner with new species of mutualistic symbiotic microorganisms (Sun et al. 2013). In its invasive range the beetle mainly attacks Chinese pine (*Pinus tabuliformis*) and sometimes Chinese white pine (*Pinus armandii*). It attacks both healthy trees and trees that have been stressed by drought, fire or root disturbance. The beetle naturally vectors different species of bluestain fungi and some of these were introduced in China together with the beetle. In addition, the beetle has picked up several native Chinese bluestain fungi and this appears to have contributed to the beetle's impact in China (Lu et al. 2009). The beetle's potential geographic range in China is much larger than its current range, suggesting there is a high risk of future range expansion (Tang et al. 2008; He et al. 2015). Chinese pine is a widely planted reforestation tree used to reduce soil erosion and further expansion of the red turpentine beetle in China will probably have severe ecological impacts.

7.4.2.3 Balsam Woolly Adelgid: An Invasive Sucking Insect Killing North American Firs

The balsam woolly adelgid (*Adelges piceae*) is an invasive piercing-sucking insect that has devastated most naturally occurring populations of the premier Christmas tree species in North America, Fraser fir (*Abies fraseri*). Since its accidental introduction into North America from Europe around 1900, the adelgid has killed thousands of hectares of Fraser fir, its main host in North America. The adelgid has also spread west across the continent and reached most areas where suitable host trees occur. All North American fir species are highly susceptible to the pest, while European firs tolerate infestation for several years with little symptoms (Newton et al. 2011). In its invasive North American range, the balsam woolly adelgid reproduces strictly through parthenogenesis and completes two or more generations per year (Arthur and Hain 1984). The adults are wingless and the only mobile life stage is the early phase of the first larval instar (the crawler), which disperses from tree to tree primarily by wind or gravity. When the crawler finds a suitable feeding site on a branch or trunk it inserts its mouthparts into the bark and remains attached at that site for the rest of its life (Balch and Carroll 1956). The formation of 'rotholz' (red wood) around feeding sites is a characteristic symptom of balsam woolly adelgid feeding in Fraser fir (Mester et al. 2016). This abnormal wood formation resembles compression wood and is considered to be a major cause of decline in infested trees (Timell 1986). Fraser fir is a specialty crop conifer and the most valuable Christmas tree species in the US. Christmas tree revenues total more than 2 billion USD annually. Both the entire natural range and the largest production region of Fraser fir are located in small rural communities in the Southern Appalachian Mountains of the southeastern US. Here, the balsam wooly adelgid has killed 80% of the mature Fraser fir trees across the very restricted natural range, reducing Fraser fir to an endangered species (White et al. 2012). Tree resistance mechanisms to infestation are not well understood but probably involve a combination of physical and chemical defenses at the infestation site (Hain et al. 1991; Newton et al. 2011). Methodologies to screen for genetic resistance in Fraser fir to the adelgid have been developed (Newton et al. 2011) and the ultimate goal is to develop tolerant or resistant Fraser fir genotypes through genetic improvement and thus support the Christmas tree industry for future generations.

7.5 Conclusions and Future Prospects

Climate change is expected to reduce forest health and amplify damage from native and invasive insect pests (Allen et al. 2010; Bentz et al. 2010). Ecological constraints tend to keep insect populations more or less stable and prevent large-scale pest eruptions. However, increasing temperatures alter species interactions and remove natural climatic barriers that have historically prevented population growth and range expansion of forest pests. Warming temperatures over the last several decades have already resulted in some of the most severe forest insect outbreaks reported in the literature. These include outbreaks of well-known pests such as mountain pine beetle, spruce budworm (*Christoneura occidentalis*), and Eurasian spruce bark beetle. In addition, new invasive forest pests have emerged, such as emerald ash borer in North America, red turpentine beetle in China, and redbay ambrosia beetle (*Xyleborus glabratus*) in the south-eastern United States. The combination of warmer temperatures, leading to increased stress and decreased resilience of forest ecosystems, and so-called naïve host trees without co-evolved defenses provide invasive species with a favorable, potentially defense-free environment. Expansion of invasive pests into novel environments may cause extirpation of other species and disruption of ecosystems in the process (Klooster et al. 2014).

Climatic and other environmental change may favor insect pests over their host trees, because insects have much shorter life cycles and can adapt more rapidly than trees to changing conditions. As human populations continue to affect the planet through climate change and homogenization of the world's biota we will increasingly see dramatic effects of interactions between insects and trees. It is therefore more important than ever to understand the mechanisms of tree resistance to herbivore attack, in order to promote tree resistance through optimized forest management and development of resistant cultivars. Natural variability in tree defense traits, as a result of co-evolutionary history between trees and insects, can provide robust defenses against forest pests. The most effective tree defense mechanisms fend off or stop insect attack despite continual exposure to a pest.

While much is known about some of the traits that contribute to tree defense, little is known regarding how these defense traits function ecologically, or how the underlying genomic mechanisms function to control tree defenses. Researchers who study tree-insect interactions face several challenges and limitations compared with those who study annual plants and model species like *Arabidopsis thaliana*, tobacco and tomato. However, these challenges also pose opportunities for the development of novel and innovative approaches to elucidate the complex interactions between forest trees and insects. Genomics tools are opening new avenues of research in notoriously difficult-to-study non-model tree species. The marriage between ecological and genomic approaches will help to streamline the identification of genetic markers that associate with complex resistance mechanisms in tree-insect interactions and rapidly increase tree health through genetic improvement. To keep pace with the rapid impacts of climate change and prepare trees for expected future climates, the application of modern genomic technologies may be crucial to the survival of forest tree ecosystems.

References

Adams AS, Currie CR, Cardoza YJ, Klepzig KD, Raffa KF (2009) Effects of symbiotic bacteria and tree chemistry on the growth and reproduction of bark beetle fungal symbionts. Can J for Res 39:1133–1147

Adams AS, Aylward FO, Adams SM, Erbilgin N, Aukema BH, Currie CR, Suen G, Raffa KF (2013) Mountain pine beetles colonizing historical and naïve host trees are associated with a bacterial community highly enriched in genes contributing to terpene metabolism. Appl Environ Microbiol 79:3468–3475

Agrawal AA, Kurashige NS (2003) A role for isothiocyanates in plant resistance against the specialist herbivore *Pieris rapae*. J Chem Ecol 29:1403–1415

Agrawal AA, Fishbein M (2006) Plant defense syndromes. Ecology 87:S132–S149

Ali JG, Agrawal AA (2012) Specialist versus generalist insect herbivores and plant defense. Trends Plant Sci 17:293–302

Allen CD, Macalady AK, Chenchouni H, Bachelet D, McDowell N, Vennetier M, … Hogg EH (2010) A global overview of drought and heat-induced tree mortality reveals emerging climate change risks for forests. For Ecol Manag 259:660–684

Arthur FH, Hain FP (1984) Seasonal history of the balsam woolly adelgid (Homoptera: Adelgidae) in natural stands and plantations of Fraser fir. J Econ Entomol 77:1154–1158

Aukema JE, McCullough DG, Von Holle B, Liebhold AM, Britton K, Frankel SJ (2010) Historical accumulation of nonindigenous forest pests in the continental United States. Bioscience 60:886–897

Balch RE, Carroll WJ (1956) The balsam woolly aphid. Canadian department of agriculture. Publication No. 977, 7 pp

Barbehenn RV (1999) Non-absorption of ingested lipophilic and amphiphilic allelechemicals by generalist grasshoppers: the role of extractive ultrafiltration by the peritrophic envelope. Arch Insect Biochem Physiol 42:130–137

Beech E, Rivers M, Oldfield S, Smith PP (2017) GlobalTreeSearch: the first complete global database of tree species and country distributions. J Sustain For 36:454–489

Bentz BJ, Régnière J, Fettig CJ, Hansen EM, Hayes JL, Hicke JA, Kelsey RG, Negrón JF, Seybold SJ (2010) Climate change and bark beetles of the western United States and Canada: direct and indirect effects. Bioscience 60:602–613

Berasategui A, Salem H (2020) Microbial determinants of folivory in insects. In: Bosch T, Hadfield M (ed) Cellular dialogues in the holobiont. CRC Press

Bernays E, Graham M (1988) On the evolution of host specificity in phytophagous arthropods. Ecology 69(886):892

Bernays EA, Cornelius M (1992) Relationship between deterrence and toxicity of plant secondary compounds for the alfalfa weevil *Hypera brunneipennis*. Entomol Exp Appl 64:289–292

Birol I, Raymond A, Jackman SD, Pleasance S, Coope R, Taylor GA, ... Jones SJM (2013) Assembling the 20 Gb white spruce (*Picea glauca*) genome from whole-genome shotgun sequencing data. Bioinformatics 29:1492–1497

Bliss S (2011) United Nations International Year of Forests 2011. Geogr Bull 43:33

Blomquist GJ, Figueroa-Teran R, Aw M, Song M, Gorzalski A, Abbott NL, ... Tittiger C (2010) Pheromone production in bark beetles. Insect Biochem Mol Biol 40:699–712

Boone CK, Keefover-Ring K, Mapes AC, Adams AS, Bohlmann J, Raffa KF (2013) Bacteria associated with a tree-killing insect reduce concentrations of plant defense compounds. J Chem Ecol 39:1003–1006

Boone CK, Aukema BH, Bohlmann J, Carroll AL, Raffa KF (2011) Efficacy of tree defense physiology varies with bark beetle population density: a basis for positive feedback in eruptive species. Can J for Res 41:1174–1188

Bryant JP, Swihart RK, Reichardt PB, Newton L (1994) Biogeography of woody plant chemical defense against snowshoe hare browsing: comparison of Alaska and eastern North America. Oikos 70:385–395

Buotte PC, Hicke JA, Preisler HK, Abatzoglou JT, Raffa KF, Logan JA (2016) Climate influences on whitebark pine mortality from mountain pine beetle in the Greater Yellowstone Ecosystem. Ecol Appl 26:2507–2524

Campbell SA, Borden JH (2009) Additive and synergistic integration of multimodal cues of both hosts and non-hosts during host selection by woodboring insects. Oikos 118:553–563

Cappaert D, McCullough DG, Poland TM, Siegert N (2005) Emerald ash borer in North America: a research and regulatory challenge. Am Entomol 51:152–165

Caporaso JG, Lauber CL, Walters WA, Berg-Lyons D, Huntley J, Fierer N, Owens SM, Betley J, Fraser L, Bauer M, Gormley N, Gilbert JA, Smith G, Knight R (2012) Ultra-high-throughput microbial community analysis on the Illumina HiSeq and MiSeq platforms. ISME J 6:1621–1624

Cardoza YJ, Klepzig KD, Raffa KF (2006) Bacteria in oral secretions of an endophytic insect inhibit antagonistic fungi. Ecol Entomol 31:636–645

Carmona D, Lajeunesse M, Johnson M (2011) Plant traits that predict resistance to herbivores. Funct Ecol 25:358–367

Ceja-Navarro J, Vega F, Karaoz U, Hao Z, Jenkins S, Lim HC, ... Brodie EL (2015) Gut microbiota mediate caffeine detoxification in the primary insect pest of coffee. Nat Commun 6:7618

Celedon JM, Yuen MMS, Chiang A, Henderson H, Reid KE, Bohlmann J (2017) Cell-type- and tissue-specific transcriptomes of the white spruce (*Picea glauca*) bark unmask fine-scale spatial patterns of constitutive and induced conifer defense. Plant J 92:710–726

Celedon JM, Bohlmann J (2019) Oleoresin defenses in conifers: chemical diversity, terpene synthases and limitations of oleoresin defense under climate change. New Phytol 224:1444–1463

Chakraborty S, Whitehill JGA, Hill AL, Opiyo SO, Cipollini D, Herms DA, Bonello P (2014) Effects of water availability on emerald ash borer larval performance and phloem phenolics of Manchurian and black ash. Plant, Cell Environ 37:1009–1021

Che-Mendoza A, Penilla RP, Rodriguez DA (2009) Insecticide resistance and glutathione S-transferases in mosquitoes: A review. Afr J Biotech 8:1386–1397

Chiu CC, Keeling CI, Bohlmann J (2017) Toxicity of pine monoterpenes to mountain pine beetle. Sci Rep 7:8858

Chiu CC, Keeling CI, Bohlmann J (2018) Monoterpenyl esters in juvenile mountain pine beetle and sex-specific release of the aggregation pheromone *trans*-verbenol. Proc Natl Acad Sci 115:3652–3657

Chiu CC, Keeling CI, Bohlmann J (2019) The cytochrome P450 CYP6DE1 catalyzes the conversion of α-pinene into the mountain pine beetle aggregation pheromone *trans*-verbenol. Sci Rep 9:1477

Chorbadjian RA, Phelan PL, Herms DA (2019) Tight insect–host phenological synchrony constrains the life-history strategy of European pine sawfly. Agric for Entomol 21:15–27

Christiansen E, Waring RH, Berryman AA (1987) Resistance of conifers to bark beetle attack: searching for general relationships. For Ecol Manage 22:89–106

Clavijo-McCormick A, Irmisch S, Reinecke A, Boeckler GA, Veit D, Reichelt M, Hansson BS, Gershenzon J, Köllner TG, Unsicker SB (2014) Herbivore-induced volatile emission in black poplar: regulation and role in attracting herbivore enemies. Plant, Cell Environ 37:1909–1923

Conrath U, Beckers GJM, Langenbach CJG, Jaskiewicz MR (2015) Priming for enhanced defense. Annu Rev Phytopathol 53:97–119

Cooper PD (2001) What physiological processes permit insects to eat Eucalyptus leaves? Austral Ecol 26:556–562

Craighead FC (1928) Interrelation of tree-killing bark beetles (*Dendroctonus*) and blue stains. J Forest 26:886–887

Cudmore TJ, Björklund N, Carroll AL, Lindgren B (2010) Climate change and range expansion of an aggressive bark beetle: evidence of higher beetle reproduction in naïve host tree populations. J Appl Ecol 47:1036–1043

Despres L, David JP, Gallet C (2007) The evolutionary ecology of insect resistance to plant chemicals. Trends Ecol Evol 22:298–307

Dethier VG (1941) Chemical factors determining the choice of food plants by *Papilio larvae*. Am Nat 75:61–73

DiGuistini S, Wang Y, Liao NY, Taylor G, Tanguay P, Feau N, … Breuil C (2011) Genome and transcriptome analyses of the mountain pine beetle-fungal symbiont *Grosmannia clavigera*, a lodgepole pine pathogen. Proceedings of the national academy of sciences of the United States of America. 108: 2504–2509

Dion E, Polin SE, Simon JC, Outreman Y (2011) Symbiont infection affects aphid defensive behaviours. Biol Let 7:743–746

Dixon RA, Chen F, Guo D, Parvathi K (2001) The biosynthesis of monolignols: a "metabolic grid", or independent pathways to guaiacyl and syringyl units? Phytochemistry 57:1069–1084

Doussard DE, Eisner T (1987) Vein cutting behavior: insect counterploy to the latex defense of plants. Science 237:898–901

Ebata T (1991) Summary report of two spruce weevil surveys in twelve plantations in the Kitimat Valley. B. C. Ministry of Forests. Victoria Int. Rep. PM-PB-69

Ehrlich PR, Raven PH (1964) Butterflies and plants: a study in coevolution. Evolution 18:586–608

Eisner T, Johnessee JS, Carrel J, Hendry LB, Meinwald J (1974) Defensive use by an insect of a plant resin. Science 184:996–999

Enayati AA, Ranson H, Hemingway J (2005) Insect glutathione transferases and insecticide resistance. Insect Mol Biol 14:3–8

Eyles A, Jones W, Riedl K, Herms DA, Cipollini D, Schwartz S, Chan K, Bonello P (2007) Comparative phloem chemistry of Manchurian (*F. mandshurica*) and two North American Ash Species (*F. americana* and *F. pennsylvanica*). J Chem Ecol 33:1430–1448

Eyles A, Bonello P, Ganley R, Mohammed C (2010) Induced resistance to pests and pathogens in trees. New Phytol 185:893–908

Feeny pp. (1970) Seasonal changes in oakleaf tannins and nutrients as a cause of spring feeding by winter-moth caterpillars. Ecology 51:656–681

Feeny PP (1976) Plant apparency and chemical defense. Recent Adv Phytochem 10:1–40

Ferrenberg S, Mitton JB (2014) Smooth bark surfaces can defend trees against insect attack: resurrecting a 'slippery' hypothesis. Funct Ecol 28:837–845

Feyereisen R (2006) Evolution of insect P450. Biochem Soc Trans 34:1252–1255

Feyereisen R (2005) Insect cytochrome P450. In: Gilbert LI, Latrou K, Gill SS (eds) Comprehensive Molecular Insect Science, vol 4. Elsevier, Oxford, UK, pp 1–77

Food and Agriculture Organization of the United Nations Environment Program (2020) The state of the world's forests (2020) Forests, biodiversity and people. Rome. https://doi.org/10.4060/ca8642en

Fox CW, Stillwell RC, Amarillo-S AR, Czesak ME, Messina FJ (2004) Genetic architecture of population differences in oviposition behaviour of the seed beetle Callosobruchus maculatus. J Evol Biol 17:1141–1151

Fraenkel GS (1959) The raison d'etre of secondary plant substances. Science 129:1466–1470

Frago E, Zytynska SE, Fatouros NE (2020) Microbial symbionts of herbivorous species across the insect tree. In Mechanisms underlying microbial symbiosis. Academic Press Inc, pp 111–159

Franceschi VR, Krekling T, Berryman AA, Christiansen E (1998) Specialized phloem parenchyma cells in Norway spruce (Pinaceae) bark are an important site of defense reactions. Am J Bot 85:601–615

Franceschi VR, Krokene P, Christiansen E, Krekling T (2005) Anatomical and chemical defences of conifer bark against bark beetles and other pests. New Phytol 167:353–376

French E, Kim BS, Iyer-Pascuzzi AS (2016) Mechanisms of quantitative disease resistance in plants. Semin Cell Dev Biol 56:201–208

Gandhi KJ, Herms DA (2010) Direct and indirect effects of alien insect herbivores on ecological processes and interactions in forests of eastern North America. Biol Invasions 12:389–405

Gatehouse JA (2002) Plant resistance towards insect herbivores: a dynamic interaction. New Phytol 156:145–169

Geib SM, Filley. TR, Hatcher PG, Hoover K, Carlson JK, del Mar Jimenez-Gasco M, ... Tien M (2008) Lignin degradation in wood-feeding insects. Proceedings of the national academy of sciences 105:12932–12937

Gershenzon J (1994) Metabolic costs of terpenoid accumulation in higher plants. J Chem Ecol 20:1281–1328

Gijzen M, Lewinsohn E, Savage TJ, Croteau RB (1993) Conifer monoterpenes—biochemistry and bark beetle chemical ecology. ACS Symp Ser 525:8–22

Gillespie JJ, Kjer KM, Duckett CN, Tallamy DW (2003) Convergent evolution of cucurbitacin feeding in spatially isolated rootworm taxa (Coleoptera: Chrysomelidae; Galerucinae, Luperini). MolPhylogenet Evol 29:161–175

Glynn C, Herms DA, Orians CM, Hansen RC, Larsson S (2007) Testing the growth-differentiation balance hypothesis: dynamic responses of willows to nutrient availability. New Phytol 176:623–634

Hain FP, Hollingsworth RG, Arthur FH, Sanchez F, Ross RK (1991) Adelgid host interactions with special reference to the balsam woolly adelgid in North America. In: Baranchikov YN, Mattson WJ, Hain FP, Payne TL (eds) Forest insect guilds: Patterns of interaction with host trees. United States department of agriculture forest service general technical report NE 153:271–287

Hammerbacher A, Schmidt A, Wadke N, Wright LP, Schneider B, Bohlmann J, ... Paetz C (2013) A common fungal associate of the spruce bark beetle metabolizes the stilbene defenses of Norway spruce. Plant Physiol 162:1324–1336

Hanley ME, Lamont BB, Fairbanks MM, Rafferty CM (2007) Plant structural traits and their role in anti-herbivore defence. Perspect Plant Ecol, Evol Syst 8:157–178

Harborne JB (1993) Introduction to ecological biochemistry. Academic Press

Hay ME (2016) Negating the plant apparency model: rigorous tests are the fuel of progress. New Phytol 210:770–771

Hall DE, Robert JA, Keeling CI, Domanski D, Quesada AL, Jancsik S, ... Bohlmann J (2011) An integrated genomic, proteomic and biochemical analysis of (+)-3-carene biosynthesis in Sitka spruce (Picea sitchensis) genotypes that are resistant or susceptible to white pine weevil. Plant J 65:936–948

He S, Ge X, Wang T, Wen J, Zong S (2015) Areas of potential suitability and survival of *Dendroctonus valens* in China under extreme climate warming scenario. Bull Entomol Res 105:477–484

Herms DA, Mattson WJ (1992) The dilemma of plants—to grow or defend. Q R Biol 67:283–335

Herms DA, McCullough DG (2014) Emerald ash borer invasion of North America: history, biology, ecology, impacts, and management. Annu Rev Entomol 59:13–30

Hilker M, Kobs C, Varama M, Schrank K (2002) Insect egg deposition induces *Pinus sylvestris* to attract egg parasitoids. J Exp Biol 205:455–461

Hill AL, Whitehill JGA, Opiyo SO, Phelan PL, Bonello P (2012) Nutritional attributes of ash (*Fraxinus* spp.) outer bark and phloem and their relationships to resistance against the emerald ash borer. Tree Physiol 32:1522–1532

Howe GA, Jander G (2008) Plant immunity to insect herbivores. Annu Rev Plant Biol 59:41–66

Howe GA, Herde M (2015) Interaction of plant defense compounds with the insect gut: new insights from genomic and molecular analyses. Curr Opin Insect Sci 9:62–68

Hwang SY, Lindroth RL (1997) Clonal variation in foliar chemistry of aspen: effects on gypsy moths and forest tent caterpillars. Oecologia 111:99–108

Janzen DH (1966) Coevolution of mutualism between ants and acacias in Central America. Evolution 20:249–275

Kandasamy D, Gershenzon J, Andersson MN, Hammerbacher A (2019) Volatile organic compounds influence the interaction of the Eurasian spruce bark beetle (*Ips typographus*) with its fungal symbionts. ISME J 13:1788–1800

Keeling CI, Bohlmann J (2006) Genes, enzymes and chemicals of terpenoid diversity in the constitutive and induced defence of conifers against insects and pathogens. New Phytol 170:657–675

Keeling CI, Yuen MM, Liao NY, Docking TR, Chan SK, Taylor GA, ... Bohlman J (2013) Draft genome of the mountain pine beetle, *Dendroctonus ponderosae* Hopkins, a major forest pest. Genome Biol 14:R27

King JN, Alfaro RI (2009) Developing Sitka spruce populations for resistance to the white pine weevil: summary of research and breeding program. (BC. Ministry of Forests and Range, R.B. ed. Victoria, BC

King JN, Alfaro RI, Lopez MG, Akker LV (2011) Resistance of Sitka spruce (*Picea sitchensis* (Bong.) Carr.) to white pine weevil (*Pissodes strobi* Peck): characterizing the bark defence mechanisms of resistant populations. Forestry 84:83–91

Kiss GK, Yanchuk AD (1991) Preliminary evaluation of genetic variation of weevil resistance in interior spruce in British Columbia. Can J Res 21:230–234

Kirisits T (2004) Fungal associates of European bark beetles with special emphasis on the Ophiostomatoid fungi. In: Lieutier F, Al E (eds) Bark and wood boring insects in living trees in Europe, a synthesis. Kluwer Academic Publishers, pp 181–235

Kleczewski NM, Herms DA, Bonello P (2010) Effects of soil type, fertilization and drought on carbon allocation to root growth and partitioning between secondary metabolism and ectomycorrhizae of *Betula papyrifera*. Tree Physiol 30:807–817

Kliebenstein DJ (2014) Quantitative genetics and genomics of plant resistance to insects. Annu Plant Rev 47:235–262

Klooster WS, Herms DA, Knight KS, Herms CP, McCullough DG, Smith A, Gandhi KJK, Cardina J (2014) Ash (*Fraxinus* spp.) mortality, regeneration, and seed bank dynamics in mixed hardwood forests following invasion by emerald ash borer (*Agrilus planipennis*). Biol Invasions 16:859–873

Kostova I, Iossifova T (2007) Chemical components of Fraxinus species. Fitoterapia 78:85–106

Krokene P (2015) Conifer defense and resistance to bark beetles. In: Vega FE (ed) Bark beetles: biology and ecology of native and invasive species. Hofstetter RW. Elsevier Academic Press, San Diego, USA, pp 177–207

Kuhn J, Pettersson EM, Feld BK, Burse A, Termonia A, Pasteels JM, Boland W (2004) Selective transport systems mediate sequestration of plant glucosides in leaf beetles: a molecular basis for adaptation and evolution. Proc Natl Acad Sci USA 101:13808–13813

Labandeira CC (1998) Early history of arthropod and vascular plant associations. Annu Rev Earth Planet Sci 26:329–377

Lee FJ, Rusch DB, Stewart FJ, Mattila HR, Newton IL (2015) Saccharide breakdown and fermentation by the honey bee gut microbiome. Environ Microbiol 17:796–815

Li X, Schuler MA, Berenbaum MR (2007) Molecular mechanisms of metabolic resistance to synthetic and natural xenobiotics. Annu Rev Entomol 52:231–253

Lindroth RL, Hwang SY (1996) Diversity, redundancy, and multiplicity in chemical defense systems of aspen. In: Romeo JT, Saunders JA, Barbosa P (eds) Phytochemical diversity and redundancy in ecological interactions. recent advances in phytochemistry, vol 30. Springer, Boston, MA

Loomis WE (1932) Growth–differentiation balance vs carbohydrate–nitrogen ratio. Proc Am Soc Hort Sci 29:240–245

Lopresti E, Karban R (2016) Chewing sandpaper: grit, plant apparency, and plant defense in sand-entrapping plants. Ecology 97:826–833

Lu M, Zhou XD, De Beer ZW, Wingfield MJ, Sun J (2009) Ophiostomatoid fungi associated with the invasive pine-infesting bark beetle, *Dendroctonus valens*, in China. Fungal Divers 38:133–145

Mageroy MH, Christiansen E, Solheim H, Borg Karlsson A.-K, Zhao T, Björklund N, ... Krokene P (2020a) Priming of inducible defenses protects Norway spruce against tree-killing bark beetles. Plant Cell Environ 43:420–430

Mageroy MH, Wilkinson SW, Tengs T, Cross H, Almvik M, Pétriacq P, ... Krokene P (2020b) Molecular underpinnings of methyl jasmonate-induced resistance in Norway spruce. Plant, Cell Environ 43:1827–1843

Maïbèche-Coisne M, Merlin C, François MC, Porcheron P, Jacquin- JE (2005) P450 and P450 reductase cDNAs from the moth *Mamestra brassicae*: cloning and expression patterns in male antennae. Gene 346:195–203

Massey FP, Hartley SE (2006) Experimental demonstration of the antiherbivore effects of silica in grasses: impacts on foliage digestibility and vole growth rates. Proc Royal Soc B Biol Sci B: Biol Sci 273:2299–2304

Massey FP, Ennos AR, Hartley SE (2007) Grasses and the resource availability hypothesis: the importance of silica-based defences. J Ecol 95:414–424

Mauch-Mani B, Baccelli I, Luna E, Flors V (2017) Defense priming: an adaptive part of induced resistance. Annu Rev Plant Biol 68:485–512

McCullough D, Katovich S (2004) Pest alert: emerald ash borer. U.S. Dept. of Agriculture, Forest Service, Northeastern Area, State & Private Forestry

Meddens AJ, Hicke JA, Ferguson CA (2012) Spatiotemporal patterns of observed bark beetle caused mortality in British Columbia and the western United States. Ecol Appl 22:1876–1891

Meinwald J, Jones TH, Eisner T, Hicks K (1977) New methylcyclopentanoid terpenes from the larval defensive secretion of a chrysomelid beetle (*Plagiodera versicolora*). Proc Natl Acad Sci USA 74:2189–2193

Mester EC, Lucia L, Frampton J, Hain FP (2016) Physico-chemical responses of Fraser fir induced by balsam woolly adelgid (Homoptera: Adelgidae) infestation. J Entomolog Sci 51:94–97

Moran NA, McCutcheon JP, Nakabachi A (2008) Genomics and evolution of heritable bacterial symbionts. Annu Rev Genet 42:165–190

Mumm R, Dicke M (2010) Variation in natural plant products and the attraction of bodyguards involved in indirect plant defense. Can J Zool 88:628–667

Nagy NE, Sikora K, Krokene P, Hietala AM, Solheim H, Fossdal CG (2014) Using laser micro-dissection and qRT-PCR to analyze cell type-specific gene expression in Norway spruce phloem. PeerJ 2:e362

Naidoo S, Külheim C, Zwart L, Mangwanda R, Oates CN, Visser EA, ... Myburg AA (2014) Uncovering the defence responses of Eucalyptus to pests and pathogens in the genomics age. Tree Physiol 34:931–943

Newton L, Frampton J, Monahan J, Goldfarb B, Hain F (2011) Two novel techniques to screen *Abies* seedlings for resistance to the balsam woolly adelgids, *Adelges piceae*. J Insect Sci 11:1–16

Niemelä P, Mannila R, Mäntsälä P (1982) Deterrent in Scots pine, *Pinus sylvestris*, influencing feeding behaviour of the larvae of *Neodiprion sertifer* (Hymenoptera, Diprionidae). Annales Entomologici Fennici 48:57–59

Painter RH (1951) Insect resistance in crop plants. Soil Sci 72:481

Pasteels JM, Duffy S, Rowell-Rahier M (1990) Toxins in chrysomelid beetles possible evolutionary sequence from de novo synthesis to derivation from food-plant chemicals. J Chem Ecol 16:211–222

Pastor V, Luna E, Mauch-Mani B, Ton J, Flors V (2013) Primed plants do not forget. Environ Exp Bot 94:46–56

Petersen RA, Zangrel AR, Berenbaum MR, Schuler MA (2001) Expression of CYP6B1 and CYP6B3 cytochrome P450 monooxygenases and furanocoumarin metabolism in different tissues of *Papilio polyxenes* (Lepidoptera: Papilionidae). Insect Biochem Mol Biol 31:679–690

Philippe RN, Bohlmann J (2007) Poplar defense against insect herbivores. Can J Bot 85:1111–1126

Poland TM, McCullough DG (2006) Emerald ash borer: Invasion of the urban forest and the threat to North America's ash resource. J Forest 104:118–124

Ploetz RC, Hulcr J, Wingfield MJ, De Beer ZW (2013) Destructive tree diseases associated with ambrosia and bark beetles: black swan events in tree pathology? Plant Dis 97:856–872

Raffa KF (2001) Mixed messages across multiple trophic levels: the ecology of bark beetle chemical communication systems. Chemoecology 11:49–65

Raffa KF, Aukema BH, Bentz BJ, Carroll AL, Hicke JA, Turner MG, Romme WH (2008) Cross-scale drivers of natural disturbances prone to anthropogenic amplification: the dynamics of bark beetle eruptions. Bioscience 58:501–517

Raffa KF, Aukema B, Bentz BJ, Carroll A, Erbilgin N, Herms DA, … Wallin KF (2009) A literal use of "Forest Health" safeguards against misuse and misapplication. J For 107:276–277

Raffa KF, Berryman AA, Simasko J, Teal W, Wong BL (1985) Effects of grand fir monoterpenes on the fir engraver, *Scolytus ventralis* (Coleoptera: Scolytidae), and its symbiotic fungus. Environ Entomol 14:552–556

Raffa KF, Mason CJ, Bonello P, Cook S, Erbilgin N, Keefover-Ring K, … Townsend PA (2017) Defence syndromes in lodgepole—whitebark pine ecosystems relate to degree of historical exposure to mountain pine beetles. Plant, Cell Environ 40:1791–1806

Rasmann S, Kollner TG, Degenhardt J, Hiltpold I, Toepfer S, Kuhlmann U, Gershenzon J, Turlings TCJ (2005) Recruitment of entomopathogenic nematodes by insect-damaged maize roots. Nature 434:732–737

Raupp MJ (1985) Effects of leaf toughness on mandibular wear of the leaf beetle. Plagiodera Versicolora. Ecol Entomol 10:73–79

Rebek EJ, Herms DA, Smitley DR (2008) Interspecific variation in resistance to emerald ash borer (Coleoptera: Buprestidae) among North American and Asian ash (*Fraxinus* spp.). Environ Entomol 37:242–246

Rhoades DF, Cates RG (1976) Toward a general theory of plant antiherbivore chemistry. In: Biochemical interaction between plants and insects. Springer, Boston, MA, pp 168–213

Ro DK, Arimura GI, Lau SY, Piers E, Bohlmann J (2005) Loblolly pine abietadienol/abietadienal oxidase PtAO (CYP720B1) is a multifunctional, multisubstrate cytochrome P450 monooxygenase. Proc Natl Acad Sci 102:8060–8065

Robert JA, Bohlmann J (2010) Behavioral and reproductive response of white pine weevil (*Pissodes strobi*) to resistant and susceptible Sitka spruce (*Picea sitchensis*). Insects 1:3–19

Robert JA, Madilao LL, White R, Yanchuk A, King J, Bohlmann J (2010) Terpenoid metabolite profiling in Sitka spruce identifies association of dehydroabietic acid, (+)-3-carene, and terpinolene with resistance against white pine weevil. Botany 88:810–820

Salminen JP, Karonen M (2011) Chemical ecology of tannins and other phenolics: we need a change in approach. Funct Ecol 25:325–338

Samsone I, Andersone U, Ievinsh G (2012) Variable effect of arthropod-induced galls on photochemistry of photosynthesis, oxidative enzyme activity and ethylene production in tree leaf tissues. Environ Exp Biol 10:15–26

Sandström J, Moran N (1999) How nutritionally imbalanced is phloem sap for aphids? Entomol Exp Appl 91:203–210

Schiebe C, Hammerbacher A, Birgersson G, Witzell J, Brodelius PE, Gershenzon J, … Schlyter F (2012) Inducibility of chemical defenses in Norway spruce bark is correlated with unsuccessful mass attacks by the spruce bark beetle. Oecologia 170:183–198

Scully ED, Geib SM, Carlson JE, Tien M, McKenna D, Hoover K (2014) Functional genomics and microbiome profiling of the Asian longhorned beetle (Anoplophora glabripennis) reveal insights into the digestive physiology and nutritional ecology of wood feeding beetles. BMC Genomics 15:1096

Shigenobu S, Watanabe H, Hattori M, Sakaki Y, Ishikawa H (2000) Genome sequence of the endocellular bacterial symbiont of aphids Buchnera sp. Nature 407:81–86

Singer MS, Stireman JO III (2005) The tri-trophic niche concept and adaptive radiation of phytophagous insects. Ecol Lett 8:1247–1255

Six D, Bracewell R (2015) Dendroctonus. In: Vega FE, Hofstetter RW (eds) Bark beetles. biology and ecology of native and invasive species (pp. 177–207). San Diego: Elsevier Academic Press.

Six DL, Wingfield MJ (2011) The role of phytopathogenicity in bark beetle-fungus symbioses: a challenge to the classic paradigm. Annu Rev Entomol 56:255–272

Stamp N (2003) Out of the quagmire of plant defenses hypotheses. Q R Biol 78:23–55

Steele CL, Katoh S, Bohlmann J, Croteau R (1998) Regulation of oleoresinosis in grand fir (Abies grandis): differential transcriptional control of monoterpene, sesquiterpene, and diterpene synthase genes in response to wounding. Plant Physiol 116:1497–1504

Steppuhn A, Baldwin IT (2007) Resistance management in a native plant: nicotine prevents herbivores from compensating for plant protease inhibitors. Ecol Lett 10:499–511

Stork NE, McBroom J, Gely C, Hamilton AJ (2015) New approaches narrow global species estimates for beetles, insects, and terrestrial arthropods. Proc Natl Acad Sci 112:7519–7523

Strauss SY, Agrawal AA (1999) The ecology and evolution of plant tolerance to herbivory. Trends Ecol Evol 14:179–185

Strauss AS, Peters S, Boland W, Burse A (2013) ABC transporter functions as a pacemaker for sequestration of plant glucosides in leaf beetles. Elife 2: e01096

Sun JH, Lu M, Gillette NE, Wingfield MJ (2013) Red turpentine beetle: innocuous native becomes invasive tree killer in China. Annu Rev Entomol 58:293–311

Tang WD, Shi J, Luo YQ (2008) Evaluation of the potential risk of Dendroctonus valens with @Risk software. For. Pest Dis 27:7–14

Timell TE (1986) Compression wood in gymnosperms, vol. 1–3. Springer-Verlag, Berlin

Trumbore S, Brando P, Hartmann H (2015) Forest health and global change. Science 349:814–818

Turlings TC, Erb M (2018) Tritrophic interactions mediated by herbivore-induced plant volatiles: mechanisms, ecological relevance, and application potential. Annu Rev Entomol 63:433–452

Vega FE, Hofstetter RW (2015) Bark beetles: biology and ecology of native and invasive species. Elsevier Academic Press, 640 pp

Wainhouse D, Cross DJ, Howell RS (1990) The role of lignin as a defence against the spruce bark beetle Dendroctonus micans: effect on larvae and adults. Oecologia 85:257–265

Walling LL (2000) The myriad plant responses to herbivores. J Plant Growth Regul 19:195–216

White PB, van de Gevel SL, Soulé PT (2012) Succession and disturbance in an endangered red spruce–Fraser fir forest in the southern Appalachian Mountains, North Carolina, USA. Endanger Species Res 18:17–25

Whitehill JGA, Popova-Butler A, Green-Church KB, Koch JL, Herms DA, Bonello P (2011) Interspecific proteomic comparisons reveal ash phloem genes potentially involved in constitutive resistance to the emerald ash borer. PLoS ONE 6:e24863

Whitehill JGA, Opiyo SO, Koch JL, Herms DA, Cipollini DF, Bonello P (2012) Interspecific comparison of constitutive ash phloem phenolic chemistry reveals ompounds unique to Manchurian ash, a species resistant to emerald ash borer. J Chem Ecol 38:499–511

Whitehill JGA, Rigsby C, Cipollini D, Herms DA, Bonello P (2014) Decreased emergence of emerald ash borer from ash treated with methyl jasmonate is associated with induction of general defense traits and the toxic phenolic compound verbascoside. Oecologia 176:1047–1059

Whitehill JGA, Henderson H, Schuetz M, Skyba O, Yuen MMS, King J, Samuels AL, … Bohlmann J (2016a) Histology and cell wall biochemistry of stone cells in the physical defence of conifers against insects. Plant, Cell Environ 39:1646–1661

Whitehill JGA, Henderson H, Strong W, Jaquish B, Bohlmann J (2016b) Function of Sitka spruce stone cells as a physical defense against white pine weevil. Plant, Cell Environ 39:2545–2556

Whitehill JGA, Yuen MMS, Henderson H, Madilao L, Kshatriya K, Bryan J, Jaquish B, Bohlmann J (2019) Function of stone cells and oleoresin terpenes in the conifer defense syndrome. New Phytol 221:1503–1517

Whitehill JGA, Bohlmann J (2019) A molecular and genomic reference system for conifer defence against insects. Plant, Cell Environ 10:2844–2859

Wilkinson SW, Magerøy MH, Sánchez AL, Smith LM, Furci L, Cotton TEA, Krokene P, Ton J (2019) Surviving in a hostile world: plant strategies to resist pests and diseases. Annu Rev Phytopathol 57:505–529

Wondji CS, Morgan J, Coetzee M, Hunt RH, Steen K, Black WC, Hemingway J, Ranson H (2007) Mapping a quantitative trait locus (QTL) conferring pyrethroid resistance in the African malaria vector *Anopheles funestus*. BMC Genomics 8:34

Wood DL (1982) The role of pheromones, kairomones, and allomones in the host selection and colonization behavior of bark beetles. Annu Rev Entomol 27:411–446

Yan ZL, Sun J, Owen DR, Zhang Z (2005) The red turpentine beetle, *Dendroctonus valens* LeConte (Scolytidae): an exotic invasive pest of pine in China. Biodivers Conserv 14:1735–1760

Yang Z, Zhang F, He Q, He G (2005) Molecular dynamics of detoxification and toxin-tolerance genes in brown planthopper (*Nilaparvata lugens* Stål., Homoptera: Delphacidae) feeding on resistant rice plants. Arch Insect Biochem Physiol 59:59–66

Zagrobelny M, Bak S, Rasmussen AV, Jørgensen B, Naumann CM, Møller BL (2004) Cyanogenic glucosides and plant–insect interactions. Phytochemistry 65:293–306

Zeneli G, Krokene P, Christiansen E, Krekling T, Gershenzon J (2006) Methyl jasmonate treatment of mature Norway spruce (*Picea abies*) trees increases the accumulation of terpenoid resin components and protects against infection by *Ceratocystis polonica*, a bark beetle-associated fungus. Tree Physiol 26:977–988

Zhang QH, Schlyter F (2004) Olfactory recognition and behavioural avoidance of angiosperm nonhost volatiles by conifer-inhabiting bark beetles. Agric for Entomol 6:1–20

Zhao T, Krokene P, Hu J, Christiansen E, Björklund N, Långström B, Solheim H, Borg-Karlson AK (2011) Induced terpene accumulation in Norway spruce inhibits bark beetle colonization in a dose-dependent manner. PLoS One 6:e26649

Zhao T, Kandasamy D, Krokene P, Chen J, Gershenzon J, Hammerbacher A (2019a) Fungal associates of the tree-killing bark beetle, *Ips typographus*, vary in virulence, ability to degrade conifer phenolics and influence bark beetle tunneling behavior. Fungal Ecol 38:71–79

Zhao T, Ganji S, Schiebe C, Bohman B, Weinstein P, Krokene P, Borg-Karlson AK, Unelius CR (2019b) Convergent evolution of semiochemicals across Kingdoms: bark beetles and their fungal symbionts. ISME J 1535–1545

Chapter 8
Insects and Forest Succession

Sean C. Thomas

8.1 Introduction—Foundations of "Succession" in Plant Ecology

There is a long-standing, even ancient, belief in Western thought that forests, particularly unmanaged forests relatively free from obvious human impacts, are never-changing; this is the connotation of the German word "urwald" or "original forest" that influenced early thinking on forests from the origins of the emerging scientific disciplines of forestry and ecology in the 1800s. However, all forests, including extant ancient forests, are in fact in a state of flux. In addition to changes due to seasonality and forest responses to vicissitudes of the environment, forests nearly always show directional changes in species composition, structure, and ecosystem processes that are termed **succession** (Box 8.1). In general, forest succession is initiated by **disturbance** (Box 8.1), defined as a (more or less) discrete event in which some or all vegetation is destroyed or removed from the system. The most common agents of forest disturbance are fire, windstorms, floods, and (very commonly) tree removal by human activities; however, animals, including insects, and microbes such as fungal pathogens, can also be important disturbance agents in many forest ecosystems. Succession may in general be viewed as the process of biotic recovery of the system following such a disturbance event.

Citation: Thomas, S.C. (2021) Insects and forest succession. In: Forest Entomology, J. Allison, editor, Springer-Verlag, New York. In press.

S. C. Thomas (✉)
Institute of Forestry and Conservation, University of Toronto, Toronto, ON, Canada
e-mail: sc.thomas@utoronto.ca

Box 8.1 Definitions of succession and disturbance

"Disturbance" and "succession" are both terms that have a long use in the ecological literature, and a correspondingly long history of debate over precise definitions. To most ecologists, "disturbance" connotes a large and sudden reduction in biomass that is associated with a discrete event, such as a fire, windstorm, or forest harvest. A definition of disturbance based on loss of biomass of primary producers has been promoted by Grime (1979, 2006), and is the most commonly cited definition. Grime argues that broader definitions include too many types of environmental perturbations to be useful: forest community responses to atmospheric pollutants or climate variation, for example, generally have little in common with changes following clearcut harvesting. Likewise, some proposed definitions of "succession" encompass any change in the structure, function, or composition of community (or ecosystem). However, such all-encompassing definitions have been widely critiqued as overly broad, including patterns and processes that range from community drift (stochastic variation in populations of individual species under stable conditions), to responses to atmospheric pollutants.

While recognizing that alternative definitions exist, the present chapter (and most of the ecological and forestry literature) adheres to the following definitions that essentially paraphrase Grime (1979; 2006):

Disturbance: an event that removes biomass.

Succession: a directional change in community structure over time.

Understanding successional changes in structure, species composition, and diversity of dominant vegetation following disturbance has been a central focus of ecology since the discipline's inception. Many early ideas and generalizations concerning succession—such as the idea of an unchanging "urwald"—have remained surprisingly influential, even when convincingly falsified. An historical approach is therefore taken here as a framework.

The earliest[1] formal studies of ecological succession focused on dune vegetation (Cowles 1899, 1901), but ecologists soon began to examine this process in forest ecosystems (Gleason 1917; Lee 1924). Due to the long lifespan of trees, changes in forest community composition driven by succession can take place over centuries to millennia. This timescale has presented a long-standing challenge to

---

[1] As an historical note of particular interest to entomologists, an earlier but strikingly similar development of theory on ecological succession was the work of Pierre Mégnin in the 1880s (Michaud et al. 2015). Mégnin, trained as a veterinarian and entomologist, was the first to systematically investigate the timing of insect colonization of human corpses, with a view toward supporting the work of forensic scientists in court cases. He described eight "squads" of colonizing insects that formed a predictable sequential series on corpses and used the term "succession" to describe this pattern (Mégnin 1894). The predictability of this pattern was then challenged in the literature by American physician Murray Motter (Motter 1898), paralleling aspects of the Clements-Gleason debate, but predating it by more than two decades (Michaud et al. 2015).

understanding mechanisms that determine patterns of forest succession, since manipulative experiments or even sequential observations at the correct temporal scale are
rarely possible. Models, ranging from simple conceptual representations to complex
simulation models, have thus played a central role in the study of forest succession.
Some of the earliest ecological computer simulation models, such as the forest "gap
models" JABOWA and FORET (Botkin et al. 1972; Shugart 1984), were specifically
aimed at elucidating mechanisms of forest succession. This focus on forest succession has continued as a central preoccupation in ecological modeling to the present
(e.g. Pacala et al. 1993; Liu and Ashton 1995; Grimm et al. 2005; Taylor et al. 2009;
Ma et al. 2022). However, early inquiry on succession relied on simpler conceptual
models that remain influential.

Historically, the works of Frederic E. Clements (1916, 1936) had great influence
on the conceptualization of the process of succession and the ecological mechanisms involved. Clements formulated two central theories. The first was the idea
that succession generally operated by means of facilitation, with colonizing species
creating conditions that lead to the success of other species. For example, early
colonizing tree species would enhance soil organic matter and nutritional status in
a manner that would enable later-successional species to successfully establish and
grow (Clements 1916). The second theory was that of the climax community, toward
which succession under a given set of soil (edaphic) and climatic conditions would
gradually converge (Clements 1936). Climax communities were hypothesized to be
stable over long time periods, showing no directional change in species composition.

Both the climax community concept and predominance of facilitation processes
in succession were hotly debated in ensuing years. Most prominently, Henry
Gleason promoted an individualistic view of succession, which proposed that ecological communities form and develop in a non-deterministic way (Gleason 1926).
Another influential ecologist, Alexander Watt, described systems in which succession appeared to by cyclic, with no set end point (Watt 1947). Frank Egler presented
evidence that species coming to dominate late in succession were generally present
early in succession, and that there could be "precedence effects" in which early
presence of plant species could strongly influence subsequent successional patterns
(Egler, 1954). Egler argued strongly against what he termed the "relay floristics"
model of Clements, and even offered a $10,000 reward to any ecologist who could
demonstrate a clear example of Clementsian succession through at least 5 stages
(Anderson 2018). The award was never collected.

The concept of a climax community likewise has been extensively critiqued,
and in modern ecology is viewed as an abstraction not actually observed in nature.
Thus, a given forest may be thought of as "late seral" (i.e. dominated by species not
typical of early stages of succession, and not undergoing rapid successional change
in species composition), but essentially no forest is a true ecological climax that
does not show directional change. The main reasons for the non-existence of true
ecological climax communities are: (1) a mis-match of current communities with
climatic conditions; (2) persistence of disturbance, including disturbances that are
"endogenous" to communities (such as treefall gaps formed following the death of

individual trees); and (3) a sufficiently short return interval for large-scale disturbance such that the community cannot reach equilibrium (Pickett and McDonnell 1989).

A host of commonly used terms and concepts attach to ecological succession (Box 8.2). It has been argued that there is substantial redundancy in terminology related to ecological succession (Pulsford et al. 2016); however, in any discussion of succession it is difficult to avoid the terminological morass. Succession has classically been described as falling into categories of primary succession and secondary succession (on "new" and "previously occupied" substrates, respectively); in secondary succession individuals and structures that derive from the pre-disturbance community are termed "biological legacies". Additional descriptors have often been applied to describe the pattern of succession, including "progressive", "retrogressive", "cyclic", and "arrested" succession. Species that initially colonize sites following disturbance are most often referred to as "pioneer" species. Although the term and concept of a "climax community" in a Clementsian sense have been discarded in modern ecology, forest communities late in succession are often termed "late-seral", and such forests are typically characterized by intrinsically generated small-scale disturbance events as individual trees senesce and die, forming gaps. The formation of such gaps, together with the process of forest regrowth at gap sites, is termed "gap-phase dynamics", and is characteristic of most late-seral forests.

Box 8.2 Forest succession concepts and terminology

Community: a set of interacting organisms in a given space and time, generally quantified as the relative abundances of these organisms.

Primary succession: succession occurring in areas lacking a prior community, such as plants colonizing newly formed geological deposits.

Secondary succession: succession occurring in areas that have a pre-existing community.

Progressive succession: succession accompanied by an increase in total biomass.

Retrogressive succession: succession accompanied by a decrease in total biomass.

Cyclic succession: succession in which species reciprocally replace each other over time.

Arrested succession: succession in which the typical progression of successional stages does not occur, often associated with anthropogenic or invasive species effects.

Pioneer species: species that are early colonizers following disturbance; synonyms include "ephemeral species", "fugitive species", and "opportunist species".

Gap phase dynamics: the process of tree death and subsequent forest regrowth characteristic of late-seral forest systems.

Initial floristics: theory that all species, including those dominating later successional stages, are present early in succession but change in abundance.

Relay floristics: theory that groups of species colonize and disappear from a given site through the course of succession, and characteristically act to make the site less suitable for themselves and more suitable for subsequent sets of species.

Biological legacy: structures or organisms that carry over from pre-disturbance communities.

Sere: successional stage.

Late-seral community: a community dominated by late-successional (non-pioneer) species.

Primary forest: forest that has not been logged.

Old-growth forest: variously defined—a common ecological definition is a late-seral forest showing gap-phase dynamics.

The term "old-growth" is somewhat problematic. It has connotations of a Clementsian climax community, and for this reason is avoided by some ecologists and foresters. In many regions there are working definitions of "old-growth forest" based on management objectives or specified in a legal framework. For example, in the province of Ontario, Canada, regulations define "old-growth" as forests with dominant trees older than 70–150 years, depending on biogeographic region and dominant tree species (Uhlig et al. 2001). From a modern ecological perspective, "old-growth" is commonly used as a synonym for a late-seral forest showing gap-phase dynamics; as noted below this is the common usage in relation to stages of stand structural development. However, recent analyses of usage emphasize that the precise meaning of "old-growth" varies widely in both the ecological and forestry literature (Wirth et al. 2009).

8.2   Successional Changes in Forest Communities—Models and Mechanisms

There are several reasons to use studies of plant ecology as a basis for understanding successional patterns in other organisms, including insects. As noted above, the historical development of thinking on succession in ecology was based almost entirely on plant communities. It is also widely accepted that plants are generally of primary importance in determining diversity of other organisms, in particular insects (Siemann et al. 1998; Castagneyrol and Jactel 2012). One would therefore generally expect that the same ecological processes that drive plant species turnover and changes in diversity through succession would be reflected in the insect community. Changes in plant species composition may or may not be the main mechanism by which vegetation affects insect communities. Recent studies have emphasized the importance of changes in forest stand structure and dynamics (as distinct from changes in plant community composition) in understanding successional patterns in forest insects. In addition, age-related changes in the morphology and physiology of individual trees themselves may have important consequences for insect communities in guilds that interact closely with live trees such as herbivores and pollinators.

Broad generalizations or "laws" of succession—of the sort sought by early plant ecologists—have remained elusive. Pluralistic reconciliations of alternative views of patterns and mechanisms were offered in the 1970s by Drury and Nisbet (1973) and Connell and Slatyer (1977), who focused on mechanisms rather than resulting patterns. The general mechanisms may be classified as involving processes of facilitation, tolerance, or inhibition. In "facilitation", species alter the environment in a way that makes it more suitable for colonization of other species. A tolerance process in succession involves progressive lowering of resource levels, and a sorting of species by their ability to tolerate low resources; models of succession based on tolerance have been developed in detail by David Tilman (1982, 1985). "Inhibition" processes involve resistance of all species to displacement, such that early colonizers persist until they have completed their life cycle. It should be noted that the meanings of the terms "mechanism" and "model" themselves have a long history of debate in relation to ecological succession (Pickett et al. 1989). Very generally, a "mechanism" is a process operating at a lower hierarchical level of organization that explains a pattern observed at a higher level. Some recent efforts to conceptually unify community ecology advocate a focus on processes and mechanisms analogous to those operating on gene frequencies in population genetics (i.e. selection, drift, immigration, and speciation: Vellend 2016); however, this focus seems to discard the study of succession entirely.

A central question that received research attention from the 1960s onwards is the development of ecological diversity (most commonly species diversity as measured by local species richness or a diversity index) through succession. An early generalization was that increases in species diversity through the course of succession were universal (Margalef 1968; Odum 1969). However, empirical data from plant communities did not generally support this claim (Drury and Nisbet 1973), though evidence that species diversity is maximized in late-seral stands was found in tropical forests (Brünig 1973). In contrast, communities with high natural disturbance rates commonly were found to show a peak in plant species diversity early in succession, as in the case of Australian schlerophyll woodland communities (Purdie and Slatyer 1976), and a number of western conifer forests (Habeck 1968; Peet 1978). Other studies have presented strong evidence for peaks in forest plant diversity at intermediate successional stages (a hump-shaped pattern through succession) in a variety of systems (e.g. Schoonmaker and McKee 1988; Sheil 2001).

An observed peak in species richness at an intermediate successional stage in tropical forest (Eggeling 1947) was used as a principle illustration in Joseph Connell's exposition of the influential intermediate-disturbance hypothesis (Connell 1978). This hypothesis states that species diversity is expected to be maximized at an intermediate intensity or frequency of disturbance: only a few species (generally pioneer species) will be able to persist under a high disturbance regime, and under very low disturbance a small number of species are expected to out-compete other species. Although commonly attributed to Connell, the main elements of the intermediate-disturbance hypothesis go back earlier (Wilkinson 1999), particularly to works by Grime (1973) and Horn (1975).

While intuitive, the intermediate-disturbance hypothesis has repeatedly been questioned on theoretical grounds (Huston 1979; Fox 2013), and is not particularly well supported empirically (Mackey and Currie, 2001; Bongers et al. 2009). In particular, the point was made early on that the response of local (alpha) diversity to disturbance is expected to vary with site productivity (Huston 1979, 2014: Fig. 8.1). Huston's demographic equilibrium theory predicts that in very low productivity systems with low growth rates any disturbance can drive species locally to extinction; in this case peak diversity is expected at low disturbance rates. In very high productivity systems with high growth rates, competitive exclusion can take place rapidly, and peak diversity is expected at high disturbance. This analysis suggests that the intermediate-disturbance hypothesis only applies at intermediate levels of productivity. Although the intermediate disturbance hypothesis was developed in part as a potential explanation for a hump-shaped successional pattern in diversity, its application to such patterns also remains somewhat ambiguous. Neither the intermediate disturbance hypothesis nor the dynamic equilibrium model makes explicit predictions regarding how much diversity is expected immediately following a disturbance event, since this largely depends on colonization and "legacy" effects that are not part of either model.

An additional hypothesis that may provide an alternative explanation for variable patterns of species diversity through succession is that diversity is maximized in the successional stages that are most frequent at the landscape scale under the prevailing disturbance regime (Denslow 1980). The gist of this argument is that the regional species pool is a function of habitat area, following from island biogeographic theory (MacArthur and Wilson 1967). Thus, regions with infrequent disturbance are expected to show maximal diversity in late-seral stands since there has been greater opportunity for immigration and speciation to add to the regional pool of species adapted to late-seral conditions. Conversely, regions with frequent disturbance, and regions with slow recovery from disturbance, are expected to accumulate a larger species pool adapted to early-successional habitats. This theoretical framework leads to a prediction that successional patterns of species diversity may show pronounced biogeographic differences as a function of the regional disturbance regime.

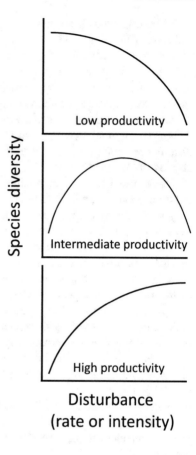

Fig. 8.1 Hypothesized relationships between species diversity and disturbance regime based on the demographic equilibrium model (Huston 1979, 2014); at intermediate levels of productivity the "intermediate disturbance hypothesis" pattern is expected

8.2.1 Forest Stand Structure and Dynamics

Successional patterns per se have predominantly been analyzed in terms of the species composition of communities (i.e. patterns of species abundance and diversity), rather than structural characteristics. However, as detailed below, there is also a long-standing applied forestry literature that has focused on stand structure rather than species composition in describing patterns of forest regrowth following a disturbance event. Stand structure is in fact often considered of primary importance in determining forest biodiversity patterns (e.g. Spies 1998; McElhinny et al. 2005). Forest structure here is generally defined in terms of patterns of macroscopic habitat elements, such as tree density and basal area, leaf area index, gap size distributions, and the amounts and decay classes of coarse woody debris, and also encompasses edaphic characteristics such as litter layer thickness, humus form, and the development of pit-and-mound topography associated with gap-phase regeneration (Spies 1998; Franklin et al. 2002).

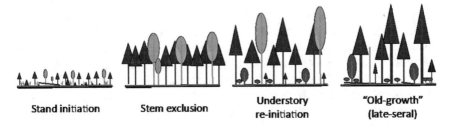

| Stand initiation | Stem exclusion | Understory re-initiation | "Old-growth" (late-seral) |

Fig. 8.2 Stages of stand development following a stand-replacing disturbance. Note presence of dense herbaceous vegetation and legacy structures during the stand initiation stage, even, closed canopy and lack of understory vegetation in the stem exclusion stage, presence of small inter-crown gaps and recruitment of shade-tolerant vegetation in the understory re-initiation stage, and uneven structure, canopy gaps, coarse wood, and patches of shade-tolerant understory vegetation in the "old-growth" stage

A four-stage scheme for forest stand development described by Oliver (1980) has been widely utilized (note that similar descriptions were commonly given in older forestry texts (e.g. Toumey and Korstian 1937), and derive from the German forestry literature of the 1800s). The four-stage scheme (Fig. 8.2) divides stand development into: (1) stand initiation, in which a new cohort of trees establishes; (2) stem exclusion, in which trees compete strongly for resources and there is high density-dependent mortality; (3) understory re-initiation, in which sufficient gaps form in the canopy to allow development of ground-layer vegetation and recruitment of shade-tolerant trees; and (4) old-growth, characterized by senescence of individual trees and gap-phase dynamics.

Recent critiques and extensions of this scheme have made a number of important refinements (Franklin et al. 2002). First "legacy" inputs from pre-disturbance stands, including dead and live trees, can critically affect stand development, particularly at the stand initiation stage. Second, the old-growth stage is an aggregate of multiple distinct stand development stages. Many forests have species that qualify as "long-lived pioneers", trees that colonize open area but that can survive for 100s to 1,000 + years (such as Douglas-fir (*Pseudotsuga menziesii*) in western North America). Late-seral forests that retain these long-lived pioneer trees are generally distinct in structure and species composition from later stages. In part due to this effect, there is commonly a peak in biomass accumulation in late-seral stands that should often be considered distinct from "old-growth" stands: the term "transition old-growth" has sometimes been used to describe such stands (Wirth et al. 2009). In some systems there is a pronounced long-term pattern of "ecosystem retrogression" with declining productivity, often accompanied by soil acidification; this is particularly well documented in boreal forests (Wardle et al. 2003) but appears to be common to many forest systems (Wardle et al. 2004). Third, there are important events and processes that may or may not correspond to the described transitions between stand development stages. For example, canopy closure commonly is used to distinguish stand initiation from stem exclusion stages; however, density-dependent mortality is not observed immediately following canopy closure. The development of gaps between individual tree crowns is a signature characteristic of the understory re-initiation stage; these canopy openings often arise through "crown shyness" effects (e.g. Fish et al. 2006) that vary greatly among tree species and in response to environmental conditions, such as the prevailing wind regime. As noted above, the term "old-growth" is also problematic in its connotation of Clementsian "climax" community, so the later stages of stand development might better be termed "late-seral" or "gap phase dynamic" stages.

In addition to the point that forest structure rather than composition may be a better predictor of community patterns of forest organisms—in particular arthropods—a focus on stand structure and dynamic stages is important for developing broad generalizations on forest succession. First, it is clear that there is high stochasticity in community composition, particularly early in succession, such that clearly defined "successional communities" do not generally exist. In contrast, there is evidence that stand structural characteristics often follow similar and predictable patterns in a wide variety of forest systems (Oliver and Larsen 1996). Stand structural patterns, in addition to being closely linked to a number of mechanisms of importance from the perspective of insect habitats (e.g. legacy structures such as coarse woody debris, canopy tree senescence, and tree gap formation), may thus also enhance comparability across studies.

A general concept of stand structure as a predictor of arthropod diversity was proposed by John Lawton in the 1980s (Lawton 1983); however, the conceptualization of forest structure differed from that presented above. Lawton focused on canopy structural complexity and did not consider coarse woody debris or edaphic factors. Lawton also predicted a continuous increase in structural complexity with stand age, whereas a stand development perspective notes that legacy structures and patchy regeneration commonly results in higher environmental heterogeneity soon after disturbance events, and low environmental heterogeneity during the stem exclusion phase. One may thus consider the hypothesis of stand structural development as a predictor of successional patterns in insect communities as an extension of, but distinct from, Lawton's plant architecture hypothesis.

8.2.2 Tree Ontogeny

Another recent perspective on potential mechanisms for forest successional patterns of particular relevance to arthropod communities is age-related changes in tree physiological and functional biology. The lifespan of individual canopy trees commonly continues through the duration of observed successional patterns; in managed forests and forests with short disturbance-return intervals, this is essentially always the case. Trees generally show large and predictable changes though ontogeny not only in structural features, but also in physiology, including large changes in leaf and woody tissue chemistry (Meinzer et al. 2011). Some tissue-level ontogenetic changes important from an arthropod perspective include: (1) increased leaf thickness and leaf mass per area (Thomas and Winner 2002); (2) reduced leaf nitrogen concentrations and a concomitant reduction in leaf photosynthetic capacity (Bond 2000); and (3) increased leaf toughness (Mason et al. 2013). The mechanisms for such changes include limitations on tree water transport that increase as trees grow (Bond 2000; Koch et al. 2004), as well as changes in allocation patterns including the effects of increasing allocation to reproductive structures as trees age (Thomas 2011). Some important traits show strongly non-linear trends, possibly as a result of reproductive allocation effects: for example, in temperate hardwoods leaf nitrogen and photosynthetic capacity show a hump-shaped pattern with a peak in younger trees (Thomas 2010). It is hypothesized that ontogenetic trends may reflect in part selective pressure for leaf herbivore defense (Boege et al. 2011; Mason and Donovan 2015). However, there appears to be no general pattern in production of herbivore defensive compounds in relation to tree age (Barton and Koricheva 2010), and indirect defenses and herbivory tolerance likewise show variable patterns (Boege et al. 2011).

Ontogenic changes in macroscopic aspects of tree structure are also common, and some of these have long been recognized to be important to arthropod habitat use. Trees add progressive layers of bark (periderm) cells produced by the cork cambium; thus, bark thickness increases with tree age, and declines from the base to the peripheral branches. Sucking insects such as scale species (Hemiptera suborder Sternorrhyncha) that feed on woody tissues must penetrate bark tissues but can benefit from reduced moisture stress in bark crevices. This tradeoff is thought to result in a peak in scale abundance on trees of intermediate size that has been seen in some systems (Wardhaugh et al. 2006). Production of large branches can result in the trapping of soil within three canopies, producing unique "canopy soil" environments that are the habitat of specialized arthropod communities in some systems (Lindo and Winchester 2006). As noted by Lawton (1983), increasing complexity of branching structure through tree ontogeny may contribute importantly to arthropod habitats. Another macroscopic pattern is age-related crown thinning, as documented in both temperate (Nock et al. 2008) and tropical (Quinn and Thomas 2015) trees. Intra-crown leaf area index of older trees declines to as little as 1/2 or 1/3 of that observed in younger trees just entering the canopy. The canopies of older trees showing crown thinning likely present a dramatically different thermal environment for canopy insects. In addition, many tree species have long-delayed reproduction, and trees generally show increased reproductive allocation through ontogeny (Thomas 2011); these patterns are certain to affect arthropods reliant on flowers or fruits as resources or habitat elements.

8.3   Key Questions on Forest Insect Succession

Forest management generally results in a replacement of late-seral forests with younger forests of simplified structure and altered tree species composition. Insects and non-insect arthropods comprise the majority of macroscopic taxa in most forest ecosystems, so an understanding of insect community changes in relation to forest stand development is essential. The mechanisms and processes involved in these

responses are likewise of central importance in developing conservation and management approaches to mitigate detrimental effects of wide-scale forest management.
In addition to forest-level successional patterns in insect communities, one expects
to find successional processes associated with aging of individual live trees, and
in structures associated with trees, such as dead wood. These patterns are both of
fundamental interest and contribute to whole-forest successional patterns important
from a management perspective. Moreover, some insects are themselves a cause of
stand-replacing disturbance events and may influence forest succession processes
via herbivory and other interactions.

The remainder of this chapter addresses the following questions: (1) How do
forest arthropod communities change in relation to stand development in terms of
species richness, overall abundance, and community composition, and what mechanisms account for these patterns? (2) Is there evidence for more than two distinct
successional stages in forest arthropod communities? (3) Does arthropod diversity
essentially track plant diversity through succession? (4) What insect groups are typically dependent on late-seral forests, and what mechanisms and processes account
for this dependence? (5) Do forest arthropod communities closely associated with
trees vary with tree size and age? (6) Given the importance of coarse woody debris
in driving many patterns in forest insect arthropods, what is the evidence for insect
succession on woody debris itself? I conclude with a brief overview of insect effects
on successional processes in forests, including insects that cause stand-replacing
disturbance events, and the effects of insects on forest succession generally.

8.3.1 Observed Successional Patterns in Forest Arthropod Assemblages

The form of the relationship between diversity and forest age is a central descriptor
of successional patterns (Fig. 8.1). However, many published studies on forest insect
succession have been based on a small number of (often only 2) stand age, successional stage, or stand development categories. Frequently studies have also lacked
true replication, making it impossible to distinguish successional patterns from stand-
to-stand variation. Table 8.1 summarizes empirical studies that have true replication
(or examined continuous variation with 12 or more sampled stands) and included
more than 2 categories and spanned at least 15 years of post-disturbance recovery in
terrestrial forest arthropods.

As has been found in syntheses aimed at testing the intermediate disturbance
hypothesis generally (MacKey and Currie 2001; Shea et al. 2004), hump-shaped
relationships as predicted by the hypothesis are not consistently observed in individual studies, though may emerge in synthesizing large data sets (Bongers et al.
2009; Yeboah and Chen 2016). Qualitative successional patterns of species richness of forest arthropod communities seem to vary considerably among studies and
specific systems (Table 8.1). Overall a somewhat greater proportion of studies found

negative rather than positive trends in species richness with stand age (33% vs. 27%); only 15% of studies exhibit a hump-shaped relationship, with a peak at intermediate stand age, while 9% of studies show a "U-shaped" pattern (Table 8.1). However, a dichotomy in patterns is apparent with respect to biome: in boreal and temperate forests most studies (68%) find a decreasing or U-shaped pattern of species richness with stand age, while in tropical forest most studies (80%) show either increasing species or hump-shaped patterns (Table 8.1).

One of the only published works to assess patterns across a full range of stand ages and development stages is that of Paquin (2008). This study provides compelling evidence for a "U-shaped" relationship between species richness and stand age in Carabid beetles in boreal forest (Fig. 8.3). Many other boreal and temperate forest studies have not had a sufficient range of stand ages or sufficient replication to possibly observe an increase in species richness among very old stands. Thus, observed negative relationships may correspond to "truncated" U-shaped patterns. The other boreal study that covers a very large age is that of Gibb et al. (2013), who note an increase in species richness mainly in the oldest stands in a long chronosequence (and who did not sample stands younger than 5 years post-harvest). Two other well-supported U-shaped patterns have also been published: a study on carabid beetles in pine plantations in Spain (Taboada et al. 2008), and a study of chrysomelid beetles in thorn forests in northern Mexico (Sánchez-Reyes et al. 2019).

In general, the patterns reported in Table 8.1 do not appear to support predictions of either the intermediate disturbance hypothesis or of the demographic equilibrium model (Huston 1979, 2014) that builds upon it. North temperate and in particular boreal forests have much lower productivity than most tropical forests, and so would be predicted to show a less pronounced decline in diversity with stand age (due to competitive exclusion effects) than tropical forests. However, precisely the opposite trend is found. Some of the best-replicated studies show U-shaped patterns of species richness through succession, which is essentially the opposite of the predicted pattern. The patterns observed are generally more consistent with mechanisms based on stand structural development. Important habitat elements such as coarse woody debris are often abundant as structural legacies in young stands, particularly after natural disturbance events such as fire and wind-throw. Coarse woody debris decays slowly in northern ecosystems, and so these legacy effects would be expected to persist for decades. The recruitment of new coarse woody debris, particularly in the form of large standing dead trees and large-dimension logs, requires that trees complete their life cycle, which may require 100 years or more. U-shaped patterns of arthropod diversity would thus be predicted as a consequence of coarse woody debris inputs and dynamics. In the tropics coarse woody debris is more ephemeral as a result of high temperatures, consistent high moisture, and the abundance of termites and other organisms that rapidly consume dead wood. Thus, legacy structures may be less likely to influence arthropod successional patterns in the tropics. Also, tropical forests likely present more structural habitat elements that consistently increase through stand development, such as those related to lianas and epiphytes. The prevailing positive trend in arthropod diversity through succession in the tropics thus also appears consistent with a stand structure mechanism.

Table 8.1 Studies examining successional patterns in forest arthropods; studies listed included assessments in 3 or more stand age categories spanning at least 15 years with true replication (or spanning a continuous age sequence with at least 12 total samples). Qualitative patterns of successional patterns in total abundance (abund.) and species richness (rich.) are described as follows: " – " and " + " indicate decline or increases with stand age or successional stage, respectively, "hump" and "U" indicates a maximum or minimum at intermediate age/stage, and "null" indicates no detectable response

Taxon	Biome	Location	Stages	Ages (y)	abund	rich	Reference
Spiders	boreal	Finland	4	0–60	–	–	Niemelä et al. 1996
Carabid beetles	boreal	Finland	4	0–60	–	–	Niemelä et al. 1996
Ants	boreal	Finland	4	0–60	–	–	Niemelä et al. 1996
Spiders	boreal	Canada	3	1–29	null	+	Buddle et al. 2000
Carabid beetles	boreal	Finland	5	5–60	null	–	Koivula et al. 2002
Carabid beetles	boreal	Canada	cont	0–341	?	U	Paquin 2008
All beetles	boreal	Sweden	cont	5–290	+	+	Gibb et al. 2013
Spiders	temp	USA	4		–	–	McIver et al. 1992
Spiders	temp	Canada	4		?	–	Brumwell et al. 1998
Carabid beetles	temp	Canada	4		?	–	Brumwell et al. 1998
Spiders	temp	USA	cont	0–15	null	null	Niwa and Peck, 2002
Carabid beetles	temp	USA	cont	0–15	null	null	Niwa and Peck 2002
Ground-dwelling beetles	temp	USA	4	5–	U	–	Heyborne et al. 2003
Butterflies	temp	Japan	4	1–	–	–	Inoue 2003
Carabid beetles	temp	Spain	5	2–80	U	U	Taboada et al. 2008
Carabid beetles	temp	New Zealand	6	1–29	–	null	Pawson et al. 2009
Orthoptera	temp	Germany	3		hump	hump	Helbing et al. 2014
Spiders	temp	Japan	cont	1–107	–	–	Haraguchi and Tayasu 2016

(continued)

Table 8.1 (continued)

Taxon	Biome	Location	Stages	Ages (y)	abund	rich	Reference
Chrysomelid beetles	sub-trop	Mexico	4	4−	+	U	Sánchez-Reyes et al. 2019
Butterflies	trop	Cameroon	4		?	+	Lawton et al. 1998
Canopy beetles	trop	Cameroon	4		?	null	Lawton et al. 1998
Canopy ants	trop	Cameroon	4		?	null	Lawton et al. 1998
Leaf litter ants	trop	Cameroon	4		?	hump	Lawton et al. 1998
Termites	trop	Cameroon	4		?	+	Lawton et al. 1998
Bees	trop	Malaysia	3	20−	+	−	Liow et al. 2001
Geometrid moths	trop	Malaysia	6		null	+	Beck et al. 2002
Butterflies	trop	Indonesia	3		?	+	Schulze et al. 2004
Dung beetles	trop	Indonesia	3		?	+	Schulze et al. 2004
Pyraloid moths	trop	Malaysia	6		?	+	Fiedler and Schulze 2004
Arctiid moths	trop	Ecuador	3		hump	hump	Hilt and Fiedler 2005
Butterflies	trop	Indonesia	4		hump	+	Vedderler et al. 2005
Geometrid moths	trop	Ecuador	3		?	hump	Nöske et al. 2008
Arctiid moths	trop	Ecuador	3		?	hump	Nöske et al. 2008
Galling insects	trop	Brazil	cont	0–21	?	hump	Fernandes et al. 2010

The attention in most studies of successional patterns in forest arthropods has been on species richness patterns and changes in species composition. Most studies have not directly reported patterns in overall arthropod abundance; however, where this is done it appears that overall arthropod abundance commonly shows similar patterns to that of species richness (Table 8.1). For example, Niemelä et al. (1996) report declines in both abundance and species richness through succession in Carabid beetles, spiders, and ants in boreal forests. Abundance patterns themselves are of interest in terms of trophic interactions, nutrient cycling, and other processes. Abundance patterns should also be taken into account in assessing species richness (Gotelli and Colwell 2001). Most recent studies have done this through use of rarefaction

Fig. 8.3 Relationship between estimated species richness of Carabid beetles (abundance-based coverage estimator, derived from analysis of species accumulation curves: Chao and Yang 1993) and stand age in naturally regenerated post-fire stands of black spruce (*Picea mariana*) sampled in western Quebec, Canada. Redrawn from Paquin (2008)

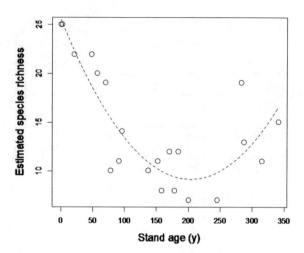

curves and related statistics (e.g. Paquin 2008). Of course, biodiversity more broadly may be assessed through numerous metrics including conventional species diversity measures that weight evenness and richness (such as Fisher's alpha, Shannon–Wiener index, Simpson index, and others: Magurran 2013), functional diversity measures (Mouchet et al. 2010), and phylogenetic diversity measures (Cadotte et al. 2010).

Additional methodological limitations pertinent to succession studies on forest arthropods bear mention. Essentially all studies involve chronosequences that substitute space for time. Some of the biases and limitations of a chronosequence approach are overcome with true replication of stands; however, chronosequence studies implicitly assume constant environmental conditions (Pickett 1988; Johnson and Miyanishi 2008). Given the long time periods involved in forest succession, there is not really an alternative; however, future studies could profitably apply emerging approaches that combine chronosequence data with direct temporal data (Damgaard 2019). The available data are also highly skewed to a few taxonomic groups. For example, most studies in temperate and boreal systems have focused on carabid beetles or spiders, both of which are readily sampled using pitfall traps. Major forest arthropod groups that have received almost no attention in terms of successional patterns include many non-insect arthropods (e.g. isopods, centipedes, millipedes, opiliones - but see Schreiner et al. 2012), and major insect groups, including Diptera, Hemiptera, and non-ant Hymenoptera.

8.3.2 Two or More Distinct Successional Stages in Forest Arthropod Communities?

In essentially all studies of forest arthropod succession, differences in community composition have been detected between post-disturbance sites and late-seral stands

(Table 8.1). In general, one finds a set of species associated with more open habitats, a set of forest species, and a gradual transition between these two groups. However, a few studies have presented evidence for a distinct mid-successional community of forest arthropods. Niemälä et al. (1996) present evidence from boreal forest in southern Finland that carabid and ant communities immediately post disturbance are more similar to late-seral communities than are communities in younger, closed-canopy stands. In a study of *Pinus sylvestris* plantations in northern Spain, Taboada et al. (2008) found that the youngest stands showed carabid beetle communities similar to surrounding open habitats, while after canopy closure (i.e. in the stem exclusion stage: Fig. 8.2), communities differed strongly in composition and were highly depauperate; older stands showed more similar species composition to natural pine forests in the region. Analyses presented by Paquin (2008) provide evidence for four distinct successional communities of carabid beetles in black spruce succession following fire: a "burned" seral community found only in the first 2 years post-fire with a set of 6 indicator species, and "regenerating", "mature", and "old growth" communities each with 2–4 distinctive characteristic species. The "regenerating" community corresponds to the progressive decline in overall carabid beetle diversity from year ~ 3–170 (Fig. 8.3).

The only temperate or boreal study included in Table 8.1 to find a hump-shaped response pattern, examining succession patterns of Orthoptera in pine woodlands in the northern Alps, also presents evidence for 3 distinct insect communities (Helbing et al. 2014). In this case, the earliest seral stage had a high proportion of bare ground, and was inferred to be poor in food resources, while the second seral stage had some tree recruitment but was essentially still open; closed-canopy forest was not found until the third stage, and this corresponded to a large decline in species richness. This study, although superficially seeming to support intermediate disturbance, thus also strongly implicates changes in forest structure as a main driver of successional patterns.

In sum, studies that have looked in detail at arthropod community patterns through succession, at least in boreal and temperate forest systems, have commonly found evidence for a distinct intermediate stage. In terms of stand development, this appears to generally correspond to the stem exclusion stage, and likely includes species that can persist under low light conditions with little understory vegetation and little coarse woody debris.

8.3.3 Relationships Between Arthropod and Vegetation Diversity Through Forest Succession

As noted earlier, it is widely accepted that there is a pervasive relationship between arthropod diversity and plant diversity. Many herbivores and seed predators have narrow host ranges; widespread specialization in forest insect communities was famously the basis for early extrapolations of global insect diversity based on host tree canopy insecticidal fogging (Erwin 1982). Siemann et al. (1998) present evidence for a general relationship between arthropod and vegetation diversity based on large-scale experimental manipulations of herbaceous plant communities. As they note, the overall relationships were significant, but with low intercepts and R^2 values (0.14 for observed total species richness), and stronger relationships between species richness of insect herbivores and higher trophic levels (predators and parasitoids). Subsequent studies have noted similar patterns (e.g. Haddad et al. 2009), and comparable effects have been seen in relation to plant genetic diversity (Johnson et al. 2006). Observational studies have indicated strong relationships between insect diversity and plant diversity, specifically in forest ecosystems (Basset et al. 2012), and in heterogeneous landscapes (Zhang et al. 2016). However, a recent experimental study that manipulated local woody plant diversity did not find effects on insect diversity (Yeeles et al. 2017).

Do changes in arthropod diversity through succession track patterns for plants? Few of the studies listed in Table 8.1 examined these relationships, however Beck et al. (2002) found a strong correlation between vegetation diversity and insect diversity in a study of geometrid moths in Malaysia, and Nöske et al. (2008) found similar results in montane forests in Ecuador. In the broader literature, a notable counterexample is a study reporting no significant relationship between geometrid moth diversity and vegetation diversity along a successional gradient on Mt. Kilimanjaro (Axmacher et al. 2004). However, in this case the oldest vegetation class was a monodominant high-elevation forest that was spatially disjunct and at higher elevation than other sites. Additional tropical studies showing relatively strong correlations between vegetation diversity and insect diversity through succession include a study of butterflies and dung beetles in Sulawesi, Indonesia (Schulze et al. 2004), and of gall-forming insects in a tropical dry forest in Mexico (Cuevas-Reyes et al. 2004). A meta-analysis on broader patterns suggests that positive correlations are generally observed between insect and plant diversity (with a pooled correlation coefficient of ~ 0.45), but that this relationship is stronger between habitats and stronger for primary consumers than secondary consumers (Castagneyrol and Jactel 2012).

8.3.4 What Insect Groups Depend on Late-Seral Forests?

Observations on general associations of arthropod groups with open vs. forested habitats are certainly as old as entomology as a science: Orthoptera, Hemiptera, and most bees and Lepidoptera are likely to be found in open areas, whereas most Isoptera, Blattoidea, and millipedes favour forest habitats. Of course, casual observations can be misleading (and biased toward the most apparent species); specific associations with late-seral forests are often less obvious, though critically important from a conservation perspective.

Studies represented in Table 8.1 may give some indication of patterns. The most important point is that essentially all studies find variable patterns within taxa, with some species associated with late-seral stands. Among broad taxonomic groups,

Fig. 8.4 *Rhysodes sulcatus*: an example of a woody-debris-dependent insect of conservation concern. This endangered saproxylic beetle is native to Eurasia, and currently extinct in much of its European range (Photo: Credit Nikolas_Rahme-Flickr14929651712_09f4855d2b_k)

those that appear to most consistently show positive relationships with stand age include most Lepidoptera and Isoptera, at least in the tropics. Consideration of this question illustrates how sparse these data are: hundreds of similar studies covering all arthropod groups would be required for an adequate assessment.

In the absence of such data, lists of threatened and endangered arthropod species provide some useful information. The most comprehensive assessments to date have been in the European Union: among non-aquatic insect groups assessed, 15% of saproxylic beetles are considered threated, compared to 9% of bees, and 9% of butterflies (Nieto et al. 2014). Eckelt et al. (2018) provide a list of 168 beetles that are strongly associated with late-seral stands in Germany. Beetle species that require large coarse woody debris in closed forest habitats appear to be among those most systematically threatened (Fig. 8.4).

8.3.5 Insect Succession Related to Tree Age and Size

Lawton (1983) noted that natural history observations suggest associations of specific insects with trees of specific age but was unable to locate any data on this phenomenon. Recent observations that there are large systematic differences in tree physiology through tree ontogeny have motivated studies on the effects of tree size/age on insect communities closely associated with trees, in particular insect herbivores. There are thus now a number of studies that allow tests for patterns of abundance of specific insects through the whole of tree ontogeny. Ontogenetic succession in myrmecophytic trees has been the subject of a number of studies. These tree species require some time to attract ants as a consequence of developmental constraints and ant dispersal limitation (e.g. Del Val and Dirzo 2003); ant inhabitants subsequently have strong effects on herbivore communities, and initial ant colonizers are commonly displaced by other species (e.g. Feldhaar et al. 2003; Fonseca and Benson 2003; Dejean et al. 2008). These studies thus provide clear examples of distinct insect successional communities that track tree age and ontogenetic stage.

Aside from studies of myrmecophytes, assessments of tree ontogeny effects on arthropod communities have focused primarily on herbivore communities. LeCorff and Marquis (1999) compared herbivore communities on understory saplings and mature trees of two oak species, finding differences in community composition and higher herbivore abundance and diversity in the understory. Other "sapling vs. mature tree" studies have yielded different results. Basset (2001) found increased herbivore abundance and diversity in mature trees of the neotropical pioneer species *Pourouma bicolor*. Jeffries et al. (2006) sampled herbivore communities from *Quercus alba* leaves across a broad chronosequence, finding an increase in the number of species per unit leaf area (from ~ 0.8 to 1.2 species/m^2 leaf sampled). Thomas et al. (2010) present data on the frequency of herbivore damage types, most of which may be traced to one or two main species, on canopy leaves of *Acer saccharum* sampled in an uneven-aged forest. These data show a positive correlation of the diversity of damage types with tree size and age (Fig. 8.5). Available data, albeit scarce, thus suggest a general trend of increasing diversity of herbivore communities with tree age (as distinct from stand age).

Sessile arthropods may have particularly strong ontogenetic associations with their hosts. As noted above, scale insect abundance commonly reaches a maximum at trees of intermediate size (Wardhaugh et al. 2006). In a tropical dry forest, Cuevas-Reyes et al. (2004) found a general tendency for increased levels of gall formation (mainly by Cecidomyid midges) on saplings than on mature trees and inferred that this may be caused by greater availability of undifferentiated meristems favorable to gall development. In contrast, maple spindle gall mite increases dramatically in

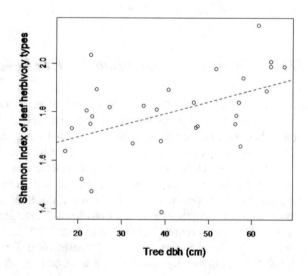

Fig. 8.5 Increase in diversity of arthropod herbivory types on canopy leaves of sugar maple (*Acer saccharum*). Linear regression line is shown (R = 0.473; P = 0.008). Data are from Thomas et al. (2010)

abundance with tree age, and galling is associated with substantial declines in leaf physiological performance (Patankar et al. 2011); a predaceous mite that invades and lays eggs within galls also tracks this pattern (Patankar et al. 2012).

8.3.6 Insect Succession on Coarse Woody Debris and Other Discrete Habitat Elements

As detailed above, the early literature on succession as an ecological process focused largely on plant communities. Nevertheless, there was at least one influential early entomological study, that of Savely (1939), who described successional patterns of arthropods on pine and oak logs in the southeast US. Logs were initially colonized by phloem-feeding taxa during the first year, in particular beetles in the families Cerambycidae, Buprestidae, and subfamily Scolytinae. These species enhanced wood decomposition by fungi, which were in turn linked with a variety of fungivorous and predaceous species that later colonized the logs (Savely 1939). Although the patterns described clearly had an affinity with prevailing ideas of Clementsian succession, Savely sought an understanding of insect succession on the basis of physical processes, with a focus on log microclimate and chemistry.

Insect succession patterns on coarse woody debris have received renewed research interest in recent years, with a focus on saproxylic beetles. In general, species with a narrow host range initially colonize, and more generalist species are found in later decay classes (Grove 2002). Varying patterns have been found with respect to diversity. Ulyshen and Hanula (2010) found the highest diversity of beetles in loblolly pine in the earliest decay class. In contrast, Hammond et al. (2004) found increasing beetle diversity through decay in poplar logs. Boulanger and Sirois (2014) describe a distinct community of beetles that colonizes standing dead trees following fire, and another than colonizes burnt trees once fallen. Ferro et al. (2012) report peak beetle diversity in mid decay class logs, with distinct communities found in early, mid, and late decay classes (Fig. 8.6).

There are a variety of other discrete (and often ephemeral) habitat elements analogous to coarse woody debris on which succession in forest arthropod communities is common. Examples include ant communities in domatia (e.g. Fonseca and Benson 2003), insects associated with decomposition of animal carcasses (e.g. Matuszewski et al. 2010), small natural ephemeral pools (phytotelmata) such as those formed by tree holes and bromeliads (Greeney 2001; Rangel et al. 2017), and larger vernal pools (Bischof et al. 2013) and animal wallows (Vanschoenwinkel et al. 2011). One might expect the successional patterns in these habitats to be affected by the local forest environment, which itself is strongly affected by stand successional status and structure. Successional patterns within these habitat elements would also be expected to contribute to overall successional patterns with stand age. These interactions have received little attention.

Fig. 8.6 Venn diagram showing species overlap of dead-wood-inhabiting beetles sampled from coarse woody debris by decay class. The area of circles is proportional to the total number of observed species. The largest distinct community occurs on mid-decay logs. Redrawn from Ferro et al. (2012)

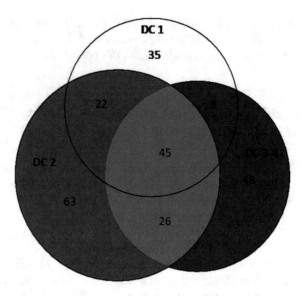

The importance of coarse woody debris as a habitat element stems from its provision of resources and effects on micro-environmental conditions over an extended period. Another forest disturbance legacy that is beginning to receive attention is charcoal generated from fire events, which has marked effects on soil properties and commonly strongly stimulates tree growth (Wardle et al. 1998; Thomas and Gale 2015). Uniquely, charcoals are exceptionally long-lived in the natural environment, potentially persisting for 1,000s or 10,000s of years, and thus are expected to remain through multiple stand-replacing disturbance events. Recent studies have addressed both recent "biochar" additions to soil (i.e. charcoals designed for use as a soil amendment), and effects of long-persistent natural chars. Although data on forest arthropods are very limited, research to date suggests the potential for large changes in soil arthropod communities associated with deposition of charcoals (Domene 2016). Recent studies also suggest unique arthropod communities associated with Amazonian "terra preta" soils defined by incorporation of chars by pre-contact Amerindians (Demetrio et al. 2019).

8.4 Effects of Insects on Forest Succession

The most dramatic and obvious effects of arthropods on forest succession processes are the relatively few species of insects that themselves can be the direct cause of stand-replacing disturbance events by killing the majority of canopy trees over a short time period. These cases are mainly restricted to boreal and north-temperate forests, and specifically include several species of Scolytine beetles—namely mountain pine beetle (*Dendroctonus ponderosae*), European spruce bark beetle (*Ips typographus*), and southern pine beetle (*Dendroctonus frontalis*), as well as three species of Lepidoptera: spongy moth (*Lymantria dispar*), spruce budworm (*Choristoneura* spp.), and eastern tent caterpillar (*Malacosoma americanum*). In addition, there are cases of invasive species that do not cause stand-replacing disturbances in their native range but can do so in their introduced range. Notable examples of include Asian long-horned beetle (*Anoplophora glabripennis*), and emerald ash borer (*Agrilus planipennis*).

The tree host ranges of these species, at least those within their native range, are relatively small. For example, mountain pine beetle essentially impacts *Pinus contorta*, but also can feed to some extent on sugar pine (*P. lambertiana*), western white pine (*P. monticola*) and ponderosa pine (*P. ponderosa*) and has recently reproduced on jack pine (*P. banksiana*) (Cullingham et al. 2011). The species acts as stand-replacing disturbance agent only because *P. contorta* forms essentially mono-dominant forests in large areas of British Colombia, Alberta, and the Western US. This raises the issue of future forest succession: is it possible that beetle-kill areas will show a complete change in species composition or possibly enter a state of arrested succession and lose forest cover entirely? This is a critically important question in view of the recent unprecedented mountain pine beetle impacts in western Canada. Although mountain pine beetle is the most extreme case, similar questions arise in essentially any case of insects as agents of stand-replacing disturbance.

Recent work on vegetation responses following complete tree mortality of lodgepole pine stands due to beetle kill suggests a large initial positive response of understory herbaceous vegetation in terms of both productivity and diversity (Pec et al. 2015). Lodgepole pine has serotinous cones and is adapted to regenerate following stand-replacing fires. In central British Colombia lodgepole pine is essentially absent from tree recruitment following beetle kill and the existing seedling bank of subalpine fir (*Abies lasiocarpa*) is the only source of tree regeneration (Astrup et al. 2008). However, higher lodgepole pine regeneration has been seen in areas of the US (Collins et al. 2011; Kayes and Tinker 2012), and in boreal forest regions where the mountain pine beetle represents a novel impact (Campbell and Antos 2015). Thus, it appears that in only some areas is there likely to be a complete change in species composition following stand-replacing mountain pine beetle outbreaks.

Given the relatively narrow host ranges of insects, it is not surprising that insects as true stand-replacing disturbance agents are essentially restricted to boreal forests and low-diversity temperate forests. However, large-scale insect outbreaks, though perhaps not true stand-replacement events, are also found in the tropics. Anderson (1961) observed stand-level defoliation, likely by a species-specific lepidopteran, in areas dominated by the dipterocarp species *Shorea albida*. This tree species forms nearly monospecific stands in peat swamp areas in Borneo. A similar example has been documented in another monodominant tropical forest in the neotropics, dominated by *Peltogyne gracilipes* (Nascimento and Proctor 1994). Dyer et al. (2012) compiled other known examples in natural tropical forests. In general, stand-level defoliation events have been reported only from low-diversity tropical forests, in particular areas where a single species dominates.

Other than extreme cases of insects causing stand-replacing disturbance events, are there more general effects of insects on the succession process in forests? It has been hypothesized that insect herbivory can act to decelerate succession (Brown 1985) by reducing overall growth and competition among plant species. Alternatively, insect herbivory might accelerate succession by herbivores having a larger effect on poorly defended early-successional species (Davidson 1993). Manipulative studies (mainly on amenable non-forest systems) have yielded variable effects depending on the system (Brown et al. 1988). It has also been hypothesized that granivory or seed predation has effects on successional processes distinct from herbivory (Davidson 1993). Seed predators generally impact large-seeded late-successional trees more so than pioneer species, and thus might be expected to favor the latter. Important insect seed predator taxa include Curculionid, Scolytid, and Bruchid beetles, Lygaeid hemipterans, Gryllid orthopterans, and members of the orders Diptera, Hymenoptera, Lepidoptera, and Thysanoptera. I am not aware of any formal test of herbivory or seed predation effects on successional patterns in forest systems. The potential importance of seed predation on tropical forest dynamics is suggested by the phenomenon of mast fruiting in the Dipterocarpaceae of Southeast Asia, thought to be an evolutionary response to seed predation pressure (Janzen 1974; Lyal and Curran 2000). Analyses of successional patterns in insect taxa important as seed predators are also lacking. One might expect large increases in diversity in these groups through forest succession, particularly an increase in species associated with large-seeded host taxa.

8.5 Conclusions

Succession has been a notoriously contentious topic from the time of Clements and Gleason to the present. In any reading of the empirical literature on insects and forest succession, it is obvious that many entomologists simply avoid broader ecological theory, being satisfied with narrow descriptions of patterns specific to a given system. Two problems arise from such narrow description. First, it is inherently important that scientific contributions form a basis for broader generalizations and test existing theory. Second, if theory is not articulated, it is often still present in the mind of the investigator in the form of unstated assumptions and bias. The uncritical assumption that old-growth forests are climax communities in the Clementsian sense is a particularly common popular misconception, as is the bias toward assuming that older forest stands must have higher species richness. The intermediate disturbance hypothesis has been the most important theoretical touchstone of studies on forest insect succession, but forest entomologists should be aware that support for this theory is generally weak, and that the foundations of the theory itself are questionable. Successional processes are almost certainly system-specific and idiosyncratic in many respects; however, the main conclusion that emerges from the present review is that forest structural development (and possibly direct effects related to tree ontogeny) is generally more useful as a framework for understanding patterns of forest insect succession than more abstract theoretical representations.

Acknowledgements I thank Sandy Smith for discussions on the topic, and the editors of this volume for constructive criticisms that helped improve the paper. I acknowledge the members of my academic "pedigree" (F.A. Bazzaz, L.C. Bliss, W.D. Billings, H.J. Oosting, W.S. Cooper, and H.C. Cowles), who largely dedicated their careers to the study of plant succession. I also thank J. Weiner, with whom I first studied much of the basics covered here, and J. Allison for entrusting a chapter in an entomology text to an unrepentant botanist.

References

Anderson JAR (1961) The destruction of *Shorea albida* forest by an unidentified insect. Emp For Rev 40:19–29

Anderson J (2018) Egler's $10,000 succession challenge. In: Finlayson CM, Everard M, Irvine K, McInnes RJ, Middleton B, Van Dam AA, Davidson NC (eds) The Wetland Book: I: Structure and function, management, and methods. Springer, Netherlands, pp 43–45

Astrup R, Coates KD, Hall E (2008) Recruitment limitation in forests: lessons from an unprecedented mountain pine beetle epidemic. For Ecol Manage 256:1743–1750

Axmacher JC, Tünte H, Schrumpf M, Müller-Hohenstein K, Lyaruu HV, Fiedler K (2004) Diverging diversity patterns of vascular plants and geometrid moths during forest regeneration on Mt Kilimanjaro, Tanzania. J Biogeogr 31:895–904

Barton KE, Koricheva J (2010) The ontogeny of plant defense and herbivory: characterizing general patterns using meta-analysis. Am Nat 175:481–493

Basset Y (2001) Communities of insect herbivores foraging on saplings versus mature trees of *Pourouma bicolor* (Cecropiaceae) in Panama. Oecologia 129:253–260

Basset Y, Cizek L, Cuénoud P, Didham RK, Guilhaumon F, Missa O et al (2012) Arthropod diversity in a tropical forest. Science 338(6113):1481–1484

Beck J, Schulze CH, Linsenmair KE, Fiedler K (2002) From forest to farmland: diversity of geometrid moths along two habitat gradients on Borneo. J Trop Ecol 17:33–51

Bischof MM, Hanson MA, Fulton MR, Kolka RK, Sebestyen SD, Butler MG (2013) Invertebrate community patterns in seasonal ponds in Minnesota, USA: Response to hydrologic and environmental variability. Wetlands 33:245–256

Boege K, Barton KE, Dirzo R (2011) Influence of tree ontogeny on plant-herbivore interactions. In: Meinzer FC, Lachenbruch B, Dawson TE (eds) Size-and age-related changes in tree structure and function. Springer, Dordrecht, pp 193–214

Bond BJ (2000) Age-related changes in photosynthesis of woody plants. Trends Plant Sci 5:349–353

Bongers F, Poorter L, Hawthorne WD, Sheil D (2009) The intermediate disturbance hypothesis applies to trop. forests, but disturbance contributes little to tree diversity. Ecol Lett 12:798–805

Botkin DB, Janak JF, Wallis JR (1972) Some ecological consequences of a computer model of forest growth. J Ecol 60:849–872

Boulanger Y, Sirois L (2014) Postfire succession of saproxylic arthropods, with emphasis on Coleoptera, in the north boreal forest of Quebec. Environ Entomol 36:128–141

Brown VK (1985) Insect herbivores and plant succession. Oikos 44:17–22

Brown VK, Jepsen M, Gibson CWD (1988) Insect herbivory: effects on early old field succession demonstrated by chemical exclusion methods. Oikos 293–302

Brumwell LJ, Craig KG, Scudder GG (1998) Litter spiders and carabid beetles in a successional Douglas-fir forest in British Columbia. Northwest Sci 72:94–95

Brünig EF (1973) Species richness and stand diversity in relation to site and succession of forests in Sarawak and Brunei (Borneo). Amazoniana: Limnologia et Oecologia Regionalis Systematis Fluminis Amazonas 4:293–320

Buddle CM, Spence JR, Langor DW (2000) Succession of boreal forest spider assemblages following wildfire and harvesting. Ecography 23:424–436

Cadotte MW, Jonathan Davies T, Regetz J, Kembel SW, Cleland E, Oakley TH (2010) Phylogenetic diversity metrics for ecological communities: integrating species richness, abundance and
evolutionary history. Ecol Lett 13:96–105
Campbell EM, Antos JA (2015) Advance regeneration and trajectories of stand development
following the mountain pine beetle outbreak in boreal forests of British Columbia. Can J for
Res 45:1327–1337
Castagneyrol B, Jactel H (2012) Unraveling plant–animal diversity relationships: a meta-regression
analysis. Ecology 93:2115–2124
Chao A, Yang MC (1993) Stopping rules and estimation for recapture debugging with unequal
failure rates. Biometrika 80:193–201
Clements FE (1916) Plant succession: an analysis of the development of vegetation (No. 242).
Carnegie Institution of Washington
Clements FE (1936) Nature and structure of the climax. J Ecol 24:252–284
Collins BJ, Rhoades CC, Hubbard RM, Battaglia MA (2011) Tree regeneration and future stand
development after bark beetle infestation and harvesting in Colorado lodgepole pine stands. For
Ecol Manage 261:2168–2175
Connell JH (1978) Diversity in tropical rain forests and coral reefs: high diversity of trees and corals
is maintained only in a nonequilibrium state. Science 199:1302–1310
Connell JH, Slatyer RO (1977) Mechanisms of succession in natural communities and their role in
community stability and organization. Am Nat 111:1119–1144
Cowles HC (1899) The ecological relations of the vegetation on the sand dunes of Lake Michigan.
Bot Gaz 27, pp 95–117, 167–202, 281–308, 361–391
Cowles HC (1901) The physiographic ecology of Chicago and vicinity; a study of the origin,
development, and classification of plant societies. Bot Gaz 31:145–182
Cuevas-Reyes P, Quesada M, Hanson P, Dirzo R, Oyama K (2004) Diversity of gall-inducing insects
in a Mexican tropical dry forest: the importance of plant species richness, life-forms, host plant
age and plant density. J Ecol 92:707–716
Cullingham CI, Cooke JE, Dang S, Davis CS, Cooke BJ, Coltman DW (2011) Mountain pine beetle
host-range expansion threatens the boreal forest. Mol Ecol 20:2157–2171
Damgaard C (2019) A critique of the space-for-time substitution practice in community ecology.
Trends Ecol Evol 34:416–421
Davidson DW (1993) The effects of herbivory and granivory on terrestrial plant succession. Oikos
68:23–35
Dejean A, Djiéto-Lordon C, Céréghino R, Leponce M (2008) Ontogenetic succession and the ant
mosaic: an empirical approach using pioneer trees. Basic Appl Ecol 9:316–323
Del Val E, Dirzo R (2003) Does ontogeny cause changes in the defensive strategies of the
myrmecophyte Cecropia peltata? Plant Ecol 169:35–41
Demetrio WC, Conrado AC, Acioli A, Ferreira AC, Bartz ML, James SW, et al (2019) A "dirty"
footprint: anthropogenic soils promote biodiversity in Amazonian rainforests. bioRxiv, 552364
Denslow JS (1980) Patterns of plant species diversity during succession under different disturbance
regimes. Oecologia 46:18–21
Domene X (2016) A critical analysis of meso-and macrofauna effects following biochar supple-
mentation. In: Ralebitso-Senior TK, Orr CH (eds) Biochar application: Essential soil microbial
ecology. Elsevier, Amsterdam, pp 268–292
Drury WH, Nisbet IC (1973) Succession. J Arnold Arbor 54:331–368
Dyer LA, Carson WP, Leigh EG (2012) Insect outbreaks in tropical forests: patterns, mechanisms,
and consequences. In: Barbosa P, Letourneau DK, Agrawal AA (eds) Insect outbreaks revisited.
Blackwell, New York, pp 219–245
Eckelt A, Müller J, Bense J, Brustel H, Bußler H, Chittaro Y, Cizek L, Frei A, Holzer E, Kadej
M, Kahlen M, Köhler F, Möller G, Mühle H, Sanchez A, Schaffrath U, Schmidl J, Smolis A,
Szallies A, Németh T, Wurst C, Thorn S, Christensen RHB, Siebold S (2018) "Primeval forest
relict beetles" of Central Europe: a set of 168 umbrella species for the protection of primeval
forest remnants. J Insect Conserv 22:15–28

Eggeling WJ (1947) Observations on the ecology of the Budongo rain forest, Uganda. J Ecol 20–87

Egler FE (1954) Vegetation science concepts I. Initial floristic composition, a factor in old-field vegetation development with 2 figs. Vegetatio 4:412–417

Erwin TL (1982) Tropical forests: their richness in Coleoptera and other arthropod species. Coleopterist's Bull 36:74–75

Feldhaar H, Fiala B, Maschwitz U (2003) Patterns of the Crematogaster-Macaranga association: the ant partner makes the difference. Insectes Soc 50:9–19

Fernandes GW, Almada ED, Carneiro MAA (2010) Gall-inducing insect species richness as indicators of forest age and health. Environ Entomol 39:1134–1140

Ferro ML, Gimmel ML, Harms KE, Carlton CE (2012) Comparison of coleoptera emergent from various decay classes of downed coarse woody debris in Great Smoky Mountains National Park, USA. Insecta Mundi 2012:1–80

Fiedler K, Schulze CH (2004) Forest modification affects diversity (but not dynamics) of speciose tropical pyraloid moth communities. Biotropica 36:615–627

Fish H, Lieffers VJ, Silins U, Hall RJ (2006) Crown shyness in lodgepole pine stands of varying stand height, density, and site index in the upper foothills of Alberta. Can J for Res 36:2104–2111

Fonseca CR, Benson WW (2003) Ontogenetic succession in Amazonian ant trees. Oikos 102:407–412

Fox JW (2013) The intermediate disturbance hypothesis should be abandoned. Trends Ecol Evol 28:86–92

Franklin JF, Spies TA, Van Pelt R, Carey AB, Thornburgh DA, Berg DR, ... Bible, K (2002) Disturbances and structural development of natural forest ecosystems with silvicultural implications, using Douglas-fir forests as an example. For Ecol Manag 155:399–423

Gibb H, Johansson T, Stenbacka F, Hjältén J (2013) Functional roles affect diversity-succession relationships for boreal beetles. PLoS ONE 8:e72764

Gleason HA (1917) The structure and development of the plant association. Bull Torrey Bot Club 44:463–481

Gleason HA (1926) The individualistic concept of the plant association. Bull Torrey Bot Club 53:7–26

Gotelli NJ, Colwell RK (2001) Quantifying biodiversity: procedures and pitfalls in the measurement and comparison of species richness. Ecol Lett 4:379–391

Greeney HF (2001) The insects of plant-held waters: a review and bibliography. J Trop Ecol 17:241–260

Grime JP (1973) Competitive exclusion in herbaceous vegetation. Nature 242:344–347

Grime JP (1979) Plant strategies and vegetation processes. John Wiley & Sons

Grime JP (2006) Plant strategies, vegetation processes, and ecosystem properties. John Wiley & Sons

Grimm V, Revilla E, Berger U, Jeltsch F, Mooij WM, Railsback SF et al (2005) Pattern-oriented modeling of agent-based complex systems: lessons from ecology. Science 310:987–991

Grove SJ (2002) Saproxylic insect ecology and the sustainable management of forests. Annu Rev Ecol Syst 33:1–23

Habeck JR (1968) Forest succession in the Glacier Park cedar-hemlock forests. Ecology 49:872–880

Haddad NM, Crutsinger GM, Gross K, Haarstad J, Knops JM, Tilman D (2009) Plant species loss decreases arthropod diversity and shifts trophic structure. Ecol Lett 12:1029–1039

Hammond HJ, Langor DW, Spence JR (2004) Saproxylic beetles (Coleoptera) using populus in boreal aspen stands of western Canada: spatiotemporal variation and conservation of assemblages. Can J for Res 34:1–19

Haraguchi TF, Tayasu I (2016) Turnover of species and guilds in shrub spider communities in a 100-year post-logging forest chronosequence. Environ Entomol 45:117–126

Helbing F, Blaeser TP, Löffler F, Fartmann T (2014) Response of Orthoptera communities to succession in alluvial pine woodlands. J Insect Conserv 18:215–224

Heyborne WH, Miller JC, Parsons GL (2003) Ground dwelling beetles and forest vegetation change over a 17-year-period, in western Oregon, USA. For Ecol Manage 179:123–134

Hilt N, Fiedler K (2005) Diversity and composition of Arctiidae moth ensembles along a successional gradient in the Ecuadorian Andes. Divers Distrib 11:387–398

Horn HS (1975) Markovian properties of forest succession. In: Cody ML, Diamond JM (eds) Ecology and evolution of communities. Belknap Press, Cambridge, MA, pp 196–211

Huston M (1979) A general hypothesis of species diversity. Am Nat 113:81–101

Huston MA (2014) Disturbance, productivity, and species diversity: empiricism vs. logic in ecological theory. Ecology 95:2382–2396

Inoue T (2003) Chronosequential change in a butterfly community after clear-cutting of deciduous forests in a cool temperate region of central Japan. Entomol Sci 6:151–163

Janzen DH (1974) Tropical blackwater rivers, animals, and mast fruiting by the Dipterocarpaceae. Biotropica 6:69–103

Jeffries JM, Marquis RJ, Forkner RE (2006) Forest age influences oak insect herbivore community structure, richness, and density. Ecol Appl 16:901–912

Johnson EA, Miyanishi K (2008) Testing the assumptions of chronosequences in succession. Ecol Lett 11:419–431

Johnson MT, Lajeunesse MJ, Agrawal AA (2006) Additive and interactive effects of plant genotypic diversity on arthropod communities and plant fitness. Ecol Lett 9:24–34

Kayes LJ, Tinker DB (2012) Forest structure and regeneration following a mountain pine beetle epidemic in southeastern Wyoming. For Ecol Manage 263:57–66

Klimes P, Idigel C, Rimandai M, Fayle TM, Janda M, Weiblen GD, Novotny V (2012) Why are there more arboreal ant species in primary than in secondary trop. forests? J Anim Ecol 81:1103–1112

Koch GW, Sillett SC, Jennings GM, Davis SD (2004) The limits to tree height. Nature 428:851–854

Koivula M, Kukkonen J, Niemelä J (2002) Boreal carabid-beetle (Coleoptera, Carabidae) assemblages along the clear-cut originated succession gradient. Biodivers Conserv 11:1269–1288

Lawton JH (1983) Plant architecture and the diversity of phytophagous insects. Annu Rev Entomol 28:23–39

Lawton JH, Bignell DE, Bolton B, Bloemers GF, Eggleton P, Hammond PM, Hodda M, Holt RD, Larsen TB, Mawdsley NA, Stork NE, Srivastava DS, Watt AD (1998) Biodiversity inventories, indicator taxa and effects of habitat modification in trop. forest. Nature 391:72–76

LeCorff J, Marquis RJ (1999) Differences between understorey and canopy in herbivore community composition and leaf quality for two oak species in Missouri. Ecol Entomol 24:46–58

Lee SC (1924) Factors controlling forest successions at Lake Itasca, Minnesota. Bot Gaz 78:129–174

Lindo Z, Winchester NN (2006) A comparison of microarthropod assemblages with emphasis on oribatid mites in canopy suspended soils and forest floors associated with ancient western redcedar trees. Pedobiologia 50:31–41

Liow LH, Sodhi NS, Elmqvist T (2001) Bee diversity along a disturbance gradient in tropical lowland forests of Southeast Asia. J Appl Ecol 38:180–192

Liu J, Ashton PS (1995) Individual-based simulation models for forest succession and management. For Ecol Manage 73:157–175

Lyal CHC, Curran LM (2000) Seed-feeding beetles of the weevil tribe Mecysolobini (Insecta: Coleoptera: Curculionidae) developing in seeds of trees in the Dipterocarpaceae. J Nat Hist 34:1743–1847

Ma L, Hurtt G, Ott L, Sahajpal R, Fisk J, Lamb R, Tang H, Flanagan S, Chini L, Chattergee A, Sullivan J (2022) Global evaluation of the ecosystem demography model (ED v3. 0). Geosci Model Dev 15:1971–1994

MacArthur RH, Wilson EO (1967) The theory of island biogeography. Princeton University Press, Princeton, NJ, USA

Mackey RL, Currie DJ (2001) The diversity-disturbance relationship: is it generally strong and peaked? Ecology 82:3479–3492

Magurran AE (2013) Measuring biological diversity. John Wiley & Sons

Margalef R (1968) Perspectives in ecological theory. Univ. of Chicago Press, Chicago

Matuszewski S, Bajerlein D, Konwerski S, Szpila K (2010) Insect succession and carrion decom+ition in selected forests of Central Europe. Part 2: composition and residency patterns of carrion fauna. Forensic Sci Int 195:42–51

Mason CM, McGaughey SE, Donovan LA (2013) Ontogeny strongly and differentially alters leaf economic and other key traits in three diverse Helianthus species. J Exp Bot 64:4089–4099

Mason CM, Donovan LA (2015) Does investment in leaf defenses drive changes in leaf economic strategy? A focus on whole-plant ontogeny. Oecologia 177:1053–1066

McElhinny C, Gibbons P, Brack C, Bauhus J (2005) Forest and woodland stand structural complexity: its definition and measurement. For Ecol Manage 218:1–24

McIver JD, Parsons GL, Moldenke AR (1992) Litter spider succession after clear-cutting in a western coniferous forest. Can J for Res 22:984–992

Mégnin JP (1894) La faune des cadavres: application de l'entomologie à la médecine légale. Encyclopédie Scientifique des Aide-Mémoire. Gauthier-Villars, Paris, France

Meinzer FC, Lachenbruch B, Dawson TE (eds) (2011) Size-and age-related changes in tree structure and function (Vol. 4). Springer Science & Business Media

Michaud JP, Schoenly KG, Moreau G (2015) Rewriting ecological succession history: did carrion ecologists get there first? Q R Biol 90:45–66

Motter MG (1898) A contribution to the study of the fauna of the grave. A study of one hundred and fifty disinterments, with some additional experimental observations. J N Y Entomol Soc 6:201–231

Mouchet MA, Villéger S, Mason NW, Mouillot D (2010) Functional diversity measures: an overview of their redundancy and their ability to discriminate community assembly rules. Funct Ecol 24:867–876

Müller J, Bußler H, Bense U, Brustel H, Flechtner G (2005) Urwald relict species—Saproxylic beetles indicating structural qualities and habitat tradition. Waldökologie Online 2:106–113

Nascimento MT, Proctor J (1994) Insect defoliation of a monodominant Amazonian rainforest. J Trop Ecol 10:633–636

Niemelä J, Haila Y, Punttila P (1996) The importance of small-scale heterogeneity in boreal forests: variation in diversity in forest-floor invertebrates across the succession gradient. Ecography 19:352–368

Nieto A, Roberts SPM, Kemp J, Rasmont P, Kuhlmann M, García Criado M, Biesmeijer JC, Bogusch P, Dathe HH, De la Rúa P, De Meulemeester T, Dehon M, Dewulf A, Ortiz-Sánchez FJ, Lhomme P, Pauly A, Potts SG, Praz C, Quaranta M, Radchenko VG, Scheuchl E, Smit J, Straka J, Terzo M, Tomozii B, Window J, Michez D (2014) European red list of bees. Publication Office of the European Union, Luxembourg

Niwa CG, Peck RW (2002) Influence of prescribed fire on carabid beetle (Carabidae) and spider (Araneae) assemblages in forest litter in southwestern Oregon. Environ Entomol 31:785–796

Nock CA, Caspersen JP, Thomas SC (2008) Large ontogenetic declines in intra-crown leaf area index in two temperate deciduous tree species. Ecology 89:744–753

Nöske NM, Hilt N, Werner FA, Brehm G, Fiedler K, Sipman HJ, Gradstein SR (2008) Disturbance effects on diversity of epiphytes and moths in a montane forest in Ecuador. Basic Appl Ecol 9:4–12

Odum EP (1969) The strategy of ecosystem development. Science 164:262–270

Oliver CD (1980) Forest development in North America following major disturbances. For Ecol Manage 3:153–168

Oliver CD, Larson BC (1996) Forest stand dynamics: updated edition. John Wiley and sons

Pacala SW, Canham CD, Silander JA Jr (1993) Forest models defined by field measurements: I. The design of a northeastern forest simulator. Can J for Res 23:1980–1988

Paquin P (2008) Carabid beetle (Coleoptera: Carabidae) diversity in the black spruce succession of eastern Canada. Biol Cons 141:261–275

Patankar R, Smith SM, Thomas SC (2011) A gall-inducing arthropod drives declines in canopy tree photosynthesis. Oecologia 167:701–709

Patankar R, Beaulieu F, Smith SM, Thomas SC (2012) The life history of a gall-inducing mite: summer phenology, predation and influence of gall morphology in a sugar maple canopy. Agric for Entomol 14:251–259

Pawson SM, Brockerhoff EG, Didham RK (2009) Native forest generalists dominate carabid assemblages along a stand age chronosequence in an exotic Pinus radiata plantation. For Ecol Manage 258:S108–S116

Pec GJ, Karst J, Sywenky AN, Cigan PW, Erbilgin N, Simard SW, Cahill JF Jr (2015) Rapid increases in forest understory diversity and productivity following a mountain pine beetle (*Dendroctonus ponderosae*) outbreak in pine forests. PLoS One 10:e0124691

Peet RK (1978) Forest vegetation of the Colorado Front Range: patterns of species diversity. Vegetatio 37:65–78

Pickett STA (1988) Space-for-time substitution as an alternative to long term studies. In: Likens GE (ed) Long-term Studies in Ecology. Springer, New York, NY, pp 110–135

Pickett STA, Collins SL, Armesto JJ (1987) Models, mechanisms and pathways of succession. Bot Rev 53:335–371

Pickett STA, Kolasa J, Armesto JJ, Collins SL (1989) The ecological concept of disturbance and its expression at various hierarchical levels. Oikos 129–136

Pickett ST, McDonnell MJ (1989) Changing perspectives in community dynamics: a theory of successional forces. Trends Ecol Evol 4:241–245

Pulsford SA, Lindenmayer DB, Driscoll DA (2016) A succession of theories: purging redundancy from disturbance theory. Biol Rev 91:148–167

Purdie RW, Slatyer RO (1976) Vegetation succession after fire in sclerophyll woodland communities in south-eastern Australia. Aust J Ecol 1:223–236

Quinn EM, Thomas SC (2015) Age-related crown thinning in tropical forest trees. Biotropica 47:320–329

Rangel JV, Araújo RE, Casotti CG, Costa LC, Kiffer WP Jr, Moretti MS (2017) Assessing the role of canopy cover on the colonization of phytotelmata by aquatic invertebrates: an experiment with the tank-bromeliad *Aechmea lingulata*. J Limnol 76:230–239

Sánchez-Reyes UJ, Niño-Maldonado S, Clark SM, Barrientos-Lozano L, Almaguer-Sierra P (2019) Successional and seasonal changes of leaf beetles and their indicator value in a fragmented low thorn forest of northeastern Mexico (Coleoptera, Chrysomelidae). ZooKeys 2019:71–103

Savely HE (1939) Ecological relations of certain animals in dead pine and oak logs. Ecol Monogr 9:321–385

Schoonmaker P, McKee A (1988) Species composition and diversity during secondary succession of coniferous forests in the western Cascade Mountains of Oregon. For Sci 34:960–979

Schreiner A, Decker P, Hannig K, Schwerk A (2012) Millipede and centipede (Myriapoda: Diplopoda, Chilopoda) assemblages in secondary succession: variance and abundance in Western German beech and coniferous forests as compared to fallow ground. Web Ecol 12:9–17

Schulze CH, Waltert M, Kessler PJA, Pitopang R, Shahabuddin VD, Muhlenberg M, Gradstein SR, Leuschner C, Steffan-Dewenter I, Tscharntke T (2004) Biodiversity indicator groups of tropical land-use systems: Comparing plants, birds, and insects. Ecol Appl 14:1321–1333

Shea K, Roxburgh SH, Rauschert ES (2004) Moving from pattern to process: coexistence mechanisms under intermediate disturbance regimes. Ecol Lett 7:491–508

Sheil D (2001) Long-term observations of rain forest succession, tree diversity and responses to disturbance. Plant Ecol 155:183–199

Shugart HH (1984) A theory of forest dynamics: the ecological implications of forest succession models. Springer, New York

Siemann E, Tilman D, Haarstad J, Ritchie M (1998) Experimental tests of the dependence of arthropod diversity on plant diversity. Am Nat 152:738–750

Spies TA (1998) Forest structure: a key to the ecosystem. Northwest Sci 72:34–36

Taboada A, Kotze DJ, Tárrega R, Salgado JM (2008) Carabids of differently aged reforested pinewoods and a natural pine forest in a historically modified landscape. Basic Appl Ecol 9:161–171

Taylor AR, Chen HY, VanDamme L (2009) A review of forest succession models and their suitability for forest management planning. For Sci 55:23–36

Thomas SC (2010) Photosynthetic capacity peaks at intermediate size in temperate deciduous trees. Tree Physiol 30:555–573

Thomas SC (2011) Age-related changes in tree growth and functional biology: the role of reproduction. In: Meinzer FC, Lachenbruch B, Dawson TE (eds) Size-and age-related changes in tree structure and function. Springer, Dordrecht, pp 33–64

Thomas SC, Gale N (2015) Biochar and forest restoration: a review and meta-analysis of tree growth responses. New For 46:931–946

Thomas SC, Sztaba A, Smith SM (2010) Herbivory patterns in mature sugar maple: variation with vertical canopy strata and tree ontogeny. Ecol Entomol 35:1–8

Thomas SC, Winner WE (2002) Photosynthetic differences between saplings and adult trees: an integration of field results by meta-analysis. Tree Physiol 22:117–127

Tilman D (1982) Resource competition and community structure. Princeton University Press, Princeton, NJ, USA

Tilman D (1985) The resource-ratio hypothesis of plant succession. Am Nat 125:827–852

Toumey JW, Korstian CF (1937) Foundations of silviculture upon an ecological basis. Wiley, New York

Uhlig P, Harris A, Craig G, Bowling C, Chambers B, Naylor B, Beemer G (2001).Old growth forest definitions for Ontario. Ont Min Nat Res, Queen's Printer for Ontario, Toronto, ON. 53 p

Ulyshen MD, Hanula JL (2010) Patterns of saproxylic beetle succession in loblolly pine. Agric for Entomol 12:187–194

Vanschoenwinkel B, Waterkeyn A, Nhiwatiwa T, Pinceel TOM, Spooren E, Geerts A et al (2011) Passive external transport of freshwater invertebrates by elephant and other mud-wallowing mammals in an African savannah habitat. Freshw Biol 56:1606–1619

Veddeler D, Schulze CH, Steffan-Dewenter I, Buchori D, Tscharntke T (2005) The contribution of tropical secondary forest fragments to the conservation of fruit-feeding butterflies: effects of isolation and age. Biodivers Conserv 14:3577–3592

Vellend M (2016) The theory of ecological communities (MPB-57) (Vol. 75). Princeton University Press, Princeton, NJ, USA

Wardhaugh CW, Blakely TJ, Greig H, Morris PD, Barnden A, Rickard S et al (2006) Vertical stratification in the spatial distribution of the beech scale insect (*Ultracoelostoma assimile*) in Nothofagus tree canopies in New Zealand. Ecol Entomol 31:185–195

Wardle DA, Hörnberg G, Zackrisson O, Kalela-Brundin M, Coomes DA (2003) Long-term effects of wildfire on ecosystem properties across an island area gradient. Science 300:972–975

Wardle DA, Walker LR, Bardgett RD (2004) Ecosystem properties and forest decline in contrasting long-term chronosequences. Science 305:509–513

Wardle DA, Zackrisson O, Nilsson MC (1998) The charcoal effect in Boreal forests: mechanisms and ecological consequences. Oecologia 115:419–426

Watt AS (1947) Pattern and process in the plant community. J Ecol 35:1–22

Wilkinson DM (1999) The disturbing history of intermediate disturbance. Oikos 84:145–147

Wirth C, Messier C, Bergeron Y, Frank D, Fankhänel A (2009) Old-growth forest definitions: a pragmatic view. In: Wirth C et al (eds) Old-Growth Forests. Springer, Berlin, pp 11–33

Yeboah D, Chen HY (2016) Diversity–disturbance relationship in forest landscapes. Landscape Ecol 31:981–987

Yeeles P, Lach L, Hobbs RJ, Wees M, Didham RK (2017) Woody plant richness does not influence invertebrate community reassembly trajectories in a tree diversity experiment. Ecology 98:500–511

Zhang K, Lin S, Ji Y, Yang C, Wang X, Yang C et al (2016) Plant diversity accurately predicts insect diversity in two tropical landscapes. Mol Ecol 25:4407–4419

Chapter 9
Foliage Feeders

Joseph Elkinton and Artemis Roehrig

9.1 Introduction

One of the most significant categories of insects that cause damage to trees are the defoliators. While many orders of insects feed on tree foliage, in this chapter we will focus on Lepidoptera, as there are so many Lepidopter larvae (caterpillars) that are known for their extensive tree damage. In this chapter we review the impact of foliage feeders on forest trees and stand composition, and the ways in which densities of these species or the defoliation they cause are monitored. We do not cover insects attacking ornamental trees in the landscape, nor do we cover insects feeding exclusively on foliage tips or buds. The species we include live and feed externally on the leaves and remove or consume leaf tissue that may or may not include leaf veins. Other species, called leaf miners, live and feed as larvae between the upper and lower surface of the leaf and produce characteristic patterns of leaf damage. Most of those species are considered pests of ornamental trees and are not included in this chapter. We provide more detail on two key species as case studies: winter moth, *Operophtera brumata* L, and spongy moth, *Lymantria dispar* L. These species are two of the most widely studied of all foliage-feeding insects attacking forest trees. Treatment of other important species such as spruce budworm, *Choristoneura fumiferana,* would produce a chapter too long for the current volume. That species, and others like it, are included in a table (Table 9.1) of the world's most forest-damaging Lepidoptera and Hymenoptera, along with key references that provide access to the most recent and important literature.

J. Elkinton (✉) · A. Roehrig
University of Massachusetts, Amherst, MA, USA
e-mail: elkinton@umass.edu

J. D. Allison et al. (eds.), *Forest Entomology and Pathology*,
https://doi.org/10.1007/978-3-031-11553-0_9

9.2 Effects of Defoliation on Forest Trees

The general public often views defoliation in terms of aesthetics and potential economic effects. Beyond simply affecting the growth and life of the defoliated trees, defoliation has many indirect effects that have implications for future defoliator population dynamics and forest nutrient cycling, in turn affecting overall forest composition.

Defoliation that removes some or all of the leaf canopy of trees has a large impact on the ability of trees to produce carbohydrates, and most studies have shown foliage loss to be directly proportional to reductions in tree growth. While defoliation can cause tree mortality, this often occurs indirectly, as defoliation increases the susceptibility of trees to secondary insects and disease, which then are the ultimate cause of tree mortality (Kulman 1971). Outbreaks of defoliators are major events in forests worldwide and may produce landscape-wide patterns of tree mortality and result in major changes in stand tree species composition.

Even if there is no current folivore outbreak, trees may still be suffering the effects of past defoliation events. For instance, a study done in Cerro Castillo National Park by Piper, Gundale and Fajardo (2015) on *Nothofagus pumilio*, a South American deciduous tree, found that natural defoliation by *Ormiscodes amphimone* (Saturniidae) did not cause tree mortality. However, defoliated trees showed significantly stunted growth in comparison to non-defoliated trees. Contrary to previous assumptions, this growth limitation could not be explained by limitations in C and N availability. Defoliation by the larvae of the invasive winter moth (*Operophtera brumata* L.) has been shown to cause a significant reduction in radial growth and latewood production of *Quercus* trees in the same year as defoliation, as well as a reduction in earlywood production the subsequent year (Simmons et al. 2014).

Many trees produce defensive compounds in their leaves, such as phenolics or tannins, to defend themselves against free-feeding insects (Feeny 1970). On the other hand, many foliage-feeding insects are well adapted to cope with these compounds in their diet. There exists a very large literature dealing with the mode of action of tannin or phenolic compounds on insect performance, and whether or not trees respond to defoliation by producing more defensive compounds (Salminen and Karonen 2011).

When it comes to tree resistance to defoliators, there are two main types of resistance: constitutive (always present) and induced (as the result of defoliation). These effects may be either direct, wherein the plant produces either mechanical or molecular herbivore deterrents, or indirect, whereby they put up defenses, chemical or otherwise, that attract defoliator predators or parasitoids (War et al. 2012).

An important molecular mechanism plants use for defoliation resistance is the production of phenolic compounds, such as tannins, which include hydrolysable tannins, proanthocyanidins, and phlorotannins. Different kinds of tannins have greater impacts on different types of herbivores. In insects, different parts of the digestive system have different pH levels, and, as a result, differently structured tannins will react and metabolize differently in different sections of the gut, as they are hydrolyzed or oxidized. Rather than tannins themselves, it is possible that tannin metabolites are what actually affect herbivores (Salminen and Karonen 2011).

Tannins may serve as an important factor in tree constitutive resistance. Although some herbivore species have adapted to feed on certain tannins, for non-adapted defoliators they can serve as a feeding deterrent. Tannins may also be important for induced defenses, as multiple studies have shown tannin production increases with insect damage. However, there are many other factors at play, and tannin concentration is affected by things such as environmental stress. There are so many different specific types of tannins produced by plants and so many potential interactions that most current studies are correlative rather than causative (Barbehenn and Constabel 2011). For instance, there have been disparate findings on the relationship between tannin content and amount of defoliation. A recent study on spongy moth defoliation on *Quercus ilex* found no relationship (Solla et al. 2016).

Haukioja (1991) reviewed studies on tree-induced resistance to insect defoliation. While in general insect growth rate declined with decreased food quality, there were very mixed results about the effect of induced responses. Some studies showed that foliage damage induced changes in present and future leaves that were detrimental to insects, while others showed no effect of induced resistance. To complicate matters, other studies mentioned in the review showed improved performance of insects that fed on defoliated trees. Haukioja's review made an important distinction between rapid and delayed induced resistance. The latter refers to changes in foliage chemistry that persist one or more years beyond the defoliation event, rather than those immediately following the defoliation in the same year. Only delayed induced resistance can cause the delayed density-dependent responses (see Chapter 7) that might cause forest insects to exhibit population cycles. Such effects have been proposed for autumnal moth (Haukioja 1991) and for larch budmoth (see Chapter 7; Baltensweiler and Fischlin 1988). In many cases it is not clear whether the changes in foliage chemistry involve defensive compounds or delayed effects on foliage that affect their nutrient quality.

White spruce (*Picea glauca*) trees resistant to defoliation by spruce budworm had different phenolic compounds present than non-resistant trees. Those phenolic compounds present in resistant trees were found to reduce fitness of spruce budworms (Delvas et al. 2011). However, as shown in a recent study, spruce budworm (*Choristoneura fumiferana* (Clem.)) that fed on resistant white spruce trees (*Picea glauca* (Moench) Voss) had greater fitness than those that fed on susceptible trees (Quezada-Garcia et al. 2015). Hodar et al. (2015) found that the chemical defenses in three species of pine were constitutive rather than induced. Several important herbivores are undeterred by these defenses, such as the pine processionary moth (*Thaumetopoea pityocampa*). Ultimately, as summarized by War et al. (2012), there is still much work needed to understand the biochemical response of induced resistance and how it is invoked by insect feeding.

9.3   Monitoring for Defoliation and Changes in Defoliator Population Densities

Defoliation has typically been mapped by aerial survey. For example, aerial maps of spruce budworm outbreaks have long been produced by the Canadian Forest Service (Fig. 9.1a). Annual defoliation maps of spongy moth in the eastern United States have been analyzed extensively to detect multi-annual cycles and spatial synchrony of spongy moth populations (Liebhold et al. 2004; Johnson et al. 2006b; Bjørnstad et al. 2008, 2010; Haynes et al. 2013, 2018a). Elkinton et al. (2014) used aerial survey maps of winter moth defoliation to estimate rates of spread of winter moth in the northeastern United States. More recently, imagery obtained from satellites or other forms of remote sensing has been used to map and analyze the expansion of defoliator outbreaks. Pasquarella et al. (2018) used Landsat imagery to portray the extent, severity and spread of spongy moth outbreak in the northeastern United States (Fig. 9.1b). Jepsen et al. (2009a) analyzed MODIS satellite data to relate winter moth defoliation to the timing of spring bud-burst in northern Fennoscandia. See reviews by Hall et al. (2006) and Chapter 19 for more detailed discussion of this topic.

Pheromone traps have often been used to map the spread of invasive species on the landscape. For example, Elkinton et al. (2010) used pheromone-baited traps to monitor the extent of the new invasion of winter moth in the northeastern United States (Fig. 9.2a) and its subsequent spatial spread (Elkinton et al. 2014). By far the most extensive use of pheromone traps anywhere in the world has been the Slow the Spread Program (Tobin and Blackburn 2007) to monitor the spread of spongy moth (Fig. 9.2b). Each year more than 100,000 traps are deployed along this invasion front. Pheromone traps are less frequently used to monitor changes in density of outbreak species in regions where they are native or widely established because such traps often fill to capacity even in low-density populations. Therefore, it is more

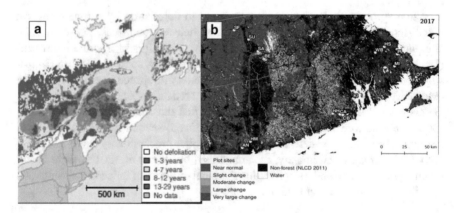

Fig. 9.1 (a) Years of defoliation by spruce budworm in eastern Canada 1954–1988 mapped by aerial survey (Williams and Birdsey 2003); (b) Defoliation by spongy moth mapped from Landsat satellite images (Pasquarella et al. 2018; Elkinton et al. 2019)

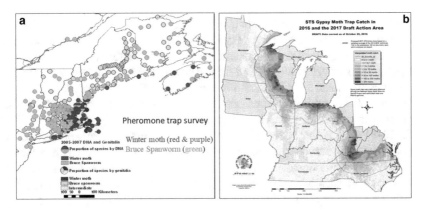

Fig. 9.2 (a) Distribution of winter moth and Bruce spanworm in pheromone-baited traps in northeastern North America in 2005–2007. Winter moths use the same pheromone compound as the native species Bruce spanworm, *Operophtera bruceata*. Identification of moths is based on male genitalia and the DNA sequence of the COI mitochondrial gene (Elkinton et al. 2010); (b) Isopleths of numbers of spongy moth males per trap captured in more than 100,000 pheromone-baited traps in 2019 from Wisconsin to North Carolina (US Forest Service Slow the Spread Annual Report 2019)

common to use sampling of other life stages, such as egg mass counts for spongy moth, to measure changes in population density. See Chapter 19 for a more thorough discussion of this topic.

9.4 Case Study 1: Winter Moth

9.4.1 Biology and Host Range

The winter moth, *Operophtera brumata* L, is a geometrid species that is native to Europe, where it is one of the most common Lepidoptera feeding on a wide range of tree species. These include oaks (*Quercus*), maples (*Acer*), birches (*Betula*) and many others (Wint 1983). It is an occasional orchard pest, because it performs extremely well on apple (*Malus*). It is also especially damaging to blueberry (*Vaccinium*) crops, because the larvae feed inside the buds, where they are inaccessible to most pesticides and destroy developing berries before the buds open. In Europe, outbreaks of winter moth have occurred on Sitka spruce (*Picea sitchensis*) (Stoakley 1985; Watt and Mcfarlane 1991), on heather (*Calluna vulgaris*) in Scotland (Kerslake et al. 1996), and on mountain birch (*Betula pubescens czereapanovii*) in Fennoscandia (Jepsen et al. 2008).

Winter moth gets its name from the fact that adults typically emerge in November or December. The females attract males with a pheromone (Roelofs et al. 1982) and, after mating, lay eggs singly on the bark of host trees and overwinter in this stage. Winter moth larvae typically hatch at or before budbreak of their host trees and then bore into the expanding buds, so much of the damage occurs before leaf expansion. Classic work by Feeny (1970) proposed that winter moth is one of a suite of early spring-feeding Lepidoptera larvae that are relatively intolerant to accumulated tannins in oak foliage. Even though there may be many larvae per bud in outbreak populations, defoliation of oak and maple in New England, at least, rarely approaches 100%, presumably because the larvae finish feeding and pupate before defoliation is complete. Given that pupation occurs before the end of May, Pepi et al. (2016) showed that winter moth larvae disperse from partially defoliated oak leaves, possibly in response to tannins or other compounds induced by defoliation. Although the typical damage caused by winter moth results in only partially defoliated leaves, this can cause lasting damage to the tree, especially when defoliation persists year after year, as it did in Nova Scotia in the 1950s (Embree 1965, 1967) and Massachusetts after 2004 (Elkinton et al. 2014). Simmons et al. (2014) showed that defoliation by winter moth caused significant decline in tree growth in red oak (*Quercus rubra* L.) in Massachusetts, as measured by growth rings in increment cores of tree stems. Embree (1967) reported that repeated defoliation by winter moth resulted in as much as 40% tree mortality in red oak stands in Nova Scotia.

9.4.2 Geographical Range

Winter moth occurs in every European country, as well as Iran and Tunisia. Early reports included the Russian Far East and Japan, but the Japanese population was redescribed as *Operophtera brunnea* (Nakajima 1991). Recent collections from the Russian Far East suggest that those populations also are closely related to *O. brunnea* (Andersen et al. unpublished). Winter moth has been introduced to four distinct locations in North America: Nova Scotia in the 1930s (Hawboldt and Cuming 1950), Oregon in the 1950s (Kimberling et al. 1986), the region around Vancouver, British Columbia in the 1970s (Gillespie et al. 1978) and in the northeastern United States in the 1990s (Elkinton et al. 2010). Recent studies of winter moth DNA (microsatellites) from these populations by Andersen et al. (2021a) indicate that all four populations represent separate introductions from different European sources. The same techniques show that European populations of winter moth arose from distinct eastern and western forested glacial refugia that existed at the height of the last Ice Age 20,000 years ago (Andersen et al. 2017). Molecular analyses also have shown that in North America winter moth readily hybridizes with a native congener Bruce spanworm, *O. bruceata*, (Elkinton et al. 2010; Havill et al. 2017), that hybridization occurs in all regions where winter moth is known to have invaded (Andersen et al. 2019a), and that, at least in the northeastern United States, the hybrid zone appears to be stable in nature, existing under a tension hybrid zone model (Andersen et al. 2022).

9.4.3 Early Ecological Studies

Winter moth is one of the most famous of all forest insects, due in large part to the classic work by Varley and Gradwell (1960, 1963, 1968, 1970) and Varley et al. (1973), who collected annual life table data on this species on four oak trees near Oxford University in England during the 1950s and 1960s. They introduced important methodology for collecting annual data on density and mortality of different life stages and how to analyze the data to detect the presence of density-dependent factors regulating density and the causes of year-to-year changes in density. Based on these studies, they concluded that winter moth densities were typically regulated at low density by a community of predators that preyed upon winter moth pupae in the leaf litter beneath the infested trees. Subsequent research suggested that pupal predation was caused mainly by staphylinid and carabid beetles (Frank 1967). Other sources of mortality, including overwintering mortality and larval mortality combined, were not density-dependent, but experienced large year-to-year variation in impact and were thus responsible for the observed changes in population density. Varley and Gradwell used the term 'key factor' to describe such mortality factors.

Varley and Gradwell (1960, 1968) believed that the main cause of overwintering mortality was the periodic failure of winter moth hatch to adequately synchronize with budburst of their principal host trees, mainly oaks (*Quercus*). These ideas have been supported by research in North America (Embree 1965) and by Jepsen et al. (2009b), who studied outbreaks of winter moth in northern Fennoscandia.

9.4.4 Pathogens

Like most outbreak species of forest Lepidoptera, winter moth larvae are killed by a nuclear polyhedrosis virus (NPV) (Wigley 1976; Raymond et al. 2002; Raymond and Hails 2007). This virus has been recovered from winter moth in North America (Burand et al. 2011; Broadley et al. 2017), but it rarely, if ever, causes a major epizootic resulting in the collapse of outbreak populations. The virus is thus different from those that occur in other forest Lepidoptera such as spongy moth, *Lymantria dispar*, (Campbell and Podgwaite 1971) or forest tent caterpillar, *Malacosoma disstria* (Cooper et al. 2003), whose outbreaks are typically terminated by these agents. Broadley et al. (2017) showed that the NPV of winter moth was closely related to, but distinct from, an NPV recovered from Bruce spanworm (*O. bruceata*), the North American congener of winter moth. These two NPV's were not cross-infective in the other species, discounting an earlier suggestion (Murdoch et al. 1985) that declines of winter moth in Nova Scotia in the 1950s might have been partially caused by infection of winter moth populations with viruses derived from Bruce spanworm.

Microsporidia are another pathogen that have been recovered from winter moth in Europe (Canning 1960; Canning et al. 1983) and were recorded by Varley et al. (1973). Broadley (2018) showed that microsporidia in North America (Donahue et al.

2019) were a major source of mortality in the rare outbreak populations of the North American congener of winter moth, Bruce spanworm, *O. bruceata*. They have not been recovered from winter moth in North America (Broadley 2018).

9.4.5   Biological Control in North America

Winter moth invaded Nova Scotia in Canada sometime before 1930 and soon caused widespread defoliation of oak forests in that region (Hawboldt and Cuming 1950). Beginning in 1954, Embree and colleagues undertook what would become one of the most famous biological control successes in forest entomology of all time (Embree 1966; Murdoch et al. 1985; Roland and Embree 1995; Kenis et al. 2017). Embree and his colleagues introduced several parasitoid species from Europe, two of which, the tachinid *Cyzenis albicans* and the ichneumonid *Agrypon flaveolatum*, began to cause high levels of mortality in winter moth populations after 4–5 years (Fig. 9.3a). By 1962, winter moth densities had declined to non-pest status, where they have remained ever since (Fig. 9.3a). Hassell (1980) presented a simulation model of *C. albicans* impact on winter moth that appears to explain why in Nova Scotia it was effective at suppressing winter moth populations, whereas it seemed to play a minor role in the population studied by Varley and Gradwell in England. The model was built on his earlier life table studies of *C. albicans* in England (Hassell 1968, 1969a, 1969b).

Similar biological control efforts were undertaken in the 1970s following an introduction of winter moth to Southwest British Columbia in Canada. Winter moth densities there soon declined following the onset of high levels of parasitism, mainly by the tachinid *C. albicans* (Roland 1986; Roland and Embree 1995). Yet another successful biological control effort was initiated by Elkinton et al. (2018, 2021)

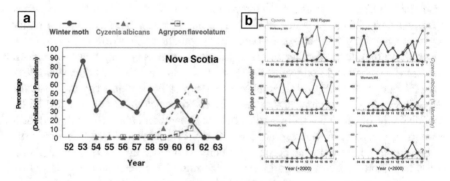

Fig. 9.3 (a) Defoliation by winter moth and percent parasitism by *C. albicans* and *Agrypon flaveolatum* in Nova Scotia in the 1950s following parasitoid release in 1954 (adapted from Embree 1965); (b) Density of winter moth pupae and percent parasitism by *C. albicans* at six widely spaced release sites in Massachusetts (Elkinton et al. 2018)

(Fig. 9.3b) against an outbreak of winter moth that appeared in the northeastern United States in the late 1990s (Fig. 9.1a) Elkinton et al. (2010). This effort was based solely on the release of the tachinid *C. albicans,* because *Agrypon flaveolatum,* the other parasitoid released in Canada, was deemed too much of a generalist and also of uncertain taxonomy. Over 14 years Elkinton and his colleagues established the fly at 41 release sites in New England and observed a substantial decline in winter moth densities (Fig. 9.3b) (Elkinton et al. 2018, 2021).

9.4.6   Population Ecology in North America

Roland (1990b) analyzed the decline of winter moth densities associated with the onset of parasitism by *C. albicans* in Nova Scotia and in British Columbia. He concluded that the decline was caused mainly by predation rather than parasitism and that the presence of *C. albicans* enhanced predation rates on winter moth pupae. He proposed several possible mechanisms for this phenomenon, which included reductions of winter moth densities to levels below which predators were saturated and caused inversely density-dependent mortality, or that parasitized pupae provided a food resource available in the spring months following the emergence of un-parasitized pupae in November and December. He further provided evidence that pupal predators caused density-dependent mortality that regulated the low-density populations of winter moth following the population decline induced by the presence of *C. albicans* (Roland 1994, 1995). Broadley et al. (2022) analyzed data from the recent biological control success in the northeast United States and confirmed Roland's findings that low-density populations of winter moth following the onset of high parasitism by *C. albicans* were regulated by density-dependent predation by a suite of pupal predators. Broadley et al. (2019) also discovered a parasitoid, *Pimpla aequalis* that consisted of two cryptic species causing density-dependent mortality of winter moth pupae. Broadley et al. (2022) found no evidence in support of Roland's findings that the presence of *C. albicans* enhanced predation on winter moth pupae.

Other research on winter moth population ecology in North America includes the life table studies of outbreak populations of winter moth in stands of red oak, *Qurecus rubra,* in Nova Scotia prior to the establishment of parasitoids (Embree 1965). Embree found that the main cause of population change in outbreak populations was synchrony of winter moth hatch with budburst, confirming similar conclusions reached by Varley et al. (1973) in England. In years where spring occurred phenologically early, hatch was well synchronized with budburst, yielding high larval survival. In contrast, in years where springtime warming came later, synchrony was poor and larval survival low. Embree's research was followed up by MacPhee et al. (1988), who studied the lower-density populations of winter moth that existed on apple trees in Nova Scotia over the decade that followed the population decline induced by *C. albicans* in the early 1960s. He found that both *C. albicans* and *A. flaveolatum* caused parasitism in the range of 10 to 20%, far lower than the values observed by Embree in high-density populations in the early 1960s. These findings reinforce the idea that *C. albicans* has its biggest impact on high-density populations of winter moth. A principal reason is that this species is attracted to defoliated trees and oviposits tiny (micro-type) eggs on partially eaten leaves (Hassell 1968, 1980; Roland 1990a; Roland et al. 1995). Winter moth becomes parasitized by *C. albicans* only when the larva consumes the egg. These eggs then hatch, and the larval fly migrates to the salivary glands of the winter moth larva, where it stays until the moth stops feeding and drops to the ground to pupate. After this, the larval fly completes development, kills the winter moth pupa and forms a puparium inside the pupal cadaver.

9.4.7   Recent European Studies

In recent years, European research has focused mainly on the outbreaks of winter moth in northern Fennoscandia (Tenow et al. 2007; Jepsen et al. 2008). Winter moth outbreaks occur approximately every 10 years in the mountain birch (*Betula pubescens czereapanovii*) forests of that region in synchrony with, but lagging 2–3 years behind, those of another well-studied geometrid, the autumnal moth, *Epirrita autumnata* (Tenow et al. 2007). Jepsen et al. (2008) showed that outbreak populations of winter moth in this region were moving to higher altitudes in response to climate change (Fig. 9.4a) and were moving into forests formerly occupied only by autumnal moth. Consecutive outbreaks of both species are threatening widespread mortality of the mountain birch forests. Vindstad et al. (2022) documented the more recent spread of winter moth into willow (*Salix*) stands in the subarctic tundra of northeastern Fennoscandia.

Jepsen et al. (2009a, 2009b) used multitemporal remotely-sensed data of leaf-out and defoliation to show that favorable synchrony of winter moth hatch with budbreak fueled the synchronous outbreak of winter moths during the increase phase of the population cycle. The spatial synchrony was reduced during the peak and declining phase of the outbreak. Analyses by Tenow et al. (2013) indicated that waves of defoliation by winter moth spread from east to west across Europe approximately every 10 years. However, subsequent analyses challenged that conclusion (Jepsen et al. 2016), and no underlying mechanism for such a phenomenon has been proposed, especially since weather systems at that latitude move from west to east and winter moth females are incapable of flight.

Vindstad et al. (2013) reported the complex of larval parasitoids attacking winter moth and autumnal moth in Norway and compared it to the complex from other sites in Western Europe. These parasitoids included a total of 18 species, including five ichneumonids, three braconids, nine tachinids and one eulophid. The majority of these species occur in winter moth in northern Fennoscandia, with the exception of the tachinids, such as *C. albicans,* which do not occur there, despite being very common elsewhere (Vindstad et al. 2013). Recent studies by Schott et al. (2010) of winter moth mortality caused by these other larval parasitoid species often showed levels of mortality exceeding 50% in northern Norway. However, they do not appear

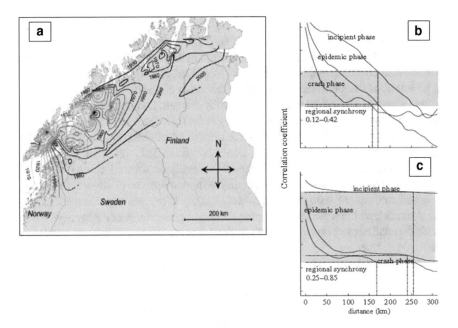

Fig. 9.4 (a) Contours connecting years of first outbreaks of winter moth in northern Fennoscandia abetted by climate change (Jepsen et al. 2008); (b) Spatial synchrony of winter moth outbreaks and; (c) spring bud-burst phenology in mountain birch forests in the incipient, epidemic and crash phases of the winter moth outbreak (Jepsen et al. 2009b)

to be responsible for the decline of outbreak populations. In contrast, Klemola et al. (2010) concluded from manipulative experiments that larval parasitoids are responsible for the decline of outbreak populations of the autumnal moth in northern Finland. Meanwhile, Schott et al. (2013) reported that outbreaks of winter moth in northern Norway are not caused by the release of winter moth populations from regulation at low density by invertebrate predation. It is evident that, despite all this research, the role of natural enemies in the dynamics of winter moth in northern Fennoscandia remains unresolved.

Other recent research has used modern molecular techniques to analyze the expansion of the winter moth's range across Europe and the European origins of winter moth in North America. Gwiazdowski et al. (2013) sequenced the CO1 barcoding gene in a world-wide study of winter moth males collected using pheromone traps and found that nearly all the sampled individuals in the four North American populations shared a single haplotype. However, this haplotype was also found in winter moths collected from 10 of the 11 sampled European countries. This study was thus unable to determine the European origins of winter moth in North America. The lack of genetic diversity revealed by Gwiazdowski et al. (2013) was surprising given the fact that female winter moths are flightless, and thus strong biogeographic patterns might be expected. In a follow-up study, Andersen et al. (2017) examined gene regions called "microsatellites" that have greater sensitivity than the CO1 barcode gene for examining the genetic structure of populations. They showed that one possible explanation for the lack of genetic diversity in Europe found by Gwiazdowski et al. (2013) is that winter moth populations in central and western Europe (Fig. 9.5) represent a blend of populations from eastern Europe and the Iberian peninsula. They argue that this pattern arose as a result of widely separated forest refugia on the Iberian peninsula and in southeastern Europe during the last glacial maximum (Fig. 9.5).

Subsequent analyses of moths collected in the Mediterranean region have identified two additional glacial refugia: one in southern Italy and another in North Africa (Andersen et al. 2019b). A follow-up analysis showed that winter moth invaded northern Scandinavia via the United Kingdom instead of alternate routes via Denmark or eastern Europe (Andersen et al. 2021b). More recently, these microsatellite markers have been used to reexamine the geographic origins of the invasive winter moth populations in North America (Andersen et al. 2021a). These analyses show

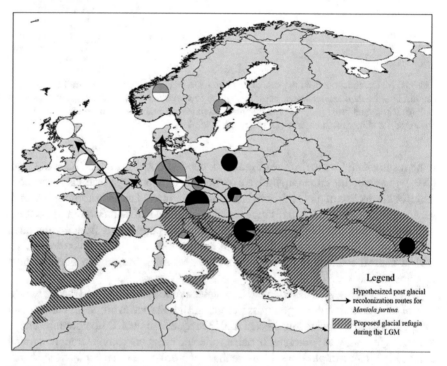

Fig. 9.5 Genetic diversity of winter moth in Europe with populations that utilized glacial refugia of the forests in southern Europe on the Iberian peninsula at the height of the last glacial maximum about 20 thousand years ago shown in white, eastern Europe shown in black, and populations that are admixed shown in grey. The populations into northern Europe represent a merger of these two populations following the retreat of the ice sheet (adapted from Andersen et al. 2017). The hash-marked lines represent the likely locations of glacial refugia during the last glacial maximum, and the arrows represent the likely post-glacial recolonization route of winter moth similar to that of another European Lepidoptera, the meadow brown, *Maniola jurtina* (adapted from Schmitt 2007)

that each one of the four North American populations of winter moth (Nova Scotia, New England, British Columbia and Oregon) are all quite distinct from one another and probably represent separate introductions (Andersen et al. 2021a). In addition, the populations from Nova Scotia, British Columbia, and New England all appear to be introduced from western Europe (likely France or Germany), while the population in Oregon appears to be introduced from somewhere in the British Isles.

Other European studies have focused on the effects of climate change on the timing of winter moth hatch in spring. Winter moth larvae have been hatching earlier and earlier as spring temperatures have become warmer over the last several decades. Although winter moth is rarely a significant defoliator in central Europe, it is an important source of food for nesting birds in the spring. Migratory birds have timed their arrival based on solar cues and in recent years have arrived too late after winter moth larvae have finished feeding and dropped to the forest floor to pupate (Visser et al. 1998). Visser and Holleman (2001) showed that warmer springs have caused winter moths to desynchronize with budbreak of oaks (*Quercus* spp.), their principal host tree, and shift to other tree species that break bud earlier. They also showed that egg hatch in spring is influenced by factors more complex than predicted by growing-degree-day models that are widely used to predict hatch of most insects in the spring. Hatch times in their model were also influenced by the number of winter days below freezing. Hibbard and Elkinton (2015) applied this model with some success to egg hatch data in North America. Salis et al. (2016) proposed a revised model, wherein developmental rate of winter moth eggs as a function of temperature increased with egg age or egg development (see also Gray, 2018). Elkinton is currently attempting to fit versions of these models for egg hatch and bud-break to data from North America. Van Dis et al. (2021) have provided detailed information on the effects of temperature on embryonic development of winter moth eggs.

9.5 Case Study 2: Spongy Moth

9.5.1 *Biology*

Spongy moth, *Lymantria dispar* L. (formerly called gypsy moth) is another major defoliator, mainly of deciduous trees, that is native to both Europe and Asia. Three subspecies have been described (Pogue and Schaefer 2007): European spongy moth (*Lymantria dispar dispar*), Asian spongy moth (*Lymantria dispar asiatica*), and Japanese spongy moth (*Lymantria dispar japonica*). Although spongy moth females have wings and the Asian subspecies tend to be capable of flight, most populations of the European subspecies *L. dispar dispar* do not fly (Keena et al. 2008). Spongy moth females mate in mid-summer and lay egg masses that contain from 100–1000 eggs on the stems of trees, rocks or other objects and cover them with their tawny brown body hairs. Larvae hatch in spring coincident with host tree budburst and develop through five (males) or six (females) larval instars until late June or early July, depending on latitude. Late-instar larvae in low-density populations seek daytime resting locations under bark flaps or on the forest floor, presumably as a defense against day active predators and parasitoids (Lance et al. 1987). Pupation typically occurs in these resting locations. Adults emerge in mid-summer. There is one generation per year.

9.5.2 Introduction to North America

European spongy moths (*L. dispar dispar*) were introduced into North America in 1868 or 1869 by Leopold Trouvelot for the purpose of various experiments. The insect escaped from his home in a suburb of Boston, Massachusetts and began to spread across the landscape. Trouvelot tried to notify local officials of the potential problem resulting from his accident, but his efforts were ignored until widespread defoliation in his neighborhood became apparent in the late 1880s. The Massachusetts state legislature allocated funds to eradicate spongy moth by mechanical removal of egg masses and applications of primitive pesticides such as lead arsenate (Spear 2005). This effort failed and spongy moth continued to spread, albeit quite slowly, since the females of the European strain of the species do not fly. Indeed, 140 years later, spongy moths are still spreading south and west in North America as shown in Fig. 9.2a and only occupy about 1/3 of their potential range (Figs. 9.6 and 9.7).

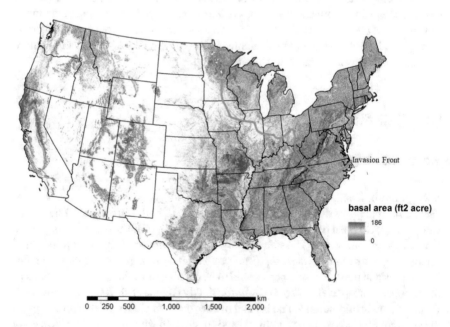

Fig. 9.6 Forest types susceptible to spongy moth invasion. Orange represents highly susceptible forest, green low susceptibility (Morin et al. 2005). Blue line indicates the current invasion front of spongy moth in N. America (see Fig. 9.2b)

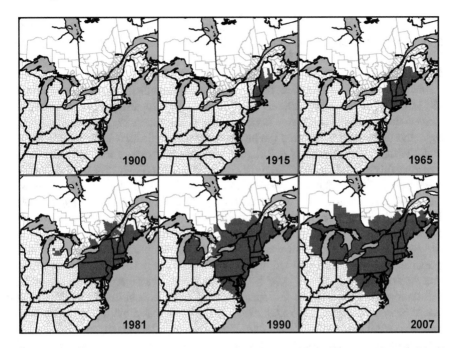

Fig. 9.7   Spread of spongy moths in northeastern North America after 1900 (Figure from Leibhold et al. 2007)

9.5.3   Host Preferences

Like winter moths, spongy moths feed on a wide range of host tree species, but perform best on oaks (*Quercus* spp), aspen (*Populus*), and birches (*Betula*) (Liebhold et al. 1995; Davidson et al. 1999). They will feed on many conifers and indeed on most tree species, especially if preferred hosts are unavailable or already defoliated. A handful of species are avoided altogether, even in stands that are otherwise completely defoliated. These species include ash (*Fraxinus spp*), silver maple (*Acer saccharinum*) and tulip poplar (*Liriodendron tulipifera*).

9.5.4   Impact on Forests and Trees

Defoliation is more frequent in forest stands that are dominated by tree species preferred by spongy moths, as described above, than in stands dominated by other tree species. In eastern North America, oaks (*Quercus*) dominate the forests in southern New England, the mid-Atlantic states and the Midwest. Aspen (*Populus*) dominated forests are often defoliated in the region around the Great Lakes (Fig. 9.6). These forests are most frequently defoliated by spongy moth and experience the greatest tree mortality (Campbell and Sloan 1977; Davidson et al. 1999).

Most hardwood trees defoliated > 50% by spongy moths will re-foliate in midsummer. However, those that fail to re-foliate at that time, or fail to re-foliate the following spring, will be killed, due to insufficient carbohydrate reserves (Kulman 1971). Defoliated trees become susceptible to attack by secondary organisms, such as the two lined chestnut borer, *Agrilus bilineatus,* or the shoestring fungus, *Armillaria* spp., and these agents are often the main causes of tree death (Campbell and Sloan 1977; Wargo 1977). Repeated defoliations in consecutive years can lead to levels of tree mortality exceeding 50% (Kegg 1973; Campbell and Sloan 1977). Other studies show less mortality following defoliation (Brown et al. 1979; Gansner et al. 1993). Campbell and Sloan (1977) analyzed the impact of spongy moth on stands from 1911 to 1931 in New England and reported that defoliation occurred most frequently on oak-dominated stands and that oaks were the most likely to die. Dominant trees survived better than ones that were subdominant or suppressed. Non-favored host trees, such as white pine and red maple, were more likely to die after one defoliation than oak trees. Morin and Liebhold (2016) analyzed the impact of spongy moth defoliation on changes in the tree species composition data collected by the USDA Forest Service between 1975 and 2010. They found that most of the stands with repeated defoliation in the northeastern USA were oak-dominated, and the effect of defoliation was to hasten the process of replacement of overstory oaks with other species such as maple (*Acer*), which are less preferred by spongy moth. Even though the volume or basal area of oak was increasing across this region due to tree growth, mortality of the younger age classes of oaks contributed to the overall decline of oaks and replacement by other species.

9.5.5   Spread of Spongy Moth

The enormous spatial detail evident in the spongy moth pheromone trap catch data (Fig. 9.2a) across the landscape, and the long time period over which spread has been monitored, have allowed investigators to study the rate of spread of spongy moths and make important contributions to the theory of spread of invasive organisms. Liebhold et al. (1992) compared historical rates of spongy moth spread (1900–1989) with predictions made using the spread model of Skellam (1951). The Skellam model consists of two components: exponential population growth defined by the parameter 'r' and diffusion analogous to molecular diffusion defined by the parameter D. The model predicts that the rate of spread V of an invasion front is constant: $V = 2\sqrt{rD}$. Liebhold et al. (1992) estimated both parameters from earlier studies of spongy moth population growth and diffusion based on dispersal of first-instar larvae that spin down on threads from tree canopies and are blown in the wind. Experimental studies of that process (Mason and McManus 1981) suggest that most such larvae spread only a few hundred meters, but a few of them spread several kilometers. The Skellam model based on these parameters predicted that spongy moth dispersal would be about 2 km/year. The spongy moth spread prior to 1966 varied between 2 and 10 km/year compared to 20.78 km/year after 1996. Liebhold et al. (1992) concluded that the discrepancy between predicted and observed spread was due to accidental human movement of spongy moth life stages which form isolated populations ahead of the advancing population front and thereby accelerate spread.

Analyses of spongy moth spread were greatly enhanced by implementation of regional grids of pheromone traps (Fig. 9.2a, 9.8a). Analyses of such data from the central Appalachians (Sharov et al. 1995, 1996, 1997) indicated a rate of spread that varied yearly and ranged from 17 to 30 km/year. These data show that clumps of small populations of spongy moths arise many kilometers in front of the infested zone (Figs. 9.2a, 9.8a), and their growth and coalescence contribute significantly to the rate of spread. These data suggest that spread of spongy moth is an excellent example of stratified dispersal (Hengeveld 1989), consisting of a short-range process governed by larval dispersal and a longer-range process governed by human transport of spongy moth egg masses. The latter process has long been understood to be a central feature of the spongy moth system. Spongy moths lay the overwintering egg masses in midsummer on backyard objects, such as lawn furniture, that are readily transported in succeeding months elsewhere in the United States. As a result, new infestations arise many kilometers from the generally infested area or indeed anywhere else in North America. Models of stratified dispersal (Shigesada and Kawasaki 1997) were fit to the spongy moth system (Sharov and Liebhold 1998a). These analyses form the theoretical basis of the spongy moth Slow the Spread Program (Sharov et al. 1997, 1998, 2002a; Sharov and Liebhold 1998a, 1998b; Tobin and Blackburn 2007) discussed below. Suppression of these incipient populations, arising ahead of the invasion front, slows the spread.

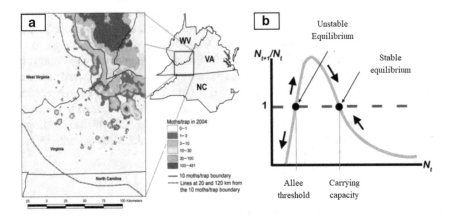

Fig. 9.8 (a) Leading edge of spongy moth infestation arising ahead of the invasion front, resulting in stratified spread and; (b) Allee effect showing population growth as a function of density. Below the horizontal dashed line populations decline; above the line they increase (from Liebhold et al. 2007)

Understanding the survival and expansion of incipient populations thus became a key feature of managing spongy moth. Such populations are governed by Allee effects (Fig. 9.8b), which express the survival or growth of populations as a function of population densities. At the very low densities characteristic of newly founded populations, survival or population growth of many species increases with population density. At higher densities, in virtually all populations survival or growth rates decline to an equilibrium that represents either the carrying capacity, or else a lower-density equilibrium maintained by natural enemies. Allee effects refer to the positive density dependence at lower densities, and they can be weak or strong (Taylor and Hastings 2005). If they are strong, then at very low densities there exists what is called the Allee threshold (Fig. 9.8b). At densities above the threshold, populations steadily increase. When populations are below the threshold, however, densities typically decline to extinction. In other words, the low-density Allee threshold is an unstable equilibrium. There are several possible causes of low-density Allee effects in spongy moth populations, including predation (see below), but probably the most common cause at the very lowest densities characteristic of incipient populations is failure to locate mates. The implication of this is that many incipient populations of spongy moth will decline to extinction on their own accord. Indeed, data suggest that this frequently occurs (Liebhold et al. 2016). Eradication of such populations with pesticides or indeed mating disruption (Sharov et al. 2002b) is entirely feasible because even if the treatment fails to kill all the spongy moths it will surely vastly lower their densities and thus hasten their natural tendency to decline to extinction.

Subsequent analyses of spongy moth spread have shown that the rate of spongy moth spread declines with the strength of Allee effects (Tobin et al. 2007, 2009), which varies in time and space across the landscape. The strength is measured by the intercept of the plot shown in Fig. 9.8b with the vertical axis; it is strongest when the intercept with the vertical axis (below the figure) is most negative. For example, Tobin (2007) reported that there were strong Allee effects and, as a result, slower spread in parts of the Midwest compared to Great Lakes or Appalachian regions.

An exciting recent finding (Tobin et al. 2014) is that spongy moth populations in North Carolina have stopped spreading, and indeed have retreated northward in recent years. Tobin et al. (2014) suggest that in that region spongy moths have exceeded temperature maximums that inhibit optimal growth and further spread to southern states, and the northward retreat may be due to climate change. These findings imply that spongy moths may never occupy southern regions of the Midwest with highly susceptible oak forests (Fig. 9.6).

9.5.6 History of Spongy Moth Control

Efforts to control spongy moth in Massachusetts began in 1890, with a large program funded by the state legislature. The program focused on an attempt to mechanically destroy spongy moth egg masses, which are present on the trunks of trees from August through April each year. In addition, there was a large effort to spray the larvae with pesticides, mainly with lead and copper arsenate. There was little or no appreciation in those days of the environmental danger posed by these toxins. Furthermore, they were largely ineffective and failed to stem the spread of the population.

In 1905, the US Department of Agriculture launched what became the most extensive worldwide effort for biological control of an invasive forest insect ever conducted. Twelve species of parasitoids became established of the 34 species that were released over several decades. Fuester et al. (2014) provide the most recent of several reviews of this effort. These included the egg parasitoid *Ooencyrtus kuvanae* (Howard) [Hymenoptera Encyrtidae]; three tachinid [Diptera] species: *Compsilura concinnata* (Meigen), *Parasetigena silvestris* (Robineau-Desvoidy), and *Blepharipa pratensis* (Meigen); a braconid *Cotesia melanoscelus* (Ratzeburg) and an ichneumonid *Phobocampe disparis* (Viereck) which attack the larval stage of spongy moth. Pupal parasitoids established were two hymenopterans: the chalcid *Brachymeria intermedia* (Ness) (Chalcidae) and the ichneumonid *Pimpla disparis* (Viereck). Of these, *O. kuvanae* and *P. disparis* were introduced from Japan, the other species from Europe. *Compsilura concinnata* was introduced to North America in 1906 and has gained some notoriety because Boettner et al. (2000) showed that it has become the dominant source of mortality on several native species of giant silk moths (Saturniidae) and is probably responsible for the decline of these species since the nineteenth century. On the other hand, Elkinton et al. (2006) showed that the same parasitoid was probably responsible for the extirpation of the invasive brown tail moth, *Euproctis chrysorrhea*, over much of its invasive range in the northeastern United States.

Unfortunately, these parasitoids did not prevent spongy moth outbreaks. Williams et al. (1992) published the only long-term data on parasitism by these species and concluded that none of them regulated spongy moth density. The results of this study confirmed the conclusions drawn by earlier investigators: that parasitoids played a limited or equivocal role in the population dynamics of spongy moth in North America (Campbell 1975; Reardon 1976; Elkinton and Liebhold 1990). In addition to parasitoids, biological control introductions included predatory beetles, such as *Calosoma sychophanta* (Weseloh 1985) and pathogens such as *Entomophaga maimaiga* from Japan (Fuester et al. 2014). That pathogen was initially collected and released in 1910 and 1911 in the Boston area but was not established (Speare and Colley 1912). The recent invasion of spongy moth populations by *E. maimaiga* in North America that began in 1989 (see below) was evidently an accidental or inadvertent introduction (Hajek 2007). *Entomophaga maimaiga* was recently established in Bulgaria from where it has spread to other European countries and has become quite common (Hajek et al. 2020). But with the notable possible exception of *E. maimaiga* after 1989, none of these introductions prevented spongy moth outbreaks.

Following World War II, the pesticide DDT became widely available. It was cheaper and more effective than any previous pesticide. In the succeeding decades, widespread aerial application of DDT was made against spongy moth. Applying pesticide by air allowed application at a landscape level, something that was never feasible or affordable from the ground. Entomologists in those days were convinced that DDT was a new tool that would solve most insect problems. By the 1960s, however, the environmental costs of DDT and related compounds were evident and were popularized by the famous book *Silent Spring* by Rachel Carson. DDT and its breakdown products persist indefinitely in the environment and accumulate in the fatty tissue of many animals. It was particularly damaging to birds, especially those at the end of long food chains, such as eagles and ospreys. DDT and other chlorinated hydrocarbon insecticides were banned in the late 1960s and 1970s. The Environmental Protection Agency was established, and laws were passed to require safety testing of all pesticides. Nevertheless, populations of birds such as eagles and ospreys took many decades to recover, a process that goes on to this day.

Meanwhile, new pesticides were developed and used against spongy moth. In the early 1980s aerial applications of carbaryl were very popular. Carbaryl gave way to diflubenzuron, an insect growth regulator. By the end of the decade the bacterial insecticide *Bacillus thuringiensis* (Bt.) became popular. Its advantage was that it affected only foliage-eating insects, and not the adult stages of their insect natural enemies. Other bacterial insecticides such as spinosad were added to the mix in subsequent decades. Thus, in the modern era, we now have much safer pesticides that affect a more narrow spectrum of target and nontarget insects. In the northeastern states large scale aerial application of pesticides largely ceased after 1990 (Fig. 9.9b), coincident with the arrival of a new fungal pathogen of spongy moth, *E. maimaiga* (see below). It appears likely that the days of aerial application of any pesticides against spongy moth in New England are finished. We now know that the spongy moth outbreaks will subside on their own, and the forests will recover, even if there is significant tree mortality. Even the modern pesticides with a narrow spectrum will kill many nontarget insects and aerial applications are too expensive to justify for the governmental agencies charged with carrying them out. Applications to individual shade trees, however, are another matter. Homeowners place high value on these trees which provide beauty and shade to their yards. If a shade tree dies, it is expensive to remove. Homeowners are thus willing to spend significant funds to protect their trees, and many tree care professionals are available to help them to do that. The small scale of such applications presumably has a limited impact on non-target species at the landscape scale.

The federal effort against spongy moth in recent years has focused on the "Slow the Spread" project (Tobin and Blackburn 2007) (Fig. 9.2a). This involves annually deploying 80,000 to 100,000 traps baited with spongy moth pheromone each year in a grid along a front that extends from Minnesota to North Carolina. The objective of this effort is to identify incipient populations arising ahead of the invasion front that facilitate spread, as described above. Efforts are thus made to suppress them and slow the overall rate of spread of spongy moth. While this effort is expensive, cost–benefit analyses have shown that it is justified (Sharov and Liebhold 1998c). To suppress isolated populations, the program mostly relies on aerial applications of pheromones in small slow-release dispensers such that spongy moth males in treated areas are unable to locate females. Consequently, many females go unmated (Sharov et al. 2002b). This approach is called mating-disruption or the confusion technique (Carde and Minks 1995). It has been widely applied against agricultural pests such as pink bollworm, *Pectinophora gossypiella,* on cotton, but this is one of the only applications that has been widely applied against a forest insect. Another

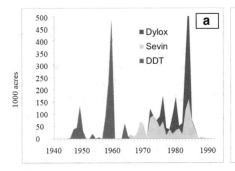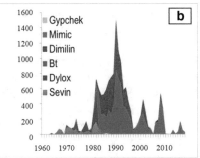

Fig. 9.9 (a) Aerial application from 1945 to 1985 of DDT, carbaryl (Sevin®) and Dylox in the northeastern United States and; (b) other more recently developed pesticides, including LdNPV (Gypchek), Mimic, diflubenzuron (Dimilin®)and Bacillus thuringiensis (Bt) after 1960 (figure courtesy of A. Liebhold)

more widely used eradication technique involves application of microbial pesticides such as *Bacillus thuringiensis (Bt)* (Hajek and Tobin 2010).

A parallel effort is used to detect and eradicate isolated populations of spongy moth that arise far from the invasion front in the western and southern United States, where spongy moth egg masses are transported inadvertently by homeowners arriving from the infested region in the east. Again, the strategy is to annually deploy networks of thousands of traps that are used to detect newly-founded populations. Following detection, these populations are eradicated, mostly using aerial applications of the microbial pesticide *Bacillus thuringiensis.* Of particular concern are populations of Asian spongy moths arriving on ships from East Asia, where the flying female spongy moths are attracted to lights associated with various ports in Asia and thus often deposit egg masses in large numbers on ships in the ports. Asian spongy moths represent a major threat to North America, because, once established, they can spread across the continent very rapidly, and they attack different tree species, including conifers (Baranchikov and Sukachev 1989). Thus, a major effort has been made to locate spongy moth egg masses on cargo and ships arriving from East Asian ports and prohibit imports of contaminated cargo. Recent theoretical studies show that eradication of incipient populations is far more feasible than originally thought (Liebhold et al. 2016).

9.5.7 Population Ecology of Spongy Moth

Robert Campbell, of the US Forest Service, in the 1960s and 1970s, led the first comprehensive research aimed at understanding the population ecology of spongy moth in North America. Campbell and Sloan (1978a) suggested that predation by small mammals, in particular the white-footed mouse, *Peromyscus leucopis,* feeding on the late larval and pupal stages, was the key to maintaining populations at low density in the years between outbreaks. Predation by birds, in contrast, was much less important. Many bird species feed to some extent on spongy moth caterpillars, but many are also deterred by the hairs on the integument.

Elkinton et al. (1996) presented results of research initiated in the 1980s at two sites in Massachusetts that confirmed the importance of small mammal predation on low-density spongy moth populations. They showed that spongy moth populations would rise when populations of white-footed mice declined. Furthermore, they showed that mouse populations fluctuate with the acorn crops, their major overwintering food source. As is true with many tree species, acorn crops vary enormously from year to year. A variety of weather conditions, such as a late spring frost or mid-summer drought, can nearly eliminate the acorn crop. They also showed that when acorn crops failed, as in the autumn of 1992 (Fig. 9.10), mouse populations had declined dramatically by the following summer, and spongy moth populations therefore increased (Fig. 9.10). All of this occurred at low spongy moth density, when they were in a non-outbreak phase (egg mass densities < 100/ha).

Somewhere above one hundred egg masses per acre, a density threshold is reached, beyond which predation by mice or other small mammals, such as shrews, declines with increasing spongy moth density. Unlike spongy moth parasitoids, changes in the density of vertebrate predators such as mice or birds are fairly constrained. Birds defend territories and so do mice. Thus, the population densities of mice rarely increase beyond about 100 mice per ha. Spongy moths, in contrast, can increase from 1 to 100 to 10,000 egg masses per ha, which is characteristic of outbreak populations. At these higher densities, mice or birds can feed all day on spongy moth and never make a dent in the population, whereas, at lower spongy moth densities, the mice may consume most of the spongy moth pupae in the forest. Therefore, as spongy moth density increases, there is decline in the percent mortality caused by mice and other generalist predators. Thus, vertebrate predators play almost no role in regulating outbreak populations. With many caterpillar species, parasitoids can regulate density and prevent outbreaks because their numbers can increase along with their hosts. Unfortunately, introduced and native parasitoids that attack spongy moth in North America do not do this effectively. Their numbers are constrained for reasons that are poorly understood, and they never cause very high levels of parasitism. So, once spongy moth densities reach a threshold in the vicinity of 100 egg masses per acre, the spongy moth population will grow inexorably over the next one or two years into an outbreak phase that results in widespread defoliation.

Outbreak populations become limited only by the availability of green foliage. Few spongy moth larvae actually starve in outbreak populations, but many fail to get sufficient food resources. As a consequence, the adults that arise from such populations are smaller and the females might lay 100 eggs per mass, instead of 600 (Campbell and Sloan 1978a). More importantly, there is a virus disease called nuclear polyhedrosis virus (*LdNPV*) that causes epidemics in these outbreak populations and may kill 99% of larvae before they reach the pupal stage (Campbell and Podgwaite 1971). Such viruses are common in outbreak populations of many insect species. Virus diseases reach epidemic proportions in outbreak populations because high caterpillar densities increase disease transmission. When the caterpillar dies from

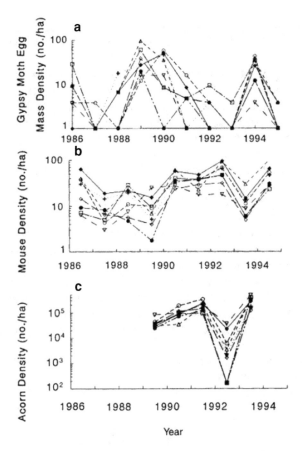

Fig. 9.10 Yearly estimates of (a) spongy moth egg masses per ha; (b) densities of white-footed mice and; (c) acorn crops at eight different plots near the Quabbin reservoir in central Massachusetts (Elkinton et al. 1996)

LdNPV, the virus causes the caterpillar cadaver to liquefy and spread virus particles over the leaf surface. Transmission occurs when a healthy caterpillar consumes virus particles released by these liquefied cadavers. Mortality from *LdNPV* starts in the early larval stages but grows exponentially in the late larval stage and peaks just before the caterpillars form pupae (Campbell and Podgwaite 1971; Murray et al. 1989). It is this epidemic that brings an end to spongy moth outbreaks and causes the populations to retreat back to low density. Therefore, outbreaks will typically last for 1 to 3 years before this population collapse happens. In the years following collapse of the outbreak, predation by small mammals resumes as the dominant force of mortality that maintains spongy moth at low density (Campbell and Sloan 1978b).

Campbell and Sloan (1978b) believed that spongy moth was a multi-equilibrium system (see Chapter 5) with a low-density equilibrium maintained by predators, mainly mice, and a high-density equilibrium wherein foliage supply and the resulting decline in fecundity, coupled with epizootics of *LdNPV*, limited further expansion of spongy moth densities and ultimately caused the collapse of outbreak populations. While it is very clear that there is indeed an upper limit to spongy moth densities, and that *LdNPV* plays a major role in the collapse of outbreaks, evidence for the low-density equilibrium remains undemonstrated. Campbell believed that predation rates by small mammals increased with spongy moth density at the lowest spongy moth densities but lacked supporting evidence. Unlike parasitoids, densities of small mammal predators do not increase in response to increased spongy moth density. Mouse densities are governed in large part by acorn crops, their principal overwintering food source. In contrast, spongy moth pupae and late instar larvae represent an extremely ephemeral food resource for mice at a time of year when they have many other things to feed on. Predation rates, if they are to increase with spongy moth density, must, in response, entail a change in foraging behavior of the predator (a Type III functional response) (Holling 1959) to increasing density of prey. In field experiments, Elkinton et al. (2004) showed that mice exhibited a Type II functional response, wherein rates of predation decline steadily as densities increase from the lowest spongy moth densities. This implies that mice cannot serve to regulate spongy moth populations at low density. This type of predation may contribute to the Allee effect in low-density spongy moth populations, as discussed above.

Dwyer et al. (2004) developed a model of spongy moth populations that combined the effects of *LdNPV* and small mammal predators. The model predicted regular outbreaks of spongy moths with an approximate 10-year periodicity. Fundamentally, this was a pathogen-driven model analogous to earlier models (e.g. Anderson and May 1981), but the addition of predators added an unstable low-density equilibrium to the system. Even a minor amount of stochasticity, however, resulted in quasi-periodic oscillations (Fig. 9.11B) that matched those of spongy moth defoliation data in New Hampshire (Fig. 9.11A) characterized by chaotic dynamics (May 1975) that make them susceptible to dynamical change with small environmental perturbations or small changes in model parameter values (Fig. 9.11C). Subsequent analyses of spongy moth defoliation data confirmed the existence of such periodicities in the spongy moth system (Bjørnstad 2000).

The Dwyer et al. (2004) model was elaborated by Bjørnstad et al. (2010) and applied to defoliation data. The revised model replaced the Type III functional response of predation with a Type II functional response, which made a low-density equilibrium caused by predators impossible. Indeed, there exists no evidence to support such an equilibrium. These analyses suggested the existence of a dominant 10-year cycle with a subdominant four-year cycle (Johnson et al. 2006a; Haynes et al. 2009a). Allstadt et al. (2013) analyzed 86 years of defoliation data, the longest available for/in North America, and concluded that population cycles appeared or disappeared four times over the duration of the spongy moth infestation in North America (Fig. 9.12B).

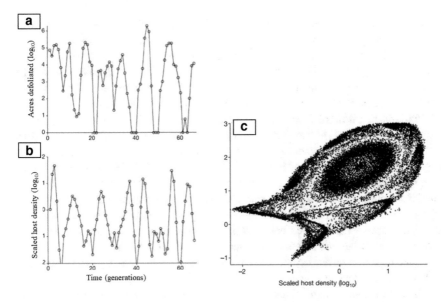

Fig. 9.11 (a) Time series of spongy moth population model of Dwyer et al. (2004) showing quasi-periodic dynamics similar to those exhibited by; (b) spongy moth defoliation in New Hampshire and; (c) a phase plot of model with stochasticity

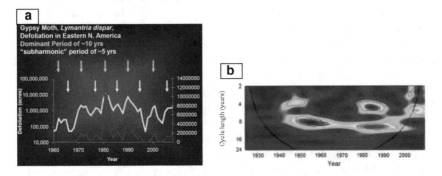

Fig. 9.12 (a) Spongy moth population dynamics model of Bjornstad et al. (2008, 2010) versus defoliation data (figure courtesy of A. Liebhold) and; (b) wavelet analysis by Allstad et al. (2013) showing changes in periodicity of spongy moth defoliation in N. America over 86 years. Vertical axis shows cycle period in years; orange/yellow colors indicate statistically significant periodicities. Only patterns above the curved black line in this figure are statistically significant

Another conspicuous feature of the spongy moth population system is that populations fluctuate in synchrony with one another across the landscape (Williams and Liebhold 1995a, 1995b; Peltonen et al. 2002; Liebhold et al. 2004; Johnson et al. 2006a, 2006b; Bjørnstad et al. 2008; Haynes et al. 2013; Allstadt et al. 2015). This phenomenon is nearly ubiquitous with most forest insects (Liebhold and Kamata

2000). Dispersal from one population to another can synchronize adjacent populations, but for spongy moth, and most other forest insects, this occurs over far too
short a distance to account for the regional synchronies observed (Peltonen et al.
2002). Instead, the standard explanation for this phenomenon involves the Moran
(1953) effect. Moran was a statistician who studied the famous snowshoe hare-
lynx predator prey oscillation in Canada. He showed that model time series of such
populations in different locations would come into synchrony with one another,
provided they were influenced by a common random factor, such as synchronous
weather. The shared weather conditions are not responsible for the oscillation, but
they do explain why snowshoe hares or forest insects typically oscillate in synchrony
with one another across much of northern Canada. The synchrony breaks down at
greater distances because weather conditions become uncorrelated at these distances.
Bjørnstad et al. (1999) developed statistical methods to detect such synchrony and
how it declines with distance between two or more populations (see Fig. 9.4b, c).
Moran's model assumed that the dynamics of spatially separated populations were all
governed by the same density-dependent processes. In fact, these dynamics undoubtedly vary somewhat in space. Peltonen et al. (2002) showed that populations with
similar but distinct dynamical parameters still exhibited spatial synchrony, as Moran
described, but the synchrony declined with distance more sharply than the synchronizing weather conditions. Haynes et al. (2009b) utilized the model of Bjørnstad et al.
(2010) and analyzed data on the spatial synchrony of spongy moths, white-footed
mice, and acorn crops in the northeastern United States. All three are synchronized
out to a distance of approximately 1000 km. They concluded that synchrony of acorn
crops was the main cause of spongy moth and mouse synchrony, as opposed to the
independent regional stochasticity (i.e. weather conditions) directly affecting each
of the latter two species. The synchrony of all three is evident on a small spatial scale
(ca 10 km) in Fig. 9.10.

In 1989, a dramatic change occurred to spongy moth populations with the accidental introduction of a fungal pathogen of spongy moth, *Entomophaga maimaiga*,
from Japan (Andreadis and Weseloh 1990; Hajek et al. 1990b). That year, the fungus
caused extensive mortality in both high and low-density populations throughout
southern New England. The following year, the infection spread over the rest of New
England and halfway across Pennsylvania (Elkinton et al. 1991). The rapid spread
was due to the fact that spongy moth cadavers killed by the fungus produce conidia
that are blown in the wind across the landscape. Subsequent research showed the
fungus depends on rainy conditions in May and June for successful transmission to
healthy larvae, and, indeed, 1989 was an especially rainy year. Beginning in 1991,
spongy moth researchers worked to spread *E. maimaiga* to Michigan (Smitley et al.
1995) and to Virginia (Hajek et al. 1996), but the fungus spread rapidly on its own,
so that by about 1996 all of the areas infested by spongy moth in the northeastern
United States were infested with the fungus (Hajek 1997, 1999). The fungus caused
a major change in status of spongy moth as a serious forest pest in New England
states. Spongy moth populations in that region declined to low density where they

have mostly remained for the last 35 years (Fig. 9.13). In contrast, spongy moth populations in areas further south, such as Pennsylvania, have continued to have periodic outbreaks despite the presence of the fungus (Morin and Liebhold 2016). Laboratory tests demonstrated that the fungus does best in cooler conditions (Hajek et al. 1990a). Temperatures in May and June in the mid-Atlantic states are much warmer than in New England.

Studies of the interaction of spongy moth fungal and viral pathogens demonstrated that *E. maimaiga* develops more quickly and outcompetes *LdNPV* when both pathogens affect the same larva (Malakar 1997; Malakar et al. 1999). The same is true for infections of *E. maimaiga* and parasitoid larvae in spongy moth larvae. Hajek et al. (2015) (Fig. 9.14a) demonstrated that *E. maimaiga* has now become the dominant mortality factor in both low and high-density populations of spongy moth. However, Liebhold et al. (2013) demonstrated that *LdNPV* still causes comparable levels of density-dependent mortality in outbreak populations in the presence of *E. maimaiga* as it had before the fungal pathogen was introduced in 1989 (Fig. 9.14b).

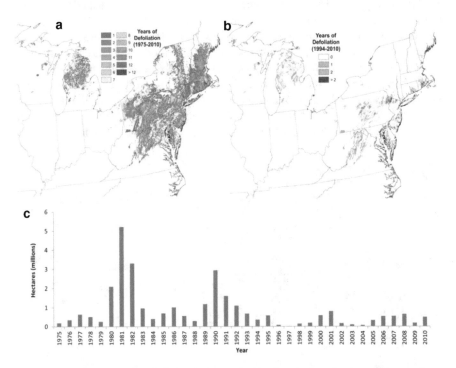

Fig. 9.13 (a) Spongy moth defoliation before and; (b) after the introduction of *Entomophaga maimaiga* in 1989 in the northeast United States (Morin and Liebhold 2016); (c) The annual hectares defoliated by spongy moth 1975–2010 in the United States

Various studies indicate that rainfall in May and June are critical to transmission of *E. maimaiga* (Hajek et al. 1990a; Hajek 1999; Reilly et al. 2014). A recent outbreak of spongy moth in New England (Fig. 9.1b; Pasquarella et al. 2018), the first widespread one since 1981, was likely caused or facilitated by three consecutive years of drought conditions in May and June beginning in 2014. Thus, rainfall has likely become a critical feature in promoting or suppressing spongy moth outbreaks. Most of the time series analyses of spongy moth defoliation data described above were applied to data collected prior to widespread establishment of *E. maimaiga*, so perhaps it is still too early to tell how it will affect the overall dynamics of spongy moth. For example, the disappearance of the population cycles after 1996 described by Allstadt et al. (2013) might be due to this major new source of mortality. Unlike the viral pathogen *LdNPV*, which only causes major epizootics in outbreak populations of spongy moth, *E. maimaiga* causes high levels of mortality in both low- and high-density populations (Hajek 1999; Fig. 9.14c). As such, it may play a significant role in preventing the onset of outbreaks in contrast to *LdNPV*. Even so, *E. maimaiga* is weakly density dependent because transmission depends on conidia that spread from nearby high-density populations (Bittner et al. 2017; Elkinton et al. 2019). Thus, *E. maimaiga* might contribute to the development of a low-density equilibrium, whose existence has not yet been demonstrated in spongy moth populations. Kyle et al. (2020) developed a population model of the impact of *E. maimaiga* on spongy moth population dynamics. Recent analyses by Liebhold et al. (2022) demonstrate that *E. maimaiga* has reduced the intensity of spongy moth outbreaks but not necessarily their frequency. Further studies and longer population time series are needed to resolve its role in low-density population dynamics of spongy moth.

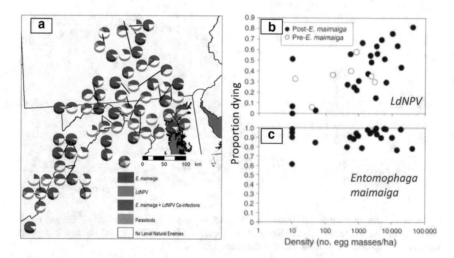

Fig. 9.14 (a) Proportions of spongy moth larvae dying from *E maimaiga, LdNPV* and parasitoids in Pennsylvania, Maryland and West Virginia (Hajek et al. 2015); (b) Mortality of spongy moth larvae in Pennsylvania from *LdNPV* and; (c) from *E maimaiga* vs. egg mass density before and after the introduction of *E maimaiga*. (Liebhold et al. 2013)

As described above, spongy moth has been exhaustively researched both from a population dynamic and from a management perspective. The extensive data on spongy moth defoliation and pheromone trap catch is almost certainly the most extensive such data for any species and has allowed researchers to make significant contributions to the general theory of population spread and eradication of invasive species. Analysis of spongy moth population data has made important contributions to the general theory of population cycles, Allee effects, and spatial synchrony of population fluctuations.

In Table 9.1, we list what we believe are the most important or damaging foliage-feeding forest insects in the world. We list the geographical range, the host tree species, and key references that give readers access to the literature on these species. We do not include the two species we have already discussed at length: winter moth, *Operophtera brumata* and spongy moth, *Lymantria dispar*.

Table 9.1 Major foliage-feeding insects: species, geographic range, hosts, and key references focusing on population ecology and impacts on forests excluding winter moth and spongy moth

Common name	Geographic range/Principal host trees	Key references
Lepidoptera		
Family: Erebidae		
Lymantria monacha Nun moth	Range: Europe, temperate Palearctic to Japan	(Bejer 1986; Skuhravý 1987; Maksimov 1999; Keena 2003; Platek 2007; Vanhanen et al. 2007; Sukovata 2010; Ilyinykh 2011; Lee et al. 2015; Nakládal and Brinkeová 2015; Faelt-Nardmann et al. 2018a, 2018b, 2018c; Hentschel et al. 2018; Melin et al. 2020)
	Hosts: *Picea abies, Pinus sylvestris*	
Euproctis chrysorrhoea Brown-tail moth	Range: mid and southern Europe into western Asia and northern Africa	(Blair 1979; Purrini 1979; Schaefer 1986; Kelly et al. 1988; Sterling and Speight 1989; Cory et al. 2000; Elkinton et al. 2006, 2008; Frago et al. 2010, 2011; Moraal and Jagers op Akkerhuis 2011; Elkinton and Boettner 2012; Frago et al. 2012; Klapwijk et al. 2013; Marques et al. 2014); Boyd et al. 2021
	Invasive range: Northeast United States into Canada (1890s); reports in China, Japan, New Guinea	
	Host trees: Polyphagous, fruit orchards, ornamental trees, most hardwoods incl. *Quercus*	
Orgyia pseudotsugata Douglas fir tussock moth	Range: western N. America into British Columbia	(Beckwith 1976; Mason 1976, 1978, 1996; Dahlsten et al. 1977; Berryman 1978; Brookes et al. 1978; Torgersen and Ryan 1981; Mason et al. 1983, 1993; Wickman et al. 1986; Alfaro et al. 1987; Mason and Torgersen 1987; Shepherd et al. 1988; Swetnam et al. 1995; Negrón et al. 2014)
	Host trees: *Pseudotsuga, Abies, P. menziesii, Abies grandis, A. concolor, A. lasiocarpa*	
Leucoma salicis Satin moth	Range: Eurasia south of the Polar circle up to northeast Siberia	(Burgess 1921; Burgess and Crossman 1927; Brown 1931; Doucette 1954; Wagner and Leonard 1979, 1980; Humphreys 1996)
	Host trees: *Salix, Populus*	

(continued)

Table 9.1 (continued)

Common name	Geographic range/Principal host trees	Key references
Family: Tortricidae		
Choristoneura fumiferana Spruce budworm, eastern spruce budworm	Range: Eastern US and Canada Host trees: *Abies balsamea, Picea glauca, P. mariana*	(Royama 1984; Crawford and Jennings 1989; Campbell 1993; Régnière and Lysyk 1995; Nealis 2003; Belle-Isle and Kneeshaw 2007; Venier and Holmes 2010; Sonia et al. 2011; Boulanger et al. 2012; Chang et al. 2012; Rhainds et al. 2012; Régnière et al. 2013; MacLean 2016; Pureswaran et al. 2016; Fuentealba et al. 2017; Royama et al. 2017; Bouchard et al. 2018a, 2018b; Drever et al. 2018; Goodbody et al. 2018; Rahimzadeh-Bajgiran et al. 2018; Legault and James 2018; Pureswaran et al. 2019; Regniere et al. 2019a, 2019b; Regniere and Nealis 2019; Lumley et al. 2020); Bhattarai et al. 2021; Berguet et al. 2021; Donovan et al. 2021; Germain et al. 2021; Maclean et al. 2019; Regniere et al. 2021; Nealis and Regniere 2021; Rhainds et al. 2022
Choristoneura freeman Western spruce budworm	Range: Western N. America (Western Canada into PNW United States) Host trees: *Pseudotsuga menziesi, Abies grandis, A. concolor,* etc	(McKnight 1974; Torgersen and Campbell 1982; Campbell et al. 1983; Anderson et al. 1987; Swetnam and Lynch 1989; Alfaro and Maclauchlan 1992; Campbell 1993; Clancy et al. 1993; Swetnam and Lynch 1993; Volney 1994; Shepherd et al. 1995; Williams and Liebhold 1995a, 1995b; Chen et al. 2001, 2002; Hummel and Agee 2003; Clancy et al. 2004; Campbell et al. 2006; Maclauchlan et al. 2006; Alfaro et al. 2014; Flower et al. 2014a, 2014b; Gilligan and Brown 2014; Nealis and Régnière 2014; Axelson et al. 2015; Meigs et al. 2015; Flower 2016; Nealis and Régnière 2016; Senf et al. 2017; Vane et al. 2017; Régnière and Nealis 2018; Santiago 2022)

(continued)

Table 9.1 (continued)

Common name	Geographic range/Principal host trees	Key references
Choristoneura pinus Jack pine budworm	Range: NE US and throughout Canada Host trees: *Pinus banksiana*	(Clancy et al. 1980; Volney and McCullough 1994; Nealis 1995; McCullough 2000; Radeloff et al. 2000; Mamet et al. 2015; Robson et al. 2015; Fahrner et al. 2016; Cadogan et al. 2018)
Tortrix viridana Green oak moth	Range: Europe, Northern Africa, western Asia Host trees: *Quercus*	(Carlisle et al. 1966; Witrowski 1975; Horstmann 1977; Du Merle 1983; Hunter 1998; Simchuk et al. 1999; Ivashov et al. 2001; Mannai et al. 2010; Ghirardo et al. 2012; Schroeder and Degen 2012; Klapwijk et al. 2013; Schroeder et al. 2015; Nedorezov 2019)
Zeiraphera griseana Larch budmoth, larch tortrix	Range: Europe, China, Korea, Japan, Russia, N. America Host trees: *Larix*	(Baltensweiler et al. 1977; Baltensweiler 1993a, 1993b; Björnstad et al. 2002; Peltonen et al. 2002; Dormont et al. 2006; Esper et al. 2007; Kress et al. 2009; Iyengar et al. 2016; Hartl-Meier et al. 2017; Saulnier et al. 2017); Büntgen et al. 2020; Liebhold et al. 2020; Rozenberg et al. 2020; Din 2021
Family: Lasiocampidae		
Malacosoma disstria Forest tent caterpillar	Range: Throughout North America Host trees: *Quercus, Liquidambar, Nyssa, Populus, Acer saccharum*	(Daniel and Myers 1995; Parry 1995; Parry et al. 1997; Rothman and Roland 1998; Cooke and Roland 2000; Parry et al. 2001; Cooke and Roland 2003; Parry et al. 2003; Wood et al. 2009; Trudeau et al. 2010; Wood et al. 2010; Charbonneau et al. 2012; Cooke et al. 2012; Moulinier et al. 2013; Hughes et al. 2015; Uelmen et al. 2016; Schowalter 2017; Haynes et al. 2018b; Lait and Hebert 2018; Plenzich and Despland 2018; Robert et al 2020; Cooke et al 2022

(continued)

Table 9.1 (continued)

Common name	Geographic range/Principal host trees	Key references
Dendrolimus superans Siberian silk moth	Range: Kazakhstan, Mongolia, China, Russia, Korea, Japan Host trees: *Larix, Picea, Pinus*	(Maeto 1991; Li et al. 2002; Kharuk et al. 2004, 2018; Kovacs et al. 2005; Kirichenko et al. 2009; Jeger et al. 2018)
Family: Geometridae		
Operophtera bruceata Bruce spanworm	Range: Southern Canada, northern USA, Alaska, Greenland Host trees: *Acer saccharum, Fagus, Populus*	(Brown 1962; Eidt et al. 1966; Ives and Cunningham 1980; Ives 1984; Troubridge and Fitzpatrick 1993; Elkinton et al. 2010; Gwiazdowski et al. 2013; Broadley et al. 2017; Havill et al. 2017); Andersen et al. 2022
Epirrita autumnata Autumnal moth	Range: Palearctic region and Middle East Host trees: *Betula*	(Ayres and MacLean 1987; Tenow and Nilssen 1990; Ruohomäki and Haukioja 1992; Tammaru et al. 1996; Ruohomäki et al. 2000; Tammaru et al. 2001; Tanhuanpaa et al. 2001; Klemola et al. 2004; Nilssen et al. 2007; Ruohola et al. 2007; Tenow et al. 2007; Jepsen et al. 2008; Klemola et al. 2008; Ruuhola et al. 2008; Klemola et al. 2009; Ammunét et al. 2010; Jepsen et al. 2011; Myers and Cory 2013; Klemola et al. 2014)
Alsophila pometaria Fall cankerworm	Range: across N. America, south to Colorado Host trees: *Ulmus, Fraxinus, Acer*	(Smith 1958; Kegg 1967; Appleby et al. 1975; Mitter and Futuyma 1977; Schneider 1980; Mitter et al. 1987; Walter et al. 2016)
Ennomos subsignaria Elm spanworm	Range: Eastern N. America spanning south to Texas and north to Alberta Host trees: *Ulmus, Malus, Betula, Acer, Quercus*	(Ciesla 1964a, 1964b; Fedde 1965; Kaya and Anderson 1974; Anderson and Kaya 1976; Fry et al. 2008, 2009; Ryall 2010)

(continued)

Table 9.1 (continued)

Common name	Geographic range/Principal host trees	Key references
Lambdina fiscellaria Hemlock looper	N. America (Pacific to Atlantic coast) south from Canada to Pennsylvania, Wisconsin and CA	(Alfaro et al. 1999; MacLean and Ebert 1999; Hébert et al. 2001; Pelletier and Piché 2003; Butt et al. 2010; Iqbal et al. 2011; Rochefort et al. 2011; Legault et al. 2012; Delisle et al. 2013, 2016; Seehausen et al. 2015; Wilson et al. 2016; Oswald et al. 2017; Sabbahi et al. 2018)
	Host trees: *Tsuga, Abies balsamea, Picea, Quercus*, other hardwoods	
Family: Notodontidae		
Heterocampa guttivitta Saddled prominent	Range: Central and eastern N. America	(Fiske and Burgess 1910; Fisher 1970; Allen 1972, 1973; Ticehurst and Allen 1973; Martinat and Allen 1987)
	Host trees: *Malus, Betula, Cornus, Corylus, Acer, Quercus, Rhus, Juglans*	
Syntypistis punctatella Beech caterpillar	Japan	(Kamata 2000, 2002)
	Host trees: *Fagus*	
Family: Coleophoridae		
Coleophora laricella Larch casebearer	Range: Central and northern Europe Introduced: N. America (nineteenth century)	(Ryan 1983, 1986, 1997; Habermann 2000; Ward and Aukema 2019; Ward et al. 2019, 2020a, 2020b)
	Host trees: *Larix decidua*; in invasive range: *L. occidentalis, L. laricina*	
Family: Thaumetopoeidae		

(continued)

Table 9.1 (continued)

Common name	Geographic range/Principal host trees	Key references
Thaumetopoea pityocampa Pine processionary caterpillar	Range: Central Asia, N. Africa, southern Europe Host trees: *Pinus, Cedrus*	(Battisti 1988; Hódar et al. 2003, 2015; Pérez-Contreras et al. 2003; Carus 2004, 2010; Battisti et al. 2005, 2006; Stastny et al. 2006; Arnaldo et al. 2010; Ronnås et al. 2010; Seixas Arnaldo et al. 2011; Jacquet et al. 2012; Van Dyck 2012; Tamburini et al. 2013; Cayuela et al. 2014; Li et al. 2015; Castagneyrol et al. 2016; Salman et al. 2016; Pimentel et al. 2017); Atvis et al. 2018; Camerero et al. 2022; Georgiev et al. 2022; Martin et al. 2022; Mirchev et al. 2021
Hymenoptera		
Family: Diprionidae		
Neodiprion sertifer European Pine sawfly	Range: Native: Europe Introduced: N. America (1925) Host trees: *Pinus, P. sylvestris, P. resinosa, P. banksiana, P. mugo*	(Austara et al. 1987; Pschorn-Walcher 1987; Kouki et al. 1998; Larsson et al. 2000; Lyytikäinen-Saarenmaa et al. 2001; Gur'yanova 2006; Kollberg et al. 2013, 2014; Bellone et al. 2017; Kosunen et al. 2016; Klapwijk and Björkman 2018; Chorbadjian et al. 2019)
Diprion similis Introduced pine sawfly	Range: central & northern Europe, Siberia, China Introduced (1914): eastern USA & Canada Host trees: *Pinus, P. strobus*	(Weber 1977; Drooz et al. 1985)
Gilpinia hercyniae European spruce sawfly	Range: Quebec, New Brunswick, NE USA Host trees: *Picea*	(Balch 1939; Dowden 1939; Neilson and Morris 1964; Wong 1972; Schopf 1989; Williams et al. 2003)
Family: Pamphiliidae		
Acantholyda erythrocephala Red-headed pine sawfly	Range: Native: Europe Introduced (1925): N. America Host trees: (Introduced range): *Pinus strobus, P. sylvestris, P. Resinosa*	(Asaro and Allen 2001; Kenis and Kloosterman 2001; Mayfield et al. 2007)

References

Alfaro RI, Maclauchlan LE (1992) A method to calculate the losses caused by western spruce budworm in uneven-aged Douglas-fir forests of British-Columbia. Forest Ecol Manag 55(1–4):295–313. https://doi.org/10.1016/0378-1127(92)90107-K

Alfaro RI, Taylor SP, Wegwitz E et al (1987) Douglas-fir tussock moth damage in British Columbia. Forest Chron 63(5):351–355. https://doi.org/10.5558/tfc63351-5

Alfaro RI, Taylor S, Brown G et al (1999) Tree mortality caused by the western hemlock looper in landscapes of central British Columbia. Forest Ecol Manag 124(2–3):285–291. https://doi.org/10.1016/S0378-1127(99)00073-0

Alfaro RI, Berg J, Axelson J (2014) Periodicity of western spruce budworm in Southern British Columbia, Canada. Forest Ecol Manag 315:72–79. https://doi.org/10.1016/j.foreco.2013.12.026

Allen DC (1972) Insect parasites of the saddled prominent, *Heterocampa guttivitta* (Lepidoptera: Notodontidae), in the northeastern United States. Can Entomol 104(10):1609–2162. https://doi.org/10.4039/Ent1041609-10

Allen DC (1973) Fecundity of the saddled prominent. *Heterocampa guttivitta*. Ann Entomol Soc Am 66(6):1181–1183.https://doi.org/10.1093/aesa/66.6.1181

Allstadt AJ, Haynes KJ, Liebhold AM et al (2013) Long-term shifts in the cyclicity of outbreaks of a forest-defoliating insect. Oecologia 172:141–151. https://doi.org/10.1007/s00442-012-2474-x

Allstadt AJ, Liebhold AM, Johnson DM et al (2015) Temporal variation in the synchrony of weather and its consequences for spatiotemporal population dynamics. Ecology 96(11):2935–2946. https://doi.org/10.1890/14-1497.1

Ammunét T, Heisswolf A, Klemola N et al (2010) Expansion of the winter moth outbreak range: no restrictive effects of competition with the resident autumnal moth. Ecol Entomol 35(1):45–52. https://doi.org/10.1111/j.1365-2311.2009.01154.x

Andersen JC, Havill NP, Caccone A et al (2017) Postglacial recolonization shaped the genetic diversity of the winter moth (*Operophtera brumata*) in Europe. Ecol Evol 7(10):3312–3323. https://doi.org/10.1002/ece3.2860

Andersen JC, Havill NP, Broadley HJ et al (2019a) Widespread hybridization among native and invasive species of *Operophtera* moths (Lepidoptera: Geometridae) in Europe and North America. Biol Ivasions 21:3383–3394. https://doi.org/10.1007/s10530-019-02054-1

Andersen JC, Havill NP, Mannai Y et al (2019b) Identification of winter moth (*Operophtera brumata*) refugia in North Africa and the Italian Peninsula during the last glacial maximum. Ecol Evol 9(24):13931–13941. https://doi.org/10.1002/ece3.5830

Andersen JC, Havill NP, Caccone A et al (2021a) Four times out of Europe: Serial invasions of the winter moth, *Operophtera brumata*, to North America. Mol Ecol 30(14):3439–3452. https://doi.org/10.1111/mec.15983

Andersen JC, Havill NP, Griffin BP et al (2021b). Northern Fennoscandia via the British Isles: evidence for a novel post-glacial recolonization route by winter moth (*Operophtera brumata*). Frontiers of Biogeography 13(1):e49581. https://doi.org/10.21425/F5FBG49581

Andersen JC, Havill NP, Boettner GH et al (2022) Real-time geographic settling of a hybrid zone between the invasive winter moth (*Operophtera brumata* L.) and the native Bruce spanworm (*O. bruceata* Hulst). Mol Ecol. https://doi.org/10.1111/mec.16349

Anderson JF, Kaya HK (1976) Parasitoids and Diseases of the Elm Spanworm. J New York Entomol S 84(3):169–177. https://www.jstor.org/stable/25009007. Accessed 18 October 2022

Anderson L, Carlson CE, Wakimoto RH (1987) Forest fire frequency and western spruce budworm outbreaks in western Montana. Forest Ecol Manag 22:251–260. https://doi.org/10.1016/0378-1127(87)90109-5

Anderson RM, May RM (1981) The population dynamics of microparasites and their invertebrate hosts. Philos T Roy Soc B 291(1054):451–524. https://doi.org/10.1098/rstb.1981.0005

Andreadis TG, Weseloh RM (1990) Discovery of *Entomophaga maimaiga* in North American gypsy moth, *Lymantria dispar*. P Natl Acad Sci USA 87(7):2461–2465. https://doi.org/10.1073/pnas.87.7.2461

Appleby JE, Bristol P, Eickhorst WE (1975) Control of fall cankerworm. J Econ Entomol 68(2):233–234. https://doi.org/10.1093/jee/68.2.233

Arnaldo PS, Chacim S, Lopes D (2010) Effects of defoliation by the pine processionary moth *Thaumetopoea pityocampa* on biomass growth of young stands of *Pinus pinaster* in northern Portugal. Iforest 3(6):159–162. https://doi.org/10.3832/ifor0553-003

Asaro C, Allen DC (2001) History of a pine false webworm (Hymenoptera: Pamphiliidae) outbreak in northern New York. Can J Forest Res 31:181–185. https://doi.org/10.1139/x00-147

Austarå Ø, Orlund A, Svendsrud A et al (1987) Growth loss and economic consequences following two years defoliation of *Pinus sylvestris* by the pine sawfly *Neodiprion sertifer* in West-Norway. Scand J Forest Res 2(1–4):111–119. https://doi.org/10.1080/02827588709382450

Avtzis DN, Petsopoulos D, Memtsas GI et al (2018) Revisiting the distribution of *Thaumetopoea pityocampa* (Lepidoptera: Notodontidae) and *T. pityocampa* ENA Clade in Greece. J Econ Entomol 111(3):1256–1260. https://doi.org/10.1093/jee/toy047

Axelson JN, Smith DJ, Daniels LD et al (2015) Multicentury reconstruction of western spruce budworm outbreaks in central British Columbia, Canada. Forest Ecol Manag 335:235–248. https://doi.org/10.1016/j.foreco.2014.10.002

Ayres MP, MacLean SF (1987) Development of birch leaves and the growth energetics of *Epirrita autumnata* (Geometridae). Ecology 68(3):558–568. https://doi.org/10.2307/1938461

Balch RE (1939) The outbreak the European spruce sawfly in Canada and some important features of its bionomics. J Econ Entomol 32(3):412–418. https://doi.org/10.1093/jee/32.3.412

Baltensweiler W (1993a) Why the larch bud-moth cycle collapsed in the subalpine larch-cembran pine forests in the year 1990 for the first time since 1850. Oecologia 94:62–66. https://doi.org/10.1007/BF00317302

Baltensweiler W (1993b) A contribution to the explanation of the larch bud moth cycle, the polymorphic fitness hypothesis. Oecologia 93:251–255. https://doi.org/10.1007/BF00317678

Baltensweiler W, Fischlin A (1988) The larch budmoth in the Alps. In: Berryman AA (ed) Dynamics of forest insect populations. Springer, MA, pp 331–351. https://doi.org/10.1007/978-1-4899-0789-9_17

Baltensweiler W, Benz G, Bovey P et al (1977) Dynamics of larch bud moth populations. Annu Rev Entomol 22:79–100. https://doi.org/10.1146/annurev.en.22.010177.000455

Baranchikov YN, Sukachev VN (1989) Ecological basis of the evolution of host relationships in Eurasian gypsy moth populations. USDA NE GTR 123:319–338

Barbehenn RV, Constabel CP (2011) Tannins in plant-herbivore interactions. Phytochemistry 72(13):1551–1565. https://doi.org/10.1016/j.phytochem.2011.01.040

Battisti A (1988) Host-plant relationships and population dynamics of the pine processionary caterpillar *Thaumetopoea pityocampa* (Denis & Schiffermuller). J App Entomol 105(1–5):393–402. https://doi.org/10.1111/j.1439-0418.1988.tb00202.x

Battisti A, Stastny M, Netherer S et al (2005) Expansion of geographic range in the pine processionary moth caused by increased winter temperatures. Ecol Appl 15(6):2084–2096. https://doi.org/10.1890/04-1903

Battisti A, Stastny M, Buffo E et al (2006) A rapid altitudinal range expansion in the pine processionary moth produced by the 2003 climatic anomaly. Global Change Biol 12(4):662–671. https://doi.org/10.1111/j.1365-2486.2006.01124.x

Beckwith RC (1976) Influence of host foliage on the Douglas-fir tussock moth. Environ Entomol 5(1):73–77. https://doi.org/10.1093/ee/5.1.73

Bejer B (1986) Outbreaks of Nun Moth (*Lymantria monacha* L) in Denmark with Remarks on their Control. Anz Schadlingskd Pfl 59:86–89. https://doi.org/10.1007/BF01903455

Belle-Isle J, Kneeshaw D (2007) A stand and landscape comparison of the effects of a spruce budworm (*Choristoneura fumiferana* (Clem.)) outbreak to the combined effects of harvesting and thinning on forest structure. Forest Ecol Manag 246(2–3):163–174. https://doi.org/10.1016/j.foreco.2007.03.038

Bellone D, Klapwijk MJ, Björkman C (2017) Habitat heterogeneity affects predation of European pine sawfly cocoons. Ecol Evol 7(24):11011–11020. https://doi.org/10.1002/ece3.3632

Berguet C, Martin M, Arseneault D et al (2021) Spatiotemporal Dynamics of 20th-Century Spruce Budworm Outbreaks in Eastern Canada: Three Distinct Patterns of Outbreak Severity. Dig Front Ecol Evol 8:544088. https://doi.org/10.3389/fevo.2020.544088

Berryman AA (1978) Population cycles of the Douglas-fir tussock moth (Lepidoptera: Lymantriidae): the time-delay hypothesis. Can Entomol 110(5):513–518. https://doi.org/10.4039/Ent110 513-5

Bhattarai R, Rahimzadeh-Bajgiran P, Weiskittel A et al (2021) Spruce budworm tree host species distribution and abundance mapping using multi-temporal Sentinel-1 and Sentinel-2 satellite imagery. ISPRS J Photogram Remote Sens 172:28–40. https://doi.org/10.1016/j.isprsjprs.2020. 11.023

Bittner TD, Hajek AE, Liebhold AM et al (2017) Modification of a pollen trap design to capture airborne conidia of *Entomophaga maimaiga* and detection by quantitative PCR. Appl Environ Microbiol 83(17):1–11. https://doi.org/10.1128/AEM.00724-17

Bjørnstad ON (2000) Cycles and synchrony: two historical 'experiments' and one experience. J Anim Ecol 69(5):869–873. https://www.jstor.org/stable/2647407. Accessed 18 October 2022

Bjørnstad ON, Ims RA, Lambin X (1999) Spatial population dynamics: analyzing patterns and processes of population synchrony. Trends Ecol Evol 14(11):427–432. https://doi.org/10.1016/ S0169-5347(99)01677-8

Bjørnstad ON, Peltonen M, Liebhold AM et al (2002) Waves of larch budmoth outbreaks in the European Alps. Science 298(5595):1020–1023. https://doi.org/10.1126/science.1075182

Bjørnstad ON, Liebhold AM, Johnson DM (2008) Transient synchronization following invasion: revisiting Moran's model and a case study. Popul Ecol 50(4):379–389. https://doi.org/10.1007/ s10144-008-0105-5

Bjørnstad ON, Robinet C, Liebhold AM (2010) Geographic variation in North American gypsy moth cycles: subharmonics, generalist predators, and spatial coupling. Ecology 91(1):106–118. https://doi.org/10.1890/08-1246.1

Blair CP (1979) Browntail moth, its caterpillar and their rash. Clin Exp Dermatol 4(2):215–222. https://doi.org/10.1111/j.1365-2230.1979.tb01621.x

Boettner GH, Elkinton JS, Boettner CJ (2000) Effects of a biological control introduction on three nontarget native species of saturniid moths. Conserv Biol 14(6):1798–1806. https://doi.org/10. 1111/j.1523-1739.2000.99193.x

Bouchard M, Régnière J, Therrien P (2018a) Bottom-up factors contribute to large-scale synchrony in spruce budworm populations. Can J for Res 48(3):277–284. https://doi.org/10.1139/cjfr-2017-0051

Bouchard M, Martel V, Régnière J et al (2018b) Do natural enemies explain fluctuations in low-density spruce budworm populations? Ecology 99(9):2047–2057. https://doi.org/10.1002/ecy. 2417

Boulanger Y, Arseneault D, Morin H et al (2012) Dendrochronological reconstruction of spruce budworm (*Choristoneura fumiferana*) outbreaks in southern Quebec for the last 400 years. Can J for Res 42(7):1264–1276. https://doi.org/10.1139/x2012-069

Boyd KS, Drummond F, Donahue C et al (2021) Factors influencing the population fluctuations of *Euproctis chrysorrhoea* (Lepidoptera: Erebidae) in Maine. Environ Entomol 50(5):1203–1216. https://doi.org/10.1093/ee/nvab060

Broadley HJ (2018) Impact of native natural enemies on populations of the invasive winter moth, (*Operophtera brumata* L) in the northeast United States. Dissertation, University of Massachusetts.https://doi.org/10.7275/12760419

Broadley HJ, Boucher M, Burand JP et al (2017) The phylogenetic relationship and cross-infection of nucleopolyhedroviruses between the invasive winter moth (*Operophtera brumata*) and its native congener, Bruce spanworm (*O. bruceata*). J Invertebr Pathol 143:61–68. https://doi.org/ 10.1016/j.jip.2016.11.016

Broadley HJ, Kula RR, Boettner GH et al (2019) Recruitment of native parasitic wasps to populations of the invasive winter moth in the northeastern United States. Biol Invasions 21:2871–2890. https://doi.org/10.1007/s10530-019-02019-4

Broadley HJ, Boettner GH, Schneider B et al (2022) Native generalist natural enemies and an introduced specialist parasitoid together control an invasive forest insect. Dig Ecol Appl. https://doi.org/10.1002/eap.2697

Brookes MH, Stark RW, Campbell RW (eds) (1978) The Douglas-fir tussock moth: a synthesis. USDA, Washington

Brown CE (1962) The life history and dispersal of the Bruce Spanworm, *Operophtera bruceata*. Can Entomol 94(10):1103–1107.https://doi.org/10.4039/Ent941103-10

Brown JH, Halliwell DB, Gould WP (1979) Gypsy moth defoliation: impact in Rhode Island forests. J for 77(1):30–32. https://doi.org/10.1093/jof/77.1.30

Brown RC (1931) Observations on the satin moth and its natural enemies in Central Europe. USDA Circular 176

Büntgen U, Liebhold A, Nievergelt D et al (2020) Return of the moth: rethinking the effect of climate on insect outbreaks. Oecologia 192:543–552. https://doi.org/10.1007/s00442-019-04585-9

Burand JP, Kim W, Welch A et al (2011) Identification of a nucleopolyhedrovirus in winter moth populations from Massachusetts. J Invertebr Pathol 108(3):217–219. https://doi.org/10.1016/j.jip.2011.08.005

Burgess AF (1921) The satin moth: an introduced enemy of poplars and willows. USDA Circular 167

Burgess AF, Crossman SS (1927) The Satin Moth, a recently introduced pest. USDA Bulletin 1469. https://www.biodiversitylibrary.org/bibliography/107926. Accessed 18 October 2022

Butt C, Quiring D, Hébert C et al (2010) Influence of balsam fir (*Abies balsamea*) budburst phenology on hemlock looper (*Lambdina fiscellaria*). Entomol Exp Appl 134(3):220–226. https://doi.org/10.1111/j.1570-7458.2009.00963.x

Cadogan BL, Scharbach RD, Krause RE et al (2018) Linkages between the phenologies of jack pine (*Pinus banksiana*) foliage and jack pine budworm (Lepidoptera: Tortricidae). Great Lakes Entomol 38(1&2):58–75. https://scholar.valpo.edu/tgle/vol38/iss1/7. Accessed 18 October 2022

Camarero JJ, Tardif J, Gazol A et al (2022) Pine processionary moth outbreaks cause longer growth legacies than drought and are linked to the North Atlantic Oscillation. Sci Total Environ 819:153041. https://doi.org/10.1016/j.scitotenv.2022.153041

Campbell R, Smith DJ, Arsenault A (2006) Multicentury history of western spruce budworm outbreaks in interior Douglas-fir forests near Kamloops, British Columbia. Can J Forest Res 36(7):1758–1769.https://doi.org/10.1139/x06-069

Campbell RW (1975) The gypsy moth and its natural enemies. USDA Bulletin 381. https://doi.org/10.22004/ag.econ.309219

Campbell RW (1993) Population dynamics of the major North American needle-eating budworms. USDA PNW-RP-463

Campbell RW, Podgwaite JD (1971) The disease complex of the gypsy moth. I. Major components. J Invertebr Pathol 18(1):101–107.https://doi.org/10.1016/0022-2011(91)90015-I

Campbell RW, Sloan RJ (1977) Forest stand responses to defoliation by the gypsy moth. For Sci 23(s2):1–34. https://academic.oup.com/forestscience/article-abstract/23/suppl_2/a0001/4675788?login=true. Accessed 18 October 2022

Campbell RW, Sloan RJ (1978a) Natural maintenance and decline of gypsy moth outbreaks. Environ Entomol 7(3):389–395. https://doi.org/10.1093/ee/7.3.389

Campbell RW, Sloan RJ (1978b) Numerical bimodality among North-American gypsy moth populations. Environ Entomol 7(5):641–646. https://doi.org/10.1093/ee/7.5.641

Campbell RW, Beckwith RC, Torgersen TR (1983) Numerical Behavior of some western spruce budworm (Lepidoptera: Tortricidae) populations in Washington and Idaho. Environ Entomol 12(5):1366. https://doi.org/10.1093/ee/12.5.1360

Canning EU (1960) Two new microsporidian parasites of the winter moth, *Operophtera brumata* (L). J Parasitol 46(6):755–763. https://doi.org/10.2307/3275526

Canning EU, Wigley PJ, Barker RJ (1983) The taxonomy of three species of microsporidia (Protozoa: Microspora) from an oakwood population of winter moths *Operophtera brumata* (L.) (Lepidoptera: Geometridae). Sys Parasitol 5:147–159. https://doi.org/10.1007/BF00049242

Cardé RT, Minks AK (1995) Control of moth pests by mating disruption: successes and constraints. Annu Rev Entomol 40:559–585. https://www.annualreviews.org/doi/abs/10.1146/annurev.en.40. 010195.003015. Accessed 18 October 2022

Carlisle A, Brown AHF, White EJ (1966) Litter fall, leaf production and the effects of defoliation by *Tortrix viridana* in a sessile oak (*Quercus petraea*) woodland. J Ecol 54(1):65–85. https://doi.org/10.2307/2257659

Carus S (2004) Impact of defoliation by the pine processionary moth (*Thaumetopoea pityocampa*) on radial, height and volume growth of Calabrian pine (*Pinus brutia*) trees in Turkey. Phytoparasitica 32:459–469. https://doi.org/10.1007/BF02980440

Carus S (2010) Effect of defoliation by the pine processionary moth (PPM) on radial, height and volume growth of Crimean pine (Pinus nigra) trees in Turkey. J Environ Biol 31(4):453–460. https://pubmed.ncbi.nlm.nih.gov/21186719/. Accessed 18 October 2022

Castagneyrol B, Jactel H, Brockerhoff EG et al (2016) Host range expansion is density dependent. Oecologia 182:779–788. https://doi.org/10.1007/s00442-016-3711-5

Cayuela L, Hernández R, Hódar JA et al (2014) Tree damage and population density relationships for the pine processionary moth: Prospects for ecological research and pest management. Forest Ecol Manag 328:319–325. https://doi.org/10.1016/j.foreco.2014.05.051

Chang WY, Lantz VA, Hennigar CR et al (2012) Economic impacts of forest pests: a case study of spruce budworm outbreaks and control in New Brunswick, Canada. Can J Forest Res 42(3):490–505.https://doi.org/10.1139/x11-190

Charbonneau D, Lorenzetti F, Doyon F et al (2012) The influence of stand and landscape characteristics on forest tent caterpillar (*Malacosoma disstria*) defoliation dynamics: the case of the 1999–2002 outbreak in northwestern Quebec. Can J Forest Res 42(10):1827–1836. https://doi.org/10.1139/x2012-126

Chen Z, Kolb TE, Clancy KM (2001) Mechanisms of Douglas-fir resistance to western spruce budworm defoliation: bud burst phenology, photosynthetic compensation and growth rate. Tree Physiol 21(16):1159–1169. https://doi.org/10.1093/treephys/21.16.1159

Chen Z, Kolb TE, Clancy KM (2002) The role of monoterpenes in resistance of Douglas fir to western spruce budworm defoliation. J Chem Ecol 28(5):897–920. https://doi.org/10.1023/A:1015297315104

Chorbadjian RA, Phelan PL, Herms DA (2019) Tight insect-host phenological synchrony constrains the life-history strategy of European pine sawfly. Agric for Entomol 21(1):15–27. https://doi.org/10.1111/afe.12299

Ciesla WM (1964a) Egg Parasites of Elm Spanworm in Southern Appalachian Mountains. J Econ Entomol 57(6):837–838. https://doi.org/10.1093/jee/57.6.837

Ciesla WM (1964b) Life history and habits of the elm spanworm, *Ennomos subsignarius*, in the southern Appalachian Mountains (Lepidoptera: Geometridae). Ann Entomol Soc Am 57(5):591–596. https://doi.org/10.1093/aesa/57.5.591

Clancy KM, Giese RL, Benjamin DM (1980) Predicting jack-pine budworm infestations in northwestern Wisconsin. Environ Entomol 9(6):743–751. https://doi.org/10.1093/ee/9.6.743

Clancy KM, Itami JK, Huebner DP (1993) Douglas-Fir nutrients and terpenes: potential resistance factors to western spruce budworm defoliation. For Sci 39(1):78–94. https://academic.oup.com/forestscience/article-abstract/39/1/78/4627128. Accessed 18 October 2022

Clancy KM, Chen Z, Kolb TE (2004) Foliar nutrients and induced susceptibility: genetic mechanisms of Douglas-fir resistance to western spruce budworm defoliation. Can J Forest Res 34(4):939–949. https://doi.org/10.1139/x03-264

Cooke BJ, Roland J (2000) Spatial analysis of large-scale patterns of forest tent caterpillar outbreaks. Ecoscience 7(4):410–422. https://doi.org/10.1080/11956860.2000.11682611

Cooke BJ, Roland J (2003) The effect of winter temperature on forest tent caterpillar (Lepidoptera: Lasiocampidae) egg survival and population dynamics in northern climates. Environ Entomol 32(2):299–311. https://doi.org/10.1603/0046-225X-32.2.299

Cooke BJ, MacQuarrie CJK, Lorenzetti F (2012) The dynamics of forest tent caterpillar outbreaks across east-central Canada. Ecography 35(5):422–435. https://doi.org/10.1111/j.1600-0587. 2011.07083.x

Cooke BJ, Sturtevant BR, Robert LE (2022) The forest tent caterpillar in Minnesota: detectability, impact, and cycling dynamics. Forests 13(4):601. https://doi.org/10.3390/f13040601

Cooper D, Cory JS, Theilmann DA et al (2003) Nucleopolyhedroviruses of forest and western tent caterpillars: cross-infectivity and evidence for activation of latent virus in high-density field populations. Ecol Entomol 28(1):41–50. https://doi.org/10.1046/j.1365-2311.2003.00474.x

Cory JS, Hirst ML, Sterling PA et al (2000) Narrow host range nucleopolyhedrovirus for control of the browntail moth (Lepidoptera: Lymantriidae). Environ Entomol 29(3):661–667. https://doi. org/10.1603/0046-225X-29.3.661

Crawford HS, Jennings DT (1989) Predation by birds on spruce budworm *Choristoneura fumiferana*: functional, numerical, and total responses. Ecology 70(1):152–163. https://doi.org/ 10.2307/1938422

Dahlsten DL, Luck RF, Schlinger EI et al (1977) Parasitoids and predators of Douglas-Fir tussock moth, *Orgyia pseudotsugata* (Lepidoptera: Lymantridae), in low to moderate populations in central California. Can Entomol 109(5):727–746. https://doi.org/10.4039/Ent109727-5

Daniel CJ, Myers JH (1995) Climate and outbreaks of the forest tent caterpillar. Ecography 18(4):353–362. https://doi.org/10.1111/j.1600-0587.1995.tb00138.x

Davidson CB, Gottschalk KW, Johnson JE (1999) Tree mortality following defoliation by the European gypsy moth (*Lymantria dispar* L.) in the United States: A Review. For Sci 45(1):74–84. https://doi.org/10.1093/forestscience/45.1.74

Delisle J, Labrecque A, Royer L et al (2013) Impact of short-term exposure to low subzero temperatures on egg hatch in the hemlock looper, *Lambdina fiscellaria*. Entomol Exp Appl 149(3):206–218.https://doi.org/10.1111/eea.12123

Delisle J, Bernier-Cardou M, Laroche G (2016) Reproductive performance of the hemlock looper, *Lambdina fiscellaria*, as a function of temperature and population origin. Entomol Exp Appl 161(3):219–231. https://doi.org/10.1111/eea.12469

Delvas N, Bauce É, Labbé C et al (2011) Phenolic compounds that confer resistance to spruce budworm. Entomol Exp Appl 141(1):35–44. https://doi.org/10.1111/j.1570-7458.2011.01161.x

Din Q (2021) Dynamics and chaos control for a novel model incorporating plant quality index and larch budmoth interaction. Chaos Solitons Fractals 153(2):111595. https://doi.org/10.1016/ j.chaos.2021.111595

Donahue KL, Broadley HJ, Elkinton JS et al (2019) Using the SSU, ITS, and Ribosomal DNA Operon arrangement to characterize two microsporidia infecting bruce spanworm, *Operophtera bruceata* (Lepidoptera: Geometridae). J Eukaryotic Microbiol 66(3):424–434. https://doi.org/10. 1111/jeu.12685

Donovan SD, MacLean DA, Zhang Y et al (2021) Evaluating annual spruce budworm defoliation using change detection of vegetation indices calculated from satellite hyperspectral imagery. Remote Sens Environ 253:112204. https://doi.org/10.1016/j.rse.2020.112204

Dormont L, Baltensweiler W, Choquet R et al (2006) Larch- and pine-feeding host races of the larch bud moth (*Zeiraphera diniana*) have cyclic and synchronous population fluctuations. Oikos 115(2):299–307. https://doi.org/10.1111/j.2006.0030-1299.15010.x

Doucette CF (1954) Recurrence of the Satin Moth in the Pacific Northwest. J Econ Entomol 47(5):939–940. https://doi.org/10.1093/jee/47.5.939

Dowden PB (1939) Present status of the European spruce sawfly, *Diprion polytomum* (Htg.), in the United States. J Econ Entomol 32(5):619–624. https://academic.oup.com/jee/article-abstract/32/ 5/619/798754?redirectedFrom=fulltext. Accessed 18 October 2022

Drever MC, Smith AC, Venier LA et al (2018) Cross-scale effects of spruce budworm outbreaks on boreal warblers in eastern Canada. Ecol Evol 8(15):7334–7345. https://doi.org/10.1002/ece3. 4244

Drooz AT, Ghent JH, Huber CM (1985) Insect parasites associated with the Introduced pine Sawfly, *Diprion-similis* (Hartig) (Hymenoptera: Diprionidae), in North-Carolina. Environ Entomol 14(4):401–403.https://doi.org/10.1093/ee/14.4.401

Du Merle P (1983) Mortality factors for the eggs of *Tortrix viridana* L (Lep. Tortricidae) 3. Regulatory action of each factor and examination of total mortality [green oak tortrix; predation, parasitism, disease; Southern France]. Agronomie 3(5):429–434. https://agris.fao.org/agris-sea rch/search.do?recordID=XE8436337. Accessed 18 October 2022

Dwyer G, Dushoff J, Yee SH (2004) The combined effects of pathogens and predators on insect outbreaks. Nature 430:341–345. https://doi.org/10.1038/nature02569

Eidt DC, Embree DG, Smith CC (1966) Distinguishing adults of the winter moth *Operophtera brumata* (L.), and Bruce spanworm *O. bruceata* (Hulst) (Lepidoptera: Geometridae). Can Entomol 98(3):258–261. https://doi.org/10.4039/Ent98258-3

Elkinton JS, Boettner GH (2012) Benefits and harm caused by the introduced generalist tachinid, *Compsilura concinnata*, in North America. Biocontrol 57:277–288. https://doi.org/10.1007/s10 526-011-9437-8

Elkinton JS, Liebhold AM (1990) Population dynamics of gypsy moth in North America. Annu Rev Entomol 35:571–596. https://doi.org/10.1146/annurev.en.35.010190.003035

Elkinton JS, Hajek AE, Boettner GH et al (1991) Distribution and apparent spread of *Entomophaga maimaiga* (Zygomycetes: Entomophthorales) in gypsy moth (Lepidoptera: Lymantriidae) populations in North America. Environ Entomol 20(6):1601–1605. https://doi.org/10.1093/ee/20.6. 1601

Elkinton JS, Healy WM, Buonaccorsi JP et al (1996) Interactions among gypsy moths, White-footed mice, and acorns. Ecology 77(8):2332–2342. https://doi.org/10.2307/2265735

Elkinton JS, Liebhold AM, Muzika RM (2004) Effects of alternative prey on predation by small mammals on gypsy moth pupae. Popul Ecol 46:171–178. https://doi.org/10.1007/s10144-004-0175-y

Elkinton JS, Parry D, Boettner GH (2006) Implicating an introduced generalist parasitoid in the invasive browntail moth's enigmatic demise. Ecology 87(10):2664–2672. https://doi.org/10.1890/ 0012-9658(2006)87[2664:IAIGPI]2.0.CO;2

Elkinton JS, Preisser E, Boettner G et al (2008) Factors influencing larval survival of the invasive Browntail moth (Lepidoptera: Lymantriidae) in relict North American populations. Environ Entomol 37(6):1429–1437. https://doi.org/10.1603/0046-225X-37.6.1429

Elkinton JS, Boettner GH, Sermac M et al (2010) Survey for Winter moth (Lepidoptera: Geometridae) in northeastern North America with pheromone-baited traps and hybridization with the native Bruce spanworm (Lepidoptera: Geometridae). Ann Entomol Soc Am 103(2):135–145. https://doi.org/10.1603/AN09118

Elkinton JS, Liebhold A, Boettner GH et al (2014) Invasion spread of *Operophtera brumata* in northeastern United States and hybridization with *O. bruceata*. Biol Invasions 16:2263–2272. https://doi.org/10.1007/s10530-014-0662-9

Elkinton JS, Boettner GH, Broadley HJ et al (2018) Biological control of the winter moth in the northeastern North America. USDA FHAAST-2018-03. 8p. https://www.fs.usda.gov/foresthea lth/publications/fhaast/index.shtml. Accessed 19 October 2022

Elkinton JS, Bittner TD, Pasquarella VJ et al (2019) Relating aerial deposition of *Entomophaga maimaiga* Conidia (Zoopagomycota: Entomophthorales) to mortality of Gypsy moth (Lepidoptera: Erebidae) larvae and nearby defoliation. Environ Entomol 48(5):1214–1222. https:// doi.org/10.1093/ee/nvz091

Elkinton JS, Boettner GH, Broadley HJ (2021) Successful biological control of winter moth, *Operophtera brumata*, in the northeastern United States. J Ecol Appl 31(5):e02326. https://doi.org/ 10.1002/eap.2326

Embree DG (1965) The population dynamics of the winter moth in Nova Scotia, 1954–1962. Mem Entomol Soc Can 97(S46):5–57. https://doi.org/10.4039/entm9746fv

Embree DG (1966) The role of introduced parasites in the control of the winter moth in Nova Scotia. Can Entomol 98(11):1159–1168. https://doi.org/10.4039/Ent981159-11

Embree DG (1967) Effects of winter moth on growth and mortality of red oak in Nova Scotia. For Sci 13(3):295–299. https://academic.oup.com/forestscience/article-abstract/13/3/295/4709561. Accessed 19 October 2022

Esper J, Büntgen U, Frank DC et al (2007) 1200 years of regular outbreaks in alpine insects. P Roy Soc B-Biol Sci 274:671–679. https://doi.org/10.1098/rspb.2006.0191

Fält-Nardmann JJJ, Ruohomäki K, Tikkanen O-P et al (2018a) Cold hardiness of *Lymantria monacha* and *L. dispar* (Lepidoptera: Erebidae) eggs to extreme winter temperatures: implications for predicting climate change impacts. Ecol Entomol 43(4):422–430. https://doi.org/10.1111/een.12515

Fält-Nardmann JJJ, Klemola T, Ruohomäki K et al (2018b) Local adaptations and phenotypic plasticity may render gypsy moth and nun moth future pests in northern European boreal forests. Can J Forest Res 48(3):265–276. https://doi.org/10.1139/cjfr-2016-0481

Fält-Nardmann JJJ, Tikkanen O-P, Ruohomäki K et al (2018c) The recent northward expansion of *Lymantria monacha* in relation to realised changes in temperatures of different seasons. Forest Ecol Manag 427:96–105. https://doi.org/10.1016/j.foreco.2018.05.053

Fahrner SJ, Albers JS, Albers MA et al (2016) Oviposition and feeding on red pine by jack pine budworm at a previously unrecorded scale. Agric for Entomol 18(3):214–222. https://doi.org/10.1111/afe.12154

Fedde GF (1965) The popular and scientific nomenclature of the Elm Spanworm, *Ennomos subsignarius* (Lepidoptera: Geometridae). Ann Entomol Soc Am 58(1):68–71. https://doi.org/10.1093/aesa/58.1.68

Feeny P (1970) Seasonal changes in oak leaf tannins and nutrients as a cause of spring feeding by winter moth caterpillars. Ecology 51(4):565–581. https://doi.org/10.2307/1934037

Fisher GT (1970) Parasites and predators of species of a saddled prominent complex at Groton, Vermont. J Econ Entomol 63(5):1613–1614. https://doi.org/10.1093/jee/63.5.1613

Fiske WF, Burgess AF (1910) The natural control of Heterocampa guttivitta. J Econ Entomol 3(5):389–394. https://doi.org/10.1093/jee/3.5.389

Flower A (2016) Three centuries of synchronous forest defoliator outbreaks in western North America. PLoS ONE 11(10):e0164737. https://doi.org/10.1371/journal.pone.0164737

Flower A, Gavin DG, Heyerdahl EK et al (2014a) Drought-triggered western spruce budworm outbreaks in the interior Pacific Northwest: A multi-century dendrochronological record. For Ecol Manag 324:16–27. https://doi.org/10.1016/j.foreco.2014.03.042

Flower A, Gavin DG, Heyerdahl EK et al (2014b) Western spruce budworm outbreaks did not increase fire risk over the last three centuries: a dendrochronological analysis of inter-disturbance synergism. PLoS ONE 9(12):e114282. https://doi.org/10.1371/journal.pone.0114282

Frago E, Guara M, Pujade-Villar J et al (2010) Winter feeding leads to a shifted phenology in the browntail moth *Euproctis chrysorrhoea* on the evergreen strawberry tree *Arbutus unedo*. Agric for Entomol 12(4):381–388. https://doi.org/10.1111/j.1461-9563.2010.00489.x

Frago E, Pujade-Villar J, Guara M et al (2011) Providing insights into browntail moth local outbreaks by combining life table data and semi-parametric statistics. Ecol Entomol 36(2):188–199. https://doi.org/10.1111/j.1365-2311.2010.01259.x

Frago E, Pujade-Villar J, Guara M et al (2012) Hyperparasitism and seasonal patterns of parasitism as potential causes of low top-down control in *Euproctis chrysorrhoea* L. (Lymantriidae). Biol Control 60(2):123–131. https://doi.org/10.1016/j.biocontrol.2011.11.013

Frank JH (1967) The insect predators of the pupal stage of the winter moth, *Operophtera brumata* (L) (Lepidoptera: Hydriomenidae). J Anim Ecol 36(2):375–389. https://doi.org/10.2307/2920

Fry HRC, Quiring DT, Ryall KL et al (2008) Relationships between elm spanworm, *Ennomos subsignaria*, juvenile density and defoliation on mature sycamore maple in an urban environment. Forest Ecol Manag 255(7):2726–2732. https://doi.org/10.1016/j.foreco.2008.01.039

Fry HRC, Quiring DT, Ryall KL et al (2009) Influence of intra-tree variation in phenology and oviposition site on the distribution and performance of *Ennomos subsignaria* on mature sycamore maple. Ecol Entomol 34(3):394–405. https://doi.org/10.1111/j.1365-2311.2009.01091.x

Fuentealba A, Pureswaran D, Bauce É et al (2017) How does synchrony with host plant affect the performance of an outbreaking insect defoliator? Oecologia 184:847–857. https://doi.org/10. 1007/s00442-017-3914-4

Fuester RW, Hajek AE, Elkinton JS et al (2014) Gypsy moth (*Lymantria dispar* L.) (Lepidoptera: Erebidae: Lymantriinae). In: Van Driesche R, Reardon R (eds) The use of classical biological control to preserve forests in North America. USDA FHET-2013 2:49–82. https://bugwoodcloud. org/resource/files/14888.pdf. Accessed 19 October 2022

Gansner DA, Arner SL, Widmann RH et al (1993) After two decades of gypsy moth, is there any oak left? North J Appl for 10(4):184–186. https://doi.org/10.1093/njaf/10.4.184

Georgiev G, Hubenov Z, Mirchev P et al (2022) New tachinid parasitoids on pine processionary moth (*Thaumetopoea pityocampa*) (Diptera: Tachinidae) in Bulgaria. Silva Balcanica 23(1):5–10. https://doi.org/10.3897/silvabalcanica.23.e81890

Germain M, Kneeshaw D, De Grandpré L et al (2021) Insectivorous songbirds as early indicators of future defoliation by spruce budworm. Landscape Ecol 36(10):3013–3027. https://doi.org/10. 1007/s10980-021-01300-z

Ghirardo A, Heller W, Fladung M et al (2012) Function of defensive volatiles in pedunculate oak (*Quercus robur*) is tricked by the moth *Tortrix viridana*. Plant Cell Environ 35(12):2192–2207. https://doi.org/10.1111/j.1365-3040.2012.02545.x

Gillespie DR, Finlayson T, Tonks NV et al (1978) Occurrence of the winter moth, *Operophtera brumata* (Lepidoptera: Geometridae), on southern Vancouver Island, British Columbia. Can Entomol 110(2):223–224.https://doi.org/10.4039/Ent110223-2

Gilligan TM, Brown JW (2014) A new name for the western spruce budworm (Lepidoptera: Tortricidae)? Can Entomol 146(6):583–589. https://doi.org/10.4039/tce.2014.17

Goodbody TRH, Coops NC, Hermosilla T et al (2018) Digital aerial photogrammetry for assessing cumulative spruce budworm defoliation and enhancing forest inventories at a landscape-level. ISPRS J Photogramm 142:1–11. https://doi.org/10.1016/j.isprsjprs.2018.05.012

Gray DR (2018) Age-dependent developmental response to temperature: an examination of the rarely tested phenomenon in two species (gypsy moth (*Lymantria dispar*) and winter moth (*Operophtera brunata*)). InSects 9(2):41. https://doi.org/10.3390/insects9020041

Gur'yanova TM (2006) Fecundity of the European pine sawfly *Neodiprion sertifer* (Hymenoptera, Diprionidae) related to cyclic outbreaks: invariance effects. Entomol Rev 86:910–921. https:// doi.org/10.1134/S0013873806080069

Gwiazdowski RA, Elkinton JS, Dewaard JR et al (2013) Phylogeographic diversity of the winter moths *Operophtera brumata* and *O. bruceata* (Lepidoptera: Geometridae) in Europe and North America. Ann Entomol Soc Am 106(2):143–151. https://doi.org/10.1603/AN12033

Habermann M (2000) The larch casebearer and its host tree: I. Population dynamics of the larch casebearer (*Coleophora laricella* Hbn.) from latent to outbreak density in the field. Forest Ecol Manag 136(1–3):11–22. https://doi.org/10.1016/S0378-1127(99)00266-2

Hajek AE (1997) Fungal and viral epizootics in gypsy moth (Lepidoptera: Lymantriidae) populations in central New York. Biol Control 10(1):58–68. https://doi.org/10.1006/bcon.1997. 0541

Hajek AE (1999) Pathology and epizootiology of *Entomophaga maimaiga* infections in forest Lepidoptera. Microbiol Mol Biol R 63(4):814–835. https://doi.org/10.1128/MMBR.63.4.814- 835.1999

Hajek AE (2007) Introduction of a fungus into North America for control of gypsy moth. In: Vincent C, Goettel MS, Lazarovits G (eds) Biological control: a global perspective: case studies from around the world, CABI, Cambridge, MA, pp 53–62. https://doi.org/10.1079/978184593 2657.0053

Hajek AE, Tobin PC (2010) Micro-managing arthropod invasions: eradication and control of invasive arthropods with microbes. Biol Invasions 12:2895–2912. https://doi.org/10.1007/s10530- 010-9735-6

Hajek AE, Carruthers RI, Soper RS (1990a) Temperature and moisture relations of sporulation and germination by *Entomophaga-maimaiga* (Zygomycetes: Entomophthoraceae), a fungal pathogen of *Lymantria dispar* (Lepidoptera: Lymantriidae). Environ Entomol 19(1):85–90. https://doi.org/10.1093/ee/19.1.85

Hajek AE, Humber RA, Elkinton JS et al (1990b) Allozyme and restriction fragment length polymorphism analyses confirm *Entomophaga maimaiga* responsible for 1989 epizootics in North American gypsy moth populations. P Natl Acad Sci USA 87(18):6979–6982. https://doi.org/10.1073/pnas.87.18.6979

Hajek AE, Elkinton JS, Witcosky JJ (1996) Introduction and spread of the fungal pathogen *Entomophaga maimaiga* (Zygomycetes: Entomophthorales) along the leading edge of gypsy moth (Lepidoptera: Lymantriidae) spread. Environ Entomol 25(5):1235–1247. https://doi.org/10.1093/ee/25.5.1235

Hajek AE, Tobin PC, Haynes KJ (2015) Replacement of a dominant viral pathogen by a fungal pathogen does not alter the collapse of a regional forest insect outbreak. Oecologia 177:785–797. https://doi.org/10.1007/s00442-014-3164-7

Hajek AE, Gardescu S, Delalibera I (2020) Summary of classical biological control introductions of entomopathogens and nematodes for insect control. Biocontrol 66:1671–2180. https://doi.org/10.1007/s10526-020-10046-7

Hall RJ, Skakun RS, Arsenault EJ (2006) Remotely sensed data in the mapping of insect defoliation. In: Wulder MA, Franklin SE (eds) Understanding forest disturbance and spatial pattern: remote sensing and GIS approaches, Taylor & Francis, Boca Raton, FL, pp 85–111. https://www.researchgate.net/publication/238730970. Accessed 19 October 2022

Hartl-Meier C, Esper J, Liebhold A et al (2017) Effects of host abundance on larch budmoth outbreaks in the European Alps. Agric for Entomol 19(4):376–387. https://doi.org/10.1111/afe.12216

Hassell MP (1968) Behavioural response of a tachinid fly (*Cyzenis albicans* (Fall) to its host, the winter moth (*Operophtera brumata* (L)). J Anim Ecol 37(3):627–639. https://doi.org/10.2307/3079

Hassell MP (1969a) A study of mortality factors acting upon *Cyzenis albicans* (Fall), a tachinid parasite of the winter moth (*Operophtera brumata* (L)). J Anim Ecol 38(2):329–339. https://doi.org/10.2307/2774

Hassell MP (1969b) A population model for the interaction between *Cyzenis albicans* (Fall.) (Tachinidae) and *Operophtera brumata* (L.) (Geometridae) at Wytham, Berkshire. J Anim Ecol 38(3):567–576. https://doi.org/10.2307/3035

Hassell MP (1980) Foraging strategies, population models and biological control—a case study. J Anim Ecol 49(2):603–628. https://doi.org/10.2307/4267

Haukioja E (1991) Induction of defenses in trees. Annu Rev Entomol 36:25–42. https://doi.org/10.1146/annurev.en.36.010191.000325

Havill NP, Elkinton J, Andersen JC et al (2017) Asymmetric hybridization between non-native winter moth, *Operophtera brumata* (Lepidoptera: Geometridae), and native Bruce spanworm, *Operophtera bruceata*, in the Northeastern United States, assessed with novel microsatellites and SNPs. Bull Entomol Res 107(2):241–250. https://doi.org/10.1017/S0007485316000857

Hawboldt LS, Cuming FG (1950) Cankerworms and European winter moth in Nova Scotia. Bi-m Progr Rep For Insect Invest Dep Agric Can 6(1):1–2. https://afc-fr.cfsnet.nfis.org/fias/pdfs/afc/atlantic_bimonthly_1950-6(1-6).pdf. Accessed 18 October 2022

Haynes KJ, Liebhold AM, Johnson DM (2009a) Spatial analysis of harmonic oscillation of gypsy moth outbreak intensity. Oecologia 159:249–256. https://doi.org/10.1007/s00442-008-1207-7

Haynes KJ, Liebhold AM, Fearer TM et al (2009b) Spatial synchrony propagates through a forest food web via consumer-resource interactions. Ecology 90(11):2974–2983. https://doi.org/10.1890/08-1709.1

Haynes KJ, Bjørnstad ON, Allstadt AJ et al (2013) Geographical variation in the spatial synchrony of a forest-defoliating insect: isolation of environmental and spatial drivers. P Roy Soc B-Biol Sci 280:20122373. https://doi.org/10.1098/rspb.2012.2373

Haynes KJ, Liebhold AM, Bjørnstad ON et al (2018a) Geographic variation in forest composition and precipitation predict the synchrony of forest insect outbreaks. Oikos 127(4):634–642. https://doi.org/10.1111/oik.04388

Haynes KJ, Tardif JC, Parry D (2018b) Drought and surface-level solar radiation predict the severity of outbreaks of a widespread defoliating insect. Ecosphere 9(8):e02387. https://doi.org/10.1002/ecs2.2387

Hébert C, Berthiaume R, Dupont A et al (2001) Population collapses in a forecasted outbreak of *Lambdina fiscellaria* (Lepidoptera: Geometridae) caused by spring egg parasitism by *Telenomus* spp. (Hymenoptera: Scelionidae). Environ Entomol 30(1):37–43. https://doi.org/10.1603/0046-225X-30.1.37

Hengeveld R (1989) Dynamics of biological invasions. Chapman and Hall, London

Hentschel R, Möller K, Wenning A et al (2018) Importance of ecological variables in explaining population dynamics of three important pine pest insects. Front Plant Sci 9:1667. https://doi.org/10.3389/fpls.2018.01667

Hibbard EL, Elkinton JS (2015) Effect of spring and winter temperatures on winter moth (Geometridae: Lepidoptera) larval eclosion in the northeastern United States. Environ Entomol 44(3):798–807. https://doi.org/10.1093/ee/nvv006

Hódar JA, Castro J, Zamora R (2003) Pine processionary caterpillar *Thaumetopoea pityocampa* as a new threat for relict Mediterranean Scots pine forests under climatic warming. Biol Conserv 110(1):123–129. https://doi.org/10.1016/S0006-3207(02)00183-0

Hódar JA, Torres-Muros L, Zamora R et al (2015) No evidence of induced defence after defoliation in three pine species against an expanding pest, the pine processionary moth. Forest Ecol Manag 356:166–172. https://doi.org/10.1016/j.foreco.2015.07.022

Holling CS (1959) The components of predation as revealed by a study of small-mammal predation of the European pine sawfly. Can Entomol 91(5):293–320. https://doi.org/10.4039/Ent91293-5

Horstmann K (1977) Wood ants (*Formica polyctena* Foerster) as mortality factors in population dynamics of the oak tortrix *Tortrix viridana* L. Z Angew Entomol 82(4):421–435. https://www.cabdirect.org/cabdirect/abstract/19770548022. Accessed 20 October 2022

Hughes JS, Cobbold CA, Haynes K et al (2015) Effects of Forest Spatial Structure on Insect Outbreaks: Insights from a Host-Parasitoid Model. Dig Am Nat 185(5):E130-152. https://doi.org/10.1086/680860

Hummel S, Agee JK (2003) Western spruce budworm defoliation effects on forest structure and potential fire behavior. Northwest Science 77(2):159–169. https://hdl.handle.net/2376/807

Humphrey N (1996) Satin moth in British Columbia. PFC CFS Forest Pest Leaflet No 38:1–4. https://www.cabdirect.org/cabdirect/abstract/19971102190. Accessed 20 October 2022

Hunter MD (1998) Interactions between *Operophtera brumata* and *Tortrix viridana* on oak: new evidence from time-series analysis. Ecol Entomol 23(2):168–173. https://doi.org/10.1046/j.1365-2311.1998.00124.x

Ilyinykh A (2011) Analysis of the causes of declines in Western Siberian outbreaks of the nun moth *Lymantria monacha*. Biocontrol 56:123–131. https://doi.org/10.1007/s10526-010-9316-8

Iqbal J, MacLean DA, Kershaw JA Jr (2011) Impacts of hemlock looper defoliation on growth and survival of balsam fir, black spruce and white birch in Newfoundland, Canada. Forest Ecol Manag 261(6):1106–1114.https://doi.org/10.1016/j.foreco.2010.12.037

Ivashov AV, Simchuk AP, Medvedkov DA (2001) Possible role of inhibitors of trypsin-like proteases in the resistance of oaks to damage by oak leafroller *Tortrix viridana* L. and gypsy moth *Lymantria dispar* L. Ecol Entomol 26:664–668. https://resjournals.onlinelibrary.wiley.com/doi/pdf/10.1046/j.1365-2311.2001.00362.x. Accessed 20 October 2022

Ives WGH (1984) *Operophtera bruceata* (Hulst), Bruce Spanworm (Lepidoptera: Geometridae). In: Kelleher JS, Hulme MA (eds) (1984) Biological control programmes against insects and weeds in Canada 1969–1980. CAB, Slough, England, pp 349–351.

Ives WGH, Cunningham JC (1980) Application of nuclear polyhedrosis virus to control Bruce spanworm (Lepidoptera: Geometridae). Can Entomol 112(7):741–744. https://doi.org/10.4039/Ent112741-7

Iyengar SV, Balakrishnan J, Kurths J (2016) Impact of climate change on larch budmoth cyclic outbreaks. Sci Rep 6:27845. https://doi.org/10.1038/srep27845

Jacquet J-S, Orazio C, Jactel H (2012) Defoliation by processionary moth significantly reduces tree growth: a quantitative review. Ann for Sci 69:857–866. https://doi.org/10.1007/s13595-012-0209-0

Jeger M, Bragard C, Caffier D et al (2018) Pest categorisation of *Dendrolimus sibiricus*. EFSA J 16(6):e05301. https://doi.org/10.2903/j.efsa.2018.5301

Jepsen JU, Hagen SB, Ims RA et al (2008) Climate change and outbreaks of the geometrids *Operophtera brumata* and *Epirrita autumnata* in subarctic birch forest: evidence of a recent outbreak range expansion. J Anim Ecol 77(2):257–264. https://doi.org/10.1111/j.1365-2656.2007.01339.x

Jepsen JU, Hagen SB, Hogda KA et al (2009a) Monitoring the spatio-temporal dynamics of geometrid moth outbreaks in birch forest using MODIS-NDVI data. Remote Sens Environ 113:1939–1947. https://scholar.google.ca/scholar?q=Monitoring+the+spatio-temporal+dynamics+of+geometrid+moth+outbreaks+in+birch+forest+using+MODIS-NDVI+data&hl=en&as_sdt=0&as_vis=1&oi=scholart. Accessed 20 October 2022

Jepsen JU, Hagen SB, Karlsen SR et al (2009b) Phase-dependent outbreak dynamics of geometrid moth linked to host plant phenology. Proc R Soc B-Biol Sci 276(1676):4119–4128. https://doi.org/10.1098/rspb.2009.1148

Jepsen JU, Kapari L, Hagen SB et al (2011) Rapid northwards expansion of a forest insect pest attributed to spring phenology matching with sub-Arctic birch. Global Change Biol 17(6):2071–2083. https://doi.org/10.1111/j.1365-2486.2010.02370.x

Jepsen JU, Vindstad OPL, Barraquand F et al (2016) Continental-scale travelling waves in forest geometrids in Europe: an evaluation of the evidence. J Anim Ecol 85(2):385–390. https://doi.org/10.1111/1365-2656.12444

Johnson DM, Liebhold AM, Bjørnstad ON (2006a) Geographical variation in the periodicity of gypsy moth outbreaks. Ecography 29(3):367–374. https://doi.org/10.1111/j.2006.0906-7590.04448.x

Johnson DM, Liebhold AM, Tobin PC et al (2006b) Allee effects and pulsed invasion by the gypsy moth. Nature 444:361–363. https://doi.org/10.1038/nature05242

Kamata N (2000) Population dynamics of the beech caterpillar, *Syntypistis punctatella*, and biotic and abiotic factors. Popul Ecol 42(3):267–278. https://doi.org/10.1007/PL00012005

Kamata N (2002) Outbreaks of forest defoliating insects in Japan, 1950–2000. Bull Entomol Res 92(2):109–117. https://doi.org/10.1079/BER2002159

Kaya HK, Anderson JF (1974) Collapse of elm spanworm outbreak in Connecticut: role of *Ooencyrtus* sp. Environ Entomol 3(4):659–663. https://doi.org/10.1093/ee/3.4.659

Keena MA (2003) Survival and development of *Lymantria monacha* (Lepidoptera: Lymantriidae) on North American and introduced Eurasian tree species. J Econ Entomol 96(1):43–52. https://doi.org/10.1093/jee/96.1.43

Keena MA, Côté MJ, Grinberg PS et al (2008) World distribution of female flight and genetic variation in *Lymantria dispar* (Lepidoptera: Lymantriidae). Environ Entomol 37(1):636–649. https://doi.org/10.1603/0046-225X(2008)37[636:WDOFFA]2.0.CO;2

Kegg JD (1967) Sampling techniques for predicting fall cankerworm defoliation. J Econ Entomol 60(3):889–890. https://doi.org/10.1093/jee/60.3.889

Kegg JD (1973) Oak mortality caused by repeated gypsy moth defoliations in New Jersey. J Econ Entomol 66(3):639–641. https://doi.org/10.1093/jee/66.3.639

Kelly PM, Sterling PH, Speight MR et al (1988) Preliminary spray trials of a nuclear polyhedrosis virus as a control agent for the brown-tail moth, *Euproctis chrysorrhoea* (L) (Lepidoptera: Lymantriidae). Bull Entomol Res 78(2):227–234. https://doi.org/10.1017/S0007485300012992

Kenis M, Kloosterman K (2001) European parasitoids of the pine false webworm (*Acantholyda erythrocephala* (L.)) and their potential for biological control in North America. USDA GTR NE-277:65–73. https://www.researchgate.net/publication/242313271_European_Parasitoids_of_the_Pine_False_Webworm_Acantholyda_erythrocephala_L_and_Their_Potential_for_Biological_Control_in_North_America

Kenis M, Hurley BP, Hajek AE et al (2017) Classical biological control of insect pests of trees: facts and figures. Biol Invasions 19:3401–3417. https://doi.org/10.1007/s10530-017-1414-4

Kerslake JE, Kruuk LEB, Hartley SE et al (1996) Winter moth (*Operophtera brumata* (Lepidoptera: Geometridae)) outbreaks on Scottish heather moorlands: effects of host plant and parasitoids on larval survival and development. Bull Entomol Res 86(2):155–164. https://doi.org/10.1017/S00 07485300052391

Kharuk VI, Ranson KJ, Kozuhovskaya AG et al (2004) NOAA/AVHRR satellite detection of Siberian silkmoth outbreaks in eastern Siberia. Int J Remote Sens 25(24):5543–5555. https:// doi.org/10.1080/01431160410001719858

Kharuk VI, Im ST, Yagunov MN (2018) Migration of the northern boundary of the Siberian silk moth. Contemp Probl Ecol 11:26–34. https://doi.org/10.1134/S1995425518010055

Kimberling DN, Miller JC, Penrose RL (1986) Distribution and parasitism of winter moth, *Operophtera brumata* (Lepidoptera: Geometridae), in western Oregon. Environ Entomol 15(5):1042–1046. https://doi.org/10.1093/ee/15.5.1042

Kirichenko NI, Baranchikov YN, Vidal S (2009) Performance of the potentially invasive Siberian moth *Dendrolimus superans sibiricus* on coniferous species in Europe. Agric for Entomol 11(3):247–254. https://doi.org/10.1111/j.1461-9563.2009.00437.x

Klapwijk MJ, Björkman C (2018) Mixed forests to mitigate risk of insect outbreaks. Scand J Forest Res 33(8):772–780. https://doi.org/10.1080/02827581.2018.1502805

Klapwijk MJ, Csóka G, Hirka A et al (2013) Forest insects and climate change: long-term trends in herbivore damage. Ecol Evol 3(12):4183–4196. https://doi.org/10.1002/ece3.717

Klemola N, Heisswolf A, Ammunét T et al (2009) Reversed impacts by specialist parasitoids and generalist predators may explain a phase lag in moth cycles: a novel hypothesis and preliminary field tests. Ann Zool Fenn 46(5):380–393. https://doi.org/10.5735/086.046.0504

Klemola N, Andersson T, Ruohomäki K et al (2010) Experimental test of parasitism hypothesis for population cycles of a forest lepidopteran. Ecology 91(9):2506–2513. https://doi.org/10.1890/09-2076.1

Klemola T, Ruohomäki K, Andersson T et al (2004) Reduction in size and fecundity of the autumnal moth, *Epirrita autumnata*, in the increase phase of a population cycle. Oecologia 141:47–56. https://doi.org/10.1007/s00442-004-1642-z

Klemola T, Andersson T, Ruohomäki K (2008) Fecundity of the autumnal moth depends on pooled geometrid abundance without a time lag: implications for cyclic population dynamics. J Anim Ecol 77(3):597–604. https://doi.org/10.1111/j.1365-2656.2008.01369.x

Klemola T, Andersson T, Ruohomäki K (2014) Delayed density-dependent parasitism of eggs and pupae as a contributor to the cyclic population dynamics of the autumnal moth. Oecologia 175:1211–1225. https://doi.org/10.1007/s00442-014-2984-9

Kollberg I, Bylund H, Schmidt A et al (2013) Multiple effects of temperature, photoperiod and food quality on the performance of a pine sawfly. Ecol Entomol 38(2):201–208. https://doi.org/10.1111/een.12005

Kollberg I, Bylund H, Huitu O et al (2014) Regulation of forest defoliating insects through small mammal predation: reconsidering the mechanisms. Oecologia 176:975–983. https://doi.org/10.1007/s00442-014-3080-x

Kosunen M, Kantola T, Starr M et al (2016) Influence of soil and topography on defoliation intensity during an extended outbreak of the common pine sawfly (*Diprion pini* L.). IForest 10(1):164–171. https://doi.org/10.3832/ifor2069-009

Kouki J, Lyytikäinen-Saarenmaa P, Henttonen H et al (1998) Cocoon predation on diprionid sawflies: the effect of forest fertility. Oecologia 116:482–488. https://doi.org/10.1007/s004420050613

Kovacs K, Ranson KJ, Kharuk VI (2005) Detecting siberian silk moth damage in central Siberia using multi-temporal MODIS data. In: International workshop on the analysis of multi-temporal remote sensing images, pp 25–29. https://doi.org/10.1109/AMTRSI.2005.1469833

Kress A, Saurer M, Büntgen U et al (2009) Summer temperature dependency of larch budmoth outbreaks revealed by Alpine tree-ring isotope chronologies. Oecologia 160:353–365. https://doi.org/10.1007/s00442-009-1290-4

Kulman HM (1971) Effects of insect defoliation on growth and mortality of trees. Ann Rev Entomol 16:289–324. https://doi.org/10.1146/annurev.en.16.010171.001445

Kyle CH, Liu J, Gallagher ME et al (2020) Stochasticity and infectious disease dynamics: density and weather effects on a fungal insect pathogen. Am Nat 195(3):504–523. https://doi.org/10.1086/707138

Lait LA, Hebert PDN (2018) Phylogeographic structure in three North American tent caterpillar species (Lepidoptera: Lasiocampidae): *Malacosoma americana, M. californica,* and *M. disstria.* Peerj 6:e4479. https://doi.org/10.7717/peerj.4479

Lance DR, Elkinton JS, Schwalbe CP (1987) Behavior of late-instar gypsy moth larvae in high and low density populations. Ecol Entomol 12(3):267–273. https://doi.org/10.1111/j.1365-2311.1987.tb01005.x

Larsson S, Ekbom B, Björkman C (2000) Influence of plant quality on pine sawfly population dynamics. Oikos 89(3):440–450. https://doi.org/10.1034/j.1600-0706.2000.890303.x

Lee K-S, Kang TH, Jeong JW et al (2015) Taxonomic review of the genus *Lymantria* (Lepidoptera: Erebidae: Lymantriinae) in Korea. Entomol Res 45(5):225–234. https://doi.org/10.1111/1748-5967.12116

Legault S, James PMA (2018) Parasitism rates of spruce budworm larvae: testing the enemy hypothesis along a gradient of forest diversity measured at different spatial scales. Environ Entomol 47(5):1083–1095. https://doi.org/10.1093/ee/nvy113

Legault S, Hébert C, Blais J et al (2012) Seasonal ecology and thermal constraints of *Telenomus* spp. (Hymenoptera: Scelionidae), egg parasitoids of the Hemlock looper (Lepidoptera: Geometridae). Environ Entomol 41(6):1290–1301. https://doi.org/10.1603/EN12129

Li L, Huanwen M, Xu S (2002) Bionomics of *Dendrolimus superans* Butler. J Inner Mongolia Agri Univ (Natural Science Edition) 23(1):101–103.

Li S, Daudin JJ, Piou D et al (2015) Periodicity and synchrony of pine processionary moth outbreaks in France. Forest Ecol Manag 354:309–317. https://doi.org/10.1016/j.foreco.2015.05.023

Liebhold A, Kamata N (2000) Population dynamics of forest-defoliating insects. Popul Ecol 42(3):205–209. https://doi.org/10.1007/PL00011999

Liebhold A, Koenig WD, Bjørnstad ON (2004) Spatial synchrony in population dynamics. Ann Rev Ecol Evol Syst 35:467–490. https://www.jstor.org/stable/30034124

Liebhold AM, Halverson JA, Elmes GA (1992) Gypsy moth invasion in North America: a quantitative analysis. J Biogeogr 19(5):513–520. https://doi.org/10.2307/2845770

Liebhold AM, Gottschalk KW, Muzika R-M et al (1995) Suitability of North American tree species to the gypsy moth: a summary of field and laboratory tests. USDA GTR NE-211. https://doi.org/10.2737/NE-GTR-211. Accessed 21 October 2022

Liebhold AM, Sharov AA, Tobin PC (2007) Population biology of gypsy moth spread. In: Tobin PC, Blackburn LM (eds) "Slow the spread: a national program to manage the gypsy moth". USDA GTR NRS-6:15–32. https://doi.org/10.2737/NRS-GTR-6. Accessed 21 October 2022

Liebhold AM, Plymale R, Elkinton JS et al (2013) Emergent fungal entomopathogen does not alter density dependence in a viral competitor. Ecology 94(6):1217–1222. https://doi.org/10.1890/12-1329.1

Liebhold AM, Berec L, Brockerhoff EG et al (2016) Eradication of invading insect populations: from concepts to applications. Ann Rev Entomol 61:335–352. https://doi.org/10.1146/annurev-ento-010715-023809

Liebhold AM, Björkman C, Roques A et al (2020) Outbreaking forest insect drives phase synchrony among sympatric folivores: Exploring potential mechanisms. Popul Ecol 62(4):372–384. https://doi.org/10.1002/1438-390X.12060

Liebhold AM, Hajek AE, Walter JA et al (2022) Historical change in the outbreak dynamics of an invading forest insect. Biol Invasions 24:879–889. https://doi.org/10.1007/s10530-021-02682-6

Lumley LM, Pouliot E, Laroche J et al (2020) Continent-wide population genomic structure and phylogeography of North America's most destructive conifer defoliator, the spruce budworm (*Choristoneura fumiferana*). Ecol Evol 10(2):914–927. https://doi.org/10.1002/ece3.5950

Lyytikäinen-Saarenmaa P, Varama M, Anderbrant O et al (2001) Predicting pine sawfly population densities and subsequent defoliation with pheromone traps. In: Integrated management and dynamics of forest defoliating insects, Proceedings, USDA GTR NE-277:108–116. https://www.fs.usda.gov/nrs/pubs/gtr/gtr_ne277.pdf. Accessed 21 October 2022

Maclauchlan LE, Brooks JE, Hodge JC (2006) Analysis of historic western spruce budworm defoliation in south central British Columbia. For Ecol Manag 226(1–3):351–356. https://doi.org/10.1016/j.foreco.2006.02.003

MacLean DA (2016) Impacts of insect outbreaks on tree mortality, productivity, and stand development. Can Entomol 148(Suppl 1):138–159. https://doi.org/10.4039/tce.2015.24

MacLean DA, Ebert P (1999) The impact of hemlock looper (*Lambdina fiscellaria fiscellaria* (Guen.)) on balsam fir and spruce in New Brunswick, Canada. Forest Ecol Manag 120(103):77–87. https://doi.org/10.1016/S0378-1127(98)00527-1

MacLean DA, Amirault P, Amos-Binks L et al (2019) Positive results of an early intervention strategy to suppress a spruce budworm outbreak after five years of trials. Forests 10(5):448. https://doi.org/10.3390/f10050448

MacPhee A, Newton A, McRae KB (1988) Population studies on the winter moth *Operophtera brumata* (L) (Lepidoptera: Geometridae) in apple orchards in Nova Scotia. Can Entomol 120(1):73–83. https://doi.org/10.4039/Ent12073-1

Maeto K (1991) Outbreaks of *Dendrolimus superans* (Butler) (Lepidoptera: Lasiocampidae) related to weather in Hokkaido. App Entomol Zool 26(2):275–277. https://doi.org/10.1303/aez.26.275

Maksimov SA (1999) On factors responsible for population outbreaks in nun moth (*Lymantria monacha* L.). Russian J Ecol 30:47–51

Malakar R, Elkinton JS, Hajek AE et al (1999) Within-host interactions of *Lymantria dispar* (Lepidoptera: Lymantriidae) nucleopolyhedrosis virus and *Entomophaga maimaiga* (Zygomycetes: Entomophthorales). J Invertebr Pathol 73(1):91–100. https://doi.org/10.1006/jipa.1998.4806

Malakar RD (1997) Interactions between two gypsy moth (*Lymantria dispar* L.) pathogens: nuclear polyhedrosis virus and *Entomophaga maimaiga* (Entomophthorales: Zygomycetes). UMASS, PhD Dissertation. https://scholarworks.umass.edu/dissertations/AAI9809364/. Accessed 21 October 2022

Mamet SD, Chun KP, Metsaranta JM et al (2015) Tree rings provide early warning signals of jack pine mortality across a moisture gradient in the southern boreal forest. Environ Res Lett 10(8):084021. https://doi.org/10.1088/1748-9326/10/8/084021

Mannai Y, Ben Jamâa ML, M'nara S et al (2010) Biology of *Tortrix viridana* (Lep., Tortricidae) in cork oak forests of North-West Tunisia 57:153–160. https://www.iobc-wprs.org/members/shop_en.cfm

Marques JF, Wang HL, Svensson GP et al (2014) Genetic divergence and evidence for sympatric host-races in the highly polyphagous brown tail moth, *Euproctis chrysorrhoea* (Lepidoptera: Erebidae). Evol Ecol 28:829–848. https://doi.org/10.1007/s10682-014-9701-3

Martin JC, Mesmin X, Buradino M et al (2022) Complex drivers of phenology in the pine processionary moth: lessons from the past. Agric for Entomol 24(2):247–259. https://doi.org/10.1111/afe.12488

Martinat PJ, Allen DC (1987) Relationship between outbreaks of saddled prominent, *Heterocampa guttivitta* (Lepidoptera: Notodontidae), and drought. Environ Entomol 16(1):246–249. https://doi.org/10.1093/ee/16.1.246

Mason CJ, McManus ML (1981) Larval dispersal of the gypsy moth. In: Doane CC, McManus ML (eds) The Gypsy moth: research toward integrated pest management. USDA APHIS TB-1584:161–202. https://handle.nal.usda.gov/10113/CAT82474520. Accessed 24 October 2022

Mason RR (1976) Life tables for a declining population of Douglas-fir tussock moth in northeastern Oregon. Ann Entomol Soc Am 69(1):948–958. https://doi.org/10.1093/aesa/69.5.948

Mason RR (1978) Synchronous patterns in an outbreak of Douglas-fir tussock moth. Environ Entomol 7(5):672–675. https://doi.org/10.1093/ee/7.5.672

Mason RR (1996) Dynamic behavior of Douglas-fir tussock moth populations in the Pacific North-west. For Sci 42(2):182–191. https://academic.oup.com/forestscience/article/42/2/182/4627302. Accessed 24 October 2022

Mason RR, Torgersen TR (1987) Dynamics of a nonoutbreak population of the Douglas-fir tussock moth (Lepidoptera, Lymantriidae) in southern Oregon. Environ Entomol 16(6):1217–1227. https://doi.org/10.1093/ee/16.6.1217

Mason RR, Torgersen TR, Wickman B et al (1983) Natural regulation of a Douglas-fir tussock moth (Lepidoptera, Lymantriidae) population in the Sierra-Nevada. Environ Entomol 12(2):587–594. https://doi.org/10.1093/ee/12.2.587

Mason RR, Scott DW, Paul HG (1993) Forecasting outbreaks of the Douglas-fir tussock moth from lower crown cocoon samples. USDA PNW-RP-460:1–12. https://doi.org/10.2737/PNW-RP-460

May RM (1975) Deterministic models with chaotic dynamics. Nature 256:165–166. https://doi.org/10.1038/256165a0

Mayfield AE III, Allen DC, Briggs RD (2007) Site and stand conditions associated with pine false webworm populations and damage in mature eastern white pine plantations. North J Appl for 24(3):168–176. https://doi.org/10.1093/njaf/24.3.168

McCullough DG (2000) A review of factors affecting the population dynamics of jack pine budworm (*Choristoneura pinus pinus* Freeman). Popul Ecol 42(3):243–256. https://doi.org/10.1007/PL0 0012003

McKnight ME (1974) Parasitoids of western spruce budworm in Colorado. Environ Entomol 3(1):186–187. https://doi.org/10.1093/ee/3.1.186

Meigs GW, Kennedy RE, Gray AN et al (2015) Spatiotemporal dynamics of recent mountain pine beetle and western spruce budworm outbreaks across the Pacific Northwest Region, USA. Forest Ecol Manag 339:71–86. https://doi.org/10.1016/j.foreco.2014.11.030

Melin M, Viiri H, Tikkanen OP et al (2020) From a rare inhabitant into a potential pest–status of the nun moth in Finland based on pheromone trapping. Silva Fennica 54(1):10262. https://doi.org/10.14214/sf.10262

Mirchev P, Georgiev G, Zaemdzhikova G et al (2021) Impact of egg parasitoids on pine proces-sionary moth *Thaumetopoea pityocampa* (Denis & Schiffermüller, 1775) (Lepidoptera: Notodon-tidae) in a new habitat. Acta Zool Bulg 73(1):131–134. https://www.researchgate.net/publication/350484967

Mitter C, Futuyma D (1977) Parthenogenesis in Fall Cankerworm, *Alsophila pometaria* (Lepi-doptera: Geometridae). Entomol Exp Appl 21(2):192–198. https://doi.org/10.1111/j.1570-7458.1977.tb02672.x

Mitter C, Neal JW, Gott KM et al (1987) A geographic comparison of pseudogamous populations of the fall cankerworm (*Alsophila pometaria*). Entomol Exp Appl 43(2):133–143. https://doi.org/10.1111/j.1570-7458.1987.tb03597.x

Moraal LG, Jagers op Akkerhuis GAJM (2011) Changing patterns in insect pests on trees in the Netherlands since 1946 in relation to human induced habitat changes and climate factors-An anal-ysis of historical data. Forest Ecol Manag 261(1):50–61.https://doi.org/10.1016/j.foreco.2010.09.024

Moran PAP (1953) The statistical analysis of the Canadian Lynx cycle. Australian J Zool 1(3):291–298. https://doi.org/10.1071/ZO9530291

Morin RS, Liebhold AM (2016) Invasive forest defoliator contributes to the impending downward trend of oak dominance in eastern North America. Forestry 89(3):284–289. https://doi.org/10.1093/forestry/cpv053

Morin RS, Liebhold AM, Luzader ER et al (2005) Mapping host-species abundance of three major exotic forest pests. USDA NE- 726:11. https://doi.org/10.2737/NE-RP-726

Moulinier J, Lorenzetti F, Bergeron Y (2013) Effects of a forest tent caterpillar outbreak on the dynamics of mixedwood boreal forests of eastern Canada. Écoscience 20(2):182–193. https://doi.org/10.2980/20-2-3588

Murdoch WW, Chesson J, Chesson PL (1985) Biological control in theory and practice. Am Nat 125(3):344–366. https://doi.org/10.1086/284347

Murray KM, Elkinton JS, Woods SA (1989) Epizootiology of gypsy moth nucleopolyhedrous virus. In: Wallner WE, McManus KA (tech coords) Proceedings, Lymantriidae: a comparison of features of New and Old World tussock moths. USDA NE-123:439–453. https://doi.org/10.2737/NE-GTR-123

Myers JH, Cory JS (2013) Population cycles in forest Lepidoptera revisited. In: Futuyma DJ (ed) The annual review of ecology, evolution, and eystematics 44:565–592. https://doi.org/10.1146/annurev-ecolsys-110512-135858

Nakajima H (1991) Two new species of the genus *Operophtera* (Lepidoptera, Geometridae) from Japan. Lep Sci 42(3): 195–205. https://doi.org/10.18984/lepid.42.3_195

Nakládal O, Brinkeová H (2015) Review of historical outbreaks of the nun moth (Lymantria monacha) with respect to host tree species. J For Sci 61(1):18–26. https://doi.org/10.17221/94/2014-JFS

Nealis V, Régnière J (2021) Ecology of outbreak populations of the western spruce budworm. Ecosphere 12(7):e03667. https://doi.org/10.1002/ecs2.3667

Nealis VG (1995) Population biology of the jack pine budworm. In: Volney WJA, Nealis VG, Howse GM et al (eds) Jack pine budworm biology and management. Proceedings of the jack pine budworm Symposium. Nat Res Can, Info Rep NOR-X-342:55–71. https://d1ied5g1xfgpx8.cloudfront.net/pdfs/12135.pdf. Accessed 24 October 2022

Nealis VG (2003) Host-plant relationships and comparative ecology of conifer-feeding budworms (*Choristoneura* spp.). In: McManus ML, Liebhold AM (eds) Proceedings ecology, survey and management of forest insects. USDA NE-311:68–71. https://www.iufro.org/download/file/4456/75/70307-krakow02.pdf#page=75

Nealis VG, Régnière J (2014) An individual-based phenology model for western spruce budworm (Lepidoptera: Tortricidae). Can Entomol 146(3):306–320. https://doi.org/10.4039/tce.2013.67

Nealis VG, Régnière J (2016) Why western spruce budworms travel so far for the winter. Ecol Entomol 41(5):633–641. https://doi.org/10.1111/een.12336

Nedorezov LV (2019) Identification of the type of population dynamics type of green oak tortrix with a generalized discrete logistic model. Bio Bull Rev 9:243–249. https://doi.org/10.1134/S2079086419030071

Negrón JF, Lynch AM, Schaupp WC Jr et al (2014) Douglas-fir tussock moth- and Douglas-fir beetle-caused mortality in a Ponderosa pine/Douglas-fir forest in the Colorado Front Range, USA. Forests 5(12):3131–3146. https://doi.org/10.3390/f5123131

Neilson MM, Morris RF (1964) Regulation of European spruce sawfly numbers in the Maritime Provinces of Canada from 1937 to 1963. Can Entomol 96(5):773–784. https://doi.org/10.4039/Ent96773-5

Nilssen AC, Tenow O, Bylund H (2007) Waves and synchrony in *Epirrita autumnata/Operophtera brumata* outbreaks. II. Sunspot activity cannot explain cyclic outbreaks. J Anim Ecol 76(2):269–275. https://doi.org/10.1111/j.1365-2656.2006.01205.x

Oswald WW, Doughty ED, Foster DR et al (2017) Evaluating the role of insects in the middle-Holocene *Tsuga* decline. J Torrey Bot Soc 144(1):35–39. https://harvardforest.fas.harvard.edu/publications/pdfs/Oswald_JTorrBot_2017.pdf

Parry D (1995) Larval and pupal parasitism of the forest tent caterpillar, *Malacosoma disstria* Hübner (Lepidoptera: Lasiocampidae), in Alberta, Canada. Can Entomol 127(6):877–893.https://doi.org/10.4039/Ent127877-6

Parry D, Spence JR, Volney WJA (1997) Responses of natural enemies to experimentally increased populations of the forest tent caterpillar, *Malacosoma disstria*. Ecol Entomol 22(1):97–108.https://doi.org/10.1046/j.1365-2311.1997.00022.x

Parry D, Goyer RA, Lenhard GJ (2001) Macrogeographic clines in fecundity, reproductive allocation, and offspring size of the forest tent caterpillar *Malacosoma disstria*. Ecol Entomol 26(3):281–291. https://doi.org/10.1046/j.1365-2311.2001.00319.x

Parry D, Herms DA, Mattson WJ (2003) Responses of an insect folivore and its parasitoids to multiyear experimental defoliation of aspen. Ecology 84(7):1768–1783. https://doi.org/10.1890/0012-9658(2003)084[1768:ROAIFA]2.0.CO;2

Pasquarella VJ, Elkinton JS, Bradley BA (2018) Extensive gypsy moth defoliation in southern New England characterized using Landsat satellite observations. Biol Invasions 20:3047–3053. https://doi.org/10.1007/s10530-018-1778-0

Pelletier G, Piché C (2003) Species of *Telenomus* (Hymenoptera: Scelionidae) associated with the hemlock looper (Lepidoptera: Geometridae) in Canada. Can Entomol 135(1):23–39. https://doi.org/10.4039/n02-035

Peltonen M, Liebhold AM, Bjørnstad ON et al (2002) Spatial synchrony in forest insect outbreaks: roles of regional stochasticity and dispersal. Ecology 83(11):3120–3129. https://doi.org/10.1890/0012-9658(2002)083[3120:SSIFIO]2.0.CO;2

Pepi AA, Broadley HJ, Elkinton JS (2016) Density-dependent effects of larval dispersal mediated by host plant quality on populations of an invasive insect. Oecologia 182: 499–509. Erratum Oecologia 185(2017):533–535. https://doi.org/10.1007/s00442-016-3689-z

Pérez-Contreras T, Soler JJ, Soler M (2003) Why do pine processionary caterpillars *Thaumetopoea pityocampa* (Lepidoptera, Thaumetopoeidae) live in large groups? An experimental study. Ann Zool Fenn 40(6):505–515. https://www.jstor.org/stable/23736507. Accessed 25 October 2022

Pimentel CS, Ferreira C, Santos M et al (2017) Spatial patterns at host and forest stand scale and population regulation of the pine processionary moth *Thaumetopoea pityocampa*. Agri for Entomol 19(2):200–209. https://doi.org/10.1111/afe.12201

Piper FI, Gundale MJ, Fajardo A (2015) Extreme defoliation reduces tree growth but not C and N storage in a winter-deciduous species. Ann Bot 115(7):1093–1103. https://doi.org/10.1093/aob/mcv038

Płatek K (2007) Variability of population density of the nun moth (*Lymantria monacha* L.) larvae in the Tuczno Forest District in the years 1996–2003. Sylwan 151(9):57–65. https://doi.org/10.26202/sylwan.2006112

Plenzich C, Despland E (2018) Host-plant mediated effects on group cohesion and mobility in a nomadic gregarious caterpillar. Behav Ecol Sociobiol 72:7p. https://doi.org/10.1007/s00265-018-2482-x

Pogue MG, Schaefer PW (2007) A review of selected species of *Lymantria* Hübner [1819] (Lepidoptera: Noctuidae: Lymantriinae) from subtropical and temperate regions of Asia, including the descriptions of three new species, some potentially invasive to North America. USDA FHTET 2006–07. https://handle.nal.usda.gov/10113/45484

Pschorn-Walcher H (1987) Interspecific competition between the principal larval parasitoids of the pine sawfly, *Neodiprion sertifer* (Geoff.) (Hym.: Diprionidae). Oecologia 73:621–625. https://doi.org/10.1007/BF00379426

Pureswaran DS, Johns R, Heard SB et al (2016) Paradigms in eastern spruce budworm (Lepidoptera: Tortricidae) population ecology: a century of debate. Environ Entomol 45(6):1333–1342. https://doi.org/10.1093/ee/nvw103

Pureswaran DS, Neau M, Marchand M et al (2019) Phenological synchrony between eastern spruce budworm and its host trees increases with warmer temperatures in the boreal forest. Ecol Evol 9(1):576–586. https://doi.org/10.1002/ece3.4779

Purrini K (1979) On natural diseases of *Euproctis chrysorrhoea* L (Lep., Lymantriidae) in Bavaria, 1977. Anz Schadlingskd Pfl 52:56–58. https://doi.org/10.1007/BF01988653

Quezada-Garcia R, Fuentealba A, Nguyen N et al (2015) Do offspring of insects feeding on defoliation-resistant trees have better biological performance when exposed to nutritionally-imbalanced food? InSects 6(1):112–121. https://doi.org/10.3390/insects6010112

Radeloff VC, Mladenoff DJ, Boyce MS (2000) The changing relation of landscape patterns and jack pine budworm populations during an outbreak. Oikos 90(3):417–430. https://doi.org/10.1034/j.1600-0706.2000.900301.x

Rahimzadeh-Bajgiran P, Weiskittel A, Kneeshaw D et al (2018) Detection of annual spruce budworm defoliation and severity classification using Landsat imagery. Forests 9(6):357. https://doi.org/10.3390/f9060357

Raymond B, Hails RS (2007) Variation in plant resource quality and the transmission and fitness of the winter moth, *Operophtera brumata* nucleopolyhedrovirus. Biol Control 41(2):237–245.https://doi.org/10.1016/j.biocontrol.2007.02.005

Raymond B, Vanbergen A, Pearce I et al (2002) Host plant species can influence the fitness of herbivore pathogens: the winter moth and its nucleopolyhedrovirus. Oecologia 131:533–541. https://doi.org/10.1007/s00442-002-0926-4

Reardon RC (1976) Parasite incidence and ecological relationships in field populations of gypsy moth larvae and pupae. Environ Entomol 5(5):981–987. https://doi.org/10.1093/ee/5.5.981

Régnière J, Lysyk TJ (1995) Population dynamics of the spruce budworm, *Choristoneura fumiferana*. In: Armstrong JA, Ives WGH (eds) Forest insect pests in Canada. NRCAN, CFS/LFC, Science and Sustainable Dev't Directorate, Ottawa, ON, pp 95–105. https://cfs.nrcan.gc.ca/publications?id=17004. Accessed 25 October 2022

Régnière J, Nealis VG (2018) Two sides of a coin: host-plant synchrony fitness trade-offs in the population dynamics of the western spruce budworm. Insect Science 25(1):117–126. https://doi.org/10.1111/1744-7917.12407

Régnière J, Nealis VG (2019) Density dependence of egg recruitment and moth dispersal in spruce budworms. Forests 10(8):706. https://doi.org/10.3390/f10080706

Régnière J, Delisle J, Pureswaran DS et al (2013) Mate-finding allee effect in spruce budworm population dynamics. Entomol Exp Appl 146(1):112–122. https://doi.org/10.1111/eea.12019

Régnière J, Cooke BJ, Béchard A et al (2019a) Dynamics and management of rising outbreak spruce budworm populations. Forests 10(9):748. https://doi.org/10.3390/f10090748

Régnière J, Delisle J, Dupont A et al (2019b) The impact of moth migration on apparent fecundity overwhelms mating disruption as a method to manage spruce budworm populations. Forests 10(9):775. https://doi.org/10.3390/f10090775

Régnière J, Venier L, Welsh D (2021) Avian predation in a declining outbreak population of the spruce budworm, *Choristoneura fumiferana* (Lepidoptera: Tortricidae). InSects 12(8):720. https://doi.org/10.3390/insects12080720

Reilly JR, Hajek AE, Liebhold AM et al (2014) Impact of *Entomophaga maimaiga* (Entomophthorales: Entomophthoraceae) on outbreak gypsy moth populations (Lepidoptera: Erebidae): the role of weather. Environ Entomol 43(3):632–641. https://doi.org/10.1603/EN13194

Rhainds M, Kettela EG, Silk PJ (2012) Thirty-five years of pheromone-based mating disruption studies with *Choristoneura fumiferana* (Clemens) (Lepidoptera: Tortricidae). Can Entomol 144(3):379–395. https://doi.org/10.4039/tce.2012.18

Rhainds M, Lavigne D, Boulanger Y et al (2022) I know it when I see it: Incidence, timing and intensity of immigration in spruce budworm. Agric for Entomol 24(2):152–166. https://doi.org/10.1111/afe.12479

Robert LE, Sturtevant BR, Kneeshaw D et al (2020) Forest landscape structure influences the cyclic-eruptive spatial dynamics of forest tent caterpillar outbreaks. Ecosphere 11(8):e03096. https://doi.org/10.1002/ecs2.3096

Robson JRM, Conciatori F, Tardif JC et al (2015) Tree-ring response of jack pine and scots pine to budworm defoliation in central Canada. Forest Ecol Manag 347:83–95. https://doi.org/10.1016/j.foreco.2015.03.018

Rochefort S, Berthiaume R, Hébert C et al (2011) Effect of temperature and host tree on cold hardiness of hemlock looper eggs along a latitudinal gradient. J Insect Physiol 57(6):751–759. https://doi.org/10.1016/j.jinsphys.2011.02.013

Roelofs WL, Hill AS, Linn CE et al (1982) Sex pheromone of the winter moth, a geometrid with unusually low temperature precopulatory responses. Science 217(4560):657–659. https://doi.org/10.1126/science.217.4560.657

Roland J (1986) Parasitism of winter moth in British Columbia during build-up of its parasitoid *Cyzenis albicans*: Attack rate on oak v. apple. J Anim Ecol 55(1):215–234. https://doi.org/10.2307/4703

Roland J (1990a) Parasitoid aggregation: chemical ecology and population dynamics. In: Mackauer M, Ehler LE (eds) Critical issues in biological control. Intercept, Andover, pp 185–211

Roland J (1990b) Interaction of parasitism and predation in the decline of winter moth in Canada. In: Watt A, Leather SR, Hunter AF (eds) Population dynamics of forest insects, Chapter 26. Intercept Ltd, Andover, Hampshire, UK, pp 289–301. https://www.academia.edu/60341405/Pop ulation_Dynamics_of_forest_insects. Accessed 25 October 2022

Roland J (1994) After the decline: what maintains low winter moth density after successful biological control? J Anim Ecol 63(2):392–398. https://doi.org/10.2307/5556

Roland J (1995) Response to Bonsall & Hassell 'Identifying density-dependent processes: a comment on the regulation of winter moth.' J Anim Ecol 64(6):785–786. https://doi.org/10. 2307/5859

Roland J, Embree DG (1995) Biological control of the winter moth. Ann Rev Entomol 40:475–492. https://doi.org/10.1146/annurev.en.40.010195.002355

Roland J, Denford KE, Jimenez L (1995) Borneol as an attractant for *Cyzenis albicans*, a tachinid parasitoid of the winter moth, *Operophtera brumata* L (Lepidoptera: Geometridae). Can Entomol 127(3):413–421. https://doi.org/10.4039/Ent127413-3

Ronnås C, Larsson S, Pitacco A et al (2010) Effects of colony size on larval performance in a processionary moth. Ecol Entomol 35(4):436–445. https://doi.org/10.1111/j.1365-2311.2010. 01199.x

Rothman LD, Roland J (1998) Forest fragmentation and colony performance of forest tent caterpillar. Ecography 21(4):383–391. https://doi.org/10.1111/j.1600-0587.1998.tb00403.x

Royama T (1984) Population dynamics of the spruce budworm *Choristoneura fumiferana*. Ecol Monogr 54(4):429–462. https://doi.org/10.2307/1942595

Royama T, Eveleigh ES, Morin JRB et al (2017) Mechanisms underlying spruce budworm outbreak processes as elucidated by a 14-year study in New Brunswick, Canada. Ecol Monogr 87(4):600–631.https://doi.org/10.1002/ecm.1270

Rozenberg P, Pâques L, Huard F et al (2020) Direct and indirect analysis of the elevational shift of larch budmoth outbreaks along an elevation gradient. Front Glob Chang 3:86. https://doi.org/10. 3389/ffgc.2020.00086

Ruohomäki K, Haukioja E (1992) Interpopulation differences in pupal size and fecundity are not associated with occurrence of outbreaks *in Epirrita autumnata* (Lepidoptera, Geometridae). Ecol Entomol 17(1):69–75. https://doi.org/10.1111/j.1365-2311.1992.tb01041.x

Ruohomäki K, Tanhuanpää M, Ayres MP et al (2000) Causes of cyclicity of *Epirrita autumnata* (Lepidoptera, Geometridae): grandiose theory and tedious practice. Popul Ecol 42(3):211–223. https://doi.org/10.1007/PL00012000

Ruuhola T, Salminen J-P, Haviola S et al (2007) Immunological memory of mountain birches: effects of phenolics on performance of the autumnal moth depend on herbivory history of trees. J Chem Ecol 33:1160–1176. https://doi.org/10.1007/s10886-007-9308-z

Ruuhola T, Yang S, Ossipov V et al (2008) Foliar oxidases as mediators of the rapidly induced resistance of mountain birch against *Epirrita autumnata*. Oecologia 154:725–730. https://doi. org/10.1007/s00442-007-0869-x

Ryall KL (2010) Effects of larval host plant species on fecundity of the generalist insect herbivore *Ennomos subsignarius* (Lepidoptera: Geometridae). Environ Entomol 39(1):121–126. https:// doi.org/10.1603/EN09117

Ryan RB (1983) Population-density and dynamics of larch casebearer (Lepidoptera: Coleophoridae) in the Blue Mountains of Oregon and Washington before the build-up of exotic parasites. Can Entomol 115(9):1095–1102. https://doi.org/10.4039/Ent1151095-9

Ryan RB (1986) Analysis of life-tables for the larch casebearer (Lepidoptera, Coleophoridae) in Oregon. Can Entomol 118(12):1255–1263. https://doi.org/10.4039/Ent1181255-12

Ryan RB (1997) Before and after evaluation of biological control of the larch casebearer (Lepidoptera: Coleophoridae) in the Blue Mountains of Oregon and Washington, 1972–1995. Environ Entomol 26(3):703–715. https://doi.org/10.1093/ee/26.3.703

Sabbahi R, Royer L, O'Hara JE et al (2018) A review of known parasitoids of hemlock looper (Lepidoptera: Geometridae) in Canada and first records of egg and larval parasitoids in Labrador forests. Can Entomol 150(4):499–510. https://doi.org/10.4039/tce.2018.27

Salis L, Lof M, van Asch M et al (2016) Modeling winter moth *Operophtera brumata* egg phenology: nonlinear effects of temperature and developmental stage on developmental rate. Oikos 125: 1772–1781. Erratum 2017 Oikos 126:1522. https://doi.org/10.1111/oik.03257; https://doi.org/10.1111/oik.04819

Salman MHR, Hellrigl K, Minerbi S et al (2016) Prolonged pupal diapause drives population dynamics of the pine processionary moth (*Thaumetopoea pityocampa*) in an outbreak expansion area. Forest Ecol Manag 361:375–381. https://doi.org/10.1016/j.foreco.2015.11.035

Salminen JP, Karonen M (2011) Chemical ecology of tannins and other phenolics: we need a change in approach. Funct Ecol 25(2):325–338. https://doi.org/10.1111/j.1365-2435.2010.01826.x

Santiago O (2022) Western spruce budworm outbreak associated with wet periods in the Colorado Front Range: a multicentury reconstruction (CSU Theses and Dissertations). https://mountainscholar.org/handle/10217/235662. Accessed 27 October 2022

Saulnier M, Roques A, Guibal F et al (2017) Spatiotemporal heterogeneity of larch budmoth outbreaks in the French Alps over the last 500 years. Can J Forest Res 47(5):667–680. https://doi.org/10.1139/cjfr-2016-0211

Schaefer PW (1986) Bibliography of the browntail moth, *Euproctis chrysorrhoea* (L) (Lepidoptera: Lymantriidae) and its natural enemies. Del Agr Exp Sta Bull: 1–66.

Schmitt T (2007) Molecular biogeography of Europe: pleistocene cycles and postglacial trends. Front Zool 4:11. https://doi.org/10.1186/1742-9994-4-11

Schneider JC (1980) The role of parthenogenesis and female aptery in microgeographic, ecological adaptation in the fall cankerworm, *Alsophila pometaria* Harris (Lepidoptera: Geometridae). Ecology 61(5):1082–1090. https://doi.org/10.2307/1936827

Schopf R (1989) Spruce needle compounds and the susceptibility of Norway spruce (*Picea abies* Karst.) to attacks by the european sawfly, *Gilpinia hercyniae* Htg (Hym., Diprionidae). J Appl Entomol 107(1–5):435–445. https://doi.org/10.1111/j.1439-0418.1989.tb00280.x

Schott T, Hagen SB, Ims RA et al (2010) Are population outbreaks in sub-arctic geometrids terminated by larval parasitoids? J Anim Ecol 79(3):701–708. https://doi.org/10.1111/j.1365-2656.2010.01673.x

Schott T, Kapari L, Hagen SB et al (2013) Predator release from invertebrate generalists does not explain geometrid moth (Lepidoptera: Geometridae) outbreaks at high altitudes. Can Entomol 145(2):184–192. https://doi.org/10.4039/tce.2012.109

Schowalter TD (2017) Biology and management of the forest tent caterpillar (Lepidoptera: Lasiocampidae). J Integr Pest Manage 8(1):24;1–10. https://doi.org/10.1093/jipm/pmx022

Schroeder H, Degen B (2012) Phylogeography of the green oak leaf roller, *Tortrix viridana* L. (Lepidoptera, Tortricidae). M D Gesell Allg Ange 18:401–404. https://www.researchgate.net/publication/292953371_Phylogeography_of_the_green_oak_leaf_roller_Tortrix_viridana_L_Lepidoptera_Tortricidae

Schroeder H, Orgel F, Fladung M (2015) Performance of the green oak leaf roller (Tortrix viridana L.) on leaves from resistant and susceptible oak genotypes. M D Gesell Allg Ange 20:265–269. https://www.openagrar.de/receive/openagrar_mods_00022329

Seehausen ML, Bauce E, Régnière J et al (2015) Short-term influence of partial cutting on hemlock looper (Lepidoptera: Geometridae) parasitism. Agric for Entomol 17(4):347–354. https://doi.org/10.1111/afe.12113

Seixas Arnaldo P, Oliveira I, Santos J et al (2011) Climate change and forest plagues: the case of the pine processionary moth in Northeastern Portugal. For Sys 20(3):508–515. https://doi.org/10.5424/fs/20112003-11394

Senf C, Campbell EM, Pflugmacher D et al (2017) A multi-scale analysis of western spruce budworm outbreak dynamics. Landscape Ecol 32:501–514. https://doi.org/10.1007/s10980-016-0460-0

Sharov AA, Liebhold AM (1998a) Bioeconomics of managing the spread of exotic pest species with barrier zones. Ecol Appl 8(3):833–845. https://doi.org/10.2307/2641270

Sharov AA, Liebhold AM (1998b) Model of slowing the spread of gypsy moth (Lepidoptera: Lymantriidae) with a barrier zone. Ecol Appl 8(4):1170–1179. https://doi.org/10.2307/2640970

Sharov AA, Roberts EA, Liebhold AM et al (1995) Gypsy moth (Lepidoptera: Lymantriidae) spread in the central Appalachians: three methods for species boundary estimation. Environ Entomol 24(6):1529–1538. https://doi.org/10.1093/ee/24.6.1529

Sharov AA, Liebhold AM, Roberts EA (1996) Spread of gypsy moth (Lepidoptera: Lymantriidae) in the central Appalachians: comparison of population boundaries obtained from male moth capture, egg mass counts, and defoliation records. Environ Entomol 25(4):783–792. https://doi.org/10.1093/ee/25.4.783

Sharov AA, Liebhold AM, Roberts EA (1997) Methods for monitoring the spread of gypsy moth (Lepidoptera: Lymantriidae) populations in the Appalachian mountains. J Econ Entomol 90(5):1259–1266. https://doi.org/10.1093/jee/90.5.1259

Sharov AA, Liebhold AM, Roberts EA (1998) Optimizing the use of barrier zones to slow the spread of gypsy moth (Lepidoptera: Lymantriidae) in North America. J Econ Entomol 91(1):165–174. https://doi.org/10.1093/jee/91.1.165

Sharov AA, Leonard D, Liebhold AM et al (2002a) "Slow the spread": a national program to contain the gypsy moth. J Forest 100(5):30–36. https://academic.oup.com/jof/article/100/5/30/4608639. Accessed 27 October 2022

Sharov AA, Leonard D, Liebhold AM et al (2002b) Evaluation of preventive treatments in low-density gypsy moth populations using pheromone traps. J Econ Entomol 95(6):1205–1215. https://doi.org/10.1603/0022-0493-95.6.1205

Shepherd RF, Bennett DD, Dale JW et al (1988) Evidence of synchronized cycles in outbreak patterns of Douglas-fir tussock moth, Orgyia pseudotsugata (McDunnough) (Lepidoptera: Lymantriidae). Mem Entomol Soc Can 120(S146):107–121. https://doi.org/10.4039/entm12014 6107-1

Shepherd RF, Gray TG, Harvey GT (1995) Geographical distribution of Choristoneura species (Lepidoptera: Tortricidae) feeding on Abies, Picea, and Pseudotsuga in western Canada and Alaska. Can Entomol 127(6):813–830. https://doi.org/10.4039/Ent127813-6

Shigesada N, Kawasaki K (1997) Biological invasions: theory and practice. Oxford University Press, UK

Simchuk AP, Ivashov AV, Companiytsev VA (1999) Genetic patterns as possible factors causing population cycles in oak leafroller moth, Tortrix viridana L. Forest Ecol Manag 113(1):35–49.https://doi.org/10.1016/S0378-1127(98)00340-5

Simmons MJ, Lee TD, Ducey MJ et al (2014) Effects of invasive winter moth defoliation on tree radial growth in eastern Massachusetts, USA. InSects 5(2):301–318. https://doi.org/10.3390/insects5020301

Skellam JG (1951) Random dispersal in theoretical populations. Biometrika 38(1/2):196–218. https://doi.org/10.2307/2332328

Skuhravý V (1987) A review of research on the nun moth (Lymantria monacha L.) conducted with pheromone traps in Czechoslovakia, 1973–1984. Anz Schadlingskd Pfl 60:96–98. https://doi.org/10.1007/BF01906038

Smith CC (1958) A note on the association of fall cankerworm (Alsophila pometaria (Harr.)) with winter moth (Operophtera brumata (Linn.)) (Lepidoptera: Geometridae). Can Entomol 90(9):538–540. https://doi.org/10.4039/Ent90538-9

Smitley DR, Bauer LS, Hajek AE et al (1995) Introduction and establishment of Entomophaga maimaiga, a fungal pathogen of gypsy moth (Lepidoptera: Lymantriidae) in Michigan. Environ Entomol 24(6):1685–1695. https://doi.org/10.1093/ee/24.6.1685

Solla A, Milanović S, Gallardo A et al (2016) Genetic determination of tannins and herbivore resistance in Quercus ilex. Tree Genet Genomes 12:117. https://doi.org/10.1007/s11295-016-1069-9

Sonia S, Morin H, Krause C (2011) Long-term spruce budworm outbreak dynamics reconstructed from subfossil trees. J Quat Sci 26(7):734–738. https://doi.org/10.1002/jqs.1492

Spear RJ (2005) The great gypsy moth war: a history of the first campaign in Massachusetts to eradicate the gypsy moth, 1890–1901. University of Massachusetts Press. https://www.jstor.org/stable/j.ctt5vk7pz

Speare AT, Colley RH (1912) The artificial use of the brown-tail fungus in Massachusetts: with practical suggestions for private experiment, and a brief note on a fungous disease of the gypsy caterpillar. Wright & Potter Printing Company, Boston. https://tile.loc.gov/storage-services/public/gdcmassbookdig/artificialuseofb00mass/artificialuseofb00mass.pdf. Accessed 31 October 2022

Stastny M, Battisti A, Petrucco-Toffolo E et al (2006) Host-plant use in the range expansion of the pine processionary moth, *Thaumetopoea pityocampa*. Ecol Entomol 31(5):481–490.https://doi.org/10.1111/j.1365-2311.2006.00807.x

Sterling PH, Speight MR (1989) Comparative mortalities of the brown-tail moth, *Euproctis chrysorrhoea* (L.) (Lepidoptera: Lymantriidae), in south-east England. Bot J Linn Soc 101(1):69–78. https://doi.org/10.1111/j.1095-8339.1989.tb00137.x

Stoakley JT (1985) Outbreaks of winter moth, *Operophthera brumata* L. (Lep, Geometridae) in young plantations of Sitka spruce in Scotland: insecticidal control and population assessment using the sex attractant pheromone. Z Angew Entomol 99(1–5):153–160. https://doi.org/10.1111/j.1439-0418.1985.tb01973.x

Sukovata L (2010) Prediction and control of the nun moth *Lymantria monacha* L. (Lepidoptera, Lymantriidae). Dissertations and Monographs 14:128. https://www.researchgate.net/publication/266970365_Prediction_and_control_of_the_nun_moth_Lymantria_monacha_L_Lepidoptera_Lymantriidae. Accessed 31 October 2022

Swetnam TW, Lynch AM (1989) A tree-ring reconstruction of western spruce budworm history in the southern Rocky Mountains. For Sci 35(4):962–986. https://www.ltrr.arizona.edu/~ellisqm/outgoing/dendroecology2014/readings/Swetnam%20and%20Lynch.1989.A%20tree-ring%20reconstruction.pdf. Accessed 31 October 2022

Swetnam TW, Lynch AM (1993) Multicentury, regional-scale patterns of western spruce budworm outbreaks. Ecol Mono 63(4):399–424. https://doi.org/10.2307/2937153

Swetnam TW, Wickman BE, Paul HG et al (1995) Historical patterns of western spruce budworm and Douglas-fir tussock moth outbreaks in the northern Blue Mountains, Oregon, since A.D. 1700. USDA PNW-RP-484:27. https://doi.org/10.2737/PNW-RP-484

Tamburini G, Marini L, Hellrigl K et al (2013) Effects of climate and density-dependent factors on population dynamics of the pine processionary moth in the Southern Alps. Clim Change 121:701–712. https://doi.org/10.1007/s10584-013-0966-2

Tammaru T, Kaitaniemi P, Ruohomäki K (1996) Realized fecundity in *Epirrita autumnata* (Lepidotera: Geometridae): relation to body size and consequences to population dynamics. Oikos 77(3):407–416. https://doi.org/10.2307/3545931

Tammaru T, Tanhuanpää M, Ruohomäki K et al (2001) Autumnal moth—why autumnal? Ecol Entomol 26(6):646–654. https://doi.org/10.1046/j.1365-2311.2001.00363.x

Tanhuanpää M, Ruohomäki K, Uusipaikka E (2001) High larval predation rate in non-outbreaking populations of a geometrid moth. Ecology 82(1):281–289. https://doi.org/10.1890/0012-9658(2001)082[0281:HLPRIN]2.0.CO;2

Taylor CM, Hastings A (2005) Allee effects in biological invasions. Ecol Lett 8(8):895–908. https://doi.org/10.1111/j.1461-0248.2005.00787.x

Tenow O, Nilssen A (1990) Egg cold hardiness and topoclimatic limitations to outbreaks of *Epirrita autumnata* in northern Fennoscandia. J Appl Ecol 27(2):723–734. https://doi.org/10.2307/2404314

Tenow O, Nilssen AC, Bylund H et al (2007) Waves and synchrony in *Epirrita autumnata/Operophtera brumata* outbreaks. I. Lagged synchrony: regionally, locally and among species. J Anim Ecol 76(2):258–268. https://www.jstor.org/stable/4539126. Accessed 31 October 2022

Tenow O, Nilssen AC, Bylund H et al (2013) Geometrid outbreak waves travel across Europe. J Anim Ecol 82(1):84–95. https://doi.org/10.1111/j.1365-2656.2012.02023.x

Ticehurst M, Allen DC (1973) Notes on biology of *Telenomus coelodasidis* (Hymenoptera: Scelionidae) and its relationship to saddled prominent, *Heterocampa guttivitta* (Lepidoptera: Notodontidae). Can Entomol 105(8):1133–1143. https://doi.org/10.4039/Ent1051133-8

Tobin PC (2007) Space-time patterns during the establishment of a nonindigenous species. Popul Ecol 49:257–263. https://doi.org/10.1007/s10144-007-0043-7

Tobin PC, Blackburn L (2007) Slow the Spread: a national program to manage the gypsy moth. USDA, Forest Service, Newtown Square, PA. NRS-GTR-6:109. https://doi.org/10.2737/NRS-GTR-6

Tobin PC, Whitmire SL, Johnson DM et al (2007) Invasion speed is affected by geographical variation in the strength of Allee effects. Ecol Lett 10(1):36–43. https://doi.org/10.1111/j.1461-0248.2006.00991.x

Tobin PC, Robinet C, Johnson DM et al (2009) The role of Allee effects in gypsy moth, *Lymantria dispar* (L.), invasions. Popul Ecol 51:373–384. https://doi.org/10.1007/s10144-009-0144-6

Tobin PC, Gray DR, Liebhold AM (2014) Supraoptimal temperatures influence the range dynamics of a non-native insect. Divers Distrib 20(7):813–823. https://doi.org/10.1111/ddi.12197

Torgersen TR, Ryan RB (1981) Field biology of *Telenomus californicus* Ashmead, an important egg parasite of Douglas-fir tussock moth. Ann Entomol Soc Am 74(2):185–186. https://doi.org/10.1093/aesa/74.2.185

Torgersen TR, Campbell RW (1982) Some effects of avian predators on the western spruce budworm *Choristoneura occidentalis* Freeman (Lepidoptera, Tortricidae) in north central Washington. Environ Entomol 11(2):429–431. https://doi.org/10.1093/ee/11.2.429

Troubridge JT, Fitzpatrick SM (1993) A revision of the North American *Operophtera* (Lepidoptera: Geometridae). Can Entomol 125(2):379–397. https://doi.org/10.4039/Ent125379-2

Trudeau M, Mauffette Y, Rochefort S et al (2010) Impact of host tree on forest tent caterpillar performance and offspring overwintering mortality. Environ Entomol 39(2):498–504. https://doi.org/10.1603/EN09139

Uelmen JA Jr, Lindroth RL, Tobin PC et al (2016) Effects of winter temperatures, spring degree-day accumulation, and insect population source on phenological synchrony between forest tent caterpillar and host trees. Forest Ecol Manag 362:241–250. https://doi.org/10.1016/j.foreco.2015.11.045

van Dis NE, van der Zee M, Hut RA et al (2021) Timing of increased temperature sensitivity coincides with nervous system development in winter moth embryos. J Exp Biol 224(17):jeb242554. https://doi.org/10.1242/jeb.242554

Van Dyck H (2012) Dispersal under global change-the case of the pine processionary moth and other insects. In: Clobert J, Baquette M, Benton TG et al (eds) Dispersal Ecology and Evolution. Oxford University Press, UK, pp 357–365. https://doi.org/10.1093/acprof:oso/9780199608898.003.0028

Vane E, Waring K, Polinko A (2017) The influence of western spruce budworm on fire in Spruce-Fir forests. Fire Ecol 13:16–33. https://doi.org/10.4996/fireecology.1301016

Vanhanen H, Veleli TO, Paivinen S et al (2007) Climate change and range shifts in two insect defoliators: gypsy moth and nun moth—a model study. Silva Fenn 41(4):621–638. https://doi.org/10.14214/sf.469

Varley GC, Gradwell GR (1960) Key factors in population studies. J Anim Ecol 29(2):399–401. https://doi.org/10.2307/2213

Varley GC, Gradwell GR (1963) Predatory insects as density dependent mortality factors. In: Moore JA (ed) Proceedings, XVI International Congress of Zoology 1:240, Washington, DC. https://catalog.hathitrust.org/api/volumes/oclc/2721561.html

Varley GC, Gradwell GR (1968) Population models for the winter moth. In: Southwood TRE (ed.) Insect abundance: symposia of the Royal Entomological Society of London 4:132–142.

Varley GC, Gradwell GR (1970) Recent advances in insect population dynamics. Ann Rev Entomol 15:1–24. https://doi.org/10.1146/annurev.en.15.010170.000245

Varley GC, Gradwell GR, Hassell MP (1973) Insect population ecology: an analytical approach. Blackwell Scientific, Oxford, England.

Venier LA, Holmes SB (2010) A review of the interaction between forest birds and eastern spruce budworm. Environ Rev 18:191–207. https://doi.org/10.1139/A10-009

Vindstad OPL, Schott T, Hagen SB et al (2013) How rapidly do invasive birch forest geometrids recruit larval parasitoids? Insights from comparison with a sympatric native geometrid. Biol Invasions 15:1573–1589. https://doi.org/10.1007/s10530-012-0393-8

Vindstad OPL, Jepsen JU, Molvig H et al (2022) A pioneering pest: the winter moth (*Operphtera brumata*) is expanding its outbreak range into low Arctic shrub tundra. Arct Sci 8(2):450–470. https://doi.org/10.1139/as-2021-0027

Visser ME, van Noordwijk AJ, Tinbergen JM et al (1998) Warmer springs lead to mistimed reproduction in great tits (*Parus major*). Proc R Soc Ser B Biol Sci 265(1408):1867–1870. https://doi.org/10.1098/rspb.1998.0514

Visser ME, Holleman LJM (2001) Warmer springs disrupt the synchrony of oak and winter moth phenology. Proc R Soc Ser B Biol Sci 268(1464):289–294. https://doi.org/10.1098/rspb.2000.1363

Volney WJA (1994) Multi-century regional western spruce budworm outbreak patterns. Trends Ecol Evol 9(2):43–45. https://doi.org/10.1016/0169-5347(94)90265-8

Volney WJA, McCullough DG (1994) Jack pine budworm population behaviour in northwestern Wisconsin. Can J Forest Res 24(3):502–510. https://doi.org/10.1139/x94-067

Wagner TL, Leonard DE (1979) Aspects of mating, oviposition, and flight in the satin moth, *Leucoma salicis* (Lepidoptera: Lymantriidae). Can Entomol 111(7):833–840. https://doi.org/10.4039/Ent111833-7

Wagner TL, Leonard DE (1980) Mortality factors of satin moth, *Leucoma salicis* [Lep.: Lymantriidae], in aspen forests in Maine. Entomophaga 25:7–16. https://doi.org/10.1007/BF02377517

Walter JA, Finch FT, Johnson DM (2016) Re-evaluating fall cankerworm management thresholds for urban and suburban forests. Agric for Entomol 18(2):145–150. https://doi.org/10.1111/afe.12147

War AR, Paulraj MG, Ahmad T et al (2012) Mechanisms of plant defense against insect herbivores. Plant Signaling Behav 7(10):1306–1320. https://doi.org/10.4161/psb.21663

Ward SF, Aukema BH (2019) Climatic synchrony and increased outbreaks in allopatric populations of an invasive defoliator. Biol Invasions 21:685–691. https://doi.org/10.1007/s10530-018-1879-9

Ward SF, Venette RC, Aukema BH (2019) Cold tolerance of the invasive larch casebearer and implications for invasion success. Agric for Entomol 21(1):88–98. https://doi.org/10.1111/afe.12311

Ward SF, Eidson EL, Kees AM et al (2020a) Allopatric populations of the invasive larch casebearer differ in cold tolerance and phenology. Ecol Entomol 45(1):56–66. https://doi.org/10.1111/een.12773

Ward SF, Aukema BH, Fei S et al (2020b) Warm temperatures increase population growth of a nonnative defoliator and inhibit demographic responses by parasitoids. Ecology 101(11):e03156. https://doi.org/10.1002/ecy.3156

Wargo PM (1977) *Armillariella mellea* and *Agrilus bilineatus* and mortality of defoliated oak trees. For Sci 23(4):485–492. https://academic.oup.com/forestscience/article-abstract/23/4/485/4675948?redirectedFrom=fulltext#no-access-message

Watt AD, McFarlane AM (1991) Winter moth on Sitka spruce: synchrony of egg hatch and budburst, and its effect on larval survival. Ecol Entomol 16(3):387–390. https://doi.org/10.1111/j.1365-2311.1991.tb00231.x

Weber BC (1977) Parasitoids of the introduced pine sawfly, *Diprion similis* (Hymenoptera: Diprionidae), in Minnesota. Can Entomol 109(3):359–364.https://doi.org/10.4039/Ent109359-3

Weseloh RM (1985) Predation by *Calosoma sycophanta* L. (Coleoptera: Carabidae): evidence for a large impact on gypsy moth, *Lymantria dispar* L. (Lepidoptera: Lymantriidae), pupae. Can Entomol 117(9):1117–1126. https://doi.org/10.4039/Ent1171117-9

Wickman BE, Seidel KW, Star GL (1986) Natural regeneration 10 years after a Douglas-fir tussock moth outbreak in northeastern Oregon. USDA PNW-RP-370:1–15. https://doi.org/10.2737/PNW-RP-370

Wigley PJ (1976) The Epizooliology of a nuclear polyhedrosis virus disease of the winter moth *Operophtera brumata* L., at Wistmanns Wood, Dartmoor. PhD thesis. University of Oxford, Oxford, UK

Williams DT, Straw NA, Day KR (2003) Defoliation of Sitka spruce by the European spruce sawfly, *Gilpinia hercyniae* (Hartig): a retrospective analysis using the needle trace method. Agric for Entomol 5(3):235–245. https://doi.org/10.1046/j.1461-9563.2003.00183.x

Williams DW, Birdsey RA (2003) Historical patterns of spruce budworm defoliation and bark beetle outbreaks in North American conifer forests: an atlas and description of digital maps. USDA NE-GTR-308:1–33. https://doi.org/10.2737/NE-GTR-308

Williams DW, Liebhold AM (1995a) Forest defoliators and climatic change: potential changes in spatial distribution of outbreaks of western spruce budworm (Lepidoptera: Tortricidae) and gypsy moth (Lepidoptera: Lymantriidae). Environ Entomol 24(1):1–9. https://doi.org/10.1093/ee/24.1.1

Williams DW, Liebhold AM (1995b) Influence of weather on the synchrony of gypsy moth (Lepidoptera: Lymantriidae) outbreaks in New England. Environ Entomol 24(5):987–995. https://doi.org/10.1093/ee/24.5.987

Williams DW, Fuester RW, Metterhouse WW et al (1992) Incidence and ecological relationships of parasitism in larval populations of *Lymantria dispar* (Lepidoptera: Lymantriidae). Biol Control 2(1):35–43. https://doi.org/10.1016/1049-9644(92)90073-M

Wilson CM, Vendettuoli JF, Orwig DA et al (2016) Impact of an invasive insect and plant defense on a native forest defoliator. InSects 7(3):45. https://doi.org/10.3390/insects7030045

Wint W (1983) The role of alternative host-plant species in the life of a polyphagous moth, *Operophtera brumata* (Lepidoptera: Geometridae). J Anim Ecol 52(2):439–450. https://doi.org/10.2307/4564

Witrowski Z (1975) Environmental regulation of population size of oak leaf roller moth (*Tortrix viridana* L.) in Niepolomice Forest. B Acad Pol Sci Biol 23(8):513–519.

Wong HR (1972) Spread of the European spruce sawfly, *Diprion hercyniae* (Hymenoptera: Diprionidae), in Manitoba. Can Entomol 104(5):755–756.https://doi.org/10.4039/Ent104755-5

Wood DM, Yanai RD, Allen DC et al (2009) Sugar maple decline after defoliation by forest tent caterpillar. J For 107(1):29–37. https://academic.oup.com/jof/article/107/1/29/4598872. Accessed 1 November 2022

Wood DM, Parry D, Yanai RD et al (2010) Forest fragmentation and duration of forest tent caterpillar (*Malacosoma disstria* Hübner) outbreaks in northern hardwood forests. Forest Ecol Manag 260(7):1193–1197. https://doi.org/10.1016/j.foreco.2010.07.011

Chapter 10
Bark Beetles

Demian F. Gomez, John J. Riggins, and Anthony I. Cognato

10.1 Introduction

In general, the term "bark beetle" most commonly applies to the weevil (Curculionidae) subfamily Scolytinae (Fig. 10.1). The Scolytinae also includes ambrosia beetles that feed on symbiotic fungi and these are addressed in Chapter 11. The lifecycle of these small (0.05–10 mm) snoutless weevils occurs almost exclusively in the interior of plant tissues. As adults and larvae, bark beetles feed on plant tissues including twigs, branches, trunks and roots, xylem, piths, fruits, and cones. Adults bore into the plant tissue and create a chamber to mate, lay eggs and for larvae to grow, pupate, and eclose as adults. Most often this plant tissue is dead or dying and bark beetles serve as primary decomposers (Stokland et al. 2012), and create pathways into the wood for other decomposers.

A minority of bark beetles kill healthy trees, although at the time of attack, these healthy trees are often experiencing stressful conditions (e.g. due to drought or lightning strikes). Populations of tree-killing bark beetles can increase in size to a level where they can overcome the resistance of healthy trees and cause mass destruction of forests resulting in tremendous economic and ecological damage. It is these few species that give bark beetles their nefarious reputation and demand the attention of forest entomologists.

This chapter introduces the reader to bark beetle natural history, diversity, evolution and management. **Natural history** is organized by feeding ecology, mating strategies, and intra– and interspecific interactions. **Evolution and diversity of bark**

D. F. Gomez (✉)
Texas A&M Forest Service, Austin, TX, USA
e-mail: demian.gomez@tfs.tamu.edu

J. J. Riggins
Mississippi State University, Starkville, MS, USA

A. I. Cognato
Michigan State University, East Lansing, MI, USA

© The Author(s) 2023
J. D. Allison et al. (eds.), *Forest Entomology and Pathology*,
https://doi.org/10.1007/978-3-031-11553-0_10

Fig. 10.1 Examples of bark beetles representing different tribes. Unlike most other weevils, scolytines lack an elongated rostrum, have oval or kidney shaped eyes, and antennae with round or conical clubs. a. *Scolytus aztecus*, b. *Chramesus crenatus*, c. *Cactopinus burjosi*, d. *Pseudips mexicanus*. Photos courtesy of Thomas Atkinson, www. barkbeetle.info

beetles includes discussion of phylogeny, timing of evolutionary events and an annotated and illustrated list of bark beetle genera important to forest entomologists. **Management and control** covers efforts to reduce losses to bark beetle destruction of forests and plant products. Finally, we present case studies, including outbreak events, which have resulted in vast economic and ecological loss.

10.2   Natural History

10.2.1   Feeding Ecology

Upon emerging from their natal host, progeny adults search for a suitable host. Dispersal flights are usually short, consisting of a few hundred meters, but some species have the potential to fly more than 30 km (Zumr 1992; Yan et al. 2005). For most beetle species, a suitable host is limited to a certain tree taxon in a suitable physiological condition for infestation. The pioneer sex, first to arrive at the host, can vary depending on the mating system of the beetle.

Host specificity ranges widely for bark beetles from a few species restricted to one tree species to some that exploit entire plant families. Most bark beetles that attack living trees exploit hosts within the Pinaceae, whereas species that breed in angiosperms are usually saprophagous. Specificity to one tree species is uncommon, occurring in approximately 1% of all bark beetles, whereas specificity to tree family is more common (Kirkendall et al. 2015). The ability to feed in both gymnosperms and angiosperms is rare and has been documented only for *Polygraphus grandiclava* (Avtzis et al. 2008).

Maturation feeding outside the maternal gallery can occur, including feeding on fresh shoots from the natal or a new host (Raffa et al. 2015). Adults usually overwinter in host material. Larvae are generally not able to survive cold weather. In some species of *Ips*, groups of individuals will bore into a tree and "roost" for the winter (Cognato 2015). *Dendroctonus frontalis* and *D. micans* are exceptions that are able to overwinter in all life-stages (Luik and Voolma 1990; Hain et al. 2011). During overwintering, beetles stop feeding and reduce their water content, accumulating compounds such as glycerol and ethylene glycol in their hemolymph to withstand freezing conditions (Gehrken 1985, 1989).

Successful colonization of the host tree depends on the population level of the available beetles to produce aggregation pheromones and on the vigor of the tree, which determines the defensive response. The great evolutionary success of conifers, for example, is directly related to their complex defense mechanisms to deter herbivores and pathogens through the production of resin (Trapp and Croteau 2001). The relationship among aggregation pheromones, conifer resin defenses and bark beetle mass attacks probably reflects the coevolution of bark beetles and their hosts (Borden 1982; Franceschi et al. 2005). Pheromones, highly important for the achievement of rapid and massive attacks, have been suggested to have originated as detoxification products of host monoterpenes (Lindgren and Raffa 2013).

Because wood is a nutritionally poor substrate, most bark beetle feeding occurs in the phloem. This tissue is a relatively thin layer, and there are different minimal requirements of phloem thickness for different bark beetle species. To increase nitrogen intake, several species feed on either fungus or fungus-infected phloem (Bleiker and Six 2007). Many bark beetles feed on fungi as well as plant tissues both as larvae and as adults. Symbiotic fungi, carried by many species in specific integument structures called mycangia or directly in the exoskeleton, are inoculated in the galleries where they grow into the host tissue (Happ et al. 1976). Females of *Dendroctonus frontalis* for example, possess mycangia in which they carry their symbiotic fungi, most commonly *Entomocorticium* (Basidiomycota) species and *Ceratocystiopsis* (Ascomycota) species, the predominant source of nutrition for the larvae (Barras and Perry 1972; Bridges 1983; Six and Wingfield 2011; Harrington et al. 2021). Other fungal species, such as the ascomycetes *Ophiostoma* spp., also alter tree condition which facilitates larval development (Barras and Taylor 1973; Goldhammer et al. 1989; Six and Wingfield 2011). These fungi, like the rest of the Ophiostomatales, are well adapted for insect dispersal, as most produce long sexual fruiting bodies with sticky spores that facilitate contact with the vector (Kirisits 2007). Interestingly, these ophiostomatalean symbionts supress wood decomposition through competitive interactions with decay fungi (Skelton et al. 2020). Moreover, it may increase feeding by subterranean termites (Little et al. 2012; Riggins et al. 2014; Clay et al. 2017).

10.2.1.1   Host Location and Acceptance

Visual and chemical cues, such as vertical silhouettes, host volatiles, and/or pheromones, are important for orientation and initial landing on the host (Person 1931; Vité and Gara 1962; Wood 1982a; Payne 1986; Saint-Germain et al. 2007). Gustatory and olfactory stimulants are important in the boring phase subsequent to host location, when the beetle determines the quality of the host in terms of nutrition and humidity (Webb and Franklin 1978). Several sensory receptors located in the antennae and mouthparts are involved in the perception and location of the host tree (Payne 1979).

Antennal sensillae are highly responsive to pheromones and host-derived volatiles, where each antennal receptor cell contains multiple sites that interact with the chemicals. Sensillae from maxillary and labial palps are also important in host selection and food discrimination, as suggested in morphological studies from *D. ponderosae* and *I. typographus* (Byers 2007). These chemicals are not only relevant in recognizing suitable hosts, but also help to avoid colonized or decaying hosts. Pine monoterpenes, such as α-pinene, are present in the oleoresin and can serve as part of the tree defense system in high concentrations. Moreover, they also function as kairomones for bark beetles, attracting them to suitable hosts, sometimes in combination with sex pheromones (Vité and Gara 1962; Wood 1982a; Payne 1986; Seybold et al. 2000,

2006). Conversely, when a large number of bark beetles are present in the host tree, a deterrent or anti-aggregation pheromone, such as verbenone, signals that the tree is no longer suitable for colonization in some species (Pitman and Vité 1969; Renwick and Vité 1970; Etxebeste and Pajares 2011). For example, in *D. frontalis*, females are initially attracted by kairomones (α-pinene) released by the host trees. Soon after the initial colonization, females release the pheromone frontalin attracting both males and females, resulting in a mass attack that overcomes tree defenses. Males later produce (+)-endo-brevicomin, an antiaggregation pheromone in high concentration (Sullivan et al. 2007).

Bark beetles can be attracted to susceptible hosts by tree volatiles (Lindelöw et al. 1992; Tunset et al. 1993), but encounter rates are also based on random alightment with no need of kairomones (Wermelinger 2004). Host location can also be influenced by abiotic effects. For example as sun-exposed trees are more likely to be attacked than trees in the shade (Lobinger and Skatulla 1996).

10.2.2   Mating Systems

Most bark beetles outbreed, but there is variation among the mating systems (Kirkendall et al. 2015). In the early colonization phase, when reproductive pairs form, conflict in the gallery entrances between conspecifics of the same sex are common in bark beetles (Kirkendall et al. 2015). In female-initiated mating systems, such as in *Dendroctonus* and *Tomicus*, male-male competition is common. Males wander and attempt to enter active galleries, but are usually blocked by already established males. Both chemical and acoustic communication are involved in gaining access to galleries and during courtship (Barr 1969; Oester et al. 1981; Ryker 1984). Females can re-emerge from the initial gallery and lay eggs in a new gallery constructed in the same host or disperse to a new tree. Eggs are commonly laid in individual niches on one or both sides of the gallery. Before re-emerging, females feed in the gallery, likely to regenerate wing muscles (Sauvard 2007).

Monogyny is the most ancestral and predominant mating system in bark beetles, and is present in almost every genus (Kirkendall 1983). In monogamous species, females typically select the host and initiate colonization. Males are subsequently attracted by female-released pheromones (Raffa et al. 2015). Exceptions exist among the Bothrosternini and Pityophthorina, where some genera are known to have male-initiated monogyny (Beaver 1973). This tends to occur with species that breed in resources where no more than one female can breed because of interbrood competition (Kirkendall 1983). A few species may have females that mate with siblings or with a newly arrived male before emergence (Bleiker et al. 2013). Depending on the species, mating occurs on the bark or in the gallery. Bigyny, where males regularly have two females, occurs in 19 genera but is most common within the Micracidini. Given that scolytines are the only insects to engage in simultaneous bigyny in nature, it has been suggested that it may be related to geometric constraints on egg tunnel construction. More than two colonizing females would decrease host real estate, resulting in increased competition among larvae and subsequently greater larval mortality due to diminished resources (Kirkendall et al. 2015).

In polygamous species, males initiate the attack, build a nuptial chamber, and are joined by several females. Harem polygyny (simultaneous polygyny) has evolved at least 12 times in Scolytinae and is found in 26 genera, being predominant in the Ipini, and common in the Corthylini and Polygraphini (Kirkendall et al. 2015). The evolutionary context of why females would join already mated males is hypothesized to be related to resource quality (Kirkendall 1983). Because resource quality is variable, some males will have high-quality resource patches to support several females, whereas other males initiate their attacks in low-quality patches that would not be able to support multiple females.

Colonial polygyny is found in a few genera and is based on having multiple males and multiple females in the same network of interconnected tunnel systems (Kirkendall et al. 2015). Colonial polygyny has been reported for *Aphanarthrum* and *Crypturgus* (Crypturgini), and *Cyrtogenius* (Dryocoetini) (Chararas 1962; Roberts 1976; Jordal 2006). Inbreeding polygyny is most common in ambrosia beetles, but also exists in several genera of bark beetles that usually do not show phloeophagous feeding habits (Kirkendall et al. 2015). The few phloeophagous inbreeders are atypical for bark beetles. Some species within *Ozopemon* (Dyocoetini), *Hypothenemus* (Cryphalini), and *Dendroctonus* (Hylurgini) breed in large chambers with larvae feeding communally (Kirkendall 1993). Partial inbreeding can also occur in *Dendroctonus micans* and *D. punctatus*, which produce small males and female-biased sex ratios (Kirkendall 1983).

Different forms of parthenogenesis are found in the Scolytinae. Arrhenotoky is the most commonly known; observed in the most successful ambrosia beetle clade, the Xyleborini. Pseudo-arrhenotoky, where daughters are sexually produced and the paternal genome is eliminated, is known from the genus *Hypothenemus*, having been demonstrated in *H. hampei* (Vega et al. 2015). Pseudogamy, where females are produced clonally, are genetic copies of their mothers, and fertilization is required but male genomes are not passed to the offspring, occurs in some species of the spruce-feeding *Ips* in North America (Lanier and Kirkendall 1986).

10.2.3   Social Behavior

Bark beetles are largely considered sub-social, with aggregated breeding and, to some extent, parental care for offspring (Jordal et al. 2011). Sub-sociality is facilitated by their subcortical lifestyle, which offers a protected abundant resource and, by inoculating it with symbiotic fungi, an easily assimilated food substrate (Kirkendall et al. 2015).

Males typically stay within galleries with females for at least days or weeks. Mate guarding, increased offspring number and survivorship, and mate attraction, have been suggested as some of the reasons for male post-copulatory residence in galleries (Kirkendall et al. 2010). For example, blocking the entry of natural enemies into the gallery would positively affect offspring survivorship. Experiments conducted for *Ips pini* suggest that the presence of males in the galleries increases the number of eggs laid by females by removing female-produced frass, and significantly reduces the number of predators in the egg galleries (Reid and Roitberg 1994). Clearing frass from egg tunnels, one of the most widespread forms of parental care, is conducted by either males or females depending on the species, using elytral declivities to push it out of the galleries (Wichmann 1967). Aggregated breeding through multiple colonization may occur without the production of pheromones, such as in species of *Hylastes* (Hylastini) or *Tomicus* (Hylurgini), where individuals are attracted by host volatiles.

10.2.4 Communication

Interactions between bark beetles and their hosts involve different stimuli such as semiochemicals (Blomquist et al. 2010). For example, feeding induces the production of aggregation pheromones that attract both sexes during a mass attack, such as ipsdienol and ipsenol in the genus *Ips*, and frontalin in some species of *Dendroctonus*. Pheromonal communication, which may have been co-opted from the detoxification of terpenes (Franceschi et al. 2005), is essential in this attraction-based system, which for some species helps to overcome tree defenses. Host colonization starts with the ability to locate a suitable host, followed by the attraction of conspecifics, and finally, as tree defenses decline and colonization proceeds, the emission of anti-aggregation pheromones to reduce competition (Wood 1982a). The same compounds produced by bark beetles to stop aggregation on a host, among other aggregation compounds, serve as kairomones and are attractive to a large number of organisms, including predators (Reeve 1997).

Acoustic signals are also important stimuli for intraspecific interactions within a host, with stridulatory organs present in one or both sexes depending on the species. Acoustic signals, commonly used by insects in the context of mating, have been associated in bark beetles with arrival announcement of the stridulating sex, or premating species recognition (Barr 1969; Oester et al. 1981; Ryker 1984). Stridulatory organs can be located on different parts of the body depending on the species, and play an important role in mate choice and male competition. For example, the elytra-abdominal stridulatory structure of *Dendroctonus valens* is capable of producing several distinct chirps, that males produce to induce female acceptance into the gallery (Lindeman and Yack 2015).

10.2.5 Interspecific Interactions

It is common in multiple species of bark beetles to feed on a common resource and therefore, there are several strategies for reducing direct competition (Raffa et al.

2015). In the broad sense, many bark beetle species achieve resource partitioning by having different host preferences. On a smaller scale, such as within a single tree, bark beetles can achieve resource partitioning by utilizing different parts of the tree. This within-tree niche partitioning by multiple species is usually not absolute and involves an opportunistic extension of the galleries in the absence of other species. For example, in the southeastern US, *Dendroctonus terebrans* can be observed at the base of the trunk in trees previously attacked by *D. frontalis* (Payne et al. 1987). Species of *Ips* will subsequently attack the higher portions of the trunk according to their size, with *I. calligraphus* colonizing larger diameters and *I. grandicollis* and *I. avulsus* colonizing smaller diameters and branches in the crown (Paine et al. 1981). This partitioning is also explained by chemical communication, as their pheromones have both intraspecific and interspecific effects on the distribution of the species across the tree (Birch 1980). Moreover, bark beetle predators and parasitoids can exploit pheromone signals to locate prey (Ayres et al. 2001).

Phenology is another form of partitioning, with differences in flight and reproductive cycles allowing some bark beetle species to occupy the same geographic range and host with minimal competition. In the case of *Ips pini*, *I. perroti*, and *I. grandicollis*, which coexist in pine forests of the north-central United States and share the same host tree, differences in flight phenology, development time, voltinism, and spatial colonization patterns reduce congeneric competition (Ayres et al. 2001). The physiological condition of the host can also partition the resource, as different colonization patterns have been observed for different bark beetle genera among trees and snags of different physiological and decomposition states (Saint-Germain et al., 2009).

Closely related species of bark beetles with similar life histories and hosts often inhabit distinct geographic regions. For example, *Tomicus piniperda* and *T. destruens*, species of great importance across the Mediterranean region, present contrasting distributions as a result of different climate demands, where *T. destruens* occurs in locations with warmer temperatures and low altitudes and *T. piniperda* occurs in locations with colder temperatures and higher altitudes (Horn et al. 2012). Another example is the distributions of *Dendroctonus terebrans*, found throughout the eastern United States from coastal New Hampshire south to Florida and west to Texas and Missouri, and *D. valens*, which occurs from Alaska to Mexico and eastward to New England, but does not occur in the southeastern United States (Mayfield and Foltz 2005). These species are morphologically and behaviorally similar, but only co-occur in a narrow zone where their ranges overlap.

Other woodborers compete with bark beetles for resources. Cerambycids for example, such as *Monochamus* spp., feed in the phloem of recently killed pine trees and are facultative intra-guild predators of larvae of other phloem feeders, influencing bark beetle population dynamics (Dodds et al. 2001; Schoeller et al. 2012). Moreover, because cerambycids are larger, competition for phloem results in a loss of resource for bark beetles (Stephen 2011). In the southeastern United States for example, *Monochamus* spp. are common after the attack of bark beetles, such as *Dendroctonus frontalis* and *Ips* spp., attracted by host volatiles and a kairomonal response to sympatric bark beetle pheromones (Allison et al. 2001; Stephen 2011). Other species of cerambycids, such as *Acanthosinus nodosus*, appear to colonize thicker phloem, acting as a potential competitor for several bark beetle species (Stephen 2011).

Symbiotic organisms are commonly associated with bark beetles, including mites, protozoa and nematodes (Hofstetter et al. 2015). Phoretic mites, of which there are more than 250 species associated with bark beetles, have diverse roles ranging from antagonistic parasites or predators of immature beetles, to mutualists that are mycophages or nematophages (Hofstetter et al. 2013, 2015). For example, some mites contribute to fungal diversity in the galleries by carrying different fungal species in a specialized structure (sporotheca) (Moser 1985). At least 57 species of phoretic mites have been recorded for *Dendroctonus frontalis*, and some of these mites have sporothecae that frequently contain spores of *Ophiostoma minus* and *Ceratocystiopsis ranaculosa* (Hofstetter et al. 2013). Because of its pathogenicity, *O. minus* has long been considered a critical mutualist of *D. frontalis*, but several observations suggest that *O. minus* is not always present in trees killed by the beetle, and, moreover, is not capable of killing mature pines (Klepzig et al. 2005). In addition, larvae of *D. frontalis* turn away from phloem colonized by *O. minus* and cannot survive in wood colonized by the fungus (Barras 1970). *Ips typographus* is associated with 38 species of phoretic mites (Hofstetter et al. 2015), which can potentially carry spores of several fungal pathogens that cause mortality to spruce trees (Hofstetter et al. 2013). Because of the lack of mycangia on *Ips* spp., mites are frequently associated with them and critical to the maintenance of fungal associations (Harrington 2005). Nematodes are also common symbiotic organisms associated with bark beetles, often with thousands of individuals in one single beetle, ranging from mutualistic, parasitic, or commensal relationships (Hofstetter et al. 2015).

10.3 Evolution and Diversity

Bark beetles began their diversification at least 120 million years ago as evidenced by a specimen from Lebanese amber (Kirejtshuk et al. 2009). This species, *Cylindrobrotus pectinatus*, resembles *Dryocoetes* but possesses a mixture of ancestral and derived traits leading the authors to place it in a unique tribe. In 100 million-year-old Burmese amber, *Microborus inertus*, represents an extant genus of bark beetle (Cognato and Grimaldi 2009). The variation of morphological features represented in these two species suggests that scolytine diversity was well-established and greater than its fossilized representation. Many species of their extant relatives feed on angiosperms and it is postulated that these Cretaceous species also fed on the burgeoning angiosperm diversity. This would explain the scarcity of scolytine specimens from Cretaceous coniferous ambers (Hulcr et al. 2015).

Bark beetles survived the impact of the great celestial object that was the demise of the dinosaurs and perhaps flourished with the abundance of stressed trees. The next window to their ancient diversity occurred 20 million years later. The scolytine fossil record is well represented in Baltic (45 million years ago) and Dominican ambers (20 million years ago). The Baltic amber fauna is represented mainly by *Hylurgops* and

Hylastes species and along with the plant diversity, suggests an ecosystem similar to the southeastern US (Grimaldi 1996). The Dominican amber fauna is represented mostly by tropical fungi feeding scolytines (except Xyleborini) however several bark beetle genera occur and suggest ties to the current Afrotropical fauna (Bright and Poinar 1994; Cognato 2013). By this point, much of the extant generic diversity was achieved.

Bark beetles spread throughout the world's forests over 120 million years after their origin, when vast distances of ocean separated most of the continents. Bark beetles likely dispersed between land masses by wind and within tree-flotsam and likely seeded multiple species radiations (Gohli et al. 2016; Cognato et al. 2018). There are currently 189 genera and ~ 4300 species of bark beetles. Their diversity is concentrated in the Old and New World tropics, representing half of the total diversity. This is not surprising given the great diversity of plants in the tropics. Likely, natural selection caused by the close association between tree host and beetle, influenced the diversification of scolytines (Gohli et al. 2017). Also, geographic isolation had a major influence on species diversification, as evidenced by bursts of radiation through time (Jordal and Cognato 2012; Gohli et al. 2017).

Introduction of DNA sequence data for phylogenetic reconstruction has resulted in major advances in the understanding of bark beetle relationships. Prior to the 1990's bark beetle phylogenies were mostly unknown given the lack of informative morphological characters considering the canalized scolytine body form (e.g. Cognato 2000). DNA sequences from just a few genes provided needed data to address vexing questions in taxonomy and evolution (Farrell et al. 2001; Cognato and Sun 2007; Jordal and Cognato 2012). Recent use of genomic data has produced the largest and most informative phylogenies to date (Gohli et al. 2017; Johnson et al. 2018). These and other phylogenies are important because they provide evolutionary based hypotheses to the organization of scolytine taxonomy and to the investigations of biological processes. For example, the taxonomy of some of the genera of Ipini was debated (e.g. Cognato 2000; Wood 2007). DNA-based phylogenies supported the recognition of *Pseudips* for *Ips mexicanus* and *I. concinnus*, placement of *I. latidens* and *I. spinifer* in *Orthotomicus*, and the inclusion of the ambrosia fungus feeding Premnobina within Ipini (Cognato and Sperling 2000; Cognato and Vogler 2001; Cognato 2013; see Fig. 9.4 in Cognato 2015). Additionally, behavioral traits can be mapped on phylogenies to identify evolutionary patterns. For example, mapping food preferences on a phylogeny, reveals evolutionary patterns and in this case, that feeding in phloem occurred prior to feeding in other plant parts or on fungi (e.g. Kirkendall et al. 2015).

Phylogenies can also be used to predict a behavior or control method for a new bark beetle pest based on its relationship to other known species. Thereby, the cost for developing management strategies for a potential pest will be reduced. For further detailed examples of bark beetle evolution and diversity see reviews of Kirkendall et al. (2015) and Hulcr et al. (2015).

10.3.1   Ten High Impact Bark Beetle Genera and Selected Case-Studies

10.3.1.1   *Conophthorus*

Conophthorus species are similar to *Pityophthorus,* but species within *Pityophthorus* are smaller. They are distinguished by the gradual transition from asperate to punctate in the pronotum and the costal margin of the declivity descending towards the apex.

There are 13 species of *Conophthorus* in the Nearctic region, from Canada to Guatemala (Alonso-Zarazaga and Lyal 2009). Both larvae and adults feed on pine cones, although some species can infest twigs and buds. Females initiate the galleries near the base of second year cones in early summer (Kirkendall 1983). As reported for several species in the genus, females attract males to the cone with the sex pheromone (+)-pityol (Miller et al. 2000). There is usually one monogamous mating pair per cone (Trudel et al. 2004). Females deposit eggs along the gallery in individual niches close to the developing seeds.

Conophthorus ponderosae is an economically important species that occurs in many *Pinus* species in western North America, ranging from Canada to Mexico (Fig. 10.3). This species can cause up to 90% cone mortality with 100% seed mortality within each cone (Bennett 2000; Smith and Hulcr 2015). It has been suggested that this species is polyphyletic and that southern populations represent a different species (Cognato et al. 2005). *Conophthorus ponderosae* can be distinguished from other *Conophthorus* species by the absence of tubercles on the declivital interstriae 1, and by the lateral convexities on the declivity.

10.3.1.2   *Dendroctonus*

The genus *Dendroctonus* is distinguished by its flattened and rounded antennal club, 5-segmented funicle, steep convex declivity, and an entire compound eye. Species can be confused with *Hylurgus* or *Tomicus,* but these genera have a conical antennal club with a 6-segmented funicle.

There are 20 described species of *Dendroctonus* distributed across the Nearctic region (18 species), and two species in the Palearctic region (Armendáriz-Toledano et al. 2015; Six and Bracewell 2015). *Dendroctonus* contains some of the major conifer-killing bark beetles in the world. Most species colonize *Pinus,* and five reproduce in *Picea, Pseudotsuga,* or *Larix.* Females initiate colonization and build a nuptial chamber, followed by a male that is attracted by sex pheromones and/or host kairomones. After mating, females lay eggs in a newly constructed gallery in the phloem. In this monogamous genus, females typically build galleries that are packed with frass. Some re-emergence and re-mating can occur, as well as sib-mating in a few species (Six and Bracewell 2015). Larvae usually feed on phloem and symbiotic fungi. Larger individuals can fly further, produce more pheromone and offspring, and have a greater overwintering success (Six and Bracewell 2015). Attack of the basal portion of a living tree by some species involves a few individuals with gregarious larval feeding that usually does not kill the tree in usual climatic conditions. However, other representatives of the genus conduct a pheromone-based mass attack that results in the death of the tree and potential massive outbreaks (Raffa et al. 2008).

The southern pine beetle, *Dendroctonus frontalis*, is the most destructive native pest of pine trees in the southeastern United States, Mexico, and Central America (Fig. 10.2) (Thatcher et al. 1980; Coulson and Klepzig 2011). During outbreaks, southern pine beetle infestations often begin in weakened or injured trees, but high beetle populations can mass-attack and kill healthy trees (Cara and Coster 1968; Hain et al. 2011). Uncontrolled infestations may grow to thousands of acres in size, persisting for multiple beetle generations, until depletion of hosts, cold temperatures, direct control, or other factors intervene (Billings 2011). Trees attacked by southern pine beetle often exhibit hundreds of pitch tubes on the outer bark. Beetles feed on phloem and bore S-shaped galleries which can girdle a tree, causing its death. This species is distinguished by its small size (2 to 3 mm) and the convex elytral declivity with the striae distinct and impressed. Males have a distinct notch in the frons and females have a transverse ridge (mycangium) along the anterior pronotum (Fig. 10.3).

Southern pine beetle outbreaks have been cyclical in occurrence, occurring on six to 12 year-intervals and generally last for two to three years after they begin. It has shown a dramatic decline in outbreak activity over much of the southeastern United States since the turn of the twenty-first century compared to previous decades (Birt

Fig. 10.2 Southern Pine Beetle (*Dendroctonus frontalis*) damage in Honduras. Photograph by Ronald Billings, US Forest Service

Fig. 10.3 From left to right, top to bottom, lateral view of *Conophthorus ponderosae, Dryocoetes confuses, Dendroctonus frontalis, Dendroctonus ponderosae, Ips typographus, Pityophthorus juglandis, Polygraphus proximus, Polygraphus ruffipenis.* Scale bar: 1.0 mm. Photographs by Demian F. Gomez, University of Florida

2011; Clarke 2012; Clarke et al. 2016; Asaro et al. 2017). The major outbreak, from 1998 to 2002 in the southern Appalachian Mountains, affected more than 400,000 hectares with an economic loss of more than US\$ 1 billion (Nowak et al. 2008; Clarke and Nowak 2009).

Female *D. frontalis* carry symbiotic fungi in their mycangia, most commonly *Entomocorticium* and *Ceratocystiopsis* species (Yuceer et al. 2011; Six and Bracewell 2015; Harrington et al. 2021). These fungi are introduced into the phloem and serve as the predominant source of nutrition for larvae. The beetles also inadvertently carry blue-stain fungi such as *Ophiostoma minus* in association with phoretic mites

(Moser 1985; Moser and Bridges 1986; Hofstetter et al. 2006). Despite the ongoing controversy over the role of these fungi in tree death, it is known that it has a limited impact compared to the actual beetle attack (Six and Wingfield 2011).

Dendroctonus ponderosae is the most destructive species of bark beetle, colonizing weak *P. ponderosa* and *P. contorta*, and producing extensive outbreaks in healthy trees facilitated by drought and warming climate (Raffa et al. 2008; Creeden et al. 2014) (Fig. 10.3). This species caused the death of more than 11 million hectares of pine trees in a 13-year period in North America, responsible of 50% of tree mortality in the western United States (Ramsfield et al. 2016). Historically distributed in western North America, it has been increasing its natural range mainly because of a warming climate through northern British Columbia towards new regions in Alberta, Canada (Robertson et al. 2009). Obligate symbionts are carried in the mycangia to provide nutritional supplementation, increasing nitrogen availability for larvae (Bleiker and Six 2007). *Grosmannia clavigera*, *Leptographium longiclavatum*, and *Ophiostoma montium* are common symbions of *D. ponderosa* (Six and Bracewell 2015). These fungi can vary within a population due to changing nutrient and moisture profiles in the host, competition among the fungi, and temperature (Six and Bentz 2007). *Dendroctonus ponderosae* can be distinguished from other *Dendroctonus* species by the absence of tubercles in the frons, the large punctures in the pronotum (larger than the distance between them), and the impressed interstria 2 on the elytral declivity.

10.3.1.3 *Dryocoetes*

Dryocoetes species can be recognized by their 5-segmented funicle, the truncated antennal club with corneous first segment, and the short, steep, and unarmed elytral declivity. This genus is similar to *Coccotrypes*, but it can be distinguished by the broad oral region and the non-aciculate frons.

There are 46 species in the genus *Dryocoetes* distributed in the Holarctic and Oriental regions, with seven species occurring in North America (Smith and Hulcr 2015). This genus is mostly phloeophagous feeding in broadleaved and conifer hosts. During colonization, males initiate the attack and build the nuptial chamber in the phloem (Furniss and Kegley 2006). Depending on the species, between 2–6 females will join and construct star-shaped egg galleries. Males remove the frass through the entrance hole after females remove it from the egg galleries. Larval galleries are short and development time may vary according to temperature and altitude, ranging from 1 to 2 years (Smith and Hulcr 2015).

Dryocoetes confusus is the most destructive species in the genus, causing severe damage mainly to *Abies lasiocarpa*, but it can also attack other firs (Garbutt 1992; Smith and Hulcr 2015) (Fig. 10.3). This species can colonize fallen trees as well as kill over-mature trees (i.e. beyond the stage of desirable or optimal development or productivity) in association with the fungal pathogen *Grosmannia dryocoetis* transmitted through a mandibular mycangia. Successfully attacked hosts also show less induced resinosis and less radial growth than unsuccessfully attacked trees, and mortality typically occurs in spots (groups of infested trees) (Bleiker et al. 2003; McMillin et al. 2003).

10.3.1.4 *Ips*

Ips species are distinguished by the three to six spines that line the lateral margin of the elytral declivity. They can be confused with *Orthotomicus* and *Pseudips*; however, in these genera, the elytral declivity is steep and the sutures of the antennal club are slightly to distinctly procurved.

There are 37 *Ips* spp. distributed throughout the Holarctic and most species diversity lies in North America (23 spp.) followed by Eurasia (14 spp.) (Cognato 2015). The adults and larvae feed and complete their life cycle under the bark of the conifer genera *Abies*, *Pinus*, *Picea*, and *Larix*. Most species are specific to *Pinus* or *Picea*, but two Asian species are specific to *Larix*. When *Ips* spp. are restricted to one or two hosts, this appears to be the result of host availability within specific geographic areas. Adult *Ips* males initiate mating by locating a suitable dead or dying host and bore into the phloem to create a nuptial chamber. Males produce aggregation pheromones while feeding, which attract conspecifics to the tree. *Ips* spp. are polygamous and 3–7 females may join the male in the nuptial chamber where they mate. The females then create a tunnel where they lay eggs in niches along the tunnel walls. The hatched larvae feed by tunneling through the phloem. The larvae complete their development in 6–8 weeks depending on temperature.

Ips typographus, the European spruce bark beetle, is the most destructive species of the genus attacking primarily *Picea*, but it can also breed in *Abies* and *Pinus* (Fig. 10.3). The species is distributed across Europe and Asia and although it usually behaves as a secondary pest attacking and killing trees under some level of stress, mass attacks on neighboring healthy trees and enormous economic losses have been reported (Wermelinger 2004). This species can be identified by the four spines on the dull elytral declivity, and the impunctate interstriae on the basal half of the elytral disc.

During the last decade of the twentieth century in Europe, storms caused severe damage to spruce forests, triggering outbreaks of *Ips typographus*. The extent of the damage was highly significant, with millions of cubic meters of spruce killed and large amount of public money spent to manage the outbreaks (Wermelinger 2004). In recent years, severe storms, windthrow events, and high temperatures, have caused the return of new outbreaks in several European countries and parts of Asia (Lausch et al. 2013; Mezei et al. 2017).

Outbreaks depend on weather, drought, storms, and the availability and susceptibility of host trees. Unmanaged forests do not necessarily have higher populations of *I. typographus*. However, after a disturbance, the populations of beetles in unmanaged forests are more likely to increase to epidemic levels (Schlyter and Lundgren 1993). Site and silvicultural characteristics, such as water availability and slope, are

related to attack probability. The maintenance of heterogeneous stands is recommended to reduce attacks in managed spruce forests, as multi-tree species forests
are often less susceptible to bark beetle attack (Wermelinger 2004). Aggregation
pheromones, biosynthesized from tree resin compounds, play a role in attracting *Ips
typographus* to suitable breeding hosts. Colonization usually occurs in windthrown
trees and large trunks are most commonly attacked.

Ips acuminatus has recently increased the frequency and intensity of outbreaks
in *Pinus sylvestris* of the south-eastern Alps (Colombari et al. 2012). In Belarus,
timber losses accounted for more than 184,000 ha in the last ten years. This species
often initiates attacks in the upper bole of mature trees and may infest twigs as
small as 2 mm in diameter. Trees are subsequently attacked by second-generation *I.
acuminatus* and by *Ips sexdentatus* in the lower part of the trunk. *Ips acuminatus* can
be identified by the three spines in the elytral declivity, of which the third is flattened
and acuminate in the male.

10.3.1.5 *Pityophthorus*

Pityophthorus can be distinguished by the pronotal asperities on the anterior half of
the pronotum and by the presence of a sclerotized septum in both antennal sutures of
the club. Species of *Pityophthorus* can be confused with the genus *Araptus*, however,
species of the latter genus do not have a sclerotized septum in the antennal club.

There are 386 species in the genus *Pityophthorus* distributed mostly in North
and Central America, but also ranging from the Palearctic to the Oriental Region
(Alonso-Zarazaga and Lyal 2009). This genus typically infests twigs and branches
from a broad range of hosts, such as conifers, woody shrubs, vines, hardwood trees,
and herbaceous plants (Bright 1981). In North America, most species develop in
Pinus, with a few colonizing *Abies, Picea, Pseudotsuga,* and *Larix*. Mating systems
vary widely in this genus from polygamy to monogamy and thelytokous parthenogenesis. In phloeophagous species, males initiate the attack and build a nuptial
chamber, joined by 3–5 females attracted by aggregation pheromones (Smith and
Hulcr 2015). Females then excavate egg galleries radiating from the central nuptial
chamber. Females of myelophagous species feed and construct galleries in the pith
of small twigs.

Most species in this genus are secondary pests and usually are not of economic
importance, with the exception of a few species that vector fungi such as the conifer
pathogen *Fusarium circinatum* or the walnut pathogen *Geosmithia morbida*. *Pityophthorus juglandis*, endemic to Mexico and the southwestern continental United States,
is the most economically important species in the genus (Fig. 10.3). *Pityophthorus
juglandis* causes black walnut tree mortality when they colonize branches and trunks
in high numbers and cankers develop around the galleries as a result of its association with *G. morbida* (Kolařík et al. 2011; Rugman-Jones et al. 2015). After 3 years,
trees show symptoms of die-back and flagging. The combination of the insect and
the fungus threatens the $500 billion black walnut industry in the eastern United
States (Newton et al. 2009). However, the pathogenicity of *G. morbida* has recently

been questioned as different strains may cause different effects, and therefore, the consequences of *P. juglandis* colonization are dependent on the pathogenicity of the *G. morbida* strain and environmental factors (Sitz et al. 2017).

10.3.1.6 *Polygraphus*

Polygraphus species are distinguished from other related genera such as *Carphoborus*, by the divided eye, the antennal club with no sutures, and the absence of a scutellum. There are 101 species distributed through the Holarctic, Oriental, and Ethiopian regions (Alonso-Zarazaga and Lyal 2009). All the species within this genus are phloeophagous, feeding mainly on Pinaceae (*Abies*, *Cedrus*, *Larix*, *Picea*, and *Pinus*) and hardwoods (Wood and Bright 1992). *Polygraphus* spp. are polygamous, with males usually initiating attack and excavating the nuptial chamber. Attracted to male aggregation pheromones, 2–4 females can join and start individual egg galleries that can reach up to 10 cm length (Smith and Hulcr 2015).

Polygraphus proximus, distributed in the eastern Palearctic region, attacks several species of *Abies*, and is one of the main factors contributing to the destruction of large areas of Siberian forests since the early 2000s (Krivets et al. 2015) (Fig. 10.3). This species typically colonizes weakened or dying trees, but when population levels are high healthy trees are attacked (Kerchev 2014). Trees usually die after 2–4 years of attack. The ophiostomatoid fungus *Grosmannia aoshimae*, is symbiotic with *P. proximus*, considered an aggressive phytopathogen (Pashenova et al. 2011) and likely contributes to tree mortality. *Polygraphus proximus* can be distinguished from other European species by the pointed antennal club, yellow legs, and elytral base slightly wider than pronotum (Pfeffer 1995).

Polygraphus rufipennis, common across the Nearctic region, is a secondary species that usually colonizes stumps, trunks, or branches of *Picea,* particularly *P. glauca*. In association with the blue stain fungus *Ophiostoma piceaperdum*, it can cause mortality to trees previously weakened by other biotic factors (Fig. 10.3). For example, *P. rufipennis* often colonizes trees weakened by *Dendroctonus rufipennis* or the spruce budworm (*Choristoneura* spp.) (Simpson 1929). This species has one generation per year, with females emerging in mid-summer to establish a second brood. *Polygraphus rufipennis* is distinguished by the stout body, the obtusely pointed antennal club, and by the densely punctured frons in females (Wood 1982b).

10.3.1.7 *Pseudohylesinus*

Pseudohylesinus species are distinguished by the scaled vestiture, the seven-segmented funicle, and the antennal club with the first segment larger. *Pseudo-hylesinus* species are similar to those of *Xylechinus* and *Hylorgupinus*, but *Pseu-dohylesinus* can be distinguished by the two-color pattern of dark and light scales covering the body.

There are 13 species of *Pseudohylesinus*, all endemic to North America, distributed from Alaska and western Canada and contiguous United States, to Mexico (Wood and Bright 1992). Species in this genus are all phloeophagous and are attracted to host compounds, feeding mostly on *Abies*, whereas a few species also develop on *Picea, Pinus, Pseudotsuga*, and *Tsuga*. Only a few severe outbreaks have been recorded for species within this genus, but the common observed damage is in discrete patches or individual trees (Carlson and Ragenovich 2012). This genus is monogamous, with females initiating the attack and boring the entrance tunnel (Bright 1969). Once the male joins, they both excavate branched egg galleries (1 or 2 ramifications), and females deposit individual eggs along the gallery and cover them with boring dust.

Pseudohylesinus granulatus, the most economically important species of the genus, is distributed from British Columbia to California and attacks mostly *Abies amabalis* (Fig. 10.4). It can kill overmature trees in association with the brown-staining fungus *Ophiostoma subannulatum*, but usually colonizes fallen trees (Carlson and Ragenovich 2012). Mortality can occur as a result of girdling from accumulated attack patches over several years (Smith and Hulcr 2015). *Pseudohylesinus granulatus* can be distinguished by the large and deep pronotal punctures and by the slender body (Wood 1982b).

10.3.1.8 *Pseudopityophthorus*

Pseudopityophthorus can be distinguished by the reduced or absent striae in the elytra, the convex elytral declivity with abundant hair-like to scale-like setae, and the septate and procurved sutures in the antennal club. Species in this genus can be confused with *Pityophthorus*, but the absence of striae in *Pseudopityophthorus* differentiate them.

There are 27 species of *Pseudopityophthorus*, distributed mostly in the Nearctic region, but some species occur in the Neotropical and eastern Palearctic region (Wood 1986; Alonso-Zarazaga and Lyal 2009). Species of *Pseudopityophthorus* are phloeophagous and mainly found on *Quercus,* although other Fagaceae have been reported as hosts. Males initiate the colonization process in this monogamous genus by excavating the entrance tunnel and a short longitudinal gallery in cut, broken, or fallen branches or trunks (Wood 2007). The female then joins and begins a transverse egg gallery in the opposite direction from the male gallery. Larval galleries are longitudinal and almost straight.

Pseudopityophthorus minutissimus (Fig. 10.4) and *P. pruinosus* (Fig. 10.4), have been implicated as vectors of the oak wilt fungal pathogen, *Bretziella fagacearum* in North America, although different roles of the beetle as a vector have been suggested for this pathogen (Berry and Britz 1966; Ambourn et al. 2006). These beetles produce two generations per year through most of the disease range. *Bretziella fagacearum* causes a vascular wilt in more than 30 species of *Quercus* and kills thousands of trees every year in urban landscapes of the United States (Tainter and Baker 1996). *Pseudopityophthorus minutissimus* can be identified by the reticulate frons, the confused

Fig. 10.4 From left to right, top to bottom, lateral view of *Pseudohylesinus granulatus*, *Pseu-dopityophthorus minutissimus*, *Pseudopityophthorus pruinosus*, *Scolytus multistriatus*, *Scolytus quadrispinosus*, *Scolytus schevyrewi*, *Scolytus ventralis*. Scale bar: 1.0 mm. Photographs by Demian F. Gomez, University of Florida

elytral punctures, and by the uniformly short and confused elytral setae (Wood 1982b). *Pseudopityophthorus pruinosus* is similar to *P. minutissimus* but is larger with larger elytral punctures, and with a row of scales on interstria 1 and 3.

10.3.1.9   *Scolytus*

The genus *Scolytus* can be distinguished by the single curved process in the outer margin of the protibiae, the flattened antennal club, the seven-segmented funicle, and by the slightly sloped elytra (Smith and Cognato 2014). This genus is similar to *Cnemonyx*, but the indistinct declivity in this genus differentiates them.

There are 213 species in the genus *Scolytus* distributed in the Holarctic, Oriental and Neotropical regions. They are phloeophagous and colonize either Pinaceae, such as *Abies, Larix, Picea, Pseudotsuga*, and *Tsuga*, or hardwoods. Host selection is usually mediated by host volatiles and severe attacks are usually stress related, commonly associated with drought or other insects (Smith and Cognato 2014). All Holarctic *Scolytus* species are monogamous and Neotropical species are bigamous or polygamous (Smith and Hulcr 2015). In monogamous species, females colonize the host and start the construction of the nuptial chamber. Males join the entrance tunnel where mating occurs. The female then excavates 1 or 2 egg galleries (depending on the species) with eggs deposited individually inside niches. Males leave the gallery after the egg gallery is complete and females typically die in the entrance hole. Maturation feeding in twigs has been reported for some species.

Scolytus multistriatus is a Palearctic species that has been introduced in the Americas, Australia, and New Zealand (Fig. 10.4). Through the production of an aggregation pheromone, females colonize stressed native and exotic *Ulmus* species. This species is the principal vector of the pathogen *Ophiostoma novo-ulmi*, that causes Dutch elm disease, responsible for the death of millions of elm trees in North America (Furniss and Carolin 1977; Bloomfield 1979). Adults, covered in fungal spores of *Ophiostoma novo-ulmi* upon emergence, inoculate the trees with the pathogen during maturation feeding in the twigs. This species can be identified by the presence of lateral teeth on ventrites 2–4 and by a median conical spine on ventrite 2 (Smith and Cognato 2014).

Scolytus quadrispinosus, a native species in North America, is one of the most destructive pests of hardwoods, in particular species of the genus *Carya* (Fig. 10.4). It usually attacks and kills single trees through mass attack and subsequent girdling of the host, but can develop outbreaks during periods of drought (Blackman 1922). Males are distinguished by the apical margin of ventrite 3 armed by three spines, ventrite 4 armed by one median tooth, and ventrite 1 apically descending (Smith and Cognato 2014). Females are distinguished by the flattened and longitudinally aciculate frons.

Scolytus schevyrewi, a Palearctic species that has been introduced in North America, colonizes stressed *Ulmus* trees and is attracted by host volatiles (Fig. 10.4). This species is a less effective vector of the Dutch elm disease pathogen than *S. multistriatus* (Jacobi et al. 2013). *Scolytus schevyrewi* resembles *S. piceae*, but it can be distinguished by the subapical carina on ventrite 5 located just before the end of the segment (Smith and Cognato 2014).

Scolytus ventralis, native to North America, attacks several *Abies* species and can cause significant mortality, being the most destructive conifer-feeding species in the genus (Fig. 10.4). This species is associated with the symbiotic fungus *Trichosporium symbioticum*, introduced by females in the gallery (Bright and Stark 1973). Development time varies from 41 to 380 days depending on latitude and elevation. Males of this species can be distinguished from females by the elevated base of ventrite 2, the surface of ventrite 2 flat, the apical margin of ventrite 2 often bearing a median denticle, and by the glabrous ventrite 2 (Smith and Cognato 2014). Females are distinguished by the weakly aciculate and strongly punctate frons and by the apical margin of ventrite 1 flush with basal margin of ventrite 2.

10.3.1.10   *Tomicus*

Tomicus species can be differentiated by the 6-segmented antennal funicle, an ovate club with straight sutures, the pronotum wider than long, and the convex declivity with interstrial granules and erect setae. This genus is similar to *Hylurgus*, but *Tomicus* can be distinguished by the shiny frons and declivity, and by the less hairy vestiture.

Tomicus is comprised of eight species distributed across the Palearctic region with one species introduced in North America (Lieutier et al. 2015). Five species occur only in Asia, one in Europe and northern Africa, and two widely distributed in Eurasia. All species are phloeophagous and usually colonize trunks or branches of weakened *Pinus* species, and one species utilizes *Picea*. Maturation feeding occurs in the shoot of healthy and vigorous pines, causing severe problems to young plantations when a large number of shoots are destroyed resulting in growth loss. This maturation feeding, revealed by the existence of entrance holes surrounded by resin, can occur in a different tree than the natal host; therefore, their life cycle would not necessarily occur in the same tree as in most scolytines. Species of this genus are monogamous and have one generation per year, with females excavating egg galleries with individual niches.

Tomicus destruens is among the most damaging pests across the Mediterranean region and attacks native and exotic pine species through attraction to several host volatiles from stressed trees, such as ethanol, α-pinene, β-myrcene, and α-terpinolene (Faccoli et al. 2008) (Fig. 10.5). *Tomicus destruens* can be distinguished by the weakly impressed elytral declivital interstriae 2 with dense and confused punctures, and by the uniformly yellow or yellow–brown antennae (Kirkendall et al. 2008).

Tomicus piniperda, the most widespread species, can colonize several *Pinus* species but prefers *P. sylvestris* (Fig. 10.5). This species has been introduced in eastern North America where it causes damage to the Christmas tree and nursery industries (Haack and Poland 2001). Host kairomones (α-pinene) and aggregation pheromones play an important role in colonization of the host (Poland et al. 2003). *Tomicus piniperda* can be distinguished by the interstria 2 strongly impressed and concave with uniseriate regularly spaced punctures on the declivity, the erect hairs

Fig. 10.5 From top to bottom, lateral view of *Tomicus destruens, Tomicus piniperda, Tomicus yunnanensis.* Scale bar: 1.0 mm. Photographs by Demian F. Gomez, University of Florida

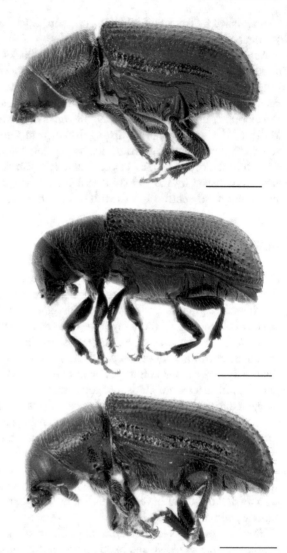

on the declivity distinctly longer than those on disc, and by the uniformly brown antennae (Kirkendall et al. 2008).

Tomicus yunnanensis, recorded only in Yunnan Province, China, has caused significant damage to more than 200,000 hectares of *Pinus yunnanensis* forests in southwest China (Liu et al. 2010) (Fig. 10.5). *Tomicus yunnanensis* can be distinguished by the interstria 2 strongly impressed and broadly convex with confused or biseriate evenly spaced punctures on the declivity, and by the uniformly yellow or yellow–brown antennae (Kirkendall et al. 2008).

10.4 Management and Control

Bark beetle epidemics are generally managed through direct and indirect control measures. Direct control involves tactics like sanitation harvests to manage current infestations, whereas the indirect approach is preventive and designed to reduce the frequency and severity of future attacks. Indirect measures involve manipulation of the stand through silvicultural practices such as thinning and prescribed burning, aimed at reducing competition among trees resulting in improved tree vigor, and selecting for favorable species composition.

In order to apply proper control strategies, monitoring and prediction programs that gather and analyze information on the extent of infestations are essential. Aerial surveys using digital mapping are commonly used for recognizing spots of infested trees that are later confirmed in the field (Fettig and Hilszczański 2015). Remote sensing techniques are becoming more commonly used to detect bark beetle outbreaks, usually relying on near-infrared (NIR) and shortwave infrared (SWIR) satellite imagery (Hais et al. 2016). Hazard prediction systems are also utilized. Some hazard rating systems are based on stand characteristics (e.g. basal area, radial growth), and others are based on bark beetle captures. The former provides an estimate of how severely a stand might be impacted if an outbreak were to occur, while the latter attempts to estimate beetle population trends. For example, a system to forecast infestation trends (increasing, static, declining) and relative population levels (high, moderate, low) of *D. frontalis* has been developed and implemented in the southeastern United States based on the captures of the pest and its major predator, the clerid *Thanasimus dubius* (Billings and Upton 2010).

Preventive measures that reduce the amount of slash material (woody debris from logging operations or forest disturbances) can help minimize populations of some bark beetles (Fettig et al. 2007). It is also important to select the appropriate tree species for the site, as well as spacing intervals that minimize tree competition. Treatments such as thinning, are recommended to enhance tree vigor, and therefore, increase forests resilience towards bark beetles.

Thinning is a silvicultural treatment with the objective to reduce stand density to improve growth and forest health (Helms 1998). Several benefits arise from thinning, such as enhanced growing space for desirable trees, increased tree vigor, reduced fire, insect, and pathogen risks, and the production of early economic benefits. Depending on the objective of the thinning, as well as the tree species involved, different practices can be used. Low thinning removes trees from smaller diameter classes, crown thinning removes mid-canopy trees, and selection thinning removes the largest trees (Fettig et al. 2007). This is a widely used method and, if conducted properly without creating physical damage to residual trees, thinning reduces bark beetle attacks and therefore, tree mortality. For southern pine beetle for example, landscape-level preventative thinning is the most economical and sustainable approach to the mitigation of epidemics (Asaro et al. 2017). In more than 10 states, the Forest Service offers the Southern Pine Beetle Assistance and Prevention Program, which promotes proactive management practices by reimbursing landowners for thinning, prescribed burning, or other management plans (Nowak et al. 2008).

The relationship between silvicultural thinning and significantly reduced tree mortality during outbreaks has been experimentally reported for the two most severe bark beetle pests in North America, *D. ponderosa* and *D. frontalis* (Fettig et al. 2007; Asaro et al. 2017). Thinning is usually conducted during periods of reduced beetle activity; however, for *D. frontalis* for example, thinning can be conducted during periods of beetle activity with limited risk if logging damage and slash material is minimized (Fettig et al. 2007). For secondary bark beetles, such as *Ips* spp., slash or damaged hosts are important for the growth of infestations, particularly in areas with high beetle populations. For *Ips typographus* for example, removal of windthrown timber is one of the most important management strategies. Moreover, these logs can act as trap trees if removed after infestation but before emergence (Göthlin et al. 2000).

In planted forests, breeding sites of these secondary bark beetles occur mostly in slash material produced by pruning and thinning, thus management of slash material is an essential tool for reducing bark beetle populations. Chipping slash residual has been proposed as a strategy to reduce breeding sites and retain biomass for nutrient cycling, however, some authors have shown that the high concentration of monoterpenes and other volatiles associated with chipping actually increases the risk of standing trees being attacked compared to scattered logs (Fettig et al. 2007). Prescribed fire, used to enhance wildlife habitat, reduce fuels, and control pests, can stress standing trees and increase susceptibility to bark beetles (Elkin and Reid 2004). Some studies have associated prescribed fires, when not properly conducted, to infestations of *D. frontalis* in the southeastern United States, but usually fire increases populations of less threatening bark beetles such as *Ips* spp. and *Dendroctonus terebrans* (Sullivan et al. 2003).

Direct control measures include insecticides, mass trapping, mating disruption, biological control, or sanitation harvests. These methods are costly, meaning that their implementation will depend on budget, equipment, and market conditions. Hence, the first step is to identify which spots are more likely to expand (Billings and Ward 1984).

Sanitation harvesting (cut-and-remove trees to remove pests) is the preferred control tactic for species like *D. frontalis* (George and Beal 1929) because it is still the most effective. Harvesting trees infested by the beetles, as well as a 15 to 30 m (1–2 tree lengths) buffer zone of uninfested trees, can stop spot growth (Billings 2011). However, sometimes salvage logging (harvest to recover some economic value) is not possible, either because of socio-political and economic hurdles (as has been the case with *D. frontalis* in the southeastern US recently), or because complex terrain in remote locations can make salvage impractical, as can be the case in the western US and Canada. For *I. typographus*, sanitation harvesting is the most effective direct management approach. However, trees need to be cut before adults emerge and logs need to be either debarked, burned or chipped before storing or removed from the forest. Debarking can be highly effective because it causes 93% mortality of the beetles (Dubbel 1993).

When cut-and-remove operations are not possible, cut-and-leave tactics are the next best option (Fig. 10.6). This control method is based on felling all freshly attacked or infested trees towards the center of the spot, in addition to a buffer zone of uninfested trees in the expanding front (Fig. 10.6), usually as wide as the average tree height (Billings and Schmidtke 2002; Fettig et al. 2007). This technique is effective because it increases solar radiation and causes less favorable microclimatic conditions for further bark beetle development, while also increasing competition with wood borers and other antagonists.

Insecticides are important control measures for some species, such a *D. ponderosae*, but the use is regulated by different agencies and approved chemicals vary among jurisdictions. Usually, insecticides are only utilized to preventatively protect unattacked or lightly attacked high value trees, such as the ones grown in urban environments or trees growing in progeny tests or seed orchards (Fettig and Hilszczański 2015). Most treatments involve spraying the tree trunk or any part that is likely to be attacked by the targeted species usually in late spring prior to adult flight. Injection of systemic insecticides to the trunk can also be used, as the product is transported throughout the tree. For example, the application of systemic pesticides, particularly emamectin benzoate, can protect high-value trees from the attack of *D. frontalis* during outbreaks (Grosman et al. 2009).

Semiochemicals, mainly used as attractants or anti-aggregation compounds in forest management, can also be employed for mass trapping, but usually these traps will not capture a significant portion of the population and catches do not necessarily correlate with high infestations (Weslien and Lindelow 1990; Dodds and Ross 2002). Moreover, some beetles attracted to these traps may infest adjacent trees causing additional mortality (Fettig and Hilszczański 2015). In some cases, such as in push–pull strategies, mass trapping devices are combined with repellents so to deter beetle attack of high quality stands or trees. Antiaggregation pheromones, such as verbenone for several species of *Dendroctonus*, are widely used to protect individual trees or forest stands. These inhibitors are usually placed as pouched release devices on individual trees before beetle flight. For *D. frontalis* for example, both male and female pheromones are used for monitoring purposes (Sullivan and Mori 2009). The female pheromone (frontalin) is deployed in multi-funnel traps, while the male pheromone ((+)-endo-brevicomin) is deployed a few meters away from the frontalin trap to significantly enhance its synergistic effect on *D. frontalis* attraction. For *I. typographus* pheromone traps baited with cis-verbenol, ipsdienol and 2-methyl-3-buten-2-ol are used to prevent attacks on living trees and for monitoring. However, catches depend on environmental and local conditions, such as temperature, sun exposure, and the presence of woody debris, slash, and susceptible trees (Lobinger 1995; Wermelinger 2004; Fettig and Hilszczański 2015).

Nonhost volatiles (NHV), released by nonhost angiosperm plants, have been shown to inhibit pheromone attraction and orientation response in several conifer bark beetle species (Byers et al. 1998; Zhang 2003). The combination of NHVs with anti-aggregation pheromones can provide potent treatments to protect trees, logs, or stands from attacks by bark beetles (Huber et al. 2001). Even though semiochemicals are widely used in bark beetle management, more studies on blends and delivery

Fig. 10.6 Top: Cut-and-leave management strategy for Southern Pine Beetle (*Dendroctonus frontalis*) damage in Honduras. Bottom: Buffer zone during direct control management for Southern Pine Beetle (*Dendroctonus frontalis*) damage in Honduras. Photographs by Ronald Billings, US Forest Service

systems are needed, as well as the performance of the semiochemicals on different hosts and beetle populations (Fettig and Hilszczański 2015).

Biological control using predators or parasitoids has been used with success to control bark beetle populations. In China, the predator *Rhizophagus grandis* (Coleoptera: Rhizophagidae) has been used to control introduced *Dendroctonus valens* in pine forests (Yang et al. 2014). Entomopathogenic fungi, mainly *Beauveria bassiana*, have been effective at causing high mortality in several bark beetle species (Whitney et al. 1984). Inoculating beetles collected in baited traps and then releasing them back into the field has been suggested (Kreutz et al. 2000), but more practical methods should be developed because of low infection rates in field trials compared to laboratory conditions.

10.4.1 Emerging Pests

10.4.1.1 *Acanthotomicus suncei* Cognato

The sweetgum inscriber, *Acanthotomicus suncei*, is a polygynous species, in which the male starts the gallery and is later joined by one to three females (Gao and Cognato 2018) (Fig. 10.7). Galleries are usually horizontal ranging from 5 to 10 cm and trees as small as 2 cm diameter can be attacked. This endemic Chinese species has been recently reported to cause severe damage to a sweetgum native to North America, *Liquidambar styraciflua*, planted as an ornamental tree in China (Gao et al. 2017) (Fig. 10.7). The outbreak occurred in nurseries and urban trees in the Shanghai area. Affected trees exude resin from wounds and branches wilt and die. As it is observed in conifer-feeding bark beetles, the accumulation of attackers eventually exhausts tree defenses and kills the tree. Outbreaks develop quickly and the extent of the damage is unknown outside the evaluated localities. Economic losses are estimated around US\$ 4 million from the loss of more than ten thousand trees. Arrival of this species to North America would be cause for concern for the health of native *L. styraciflua*. A recent economic analysis suggests a potential economic loss of US\$ 150 million to US forest industries (Susaeta et al. 2017).

10.4.1.2 *Cyrtogenius luteus* (Blandford)

Cyrtogenius luteus is an Asian bark beetle that attacks stressed or dying trees, with no economic significance recorded in its native range (Fig. 10.7). It is a polygynous species that flies mainly in summer, but colonization has also been observed in spring (Gómez et al. 2017). Irregular star-shaped galleries are bored in the phloem and eggs are laid in niches on the side (Gómez et al. 2012). Larvae will bore irregular galleries and after pupation, adults will emerge through individual exit holes. Since 2009, it has been recorded in South America (Uruguay) and Europe (Italy) (Faccoli et al. 2012; Gómez et al. 2012). More recently, it has also been reported from southern

Fig. 10.7 From left to right, top to bottom, lateral view of *Acanthotomicus suncei* (photograph by Demian F. Gomez, University of Florida), commercial nursery of sweetgum attacked by *Acanthotomicus suncei* in Shanghai, China (photograph by You Li, University of Florida), lateral view of *Cyrtogenius luteus* (photograph by Demian F. Gomez, University of Florida), lateral view of *Dendroctonus valens* (photograph by Demian F. Gomez, University of Florida). Scale bars: 1.0 mm

Brazil (Flechtmann and Atkinson 2018), where it occurs since 2006. In Italy, it has been mostly recorded from traps and no economic damage has been reported. In South America, where commercial forestry has been increasing exponentially in the last two decades, *C. luteus* is usually associated with *Pinus taeda*, the most common planted pine tree species in Brazil and Uruguay. However, observations from Brazil suggest that this species might be colonizing the native Brazilian conifer *Araucaria angustifolia*, as it has only been recovered from traps deployed 30 km away from the closest pine plantation (Flechtmann and Atkinson 2018). Even though *C. luteus* appears to behave as a secondary pest in Asia, attacking only dying trees, several infested apparently healthy pine stands have been reported in Uruguay (Gómez et al. 2012). In commercial plantations of *P. taeda*, 80% of the stand is affected with losses up to 20 hectares (Fig. 10.7). However, this observation was made after significant drought periods in dense stands. In Brazil and Italy, no significant damage to live trees has been recorded (Faccoli et al. 2012; Flechtmann and Atkinson 2018).

10.4.1.3 *Dendroctonus valens* LeConte

Dendroctonus valens is widely distributed in North and Central America, ranging from Canada to the western United States, Mexico, Guatemala, and Honduras (Fig. 10.7). It is rarely a problem in its native range, but was introduced into China where it has become a pest. After its first detection in the Shanxi Province in northern China in 1998, it has been spreading to adjacent provinces causing unprecedented tree mortality to *Pinus tabuliformis* (Yan et al. 2005). This beetle species has the broadest host range within the genus (Six and Bracewell 2015) and usually reproduces in living trees, but is highly attracted to injured, weakened, and dying trees (Fettig et al. 2004).

10.4.2 Bark Beetle Management in a Changing World

From a landscape perspective, the abundance and distribution of susceptible hosts play an important role in the distribution of bark beetles. Outbreaks occur when favorable environmental and host conditions occur. Silvicultural treatments that increase forest resilience may become even more important to stave off pest problems as climate change and invasive species introductions continue. Insects are attracted to highly concentrated patches of their hosts (Root 1973), and large forested areas with little heterogeneity make certain regions highly susceptible to outbreaks. As a result, the spatial arrangement of stands of similar age and species is relevant to reducing levels of tree mortality (Samman and Logan 2000; Jactel and Brockerhoff 2007). However, this does not mean that desirable forest conditions are free of disturbances. Forests can be both productive and sustainable, but this condition in a forest ecosystem also involves dead and dying trees. From an ecological perspective, healthy amounts of insects and pathogens are needed to keep a baseline tree mortality (Castello and Teale 2011). Beyond this baseline the impacts of insects can cause mortality with more negative consequences.

Forest insects and pathogens are seen as problems when they interfere with management objectives, but the conditions that favor insect or disease problems are usually the result of past or present human activity, such as method of harvesting, and spatial and temporal patterns in tree size, tree species, among others. For many eruptive forest insects, the existing knowledge on the drivers of outbreak eruptions and crashes is insufficient to face current challenges. Biotic variables that affect bark beetle population dynamics need to be compiled, and hypotheses on their role and their interaction with anthropogenic change need to be developed (Biedermann et al. 2019).

For severe outbreaks to occur, there must be several years of favorable weather that enhance population growth, and an abundance of susceptible trees. Increasingly, climate change is playing a substantial role in these interactions. Recent examples of drought-related tree mortality suggest that all forest types are vulnerable to climate change (Allen et al. 2010). Moreover, outbreaks of bark beetles and other insect pests are increasing in severity and frequency. Climatic changes are predicted to significantly affect the frequency and severity of disturbances, as higher latitudes and elevations will be more susceptible to bark beetle outbreaks and the resulting tree mortality in the next decades (Bentz et al. 2010).

Market forces also play a significant role in bark beetle management. For example, during the SPB outbreak of 2012 in Missisipi, cut-and-leave was the primary suppression method for 407 hectares (201 spots), whereas less than 12 hectares were treated with cut-and-remove (Meeker 2013).

Despite the effectiveness of management strategies, changing forest structure to improve resiliency is perhaps the best long-term plan for coping with climate change. Regional and international networks should support countries to increase local knowledge and forest management capacity. Cooperation among forest scientists, landowners, and governmental stakeholders is key, and will ultimately help with developing long-term and evidence-based solutions to manage outbreaks of the bark beetles (Biedermann et al. 2019). Bark beetle outbreaks will keep increasing as long as susceptible forests and favorable climatic conditions coincide.

References

Allen CD, Macalady AK, Chenchouni H, Bachelet D, McDowell N, Vennetier M, Kitzberger T, Rigling A, Breshears DD, Hogg EH, Gonzalez P, Fensham R, Zhang Z, Castro J, Demidova N, Lim JH, Allard G, Running SW, Semerci A, Cobb N (2010) A global overview of drought and heat-induced tree mortality reveals emerging climate change risks for forests. For Ecol Manage 259:660–684

Allison JD, Borden JH, Mcintosh RL, De Groot P, Gries R (2001) Kairomonal response by four Monochamus species (Coleoptera: Cerambycidae) to bark beetle pheromones. J Chem Ecol 27:633–646

Alonso-Zarazaga MA, Lyal CHC (2009) A catalogue of family and genus group names in scolytinae and platypodinae with nomenclatural remarks (Coleoptera: Curculionidae). Zootaxa 134:1–134

Ambourn AK, Juzwik J, Eggers JE (2006) Flight periodicities, phoresy rates, and levels of Pseudopityophthorus minutissimus branch colonization in oak wilt centers. For Sci 52:243–250

Armendáriz-Toledano F, Niño A, Sullivan BT, Kirkendall LR, Zúñiga G (2015) A new species of bark beetle, Dendroctonus mesoamericanus sp. nov. (Curculionidae: Scolytinae), in Southern Mexico and Central America. Ann Entomol Soc Am 108:403–414

Asaro C, Nowak JT, Elledge A (2017) Why have southern pine beetle outbreaks declined in the southeastern U.S. with the expansion of intensive pine silviculture? a brief review of hypotheses. For Ecol Manage 391:338–348

Avtzis D, Knížek M, Hellrigl K, Stauffer C (2008) polygraphus grandiclava (coleoptera: curculionidae) collected from pine and cherry trees: a phylogenetic analysis. Eur J Entomol 105:789–792

Ayres BD, Ayres MP, Abrahamson MD, Teale SA (2001) Resource partitioning and overlap in three sympatric species of Ips bark beetles (Coleoptera: Scolytidae). Oecologia 128:443–453

Barr BA (1969) Sound production in Scolytidae (Coleoptera) with emphasis on the genus Ips. Can Entomol 101:636–672

Barras SJ (1970) Antagonism between Dendroctonus frontalis and the Fungus Ceratocystis minor. Ann Entomol Soc Am 63:1187–1190

Barras SJ, Perry T (1972) Fungal Symbionts in the Prothoracic Mycangium of Dendroctonus frontalis (Coleoptera: Scolytidae). Zeitschrift für Angew. Entomol 71:95–104

Barras SJ, Taylor JJ (1973) Varietal Ceratocystis minor identified from mycangium of Dendroctonus frontalis. Mycopathol Mycol Appl 50:293–305

Beaver RA (1973) Biological studies of Brazilian Scolytidae and Platypodidae (Coleoptera): II. The tribe Bothrosternini. Sao Paulo Dep Zool Pap Avulsos 227–236

Bennett R (2000) Management of cone beetles (Conophthorus ponderosae, Scolytidae) in blister rust resistant western white pine seed orchards in British Columbia. Seed Seedlings Ext Top 12:16–18

Bentz BJ, Rgnire J, Fettig CJ, Hansen EM, Hayes JL, Hicke JA, Kelsey RG, Negron JF, Seybold SJ (2010) Climate change and bark beetles of the western United States and Canada: direct and indirect effects. Bioscience 60:602–613

Berry FH, Britz TW (1966) Small oak bark beetle: a potential vector of oak wilt. Plant Dis Rep 50:45–49

Biedermann PHW, Müller J, Grégoire JC, Gruppe A, Hagge J, Hammerbacher A, Hofstetter RW, Kandasamy D, Kolarik M, Kostovcik M, Krokene P, Sallé A, Six DL, Turrini T, Vanderpool D, Wingfield MJ, Bässler C (2019) Bark Beetle population dynamics in the anthropocene: challenges and solutions. Trends Ecol Evol 34:914–924

Billings RF (2011) Mechanical control of southern pine beetle infestations. In: South Pine Beetle II Gen Tech Rep SRS-140, pp 399–413

Billings RF, Schmidtke P (2002) Central America Southern Pine Beetle/Fire Management Assessment. USDA Foreign Agric Serv Coop Dev 19

Billings RF, Upton WW (2010) A methodology for assessing annual risk of southern pine beetle outbreaks across the southern region using pheromone traps. In: Adv Threat Assess Their Appl to For Rangel Manag, pp 73–85

Billings RF, Ward JD (1984) How to conduct a Southern pine beetle aerial detection survey. Circ For Serv

Birch M (1980) Olfactory discrimation among four sympatric species of bark beetles: a mechanism of resource partitioning. J Chem Ecol 6:395

Birt AG (2011) Risk assessment for the Southern Pine Beetle. In: South Pine Beetle II, pp 299–316

Blackman (1922) Mississippi bark beetles. Mississippi Agric Exp Station Tech Bull 11:130

Bleiker KP, Heron RJ, Braithwaite EC, Smith GD (2013) Preemergence mating in the mass-attacking bark beetle, Dendroctonus ponderosae (Coleoptera: Curculionidae). Can Entomol 145:12–19

Bleiker KP, Lindgren BS, Maclauchlan LE (2003) Characteristics of subalpine fir susceptible to attack by western balsam bark beetle (Coleoptera: Scolytidae). Can J for Res 33:1538–1543

Bleiker KP, Six DL (2007) Dietary benefits of fungal associates to an eruptive herbivore: potential implications of multiple associates on host population dynamics. Environ Entomol 36:1384–1396

Blomquist GJ, Figueroa-Teran R, Aw M, Song M, Gorzalski A, Abbott NL, Chang E, Tittiger C (2010) Pheromone production in bark beetles. Insect Biochem Mol Biol 40:699–712

Bloomfield H (1979) Elms for always. Am For 85:24–26

Borden J (1982) Aggregation pheromones. In: Bark Beetles North Am Conifers, pp 74–139

Bridges JR (1983) Mycangial Fungi of Dendroctonus frontalis (Coleoptera: Scolytidae) and thier relationship to Beetle population trends. Environ Entomol 12:858–861

Bright DE (1969) Biology and taxonomy of bark beetle species in the genus Pseudohylesinus Swaine (Curculionidae: Scolytinae). Univ Calif Publ Entomol 54:1–46

Bright DE (1981) Taxonomic monograph of the genus Pityophthorus Eichhoff in North and Central America (Coleoptera: Scolytidae). Mem Entomol Soc Canada 113:1–378

Bright DE, Poinar GO (1994) Scolytidae and Platypodidae (Coleoptera) from Dominican Republic Amber. Ann Entomol Soc Am 87:170–194

Bright DE, Stark RW (1973) The bark and ambrosia beetles of California (Coleoptera: Scolytidae and Platypodidae). Bull Calif Insect Surv 16:1–169

Byers JA (2007) Chemical ecology of Bark Beetles in a complex olfactory landscape. Bark Wood boring insects living trees Eur a synth. Springer, Netherlands, pp 89–134

Byers JA, Zhang QH, Schlyter F, Birgersson G (1998) Volatiles from nonhost birch trees inhibit pheromone response in spruce bark beetles. Naturwissenschaften 85:557–561

Cara R, Coster J (1968) Studies on the attack behavior of the southern pine beetle III. Sequence of tree infestation within stands Contrib from Boyce Thompson Inst 24:77–85

Carlson D, Ragenovich I (2012) Silver fir beetle and fir root bark beetle. United States Dep Agric For Serv US For Insect Dis Leafl 60

Castello J, Teale S (2011) Forest health: an integrated perspective. Manag Environ Qual an Int J 22:61–63

Chararas C (1962) Etude biologique des Scolytides des Conifères. Lechevalier 556

Clarke S (2012) Implications of population phases on the integrated pest management of the Southern Pine Beetle, Dendroctonus frontalis. J Integr Pest Manag 3:F1–F7

Clarke SR, Nowak JT (2009) Southern forest insect & disease leaflet. USDA For Serv Pacific Northwest Reg Portland, OR

Clarke SR, Riggins JJ, Stephen FM (2016) Forest management and southern pine beetle outbreaks: a historical perspective. For Sci

Clay NA, Little N, Riggins JJ (2017) Inoculation of ophiostomatoid fungi in loblolly pine trees increases the presence of subterranean termites in fungal lesions. Arthropod Plant Interact 11:213–219

Cognato AI (2000) Phylogenetic analysis reveals new genus of Ipini bark beetle (Scolytidae). Ann Entomol Soc Am 93:362–366

Cognato AI (2013) Electroborus brighti: the first Hylesinini bark beetle described from Dominican amber (Coleoptera: Curculionidae: Scolytinae). Can Entomol 145:501–508

Cognato AI (2015) Biology, systematics, and evolution of Ips. In: Bark Beetles Biol Ecol Nativ Invasive Species, pp 351–370

Cognato AI, Gillette NE, Bolaños RC, FaH S (2005) Mitochondrial phylogeny of pine cone beetles (Scolytinae, Conophthorus) and their affiliation with geographic area and host. Mol Phylogenet Evol 36:494–508

Cognato AI, Grimaldi D (2009) 100 Million years of morphological conservation in bark beetles (Coleoptera: Curculionidae: Scolytinae). Syst Entomol 34:93–100

Cognato AI, Jordal BH, Rubinoff D (2018) Ancient "Wanderlust" leads to diversification of Endemic Hawaiian Xyleborus Species (Coleoptera: Curculionidae: Scolytinae). Insect Syst Divers 2:1–9

Cognato AI, Sperling FAH (2000) Phylogeny of Ips DeGeer species (Coleoptera: Scolytidae) inferred from mitochondrial cytochrome oxidase I DNA sequence. Mol Phylogenet Evol 14:445–460

Cognato AI, Sun JH (2007) DNA based cladograms augment the discovery of a new Ips species from China (Coleoptera: Curculionidae: Scolytinae). Cladistics 23:539–551

Cognato AI, Vogler AP (2001) Exploring data interaction and nucleotide alignment in a multiple gene analysis of Ips (Coleoptera: Scolytinae). Syst Biol 50:758–780

Colombari F, Battisti A, Schroeder LM, Faccoli M (2012) Life-history traits promoting outbreaks of the pine bark beetle Ips acuminatus (Coleoptera: Curculionidae, Scolytinae) in the south-eastern Alps. Eur J for Res 131:553–561

Coulson RN, Klepzig K (2011) Southern Pine Beetle II, Gen Tech Rep SRS-140 Asheville, NC US Dep Agric For Serv South Res Station

Creeden EP, Hicke JA, Buotte PC (2014) Climate, weather, and recent mountain pine beetle outbreaks in the western United States. For Ecol Manage 312:239–251

Dodds KJ, Graber C, Stephen FM (2001) Facultative intraguild predation by larval Cerambycidae (Coleoptera) on bark beetle larvae (Coleoptera: Scolytidae). Environ Entomol 30:17–22

Dodds KJ, Ross DW (2002) Sampling range and range of attraction of Dendroctonus pseudotsugae pheromone-baited traps. Can Entomol 134:343–355

Dubbel V (1993) Überlebensrate von Fichtenborkenkäfern bei maschineller Entrindung. Allg Forst Z Waldwirtsch Umweltvorsorge 359–360

Elkin CM, Reid ML (2004) Attack and reproductive success of mountain pine beetles (Coleoptera: Scolytidae) in fire-damaged lodgepole pines. Environ Entomol 33:1070–1080

Etxebeste I, Pajares JA (2011) Verbenone protects pine trees from colonization by the six-toothed pine bark beetle, Ips sexdentatus Boern (Coleoptera: Scolytinae). J Appl Entomol 135:258–268

Faccoli M, Anfora G, Tasin M (2008) Responses of the mediterranean pine shoot beetle Tomicus destruens (Wollaston) to pine shoot and bark volatiles. J Chem Ecol 34:1162–1169

Faccoli M, Simonato M, Toffolo EP (2012) First record of Cyrtogenius strohmeyer in Europe, with a key to the European genera of the tribe Dryocoetini (Coleoptera: Curculionidae, Scolytinae). Zootaxa 35:27–35

Farrell BD, Sequeira AS, O'Meara BC, Normark BB, Chung JH, Jordal BH (2001) The evolution of agriculture in beetles (Curculionidae: Scolytinae and Platypodinae). Evolution 55:2011–2027

Fettig CJ, Hilszczański J (2015) Management Strategies for Bark Beetles in Conifer Forests. In: Bark Beetles Biol Ecol Nativ Invasive Species. Elsevier Inc, pp 555–584

Fettig CJ, Klepzig KD, Billings RF, Munson AS, Nebeker TE, Negrón JF, Nowak JT (2007) The effectiveness of vegetation management practices for prevention and control of bark beetle infestations in coniferous forests of the western and southern United States. For Ecol Manage 238:24–53

Fettig CJ, Shea PJ, Borys RR (2004) Seasonal flight patterns of four bark beetle species (Coleoptera: Scolytidae) along a latitudinal gradient in California. Pan-Pac Entomol 80:4–17

Flechtmann CAH, Atkinson TH (2018) Oldest record of Cyrtogenius luteus (Blandford) (Coleoptera: Curculionidae: Scolytinae) from South America with notes on its distribution in Brazil. Insecta Mundi 0645:1–4

Franceschi VR, Krokene P, Christiansen E, Krekling T (2005) Anatomical and chemical defenses of conifer bark against bark beetles and other pests. New Phytol 167:353–376

Furniss MM, Kegley SJ (2006) Observations on the biology of Dryocoetes betulae (Coleoptera: Curculionidae) in paper birch in northern Idaho. Environ Entomol 35:907–911

Furniss RL, Carolin VM (1977) Western forest insects

Gao L, Cognato AI (2018) Acanthotomicus suncei, a new sweetgum tree pest in China (Coleoptera: Curculionidae: Scolytinae: Ipini). Zootaxa 4471:595–599

Gao L, Li Y, Xu Y, Hulcr J, Cognato AI, Wang JG, Ju RT (2017) Acanthotomicus sp. (Coleoptera: Curculionidae: Scolytinae), a new destructive insect pest of North American Sweetgum Liquidambar styraciflua in China. J Econ Entomol 110:1592–1595

Garbutt RW (1992) Western balsam bark beetle. For Pest Leafl No 64

Gehrken U (1985) Physiology of diapause in the adult bark beetle, Ips acuminatus Gyll, studied in relation to cold hardiness. J Insect Physiol 31:909–916

Gehrken U (1989) Supercooling and thermal hysteresis in the adult bark beetle, Ips acuminatus Gyll. J Insect Physiol 35:347–352

George RA, Beal J (1929) The southern pine beetle: a serious enemy of pines in the south

Gohli J, Kirkendall LR, Smith SM, Cognato AI, Hulcr J, Jordal BH (2017) Biological factors contributing to bark and ambrosia beetle species diversification. Evolution (NY) 71:1258–1272

Gohli J, Selvarajah T, Kirkendall LR, Jordal BH (2016) Globally distributed Xyleborus species reveal recurrent intercontinental dispersal in a landscape of ancient worldwide distributions Phylogenetics and phylogeography. BMC Evol Biol 16

Goldhammer DS, Stephen FM, Paine TD (1989) Average radial growth rate and chlamydospore production of Ceratocystis minor, Ceratocystis minor var. barrasii, and SJB 122 in culture. Can J Bot 67:3498–3505

Gómez D, Hirigoyen A, Balmelli G, Viera C, Martínez G (2017) Patterns in flight phenologies of bark beetles (Coleoptera: Scolytinae) in commercial pine tree plantations in Uruguay. Bosque 38:47–53

Gómez D, Martinez G, Beaver R (2012) First record of Cyrtogenius luteus (Blandford) (Coleoptera: Curculionidae: Scolytinae) in the Americas and its distribution in Uruguay. Coleopt Bull 66:362–364

Göthlin E, Schroeder LM, Lindelöw A (2000) Attacks by Ips typographus and Pityogenes chalcographus on Windthrown Spruces (Picea abies) during the two years following a storm felling. Scand J for Res 15:542–549

Grimaldi D (1996) Amber: window to the past. Am Museum Nat Hist

Grosman DM, Clarke SR, Upton WW (2009) Efficacy of two systemic insecticides injected into loblolly pine for protection against southern pine bark beetles (Coleoptera: Curculionidae). J Econ Entomol 102:1062–1069

Haack RRA, Poland TMT (2001) Evolving management strategies for a recently discovered exotic forest pest: the pine shoot beetle, Tomicus piniperda (Coleoptera). Biol Invasions 3:307–322

Hain FP, Duehl AJ, Gardner MJ, Payne TL (2011) Natural history of the Southern Pine Beetle. In: Coulson RN, Klepzig KD (eds) South Pine Beetle II Gen Tech Rep SRS-140. US Department of Agriculture Forest Service, Southern Research Station, Asheville, NC, pp 13–24

Hais M, Wild J, Berec L, Brůna J, Kennedy R, Braaten J, Brož Z (2016) Landsat imagery spectral trajectories-important variables for spatially predicting the risks of bark beetle disturbance. Remote Sens 8

Happ GM, Happ CM, French JRJ (1976) Ultrastructure of the mesonotal mycangium of an ambrosia beetle, Xyleborus Dispar (F.) (Coleoptera: Scolytidae). Int J Insect Morphol Embryol 5:381–391

Harrington TC (2005) Ecology and evolution of mycophagous bark beetles and their fungal partners. In: Vega FE, Blackwell M (eds) Ecol Evol Adv Insect-Fungal Assoc. Oxford University Press, pp 257–291

Harrington TC, Batzer JC, McNew DL (2021) Corticioid basidiomycetes associated with bark beetles, including seven new Entomocorticium species from North America and Cylindrobasidium ipidophilum, comb. nov. Antonie van Leeuwenhoek, Int J Gen Mol Microbiol

Helms JA (1998) The dictionary of forestry. Soc Am For

Hofstetter RW, Cronin JT, Klepzig KD, Moser JC, Ayres MP (2006) Antagonisms, mutualisms and commensalisms affect outbreak dynamics of the southern pine beetle. Oecologia 147:679–691

Hofstetter RW, Dinkins-Bookwalter J, Davis TS, Klepzig KD (2015) Symbiotic associations of Bark Beetles. In: Vega FE, Hofstetter RW (eds) Bark Beetles Biol Ecol Nativ Invasive Species. Academic Press, pp 209–245

Hofstetter RW, Moser JC, Blomquist SR (2013) Mites associated with bark beetles and their hyperphoretic ophiostomatoid fungi. In: Seifert KA, de Beer ZW, Wingfield MJ (eds) Ophiostomatoid Fungi - Expand Front CBS Biodiversity Series, pp 165–176

Horn A, Kerdelhué C, Lieutier F, Rossi JP (2012) Predicting the distribution of the two bark beetles Tomicus destruens and Tomicus piniperda in Europe and the Mediterranean region. Agric for Entomol 14:358–366

Huber DPW, Borden JH, Stastny M (2001) Response of the pine engraver, Ips pini (Say) (Coleoptera: Scolytidae), to conophthorin and other angiosperm bark volatiles in the avoidance of non-hosts. Agric for Entomol 3:225–232

Hulcr J, Atkinson TH, Cognato AI, Jordal BH, McKenna DD (2015) Morphology, taxonomy, and phylogenetics of Bark Beetles. In: Bark Beetles Biol Ecol Nativ Invasive Species, pp 41–84

Jacobi WR, Koski RD, Negron JF (2013) Dutch elm disease pathogen transmission by the banded elm bark beetle scolytus schevyrewi. For Pathol 43:232–237

Jactel H, Brockerhoff EG (2007) Tree diversity reduces herbivory by forest insects. Ecol Lett 10:835–848

Johnson AJ, McKenna DD, Jordal BH, Cognato AI, Smith SM, Lemmon AR, Lemmon EM, Hulcr J (2018) Phylogenomics clarifies repeated evolutionary origins of inbreeding and fungus farming in bark beetles (Curculionidae, Scolytinae). Mol Phylogenet Evol 127:229–238

Jordal BH (2006) Community structure and reproductive biology of bark beetles (Coleoptera: Scolytinae) associated with Macaronesian Euphorbia shrubs. Eur J Entomol 103:71–80

Jordal BH, Cognato AI (2012) Molecular phylogeny of bark and ambrosia beetles reveals multiple origins of fungus farming during periods of global warming. BMC Evol Biol 12:133

Jordal BH, Sequeira AS, Cognato AI (2011) The age and phylogeny of wood boring weevils and the origin of subsociality. Mol Phylogenet Evol 59:708–724

Kerchev IA (2014) On monogyny of the four-eyed fir bark beetle Polygraphus proximus Blandf. (Coleoptera, Curculionidae: Scolytinae) and its reproductive behavior. Entomol Rev 94:1059–1066

Kirejtshuk AG, Azar D, Beaver RA, Mandelshtam MY, Nel A (2009) The most ancient bark beetle known: a new tribe, genus and species from Lebanese amber (Coleoptera, Curculionidae, Scolytinae). Syst Entomol 34:101–112

Kirisits T (2007) Fungal associates of European Bark Beetles with special emphasis on the Ophiostomatoid Fungi. Bark Wood boring insects living trees Eur a Synth. Springer, Netherlands, pp 181–236

Kirkendall LR (1983) The evolution of mating systems in bark and ambrosia beetles (Coleoptera: Scolytidae and Platypodidae). Zool J Linn Soc 77:293–352

Kirkendall LR (1993) Ecology and evolution of biased sex ratios in Bark and Ambrosia Beetles. In: Evol Divers Sex Ratio, pp 235–345

Kirkendall LR, Biedermann PHW, Jordal BH (2015) Evolution and diversity of Bark and Ambrosia Beetles. In: Bark Beetles Biol Ecol Nativ Invasive Species, pp 85–156

Kirkendall LR, Faccoli M, Ye HUI (2008) Description of the Yunnan shoot borer, Tomicus yunnanensis Kirkendall & Faccoli sp. n. (Curculionidae, Scolytinae), an unusually aggressive pine shoot beetle from southern China, with a key to the species of Tomicus. Zootaxa 1839:25–39

Kirkendall LR, Kent DS, Raffa KF (2010) Interactions among males, females and offspring in bark and ambrosia beetles: the significance of living in tunnels for the evolution of social behavior. In: Evol Soc Behav Insects Arachn, pp 181–215

Klepzig KD, Robison DJ, Fowler G, Minchin PR, Hain FP, Allen HL (2005) Effects of mass inoculation on induced oleoresin response in intensively managed loblolly pine. Tree Physiol 25:681–688

Kolařík M, Freeland E, Utley C, Tisserat N (2011) Geosmithia morbida sp. nov., a new phytopathogenic species living in symbiosis with the walnut twig beetle (Pityophthorus juglandis) on Juglans in USA. Mycologia 103:325–332

Kreutz J, Zimmermann G, Marohn H, Vaupel O, Mosbacher G (2000) Preliminary investigations on the use of Beauveria bassiana (Bals.) Vuill and other control methods against the bark beetle Ips typographus (Coleoptera, Scolytidae) in the field. Mitteilungen Der Dtsch Gesellschaft Fur Allg Und Angew Entomol 12:119–125

Krivets SA, Bisirova EM, Kerchev IA, Pats EN, Chernova NA (2015) Transformation of taiga ecosystems in the Western Siberian invasion focus of four-eyed fir bark beetle Polygraphus proximus Blandford (Coleoptera: Curculionidae, Scolytinae). Russ J Biol Invasions 6:94–108

Lanier GN, Kirkendall LR (1986) Karyology of pseudogamous Ips bark beetles. Hereditas 105:87–96

Lausch A, Heurich M, Gordalla D, Dobner HJ, Gwillym-Margianto S, Salbach C (2013) Forecasting potential bark beetle outbreaks based on spruce forest vitality using hyperspectral remote-sensing techniques at different scales. For Ecol Manage 308:76–89

Lieutier F, Långström B, Faccoli M (2015) The Genus Tomicus. In: Bark Beetles Biol Ecol Nativ Invasive Species, Elsevier Inc, pp 371–426

Lindelöw Å, Risberg B, Sjödin K (1992) Attraction during flight of scolytids and other bark- and wood-dwelling beetles to volatiles from fresh and stored spruce wood. Can J for Res 22:224–228

Lindeman AA, Yack JE (2015) What is the password? Female bark beetles (Scolytinae) grant males access to their galleries based on courtship song. Behav Processes 115:123–131

Lindgren BS, Raffa KF (2013) Evolution of tree killing in bark beetles (Coleoptera: Curculionidae): trade-offs between the maddening crowds and a sticky situation. Can Entomol 145:471–495

Little NS, Riggins JJ, Schultz TP, Londo AJ, Ulyshen MD (2012) Feeding Preference of Native Subterranean Termites (Isoptera: Rhinotermitidae: Reticulitermes) for Wood Containing Bark Beetle Pheromones and Blue-Stain Fungi. J Insect Behav 25:197–206

Liu H, Zhang Z, Ye H, Wang H, Clarke SR, Jun J (2010) Response of Tomicus yunnanensis (Coleoptera: Scolytinae) to infested and uninfested Pinus yunnanensis bolts. J Econ Entomol 103:95–100

Lobinger G (1995) Einsatzmöglichkeiten von Borkenkäferfallen. Allg Forst Z Waldwirtsch Umweltvorsorge

Lobinger G, Skatulla U (1996) Untersuchungen zum Einfluß von Sonnenlicht auf das Schwärmverhalten von BorkenkäfernInfluencing the flight behaviour of bark beetles by light conditions. Anzeiger Für Schädlingskd Pflanzenschutz Umweltschutz 69:183–185

Luik A, Voolma K (1990) Hibernation peculiarities and cold-hardiness of the great spruce bark beetle. Dendroctonus Micans Kug Eesti Tead Akad Toim Biol 39:214–218

Mayfield AE, Foltz JL (2005) The Black Turpentine Beetle, Dendroctonus terebrans (Olivier) (Coleoptera: Curculionidae: Scolytinae). Fla Dep Agric Consum Serv Div Plant Ind

McMillin JD, Allen KK, Long DF, Harris JL, Negrón JF (2003) Effects of Western Balsam Bark Beetle on Spruce-Fir Forests of North-Central Wyoming. West J Appl for 18:259–266

Meeker JR (2013) Forest health evaluation of southern pine beetle activity on the National Forests in Mississippi

Mezei P, Jakuš R, Pennerstorfer J, Havašová M, Škvarenina J, Ferenčík J, Slivinský J, Bičárová S, Bilčík D, Blaženec M, Netherer S (2017) Storms, temperature maxima and the Eurasian spruce bark beetle Ips typographus—an infernal trio in Norway spruce forests of the Central European High Tatra Mountains. Agric for Meteorol 242:85–95

Miller DR, Pierce HD, De Groot P, Jeans-Williams N, Bennett R, Borden JH (2000) Sex pheromone of Conophthorus ponderosae (Coleoptera: Scolytidae) in a coastal stand of western white pine (Pinaceae). Can Entomol 132:243–245

Moser JC (1985) Use of sporothecae by phoretic Tarsonemus mites to transport ascospores of coniferous bluestain fungi. Trans – Br Mycol Soc 84:750–753

Moser JC, Bridges JR (1986) Tarsonemus (Acarina: Tarsonemidae) mites phoretic on the southern pine beetle (Coleoptera: Scolytidae): attachment sites and number of bluestain (Ascomycetes: Ophiostomataceae) ascopores carried. Proc Entomol Soc Washingt 88:297–299

Newton L, Fowler G, Neeley A, Schall R, Takeuchi Y (2009) Pathway assessment: Geosmithia sp. and Pityophthorus juglandis Blackman movement from the western into the eastern United States. US Dep Agric Anim Plant Heal Insp Serv

Nowak J, Astro C, Klepzig K, Billings R (2008) The Southern Pine Beetle prevention initiative: working for healthier forests. J for 106:261–267

Oester PT, Rudinsky JA, Ryker LC (1981) Olfactory and acoustic behavior of Pseudohylesinus nebulosus (Coleoptera: Scolytidae) on douglas-fir bark. Can Entomol 113:645–650

Paine TD, Birch MC, Švihra P (1981) Niche breadth and resource partitioning by four sympatric species of bark beetles (Coleoptera: Scolytidae). Oecologia 48:1–6

Pashenova N, Baranchikov Y, Petko V (2011) Aggressive Ophiostoma species of fungi isolated from galleries of Polygraphus proximus. Zashita i Karantin Rastenii 6:31–32

Payne T (1979) Pheromone and Host Odor Perception in Bark Beetles. Neurotoxicology Insectic Pheromones. Springer, US, pp 27–57

Payne T (1986) Olfaction and vision in host finding by a bark beetle. In: Payne T, Birch M, Kennedy C (eds) Mech Insect Olfaction

Payne T, Billings RF, Delorme JD, Andryszak NA, Bartels J, Francke W, Vité JP (1987) Kairomonal-pheromonal system in the black turpentine beetle, Dendroctonus terebrans (Ol.). J Appl Entomol 103:15–22

Person H (1931) Theory in explanation of the selection of certain trees by the Western Pine Beetle. J for 29:696–699

Pfeffer A (1995) Zentral-und westpaläarktische Borken-und Kernkäfer: (Coloptera: Scolytidae, Platypodidae). Pro Entomol

Pitman GB, Vité JP (1969) Aggregation behavior of dendroctonus ponderosae (coleoptera: Scolytidae) in response to chemical messengers. Can Entomol 101:143–149

Poland TM, De Groot P, Burke S, Wakarchuk D, Haack RA, Nott R, Scarr T (2003) Development of an improved attractive lure for the pine shoot beetle, Tomicus piniperda (Coleoptera: Scolytidae). Agric for Entomol 5:293–300

Raffa KF, Aukema BH, Bentz BJ, Carroll AL, Hicke JA, Turner MG, Romme WH (2008) Cross-scale drivers of natural disturbances prone to anthropogenic amplification: the dynamics of Bark Beetle Eruptions. Bioscience 58:501–517

Raffa KF, Grégoire JC, Lindgren BS (2015) Natural history and ecology of Bark Beetles. In: Vega FE, Hofstetter RW (eds) Bark Beetles Biol Ecol Nativ Invasive Species. Academic Press, San Diego, pp 1–40

Ramsfield TD, Bentz BJ, Faccoli M, Jactel H, Brockerhoff EG (2016) Forest health in a changing world: effects of globalization and climate change on forest insect and pathogen impacts. Forestry 89:245–252

Reeve JD (1997) Predation and bark beetle dynamics. Oecologia 112:48–54

Reid ML, Roitberg BD (1994) Benefits of prolonged male residence with mates and brood in Pine Engravers (Coleoptera: Scolytidae). Oikos 70:140

Renwick JAA, Vité JP (1970) Systems of chemical communication in Dendroctonus. Contrib Boyce Thompson Inst Plant Res 24:283–292

Riggins JJ, Little NS, Eckhardt LG (2014) Correlation between infection by ophiostomatoid fungi and the presence of subterranean termites in loblolly pine (Pinus taeda L.) roots. Agric for Entomol 16:260–264

Roberts H (1976) Observations on the biology of some tropical rain forest Scolytidae (Coleoptera) from Fiji: I. Subfamilies—Hylesininae, Ipinae (excluding xyleborini). Bull Entomol Res 66:373–388

Robertson C, Nelson TA, Jelinski DE, Wulder MA, Boots B (2009) Spatial-temporal analysis of species range expansion: the case of the mountain pine beetle, Dendroctonus ponderosae. J Biogeogr 36:1446–1458

Root RB (1973) Organization of a plant-arthropod association in simple and diverse habitats: the fauna of collards (Brassica Oleracea). Ecol Monogr 43:95–124

Rugman-Jones PF, Seybold SJ, Graves AD, Stouthamer R (2015) Phylogeography of the walnut twig beetle, Pityophthorus juglandis, the vector of thousand cankers disease in North American walnut trees. PLoS One 10

Ryker LC (1984) Acoustic and chemical signals in the life cycle of a Beetle. Sci Am 250:112–123

Saint-Germain M, Buddle CM, Drapeau P (2007) Primary attraction and random landing in host-selection by wood-feeding insects: a matter of scale? Agric for Entomol 9:227–235

Saint-Germain M, Drapeau P, Buddle CM (2009) Landing patterns of phloem- and wood-feeding Coleoptera on black spruce of different physiological and decay states. Environ Entomol 38:797–802

Samman S, Logan J (2000) Assessment and response to bark beetle outbreaks in the Rocky Mountain area, Rep to Congr from For Heal Prot Washingt Off For Serv USDA

Sauvard D (2007) General biology of Bark Beetles. In: Lieutier F, Day KR, Battisti A, Gregoire JC, Evans HF (eds) Bark Wood boring insects living trees Eur a Synth. Springer, Netherlands, Dordrecht, pp 63–88

Schlyter F, Lundgren U (1993) Distribution of a bark beetle and its predator within and outside old growth forest reserves: no increase of hazard near reserves. Scand J for Res 8:246–256

Schoeller EN, Husseneder C, Allison JD (2012) Molecular evidence of facultative intraguild predation by Monochamus titillator larvae (Coleoptera: Cerambycidae) on members of the southern pine beetle guild. Naturwissenschaften 99:913–924

Seybold S, Bohlmann J, Raffa K (2000) Biosynthesis of coniferophagous bark beetle pheromones and conifer isoprenoids: evolutionary perspective and synthesis. Can Entomol

Seybold S, Huber D, Lee J, Graves A, Bohlmann J (2006) Pine monoterpenes and pine bark beetles: a marriage of convenience for defense and chemical communication. Phytochem Rev 5:143–178

Simpson LJ (1929) The seasonal history of Polygraphus rufipennis Kirby. Can Entomol 61:146–151

Sitz RA, Luna EK, Caballero JI, Tisserat NA, Cranshaw WS, Stewart JE (2017) Virulence of genetically distinct Geosmithia morbida Isolates to Black Walnut and their response to Coinoculation with Fusarium solani. Plant Dis 101:116–120

Six DL, Bentz BJ (2007) Temperature determines symbiont abundance in a multipartite bark beetle-fungus ectosymbiosis. Microb Ecol 54:112–118

Six DL, Bracewell R (2015) Dendroctonus. In: Vega FE, Hofstetter RW (eds) Bark Beetles Biol Ecol Nativ Invasive Species. Academic Press, pp 305–350

Six DL, Wingfield MJ (2011) The role of Phytopathogenicity in Bark Beetle–fungus symbioses: a challenge to the classic paradigm. Annu Rev Entomol 56:255–272

Skelton J, Loyd A, Smith JA, Blanchette RA, Held BW, Hulcr J (2020) Fungal symbionts of bark and ambrosia beetles can suppress decomposition of pine sapwood by competing with wood-decay fungi. Fungal Ecol 45:100926

Smith SM, Cognato AI (2014) A taxonomic monograph of nearctic Scolytus Geoffroy (Coleoptera, Curculionidae, Scolytinae). Zookeys

Smith SM, Hulcr J (2015) Scolytus and other economically important Bark and Ambrosia Beetles. In: Vega FE, Hofstetter RW (eds) Bark Beetles Biol Ecol Nativ Invasive Species. Academic Press, San Diego, pp 495–531

Stephen FM (2011) Southern Pine Beetle competitors. In: Coulson RN, Klepzig KD (eds) South Pine Beetle II US Department of Agriculture Forest Service, Southern Research Station, Asheville, NC, pp 183–198

Stokland JN, Siitonen J, Jonsson BG (2012) Biodiversity in dead wood. Cambridge University Press, Biodivers. Dead Wood

Sullivan BT, Fettig CJ, Otrosina WJ, Dalusky MJ, Berisford CW (2003) Association between severity of prescribed burns and subsequent activity of conifer-infesting beetles in stands of longleaf pine. For Ecol Manage 185:327–340

Sullivan BT, Mori K (2009) Spatial displacement of release point can enhance activity of an attractant pheromone synergist of a bark beetle. J Chem Ecol 35:1222–1233

Sullivan BT, Shepherd WP, Pureswaran DS, Tashiro T, Mori K (2007) Evidence that (+)-endo-brevicomin is a male-produced component of the southern pine beetle aggregation pheromone. J Chem Ecol 33:1510–1527

Susaeta A, Soto JR, Adams DC, Hulcr J (2017) Expected timber-based economic impacts of a Wood-Boring Beetle (Acanthotomicus Sp) that kills American Sweetgum. J Econ Entomol 110:1942–1945

Tainter F, Baker F (1996) Principles of forest pathology, Choice Rev, Online

Thatcher RC, Searcy JL, Coster JE, Hertel GD (1980) The Southern Pine Beetle. USDA, expanded Southern Pine Beetle research and application program. Forest Service. Science and Education Administration, Pineville, LA. Technical Bulletin 1631

Trapp S, Croteau R (2001) Defensive resin biosynthesis in conifers. Annu Rev Plant Physiol Plant Mol Biol 52:689–724

Trudel R, Guertin C, De Groot P (2004) Use of pityol to reduce damage by the white pine cone beetle, Conophthorus coniperda (Coleoptera, Scolytidae) in seed orchards. J Appl Entomol 128:403–406

Tunset K, Nilssen AC, Andersen J (1993) Primary attraction in host recognition of coniferous bark beetles and bark weevils (Coleoptera, Scolytidae and Curculionidae). J Appl Entomol 115:155–169

Vega FE, Infante F, Johnson AJ (2015) The Genus Hypothenemus, with Emphasis on H. hampei, the Coffee Berry Borer. In: Bark Beetles Biol Ecol Nativ Invasive Species, pp 427–494

Vité JP, Gara RI (1962) Volatile attractants from ponderosa pine attacked by bark beetles (Coleoptera: Scolytidae). Contrib Boyce Thompson Inst 21:1–273

Webb JW, Franklin RT (1978) Influence of Phloem moisture on brood development of the Southern Pine Beetle (Coleoptera: Scolytidae). Environ Entomol 7:405–410

Wermelinger B (2004) Ecology and management of the spruce bark beetle Ips typographus - a review of recent research. For Ecol Manage 202:67–82

Weslien J, Lindelow A (1990) Recapture of marked spruce bark beetles (Ips typographus) in pheromone traps using area-wide mass trapping. Can J for Res 20:1786–1790

Whitney HS, Ritchie DC, Borden JH, Stock AJ (1984) The fungus beauveria bassiana (Deuteromycotina: Hyphomycetaceae) in the western balsam bark beetle, dryocoetes confusus (Coleoptera: Scolytidae). Can Entomol 116:1419–1424

Wichmann HE (1967) Die Wirkungsbreite des Ausstoßreflexes bei Borkenkäfernreflexes bei Borkenkäfern. Anzeiger Für Schädlingskd 40:184–187

Wood DL (1982a) The role of Pheromones, Kairomones, and Allomones in the host selection and colonization behavior of Bark Beetles. Annu Rev Entomol 27:411–446

Wood SL (1982b) The Bark and Ambrosia Beetles of North and Central America (Coleoptera: Scolytidae), a taxonomic monograph. Gt Basin Nat Mem 6:1–1359

Wood SL (1986) A reclassification of the genera of Scolytidae (Coleoptera). Gt Basin Nat Mem 10:1–126

Wood SL (2007) Bark and ambrosia beetles of South America (Coleoptera, Scolytidae). Monte L Bean Life Sci Museum Brigham Young University, Provo

Wood SL, Bright DE (1992) A catalog of Scolytidae and Platypodidae (Coleoptera), Part 2: taxonomic index, volumes A and B. J Chem Inf Model 13:1–1553

Yan Z, Sun J, Don O, Zhang Z (2005) The red turpentine beetle, Dendroctonus valens LeConte (Scolytidae): an exotic invasive pest of pine in China. Biodivers Conserv

Yang ZQ, Wang XY, Zhang YN (2014) Recent advances in biological control of important native and invasive forest pests in China. Biol Control 68:117–128

Yuceer C, Hsu CY, Erbilgin N, Klepzig KD (2011) Ultrastructure of the mycangium of the southern pine beetle, Dendroctonus frontalis (Coleoptera: Curculionidae, Scolytinae): complex morphology for complex interactions. Acta Zool 92:216–224

Zhang QH (2003) Interruption of aggregation pheromone in Ips typographus (L.) (Coleoptera, Scolytidae) by non-host bark volatiles. Agric for Entomol 5:145–153

Zumr V (1992) Dispersal of the spruce bark beetle Ips typographus (L.) (Coleoptera, Scolytidae) in spruce woods. J Appl Entomol 114:348–352

Chapter 11
Ambrosia Beetles

Jiri Hulcr and James Skelton

11.1 Ambrosia Beetle Biology

11.1.1 Taxonomic Identity

The term "ambrosia beetles" refers to an ecological strategy shared by thousands of species of wood-boring weevils from multiple lineages, rather than a single taxonomic group. Most ambrosia beetle groups evolved from within the bark beetles (Curculionidae: Scolytinae), which are a diverse group of weevils which bore into trees and whose progeny develop by feeding on the host tree tissue. Ambrosia beetles do not consume the tree tissue; instead, they introduce symbiotic fungi into their tunnels, which comprise the majority or entirety of the ambrosia beetle diet. Ambrosia fungus farming has evolved at least sixteen times within bark beetles (Johnson et al. 2018) (Fig. 11.1).

There are over 3,000 species of ambrosia beetles (Hulcr et al. 2015), making them far more species-rich than other fungus-farming insect groups, such as the fungus farming ants, termites, and wood wasps. It has been suggested that the diversity of ambrosia beetles is derived from the ecological success of the fungus-farming strategy. However, only a few ambrosia beetle lineages are particularly diverse, and in those lineages, other factors likely contribute to their high diversity. For example, the rapid and extensive diversification of Xyleborini may be better explained by their haplo-diploid genetic system rather than fungus farming (Gohli et al. 2017). Several other ambrosia lineages are diverse because they are old and not because they are speciating faster than other weevil groups, such as the Platypodinae. They

J. Hulcr (✉)
School of Forest, Fisheries, and Geomatics Sciences, University of Florida, Gainesville, FL, USA
e-mail: hulcr@ufl.edu

J. Skelton
Biology Department, William and Mary, Williamsburg, VA, USA

© The Author(s) 2023
J. D. Allison et al. (eds.), *Forest Entomology and Pathology*,
https://doi.org/10.1007/978-3-031-11553-0_11

Coptodryas pubifer

Corthyloxiphus sp.

Fig. 11.1 Ambrosia beetle galleries. Left, a cavity of *Coptodryas pubifer* (Xyleborini) in Sabah, Malaysia; new adults, larvae and the white fungal growth, the top right specimen is the haploid male. Right, an unidentified *Corthyloxiphus* in Ecuador: the male (diploid), and larval chambers with individual larvae and fungus. Other types of gallery arrangements exist. Photos: J. Hulcr. *Corthyloxiphus* was identified by Sarah M. Smith

now comprise approximately 1,400 species and are estimated to have been farming fungi for over 100 million years (Jordal and Cognato 2012; Poinar Jr and Vega 2018; Vanderpool et al. 2018).

The most practical biological unit for classification and discussion of ambrosia symbioses is not any single taxonomic level, such as species or genus. Instead, it is better to use the concept of evolutionary symbiotic unit because both the beetle and the fungus partners have been coevolving and speciating together. The coevolutionary unit represents an independent event of an evolutionary beetle-fungus association and includes its evolutionary offshoots—beetle and fungus species or genera that retain that association.

11.1.2   Relationships with Fungi

Ambrosia symbioses are most often considered reciprocally obligate mutualisms. The beetles depend on their fungi as a food source and the fungi depend on the beetles for dispersal to new trees. It is likely that at least some ambrosia fungi have retained the ability to disperse by other means such as fruiting bodies that eject spores. At least some ambrosia fungi have retained the ability to produce sexual stages (Musvuugwa et al. 2015; Mayers et al. 2017; Jusino et al. 2020), but whether they are also able to disperse independently of the beetles is not known. This capacity is known in fungal associates of other insects, including fungus growing termites (Johnson et al. 1981) and siricid woodwasps (Talbot 1977).

Ambrosia fungi originated from at least seven separate fungal clades (Alamouti et al. 2009; Hulcr and Stelinski 2017). Improved systematics and phylogenetic sampling continue to reveal more independent evolutionary origins of ambrosia fungi within well-known ambrosia fungus taxa because many are polyphyletic, particularly within the Ophiostomatales (Vanderpool et al. 2018; de Beer et al. 2022) and Cera-tocystidaceae (Mayers et al. 2015). Additionally, the increasing research interest and DNA-based studies continue to uncover a rich diversity of ambrosia fungi that are not directly related to previously known ambrosia fungi, many of which have remained unnoticed until recently (Bateman et al. 2016; Li et al. 2017).

As a result of their diverse origins, ambrosia fungi inherited various ecological strategies. Despite the shared strategy of symbiosis with insect vectors, the metabolic profiles of these fungi do not seem to be convergent. Instead, both the substrate use and the metabolic products (the beetle food) of each ambrosia fungus clade are more similar to closely related free-living fungi than to other ambrosia clades (Huang et al. 2019, 2020). Some newly discovered symbiotic fungi have metabolic capabilities and ecological strategies that were previously unknown from ambrosia fungi. For example, beetles in the genera *Ambrosiodmus* and *Ambrosiophilus* (beetle genera in the scolytine tribe Xyleborini) farm the basidiomycete genus *Irpex* (formerly *Flavodon*). Fungi in this genus are exceptional among ambrosia fungi because they are truly lignicolous and degrade the structural components of wood (Kasson et al. 2016; Jusino et al. 2020). This is in contrast to other ambrosia fungi which extract labile resources within the wood but do not decompose the wood itself.

The dichotomy between bark and ambrosia beetles is convenient, but imperfect. Many scolytine species blur the boundary between the phloem-feeding (phloephagous) bark beetles and fungus-feeding (mycetophagous) ambrosia beetles. While most bark beetles feed within bark and phloem, many species also consume wood, seeds, herbaceous plant tissue, and tissues with varying amounts of fungi. In fact, some of the best-known forest pests, such as species in the genera *Dendroctonus, Ips* and *Tomicus,* are phloeomycetophagous. This means that the larvae develop in phloem but eat primarily fungal mutualists, similar to ambrosia beetles. Similar to true ambrosia beetles, the adults of some of these phloeomycetophages even have mycangia for transporting specific fungal mutualists to new trees. Furthermore, not all ambrosia beetles are strictly fungivores. Some entire genera are xylomycetophagous: the larvae chew and ingest a mixture of wood and the mycelium of a fungal mutualist (Roeper 1995).

Phloeomycetophagous bark beetles and xylomycetophagous ambrosia beetles show similar specificity to their fungi. For example, the pine-inhabiting phloeomyce-tophages in North America are associated with only a few species of highly derived species of *Entomocorticium, Ophiostoma, Grossmania,* and/or *Ceratocystiopsis* (Harrington 2005). Similarly in Europe, the phloeomycetophagous species of *Ips acuminatus* and *Tomicus minor* are each primarily associated with a single species of *Ophiostoma* (Francke-Grosmann 1967; Seifert et al. 2013).

Another fungus-related strategy among bark beetles is sapromycetophagy: consuming degraded plant tissues rich in various fungi. A number of pygmy borers in the genus *Hypothenemus,* for example, occupy twigs and branches pre-colonized by fungi. Their larvae do not drill individual tunnels but instead develop in extensive communal spaces lined with mycelium. What exactly they consume, and whether there is any specificity to this beetle-fungus relationship, remains unexplored.

11.2   Who Is the Host and Why Does It Matter?

In many symbioses, the roles of host and symbiont are often obvious. The larger organism hosts the symbionts, which are usually smaller and more numerous. The host bears the brunt of interacting with the environment, while the symbionts experience only a subset of environmental factors. This environmental shielding and reduced population size affects symbiont evolution. For example, microbial endosymbionts often evolve reduced genetic complexity (Moran and Wernegreen 2000; McCutcheon and Moran 2011).

Ambrosia symbioses are different because the role of host and symbiont alternates throughout their shared life cycle. During dispersal, the beetle is the host to its fungal symbiont. The fungus is sheltered and nourished within the beetle's body in the mycangium. However, once a dispersing ambrosia beetle establishes a new gallery, the ambrosia fungi are released from the mycangium into the wood. At this point, they must colonize resources, sequester energy and nutrients, compete with other microbes, and resist or detoxify plant-produced defensive chemicals. Meanwhile, larvae and newly emerged adult ambrosia beetles feed primarily or exclusively on their fungal symbionts within the stable and protected environment of the fungi-laden gallery. At this stage, ambrosia fungi arguably act as the host to their beetles because the fungi bear the burden of interacting with a variable and often hostile environment. This stage comprises the majority of the life cycle of this symbiosis, and therefore incorporating a fungus-centric view with an entomological perspective will improve our understanding of the biology of the ambrosia symbiosis.

11.2.1   Biology of the Coevolutionary Units is Dictated by the Fungus

There is a tendency for research on agricultural symbioses to focus on the "farmers" as the dominating partner and to expect that their crops are passive or enslaved participants in the symbiosis. However, in insect/fungus farming, and even human agriculture, there is evidence that crops also exert significant selection on their farmers, especially during the early stages of the evolution of agricultural symbioses (Schultz et al. 2005). Support for this view comes from comparative studies of ambrosia fungi and closely related non-ambrosia fungi. The ancestors of various ambrosial lineages had distinct metabolic abilities and ecological niches, and each ambrosia lineage has retained the metabolic capacities of its recent ancestors (Huang et al. 2019). The inherited metabolic capacity of beetle-associated fungi is correlated with the diversity and taxonomic composition of the trees that the fungi utilize (Veselská et al. 2019), suggesting that the ecological niche breadth of an ambrosia beetle may be constrained by the niche of its fungal symbiont.

The contemporary ecology of each coevolutionary ambrosia beetle/fungal symbiont unit seems to be predicted more by the ancestral ecology of the fungus than by the ancestral ecology of the beetle. For example, beetle taxa that farm *Ambrosiella* and the closely related genus *Meredithiella* utilize substrates that are prone to drying such as twigs and smaller branches. These beetles include unrelated groups, such as the *Xylosandrus* clade within Xyleborini, *Corthylus* spp., *Scolytoplatypus*, all of which independently evolved the colonization of twigs or branches, but rarely trunks. Conversely, essentially all the ambrosia beetles that colonize the bases of tree trunks, which remain moist, have fungal symbionts in the genus *Raffaelea* (sensu lato) and related Ophiostomatales. These beetle clades—including multiple Xyleborini genera, Platypodinae, corthyline genera *Monarthrum* and *Gnathotrichus*, and Premnobiina—are not closely related. Even within the hyper-diverse tribe of ambrosia beetles Xyleborini, there are several separately derived coevolutionary units, and the beetle members of each unit typically follow the ecological strategy of the fungus, not the ancestral strategy of the beetles. This diversity of fungal traits and the resulting ecological variability among the fungus-beetle coevolutionary units suggests that there are many functionally diverse ambrosia symbioses, rather than a single convergent type.

In addition to influencing a beetle's ecological niche, the ecology of ambrosia fungi may also be tied to mating systems, and perhaps facilitate the evolution of sociality. This is the case especially in systems where the growth of the fungus garden lasts long enough to support multiple overlapping generations of the beetles. Most ambrosia gardens are short lived, because almost all ambrosia fungi stem from lineages of saprotrophic or plant-pathogenic ascomycetes. These fungi typically lack the ability to degrade the lignin-containing structural components that comprise the majority of wood biomass. Instead they rely on more readily digestible resources such as sugars and amino acids, but those are abundant only in living trees or fresh dead wood. This forces each new generation of most ambrosia beetles to seek new substrate, largely preventing the overlap of generations. However, there are notable exceptions. *Ambrosiodmus* and *Ambrosiophilus* beetles are the only ambrosia beetles currently known to farm ambrosia fungi capable of degrading lignin, the compound that makes wood remarkably difficult to enzymatically degrade. By partnering with *Irpex subulatus,* a true wood degrading basidiomycete, *Ambrosiophilus* and *Ambrosiodmus* can remain in the same log longer as it is slowly decomposed, and display signs of sub-social arrangement such as overlapping generations (Kasson et al. 2016). Similar delay of dispersal and acceleration of reproduction was documented in other ambrosia beetles which live in environments where the ambrosia garden is long-lasting, such as *Xyleborinus* and *Austroplatypus* (Kent and Simpson 1992; Biedermann and Taborsky 2011).

11.2.2 Mycangia

A mycangium can be one of various anatomical structures that maintains living fungal propagules in dormant and/or dispersing adult beetles. These structures are key adaptations that are essential to the evolution and maintenance of ambrosia symbioses (Mayers et al. 2022). Mycangia facilitate the persistence of associations between beetle and fungal lineages across generations. Mycangia provide an interface for the discrimination of mutualistic versus non-mutualistic fungi and a bottleneck, through which antagonistic fungal parasites and competitors are purged (Skelton et al. 2019a). Mycangia vary among beetle lineages in their size, complexity, anatomical location, sex associations, and specificity to fungal species. There is a rich literature describing the detailed morphology of these structures, identifying their fungal and bacterial contents (Hulcr and Stelinski 2017), and devising classifications according to their anatomy and complexity (Six 2003).

More than just passive containers for fungal spores, bark and ambrosia beetle mycangia support the growth of fungi during strategic moments of beetle development (Francke-Grosmann 1956; Batra 1963; Kajimura and Hijii 1992). While some mycangia are fixed structures, some are dynamic. The mesothoracic mycangium in the genera *Xylosandrus*, *Anisandrus,* and relatives are flattened in young adults, inflate with fungal matter after symbiont uptake and during dispersal, and deflate again after the new garden is established (Li et al. 2018b). These observations suggest that maintaining a mycangium that is full of active fungal tissue is a costly investment for these beetles and that selection favors precise timing and control over fungal growth.

Mycangia provide a mechanism to promote specificity in beetle-fungus relationships. Beetles in the genus *Xylosandrus* farm fungi in the genus *Ambrosiella*. Their mycangium is able to accept several species of *Ambrosiella* in no-choice situations. However, the probability of uptake of any *Ambrosiella* by the new generation of beetles is lower for species that are not the specific coevolved symbiont, and minimal for non-*Ambrosiella* genera (Skelton et al. 2019a). Likewise in *Xyleborus* and Platypodinae, the mycangium can transfer multiple species of *Raffaelea, Harringtonia* or *Dryadomyces*, but routinely only one species is numerically dominant in the mycangia of each beetle species, and other fungal genera are vectored in lesser abundances and low frequencies (Carrillo et al. 2014; Li et al. 2018a). Further experimental work is needed to determine if the dominance of particular species is enforced by selectivity of the beetles' mycangia, or additional/alternative mechanisms such as beetle behavior, substrate choice, or fungal competition.

In some beetles, the ambrosia farming lifestyle is evident from galleries lined with luxuriant fungal growth, yet the presence of a mycangium is yet to be confirmed [e.g. *Sueus* and several Platypodinae, (Li et al. 2020)]. Other beetles possess small external structures that frequently hold a few fungal cells, but their function is uncertain. Such "pit mycangia" are common in Platypodinae and have been proposed for other beetle taxa. In *Xyleborinus* and some *Xyleborus* (Scolytinae, Xyleborini), the putative elytral mycangia are also very small, and symbiotic fungi have been isolated from other body parts (Biedermann et al. 2013). It remains unclear whether these pit and elytral mycangia are truly co-evolved adaptations for bearing the propagules of fungal mutualists, or if they are simply anatomical features with an as yet unknown function and a coincidental tendency to collect spores. Uniquely, *Xyloterinus* and some *Euwallacea* beetles appear to have two types of mycangia, each occupied by a different fungus (Abrahamson and Norris 1966; Mayers et al. 2020; Spahr et al. 2020).

There are no known instances in which an obligate dependence on fungus farming has been secondarily lost in bark and ambrosia beetles. However, several lineages of ambrosia beetles have secondarily lost their mycangium in favor of a "mycocleptic" strategy. These beetles bore their galleries adjacent to the galleries of mycangium-bearing ambrosia beetles. The mycoclept's offspring feed on the parasitized fungal gardens as the fungi extend into the gallery of the mycoclept (Hulcr and Cognato 2010). Mycoclepts are perfect examples of evolutionary cheaters because they exploit the ambrosia mutualism by benefiting from the nutritional spores produced by ambrosia fungi, while they do not reciprocate by facilitating the dispersal of the ambrosia fungus (Skelton et al. 2019a).

11.2.3 Relationships with Trees

Ambrosia beetles are often said to colonize "stressed, dead or dying" trees. However, it is important to discriminate among these types of resources. From the beetle and fungus perspective, plant tissues that are stressed but still alive present a much different environment than tissues which are dead or nearly so. Grouping them together obscures significant differences between the ecology of beetles and fungi that are able to colonize stressed but living trees, and those that only colonize trees which will not recover. The ability of some species to colonize stressed live tissue explains their tendencies to become forest or silvicultural pests.

Only very few ambrosia beetle species and their fungal associates are able to colonize healthy living trees. Some beetles, such as the black twig borer *Xylosandrus compactus,* attack only the twigs of healthy trees causing the end of the twig to die, but they typically cause no serious harm to the tree unless they are present in very high abundance. There is only a single case in which an ambrosia beetle causes tree mortality by infecting the tree with a systemically pathogenic fungus: the redbay ambrosia beetle *Xyleborus glabratus* which carries the systemic laurel pathogen *Harringtonia lauricola*. Although these pest species are more studied and better known than harmless species, they do not represent the typical ambrosia ecological strategy.

Trees that are stressed while still alive are attractive to several other groups of specialized ambrosia beetles, including the common and widely introduced species *Cnestus mutilatus, Xylosandrus crassiusculus*, and *Xylosandrus germanus*. As a result, these beetles are prominent pests on intensively managed trees such as in nurseries and young orchards. The remaining ambrosia beetle species are not known to colonize healthy living trees, and only rarely colonize stressed trees. Instead, most ambrosia beetles seek freshly dead trees which no longer possess functioning defense mechanisms.

Several groups of ambrosia beetles prefer to colonize wood tissue already pre-infested by their respective fungus, instead of seeking new hosts. For example, the tea shot-hole borer *Euwallacea fornicatus* establishes large colonies by re-infesting the same portions of trees, and only when the particular tree part is no longer suitable to support the fungus, the emerging beetles take flight and seek new hosts (Mendel et al. 2017). Similarly, beetles associated with the wood decaying ambrosia fungus *Irpex* (*Ambrosiodmus, Ambrosiophilus*) are often found colonizing tree parts infected with *Irpex* inoculated by previously colonized beetles of either genus (Kasson et al. 2016; Li et al. 2017).

The question of taxonomic host tree specificity of ambrosia fungi and beetles is not yet fully resolved. On one hand, a seemingly unlimited taxonomic range of tree hosts is sometimes reported (Beaver 1979; Hulcr et al. 2007), however these analyses are typically based predominantly on beetles that associate with the polyphagous species of the polyphyletic fungal genus *Raffaelea*. There are many observations suggesting that other groups of ambrosia beetles and fungi display preferences for particular host tree families. For example, in Asia there are entire genera of Xyleborini specific to dipterocarps and species from various genera specific to Lauraceae (including the pestiferous *X. glabratus* associated with *H. lauricola*). In North America there are several phloeomycetophagous semi-ambrosial beetles that farm *Entomocorticium* and only colonize trees in the Pinaceae (Harrington 2005). There are even species that are specific to certain host tree species, such as some *Corthylus* (Roeper et al. 1987). This pattern further supports the notion that there is not one, but many different types of ambrosia symbiosis and a corresponding diversity of ecologies.

11.2.4   Host Selection and Chemical Ecology

While host searching behavior has been well studied in several important species of bark beetles, it has only recently been studied in ambrosia beetles. Just as in bark beetles, several main sources of volatile chemicals are important for ambrosia beetles: host volatiles (primary attractants), non-host volatiles that are typically repellent, volatiles generated by decay or the organisms associated with decay (secondary attractants), and pheromones produced by other scolytine beetles. Pheromones exist in ambrosia beetle groups that reproduce via regular outcrossing, such as the Platypodinae (Gonzalez-Audino et al., 2005) and Xyloterini (Macconnell et al. 1977).

Primary host attraction is most important in ambrosia beetle species that attack living trees or that are specific to certain host groups. Such host-specific species are rare among ambrosia beetles, but a well-studied example is the redbay ambrosia beetle *X. glabratus*. The beetle responds to sesquiterpenes and other compounds that are characteristic for Lauraceae, its host family (Kendra et al. 2014). On the contrary, it is repelled by volatiles from other kinds of trees, as well as volatiles from the leaves of Lauraceae, indicating healthy and unsuitable host (it is, however, attracted to volatiles released from the wood) (Hughes et al. 2017). Interestingly, the effect of the most commonly used attractant of aggressive ambrosia beetles— ethanol—on *X. glabratus* is ambiguous, and may even be a repellent (Kendra et al. 2014).

Ambrosia beetle ecology differs from that of other wood borers primarily in the reliance on fungi, and that has implications also for their chemical ecology. Fungus-produced volatiles are attractive to ambrosia beetles. In some instances, vectors are most strongly attracted by volatiles from their respective symbionts (Hulcr et al. 2011). More general fungal volatiles, such as quercivorol, are attractive to a broad diversity of ambrosia beetles or serve as synergists for other volatiles (Cooperband et al. 2017).

Some volatiles seem to be attractive to many different ambrosia beetles. For example, the aforementioned ethanol and quercivorol, byproducts of plant stress and of fungal metabolism respectively, are attractive to many unrelated beetle species (Kamata et al. 2008; Ranger et al. 2010; Kendra et al. 2017). The repeated use of the same compounds in related species, or at least the use of derivatives of the same chemical structures, and the enrichment of the information content by synergy with host volatiles, has been termed semiochemical parsimony, and has been also shown in other wood boring beetles (Hanks and Millar 2013).

An important group of ambrosia beetles, the Xyleborini, is exceptional in its lack of aggregation pheromones. All species in this tribe reproduce almost entirely via inbreeding paired with haplo-diploidy. The haploid males, which are smaller than the females, flightless, and probably blind, mostly stay in their native galleries and mate with their sisters. Because the dispersing females are already mated when they arrive at a new tree and therefore do not need to attract a male, the group does not use any long-distance pheromones. Short-distance or contact pheromones are produced at least by the genus *Euwallacea* (Cooperband et al. 2017), but their practical application as long-distance pest attractants is unlikely due to low volatility.

11.3 Economic Significance

Ambrosia beetles and fungi are ecologically diverse, and the pestiferous species are no exception. Here we introduce multiple examples, especially those that display different types of damage. Many other interesting and important ambrosia pest groups exist but could not be covered here, including tropical and temperate pinhole borers and many species that cause damage to trees stressed by climate or management.

11.3.1 Ambrosia Beetle Pests in Dead Trees

Before the contemporary era of global biotic homogenization, ambrosia beetles were known mostly for lumber damage, the result of many tunnels at timber loading sites in logged forests. Such damage is typically regional, and unrelated to the health of living trees.

11.3.1.1 Trypodendron

Distributed throughout almost the entire Northern hemisphere, the genus *Trypodendron* defies many standard narratives about ambrosia beetles. Despite its wide distribution, the genus is rather species-poor, especially compared to the hyper-diverse Xyleborini or Platypodinae. *Trypodendron* species do not kill trees, but their massive colonization of freshly cut conifer trees causes many perforations in the wood and their associated fungi cause staining around the beetle galleries. This results in a significant reduction of the monetary value of lumber. Such damage to cut lumber may exceed the financial losses caused by the tree-killing bark beetles (Lindgren and Fraser 1994).

Trypodendron damage garnered significant research attention, and consequently a considerable number of management methods that are now available for truly Integrated Pest Management (IPM) of this pest. A simple yet significant mitigation of impact can be achieved by the timing of logging and exposure of logs because *Trypodendron* spp. are distinctly seasonal (Dyer and Chapman 1965). Application of non-host volatiles, such as pine extracts, onto spruce logs achieves up to 85% protection against *T. lineatum* (Dubbel 1992). In addition, *Trypodendron* species are also highly responsive to the genus-specific pheromone lineatin, which is therefore widely used in monitoring. In heavy infestations, lineatin baited intercept traps can also trap-out significant numbers of beetles from the vicinity of the logs (Lindgren and Fraser 1994). The use of semiochemicals for the control of *Trypodendron lineatum* has been one of the most successful examples of this control technique.

11.3.2 Global Change-Induced Damage by Ambrosia Beetles

The health of the world's trees and forests is increasingly affected by many stressors. The two pressures most related to the spread of ambrosia beetles is the spread of planted monocultures and global climate change. In intensively managed nurseries and orchards, trees may experience multiple stressors, including poor matches to the local soil types, excessive or insufficient water regimes, and novel pathogens. Such stresses may not be apparent to a human observer, but some ambrosia beetles have evolved to be exquisitely sensitive to the semiochemical signature of a stressed tree (Ranger et al. 2010). As examples below demonstrate, from tropical plantations to temperate nurseries, ambrosia beetles attack managed trees emitting trace amounts of stress-related chemicals, while similar attacks are not reported from nearby natural vegetation. Perhaps rather than thinking about all ambrosia beetles strictly as pests and attempting to manage them as such, it may be more appropriate for tree managers to consider many ambrosia beetle species as reliable indicators of underlying poor tree health that should be improved. That increased insect activity is a symptom of poor tree health is one of the foundational elements of modern forest entomology (Manion 1981).

11.3.2.1 Tree Stress Responders: Xylosandrus Spp.

In the U.S. and increasingly in Europe, the invasive *Xylosandrus crassiusculus* and *X. germanus* are examples of ambrosia beetles sensitive to tree stress-related volatiles, primarily ethanol (Ranger et al. 2015). Ethanol production is triggered most often by damaged roots, for example due to frost, lack of oxygen due to saturation of soil with water, or an internal pathogen. Such stressors are common in actively managed nurseries and orchards, and therefore non-native *Xylosandrus* species are becoming notable pests in such environments.

It is important to recognize that the beetles are not the cause of the tree stress but a sign of other stressors. Focusing management on the beetle is likely going to be less effective than ameliorating the underlying causes. Ambrosia beetles are abundant throughout the landscape, difficult to monitor and even more difficult to manage. In contrast, the growing conditions and health of tree crops can be much more easily monitored and managed. If growers maintain healthy trees and optimal growing conditions, the ubiquitous ambrosia beetles will be mostly inconsequential.

Xylosandrus crassiusculus is also increasingly posing a problem to industries that process hardwood lumber. The rapidly reproducing and polyphagous beetles can colonize untreated timber in high numbers, causing extensive perforation and staining of the wood. This necessitates a much shorter turnaround of such inventory.

The third invasive and damaging species of this genus is *Xylosandrus compactus*. This minute ambrosia beetle is specialized on small living twigs, causing dieback of branch tips. Heavy infestations can cause disfiguration of trees and death of seedlings. The origin of this beetle is South East Asia (Urvois et al. 2021), and it is an increasingly common pest throughout warm regions of the US and Europe, and in coffee growing regions globally where it causes significant damage to the coffee crop (Ngoan et al. 1976; Greco and Wright 2015; Vannini et al. 2017).

11.3.3 Tree-Killing Invasive Species

The killing of mature healthy trees is very rare among ambrosia beetles. Some unusual native species, such as *Corthylus punctatissimus*, naturally colonize small tree seedlings which consequently die (Roeper et al. 1987). However, the majority of ambrosia beetles that cause extensive tree mortality are non-native species, which have not coevolved with these trees.

11.3.3.1 Xyleborus glabratus

The most dramatic and unusual case of widespread ambrosia beetle-induced tree mortality is the case of Laurel wilt, a deadly disease of susceptible trees in the Lauraceae caused by the fungus *H. lauricola* which is vectored by the ambrosia beetle *X. glabratus*. Laurel wilt is most prevalent in the Southeastern US which used to have high densities of susceptible Lauraceae trees. So far, the disease has had the greatest economic impact in avocado groves in South Florida (Evans et al. 2010). Ecologically, the most affected ecosystem has been the forest understory across the Southeastern US where mature individuals of several *Persea* and related lauraceous genera have been nearly eradicated. Only a fraction of the former population survives, which has consequences for many other members of the ecosystems, from insect herbivores to endangered plant pollinators (Hughes et al. 2015).

Large ecological impacts have occurred due to Laurel wilt in the Florida Everglades, where the void left by the deaths of millions of *Persea* is being filled by invasive plants (Rodgers et al. 2014). Laurel wilt also occurs in Asia, but with much lesser intensity (Hulcr et al. 2017). The greatest threat may yet be realized. Lauraceae and avocado are much more important ecologically, economically and culturally in South and Central America (Lira-Noriega et al. 2018). If Laurel wilt spreads to these regions, the effects could be catastrophic.

The ecology of the *X. glabratus* and *H. lauricola* mutualism in non-native regions is unusual in several respects. For instance, the vector beetle searches for live host trees by following specific sesquiterpenes, a behavior not known in other ambrosia beetles (Kendra et al. 2011). Similarly, the disease has unusual etiology. The prevailing hypothesis about the initial infection posits that the first beetle colonizes the living and healthy tree in error, or perhaps as a trial, and either leaves or dies within the tree (Martini et al. 2017). Should this be confirmed as the main mode of action of the disease spread, it is truly a unique situation as the pioneer beetle derives no fitness benefit and the behavior is not adaptive. The fungus-tree interaction is also unusual. Unlike the localized infections caused by other ambrosia fungi in living trees, *H. lauricola* rapidly spreads as a systemic infection, triggering extensive formation of tyloses in tracheids and vessels, diminishing the water conductivity of the xylem (Inch et al. 2012).

11.3.3.2 Euwallacea

The genus *Euwallacea* includes species that span the entire range from primary pests (attacking living, healthy trees) to saprophages (living in decaying wood). From the tree health management perspective, species that attack living trees are important, and those include the *E. interjectus*, *E. destruens*, *E. fornicatus*, *E. kuroshio*, and *E. perbrevis*. All *Euwallacea* species are primarily associated with *Fusarium* species from the specialized ambrosial *Fusarium* clade, collectively referred to as the "AFC" (O'Donnell et al. 2016). The spread of *E. fornicatus* to various regions around the world and the damage that has followed appear to suggest that this species is much more invasive and damaging than the remaining ones (Hulcr et al. 2017; Smith et al. 2019).

Observations from native regions suggest that some of these small *Euwallacea* species are able to colonize specific parts of healthy trees, such as branch joints (Hulcr et al. 2017). However, the greatest damage is typically observed in managed situations such as urban landscape vegetation or avocado groves. Some of the greatest damage in natural systems appear to be associated with tree stress including flooding and pollution (Boland and Woodward 2019). The greater impact in managed plantations is manifested in both the invaded and the native regions. The tea shot hole borer (*E. perbrevis*) has been known to cause losses in tea plantations in Asia where it is native (Hazarika et al. 2009), but there are no reports of damage from non-agricultural habitats. Also in the invaded regions such as South Africa or Israel, the *E. fornicatus* infestation has been documented mostly in managed urban vegetation or managed settings (Mendel et al. 2012; Paap et al. 2018). The various stressors that may trigger colonization by *E. fornicatus*, or that may facilitate development of the infestation, may also include unapparent tree disease. Attacks on living trees that have been pre-infested by a pathogen have been well documented for *E. validus* and *E. interjectus* (Kajii et al. 2013; Kasson et al. 2013).

The pattern in *Euwallacea* damage suggests a distinct role of tree stress as a predis-posing factor (Wang et al. 2021). However, there are also cases where the invasive beetles cause damage in naturally growing native vegetation, including increasingly in South Florida (Owens et al. 2018) and South Africa (Paap et al. 2018). Therefore, it may be too early to estimate the full impact of the *Euwallacea* global invasion.

11.3.4 Ambrosia Beetle Colonization Is a Sign of Tree Disease, not Its Cause

Plant pathologists have long understood the tripartite balance between a pathogen, the host, and the well-being of the host as the so called "disease triangle". In other words, for a disease to occur, the three elements must be in place: the pathogen has to be present in the susceptible host and the environment has to be conducive to disease development. In ambrosia beetle management, the role of the environment and the pre-existing conditions of the trees has not yet been broadly appreciated.

In ambrosia beetle systems where environment has been studied, it is often tree stress that determines the impact of these beetles (Ranger et al. 2010; Boland and Woodward 2019). Also, in the case of the closely related phloem-feeding bark beetles, tree stress is often required for the bark beetles to arrive and facilitate tree death (Wallace 1859; Stephenson et al. 2019). Therefore, we recommend that, when tree disease or death is being diagnosed and when ambrosia beetles are involved, the default assumption is that beetle colonization is a part of multiple interacting negative factors, unless the beetles are explicitly determined to be the primary cause of the problem. Correct determination of the cause of plant diseases is the basis of plant pathology, and the most effective path towards a solution (Leach 1940).

11.4   Questions for Further Research

11.4.1   Defense Against Invasive Ambrosia Beetles

Invasive exotic pests and diseases are causing increasing tree mortality around the world. In the past several decades, governments and agencies have been mobilizing a range of solutions to improve national biosecurity, which follow two types of approaches. One approach relies on closing pathways for all new invasions, such as certification of pest-free status of goods and packaging, inspections, and quarantine (Hulme 2009). The second approach is focused on early detection of, and rapid response to, specific exotic species that may cause harm (Kenis et al. 2018; Rabaglia et al. 2019). The two approaches are complementary. While pathway limitations are sometimes perceived to be more effective, their implementation is more likely to impede trade and are thus politically complex. A focus on responses to individual exotic species requires nimble action that is often difficult to mobilize but is much more acceptable to agencies that need to balance pressure from trade organizations and biosecurity, such as the USDA APHIS.

Both approaches are dependent on data. As knowledge about the ecology of individual bark and ambrosia beetle species is growing, we are increasingly able to predict pathways of introductions, and species that are likely to pose harm when introduced to new regions. Species that are likely to cause harm are characterized by two features: the ability to invade and thrive in new habitats, and a propensity for negative impacts on plant commodities.

In the case of ambrosia beetles, pre-invasion assessment is becoming feasible because the features predisposing some species to invasions as well as to damage are becoming increasingly understood (Li et al. 2022). Successful spread and establishment in new regions are facilitated by the fact that the majority of the life-cycle is spent in a concealed habitat and that many species are capable of inbreeding without reduced fitness (Jordal et al. 2001). The capacity for repeated inbreeding allows even minute populations to grow, while in most other outcrossing organisms, repeated inbreeding often leads to expression of recessive deleterious features. Predisposition to actual damage by ambrosia beetles is less clear, but it appears to be determined by specificity to the commodity in question, and the ability to colonize living tissues (Hulcr et al. 2017).

11.4.2   Ecological Significance

The sheer abundance of some ambrosia beetles, such as the various Platypodinae or *Xyleborus* in several regions of the world makes these beetles among the most common insects in the forest. To the best of our knowledge however, these numbers have never been quantified, and their impact on ecosystem processes such as ecosystem-scale wood decay remain unclear.

Bark and ambrosia beetles are often the first colonizers of dead and dying trees in most forest ecosystems, and as such, they are likely to play an important role in the recycling of the world's forest biomass and the release of carbon from decaying wood (Luyssaert et al. 2007; Le Quéré et al. 2013; Dossa et al. 2018). Living trees are the largest terrestrial sink for atmospheric carbon dioxide. After a tree dies, however, most of the carbon stored in its tissues is released to the soil and the atmosphere as the metabolic waste of fungal decomposers (Chambers et al. 2001). The rate of carbon release through wood decomposition is to a large degree determined by the identity, diversity, and sequence of fungal colonists (Fukami et al. 2010). Many of the saprotrophic fungi in wood, and in some cases most of the fungi, are introduced by bark and ambrosia beetles (Strid et al. 2014; Skelton et al. 2019b). Thus, by initiating fungal community assembly in recently dead wood, bark and ambrosia beetles are likely to have pervasive influence on wood decay rates, and that influence likely depends on the fungi they carry.

Contrary to the popular belief that they facilitate wood decomposition, new evidence suggests many ambrosia beetles could have the opposite effect. Relatively few fungi can degrade lignocellulose, the main structural component of wood. This process requires highly specialized enzymatic pathways. With the exception of the recently discovered ambrosial *Irpex* and perhaps the basidiomycete associates of some pine-infesting bark beetles [i.e. *Entomocorticium;* (Valiev et al. 2009), but see (Whitney et al. 1987)], no other fungi commonly associated with bark and ambrosia beetles are currently known to have this ability. Instead, most beetle associates depend on the scarcer but more labile resources present in fresh wood, such as sugars and nitrogenous compounds (Licht and Biedermann 2012; Huang et al. 2019).

Recent field and laboratory experiments have shown that some beetle-associated fungi exclude, or compete with, true wood-degrading fungi for labile resources, resulting in decreased decay rates during the early stages of decomposition (Skelton et al. 2019b, 2020). Thus, ambrosia beetles may actually slow carbon release from forest biomass by assembling saprotroph communities that do not decay wood, but instead compete with or exclude decay fungi. Whether these effects persist over the entire decomposition process and ultimately result in increased carbon burial in forest soils is currently unknown. The ecological impacts of widespread introductions and rapid increases in certain beetles that do vector aggressive decay fungi, and which displace native fungi, are also currently unknown and deserve future study (Hulcr et al. 2021; Jusino et al. 2020).

11.4.3 *Pests of the Future*

Eradicating established invasive ambrosia beetles is virtually impossible. Classical biological control has not yet been shown to work in ambrosia beetle pests. Likewise, the biology of the haplo-diploid and inbred ambrosia beetles precludes the effectiveness of some biotechnological applications such as gene drive. We see three options as most promising for forest and tree health protection against invasive ambrosia pests.

First, preventing future invasions is key. While ambrosia beetles include many global "tourist" species (Gohli et al. 2016), rather few of them become true pests. Most of the damage attributed to invasive ambrosia beetles is actually caused by a few species, namely *X. glabratus*, *E. fornicatus*, and *Xylosandrus* spp. To allow agencies to focus on the pests that are likely to cause impact, and lessen focus on harmless species, it may be worth developing a formal pre-invasion assessment of the likely future pests.

Second, tree management needs to be adapted to the new pests. Fortunately, such adaptation may be within reach. In nurseries, defense against *Xylosandrus* stem borers may require not much more than more efficient water management (Ranger et al. 2016). In orchards affected by *E. fornicatus* and *E. kuroshio*, removal of the hyper-infested tree branches is sufficient to prevent escalation of the pest impact (Mendel et al. 2017).

Third, in cases where tree deaths result from a biotic interaction that is known and characterized, resistance breeding may be a valuable tool for tree protection. In laurel wilt-stricken *Persea*, for example, a certain percentage of the tree population survives, either via resistance to the pathogen or by being undetectable to the vector. Such resistance can be harnessed and resistant populations of these trees are now grown, composed of genotypes from multiple locations (Hughes et al. 2015). In highly valued species, resistance development by biotechnology is also plausible. For pathosystems involving fungi, such as the ambrosia beetle-fungus symbioses, known anti-fungal heritable defense can be deployed, such as has already been used in the protection of trees against invasive fungal diseases (Newhouse et al. 2014).

References

Abrahamson LP, Norris DM (1966) Symbiotic interrelationships between microbes and ambrosia fungi I. The organs of microbial transport and perpetuation of *Xyloterinus politus*. Ann Entmological Soc Am 59:877–880

Alamouti SM, Tsui CKM, Breuil C (2009) Multigene phylogeny of filamentous ambrosia fungi associated with ambrosia and bark beetles. Mycol Res 113:822–835

Bateman C, Huang YT, Simmons D, Kasson MT, Stanley E, Hulcr J (2016) Ambrosia beetle *Premnobius cavipennis* (Scolytinae: Ipini) carries highly divergent ascomycotan ambrosia fungus, *Afroraffaelea ambrosiae* gen. nov. sp. nov. (Ophiostomatales). Fungal Ecol 25:41–49

Batra LR (1963) Ecology of ambrosia fungi and their dissemination by beetles. Trans Kansas Acad Sci 66:213–236

Beaver RA (1979) Host specificity of temperate and tropical animals. Nature 281:139–141

Biedermann PHW, Klepzig KD, Taborsky M, Six DL (2013) Abundance and dynamics of filamentous fungi in the complex ambrosia gardens of the primitively eusocial beetle *Xyleborinus saxesenii* Ratzeburg (Coleoptera: Curculionidae, Scolytinae). FEMS Microbiol Ecol 83:711–723

Biedermann PHW, Taborsky M (2011) Larval helpers and age polyethism in ambrosia beetles. Proc Natl Acad Sci 108:17064–17069

Boland JM, Woodward DL (2019) Impacts of the invasive shot hole borer (*Euwallacea kuroshio*) are linked to sewage pollution in southern California: the Enriched Tree Hypothesis. PeerJ 7:e6812

Carrillo D, Duncan RE, Ploetz JN, Campbell AF, Ploetz RC, Peña JE (2014) Lateral transfer of a phytopathogenic symbiont among native and exotic ambrosia beetles. Plant Pathol 63:54–62

Chambers JQ, Schimel JP, Nobre AD (2001) Respiration from coarse wood litter in central Amazon forests. Biogeochemistry 52:115–131

Cooperband MF, Cosse AA, Jones TH, Carrillo D, Cleary K, Canlas I, Stouthamer R (2017) Pheromones of three ambrosia beetles in the *Euwallacea fornicatus* species complex: ratios and preferences. PeerJ 5:e3957

de Beer ZW, Procter M, Wingfield MJ, Marincowitz S, Duong TA (2022) Generic boundaries in the Ophiostomatales reconsidered and revised. Stud Mycol 101:57–120

Dossa GG, Schaefer D, Zhang JL, Tao JP, Cao KF, Corlett RT, Cunningham AB, Xu JC, Cornelissen JH, Harrison RD (2018) The cover uncovered: Bark control over wood decomposition. J Ecol 106:2147–2160

Dubbel V (1992) The effectiveness of pine oil as a repellent against the striped ambrosia beetle *Trypodendron lineatum* (Coleoptera, Scolytidae). J Appl Entomol-Z Fur Angew Entomol 114:91–97

Dyer E, Chapman J (1965) Flight and attack of the ambrosia beetle, *Trypodendron lineatum* (Oliv.) in relation to felling date of logs. Can Entomol 97:42–57

Evans EA, Crane J, Hodges A, Osborne JL (2010) Potential economic impact of laurel wilt disease on the Florida avocado industry. HortTechnology 20:234–238

Francke-Grosmann H (1956) Hautdrüsen als Träger der Pilz-Symbiose bei Ambrosia-Käfern. Z Für Morphol Und Oekologie Tiere 45:275–308

Francke-Grosmann H (1967) Ectosymbiosis in wood-inhabiting insects. In: Henry SM (ed) Symbiosis. Academic Press, New York, pp 141–206

Fukami T, Dickie IA, Paula Wilkie J, Paulus BC, Park D, Roberts A, Buchanan PK, Allen RB (2010) Assembly history dictates ecosystem functioning: evidence from wood decomposer communities. Ecol Lett 13:675–684

Gohli J, Kirkendall LR, Smith SM, Cognato AI, Hulcr J, Jordal BH (2017) Biological factors contributing to bark and ambrosia beetle species diversification. Evolution 71:1258–1272

Gohli J, Selvarajah T, Kirkendall LR, Jordal BH (2016) Globally distributed *Xyleborus* species reveal recurrent intercontinental dispersal in a landscape of ancient worldwide distributions. BMC Evol Biol 16:37

Gonzalez-Audino P, Villaverde R, Alfaro R, Zerba E (2005) Identification of volatile emissions from Platypus mutatus (= sulcatus) (Coleoptera: Platypodidae) and their behavioral activity. J Econ Entomol 98:1506–1509

Greco EB, Wright MG (2015) Ecology, biology, and management of *Xylosandrus* compactus (Coleoptera: Curculionidae: Scolytinae) with emphasis on coffee in Hawaii. J Integr Pest Manag 6:7

Hanks LM, Millar JG (2013) Field bioassays of cerambycid pheromones reveal widespread parsimony of pheromone structures, enhancement by host plant volatiles, and antagonism by components from heterospecifics. Chemoecology 23:21–44

Harrington TC (2005) Ecology and evolution of mycophagous bark beetles and their fungal partners. In: Vega FE, Blackwell M (eds) Insect-Fungal Associations. Oxford University Press, New York, pp 257–291

Hazarika LK, Bhuyan M, Hazarika BN (2009) Insect pests of tea and their management. Annu Rev Entomol 54:267–284

Huang Y-T, Skelton J, Hulcr J (2019) Multiple evolutionary origins lead to diversity in the metabolic profiles of ambrosia fungi. Fungal Ecol 38:80–88

Huang Y-T, Skelton J, Hulcr J (2020) Lipids and small metabolites provisioned by ambrosia fungi to symbiotic beetles are phylogeny-dependent, not convergent. ISME J 14:1089–1099

Hughes MA, Martini X, Kuhns E, Colee J, Mafra-Neto A, Stelinski LL, Smith JA (2017) Evaluation of repellents for the redbay ambrosia beetle, Xyleborus glabratus, vector of the laurel wilt pathogen. J Appl Entomol 141:653–664

Hughes MA, Smith JA, Ploetz RC, Kendra PE, Mayfield AE, Hanula JL, Hulcr J, Stelinski L, Cameron S, Riggins JJ, Carrillo D, Rabaglia RJ, Eickwort JM, Pernas T (2015) Recovery plan for laurel wilt on redbay and other forest species caused by *Raffaelea lauricola* and disseminated by *Xyleborus glabratus*. Plant Health Prog 16:173–210

Hulcr J, Atkinson TH, Cognato AI, Jordal BH, McKenna DD (2015) Morphology, taxonomy and phylogenetics of Bark Beetles. In: Vega FE, Hofstetter RW (eds) Bark Beetles. Elsevier, pp 41–84

Hulcr J, Black A, Prior K, Chen CY, Li HF (2017) Studies of ambrosia beetles (Coleoptera: Curculionidae) in their native ranges help predict invasion impact. Fla Entomol 100:257–261

Hulcr J, Cognato AI (2010) Repeated evolution of crop theft in fungus-farming ambrosia beetles. Evolution 64:3205–3212

Hulcr J, Mann R, Stelinski LL (2011) The scent of a partner: ambrosia beetles are attracted to volatiles from their fungal symbiont. J Chem Ecol 37:1374–1377

Hulcr J, Mogia M, Isua B, Novotny V (2007) Host specificity of ambrosia and bark beetles (Coleoptera, Curculionidae: Scolytinae and Platypodinae) in a New Guinea rain forest. Ecol Entomol 32:762–772

Hulcr J, Gomez DF, Skelton J, Johnson AJ, Adams S, Li Y, Jusino MA, Smith ME (2021) Invasion of an inconspicuous ambrosia beetle and fungus may affect wood decay in Southeastern North America. Biol Invasions 23:1–9

Hulcr J, Stelinski LL (2017) The ambrosia symbiosis: from evolutionary ecology to practical management. Annu Rev Entomol 62:285–303

Hulme PE (2009) Trade, transport and trouble: managing invasive species pathways in an era of globalization. J Appl Ecol 46:10–18

Inch S, Ploetz R, Held B, Blanchette R (2012) Histological and anatomical responses in avocado, Persea americana, induced by the vascular wilt pathogen, Raffaelea lauricola. Botany 90:627–635

Johnson AJ, McKenna DD, Jordal BH, Cognato AI, Smith SM, Lemmon AR, Lemmon ELM, Hulcr J (2018) Phylogenomics clarifies repeated evolutionary origins of inbreeding and fungus farming in bark beetles (Curculionidae, Scolytinae). Mol Phylogenetics Evol 127:229–238

Johnson RA, Thomas RJ, Wood TG, Swift MJ (1981) The inoculation of the fungus comb in newly founded colonies of some species of the Macrotermitinae (Isoptera) from Nigeria. J Nat Hist 15:751–756

Jordal B, Beaver RA, Kirkendall LR (2001) Breaking taboos in the tropics: incest promotes colonization by wood-boring beetles. Glob Ecol Biogeogr 10:345–357

Jordal BH, Cognato AI (2012) Molecular phylogeny of bark and ambrosia beetles reveals multiple origins of fungus farming during periods of global warming. BMC Evol Biol 12

Jusino MA, Skelton J, Chen C, Hulcr J, Smith ME (2020) Sexual reproduction and saprotrophic dominance by the ambrosial fungus *Flavodon subulatus* (= *Flavodon ambrosius*). Fungal Ecol 47:100979

Kajii C, Morita T, Jikumaru S, Kajimura H, Yamaoka Y, Kuroda K (2013) Xylem dysfunction in Ficus Carica infected with wilt Fungus Ceratocystis Ficicola and the role of the vector Beetle Euwallacea Interjectus. Iawa J 34:301–312

Kajimura H, Hijii N (1992) Dynamics of the fungal symbionts in the gallery system and the mycangia of the ambrosia beetle, Xylosandrus mutilatus (Blandford) (Coleoptera: Scolytidae) in relation to its life history. Ecol Res 7:107–117

Kamata N, Esaki K, Mori K, Takemoto H, Mitsunaga T, Honda H (2008) Field trap test for bioassay of synthetic (1S,4R)-4-isopropyl-1-methyl-2-cyclohexen-1-ol as an aggregation pheromone of *Platypus quercivorus* (Coleoptera: Platipodidae). J for Res 13:122–126

Kasson MT, Davis MD, Davis DD (2013) The Invasive Ailanthus altissima in Pennsylvania: a case study elucidating species introduction, migration, invasion, and growth patterns in the Northeastern US. Northeast Nat 20:1–60

Kasson MT, Wickert KL, Stauder CM, Macias AM, Berger MC, Simmons DR, Short DPG, DeVallance DB, Hulcr J (2016) Mutualism with aggressive wood-degrading Flavodon ambrosius (Polyporales) facilitates niche expansion and communal social structure in Ambrosiophilus ambrosia beetles. Fungal Ecol 23:86–96

Kendra PE, Montgomery WS, Niogret J, Peña JE, Capinera JL, Brar G, Epsky ND, Heath RR (2011) Attraction of redbay ambrosia beetle (Coleoptera: Curculionidae: Scolytinae) to avocado, lychee, and essential oil lures. Fungal Ecol 37:932–942

Kendra PE, Montgomery WS, Niogret J, Schnell EQ, Deyrup MA, Epsky ND (2014) Evaluation of seven essential oils identifies cubeb oil as most effective attractant for detection of Xyleborus glabratus. J Pest Sci 87:681–689

Kendra PE, Owens D, Montgomery WS, Narvaez TI, Bauchan GR, Schnell EQ, Tabanca N, Carrillo D (2017) Alpha-Copaene is an attractant, synergistic with quercivorol, for improved detection of Euwallacea nr. fornicatus (Coleoptera: Curculionidae: Scolytinae). PLoS One 12

Kenis M, Li H, Fan JT, Courtial B, Auger-Rozenberg M-A, Yart A, Eschen R, Roques A (2018) Sentinel nurseries to assess the phytosanitary risks from insect pests on importations of live plants. Sci Rep 8:11217

Kent DS, Simpson JA (1992) Eusociality in the beetle *Austroplatypus incompertus* (Coleoptera: Curculionidae). Naturwissenschaften 79:86–87

Le Quéré C, Andres R, Boden T, Conway T, Houghton R, House J, Marland G, Peters G, Van der Werf G, Ahlström A (2013) The global carbon budget 1959–2011. Earth Syst Sci Data 5:165–185

Leach JG (1940) Insect transmission of plant diseases. McGraw-Hill Book Company

Li Y, Bateman CC, Skelton J, Jusino MA, Nolen ZJ, Simmons DR, Hulcr J (2017) Wood decay fungus *Flavodon ambrosius* (Basidiomycota: Polyporales) is widely farmed by two genera of ambrosia beetles. Fungal Biol 121:984–989

Li Y, Huang Y-T, Kasson MT, Macias AM, Skelton J, Carlson CS, Yin M, Hulcr J (2018a) Specific and promiscuous ophiostomatalean fungi associated with Platypodinae ambrosia beetles in the southeastern United States. Fungal Ecol 35:42–50

Li Y, Ruan YY, Stanley EL, Skelton J, Hulcr J (2018b) Plasticity of mycangia in Xylosandrus ambrosia beetles. Insect Sci 26:732–742

Li Y, Skelton J, Adams S, Hattori Y, Smith ME, Hulcr J (2020) The ambrosia beetle *Sueus niisimai* (Scolytinae: Hyorrhynchini) is associated with the canker disease fungus *Diatrypella japonica* (Xylariales). Plant Disease (in press)

Li Y, Bateman C, Skelton J, Wang B, Black A, Huang Y-T, Gonzalez A, Jusino MA, Nolen ZJ, Freeman S, Mendel Z, Kolařík M, Knížek M, Park J-H, Sittichaya W, Pham T-H, Ito S-I, Torii M, Gao L, Johnson AJ, Lu M, Sun J, Zhang Z, Adams DC, Hulcr J (2022) Preinvasion assessment of exotic bark beetle-vectored fungi to detect tree-killing pathogens. Phytopathology 112:261–270

Licht HHD, Biedermann PHW (2012) Patterns of functional enzyme activity in fungus farming ambrosia beetles. Front Zool 9:13

Lindgren B, Fraser R (1994) Control of ambrosia beetle damage by mass trapping at dryland log sorting area in British Columbia. For Chron 70:159–163

Lira-Noriega A, Soberon J, Equihua J (2018) Potential invasion of exotic ambrosia beetles *Xyleborus glabratus* and *Euwallacea* sp in Mexico: a major threat for native and cultivated forest ecosystems. Sci Rep 8:1–13

Luyssaert S, Inglima I, Jung M, Richardson AD, Reichstein M, Papale D, Piao S, Schulze ED, Wingate L, Matteucci G (2007) CO2 balance of boreal, temperate, and tropical forests derived from a global database. Glob Change Biol 13:2509–2537

Macconnell JG, Borden JH, Silverstein RM, Stokkink E (1977) Isolation and tentative identification of lineatin, a pheromone from the frass of Trypodendron lineatum (Coleoptera: Scolytidae). J Chem Ecol 3:549–561

Manion PD (1981) Tree disease concepts. Prentice-Hall, Englewood Cliffs, NJ

Martini X, Hughes MA, Killiny N, George J, Lapointe SL, Smith JA, Stelinski LL (2017) The fungus Raffaelea lauricola modifies behavior of its symbiont and vector, the redbay ambrosia beetle (Xyleborus Glabratus), by altering host plant volatile production. J Chem Ecol 43:519–531

Mayers CG, Harrington TC, Ranger CM (2017) First report of a sexual state in an ambrosia fungus: *Ambrosiella cleistominuta* sp nov associated with the ambrosia beetle *Anisandrus maiche*. Botany 95:503–512

Mayers CG, Harrington TC, McNew DL, Roeper RA, Biedermann P, Masuya H, Bateman C (2020) Four mycangium types and four genera of ambrosia fungi suggest a complex history of fungus farming in the ambrosia beetle tribe Xyloterini. Mycologia (in press)

Mayers CG, McNew DL, Harrington TC, Roeper RA, Fraedrich SW, Biedermann PHW, Castrillo LA, Reed SE (2015) Three genera in the Ceratocystidaceae are the respective symbionts of three independent lineages of ambrosia beetles with large, complex mycangia. Fungal Biol 119:1075–1092

Mayers CG, Harrington TC, Biedermann PH (2022) Mycangia define the diverse ambrosia beetle-fungus symbiosis. The convergent evolution of agriculture in humans and insects. MIT Press, Cambridge, MA, pp 1–38

McCutcheon JP, Moran NA (2011) Extreme genome reduction in symbiotic bacteria. Nat Rev Microbiol 10:13–26

Mendel Z, Protasov A, Maoz Y, Maymon M, Miller G, Elazar M, Freeman S (2017) The role of Euwallacea nr fornicatus (Coleoptera: Scolytinae) in the wilt syndrome of avocado trees in Israel. Phytoparasitica 45:341–359

Mendel Z, Protasov A, Sharon M, Zveibil A, Yehuda SB, O'Donnell K, Rabaglia R, Wysoki M, Freeman S (2012) An Asian ambrosia beetle *Euwallacea fornicatus* and its novel symbiotic fungus *Fusarium* sp pose a serious threat to the Israeli avocado industry. Phytoparasitica 40:235–238

Moran NA, Wernegreen JJ (2000) Lifestyle evolution in symbiotic bacteria: insights from genomics. Trends Ecol Evol 15:321–326

Musvuugwa T, de Beer ZW, Duong T, Dreyer LL, Oberlander KC, Roets F (2015) New species of Ophiostomatales from Scolytinae and Platypodinae beetles in the Cape Floristic Region, including the discovery of the sexual state of *Raffaelea*. Antonie Leeuwenhoek 108:933–950

Newhouse AE, Polin-McGuigan LD, Baier KA, Valletta KER, Rottmann WH, Tschaplinski TJ, Maynard CA, Powell WA (2014) Transgenic American chestnuts show enhanced blight resistance and transmit the trait to T1 progeny. Plant Sci 228:88–97

Ngoan ND, Wilkinson RC, Short DE, Moses CS, Mangold JR (1976) Biology of an introduced ambrosia beetle, *Xylosandrus compactus*, (Coleoptera, Scolytidae) in Florida. Ann Entomol Soc Am 69:872–876

O'Donnell K, Libeskind-Hadas R, Hulcr J, Bateman C, Kasson MT, Ploetz RC, Konkol JL, Ploetz JN, Carrillo D, Campbell A, Duncan RE, Liyanage PNH, Eskalen A, Lynch SC, Geiser DM, Freeman S, Mendel Z, Sharon M, Aoki T, Cosse AA, Rooney AP (2016) Invasive Asian *Fusarium* - *Euwallacea* ambrosia beetle mutualists pose a serious threat to forests, urban landscapes and the avocado industry. Phytoparasitica 44:435–442

Owens D, Cruz LF, Montgomery WS, Narvaez TI, Schnell EQ, Tabancal N, Duncan RE, Carrillo D, Kendra PE (2018) Host range expansion and increasing damage potential of *Euwallacea* nr. *fornicatus* (Coleoptera: Curculionidae) in Florida. Fla Entomol 101:229–236

Paap T, de Beer ZW, Migliorini D, Nel WJ, Wingfield MJ (2018) The polyphagous shot hole borer (PSHB) and its fungal symbiont *Fusarium euwallaceae*: a new invasion in South Africa. Australas Plant Pathol 47:231–237

Poinar GO Jr, Vega FE (2018) A mid-Cretaceous ambrosia fungus, *Paleoambrosia entomophila* gen nov et sp nov (Ascomycota: Ophiostomatales) in Burmese (Myanmar) amber, and evidence for a femoral mycangium. Fungal Biol 122:1159–1162

Rabaglia RJ, Cognato AI, Hoebeke ER, Johnson CW, LaBonte JR, Carter ME, Vlach JJ (2019) Early detection and rapid response: a ten-year summary of the U.S. Forest Service Program of Surveillance for Non-Native Bark and Ambrosia Beetles. Am Entomol 65:29–42

Ranger C, Reding M, Persad A, Herms D (2010) Ability of stress-related volatiles to attract and induce attacks by *Xylosandrus germanus* (Coleoptera: Curculionidae, Scolytinae) and other ambrosia beetles. Agric for Entomol 12:177–185

Ranger CM, Reding ME, Schultz PB, Oliver JB, Frank SD, Addesso KM, Chong JH, Sampson B, Werle C, Gill S, Krause C (2016) Biology, ecology, and management of nonnative ambrosia beetles (Coleoptera: Curculionidae: Scolytinae) in ornamental plant nurseries. J Integr Pest Manag 7:9

Ranger CM, Schultz PB, Frank SD, Chong JH, Reding ME (2015) Non-native Ambrosia Beetles as opportunistic exploiters of living but weakened trees. PLoS One 10

Rodgers L, Derksen A, Pernas T (2014) Expansion and impact of Laurel Wilt in the Florida Everglades. Fla Entomol 97:1247–1250

Roeper RA (1995) Patterns of mycetophagy in Michigan ambrosia beetles. Mich Acad 26:153–161

Roeper RA, Palik BJ, Zestos DV, Hesch PG, Larsen CD (1987) Observations of the habits of *Corthylus punctatissimus* (Coleoptera, Scolytidae) infesting maple saplings in central Michigan. Gt Lakes Entomol 20:173–176

Schultz TR, Mueller UG, Currie CR, Rehner SA (2005) Reciprocal illumination a comparison of agriculture in humans and in fungus-groing ants. Insect–Fungal Assoc: Ecol Evol 149

Seifert KA, de Beer ZW, Wingfield MJ (2013) The ophiostomatoid fungi: expanding frontiers. CBS-KNAW fungal biodiversity centre

Six DL (2003) Bark beetle-fungus symbioses. In: Bourtzis K, Miller TA (eds) Insect symbiosis. CRC Press, New York, pp 97–114

Skelton J, Johnson AJ, Jusino MA, Bateman CC, Li Y, Hulcr J (2019a) A selective fungal transport organ (mycangium) maintains coarse phylogenetic congruence between fungus-farming ambrosia beetles and their symbionts. Proc R Soc B 286:20182127

Skelton J, Jusino MA, Carlson PS, Smith K, Banik MT, Lindner DL, Palmer JM, Hulcr J (2019b) Relationships among wood-boring beetles, fungi, and the decomposition of forest biomass. Mol Ecol 28:4971–4986

Skelton J, Loyd A, Smith JA, Blanchette RA, Held BW, Hulcr J (2020) Fungal symbionts of bark and ambrosia beetles can suppress decomposition of pine sapwood by competing with wood-decay fungi. Fungal Ecol 45:100926

Smith SM, Gomez DF, Beaver RA, Hulcr J, Cognato AI (2019) Reassessment of the species in the *Euwallacea fornicatus* (Coleoptera: Curculionidae: Scolytinae) complex after the rediscovery of the "lost" type specimen. Insects 10:261

Spahr E, Kasson MT, Kijimoto T (2020) Micro-computed tomography permits enhanced visualization of mycangia across development and between sexes in Euwallacea ambrosia beetles. PLoS One 15:e0236653

Stephenson NL, Das AJ, Ampersee NJ, Bulaon BM, Yee JL (2019) Which trees die during drought? The key role of insect host-tree selection. J Ecol 107:2383–2401

Strid Y, Schroeder M, Lindahl B, Ihrmark K, Stenlid J (2014) Bark beetles have a decisive impact on fungal communities in Norway spruce stem sections. Fungal Ecol 7:47–58

Talbot PHB (1977) The *Sirex-Amylostereum-Pinus* association. Annu Rev Phytopathol 15:41–54

Urvois T, Perrier C, Roques A, Sauné L, Courtin C, Li Y, Johnson AJ, Hulcr J, Auger-Rozenberg M-A, Kerdelhué C (2021) A first inference of the phylogeography of the worldwide invader *Xylosandrus compactus*. J Pest Sci (in press)

Valiev A, Ogel ZB, Klepzig KD (2009) Analysis of cellulase and polyphenol oxidase production by southern pine beetle associated fungi. Symbiosis 49:37–42

Vanderpool D, Bracewell RR, McCutcheon JP (2018) Know your farmer: ancient origins and multiple independent domestications of ambrosia beetle fungal cultivars. Mol Ecol 27:2077–2094

Vannini A, Contarini M, Faccoli M, Valle MD, Rodriguez CM, Mazzetto T, Guarneri D, Vettraino AM, Speranza S (2017) First report of the ambrosia beetle *Xylosandrus compactus* and associated fungi in the Mediterranean maquis in Italy, and new host–pest associations. EPPO Bull 47:100–103

Veselská T, Skelton J, Kostovcik M, Hulcr J, Baldrian P, Chudíčková M, Cajthaml T, Vojtová T, Garcia-Fraile P, Kolařík M (2019) Adaptive traits of bark and ambrosia beetle-associated fungi. Fungal Ecol 41:165–176

Wallace A (1859) Note on the habits of Scolytidae and Bostrichidae. Trans Entomol Soc Lond 5:218–220

Wang Z, Li Y, Ernstsons AS, Sun R, Hulcr J, Gao J (2021) The infestation and habitat of the ambrosia beetle Euwallacea interjectus (Coleoptera: Curculionidae: Scolytinae) in the riparian zone of Shanghai, China. Agric For Entomol. https://doi.org/10.1111/afe.12405

Whitney H, Bandoni R, Oberwinkler F (1987) *Entomocorticium dendroctoni* gen. et sp. nov. (Basidiomycotina), a possible nutritional symbiote of the mountain pine beetle in lodgepole pine in British Columbia. Can J Bot 65:95–102

Chapter 12
Woodborers in Forest Stands

Kevin J. Dodds, Jon Sweeney, and Jeremy D. Allison

12.1 Introduction

The term woodborer is used to describe a polyphyletic group of insects that primarily inhabit the wood of angiosperm and conifer trees in various stages of decay. In the broadest sense, this term includes any insect that inhabits tissues of living woody plants or wood at any stage of the decay process. Common wood associates include Coleoptera (beetles), Hymenoptera (ants, wasps), Lepidoptera (moths), Diptera (flies) and Blattodea (termites and cockroaches). For this chapter, however, we focus on woodborer families that represent the majority of both ecologically and economically important species worldwide. These will include members of two beetle families (Buprestidae, Cerambycidae) as well as woodwasps (Hymenoptera: Siricidae) (Fig. 12.1). Another woodborer group, ambrosia beetles (Coleoptera: Curculionidae: Scolytinae) are covered in depth in Chapter 11. Finally, while there is some overlap in pests of urban and natural forests, this chapter will focus on woodborers of natural and managed forested ecosystems.

While woodborers have gained notoriety based on invasion success of a few species, such as the emerald ash borer (*Agrilus planipennis* Fairmaire) in

K. J. Dodds (✉)
U.S. Forest Service, Eastern Region, State, Private, and Tribal Forestry, Durham, NH, USA
e-mail: kevin.j.dodds@usda.gov

J. Sweeney
Natural Resources Canada, Canadian Forest Service, Atlantic Forestry Centre, Fredericton, NB, Canada

J. D. Allison
Natural Resources Canada, Canadian Forest Service, Great Lakes Forestry Centre, Sault Ste. Marie, ON, Canada
e-mail: Jeremy.Allison@NRCan-RNCan.gc.ca; jeremy.allison@canada.ca

Department of Zoology and Entomology, Forestry and Agricultural Biotechnology Institute, University of Pretoria, Pretoria, Gauteng, South Africa

© The Author(s) 2023   361
J. D. Allison et al. (eds.), *Forest Entomology and Pathology*,
https://doi.org/10.1007/978-3-031-11553-0_12

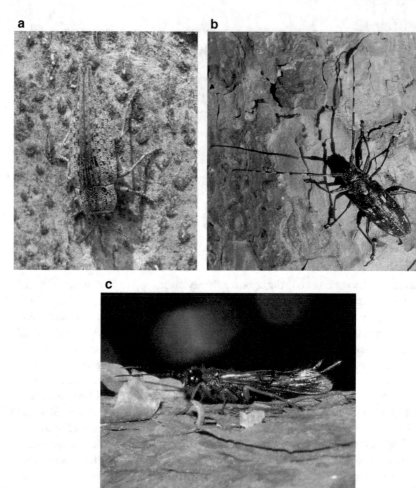

Fig. 12.1 Examples of common woodborers, including **a** *Dicerca divaricata* (Buprestidae), **b** *Monochamus scutellatus* (Cerambycidae), and **c** *Sirex noctilio* (Siricidae). Photo credit: Kevin Dodds

North America and Russia, Asian longhorned beetle [*Anoplophora glabripennis* (Motschulsky)] in North America and Europe, and Sirex woodwasp (*Sirex noctilio* F.) throughout much of the Southern Hemisphere, the majority of insects in these families provide important ecosystem services and rarely develop into epidemic populations that cause economic losses or severe ecological impacts. Most of these species inhabit dead woody material, with the exception being species colonizing and sometimes killing living, healthy trees. Woodborers are cornerstones of decay processes through material fragmentation, introduction of fungi, and wood digestion (Edmonds and Eglitis 1989; Martius 1997; Hadfield and Magelssen 2006; Parker et al. 2006; Ulyshen 2016). They create and/or facilitate access to habitat for other species (Georgiev et al. 2004; Buse et al. 2008) and are important components of forest food webs (Murphy and Lehnhausen 1998; Hunt 2000).

The Buprestidae and Cerambycidae represent diverse families that can be found in all woody plant parts and most wood decay stages. Worldwide, there are between 12,000 and 15,000 species of Buprestidae (Bright 1987; Bellamy 2002; Evans et al. 2004), while there are approximately 36,300 cerambycid species (Monné et al. 2017). Siricidae have much less diversity in the family compared to buprestids and cerambycids, with ~122 species worldwide from 10 genera (Schiff et al. 2012).

12.2 Natural History/Ecology of Woodborers

12.2.1 Woodborer Habitat

Some cerambycids and buprestids are found in vines and herbaceous plants (Bellamy and Nelson 2002), however, the majority, along with siricids, are found in hardwood and conifer tree tissues. Collectively, these insects inhabit all vertical portions of trees, from the roots up to small twigs in crowns, and even within leaves (Hespenheide 1991; Bellamy and Nelson 2002). Horizontally in wood, all tissues from the outer bark to heartwood are also colonized by woodborers during some portion of the decay process, with insects and their associates capable of gaining nutrition from even seemingly poor habitat (Haack and Slansky 1987). Woodborers spend most of their lives developing within host material, then emerge to locate hosts, mate, and reproduce. Eggs are laid on or within specific plant tissues on which early instar larvae establish and feed. Some species may feed sequentially on different tissues in later instars as development progresses (Donley and Acciavatti 1980; Hu et al. 2009).

Generically, woodborers are often referred to by the plant tissues or tree portions on which they feed, such as phloem, sapwood, heartwood, root and bole borers, or twig girdlers. In regard to nutrition, phloeophagous species gain all their nutrition from the phloem/cambium layer, but some may also enter the sapwood for further feeding and/or pupation (e.g. *Monochamus* Guérin spp., *Anoplophora* Hope spp.). Xylophagous species generally gain most of their nutrition from sapwood and/or heartwood and are found deeper inside trees. However, some of these species may briefly feed in the phloem. Aside from the outer bark that is of limited nutritional value, nutritional quality diminishes from the bark of trees inwards to the heartwood (Haack and Slansky 1987). Outside of the phloem/cambium layer, tissues are dominated by cellulose, hemicellulose, and lignin, all compounds that are more difficult to digest and require specialized enzymes to aid in acquisition (Stokland 2012). Woodborers developing within these tissues may take longer to develop (Haack and Slansky 1987). In addition to vertical and horizontal feeding on trees, there is a temporal aspect to food resources where woodborers are often associated with specific stages of tree death or wood decay (Howden and Vogt 1951; Saint-Germain et al. 2007; Ulyshen and Hanula 2010; Ferro et al. 2012).

Both spatial and temporal partitioning occurs with woodborers that utilize the same habitat. For example, succession and resource partitioning, similar to what has been observed in conifer inhabiting bark beetles (Paine et al. 1981; Ayres et al. 2001) likely occurs among woodborers in dying or recently dead conifers. On available stressed or dying trees, or fresh stumps and windfall, woodborer genera such as *Tetropium* Kirby and *Asemum* Eschscholtz may colonize lower bole positions (Lowell et al. 1992), while genera such as *Monochamus*, *Sirex*, and *Xylotrechus* Chevrolat colonize mid- and upper-bole positions. Other genera of buprestids and cerambycids also colonize the crowns. Horizontal partitioning can occur simultaneously to vertical partitioning, with some cerambycids, buprestids, and siricids in the sapwood, while other species of buprestids and cerambycids feed primarily in the phloem and occur only shallowly in the sapwood.

Temporally, phloeophagous woodborers arrive early where some may compete with bark beetles colonizing the same material (Dodds and Stephen 2002). Some of these woodborers, like *Monochamus* spp., utilize kairomones (e.g. host volatiles, bark beetle pheromones) to locate freshly killed or stressed trees quickly (Allison et al. 2001; Miller 2006; Miller et al. 2011). Species that specialize on sapwood or heartwood may arrive later. Their colonization period may be longer as their habitat is less ephemeral and remains suitable longer after tree death. As trees begin to decay, species such as *Orthosoma brunneum* (Forster) that specialize on more decayed material arrive and colonize the trees or logs (Craighead 1950).

12.2.2 Live Tree Inhabitants

With the exception of invasive species, it is rare for woodborers to kill healthy living trees. However, living trees do provide habitat for cerambycid and buprestid species. Tree roots, boles, crowns, and leaves provide habitat for specialized species that can tolerate or avoid host defenses. For example, some *Prionus* F. species colonize roots of living host trees (Duffy 1946; Benham and Farrar 1976). Bole specialists, like the sugar maple borer [*Glycobius speciosus* (Say)] and locust borer [*Megacyllene robiniae* (Forst.)], colonize living trees and cause damage through their feeding activities (MacAloney 1971) (Fig. 12.2a). The buprestid *Coraebus undatus* (F.) colonizes the boles of living cork oak trees and can negatively impact cork harvesting (Jiménez et al. 2012). In some cases, these trees may be slow growing or under some other form of stress that allows the establishment of these woodborers (Newton and Allen 1982; O'Leary et al. 2003). Feeding damage by woodborers often causes stem failure, or further degrade of tree health that eventually results in tree mortality (Galford 1984).

Another woodborer guild of live tree inhabitants specializes in colonizing crowns of trees and includes twig girdlers, twig pruners, and leaf-mining species. Twig girdlers such as *Oncideres cingulata* (Say) can damage >40% of twigs (Forcella 1984), and as a result can reduce timber quality and height growth in hickory a   b

Fig. 12.2 Two examples of woodborer damage to living trees: **a** Damage caused to a living sugar maple (*Acer saccharum*) by the sugar maple borer, *Glycobius speciosus* in central New York, USA. **b** Wound periderm surrounding an unsuccessful attempt at colonization by *Agrilus planipennis* on stem of a *Fraxinus mandshurica* tree in Jilin province, China. The typical sinusoidal larval gallery is apparent but lack of an exit hole indicates the larva did not complete development. Photo credits: **a** Kevin Dodds; **b** Jon Sweeney

(Kennedy et al. 1961). Even though trees attacked by twig girdlers may appear healthy, there is some evidence that these beetles are attracted to stressed trees (Ansley et al. 1990). Twig pruners, such as *Anelaphus villosus* (F.) that colonizes various hardwoods, have larval stages that feed within branches, effectively killing those sections of trees. Similarly, some buprestids, such as *Agrilus arcuatus* Say in hickories, colonize and kill branches of healthy trees (Brooks 1926). Some species of buprestids are leafminers and do not bore in wood at all (Weiss and Nicolay 1919; Bellamy 2002; Queiroz 2002).

While few woodborers use living trees for larval development, many species use living plant material as an adult food resource. Adult feeding in woodborers is common in cerambycids and buprestids (Bright 1987; Hanks 1999; Bellamy 2002; Haack 2017) but does not occur in siricids. Many buprestids and cerambycids feed as adults and need a period of maturation feeding before mating and oviposition can occur (Linsley 1961; Hanks 1999; Poland and McCullough 2006; Lopez and Hoddle 2014). The primary source of nutrition for woodborer adults that do feed is plant material, including phloem tissue, floral resources (nectar, pollen, etc.), thin bark tissue, and leaves and needles (Linsley 1959; Hanks 1999).

12.2.3   Generic Life Cycle

The following is a generalized life cycle and given the size of the guild, it is not surprising that exceptions exist. The woodborers covered in this chapter are all holometabolous insects. Sexual reproduction is typical for buprestids and cerambycids, while siricids are parthenogenetic and can lay viable eggs (males) without mating. Fertilized eggs are necessary to produce female brood. Mating generally occurs on the host plant for most woodborers, with females laying eggs on the bark or under bark scales, in the phloem/cambium region or directly into the sapwood. In most species, males are not present when the female oviposits. However, post-copulation mate guarding does occur in some species (Hughes 1979; Hanks et al. 1996a; Wang and Zeng 2004; GodÍnez-Aguilar et al. 2009). Developing larvae of all woodborers either feed directly on plant tissue, or on associated fungi that females inoculate into trees (Madden 1981; but see Thompson et al. 2013), or on plant material that has been partially digested and broken down by associated organisms (Adams et al. 2011; Thompson et al. 2014). Larvae may go through as few as 3 or as many as 15 larval instars before they pupate; the number of instars varies both intra- and interspecifically and can be affected by temperature, photoperiod, and food quantity and quality (Esperk et al. 2007). Most species developing in the phloem tissue pupate in this region (Ness 1920) or go into the sapwood to pupate (Webb 1910) but some in the Lepturinae subfamily leave the larval host and pupate in the soil (Iwata et al. 2004). Most sapwood colonizers pupate within the same region where larval development occurs. After successful pupation, newly formed adults chew through the sapwood and/or bark to emerge and disperse from host trees.

12.2.4   Importance of Symbionts

There has been a longstanding understanding that symbionts are important in the nutrition of wood-feeding insects (Graham 1967) and molecular techniques are illuminating the diversity and function of these relationships (Grünwald et al. 2010). Common symbionts of cerambycids include bacteria, fungi, and yeasts (Douglas 1989; Schloss et al. 2006; Grünwald et al. 2010; Calderon and Berkov 2012). These symbionts can aid insects in several ways, but a primary role is the conversion of difficult-to-digest plant material (lignin, cellulose, hemicellulose) into useable nutrients (Delalibera et al. 2005). Some buprestids can digest cellulose (Martin 1991), but less is known about symbionts in this family. Siricids have an obligate symbiotic relationship with associated white-rot fungi (Gilbertson 1984). Cerambycids and siricids also ingest fungal enzymes that help break down wood (Kukor and Martin 1983, 1986; Kukor et al. 1988).

12.3 Population Regulation

Woodborer populations are affected by many abiotic and biotic factors and their interactions. Temperature, rainfall, and other weather variables affect woodborers directly (e.g. development rate, overwintering survival, foraging activity) as well as indirectly through the host plant (e.g. trees stressed by drought, flooding, wind storms or disturbances are often more susceptible to woodborer colonization) (Juutinen 1955; Hanks et al. 1999) and their impact on symbionts. The relative impact of these factors on woodborer populations varies among species according to their life histories, and within species, both temporally and spatially. We provide some examples of how climate, fire, and other disturbances affect the distribution and abundance of woodborers. We then discuss the influence of biotic factors on woodborer populations, including bottom-up effects like host tree availability and host defenses, intra- and interspecific competition, and top-down effects like parasitoids, predators, and pathogens.

12.3.1 Abiotic Factors

12.3.1.1 Climate

Each species has optimum temperatures for development and activity as well as minimum and maximum lethal temperatures and these play a large role in determining its geographic range. These temperature optima and limits may vary depending on the life stage and season, especially in temperate climates (Wellington 1954). There are also minimum and maximum threshold temperatures for development of each life stage and a minimum number of heat units (e.g. degree-hours or degree-days = accumulated time between the minimum and maximum threshold temperatures) required to complete development. For example, emerald ash borer larvae need at least 150 frost-free days for feeding (Wei et al. 2007) and have a 2-year life cycle in the most northern province of Heilongjiang in China (Yu 1992), a 1-year life cycle in the more southern Liaoning Province (Zhao et al. 2004) and a 1–2 year life cycle at intermediate latitudes in Jilin province (Wei et al. 2007) and the USA (Tluczek et al. 2011).

Rate of egg and larval development (Schimitschek 1929) as well as adult woodborer activity (Sánchez and Keena 2013) normally increases with temperature above the minimum threshold until temperatures exceed the optimum, beyond which development rate and survival are reduced (Keena and Moore 2010). Temperatures experienced by woodborer larvae in the microclimate under the bark of host trees often differ from ambient air temperatures, and this can affect predictions of overwintering mortality and development rates (Bolstad et al. 1997). For example, the minimum daily temperatures measured under the bark of ash trees were significantly warmer than those measured in the air (Vermunt et al. 2012). Although temperature is a dominant factor affecting woodborer development rate, host condition also affects development rate, i.e. healthy vs. stressed or moribund, as discussed in Sect. 12.2.1.

Upper lethal temperature thresholds vary with species, life stage, and duration of exposure. For example, brown spruce longhorn beetle, *Tetropium fuscum* F., adults died after 30 min exposure to 40 °C and 15 min exposure to 45 °C, whereas mortality of pre-pupal larvae required 30 min exposure to 50 °C or 15 min exposure to 55 °C (Mushrow et al. 2004). Larvae of the emerald ash borer, on the other hand, have survived 30 min exposures to 60 °C (Myers et al. 2009). However, few life stages of woodboring species appear to survive exposure to temperatures >55 °C for 30 min (Pawson et al. 2019) and thus, heat treatment is a common phytosanitary treatment for solid wood packaging used for international shipping of goods. The International Standard for Phytosanitary Measures 15 (ISPM 15) requires that wood packaging be either fumigated or heated to 55 °C for 30 min to reduce the risk that it contains live woodborers (Humble 2010).

In temperate regions, overwintering success is a critical factor affecting the potential geographic range of woodborer populations. Cold hardiness is the capacity of insects to survive exposure to cold temperatures and it varies with species, developmental stage, season, intensity, frequency and duration of exposure, and nutritional status (Lee 1989; Marshall and Sinclair 2015). Some insects avoid freezing and enhance their cold hardiness by increasing the concentration of cryoprotectants (e.g. glycerol, glycogen) in the hemolymph (Danks 2000). The supercooling point (SCP) is the temperature at which ice crystals form in the hemolymph and is a useful index of cold hardiness. In general, the lower the SCP, the greater the cold hardiness. The SCP may vary significantly among species, among different geographic populations within species, and among individuals within populations (Feng et al. 2014). Cold hardiness also varies with time of year, e.g. the SCP of Japanese sawyer beetle larvae (*Monochamus alternatus* Hope) ranged from −6 °C in the summer to −15 °C in the winter (Ma et al. 2006). If minimum winter temperatures increase because of climate change, then distributions of woodborer populations may shift northwards, similar to what has been documented in bark beetles (Lesk et al. 2017). In addition to affecting development and survival of immature life stages, temperature affects adult activity and flight in wood boring beetles, e.g. in a mark-release-recapture study of the Eucalyptus longhorned beetle, *Phoracantha semipunctata* (F.), Hanks et al. (1998) concluded that adult dispersal flights declined sharply as air temperature dropped below about 22 °C.

12.3.1.2 Natural Disturbances

Natural disturbances can significantly increase populations of some woodborer species by greatly increasing the volume of weakened or freshly-felled host trees suitable for colonization (Gandhi et al. 2007). Haack et al. (2017) list many cerambycid genera whose populations increase following disturbances like drought, ice and windstorms, and fire, due to increased availability of stressed host trees. Infestations of *Tetropium* spp. and *Monochamus* spp. increased in spruce forests weakened by windstorms in eastern North America, and their damage far exceeded that of the spruce beetle, *Dendroctonus rufipennis* (Kirby), which often erupts following severe wind events (Gardiner 1975). Both drought and flooding can increase tree moisture stress and susceptibility to woodborer colonization (Craighead 1937, 1950; Mattson and Haack 1987). Larval survival and damage by the locust borer, *M. robiniae*, increased during drought conditions (Craighead 1937). Drought is considered to be an important factor associated with the unprecedented outbreak of red oak borer, *Enaphalodes rufulus* (Haldeman), in red oak forests in Arkansas from 1999 to 2003 (Stephen et al. 2001; Haavik and Stephen 2010; Haavik et al. 2012b).

Fire can significantly affect woodborer populations by changing the distribution and abundance of suitable host trees and can directly suppress woodborer populations by destroying brood in infested trees. Felling and burning of infested trees in winter is sometimes used in sanitation control of satellite infestations of invasive woodborers, e.g. the brown spruce longhorn beetle in Nova Scotia, Canada (Fig. 12.3). Depending on the severity and extent of a forest fire, and the species of woodborer, fire can have positive and negative effects on host availability. For example, conversion of a mature forest to an early successional stage by a severe fire will reduce host availability for many years for woodborer species that favor large diameter, mature trees or other structural components associated with these forests. On the other hand, many species of woodborers prefer to colonize trees weakened or freshly killed by fire, so fires may greatly increase host availability and increase populations of these woodborers (Costello et al. 2013). Females of the longhorn beetle *Arhopalus ferus* (Mulsant) prefer to lay eggs on trees that have been damaged by fire (Hosking and Bain 1977) as do those of the buprestid, *Melanophila acuminata* (DeGeer) (Linsley 1943). The latter species has infrared-sensitive pit organs on the underside of their metathorax (Evans 1964, 1966) and it has been suggested that these are used to detect infrared radiation from forest fires as far away as 50 km (Linsley 1943) (Fig. 12.4).

12.3.2   Biotic Factors

12.3.2.1   Host Availability

Woodborer populations and their distribution on the landscape are significantly affected by the availability of suitable hosts, i.e. those in which broods can be success-fully produced (Haavik et al. 2016), and this is affected by abiotic factors (discussed above) as well as biotic factors such as inter- and intraspecific competition (see Sect. 12.3.2.3). Host availability especially affects woodborers that specialize on one or few host species or genera and/or ephemeral host conditions. For example, *Tetropium* spp. and *Monochamus* spp. typically colonize stressed, dying or recently dead trees, and their populations have increased in conifer stands weakened by defoliator outbreaks (Basham and Belyea 1960; Haack 2017).

Some woodborer species are polyphagous while others breed in a single plant genus or species. For example, the linden borer, *Saperda vestita* Say, breeds only

Fig. 12.3 Sanitation burn of red spruce trees and stumps suspected of infestation by the brown spruce longhorn beetle, *Tetropium fuscum*, at a satellite infestation near Glenholme, Nova Scotia. Photo credit: Wayne MacKinnon, Natural Resources Canada, Canadian Forest Service

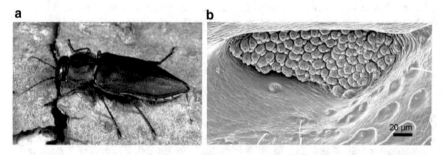

Fig. 12.4 **a** Adult of the "fire-loving" jewel beetle *Melanophila acuminata*; **b** Scanning electron microscope image of one of two infrared (IR) pit organs located between the base of the middle legs on the underside of the beetle's thorax. Each IR pit organ has about 70 hemi-spherical IR sensilla. Reproduced with permission from Schmitz et al. (2009)

in dead and dying linden trees, *Tilia* L. spp. (Yanega 1996), whereas *Neoclytus acuminatus* (F.) breeds in at least 26 genera of broadleaf trees (Haack 2017). Larval feeding by most of the >3000 described species of jewel beetles in the genus *Agrilus* is restricted to a single genus or family of host plants, but there are several exceptions, e.g. *Agrilus viridis* (L.) will colonize many genera such as *Betula* L., *Salix* L., and *Fagus* L. (Jendek and Poláková 2014).

Typically, species that feed in healthy trees tend to specialize on one or few host genera whereas those that feed in dead hosts tend to be polyphagous (Hanks 1999). But there are exceptions like *Anoplophora chinensis* (Forster) whose larvae can complete development in live healthy plants from at least 13 different genera (Sjöman et al. 2014). Even for polyphagous woodborers, there are differences among host species in terms of preference (by ovipositing females) and performance (survival and reproduction of offspring). For example, *A. glabripennis* has been recorded from 24 tree genera (Sjöman et al. 2014) but extensive surveys of infestations in Toronto, Canada (Turgeon et al. 2016) and Chicago, USA (Haack et al. 2006) found that *Acer* L. and *Ulmus* L. were clearly preferred to other tree species. Similarly, *A. glabripennis* has been recorded in seven genera in Northern Italy but 98% of infested trees belonged to only four genera (*Acer, Ulmus, Salix, Betula*) and both oviposition and larval survival was greatest on *Acer* (Faccoli and Favaro 2016). There are also differences in host preference or performance within genera. For example, oviposition and reproductive success of *A. glabripennis* were greater on *Acer rubrum* L. than on *A. platanoides* L. or *A. saccharum* Marsh. (Dodds et al. 2014b), emerald ash borer females lay significantly more eggs on highly susceptible North American ash species than on the more resistant Manchurian ash (*Fraxinus mandshurica* Rupr.) (Rigsby et al. 2014), and reproductive potential of *Monochamus galloprovincialis* Olivier was greater on *Pinus sylvestris* L. than on *Pinus nigra* Arnold (Akbulut 2009).

Host suitability also varies within tree species according to variables such as tree vigor, diameter, and bark thickness. Host condition (e.g. healthy *versus* stressed live trees, recently dead *versus* partially decomposed) affects the preference and/or performance of many wood boring species (Haack 2017). For example, colonization success of *P. semipunctata* larvae in *Eucalyptus* was significantly greater in fresh logs and moisture-stressed trees than in healthy trees (Hanks et al. 1991). As trees die and advance through stages of decay, there are successional changes in the woodborer community following changes in host condition and suitability for different species (Haack 2017).

Variability in host quality combined with the inability of woodborer larvae to move from the brood host selected by females may be responsible for the large intraspecific variation in adult body size (Andersen 1983). According to the preference-performance hypothesis (Jaenike 1978), females should preferentially oviposit in hosts that optimize offspring fitness. Results of some woodborer studies have supported this hypothesis and others have not. Survival and development rate of brown spruce longhorn beetles were greater in stressed than in healthy spruce (Flaherty et al. 2013a) and females landed 10 times more frequently and laid 3 times as many eggs on stressed trees than on healthy trees (Flaherty et al. 2013b). However, Hanks et al. (1993) found that survival of eucalyptus longhorn beetles in field trials was actually lower in preferred hosts, due to high larval densities and intense intraspecific competition; brood survival was greater in the preferred hosts only when larval densities were kept artificially low in laboratory studies.

12.3.2.2 Host Defenses

Tree defenses may be constitutive (always present) or induced (e.g. by herbivore feeding or fungal infection) and both types can significantly reduce survival and colonization success of herbivores, including woodborers (Raffa 1991; Phillips and Croteau 1999). Constitutive and induced resins (complex mixtures of phenolics and terpenoids) may prevent establishment of early instar larvae physically by drowning them or chemically by reducing food digestibility. Drowning of early instar larvae in host oleoresin is a major mortality factor in *T. fuscum* (Juutinen 1955), *Semanotus japonicas* Lacordaire (Shibata 1987, 2000; Kato 2005), and other woodborers that attack live but weakened hosts.

Trees may also increase toxin concentrations at the site of feeding and surround larvae with tougher, less digestible wound periderm tissue (Lieutier et al. 1991). Establishment and survival of early instar buprestid larvae in healthy trees is usually low due to callus formation (Evans et al. 2004; Chakraborty et al. 2014) (Fig. 12.2b). When trees are stressed, these defenses are reduced and larval establishment, colonization success, and woodborer populations increase. For example, incipient root rot in *Eucalyptus* was correlated with attack by the bullseye borer, *Phoracantha acanthocera* (Macleay) (Farr et al. 2000), extensive areas of *P. sylvestris* weakened by root rots were infested and killed by *Phaenops cyanea* (F.) in Germany in the late 1960s (Evans et al. 2004), and oaks undergoing temporary periods of stress from defoliation may be colonized and killed by *Agrilus bigutattus* (F.) in Europe (Moraal and Hilszczanski 2000) and *Agrilus bilineatus* (Weber) in North America (Dunbar and Stephens 1975).

Tree defenses are also less effective at preventing woodborer colonization of naïve hosts, i.e. tree species that have not coevolved with a woodborer species introduced to a new range. A good example of this phenomenon is the devastating mortality of North American ash, *Fraxinus* L. spp., caused by the exotic invasive emerald ash borer compared to the relatively benign effect of this insect on *Fraxinus* spp. in its native range (Poland and McCullough 2006; Herms and McCullough 2014). Similarly, "evolutionary naïve" Eurasian species of birch are far more susceptible to colonization and mortality by the Nearctic bronze birch borer, *Agrilus anxius* Gory, than are North American species of birch (Muilenburg and Herms 2012).

Development rate and survival of woodborer larvae that normally attack weakened hosts (e.g. brown spruce longhorn beetle, emerald ash borer) is lower in healthy trees than in stressed trees, likely due to differences in defensive compounds or host nutrients (Flaherty et al. 2011, 2013a; Tluczek et al. 2011). Growth rate of *Hylotrupes bajulus* L. larvae was negatively correlated with increases in secondary carbon-based compounds in *P. sylvestris* (Heijari et al. 2008). Low host nutritional quality and low moisture content can also prolong the development time of cerambycids and buprestids, with several cases where adults emerged from finished wood products up to 40 years after the presumed oviposition (Duffy 1953; Haack 2017).

Resistance of *Eucalyptus* L'Hér. to colonization by *P. semipunctata* is related to bark moisture content (Hanks et al. 1991) and resistance of *Populus tomentosa* Carr. to colonization by Asian longhorned beetle is related to bark glycoside and phenolic acid content (Wang et al. 1995). The maintenance of healthy, vigorous trees is the best defense against attack by most species of cerambycids and buprestids (Evans et al. 2004). In addition to "bottom-up" factors like host availability and host defenses, woodborer populations are also regulated by "top-down" factors, i.e. natural enemies like parasitoids and predators, and these are discussed in Sect. 12.3.2.4.

12.3.2.3 Competition

Woodborers must compete for limited food and space with conspecifics as well as other species of woodborers and other insects and microorganisms that exploit the same host species and tissues. For example, 27 different species of longhorn beetles, plus a few species of buprestids, curculionids and other beetles were recorded co-inhabiting branches and small saplings of *Leucaena pulverulenta* (Schlect.) Benth. that had been girdled by the twig-girdler, *Oncideres pustulata* LeConte (Hovore and Penrose 1982).

Woodborers may be subject to indirect or exploitative competition, in which larvae that establish later have less food or space for development than earlier colonists (Ikeda 1979), or direct competition, i.e. cannibalism or intra-guild predation (Rose 1957; Anbutsu and Togashi 1997b; Dodds et al. 2001; Ware and Stephen 2006), or both (Powell 1982; Shibata 1987). Lower survival of brown spruce longhorned beetle in cut logs than girdled trees was partially attributed to interspecific competition with other species of phloem-feeding insects which were more numerous in cut logs than girdled trees (Flaherty et al. 2011). The impact of cannibalism on woodborer survival increases with larval densities (Richardson et al. 2010) and later colonists (i.e. smaller larvae) are usually the victims (Anbutsu and Togashi 1997b). Intraspecific competition resulting from overcrowding can be a major mortality factor of *P. semipunctata* (Powell 1982; Way et al. 1992; Hanks et al. 2005) and *Monochamus* spp. (Shibata 1987; Dodds et al. 2001; Akbulut et al. 2008). Larvae of the red oak borer will sometimes cannibalize one another (Ware and Stephen 2006) but subsequent life table studies indicated that intraspecific competition was not an important mortality factor (Haavik et al. 2012a).

Another form of intraspecific competition is when polygamous male cerambycid species compete with other conspecific males for access to females for mating, e.g. larger males of *Glenea cantor* (F.) have greater mating success than smaller males (Lu et al. 2013). Mate guarding, in which the male remains in copula or stays close to the female after copulation to prevent copulation with other males, occurs in several species of cerambycids (Fig. 12.5).

Fig. 12.5 Pair of *Moechtypa diphysis* adults on the stem of *Quercus mandshurica.* Note the pair are not in copula but the male remained mounted on the female for a prolonged period, possibly as a form of mate guarding to prevent her from mating with other males. Photo credit: Jon Sweeney

12.3.2.4   Natural Enemies

Parasitoids

Woodborers are parasitized by many species, mainly wasps (Hymenoptera), particularly the families Ichneumonidae and Braconidae, but also flies (Diptera: Tachinidae) and beetles (Coleoptera: Bothrideridae). Most woodborer parasitoids attack host larvae but some species exploit eggs and pupae (Yu et al. 2016). Some species are ectoparasitoids that feed externally on hosts while others are endoparasitoids that feed internally. Parasitoids can also be classified as idiobionts that kill or paralyze their host immediately following oviposition or koinobionts that allow their host to continue developing and consume it at a later stage (Askew and Shaw 1986). Koinobionts tend to have a narrower host range than idiobionts (Spradbery 1968) possibly because they have had to evolve defenses against host immune systems (Gauld 1988).

Table 12.1 lists some parasitoid genera recorded from cerambycids, buprestids and siricids, along with some features of their biology. Due to the cryptic nature of most woodborers, obtaining accurate host records of parasitoid species is not straightforward, but the associations of parasitoid genera with woodborer families in Table 12.1 may be considered accurate. In simple collections of parasitoids and woodborers that emerge from the same log or tree, it is generally not possible to know from which woodborer species the parasitoids emerged when more than one potential host woodborer species emerges. Molecular techniques have been used to a limited degree to associate emerging parasitoids from trees with more than one brood species (Foelker et al. 2016). Unequivocal woodborer species-parasitoid associations have generally been determined either by manipulative experiments that expose a single woodborer species to parasitoids or by isolating individual woodborer larvae from infested trees and then recording parasitoids that emerge.

Apart from a few pest species like *P. semipunctata*, emerald ash borer, *S. noctilio*, Asian longhorned beetle, and *M. galloprovincialis*, natural enemies of woodborers have not been the subject of much research, and few studies have documented their

Table 12.1 Partial list of parasitoid genera recorded from woodborers in coniferous (C) and broadleaf (B) trees and some of their reported characteristics: EC = ectoparasitoids, EN = endoparasitoids, I = idiobionts, K = koinobionts, E = egg parasitoids, G = generalists, S = specialists (known from only one or few host genera)

Order/family/subfamily/genus	Cerambycidae				Buprestidae			Siricidae			
Hymenoptera											
Ichneumonidae											
Campopleginae	EN		K	S							
Rhimphoctona Foerster	EN	C	K	S							
Cryptinae											
Cryptus Fabricius					EC	B					
Echthrus Gravenhorst	EC	C	I								
Pimplinae											
Dolichomitus Smith	EN	BC	I	G	EN	B	I				
Poemeniinae											
Neoxorides Clement	EC	BC	I		EC	BC	I				
Pseudorhyssa Merill[a]								EC	C	I	S
Rhyssinae											
Megarhyssa Ashmead								EC	BC	I	S
Rhyssa Gravenhorst								EC	BC	I	S
Xoridinae											
Ischnocerus Agassiz	EC	BC	I		EC	B	I				

(continued)

Table 12.1 (continued)

Order/family/subfamily/genus	Cerambycidae			Buprestidae					Siricidae		
Odontocolon Cushman	EC	C	I								
Xorides Latreille	EC	C	I	EC	BC	I					
Aulacidae											
Aulacinae											
Pristaulacus Keiffer	EN	C		EN							
Stephanidae											
Stephaninae											
Foenatopus Smith					B						
Schletteriinae											
Schlettererius Ashmead									EC	C	I
Braconidae											
Doryctinae											
Callihormius Ashmead	EC	B	I	EC	B						
Doryctes Haliday	EC	BC	I	EC	B						
Ecphylus Foerster	EC		I	EC	B						
Heterospilus Haliday	EC	BC	I	EC	B	I		G			
Jarra Marsh & Austin	EC	B	I								
Leluthia Cameron	EC		I	EC	B			G			

(continued)

Table 12.1 (continued)

Order/family/subfamily/genus	Cerambycidae				Buprestidae				Siricidae
Ontsira Cameron	EC	B	I						
Pareucorystes Tobias	EC		I		EC	B			
Polystenus Foerster	EC				EC	B	I		
Rhoptrocentrus Marshall	EC	B	I	G					
Spathius Nees	EC	B	I		EC	BC	I		
Syngaster Brullé	EC	B	I						
Braconinae									
Atanycolus Foerster	EC	BC	I		EC	BC	I		
Coeloides Wesmael	EC	BC	I		EC	C	I		
Cyanopterus Haliday	EC	C	I						
Digonogastra Viereck	EC	C	I						
Iphiaulax Foerster	EC	BC	I	G	EC	B	I	G	
Megalommum Szepligeti	EC	B	I						
Monogonogastra Viereck					EC	B	I		
Helconinae									
Helcon Nees	EN	BC	K						
Helconidea Viereck	EN	C	K						
Cenocoelius Westwood	EN	B	K		EN	B	K		

(continued)

Table 12.1 (continued)

Order/family/subfamily/genus	Cerambycidae				Buprestidae				Siricidae
	EN	BC	K	S	EN	B	K	S	
Wroughtonia Cameron	EN			S	EN	B			
Eulophidae									
Entedoninae									
Entedon Dalman					EN	B	K		
Pediobius Walker					EN	B			
Euderinae									
Euderus Haliday	EN	B			EN	B			
Tetrastichinae									
Aprostocetus Westwood	EN	B							
Baryscapus Foerster					EN	B			
Quadrastichus Girault					EN	B			
Tetrastichus Haliday					EN	B	K	S	
Encyrtidae									
Encyrtinae									
Avetianella Trjapitzin					E	B	I		
Ooencrytus Ashmead					E	B	I		
Oobius Trjapitzin		B	I	S	E	B	I	S	
Orianos Noyes					E	B	I		

(continued)

Table 12.1 (continued)

Order/family/subfamily/genus	Cerambycidae	Buprestidae				Siricidae
Eupelmidae						
Calosotinae						
Balcha Walker		EC	B		I	
Calosota Curtis		EC	B			
Eupelminae						
Eupelmus Dalman		E	B			
Neanastatinae						
Metapelma Westwood			B			
Pteromalidae						
Clyonyminae						
Cleonymus Latreille	B					
Oodera Westwood			B			
Pteromalinae						
Trigonoderus Westwood		EC	B			
Zatropis Crawford			B			
Chalcididae						
Chalcidinae						
Phasgonophora Westwood			B	K		S

(continued)

Table 12.1 (continued)

Order/family/subfamily/genus	Cerambycidae				Buprestidae				Siricidae			
Ibaliidae												
Ibaliinae												
Ibalia Latreille									EN	BC	K	S
Bethlyidae												
Epyrinae												
Sclerodermus Latreille	EC	B	I	G	EC	B	I	G				
Diptera												
Tachinidae												
Dexiinae												
Billaea Robineau-Desvoidy	EN	BC										
Coleoptera												
Bothrideridae												
Dastarcus Walker	EC	B		G				G				

Note [a] kleptoparasite on *Rhyssa* spp.

Adapted from Kenis and Hilszczanski (2004), Schimitschek (1929), Finlayson (1969), Raske (1973), Wanjala and Khaemba (1987), Duan et al. (2009, 2016), Kenis and Hilszczanski (2004), Naves et al. (2005), Petrice et al. (2009), Li et al. (2010), Flaherty et al. (2011), Van Achterberg and Mehrnejad (2011), Coyle and Gandhi (2012), Schiff et al. (2012), Taylor et al. (2012), Roscoe et al. (2016), Triapitsyn et al. (2015), Golec et al. (2016), Jennings et al. (2017), Yu et al. (2016), and Paine (2017) impact on woodborer population dynamics (Paine 2017). A multi-year life table study of the red oak borer concluded that competition and natural enemies contributed very little to mortality during the crash of the outbreak, and that tree defenses were more likely responsible (Haavik et al. 2012a). Similar results have been reported with the woodwasp *S. noctilio* (Haavik et al. 2015), although factorial exclusion trials suggest that biotic factors (e.g. competitors and their associated fungi, and natural enemies) may also be important in parts of the range (Haavik et al. 2020). However, relatively high parasitism rates have been recorded in some species, e.g. 20–75% mortality of *Tetropium gabrieli* Weise and *T. fuscum* in Europe (Schimitschek 1929; Juutinen 1955) and 22–28% mortality of *S. noctilio* (Long et al. 2009; Zylstra and Mastro 2012), suggesting parasitoids may be important in regulating populations of some woodborer species. Further evidence for this comes from successful biological control programs that are discussed in Sect. 12.7.2.3.

An interesting question is how parasitoids locate cryptic woodborer hosts beneath the bark and wood of a tree. Increased parasitism of bark beetles in stressed trees *vs.* healthy trees suggests that parasitoids may use volatiles emitted from stressed trees as olfactory cues associated with their hosts (Sullivan et al. 1997). Percent parasitism of *Tetropium* spp. (Flaherty et al. 2013a) and *Semanotus japonicus* (Shibata 2000) was greater in stressed trees than in healthy trees. After landing on a tree, some parasitoids use auditory cues to locate their hosts. Ichneumonid wasps in the Cryptini tribe have hammer-like structures on their antennae that they use to echo-locate wood boring larvae and pupae of both cerambycids and buprestids (Laurenne et al. 2009), and the braconid, *Syngaster lepidus* Brullé, uses chordotonal organs to detect the vibrational cues of *P. semipunctata* larvae feeding under the bark (Joyce et al. 2011). Aspects of a woodborer's host tree can affect the foraging success of its parasitoids. For example, the ovipositor of *Tetrastichus planipennisi* Yang cannot penetrate >3.2 mm of bark so its effectiveness against emerald ash borer is restricted to smaller diameter trees (Abell et al. 2012). On the other hand, parasitism of *S. noctilio* by *Ibalia leucospoides ensiger* Norton peaked at bole diameters of 15 cm but was not affected by bark thickness (Eager et al. 2011).

Predators

Woodborers are attacked by a variety of predators (vertebrate and invertebrate), including beetles in the families Cleridae (e.g. *Thanasimus dubius* (Fabricius)), Trogossitidae (e.g. *Trogossita japonica* Reitter), and Elateridae (e.g. *Athous subfuscus* Müller), flies in the families Asilidae (e.g. *Laphria gibbosa* (L.)), Lonchaeidae (e.g. *Lonchae chorea* (F.)), Odiniidae (e.g. *Odinia xanthocera* Collin), Pallopteridae (e.g. *Palloptera usta* (Meighen)), crabronid wasps (e.g. *Cerceris fumipennis* (Say)), lacewings (e.g. *Raphidia xanthostigmus* Schummel), and earwigs (e.g. *Forficula auricularia* L.) (Kenis and Hilszczanski 2004). Ants prey on eggs of *P. semipunctata* (Way et al. 1992) and the red oak borer (Muilenburg et al. 2008). Woodpeckers (Piciformes: Picidae) are common and important predators of woodborers. Cerambycid larvae are the preferred food of woodpeckers (Pechacek and

Fig. 12.6 **a** Three-toed woodpecker, *Picoides tridactylus* is a common predator of woodboring larvae in Europe. **b** Woodpeckers often excavate deeply into trees to find woodboring larvae. Photo credits: **a** Dariusz Graszka-Petrykowski; **b** Kevin Dodds

Kristin 2004) and their availability is critical to the reproductive success of the three-toed woodpecker, *Picoides tridactylus* L. in Europe (Fayt 2003) (Fig. 12.6). Woodpeckers mainly consume mature larvae and pupae and predation rates often increase as larval density increases (McCann and Harman 2003; Lindell et al. 2008; Flaherty et al. 2011) but not always [e.g. woodpecker-caused mortality of *P. semipunctata* decreased with increasing larval density in trap logs (Mendel et al. 1984)]. Woodborer mortality from woodpeckers can be considerable, e.g. woodpeckers have been reported to consume 65% of oak branch borer, *Goes debilis* LeConte larvae (Solomon 1977) and 32–42% of emerald ash borer larvae (Duan et al. 2012).

Pathogens

Woodborers may be infected and killed by various pathogens like nematodes [e.g. *Steinernema carpocapsae* (Weiser)] and fungi [e.g. *Beauveria bassiana* (Bals.-Criv.), Vuill.] (Morales-Rodríguez et al. 2015; Liu et al. 2016). *Beauveria bassiana* caused significant natural mortality of the pine sawyer, *M. galloprovincialis* (Naves et al. 2008). *Beauveria pseudobassiana* Rehner & Humber, isolated from natural populations of the pine sawyer in Spain was highly virulent in lab tests, killing 100% of adults and significantly reducing adult lifespan and number of eggs laid, both via direct contact and by mating with infected beetles, i.e. horizontal transmission (Álvarez-Baz et al. 2015).

Entomopathogenic fungi such as *Beauveria* spp. and *Metarhizium anisopliae* (Metsch.) Sorok. have shown potential for applied control of woodborers. For example, direct application of aqueous suspensions of *B. bassiana* conidia (or mitospores) to ash trees infested with emerald ash borer significantly reduced larval densities and the number of emerging adults in the next generation (Liu and Bauer 2008). Wrapping non-woven fabric strips impregnated with fungal conidia around host tree trunks was effective at infecting and killing *M. alternatus* Hope (Shimazu and Sato 1995; Shimazu 2004) and reducing longevity and fecundity of *A. glabripennis* (Dubois et al. 2004). Another method that has been tested for woodborer control is auto-dissemination, in which the target species is attracted to a trap baited with pheromone and/or host volatiles where it receives a dose of fungal conidia and is allowed to escape and horizontally transmit the pathogen within the local population (Klein and Lacey 1999; Lyons et al. 2012; Francardi et al. 2013; Sweeney et al. 2013; Álvarez-Baz et al. 2015; Srei et al. 2020).

One of the most interesting and successful examples of microbial control of woodborers is the use of the nematode, *Deladenus siridicola* Bedding for control of *S. noctilio*. The nematode does not kill the woodwasp but infected insects become sterile and the female spreads the nematode from tree to tree when depositing nematodes instead of eggs (Bedding and Akhurst 1974). For a more complete story on the woodwasp and its control by the nematode, see Chapter 17.

12.4 Ecological Roles

The vast majority of woodborers serve important ecological functions while inhabiting dead or stressed materials and provide critical services that benefit forested ecosystems. Important contributions from woodborers include facilitating nutrient cycling (Edmonds and Eglitis 1989; Cobb et al. 2010), influencing forest structure (Feller and McKee 1999), creating habitat (Buse et al. 2008), and providing food for predaceous invertebrates and vertebrates.

12.4.1 Nutrient Cycling

Saproxylic woodborers are an important group of insects that help drive nutrient cycling in forested environments through the breakdown of dead wood (Fig. 12.7). Woodborers are some of the earliest arriving insects at stressed trees and dead wood (Savely 1939; Saint-Germain et al. 2007) and a successive community of these species colonize wood throughout the decay process (Graham 1925; Howden and Vogt 1951; Stokland and Siitonen 2012). Through their feeding and tunneling behavior, woodborers begin the process of fragmentation and nutrient cycling as well as exposing wood to other organisms, such as decay fungi, which are also important decomposers (Harmon et al. 1986; Edmonds and Eglitis 1989; Hadfield and Magelssen 2006; Parker et al. 2006). Woody debris is an important forest structure and can contain large sources of nutrients (Harmon et al. 1986). The return of nutrients to the soil through decomposition of dead wood is a critical ecosystem service and one that is aided by woodborers and other organisms (Edmonds and Eglitis 1989; Ulyshen 2016).

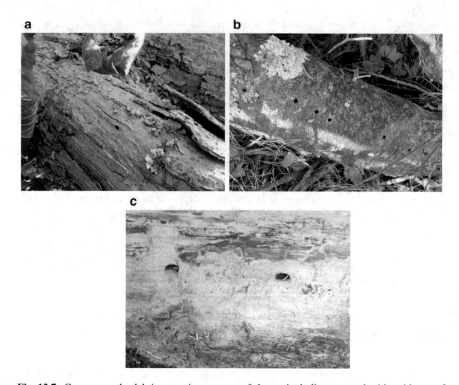

Fig. 12.7 Coarse woody debris at various stages of decay, including **a** wood with evidence of Buprestidae emergence and decay fungi, **b** Siricidae emergence, and **c** *Monochamus* sp. sapwood entrance holes. Photo credit: Kevin Dodds

While the relationship between woodborers and wood decay and nutrient cycling is well known, few studies have attempted to quantify this relationship. *Monochamus scutellatus* (Say) was an important contributor to Douglas fir [*Pseudotsuga menziesii* (Mirbel) Franco] log decay, most likely through providing pathways for decay fungi into larger diameter logs (Edmonds and Eglitis 1989). At small scales, *M. scutellatus* larval activity can influence total carbon and nitrogen in soil around infested logs (Cobb et al. 2010). Cerambycids were also a factor in decay of both deciduous and coniferous standing snags (Angers et al. 2011).

12.4.2   Forest Structure

Forest structural diversity, including standing snags and downed wood of various decay classes, is an important component of natural forests and an important reservoir of organic matter and forest nutrients (Harmon et al. 1986). Insects, such as woodborers and bark beetles, play a critical role in the creation of these structural components either through directly killing trees, or colonizing this material and facilitating decay, thus creating additional habitat for saproxylic (Buse et al. 2008) and other organisms. Woodborers can drive changes in the structure (crown characteristics, bark attachment) and physical properties (wood density, moisture) of coarse woody debris. These changes to coarse woody debris are critical to maintaining biological diversity and help support healthy forest ecosystems (Harmon et al. 1986; Jia-bing et al. 2005).

In addition to the obvious contributions that woodborers make to forest structure through the breakdown and decay of standing and downed wood, they also influence stand structure through impacts on living trees. For example, the cerambycid, *Elaphidion mimeticum* Shaeffer is important for creating small-scale gaps in mangrove forests and subsequently promoting understory regeneration (Feller and McKee 1999). Black locust (*Robinia pseudoacacia* L.), an early successional tree species in North America, can be killed from successive years of *M. robiniae* attacks. Through this mortality, canopy gaps are created that allow more shade-intolerant tree species to become established (Boring and Swank 1984). Red oak borer in combination with other factors, can kill overstory trees that results in changes to residual forest structure (Heitzman et al. 2007; Haavik et al. 2012b). Stand structure in pine stands is also influenced by attack patterns of woodborers. For example, *S. noctilio* preferentially attacks and colonizes smaller suppressed trees (McKimm and Walls 1980; Dodds et al. 2010a). Species such as *Plectodera scalator* F. that colonize lower bole and root positions on trees can structurally weaken stems to a point where they break. Through these actions, *P. scalator* may influence the spatial structure of stands and regeneration.

The behavior of branch girdlers can also have an impact on forest structure. The cerambycid *Oncideres rhodosticta* Bates influences crown architecture of honey mesquite (*Prosopis glandulosa* Torr.) through its stem girdling behavior (Martínez et al. 2009) and the resultant branch-heavy crowns may be a critical factor in increased desertification in parts of the Chihuahuan Desert (Duval and Whitford 2008). Similarly, twig girdlers influenced understory crown architecture of *Dicorynia guianensis* Amshoff. (Caraglio et al. 2001). *Oncideres humeralis* Thomson influenced forest composition and structure in a Brazilian forest through species-specific tree attacks that likely allowed other tree species to respond to increased resources (Romero et al. 2005). Through their actions, these species can also influence the invertebrate community in these stands (Calderón-Cortés et al. 2011) and provide habitat for many other species (Lemes et al. 2015).

12.4.3 Ecosystem Services

Woodborer larvae represent a relatively large source of nutrition for animals foraging in wood. Woodpeckers are commonly seen foraging on dead standing or downed trees, and woodborers are a common prey item taken (Hanula et al. 1995; Murphy and Lehnhausen 1998; McCann and Harman 2003; Nappi et al. 2015). Crows have also been reported to use twigs to extract larvae from wood (Hunt 2000). Adult ceram-bycids are reported as prey for bats (Medellín 1988), owls (Haw et al. 2001), pitcher plants (Cresswell 1991), lizards (Vitt and Cooper 1986) and passerines (Tryjanowski et al. 2003). Predaceous insects attack all stages of woodborers and are important factors in population regulation as previously discussed in section "Predators".

Pollination is an important ecosystem service carried out by a diverse group of insects that includes woodborers. Cerambycids, especially from the subfamily Lepturinae, feed on pollen as adults (Linsley and Chemsak 1972) and are frequently found with pollen on their integument (Willemstein 1987). Buprestids have also been commonly associated with plants as pollinators (Williams and Williams 1983). Siricids, however, with their lack of adult feeding, are not known to serve a role in pollination.

Because of their impacts on plants, several woodborers have been introduced into new environments as biological control agents against unwanted invasive plants. Several species have been introduced targeting the invasive plant, *Lantana camara* (L.), including the cerambycids *Plagiohammis spinipennis* (Thomsom) in Hawaii (Broughton 2000) and *Aerenicopsis championi* Bates in Australia (Palmer et al. 2000). Australia has successfully introduced other woodborer species for invasive plant management including the cerambycids *Alcidion cereicola* Fisher targeting *Harrisia* Britton cactus (McFadyen and Fidalgo 1976) and *Megacyllene mellyi* (Chevrolat) for *Baccharis halimifolia* L. management (McFadyen 1983), and the buprestid *Hylaeogena jureceki* Obenberger targeting cats claw creeper (Dhileepan et al. 2013). South Africa has also released *A. cereicola* for *Harrisia* cactus management (Klein 1999) and *H. jureceki* for cats claw creeper (King et al. 2011). North American woodborer releases have included *Oberea erythro-cephala* (Schrank) (Cerambycidae) targeting leafy spurge (Rees et al. 1986), and the buprestids *Sphenoptea jugoslavica* Obenb. and *Agrilus hyperici* (Creutzer) for knapweed (Powell and Myers 1988; Harris and Shorthouse 1996) and St. Johns wort (Campbell and McCaffrey 1991), respectively.

12.4.4   Woodborer Conservation

Some species of woodborers are rarely observed and may be in danger of extirpation due to loss of suitable habitat and hosts. Many species feed in dead and decaying heartwood of ancient "veteran" trees and these trees have become increasingly rare in Europe (Nieto and Alexander 2010). Forestry practices that leave less dead wood in the forest have resulted in declining populations of some wood boring beetles, e.g. *Cerambycx cerdo* L. is listed as "near threatened" in Europe (Evans et al. 2004). Similarly, populations of species that depend on old growth forests or which feed in large diameter wood may decline as the area of old growth forest declines; more than 80% of land in Europe is under some form of direct management (Anonymous 2007). Risk of longhorn beetle extinction increases with larval host plant specialization and length of generation time (Jeppsson and Forslund 2014). The hoptree borer, *Prays atomocella* (Dyar) (Lepidoptera: Praydidae) is listed as an endangered species in Canada because its sole larval host is the "common" hoptree which is limited to a very narrow range in southwestern Ontario (Harris 2018; Anonymous 2020).

The International Union for Conservation of Nature (IUCN) assessed 431 species of saproxylic insects in Europe (of which 153 species were cerambycids and 1 species was a buprestid) and designated 2, 27 and 17 as critically endangered, endangered or vulnerable, respectively (Nieto and Alexander 2010). More than half of these species are endemic to Europe and found nowhere else in the world. The drivers of this decline are habitat loss due to forest harvesting and a general decrease in old growth "veteran trees" on the landscape (Nieto and Alexander 2010). Other threats include agricultural and urban expansion, forest fires and climate change. It is more than a little ironic that invasive woodborers may threaten populations of native woodborers and other arthropods. The community of arthropods on ash trees in the state of Maryland, USA, included 13 orders, 60 families and 41 genera (Jennings et al. 2017) and the decimation of North American *Fraxinus* species by the invasive emerald ash borer may threaten woodborers and other herbivores tightly associated with ash (Herms and McCullough 2014).

The IUCN identifies species at risk of extinction (so called Red Lists) and promotes their conservation by increasing public awareness and conserving wildlife habitat (Rodrigues et al. 2006). When it comes to woodborers and other insects, Red Lists often reflect a lack of knowledge of species range and population trends rather than actual extinction risk (Cardoso et al. 2012). According to the European Red List, 14% of saproxylic beetles have declining populations but the trend is unknown for more than half of the species (57%) on the list (Nieto and Alexander 2010).

In an effort to conserve species that rely on old growth forests, some countries have forest management regulations in place that mandate conservation of coarse woody debris, snags, and dead wood in the forest. Many countries are signatories to the 1979 Bern Convention on the Conservation of European Wildlife and Natural habitats and the 1992 Convention on Biological Diversity, which provide official impetus for conserving wildlife biodiversity, including woodboring insects. Each member state is required to identify threatened species and their respective habitats, and then develop management plans to protect these natural areas. In Europe, this makes up the Natura 2000 network, a coordinated network of protected areas home to rare and threatened species that makes up 18% of the European Union's land base and 10% of marine territory (European Commission, Directorate-General for Environment and Sundseth 2021). While the goal of conserving biodiversity is valid, the effectiveness of the Natura 2000 network for conserving saproxylic beetles has been questioned (D'Amen et al. 2013). In Canada, the Committee on the Status of Endangered Wildlife in Canada (COSEWIC) meets twice a year to review and assess the status of wildlife species, including arthropods, and submits an annual report to the federal Minister of Environment and Climate Change. Species listed as extirpated, endangered, threatened or of special concern are considered for legal protection and management under the *Species at Risk Act* (SARA). Only two wood boring insects are currently listed as endangered in Canada: the hoptree borer, *Prays atomocella* (Dyar) (Lepidoptera: Praydidae) and the Aweme borer, *Papaipema aweme* (Lyman)

(Lepidoptera: Noctuidae) (Anonymous 2020). Ultimately, woodborer species rich-ness depends on the quantity and diversity of living and dead wood in the forest, forest size and fragmentation, and management practices.

12.5 Chemical Ecology

As discussed above, adult woodborers usually live a few days to a few weeks and are host specific both in terms of the species and physiological condition of the host. This specificity can result in a heterogeneous spatial and temporal distribution of suitable hosts across the landscape and variance in larval performance in hosts. For example larval survival, developmental time, and adult size are all affected by host quality in *P. semipunctata* (Hanks et al. 1993, 1995). In addition to this variance in host quality, due to their short life-span, a delay in mate or host location of only a few days can have significant fitness consequences. Cumulatively, these factors generate selection for rapid host and mate location. Not surprisingly, most adult woodborers typically have highly developed sensory systems while immature life stages do not.

The dominant modality that woodborers use to obtain information about their biotic and abiotic environment is olfaction. The advantages of olfaction include: (i) the availability of a large number of "channels" due to the diversity of chemicals woodborers and their host plants can synthesize and that woodborers can perceive. As a result, chemical signals and cues can have high information content and be highly specific; (ii) volatile chemicals can be transmitted over large distances and around obstacles; and (iii) woodborers can perceive and discriminate among chemical cues and signals with high levels of sensitivity and precision. The disadvantages of olfaction include the fact that: (i) they cannot be transmitted quickly over large distances; (ii) the primary direction of transmission is determined by wind direction; and (iii) they require complex behaviors [e.g. optomotor anemotaxis, see Cardé (2016)] to locate the odor source.

Interest in and our knowledge of the chemical ecology of woodboring insects has increased dramatically in the past 20 years. For example, while fewer than 10 attractant pheromones were known for the family Cerambycidae in 2004 (Allison et al. 2004), approximately a decade later pheromones and likely pheromones (i.e. attraction observed in the field but production and release not yet demonstrated) were known for more than 100 species of Cerambycidae (Hanks and Millar 2016). This increase has been driven by the realization that woodborers can have significant economic impacts (particularly in plantation and urban trees), recognition that they are among the most common and damaging exotic insects (Brockerhoff et al. 2006; Haack 2006) and increased awareness of the importance of the ecosystem services that they deliver.

In general, woodboring insects use volatile sex pheromones to mediate mate loca-tion over large distances and low volatility cuticular hydrocarbons for mate recog-nition at close range (Allison et al. 2004; Hanks and Millar 2016; Millar and Hanks

2017; Silk et al. 2019). A large number of studies have reported the identification of pheromones and the demonstration of attraction to known pheromones and their analogues in the Cerambycidae and we refer readers to the review by Millar and Hanks (2017) for an excellent synthesis of this literature. In brief, two general patterns of volatile pheromone use in the Cerambycidae have emerged: (i) male-produced pheromones are released in large quantities, attract both sexes and occur in the subfamilies Cerambycinae, Lamiinae and Spondylidinae; and (ii) female-produced pheromones are released in small quantities, only attract males and occur in the subfamilies Prioninae and Lepturinae. Volatile pheromones are only known for a single species in the Buprestidae and Siricidae, the emerald ash borer (Silk et al. 2011) and *S. noctilio* (Cooperband et al. 2012), respectively. In the emerald ash borer, it is a female-produced pheromone and male response to it is synergized by host volatiles [synergy of the response to pheromone by host volatiles has also been reported in the Cerambycidae (Allison et al. 2012; Millar and Hanks 2017)]. The putative pheromone in *S. noctilio* is male-produced and behavioral activity has been demonstrated in laboratory trials (Cooperband et al. 2012; Guignard et al. 2020) but field trials did not observe activity (Hurley et al. 2015).

Although the active space of these attractant pheromones has not been quantified empirically for any woodborer, mark-release-recapture trials with several species of Cerambycidae suggest they may range from ca. 50 to 500 m (Maki et al. 2011; Torres-Vila et al. 2013, 2015). These estimates are consistent with research in moths which suggests that attraction likely occurs over a distance of a few meters to a maximum of a few hundred meters (Cardé 2016). Often the release of volatile pheromones is sex-specific, occurs from specific habitats (e.g. host material) (Hanks 1999) and is facilitated by "calling" behaviors (Lacey et al. 2007).

For some of these species, females have been observed to deposit nonvolatile compounds while walking that males use to locate females [i.e. trail pheromones (Hoover et al. 2014)]. Alternatively, in some species males form leks and females may be attracted to these by visual (Allison et al. 2021) and olfactory stimuli (Cooperband et al. 2012; but see Hurley et al. 2015). These mechanisms (sex and trail pheromones, leks) bring the sexes into close proximity but in many woodborers mate recognition appears to be mediated by contact pheromones (Allison et al. 2004; Millar and Hanks 2017; Silk et al. 2019). In these species, males do not appear to recognize females until their antennae contact the female cuticle and they detect cuticular hydrocarbons. After contact males often begin a sequence of characteristic behaviors that culminate in copulation (Hanks et al. 1996a).

Due to the heterogeneous distribution of suitable hosts in space and time, wood-boring insects are expected to experience strong selection to rapidly locate available host material. For woodborers, oviposition can generally be considered as two separate and sequential events: host location and host acceptance. Host location is generally thought to occur first and be initiated from a distance (i.e. before landing on the host plant), whereas host acceptance does not occur until the host has been contacted. Both host location and acceptance are mediated, at least in part, by chemicals. Meurer-Grimes and Tavakilian (1997) evaluated the phytochemistry and diversity of Cerambycidae associated with 51 species of Leguminosae. The host plants of cerambycid guilds (species sharing host plants) were taxonomically related and had similar phytochemistry. In contrast guild members were not usually related suggesting that host location and/or acceptance are mediated by phytochemicals.

In support of the hypothesis that host location and acceptance are chemically mediated, numerous studies have identified primary attractants for woodborers including floral, smoke, trunk and leaf volatiles (Allison et al. 2004; Hanks and Millar 2016; Millar and Hanks 2017; Silk et al. 2019). Some woodborers from the family Cerambycidae overlap temporally in host trees with bark beetles and are attracted by bark beetle pheromones (Allison et al. 2001, 2013). In addition to competing with bark beetle larvae for limited host tissues, larvae of these woodborers are also facultative intraguild predators of bark beetle larvae (Dodds et al. 2001; Schoeller et al. 2012). The current paradigm for host selection by phytophagous insects argues that to optimize foraging efficiency all available cues and signals should be used. Although most studies have focused on the role of attractive semiochemicals in host location and acceptance, a few studies have demonstrated that woodborers (Coleoptera: Cerambycidae) respond to repellent non-host volatiles to avoid non-host trees (Aojin and Qing'an 1998; Suckling et al. 2001; Morewood et al. 2003).

Semiochemicals produced by the host plant and conspecifics influence female oviposition behavior. The woodborer *H. bajulus* preferentially oviposits in wood infested with larval conspecifics and several monoterpenoids identified in larval frass appear to stimulate oviposition in females (Evans and Higgs 1975; Higgs and Evans 1978; Fettköther et al. 2000). In other woodborers, the presence of conspecifics reduces oviposition (Wang et al. 1990; Anbutsu and Togashi 1996, 1997a, 2000; Peddle et al. 2002). Treatment of host material with larval frass or extracts of larval frass reduced oviposition by *M. alternatus* (Anbutsu and Togashi 2002), suggesting that semiochemicals in larval frass mediate the effect. Some woodborers deposit a jelly-like substance over their eggs (Anbutsu and Togashi 1996, 1997a, 2000; Peddle et al. 2002) and females palpate the bark surface before oviposition. It has been hypothesized that semiochemicals in the material deposited over the eggs mediates the recognition and avoidance of host material already infested with conspecifics. To date, the role of phytochemicals in the induction of oviposition have only been studied in *M. alternatus*. In this species chemicals in the inner bark of the host *Pinus densiflora* Siebold & Zucc. have been demonstrated to induce oviposition in females (Yamasaki et al. 1989; Islam et al. 1997; Sato et al. 1999a, 1999b).

12.6 Economically Important Species

Most woodborers develop in dead or stressed trees, or downed wood, and provide important ecosystem services that contribute to healthy forest ecosystems. The adults of some species oviposit in stressed trees (e.g. fire, drought, and storm damaged; defoliated) and fewer in apparently healthy trees (Craighead 1950; Keen 1952; Solomon 1995). The associated larval feeding and development can result in mortality in both classes of trees. Although some woodborers are significant pests of woody plants in their native range, exotic species are often among the most damaging species, especially in terms of causing direct tree mortality. For example, in the United States annual costs of tree removal, replacement and treatment due to invasive phloem and woodborers are estimated to be approximately $1.7 billion USD, of which 50% is a result of the emerald ash borer (Aukema et al. 2011).

The significant economic and ecological impacts of woodborers is of concern given the increase in the number of introductions outside of their native ranges in recent years (Haack 2006; Aukema et al. 2011). As mentioned above, woodborer larvae feed cryptically within phloem and xylem tissues and development takes months to years. These traits make woodborers ideally suited for movement outside of their native ranges in wood products, wood packaging material, dunnage and nursery stock. Additionally, many species attack low quality, stressed hosts and this type of wood is often used for wood packaging and dunnage in container shipping. It is therefore not surprising that the increase in introduction of woodborers outside of their native ranges is coincident with increased movement of goods in container shipping (Haack 2006; Aukema et al. 2011).

In addition to the direct impacts on tree health, larval development and feeding and the associated invasion by fungi can result in degrade losses to wood products. Few studies have quantified these losses but degrade affecting as much as two-thirds of the inventory in log yards have been reported (Becker 1966), as well as monetary losses of 35% to logs infested by woodborers (Becker and Abbott 1960). Woodborers can also negatively affect trees by contributing to disease transmission that leads to increased stress or mortality. In North America, several species of Cerambycidae are known vectors of the tree-killing nematode, *Bursaphelenchus xylophilus* (Steiner & Buhrer) Nickle (Linit 1988; Vallentgoed 1991), and this disease has been particularly problematic in Japan where it is transmitted by *M. alternatus* (Mamiya 1988). Woodborers have also been implicated in the transmission of several fungal pathogens including Dutch elm disease, chestnut blight, dieback of balsam-fir, oak wilt and hypoxylon canker on aspen (Donley 1959; Linsley 1961; Nord and Knight 1972; Ostry and Anderson 1995).

By far, the most important economic genus of Buprestidae is *Agrilus*. This genus contains over 3000 species, most of which inhabit angiosperms (Chamorro et al. 2015). Some native species are problematic on stressed trees in North America and Europe or are able to colonize non-native host trees common in more urban settings. Species such as the bronze birch borer (*A. anxius*), twolined chestnut borer (*A. bilineatus*), oak buprestid beetle [*Agrilus biguttatus* (F.)] and bronze poplar borer (*Agrilus liragus* Barter & Brown) can transition into primary tree killers given favorable environmental conditions (Barter 1957; Haack and Benjamin 1982; Dunn et al. 1986; Moraal and Hilszczanski 2000; Vansteenkiste et al. 2004) (Fig. 12.8). Exotic *Agrilus* spp. have been much more aggressive than their native counterparts in urban and forested settings in newly invaded areas. For example, the emerald ash borer has successfully invaded and spread into a large portion of North America (Herms and McCullough 2014), while also establishing and spreading in parts of Russia (Orlova-Bienkowskaja 2014). Native to eastern Asia, this species is a pest of ash throughout its introduced range. Indigenous exotic species (sensu Dodds et al. 2010b) have also been problematic in introduced areas, including the goldspotted oak borer (*Agrilus auroguttatus* Schaeffer) and soapberry borer (*Agrilus prionurus* Chevrolat) in forests in California and Texas, respectively (Coleman and Seybold 2008; Billings et al. 2014). These are native species that were previously isolated from areas where they have inadvertently been introduced.

Several genera of Cerambycidae can have significant economic or ecological importance. *Monochamus* species are secondary species colonizing weakened or recently dead material (Baker 1972), but through their maturation feeding can transfer the pinewood nematode to pine trees (Linit 1988). Pinewood nematode has caused serious tree losses in East Asia and Portugal, and threatens European pines (Mamiya 1988; Mota et al. 1999; Shin 2008; Zhao 2008; Robertson et al. 2011). Brown spruce longhorned beetle, a European species that was introduced into maritime Canada, has caused mortality in spruce stands (Smith and Hurley 2000). Asian longhorned beetle and citrus longhorned beetle [*Anoplophora chinensis* (Forster)], both native to Asia, have been repeatedly introduced in North America and Europe where they have successfully established multiple times (Haack et al. 2010; Meng et al. 2015).

Fig. 12.8 Damage caused by twolined chestnut borer, *A. bilineatus*, in the eastern US. Trees **a** stressed by drought and *Lymantria dispar* defoliation were killed by the buprestid, while logs **b** with damage from high densities of larvae **c** were salvage logged. Photo credit: Kevin Dodds

Worldwide, the most well-known siricid is *S. noctilio*. This species has been a pest of pine plantations since the early 1900s when it was detected in New Zealand (Bain et al. 2012) and later spread to other parts of the Southern Hemisphere (Slippers et al. 2002). Because the majority of siricids are associated with dead wood, very few species have caused economic losses. *Sirex noctilio* uses a phytotoxic venom (Bordeaux et al. 2014) to help it overcome host tree defenses and colonize trees most siricids cannot occupy.

12.7 Management of Woodborers

Management to reduce populations of most woodborers is unnecessary. However, management is required for some invasive species that damage and kill live trees, or if degradation to standing salvageable trees or stored wood products by native species is a concern. Where management is necessary in forested environments, approaches taken include silvicultural treatments, aggressive tree removal, and biological control efforts with bacteria, fungi, nematodes, and other insects.

12.7.1 Native Species

For native woodborers, there are few circumstances where population management is necessary. In situations where woodborer populations are building in a forest, it is generally in association with some form of abiotic or biotic disturbance that is predisposing trees to attack by secondary insects, including woodborers. Once the disturbance has subsided, or susceptible trees have been eliminated, woodborer attacks on trees rapidly diminish because of improved vigor of residual trees. Maintaining healthy forests with suitable stocking for a given site will reduce the number of susceptible trees that could be colonized by woodborers or act as sources for initial population outbreaks.

Because some woodborers respond to recently dead trees after a large-scale disturbance (Amman and Ryan 1991) and mine through wood, they can result in degrade losses, especially related to timber salvage after a disturbance. *Monochamus* species, in particular, cause rapid decline in wood quality of various conifer species after a disturbance (Richmond and Lejeune 1945; Gardiner 1957, 1975; Prebble and Gardiner 1958) because they are attracted to injured or recently dead trees where females oviposit and larvae mine into sapwood. Timely salvage and storage practices that minimize exposure to peak woodborer populations (Post and Werner 1988) can reduce the chance of excessive woodborer damage in logs destined for markets.

12.7.2   Invasive Species

Limiting the introduction of invasive species is an important first step to keeping damaging woodborers out of new environments. Strong legislation that focuses on preventative measures for limiting the introduction of these species or curtailing their spread once established can help reduce the impacts of these organisms. Once established and causing damage, various techniques have been implemented to attempt eradication and/or management of invasive woodborers. In some cases, well-developed integrated pest management plans have been developed through decades of research (Haugen et al. 1990), and in other cases, the development of management plans continues, even long after establishment (Herms and McCullough 2014). Common components involved with invasive woodborer management in forests include silvicultural treatments (Dodds et al. 2014a), tree removal (Hérard et al. 2006; Herms and McCullough 2014), biological control (Bedding 2009; Collett and Elms 2009), and restrictions on wood movement (USDA-APHIS 2010). Chemical insecticides are rarely implemented in invasive woodborer management within forested environments, although they can be important components of managing these species in urban forests.

12.7.2.1   Silvicultural Treatments

For invasive woodborer species that behave similarly to secondary species, or species that target specific trees (e.g. trees of certain species, sizes, vigor, or crown class), forest management may provide a solution for eliminating or reducing the effects of these insects. An example of silvicultural treatments reducing the impact of an invasive species is *S. noctilio*. Early observations of *S. noctilio* behavior suggested this woodborer was targeting weakened trees growing under overstocked stand conditions (Morgan and Stewart 1966). Consequently, silvicultural options that promoted optimal growing conditions in younger stands and targeted suppressed trees during thinning in older stands (Neumann et al. 1987) have successfully reduced the impact of *S. noctilio* in pine stands (Dodds et al. 2014a).

Unfortunately, most invasive woodborer species do not concentrate attacks on specific age, size, or canopy classes in forests. Therefore, it is not possible to target specific trees for removal based on any of these characteristics. Most invasive woodborers have either a wide host breadth, attack trees of all size classes, or do both, making silvicultural options ineffective (Dodds and Orwig 2011).

12.7.2.2   Tree Removal

Attempts to eradicate woodborers from urban forests often involve large-scale tree removal efforts. These removals can target infested trees only, and in some cases infested and adjacent non-infested host trees (Turgeon et al. 2007; Straw et al. 2015).

The use of large-scale tree removals in forested settings have been limited because of logistical challenges as well as questions as to effectiveness (Herms and McCullough 2014). For example, a six mile wide ash-free zone was created in southern Ontario in an attempt to stop the spread of emerald ash borer, but after completion, the beetle was found already established behind the zone (i.e. in the area the zone was designed to prevent emerald ash borer from invading) (Poland and McCullough 2006). Preemptive salvage logging of host species, such as ash in North America, has been conducted in some situations. Removal of Asian longhorned beetle infested trees and non-infested host trees has been conducted in smaller forested stands in North America (Dodds and Orwig 2011; Dodds et al. 2014b), Europe (Krehan 2008) and Great Britain (Straw et al. 2015) (Fig. 12.9). Woodborer dispersal behavior and initial distribution upon detection are generally the deciding factor for determining if eradication through tree removal is a feasible option for a given species. Asian longhorned beetle eradication has been successful because the adult beetles often reattack natal host trees and generally do not disperse long distances (Smith et al. 2004), allowing for more containment of infestations. Invasive species that are more widely dispersed upon detection, like *S. noctilio* in North America, are most often beyond the point where eradication would be feasible or cost effective.

Fig. 12.9 Asian longhorned beetle infested trees cut as part of an eradication program in Massachusetts, USA. Photo credit: Kevin Dodds

12.7.2.3 Biological Control

Both classical (introducing biological agents from other regions into new environments to control damaging invasive species) and augmentative biological (increasing native biological agents to control damaging invasive species) control of woodborers has been attempted for several species. Classical biological control using parasitic nematodes and wasps (Ichneumonidae, Ibaliidae) to manage *S. noctilio* populations has been implemented throughout the Southern Hemisphere (Hurley et al. 2007). These species have been important components of integrated pest management plans that also include silvicultural treatments for *S. noctilio* and have been responsible for keeping populations below damaging levels in many places. Classical biological control using an egg parasitoid, *Avetianella longoi* (Hymenoptera: Encyrtidae), with parasitism rates sometimes >90%, has also been helpful for reducing *Eucalyptus* mortality and damage from *P. semipunctata* in California, USA (Hanks et al. 1996b). However, biological control against *P. recurva* using *A. longoi* has not been effective (Luhring et al. 2000).

Because some invasive woodborers are congeners of native species and colonize the same habitat, there is often overlap in population regulation factors, providing opportunities for augmentative biological control if populations become economically or ecologically problematic. Native parasitoids that attack North American siricids have been documented attacking *S. noctilio* in these same forests (Ryan et al. 2012; Standley et al. 2012; Zylstra and Mastro 2012) (Fig. 12.10). Similarly, native parasitoids have also been found attacking the invasive brown spruce longhorned beetle (Flaherty et al. 2013a) and emerald ash borer (Gaudon and Smith 2020) in North America. Asian longhorned beetle and *A. chinensis* have been colonized by parasitoids native to the invaded region as well (Brabbs et al. 2015; Duan et al. 2016).

Purely augmentative biological control using native natural enemies on the native cerambycid *Massicus raddei* (Blessig) that causes damage to oak and chestnut species has been attempted in China. The parasitic wasp *Sclerodermus pupariae* Yang *et*

Fig. 12.10 Two rhyssine (Ichneumonidae) parasitoids search for hosts on a *Sirex noctilio* infested Scots pine in New York, USA. Photo credit: Kevin Dodds

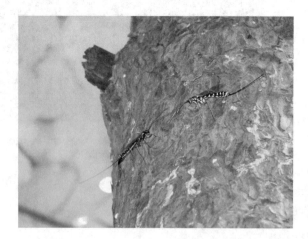

Yao (Hymenoptera: Bethylidae) and the beetle *Dastarcus helophoroides* (Fairmaire) (Coleoptera: Bothrideridae) have been used as biological control agents (Yang et al. 2014). Both are potential management tools for reducing the impact of *M. raddei* on native trees.

The combination of augmentative and classical biological control could be beneficial for reducing woodborer populations. This approach is currently being developed for emerald ash borer in North America. Four hymenopteran species have been approved as biological control agents for release in the U.S., including *Oobius agrili* Zhang and Huang (Encyrtidae), *Spathius agrili* Yang (Braconidae), *Tetrastichus planipennisi* Yang (Eulophidae), and *Spathius galinae* Belokobylskij & Strazanac (Braconidae) (Gould et al. 2015; Duan et al. 2019). These species are native to China, Russia, and Korea and attack eggs (*O. agrili*) or larvae (*S. agrili, T. planipennisi, S. galinae*). Native species, including *Phasgonophora sulcata* Westwood and *Atanycolus* Foerster spp. can be used as augmentative biocontrol agents (Gaudon and Smith 2020). The fungus *B. bassiana*, a native species, has also been tested to manage EAB (Lyons et al. 2012) and ALB populations (Dubois et al. 2004).

12.7.2.4 Chemical Control

Similar to tree removal, chemical control options have been used successfully on woodborers in urban forests, but their utility in natural or managed forests is limited. Several compounds are available, primarily for invasive species control, including systemic and contact insecticides. Compounds such as imidacloprid, cypermethrin, and emamectin benzoate have been used on invasive woodborers in urban settings (Hu et al. 2009). However, because of the cost associated with treatments, logistical challenges, and environmental concerns, these compounds are not seen as valid options for managing woodborers in forested environments.

12.7.2.5 Cultural Control

An important factor to limiting the spread of invasive woodborers is restricting movement of host material that may be infested by these insects. Wood products or wood packing material are often moved large distances, both intra- and inter-continentally. This material, combined with the cryptic nature of woodborers, provides an efficient pathway of introduction for woodborers and associated organisms (Mamiya 1988) into new environments. Wood packing is often colonized by woodborers that are then transported in the egg, larval, or pupal stage and emerge in the new environment to seek hosts unless wood is properly handled at the destination. The most damaging woodborers worldwide, including emerald ash borer, Asian longhorned beetle, and *S. noctilio*, are all believed to be introduced via whole logs or wood packing material.

Firewood is a documented pathway for movement of invasive species in North America and likely elsewhere when this material is used for heating or recreation (USDA-APHIS 2010). It has been linked to the spread of emerald ash borer across

Fig. 12.11 Disturbances such as windstorms can leave large volumes of downed trees **a** that are quickly colonized by woodborers and other insects. This material is often cut into firewood **b** that can provide a pathway for insects into new environments. Photo credit: Kevin Dodds.

portions of eastern North America (Cappaert et al. 2005) and may have been the pathway for *Agrilus prionurus* into Texas from Mexico (Haack 2006) and *A. auroguttatus* into California (Coleman and Seybold 2011). Trees harvested for firewood are often recently dead and already infested with woodborers or at a stage of decay that makes them suitable for colonization (Fig. 12.11). Seasoning of trees or cut firewood also provides opportunity for insects to colonize this material. Firewood can host large communities of wood-inhabiting insects, including buprestids, cerambycids and siricids (Dodds et al. 2017) that can then be moved large distances often for recreational camping (Jacobi et al. 2011; Koch et al. 2014). If this wood is infested with invasive species, introductions into new environments can occur.

12.8   Summary

Woodborers are an ecologically important guild in forested ecosystems. They contribute to various ecological processes including nutrient cycling, forest succession, and are important components of food webs. Woodborers colonize all parts of dead or living trees, while generally causing little impact on overall tree health. Most Buprestidae, Cerambycidae, and Siricidae encountered in native and managed forests are secondary species that rarely kill trees, however, important invasive species like Asian longhorned beetle, emerald ash borer, and *S. noctilio* can have broad ranging impacts on urban, managed, and natural forests. At times, management of woodborers is necessary and includes preventative silvicultural treatment, tree removal, biological control, chemical control, and cultural methods to reduce movement of infested materials.

   Various factors effect woodborer populations, including both abiotic (i.e. climate, fire, and other natural disturbances) and biotic factors (i.e. host suitability, natural enemies, and competition). Biotic factors such as parasitic nematodes, fungi, and parasitoids have been important in management of some woodborers, including the invasive emerald ash borer and *S. noctilio*. In some forest types and regions, some woodborer species are considered threatened or endangered, primarily from habitat loss due to forest fragmentation, conversion, and loss of old trees.

References

Abell KJ, Duan JJ, Bauer L, Lelito JP, Van Driesche RG (2012) The effect of bark thickness on host partitioning between *Tetrastichus planipennisi* (Hymen: Eulophidae) and *Atanycolus* spp. (Hymen: Braconidae), two parasitoids of emerald ash borer (Coleop: Buprestidae). Biol Control 63:320–325

Adams AS, Jordan MS, Adams SM, Suen G, Goodwin LA, Davenport KW, Currie CR, Raffa KF (2011) Cellulose-degrading bacteria associated with the invasive woodwasp *Sirex noctilio*. ISME J 5:1323–1331

Akbulut S (2009) Comparison of the reproductive potential of *Monochamus galloprovincialis* on two pine species under laboratory conditions. Phytoparasitica 37:125–135

Akbulut S, Keten A, Stamps WT (2008) Population dynamics of *Monochamus galloprovincialis* Olivier (Coleoptera: Cerambycidae) in two pine species under laboratory conditions. J Pest Sci 81:115–121

Allison JD, Borden JH, McIntosh RL, de Groot P, Gries R (2001) Kairomonal response by four *Monochamus* species (Coleoptera: Cerambycidae) to bark beetle pheromones. J Chem Ecol 27:633–646

Allison JD, Borden JH, Seybold SJ (2004) A review of the chemical ecology of the Cerambycidae (Coleoptera). Chemoecology 14:123–150

Allison JD, McKenney JL, Millar JG, McElfresh JS, Mitchell RF, Hanks LM (2012) Response of the woodborers *Monochamus carolinensis* and *Monochamus titillator* (Coleoptera: Cerambycidae) to known cerambycid pheromones in the presence and absence of the host plant volatile alpha-pinene. Environ Entomol 41:1587–1596

Allison JD, McKenney JL, Miller DR, Gimmel ML (2013) Kairomonal responses of natural enemies and associates of the southern *Ips* (Coleoptera: Curculionidae: Scolytinae) to ipsdienol, ipsenol and cis-verbenol. J Insect Behav 26:321–335

Allison JD, Slippers B, Bouwer M, Hurley BP (2021) Simulated leks increase the capture of female *Sirex noctilio* in the absence of host volatiles. Int J Pest Manag 67:58–64

Álvarez-Baz G, Fernández-Bravo M, Pajares J, Quesada-Moraga E (2015) Potential of native *Beauveria pseudobassiana* strain for biological control of pine wood nematode vector *Monochamus galloprovincialis*. J Invertebr Pathol 132:48–56

Amman GD, Ryan KC (1991) Insect infestation of fire-injured trees in the Greater Yellowstone area. Res. Note INT-398. U.S. Department of Agriculture, Forest Service

Anbutsu H, Togashi K (1996) Deterred oviposition of *Monochamus alternatus* (Coleoptera: Cerambycidae) on *Pinus densiflora* bolts from oviposition scars containing eggs or larvae. Appl Entomol Zool 31:481–488

Anbutsu H, Togashi K (1997a) Oviposition behavior and response to the oviposition scars occupied by eggs in *Monochamus saltuarius* (Coleoptera: Cerambycidae). Appl Entomol Zool 32:541–549

Anbutsu H, Togashi K (1997b) Effects of spatio-temporal intervals between newly-hatched larvae on larval survival and development in *Monochamus alternatus* (Coleoptera: Cerambycidae). Res Popul Ecol 39:181–189

Anbutsu H, Togashi K (2000) Deterred oviposition response of *Monochamus alternatus* (Coleoptera: Cerambycidae) to oviposition scars occupied by eggs. Agric for Entomol 2:217–223

Anbutsu H, Togashi K (2002) Oviposition deterrence associated with larval frass of the Japanese pine sawyer, *Monochamus alternatus* (Coleoptera: Cerambycidae). J Insect Physiol 48:459–465

Andersen J (1983) Intrapopulation size variation of free-living and tree-boring Coleoptera. Can Entomol 115:1453–1464

Angers VA, Drapeau P, Bergeron Y (2011) Mineralization rates and factors influencing snag decay in four North American boreal tree species. Can J For Res 42:157–166

Anonymous (2007) Halting the loss of biodiversity by 2010: proposal for a first set of indicators to monitor progress in Europe. EE Agency EEA Technical Report No 11.

Anonymous (2020) Species at risk act. https://laws.justice.gc.ca/eng/acts/S-15.3/page-17.html#h-435647. Accessed on 14 Jan 2021

Ansley RJ, Meadors CH, Jacoby PW (1990) Preferential attraction of the twig girdler, *Oncideres cingulata texana* Horn, to moisture-stressed mesquite. Southwest Entomol 15:469–474

Aojin Y, Qing'an T (1998) Repellency effects of essential oil derived from Eucalyptus leaf against three species of sawyers (Abstract in English). J Nanjing For Univ 22:87–90

Askew RR, Shaw MR (1986) Parasitoid communities: their size, structure and development. In: Waage JK, Greathead DJ (eds) Insect parasitoids. Academic Press, London

Aukema JE, Leung B, Kovacs K, Chivers C, Britton KO, Englin J, Frankel SJ, Haight RG, Holmes TP, Liebhold AM, McCullough DG, Von Holle B (2011) Economic impacts of non-native forest insects in the continental United States. PLoS ONE 6:e24587

Ayres BD, Ayres MP, Abrahamson MD, Teale SA (2001) Resource partitioning and overlap in three sympatric species of *Ips* bark beetles (Coleoptera: Scolytidae). Oecologia 128:443–453

Bain J, Sopow SL, Bulman LS (2012) The Sirex woodwasp in New Zealand: history and current status. In: Slippers B, de Groot P, Wingfield MJ (eds) The Sirex woodwasp and its fungal symbiont. Springer, Dordrecht, pp 167–173

Baker WL (1972) Eastern forest insects. Miscellaneous Publications No 1175. U.S. Department of Agriculture, Forest Service, Washington, DC

Barter GW (1957) Studies of the bronze birch borer, *Agrilus anxius* Gory, in New Brunswick. Can Entomol 89:12–36

Basham JT, Belyea RM (1960) Death and deterioration of balsam fir weakened by spruce budworm defoliation in Ontario—part III. The deterioration of dead trees. For Sci 6:78–96

Becker WB (1966) Worm-hole free lumber salvaged from borer-damaged pine logs. J For Res 64:126–128

Becker WB, Abbott HG (1960) Lumber downgrading losses due to insect damage to small unseasoned pine sawlogs. J For 58:46–47

Bedding RA (2009) Controlling the pine-killing woodwasp, *Sirex noctilio*, with nematodes. In: Hajek AE, Glare TR, O'Callaghan M (eds) Use of microbes for control and eradication of invasive arthropods. Springer, Dordrecht, pp 213–235

Bedding RA, Akhurst RJ (1974) Use of the nematode *Deladenus siricidicola* in the biological control of *Sirex noctilio* in Australia. J Aust Entomol Soc 13:129–135

Bellamy CL (2002) Zoological catalogue of Australia 29.5, *Coleoptera: Buprestoidea*. CSIRO Publishing, Collingwood, VIC

Bellamy CL, Nelson GH (2002) Buprestidae Leach 1815. In: Arnett RH Jr, Thomas MC, Skelley PE, Frank JH (eds) American beetles, Volume II polyphaga: Scarabaeoidea through Curculionoidea. CRC Press LLC, Boca Raton, FL, pp 98–112

Benham GS, Farrar RJ (1976) Notes on the biology of *Prionus laticollis* (Coleoptera: Cerambycidae). Can Entomol 108:569–576

Billings RF, Grosman DM, Pase HA (2014) Soapberry borer, *Agrilus prionurus* (Coleoptera: Buprestidae): an exotic pest threatens western soapberry in Texas. Southeast Nat 13:105–116

Bolstad PV, Bentz BJ, Logan JA (1997) Modelling micro-habitat temperature for *Dendroctonus ponderosae* (Coleoptera: Scolytidae). Ecol Model 94:287–297

Bordeaux JM, Lorenz WW, Johnson D, Badgett MJ, Glushka J, Orlando R, Dean JFD (2014) Noctilisin, a venom glycopeptide of *Sirex noctilio* (Hymenoptera: Siricidae), causes needle wilt and defense gene responses in pines. J Econ Entomol 107:1931–1945

Boring LR, Swank WT (1984) The role of black locust (*Robinia pseudoacacia*) in forest succession. J Ecol 72:749–766

Brabbs T, Collins D, Hérard F, Maspero M, Eyre D (2015) Prospects for the use of biological control agents against *Anoplophora* in Europe. Pest Manag Sci 71:7–14

Bright DE (1987) The metallic wood-boring beetles of Canada and Alaska, Coleoptera: Buprestidae. The insects and arachnids of Canada, part 15. Canadian Government Publishing Centre, Ottawa, ON

Brockerhoff EG, Bain J, Kimberley M, Knížek M (2006) Interception frequency of exotic bark and ambrosia beetles (Coleoptera: Scolytinae) and relationship with establishment in New Zealand and worldwide. Can J For Res 36:289–298

Brooks FE (1926) Life history of the hickory spiral borer, *Agrilus arcuatus* Say. J Agric Res 33:331–338

Broughton S (2000) Review and evaluation of lantana biocontrol programs. Biol Control 17:272–286

Buse J, Ranius T, Assmann T (2008) An endangered longhorn beetle associated with old oaks and its possible role as an ecosystem engineer. Conserv Biol 22:329–337

Calderon O, Berkov A (2012) Midgut and fat body bacteriocytes in neotropical cerambycid beetles (Coleoptera: Cerambycidae). Environ Entomol 41:108–117

Calderón-Cortés N, Quesada M, Escalera-Vázquez LH (2011) Insects as stem engineers: interactions mediated by the twig-girdler *Oncideres albomarginata chamela* enhance arthropod diversity. PLoS ONE 6:e19083

Campbell CL, McCaffrey JP (1991) Population trends, seasonal phenology, and impact of *Chrysolina quadrigemina*, *C. hyperici* (Coleoptera: Chrysomelidae), and *Agrilus hyperici* (Coleoptera: Buprestidae) associated with *Hypericum perforatum* in northern Idaho. Environ Entomol 20:303–315

Cappaert D, McCullough DG, Poland TM, Siegert NW (2005) Emerald ash borer in North America: a research and regulatory challenge. Am Entomol 51:152–165

Caraglio Y, Nicolini E, Petronelli P (2001) Observations on the links between the architecture of a tree (*Dicorynia guianensis* Amshoff) and Cerambycidae activity in French Guiana. J Trop Ecol 17:459–463

Cardé RT (2016) Moth navigation along pheromone plumes. In: Allison JD, Cardé RT (eds) Pheromone communication in moths: evolution, behavior and application. University of California Press, Oakland, CA, pp 173–189

Cardoso P, Borges PAV, Triantis KA, Ferrández MA, Martín JL (2012) The underrepresentation and misrepresentation of invertebrates in the IUCN Red List. Biol Cons 149:147–148

Chakraborty S, Whitehill JGA, Hill AL, Opiyo SO, Cipollini D, Herms DA, Bonello P (2014) Effects of water availability on emerald ash borer larval performance and phloem phenolics of Manchurian and black ash. Plant Cell Environ 37:1009–1021

Chamorro ML, Jendek E, Haack RA, Petrice TR, Woodley NE, Konstantinov AS, Volkovitsh MG, Yang X-K, Grebennikov VV, Lingafelter SW (2015) Illustrated guide to the emerald ash borer *Agrilus planipennis* Fairmaire and related species (Coleoptera, Buprestidae). Pensoft, Sofia

Cobb TP, Hannam KD, Kishchuk BE, Langor DW, Quideau SA, Spence JR (2010) Wood-feeding beetles and soil nutrient cycling in burned forests: implications of post-fire salvage logging. Agric For Entomol 12:9–18

Coleman TW, Seybold SJ (2008) Previously unrecorded damage to oak, *Quercus* spp., in southern California by the goldspotted oak borer, *Agrilus coxalis* Waterhouse (Coleoptera: Buprestidae). Pan-Pac Entomol 84:288–300

Coleman TW, Seybold SJ (2011) Collection history and comparison of the interactions of the goldspotted oak borer, *Agrilus auroguttatus* Schaeffer (Coleoptera: Buprestidae), with host oaks in southern California and southeastern Arizona, U.S.A. Coleopt Bull 65:93–108

Collett NG, Elms S (2009) The control of Sirex wood wasp using biological control agents in Victoria, Australia. Agric For Entomol 11:283–294

Cooperband MF, Boroczky K, Hartness A, Jones TH, Zylstra KE, Tumlinson JH, Mastro VC (2012) Male-produced pheromone in the European woodwasp, *Sirex noctilio*. J Chem Ecol 38:52–62

Costello SL, Jacobi WR, Negron JF (2013) Emergence of Buprestidae, Cerambycidae, and Scolytinae (Coleoptera) from mountain pine beetle-killed and fire-killed ponderosa pines in the Black Hills, South Dakota, USA. Coleopt Bull 67:149–154

Coyle DR, Gandhi KJK (2012) The ecology, behavior, and biological control potential of hymenopteran parasitoids of woodwasps (Hymenoptera: Siricidae) in North America. Environ Entomol 41:731–749

Craighead FC (1937) Locust borer and drought. J For 35:792–793

Craighead FC (1950) Insect enemies of Eastern Forests. Misc. Publ. 657. U.S. Department of Agriculture, Washington, DC

Cresswell JE (1991) Capture rates and composition of insect prey of the pitcher plant *Sarracenia purpurea*. Am Midl Nat 125:1–9

D'Amen M, Bombi P, Campanaro A, Zapponi L, Bologna MA, Mason F (2013) Protected areas and insect conservation: questioning the effectiveness of Natura 2000 network for saproxylic beetles in Italy. Anim Conserv 16:370–378

Danks HV (2000) Insect cold hardiness: a Canadian perspective. Cryo-Letters 21:297–308

Delalibera I, Handelsman J, Raffa KF (2005) Contrasts in cellulolytic activities of gut microorganisms between the wood borer, *Saperda vestita* (Coleoptera: Cerambycidae), and the bark beetles, *Ips pini* and *Dendroctonus frontalis* (Coleoptera: Curculionidae). Environ Entomol 34:541–547

Dhileepan K, Taylor DBJ, Lockett C, Treviño M (2013) Cat's claw creeper leaf-mining jewel beetle *Hylaeogena jureceki* Obenberger (Coleoptera: Buprestidae), a host-specific biological control agent for *Dolichandra unguis-cati* (Bignoniaceae) in Australia. Aust J Entomol 52:175–181

Dodds KJ, Orwig DA (2011) An invasive urban forest pest invades natural environments—Asian longhorned beetle in northeastern US hardwood forests. Can J For Res 41:1729–1742

Dodds KJ, Stephen FM (2002) Arrival and species complex of Cerambycidae on pines attacked by southern pine beetle. J Entomol Sci 37:272–274

Dodds KJ, Graber C, Stephen FM (2001) Facultative intraguild predation by larval Cerambycidae (Coleoptera) on bark beetle larvae (Coleoptera: Scolytidae). Environ Entomol 30:17–22

Dodds KJ, de Groot P, Orwig DA (2010a) The impact of *Sirex noctilio* in *Pinus resinosa* and *Pinus sylvestris* stands in New York and Ontario. Can J For Res 40:212–223

Dodds KJ, Gilmore DW, Seybold SJ (2010b) Assessing the threat posed by indigenous exotics: a case study of two North American bark beetle species. Ann Entomol Soc Am 103:39–49

Dodds KJ, Cooke RR, Hanavan RP (2014a) The effects of silvicultural treatment on *Sirex noctilio* attacks and tree health in northeastern United States. Forests 5:2810–2824

Dodds KJ, Hull-Sanders HM, Siegert NW, Bohne MJ (2014b) Colonization of three maple species by Asian longhorned beetle, *Anoplophora glabripennis*, in two mixed-hardwood forest stands. Insects 5:105–119

Dodds KJ, Hanavan RP, DiGirolomo MF (2017) Firewood collected after a catastrophic wind event: the bark beetle (Scolytinae) and woodborer (Buprestidae, Cerambycidae) community present over a 3-year period. Agric For Entomol 19:309–320

Donley DE (1959) Studies of wood boring insects as vectors of the oak wilt fungus. PhD, Ohio State University

Donley DE, Acciavatti RE (1980) Red oak borer. Forest Insect & Disease Leaflet 163. U.S. Government Printing Office, Washington, DC, 7 pp

Douglas AE (1989) Mycetocyte symbiosis in insects. Biol Rev Camb Philos Soc 64:409–434

Duan JJ, Fuester RW, Wildonger J, Taylor PB, Barth S, Spichiger SE (2009) Parasitoids attacking the emerald ash borer (Coleoptera: Buprestidae) in western Pennsylvania. Fla Entomol 92:588–592

Duan JJ, Bauer LS, Abell KJ, van Driesche R (2012) Population responses of hymenopteran parasitoids to the emerald ash borer (Coleoptera: Buprestidae) in recently invaded areas in north central United States. Biocontrol 57:199–209

Duan JJ, Aparicio E, Tatman D, Smith MT, Luster DG (2016) Potential new associations of North American parasitoids with the invasive Asian longhorned beetle (Coleoptera: Cerambycidae) for biological control. J Econ Entomol 109:699–704

Duan JJ, Van Driesche RG, Crandall RS, Schmude JM, Rutledge CE, Slager BH, Gould JR, Elkinton JS (2019) Establishment and early impact of *Spathius galinae* (Hymenoptera: Braconidae) on emerald ash borer (Coleoptera: Buprestidae) in the northeastern United States. J Econ Entomol 112:2121–2130

Dubois T, Hajek AE, Hu JF, Li ZH (2004) Evaluating the efficiency of entomopathogenic fungi against the Asian longhorned beetle, *Anoplophora glabripennis* (Coleoptera: Cerambycidae), by using cages in the field. Environ Entomol 33:62–74

Duffy EAJ (1946) A contribution towards the biology of *Prionus coriarius* L. (Coleoptera: Cerambycidae). Trans R Entomol Soc Lond 97:419–442

Duffy EAJ (1953) A monograph of the immature stages of British and imported timber beetles (Cerambycidae). British Museum (Natural History), London

Dunbar DM, Stephens GR (1975) Association of twolined chestnut borer and shoestring fungus with mortality of defoliated oak in Connecticut. For Sci 21:169–174

Dunn JP, Kimmerer TW, Nordin GL (1986) The role of host tree condition in attack of white oaks by the twolined chestnut borer, *Agrilus bilineatus* (Weber) (Coleoptera: Buprestidae). Oecologia 70:596–600

Duval BD, Whitford WG (2008) Resource regulation by a twig-girdling beetle has implications for desertification. Ecol Entomol 33:161–166

Eager PT, Allen DC, Frair JL, Fierke MK (2011) Within-tree distributions of the *Sirex noctilio*-parasitoid complex and development of an optimal sampling scheme. Environ Entomol 40:1266–1275

Edmonds RL, Eglitis A (1989) The role of the Douglas-fir beetle and wood borers in the decomposition of and nutrient release from Douglas-fir logs. Can J For Res 19:853–859

Esperk T, Tammaru T, Nylin S (2007) Intraspecific variability in number of larval instars in insects. J Econ Entomol 100:627–645

European Commission, Directorate-General for Environment, Sundseth K (2021) The state of nature in the EU: conservation status and trends of species and habitats protected by the EU nature directives 2013–2018. Publications Office, 39 pp. https://data.europa.eu/doi/10.2779/5120

Evans WG (1964) Infra-red receptors in *Melanophila acuminata* DeGeer. Nature 202:211

Evans WG (1966) Perception of infrared radiation from forest fires by *Melanophila acuminata* DeGeer (Buprestidae, Coleoptera). Ecology 47:1061–1065

Evans DA, Higgs MD (1975) Mono-oxygenated monoterpenes from the frass of the wood-boring beetle *Hylotrupes bajulus*. Tetrahedron Lett 41:3585–3586

Evans HF, Moraal LG, Pajares JA (2004) Biology, ecology and economic importance of Buprestidae and Cerambycidae. In: Lieutier F, Day KR, Battisti A, Grégoire J-C, Evans HF (eds) Bark and wood boring insects in living trees in Europe, a synthesis. Springer, Dordrecht, pp 447–474

Faccoli M, Favaro R (2016) Host preference and host colonization of the Asian long-horned beetle, *Anoplophora glabripennis* (Coleoptera Cerambycidae), in Southern Europe. Bull Entomol Res 106:359–367

Farr JD, Dick SG, Williams MR, Wheeler IB (2000) Incidence of bullseye borer (*Phoracantha acanthocera*, (Macleay) Cerambycidae) in 20–35 year old regrowth karri in the south west of Western Australia. Aust For 63:107–123

Fayt P (2003) Insect prey population changes in habitats with declining vs. stable three-toed woodpecker *Picoides tridactylus* populations. Ornis Fenn 80:182–192

Feller IC, McKee KL (1999) Small gap creation in Belizean mangrove forests by a wood–boring insect. Biotropica 31:607–617

Feng YQ, Xu LL, Tian B, Tao J, Wang JL, Zong SX (2014) Cold hardiness of Asian longhorned beetle (Coleoptera: Cerambycidae) larvae in different populations. Environ Entomol 43:1419–1426

Ferro ML, Gimmel ML, Harms KE, Carlton CE (2012) Comparison of Coleoptera emergent from various decay classes of downed coarse woody debris in Great Smoky Mountains National Park, USA. Insecta Mundi 0260:1–80

Fettköther R, Reddy GVP, Noldt U, Dettner K (2000) Effect of host and larval frass volatiles on behavioural response of the old house borer, *Hylotrupes bajulus* (L.) (Coleoptera: Cerambycidae), in a wind tunnel bioassay. Chemoecology 10:1–10

Finlayson T (1969) Final-instar larvae of two hymenopterous parasites of a wood-boring beetle, *Tetropium velutinum* LeConte (Coleoptera: Cerambycidae). J Entomol Soc Br C 66:62–65

Flaherty L, Sweeney JD, Pureswaran D, Quiring DT (2011) Influence of host tree condition on the performance of *Tetropium fuscum* (Coleoptera: Cerambycidae). Environ Entomol 40:1200–1209

Flaherty L, Quiring D, Pureswaran D, Sweeney J (2013a) Evaluating seasonal variation in bottom-up and top-down forces and their impact on an exotic wood borer, *Tetropium fuscum* (Coleoptera: Cerambycidae). Environ Entomol 42:957–966

Flaherty L, Quiring D, Pureswaran D, Sweeney J (2013b) Preference of an exotic wood borer for stressed trees is more attributable to pre-alighting than post-alighting behaviour. Ecol Entomol 38:546–552

Foelker CJ, Standley CR, Fierke MK, Parry D, Whipps CM (2016) Host tissue identification for cryptic hymenopteran parasitoids associated with *Sirex noctilio*. Agric For Entomol 18:91–94

Forcella F (1984) Trees size and density affect twig-girdling intensity of *Oncideres cingulata* (Say) (Coleoptera: Cerambycidae). Coleopt Bull 38:37–42

Francardi V, Benvenuti C, Barzanti GP, Roversi PF (2013) Autocontamination trap with ento-mopathogenic fungi: a possible strategy in the control of *Rhynchophorus ferrugineus* (Olivier) (Coleoptera Curculionidae). Redia 96:57–67

Galford JR (1984) The locust borer. Forest Insect & Disease Leaflet 71. USDA Forest Service, 6 pp

Gandhi KJK, Gilmore DW, Katovich SA, Mattson WJ, Spence JR, Seybold SJ (2007) Physical effects of weather events on the abundance and diversity of insects in North American forests. Environ Rev 15:113–152

Gardiner LM (1957) Deterioration of fire-killed pine in Ontario and the causal wood-boring beetles. Can Entomol 89:241–263

Gardiner LM (1975) Insect attack and value loss in wind-damaged spruce and jack pine stands in northern Ontario. Can J For Res 5:387–398

Gaudon JM, Smith SM (2020) Augmentation of native North American natural enemies for the biological control of the introduced emerald ash borer in central Canada. Biocontrol 65:71–79

Gauld ID (1988) Evolutionary patterns of host utilization by ichneumonoid parasitoids (Hymenoptera: Ichneumonidae and Braconidae). Biol J Lin Soc 35:351–377

Georgiev G, Ljubomirov T, Raikova M, Ivanov K, Sakalian V (2004) Insect inhabitants of old larval galleries of *Saperda populnea* (L.) (Coleoptera: Cerambycidae) in Bulgaria. J Pest Sci 77:235–243

Gilbertson RL (1984) Relationship between insects and wood-rotting basidiomycetes. In: Wheeler Q, Blackwell M (eds) Fungus-insect relationships: perspectives. Columbia University Press, New York, pp 130–165

Godínez-Aguilar JL, Macías-Sámano JE, Morón-Ríos A (2009) Notes on biology and sexual behavior of *Tetrasarus plato* Bates (Coleoptera: Cerambycidae), a tropical longhorn beetle in coffee plantations in Chiapas, Mexico. Coleopt Bull 63:311–318

Golec JR, Duan JJ, Aparicio E, Hough-Goldstein J (2016) Life history, reproductive biology, and larval development of *Ontsira mellipes* (Hymenoptera: Braconidae), a newly associated parasitoid of the invasive Asian longhorned beetle (Coleoptera: Cerambycidae). J Econ Entomol 109:1545–1554

Gould JR, Bauer LS, Duan JJ, Williams DC, Liu H (2015) History of emerald ash borer biological control. In: Van Driesche RG, Reardon RC (eds) Biology and control of emerald ash borer. USDA Forest Service, Morgantown, WV, pp 83–95

Graham SA (1925) The felled tree trunk as an ecological unit. Ecology 6:397–411

Graham K (1967) Fungal-insect mutualism in trees and timber. Annu Rev Entomol 12:105–126

Grünwald S, Pilhofer M, Höll W (2010) Microbial associations in gut systems of wood- and bark-inhabiting longhorned beetles [Coleoptera: Cerambycidae]. Syst Appl Microbiol 33:25–34

Guignard Q, Bouwer M, Slippers B, Allison JD (2020) Biology of a putative male aggregation-sex pheromone in *Sirex noctilio* (Hymenoptera: Siricidae). PLoS ONE 15:e0244943

Haack RA (2006) Exotic bark- and wood-boring Coleoptera in the United States: recent establishments and interceptions. Can J For Res 36:269–288

Haack RA (2017) Feeding biology of cerambycids. In: Wang Q (ed) Cerambycidae of the world: biology and pest management. Boca Raton, FL: CRC Press, Taylor & Francis Group, pp 105–132

Haack RA, Benjamin DM (1982) The biology and ecology of the twolined chestnut borer, *Agrilus bilineatus* (Coleoptera: Buprestidae), on oaks, *Quercus* spp., in Wisconsin. Can Entomol 114:385–396

Haack RA, Slansky F Jr (1987) Nutritional ecology of wood-feeding Coleoptera, Lepidoptera, and Hymenoptera. In: Slansky F, Rodriguez JG (eds) Nutritional ecology of insects, mites, and spiders. Wiley, New York, pp 449–486

Haack RA, Bauer LS, Gao R, McCarthy JJ, Miller DL, Petrice TR, Poland TM (2006) *Anoplophora glabripennis* within-tree distribution, seasonal development, and host suitability in China and Chicago. Gt Lakes Entomol 39:169–183

Haack RA, Hérard F, Sun J, Turgeon JJ (2010) Managing invasive populations of Asian longhorned beetle and citrus longhorned beetle: a worldwide perspective. Annu Rev Entomol 55:521–546

Haack RA, Keena MA, Eyre D (2017) Life history and population dynamics of cerambycids. In: Wang Q (ed) Cerambycidae of the world: biology and pest management. Contemporary topics in entomology series. Boca Raton, FL: Taylor & Francis, pp 71–103

Haavik LJ, Stephen FM (2010) Historical dynamics of a native cerambycid, *Enaphalodes rufulus*, in relation to climate in the Ozark and Ouachita Mountains of Arkansas. Ecol Entomol 35:673–683

Haavik LJ, Crook DJ, Fierke MK, Galligan LD, Stephen FM (2012a) Partial life tables from three generations of *Enaphalodes rufulus* (Coleoptera: Cerambycidae). Environ Entomol 41:1311–1321

Haavik LJ, Jones JS, Galligan LD, Guldin JM, Stephen FM (2012b) Oak decline and red oak borer outbreak: impact in upland oak-hickory forests of Arkansas, USA. For: Int J For Res 85:341–352

Haavik LJ, Dodds KJ, Allison JD (2015) Do native insects and associated fungi limit non-native woodwasp, *Sirex noctilio*, survival in a newly invaded environment? PLoS ONE 10:e0138516

Haavik LJ, Dodds KJ, Ryan K, Allison JD (2016) Evidence that the availability of suitable pine limits non-native *Sirex noctilio* in Ontario. Agric For Entomol 18:357–366

Haavik LJ, Slippers B, Hurley BP, Dodds KJ, Scarr T, Turgeon JJ, Allison JD (2020) Influence of the community of associates on *Sirex noctilio* brood production is contextual. Ecol Entomol 45:456–465

Hadfield J, Magelssen R (2006) Wood changes in fire-killed tree species in eastern Washington. USDA Forest Service. Wenatchee, Washington, DC, 53 pp

Hanks LM (1999) Influence of the larval host plant on reproductive strategies of cerambycid beetles. Annu Rev Entomol 44:483–505

Hanks LM, Millar JG (2016) Sex and aggregation-sex pheromones of cerambycid beetles: basic science and practical applications. J Chem Ecol 42:631–654

Hanks LM, Paine TD, Millar JG (1991) Mechanisms of resistance in *Eucalyptus* against larvae of the Eucalyptus longhorned borer (Coleoptera, Cerambycidae). Environ Entomol 20:1583–1588

Hanks LM, Paine TD, Millar JG (1993) Host species preference and larval performance in the wood-boring beetle *Phoracantha semipunctata* F. Oecologia 95:22–29

Hanks LM, Millar JG, Paine TD (1995) Biological constraints on host-range expansion by the wood-boring beetle *Phoracantha semipunctata* (Coleoptera, Cerambycidae). Ann Entomol Soc Am 88:183–188

Hanks LM, Millar JG, Paine TD (1996a) Mating behavior of the eucalyptus longhorned borer (Coleoptera: Cerambycidae) and the adaptive significance of long "horns". J Insect Behav 9:383–393

Hanks LM, Paine TD, Millar JG (1996b) Tiny wasp helps protect eucalypts from eucalyptus longhorned borer. Calif Agric 50:14–16

Hanks LM, Millar JG, Paine TD (1998) Dispersal of the eucalyptus longhorned borer (Coleoptera: Cerambycidae) in urban landscapes. Environ Entomol 27:1418–1424

Hanks LM, Paine TD, Millar JG, Campbell CD, Schuch UK (1999) Water relations of host trees and resistance to the phloem-boring beetle *Phoracantha semipunctata* F (Coleoptera: Cerambycidae). Oecologia 119:400–407

Hanks LM, Paine TD, Millar JG (2005) Influence of the larval environment on performance and adult body size of the wood-boring beetle *Phoracantha semipunctata*. Entomol Exp Appl 114:25–34

Hanula JL, Franzreb KE, Kathleen EF (1995) Arthropod prey of nestling red-cockaded woodpeckers in the upper coastal plain of South Carolina. Wilson Bull 107:485–495

Harmon ME, Franklin JF, Swanson FJ, Sollins P, Gregory SV, Lattin JD, Anderson NH, Cline SP, Aumen NG, Sedell JR, Lienkaemper GW, Cromack K, Cummins KW (1986) Ecology of coarse woody debris in temperate ecosystems. Adv Ecol Res 15:133–302

Harris AG (2018) Recovery strategy for the hoptree borer (*Prays atomocella*) in Ontario. Ontario Recovery strategy series. Prepared for the Ministry of the Environment, Conservation and Parks, Peterborough, ON, iv + 19 pp. https://www.ontario.ca/page/hoptree-borer-recovery-strategy. Accessed on 14 Jan 2021

Harris P, Shorthouse JD (1996) Effectiveness of gall inducers in weed biological control. Can Entomol 128:1021–1055

Haugen DA, Bedding RA, Underdown MG, Neumann FG (1990) National strategy for control of *Sirex noctilio* in Australia. Australian Forest Grower, Special Liftout Section No 13, 8 pp

Haw JM, Clout MN, Powlesland RG (2001) Diet of moreporks (*Ninox novaeseelandiae*) in Pureora Forest determined from prey remains in regurgitated pellets. N Z J Ecol 25:61–67

Heijari J, Nerg A-M, Kainulainen P, Noldt U, Levula T, Raitio H, Holopainen JK (2008) Effect of long-term forest fertilization on Scots pine xylem quality and wood borer performance. J Chem Ecol 34:26–31

Heitzman E, Grell A, Spetich M, Starkey D (2007) Changes in forest structure associated with oak decline in severely impacted areas of northern Arkansas. South J Appl For 31:17–22

Hérard F, Ciampitti M, Maspero M, Krehan H, Benker U, Boegel C, Schrage R, Bouhot-Delduc L, Bialooki P (2006) *Anoplophora* species in Europe: infestations and management processes. EPPO Bull 36:470–474

Herms DA, McCullough DG (2014) Emerald ash borer invasion of North America: history, biology, ecology, impacts, and management. Annu Rev Entomol 59:13–30

Hespenheide HA (1991) Bionomics of leaf-mining insects. Annu Rev Entomol 36:535–560

Higgs MD, Evans DA (1978) Chemical mediators in the oviposition behavior of the house longhorn beetle, *Hylotrupes bajulus*. Experientia 34:46–47

Hoover K, Keena M, Nehme M, Wang SF, Meng P, Zhang AJ (2014) Sex-specific trail pheromone mediates complex mate finding behavior in *Anoplophora glabripennis*. J Chem Ecol 40:169–180

Hosking GP, Bain J (1977) *Arhopalus ferus* (Coleoptera: Cerambycidae); its biology in New Zealand. NZ J For Sci 7:3–15

Hovore FT, Penrose RL (1982) Notes on Cerambycidae co-inhabiting girdles of *Oncideres pustulata* LeConte (Coleoptera, Cerambycidae). Southwestern Naturalist 27:23–27

Howden HF, Vogt GB (1951) Insect communities of standing dead pine (*Pinus virginiana* Mill). Ann Entomol Soc Am 44:581–595

Hu J, Angell S, Schuetz S, Luo Y, Hajek AE (2009) Ecology and management of exotic and endemic Asian longhorned beetle *Anoplophora glabripennis*. Agric For Entomol 11:359–375

Hughes AL (1979) Reproductive behavior and sexual dimorphism in the white-spotted sawyer *Monochamus scutellatus* (Say). Coleopt Bull 33:45–47

Humble L (2010) Pest risk analysis and invasion pathways—insects and wood packing revisited: what have we learned? N Z J For Sci 40(Suppl.):57–72

Hunt GR (2000) Tool use by the New Caledonian Crow *Corvus moneduloides* to obtain Cerambycidae from dead wood. Emu—Austral Ornithol 100:109–114

Hurley BP, Slippers B, Wingfield MJ (2007) A comparison of control results for the alien invasive woodwasp, *Sirex noctilio*, in the southern hemisphere. Agric For Entomol 9:159–171

Hurley BP, Garnas J, Cooperband MF (2015) Assessing trap and lure effectiveness for the monitoring of *Sirex noctilio*. Agric For Entomol 17:64–70

Ikeda K (1979) Consumption and food utilization by individual larvae and the population of a wood borer *Phymatodes maaki* Kraatz (Coleoptera, Cerambycidae). Oecologia 40:287–298

Islam SQ, Ichiryu J, Sato M, Yamasaki T (1997) D-catechin: an oviposition stimulant for the cerambycid beetle, *Monochamus alternatus*, from *Pinus densiflora*. J Pestic Sci 22:338–341

Iwata R, Hirayama Y, Shimura H, Ueda M (2004) Twig foraging and soil-burrowing behaviors in larvae of *Dinoptera minuta* (Gebler) (Coleoptera: Cerambycidae). Coleopt Bull 58:399–408

Jacobi WR, Goodrich BA, Cleaver CM (2011) Firewood transport by national and state park campers: a risk for native or exotic tree pest movement. Arboric Urban For 37:126–138

Jaenike J (1978) On optimal oviposition behavior in phytophagous insects. Theor Popul Biol 14:350–356

Jendek E, Poláková J (2014) Host plants of the world *Agrilus* (Coleoptera: Buprestidae): a critical review. Springer, Cham, 706 pp

Jennings DE, Duan JJ, Bean D, Rice KA, Williams GL, Bell SK, Shurtleff AS, Shrewsbury PM (2017) Effects of the emerald ash borer invasion on the community composition of arthropods associated with ash tree boles in Maryland, U.S.A. Agric For Entomol 19:122–129

Jeppsson T, Forslund P (2014) Species' traits explain differences in Red List status and long-term population trends in longhorn beetles. Anim Conserv 17:332–341

Jia-bing W, De-xin G, Shi-jie H, Mi Z, Chang-jie J (2005) Ecological functions of coarse woody debris in forest ecosystem. J For Res 16:247–252

Jiménez A, Gallardo A, Antonietty CA, Villagrán M, Ocete ME, Soria FJ (2012) Distribution of *Coraebus undatus* (Coleoptera: Buprestidae) in cork oak forests of southern Spain. Int J Pest Manag 58:281–288

Joyce AL, Millar JG, Gill JS, Singh M, Tanner D, Paine TD (2011) Do acoustic cues mediate host finding by *Syngaster lepidus* (Hymenoptera: Braconidae)? Biocontrol 56:145–153

Juutinen P (1955) Zur Biologie und forstlichen Bedeutung der Fichtenböcke (*Tetropium* Kirby) in Finnland. Acta Entomol Fenn 11:1–112

Kato K (2005) Factors enabling *Epinotia granitalis* (Lepidoptera: Tortricidae) overwintered larvae to escape from oleoresin mortality in *Cryptomeria japonica* trees in comparison with *Semanotus japonicus* (Coleoptera: Cerambycidae). J For Res 10:205–210

Keen FP (1952) Insect enemies of western forests. Misc Publ No 273. U.S. Department of Agriculture, 280 pp

Keena MA, Moore PM (2010) Effects of temperature on *Anoplophora glabripennis* (Coleoptera: Cerambycidae) larvae and pupae. Environ Entomol 39:1323–1335

Kenis M, Hilszczanski J (2004) Natural enemies of Cerambycidae and Buprestidae infesting living trees. In: Lietier F, Day KR, Battisti A, Grégoire JC, Evans HF (eds) Bark and wood boring insects in living trees in Europe, a synthesis. Springer, Dordrecht, pp 475–498

Kennedy HE, Solomon JD, Krinard RM (1961) Twig girdler, *Oncideres cingulata* (Say), attacks terminals of plantation-managed pecans. USDA Forest Service, Southern Forest Experiment Station, New Orleans, LA, 4 pp

King AM, Williams HE, Madire LG (2011) Biological control of cat's claw creeper, *Macfadyena unguis-cati* (L.) A.H.Gentry (Bignoniaceae), in South Africa. Afr Entomol 19:366–377

Klein H (1999) Biological control of three cactaceous weeds, *Pereskia aculeata* Miller, *Harrisia martinii* (Labouret) Britton and *Cereus jamacaru* De Candolle in South Africa. Afr Entomol 1:3–14

Klein MG, Lacey LA (1999) An attractant trap for autodissemination of entomopathogenic fungi into populations of the Japanese beetle *Popillia japonica* (Coleoptera: Scarabaeidae). Biocontrol Sci Tech 9:151–158

Koch FH, Yemshanov D, Haack RA, Magarey RD (2014) Using a network model to assess risk of forest pest spread via recreational travel. PLoS ONE 9:e102105

Krehan H (2008) Asian longhorned beetle *Anopolophora glabripennis* (ALB)—eradication program in Braunau (Austria) in 2007. Forstschutz Aktuell 44:27–29

Kukor JJ, Martin MM (1983) Acquisition of digestive enzymes by siricid woodwasps from their fungal symbiont. Science 220:1161–1163

Kukor JJ, Martin MM (1986) Cellulose digestion in *Monochamus marmorator* Kby. (Coleoptera: Cerambycidae): role of acquired fungal enzymes. J Chem Ecol 12:1057–1070

Kukor JJ, Cowan DP, Martin MM (1988) The role of ingested fungal enzymes in cellulose digestion in the larvae of cerambycid beetles. Physiol Zool 61:364–371

Lacey ES, Ray AM, Hanks LM (2007) Calling behavior of the cerambycid beetle *Neoclytus acuminatus acuminatus* (F.). J Insect Behav 20:117–128

Laurenne N, Karatolos N, Quicke DLJ (2009) Hammering homoplasy: multiple gains and losses of vibrational sounding in cryptine wasps (Insecta: Hymenoptera: Ichneumonidae). Biol J Lin Soc 96:82–102

Lee RE (1989) Insect cold-hardiness—to freeze or not to freeze. Bioscience 39:308–313

Lemes PG, Cordeiro G, Jorge IR, Anjos ND, Zanuncio JC (2015) Cerambycidae and other Coleoptera associated with branches girdled by *Oncideres saga* Dalman (Coleoptera: Cerambycidae: Lamiinae: Onciderini). Coleopt Bull 69:159–166

Lesk C, Coffel E, D'Amato AW, Dodds K, Horton R (2017) Threats to North American forests from southern pine beetle with warming winters. Nat Clim Chang 7:713–717

Lieutier F, Yart A, Jayallemand C, Delorme L (1991) Preliminary investigations on phenolics as a response of Scots pine phloem to attacks by bark beetles and associated fungi. Eur J For Pathol 21:354–364

Li L, Wei W, Liu Z, Sun J (2010) Host adaptation of a gregarious parasitoid *Sclerodermus harmandi* in artificial rearing. Biocontrol 55:465–472

Lindell CA, McCullough DG, Cappaert D, Apostolou NM, Roth MB (2008) Factors influencing woodpecker predation on emerald ash borer. Am Midl Nat 159:434–444

Linit MJ (1988) Nematode-vector relationships in the pine wilt disease system. J Nematol 20:227–235

Linsley EG (1943) Attraction of *Melanophila* beetles by fire and smoke. J Econ Entomol 36:341–342

Linsley EG (1959) Ecology of Cerambycidae. Annu Rev Entomol 4:99–138

Linsley EG (1961) The Cerambycidae of North America, part I. Introduction. Univ Calif Publ Entomol 18:1–135

Linsley EG, Chemsak JA (1972) Cerambycidae of North America, Part VI, No. 1. Taxonomy and classification of the subfamily Lepturinae. Univ Calif Publ Entomol 69:1–138

Liu H, Bauer LS (2008) Microbial control of *Agrilus planipennis* (Coleoptera: Buprestidae) with *Beauveria bassiana* strain GHA: field applications. Biocontrol Sci Tech 18:557–571

Liu H, Bauer LS, Zhao T, Gao R, Poland TM (2016) Seasonal abundance and development of the Asian longhorned beetle and natural enemy prevalence in different forest types in China. Biol Control 103:154–164

Long SJ, Williams DW, Hajek AE (2009) *Sirex* species (Hymenoptera: Siricidae) and their parasitoids in *Pinus sylvestris* in eastern North America. Can Entomol 141:153–157

Lopez VM, Hoddle MS (2014) Effects of body size, diet, and mating on the fecundity and longevity of the goldspotted oak borer (Coleoptera: Buprestidae). Ann Entomol Soc Am 107:539–548

Lowell EC, Willits SA, Krahmer RL (1992) Deterioration of fire-killed and fire-damaged timber in the western United States. USDA Forest Service, Portland, OR, 27 pp

Lu W, Wang Q, Tian M, Xu J, Lv J, Qin A (2013) Mating behavior and sexual selection in a polygamous beetle. Curr Zool 59:257–264

Luhring KA, Paine TD, Millar JG, Hanks LM (2000) Suitability of the eggs of two species of eucalyptus longhorned borers (*Phoracantha recurva* and *P. semipunctata*) as hosts for the encyrtid parasitoid *Avetianella longoi*. Biol Control 19:95–104

Lyons DB, Lavallée R, Kyei-Poku G, Van Frankenhuyzen K, Johny S, Guertin C, Francese JA, Jones GC, Blais M (2012) Towards the development of an autocontamination trap system to manage populations of emerald ash borer (Coleoptera: Buprestidae) with the native entomopathogenic fungus, *Beauveria bassiana*. J Econ Entomol 105:1929–1939

Ma RY, Hao SG, Tian J, Sun JH, Kang L (2006) Seasonal variation in cold-hardiness of the Japanese pine sawyer *Monochamus alternatus* (Coleoptera: Cerambycidae). Environ Entomol 35:881–886

MacAloney HJ (1971) Sugar maple borer. Forest Pest Leaflet 108. U.S. Government Printing Office, Washington, DC, 4 pp

Madden JL (1981) Egg and larval development in the woodwasp, *Sirex noctilio* F. Aust J Zool 29:493–506

Maki EC, Millar JG, Rodstein J, Hanks LM, Barbour JD (2011) Evaluation of mass trapping and mating disruption for managing *Prionus californicus* (Coleoptera: Cerambycidae) in hop production yards. J Econ Entomol 104:933–938

Mamiya Y (1988) History of pine wilt disease in Japan. J Nematol 20:219–226

Marshall KE, Sinclair BJ (2015) The relative importance of number, duration and intensity of cold stress events in determining survival and energetics of an overwintering insect. Funct Ecol 29:357–366

Martin MM (1991) The evolution of cellulose digestion in insects. Philos Trans R Soc Lond B 333:281–288

Martínez AJ, López-Portillo J, Eben A, Golubov J (2009) Cerambycid girdling and water stress modify mesquite architecture and reproduction. Popul Ecol 51:533–541

Martius C (1997) Decomposition of wood. In: Junk WJ (ed) The central Amazon floodplain: ecology of a pulsing system. Springer-Verlag, Heidelberg, pp 267–276

Mattson WJ, Haack RA (1987) The role of drought in outbreaks of plant-eating insects. Bioscience 37:110–118

McCann JM, Harman DM (2003) Avian predation on immature stages of the locust borer, *Megacyllene robiniae* (Forster) (Coleoptera: Cerambycidae). Proc Entomol Soc Wash 105:970–981

McFadyen PJ (1983) Host specificity and biology of *Megacyllene mellyi* [Col.: Cerambycidae] introduced into Australia for the biological control of *Baccharis halimifolia* [Compositae]. Entomophaga 28:65–71

McFadyen RE, Fidalgo AP (1976) Investigations on *Alcidion cereicola* [Col.: Cerambycidae] a potential agent for the biological control of *Eriocereus martinii* [Cactaceae] in Australia. Entomophaga 21:103–111

McKimm RJ, Walls JW (1980) A survey of damage caused by the Sirex woodwasp in the radiata pine plantations at Delatite, north-eastern Victoria, between 1972–1979. For Comm Vic For Tech Pap 28:3–11

Medellín RA (1988) Prey of *Chrotopterus auritus*, with notes on feeding behavior. J Mammal 69:841–844

Mendel Z, Golan Y, Madar Z (1984) Natural control of the eucalyptus borer, *Phoracantha semipunctata* (F.) (Coleoptera: Cerambycidae), by the Syrian woodpecker. Bull Entomol Res 74:121–127

Meng PS, Hoover K, Keena MA (2015) Asian longhorned beetle (Coleoptera: Cerambycidae), an introduced pest of maple and other hardwood trees in North America and Europe. J Integr Pest Manag 6. https://doi.org/10.1093/jipm/pmv003

Meurer-Grimes B, Tavakilian G (1997) Chemistry of cerambycid host plants. Part I: survey of leguminosae—a study in adaptive radiation. Bot Rev 63:356–394

Millar JG, Hanks LM (2017) Chemical ecology of Cerambycids. In: Wang Q (ed) Cerambycidae of the world: biology and pest management. CRC Press, New York, pp 161–208

Miller DR (2006) Ethanol and (-)-α-pinene: attractant kairomones for some large wood-boring beetles in southeastern USA. J Chem Ecol 32:779–794

Miller DR, Asaro C, Crowe CM, Duerr DA (2011) Bark beetle pheromones and pine volatiles: attractant kairomone lure blend for longhorn beetles (Cerambycidae) in pine stands of the southeastern United States. J Econ Entomol 104:1245–1257

Monné ML, Monné MA, Wang Q (2017) General morphology, classification, and biology of Cerambycidae. In: Wang Q (ed) Cerambycidae of the world: biology and pest management. Taylor & Francis Group, Boca Raton, FL, pp 1–70

Moraal LG, Hilszczanski J (2000) The oak buprestid beetle, *Agrilus biguttatus* (F.) (Col., Buprestidae), a recent factor in oak decline in Europe. J Pest Sci 73:134–138

Morales-Rodríguez C, Sánchez-González Á, Conejo-Rodríguez Y, Torres-Vila LM (2015) First record of *Beauveria bassiana* (Ascomycota: Clavicipitaceae) infecting *Cerambyx welensii* (Coleoptera: Cerambycidae) and pathogenicity tests using a new bioassay method. Biocontrol Sci Tech 25:1213–1219

Morewood WD, Simmonds KE, Gries R, Allison JD, Borden JH (2003) Disruption by conophthorin of the kairomonal response of sawyer beetles to bark beetle pheromones. J Chem Ecol 29:2115–2129

Morgan FD, Stewart NC (1966) The biology and behaviour of the woodwasp *Sirex noctilio* F. in New Zealand. Trans R Soc N Z 7:195–204

Mota M, Braasch H, Bravo MA, Penas AC, Burgermeister W, Metge K, Sousa E (1999) First report of *Bursaphelenchus xylophilus* in Portugal and in Europe. Nematology 1:727–734

Muilenburg VL, Herms DA (2012) A review of bronze birch borer (Coleoptera: Buprestidae) life history, ecology, and management. Environ Entomol 41:1372–1385

Muilenburg VL, Goggin FL, Hebert SL, Jia L, Stephen FM (2008) Ant predation on red oak borer confirmed by field observation and molecular gut-content analysis. Agric For Entomol 10:205–213

Murphy EC, Lehnhausen WA (1998) Density and foraging ecology of woodpeckers following a stand-replacement fire. J Wildl Manag 62:1359–1372

Mushrow L, Morrison A, Sweeney J, Quiring D (2004) Heat as a phytosanitary treatment for the brown spruce longhorn beetle. For Chron 80:224–228

Myers SW, Fraser I, Mastro VC (2009) Evaluation of heat treatment schedules for emerald ash borer (Coleoptera: Buprestidae). J Econ Entomol 102:2048–2055

Nappi A, Drapeau P, Leduc A (2015) How important is dead wood for woodpeckers foraging in eastern North American boreal forests? For Ecol Manage 346:10–21

Naves PM, Kenis M, Sousa E (2005) Parasitoids associated with *Monochamus galloprovincialis* (Oliv.) (Coleoptera: Cerambycidae) within the pine wilt nematode-affected zone in Portugal. J Pest Sci 78:57–62

Naves PM, Sousa E, Rodrigues JM (2008) Biology of *Monochamus galloprovincialis* (Coleoptera, Cerambycidae) in the pine wilt disease affected zone, southern Portugal. Silva Lusit 16:133–148

Ness WN (1920) The ribbed pine-borer. Cornell University, Ithaca, NY, pp 367–381

Neumann FG, Morey JL, McKimm RJ (1987) The Sirex wasp in Victoria. Bulletin 29. Department of Conservation, Forests and Lands, Melbourne, VIC, 41 pp

Newton WG, Allen DC (1982) Characteristics of trees damaged by sugar maple borer, *Glycobius speciosus*. Can J For Res 12:738–744

Nieto A, Alexander KNA (2010) European Red List of saproxylic beetles. Publications Office of the European Union, Luxembourg

Nord JC, Knight FB (1972) The importance of *Saperda inornata* and *Oberea schaumii* (Coleoptera: Cerambycidae) galleries as infection courts of *Hypoxylon pruinatum* in trembling aspen, *Populus Tremuloides*. Gt Lakes Entomol 5:87–92

O'Leary K, Hurley JE, Mackay W, Sweeney J (2003) Radial growth rate and susceptibility of *Picea rubens* Sarg. to *Tetropium fuscum* (Fabr.). In: McManus ML, Liebhold AM (eds) Proceedings: ecology, survey and management of forest insects. USDA Forest Service, Northern Research Station Gen. Tech. Rep. NE-311, Newtown Square, PA, pp 107–114

Orlova-Bienkowskaja MJ (2014) European range of the emerald ash borer *Agrilus planipennis* (Coleoptera: Buprestidae) is expanding: the pest destroys ashes in the northwest of Moscow oblast and in part of Tver oblast. Russ J Biol Invasions 5:32–37

Ostry ME, Anderson NA (1995) Infection of *Populus tremuloides* by *Hypoxylon mammatum* ascospores through *Saperda inornata* galls. Can J For Res 25:813–816

Paine TD (2017) Natural enemies and biological control of cerambycid pests. In: Cerambycidae of the world: biology and pest management. Boca Raton: CRC Press, Taylor & Francis Group, pp 291–304

Paine TD, Birch MC, Švihra P (1981) Niche breadth and resource partitioning by four sympatric species of bark beetles (Coleoptera: Scolytidae). Oecologia 48:1–6

Palmer WA, Willson BW, Pullen KR (2000) Introduction, rearing, and host range of *Aerenicopsis championi* Bates (Coleoptera: Cerambycidae) for the biological control of *Lantana camara* L. in Australia. Biol Control 17:227–233

Parker TJ, Clancy KM, Mathiasen RL (2006) Interactions among fire, insects and pathogens in coniferous forests of the interior western United States and Canada. Agric For Entomol 8:167–189

Pawson SM, Bader MKF, Brockerhoff EG, Heffernan WJB, Kerr JL, O'Connor B (2019) Quantifying the thermal tolerance of wood borers and bark beetles for the development of Joule heating as a novel phytosanitary treatment of pine logs. J Pest Sci 92:157–171

Pechacek P, Kristin A (2004) Comparative diets of adult and young three-toed woodpeckers in a European alpine forest community. J Wildl Manag 68:683–693

Peddle S, de Groot P, Smith S (2002) Oviposition behaviour and response of *Monochamus scutellatus* (Coleoptera: Scolytidae) to conspecific eggs and larvae. Agric For Entomol 4:217–222

Petrice TR, Haack RA, Strazanac JS, Lelito JP (2009) Biology and larval morphology of *Agrilus subcinctus* (Coleoptera: Buprestidae), with comparisons to the emerald ash borer, *Agrilus Planipennis*. Gt Lakes Entomol 42:173–184

Phillips MA, Croteau RB (1999) Resin-based defenses in conifers. Trends Plant Sci 4:184–190

Poland TM, McCullough DG (2006) Emerald ash borer: invasion of the urban forest and the threat to North America's ash resource. J For 104:118–124

Post KE, Werner RA (1988) Wood borer distribution and damage in decked white spruce logs. North J Appl For 5:49–51

Powell W (1982) Age-specific life table data for the *Eucalyptus* boring beetle, *Phoracantha semipunctata* (F) (Coleoptera, Cerambycidae), in Malawi. Bull Entomol Res 72:645–653

Powell RD, Myers JH (1988) The effect of *Sphenoptera jugoslavica* Obenb. (Col., Buprestidae) on its host plant *Centaurea diffusa* Lam. (Compositae). J Appl Entomol 106:25–45

Prebble ML, Gardiner LM (1958) Degrade and value loss in fire-killed pine in the Mississagi area of Ontario. For Chron 34:139–158

Queiroz JM (2002) Distribution, survivorship and mortality sources in immature stages of the neotropical leaf miner *Pachyschelus coeruleipennis* Kerremans (Coleoptera: Buprestidae). Braz J Biol 62:69–76

Raffa KF (1991) Induced defensive reactions in conifer-bark beetle systems. In: Tallamy DW, Raupp MJ (eds) Phytochemical induction by herbivores. Academic Press, New York, pp 245–276

Raske AG (1973) Notes on the biology of *Tetropium parvulum* (Coleoptera: Cerambycidae) in Alberta. Can Entomol 105:757–760

Rees NE, Pemberton RW, Rizza A, Pecora P (1986) First recovery of *Oberea erythrocephala* on the leafy spurge complex in the United States. Weed Sci 34:395–397

Richardson ML, Mitchell RF, Reagel PF, Hanks LM (2010) Causes and consequences of cannibalism in noncarnivorous insects. Annu Rev Entomol 55:39–53

Richmond HA, Lejeune RR (1945) The deterioration of fire-killed white spruce by wood-boring insects in northern Saskatchewan. For Chron 21:168–192

Rigsby CM, Muilenburg V, Tarpey T, Herms DA, Cipollini D (2014) Oviposition preferences of *Agrilus planipennis* (Coleoptera: Buprestidae) for different ash species support the mother knows best hypothesis. Ann Entomol Soc Am 107:773–781

Robertson L, Cobacho AS, Escuer M, Santiago MR, Esparrago G, Abelleira A, Navas A (2011) Incidence of the pinewood nematode *Bursaphelenchus xylophilus* Steiner & Buhrer, 1934 (Nickle, 1970) in Spain. Nematology 13:755–757

Rodrigues ASL, Pilgrim JD, Lamoreux JF, Hoffmann M, Brooks TM (2006) The value of the IUCN Red List for conservation. Trends Ecol Evol 21:71–76

Romero GQ, Vasconcellos-Neto J, Paulino Neto HF (2005) The effects of the wood-boring *Oncideres humeralis* (Coleoptera, Cerambycidae) on the number and size structure of its host-plants in south-east Brazil. J Trop Ecol 21:233–236

Roscoe LE, Lyons DB, Smith SM (2016) Observations on the life-history traits of the North American parasitoid *Phasgonophora sulcata* Westwood (Hymenoptera: Chalcididae) attacking *Agrilus planipennis* (Coleoptera: Buprestidae) in Ontario, Canada. Can Entomol 148:294–306

Rose AH (1957) Some notes on the biology of *Monochamus scutellatus* (Say) (Coleoptera: Cerambycidae). Can Entomol 89:547–553

Ryan K, de Groot P, Nott RW, Drabble S, Ochoa I, Davis C, Smith SM, Turgeon JJ (2012) Natural enemies associated with *Sirex noctilio* (Hymenoptera: Siricidae) and *S. nigricornis* in Ontario, Canada. Environ Entomol 41:289–297

Saint-Germain M, Drapeau P, Buddle CM (2007) Host-use patterns of saproxylic phloeophagous and xylophagous Coleoptera adults and larvae along the decay gradient in standing dead black spruce and aspen. Ecography 30:737–748

Sánchez V, Keena MA (2013) Development of the teneral adult *Anoplophora glabripennis* (Coleoptera: Cerambycidae): time to initiate and completely bore out of maple wood. Environ Entomol 42:1–6

Sato M, Islam SQ, Awata S, Yamasaki T (1999a) Flavanonol glucoside and proanthocyanidins: oviposition stimulants for the cerambycid beetle, *Monochamus alternatus*. J Pestic Sci 24:123–129

Sato M, Islam SQ, Yamasaki T (1999b) Glycosides of a phenylpropanoid and neolignans: oviposition stimulants in pine inner bark for cerambycid beetle, *Monochamus alternatus*. J Pestic Sci 24:397–400

Savely HE Jr (1939) Ecological relations of certain animals in dead pine and oak logs. Ecol Monogr 9:321–385

Schiff NM, Goulet H, Smith DR, Boudreault C, Wilson AD, Scheffler BE (2012) Siricidae (Hymenoptera: Symphyta: Siricoidea) of the western hemisphere. Can J Arthropod Identif 21:1–305

Schimitschek E (1929) *Tetropium gabrieli* Weise und *Tetropium fuscum* F: Ein Beitrag zu ihrer Lebensgeschichte und Lebensgemeinschaft. Z Für Angew Entomol 15:229–234

Schloss PD, Delalibera I, Handelsman J, Raffa KF (2006) Bacteria associated with the guts of two wood-boring beetles: *Anoplophora glabripennis* and *Saperda vestita* (Cerambycidae). Environ Entomol 35:625–629

Schmitz H, Norkus V, Hess N, Bousack H (2009) The infrared sensilla in the beetle *Melanophila acuminata* as model for new infrared sensors. In: Rodriguez-Vázquez AB, Carmona-Galán RA, Liñán-Cembrano G (eds) Proceedings of SPIE, vol. 7365, Bioengineered and Bioinspired Systems IV, 73650A

Schoeller EN, Husseneder C, Allison JD (2012) Molecular evidence of facultative intraguild predation by *Monochamus titillator* larvae (Coleoptera: Cerambycidae) on members of the southern pine beetle guild. Naturwissenschaften 99:913–924

Shibata E (2000) Bark borer *Semanotus japonicus* (Col., Cerambycidae) utilization of Japanese cedar *Cryptomeria japonica*: a delicate balance between a primary and secondary insect. J Appl Entomol 124:279–285

Shibata E (1987) Oviposition schedules, survivorship curves, and mortality factors within trees of two cerambycid beetles (Coleoptera: Cerambycidae), the Japanese pine sawyer, *Monochamus alternatus* Hope, and sugi bark borer, *Semanotus japonicus* Lacordaire. Res Popul Ecol 29:347–367

Shimazu M (2004) A novel technique to inoculate conidia of entomopathogenic fungi and its application for investigation of susceptibility of the Japanese pine sawyer, *Monochamus alternatus*, to *Beauvéria bassiana*. Appl Entomol Zool 39:485–490

Shimazu M, Sato H (1995) Microbial control of *Monochamus alternatus* Hope (Coleoptera: Cerambycidae) by application of nonwoven fabric strips with *Beauveria bassiana* (Deuteromycotina: Hyphomycetes) on infested tree trunks. Appl Entomol Zool 30:207–213

Shin S-C (2008) Pine wilt disease in Korea. In: Zhao BG, Futai K, Sutherland JR, Takeuchi Y (eds) Pine wilt disease. Springer, Tokyo, pp 26–32

Silk PJ, Ryall K, Mayo P, Lemay MA, Grant G, Crook D, Cosse A, Fraser I, Sweeney JD, Lyons DB, Pitt D, Scarr T, Magee D (2011) Evidence for a volatile pheromone in *Agrilus planipennis* Fairmaire (Coleoptera: Buprestidae) that increases attraction to a host foliar volatile. Environ Entomol 40:904–916

Silk PJ, Mayo P, Ryall K, Roscoe L (2019) Semiochemical and communication ecology of the emerald ash borer, *Agrilus planipennis* (Coleoptera: Buprestidae). Insects 10:323

Sjöman H, Östberg J, Nilsson J (2014) Review of host trees for the wood-boring beetles *Anoplophora glabripennis* and *Anoplophora chinensis*: an urban forest perspective. Agric Urban For 40:143–164

Slippers B, Wingfield BD, Coutinho TA, Wingfield MJ (2002) DNA sequence and RFLP data reflect geographical spread and relationships of *Amylostereum areolatum* and its insect vectors. Mol Ecol 11:1845–1854

Smith G, Hurley JE (2000) First North American record of the palearctic species *Tetropium fuscum* (Fabricius) (Coleoptera: Cerambycidae). Coleopt Bull 54:540

Smith MT, Tobin PC, Bancroft J, Li GH, Gao RT (2004) Dispersal and spatiotemporal dynamics of Asian longhorned beetle (Coleoptera: Cerambycidae) in China. Environ Entomol 33:435–442

Solomon JD (1977) Biology and habits of the oak branch borer (*Goes debilis*). Ann Entomol Soc Am 70:57–59

Solomon JD (1995) Guide to insect borers of North American broadleaf trees and shrubs. Agriculture Handbook No 706. U.S. Department of Agriculture Forest Service

Spradbery JP (1968) A technique for artificially culturing ichneumonid parasites of woodwasps (Hymenoptera: Siricidae). Entomol Exp Appl 11:257–260

Srei N, Guertin C, Lavallée R, Lajoie M, Brousseau C, Bergevin R, Miller F, McMillin K, Trudel R (2020) Microbial control of the emerald ash borer (Coleoptera: Buprestidae) using *Beauveria bassiana* (Hypocreales: Cordycipitaceae) by the means of an autodissemination device. J Econ Entomol 113:2657–2665

Standley CR, Hoebeke ER, Parry D, Allen DC, Fierke MK (2012) Detection and identification of two new native hymenopteran parasitoids associated with the exotic *Sirex noctilio* in North America. Proc Entomol Soc Wash 114:238–249

Stephen FM, Salisbury VB, Oliveria FL (2001) Red oak borer, *Enaphalodes rufulus* (Coleoptera: Cerambycidae), in the Ozark Mountains of Arkansas, U.S.A.: an unexpected and remarkable forest disturbance. Integr Pest Manag Rev 6:247–252

Stokland JN (2012) The saproxylic food web. In: Stokland JN, Siitonen J, Jonsson BG (eds) Biodiversity in dead wood. Cambridge University Press, New York, pp 29–57

Stokland JN, Siitonen J (2012) Mortality factors and decay succession. In: Stokland JN, Siitonen J, Jonsson BG (eds) Biodiversity in dead wood. Cambridge University Press, New York, pp 110–149

Straw NA, Fielding NJ, Tilbury C, Williams DT, Inward D (2015) Host plant selection and resource utilisation by Asian longhorn beetle *Anoplophora glabripennis* (Coleoptera: Cerambycidae) in southern England. Forestry 88:84–95

Suckling DM, Gibb AR, Daly JM, Chen D, Brockerhoff EG (2001) Behavioral and electro-physiological responses of *Arhopalus tristis* to burnt pine and other stimuli. J Chem Ecol 27:1091–1104

Sullivan BT, Berisford CW, Dalusky MJ (1997) Field response of southern pine beetle parasitoids to some natural attractants. J Chem Ecol 23:837–856

Sweeney J, Silk PJ, Hughes C, Lavallée R, Blais M, Guertin C (2013) Auto-dissemination of *Beauveria bassiana* for control of brown spruce longhorned beetle, *Tetropium fuscum* (F.), Coleoptera: Cerambycidae. In: McManus KA, Gottschalk, KW, 24th USDA interagency research forum on invasive species. USDA Forest Health Technology Enterprise Team FHTET 13-01, Fort Collins, CO, 98 pp

Taylor PB, Duan JJ, Fuester RW, Hoddle M, Van Driesche R (2012) Parasitoid guilds of *Agrilus* woodborers (Coleoptera: Buprestidae): their diversity and potential for use in biological control. Psyche 2012:813929

Thompson BM, Grebenok RJ, Behmer ST, Gruner DS (2013) Microbial symbionts shape the sterol profile of the xylem-feeding woodwasp, *Sirex noctilio*. J Chem Ecol 39:129–139

Thompson BM, Bodart J, McEwen C, Gruner DS (2014) Adaptations for symbiont-mediated external digestion in *Sirex noctilio* (Hymenoptera: Siricidae). Ann Entomol Soc Am 107:453–460

Tluczek AR, McCullough DG, Poland TM (2011) Influence of host stress on emerald ash borer (Coleoptera: Buprestidae) adult density, development, and distribution in *Fraxinus pennsylvanica* trees. Environ Entomol 40:357–366

Torres-Vila LM, Sanchez-Gonzalez A, Merino-Martinez J, Ponce-Escudero F, Conejo-Rodriguez Y, Martin-Vertedor D, Ferrero-Garcia JJ (2013) Mark-recapture of *Cerambyx welensii* in dehesa woodlands: dispersal behaviour, population density, and mass trapping efficiency with low trap densities. Entomol Exp Appl 149:273–281

Torres-Vila LM, Zugasti C, De-Juan JM, Oliva MJ, Montero C, Mendiola FJ, Conejo Y, Sanchez A, Fernandez F, Ponce F, Esparrago G (2015) Mark-recapture of *Monochamus galloprovincialis* with semiochemical-baited traps: population density, attraction distance, flight behaviour and mass trapping efficiency. Forestry 88:224–236

Triapitsyn SV, Petrice TR, Gates MW, Bauer LS (2015) Two new species of *Oobius* Trjapitzin (Hymenoptera, Encyrtidae) egg parasitoids of *Agrilus* spp. (Coleoptera, Buprestidae) from the USA, including a key and taxonomic notes on other congeneric Nearctic taxa. ZooKeys 498:29–50

Tryjanowski P, Karg MK, Karg J (2003) Diet composition and prey choice by the red-backed shrike *Lanius collurio* in western Poland. Belg J Zool 133:157–162

Turgeon JJ, Ric J, de Groot P, Gasman B, Orr M, Doyle J, Smith MT, Dumouchel L, Scarr T (2007) Détection des signes et des symptômes d'attaque par le longicorne étoilé: guide de formation. Service canadien des forêts, Ressources naturelles Canada, Ottawa, ON

Turgeon JJ, Jones C, Smith MT, Orr M, Scarr TA, Gasman B (2016) Records of unsuccessful attack by *Anoplophora glabripennis* (Coleoptera: Cerambycidae) on broadleaf trees of questionable suitability in Canada. Can Entomol 148:569–578

Ulyshen MD (2016) Wood decomposition as influenced by invertebrates. Biol Rev 91:70–85

Ulyshen MD, Hanula JL (2010) Patterns of saproxylic beetle succession in loblolly pine. Agric For Entomol 12:187–194

USDA-APHIS (2010) Risk assessment of the movement of firewood within the United States. USDA Animal and Plant Health Services, Raleigh, NC

Vallentgoed J (1991) Some important woodborers related to export restrictions. Forest Pest Leaflet No 74. Natural Resources Canada, Canadian Forest Services, Ottawa, ON

van Achterberg C, Mehrnejad MR (2011) A new species of *Megalommum* Szépligeti (Hymenoptera, Braconidae, Braconinae); a parasitoid of the pistachio longhorn beetle (*Calchaenesthes pistacivora* Holzschuh; Coleoptera, Cerambycidae) in Iran. ZooKeys 112:21–38

Vansteenkiste D, Tirry L, Acker JV, Stevens M (2004) Predispositions and symptoms of *Agrilus* borer attack in declining oak trees. Ann For Sci 61:815–823

Vermunt B, Cuddington K, Sobek-Swant S, Crosthwaite JC, Lyons DB, Sinclair BJ (2012) Temperatures experienced by wood-boring beetles in the under-bark microclimate. For Ecol Manage 269:149–157

Vitt LJ, Cooper WE (1986) Foraging and diet of a diurnal predator (*Eumeces laticeps*) feeding on hidden prey. J Herpetol 20:408–415

Wang Q, Zeng W (2004) Sexual selection and male aggression of *Nadezhdiella cantori* (Hope) (Coleoptera: Cerambycidae: Cerambycinae) in relation to body size. Environ Entomol 33:657–661

Wang Q, Zeng WY, Li JS (1990) Reproductive behavior of *Paraglenea fortunei* (Coleoptera, Cerambycidae). Ann Entomol Soc Am 83:860–866

Wang R, Ju G, Qin X (1995) Study on the chemicals in bark of *Populus tomentosa* Carr. resistant to *Anophlophora glabripennis* Motsh. Sci Silvae Sin 31:184–188

Wanjala FME, Khaemba BM (1987) The biology and behaviour of *Iphiaulax varipalpis* (Hymenoptera: Braconidae) as a parasite of *Dirphya nigricornis* (Coleoptera: Cerambycidae). Entomophaga 32:281–289

Ware VL, Stephen FM (2006) Facultative intraguild predation of red oak borer larvae (Coleoptera: Cerambycidae). Environ Entomol 35:443–447

Way MJ, Cammell ME, Paiva MR (1992) Studies on egg predation by ants (Hymenoptera, Formicidae) especially on the eucalyptus borer *Phoracantha semipunctata* (Coleoptera, Cerambycidae) in Portugal. Bull Entomol Res 82:425–432

Webb JL (1910) Injuries to forests and forest products by roundheaded borers. U.S. Department of Agriculture, Washington, DC, pp 341–358

Wei X, Wu Y, Reardon R, Sun TH, Lu M, Sun JH (2007) Biology and damage traits of emerald ash borer (*Agrilus planipennis* Fairmaire) in China. Insect Science 14:367–373

Weiss HB, Nicolay AS (1919) Notes on the life-history and early stages of *Brachys ovatus* Web., and *Brachys aerosus* Melsh. Can Entomol 51:86–88

Wellington WG (1954) Weather and climate in forest entomlogy. In: Sargent F, Stone RG (eds) Recent studies in bioclimatology, meterorological monographs. American Meteorological Society, Boston, MA

Willemstein SC (1987) An evolutionary basis for pollination ecology. Leiden botanical series. E.J. Brill, Leiden

Williams GA, Williams T (1983) List of Buprestidae (Coleoptera) of the Sydney basin, New South Wales, with adult plant records and biological notes on food plant associations. Aust Entomol 9:81–93

Yamasaki T, Sakai M, Miyawaki S (1989) Oviposition stimulants for the beetle, *Monochamus alternatus* Hope, in inner bark of pine. J Chem Ecol 15:507–516

Yanega D (1996) Field guide to northeastern longhorned beetles (Coleoptera: Cerambycidae). Illinois Natural History Museum, Champaign

Yang ZQ, Wang XY, Zhang YN (2014) Recent advances in biological control of important native and invasive forest pests in China. Biol Control 68:117–128

Yu C (1992) *Agrilus marcopoli* Obenberger. Forest insects of China, 2nd edn. China Forestry Publishing. Beijing

Yu DSK, van Achterberg C, Horstmann K (2016) Taxapad 2016—world Ichneumonoidea 2015 database. Nepean, ON

Zhao BG (2008) Pine wilt disease in China. In: Zhao BG, Futai K, Sutherland JR, Takeuchi Y (eds) Pine wilt disease. Springer, Tokyo, pp 18–25

Zhao T, Gao R, Liu H, Bauer LS, Sun L (2004) Host range of emerald ash borer, *Agrilus planipennis* Fairmaire, its damage and the countermeasures. Acta Entomol Sin 48:594–599

Zylstra KE, Mastro VC (2012) Common mortality factors of woodwasp larvae in three northeastern United States host species. J Insect Sci 12:1–8

Chapter 13
Sap-Sucking Forest Pests

Manuela Branco, José Carlos Franco, and Zvi Mendel

13.1 Introduction

Piercing-and-sucking insects are distinguished by their specialized mouthparts, with stylets adapted to penetrate and suck fluids from plant or animal tissues. These insects are primarily hemipteroids (e.g. Psocodea, Hemiptera, and Thysanoptera), Siphonaptera (fleas), and nematocerous Diptera. Among the piercing-and-sucking hemipteroids, the sucking lice (Psocodea, Anoplura) are obligate ecotoparasites, feeding on mammal or bird blood. Most of the hemipteran species are phytophagous. The non-phytophagous species all belong to the sub-order Heteroptera and are predators, scavengers, a few are blood-feeders and some are necrophages. Thrips (Thysanoptera) include mycetophagous species (about 50%), phytophagous and a few predatory species. Non-hemipteroid piercing-and-sucking insects are mostly insectivores or blood-feeders, and less commonly feed on fungi or algae (Gullan and Cranston 2014; Labandeira and Phillips 1996; Morse and Hoddle 2006).

Two feeding strategies exist within the phytophagous hemipterans: (1) salivary sheath feeding, whereby individuals feed on contents of plant vascular tissue, i.e. phloem or xylem (most sap feeders); or (2) cell rupture feeding, i.e. mesophyll feeders. Sap-sucking insects are salivary sheath feeders and can be further differentiated in two feeding guilds: (1) phloem-feeders, most of the species of the suborder Sternorrhyncha, including aphids (Aphidoidea), whiteflies (Aleyrodoidea), scale insects (Coccoidea), and psyllids (Psylloidea); many Auchenorrhyncha, including treehoppers and many leafhoppers (Membracoidea: Membracidae, most Cicadellidae), and planthoppers (e.g. Fulgoroidea: Cixiidae,

M. Branco (✉) · J. C. Franco
Forest Research Center, School of Agriculture, University of Lisbon, Lisbon, Portugal
e-mail: mrbranco@isa.ulisboa.pt

Z. Mendel
Department of Entomology, Agriculture Research Organization, The Volcani Center, Bet Dagan, Israel

© The Author(s) 2023
J. D. Allison et al. (eds.), *Forest Entomology and Pathology*,
https://doi.org/10.1007/978-3-031-11553-0_13

Delphacidae, Flatidae, Fulgoridae), and some Heteroptera; and (2) xylem-feeders, such as cicadas (Cicadoidea), spittlebugs (Cercopoidea), and sharpshooter leafhoppers (Cicadellinae), all belonging to the suborder Auchenorrhyncha. Phytophagous thrips feed on the content of individual epidermal or mesophyll cells (i.e. mesophyll feeders) (Bennett and Moran 2013; Chuche et al. 2017a; Douglas 2006; Labandeira and Phillips 1996; Redak et al. 2004).

Many sap-sucking insect species are of major economic importance, because they often cause plant stress, distortion, shoot stunting, and gall formation, or transmit plant pathogens (Baumann 2005; Gullan and Cranston 2014). In this chapter, we address the evolution and diversity of sap-sucking insects of forest trees, with an emphasis on the two major groups, aphids and scale insects. We present their biology and ecology. Particular emphasis is given to their highly specialized feeding mode and biotic interactions. Finally, we discuss sap-sucking forest pests and their management. In Figs. 13.1a–l and 13.2a–k, we provide images of example species and some of the aspects mentioned in the chapter.

13.2 Diversity and Biology of Sap-Sucking Insects with Emphasis on Importance for Forestry

13.2.1 Background

Insects evolved on land in the Ordovician, about 480 million years ago (Misof et al. 2014). Approximately 80 million years later, in the Devonian period, one lineage of insects evolved flight. Hemipteran insects, with their defining trait of piercing-sucking stylet mouthparts (Fig. 13.1d), probably arose in the Carboniferous period, about 300 million years ago. The order Hemiptera is divided into several monophyletic branches, including the Auchenorrhyncha (cicadas, spittlebugs, leafhoppers, treehoppers), the Sternorrhyncha (scale insects, psyllids, whiteflies, aphids) and the Heteroptera (true bugs) (Song et al. 2012). The oldest fossils of aphids, coccoids and Heteroptera are from the Triassic period, about 220 million years ago (Hong et al. 2009). Table 13.1 displays the different major groups of Hemiptera and the significance of their members as forest pests.

13.2.2 Aphids: Aphidomorpha

The entire Aphidomorpha infraorder has now been fully catalogued: 5,218 valid extant and 314 valid extinct species (Favret et al. 2016). Aphids form a distinctive insect clade that features considerable variability in their biological traits. For example there are many distinct, yet genetically identical, forms of females during the life cycle (polyphenism), alternation of sexual and asexual reproduction and seasonal

Fig. 13.1 **a** *Paleococcus fuscipennis* mating; **b** *Thaumastocoris peregrinus* development stages; **c** *Matsucoccus feytaudi* male; **d** Mouth parts of *T. peregrinus*; **e** and **h** Ants tending lerps of *G. brimblecombei*, *Crematogaster scutellaris* and *Plagiolepis pygmaea*, respectively; **f** *Elatophilus hebraicus* (Anthocoridae) feeding on *Matsucoccus jospehi* while mating; **g** Larviform nymphs of *M. feytaudi* eclosing from egg mass. **i** A whitefly (Aleyrodidae); **j** A soft scale, *Kermes echinatus*; **k** and **l** Adult and larvae of *Iberorhyzobius rondensis* (Coccinelidae), monophagous predator of *M. feytaudi*. Photo credits: 13.1a, b, f, j © Alex Protasov; 13.1d, i © Zvi Mendel; 13.1e, h © Vera Zina; 13.1c, g, k, l © Manuela Branco

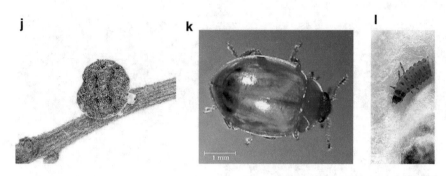

Fig. 13.1 (continued)

alternation between unrelated groups of host plants. These traits vary among species, reflecting evolutionary histories of biogeographical expansions and contractions and co-diversification with plant hosts. Many aspects of phylogenetic relationships among aphids remain unresolved, and several evolutionary questions unanswered (Nováková et al. 2013).

Extant Aphidomorpha are divided in two superfamilies. The superfamily Aphidoidea includes the true aphids (Aphididae). The superfamily Phylloxeroidea includes the adelgids (Adelgidae), feeding on conifers, and phylloxerids (Phylloxeridae), whose members develop on broadleaf trees. Adelgidae, unlike true aphids, have no tail-like cauda and no cornicles. All three families have winged and wingless forms. Winged forms are produced for dispersal in asexual generations, and for migration between the secondary and primary host plants (see Sect. 13.3.1). In these species, the sexual morphs mate on the primary host, producing overwintering eggs.

Adelgids are often known as "woolly conifer aphids" or "woolly adelgids". The family is composed of species associated with pine, spruce and other Pinaceae genera (Table 13.2, Fig. 13.2i). The most common classification system recognizes 8 genera (*Adelges, Aphrastasia, Cholodkovskya, Dreyfusia, Eopineus, Gilletteella, Pineus* and *Sacchiphantes*), including about 50 species. All of them are native to the northern hemisphere, although some have been spread to the southern hemisphere as invasive species. Adelgids exhibit cyclical parthenogenesis and are oviparous (Havill and Foottit 2007). They exhibit a two-year life cycle, with some species alternating hosts between spruce (*Picea*) and other Pinaceae (*Abies, Larix, Pinus, Pseudotsuga, Tsuga*).

The phylloxerids include 75 described species within two subfamilies (Phylloxerininae and Phylloxerinae) and 11 genera (Blackman and Eastop 1994). They have a worldwide distribution but are more diverse in temperate climates where they likely originated. Their adaptation to tropical habitats is probably secondary. They feed on leaves and roots and induce galls at their feeding sites. Most phylloxerids feed on Juglandaceae or Fagaceae. Still, they are not considered significant pests of forest trees. In Israel, *Phylloxera quercus* occurs on oak trees (mainly *Quercus calliprinos*) in very low densities and may be often found on shoots developed after a major

Fig. 13.2 a and **b** Predators (Anthocoridae and Hemerobiidae) and males of *Matsucoccus feytaudi* attracted to lures with the sex pheromone of the bast scale; **c** Band trap to collect ovipositing females of *M. feytaudi*; **d** Honeydew exploited by ants; **e** Soft scales (Coccidae) and sooty mold on *Ficus sycamorus*; **f** Infestation of *Glycaspis brimblecombei* on *Eucalyptus camaldulensis*; **g** *Ultracoelostoma* sp. canal tube with a drop of honeydew and sooty mold on *Nothofagus*; **h** Honey production in Crete from *Marchalina helenica*; **i** Forest damage caused by *Pineus pini* on *Pinus pinea*; **j** Crystallized honeydew excreted by *Cinara palestina*; **k** Large colony of *Cinara cedri*. Photo credits: 13.2a, e, h, i, j, k © Zvi Mendel; 13.2d, f © Vera Zina; 13.2b, c, g © Manuela Branco

Table 13.1 General economic relevance as forest pests of Hemiptera, by suborder and superfamily

Suborder	Superfamily or family	Common names	Significance as forest pests
Auchenorrhyncha	Cercopoidea	Spittlebugs, froghoppers	Low
	Cicadoidea	Cicadas	Low
	Membracoidea	Leafhoppers and treehoppers	Low
	Fulgoroidea	Plant-hoppers	Low
Coleorrhyncha	Peloridiidae	Moss bugs	None
Heteroptera	eight infraorders	True bugs	Low
Sternorrhyncha	Aleyrodoidea	Whiteflies	Low
	Aphidoidea	Aphids	High
	Coccoidea	Scale insects	High
	Phylloxeroidea	Adelgids, phylloxerids	High
	Psylloidea	Jumping plant lice, psyllids	high

Table 13.2 Aphidomorpha as related to major forest tree genera

Family	*Acacia*	*Eucalyptus*	*Quercus*	*Pinus*
Adelgidae	None	None	None	Several species mainly of the genus *Pineus*
Phylloxeridae	None	None	Several species mainly of the genus *Phylloxera*	None
Aphididae	Nonspecific, polyphagous species are often recorded	Nonspecific, polyphagous species were rarely recorded	225 species, from six genera, *Hoplocallis, Lachnus, Myzocallis, Stomaphis, Thelaxes, Tuberculatus*	Members of the Lachninae, mainly the genera *Cinara, Eulachnus* and *Schizolachnus*

fire (Table 13.2). However, like the grape phylloxera *Daktulsphaira vitifoliae*, the majority of the population occurs on the roots.

The true aphids are a very large insect family, including several thousand species, many of which are known as serious plant pests (Table 13.2). The oldest aphid fossils are from the Triassic (at least 220–210 mya) but aphids may have originated in the Permian. Phylogenetic analysis of molecular data suggests that aphids underwent a rapid radiation into the current tribes after switching from gymnosperms to angiosperms, sometime during the Upper Cretaceous. Furthermore, the ancestral aphid probably had a simple life cycle with host alternation evolving independently in each of the subfamilies (Gullan and Martin 2009).

Two genera of the subfamily Lachninae, i.e. *Cinara*, the conifer aphids or giant conifer aphids, and *Eulachnus*, are well known as forest pests. *Cinara* species ($\approx$150) are widespread in the Northern Hemisphere. Many species are native to North America, but there are also 55 species found in Europe and Asia. They are specialized on members of the families Pinaceae and Cupressaceae. *Cinara* spp. do not alternate hosts. The genus *Eulachnus* comprises about 17 species, all of which live on pine needles. They are cryptic when feeding, but very active when disturbed. The best-known species show preferences for certain *Pinus* spp., but none is strictly confined to one species. Several introduced *Cinara* spp. have become serious pests of forest plantations. For example, *Cinara cedri* (Fig. 13.2k) and *Cinara laportai* attack true cedars. *Cinara cupressi* causes damage to cypress trees. This aphid of unclear origin is an invasive species in Africa and Europe, South America and in the Middle East.

13.2.3 Jumping Plant Lice: Psylloidea

Psylloidea is a superfamily of true bugs, including the jumping plant lice, recently classified in eight families (Aphalaridae, Calophyidae, Carsidaridae, Homotomidae, Liviidae, Phacopteronidae, Psyllidae and Triozidae). There are about 3,000–3,500 described species. They are common worldwide, but most diverse in tropical and subtropical areas. Most Australian psyllids belong to the subfamilies Spondyliaspidinae (Aphalaridae) and Acizziinae (Psyllidae). The former is largely associated with eucalypts and the latter with acacias (Carver et al. 1991). Members of the Psylloidea also include many gall-inducing species, which are narrowly host-specific, and are most species-rich in the tropics and south temperate regions (Burckhardt 2005).

Psyllids reproduce sexually and mature through five nymphal stages. Unlike aphids, psyllids insert their eggs into host plant tissue (Hodkinson 1974). They are phloem feeders and produce honeydew. Many are known vectors of plant diseases. Of major economic importance are those vectoring *Liberibacter* and *Phytoplasma* species, the causal agents of serious plant diseases. They generally have narrow host ranges and are restricted almost exclusively to perennial dicotyledonous plants. Within species, nymphs usually have a more restricted host range than adults. At low densities, nymphal survival is enhanced by group feeding similar to many Heteroptera. Many species occur in dry or semi dry areas and the immature stages exhibit morphological and behavioral adaptations to resist desiccation. This is for example seen in the circular lerp of the red gum lerp psyllid, *Glycaspis brimblecombei*, which may also protect them from predators (Fig. 13.1e, h).

Psyllids are not known from conifers, but may overwinter on them (e.g. Čermák and Lauterer 2008). Also, they are not common on oak trees and none are known as pests of oaks. Conversely, psyllids feeding on eucalypts are among the most devastating insect pest groups in Australia, affecting both native forests and plantations.

Psyllids of several different families have become economically important invasive pests. Several Aphalaridae species, originating from Australia, have been introduced into other continents, where they become important pests, causing severe damage in eucalypt plantations (Hurley et al. 2016). Most of these species infest the river red gum, *Eucalyptus camaldulensis*. One of the most widely distributed psyllid species is the red gum lerp psyllid *G. brimblecombei* (Figs. 13.1e and 13.2f). The psyllid *Acizzia jamatonica*, native to China, has been reported from Europe and North America as a pest of *Albizia julibrissin*. *Acizzia uncatoides*, native to Australia, develops on many ornamental *Acacia* and *Albizia* species outside Australia. In Hawaii, it occurs on the native *Acacia koa*. *Calophya schini* (Calophyidae) is a leaf galling psyllid that feeds exclusively on Peruvian pepper tree, *Schinus molle* (Anacardiaceae). This psyllid is now present in California, Mexico, Portugal, South Africa, Ethiopia and Kenya (e,g. Overholt et al. 2013; Zina et al. 2012). The psyllid *Macrohomotoma gladiata* (Homotomidae) is a new insect pest of *Ficus microcarpa* originating from Asia, which has recently been found in Spain (Alicante) on urban trees (Mifsud and Porcelli 2012) and is now widely spread in East Mediterranean.

13.2.4   Scale Insects: Coccoidea

The Coccoidea is one of the four superfamilies of the monophyletic suborder Sternorrhyncha (e.g. Gullan and Martin 2009), with 49 families presently recognized (Ben-Dov et al. 2014). The early evolution of Coccoidea must have occurred during the early to mid-Mesozoic, as a sister group of the Aphidoidea (Hennig 1981). Almost all the main lineages of modern coccoids have been identified from Tertiary amber, but relatively few earlier fossils are known (Koteja 1986, 1990). The morphology of these early fossils suggests that some groups of the plesiomorphic Margarodidae sensu lato had reached their contemporary organization by the Lower Cretaceous (Koteja 1990). Miller (1984) used the aphids as outgroup and showed that the earliest scale insects were margarodid-like and that Margarodidae and then the Ortheziidae are successively sister to the remainder of the Coccoidea.

Traditionally, based on the possession of abdominal spiracles, the scale insects were separated between Archaeococcoids and Neococcoids, with the latter characterized by features such as the loss of abdominal spiracles (Koteja 1990). The Archaeococcoids comprise several families, such as Monophlebidae, Margarodidae, Orthezidae and Matsucoccidae. The Neococcoids comprise most of the currently recognized families and species of scale insects (e.g. Kosztarab 1982). Its major groups probably evolved in conjunction with the angiosperms. Almost all neococcoid fossils are from Eocene or younger deposits. Yet, the neococcoid radiation must have begun much earlier because all of the major families (Coccidae, Diaspididae, Eriococcidae and Pseudococcidae) were already present in the Eocene. The soft scales (Coccidae) consist of four major subgroups, the Ceroplastinae, Coccini, Pulvinariini and Saissetiini (Miller and Hodgson 1997). The armored scales (Diaspididae) are divided in several subfamilies and tribes with the most important being Aspidiotini and the Diaspidini. The felt scales (Eriococcidae) are not a single monophyletic group, but a complex of several different groups and several families (Cook et al. 2002). The mealybugs (Pseudococcidae) are currently separated in two subfamilies, Phenacoccinae and Pseudococcinae (Danzig and Gavrilov-Zimin 2015).

There is a very marked sexual dimorphism of adult male and female scale insects (Fig. 13.1a, c). This sexual dimorphism is established by divergent postembryonic developmental pathways after the first-instar nymph, possibly regulated by growth hormones (Vea et al. 2016). Danzig (1980) suggested that neoteny shortens female development time on coccids, resulting in nymphal morphology in adult females. This is not unique to scale insects. The maintenance of juvenile features in adults, through neoteny, has evolved independently at least six times in insects. Mature adult female scale insects generally have a large body relative to nymphs or teneral females (Ben-Dov 1990) and often have very high fecundity (McKenzie 1967).

The inability to fly has been suggested to be adaptive for female, but not male scale insects, because more resources can be allocated to egg production (Roff 1990). The adult scale insect male morphology is adapted for flight and finding females. The male mesothoracic wings have reduced venation, whereas the metathoracic ones are lost or reduced enabling the male to control its equilibrium in flight. The mouthparts in males have become nonfunctional (e.g. Afifi 1968; Gullan and Kosztarab 1997; Kawecki 1964). Waxy caudal filaments in males are known in several coccoid families (Fig. 13.1c) (Afifi 1968; Giliomee 1967). Duelli (1985) suggested that these filaments assist in stabilizing flight. The elongation of the wax caudal filaments was shown to be correlated with sexual maturation of adult males in mealybugs (Mendel et al. 2011). Because of extreme sexual dimorphism in scale insect, the conspicuous and longer-lived adult females are used in taxonomy. For many coccoid genera and species, the male is unknown.

Modern scale insects are all plant feeders, using their stylet-like mouthparts to suck sap from the phloem or parenchyma cells. The first scale insects probably fed on proto-angiosperms, gymnosperms or lower plants, or on fungi and bacteria (Koteja 1986). Vea and Grimaldi (2016) suggested that most major lineages of coccoids shifted from gymnosperms onto angiosperms, when the latter became diverse and abundant in the mid- to late Cretaceous. Alternatively, the ancestral scale insects may have fed on the contents of individual cells from roots, rotting plants, or fungal hyphae. Koteja (1986) has hypothesized that the leaf-litter layer is the primary habitat of coccoids and that feeding on above-ground plant parts is a secondary adaptation.

All forest tree species are infested by scale insects. Figure 13.3 compares the frequency of occurrence of species belonging to eight families, on four tree genera from different botanical families: *Pinus* (with ~111 tree species and 182 scale species) and *Quercus* (~600 tree species and 227 scale species), of Laurasia origin; *Eucalyptus* (~700 tree species and 273 scale species) and *Acacia* (~980 tree species and 308 scale species), of Gondwanaland origin. These tree genera are naturally distributed over large areas and comprise several tree species, which are among the most economically and environmentally important globally.

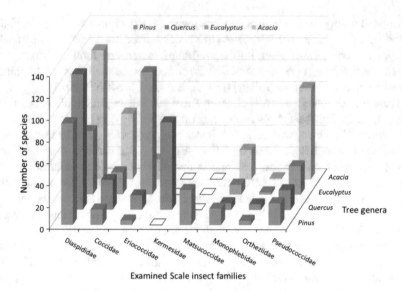

Fig. 13.3 Species frequency distribution of eight scale insect families among four genera of forest trees. The data were retrieved from Scale Net (García Morales et al. 2016)

It is interesting to note that some scale insects are primarily associated with specific tree genera. For example, Eriococcidae is the largest family of scale insects on *Eucalyptus* (~49%), but this family has limited representation on other tree genera (~2–6%); members of the family Kermesidae are only able to develop on Fagaceae (Fig. 13.1i), predominantly the genus *Quercus* (e.g. Spodek et al. 2013), while Matsucoccidae are associated only with the genus *Pinus* (e.g. Foldi 2004). Conversely, other families are evenly distributed across tree taxa. In particular Diaspididae, the largest family of scale insects, is the dominant scale insect family of *Pinus, Quercus* and *Acacia* (~38–51% of the species), and the second most species-rich family in *Eucalyptus*. Coccidae is also present in all the four tree genera, but is most common on *Acacia* (Fig. 13.3).

13.2.5 Other Hemipteran Superfamilies and Their Importance in Forestry

13.2.5.1 Auchenorrhyncha: Cicadomorpha

Frog hoppers and spittlebugs (Cercopoidea) are common insects on eucalypts. Frog hopper nymphs construct tubes which are attached to stems and twigs. The nymphs live inside these tubes where they are protected from desiccation and also to some extent from parasites and predators. Spittlebug nymphs live within a white, frothy secretion that resembles spittle (hence the name). As with the froghopper tubes, the spittle protects the nymphs from desiccation and from parasitoids and predators. The nymphs take shelter and feed beneath this froth. Members of the genus *Aphrophora* are known as minor pests of several trees in North America and Europe (Floren and Schmidl 2009).

Generally, cicadas are not significant forest pests. Nymphs live in the soil and feed on roots while adults feed on above-ground parts of plants, but this seems to have little effect on plant growth. During oviposition, females pierce plant tissues with their ovipositor to lay eggs and this can result in structural damage. For example, the cuts made by *Amphipsalta* spp. during oviposition may weaken twigs and branches sufficiently that they break in high winds. Such broken branches on conifers show up as reddish "flags" in the canopy when the foliage dies (Kay 1980). Open cuts also provide points of entry for pathogens and wood-boring insects. Often the cuts heal over making the twigs gnarled in appearance.

Leaf hoppers, family Cicadellidae (whose hind legs are modified for jumping) and treehoppers and thorn bugs, family Membracidae, are common in various types of forest all over the world, but not considered important forest pests. For example, aggregations of *Oxyrhachis versicolor* are conspicuous on *Tamarix* trees in Israel, but result in no apparent damage to the trees. Leafhoppers are commonly associated with broadleaved forest trees (Beirne 1956). In Costa Rica, some species have been reported to cause minor damage to native broadleaf trees, as well as to introduced eucalypts (e.g. *Macumolla ventralis* and *Graphocephala coccinea*) (Gamboa 2007). Leafhoppers gained particular attention as vectors of the plant pathogenic bacterium *Xylella fastidiosa*. In particular, sharpshooter leafhoppers (Cicadellinae) are the best-studied vectors of *X. fastidiosa* (Cornara et al. 2019). For example, the glassy-winged sharpshooter, *Homalodisca vitripennis*, an invasive species in California affecting a wide host range of trees, is an efficient vector of *X. fastidiosa* (Almeida and Nuney 2015).

13.2.5.2 Auchenorrhyncha: Fulgoroidea

The planthopper superfamily Fulgoroidea comprises approximately 20 described insect families, depending on which classification is followed, and includes a diverse group of phytophagous or fungivorous insects, exceeding 12,500 species. At least 160 species in 16 families are recorded as pests, including some of major economic importance, such as the brown planthopper, *Nilaparvata lugens* on rice. Planthoppers are vectors of viral and bacterial (including phytoplasma) agents causing plant diseases. The Tropiduchidae comprises 652 described species, about 4.9% of all Fulgoromorpha. Most species feed on shrubs and trees, and some are crop pests. Their association with host plants is quite diverse, including 21 plant orders. Still, few or no species in most of the families cause economic damage to forests.

13.2.5.3   Sternorrhyncha: Aleyrodoidea

Whiteflies are almost entirely leaf feeders (Fig. 13.1i). In recent years, whitefly pests have become a major problem in agriculture, almost worldwide, but as forest insects are of little concern (Nair 2001). In recent years the *Ficus* whitefly, *Singhiella simplex* (Aleyrodidae) has become a pest of *Ficus* trees in North America and the Mediterranean. Primarily a tropical group, pest species are found in the warmer parts of the world. In temperate areas, several species are serious pests in glasshouses, but they do not pose any risk to forest trees (Martin et al. 2000).

13.2.5.4   True Bugs: Heteroptera

This suborder is highly diverse, although few species are considered as important pests of forest trees. It is interesting to mention two invasive species in Europe and the Mediterranean. One is the bronze bug *Thaumastocoris peregrinus* (Thaumasto-coridae), native in Western Australia affecting *Eucalyptus* (Fig. 13.1b). Infestations are noted by the reddening of eucalypt canopy leaves and loss of leaves, leading to canopy thinning and occasionally branch dieback or tree mortality (Nadel et al. 2010). Another pest is the Sycamore Lace Bug *Corythucha ciliata* (Heteroptera: Tingidae). This North American species, was introduced in Europe in the 1960s and first found in China in 2006, where it has major impacts on *Platanus* tree health in urban parks and street trees (Maceljski 1986; Ju et al. 2009).

13.3   Biology and Ecology of Sap-Sucking Insects

13.3.1   General Models of Life History and Seasonal History

The life cycle of sap-sucking insects may include two or three nymphal stages (e.g. female scale insects), four (e.g. male scale insects, aphids, whiteflies) or five (e.g. jumping plant lice, Auchenorrhyncha and Heteroptera (Fig. 13.1b) (Dietrich 2003; Gullan and Martin 2009). In some cases, such as in whiteflies and scales insects, the immature stages are larviform, much different from adults (Fig. 13.1g), followed by a non-feeding pupal instar. This developmental pattern is similar to holometabolic insects (Gullan and Cranston 2014). The number of complete generations within a year varies among species and with climatic conditions. Univoltine species (one generation per year) are common in cold temperate regions and multivoltine species (more than two generations) in warmer climates (Dietrich 2003; Gullan and Martin 2009). Some species (e.g. the psyllid *Strophingia ericae*) may take more than one year to complete one generation (Hodkinson 2009). *Magicicada* (Cicadidade) species have periodical life cycles that last 13 or 17 years (Grant 2005).

Polyphenism, i.e. phenotypic differences determined by environmental conditions, occurs in some sap feeders. Some aphids show very complex polyphenisms. Parthenogenetic females may have up to eight different phenotypes, and sexual forms are polymorphic. Phenotypic differences may include morphological and physiological aspects, as well as fecundity, timing and size of progeny, developmental time, longevity, and host-plant selection. Photoperiod, temperature and maternal effects are among the environmental cues triggering the development of different aphid morphs (Gullan and Cranston 2014). Adelgids, phylloxerids, and aphids present a complex, polymorphic life cycle with cyclical parthenogenesis and host alternation. Some species are holocyclic, meaning they produce both asexual and sexual generations, while others are anholocyclic, producing only asexual generations. The typical adelgid holocycle takes two years to complete and involves five generations: three are completed on the primary host (spruce, *Picea* spp.), with sexual reproduction and gall formation; and the other two are asexual generations and occur on the secondary host (*Abies, Larix, Pseudotsuga, Tsuga* or *Pinus*) (Havill and Foottit 2007).

13.3.2   Feeding Ecology

Hemipteran insects modified mouth parts include a slender beak-like labium, within which there are two pairs of long stylets (two outer mandibles and two inner maxillae), forming a bundle (Fig. 13.1d). Only the stylets penetrate into the plant tissues for piercing-and-sucking. The maxillary stylets form in their inner surface both the salivary canal and the food canal. Through these channels the insect injects saliva into the plant, or sucks up plant sap into their gut, respectively. Commonly, in sap-feeding insects, the stylets pathway up to the vascular tissue is mostly intercellular, following an apoplastic transit, i.e. following a route from cell wall to cell wall, not entering the cytoplasm. Eventually, some intracellular punctures occur to assess cell content, for host acceptance and setting the position of the stylets inside the plant.

As salivary sheath-feeding insects, sap feeders produce two types of saliva. A gelling lipoproteinaceous saliva is secreted during the penetration process of stylets, which forms a lubricating and hardening sheath around them. This gelling saliva remains within plant tissues after stylet withdrawal. Additionally, a watery saliva, is directly injected in the vascular tissue before sap uptake. This watery saliva may interfere with host-plant defense responses, for example, through proteins that are involved in the detoxification of phenols or interact directly with plant defense signaling (Giordanengo et al. 2010; Gullan and Cranston 2014; Kingsolver and Daniel 1995; Will et al. 2013).

Phloem and xylem tissues have very different characteristics and consequently sap-feeding insects usually specialize on one or the other (Labandeira and Phillips 1996; see Table 13.3). However, phloem-feeding insects may occasionally ingest xylem-sap, possibly for regulating osmotic potential (Pompon et al. 2011). Phloem sap is a rich source of carbon and energy (i.e. sugars), also providing nitrogen (mostly as free amino acids). Also phloem sap usually has no toxins and feeding deterrents.

Table 13.3 Main characteristics of phloem and xylem (based on Chuche et al. 2017b; Dinant et al. 2010; Kehr 2006; Lucas et al. 2013; Redak et al. 2004)

Type	Phloem	Xylem
Sap flow	From source (e.g. leaves) to sink tissues (e.g. roots, growing shoots, and fruits)	From roots to aboveground tissues
Sap pressure	Positive	Negative
Compounds transported	Water, minerals (especially K), amino acids, organic acids, sugars (e.g. sucrose, raffinose, polyols), information molecules (e.g. phytohormones, proteins, RNAs)	Water, minerals, little organic nutrients (e.g. amino acids, organic acids, sugars), information molecules (e.g. phytohormones)
Main functions	Allocating photoassimilates, and information molecules	Transporting water, minerals, and information molecules

However, sap-sucking insects need to overcome two major nutritional problems to feed on phloem-sap: (1) nitrogen quality in phloem-sap is low (the ratio between essential and non-essential amino acids in phloem-sap is 1:4–1:20); and (2) phloem sap has a very high concentration of sugars (i.e. an osmotic pressure 2–5 times higher than that in insect's hemolymph) (Douglas 2006).

Phloem feeders excrete the excess carbohydrates from their unbalanced diet in the form of a sugary fluid, honeydew (Baumann 2005) (Fig. 13.2d, g). Xylem sap is often less nutritional than phloem sap, containing low concentrations of carbon and nitrogen compounds and is under negative pressure. Xylem-feeders compensate this constraint with high ingestion rates and generally have larger bodies than phloem-feeders. Also due to the relatively higher metabolic costs of xylem-sap extraction (Chuche et al. 2017b). In fact, the suction pressure needed for sucking xylem sap decreases with food canal width, which is directly correlated to body size of the insect (Novotny and Wilson 1997).

As both phloem and xylem are unbalanced food sources, sap-feeding insects rely on symbiotic bacteria to provide them the essential nutrients lacking in their diet (Bennett and Moran 2013). Primary endosymbionts (P-endosymbionts) are obligate mutualistic bacteria, localized within hemipteran-host polyploid cells (the bacteriocytes), which normally aggregate into a specialized organ, the bacteriome. P-endosymbionts are essential for host survival and reproduction, and are present in all individuals of the host population (Table 13.4). They present an extreme genomic reduction as a result of vertical transmission, from reproductive females to progeny, and living inside bacteriocytes.

In addition to these nutritional primary symbionts, sap-sucking insects may also contain one or more facultative or secondary symbionts (S-symbionts). In general these symbionts are not necessary for host development and reproduction. S-symbionts may inhabit a variety of tissues other than bacteriocytes, often do not infect all individuals within host populations, and can be horizontally transmitted among hosts. These symbionts are known to manipulate host reproduction and provide their

Table 13.4 Examples of P-endosymbionts found in sap-feeding insects (based on Baumann 2005; Morrow et al. 2017)

Host insect taxa	Studied host food source	P-endosymbiont
Sternorrhyncha		
– Psylloidea	Phloem sap	*Carsonella rudii* (Gammaproteobacteria)
– Aleyrodoidea	Phloem sap	*Portiera aleydodidarum* (Gammaproteobacteria)
– Aphidoidea	Phloem sap	*Buchnera aphidicola* (Gammaproteobacteria)
– Pseudococcidae	Phloem sap	*Tremblaya princeps* (Betaproteobacteria)
Auchenorrhyncha	Phloem or Xylem sap	*Sulcia muelleri* (Bacteroidetes) and co-primary symbionts from different bacterial division, e.g. *Hodgkinia* (Alphaproteobacteria), *Vidania, Nasuia,* and *Zinderia* (Betaproteobacteria), *Baumannia,* and *Purcelliella* (Gammaproteobacteria)

hosts with a range of adaptive ecological traits. These include increased host-plant range, efficiency of plant pathogen transmission, and greater resistance to biotic (e.g. parasitoids) or abiotic (e.g. temperature, insecticides) environmental stress (Baumann 2005; Chuche et al. 2017a; López-Madrigal and Gil 2017; Oliver et al. 2010).

Examples of secondary symbionts found in aphids include *Serratia* (47% of the studied aphid species), *Wolbachia* (43%), *Hamiltonella* (34%), *Regiella* (33%), *Rickettsia* (29%), X-type (14%), *Spiroplasma* (13%), and *Arsenophonus* (9%) (Zytynska and Weisser 2016). A peculiar symbiotic organization was observed in the citrus mealybug *Planococcus citri*, in which each cell of the P-endosymbiont *Tremblaya princeps* harbors several cells of the S-symbiont *Moranella endobia*, representing the first known case of prokaryote-prokaryote endocelullar symbiosis (López-Madrigal et al. 2013).

13.3.3 Reproductive Strategies

Sexual reproduction and oviparity are the most common modes of reproduction in sap-sucking insects (Dietrich 2003; Gullan and Martin 2009). However, other reproductive strategies can be observed in this insect guild (Table 13.5). Different types of parthenogenesis, i.e. apomixis, automixis, and pseudogamy, are known for example in aphids, whiteflies, scale insects and plant hoppers. Examples of mixed systems, including different types of alternation between sexual reproduction and parthenogenesis (e.g. facultative and cyclic parthenogenesis), have been described in adelgids, phylloxerans, aphids and other hemipterans. Few species, such as *Icerya* spp. are hermaphrodite (Gullan and Martin 2009; Ross et al. 2010).

Table 13.5 Genetic systems of sap-sucking insects (based on Normark 2003)

Genetic system			Examples
1. Obligate amphimixis (sexual reproduction) Females receive one haploid genome from each parent	**Diplodiploidy** Males receive one haploid genome from each parent		Most insect species
	Haplodiploidy Only mother's genome is transmitted to males	**Arrhenotoky** Males develop from unfertilized eggs inheriting only a haploid genome from the mother	Whiteflies (Aleyrododoidea), and iceryine scales (Margarodidae)
		Paternal genome elimination Males develop from a diploid zygote, but transmit only mother's genome	Most scale insects (Coccoidea)
2. Thelytoky (parthenogenesis) Only the maternal genome is transmitted by females; only daughters are produced	**Thelytokous parthenogenesis** Absence of mating and males	**Apomixis** Ameiotic parthenogenesis	Some aphid species (Aphididae), scale insects (Margarodidae, Coccidae, Pseudococcidae, Diaspididae) and Delphacidae
		Automixis Meiotic parthenogenesis	Some species of whiteflies, and scale insects (Margarodidae, Pseudococcidae, Coccidae, Diaspididae)

(continued)

Table 13.5 (continued)

Genetic system		Examples
	Sperm-dependent thelytoky Mating with males of a related population is necessary to initiate development	**Pseudogamy** The sperm nucleus does not fuse with the egg nucleus. Only maternal genes are transmitted	Known in Delphacidae
3. Mixed systems Regular or irregular alternation between amphimixis and thelytoky	**Thelytoky alternating with haplodiploidy**	**Facultative haplodiploidy-thelytoky** Facultative parthenogenesis	Known in few Hemiptera
	Thelytoky alternating with diplodiploidy	**Cyclic diplidiploidy-thelytoky** Cyclic parthenogenesis	Adelgids (Adelgidae), phylloxerans (Phylloxeridae), and aphids (Aphididae)

13.3.4   Insect-Plant Interactions

Sap sucking insects may be classified in different feeding groups, based on the part of the host plant they feed on: (1) shoots and tips, e.g. the Cooley spruce gall adelgid, *Adelges cooley*; (2) foliage, e.g. the green spruce aphid, *Elatobium abietinum*; (3) trunk and branches, e.g. maritime pine bast scale, *Matsucoccus feytaudi*, Beech scale insect *Ultracoelostoma* spp. (Fig. 13.2g), and (4) roots, e.g. spruce root aphid, *Pachypappa termulae* (Foldi 2004; Wood and Storer 2003).

Host range is variable among sap sucking insects. Most Auchenorrhyncha are apparently very specific, feeding in one single plant genus or species. However, many Auchenorrhyncha, especially xylem-feeders can feed and develop on several alternate plant species if the preferred host is not present. Phloem- and mesophyll-feeders tend to be more host-specific than xylem-feeders, with many species limited to host plants from a single family, genus, or species (Dietrich 2003).

In the Sternorrhyncha, host range varies among taxa. Adelgids are host specific, as each species survives and reproduces only on trees from a single genus, for both primary and secondary hosts (Havill and Foottit 2007). Most species of plant jumping lice are also host specialists as nymphs, with many restricted to a single plant genus or species, and often to certain host parts (e.g. leaves, young shoots), or growth stages. In the case of whiteflies, most are apparently oligophagous, with a few monophagous. Host plant specificity in scale insects ranges from monophagous to polyphagous. Most aphids are monoecious, meaning that development occurs on one or a few closely related host plants; however, about 10% of species are heteroecious, i.e. with host alternation. As a result, most aphids are host specific. This property is conspicuous in aphids developing on forest trees. An aphid genus is usually associated with a single host-plant family and species with a plant genus or species. The primary and secondary host of heteroecious aphids are usually unrelated and host specificity is higher in the case of the primary than in the secondary host (Gullan and Martin 2009).

Several lineages of sap sucking insects induce the formation of plant galls. Plant galls, or cecidia, are abnormal growths of plant tissue, involving cell proliferation (hyperplasy) and enlargement (hypertrophy). This abnormal tissue growth results in the development of characteristic gall structures, which are specific to a certain gall making organism (Schick and Dahlsten 2003). Gall makers evolved independently in the Hemiptera, primarily within the Sternorrhyncha, including the aphids, adelgids, phylloxerids, woolly aphids (Eriosomatidae) (Wool 2005), scale insects (Asterolecaniidae, Coccidae, Diaspididae, Eriococcidae, Kermidae) (Gullan et al. 2005), psyllids (Psyllidae), and few Auchenorrhyncha (Cercopidae, Cicadellidae) (Burckhardt 2005). Galls behave as physiological sinks in the host plant, sequestering nutrients used by developing insects inside them, as well as a defensive refuge against the natural enemies of gall makers (Schick and Dahlsten 2003; see also Chapter 14).

13.4 Associated Organisms

13.4.1 Natural Enemies

Although sap sucking insects are prey of some insectivorous vertebrates (e.g. birds and lizards), they are primarily predated by invertebrates (Table 13.6). This diverse assemblage of predators includes spiders, and insects from different orders [e.g. assassin and minute pirate bugs (Hemiptera: Reduviidae and Anthocoridae) (Fig. 13.1f), ladybirds (Coleoptera: Coccinellidae) (Fig. 13.1k), green and brown lacewings (Neuroptera: Chrysopidae and Hemerobiidae), ants (Hymenoptera: Formicidae) (but see Sect. 13.4.2), wasps (Hymenoptera: Vespidae), and predatory flies (Diptera: Cecidomyiidae, Chamaemyiidae and Asilidae)]. Parasitoids in the families Dryinidae (Hymenoptera, Chrysidoidea), Encyrtidae, Eulophidae, Aphelinidae (Hymenoptera, Chalcidoicdea), and Braconidae (Hymenoptera, Ichneumonoidea), are also natural enemies of sap suckers.

Due to their feeding habits, sap feeders are usually not affected by entomopathogenic viruses, bacteria, protozoa, or nematodes, as these entomopathogens are not common in the plant vascular system. Entomopathogenic fungi are their most important pathogens. Unlike other entomopathogens, fungi usually actively penetrate

Table 13.6 Natural enemies and other biotic factors of mortality for the major groups of Hemipteran forest pests

Hemipteran pest Superfamily	Family	Major biotic factor	1–3 major predator groups	1–2 major parasitoid groups
Aphidoidea	Aphididae	Both plant resistance and natural enemies	Coccinellidae	Aphidiinae
Phylloxeroidea	Adelgidae	Plant resistance	Coccinellidae, Chamaemyiidae, Derodontidae	–
	Phylloxeridae	Plant resistance	Coccinellidae	–
Coccoidea	Coccidae	Natural enemies	Coccinellidae	Encyrtidae
	Diaspididae	Natural enemies	Coccinellidae	Aphelinidae
	Matsucoccidae	Plant resistance	Anthocoridae	–
	Monophlebidae	Natural enemies	Coccinellidae	Cryptochetidae
	Pseudococcidae	Natural enemies	Coccinellidae, Hemerobiidae	Encyrtidae
Psylloidea	Phacopteronidae Aphalaridae Liviidae	Both plant resistance and natural enemies	Dominant groups are related to different groups and environments[a]	Eulophidae Encyrtidae

Note [a] Anthocoridae, Miridae, Coccinellidae, Chrysopidae the insect cuticle and do not need to be ingested by insects for infection to occur. Populations of sap sucking insects, particularly aphids and leafhoppers, may suffer epizootics caused by fungi (Dietrich 2003; Federici 2003; Gullan and Martin 2009).

13.4.2 Interaction with Ants

Honeydew produced by many Sternorrhyncha is a food source for different animal species, including many insects (e.g. flies, wasps, bees, ants, beetles, lacewings, butterflies, and moths), and nectar feeding birds and bats. Ant-tending behavior, which consists of collecting honeydew droplets directly from the anus of sap-sucking insects, is common among ants, especially in Dolichoderinae and Formicinae (Douglas 2006) (Figs. 13.1e, h and 13.2d). This ant behavior is linked to a food-for-protection type of mutualism (Styrsky and Eubanks 2007). A food resource, the sugar-rich honeydew excreted by tended sap-sucking insects, is traded for a service delivered by ants, the protection of hemipterans from predators and parasitoids (Way 1963). In the presence of honeydew-producing hemipterans, increased ant predation of other herbivores may indirectly benefit host plants if the amount of damage originated by those herbivores is greater than that inflicted by ant-tended hemipterans (Fig. 13.4). Conversely, tending ants may enhance the negative effects of sap-sucking insects on plants (e.g. reduced plant growth, transmission of plant pathogens), by protecting them from their natural enemies, and by increasing their feeding rate, fecundity and dispersal (Styrsky and Eubanks 2007; Vandegehuchtea et al. 2017 and references therein).

Styrsky and Eubanks (2007) reviewed the literature on the influence of ant–hemipteran interactions on arthropod communities and their host plants and reported that these interactions have mostly negative effects on the abundance and species richness of different herbivore and predator guilds. These authors also observed that in about 73% of the studies plants indirectly benefited from those interactions, as a result of increased predation of other more damaging herbivores by hemipteran-tending ants.

Fig. 13.4 Interactions among honeydew-producing hemipterans, ants and host-plants. Arrows indicate the direction of effects, positive (+) or negative (−), whereas solid arrows indicate direct effects, and dashed ones indirect effects. Redrawn from Styrsky and Eubanks (2007)

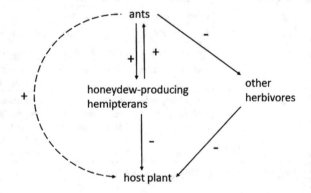

13.4.3 Bees and Honey Production from Honeydew

Honeybees are among the insects using the honeydew excreted by sap feeders as a food resource. Honeydew honeys are well known and valued in Europe and New Zealand (Table 13.7), but are also produced in North America (e.g. white cedar honey, from honeydew of the scale insect *Xylococculus macrocarpae* on *Calocedrus decurrens*) and South America (e.g. from honeydew of the scale insects *Stigmacoccus asper* on *Quercus humboldtii*, and *Tachardiella* sp. on *Mimosa scabrella*) (Azevedo et al. 2017; Chamorro et al. 2013). Originated from different plant species, mostly conifers, such as fir, spruce and pines, but also some broadleaf trees, honeydew honeys are highly valued in many European countries and are marketed with specific designations (Oddo et al. 2004). In some countries, such as Greece and Turkey, honeydew honey may represent more than 65% of total honey production, most of it produced from honeydew excreted by *Marchalina hellenica* (Marchalinidae) in pine forests (Santas 1983) (Fig. 13.2h). This species has been deliberately introduced by beekeepers, for producing honeydew, in pine forests in Crete island, Greece, where it became a serious problem. Similarly, the invasive species in Europe, the citrus flatid planthopper, *Metalfa pruinosa*, is highly appreciated by beekeepers for the honeydew with high economic value (Preda and Skolka 2011).

13.4.4 Hemiptera as Vectors of Microorganisms

As a result of their feeding habits, sap-sucking insects interact with plant pathogens which colonize the plant vascular system, such as viruses and bacteria, functioning as vectors (Perilla-Henao and Casteel 2016). Vector-borne bacteria are primarily transmitted by Auchenorrhyncha insects, including species from the superfamilies Membracoidea, Cercopoidea, and Fulgoroidea. Although less common, some Sternorrhyncha (e.g. psyllids) are also important vectors of phytopathogenic bacteria. Sap-feeding vectors of plant viruses have been reported in Auchenorrhyncha (Fulgoroidea and Membracoidea), and Sternorrhyncha (Aphidoidea, Aleyrodoidea, and Coccoidea) (Chuche et al. 2017a; Cornara et al. 2019; Perilla-Henao and Casteel 2016).

Four main types of transmission relationship between vectors and plant pathogens have been defined: (1) non-persistent; (2) semi-persistent; (3) persistent, non-propagative; and (4) persistent, propagative (Table 13.8). These types of transmission relationships were defined based on the following parameters: (1) the time needed for the acquisition and inoculation of the pathogen by the vector; (2) the retention time of the pathogen within the vector; (3) whether or not the pathogen circulates within the vector and (4) the ability of the pathogen to reproduce within the vector.

Non-persistent, non-circulative transmission of plant viruses has been only observed among viruses transmitted by aphids, whereas semi-persistent, non-circulative transmission is known in aphid, whitefly and leafhopper-transmitted

Table 13.7 Examples of honeydew honeys (based on Chamorro et al. 2013; Crane and Walker 1985; Crozier 1981; Honey Traveler 2017; Santas 1983)

Designation	Host plants	Honeydew producing insects[a]	Countries
Silver fir honeydew honey	*Abies alba*	[b]*Cinara* spp.	Germany, Italy
Greek fir honey	*Abies cephalonica*	[c]*Physokermes hemicryphus*	Greece
Oak-tree honeydew honey	*Quercus* spp.	[b]*Tuberculatus annulatus, T. borealis*	Bulgaria, Croatia, France, Greece, Italy, Serbia, Slovakia
Pine honeydew honey	*Pinus* spp.	[c]*Marchalina hellenica*	Greece, Turkey
Spruce honeydew honey	*Picea abies*	[b]*Cinara costata, C. piceae, Physokermes hemicryphus*	Slovenia, Czech Republic, Italy
Melcalfa pruinosa honeydew honey	Many wild, ornamental and cultivated plants	[d]*Melcalfa pruinosa*	France, Italy
Beech honeydew honey	*Nothofagus fusca*	[c]*Ultracoelostoma assimile, U. brittini*	New Zealand
Willow-tree honeydew honey	*Salix* spp.	[b]*Tuberolachnus salignus*	Croatia, Denmark, Italy, Lithuania, Norway, Spain, Sweden

Note [a]Associated honeydew producing insects: [b]Aphids (Aphidoidea); [c]Scale insects (Coccoidea); and [d]Planthoppers (Fulgoroidea)

viruses. Luteoviruses are an example of circulative, non-propagative viruses, which are transmitted by aphid vectors, whereas the genera *Fijivirus, Phytoreovirus, and Oryzavirus* of Reoviridae are circulative, propagative viruses transmitted by planthoppers or leafhoppers vectors (Whitfield et al. 2015).

The only known xylem-limited pathogenic bacteria, *Xylella fastidiosa* (class Gammaproteobacteria) is transmitted in a non-circulative mode by different sap-feeding insect vectors from Membracoidea, Cercopoidea, and Cicadoidea. All phloem-limited vector-borne bacteria, including phytoplasmas (class Mollicutes), liberibacters (class Alphaproteobacteria), and spiroplasmas (class Mollicutes), are apparently circulative, propagative, colonizing both the plant host and the insect vector intracellularly. The few phytopathogenic spiroplasmas are transmitted by leafhoppers, which are also the main vectors of the many plant diseases caused by phytoplasmas. The few species of phloem-limited and phytopathogenic bacteria of the genus *Liberibacter* are vectored by psyllids (Perilla-Henao and Casteel 2016).

Although vectors usually acquire viruses by feeding on infected plants, some propagative viruses can be transmitted transovarially, from female insect vector to the progeny. Vector specificity in plant pathogen transmission may vary greatly among plant pathogens. For example, some viruses are transmitted by only one insect vector,

Table 13.8 Types of transmission relationships between insect vectors and plant pathogens (based on Chuche et al. 2017a; Perilla-Henao and Casteel 2016; Whitfield et al. 2015)

Persistence	Location	Replication	Latent period	Acquisition time	Retention time	Retention sites
Non-persistent	Non-circulative	–	No	Minutes	<12 h	Stylet or foregut
Semi-persistent	Non-circulative	–	No	Hours-days	Hours-days; infectivity lost after molt	Alimentary canal or gut lumen
Persistent	Circulative	Non-propagative (pathogen does not multiply in the vector)	Hours	Hours-days	Until insect dead	Salivary glands
Persistent	Circulative	Propagative (pathogen multiplies in the vector)	Days-weeks	Hours-days	Until insect dead	Salivary glands

whereas other viruses are less vector specific and may be transmitted by insect species from a single family or subfamily (Purcell 2003).

13.5 Hemipteran Sap Suckers as Forest Pests: Damage, Management and Control

13.5.1 Damage

Despite the frequent occurrence and the high species richness of sap suckers on forest trees, in particular aphids, psyllids and scale insects, they have seldom been considered serious pests in natural or planted forests, in their native ranges. Sap suckers feeding in their native forest areas are usually regulated at low population densities by natural enemies. Invasive hemipteran sap suckers from several families, including Psyllidae, Adelgidae, Coccidae, Pseudococcidae, Diaspididae, Matsucoccidae and Monophlebidae have established on forest trees in all major forest areas globally and several of these species have become key forest pests. These species represent a minority of sap suckers associated with forest trees. For example, 191 scale insect species are associated with pines worldwide (García Morales et al. 2016), but only a few have been reported to cause serious damage on pines, leading to intensive defoliation and tree death. All these species had been introduced outside their native range and became established and spread as invasive species. Effectively, scale insects are frequent invaders in forest and agriculture areas. Illustratively, alien scale insect species represent an important component of the European entomofauna, accounting for about 30% of the total scale fauna in Europe (Pellizzari and Germain 2010). Some examples of well-known invasive scale insects are, the maritime pine blast scale *M. feytaudi*, which is native to southwestern Europe and invasive in Corsica and Italy (Sciarretta et al. 2016); the Israeli bast scale *Matsucoccus josephi* is native to Cyprus and invasive in Israel (Mendel et al. 2016); and *Matsuccocus matsumarae* is native to Japan and invasive in North America and China (McClure 1986). Other examples of important sap suckers on pine are *Palaeococcus fuscipennis* (Monophlebidae), native to southern Europe and invasive in Israel, and the margarodid *M. hellenica*, native to Turkey and Greece and invasive in Italy and Australia (Mendel et al. 2016; Nahrung et al. 2016). Similarly, for the genus *Eucalyptus* a number of sap suckers native to Australia have been established worldwide and cause serious damage. In fact, 40% of invasive species affecting *Eucalyptus* outside Australia are sap suckers. These include 13 species of psyllids, two scale insects, an Eriococcidae and a Diaspididae, a whitefly and the bronze bug, *T. peregrinus* (Hurley et al. 2016). In their native range, these species, like many other sap suckers feeding on *Eucalyptus*, exist mostly at endemic population levels and rarely reach outbreak population levels.

Nevertheless, outbreak of populations in its native range also occur. They are often associated with tree physiological status, weather conditions favorable for sap sucker population growth, or disruption of natural enemy populations. For example, high infestations of eucalyptus psyllids, like the red gum lerp psyllid *G. brimblecombei*, in its native range in Australia, are mainly reported in urban parks or forests where trees are under some kind of physiological stress (Stone 1996). In Europe, native aphids in conifers, such as *Cinara* spp. and *Pineus pini*, or on oaks e.g. *Lachnus* spp. may build up their populations in years with favorable weather conditions. Such outbreaks occur mostly in spring and summer, facilitated by the high fecundity of females and parthenogenetic generations. Normally, these outbreak populations decline as a result of increases in natural enemy populations or unfavorable weather conditions, such as low temperatures. During outbreaks, some of these species produce honeydew in great abundance, a valuable resource for beekeepers, for the production of high value forest honeys (see Table 13.7). Site conditions and silvicultural treatments that result in poor tree health can also promote outbreaks of sap sucking insects. Intense irrigation and fertilization, as well as tree stress, promote tree physiological conditions that may favor such outbreaks. For these reasons, trees in urban and nursery settings are more likely to suffer from intense attacks of sap suckers than their conspecifics in natural forests. In their native range, the importance of sap suckers also frequently increases when host trees are planted in non-natural habitats (urban parks and street trees) due to tree physiological stress and the lack of natural enemies in these fragmented habitats. For example, psyllid outbreaks on eucalyptus in their native range in Australia were often observed on roadsides, farmlands, grazing areas and sewage-irrigated sites (e.g. Collett 2001). Drought-stressed *Eucalyptus* and *Acacia* are often more susceptible to psyllids.

Since individuals are usually small, inconspicuous, frequently hiding in crevices of the bark, they are easily spread unknowingly by human activities and wind. When populations are at low densities, individual insects, particularly eggs and young nymphs, are difficult to detect even for experienced eyes. Nursery trees and seedlings, branches and leaves are all plant materials that may easily transport sap suckers. Dissemination by wind occurs mainly during the early developmental stages. Scale insects typically display wind dispersal behavior during the early non-sedentary young nymphs, termed crawlers. For example, wind dispersal of first instar nymphs of the pine blast scale, *Matsucoccus* spp. facilitates colonization of new areas. Similarly, the elongate hemlock scale *Fiorinia externa* (Diaspididae) is often dispersed long distances by wind (McClure 1979). Young nymphs of aphids may also disperse up to several thousand meters by wind, as has been observed in the hemlock woolly adelgid *Adelges tsugae* (McClure 1990). Sap suckers with winged adult females (e.g. psyllids and aphids) may also disperse by flight. In some species, eggs and nymphs may be spread by animals, such as birds or mammals (McClure 1990).

Whereas at low densities sap suckers cause only minor or no harm to individual trees, at high densities they may cause extensive leaf necrosis, discoloration and defoliation triggering reduced tree growth, and increasing susceptibility to tree pathogens and other insect pests. In some cases, such as in *C. laportei* infesting cedars, sap suckers may kill twigs and branches of infested trees. Ultimately, the activity of sap suckers may result in tree death and high enough levels of tree mortality, to result in forest decline as observed following *M. josephi* attacks in Aleppo pine in Israel (Mendel et al. 2016) or *A. tsugae* declining hemlock stands in eastern North America

(Ellison et al. 2018). Some sap suckers become particularly problematic in nurseries or greenhouse conditions. This is the case of the oak phylloxera in Israel.

At high densities sap suckers can pose serious problems in urban settings. This is due to aesthetic concerns caused by foliage discoloration, deformation and defoliation, as well as the intense honeydew production and subsequent growth of sooty mold fungi (Fig. 13.2e, i, j). Under these circumstances, the implementation of control tactics is required.

13.5.2 Pest Management

Sap suckers are often difficult to control. Due to their small size and cryptic feeding habits, they are difficult to detect at low population levels. Their high dispersal capacity facilitates the colonization of new hosts. Furthermore, many of these species are protected by cuticular waxes, rendering contact insecticide treatments less successful or ineffective. The implementation of integrated pest management (IPM) strategies, including monitoring, biological control, cultural practices, the use of behavior modifying chemicals, and the injection of systemic insecticides in urban areas, is essential.

13.5.2.1 Behavior Modifying Signals

Behaviors elicited by chemical stimuli (e.g. host plant volatiles (HPV) or pheromones) have been studied for a few sap sucker species feeding on forest trees. These compounds can be used in insect monitoring and different pest management tactics, such as mating disruption, mass trapping, and lure and kill applications, among others.

Sex pheromones of many insects are very powerful attractants, as males may be attracted from long distances to lures impregnated with few micrograms of pheromone (Fig. 13.2a, b). For example, males of *Matsucoccus* spp. and *P. citri* are attracted to the female sex pheromone from distances up to several hundred meters, although the level of captures per trap decreases with the distance between the source population and the lures (Branco et al. 2006b).

Sex pheromones have been identified for 32 scale insect species (Franco et al. 2022) and several aphid species (Dewhirst et al. 2010; Pickett et al. 2013). No sex pheromones have been identified to date for forest psyllids or adelgids. There is evidence that psyllids and other sap suckers, such as Cicadellidae, use vibrational signals in sex communication (Čokl and Millar 2009). Although aphids frequently reproduce asexually, in holocyclic species there are populations that reproduce sexually in part of their life cycle. Sex pheromones have been identified for a number of aphid species, all comprising compounds of the group of cyclopentanoid nepetalactones (Dawson et al. 1996). These compounds can act in synergy with plant volatiles. For example, bird-cherry aphid, *Rhopalosiphum padi*, male response to nepetalactol is synergized by *Prunus padus* leaf extract. To date, applications of aphid sex pheromones to control or monitor aphid pests have not been developed.

Among scale insects, diaspidid and pseudococcid pheromones are mostly terpenoid derivatives, whereas those of Matsucoccidae are ketones (Zou and Millar 2015). Only a minority of the sex pheromone compounds identified for sap suckers are commercially available, and with one or two exceptions, their use in pest management is still limited, particularly in forestry.

Mating disruption attempts to prevent males from locating females by releasing high concentrations of synthetic sex pheromone in the field. Males may then be unable to locate females and females may remain unmated. A revision of the mechanisms of mating disruption is provided by Evenden (2016). Franco et al. (2022) provided an overview of the current knowledge on mating disruption of scale pests. This technique only applies to populations that reproduce sexually. The success of mating disruption depends on the reproductive behavior of the species, the size of the treatment area and the habitat adjacent to the treatment area. For small-scale plots, immigrant gravid females, coming from adjacent habitats or from alternate host plants supporting populations outside the treatment area, can render the method inefficient. Nevertheless, this is not a problem in the case of scale insects, as the females are wingless insects (Franco et al. 2022). Further, the method is usually more efficient at low population density levels for which the probability of mating will be lower. In theory, the method could be used to increase the frequency of Allee effects, facilitating local extinction of non-native species, when populations are recently established and at very low population levels (Tobin et al. 2011; see also Chapter 18).

Whereas mating disruption has been widely investigated in Lepidoptera, only a few attempts have been made to manage hemipteran pests with mating disruption. Based on the data available on Pherobase (El-Sayed 2018), mating disruption has been investigated in about 127 Lepidoptera species, representing 86% of total species in which mating disruption was tested. However, only nine hemipteran sap suckers were targeted for the same purpose. To note, in particular three scale insects, the vine mealybug *Planococcus ficus* (Walton et al. 2006), the California red scale *Aonidiella aurantii* (Rice et al. 1997), and the San Jose scale *Comstockaspis perniciosa* (=*Quadraspidiotus perniciosus*) (Bar Zakay et al. 1987). Commercial formulations are currently available for mating disruption of the first two species, in vineyards and citrus orchards, respectively (Cocco et al. 2014; Vacas et al. 2014). The scientific, technological and practical developments in mating disruption of scale insects, as well as future prospects were recently reviewed by Franco et al. (2022). There are no examples in the literature of mating disruption to control sap suckers in forest plantations.

Disruption or manipulation of acoustic signals is a new possibility for pest management of sap sucking insects. Using artificial signals to interfere with acoustic communication between male and female sap suckers has been suggested as a mating disruption tactic to control stink bugs (e.g. *Nezara viridula, Euschistus heros*) (Čokl and Millar 2009), the Asian citrus psyllid *Diaphorina citri* (Lujo et al. 2016), and the glassy-winged sharpshooter *Homalodisca vitripennis* (Mazzoni et al. 2017).

As a pest management tactic, mass trapping attempts to remove a large portion of the target population by capturing individuals in traps, usually baited with kairomones or pheromones, and eliminating them. A weakness of using sex pheromones is that only males are normally attracted. This hinders the efficiency of the method because even a small fraction of surviving males may be enough to ensure that most females are mated, particularly in polygynous species. Mass trapping of mealybugs and scale insects using sex pheromone was tested for *P. citri* (Franco et al. 2003) and *M. feytaudi* (Binazzi et al. 2002). In both cases, the results did not prove the efficiency of the method. This was partly attributed to the small study areas and the attraction of males from the surrounding fields. In Italy, a management program against the pine bast scale *M. feytaudi*, combining the use of mass trapping and silvicultural interventions, supposedly delayed the loss of a *P. pinaster* forest caused by the pine bast scale (Sciarretta et al. 2016). One consideration is the cost-effectiveness of the technique, even if there is a substantial reduction of the population, the costs may not compensate for the expected benefits.

Mass trapping with aggregation pheromones (which cause the increase of insect density, usually of both males and females, in the vicinity of pheromone source) is theoretically possible, and could be much more powerful than the use of sex pheromones. Few aggregation pheromones are known for forest sap suckers, most from true bugs (Heteroptera). For example, adult males and nymphs of *T. peregrinus* display an aggregation behavior induced by a male specific pheromone, whose major component is 3-methylbut-2-enyl butanoate (González et al. 2012).

The lure and kill technique is similar to mass trapping, but instead of trapping the insects, individuals are attracted to a semiochemical-baited lure and then killed by exposure to an insecticide. Although limited by the small surface of lure devices, there may be environmental concerns that limit the use of insecticides in forest settings. To date, there are no examples of lure and kill tactics applied in forestry (El-Sayed 2018).

The dispersal behavior induced in aphids by the alarm pheromone (E)-β-farnesene has been commercially exploited for the control of aphids, by combining its application with an insecticide or entomopathogenic fungi. The induction of dispersal behavior by the application of alarm pheromone is expected to increase the probability of the aphids contacting with the insecticide, thus reducing the effective dosage of the required toxicant. Nevertheless, the efficacy of this control tactic, as well as its cost effectiveness are unclear (Čokl and Millar 2009). An innovative application of the alarm pheromone was recently developed for pest management of aphids. A hexaploid variety of wheat was genetically engineered to release (E)-β-farnesene. Laboratory tests showed that three different aphid species were repelled and the foraging behavior of an aphid parasitoid was enhanced (Bruce et al. 2015).

The natural enemies of a number of sap sucker pests are attracted to the pheromones of their prey (i.e. the pheromones act as kairomones for the natural enemies). In these cases, the pheromone could be used to attract natural enemies and potentially increase predation rates and reduce pest population densities and ultimately crop damage. For example, attraction of parasitic wasps of the citrus mealybug (Franco et al. 2008, 2011) and of different predators of pine blast scales, including brown lacewings (Hemerobiidae), flower bugs (Anthocoridae), ladybirds (Coccinellidae), and flower beetles (Dasytidae) (Branco et al. 2006a; Mendel et al. 2003) has been reported. It is interesting that for most of these predators, both the adult and larval stages were found to be attracted to the pheromone of the prey (Branco et al. 2006c).

Volatile cues emitted from plants and attractive to sap suckers may theoretically be used for monitoring or trapping these insect pests. These include several structural categories of volatiles, such as alcohols, aldehydes, ketones, esters and terpenes. Traps baited with these compounds must compete with volatiles released by host plants and consequently often have low capture rates. In some cases, interactions with other organisms vectored by sap suckers may alter the volatile profile of host trees and increase their attractiveness to vectors. For example, apple trees, infected by the pathogen *Phytoplasma mali* emit higher amounts of the sesquiterpene β-caryophyllene, and are highly attractive to the vector, *Cacopsylla picta* (Psyllidae) (Mayer et al. 2011). Similarly, citrus trees infected by the bacterial pathogen *Liberibacter asiaticus* release odors specifically attracting its psyllid vector *Diaphorina citri* (Mann et al. 2012). These compounds could potentially be used to control vector populations. So far there are no known examples of this phenomenon with forest sap suckers.

13.5.2.2 Monitoring

Monitoring programs normally target a single species as the deployment protocols of monitoring tools are often species-specific (e.g. habitat, part of the plant affected and distinctive symptoms). Visual surveys are recommended for urban trees and plants in nurseries, and for sap sucker species whose symptoms are easily discernible. Band traps may be used to intercept individuals moving along the tree trunk, such as adult females of the pine bast scale, while they are searching for sites suitable for oviposition (Fig. 13.2c). Interception sticky traps may be used to detect crawlers, when dispersing by wind, or winged adults stages, such as for psyllids. For example, sticky traps were used to monitor the blue gum psyllid, *Ctenarytaina eucalypti* and the red gum psyllid *G. brimblecombei* in California (Dahlsten et al. 2003). The performance of sticky traps for monitoring can vary depending on trap color and position in the canopy, as observed for the green spruce aphid *Elatobium abietinum*, in Sitka spruce (Straw et al. 2011). Therefore, trap design and deployment protocol must be optimized for each species-habitat combination.

Monitoring is probably the most frequent use of sex pheromones in IPM. Pheromone-baited traps can catch individuals when populations are at extremely low levels. Therefore, this highly specific method allows detection of population growth before outbreaks or in the early phases of population establishment. Consequently, managers may apply control methods in an early stage. This allow for timely application of treatment, before populations reach hard-to-control proportions and damage has already occurred. It also avoids increased costs of treatments. As for sticky traps, protocols for deployment of pheromone-baited traps need to be optimized for each species, namely regarding, trap design, size and dosage (e.g. Branco et al. 2004). Pheromone-baited traps are often used to follow seasonal activity of flying males of a given species. Among other applications, this allows managers to anticipate treatment periods and follow generations through seasons. For spreading populations, such as invasive species, monitoring by pheromones allows the tracking of their dispersion rate and range.

13.5.2.3 Biological Control

Many sap sucker pests in forest ecosystems are invasive species or native species whose natural enemies have been locally disrupted. Sap suckers are usually well regulated by natural enemies in their native habitats. Therefore, in many cases classical biological control is the optimal solution for invasive pest species. For example, outbreaks of *P. fuscipennis* (Monophlebidae) in pine stands in Israel came to an end following the introduction of natural enemies from Spain, such as *Novius cruentatus* (Coleoptera, Coccinellidae) and *Cryptochaetum jorgpastori* (Diptera, Cryptochaetidae) (Mendel et al. 1998). Similarly, outbreaks of *Cinara* spp. in the Mediterranean and South Africa were controlled by introduction of specific parasitic wasps. The horse chestnut scale, *Pulvinaria regalis* (Coccidae), native to Asia and first detected in Europe in 1968 has become a serious pest of horse chestnuts, as well as many other tree species in Europe (Trierweiler and Balder 2005). Efforts are underway to develop a biological control program using the parasitiod *Coccophagus lycimnia* (Hymenoptera: Aphellinidae), although this parasitoid is already present in Europe and its origin is unknown. Recent invasion of Victoria and South Australia by the giant pine scale *M. hellenica* has resulted in serious damage to several introduced pine species (Nahrung et al. 2016). *Neoleucopis kartliana* (Diptera: Chamaemyiidae), which has already been used to control *M. hellenica* on the Italian island of Ischia, may be the solution for the scale invasion in South Australia (Avtzis et al. 2020).

In some cases, classical biological control alone does not provide adequate control. The Hemlock woolly adelgid (HWA) *A. tsugae*, native to East Asia, is one of the most damaging agents on hemlock and spruce trees (*Tsuga* spp., *Picea* spp.) in North America. It was unintentionally introduced in the 1950s in the eastern states, spreading to nearby states (McClure et al. 2000). The biological control agent *Sasajiscymnus tsugae* (=*Pseudoscymnus tsugae*) (Coleoptera, Coccinelidae) was introduced in North America from Japan in the 1990s. In its natural range, the ladybeetle was considered very prey-specific, keeping the adelgid under control. More than 100,000 adult beetles were released in eastern North America in highly affected forests. Although adelgid densities were reduced in treated areas (McClure and Cheah 1998; McClure et al. 2000), 20 years later the Hemlock woolly adelgid was still the most important threat to the native *Tsuga* species in the eastern USA (Letheren et al. 2017). Additionally, three species of ladybeetles of the genus *Scymnus* were first found in China and introduced in USA for studies under quarantine conditions. However, due to mass rearing difficulties, only two species were released in limited numbers and to date there have been no field recoveries of these species (Havill et al. 2014). HWA is also invasive in western North America. Species of *Laricobius* are specialist predators of adelgids and predation by the native beetle *Laricobius nigrinus* (Coleoptera: Derodontidae) was thought to limit HWA in western North America. Lamb et al. (2006) proposed its release as a biocontrol agent for eastern HWA populations. Additional species of this genus from both North America and Japan are currently under study for the biocontrol of HWA. A difficulty of classical biological control results from finding adequate natural enemies on the insect pest native range. An example is the Beech scale, *Cryptococcus fagisuga* (Eriococcidae), invasive in North America. Phylogenetic analysis of this eriococcid suggested that its natural range covers the areas of northeastern Greece, the Black Sea drainage basin, the Caucasus Mountains, and northern Iran (Gwiazdowski et al. 2006). But so far efficient natural enemies have not been found.

Another example of the variability of the success of biological control is seen with the *Eucalyptus* psyllids. Biological control of the blue gum psyllid *C. eucalypti* by the Australian parasitoid *Psyllaephagus pilosus* was achieved in the first year after its introduction in Europe (Chauzat et al. 2002). However, the control of red gum lerp psyllid *G. brimblecombei*, by another Australian parasitoid *Psyllaephagus bliteus* was only partially successful both in California and Europe (Dahlsten et al. 2003; Boavida et al. 2016).

Native beneficial organisms may also prey upon non-native forest pests and contribute to regulating their populations (Table 13.6). For example, the predator *Elatophilus nigricornis* exerts some control upon the non-native pine bast scale *M. feytaudi* in Corsica. This control is thought to be more effective in mixed forests, where this native natural enemy feeds on a congeneric native scale insect *Matsucoccus pini* and consequently is expected to have more stable and persistent populations (Jactel et al. 2006). The leaf galling psyllid *C. schini* feeds on the Peruvian pepper tree, and is invasive in Kenya, where it is heavily parasitized by native eulophid parasitoids, which probably switched from native psyllids developing on Tamarisk (Overholt et al. 2013).

In summary, although biological control is usually a long lasting sustainable solution to control invasive sap suckers it is not always successful. Biological control may not be successful because the agent selected may not be appropriate for the release site. Some natural enemies require several years to establish and build up their populations until effects on target populations are realized. Even after several years of establishment, biocontrol agents may not control pest populations enough to avoid damages. Since landowners wish for inexpensive solutions, quick results and in the short-term, complementary tactics may be needed (e.g. cultural, silvicultural or chemical tactics) until biological control agents are established and providing adequate control.

13.5.2.4 Cultural, Chemical and Physical Methods

Cultural and physical methods may solve problems with temporary outbreaks of sap suckers. Mechanical cleaning of infested trees can reduce populations of sap suckers and reduce damage. Common practices are washing infested branches with soaped water or pruning. These tactics are labor-intensive and costly. Therefore, they are mainly applied in arboretum, parks or urban settings. If these actions are applied regularly, they may reduce sap sucker populations and allow natural enemies to regulate the pest populations. These techniques are mainly applied to individual high-value trees and often the objective is not the control of sap sucker populations but rather protecting the aesthetics of the host plant, or in cases when the scale population spoils the surroundings with honeydew, or causing nuisance high male flight. For example, control of the horse chestnut scale *P. regalis* in Europe is mostly done by mechanical cleaning in urban areas (Speight et al. 1998). In circumstances where populations are spatially delimited such as nurseries, colored sticky traps may exert some control on psyilld, cicada and leafhopper populations.

Chemical control measures may be used for treating individual street or park trees. Some environmentally friendly methods, such as non-toxic insecticidal soap, plant extracts or horticultural oils are available. In some cases, systemic insecticides may be used with good results to control aphids, psyllids and scale insects. The use of these products in forests is forbidden in many countries and their use is heavily restricted in nurseries. In urban settings, insecticide application might be practiced by soil drenching or trunk injections. The insecticide is transported systemically through the tree vascular system to the foliage or other plant parts where it may kill infesting sap suckers. With soil treatments, care needs to be taken to avoid contamination of water bodies. Of concern is the fact that other insect communities on the tree may be affected by systemic insecticides.

In some cases, host plant resistance plays a major role in tree health. This phenomenon is well known among several *Matsucoccus* spp., which devastated large pine areas as invasive species, for which pine provenance is a major factor in tree susceptibility (e.g. Mendel 1984; Mendel et al. 2016). Similarly, for HWA host plant resistance plays a major role on population densities of this adelgid, both in the native and introduced areas (Havill et al. 2014). The presence of resistant species or provenances may then help to control the problem. However, lack of resistance is usually a major management challenge and often remains an unresolved situation (e.g. some species of Matsucoccidae, Adelgidae and Eriococcidae). If resistant genetic materials are not available, cultural and silvicultural measures may be suggested as a solution to reduce damage, at least until other, more long-lasting solutions such as biological control, can be developed.

13.6 Conclusions

Sap sucking insects, are characterized by their specialized feeding mode and are quite species diverse in forest trees. By producing honeydew, sap suckers also establish interactions with other forest species, including both vertebrates and invertebrates, which use this sugar resource. The general effect of sucking insects on forest trees is by far much less conspicuous than other major insect groups, like bark beetles and defoliators.

Sucking insects often become major pests under two scenarios: (1) planting of highly susceptible trees outside their native ranges; and (2) introduction or natural spread of these hemipterans outside their native ranges. Invasions by non-native sap suckers often occur without their principal natural enemies. In several cases, the negative effects of these invasive species on tree health have been mitigated by the introduction of natural enemies. On the other hand, range expansion may result in interactions between sap suckers and host trees lacking resistance. These cases may benefit from tree breeding and selection programs. In their native range, damage by sap suckers also increases when the trees are under physiological stress or top-down effects of their natural enemies are disrupted. Understanding the ecology of sap suckers and the factors that promote outbreaks is essential for developing effective control strategies.

Acknowledgements We thank Alex Protasov and Vera Zina for allowing the use of some photos. Thanks are also due to the editors and reviewers for their comments and suggestions that greatly helped us improving the manuscript. This work was funded by the Forest Research Centre (CEF), a research unit funded by Fundação para a Ciência e a Tecnologia I.P. (FCT), Portugal (UIDB/00239/2020), and by the project "Technical cooperation, knowledge sharing and training in the area of forest and agricultural entomology and integrated pest management" funded by KKL – JNF Israel.

References

Afifi S (1968) Morphology and taxonomy of the adult males of the families Pseudococcidae and Eriococcidae (Homoptera: Coccoidea). Bull Br Mus Nat Hist Entomol Suppl 13:1–21

Almeida RP, Nuney LL (2015) How do plant diseases caused by *Xylella fastidiosa* emerge? Plant Dis 99:1457–1467

Avtzis DN, Lubanga UK, Lefoe GK, Kwong RM, Eleftheriadou N, Andreadi A, Elms S, Shaw R, Kenis M (2020) Prospects for classical biological control of Marchalina hellenica in Australia. Biocontrol 65:413–423

Azevedo MS, Seraglio SKT, Rocha G, Balderas CB, Piovezan M, Gonzaga LV, de Barcellos Falkenberg D, Fett R, de Oliveira MAL, Costa ACO (2017) Free amino acid determination by GC-MS combined with a chemometric approach for geographical classification of bracatinga honeydew honey (*Mimosa scabrella* Bentham). Food Control 78:383–392

Bar Zakay I, Peleg BA, Hefetz A (1987) Mating disruption of the California red scale *Aonidiella aurantii*. Hassadeh 69:1228–1231

Baumann P (2005) Biology of bacteriocyte-associated endosymbionts of plant sap-sucking insects. Annu Rev Microbiol 59:155–189

Beirne BP (1956) Leafhoppers (Homoptera: Cicadellidae) of Canada and Alaska. Can Entomol 86:548–553

Ben-Dov Y (1990) The adult female. In: Rosen D (ed) World crop pests, vol. 4A, armored scale insects: their biology, natural enemies and control. Elsevier, Amsterdam, pp 5–20

Ben-Dov Y, Miller DR, Gibson GAP (2014) ScaleNet: a database of the scale insects of the world. http://www.sel.barc.usda.gov/SCALENET/scalenet.htm

Bennett GM, Moran NA (2013) Small, smaller, smallest: the origins and evolution of ancient dual symbioses in a phloem-feeding insect. Genome Biol Evol 5:1675–1688

Binazzi A, Pennacchio F, Francardi V (2002) The use of sex pheromones of *Matsucoccus* species (Homoptera Margarodidae) for monitoring and mass trapping of *M. feytaudi* Ducasse and for kairomonal attraction of its natural enemies in Italy. Redia 85:155–171

Blackman RL, Eastop VF (1994) Aphids on the world's trees. An identification and information guide. CAB International, Wallingford, 1024 pp

Boavida C, Garcia A, Branco M (2016) How effective is *Psyllaephagus bliteus* (Hymenoptera: Encyrtidae) in controlling *Glycaspis brimblecombei* (Hemiptera: Psylloidea)? Biol Control 99:1–7

Branco M, Jactel H, Silva EB, Binazzi A, Mendel Z (2004) Effect of trap design, trap size and pheromone dose on male capture of two pine bast scales species (Hemiptera: Matsucoccidae): implications for monitoring and mass-trapping. Agric For Entomol 6:233–239

Branco M, Franco JC, Dunkelblum E, Assael F, Protasov A, Ofer D, Mendel Z (2006a) A common mode of attraction of larvae and adults of insect predators to the sex pheromone of their prey (Hemiptera: Matsucoccidae). Bull Entomol Res 96:179–185

Branco M, Jactel H, Franco JC, Mendel Z (2006b) Modelling response of insect trap captures to pheromone dose. Ecol Model 197:247–257

Branco M, Lettere M, Franco JC, Binazzi A, Jactel H (2006c) Kairomonal response of predators to three pine bast scale sex pheromones. J Chem Ecol 32:1577

Bruce TJA, Aradottir GI, Smart LE, Martin JL, Caulfield JC, Doherty A, Sparks CA, Woodcock CM, Birkett MA, Napier JA, Jones HD, Pickett JA (2015) The first crop plant genetically engineered to release an insect pheromone for defence. Sci Rep 5:11183

Burckhardt D (2005) Biology, ecology, and evolution of gall-inducing psyllids (Hemiptera: Psylloidea). In: Raman A, Schaefer CW, Withers TM (eds) Biology, ecology, and evolution of gall-inducing arthropods. Science Publishers, Enfield, NH and Plymouth, pp 143–157

Carver M, Gross GF, Woodward TE (1991) Hemiptera. In: CSIRO division of entomology, Australia—the insects of Australia, 2nd ed. Melbourne University Press, pp 429–509

Čermák V, Lauterer PC (2008) Overwintering of psyllids in South Moravia (Czech Republic) with respect to the vectors of the apple proliferation cluster phytoplasmas. Bull Insectol 61(1):147–148

Chamorro FJ, Nates-Parra G, Kondo T (2013) Mielato de Stigmacoccus asper (Hemiptera: Stigmacoccidae): Recurso melífero de bosques de roble en Colombia. Rev Colomb Entomol 39:61–70

Chauzat MP, Purvis G, Dunne R (2002) Release and establishment of a biological control agent, *Psyllaephagus pilosus* for eucalyptus psyllid (*Ctenarytaina eucalypti*) in Ireland. Ann Appl Biol 141(3):293–304

Chuche J, Auricau-Bouvery N, Danet JL, Thiéry D (2017a) Use the insiders: could insect facultative symbionts control vector-borne plant diseases? J Pest Sci 90:51–68

Chuche J, Sauvion N, Thiéry D (2017b) Mixed xylem and phloem sap ingestion in sheath-feeders as normal dietary behavior: evidence from the leafhopper *Scaphoideus titanus*. J Insect Physiol 102:62–72

Cocco A, Lentini A, Serra G (2014) Mating disruption of *Planococcus ficus* (Hemiptera: Pseudococcidae) in vineyards using reservoir pheromone dispensers. J Insect Sci 14:144

Čokl AA, Millar JG (2009) Manipulation of insect signaling for monitoring and control of pest insects. In: Ishaaya I, Horowitz AR (eds) Biorational control of arthropod pests: application and resistance management. Springer, Dordrecht, pp 279–316

Collett N (2001) Biology and control of psyllids, and the possible causes for defoliation of *Eucalyptus camaldulensis* Dehnh. (river red gum) in south-eastern Australia—a review. J Aust For 64:88–95

Cook L, Gullan P, Trueman H (2002) A preliminary phylogeny of the scale insects (Hemiptera: Sternorrhyncha: Coccoidea) based on nuclear small-subunit ribosomal DNA. Mol Phylogenet Evol 25(1):3–52

Cornara D, Morente M, Markheiser A, Bodino N, Tsai C-W, Fereres A, Redak RA, Perring TM, Spotti Lopes JR (2019) An overview on the worldwide vectors of *Xylella fastidiosa*. Entomol Gen 39:157–181

Crane E, Walker P (1985) Important honeydew sources and their honeys. Bee World 66:105–112

Crozier LR (1981) Beech honeydew: forest produce. N Z J For 26:200–209

Dahlsten DL, Rowney DL, Robb KL, Downer JA, Shaw DA, Kabashima JN (2003) Biological control of introduced psyllids on Eucalyptus. In: Proceedings of 1st international symposium biological control of arthropods, pp 356–361

Danzig EM (1980) Coccids of the Far East of the USSR (Homoptera, Coccinea) with an analysis of the phylogeny of coccids of the world fauna. Opredeliteli Po Faune SSR 124:1–367

Danzig EM, Gavrilov-Zimin IA (2015) Palaearctic mealybugs (Homoptera: Coccinea: Pseudococcidae). Part 2. Subfamily Pseudococcinae. ZIN RAS, St. Petersburg, 619 pp [Fauna of Russia and neighbouring countries. New series, No. 149. Insecta: Hemiptera: Arthroidignatha]

Dawson GW, Pickett JA, Smiley DW (1996) The aphid sex pheromone cyclopentanoids: synthesis in the elucidation of structure and biosynthetic pathways. Bioorg Med Chem 4(3):351–361

Dewhirst SY, Pickett JA, Hardie J (2010) Aphid pheromones. In: Litwack G (ed) Vitamins and hormones: pheromones, vol 8. Academic Press, London, pp 551–574

Dietrich CH (2003) Auchenorrhyncha (cicadas, spittlebugs, leafhoppers, treehoppers, and planthoppers). In: Resh VH, Cardé RT (eds) Encyclopedia of insects. Academic Press, Amsterdam, pp 66–74

Dinant S, Bonnemain J, Girousse C, Kehr J (2010) Complexité de la séve phloemienne et impact sur l'alimentation des pucerons. CR Biol 333:504–515

Douglas AE (2006) Phloem-sap feeding by animals: problems and solutions. J Exp Bot 57:747–754

Duelli P (1985) A new functional interpretation of the visual system of male scale insects (Coccida, Homoptera). Experientia 41:1036

Ellison AM, Orwig DA, Fitzpatrick MC, Preisser EL (2018) The past, present, and future of the hemlock woolly adelgid (Adelges tsugae) and its ecological interactions with eastern hemlock (Tsuga canadensis) forests. Insects 9(4):172. https://doi.org/10.3390/insects9040172

El-Sayed AM (2018) The pherobase: database of pheromones and semiochemicals. http://www.pherobase.com. Accessed on 19 Aug 2019

Evenden M (2016) Mating disruption of moth pests in integrated pest management. In: Allison JD, Carde RT (eds) Pheromone communication in moths. Evolution, behavior, and application. University of California Press, Oakland, CA, pp 365–393

Federici BA (2003) Pathogens of insects. In: Resh VH, Cardé RT (eds) Encyclopedia of insects. Academic Press, Amsterdam, pp 856–865

Foldi M (2004) The Matsucoccidae in the Mediterranean basin with a world list of species (Hemiptera: Sternorrhyncha: Coccoidea). Ann Société Entomol Fr 40(2):145–168

Favret C, Blackman RL, Miller GL, Victor B (2016) Catalog of the phylloxerids of the world (Hemiptera, Phylloxeridae). ZooKeys 629:83–101

Floren A, Schmidl J (2009) Canopy arthropod research in Europe: basic and applied studies from the high frontier. 2.4.2 Auchenorrhyncha: Cicadoidea and Membracoidea. Bioform, Germany, 576 pp

Franco JC, Gross S, Silva EB, Suma P, Russo A, Mendel Z (2003) Is mass-trapping a feasible management tactic of the citrus mealybug in citrus orchards? An Do Inst Super Agron 49:353–367

Franco JC, Silva EB, Cortegano E, Campos L, Branco M, Zada A, Mendel Z (2008) Kairomonal response of the parasitoid *Anagyrus* spec. nov. near *pseudococci* to the sex pheromone of the vine mealybug. Entomol Exp Appl 126(2):122–130

Franco JC, Silva EB, Fortuna T, Cortegano E, Branco M, Suma P, La Torre I, Russo A, Elyahu M, Protasov A, Levi-Zada, A (2011) Vine mealybug sex pheromone increases citrus mealybug parasitism by *Anagyrus* sp. near *pseudococci* (Girault). Biol Control 58(3):230–238

Franco JC, Cocco A, Lucchi A, Mendel Z, Suma P, Vacas S, Mansour R, Navarro-Llopis V (2022) Scientific and technological developments in mating disruption of scale insects. Entomol Gen. https://doi.org/10.1127/entomologia/2021/1220

Gamboa MA (2007) Plagas y enfermedades forestales en Costa Rica. Rev For Mes Kurú 4(11):1–69

García Morales M, Denno BD, Miller DR, Miller GL, Ben-Dov Y, Hardy NB (2016) ScaleNet: a literature-based model of scale insect biology and systematics. Database. http://scalenet.info. Accessed on 14 Oct 2017

Giliomee J (1967) Morphology and taxonomy of the adult males of the family Coccidae (Homoptera: Coccoidea). Bull Br Mus Nat Hist Entomol Suppl 7:1–168

Giordanengo P, Brunissen L, Rusterucci C, Vincent C, Van Bel A, Dinant S, Girousse C, Faucher M, Bonnemain JL (2010) Compatible plant-aphid interactions: how aphids manipulate plant responses. Comptes Rendus - Biol 333:516–523

González A, Calvo MV, Cal V, Hernández V, Doño F, Alves L, Gamenara D, Rossini C, Martínez G (2012) A male aggregation pheromone in the bronze bug, *Thaumastocoris peregrinus* (Thaumastocoridae). Psyche 2012:1–7

Grant PR (2005) The priming of periodical cicada life cycles. Trends Ecol Evol 20(4):169–174

Gullan PJ, Cranston PS (2014) The insects: an outline of entomology. Wiley, Chichester

Gullan PJ, Kosztarab M (1997) Adaptations in scale insects. Annu Rev Entomol 42:23–50

Gullan PJ, Martin JH (2009) Sternorrhyncha: (jumping plant-lice, whiteflies, aphids, and scale insects). In: Encyclopedia of insects, 2nd ed. Academic Press, San Diego, pp 957–967

Gullan PJ, Miller D, Cook L (2005) Gall-inducing scale insects (Hemiptera: Sternorrhyncha: Coccoidea). In: Raman A, Schaefer C, Withers T (eds) Biology, ecology and evolution of gall-inducing arthropods. Science Publishers, Enfield, NH and Plymouth, pp 159–229

Gwiazdowski RA, Van Driesche RG, Desnoyers A, Lyon S, Wu S-A, Kamata N, Normark BB (2006) Possible geographic origin of beech scale, *Cryptococcus fagisuga* (Hemiptera: Eriococcidae), an invasive pest in North America. Biol Control 39(1):9–18. ISSN 1049-9644. https://doi.org/10.1016/j.biocontrol.2006.04.009

Havill NP, Foottit RG (2007) Biology and evolution of adelgidae. Annu Rev Entomol 52:325–349

Havill NP, Vieira LC, Salom SM (2014) Biology and control of hemlock woolly adelgid. FHTET-2014-05. US Department of Agriculture, Forest Service, Forest Health Technology Enterprise Team, Morgantown, WV, pp 1–21

Hennig W (1981) Insect phylogeny. Wiley, Chichester, 514 pp

Hodkinson ID (1974) The biology of the Psylloidea (Homoptera): a review. Bull Entomol Res 64(2):325–338

Hodkinson ID (2009) Life cycle variation and adaptation in jumping plant lice (Insecta: Hemiptera: Psylloidea): a global synthesis. J Nat Hist 43:65–179

Honey Traveler (2017) Honeydew honey or forest honeys. http://www.honeytraveler.com/types-of-honey/honeydew-forest-honey/. Accessed on 20 Nov 2017

Hong Y, Zhang Z, Guo X, Heie OE (2009) A new species representing the oldest aphid (Hemiptera, Aphidomorpha) from the Middle Triassic of China. J Paleontol 83(5):826–831

Hurley BP, Garnas J, Wingfield MJ, Branco M, Richardson DM, Slippers B (2016) Increasing numbers and intercontinental spread of invasive insects on eucalypts. Biol Invasions 18(4):921–933

Jactel H, Menassieu P, Vetillard F, Gaulier A, Samalens J, Brockerhoff E (2006) Tree species diversity reduces the invasibility of maritime pine stands by the bast scale, *Matsucoccus feytaudi* (Homoptera: Margarodidae). Can J For Res 36:314–323

Ju R, Li Y, Wang F, Du Y (2009) Spread of and damage by an exotic lacebug, Corythuca ciliata (Say, 1832) (Hemiptera: Tingidae), in China. Entomol News 120(4):409–414

Kawecki Z (1964) On the suitable term for the second pair of wings in male scale insects (Homoptera—Coccoidea). Frustula Entomol 7:1–4

Kay MK (1980) *Amphipsalta zelandica* (Boisduval), *Amphipsalta cingulata* (Fabricius), *Amphipsalta strepitans* (Kirkaldy) (Hemiptera: Cicadidae). Large cicadas. New Zealand Forest Service, Forest and Timber Insects in New Zealand No. 44

Kehr J (2006) Phloem sap proteins: their identities and potential roles in the interaction between plants and phloem-feeding insects. J Exp Bot 57:767–774

Kingsolver JG, Daniel TL (1995) Mechanisms of food handling by fluid-feeding insects. In: Chapman RF, de Boer G (eds) Regulatory mechanisms in insect feeding. Chapman & Hall, New York, pp 32–73

Kosztarab M (1982) Homoptera. In: Parker SP (ed) Synopsis and classification of living organisms. New York: McGraw-Hill, 1232 pp, pp 447–470

Koteja J (1986) Current state of coccid paleontology. Boll Del Lab Di Entomol Agrar Filippo Silvestri 43(suppl.):29–34

Koteja J (1990) Paleontology. In: Rosen D (ed) World crop pests, vol. 4A, armored scale insects: their biology, natural enemies and control. Elsevier, Amsterdam, 384 pp, pp 149–163

Labandeira CC, Phillips TOML (1996) Insect fluid-feeding on Upper Pennsylvanian tree ferns (Palaeodictyoptera, Marattiales) and the early history of the piercing-and-sucking functional feeding group. Ann Entomol Soc Am 89:157–183

Lamb AB, Salom SM, Kok LT, Mausel DL (2006) Confined field release of *Laricobius nigrinus* (Coleoptera: Derodontidae), a predator of the hemlock woolly adelgid, Adelges tsugae (Hemiptera: Adelgida) in Virginia. Can J For Res 36(2):369–375

Letheren A, Hill S, Salie J, Parkman J, Chen J (2017) A Little bug with a big bite: impact of hemlock woolly adelgid infestations on forest ecosystems in the eastern USA and potential control strategies. Int J Environ Res Public Health 14(4):438

López-Madrigal S, Gil R (2017) Et tu, brute? Not even intracellular mutualistic symbionts escape horizontal gene transfer. Genes 8:1–16

López-Madrigal S, Latorre A, Porcar M, Moya A, Gil R (2013) Mealybugs nested endosymbiosis: going into the "matryoshka" system in *Planococcus citri* in depth. BMC Microbiol 13:74

Lucas WJ, Groover A, Lichtenberger R, Furuta K, Yadav SR, Helariutta Y, He XQ, Fukuda H, Kang J, Brady SM, Patrick JW, Sperry J, Yoshida A, López-Millán AF, Grusak MA, Kachroo P (2013) The plant vascular system: evolution, development and functions. J Integr Plant Biol 55:294–388

Lujo S, Hartman E, Norton K, Pregmon EA, Rohde BB, Mankin RW (2016) Disrupting mating behavior of *Diaphorina citri* (Liviidae). J Econ Entomol 109:2373–2379

Maceljski M (1986) Current status of *Corythuca ciliata* in Europe. EPPO Bull 16(4):621–624

Mann RS, Ali JG, Hermann SL, Tiwari S, Pelz-Stelinski KS, Alborn HT, Stelinski LL (2012) Induced release of a plant-defense volatile 'deceptively' attracts insect vectors to plants infected with a bacterial pathogen. PLoS Pathog 8(3):e1002610

Martin JH, Mifsud D, Rapisarda C (2000) The whiteflies (Hemiptera: Aleyrodidae) of Europe and the Mediterranean Basin. Bull Entomol Res 90:407–448, 407

Mayer CJ, Vilcinskas A, Gross J (2011) Chemically mediated multitrophic interactions in a plant–insect vector–phytoplasma system compared with a partially nonvector species. Agric For Entomol 13:25–35

Mazzoni V, Gordon SD, Nieri R, Krugner R (2017) Design of a candidate vibrational signal for mating disruption against the glassy-winged sharpshooter, *Homalodisca Vitripennis*. Pest Manag Sci 73(11):2328–2333

McClure MS (1979) Self-regulation in populations of the elongate hemlock scale, Fiorinia externa (Homoptera: Diaspididae). Oecologia 39(1):25–36

McClure MS (1986) Role of predators in regulation of endemic populations of Matsucoccus matsumurae (Homoptera: Margarodidae) in Japan. Environ Entomol 15(4):976–983

McClure MS (1990) Role of wind, birds, deer, and humans in the dispersal of hemlock woolly adelgid (Homoptera: Adelgidae). Environ Entomol 19(1):36–43

McClure MS, Cheah CASJ (1998) Released Japanese ladybugs are multiplying and killing hemlock woolly adelgids. Front Plant Sci 50(2):6–8

McClure MS, Cheah CASJ, Tigner TC (2000) Is *Pseudoscymnus tsugae* the solution to the hemlock woolly adelgid problem? An early perspective. In: McManus KA, Shields KS, Souto DR (eds) Proceedings: symposium on sustainable management of hemlock ecosystems in eastern North America, Durham, New Hampshire, June 22–24, 1999. General Technical Report 267. USDA, Newtown Square, PA, pp 89–96

McKenzie H (1967) Mealybugs of California with taxonomy, biology and control of North American species. University of California Press, Berkeley and Los Angeles, 526 pp

Mendel Z (1984) Provenance as a factor in susceptibility of *Pinus halepensis* to *Matsucoccus josephi* (homoptera: Margarodidae). For Ecol Manage 9(4):259–266

Mendel Z, Assael F, Zeidan S, Zehavi A (1998) Classical Biological Control of *Palaeococcus fuscipennis* (Burmeister) (Homoptera: Margarodidae) in Israel. Biol Control 12:151–157

Mendel Z, Dunkelblum E, Branco M, Franco JC, Kurosawa S, Mori K (2003) Synthesis and structure-activity relationship of diene modified analogs of Matsucoccus sex pheromones. Naturwissenschaften 90(7):313–317

Mendel Z, Protasov A, Jasrotia P, Silva E, Levi-Zada A, Franco J (2011) Sexual maturation and aging of adult male mealybug (Hemiptera; Pseudococcidae). Bull Entomol Res 15:1–10

Mendel Z, Branco M, Battisti A (2016) Invasive sap-sucker insects in the Mediterranean Basin. In: Insects and diseases of Mediterranean forest systems. Springer, Cham, pp 261–291

Mifsud D, Porcelli F (2012) The psyllid *Macrohomotoma gladiata* Kuwayama, 1908 (Hemiptera: Psylloidea: Homotomidae): a Ficus pest recently introduced in the EPPO region. EPPO Bull 42:161–164

Miller DR (1984) Phylogeny and classification of the Margarodidae and related groups (Homoptera: Coccoidea). In: Kaszab Z (ed) Proceedings of the 10th International Symposium of Central European Entomofaunistics (SIEEC), Budapest, Hungary, 1983. Muzsak Public-Educ., Budapest, 420 pp, pp 321–324

Miller DR, Hodgson CJ (1997) 1.1.3.7 phylogeny. In: World crop pests, vol. 7. Elsevier, Amsterdam, pp 229–250

Misof B, Liu S, Meusemann K, Peters RS., Donath A et al (2014) Phylogenomics resolves the timing and pattern of insect evolution. Science 346(6210)(7):763–767

Morrow JL, Hall AAG, Riegler M (2017) Symbionts in waiting: the dynamics of incipient endosymbiont complementation and replacement in minimal bacterial communities of psyllids. Microbiome 5:58

Morse JG, Hoddle MS (2006) Invasion biology of thrips. Annu Rev Entomol 51:67–89

Nadel RL, Slippers B, Scholes MC, Lawson SA, Noack AE, Wilcken CF, Bouvet JP, Wingfield MJ (2010) DNA bar-coding reveals source and patterns of *Thaumastocoris peregrinus* invasions in South Africa and South America. Biol Invasions 12(5):1067–1077

Nahrung HF, Loch AD, Matsuki M (2016) Invasive insects in Mediterranean forest systems: Australia. In: Insects and diseases of Mediterranean forest systems. Springer, Cham, pp 475–498

Nair KSS (2001) Exotic forests pest outbreaks in tropical forest plantations: is there a greater threat. https://books.google.co.il/books?isbn=9798764870

Normark B (2003) The evolution of alternative genetic systems in insects. Annu Rev Entomol 48:397–423

Nováková E, Hypša V, Klein J, Foottit RG, von Dohlen CD, Moran NA (2013) Reconstructing the phylogeny of aphids (Hemiptera:Aphididae) using DNA of the obligate symbiont *Buchnera aphidicola*. Mol Phylogenet Evol 68:42–54

Novotny V, Wilson MR (1997) Why are there no small species among xylem-sucking insects? Evol Ecol 11:419–437

Oddo LP, Piana L, Bogdanov S, Bentabol A, Gotsiou P, Kerkvliet J, Martin P, Morlot M, Valbuena AO, Ruoff K, Von Der Ohe K (2004) Botanical species giving unifloral honey in Europe. Apidologie 35(Suppl. 1):S82–S93

Oliver KM, Degnan PH, Burke GR, Moran NA (2010) Facultative symbionts in aphids and the horizontal transfer of ecologically important traits. Annu Rev Entomol 55:247–266

Overholt WA, Copeland RS, Halbert SE (2013) First record of *Calophya schini* (Hemiptera: Calophyidae) in Ethiopia and Kenya. Int J Trop Insect Sci 33(4):291–293

Perilla-Henao LM, Casteel CL (2016) Vector-borne bacterial plant pathogens: interactions with hemipteran insects and plants. Front Plant Sci 7:1–15

Pellizzari G, Germain J-F (2010) Scales (Hemiptera, Superfamily Coccoidea). Chapter 9.3. In: Roques A et al (eds) Alien terrestrial arthropods of Europe. BioRisk 4(1):475–510. https://doi.org/10.3897/biorisk.4.45

Pickett JA, Allemann RK, Birkett MA (2013) The semiochemistry of aphids. Nat Prod Rep 30:1277–1283

Pompon J, Quiring D, Goyer C, Giordanengo P, Pelletier Y (2011) A phloem-sap feeder mixes phloem and xylem sap to regulate osmotic potential. J Insect Physiol 57:1317–1322

Preda C, Skolka M (2011) Range expansion of *Metcalfa pruinosa* (Homoptera: Fulgoroidea) in southeastern Europe. Ecol Balk 3(1):79–87

Purcell AH (2003) Plant diseases and insects. In: Resh VH, Cardé RT (eds) Encyclopedia of insects. Academic Press, Amsterdam, pp 907–912

Redak RA, Purcell AH, Lopes JRS, Blua MJ, Mizell RF III, Andersen PC (2004) The biology of xylem fluid–feeding insect vectors of *Xylella fastidiosa* and their relation to disease epidemiology. Annu Rev Entomol 49:243–270

Rice RE, Atterholt CA, Delwiche MJ, Jones RA (1997) Efficacy of mating disruption pheromones in paraffin emulsion dispensers. IOBC WPRS Bull 20:151–162

Roff D (1990) The evolution of flightlessness in insects. Ecol Monogr 60:389–421

Ross L, Pen I, Shuker DM (2010) Genomic conflict in scale insects: the causes and consequences of bizarre genetic systems. Biol Rev 85:807–828

Santas LA (1983) Insects producing honeydew exploited by bees in Greece. Apidologie 14:93–103

Schick KN, Dahlsten DL (2003) Gallmaking insects. In: Resh VH, Cardé RT (eds) Encyclopedia of insects. Academic Press, Amsterdam, pp 464–466

Sciarretta A, Marziali L, Squarcini M, Marianelli L, Benassai D, Logli F, Roversi P (2016) Adaptive management of invasive pests in natural protected areas: the case of Matsucoccus feytaudi in Central Italy. Bull Entomol Res 106(1):9–18

Song N, Liang A-P, Bu C-P (2012) A molecular phylogeny of Hemiptera inferred from mitochondrial genome sequences. PLoS ONE 7(11):e48778

Speight MR, Hails RS, Gilbert M, Foggo A (1998) Horse chestnut scale (*Pulvinaria regalis*) (Homoptera: Coccidae) and urban host tree environment. Ecology 79:1503–1513

Spodek M, Ben-Dov Y, Mendel Z (2013) The scale insects (Hemiptera: Coccoidea) of oak trees (Fagaceae: Quercus spp.) in Israel. Isr J Entomol 43:95–124

Stone, C (1996) The role of psyllids (Hemiptera: Psyllidae) and bell miners (*Manorina melanophrys*) in canopy dieback of Sydney blue gum (*Eucalyptus saligna* Sm.). Austral Ecol 21(4):450–458

Straw NA, Williams DT, Green G (2011) Influence of sticky trap color and height above ground on capture of Alate *Elatobium abietinum* (Hemiptera: Aphididae) in sitka spruce plantations. Environ Entomol 40(1):120–125

Styrsky JD, Eubanks MD (2007) Ecological consequences of interactions between ants and honeydew-producing insects. Proc R Soc B: Biol Sci 274:151–164

Tobin PC, Berec L, Liebhold AM (2011) Exploiting Allee effects for managing biological invasions. Ecol Lett 14(6):615–624

Trierweiler T, Balder H (2005) Spread of horse chestnut scale (*Pulvinaria regalis*) in Germany. In: Symposium article. Plant protection and plant health in Europe. Introduction and spread of invasive species. Humboldt University, Berlin, Germany

Vacas S, Alfaro C, Primo J, Navarro-Llopis V (2014) Deployment of mating disruption dispensers before and after first seasonal male flights for the control of *Aonidiella aurantii* in citrus. J Pest Sci 88:321–329

Vandegehuchtea ML, Wermelingerc B, Fraefeld M, Baltensweilerd A, Düggelind C, Brändlid U-B, Freitage A, Bernasconie C, Cherixf D, Rischa AC (2017) Distribution and habitat requirements of red wood ants in Switzerland: implications for conservation. Biol Conserv 212:366–375

Vea I, Grimaldi D (2016) Putting scales into evolutionary time: the divergence of major scale insect lineages (Hemiptera) predates the radiation of modern angiosperm hosts. Sci Rep 6:23487

Vea I, Tanaka S, Shiotsuki T, Jouraku A, Tanaka T, Minakuchi C (2016) Differential juvenile hormone variations in scale insect extreme sexual dimorphism. PLoS ONE 11(2):e0149459

Walton VM, Daane KM, Bentley WJ, Millar JG, Larsen TE, Malakar-Kuenen R (2006) Pheromone-based mating disruption of *Planococcus ficus* (Hemiptera: Pseudococcidae) in California vineyards. J Econ Entomol 99(4):1280–1290

Way MJ (1963) Mutualism between ants and honeydew-producing Homoptera. Annu Rev Entomol 8:307–344

Whitfield AE, Falk BW, Rotenberg D (2015) Insect vector-mediated transmission of plant viruses. Virology 479:278–289

Will T, Furch A, Zimmermann M (2013) How phloem-feeding insects face the challenge of phloem-located defenses. Front Plant Sci 4:336

Wool DL (2005) Gall-inducing aphids: biology, ecology, and evolution. In: Raman A, Schaefer C, Withers T (eds) Biology, ecology, evolution of gall-inducing arthropods. Science Publishers, Enfield, NH and Plymouth, pp 73–132

Wood DL, Storer AJ (2003) Forest habitats. In: Resh VH, Cardé RT (eds) Encyclopedia of insects. Academic Press, Amsterdam, pp 442–454

Zina V, Lima A, Caetano F, Silva EB, Ramos AP, Franco JC (2012) First record of the pepper tree psyllid, *Calophya schini* Tuthill (Hemiptera, Calophyidae), in the Palaearctic region. Phytoparasitica 40:127–130

Zou Y, Millar JG (2015) Chemistry of the pheromones of mealybug and scale insects. Nat Prod Rep 32(7):1067–1113

Zytynska SE, Weisser WW (2016) The natural occurrence of secondary bacterial symbionts in aphids. Ecol Entomol 41:13–26

Chapter 14
Gall Formers

Brett P. Hurley, Gudrun Dittrich-Schröder, and Caitlin R. Gevers

14.1 Introduction

Gall formers are among the most highly evolved herbivores. Several organisms induce gall formation including viruses, bacteria, mites and nematodes. Insects are one of the most dominant gall-forming groups, with estimates ranging from 21,000 to 211,000 species (Ciesla 2011). Within the insects, gall formers have evolved independently in the Coleoptera, Diptera, Hemiptera, Hymenoptera, Lepidoptera and Thysanoptera. Galls are induced by the gall-forming insect, where specific metabolic interactions result in differentiation of the plant tissue and the consequent abnormal growths are referred to as galls. Through this manipulation of the plant's growth, the insect obtains food and shelter. Galls can vary greatly in size and shape, from pits or folds to the 'oak apples' of some cynipid species. Gall forming insects often display fascinating and complex biologies, including host alternation and cyclical parthenogenesis. However, the biology of many gall formers is poorly understood, in part because of their cryptic habit of living primarily within the gall.

Historically, gall formers of trees have often been reported as pests of little economic importance. However, this has changed with the introduction and spread of a number of invasive non-native gall forming species, some of which have become pests of serious economic importance. Thus, for some forestry tree species like eucalypts this group of insects has now become one of the most important groups of insect pests (Dittrich-Schröder et al. 2020). The importance of gall formers as pests of forestry trees will increase in the future with the increased movement of these insects around the world.

In this chapter, we discuss insect gall formers of forestry trees. We examine the natural history and ecology, as well as the evolution and diversity, of these fascinating

B. P. Hurley (✉) · G. Dittrich-Schröder · C. R. Gevers
Department of Zoology and Entomology, Forestry and Agricultural Biotechnology Institute (FABI), University of Pretoria, Pretoria, South Africa
e-mail: brett.hurley@fabi.up.ac.za

© The Author(s) 2023
J. D. Allison et al. (eds.), *Forest Entomology and Pathology*,
https://doi.org/10.1007/978-3-031-11553-0_14

insects. We provide a list of the gall forming insects associated with forestry trees and discuss management strategies, with a focus on the species most relevant to forestry. In addition, we use case studies to discuss three of the important gall forming species, to provide more specific details of the biology, spread and management of these insects.

14.2 Natural History and Ecology

As a group gall forming insects are polyphyletic and include a variety of orders and families, and consequently represent a very diverse range of life histories. As it is not possible to cover the details of all the different gall formers in this chapter, we focus on three key aspects related to their natural history and ecology; namely gall formation, reproductive strategies and the gall community.

14.2.1 Gall Formation

Insect-induced plant galls are the product of a highly specialised and unique type of insect-plant interaction. Most gall forming insects have a high fidelity to a specific host genus or even species, and thus do not attack an extensive range of host plants (Csóka et al. 2017). In addition, most gall forming insects only attack specific plant organs of their host, for example the flowers, fruits, buds, shoots or leaves, and these organs must be in the correct development or phenological stage. Gall shape is generally consistent within a species, but not between related taxa, and thus gall morphology can be used to assist species identification (Raman 2011). As opposed to the abnormal growths induced by fungi and bacteria, 90% of galls induced by insects show bilateral or radial symmetry (Raman 2007).

Plant galls are an incredible example of a modified natural structure caused by messages from a foreign organism. Initiation of a gall occurs when the plant is either exposed to an accessory gland secretion during oviposition (e.g. cynipids, sawflies and some beetles), or a salivary chemical from the feeding of first-instar larvae (e.g. cecidomyiids, coccids and aphids) (Rohfritsch 1992). Subsequently the exposed tissues no longer experience normal growth but, instead, the cells are physiologically modified through differentiation and hypertrophy (increase in cell size), resulting in formation of the inner-gall tissue and the outer gall tissue (cortical parenchyma) (Harper et al. 2004; Klein 2009). The growth stage of the gall consists mainly of hyperplasia (cell expansion), which allows the gall and the inner chamber to grow. Hyperplasia results in the formation of nutritive tissue that the larvae are able to manipulate into a suitable food source (Rohfritsch 1992; Harper et al. 2004; Klein 2009). The larva is then able to feed on the inner-gall tissue for the duration of its development, thereby decreasing the number of gall cell layers as the larva increases in size.

The gall reaches maturity once cell proliferation ceases. Major physiological and chemical changes in the gall tissues, such as when the flow of water and sap to the gall stops, results in gall dehiscence (Rohfritsch 1992). These changes are correlated with the development of the gall former and facilitate its exit from the gall. Although there is limited information on the communication between gall inducer and host plant, it has become evident that constant stimuli by the gall inducer is required to retain gall formation (Stone et al. 2002; Harper et al. 2004). If a natural enemy kills the gall inducer, gall development will cease.

14.2.2 Reproductive Strategies

Gall forming insects are present in several orders and families, and, not surprisingly, there are a range of reproductive strategies associated with these insects, from sexual reproduction to parthenogenesis, where males are either lacking or very scarce (see Case Study—*Leptocybe invasa*). Perhaps one of the most interesting reproductive strategies, present in a number of gall forming insects in the Hemiptera (Adelgidae and Pemphigidae) and Hymenoptera (Cynipidae), is that of cyclical parthenogenesis. Cyclical parthenogenesis involves the alternation of sexual and asexual reproduction. This reproductive mode has been described in over 15,000 species (Stone et al. 2008). The alternation of reproductive modes within a year is often associated with differences in morphology and ecology among generations. The differences between these generations can be so great that some gall formers were originally identified as different species and even genera (Felt 1940). In some cases, the different reproductive modes are associated with different host plants (see Case Studies—*Adelges cooleyi* and *Andricus* spp).

Oak and sycamore gall wasps (Cynipidae) are unique as they produce their asexual and sexual generations in strict alternation (Stone et al. 2002, 2008; Atkinson et al. 2003). This is unlike other gall wasps, which reproduce asexually with the occasional sexual generation in response to environmental changes. Many of the cynipids exhibit obligate alternating sexual and asexual generations (Hood and Ott 2011). These generations differ by the host plant used, the plant part galled, morphology and size of the gall, the number of siblings in each gall and the adult body size (Stone et al. 2002; Rokas et al. 2003; Stone and Schönrogge 2003; Hood and Ott 2011; Schönrogge et al. 2012). A common difference observed between sexual and asexual generations is that asexual galls are large and complex while the sexual galls are simple, often cryptic and usually much smaller.

14.2.3 Gall Community

Galls can contain multitrophic, closed and complex communities with a number of different inhabitants, including the gall inducer, parasitoids, hyperparasitoids, and inquilines (Table 14.1). Parasitoids may feed exclusively on the host gall former, but in some parasitoid species the larvae initially feed on the gall tissue and then switch to feed on the host, while in other species the larvae start by feeding on the host and later feed on the gall tissue in order to complete their development (Roskam 1992; Klein 2009). There is a great diversity of parasitoids of gall formers, and in some cases there are different cohorts of parasitoids of the sexual and asexual galls of the same species (Table 14.1).

Inquilines inhabit galls of other insects to obtain food and shelter. They are incapable of inducing galls of their own but, as with the parasitoids of gall insects, they are highly specialised to gain access to existing galls. Unlike parasitoids, however, inquilines do not feed on the developing gall insect but obtain their nutrition from the gall tissues (Brooks and Shorthouse 1998). The gall inducer and inquiline can sometimes co-exist and partition the gall resources, resulting in the successful development of both species. In other cases, the inquiline may cause the death of the gall inducer because it either develops and feeds much faster than the gall inducer, causing it to starve, or it manipulates the gall, creating additional chambers and tissues of its own (endohalls), which cause the gall inducer to die due to insufficient space or crushing (Brooks and Shorthouse 1998; Ferraz and Monteiro 2003; Klein 2009). Although some gall communities are relatively well studied, in many cases the taxonomy and role of the different gall inhabitants is not well resolved.

14.3 Evolution and Diversity

The ability to form galls has evolved many times in phytophagous insects and has been recorded from six insect orders (Cook and Gullan 2004; Hardy and Cook 2010). More than 13,000 insect species have been described as gall inducers (Raman et al. 2005); however, the two families with the greatest number of species are the Cecidomyiidae (Diptera) and Cynipidae (Hymenoptera), with each family comprising approximately 1400 described species (Dreger-Jauffret and Shorthouse 1992; Ronquist and Liljeblad 2001; Ronquist et al. 2015) (Fig. 14.1).

The complex relationship between the gall former and host plant is considered to be an advanced association (Shorthouse et al. 2005). Gall forming insects have been shown to be highly host and tissue specific, showing significant phenotypical specificity and are significantly more host-specific than closely related non gall-forming species (Hardy and Cook 2010). An example, illustrating the phenotypic specificity, are the cynipids associated with oak, which are represented by more than 1000 described species, with each species having a characteristic gall structure (Stone and Cook 1998; Ronquist and Liljeblad 2001). Evolution of gall formation is best-described using cynipids due to the extensive work conducted on this family. The ancestral origin of gall-forming cynipids is thought to be the Palearctic (Ronquist et al. 2015). Some of the oldest records of studies on the evolution of gall formation, date back to the 1920's. In these studies, Alfred Kinsey used morphological and biological information to explain the relationships between gall formers and their

Table 14.1 Examples of inquilines and parasitoids of gall insects that infest forest trees

Species	Inquilines	Parasitoids
Hymenoptera: Cynipidae		
Andricus burgundus	*Synophrus politus*	*Mesopolobus dubius, M. xanthocerus, M. mediterraneus, Ormocerus vernalis, Aulogymnus gallaru, A. testaceoviridis, Aprostocetus* sp. 1, *Tetrastichus* sp. 1, *Torymus auratus, Macroneura vesicularis*
Andricus corruptrix	Sexual generation: *Synergus reinhardii* & *S. umbraculus*	Sexual generation: *Mesopolobus dubius, M. fuscipes, M. tibialis* and *M. xanthocerus*
Andricus grossulariae	*Ceroptres cerri* (Cynipidae)	
Andricus kollari	Sexual generation: *Synergus reinhardii, S. umbraculus*	Sexual generation: *Mesopolobus dubius, M. fuscipes, M. tibialis* and *M. xanthocerus*
Andricus lignicola	Sexual generation: *Synergus reinhardii, S. umbraculus*	Sexual generation: *Mesopolobus dubius, M. fuscipes, M. tibialis* and *M. xanthocerus*
Andricus quercuscalisis (Knopper gall wasp)	Asexual galls: Synergus gallaepomiformis, S. umbraculus; Sexual galls: *Synergus reinhardii, S. umbraculus*	Asexual galls: *Sycophila biguttata, Mosopolobus amaenus, Megastigmus stigmatizana, Gelis formicarcius*. Sexual galls: *Mesopolobus dubius, M. fuscipes, M. tibialis* and *M. xanthocerus*
Dryocosmus kuriphilus (Chestnut gall wasp)		*Torymus sinensis*
Neuroterus saltatorius (Jumping oak gall wasp)		*Ormyrus distinctus, Amphidocius schickae, Mesopolobus longicausae, Dibrachys cavus, Amphidociusn.* sp., *Aprostocetus pattersonae, Aprostocetus verrucarii, Aprostocetus* n. sp., *Brasema* sp.
Hymenoptera: Eulophidae		
Leptocybe invasa (bluegum chalcid)		*Selitrichodes neseri, Quadrastichus mendeli, Selitrichodes kryceri*

(continued)

Table 14.1 (continued)

Species	Inquilines	Parasitoids
Ophelimus maskelli (eucalypt gall wasp)		*Closterocerus chamaeleon*
Quadrastichodella nova		*Leprosa milga*
Diptera: Cecidomyiidae		
Obolodiplosis robiniae		*Platygaster robiniae*
Thecodiplosis jaonensis (pine needle gall midge)		*Inostemma matsutama, I. seoulis*

References: Smith (1954), Claridge (1962), Hutchinson (1974), Payne (1978), Moriya et al. (1989b), Schönrogge et al. (1996), Walker et al. (2002), Ciesla (2011), and Schönrogge et al. (2012)

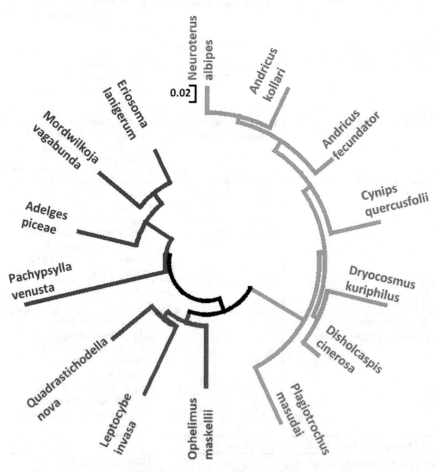

Fig. 14.1 A circular neighbour joining tree showing the relatedness of gall forming insects in forests. Cytochrome b sequences available for gall-forming insects associated with forest trees were downloaded from GenBank and used to generate the tree. Colours correspond to the insect family, namely Adelgidae, Aphididae and Psyllidae (*blue*); Cynipidae (*green*); and Eulophidae (*red*)

hosts (Ronquist et al. 2015). Kinsey (1920) hypothesized that gall formers initially utilized herbs as their hosts and inhabited the plant tissue rather than induced galls. These galls subsequently evolved and increased in complexity with for example multi-chambered galls, considered structurally simple, evolving into single chambered galls, considered structurally complex (Ronquist et al. 2015). This increased complexity over evolutionary time has also been observed in aphids, thrips and sawflies (Fukatsu et al. 1994; Nyman et al. 1998; Morris et al. 1999). Other authors such as Malyshev (1968) have argued that early cynipids were associated with oaks and first formed galls on buds or seeds. However, Roskam (1992) suggested that recent radiation events in the Oligocene led to the radiation of the cynipids from the family of flowering plants, the Asteraceae.

The three main advantages of development within galls are the availability of nutrition, a constant microclimate, and protection against parasitoids and herbivores due to physical and/or chemical features of the gall (Stone and Schönrogge 2003; Ronquist et al. 2015). It is generally accepted that the evolution of gall formers has largely been driven by selection pressure from natural enemies (Stone and Cook 1998) and to a lesser extent mistakes during oviposition/host selection by female gall formers (Price 2005). High mortality levels of gall formers, caused by parasitoids, has been suggested to drive the evolution of various morphological gall characteristics, such as gall size, thickness of gall exterior and spiny surfaces, in an attempt to decrease mortality levels (Inbar et al. 2004; Hardy and Cook 2010). This top-down force could be one of numerous factors driving variation in gall formation (Nosil and Crespi 2006). Parasitoids may parasitize gall formers on a preferred plant host, leading to the use of a new plant host by the gall former, in an attempt to evade mortality thereby resulting in variation in gall formation (Rott and Godfray 2000; Singer and Stireman 2005). Studies have shown that most often there is a correlation between gall morphology and relatedness of the gall former, however there are a few exceptions (Inbar et al. 2004).

Phylogenetic analyses suggest that most of the diversity making up the gall forming species, within for example the Cecidomyiidae and Cynipoidea, is due to a radiation event (the rapid splitting of a lineage into two distinct lineages, due to certain conditions which result in a new feature, permitting the lineage to access a new niche), rather than the independent acquisition of the gall forming trait (Ronquist and Liljeblad 2001). This conclusion is based on the occurrence of many gall forming species in only a few insect families, rather than gall forming species distributed amongst many insect families (Cook and Gullan 2004; Vardal 2004). Some groups of gall formers, such as the Cynipdae contain many species whereas others such as the Aphididae contain far less, possibly indicating that a large degree of speciation occurred within the Cynipidae at a rapid rate (Price 2005). Consequently, such groups are difficult to separate at a taxonomic level, as morphological characteristics are not distinctive enough to separate specimens into different genera.

Diversification of gall forming insects has been thought to be influenced and driven by individual phenotypic traits (e.g. reproductive rate, dispersal ability, body size, longevity, ecological constraint, sexual selection), species-specific traits (e.g. population size, abundance, geographic range), and availability of sufficient suitable hosts (Hardy and Cook 2010). The most likely factor contributing to speciation within gall-formers is the acquisition of a new host or plant organ (Medonça 2001). Due to the specificity of the relationship between the gall former to its host plant, the shift of a gall former to a new host is very rarely observed and rarely successful (Csóka et al. 2017). For this reason, no gall former has been recorded as utilising a wide range of host plants (Ronquist et al. 2015; Csóka et al. 2017).

Interestingly, within the Hymenoptera, the family Eulophidae contains species with diverse biologies, ranging from predominantly parasitoids to gall formers and seed feeders (Gauthier et al. 2000). The Eulophidae represents the largest group of parasitoids within the Chalcidoidea and contains over 4000 described species, including many that have been successfully used for biological control (Noyes 1998; Gauthier et al. 2000). The origin of parasitism is thought to have arisen from the group ancestral to the Orussoidea and Apocrita (Whitfield 1998). Early parasitoids evolved from hymenopterans feeding on both wood-boring larvae and to a lesser extent on fungi.

14.4 Gall Formers of Forest Trees

Gall forming insects are present on all continents except Antarctica and infest many forest tree species including both hardwood and softwood species. In some cases these infestations are not considered to significantly affect the health of the trees, whereas in other cases these insects are reported as serious pests. The majority have been reported from the northern hemisphere, but it is unclear whether this is a true reflection of higher diversity or if it is due to differences in reporting and research on these insects between the two hemispheres. We provide a summary of gall forming insects known to infest forest trees in Table 14.2, focusing on those that are considered as pests and/or which have been well studied.

Although gall forming insects that infest forest tree species are present in a number of insect orders and families, most of these insects are from the Adelgidae (Hemiptera), the Cecidomyiidae (Diptera), and the Cynipidae and Eulophidae (Hymenoptera) (Table 14.2). The Cecidomyiidae gall formers are associated with a range of host trees and gall types, including artichoke-shaped galls of some of the *Dasineura* species, spindle shaped galls at the base of pine needles, and galls which at maturity resemble a small flower on the branches. Nearly all of the adelgid gall formers belong to the genus *Adelges*. Several of these species have been shown to be holocyclic, where the primary host is always *Picea* and the secondary host is *Abies*, *Larix*, *Pseudostuga*, *Tsuga* or *Pinus*. These insects cause damage at the non-galling (asexual) stage, whereas the galling stage is not considered problematic. The galls resemble small pineapples.

Most of the Cynipidae are gall formers on oak species and associated with visually striking galls, sometimes with the appearance of apples. The oak gall wasps are also known to have complex communities of parasitoids and inquilines. An important cynipid that is not associated with oaks is the chestnut gall wasp, *Dryocosmus*

Table 14.2 Gall-forming insects of forest trees. This list is based on an extensive search of the available literature. It is not exhaustive, but provides an indication of the diversity of these insects. For 'Known Distribution', I = introduced range and N = native range (not known for all cases)

Species	Known distribution	Host	Gall type	References
Coleoptera: Cerambycidae				
Podapion gallicola (Pine gall weevil)	North America	*Pinus*	Spindle shaped stem galls	Drooz (1985), and Rose and Lindquist (1999)
Saperda inornata (Poplar gall saperda)	North America	*Populus* spp.	Globose galls on twigs and stems	Browne (1968), and Ciesla (2011)
Saperda concolor	North America	*Populus* spp.	Globose stem galls	Ciesla (2011)
Saperda pupulnea (Small poplar borer)	Europe (N); China (N); North America (I); Canada (I)	*Populus* spp. especially on *P. nigra* & *P. tremula* in Turkey & *Salix* spp. (willows)	Globose stem galls	Ciesla (2011)
Diptera: Agromyzidae				
Hexomyza (=*Melanagromyza*) *schineri* (Poplar twig gall fly)	Northern Hemisphere	*Populus* spp. and *Salix* spp.	Spherical galls on new shoots which continue to grow as the tree grows	Browne (1968), and Ciesla (2011)
Diptera: Cecidomyiidae				
Contarinia pseudotsugae (Douglas-fir needle midge)	North America (I); Europe (I)	*Pseudotsuga menziesii* (Douglas-fir)	Yellowish galls on the needles	Bulaon (2005)
Dasineura gleditchiae (Honey locust gall midge)	North America (I); Europe (I)	*Gleditsia triacanthos* (Honey locust)	Pod-like galls on Individual leaflets; each gall contains 1–5 larvae	Molnár et al. (2009), and Csóka et al. (2017)
Dasineura kellneri	Central Europe	*Larix decidua*	Artichoke shaped galls on stems and flowering buds	Isaev et al. (1988), and Ciesla (2011)

(continued)

Table 14.2 (continued)

Species	Known distribution	Host	Gall type	References
Dasineura nipponica	Japan	*Larix kaempferi*	Artichoke shaped galls on stems and flowering buds	Isaev et al. (1988), and Ciesla (2011)
Dasineura rozhkovi	Europe; Asia	*Larix czekanowski, L. gmelinii & L sibirica*	Artichoke shaped galls on stems and flowering buds	Isaev et al. (1988), and Ciesla (2011)
Dasineura strobilas	Australia	*Leptospermum laevigatum* (Coastal tea tree) (Myrtaceae)	Artichoke shaped galls on stems and flowering buds	Isaev et al. (1988), and Ciesla (2011)
Dasineura verae	Europe; Asia	*L. gmelinii & L sibirica*	Artichoke shaped galls on stems and flowering buds	Isaev et al. (1988), and Ciesla (2011)
Obolodiplosis robiniae	Europe; South Korea; Japan; Russia	*Robinia pseudoacacia* (North American Black Locust)	Downward-folded swellings	Molnár et al. (2009), and Csóka et al. (2017)
Oligotrophus apicis	Central America	*Juniperus* spp.	Blue/green apical conical galls resembling a small flower which turns red/brown as it matures on branches	Ciesla (2011)
Oligotrophus betheli (Juniper tip midge)	North America (N); British Columbia (I); Canada (I)	*Juniperus* spp. *J. occidentalis, J. osteosperma, J. scopulorum & J. virginiana*	Blue/green apical conical galls resembling a small flower which turns red/brown as it matures on branches	Ciesla (2011)
Oligotrophus gemmarum	Europe	*Juniperus* spp.	Blue/green apical conical galls resembling a small flower which turns red/brown as it matures on branches	Ciesla (2011)

(continued)

Table 14.2 (continued)

Species	Known distribution	Host	Gall type	References
Oligotrophus juniperinus	Europe	*Juniperus* spp.	Blue/green apical conical galls resembling a small flower which turns red/brown as it matures on branches	Ciesla (2011)
Oligotrophus nezu	Japan	*Juniperus* spp.	Blue/green apical conical galls resembling a small flower which turns red/brown as it matures on branches	Ciesla (2011)
Oligotrophus panteli	Europe	*Juniperus* spp.	Blue/green apical conical galls resembling a small flower which turns red/brown as it matures on branches	Ciesla (2011)
Paradiplosis tumifex (Balsam gall midge)	Canada (I); North America (N)	*Abie balsamea* (Balsam fir) and *Abies fraseri* (Fraser fir)	Oval galls on the needles	Drooz (1985), and Ciesla (2011)
Taxodiomyia cupressiananassa (Cypress twig gall midge)	North America; Central America	*Taxodium distichum* (Bald cypress)	Oval galls on terminal portion of branchlets which resemble pineapples at maturity; pink in colour which changes to green/white	Chen and Applybyby (1984), and Ciesla (2011)
Thecodiplosis jaonensis (Pine needle gall midge)	Japan (N); South Korea (I); North Korea (I)	*Pinus densiflora* & *P. thunbergii*	Spindle shaped galls on the base of needles	Ciesla (2011)

(continued)

Table 14.2 (continued)

Species	Known distribution	Host	Gall type	References
Pinyonia edulicola (Pinon spindle gall midge)	North America	*Pinus edulis* (Pinon pine)	Spindle shaped galls base of current years needles	Ciesla (2011)
Rhabdophaga rosaria	Europe	*Salix* spp. *Salix alba*	Pine cone shaped gall	Kollár (2007), and Ciesla (2011)
Rhabdophaga strobiloides (Willow pine cone gall)	North America (N); South America (I); Central America (I)	*Salix* spp.	Pine cone shaped bud gall (25 mm)	Kollár (2007), and Ciesla (2011)
Hemiptera: Adelgidae				
Adelges abietis (Eastern spruce gall adelgid)	Europe (N); North America (I)	*Picea* spp.	Spherical to small pineapple/conelike galls (10–18 mm long) on the base of new twigs, size and shape varies depending on hosts	Carter (1971), and Ciesla (2011)
Adelges cooleyi (Cooley spruce gall adelgid)	North America (N); Europe (I)	*Picea* spp. *Picea sitchensis* (Sitka spruce)(primary host where galls induced); *Pseudotsuga* species (*Pseudotsuga menziesii* - Douglas fir)	Small light green to dark purple pineapple shaped galls on shoots and needles	Havill and Foottit (2007), and Csóka et al. (2017)
Adelges glandulae	China	*Picea* (primary host) & *Abies* (secondary host)	Leaf cone-like gall without associated needles	Zhang et al. (1980), and Havill and Foottit (2007)
Adelges isedakii	Japan	*Picea* (primary host) & *Larix* (secondary host)	Cone-like gall	Havill and Foottit (2007), and Ciesla (2011)

(continued)

Table 14.2 (continued)

Species	Known distribution	Host	Gall type	References
Adelges japonicus	Japan	*Picea* (primary host) & *Abies* (secondary host)	Cone-like gall	Ciesla (2011)
Adelges knucheli	South Asia	*Picea* (primary host) & *Larix kaempferi* (secondary host)	Cone-like gall	Ciesla (2011)
Adelges lapponicus	Europe	*Picea* (primary host) & *Larix decidua* (secondary host)	Cone-like gall	Ciesla (2011)
Adelges lariciatus	Eastern North America (N); Canada (I)	*Picea* (primary host)	Buds and cones	Ciesla (2011)
Adelges laricis	Europe	*Picea* (primary host) & *Larix laricina* & *L. lyallii* (secondary host)	Small pineapple shaped galls of shoots and needles	Carter (1971)
Adelges merkeri	Asia Minor (N); Europe (I)	*Abies* (secondary host)	Cone-like gall	Havill and Foottit (2007), and Csóka et al. (2017)
Adelges nebrodensis	Europe	*Picea* (primary host) & *Abies* (secondary host)	Cone-like gall	Ciesla (2011)
Adelges nordmanniana	North America (N); Europe (I); Australia (I); New Zealand (I)	*Picea* spp. (primary) & *Abies nordmanniana* (secondary host)	Cone like galls	Ciesla (2011), and Csóka et al. (2017)
Adelges normannianae	North America (N); Europe (I)	*Picea orientalis* (primary host); *Albies* (secondary host)	Cone-like galls of shoots and needles on primary host; curling and thickening of needles (similar to bottlebrush) on secondary host	Carter (1971)

(continued)

Table 14.2 (continued)

Species	Known distribution	Host	Gall type	References
Adelges pectinata	Europe, China and Japan	*Picea* (primary host) & *Abies* (secondary host)		Ciesla (2011)
Adelges piceae (Balsam woolly adelgid)	Central Europe (I); North America (I)	*Abies* (true firs): *A. balsamea* and *A.fraseri* (Balsam and Fraser)	Swellings and thickening of shoots and needles	Carter (1971), and Ciesla (2011)
Adelges prelli	North America (N); Europe (I)	*Picea* spp. (primary) & *Pinus* (secondary host)	Cone-like gall	Ciesla (2011), and Csóka et al. (2017)
Adelges tardoides	Europe	*Picea* (primary host) & *Larix* (secondary host)	Cone-like gall	Ciesla (2011)
Adelges torii	Japan	*Picea* (primary host) & *Larix* (secondary host)	Cone-like gall	Ciesla (2011)
Adelges tsugae (Hemlock woolly adelgid)	Asia and western North America (N); eastern North America (I)	*Picea* (Primary host in native range); *Tsuga* (secondary host)	Small pineapple shaped galls of shoots and needles	Ciesla (2011)
Adelges viridis	Europe	*Picea* (primary host) & *Larix decidua* (secondary host)	Cone-like gall	Ciesla (2011)
Pineus pinifoliae (Pine leaf adelgid/chermes)	North America (N)	*Picea* spp.	Loose, terminal and leafy cone-shaped galls on young shoots	Drooz (1985)
Hemiptera: Aphididae				
Eriosoma lanigerum (Woolly apple aphid)	North America (N); South Africa (I); Australia (I)	*Malus pulima* (Apple) & *Pyrus* (pear)	Root and swollen and twisted twig galls	Nicholas et al. (2005)

(continued)

Table 14.2 (continued)

Species	Known distribution	Host	Gall type	References
Mordvilkoja vagabunda (Poplar vagabond aphid)	North America (N); Canada (I); Turkey (I)	*Populus* (primary host); *Lythrum* spp. (Loosestrife) (alternative host)	Large and convoluted green or brown galls (5–7 cm) on the tips of shoots	Ciesla (2011)
Pachypsylla celtidisgemma (Budgall psyllid)	North America	*Celtis occidentalis* (Hackberry)	Small round bud galls	Riley (1885)
Pachypsylla venusta (Petiolegall psyllid)	North America	*Celtis occidentalis* (Hackberry)	Woody subspherical petiole galls	Johnson and Lyon (1988)
Pachypsylla celtidisvesicula (Hackberry blister gall maker)	North America	*Celtis occidentalis* (Hackberry)	Blister-like galls (3–4 mm diameter and slightly raised) on the leaf surface	Johnson and Lyon (1988)
Pachypsylla celtidismamma (Hackberry nipple gall maker)	North America	*Celtis* sp. (Hackberry) especially *Celtis occidentalis*	Small round yellow to yellow-green galls (4 mm diameter and 6 mm high)	Drooz (1985), and Ciesla (2011)
Pemphigus populivenae (Sugarbeet root aphid)	Europe; North America	*Celtis laevigata* (Sugarbeets)	Elongated root galls	Foottit et al. (2010)
Hemiptera: Homotomidae				
Phytolyma fusca	Tropical Africa	*Milicia* sp. especially *M. excelsia*	Leaf galls are globular in shape; galls on the stems and shoots are oblong in shape	Wagner et al. (2008), and Ciesla (2011)
Phytolyma lata (Iroko gall fly)	Tropical Africa	*Milicia* sp. especially *M. regia*	Shoot, bud and leaf galls	Hollis (1973), and Cobbinah (1986)
Phytolyma tuberculate	Tropical Africa	*Milicia* sp. especially *M. excelsia*		Hollis (1973)
Hymenoptera: Cynipidae				
Amphibolips californicus	North America	*Quercus alba* (White oak)	Apple galls	

(continued)

Table 14.2 (continued)

Species	Known distribution	Host	Gall type	References
Amphibolips confluenta (Oak apple gall wasp)	Australia (N); North America (I)	*Quercus* spp. (Oaks) *Q. coccinea, Q. rubra & Q. velutina*	Large green apple galls which are quite robust; become brown as they mature	Drooz (1985)
Andricus quercuscalifornicus (=*californicus*) (California gall wasp)	North America	*Quercus garryana*	Apple gall (asexual)	Drooz (1985)
Andricus corruptrix	South Eastern Europe (N); Britain & British Isles (I)	*Quercus robur; Quercus cerris & Quercus petraea*	Single cone-shaped gall with a darkened point (sexual); irregular shaped, smooth gall with 3–5 lobed tip (asexual)	Schönrogge et al. (1996, 2012)
Andricus lucidus	South Eastern Europe (N); Britain & British Isles (I)	Sexual: *Quercus cerris*. Asexual: Q.robur & Q. petreae	Groups of bright yellow-green apple galls	Schönrogge et al. (2012)
Andricus grossulariae	South Eastern Europe (N); Britain & British Isles (I)	Sexual: *Quercus cerris*. Asexual: Q.robur & Q. petreae	Unilocular deep purple rounded galls (usually clustered) on catkin (bisexual); green-purple galls which are spined (asexual)	Schönrogge et al. (1996, 2012)
Andricus kollari	South Eastern Europe (N); Britain & British Isles (I)	*Quercus robur; Quercus petraea & Quercus cerris*	1–8 thin pointed galls per bud and occasionally have tufts of hair on the upper part (bisexual); hard, round and smooth small red-brown globular bumps on the bud (marble gall) (asexual)	Schönrogge et al. (1996, 2012)

(continued)

Table 14.2 (continued)

Species	Known distribution	Host	Gall type	References
Andricus lignicola	South Eastern Europe (N); Britain & British Isles (I)	*Quercus robur, Quercus cerris & Quercus petraea*	1–8 golden/brown galls with plunt rounded tip on a bud (bisexual); leathery brown globular bud gall (asexual)	Schönrogge et al. (1996, 2012)
Andricus lucidus (Hedgehog gall)	South Eastern Europe (N); Britain & British Isles (I)	*Quercus robur, Quercus cerris & Quercus petraea*	Bright green galls aggregated together (bisexual); rounded gall covered in spines (hedgehog gall) (asexual)	Chinery (2011)
Andricus quercuscalicis (Knopper gall wasp)	South Eastern Europe (N); Britain & British Isles (I)	*Quercus robur* (English Oak) & *Quercus cerris* (Turkey Oak)	Slender pointed single galls on the male flowers of Turkey Oak (bisexual); irregular gall which appears around the acorns of English Oak (knopper gall) (asexual)	Schönrogge et al. (1996, 2012)
Aphelonyx cerricola	South Eastern Europe (N); Britain & British Isles (I)	*Quercus cerris*	Large irregularly shaped galls which appear green and velvety and become brown and woody; neighbouring galls often fuse together	Crawley (1997), and Schönrogge et al. (2012)
Biorhiza pallida	Europe	*Quercus* spp. (Oaks)	Spongy apple bud gall (bisexual); spherical root galls (asexual)	Drooz (1985)
Callirhytis tumifica	North America (N); Europe (I)	*Quercus rubra* (planted in other countries as plantation trees), *Q. palustris, Q. coccinea, Q. schumardi,* etc. (North American red oaks)	Acorn (asexual); leaves (bisexual)	Csóka et al. (2017)

(continued)

Table 14.2 (continued)

Species	Known distribution	Host	Gall type	References
Callirhytis cornigera (Horned oak wasp)	Canada; North America	*Quercus palustris* (Pin Oak)	Golf ball sized woody stem galls with horn-like protrusions (asexual) and small blister-like leaf galls (bisexual)	Felt (1917)
Callirhytis quercusclaviger	Canada; North America	*Quercus laurifolia* (Laurel Oak)	Woody stemmed potato galls bearing spines (Spine bearing Potato Gall)	Felt (1917)
Callirhytis quercuspomiformis (=*pomiformis*) (live oak apple gall wasp)	Great Britain; North America	*Quercus agrifolia* (California live oak)	Large apple type galls with a gnarled surface (asexual); leaf galls (bisexual)	
Cycloneuroterus hisashii	Japan	*Quercus* (*Cyclobalanopsis*) *glauca*	Globular bud gall	Ide et al. (2012)
Cycloneuroterus arakashiphagus	Japan	*Quercus* (*Cyclobalanopsis*) *sessilifolia*; *Quercus* (*Cyclobalanopsis*) *glauca*	Clustered pale green/yellowish brown oval bud gall (asexual); leaf galls (sexual)	Ide et al. (2012)
Cycloneuroterus akagashiphilus	Japan	*Quercus* (*Cyclobalanopsis*) *acuta*	Oval gall on margin/apex of young leaves	Ide et al. (2012)
Cycloneuroterus fortuitusus	Japan; Taiwan	*Quercus* (*Cyclobalanopsis*) *sessilifolia*; *Quercus* (*Cyclobalanopsis*) *glauca*	Oval galls at the base of young shoots and buds	Ide et al. (2012)
Cycloneuroterus wangi	South eastern China; Japan	*Quercus* (*Cyclobalanopsis*) *sessilifolia*	Bud galls	Mendel et al. (2004)
Cynips longiventris	Europe; Asia	*Quercus* spp.		

(continued)

Table 14.2 (continued)

Species	Known distribution	Host	Gall type	References
Cynips divisa	Europe; Asia	*Quercus* spp.	Thin walled round pall yellow galls which have red-brown markings (asexual); bright yellow–brown smooth gall (red-wart gall) (bisexual)	Browne (1968)
Cynips insana		*Quercus infectoria*		Fagan (1918), and Felt (1940)
Cynips quercusfolii (oak bud wasp)	Asia	*Quercus* spp. most commonly found on *Q. petraea* and *Q. robur*	Small, ovoid galls which begin as a deep red/purple colour and turn black as the gall matures (violet-egg gall) (bisexual); spherical yellow/green–brown hard and brittle leaf galls (asexual)	Ciesla (2011)
Disholcaspis cinerosa	North America	*Quercus fusiforme* & *Q. virginiana*	Sexual generation: inconspicuous galls on the bud; Asexual generation: large ranging from 3 to 25 mm on the stem	Ciesla (2011)
Disholcaspis quercusmamma (rough oak bullet gall wasp)	Central America	*Quercus bicolor* (Swamp white oak) & *Q. macrocarpa* (Burr Oak)	Round woody galls with a point at the apex these galls change to red/brown colour as maturation occurs; galls remain on the branch for several years after the insect has left	Ciesla (2011)

(continued)

Table 14.2 (continued)

Species	Known distribution	Host	Gall type	References
Dryocosmus kuriphilus (Chestnut gall wasp)	China (N); Japan (I); North America (I); Europe (I)	*Castanea* spp. China: *C. mollissim.* Korea & Japan: *C. crenata* & *C. crenata* x *C. mollissim.* USA: *C. mollissima.* Italy: *C. denata* (native to Italy)	Rose-coloured bud gall	Moriya et al. (1989a), and Csóka et al. (2017)
Neuroterus saliens	Britain & British Isles	Sexual: *Quercus cerris.* Asexual: *Q.robur* & *Q. petreae*	Uniocular green–brown pimple-like leaf galls (asexual); red multi-ocular acorn galls with disc-like eruptions (bisexual)	Schönrogge et al. (2012), and Csóka et al. (2017)
Neuroterus saltatorius (Jumping oak gall wasp)	North America	White Oaks: *Quercus lobata* (Valley oak), *Quercus douglasii* (blue oak), *Quercus garryana* (Oregon or Garry oak), *Quercus dumosa* (California scrub oak), *Quercus arizonica* (Arizona white oak), *Quercus virginiana* (Live oak)	Sexual: non-detachable cluster galls, galls are originally green and turn brown as they mature; asexual: small mustard coloured sub-spherical galls, single chamber and detachable from leaf	Csóka et al. (2017)
Plagiotrochus coriaceus/Andricus pseudococcus	Britain & British Isles	*Quercus ilex*	Uniocular oval leaf gall	Schönrogge et al. (2012)
Plagiotrochus quercuscalis	Britain & British Isles	*Quercus ilex*	Bud gall	Hancey and Hancey (2004), and Schönrogge et al. (2012)

(continued)

Table 14.2 (continued)

Species	Known distribution	Host	Gall type	References
Plagiotrochus amenti	southwest Europe (N) and northwest Africa (N); Europe (I); North America (I); South America (I)	*Quercus suber* (Cork oak)	Sexual generation: catkins & shoots. Asexual generation: underneath the bark of 2–3 year old twigs	Csóka et al. (2017)
Plagiotrochus australis	United Kingdom	*Quercus ilex*	Sexual generation: galls in the leafblade. Asexual: galls beneath the bark of twigs	Schönrogge et al. (2012)
Plagiotrochus masudai	Taiwan	*Quercus glauca*	Bud galls	
Trichoteras vaccinifoliae	North America	*Quercus chrysolepis* (canyon live oak)	Oak apple gall (30 mm) with small red dots	
Hymenoptera: Eulophidae				
Epichrysocharis burwelli	Asia; North America; South America	*Eucalyptus*	Numerous blister galls on the leaves	Schauff and Garrison (2000)
Leptocybe invasa (Blue gum chalcid) - Haplogroup 1/Lineage A	Australia (N); Mediterranean (I); south eastern Europe (I); Asia (I); southern Asia (I); South America(I)	*Eucalyptus*	Rounded pink - red protrusions on the midrib of leaf, petiole, stem, and young shoots	Csóka et al. (2017), Nugnes et al. 2015, and Dittrich-Schröder et al. 2018, 2020)
Leptocybe invasa (Blue gum chalcid) - Haplogroup 2/Lineage B	Australia (N); Asia(I); Africa (I); South East Asia (I)	*Eucalyptus*	Rounded pink - red protrusions on the midrib of leaf, petiole, stem, and young shoots	Nugnes et al. (2015), and Dittrich-Schröder et al. (2018, 2020)
Moona spermophaga	Australia (N); Africa(I); South America(I)	*Corymbia maculata* (spotted gum)	Seed galls	Kim et al. (2005)

(continued)

Table 14.2 (continued)

Species	Known distribution	Host	Gall type	References
Ophelimus maskelli	Australia (N); Africa (I), Asia (I), Europe (I), North America (I)	*Eucalyptus*	Blister-type galls on the leaf midrib and young shoots	Csóka et al. (2017)
Ophelimus eucalypti (Eucalyptus gall wasp)	Australia (N); New Zealand (I); Mediterranean (I): Europe (I); Middle East (I) and Africa (I)	*Eucalyptus* spp. In New Zealand: *E. botryoides, E. deanei, E. grandis, E. saligna, E. camuldulensis* and *E. globulus*	Female larvae produce a circular protruding galls on the leaves; male larvae produce a pit gall	Withers et al. (2000), and Raman and Withers (2003)
Ophelimus migdanorum	China; South America (I)	*Eucalyptus*	Blister-type galls on the leaf surface and stems	Molina-Mercader et al. (2019)
Ophelimus mediterraneus	Australia (N); Europe (I); Mediterranean (I)	*Eucalyptus*	Brown coloured ellipsoid shaped galls on the upper leaf surface	Borowiec et al. (2018)
Quadrastichus erythrinae (Erythrina gall wasp)	Africa (N: some uncertainty); China (I); North America (I); South Asia (I); South East Asia (I); Taiwan (I)	*Erythrina* (tiger's claw & Indian coral)	Tiny green globular galls in large numbers on the leaflets and petioles	Csóka et al. (2017)
Quadrastichodella nova	Africa; Europe; South America;North America	*Eucalyptus*	Fully developed flowers and flower buds	Klein et al. (2015)
Selitrichodes globulus (Bluegum gall wasp)	Australia (N); North America (I)	*Eucalyptus globulus* (Blue gum)	Multiple small brown galls on the branches and sometimes the leaves	La Salle et al. (2009)
Hymenoptera: Eurytomidae				
Eurytoma tumoris	North America	Scot's pine Christmas tree	Terminal and lateral branches	Smith (1968), and Ciesla (2011)

(continued)

Table 14.2 (continued)

Species	Known distribution	Host	Gall type	References
Hymenoptera: Pteromalidae				
Nambouria xanthops	New Zealand	*Eucalyptus*	Irregular shaped gall resembles a comb on the leaves	Berry and Withers (2002)
Hymenoptera: Tenthredinidae				
Euura shibayanagii	Japan	*Salix japonica* (exclusively on this species)	Bud galls have a swollen appearance; stem galls tend to be narrow and bulbous; colouration varies from white, dark green, red or red-brown	Ciesla (2011)
Euura mucronata	North Africa; North America; Southeast Asia	*Salix* spp. (at least 30 species)	Bud galls have a swollen appearance; stem galls tend to be narrow and bulbous; colouration varies from white, dark green, red or red-brown	Nyman (2002)
Euura exigue	North America	*Salix* spp. (Willows)	Elongated stem galls	Bugbee (1970)
Pontania proxima (Willow red gall sawfly)	Australia; Europe; India and North America	*Salix* spp. (willows)	Conspicuous, red, round and often occur in clusters on the leaves	Smith (1968), and Ciesla (2011)
Lepidoptera: Olethreutidae				
Proteotera willingana (Boxelder twig borer)	North America	Maple	Spindle-shaped galls on dormant leaf buds	Kearfott (1904)

kuriphilus. This insect is native to China, but has been introduced into other parts of Asia, North America and Europe where it infests chestnut trees and can cause substantial damage. Of the Eulophidae, the best-known species are *Leptocybe invasa, Ophelimus maskeli* and *O. eucalypti*, all of which are invasive pests of *Eucalyptus* (Dittrich-Schröder et al. 2020).

Gall forming insects of forest tree species are less common in the orders Coleoptera and Lepidoptera. Examples include the poplar gall *Saperda inornata* (Coleoptera: Ceramycidae) and the boxelder twig borer, *Proteotera willingana* (Lepidoptera: Olethreutidae), which infest poplar and maple in North America, respectively (Table 14.2). Other families which include gall forming insects on forest trees include Agromyzidae in the Diptera; Aphidae and Homotomidae in the Hemiptera; Eurytomidae, Pteromalidae and Tenthredinidae in the Hymenoptera.

14.5 Management

In cases where gall insects become serious pests on forest trees, a number of management options exist. Chemical control is of limited use as the majority of the insect's life cycle is protected within the gall. Systemic insecticides may be effective in a small number of cases. For example, stem injections of individual trees are used to control the hemlock woolly adelgid, *Adelges tsugae* (Csóka et al. 2017). However, the use of systemic insecticides is generally not a financially feasible option for large areas of forest (i.e. in natural or plantation forests).

Natural enemies of gall forming insects can assist to maintain populations at low levels. Although the natural enemies of many of the gall forming insects are not known, where they have been studied, gall formers have often been found to have a number of parasitoids (Table 14.1). For invasive gall insects a classical biological control (CBC) approach has often been used, as is the case with the eucalypt-infesting gall wasps, *L. invasa* and *O. maskelli* (Protasov et al. 2007; Kim et al. 2008; Dittrich-Schröder et al. 2014). There are no known parasitoids of adelgid gall formers, and thus biological control programmes for these gall formers have focused on pathogens and predators.

Host genetic selection is a management approach that is often used in plantation forests. Here, more resistant species or genotypes are planted, replacing those that are more susceptible. This approach can include in field or nursery screening trials (Dittrich-Schröder et al. 2012) in addition to research elucidating the mechanisms behind host resistance (Oates et al. 2015, 2016). This approach can be very successful to control gall insects due to their high host specificity.

It is important to note that most gall forming insects that infest forest trees are not considered pests, and of those that are considered pests, many are not of economic concern and thus do not require management. This is partly because most gall forming insects have not established outside their native range, and within their native range their impact is limited by bottom-up (e.g. host resistance) and top-down (e.g. natural enemies) factors. For example, although there has been much study on the parasitoid community of the oak gall wasps, *Andricus* spp. (Stone and Sunnucks 1993; Stone et al. 1995) (see Case Study 14.6.2), biological control programmes for these wasps has not been necessary. Native gall forming insects can still become serious pests, but it is often the non-native introduced species that require management intervention. This has been the case with the gall wasps of *Eucalyptus*, such as *Leptocybe invasa* (see Case Study 14.6.3).

14.6 Case Studies

14.6.1 Adelges cooleyi *(Gillette 1907), Cooley Spruce Gall Adelgid (Adelgidae, Hemiptera)*

The Cooley Spruce Gall Adelgid (Fig. 14.2) is native to Western North America and reproduces by cyclical parthenogenesis (Havill and Foottit 2007). Cyclical partheno-genesis may be defined as "several rounds of clonal reproduction followed by a sexual event" (Rouger et al. 2016). The duration of the entire life cycle is two years. This species is holocyclic, which refers to the occurrence of sexual reproduction in at least one of the generations, whereas reproduction during the other genera-tions is parthenogenic. A characteristic of holocyclic species is the use of alternate hosts to complete their life cycle. *Adelges cooleyi* has five distinct generations of which three occur on its primary host, spruce (*Picea sitchensis, Picea pungens* and *Picea engelmanii*), and two occur on its secondary hosts Douglas fir and the big-cone Douglas fir (*Pseudotsuga menziesii* and *Pseudotsuga macrocarpa*). Holocyclic species form galls on their primary host and utilise their secondary host to support the parthenogenetic generations (Havill and Foottit 2007).

Winged adelgids, also known as sexupare, move from their secondary host to their primary host where they lay a single clutch of eggs. Thereafter, the sexupare die and the eggs are sheltered by their wings until they hatch. Male and female offspring are wingless and feed at the emergence site until the end of the fourth instar (Havill and Foottit 2007). Moulting occurs and the adult generation move towards the centre of the tree where they mate (Havill and Foottit 2007). Thereafter a large single egg is laid which gives rise to a wingless form, known as a fundatrix. The fundatrix selects a bud where she will overwinter until spring. At the onset of spring she feeds on the sap of the bud thereby initiating gall formation. Once the fundatrix is mature she produces a cluster of eggs that hatch into brown nymphs. These nymphs or gallicolae crawl into and feed inside the gall. The size of the gall is an indication of the number of nymphs present, with large galls containing more nymphs than smaller galls (Sopow and Quiring 2001). At the onset of summer the gall opens, allowing the gallicolae to emerge and moult into adults (Havill and Foottit 2007). This generation is winged allowing the gallicolae to disperse to their secondary host to lay eggs. The offspring emerging from these eggs are referred to as exules and are parthenogenic females. A portion of these exules, the progredientes, reproduce by

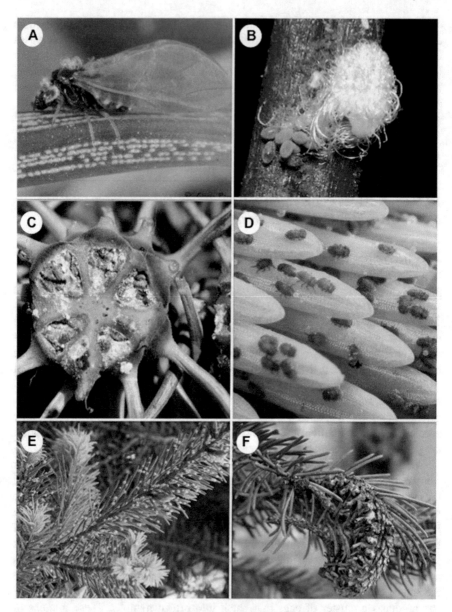

Fig. 14.2 The Cooley Spruce Adelgid, *Adelges cooleyi*: **A** winged female; **B** adult sistens with egg mass; **C** gallicolae feeding inside gall chambers; **D** nymphs; **E** Douglas fir infested with *Adelges cooleyi*; **F** old hardened gall case. Photo credits: **A** © Image, copyright Claude Pilon/Les Hemipteres du Quebec – pucerons all rights reserved, **B** © http://influentialpoints.com/Gallery/Adelges_cool eyi_Douglas_fir_adelgid_Cooley_spruce_gall_adelgid.htm#identi, **C** © courtesy of Whitney Cranshaw, Colorado State University / copyright Bugwood.org, **D** and **E** © D. Manastyrski/Cone and Seed Insect Pest Leaflet No. 14, **F** © Courtesy Ies, licensed under the Creative Commons Attribution 3.0 United States license/https://creativecommons.org/licenses/by/3.0/us/legalcode

parthenogenesis whereas others, the sistentes, overwinter on their secondary host. After overwintering, sistentes start feeding in early spring and produce waxy hair-like protrusions. They lay eggs which give rise to both winged and wingless offspring. The wingless generation remains on the secondary host whereas the winged generation return to their primary host where they produce both male and female offspring.

Damage on the primary host, spruce, is mainly aesthetic however severe gall-formation on the branch tips may lead to mortality of the growing tips. The feeding activities of *A. cooleyi* nymphs on the secondary host, Douglas fir and big-cone Douglas fir, cause needle discolouration and may cause the dropping of needles. As a result of the honey dew excreted by the feeding nymphs, black sooty mould may develop. Large numbers of adelgids may lead to a reduction in seed production.

Natural predators, such as lacewings, assassin bugs and lady beetles, are an important component of adelgid control. The effectiveness of insecticidal control is limited due to the waxy hair-like covering protecting many life stages of the insect. Short periods in the life cycle of adelgids are suitable for use of insecticidal soap, such as the emergence of nymphs from the eggs. Management approaches include ensuring that both primary and secondary hosts are not planted in close proximity.

14.6.2   Andricus spp., Andricus Gall Wasps (Cynipidae, Hymenoptera)

The oak gall wasps belong to the tribe Cynipini (Cynipidae: Hymenoptera) and are characterized by their heterogonic (cyclically parthenogenic) life cycle. Each species produces an asexual and bisexual generation that have morphologically distinct galls associated with the plant family Fagaceae, specifically on *Quercus* (oaks) (Cook et al. 2002). There are approximately 1000 known species of oak gall wasps, of which the *Andricus* genus (Fig. 14.3) is one of the largest and most ecologically diverse (Cook et al. 2002). Species from this genus exhibit a lifecycle that involves two generations per year: sexual in spring and asexual (agamic) in autumn (Fig. 14.4). Each generation is specific with regards to the host species and the plant organ it attacks. Thus cynipids such as *Andricus*, which have a heterogonic life cycle, are restricted to the areas which contain both its hosts; this prerequisite is important in determining their patterns of global distribution (Stone et al. 2002).

There are four host-altering gall wasps from the genus *Andricus* (*Andricus corruptrix, A. kollari, A. lignicola* and *A. quercuscalicis*) that are native to south-eastern Europe but have become invasive in Britain since 1934. The sexual generation of these species occurs in spring on *Quercus cerris* (Turkey oak) and the asexual generation in autumn on *Q. robur* (English oak) and *Q. petreae* (sessile oak) (Schönrogge et al. 1998). *Quercus cerris* is a necessary secondary host for these gall wasps and their invasion was likely facilitated by its introduction into Britain (Schönrogge et al. 1998). *Quercus cerris* has been planted further north and west from its native range and this has created patches where it co-occurs with other *Quercus* species native

Fig. 14.3 The oak gall wasps, *Andricus* spp. **A** agamic gall induced by *A. quercuscalicis* (Knopper gall); **B** agamic gall induced by *A. kollari*; **C** agamic gall induced by *A. lignicola*; **D** agamic gall induced by *A. corruptrix*. Photo credits: **A**, **B** and **C** © David Fenwick/http://www.aphotofauna. com; **D** © Saxifraga – Frits Bink

to those areas (Stone et al. 2002). As a consequence of anthropogenic activity and the ability of *Q. cerris* to self-seed, it is likely that the area containing the necessary hosts to support the spread of *Andricus* wasps will increase (Stone and Sunnucks 1993; Walker et al. 2002).

As mentioned above, heterogonic life cycles are used by gall formers from many insect families. This mode of reproduction consists primarily of an annual agamic generation with a single sexual generation prompted by environmental cues (Stone et al. 2002). The heterogonic reproduction of the Cynipid wasps in general is unusual because of the strict alternation between sexual-agamic generations. As mentioned above, the gall which contains the sexual generation in oak gall wasps develops during the spring on *Quercus cerris*. After emergence from these galls, females mate and then oviposit on either *Q. robur* or *Q. petreae*. The resulting galls produce the asexual generation. In autumn, these females emerge and oviposit eggs on *Q. cerris,* which are dormant until spring the following year and the cycle begins again (Fig. 14.4). Deviations occur from this general pattern depending on the species and environmental conditions (Stone et al. 2002). The life-cycle of *A. kollari* alters depending on where it is geographically situated, in southern Europe it follows an annual life cycle, whereas in northern Scotland its development takes two years.

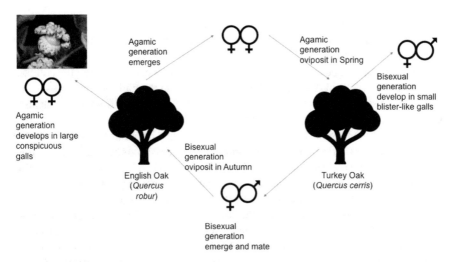

Fig. 14.4 An annotated diagram illustrating the heterogonic life cycle of the four invasive *Andricus* wasps (*Andricus corruptrix, A. kollari, A. lignicola* and *A. quercuscalicis*) described in this case study. This is a generalized depiction of the life-cycle, each species have slight deviations from the general rule, such as the organ choice. The agamic female wasp oviposition occurs in spring, the bisexual generation develops in the small blister-like galls. The bisexual generation mates and oviposits in Autumn, where the agamic females develop in large conspicuous galls. They alter between two hosts. Inset image of the *Andricus quercuscalicis* (Knopper) agamic gall. Photo credit: David Fenwick/http://www.aphotofauna.com). The Diagram was constructed by one of the authors, C. Gevers

Andricus corruptrix, A. lignicola and *A. kollari* induce galls on buds in both their sexual and agamic generations. In spring, the sexual galls on *Quercus cerris* are minute galls that are virtually unnoticeable as they barely protrude from the scales of the bud. In contrast, the agamic generation produces large conspicuous galls on either *Q. robur* or *Q. petreae* during the autumn months (Schönrogge et al. 2000). *Andricus quercuscalicis* is slightly different as the sexual generation induces small galls on the catkins of *Q. cerris* and the asexual generation produces a yellowish-green rigid gall on the acorns (knopper gall) of *Q. robur* in autumn, hence its common name of the knopper gall wasp (Schönrogge et al. 2000). The adults of the sexual generation are generally much smaller than the adults of the asexual generation, and produce fewer eggs, only 70–80 eggs in comparison to the 800–1000 eggs produced by the asexual generation (Hails and Crawley 1991).

The communities associated with these four *Andricus* species have been studied in detail (Stone and Sunnucks 1993; Rokas et al. 2003) and are fascinating in that the suite of natural enemies attracted to the sexual and asexual generations differ (Stone et al. 1995). It has been suggested that the alteration of generations in the Cyinipidae wasp may have allowed them to escape their natural enemies by "partitioning the host space" for the parasitoids (Stone et al. 1995). It is interesting that in general these four wasps share parasitoids and inquilines. For example, they share the parasitoids *Mesopolobus dubius* (except in *A. kollari*), *M. fuscipes* and *M. tibialis* (except in

A. corruptrix) in their sexual galls. However, there are parasitoids, for example, *Megastigmus stigmatizan,* which has currently only been seen to emerge from the agamic gall of *Andricus kollari* (Hayward and Stone 2005).

The appearance and structure of the gall is imperative in determining the community of insects (natural enemies) able to gain access to the gall. With regards to the *Andricus* wasps, the galls for each generation are noticeably different. The agamic galls have a longer developmental time (approximately four months), the size is almost 1000 times larger than the sexual galls, and they have thick woody walls and harbour inquilines and a more diverse community of parasitoids and inquilines (Stone et al. 1995). Conversely, the significantly smaller and thin walled sexual galls develop in approximately three weeks and do not appear to attract inquilines as effectively, thus attracting natural enemies that are predominantly parasitoids.

In addition to gall morphology, the host tree can also influence the parasitoid community. *Quercus robur,* which houses the agamic galls *of Andricus,* has been present in the invaded range for thousands of years and attracted other gall forming cynipids besides *Andricus,* indicating that there is already a rich diversity of parasitoids and inquilines present in the invaded range (Schönrogge et al. 2000). In contrast, *Q. cerris* has been invasive for fewer than 500 years, which has caused it to have a more patchy and random distribution (Stone and Sunnucks 1993; Stone et al. 1995). It is hypothesized that the more recent invasion of *Q. cerris* has likely affected the availability of parasitoids that are able to attack both the sexual and agamic galls as the sexual generation only develops on this host.

14.6.3 *Leptocybe invasa (Fisher & LaSalle), Bluegum Chalcid (Eulophidae, Hymenoptera)*

Leptocybe invasa, commonly referred to as the bluegum chalcid, is a wasp in the family Eulophidae that induces galls on *Eucalyptus* trees (Fig. 14.5). Although native to Australia, the first report, and subsequent description, of this insect was in 2000 when it was detected outside its native range, in Israel (Mendel et al. 2004). Subsequent to its detection in 2000, *L. invasa* has spread to Europe, Asia, Africa, South and North America, as well as New Zealand. By means of molecular markers it was shown that the global distribution of *L. invasa* in fact represents two different lineages, possibly different species (Nugnes et al. 2015). The lineage originally introduced into Israel (Lineage A) is also present in Europe, the Americas, eastern and southern Africa and parts of Asia, while Lineage B is present in Asia, Ghana and South Africa (Dittrich-Schröder et al. 2018). However, the exact distribution of these two lineages is only just being explored and is also expected to change over time due to the continued natural and human-assisted movement of the insect. Countries that contain both lineages could lead to genetic admixture (Dittrich-Schröder et al. 2018). Interestingly, to date (as of 2018), Lineage A, from which *L. invasa* was described, has not yet been found in Australia, the assumed native range.

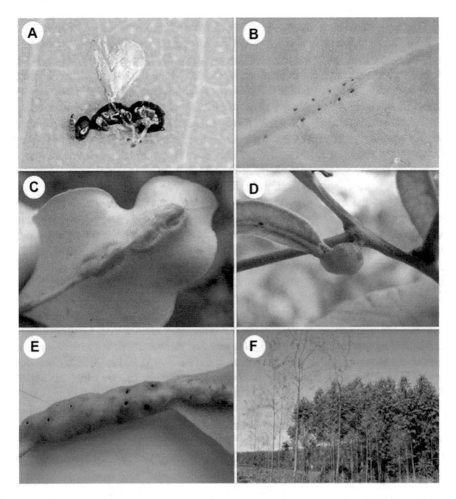

Fig. 14.5 The bluegum chalcid, *Leptocybe invasa*. **A** adult female wasp; **B** oviposition scars on midrib of leaf; **C** gall on midrib of leaf; **D** gall on petiole of leaf; **E** gall showing adult emergence holes; **F** damage from *L. invasa* (trees in foreground are a susceptible *Eucalyptus* clone, whereas trees in background are a resistant *Eucalyptus* clone)

Gall formation is induced when female *L. invasa* oviposits in the host plant. Oviposition sites include the stems, midribs and petioles of young leaves. Eggs are laid 0.3–0.5 mm apart in a line. A spherical 'bump-shaped' gall is formed and galls often fuse on a leaf (Fig. 14.5). Mendel et al. (2004) recognized five stages of galls, where the colour changes from green to pink to red as the gall matures, and light brown on adult emergence; however, the gall colour can also be influenced by exposure to the sun. The larvae develop within separate cavities within the gall. The exact number of larvae on one gall or leaf is highly variable, ranging from 1 to over 60 (Mendel et al. 2004).

Emerging adults are about 1.1–1.4 mm in length, although the male is often smaller in size. The head and body are brown with a blue to green metallic shine, and the legs are yellow (except last tarsal segment, which is brown) (Mendel et al. 2004) When *L. invasa* was first recorded in Israel no males were found in the population and the wasp was thought to reproduce thelytokously. However, males were subsequently found in Asia and other invaded areas, suggesting occasional sexual reproduction. Endosymbiotic bacteria in the genus *Rickettsia* are suggested to induce thelytokous parthenogenesis in *L. invasa* populations (Nugnes et al. 2015). Laboratory studies have stated a mean potential female fecundity of 158 eggs/female, and mean development time from oviposition to emergence of 132 days (at room temperature) (Mendel et al. 2004; Sangtongpraow et al. 2011).

Galling on the *Eucalyptus* hosts results in malformation and stunted growth of the plant and in severe cases tree death. In its invaded range, *L. invasa* has resulted in substantial losses in eucalypt plantations and is considered one of the most serious insect pests of *Eucalyptus*. One of the main management approaches is host selection; *L. invasa* has a relatively narrow host range, which allows more resistant species and hybrids to be selected for planting (Javaregowda and Prabhu 2010; Nyeko et al. 2010; Dittrich-Schröder et al. 2012). However, this approach requires the continual development of new species and clones due to the continual arrival of new insect pests with differing host preference. Biological control is the other main management approach. Biological agents released include *Quadrastichus mendeli*, *Selitrichodes kryceri*, *S. neseri* and *Megastigmus* species (Kim et al. 2008; Kelly et al. 2012; Dittrich-Schröder et al. 2014). There have also been a number of *Megastigmus* species which have been found to be associated with *L. invasa* in the invaded range (Sangtongpraow and Charernsom 2013); some of these are thought to be parasitoids of *L. invasa,* but in other cases, the role of these species and their interaction with introduced biological control agents is uncertain (Gevers et al. 2021). Systemic insecticides have been reported to be effective in some cases (Jhala et al. 2010), but their use is often limited due to high costs and restrictions from forestry certification bodies.

References

Atkinson RJ, Brown GS, Stone GN (2003) Skewed sex ratio and multiple founding in galls of the oak gall wasp *Biorhiza pallida.* Ecol Entomol 28:14–24

Berry JA, Withers TM (2002) New gall-inducing species of ormocerine pteromalid (Hymenoptera: Pteromalidae: Ormocerinae) described from New Zealand. Aust J Entomol 41:18–22

Borowlec N, La Salle J, Brancaccio L, Thon M, Warot S, Branco M, Ris N, Malausa J, Burks R (2018) *Ophelimus mediterraneus* sp. n. (Hymenoptera, Eulophidae): a new *Eucalyptus* gall wasp in the Mediterranean region. Bull Entomol Res 109:678–694

Brooks SE, Shorthouse JD (1998) Developmental morphology of stem galls of *Diplolepis nodulosa* (Hymenoptera: Cynipidae) and those modified by the inquiline *Periclistus pirata* (Hymenoptera: Cynipidae) on *Rosa blanda* (Rosaceae). Can J Bot 76:365–381

Browne FG (1968) Pests and diseases of forest plantation trees: an annotated list of the principal species occurring in the British Commonwealth. Clarendon Press, Oxford

Bugbee RE (1970) Descriptions of two new species from *Euura* galls and redescriptions of three Ashmead species of the genus *Eurytoma* (Hymenoptera: Eurytomidae). Ann Entomol Soc Am 63:433–437

Bulaon B (2005) Management guide for Douglas-fir needle midges. Forest Health Protection and State Forestry Organizations, USA

Carter CI (1971) Conifer woolly aphids (Adelgidae) in Britain, vol 42. Forestry Commission, England

Chen C, Applybyby J (1984) Biology of the cypress twig gall midge, *Taxodiomyia cupressiananassa* (Diptera: Cecidomyiidae), in central Illinois. Ann Entomol Soc Am 77:203–207

Chinery M (2011) Britain's plant galls: a photographic guide. ABC Print, Hampshire

Ciesla WM (2011) Forest entomology: a global perspective. John Wiley & Sons, West Sussex

Claridge MF (1962) *Andricus quercuscalicis* (Burgsdorf) in Britain (Hymenoptera: Cyniipdae). The Entomologist 95:60–61

Cobbinah JR (1986) Factors affecting the distribution and abundance of Phytolyma lata (Homoptera: Psyllidae). Int J Insect Sci 7:111–115

Cook LG, Gullan PJ (2004) The gall-inducing habit has evolved multiple times among the eriococcid scale insects (Sternorrhyncha: Coccoidea: Eriococcidae). Biol J Linn Soc 83:441–452

Cook JM, Rokas A, Pagel M, Stone GN (2002) Evolutionary shifts between host oak sections and host-plant organs in *Andricus* gallwasps. Evolution 56:1821–1830

Crawley MJ (1997) *Aphelonyx cerricola* Giraud (Hym., Cynipidae), an alien gall-former new to Britain. Entomol Mon Mag 133:61

Csóka G, Stone GN, Melika G (2017) Non-native gall-inducing insects on forest trees: a global review. Biol Invasions 19:3161–3181

Dittrich-Schröder G, Harney M, Neser S, Joffe T, Bush S, Hurley BP, Wingfield MJ, Slippers B (2014) Biology and host preference of *Selitrichodes neseri*: A potential biological control agent of the Eucalyptus gall wasp, *Leptocybe invasa*. Biol Control 78:33–41

Dittrich-Schröder G, Hoareau TB, Hurley BP, Wingfield MJ, Lawson S, Nahrung HF, Slippers B (2018) Population genetic analyses of complex insect invasions in managed landscapes: A *Leptocybe invasa* (Hymenoptera) case study. Biol Control 20:2395–2420

Dittrich-Schröder G, Hurley BP, Wingfield MJ, Nahrung HF, Slippers B (2020) Invasive gall-forming wasps that threaten non-native plantations-grown *Eucalyptus*: diversity and invasion patterns. Agric For Entomol 22:285–297

Dittrich-Schröder G, Wingfield MJ, Hurley BP, Slippers B (2012) Diversity in *Eucalyptus* susceptibility to the gall-forming wasp *Leptocybe invasa*. Agric For Entomol 14:419–427

Dreger-Jauffret F, Shorthouse JD (1992) Diversity of gall-inducing insects and their galls. In: Shorthouse JD, Rohfritsch O (eds) Biology of insect-induced galls. Oxford University Press, UK

Drooz AT (1985) Insects of eastern forests. USDA Forest Service

Fagan MM (1918) The uses of insect galls. Am Nat 52:155–176

Felt EP (1917) Key to American insect galls. New York State Museum Bulletin No. 200

Felt EP (1940) Plant galls and gall makers. Comstock Publishing, Ithaca, NY

Ferraz FFF, Monteiro RF (2003) Complex interactions envolving a gall midge *Myrciamyia maricaensis* Maia (Diptera, Cecidomyiidae), phytophagous modifiers and parasitoids. Rev Bras Zool 20:433–437

Foottit RG, Floate K, Maw E (2010) Molecular evidence for sympatric taxa within Pemphigus betae (Hemipetra: Aphididae: Eriosomantinae). Can Entomol 142:344–353

Fukatsu T, Aoki S, Kurosu U, Ishikawa H (1994) Phylogeny of Cerataphidini aphids revealed by their symbiotic microorganisms and basic structure of their galls: implications for host-symbiont coevolution and evolution of sterile soldier castes. Zool Sci 11:613–623

Gauthier N, La Salle J, Quicke DLJ, Godfray HCJ (2000) Phylogeny of Eulophidae (Hymenoptera: Chalcidoidea) with a reclassification of Eulophinae and the recognition that Elasmidae are derived eulophids. Syst Entomol 25:521–539

Gevers CR, Dittrich-Schröder G, Slippers B, Hurley BP (2021) Interactions between hymenopteran species associated with gall-forming wasps: the *Leptocybe invasa* community as a case study. Agric For Entomol 23:146–153

Hails RS, Crawley MJ (1991) The population dynamics of an alien Insect: *Andricus quercuscalicis* (Hymenoptera: Cynipidae). J Anim Ecol 60:545–561

Hancey R, Hancey B (2004) First British records of *Plagiotrochus quercusilicis*. Cecidology 19:98

Hardy NB, Cook LG (2010) Gall-induction in insects: evolutionary dead-end or speciation driver? BMC Evol Biol 10:257

Harper LJ, Schönrogge K, Lim KY, Francis P, Lichtenstein CP (2004) Cynipid galls: insect-induced modifications of plant development create novel plant organs. Plant Cell Environ 27:327–335

Havill NP, Foottit RG (2007) Biology and evolution of adelgidae. Annu Rev Entomol 52:325–349

Hayward A, Stone GN (2005) Oak gall wasp communities: evolution and ecology. Basic Appl Ecol 6:435–443

Hollis D (1973) African gall bugs of the genus *Phytolyma* (Hemiptera, Psylloidea). Bull Entomol Res 63:143–154

Hood GR, Ott JR (2011) Generational shape shifting: changes in egg shape and size between sexual and asexual generations of a cyclically parthenogenic gall former. Entomol Exp Appl 141:88–96

Hutchinson MH (1974) *Andricus lignicolus* (Hartig) (Hymnoptera. Cynipidae) in S.E. England, a species new to Britain. Entomologist's Rec 86:158–159

Ide T, Wachi N, Abe Y (2012) Three new species and a new record of *Cycloneuroterus* (Hymenoptera: Cynipidae: Cynipini) inducing galls on *Cyclobalanopsis* in Japan. Ann Entomol Soc Am 105:539–549

Inbar M, Wink M, Wool D (2004) The evolution of host plant manipulation by insects: molecular and ecological evidence from gall-forming aphids on Pistacia. Mol Phylogenet Evol 32:504–511

Isaev AS, Barachikov YN, Malutina VS (1988) The larch gall midge in seed orchards of south Siberia. In: Berryman AA (ed) Dynamics of forest insect populations: patterns, causes, and implications. Plenum Press, New York, pp 19–44

Javaregowda J, Prabhu ST (2010) Susceptibility of *Eucalyptus* species and clones to gall wasp, *Leptocybe invasa* Fisher and La Salle (Eulophidae: Hymenoptera) in Karnataka. Karnataka J Agric Sci 23:220–221

Jhala RC, Patel MG, Vaghela NM (2010) Effectiveness of insecticides against blue gum chalcid, *Leptocybe invasa* Fisher & La Salle (Hymenoptera: Eulophidae), infesting eucalyptus seedlings in middle Gujarat, India. Karnataka J Agric Sci 23:84–86

Kearfort WD (1904) A new Proteopteryx. Can Entomol 36:306

Kelly J, La Salle J, Harney M, Dittrich-Schröder G, Hurley BP (2012) *Selitrichodes neseri* n.sp., a new parasitoid of the eucalyptus gall wasp *Leptocybe invasa* Fisher & La Salle (Hymenoptera: Eulophidae: Tetrastichinae). Zootaxa 3333:50–57

Kim I-K, McDonald M, La Salle J (2005) *Moona*, a new genus of tetrastichiniae gall inducers (Hymenotpera: Eulophidae) on seeds of *Corymbia* (Myrtaceae) in Australia. Zootaxa 989:1–10

Kim I-K, Mendel Z, Protasov A, Blumberg D, La Salle J (2008) Taxonomy, biology, and efficacy of two Australian parasitoids of the eucalyptus gall wasp, *Leptocybe invasa* Fisher & La Salle (Hymenoptera: Eulophidae: Tetrastichinae). Zootaxa 1910:1–20

Kinsey AC (1920) Phylogeny of cynipid genera and biological characteristics. Bull Am MusM Nat Hist 42:357–402

Klein H (2009) Wasps (Hymenoptera: Chalcidoidea) associated with galls in seed-capsules of *Eucalyptus camaldulensis* (Myrtaceae) in South Africa: Species composition, trophic relationships and effects. Masters, University of Cape Town

Klein H, Hoffmann JH, Neser S, Dittrich-Schröder G (2015) Evidence that *Quadrastichodella nova* (Hymenoptera: Eulophidae) is the only gall inducer among four hymenopteran species associated with seed capsules of *Eucalyptus camaldulensis* (Myrtaceae) in South Africa. Afr Entomol 23:207–223

Kollár J (2007) The harmful entomofauna of woody plants in Slovakia. Acta Entomol Serbica 12:67–69

Johnson WT, Lyon HH (1988) Insects that feed on trees and shrubs. Cornell University Press, Ithaca, New York

La Salle J, Arakelian G, Garrison RW, Gates MW (2009) A new species of invasive gall wasp (Hymenoptera: Eulophidae: Tetrastichinae) on blue gum (*Eucalyptus globulus*) in California. Zootaxa 2121:35–43

Malyshev SI (1968) Genesis of the Hymenoptera and the phases of their evolution. Methuen, London

Medonça MDJ (2001) Galling insect diversity patterns: the resource synchronisation hypothesis. Oikos 95:171–176

Mendel Z, Protasov A, Fisher N, La Salle J (2004) Taxonomy and biology of *Leptocybe invasa* gen. & sp. n. (Hymenoptera: Eulophidae), an invasive gall inducer on *Eucalyptus*. Aust J Entomol 43:101–113

Molina-Mercader G, Angulo AO, Olivares TS, Sanfentees E, Castillio-Salazar M, Rojas E, Toro-Núnez O, Benítez HA, Hasbún R (2019) *Ophelimus megdanorum* Molina-Mercader sp. nov. (Hymenoptera: Eulophidae): application of integrative taxonomy for disentangling a polyphenism case in *Eucalyptus glubulus* Labil forest in Chile. Forests 10:720

Molnár B, Boddum T, Szöcs G, Hilbur Y (2009) Occurrence of two pest gall midges, *Oboloiplosis robiniae* (Haldeman) and *Dasineura gleditchiae* (Osten Sacken) (Diptera: Cecidomyiidae) on ornamental trees in Sweden. Entomolologica Tidskr 130:113–120

Moriya S, Inoeum K, Ôtake A, Shiga M, Mabuchi M (1989a) Decline of chestnut gall wasp population, *Dryocosmus kuriphilius* Yasumatsu (Hymenoptera: Cynipidae) after the establishment of *Torymus sinensi* Kamijo (Hymenoptera: Torymidae). Appl Entomol Zool 24:231–233

Moriya S, Inoue K, Otake A, Shiga M, Mabuchi M (1989b) Decline of the chestnut gall wasp population, *Dryocosmus kurphilus* Yasumatsu (Hymenoptera: Cynipidae) after the establishment of *Torymus sinensis* Kamijo (Hymenoptera: Torymidae). Appl Entomol Zool 24:231–233

Morris DC, Mound LA, Schwarz MP, Crespi B (1999) Morphological phylogenetics of Australian gall-inducing thrips and their allies: the evolution of host-plant affiliations, domicile use and social behaviour. Syst Entomol 24:289–299

Nicholas AH, Spooner-Hart RN, Vickers RA (2005) Abundance and natural control of the woolly aphid *Eriosoma lanigerum* in an Australian apple orchard IPM program. Biocontrol 50:271–291

Nosil P, Crespi BJ (2006) Experimental evidence that predation promotes divergence in adaptive radiation. Proc Acad Nat Sci USA 103:9090–9095

Noyes JS (1998) Catalogue of the Chalcidoidea of the world. Electronic Publication, Amsterdam

Nugnes F, Gebiola M, Monti MM, Gualtieri L, Giorgini M, Wang J, Bernardo U (2015) Genetic diversity of the invasive gall wasp *Leptocybe invasa* (Hymenoptera: Eulophidae) and of its *Rickettsia* endosymbiont, and associated sex-ratio differences. PLoS ONE 10:e0124660

Nyeko P, Mutitu KE, Otieno BO, Ngae GN, Day RK (2010) Variations in *Leptocybe invasa* (Hymenoptera: Eulophidae) population intensity and infestation on eucalyptus germplasms in Uganda and Kenya. Int J Pest Manage 56:137–144

Nyman T (2002) The willow bud galler *Euura mucronate* Hartig (Hymenoptera; Tenthredinidae): one polyphage or many polyphages? Heredity 88:288–295

Nyman T, Roininen H, Vuorinen J (1998) Evolution of different gall types in willow-feedgin sawflies (Hymenoptera: Tenthredinidae). Evolution 52:465–474

Oates CN, Denby KJ, Myburg AA, Slippers B, Naidoo S (2016) Insect gallers and their plant hosts: from omics data to systems biology. Int J Mol Sci 17:1891

Oates CN, Külheim C, Myburg AA, Slippers B, Naidoo S (2015) The transcriptome and terpene profile of *Eucalyptus grandis* reveals mechanisms of defence against the insect pest, *Leptocybe invasa*. Plant Cell Physiol 56:1418–1428

Payne JA (1978) Oriental chesnut gall wasp: new nut pest in North America. Georgia, USA

Price PW (2005) Adaptive radiation of gall-inducing insects. Basic Appl Ecol 6:413–421

Protasov A, Blumberg D, Brand D, La Salle J, Mendel Z (2007) Biological control of the eucalyptus gall wasp *Ophelimus maskelli* (Ashmead): taxonomy and biology of the parasitoid species *Closterocerus chamaeleon* (Girault), with information on its establishment in Israel. Biol Control 42:196–206

Raman A (2007) Insect-induced plant galls of India: unresolved questions. Curr Sci 92:748–757

Raman A (2011) Morphogenesis of insect-induced plant galls: facts and questions. Flora 206:517–533

Raman A, Schaeffer CW, Withers TM (2005) Biology, ecology and evolution of gall-inducing arthropods. Science Publishers Inc., Enfield

Raman A, Withers TM (2003) Oviposition by introduced *Ophelimus eucalypti* (Hymenoptera: Eulophidae) and morphogenesis of female-induced galls on *Eucalyptus saligna* (Myrtaceae) in New Zealand. Bull Entomol Res 93:55–63

Riley CV (1885) Notes on North American Psyllidae. Proc Biol Soc Wash 1882–1884(2):67–79

Rohfritsch O (1992) Patterns in gall development. In: Shorthouse JD, Rohfritsch O (eds) Biology of insect-induced galls. Oxford University Press, New York, pp 60–86

Rokas A, Atkinson RJ, Webster L, Csoka G, Stone GN (2003) Out of Anatolia: longitudinal gradients in genetic diversity support an eastern origin for a circum-Mediterranean oak gallwasp *Andricus quercustozae*. Mol Ecol 12:2153–2174

Ronquist F, Liljeblad J (2001) Evolution of the gall wasp-host plant association. Evolution 55:2503–2522

Ronquist F, Nieves-Aldrey JL, Buffington ML, Liu Z, Liljeblad J, Nylander JA (2015) Phylogeny, evolution and classification of gall wasps: the plot thickens. PLoS ONE 10:e0123301

Rose AH, Lindquist OH (1999) Insects of eastern pines, vol 1313

Roskam HC (1992) Evolution of the gall-inducing guild. In: Shorthouse JD, Rohfritsch O (eds) Biology of insect-induced galls. University Press, Oxford

Rott AS, Godfray HCJ (2000) The structure of a leafmine-parasitoid community. J Anim Ecol 69:274–289

Rouger R, Reichel K, Masson JP, Stoeckel S (2016) Effects of complex life cycles on genetic diversity: cyclical parthenogenesis. Heredity 117:336–347

Sangtongpraow B, Charernsom K (2013) Evaluation of parasitism capacity of *Megastigmus thitipornae* Doğanlar & Hassan (Hymenoptera: Torymidae), the local parasitoid of eucalyptus gall wasp, *Leptocybe invasa* Fisher & La Salle (Hymenoptera: Eulophidae). Kasetsart J 47:191–204

Sangtongpraow B, Charernsom K, Siripatanadilok S (2011) Longevity, fecundity and development time of eucalyptus gall wasp, *Leptocybe invasa* Fisher & La Salle (Hymenoptera: Eulophidae) in Kanchanaburi province, Thailand. Thai J Agric Sci 44:155–163

Schauff ME, Garrison R (2000) An introduced species of *Epichrysocharis* (Hymenoptera: Eulophidae) producing galls on *Eucalyptus* in California with notes on the described species and placement of genus. J Hymenoptera Res 9:176–181

Schönrogge K, Begg T, Williams R, Melika G, Randle Z, Stone GN (2012) Range expansion and enemy recruitment by eight alien gall wasp species in Britain. Insect Conser Divers 5:298–311

Schönrogge K, Stone GN, Crawley MJ (1996) Alien herbivores and native parasitoids: rapid developments and structure of the parasitoid and inquiline complex in an invading gall wasp *Andricus quercuscalicis* (Hymenoptera: Cynipidae). Ecol Entomol 21:71–80

Schönrogge K, Walker P, Crawley MJ (1998) Invaders on the move: parasitism in the sexual galls of four alien gall wasps in Britain (Hymenoptera: Cynipidae). Proc R Soc B: Biol Sci 265:1643–1650

Schönrogge K, Walker P, Crawley MJ (2000) Parasitoid and inquiline attack in the galls of four alien, cynipid gall wasps: host switches and the effect on parasitoid sex ratios. Ecol Entomol 25:208–219

Shorthouse JD, Wool D, Raman A (2005) Gall-inducing insects—nature's most sophisticated herbivores. Basic Appl Ecol 6:407–411

Singer MS, Stireman JO (2005) The tri-trophic niche concept and adaptive radiation of phytophagous insects. Ecol Lett 8:1247–1255

Smith F (1954) A new British Cynips and the galls made thereby. Trans Proc Entomol Soc Lond 3:35

Smith EL (1968) Biosystematics and morphology of Symphyta: I. Stem galling *Euura* of the California region, and a new female nomenclature. Ann Entomol Soc Am 61:1389–1407

Sopow SL, Quiring DT (2001) Is gall size a good indicator of adelgid fitness? Entomol Exp Appl 99:267–271

Stone GN, Atkinson RJ, Rokas A, Aldrey JL, Melika G, Acs Z, Csoka G, Hayward A et al (2008) Evidence for widespread cryptic sexual generations in apparently purely asexual *Andricus* gallwasps. Mol Ecol 17:652–665

Stone GN, Cook JM (1998) The structure of cynipid oak galls: patterns in the evolution of an extended phenotype. Proc R Soc Lond Ser B: Biol Sci 265:979–988

Stone GN, Schönrogge K (2003) The adaptive significance of insect gall morphology. Trends Ecol Evol 18:512–522

Stone GN, Schönrogge K, Atkinson RJ, Bellido D, Pujade-Villar J (2002) The population biology of oak gall wasps (Hymenoptera: Cynipidae). Annu Rev Entomol 47:633–668

Stone GN, Schönrogge K, Crawley MJ, Fraser S (1995) Geographic and between-generation variation in the parasitoid communities associated with an invading gallwasp, *Andricus quercuscalicis* (Hymenoptera: Cynipidae). Oecologia 104:207–217

Stone GN, Sunnucks P (1993) Genetic consequences of an invasion through a patchy environment— the cynipid gallwasp *Andricus quercuscalicis* (Hymenoptera: Cynipidae). Mol Ecol 2:251–268

Vardal H (2004) From parasitoids to gall inducers and inquilines: morphological evolution in cynipid wasps. Uppsala University

Wagner MR, Cobbinah JR, Bosu PP (2008) Forest entomology in West Tropical Africa: forest insects of Ghana. Springer, Dordrecht, Netherlands

Walker P, Leather SR, Crawley MJ (2002) Differential rates of invasion in three related alien oak gall wasps (Cynipidae: Hymenoptera). Divers Distrib 8:335–349

Whitfield JB (1998) Phylogeny and evolution of host-parasitoid interactions in Hymenoptera. Annu Rev Entomol 43:129–151

Withers TM, Raman A, Berry JA (2000) Host range and biology of *Ophelimus eucalypti* (Gahan) (Hym: Eulophidae), a pest of New Zealand eucalypts. NZ Plant Protect 53:339–344

Zhang G, Zhong T, Tian Z (1980) Two new species and a new subspecies of Adelgidae from Sichuan, China (Homoptera: Adelgidae). Zool Res 1:381–388

Chapter 15
Tip, Shoot, Root, and Regeneration Pests

David R. Coyle

15.1 Introduction

Actively growing tree tissues, such as branch and shoot tips and fine roots, are high in nutritive value and generally have comparatively lower amounts of defensive compounds than older tissues. Many arthropods have evolved to feed on or in these nutritious tree tissues, and most of these herbivores consume a relatively small amount of living tissue or fluids. These particular tissues often lack the physical or chemical defenses present in other parts of the tree, and can be easier to access by herbivores.

While removal of any living tissue or fluid has some effect on the host tree, impacts on the overall health of the tree can be highly variable, ranging from negligible to tree death. The magnitude of these impacts depends on host tree vigor, tree age or size, the amount of material consumed or removed, and the location of the damage. For example, adult *Hylobius abietis* consume the phloem from small diameter branches in the crowns of mature conifer trees in Europe, and this feeding can kill branch tips (Örlander et al. 2000). On a mature tree, however, the loss of some branch tips is not likely to be detrimental to that tree's overall health. Likewise, despite high branch tip mortality from periodical cicada (*Magicicada*) oviposition damage, branch and stem diameter growth of maples (*Acer*), dogwoods (*Cornus*), and redbuds (*Cercis*) trees in Indiana, U.S. was not affected (Flory and Mattingly 2008). Twig girdler damage (Fig. 15.1) can eliminate apical dominance and lead to a high rate of lateral bud development on branches, causing a change in tree structure (Martínez et al. 2009). Adult feeding by the weevil *Cylindrocopturus eatoni*, which occurs in Coastal Western North America, is known to injure young Ponderosa (*Pinus ponderosa*) and Jeffrey (*Pinus jeffreyi*) pine trees. Adults emerge in mid-summer and begin feeding on the bark of small branches, and late summer feeding by larvae on the phloem of

D. R. Coyle (✉)
Department of Forestry and Environmental Conservation, Clemson University, Clemson, SC, USA
e-mail: dcoyle@clemson.edu

© The Author(s) 2023
J. D. Allison et al. (eds.), *Forest Entomology and Pathology*,
https://doi.org/10.1007/978-3-031-11553-0_15

Fig. 15.1 Twig girdling damage by *Oncideres cingulata* on pecan (*Carya illinoinsis*) results in the loss of branch ends, which can alter the structure of host trees (Photo credit: Clemson University – USDA Cooperative Extension Slide Series, Bugwood.org)

twigs and shoots can cause branch mortality, tree deformation, and death of smaller trees (Eaton 1942; Furniss 1942).

In forests managed for commercial production, insect feeding and damage on tips, roots, and shoots can kill terminal leaders and negatively impact tree growth and form, resulting in volume and economic losses. The two most common groups of insects in these situations are tip moths and root weevils. *Rhyacionia* and *Dioryctria* tip moths impact tree branches or terminal tips in conifers, while root weevils (of which there are many species) feed on branches, shoots, or root tissues in conifers and hardwoods. Many of these insects are colloquially referred to as "regeneration pests" because they often damage seedlings or recently transplanted trees and, at times, significantly impact the regeneration process. It is important to note that the same herbivorous species may act as a regeneration pest in some situations (i.e. if it attacks and damages young tree seedlings) and not in others. Other species damage mature trees and can lead to tree health declines and contribute to mortality. The biology, ecology, and management of many of these species has been well studied.

15.2 Similarities and Differences Between Tip Moths and Root Weevils

While superficially different, these two groups of insects share several commonalities. Both tip moths and root weevils come from highly speciose taxonomic groups. There are over 10,300 species in the Family Tortricidae (Order Lepidoptera, Gilligan et al. 2018), many of which are small, usually with a <3 cm wingspan, and having brownish coloration as adults. The genus *Rhyacionia* contains many horticultural, agricultural, or forestry pests, and larvae exhibit a wide variety of feeding strategies, including leaf rollers, gall makers, fruit, root, or shoot borers, and seed or flower feeders (Gilligan et al. 2018). The lepidopteran family Pyralidae contains nearly 6,000 species, many of which are small to mid-sized (wingspans <4 cm) agricultural pests that are variably colored adults (Regier et al. 2012). Within this family the genus *Dioryctria* contains 79 recognized species, most of which impact cones or seeds (Whitehouse et al. 2011) though several species do cause damage to branches and shoots of conifers (Roe et al. 2011).

Within the Order Coleoptera, the Family Curculionidae (the "true" weevils) is the largest, with over 77,000 species (GBIF 2021). This family contains many insects that attack trees, including the bark beetles (Chapter 10) and ambrosia beetles (Chapter 11). Until recently, bark and ambrosia beetles were considered a separate Family, even though there was substantial evolutionary evidence to the contrary (Jordal et al. 2011), and bark and ambrosia beetles are now a subfamily within the Curculionidae. Several weevil species impact trees in forests, nurseries, urban and suburban landscapes, and natural or managed landscapes.

Both tip moths and root weevils have holometabolous life cycles, going through four morphologically distinct developmental stages: egg, larva, pupa, and adult. Adults are mobile—*Rhyacionia* and *Dioryctria* moths are active fliers, and most (but not all) root weevil adults can fly. Both tip moth and root weevil larvae are relatively immobile—*Rhyacionia* and *Dioryctria* larvae are confined to the shoot or meristem tip on which the female oviposited, and root weevil larvae either live in a tunnel under the bark or in the soil where they feed on fine roots (hence the common name). In all cases, the success of the individual larva—and of the species—depends largely on the ability of the adult female to choose an oviposition site that will be favorable to the offspring. This is known as the "mother knows best" hypothesis (Scheirs et al. 2000; Mayhew 2001) in that the mother chooses a location for her offspring that will allow them the best chance at success (i.e. reaching maturity and reproducing). Individuals who select the best oviposition sites produce offspring that survive and pass on those genetics; those that do not select favorable oviposition sites will be less likely to pass their genes on to subsequent generations.

There are significant ecological and biological differences between these two insect groups, primarily in terms of life cycle duration, location where the feeding damage occurs, and which life stages cause this damage. Adult *Rhyacionia* moths live only a few weeks (e.g. Friend and West 1933; Asaro and Berisford 2001b), which is just long enough to mate and oviposit. Eggs are laid on pine needles, and larvae first bore into needles, later entering the growing lateral and terminal shoots (e.g. Stevens 1966; Jennings 1975). Larvae seldom grow more than several mm in size, but their feeding on vascular tissue causes tip mortality. *Dioryctria* moths have between one and several generations a year (e.g. Butcher and Carlson 1962; Neunzig et al. 1964). Like *Rhyacionia*, larvae feed in tree tissues, most often cones but also shoot tips; this feeding can cause damage and mortality to branches, deterioration of tree form, and can negatively impact stand value (Neunzig et al. 1964; Speight and Speechly 1982; Hainze and Benjamin 1984).

While the loss of a single growing tip is not detrimental to a mature tree's health, severe damage can occur when high pest populations infest younger trees. For instance, significant mortality of newly planted Caribbean pine (*Pinus caribaea*) resulted from shoot feeding by *Dioryctria* larvae in the Philippines (Speight and Speechly 1982) and *Dioryctria resinosella* larval feeding killed nearly a third of current year leaders on young red pine (*Pinus resinosa*) in Maine, U.S. (Patterson et al. 1983). While some of these trees had a secondary leader assume dominance, 16% assumed a forked growth form and were permanently damaged. An outbreak of the pine tip moth *Rhyacionia leptotubula* damaged 40% of growing terminals in China and caused significant long-term damage to Armand pine (*Pinus armandii*) growth and form (Yang et al. 2012). After feeding is complete, larvae pupate in the now dead shoot. In some cases, trees that were previously attacked by the Nantucket pine tip moth (*Rhyacionia frustrana*) become predisposed to additional, subsequent *R. frustrana* attacks (Coody et al. 2000). Multiple moth generations can occur during a single growing season (Powell and Miller 1978).

In contrast to shoot moths, the entire life cycle of most root weevils may take two years or more. Adults can live up to several months (Wen et al. 2004; Son and Lewis 2005) and are active feeders, sometimes causing significant damage. For example, the cypress weevil (*Eudociminus mannerheimii*) primarily impacts weakened or damaged cypress and related trees in the Family Cupressaceae. It has been reported in much of the eastern U.S. (Skvarla et al. 2015) and Central Mexico (Sánchez-Martínez et al. 2010). Confirmed and reported instances of this weevil damaging trees span the range of small diameter nursery stock (Mayfield 2017), urban landscape trees (Skvarla et al. 2015), and natural riparian areas (Sánchez-Martínez et al. 2010). This weevil damages and kills trees by infesting the stems and tunneling under the bark, consuming the phloem, eventually killing the tree.

Root weevil larvae are whitish and grublike, though not C-shaped like a typical soil-dwelling grub. Larvae feed on fine roots or within a gallery created in the phloem of the root or shoot. Like tip moths, larval damage can be significant. For instance, a weevil in the genus *Aclees*, previously undocumented as a forestry pest in its home range, became a new pest of planted Spanish cedar (*Cedrela odorata*) when this tree was planted in commercial plantations in Vietnam (Thu et al. 2010). This tree is highly valued for furniture and is native to tropical areas of the Americas. But, when planted in Vietnam as a non-native tree species, the native *Aclees* weevil became extremely prevalent in these plantings, with infestation rates of 80–100% and high damage rates on inspected trees. This particular system is an excellent example of a native pest, which was of so little economic consequence that it had not yet been identified to species, becoming a serious concern for growers due to forest management practices (i.e. installing plantations of non-native Spanish cedar trees). Pupation of root weevil larvae occurs in the ground or in small "pupal chambers" inside the tissue on which the larva fed. The amount of time a root weevil spends in each of these life stages varies greatly, and depends on the insect species, climate, local conditions, and host.

Unlike tip moths, whose larval stage is the only life stage to feed on tree tissue, feeding by both larval and adult root weevils can damage trees, but the amount and severity of damage varies greatly depending on the insect species. In most cases adults and larvae feed on different tree tissues. For instance, several non-native root weevil species (primarily *Phyllobius oblongus*, *Sciaphilus asperatus*, *Barypeithes pellucidus*, and *Polydrusus formosus*, which is also known as *P. sericeus*, plus a few other less commonly encountered species) have established in hardwood forest stands throughout the Western Great Lakes Region of North America (Coyle et al. 2008b) (Fig. 15.2). Adults are only present for several weeks in early summer, occasionally causing extensive defoliation of sugar maple (*Acer saccharum*), yellow birch (*Betula alleghaniensis*), American basswood (*Tilia americana*), hop-hornbeam (*Ostrya virginiana*), and *Rubus* in the forest understory (Coyle et al. 2008b). This adult defoliation, coupled with fine root herbivory by larvae – densities of which can exceed 1000/m^2 – can lead to seedling mortality (Pinski et al. 2005; Coyle et al. 2008b, 2014).

Fig. 15.2 *Polydrusus sericeus* (**a**) and *Phyllobius oblongus* (**b**) adults with characteristic feeding on leaf margins of *Ostrya virginiana* (**c**) and *Acer saccaharum* (**d**) in northern hardwood forests of the Great Lakes Region of North America (Photo credits: Steven Katovich, Bugwood.org [a]; György Csóka, Hungary Forest Research Institute, Bugwood.org [b]; David Coyle, Clemson University [c and d])

15.3   Management Strategies for Tip, Shoot, Root, and Regeneration Pests in Forest Systems

Determining when and where to dedicate resources towards pest management is one of the primary decisions facing forestry professionals around the world. This decision is, in part, dictated by geography and socioeconomic factors. In places where forestry is a major industrial activity and component of the economy, keeping forest stands pest-free and healthy requires, and is given, more emphasis and resources.

Whether trees are in natural or unmanaged uneven-aged forests, or managed and even-aged forests; tip, shoot, root, and regeneration pests are usually not problematic or controlled in healthy older or larger trees, as they rarely cause widescale measurable damage. However, several types of tip, shoot, and root feeding insects, specifically root weevils, can and do feed on tissues of stressed trees. For instance, several genera of root weevils in the southeastern U.S. oviposit in dying or dead pines (Matusick et al. 2013) and the larvae feed on and develop in dying or dead roots or stumps (reviewed by Coyle et al. 2015). These insects play an important ecological role in that their feeding helps break down woody tissue. Further, by creating their feeding galleries they help several species of fungi proliferate inside the tree tissues, thus aiding in wood decomposition.

Younger trees in any forest situation are at a greater risk of damage from tip, shoot, and root pests. Managing these pests in mixed species or uneven-aged stands is often logistically difficult and economically unfeasible due to the heterogeneity of the system; conversely, in managed or planted forests control of tip, shoot, root, and regeneration pests is often a part of the overall management plan. In some cases, even-aged managed stands are at a greater risk of pest pressure. For example, uneven-aged Norway spruce (*Picea abies*) or Scots pine (*Pinus sylvestris*) stands were at a lower risk for *Hylobius abietis* damage than even-aged stands in Scandinavia (Nevalainen 2017). In contrast, even-aged stands of shortleaf pine (*Pinus echinata*) had lower levels of damage from the Nantucket pine tip moth and several species of *Hylobius* weevils than uneven-aged stands (Land and Rieske 2006). Although these studies report different effects of stand structure, both support the idea that regardless of location, smaller trees are more susceptible to damage by this group of herbivores than are larger trees.

In heavily managed systems, such as seedling nurseries or intensively managed production forests, there are often established protocols to control damage from tip, shoot, root, and regeneration pests (e.g. Coyle et al. 2005; Cram et al. 2012). These pest management strategies may include pesticides, cultural or silvicultural treatments, the use of natural enemies, or a combination of tools. Different forest systems around the world have developed methods to manage tip, shoot, root, and regeneration pests, but these are often tailored to a particular pest and tree species. Pest behavior, host preferences and choices, and different forest types and forestry management tactics around the world all influence pest management strategies.

The next portion of this chapter will highlight the biology, ecology, and management of several arthropod species that impact tips, shoots, roots, or seedlings in forest systems. Each case study will focus on a particular organism or group of organisms, and discuss how these pests can impact tree growth, form, and productivity.

15.3.1 Case Study: Rhyacionia Tip Moths

There are 44 species of *Rhyacionia* tip moths worldwide (Gilligan et al. 2018) of which several species are found in North America (Miller 1967; Dickerson and Kearby 1972; Bell 1993). All tip moths attack pine (*Pinus*) species (Powell and Miller 1978). Although adults do not feed, young larvae mine needles and older instars feed inside shoot tissue. Several *Rhyacionia* species are known to cause significant damage under certain conditions.

Rhyacionia frustrana is an important pest of planted and natural pine in North and Central America and several Caribbean Islands (Powell and Miller 1978; Ford 1986; Asaro et al. 2003). This moth has up to five generations annually in the U.S., with adults emerging as early as March in some areas (Fettig et al. 2000, 2003). The life cycle of *R. frustrana* is roughly synchronized so that oviposition occurs with each new flush of growth on host trees (Berisford 1988). Preferred hosts in the southern U.S. are loblolly (*Pinus taeda*), shortleaf, and Virginia pine (*P. virginiana*) (Yates 1966; Nowak et al. 2010). Adults in early generations typically emerge synchronously during a growing season, though this synchronicity can be reduced in generations later in the year (e.g. Gargiullo et al. 1985). After mating, females use volatile host terpenoids to locate oviposition sites (Asaro et al. 2004) and lay eggs on needles or shoots. After hatching, larvae bore into needles (later moving to buds or shoots), buds, or shoots, and feed inside these tissues (Asaro et al. 2003). This feeding kills the bud or shoot tip, and in response to this the tree produces additional shoots. This results in a severely forked branch or stem (Fig. 15.3). Feeding can also kill developing cones (Yates and Ebel 1972). Damage can be variable during the year, as some studies show an increasing level of tree shoot mortality as the growing season progresses (e.g. Nowak and Berisford 2000; Coyle et al. 2003) while in others the highest damage levels occurred early in the growing season (Yates 1966; Miller and Stephen 1983). Larvae pupate inside the shoot, and the entire life cycle takes just a few weeks (Gargiullo and Berisford 1983; Haugen and Stephen 1984).

Newly planted or young pine stands are considered most susceptible (Asaro et al. 2003), as *R. frustrana* attack rates decrease as trees grow taller (White et al. 1984; Sun et al. 1998). Because damage rarely kills trees, a common dogma was that trees would "outgrow" *R. frustrana* damage by the end of the harvest cycle. However, long-term studies show a lasting impact of *R. frustrana* damage on tree productivity, as loblolly pine stands in North Carolina and Georgia that received tip moth control early in the rotation had greater stem volume and better stem form after 20 and 15 years, respectively (Berisford et al. 2013). Neither fertilization, irrigation, nor vegetative competition control consistently impacts *R. frustrana* populations (Asaro et al. 2003; Coyle et al. 2003; Nowak and Berisford 2010), likely due to the highly variable *R. frustrana* and natural enemy populations on the landscape.

Fig. 15.3 Adult *Rhyacionia frustrana* (**a**) oviposit on *Pinus* spp., where the larvae feed inside a shoot tip eventually causing shoot mortality (**b**). Repeated attacks can result in heavy terminal mortality and shoot production by the tree, resulting in a stunted, bushy tree (**c**) (Photo credits: James A. Richmond, USDA Forest Service, Bugwood.org [a]; David Coyle, Clemson University [b]; Terry Price, Georgia Forestry Commission, Bugwood.org [c])

Management strategies for *R. frustrana* have centered around pheromone trapping and insecticide application (e.g. Donley 1960). Pheromone components for *R. frustrana* have been identified and isolated (Hill et al. 1981) and successfully used in monitoring programs to predict infestation levels (Asaro and Berisford 2001a). For years, optimal insecticide periods for foliar insecticide applications were based on moth phenology (Berisford et al. 1984) or pheromone trapping, and detailed maps helped advise pine growers when to spray for *R. frustrana* based on degree-day accumulation (Malinoski and Paine 1988; Fettig et al. 2000, 2003). Recent advances in systemic insecticide technology have provided growers the option of applying treatments directly to seedlings at planting (or planting pre-treated seedlings), and this management method has resulted in substantial reductions in *R. frustrana* damage and, consequently, increases in tree growth and biomass accumulation (King et al. 2014).

Several other *Rhyacionia* species cause occasional damage in pine plantations around the world. The European pine shoot moth (*Rhyacionia bouliana*), long regarded as a pine pest in Europe (Friend and West 1933), has established in North America where it is primarily a pest of ornamental pines and Christmas trees. It primarily impacts red pine, Scots pine, white pine (*P. strobus*), jack pine

(*P. banksiana*), and Austrian pine (*P. nigra*) (Butcher and Haynes 1960). However, *R. bouliana* will occasionally damage lodgepole pine (*P. contorta*) seed orchards (Heeley et al. 2003) and red pine plantations, although trees may eventually outgrow this damage, which typically occurs early in the rotation (Miller et al. 1978). This pest is also known to cause significant damage in planted pine stands in South America, especially in loblolly pine, radiata pine (*P. radiata*), and slash pine (*P. elliottii*) (Eglitis and Gara 1974; Ide and Lanfranco 1996). As with most non-native species, *R. bouliana* populations tend to be higher in the invaded range (i.e. North America) than its native range (i.e. Europe) (Miller 1962). Additional reports of *Rhyacionia* species as occasional pests of pines appear throughout the literature. For instance, *R. duplana* occasionally causes shoot mortality and damage to Japanese black pine (*P. thunbergia*) in Japan (Kanamitsu 1965; Saito 1969), *R. leptobula* is a serious pest of Yunnan pine (*P. yunnanensis*) and Armand pine (*P. armandii*) in China (Huang 1987), and *R. neomexicana* is an important pest of ponderosa pine in the southwestern U.S. (Jennings 1975).

15.3.2 Case Study: Otiorhynchus Root Weevils

Tree seedlings are particularly susceptible to herbivory because they do not have large carbohydrate reserves from which to draw upon to help regenerate tissues and alleviate stress resulting from foliage loss. In forest nurseries, millions of tree seedlings can be present in one area, representing a highly concentrated food resource for pests. Weevils in the genus *Otiorhynchus*, native to Europe but now found in most of North America, parts of Australia, New Zealand, and Japan (Nielsen 1989), can be significant nursery pests. Adults are nocturnal and feed on the foliage of many woody plant species, though this leaf "notching" does little damage to the plant (e.g. Löf et al. 2004). Eggs are laid near the base of the seedling and larvae feed on the fine roots and lower stem, often causing significant damage (La Lone and Clarke 1981; Halldorsson et al. 2000) (Fig. 15.4). Adult *Otiorhynchus* weevils do not fly, and after emergence do not disperse far from where they developed as larvae (Moorhouse et al. 1992; Brandt et al. 1995).

The most destructive *Otiorhynchus* weevils in the northern U.S. and Canada are *O. sulcatus*, *O. ovatus*, and *O. rugusostriatus*. In northern Europe, *O. rugifrons*, *O. singularis*, *O. arcticus*, and *O. nodosus* are also known to impact young nursery-grown or planted seedlings (Halldorsson et al. 2000). Both conifer and hardwood seedlings may be impacted by *Otiorhynchus* weevils and, in some cases, damage can be severe. For instance, McDaniel (1932) detailed an infestation of *O. ovatus* in central Michigan, U.S., and noted that larval weevil feeding resulted in 33% of the seedlings being unusable in 1929, to 67% unusable in 1930, and in 1931 the entire crop of Norway and white spruce (*Picea glauca*), white pine, red pine, and ponderosa pine, and white cedar (*Thuja occidentalis*) were completely destroyed. Nearly 35% of Russian larch seedlings (*Larix sibirica*) were killed by *Otiorhynchus* feeding over a 3-year period in southern Iceland (Halldorsson et al. 2000).

Fig. 15.4 Adult *Otiorhynchus* weevils, like this *Otiorhynchus rugostriatus* (**a**), are black, stout-bodied, and flightless. Adult feeding appears as leaf notches, usually on the edges of leaves, and does very little harm to the plant (**b**). Larvae are white and grublike (**c**) and consume fine roots and the phloem of larger roots (**d**) (Photo credit: Matt Bertone, North Carolina State University [a]; Whitney Cranshaw, Colorado State University, Bugwood.org [b]; Michael Reding, USDA Agricultural Research Service, Bugwood.org [c]; David Gent, USDA Agricultural Research Service, Bugwood.org [d])

In nurseries, management of regeneration weevils is commonly done via fumigation (i.e. insecticide) treatments, or occasionally with entomopathogenic nematodes (parasitic nematodes that live in soil and feed on weevil larvae). Other management techniques include practicing clean cultivation, allowing infested areas to remain fallow, and rotating transplant beds so weevils do not have adequate host material (Cram et al. 2012). The use of biological control, specifically soil fungi, to manage *Otiorhynchus* weevils in nurseries is being studied in Iceland (Oddsdottir et al. 2010) and thus far results seem promising—the addition of several types of soil fungi resulted in significant *Otiorhynchus* larval mortality. The cryptic nature of these beetles makes their detection difficult, and this can directly impact the efficacy of any integrated pest management program put in place to control populations. However, developing chemical attractants for *Otiorhynchus* weevils is possible (van Tol et al. 2012). Assuming that this technology is fully developed, growers could better time chemical applications and increase their management efficacy.

15.3.3   Case Study: *Hylobius abietis*

Over 1 billion ha of natural and plantation forest land occurs in Europe and Asia, of which spruces (*Picea*) and pines are major components of the native flora and local timber industry (FAO 2020). This area, specifically the Scandinavian countries and Russian Federation, is a major producer of forest products. Consequently, adequate management for pests that impact the health of these forests is essential.

The large pine weevil (*Hylobius abietis*), is one of the most serious pests of conifers in this region (Fig. 15.5). While Scots pine and Norway spruce are preferred hosts, *H. abietis* also feeds on other conifers and some hardwoods (e.g. silver birch [*Betula pendula*] and common beech [*Fagus sylvatica*]). Adults feed on the tender shoot tissue of young trees, causing stem girdling that can result in damage and mortality on over 80% of young seedlings in some areas (Gourov 2000; Hannerz et al. 2002; López-Villamor et al. 2019; Hardy et al. 2020). After mating, females deposit their eggs in the vicinity of or in notches chewed on the bark of roots of weakened or mature trees or stumps just below the soil surface. Larvae create galleries under the bark, feeding on the living tissue and disrupting nutrient transport in the tree, and pupate in chambers under the bark. Larval development and adult emergence often occurs within the same year, however, larvae may take up to five years to develop in some cases (Leather et al. 1999). Adults can live up to four years, and upon emergence require several weeks of feeding before becoming reproductively mature (Leather et al. 1999). Adults are attracted to volatiles emitted by cut stumps (Lindelöw et al. 1993) where they mate and oviposit.

Scientists have worked for decades evaluating different management strategies for *H. abietis*. Various types of physical barriers have been evaluated (e.g. Lindström et al. 1986; Eidmann and Von Sydow 1989; Nordlander et al. 2011; Lalík et al. 2020). While many are as effective as insecticides, not all are likely to be commercially viable due to cost and/or time necessary to install. Insecticides (including permethrin, cypermethrin, imidacloprid, and others, Nordlander et al. 2011; Willoughby et al. 2020) have been used as a management tool for *H. abietis*, many of which can reduce weevil feeding damage to acceptable levels. However, there is evidence that adult *H. abietis* can detect insecticides in woody tissue and actively avoid that tissue (Rose et al. 2005), calling into question the effectiveness of some insecticide treatments. Further, adult weevils may live for weeks after ingesting insecticides, during which time oviposition or new feeding damage could occur (Rose et al. 2005). However, the availability of insecticides is not perpetual (e.g. permethrin was banned from use in Europe in 2003). As the scientific community learns more about the impacts of different active ingredients, we use this knowledge to help make decisions on how these chemicals should and can be used, with human and environmental safety in mind. Sometimes, chemical formulations that were once thought to be safe turn out to be unsafe after new information is gathered. In cases such as these (e.g. permethrin), the insecticide may cease to be available for use.

We cannot simply rely on chemical control of forest pests. Hence, for pests such as *H. abietis*, much research effort has also gone towards silvicultural techniques and

Fig. 15.5 Feeding by *Hylobius abietis* adults (**a**) can girdle conifer seedlings throughout Europe and Asia, resulting in seedling mortality (**b**). While adult damage is the most dramatic and impactful, larvae feed on roots (**c**) and can cause damage to already stressed trees (Photo credit: Jean-Paul Grandjean, Office National des Forêts, Bugwood.org [a]; György Csóka, Hungary Forest Research Institute, Bugwood.org [b]; Petr Srutka, Czech University of Agriculture, Bugwood.org [c])

residual stump management with a focus on determining the relationship between *H. abietis* and its host material (i.e. cut stumps, in-ground roots, and slash). Adult *H. abietis* are significantly more abundant in areas with white spruce stumps (Rahman et al. 2015; Piri et al. 2020) and increased incidence of seedling damage by adult feeding is expected and occurs in areas where seedlings are planted in close proximity to stumps (Piri et al. 2020). Site preparation, in particular the removal of stumps, which reduces volatiles that attract adults and oviposition and larval feeding sites, significantly reduces the risk of *H. abietis* damage to seedlings (Rahman et al. 2018; Wallertz et al. 2018). And, while commercial stump removal does not remove all roots from the site, Rahman et al. (2018) showed that waiting two years after harvest to remove stumps reduced larval *H. abietis* feeding and densities by 50%. *Hylobius abietis* damage was also positively correlated with the amount of slash remaining on the ground, which declined as time after harvest increased (López-Villamor et al. 2019). Planting seedlings in the summer resulted in less *H. abietis* feeding damage than planting in other seasons (Wallertz et al. 2016; Nordlander et al. 2017a). Thus, timing for certain silvicultural management strategies for *H. abietis* is critical.

Many other management methods have been evaluated with varying levels of success. Methyl jasmonate is a phytohormone present in plants that is involved in plant defenses. Exogenous application of methyl jasmonate has been shown to reduce *H. abietis* damage on maritime pine (*P. pinaster*), radiata pine, white spruce, and Scots pine seedlings in the field for two years (Zas et al. 2014). However, application of methyl jasmonate often results in growth reductions (Heijari et al. 2005), thus a trade-off between protection from *H. abietis* damage and tree growth occurs. Entomopathogenic fungi (e.g. *Metarhizium burnneum* and *Beauveria* spp.) and nematodes (*Steinernema carpocapsae* and *Heterorhabditis downesi*) can persist in the soil for up to two years and provide significant (>85%) control of *H. abietis* larvae (McNamara et al. 2018), and efficacy of different nematode species can be maximized by applying nematodes in specific ways, either directly on the stump or into the soil (Kapranas et al. 2017). Silvicultural practices such as fertilization (Zas et al. 2006) and prescribed fire (Pitkänen et al. 2008) led to increased *H. abietis* captures and damage. Further, there is significant genetic variation in *P. abies* seedling resistance to *H. abietis* feeding and mortality (Zas et al. 2017), and selection for resistant or tolerant families may be a viable management method in the future.

Decades of research has contributed to a solid understanding of factors that impact *H. abietis* damage, and because of these efforts our ability to predict where damage might occur is improving. The age of the clearcut, amount of mineral soil exposed on the ground, seedling size, and temperature are all factors that can help predict where and when *H. abietis* damage will occur (Louranen et al. 2017; Nordlander et al. 2017b). No one management method is completely effective or sustainable (Eidmann 1979), and an integrated pest management strategy for *H. abietis* is necessary for successful conifer production in the region.

15.3.4   Case Study: Root Weevil Complex in the Southeastern United States Pine Forests

The most common tree species grown for commercial purposes in the southeastern U.S. are the southern pines, including loblolly, slash, longleaf, and shortleaf pine. These pines are economically and ecologically important to this region, contributing significantly to the economy and comprising some of the most biodiverse places in the world (Aruna et al. 1997; Noss et al. 2015).

Several species of root-feeding weevils, including *Hylastes salebrosus*, *H. tenuis*, *H. porculus*, *Dendroctonus terebrans*, *Hylobius pales*, and *Pachylobius picivorous* are common in pine stands throughout the southeastern U.S. (Eckhardt et al. 2007; Zanzot et al. 2010; Coyle et al. 2015). *Hylobius* and *Pachylobius* weevils are commonly called "regeneration weevils" as they can severely impact young conifer plantings by feeding on the bark and phloem of pine seedlings (Fig. 15.6). On mature trees, adults will occasionally feed on the bark and cambium of twigs, causing a "flagging" where the tip of a branch dies and turns brown. Adults of these species are attracted to recently cut pine stumps, where they breed. Eggs are laid on the roots of the stump, and larvae feed on the root tissue. The impact of this damage is negligible to a mature tree (though an exception is with Christmas tree growers, where the loss of branches can decrease the value of trees; financial losses of nearly 20% can occur in some cases [Corneil and Wilson 1986]). The majority of the damage these weevils cause is by feeding on the bark and cambium of seedlings, either in the nursery or in newly planted forest stands. This feeding often kills the seedling and can cause significant economic damage to the landowner (Thatcher 1960; Lynch and Hedden 1984).

In the late 2000s and early 2010s, reports of pines dying in several parts of Alabama and Georgia concerned landowners and foresters and prompted scientists to take a closer look at the situation. In most cases, the dying trees involved older pines, with symptoms including yellowing needles and branch dieback. This phenomenon was called "southern pine decline" and was somewhat controversial (Coyle et al. 2015). While the aforementioned weevils were commonly associated with pine mortality, they were subsequently ruled out as the primary cause. It has been suggested that management and environmental conditions have a much greater impact on tree health (Coyle et al. 2020), as these weevils are secondary herbivores and are attracted to weakened or dying trees (Helbig et al. 2016). For example, *Hylobius* spp. adults are attracted to several volatile chemicals, specifically ethanol and monoterpenes (e.g. turpentine) (Siegfried 1987; Rieske and Raffa 1991). These volatiles are released by dying or recently dead pine trees and can be common in nurseries or where harvests have recently occurred (Fox and Hill 1973). The black turpentine beetle (*Dendroctonus terebrans*) typically attacks the lower bole of pine trees, especially trees injured by fire, machines, construction damage, or stressed by drought.

Since these weevils feed on stressed, weakened, dying, or recently dead pine trees, the fact that they are commonly captured in areas with declining pine trees is not surprising—in fact, it is to be expected as this is suitable habitat for mating and oviposition. These weevils do not kill living, healthy, mature trees. Southern pine

Fig. 15.6 *Hylobius pales* (**a**) and *Pachylobius picivorous* (**b**) are significant pests of young pines and Christmas tree plantations in North America. Adult feeding removes phloem, resulting in seedling mortality (**c**), or can kill branches leading to large reductions in tree value (**d**). Larvae feed in stumps and large roots of weak, dying, or dead trees (Photo credits: Robert Anderson, USDA Forest Service, Bugwood.org [a, b]; Lacy Hyche, Auburn University, Bugwood.org [c]; Eric Day, Virginia Polytechnic University, Bugwood.org [d])

decline is a combination of many factors, including management, soil characteristics, weather and climate, and tree species. Planting the correct tree species on the appropriate site, maintaining proper basal area, and controlling competing vegetation is essential for pine growth in the southeastern U.S. The observed pine decline in the southeastern U.S is likely due to mismanagement of one or more of these factors and the root weevil complex associated with these declining pines is unlikely to be the cause of the problem.

15.4 Other Arthropods Affecting Tips, Shoots, and Roots of Trees

Given the great diversity of the Curculionidae, it is not surprising that many weevil species impact trees worldwide. And, while their damage is often negligible, or extremely limited in time or space, there are occasional occurrences when a typically non-impactful species causes measurable damage. For instance, although minor twig damage in Douglas-fir (*Pseudotsuga menziesii*) can occur from *Cylindrocopturus furnissi* when populations reach high densities (Douglas et al. 2013), these weevils are rarely noticed. Damage by *C. furnissi* often appears as scattered branch mortality on mature trees—hardly enough to cause any negative impacts (Furniss 1942). But weevil densities can increase rapidly, especially when trees are stressed by factors such as drought, at which point weevil damage can severely deform or even kill trees. The Norway spruce weevil (*Pissodes harcyniae*), can be an occasional pest of stressed Norway spruce in central and northern Europe, but relatively little is known about this pest's life history (Kolk and Starzyk 1996). Damage to expanding terminal shoots of lodgepole pine can occur by the lodgepole terminal weevil (*Pissodes terminalis*); this damage can cause severe forking of the stem, but delaying the first thinning of the forest stand can help manage weevil populations (Maclauchlan and Borden 1996). The elephant weevil (*Orthorhinus cylindrirostris*) is primarily a pest of vines but can occasionally impact *Eucalyptus*, *Acacia*, and *Castanospermum* trees (Froggatt 1900; Hely et al. 1982). Further, many herbivores that feed on tree branch tips, shoots, and roots are opportunistic secondary pests. Developing effective management plans for these species is not only challenging, but the need is often unanticipated. For example, the Warren root collar weevil (*Hylobius warreni*) is an insect native to Canada normally found in low populations throughout the boreal forest. However, after the mountain pine beetle (*Dendroctonus ponderosae*) outbreak in western North America in the 1990s and early 2000s, *H. warreni* populations increased and caused considerable damage on young, replanted lodgepole pine in areas impacted by the outbreak (Robert and Lindgren 2006). As *H. warreni* adults do not fly, it appears the adults migrated via walking to replanted areas from older forests in search of food (Klingenberg et al. 2010). These are just a few examples of Curculionidae species that are occasional root and shoot pests of trees.

Periodical cicada emergence occurs every 17 years and is a visually spectacular event where millions of larval cicadas synchronously crawl out of the soil and emerge as adults. Found only in eastern North America (Cooley et al. 2009) *Magicicada* spp. oviposit near the ends of hardwood tree branches, in the process killing the terminal end of the branch (Fig. 15.7). Damage to host trees can be highly variable and dependent upon tree species (Cook et al. 2001; Cook and Holt 2002), with branch mortality usually less than 30% (Miller and Crowley 1998). While younger trees are more susceptible, damage from cicada oviposition generally doesn't result in long-lasting impacts on larger host trees (Miller and Crowley 1998; Cook and Holt 2002; Flory and Mattingly 2008).

Fig. 15.7 Despite high levels of terminal mortality (**a**), oviposition damage (**b**) from periodical cicadas (*Magicidada septendicim*) (**c**) rarely results in long-lasting damage on trees (Photo credits: Pennsylvania Department of Conservation and Natural Resources, Bugwood.org [a]; David Coyle, Clemson University [b]; Susan Ellis, Bugwood.org (c))

Defoliators can also cause tree shoot mortality, particularly when pest populations are high and their preferred leaf material has been exhausted. For example, two defoliators of eastern cottonwood (*Populus deltoides*) trees can, where high populations occur or where host conditions are conducive to infestation, cause substantial defoliation of tree terminals and terminal mortality (Fig. 15.8). In the southern U.S.,

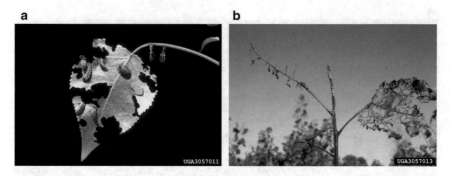

Fig. 15.8 Adult and larval cottonwood leaf beetles, *Chrysomela scripta* (**a**), feed on foliage of *Populus* trees. In some cases, feeding on tender shoot tissue will result in terminal mortality (**b**) (Photo credits: James Solomon, USDA Forest Service, Bugwood.org)

the cottonwood leafcurl mite (*Tetra lobulifera*) can cause severe leaf loss and occasional terminal mortality in plantation grown cottonwood trees, particularly those that exhibit rapid growth rates (Coyle 2002). The cottonwood leaf beetle (*Chrysomela scripta*) will also kill cottonwood terminals, as both adults and larvae will feed on the tender terminal tissue if preferred leaf tissue is no longer available (Coyle et al. 2002). Terminal mortality rarely kills trees, but it can have long-lasting impacts on tree form, causing increased branching and reduced stem growth, which can negatively impact tree value. In some cases, certain *P. deltoides* clones had nearly 40% more branch biomass when subjected to intense *C. scripta* defoliation compared to trees protected from defoliation (Coyle et al. 2008a).

Root herbivory can have a significant impact on many tree species, particularly when herbivores feed on ephemeral tissues such as lower order roots (Hunter 2008). While difficult to measure due to the cryptic nature of belowground feeding and the spatially variable density of root feeding fauna (e.g. Coyle et al. 2008b), it is possible to estimate impacts based on measurements of pest or root density. One technique is to compare tree root growth responses in areas where root feeding fauna have been removed (often via chemical applications) to areas where root herbivores have not been removed. For example, chemical removal of 95% of root herbivores in Manchurian ash (*Fraxinus mandshurica*) and Gmelin larch (*Larix gmelinii*) plantations in China resulted in first and second order root biomass increases of >42% in Manchurian ash and >53% in Gmelin larch (Sun et al. 2011). The impacts can be even more profound in seedlings, which have less capacity to cope with tissue loss. In Belarus, Kozel et al. (2017) exposed Scots pine, white spruce, and silver birch (*Betula pendula*) seedlings to root herbivory by larval cockchafers (*Melolontha melolontha*). In less than two months, scarab larvae consumed up to 73% of fine root biomass. These studies indicate that root herbivory—while not obvious—can have significant impacts on tree growth, health, and productivity.

15.5 Conclusions

Tree damage by shoot and root feeding pests will continue to require management efforts in forest stands worldwide. Early detection of forest pests is key to effective management, and recent advances in digital imagery and audio application and acquisition may be used to improve pest management strategies in forest systems. While detection of forest pest damage is possible via landscape-level imagery (e.g. Moderate Resolution Imaging Spectroradiometer, or MODIS), increased resolution of images is necessary before this technology becomes reliably usable from a practical standpoint (Gomez et al. 2020). However, images captured by unmanned aerial vehicles (UAVs, also called drones) appear to have more promise as they are able to obtain more usable imagery (Klouček et al. 2019). Capturing images with UAVs is more labor intensive than using widely available digital imagery, and it requires the operator to be appropriately licensed, but it is clear that both of these technologies hold great promise for the detection of forest pest damage. Further, acoustic sensor technologies that are being developed for pest detection in individual trees (e.g. Ashry et al. 2020) may eventually be useful for larger-scale forest pest detection and management.

Proper silvicultural techniques and chemical management are and will remain effective management tools for many forest pests. But, as more insect species develop resistance to insecticides and the cost of insecticide development continues to rise (Sparks 2013), the forest industry needs to look to new technologies for pest management. One such technology is tree resistance breeding (Showalter et al. 2018; Naidoo et al. 2019). Breeding trees for resistance to pests, either through traditional breeding methods or genetic engineering, has great potential for improving pest management in forestry. In fact, fusiform rust [*Cronatrium quercuum* (Berk.) Miyabe ex Shirai f. sp. *fusiforme*] management on pines in the southern U.S. via resistance breeding has been very successful (Schmidt 2003), and similar strategies could be applied tip, root, and shoot feeding pests.

A major unknown in forest pest management is how climate change will impact arthropod-plant interactions and, consequently, pest management strategies. While generalities exist, there is a great deal of species-specific and system-specific heterogeneity in terms of pest responses to climate change (Pureswaran et al. 2018; Jactel et al. 2019). For example, increasing temperatures are expected to result in increased fecundity and damage by *H. abietis* (Inward et al. 2012) and also changes in voltinism, which may require more localized management strategies (Wainhouse et al. 2014). Further, as climate change impacts are not restricted to temperature changes (e.g. weather unpredictability is expected to increase) it is difficult to predict what additional impacts may affect forest pest management. Ultimately, a robust integrated pest management strategy is necessary for any economically important forest pest.

References

Aruna PB, Cubbage F, Lee KJ, Redmond C (1997) Regional economic contributions of the forest-based industries in the South. For Prod J 47:35–45

Asaro C, Berisford CW (2001a) Predicting infestation levels of the Nantucket pine tip moth (Lepidoptera: Tortricidae) using pheromone traps. Environ Entomol 30:776–784

Asaro C, Berisford CW (2001b) Seasonal changes in adult longevity and pupal weight of the Nantucket pine tip moth (Lepidoptera: Tortricidae) with implications for interpreting pheromone trap catch. Environ Entomol 30:999–1005

Asaro C, Fettig CJ, McCravy KW, Nowak JT, Berisford CW (2003) The Nantucket pine tip moth (Lepidoptera: Tortricidae): a literature review with management implications. J Entomol Sci 38:1–40

Asaro C, Sullivan BT, Dalusky MJ, Berisford CW (2004) Volatiles associated with preferred and nonpreferred hosts of the Nantucket pine tip moth, *Rhyacionia frustrana*. J Chem Ecol 30:977–989

Ashry I, Mao Y, Al-Fehaid Y, Al-Shawaf A, Al-Bagshi M, Al-Brahim S, Ng TK, Ooi BS (2020) Early detection of red palm weevil using distributed optical sensor. Sci Rep 10:3155. https://doi.org/10.1038/s41598-020-60171-7

Bell CS (1993) The distribution, larval survival, and impact of a tip moth guild (Lepidoptera: Tortricidae: *Rhyacionia* species) in the Northern Plains. M.S. Thesis, Oregon State University, Corvallis, OR, 125 p

Berisford CW (1988) The Nantucket pine tip moth. In: Berryman AA (ed) Dynamics of forest insect populations: patterns, causes, implications. Plenum Publishing Corp., New York, pp 141–161

Berisford CW, Garguillo PM, Canalos CG (1984) Optimum timing for insecticidal control of the Nantucket pine tip moth (Lepidoptera: Tortricidae). J Econ Entomol 77:174–177

Berisford CW, Cameron RS, Godbee JF Jr, Jones JM, Dalusky MJ, Seckinger JO (2013) Long-term effects of pine tip moth (Lepidoptera: Tortricidae) control, vegetation control, and fertilization on growth and yield of loblolly pine, *Pinus taeda* L. J Entomol Sci 48:23–35

Brandt JP, Smith SM, Hubbes M (1995) Bionomics of strawberry root weevil adults, *Otiorhynchus ovatus* (L.) (Coleoptera: Curculionidae), on young ornamental conifer trees in southern Ontario. Can Entomol 127:595–604

Butcher JW, Carlson RB (1962) Zimmerman pine tip moth biology and control. J Econ Entomol 55:668–671

Butcher JW, Haynes DL (1960) Influence of timing and insect biology on the effectiveness of insecticides applied for control of European pine shoot moth, *Rhyacionia buoliana*. J Econ Entomol 53:349–354

Coody AR, Fettig CJ, Nowak JT, Berisford CW (2000) Previous tip moth infestation predisposes trees to heavier attacks in subsequent generations. J Entomol Sci 35:83–85

Cook WM, Holt RD (2002) Periodical cicada (*Magicicada cassini*) oviposition damage: visually impressive yet dynamically irrelevant. Am Midl Nat 147:214–224

Cook WM, Holt RD, Yao J (2001) Spatial variability in oviposition damage by periodical cicadas in a fragmented landscape. Oecologia 127:51–61

Cooley JR, Kritsky G, Edwards MJ, Zyla JD, Marshall DC, Hill KBR, Krauss R, Simon C (2009) The distribution of periodical cicada Brood X in 2004. Am Entomol 55:106–112

Corneil JA, Wilson LF (1986) Impact of feeding by adult pales weevil (Coleoptera: Curculionidae) on Christmas tree stands on southeastern Michigan. J Econ Entomol 79:192–196

Coyle DR (2002) Effects of clone, silvicultural, and miticide treatments on cottonwood leafcurl mite (Acari: Eriophyidae) damage in plantation *Populus*. Environ Entomol 31:1000–1008

Coyle DR, McMillin JD, Hall RB, Hart ER (2002) Cottonwood leaf beetle (Coleoptera: Chrysomelidae) defoliation impact on *Populus* growth and above-ground volume in a short-rotation woody crop plantation. Agric For Entomol 4:293–300

Coyle DR, Nowak JT, Fettig CJ (2003) Irrigation and fertilization effects on Nantucket pine tip moth (Lepidoptera: Tortricidae) damage levels and pupal weight in an intensively-managed pine plantation. J Entomol Sci 38:621–630

Coyle DR, Nebeker TE, Hart ER, Mattson WJ Jr (2005) Biology and management of insect pests in North American intensively-managed hardwood forest systems. Annu Rev Entomol 50:1–29

Coyle DR, Hart ER, McMillin JD, Rule LC, Hall RB (2008a) Effects of repeated cottonwood leaf beetle defoliation on *Populus* growth and economic value over an 8-year harvest rotation. For Ecol Manage 255:3365–3373

Coyle DR, Mattson WJ, Raffa KF (2008b) Invasive root feeding insects in natural forest ecosystems of North America. Chapter 8, pp 134–151. In: Johnson SN, Murray PJ (eds) Root feeders: an ecosystem perspective. CABI Press, Oxfordshire, UK, 230 p

Coyle DR, Mattson WJ, Friend AL, Raffa KF (2014) Effects of an invasive herbivore at the single plant scale do not extend to population-scale seedling dynamics. Can J For Res 44:8–16

Coyle DR, Klepzig KD, Koch FH, Morris LA, Nowak JT, Oak SW, Otrosina WJ, Smith WD, Gandhi KJK (2015) A review of southern pine decline in North America. For Ecol Manage 349:134–148

Coyle DR, Barnes BF, Klepzig KD, Koch FH, Morris LA, Nowak JT, Otrosina WJ, Smith WD, Gandhi KJK (2020) Abiotic and biotic factors affecting loblolly pine health in the southeastern U.S. For Sci 66:145–156

Cram MM, Frank MS, Mallams, KM tech. coords (2012) Forest nursery pests. United States Department of Agriculture Forest Service, Agriculture Handbook No. 680. 212 p

Dickerson WA, Kearby WH (1972) The identification and distribution of the tip moths of the genus *Rhyacionia* (Lepidoptera: Olethreutiday) in Missouri. J Kansas Entomol Soc 45:542–551

Donley DE (1960) Field testing of insecticides for control of the Nantucket pine tip moth, *Rhyacionia frustrana*, and the European pine shoot moth, *R. bouliana*. J Econ Entomol 53:365–367

Douglas H, Bouchard P, Anderson RS, de Tonnancour P, Vigneault R, Webster RP (2013) New Curculionoidea (Coleoptera) records for Canada. ZooKeys 309:13–48

Eaton CB (1942) Biology of the weevil *Cylindrocopturus eatoni* Buchanan, injurious to ponderosa and Jeffrey pine reproduction. J Econ Entomol 35:20–25

Eckhardt LG, Weber AM, Menard RD, Jones JP, Hess NJ (2007) Association of an insect-fungal complex with loblolly pine decline in central Alabama. For Sci 35:84–92

Eglitis A, Gara RI (1974) Algunal observaciones sobre la polilla del brote *Rhyacionia buoliana* (Den. & Schiff.) en Argentina. Rev Chil Ent 8:71–81

Eidmann HH (1979) Integrated management of pine weevil (*Hylobius abietis* L.) populations in Sweden, pp 103–109. In: Current topics in forest entomology: selected papers from the XVth International Congress of Entomology, Gen. Tech. Rep. WO-8, Washington, DC, 174 p

Eidmann HH, Von Sydow F (1989) Stockings for protection of containerized conifer seedlings against pine weevil (*Hylobius abietis* L.) damage. Scand J for Res 4:537–547

Food and Agricultural Organization of the United Nations (FAO) (2020) Global Forest Resources Assessment 2020: main report. Rome, Italy. https://doi.org/10.4060/ca9825en. 184 p

Fettig CJ, Dalusky MJ, Berisford CW (2000) Nantucket pine tip moth phenology and timing of insecticide spray applications in seven southeastern states. Res. Pap. SRS-18. U.S. Department of Agriculture, Forest Service, Southern Research Station, Asheville, NC, 21 p

Fettig CJ, Nowak JT, Grossman DM, Berisford CW (2003) Nantucket pine tip moth phenology and timing of insecticide spray applications in the Western Gulf Region. Res. Pap. SRS-32. U.S. Department of Agriculture, Forest Service, Southern Research Station, Asheville, NC, 13 p

Flory SL, Mattingly WB (2008) Response of host pants to periodical cicada oviposition damage. Oecologia 156:649–656

Ford LB (1986) The impact of *Rhyacionia frustrana* in Costa Rica. Turrialba 36:559–561

Fox RC, Hill TM (1973) The relative attraction of burned and cutover pine areas to the pine seedling weevils *Hylobius pales* and *Pachylobius picivorous*. Ann Entomol Soc Am 66:52–54

Friend RB, West, AS, Jr (1933) The European pine shoot moth (*Rhyacionia bouliana* Schiff.): with special reference to its occurrence in the Eli Whitney Forest. Yale School of Forestry Bulletin 37. 65 pages + plates.

Froggatt WW (1900) The reappearance of the elephant beetle *Orthorhinus cylindrirostris,* Fab. The Agricultural Gazette of New South Wales 11:847–851

Furniss RL (1942) Biology of *Cylindrocopturus furnissi* Buchanan on Douglas-Fir. J Econ Entomol 35:853–859

Gargiullo PM, Berisford CW (1983) Life tables for the Nantucket pine tip moth, *Rhycaionia frustrana* (Comstock), and the pitch pine tip moth, *Rhyacionia rigidana* (Fernald) (Lepidoptera: Tortricidae). Environ Entomol 12:1391–1402

Gargiullo PM, Berisford CW, Godbee JF Jr (1985) Prediction of optimal timing for chemical control of the Nantucket pine tip moth, *Rhyacionia frustrana* (Comstock) (Lepidoptera: Tortricidae), in the Southeastern Coastal Plain. J Econ Entomol 78:148–154

Gilligan TM, Baixeras J, Brown JW (2018) T@RTS: Online World Catalogue of the Tortricidae (Ver. 4.0). http://www.tortricid.net/catalogue.asp

Global Biodiversity Information Facility Secretariat (GBIF) (2021) GBIF backbone taxonomy: Curculionidae. Checklist dataset https://doi.org/10.15468/39omei accessed via GBIF.org on 2022-04-18

Gomez DF, Ritger HMW, Pearce C, Eickwort J, Hulcr J (2020) Ability of remote sensing systems to detect bark beetle spots in the southeastern US. Forests 11:1167. https://doi.org/10.3390/f11111167

Gourov AV (2000) *Hylobius* species (Coleoptera: Curculionidae) from Siberia and the distribution patterns of adults feeding on Scots pine stands. Entomol Fennica 11:57–66

Hainze JH, Benjamin DM (1984) Impact of the red pine shoot moth, *Dioryctria resinosella* (Lepidoptera: Pyralidae), on height and radial growth in Wisconsin red pine plantations. J Econ Entomol 77:36–42

Halldorsson G, Sverrisson H, Eyjolfsdottir GG, Oddsdottir ES (2000) Ectomycorrhizae reduce damage to Russian larch by *Otiorhynchus* larvae. Scand J for Res 15:354–358

Hannerz M, Thorsén Å, Mattsson S, Weslien J (2002) Pine weevil (*Hylobius abietis*) damage to cuttings and seedlings of Norway spruce. For Ecol Manag 160:11–17

Hardy C, Sayyed I, Leslie AD, Dittrich ADK (2020) Effectiveness of insecticides, physical barriers and size of plating stock against damage by the pine weevil (*Hylobius abietis*). Crop Protection 137: article 105307. 9 p

Haugen DA, Stephen FM (1984) Development rates of Nantucket pine tip moth, *Rhyacionia frustrana* (Comstock) (Lepidoptera: Tortricidae), life stages in relation to temperature. Environ Entomol 13:56–60

Heeley T, Alfaro RI, Humble L (2003) Distribution and life cycle of *Rhyacionia buoliana* (Lepidoptera: Tortricidae) in the interior of British Columbia. J Entomol Soc Brit Columbia 100:19–25

Heijari J, Nerg A-M, Kainulainen P, Viiri H, Vourinen M, Holopainen JK (2005) Application of methyl jasmonate reduces growth but increases chemical defence and resistance against *Hylobius abietis* in Scots pine seedlings. Entomol Exp Appl 115:117–124

Helbig CE, Coyle DR, Klepzig KD, Nowak JT, Gandhi KJK (2016) Colonization dynamics of subcortical insects on forest sites with relatively stressed and unstressed loblolly pine trees. J Econ Entomol 109:1729–1740

Hely PC, Pasfield G, Gellatley JG (1982) Insect pests of fruit and vegetables in NSW. The Australian Landscape Incarta Press, Clayton, Victoria, Australia

Hill AS, Berisford CW, Brady UE, Roelofs WL (1981) Nantucket pine tip moth, *Rhyacionia frustrana*: identification of two sex pheromone components. J Chem Ecol 7:517–528

Huang FS (1987) Forest insects of Yunnan. Yunnan Science and Technology Press, Kunming, China

Hunter MD (2008) Root herbivory in forest systems. Chapter 5, pp. 68–95. In: Johnson SN, Murray, PJ (eds) Root feeders: an ecosystem perspective. CABI Press, Oxfordshire, UK, 230 p

Ide S, Lanfranco D (1996) Evolución de los defectos fustales producidos por *Rhyacionia buoliana* en Chile: un ejemplo en la Décima Región. Bosque 17:15–19

Inward DJG, Wainhouse D, Peace A (2012) The effect of temperature on the development and life cycle regulation of the pine weevil *Hylobius abietis* and the potential impacts of climate change. Agric For Entomol 14:348–357

Jactel H, Koricheva J, Castagneyrol B (2019) Responses of forest insect pests to climate change: not so simple. Curr Opin Insect Sci 35:103–108

Jennings DT (1975) Life history and habits of the southwestern pine tip moth, *Rhyacionia neomexicana* (Dyar) (Lepidoptera: Olethreutidae). Ann Entomol Soc Am 68:597–606

Jordal BH, Sequeira AS, Cognato AI (2011) The age and phylogeny of wood boring weevils and the origin of subsociality. Mol Phylogenet Evol 59:708–724

Kanamitsu K (1965) Length and height position of pine shoots in relation to attack by the shoot moths. J Japanese For Soc 47:97–100

Kapranas A, Malone B, Quinn S, Tuama PO, Peters A, Griffin CT (2017) Optimizing the application method of entomopathogenic nematode suspension for biological control of large pine weevil *Hylobius abietis*. BioControl 62: 659–667

King JS, Kelly AM, Rees R (2014) Systemic control of Nantucket pine tip moth (*Rhyacionia frustrana* Scudder in Comstock, 1880) enhances seedling vigor, plantation establishment, and early stand-level productivity in *Pinus taeda* L. For Sci 60:97–108

Klingenberg MD, Lindgren BS, Gillingham MP, Aukema BH (2010) Management response to one insect pest may increase vulnerability to another. J Appl Ecol 47:566–574

Klouček T, Komárek J, Surový P, Hrach K, Janata P, Vašíček B (2019) The use of UAV mounted sensors for precise detection of bark beetle infestation. Remote Sens 11:1561. https://doi.org/10.3390/rs11131561

Kolk, A, Starzyk JR (1996) The atlas of forest insect pests. The Polish Forest Research Institute. Multico Warszawa, 705 pp

Kozel AV, Zvereva EL, Kozlov MV (2017) Impacts of root herbivory on seedlings of three species of boreal forest trees. Appl Soil Ecol 117–118:203–207

La Lone RS, Clarke, RG Larval development of *Otiorhynchus sulcatus* (Coleoptera: Curculionidae) and effect of larval density on larval mortality and injury to Rhododendron. Environ Entomol 10:190–191

Lalík M, Galko J, Nikolov C, Rell S, Kunca A, Modlinger R, Holuša J (2020) Non-pesicide alternatives for reducing feeding damage caused by the large pine weevil (*Hylobius abietis* L.). Ann Appl Biol 177:132–142

Land AD, Rieske LK (2006) Interactions among prescribed fire, herbivore pressure and short-leaf pine (*Pinus echinata*) regeneration following southern pine beetle (*Dendroctonus frontalis*) mortality. For Ecol Manage 235:260–269

Leather SR, Day KR, Salisbury AN (1999) The biology and ecology of the large pine weevil, *Hylobius abietis* (Coleoptera: Curculionidae): A problem of dispersal? Bull Entomol Res 89:3–16

Lindelöw Á, Eidmann HH, Nordenhem H (1993) Response on the ground of bark beetle and weevil species colonizing conifer stumps and roots to terpenes and ethanol. J Chem Ecol 19:1393–1403l

Lindström A, Hellqvist C, Gyldberg B, Långström B, Mattsson A (1986) Field performance of a protective collar against damage by *Hylobius abietis*. Scand J for Res 1:3–15

Löf M, Isacsson G, Rhydberg D, Welander TN (2004) Herbivory by the pine weevil (*Hylobius abietis* L.) and short-snouted weevils (*Strophosoma melanogrammum* Forst. And *Otiorhynchus scaber* L.) during the conversion of a wind-thrown Norway spruce forest into a mixed-species plantation. For Ecol Manage 190:281–290

López-Villamor A, Carreño S, López-Goldar X, Suárez-Vidal E, Sampedro L, Nordlander G, Björklund N, Zas R (2019) Risk of damage by the pine weevil *Hylobius abietis* in southern Europe: effects of silvicultural and landscape factors. For Ecol Manag 444:290–298

Louranen J, Viiri H, Sainoja M, Poteri M, Lappi J (2017) Predicting pine weevil risk: effects of site, planting spot and seedling level factors on weevil feeding and mortality of Norway spruce seedlings. For Ecol Manag 389:260–271

Lynch AM, Hedden RL (1984) Relation between early- and late-season loblolly pine seedling mortality from pales and pitcheating weevil attack in southeast Oklahoma. Southern J Appl for 8:172–176

Maclauchlan LE, Borden JH (1996) Spatial dynamics and impacts of *Pissodes terminalis* (Coleoptera: Curculionidae) in regenerating stands of lodgepole pine. For Ecol Manage 82:103–116

Malinoski MK, Paine TD (1988) A degree-day model to predict Nantucket pine tip moth, *Rhyacionia frustrana* (Comstock) (Lepidoptera: Tortricidae), flights in southern California. Environ Entomol 17:75–79

Martínez AJ, López-Portillo J, Eben A, Golubov J (2009) Cerambycid girdling and water stress modify mesquite architecture and reproduction. Popul Ecol 51:533–541

Matusick G, Menard RD, Zeng Y, Eckhardt LG (2013) Root-inhabiting bark beetles (Coleoptera: Curculionidae) and their fungal associates breeding in dying loblolly pine in Alabama. Fla Entomol 96:238–241

Mayfield AE III (2017) Cypress weevil, *Eudociminus mannerheimii* (Boheman) (Coleoptera: Curculionidae). UF/IFAS, EENY 360, 4 p

Mayhew PJ (2001) Herbivore host choice and optimal bad motherhood. Trends Ecol Evol 16:165–167

McDaniel EI (1932) The strawberry root weevil, *Brachyrhinus* (*Otiorhynchus*) *ovatus* as a conifer pest. J Econ Entomol 25:841–843

McNamara L, Kapranas A, Williams CD, O'Tuama P, Kavanaugh K, Griffin CT (2018) Efficacy of entomopathogenic fungi against large pine weevil, *Hylobius abietis*, and their additive effects when combined with entomopathogenic nematodes. J Pest Sci 91:1407–1419

Miller WE (1962) Differential population levels of the European pine shoot moth, *Rhyacionia buoliana*, between Europe and North America. Ann Entomol Soc Am 55:672–675

Miller WE (1967) Taxonomic review of the *Rhyacionia frustrana* group of pine-tip moths, with description of a new species. Can Entomol 99:590–596

Miller F, Crowley W (1998) Effect of periodical cicada ovipositional injury on woody plants. J Arboric 24:248–253

Miller FD Jr, Stephen FM (1983) Effects of competing vegetation on Nantucket pine tip moth (Lepidoptera: Tortricidae) populations in loblolly pine plantations in Arkansas. Environ Entomol 12:101–105

Miller WE, Wambach RF, Anfang RA (1978) Effect of past European pine shoot moth infestations on volume yield of pole-sized red pine. For Sci 24:543–550

Moorhouse ER, Charnley AK, Gillespie AT (1992) A review of the biology and control of the vine weevil, *Otiorhynchus sulcatus* (Coleoptera: Curculionidae). Ann Appl Biol 121:431–454

Naidoo S, Slippers B, Plett JM, Coles D, Oates CN (2019) The road to resistance in forest trees. Front Plant Sci 10:273. https://doi.org/10.3389/fpls.2019.00273

Neunzig HH, Cashatt ED, Matuza GA (1964) Observations on the biology of four species of *Dioryctria* in North Carolina (Lepidoptera: Phycitidae). Ann Entomol Soc Am 57:317–321

Nevalainen S (2017) Comparison of damage risks in even- and uneven-aged forestry in Finland. Silva Fennica 51(3), article 1741, 28 p. https://doi.org/10.14214/sf.1741

Nielsen DG (1989) Minimizing *Otiorhynchus* root weevil impact in conifer nurseries, pp 71–79. In: Alfaro RI, Glover SG (eds) Proceedings of Conference: Insects affecting reforestation: biology and damage. Forestry Canada, Pacific and Yukon Region, Victoria, British Columbia

Nordlander G, Hellqvist C, Johansson K, Nordenhem H (2011) Regeneration of European boreal forests: effectiveness of measures against seedling mortality caused by the pine weevil *Hylobius abietis*. For Ecol Manag 26:2354–2363

Nordlander G, Hellqvist C, Hjelm K (2017a) Replanting conifer seedlings after pine weevil emigration in spring decreases feeding damage and seedling mortality. Scand J For Res 32:60–67

Nordlander G, Mason EG, Hjelm K, Nordenhem H, Hellqvist C (2017b) Influence of climate and forest management on damage risk by the pine weevil *Hylobius abietis* in Northern Sweden. Silva Fennica 51(5), article id 7751, 20 p. https://doi.org/10.14214/sf.7751

Noss RF, Platt WJ, Sorrie BA, Weakley AS, Means DB, Costanza J, Peet RK (2015) How global biodiversity hotspots may go unrecognized: lessons from the North American Coastal Plain. Divers Distrib 21:236–244

Nowak JT, Berisford CW (2000) Effects of intensive forest management practices on insect infestation levels and loblolly pine growth. J Econ Entomol 93:336–341

Nowak JT, Asaro C, Fettig CJ, McCravy KW (2010) Nantucket pine tip moth. Forest Insect and Disease Leaflet 70. U.S. Department of Agriculture, Forest Service, Washington, DC, 8 p

Oddsdottir ES, Eilenberg J, Sen R, Halldorsson G (2010) The effects of insect pathogenic soil fungi and ectomycorrhizal inoculation of birch seedlings on the survival of *Otiorhynchus* larvae. Agric For Entomol 12:319–324

Örlander G, Nordlander G, Wallertz K, Nordenhem H (2000) Feeding in the crowns of Scots pine trees by the pine weevil *Hylobius abietis*. Scand J for Res 15:194–201

Patterson GS, Tracy RA, Osgood EA (1983) The biology and ecology of *Dioryctria resinosella* Mutuura (Lepidoptera:Pyralidae) on young red pine in Maine. Maine Agricultural Experiment Station Technical Bulletin 110, 28 p

Pinski RA, Mattson WJ, Raffa KF (2005) Composition and seasonal phenology of a nonindigenous root-feeding weevil (Coleoptera: Curculionidae) complex in northern hardwood forests in the Great Lakes Region. Environ Entomol 34:298–307

Piri T, Viiri H, Hyvonen J (2020) Does stump removal reduce pine weevil and other damage n Norway spruce regenerations?—Results of a 12-year monitoring period. For Ecol Manage 465:118098

Pitkänen A, Kouki J, Viiri H, Martikainen P (2008) Effects of controlled forest burning and intensity of timber harvesting on the occurrence of pine weevils, *Hylobius* spp., in regeneration areas. For Ecol Manage 255:522–529

Powell JA, Miller WE (1978) Nearctic pine tip moths of the genus *Rhyacionia*: biosystematics review (Lepidoptera: Tortricidae, Olethreutinae). U.S. Department of Agriculture, Agricultural Handbook 514. 51 p + illustrations

Pureswaran DS, Roques A, Barristi A (2018) Forest insects and climate change. Curr For Rep 4:35–50

Rahman A, Viiri H, Pelkonen P, Khanam T (2015) Have stump piles any effect on the pine weevil (*Hylobius abietis* L.) incidence and seedling damage? Global Ecol Cons 3:424–432

Rahman A, Viiri H, Tikkanen O-P (2018) Is stump removal for bioenergy production effective in reducing pine weevil (*Hylobius abietis*) and *Hylastes* spp. breeding and feeding activities at regeneration sites? For Ecol Manage 424:184–190

Regier JC, Mitter C, Solis MA, Hayden JE, Landry B, Nuss M, Simonsen TJ, Yen S-H, Zwick A, Cummings MP (2012) A molecular phylogeny for the pyraloid moths (Lepidoptera: Pyraloidea) and its implications for higher-level classification. Syst Entomol 37:635–656

Rieske LK, Raffa KF (1991) Effects of varying ethanol and turpentine levels on attraction of two pine root weevil species, *Hylobius pales* and *Pachylobius picivorus* (Coleoptera: Curculionidae). Environ Entomol 20:48–52

Robert JA, Lindgren BS (2006) Relationships between root form and growth, stability, and mortality in planted versus naturally regenerated lodgepole pine in north-central British Columbia. Can J For Res 36:2642–2653

Roe AD, Miller DR, Weller SD (2011) Complexity in *Dioryctria zimmermani* species group: incongruence between species limits and molecular diversity. Ann Entomol Soc Am 104:1207–1220

Rose D, Leather SR, Matthews GA (2005) Recognition and avoidance of insecticide-treated Scots pine (*Pinus sylvestris*) by *Hylobius abietis* (Coleoptera: Curculionidae): Implications for pest management strategies. Agric For Entomol 7:187–191

Saito A (1969) Damage analysis of fertilized young *Pinus thunbergii* forests injured by the shoot moth. J Japanese For Soc 51:212–214

Sánchez-Martínez G, Equihua-Martinez A, Gonzalez-Gaona E, Jones RW (2010) First record of *Eudociminus mannerheimii* (Boheman) (Coleoptera: Curculionidae) attacking *Taxodium mucronatum* Ten. (Cupressaceae) in Jalisco, Mexico. Coleop Bull 64:96–97

Scheirs J, De Bruyn L, Verhagen R (2000) Optimization of adult performance determines host choice in a grass miner. Proc R Soc Lond B 267:2065–2069

Schmidt RA (2003) Fusiform rust of southern pines: a major success for forest diseast management. Phytopathology 93:1048–1051

Showalter DN, Raffa KF, Sniezko RA, Herms DA, Liebhold AM, Smith JA, Bonello P (2018) Strategic development of tree resistance against forest pathogen and insect invasions in defense-free space. Front Ecol Evol 6:124. https://doi.org/10.3389/fevo.2018.00124

Siegfried BD (1987) In-flight responses of the pales weevil, *Hylobius pales* (Coleoptera: Curculionidae) on monoterpene constituents of southern pine gum turpentine. Fla Entomol 70:97–102

Skvarla MJ, Bertone MA, Fisher JR, Dowling APG (2015) New information about the cypress weevil, *Eudociminus mannerheimii* (Boheman, 1836) (Coleoptera: Curculionidae: Molytinae): Redescription, range expansion, new host records, and report as a possible causative agent of tree mortality. Coleop Bull 69:751–757

Son Y, Lewis EE (2005) Effects of temperature on the reproductive life history of the black vine weeil, *Otiorhynchus sulcatus*. Entomol Exp Appl 114:15–24

Sparks TC (2013) Insecticide discovery: an evaluation and analysis. Pest Biochem Physiol 107:8–17

Speight MR, Speechly HT (1982) Pine shoot moths in S.E. Asia I. distribution, biology and impact. Commonwealth For Rev 61:121–134

Stevens RE (1966) The Ponderosa pine tip moth, *Rhyacionia zozana*, in California (Lepidoptera: Olethreutidae). Ann Entomol Soc Am 59:186–192

Sun J, Kulhavy DL, Yan S-C (1998) Prediction models of Nantucket pine tip moth, *Rhyacionia frustrana* (Comstock) (Lep., Tortricidae) infestation using soil and tree factors. J Appl Ent 122:1–3

Sun Y, Gu J, Zhuang H, Guo D, Wang Z (2011) Lower order roots more palatable to herbivores: a case study with two temperate tree species. Plant Soil 347:351–361

Thatcher EC (1960) Influence of the pitch-eating weevil on pine regeneration in east Texas. For Sci 6:354–361

Thu PQ, Quang DN, Dell B (2010) Threat to cedar, *Cedrela odorata*, plantations in Vietnam by the weevil, *Aclees* sp. J Insect Sci 10, Article 192, available online: insectscience.org/10.192

van Tol RWHM, Bruck DJ, Griepink FC, De Kogel WJ (2012) Field attraction of the vine weevil *Otiorhynchus sulcatus* to kairomones. J Econ Entomol 105:169–175

Wainhouse D, Inward DJG, Morgan G (2014) Modelling geographical variation in voltinism of *Hylobius abietis* under climate change and implications for management. Agric for Entomol 16:136–146

Wallertz K, Hanssen KH, Hjelm K, Fløistad IS (2016) Effects of planting time one pine weevil (*Hylobius abietis*) damage to Norway spruce seedlings. Scand J For Res 31:262–270

Wallertz K, Björklund N, Hjelm K, Petersson M, Sundblad L-G (2018) Comparison of different site preparation techniques: quality of planting spots, seedling growth and pine weevil damage. New For 49:705–722

Wen X, Kuang Y, Shi M, Li H, Luo Y, Deng R (2004) Biology of *Hylobitelus xiaoi* (Coleoptera: Curculionidae), a new pest of slash pine, *Pinus elliottii*. J Econ Entomol 97:1958–1964

White MN, Kulhavy DL, Conner RN (1984) Nantucket pine tip moth (Lepidoptera: Tortricidae) infestation rates related to site and stand characteristics in Nacogdoches County, Texas. Environ Entomol 13:1598–1601

Whitehouse CM, Roe AD, Strong WB, Evenden ML, Sperling FAH (2011) Biology and management of North American cone-feeding *Dioryctria* species. Can Entomol 143:1–34

Willoughby IH, Moore R, Moffat AJ, Forster J, Sayyed I, Leslie K (2020) Are there viable chemical and non-chemical alternatives to the use of conventional insecticides for the protection of young trees from damage by the large pine weevil *Hylobius abietis* L. in UK forestry? Forestry 93:694–712

Yang S, Ma M, Li Q, Chai S (2012) Distribution and damage of *Rhyacionia leptotubula* in Northeastern Yunnan. For Pest Dis 31:20–21

Yates HOIII (1966) Susceptibility of loblolly and slash pines to *Rhyacionia* spp. oviposition, injury, and damage. J Econ Entomol 59:1461–1464

Yates HO III, Ebel BH (1972) Shortleaf pine conelet loss caused by the Nantucket pine tip moth, *Rhyacionia frustrana* (Lepidoptera: Olethreutidae). Ann Entomol Soc Am 65:100–104

Zanzot JW, Matusick G, Eckhardt LG (2010) Ecology of root-feeding insects and their associated fungi on longleaf pine in Georgia. J Econ Entomol 39:415–423

Zass L, Sampedro E, Prada MJL, Fernández-López J (2006) Fertilization increases *Hylobius abietis* L. damage in *Pinus pinaster* Ait. seedings. For Ecol Manage 222:137–144

Zas R, Björklund N, Nordlander G, Cendan C, Hellqvist C, Sampedro L (2014) Exploiting jasmonate induced responses for field protection of conifer seedlings against a major forest pest, *Hylobius abietis*. For Ecol Manage 313:212–223

Zas, R, Björklund N, Sampedro L, Hellqvist C, Karlsson B, Jansson S, Nordlander G (2017) Genetic variation in resistance to Norway spruce seedlings to damage by the pine weevil *Hylobius abietis*. Tree Genet Genom 13:111, 12 p. https://doi.org/10.1007/s11295-017-1193-1

Chapter 16
Insects of Reproductive Structures

Ward B. Strong, Alex C. Mangini, and Jean-Noel Candau

16.1 Introduction

The insects that feed on reproductive structures of forest trees are not only economically important, they are fascinating examples of the ability of insects to adapt to and
exploit the many niches available in forest ecosystems. Cones, fruits, seeds, nuts,
catkins and pollen are rich food sources available to insect herbivores (Sallabanks
and Courtney 1992; Turgeon et al. 1994). These reproductive structures are qualitatively different from vegetative parts of the tree. Their food quality is high relative
to leaves, needles, wood and bark (see Sect. 16.3.1.1). Cones, fruits and seeds are
discrete packages, often very small and are only present on the tree for a short time.
Cone and fruit production are much less predictable through time than other plant
structures (Janzen 1971; Crawley 2000). Insects consuming these tissues are forced
to adapt to these constraints. In this sense, insects feeding on reproductive structures often behave more like predators than herbivores, searching out and exploiting
multiple structures for an individual insect to develop (Janzen 1971; Mattson 1971;
Shea 1989). In this chapter we will refer to this group as reproductive structure
herbivores.

Fruit, cone and seed feeders have a long evolutionary association with their
hosts. As far back as the Late Pennsylvanian Epoch (300 Mya) there is evidence of
seed herbivory. The Molteno Formation in South Africa has yielded fossil evidence

W. B. Strong (✉)
British Columbia Ministry of Forests, Lands, Natural Resource Operations, and Rural
Development (retired), Vernon, BC, Canada
e-mail: ward.strong@shaw.ca

A. C. Mangini
Southern Region, Forest Health Protection, USDA Forest Service, Pineville, LA, USA

J.-N. Candau
Natural Resources Canada, Canadian Forest Service, Great Lakes Forestry Centre, Sault Ste.
Marie, ON, Canada

© The Author(s) 2023   523
J. D. Allison et al. (eds.), *Forest Entomology and Pathology*,
https://doi.org/10.1007/978-3-031-11553-0_16

of heteropteran herbivory scars on seeds of several plant genera from the Late Triassic Period (250–200 Mya) (Labandeira 2006). These scars are similar to damage produced by present-day pine seed bugs (Coreidae: *Leptoglossus*) (Krugman and Koerber 1969). Jurassic Period sawflies (Xyelidae) fed on pollen (Labandeira 2006), as do present-day xyelids (Burdick 1961). In contrast, some cone herbivores are geologically recent. Cone beetles (Scolytinae: *Conophthorus*) began diverging from their sister taxon (*Pityophthorus*) in the Early Pliocene (4 Mya). Later glaciation in the Pleistocene may have caused separation of host *Pinus* species ranges which facilitated the evolution of the thirteen extant cone-feeding *Conophthorus* species in North America (Cognato et al. 2005).

Insects attacking conifer reproductive structures are found in the orders Hemiptera, Thysanoptera, Coleoptera, Hymenoptera, Lepidoptera, and Diptera (Hedlin et al. 1981). The Coleoptera, Hymenoptera and Lepidoptera contain the majority of angiosperm fruit and seed consumers (Sallabanks and Courtney 1992). Diversity of genera and species of insects of reproductive structures is notably greater for gymnosperms compared to angiosperms; the more structurally complex conifer seed cones may result in more niche availability (Boivin and Auger-Rozenberg 2016). Species diversity is similar for Western Europe, North America and the Mediterranean Basin (Turgeon et al. 1994; Boivin and Auger-Rozenberg 2016) (see Sect. 16.4.1). Seed losses due to herbivory are often higher in temperate deciduous forests (e.g. 80% loss of acorns due to acorn weevils) than in tropical systems (e.g. 37.8% loss of acacia seeds due to bruchids) (Hulme and Benkman 2002).

Thirteen families feed on cones of conifers in Western Europe; however, only 30 genera have been recorded suggesting specialization for reproductive structure herbivory by a limited number of genera and species (Roques 1991). Specialization also occurs in angiosperms. Larvae of all Bruchidae (Coleoptera) feed and develop primarily within legume seeds, including many leguminous trees and shrubs (Southgate 1979; Derbel et al. 2007).

Insects of reproductive structures typically do not cause unpredictable and catastrophic ecological damage to natural forest stands. Unlike bark beetle and defoliator species, whose eruptive population dynamics can cause major disturbance at the landscape level; cone and seed insects are closely linked to the seasonal phenology of their hosts and their role in forest ecosystems is more subtle, though still important (Turgeon et al. 1994; Boivin and Auger-Rozenberg 2016). Their population dynamics are tied to the periodicity of their host cone or fruit crops (Shea 1989). Generally, fruit and seed herbivory is inversely proportional to crop size (Shea 1989; Turgeon et al. 1994). The larger the seed crop size relative to the size of the herbivore population, the greater the probability that an individual seed escapes herbivory; conversely, the higher the population of seed herbivores relative to the crop size, the more likely an individual seed will be consumed (see Sects. 16.3.3.2, 16.3.4). For example, in white fir, *Abies concolor* (Gord. & Glend.) Lindl. ex Hildebr., smaller crops led to an increase in insect-infested cones (Shea 1989). A bumper crop in natural stands of shortleaf pine, *Pinus echinata* Mill., resulted in a higher proportion of healthy seeds and reduced seed herbivore damage (Mangini et al. 2004). Furthermore, greater insect-caused damage can be expected for the season following a mast year (Boivin and Auger-Rozenberg 2016) (see masting in Sect. 16.3.3.2).

Economically, insects of reproductive structures can limit production of human food and feed products (fruit, nuts, acorns) and impact broader agroforestry services such as carbon sequestration, soil enrichment and biodiversity conservation (Jose 2009). The western conifer seed bug,[1] *Leptoglossus occidentalis* Heidemann, introduced from North America into Italy, is now a serious pest of Italian stone pine, *Pinus pinea* L.; in Tuscany, edible nut collection is no longer profitable because of severe damage caused by this exotic insect (Bracalini et al. 2013; Lesieur et al 2019). Insects destroy holm oak acorns, *Quercus ilex* subsp. *Ballota* (Desf.), in the savanna-like ecosystems of southwestern Spain before they can mature and fall to the ground to be consumed by the endemic Iberian pigs that local farmers use to produce the highly prized hams known as Jamón Ibérico de Ballota (Leiva and Fernández-Alés 2005).

Insects of reproductive structures can have a profound impact on forest ecology because they affect host tree reproduction and demography and can influence the evolution of the host (Boivin and Auger-Rozenberg 2016) (see Sect. 16.3.4). Threatened tree species are particularly susceptible to insects of reproductive structures. The endangered *Juniperus cedrus* Webb and Berthel., endemic to the Canary Islands, suffers seed loss from several seed herbivores (Guido and Roques 1996). Other insects are a major problem in managed trees, primarily in seed orchards and seed collection areas (Coulson and Witter 1984; Turgeon et al. 1994; Boivin and Auger-Rozenberg 2016).

Conifer seed and cone insects were first studied by John M. Miller, Bureau of Entomology, U.S. Department of Agriculture, during 1913–1917 in the western United States (Keen 1958). This work was of little concern until the 1950s when applied tree breeding programs began producing genetically improved seeds for reforestation (Coulson and Witter 1984). Seed orchards were established in Europe and North America to mass-produce genetically superior seed (Zobel and Talbert 1984). Insect damage to cones and seeds quickly became a major factor in the production of costly genetically improved seed (Keen 1958). As a result, seed orchard pest management programs have been developed for many areas and tree species around the world (see Sect. 16.5).

Our objective in this chapter is to introduce the important families, genera and species of insects that feed on reproductive structures by discussing their behavior, ecology and evolution. Our emphasis is a functional description rather than a taxonomic listing of insects of importance; see Ciesla (2011) for a taxonomic treatment. We focus on feeding on reproductive structures prior to propagule dispersal, though post-dispersal herbivory can have important evolutionary consequences (Hulme 1998). We also primarily discuss feeding on female reproductive structures rather than male. The bulk of studies found on this topic are from temperate forests rather

---

[1] Insect common names are names approved by either the Canadian Entomological Society or the Entomological Society of America or both.

than tropical, partly because temperate are more heavily studied, and partly because they are used more in sustainable, regenerative forestry than tropical forestry systems.

16.2 Types of Herbivory—Ways that Insects Exploit Reproductive Structures

Various guild classifications have been developed for insect herbivory of tree reproductive structures (Hawkins and MacMahon 1989). Roques (1991) and Turgeon et al. (1994) specified guilds and terminology that accord with the feeding behaviors of conifer-infesting insects. Boivin and Auger-Rozenberg (2016) redefine these guilds to include both angiosperm and gymnosperm herbivores. More focused guilds have been defined for insects feeding on acorns (Fukumoto and Kajimura 2001) and fir cones (Shea 1989). Our approach will be to discuss the modes of feeding on cones, fruits, seeds, pollen and catkins; guilds will become apparent as the types of herbivory are discussed.

16.2.1 Inflorescence Feeders

Many insects are restricted to the consumption of the strobili and conelets of conifers or the buds, catkins, and flowers[2] of angiosperm trees. Others that may infest the inflorescence but complete development after pollination are discussed here.

Some insect species are incidental feeders on inflorescences. Larvae of these species are defoliators by habit but consume reproductive structures when available. Budworms, *Choristoneura* spp. (Lepidoptera: Tortricidae), can defoliate huge areas of spruce and other boreal conifers. In North America, when populations are high, budworms will also feed on strobili and young conifer cones. The eastern spruce budworm, *C. fumiferana* (Clemens) readily consumes buds of balsam fir, eastern hemlock and other species; it can impact cone production for several years during outbreaks (Hedlin et al. 1981). The western spruce budworm, *C. freemani* Razowski (formerly *C. occidentalis* Freeman), will feed on young succulent strobili and conelets of spruce and Douglas-fir, *Pseudotsuga menziesii* (Mirbel) Franco, in the spring before the needles flush.

---

[2] Terminology of reproductive structures is that of Bonner and Karrfalt (2008) with modification. An inflorescence is a bud, catkin, flower, or strobilus early in development through pollen release (male) or pollination (female). A strobilus (plural strobili) is the cone-like male or female fruiting body, composed of bracts or scales, of gymnosperms. The female strobilus becomes a cone. A conelet is a young female cone or, for pines, a first-year female cone. A catkin is a dehiscent male flower spike of an angiosperm. The term fruit includes all types (achene, berry, drupe, samara and so on) exclusive of nuts. A nut is a one-seeded fruit with a woody or leathery pericarp (as in *Quercus*), or a fruit partially or wholly encased in an involucre or husk (as in *Carya* and *Corylus*).

Adults can also be incidental feeders. Scarab beetles, *Phyllophaga* spp. (Coleoptera: Scarabaeidae), feed on emerging female strobili in hard pine seed orchards in the southern United States. Feeding beetles damage the female strobili as they enlarge and become receptive to pollen; damaged strobili often die (Ebel et al. 1980).

Other inflorescence herbivores are more intimately tied to the biology of their host species. The looper, *Nemoria arizonica* (Groté) (Lepidoptera: Geometridae), displays larval developmental polymorphism. There are two broods, spring and summer. Larvae of the spring brood feed on oak catkins and their morphology mimics their food, which reduces bird predation. The summer brood larvae develop after the catkins have fallen from the trees. These larvae mimic first-year oak twigs protecting them from avian predation. A randomized diet, temperature, and photoperiod trial demonstrated that the polymorphism is diet-based (Greene 1989).

The small, primitive sawflies in the genus *Xyela* feed on the male strobili of *Pinus* (Ebel et al. 1980). Thirty-two species are known world-wide and fifteen occur in North America (Burdick 1961; Smith 1978, 1979). Adult emergence coincides with the expansion of strobili in the spring. Females oviposit on the expanding strobili. The early instar larvae feed on pollen within the pollen sacs. Mature larvae have been found feeding in the strobili axes of loblolly pine, *Pinus taeda* L. (Mangini unpubl.). Larvae fall to the ground as pollen is shed. They remain as prepupae in the soil and emerge in one or two seasons (Hedlin et al. 1981). Catkin sawflies do not reduce the pollen crop (Hedlin et al. 1981); however, they are nuisances during pollen processing for breeding work, often emerging in huge numbers in pollen drying rooms.

The incidental feeding of budworms and scarabs has little in common with *Xyela* species, whose feeding is tightly coupled with the phenology of their hosts. Inflorescence feeders do not meet the definition of a guild as defined by Root (1967). Inflorescences are the "same class" of resource; but *Xyela* and *Nemoria* exploit them differently than the budworms and scarabs.

16.2.2   Cone or Fruit Feeders

These insects feed internally on the cone or fruit tissues and seeds. These herbivores constitute a major portion of the insects of reproductive structures. Most are obligate internal feeders, having no ability to feed on plant parts other than reproductive structures. They also tend to have higher host specificity than the inflorescence feeder guilds, perhaps as a consequence of their intimate relationship with the substrate. The group can be broken into two categories, coarse internal feeders and determinant internal feeders.

16.2.2.1   Coarse Internal Feeders

These insects indiscriminately consume all internal parts of the developing cone or fruit including the seeds. In most instances this damage is caused by the larval stages; but adults and larvae can both feed during the life cycle of some species. Typically, fruits and cones are infested early in their development; however, some insects attack after cones are nearly mature.

Conifer Insects

Coleoptera. The only Coleoptera in this guild are the cone beetles, *Conophthorus* spp. (Curculionidae: Scolytinae). These are among the most destructive insects infesting pine cones in North America. All but one of the thirteen species infest developing cones; one species, *C. banksianae* McPherson, attacks shoots of jack pine (Coulson and Witter 1984; Ciesla 2011). *Conophthorus* adults are small (2.5–4 mm) dark brown to black beetles. Larvae are C-shaped with brown head capsules. The life histories of species are similar. In late spring, the adult female bores into second-year cones at the base or through the cone stalk. The male follows and after mating, the female makes an egg gallery along the axis of the cone. This girdles the cone and it quickly dies. Eggs laid along the gallery hatch and the larvae feed on the cone tissue and seeds, leaving the cone filled with frass and cone tissue. Brood adults typically overwinter in their host cone (Kinzer et al. 1972; Hedlin et al. 1981). Often, adults will emerge to feed and overwinter in shoots or conelets (Hedlin et al. 1981; Ciesla 2011). The death of the cone is necessary for successful brood development; seed loss is complete even if the larvae do not completely consume the cone. This loss can impact natural regeneration in pine stands (Graber 1964; Kinzer et al. 1972).

Lepidoptera. Coarse internal feeders are well represented by species in the Tortricidae and Pyralidae. Of the Tortricidae, several species of *Eucosma*, known as cone borers, feed in cones and can decrease seed yields in North America (Hedlin et al. 1981; Ciesla 2011). These include *E. cocana* Kearfott on shortleaf pine, *E. rescissoriana* Heinrich on western white pine, and *E. tocullionana* Heinrich, the white pine cone borer, on eastern white pine, *Pinus strobus* L. Life histories vary (Coulson and Witter 1984); however, larvae of all species consume cone contents and leave tightly packed frass and larvae pupate in the ground (Ollieu and Schenk 1966; Hedlin et al. 1981; de Groot 1998). Late-instar *E. tocullionana* often move to fresh cones to complete development. Occasionally, cones are host to both the white pine cone borer and *Conophthrous coniperda* (Schwart), the white pine cone beetle. In Ontario, the beetle, which feeds earlier (mid-May to June) than the borer (mid-June to August), often kills the cone before the borer larvae can enter the cone, giving the beetle a competitive advantage (de Groot 1998).

The coneworm genus *Dioryctria* (Pyralidae: Phycitinae) is by far the most important Lepidoptera that feed on conifer cones and seeds (Hedlin et al. 1981; Whitehouse et al. 2011). Of the 79 species described, perhaps half that number are cone-feeders. They are distributed throughout the Holarctic region; hosts are mainly in the Pinaceae with two species infesting Cupressaceae (Yates 1986; Whitehouse et al. 2011). Adults are small to medium-sized moths with somewhat narrow forewings bearing characteristic crossbands and patches of contrasting colors (Hedlin et al. 1981). Larvae have well-sclerotized head capsules and prothoracic shields, well-developed prolegs and long setae on each segment (Keen 1958; Leidy and Neunzig 1989). Life cycles vary; however, larvae of all species feed internally on the conelets and cones (Hedlin et al. 1981; Coulson and Witter 1984; Whitehouse et al. 2011). Usually, the entire content of the cone is consumed, leaving only coarse frass and webbing within. External evidence of infestation typically manifests as frass and webbing at point of larval entrance. Larvae may infest more than one cone. Depending on host, the dead cones can be distorted or may disintegrate prematurely; the latter, occurring often to pine conelets, can result in inaccurate estimates of damage at cone harvest (DeBarr 1974; Fatzinger et al. 1980).

Some *Dioryctria* are host-specific; others are polyphagous (Roux-Morabito et al. 2008). Pestiferous coneworm species tend to be polyphages, feeding across genera as well as on multiple species in a host genus (Whitehouse et al. 2011). The fir coneworm, *Dioryctria abietivorella* Groté, widely distributed from Alaska to Mexico, throughout Canada and the eastern US, feeds on cones of most Pinaceae in its range (Hedlin et al. 1981; Whitehouse et al. 2011). This insect is a significant pest of white spruce, *Picea glauca* (Moench) Voss, and Douglas-fir seed orchards (Trudel et al. 1999; Roe et al. 2006). Entire cone clusters can be killed and left covered with frass and webbing; larvae even continue to feed in harvested cones stored before seed extraction. The southern pine coneworm, *Dioryctria amatella* (Hulst) infests southern hard pine species in the United States from Texas to Virginia (Coulson and Franklin 1970; Ebel et al. 1980). Larvae can infest strobili, conelets, cones, shoots, rust galls and even wounds in hosts (Hedlin et al. 1981). The life cycle varies by host; in spring overwintering larvae enter shoots of longleaf pine, *Pinus palustris* Miller, or fusiform rust galls on loblolly pine, *Pinus taeda* L. Subsequent generations feed on conelets or cones (Coulson and Franklin 1970). It is a major seed orchard pest in its range (Ebel et al. 1980). *Dioryctria abietella* Denis and Shiffermüeller, the spruce coneworm of Europe, ranges across the Palearctic Region (Knölke 2007) and feeds on species of fir, larch, spruce and pine. It is one of the most important pests of conifer cones in Europe (Roux-Morabito et al. 2008); in Fennoscandia, it is the primary impediment to seed production of Norway spruce, *Picea abies* (L.) Karsten (Rosenberg et al. 2015).

The coarse internal feeders are similar to the white fir "cone and seed mining guild" of Shea (1989) where "Larvae … feed throughout the cone as it develops causing damage to seeds, scales and other cone structures." Larvae of lepidopterans are the major herbivores along with cone beetle adults and larvae in North America. Ecologically, the guild allows us to assess the impact of variable cone crop size on interspecies competition (Shea 1989). For example, the white pine cone beetle, by emerging earlier and killing the cone, prevents the white pine cone borer from attacking the cone (de Groot 1998).

Hardwood Insects

Coleoptera. The acorn and nut weevils, *Curculio* spp. (Curculionidae: Curculioninae), are distributed throughout the world (Hughes and Vogler 2004); most species feed on oak acorns (Drooz 1985); but some consume nuts of hickory, chestnut and birch trees (Williams 1989; Ciesla 2011). Adults possess a long, slender rostrum with tiny mouthparts at the tip; the distinctive snout can be as long as the body or longer (Triplehorn and Johnson 2005). Life histories are much alike for most species. The adult female uses her snout to chew into a developing acorn or nut and then, with her extensile ovipositor, deposits eggs into the nutritious kernel of the nut. The developing larvae feed on the nutmeat, typically consuming it until only frass remains within the husk. Infested acorns often drop prematurely. At maturity, larvae leave the fruit and move to the soil where they remain dormant; pupation and emergence are delayed for one or two years, sometimes up to five years (Drooz 1985).

In North America, several *Curculio* species infest oak acorns. Some, including *C. pardalis* (Chittenden) and *C. proboscideus* Fabricius, have a broad host range; *C. sulcatulus* (Casey) feeds on almost all oak species. In contrast, *C. fulvus* Chittenden is found only on live oak, *Quercus virginiana* Miller (Drooz 1985). Larval feeding can destroy significant portions of nut crops to the detriment of natural oak regeneration and wildlife relying on acorns for food (Gibson 1982). The pecan weevil, *C. caryae* (Horn), feeds on nuts of hickory (*Carya*) species and is a major pest of commercial pecan, *Carya illinoiensis* (Wangenheim) K. Koch, orchards. It causes premature drop of fruits and deformed inedible nuts. The hazelnut weevil, *C. nucum* L. infests hazelnuts in Europe and Asia (AliNiazee 1997). The chestnut weevils of Europe and the Near East, *C. elephas* (Gyllenhal) and *C. propinquus* (Desbr.) feed on acorns and chestnuts. In Mediterranean woodlands, these species cause premature drop of acorns of holm oak and cork oak, *Quercus suber* L. (Cañellas et al. 2007). In France, Italy and the Near East, these two species are key pests of chestnuts, *Castanea sativa* Miller (Paparatti and Speranza 2004). As with their Nearctic counterparts, a portion of the population remain in the soil in extended diapause for one or more years which complicates management efforts (Soula and Menu 2005).

Conotrachelus is a genus of weevils in North America that feeds primarily on oaks and hickories. Three species are common on oaks, *Con.*[3] *carinifer* Casey, *Con. naso* LeConte, and *Con. posticatus* Boheman. Acorns of all oak species are attacked. Life histories are similar to *Curculio*. Adults emerge in late summer, larvae feed inside the nuts, then move to the ground, pupate and remain in pupal cells over the winter. However, *Conotrachelus* species typically cannot penetrate the acorn shell of a sound acorn; oviposition occurs in damaged or previously infested acorns (Gibson 1982). They can attack and oviposit on healthy fruits of hickories (Boucher and Sork 1979).

---

[3] In this section, to avoid confusion, we use the following genera abbreviations for repeated scientific names: *C.* for *Curculio*, *Con.* for *Conotrachelus*, and *Cyd.* for *Cydia*.

Lepidoptera. The large Holarctic genus *Cydia* (Tortricidae) contains numerous species of seed and fruit herbivores of economic importance to forestry and horticulture (Ciesla 2011). Many species infest conifer cones (see Sect. 16.2.2.2.1). Several species are coarse internal feeders of *Carya*, *Quercus* and *Fagus* nuts (Drooz 1985; Boivin and Auger-Rozenberg 2016). Native to North America, *C. latiferreana* (Walsingham), the filbertworm, is a key pest of oak acorns and other tree nuts, particularly the European hazel, *Corylus avellana* L., cultivated commercially in the Willamette Valley of Oregon (AliNiazee 1983). Adults are small moths that emerge in June. Females oviposit on leaves near the fruit clusters. Young larvae enter the developing nut at the hilum (where the nut attaches to its husk). They penetrate and feed on the kernel as they develop. Infested nuts drop prematurely, and the mature larvae move into the ground where they overwinter in silken chambers (AliNiazee 1997). Life cycles of other nut-infesting *Cydia* are similar to that of the filbertworm (Debouzie et al. 1996; Speranza 1999; Jimenez-Pino et al. 2011). *Cydia fagiglandana* (Zeller) and *C. splendana* (Hübner) infest chestnuts in commercial orchards throughout Europe and the Near East (Speranza 1999; Brown and Komai 2008). In the Far East, *C. glandicolana* (Danilevsky) feeds on chestnuts in China and is found on acorns in Japan; *C. kurokoi* (Amsel) occurs in China, Korea and Japan and is a common pest of chestnuts in Japan (Brown and Komai 2008). Another European tortricid, *Pammene fasciana* L., called the "early chestnut tortrix", causes early drop of chestnut fruits; its impact is much less than the *Cydia* species (Speranza 1999; Pedrazzoli et al. 2012).

Coleopteran and lepidopteran secondary pests can make use of the damage created by the primary pests mentioned above. *Blastobasis glandulella* (Riley) (Coleophoridae) is native to the hardwood forests of eastern and central North America. Larvae enter acorns and hickory nuts through holes made by other insects (Drooz 1985) and feed on the remaining contents of the nut, often destroying nuts otherwise capable of germination (Gibson 1971).

In the deciduous forests of eastern North America, *Curculio* spp. and *Cyd. latiferranea* are primary attackers of acorns with *Conotracheles* species and *B. gladulella* acting as secondary scavengers of infested nuts (Gibson 1964, 1971). Other associations of *Curculio* and *Cydia* species occur on oaks in British Columbia and California (Lewis 1992; Rohlfs 1999; Dunning et al. 2002), Europe (Branco et al. 2002; Leiva and Fernández-Alés 2005; Csóka and Hirka 2006) and Asia (Fukumoto and Kajimura 2001; Maeto and Ozaki 2003). In Europe, *C. elephas* and *C. propinquus*, *Cyd. fagiglandana* and *Cyd. splendana*, in various combinations, consume developing hazelnuts and chestnuts (Debouzie et al. 1996; AliNiazee 1997; Speranza 1999) while the guild is represented by several *Curculio* species and *Cyd. gladicolana* in Asia (Fukumoto and Kajimura 2001).

16.2.2.2 Determinant Internal Feeders

The determinant internal feeders follow a definite feeding pattern as they consume the cone or fruit. Typically, this involves the larva of a particular species finding its way into the reproductive structure and then moving to a specific portion of the fruit, usually the seed. Lepidoptera and Diptera are the two dominant orders in terms of the number of species in this category.

Cone Tunnel Makers

Lepidoptera. The genus *Cydia* (Tortricidae) contains several species commonly called seedworms (Hedlin et al. 1981). This genus also includes the filbertworm and other nut consumers (see Sect. 16.2.2.1.2); however, the seedworms have decidedly different hosts and life cycles. Seedworms are found primarily on pine, spruce and fir and are widely distributed across North America and Eurasia (Cibrián-Tovar et al. 1986; Yates 1986; Shin et al. 2018). The adults have a 10–20 mm wingspan; forewings are, with some exceptions, metallic gray with distinct silver crossbands. The creamy white larvae have a shiny brown head capsule. Oviposition behavior of females is synchronized with the host species. Females lay eggs on conelets shortly after pollination in fir and spruce. On pines, with a two-year cone cycle, eggs are deposited in the spring on second-year cones near the cone scale spine (Tripp 1954). The first-instar larva bores into the cone and tunnels between cone scales. It enters the seed and consumes it, leaving it full of frass. The larva repeats this for successive seeds as it develops through 4–5 instars depending on species. The mature larva bores into the cone axis where it overwinters. While overwintering, it tunnels back to a seed and cuts an exit hole then returns to the axis tunnel for pupation. Pupation occurs in spring; the larva forces its way through the exit hole in the seed, pupates in the cone and the moth emerges between opened scales of the cone. Some larvae may diapause for a year (Hedlin et al. 1981).

Seedworms reduce healthy seed yield by directly consuming seeds and can destroy a substantial portion of a seed crop (Hedlin 1967; Bakke 1970). In North America, the important species can be sorted geographically. Species in the southern United States include *C. anaranjada* (Miller), the slash pine seedworm. This unusual species is host-specific to slash pine, *Pinus elliottii* Engelm., and the adults are orange with white crossbands. The pupa has spines on the abdominal segments that help it escape the seed. The longleaf pine seedworm, *C. ingens* (Heinrich) is common on longleaf pine, *P. palustris* Miller, and favors cones on the lower crown. Four seeds per cone are killed per larva, one per instar (Merkel 1963; Coyne 1968). The eastern pine seedworm, *C. toreuta* (Groté), found throughout eastern North America infests southern pines and jack pine, *Pinus banksiana* Lamb., in the Midwest where up to 50% of larvae enter diapause. Moths emerge from extended diapause in large numbers the year after a poor cone crop. Factors that reduce the numbers of first year cones may also increase the percentage of *C. toreuta* larvae undergoing extended diapause (Kraft 1968). In western North America, the ponderosa pine seedworm, *C. piperana* (Kearfott) is a pest of ponderosa pine, *Pinus ponderosa* Dougl. ex Laws., and sometimes destroys 50% of the crop. Infested seeds are fused together and to cone scales by silken feeding tunnels (Hedlin 1967). The spruce seed moth, *C. strobilella* (L.), is a significant pest of spruce and is a Holarctic species. In North America, it destroys seeds of all spruce species (Tripp 1954). In Europe, it is a pest of Norway, white and black spruce. A cold period is required for adult emergence (Bakke 1970). There appears to be a difference in the pheromone components between North American and Swedish populations, suggesting that the two populations may be separate species (Wang et al. 2010; Svensson et al. 2012). Several species are pests of conifers in Asia (Shin et al. 2018).

Diptera—Anthomyiidae. Often called cone maggots, species in the Holarctic genus *Strobilomyia* (Brachycera: Anthomyiidae) feed as larvae in the cones of spruce, fir and larch in boreal and montane habitats. Michelsen (1988) erected the new genus *Strobilomyia* for the monophyletic cone- and seed-feeding anthomyiids formerly placed in *Hylemya* or *Lasiomma*. Twenty species have been described; most species are pests of *Larix* with a few found on *Abies*, *Picea* and *Tsuga*. Adults are moderately hairy, small flies that resemble house flies. Arista of antennae are plumose. The larvae are typical brachycerine maggots with visible mouth hooks and tubercles on the posterior. The pupae occur in puparia formed from the last larval cuticles (Hedlin et al. 1981; Triplehorn and Johnson 2005).

Strobilomyia earn their place as determinant internal feeders by the intricate feeding pattern the larva makes as it feeds within a developing cone. Females oviposit on or near the conelet and the first instar remains in the egg. The second and third instars make a spiral feeding tunnel around the cone axis, consuming seeds and cone tissue as they tunnel. Some species, such as *S. laricis* Michelsen, move through the cone axis as they complete their development (Roques et al. 1984; Sachet et al. 2006, 2009). The mature third-instar larva drops to the ground during moist weather, usually in mid-summer. It forms a puparium in the ground litter where it overwinters. The small cones of larch typically host one larva while the larger cones of fir and spruce may have several. There is only one generation per year, consonant with the yearly cone cycle of the host trees (Michelsen 1988). A portion of the population enters extended diapause. In the Alps, prolonged diapause of *S. anthracina* (Czerny) coincided with poor cone crops of Norway spruce, *Picea abies* (L.) Karst. Diapause was initiated prior to onset of winter and may be correlated with a lack of available oviposition sites (Turgeon et al. 1994; Brockerhoff and Kenis 1996) (see Sect. 16.4.4).

Strobilomyia inhibit normal development of cones; larvae are robust feeders and can impact seed production in orchards (Roques et al. 1984; Michelsen 1988; Sweeney and Turgeon et al. 1994). In the western United States, there may be four or five *S. abietis* larvae in a fir cone. Cones with multiple larvae may die in early summer; all their potential seed is lost. Up to 30 percent loss of seed has been recorded. In North American spruce, *S. neanthracina* Michelsen, is a major pest, sometimes destroying entire seed crops (Hedlin et al. 1981). The number of seeds eaten per cone is positively correlated to the size of the cone; this impacts seed orchards where large cones are desired (Fidgen et al. 1998). In France, European larch, *Larix decidua* Mill., hosts *S. laricicola* (Karl) and *S. melania* (Ackland), which together can result in 50–60% damage to cone crops and an impact on natural regeneration in the French Alps (Roques et al. 1984). Other species cause similar damage to *Larix* throughout Eurasia (Michelsen 1988; Roques et al. 1996).

Cone and Fruit Galling Insects

Conifer Galling Insects. A gall is an abnormal, localized growth of plant tissue caused by the parasitic activity of another organism (Redfern and Shirley 2002). Insects and mites induce and inhabit galls; they gain protection from hygrothermic stress and access to enhanced nutritional resources (Price et al. 1987). Although not abundant among insects of reproductive structures, several gall-makers are important pests.

The family Cecidomyiidae (Diptera: Nematocera) is a large family (>6000 species) containing many destructive agricultural pests. It is also the largest gall-making group of arthropods (Gagné and Jaschhof 2014). Strangely, few species seem to be important on conifers. The Douglas-fir cone gall midge, *Contarinia oregonensis* Foote, is found throughout the range of Douglas-fir, *Pseudotsuga menziesii* (Mirbel) Franco, from central British Columbia down through north-central Mexico and throughout the Rocky Mountains. It is perhaps the most significant cone pest of Douglas-fir in the Pacific-Northwest (Hedlin 1961; Hedlin et al. 1981). The adults, typical midges, are fragile and tiny, only 3–4 mm long, with spindly legs. Eggs are about the width of a Douglas-fir pollen grain, but several times longer. The grub-like headless larvae are white in early instars, gradually becoming orange. Pupae are dark orange (Hedlin 1961). Adults emerge in spring when Douglas-fir flowers are open for pollination; emergence is closely tied to host phenology. The female deposits eggs at the base of the opened cone scale. Larvae tunnel into the cone scale and cause a gall to form near the ovules, each larva in a separate cell where it feeds on gall tissue. In autumn, during wet weather, the larva drops to the ground and pupates in a delicate cocoon in the litter, often in a dead male Douglas-fir strobilus. Pupation occurs in early spring. A portion of the population enters diapause for one or more years (Hedlin 1961).

Damage occurs as seeds are fused to the cone scale by the galls formed near the seeds. There are usually multiple galls in a cone scale. When large numbers of larvae are present, the scales die, and all seeds are lost. At times, hundreds of larvae can be found in a single cone. Damage becomes visible in July and August as scales die and turn red (Hedlin et al. 1981).

Hardwood Galling Insects. The cecidomyiids and gall wasps in Cynipidae (Hymenoptera) cause galls in hardwood trees, typically on leaves but also on flowers and fruits. On oak acorns, the cynipid genus *Callirhytis* causes galls on oak acorns. In North America, *C. fructuosa* Weld forms hard, lignified "stone" galls within the acorn. *Callirhytis operator* (Osten Saken) forms a "pip" gall in the side of the acorn shell. The pip gall causes the acorn to drop prematurely while the stone galls destroy the seed contents (Gibson 1982). Eurasian species cause similar galls in acorns (Csóka and Hirka 2006). A full treatment of the vast array of species and habits of hardwood gall feeders is beyond the scope of this work; literature compendia include Melika and Abrahamson (2002), Abe et al. (2007) (Cynipidae) and Gagné and Jaschhof (2014) (Cecidomyiidae).

Cone Scale-Feeding Insects

Midge larvae (Diptera: Cecidomyiidae) are common on or in conifer cone scales and bracts. Species in the genera *Asynapta, Camptomyia, Cecidomyia, Contarinia, Dasineura, Kaltenbachiola,* and *Resseliella* feed between the cone scales of the developing cone. Most are not serious pests but can impede the normal development of the cone when they induce resin that fuses cone scales together (Hedlin et al. 1981). *Asynapta hopkinsi* Felt is widely distributed in North America on fir, pine and spruce and feeds on resin exuded between cone scales (Hedlin et al. 1981). *Cecidomyia bisetosa* Gagné deforms cones of slash pine in the southern United States (Ebel et al. 1980). The Douglas-fir cone scale midge, *Contarinia washingtonensis* Johnson, occurs in western North America. It resembles *C. oregonensis*; however, it does not cause galls. Eggs are laid beneath cone bracts in early summer. Larvae make longitudinal tunnels under the surface of scales. Larvae drop to the ground and overwinter in cocoons in the litter. It can be abundant but does not cause significant damage (Johnson 1963). Species of *Kaltenbachiola* and *Resseliella* are common in Europe (Skrzypczyńska 1985, 1998).

16.2.3   Seed Feeders

Insects that feed within the seed represent the most specialized of the insects of reproductive structures. The life cycles or morphology of these species allows individuals to breach the developing seedcoats. Once access to the seed has been gained, the insect consumes the inner contents consisting of the embryo and either the megagametophyte (gymnosperms) or the endosperm (angiosperms) (Bonner and Karrfalt 2008). Access to the seed may be external (exophytic) by means of specialized mouthparts or internal (endophytic) by means of a life stage, usually the larva, physically entering the seed. The obligate seed-feeding insects tend to be even more specialized and host-specific than the Cone and Fruit Feeders, again because of their intimate association with the host plant.

16.2.3.1   External (Exophytic) Seed Feeding Insects

The seed-feeding true bugs (Hemiptera: Heteroptera) feed on the nutritious seeds hidden within cones and fruits; yet all life stages occur outside the cone or fruit. Only their piercing-sucking mouthparts invade their hosts. The mouthparts form a bundle of needle-like stylets within a segmented sheath. When feeding, the bug injects saliva through a duct in the stylet. The saliva liquefies the contents of the seed. The bug sucks out the liquid through the stylet. Typically, there is little external evidence of feeding damage (Hedlin et al. 1981). Only a handful of species are involved in each of the families Coreidae, Lygaeidae, Miridae, and Scutelleridae.

Coreidae—Leaf-footed Bugs. This family of medium to large bugs are elongate and usually dark colored. The common name comes from the expanded and leaf-like hind tibiae common to most species. They have scent glands that give off a strong odor when the bugs are disturbed. Most are plant-feeders (Drooz 1985; Triplehorn and Johnson 2005). Two species damage conifer seeds in North America, the leaf-footed pine seed bug, *Leptoglossus corculus* (Say), and the western conifer seed bug, *Leptoglossus occidentalis* Heidemann. Both are similar in appearance and habits. Adults are 15–18 mm long, with a narrow head and prominent eyes. Characteristic narrow white zigzag cross-bands are on the forewings. Nymphs are similar in color to the adults. Eggs are semi-cylindrical (Hedlin et al. 1981). Adults overwinter in sheltered locations, often in buildings.[4] On emerging in spring, they feed on male strobili. Adults are good fliers and make a distinctive buzzing sound when in flight. Females lay eggs on needles throughout the spring and summer. First-instar nymphs do not feed. Second-instar nymphs feed on developing ovules in conelets. Older nymphs and adults feed through the summer on seeds in maturing cones (Hedlin et al. 1981). *Leptoglossus occidentalis* typically has one generation per year; *L. corculus* may have several generations (Hedlin et al. 1981). The bug pushes its mouthparts through the scale to the seed. Saliva softens the seed coat, which is then punctured, and the bug sucks out the contents after the saliva liquefies the tissues. The puncture is marked by a minute hole in the center of a spot of discolored tissue (Koerber 1963).

Leptoglossus occidentalis is native to western North America but has extended its range eastward into the Midwest (McPherson et al. 1990). First reported by Koerber (1963), it is a major seed orchard pest in the Pacific Northwest and Rocky Mountains. It feeds on a wide range of conifers but is of most concern on Douglas-fir and pines, causing extensive seed loss (Bates et al. 2000; Strong 2015). Feeding before the seed coat hardens causes fusion of the seed to the cone scale, feeding after hardening results in an empty seed. Feeding on conelets by second-instar nymphs causes conelet abortion (Krugman and Koerber 1969; Connelly and Schowalter 1991; Bates et al. 2002,). Management of *L. occidentalis* is difficult. Separating natural abortion from bug-caused abortion makes damage assessment difficult (Bates et al. 2000). Damage occurs throughout the season and varies with host phenology (Strong 2006). Bugs are difficult to monitor, prompting studies on their communication (Takács et al. 2008) and host location cues (Takács et al. 2009) in an effort to exploit their behavior for management.

In 1999, *L. occidentalis* was discovered in Italy and has since become a major invasive pest, spreading throughout Europe from Portugal to Turkey (Roversi et al. 2011). It is the exemplary invasive insect. Highly mobile themselves, adults readily hitch rides with humans, and they are physiologically labile, easily adapting to new hosts and habitats (Tamburini et al. 2012). In Italy, *L. occidentalis* has dramatically reduced commercial pine nut production from and reduced regeneration of Italian stone pine, *Pinus pinea* L. (Bracalini et al. 2013; Lesieur et al. 2014). It has been

---

[4] On occasion *Leptoglossus* and other true bugs seeking shelter will invade homes in huge numbers causing major annoyance to homeowners. *Leptoglossus occidentalis* adults have even caused home damage by piercing plastic plumbing with their sturdy mouthparts (Bates 2005).

implicated as a potential vector of *Diplodia sapinea* (Fr.) Fuckel, the causal agent of a tip blight in European pines (Luchi et al. 2012).

Leptoglossus corculus, found throughout the eastern and southern United States, attacks most pine species in its range. Similar in appearance to *L. occidentalis*; it can be separated by its dilations on the hind tibiae which extend nearly to the apex, the dilations are much shorter in the latter (Allen 1969). The leaf-footed pine seed bug was long overlooked as a significant pest because its damage was not obvious (DeBarr 1970). It is one of the most destructive insects in southern pine seed orchards (Hedlin et al. 1981). Second-instar nymphs feed on conelet ovules and cause conelet abortion (DeBarr and Ebel 1974; DeBarr and Kormanik 1975). Later-instar nymphs and adults feed on seeds in second-year cones. As with its western counterpart, actual damage is difficult to estimate because seeds damaged before the seed coat hardens are often overlooked (DeBarr and Ebel 1973). In southern pine orchards, *L. corculus* occurs in combination with another seed-feeding bug, *Tetyra bipunctata* (Herrich-Schäffer) (Scutelleridae) (DeBarr 1967).

Scutelleridae – Shield-backed Bugs. The Scutelleridae are similar to stink bugs (Pentatomidae) but distinguished by the scutellum which extends over most of the abdomen like a shield (Triplehorn and Johnson 2005). *Tetyra bipunctata*, the shield-backed pine seed bug, occurs on all pines in the eastern United States. Adults are robust brown to dark brown insects with dark pits on the scutellum and 11–15 mm long. Nymphs are oval, flattened and grey to red-brown in color. Eggs are spherical and green changing to red as they mature. Adults overwinter under duff at the soil surface and emerge in April. Eggs are laid on needles and cones from late July through September. First-instar nymphs do not feed, later-stage nymphs and adults feed on seeds of second-year cones (Hedlin et al. 1981). After oviposition in spring, the adults enter an obligate dormancy—adults do not feed before mid-summer. *Tetyra bipuncata* can occur in large numbers but is not considered a major pest in southern orchards. Its single yearly generation and obligate diapause limit its feeding to mid-summer through fall. In contrast, *L. corculus* feeds from early spring through cone harvest, it has several generations and second-instar nymphs through adults can destroy seeds (DeBarr and Ebel 1973; Cameron 1981).

Lygaeidae. Species in several genera of Lygaeidae feed on cones and fruits. *Belonochilus numenius* (Say), introduced from North America, feeds on the fruiting heads of *Platanus* species in Europe and the Mediterranean Basin (Gessé et al. 2009). Four species of *Orsillus* feed on cones and seeds of Cupressaceae in Europe and can reduce seed yields (Dioli 1991; Rouault et al. 2005). *Orsillus depressus* Mulsant and Rey feeds on native and exotic species of *Juniperus* and *Cedrus* and is common in the Iberian Peninsula (Ciesla 2011). *Orsillus maculatus* (Feiber) is a pest of *Cupressus sempervirens* L. cones (Ciesla 2011) (see Sect. 16.3.1.4). *Orsillus* may be a potential vector of *Seridium cardinale* Sutton and Gibson, the causal agent of cypress bark canker disease (Rouault et al. 2005). The elm seed bug, *Arocatus melanocephalus* Fabricius (Hemiptera: Heteroptera: Lygaeidae) is native to Europe and widely distributed in Central and Southern Europe (Ferracini and Alma 2008); it was reported as an invasive in China in 2013 (Gao et al. 2013) and subsequently in North America (Idaho State Department of Agriculture 2013; Acheampong and Strong 2016).

Miridae. The mirid *Platylygus luridus* (Reuter) feeds on ovules of jack pine conelets and causes their abortion (Rauf et al. 1984). The birch catkin bug *Kleidocerys resedae* feeds on reproductive structures of birch as well as seeds of many other species (Wheeler 1976), but it is not known to cause economic damage.

Rhopalidae. In North America, the boxelder bug, *Boisea trivittata* (Say), and the western boxelder bug, *Boisea rubrolineata* (Barber), feed on ash and maple samaras, primarily on boxelder, *Acer negundo* L. in summer and fall. The boxelder bug occurs in the East and extends west to Montana and Alberta; its western counterpart occurs from British Columbia to Texas. Both species are similar in appearance and habitat and have little impact on host trees but are nuisances when they enter houses for overwintering (Tinker 1952; Ciesla 2011).

16.2.3.2   Internal (Endophytic) Seed Feeding Insects

Internal seed-feeding insects are those which must complete one or more life stages, almost always the larva and pupa, within the seed, consuming the seed contents as they develop (Turgeon et al. 1994). Internal seed feeders are considered the most specialized consumers of reproductive structures (Roques 1991). Species in Coleoptera, Hymenoptera and Diptera are involved.

Hymenoptera. The seed chalcid wasp genus *Megastigmus* (Chalcidoidea: Torymidae) is by far the largest group of conifer internal seed feeders (Grissell 1999). This group contains 41 species of seed insects associated with the Pinaceae, Taxodiaceae and Cupressaceae. Geographically, they seem to be restricted to the Holarctic region (Auger-Rozenberg et al. 2006). These pests are generally highly species-specific, but some can infest several members of the same genus. *Megastigmus* species have one generation per year and are most abundant on conifers with yearly cone cycles; but some species infest *Pinus* (Kinzer et al. 1972; Hedlin et al. 1981; Cibrián-Tovar et al. 1986).

Megastigmus adults are small- to moderate-sized (3–5 mm in length) antlike wasps with elongate, enlarged hind coxae, laterally compressed abdomens, and a long ovipositor in females. The forewing has a large, dilated darkened spot (stigma) in its anterior margin. Adults are variable in color with patterns of black, brown and yellow (Keen 1958; Hedlin et al. 1981). Eggs are spindle-shaped. Larvae are legless and strongly curved or arched, giving a c-shaped appearance. Pupae are exarate. Both larvae and pupae occur within the seeds and can only be seen by dissection or radiography (Hedlin 1956; Skrzypczyńska 1978). Typically, the overwintered adult emerges in spring and early summer from seed that has been shed from the tree. A small circular hole is cut in the seed coat by the emerging adult. After mating, the female lays her eggs directly inside the host ovules by inserting the ovipositor through the scales of the young cones. All the immature stages will then develop inside the seed. Pupation occurs in spring.

Megastigmus, as a group, is a major seed destroyer of conifers. Species consistently cause damage in managed seed orchards throughout the Holarctic. Because of the high level of specialization of these seed parasitoid species and their intimate relation with their hosts, seed chalcids have evolved a series of adaptive traits to cope with the wide spatial and temporal fluctuations of conifer seed production. These adaptations include extended diapause (Suez et al. 2013), parthenogenetic reproduction (Boivin et al. 2014), effective dispersal (Jarry et al. 1997; Lander et al. 2014), and the ability to modify the physiology of seed development in ways similar to galling insects (von Aderkas et al. 2005). The ease of invasive introduction and the peculiarities of the life cycle have made seed chalcids the subject of much genetic and behavioral research (Boivin et al. 2017) (see Sects. 16.3.1.4, 16.3.3.1, 16.4.3, 16.4.4).

The Douglas-fir seed chalcid, *Megastigmus spermotrophus* Wachtl, native to the natural range of Douglas-fir, *Pseudotsuga menziesii*, has been introduced into areas where Douglas-fir is grown commercially including Great Britain, Western Europe and New Zealand (Hussey 1955; Mailleux et al. 2008). In the Pacific Northwest, it is a major seed pest in orchards (Hedlin et al. 1981). *Megastimus albifrons* Walker is one of the rare seed chalcids attacking *Pinus*; hosts are ponderosa pine, *Pinus ponderosa*, and several *Pinus* species in Mexico. It has the typical one-year life cycle. Seeds often remain in cones and the adults must escape the seed and tunnel through the scale to disperse (Kinzer et al. 1972; Cibrián-Tovar et al. 1986). The spruce seed chalcid, *M. atedius* Walker and the fir seed chalcid, *M. pinus* Parfitt attack *Picea* and *Abies* species, respectively, and are native to North America but established in Europe. Numerous species are native to Europe and Asia including *M. pictus* (Förster) on Eurasian species of *Larix* (Roques et al. 1995; Roques and Skrzypczyńska 2003).

Coleoptera. *Lignyodes bischoffi* (Blatchley) and *L. helvola* (LeConte) (Curculionidae) are weevils native to eastern North America. They feed on seeds of *Fraxinus*. Females oviposit on seeds, the larvae consume the seed, and overwinter in the seed or on the ground (Barger and Davidson 1967). *Lignyodes bischoffi* has been introduced into Central Europe (Gosik et al. 2001).

Diptera. Species of *Earomyia* (Brachycera: Lonchaeidae), called seed maggots, are found primarily on *Abies* but also occur in, *Larix*, *Picea*, *Pseudotsuga*, and *Tsuga* (McAlpine 1956). Adults are small shiny black flies with wings longer than the abdomen, which is flattened. The larvae are typical brachycerine maggots. Females oviposit on cone scales in late spring. Newly hatched larvae enter the cone and then penetrate the seeds to feed. In fall, the mature larvae drop to the ground, form puparia, and overwinter. Some may delay emergence for a year or two (Keen 1958). In North America, *Earomyia abietum* McAlpine, *E. brevistylata* McAlpine, and *E. longistylata* McAlpine infest only *Abies*; *E. aquilona* McAlpine infests *Abies*, *Larix* and *Pseudotsuga* and *E. barbara* McAlpine is found on those genera and *Picea* and *Tsuga* (McAlpine 1956). In Europe, *E. impossible* is common on *Abies* (Skrzypczyńska 1998). Seed maggots often occur in large numbers but are not usually significant pests (Hedlin et al. 1981).

Gall midges in the genus *Semudobia* form galls within the developing seeds of birch. Several species attack birch in the palearctic, nearctic, or nolarctic (Roskam

1977). One species also galls the catkin scales. *Semudobia* larvae are the basis of a suite of inquilines, chalcid parasitoids, and predators (Roskam 2013).

16.2.4 Tropical Ecosystem Herbivores of Reproductive Structures

Tropical forests are characterized by rich biodiversity in both host trees and their insects. Masting is important in tropical forest ecosystems (Herrera et al. 1998). For example, in Sarawak, Malaysia, community-wide masting or "general reproduction" occurs at intervals of 2–10 years. At general reproduction, most plant species in the community flower synchronously over a period of three to six months followed by mass dispersal of seeds (Asano et al. 2016). Seed herbivore satiation is the accepted explanation for such masting; the abundance of seed overwhelms herbivore consumption leaving a large seed crop for regeneration while in non-mast years, herbivores are starved (see Sect. 16.3.3.2) (Hosaka et al. 2009; Linhart et al. 2014; Asano et al. 2016).

Tropical tree seed herbivores consist primarily of species of Coleoptera, Lepidoptera, Hemiptera and Hymenoptera. Weevils (Curculionidae, Nanophyidae) and Lepidoptera (Pyralidae, Tortricidae) consume seeds of dipterocarps (Dipterocarpaceae), primary components of rainforests of Southeast Asia (Hosaka et al. 2017; Lyal and Curran 2000). Australian eucalypts (Myrtaceae), sheoak (Casuarinaceae), and tea tree (Myrtaceae) host anobiid beetles and chalcidoid wasps (Andersen and New 1987). Curculionoidea infest seeds of baobab (Malvaceae) and mangrove (Rhizophoraceae). Bruchid beetle larvae infest seeds of Amazonian palms (Arecaeae) (Silvius and Fragoso 2002) and are major consumers of tropical acacia (Fabaceae) seeds (Janzen 1969; Peguero et al. 2014). Bugs (Lygaeidae) feed on seeds of figs (Moraceae) (Slater 1972). Much remains to be learned about tropical seed herbivores (Basset et al. 2019).

16.3 Reproductive Structures as Habitat, and Evolutionary Consequences for the Host

Reproductive structures offer a unique habitat that is both nutrient-rich and well-defended. Herbivores have devised many mechanisms to exploit this resource, while hosts have evolved means to limit herbivory. This section explores these relationships, and places them in an evolutionary context.

16.3.1 Reproductive Structures Nutritive Value and Host Defenses

16.3.1.1 Nutritive Value

Most plants invest substantial resources in seed development, quantitatively to compensate for pre- and post-dispersal losses, and qualitatively to provide their embryos with enough reserves to germinate successfully. Consequently, seeds and reproductive structures, in general, are a greater source of carbohydrates and proteins than most other plant parts (Hulme and Benkman 2002) (Fig. 16.1). Seeds are also long lasting, offering long-term storage of nutrients that seed herbivores can exploit. Fruit and other non-seed reproductive structures can also be rich resources, but are more transient, and can be considered similar to dung or carrion (Lukasic and Johnson 2007). Because of lower defenses and increased transience, non-seed reproductive structures often undergo a distinct succession of species utilizing them, much like dung or carrion. Nutrient extraction from seeds can be improved with bacterial symbionts, such as *Burkhoderia* in *Megastigmus*, which aides in nitrogen recycling and nutrient breakdown of the megagametophyte (Paulson et al. 2014).

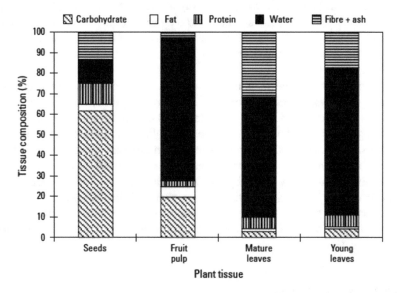

Fig. 16.1 Comparison between the average composition of seeds and other plant parts. Seeds provide a concentrated nutrient source that is particularly rich in carbohydrates. From Hulme and Benkman 2002

16.3.1.2   Chemical Defenses

Plant chemical defense strategies against herbivory are ubiquitous and explain many plant–insect interactions. While a large amount of information has been gathered on chemical defenses of vegetative parts, much less is generally known with regards to reproductive structures. In practice, it has long been established that a wide variety of toxins and repellents occur in seeds. Some tree seeds, such as those of the neem tree, *Azadirachta indica* A. Juss, even provide extracts used as insecticides or insect repellents.

Seed chemical defenses include tannins that interfere with protein absorption, cyanide precursors, enzymes inhibitors and phytohaemaglutinins that disrupt enzyme functions (Bell 1978). Defense compounds are often complex molecules that require significant energy and metabolites to synthesize. As such, their physiological cost is often significant. They also need to be stored in the seeds, thus limiting the space available for reserves destined to the embryo and the seedling. This limitation is sometimes mitigated by synthesizing toxins that can be metabolized and used by the seedling later (Harborne 1993).

Defense chemicals can also be synthesized as an induced response to herbivory. For example, infested seeds of *Mimosa bimucronata* (DC.) Kuntze have higher phenolic contents than non-infested seeds, suggesting that this is an induced defense (Kestring et al. 2009). Another strategy is to only protect the part of the seed that is important. In subtropical oaks, a higher concentration of tannins at the apical half of the acorns, near the embryo, increases resistance to seed herbivory (Xiao et al. 2007).

Herbivores have evolved detoxification and sequestration mechanisms to deal with host defenses. Physical defenses, once breached by specialists, can allow access by generalists (see Sect. 16.4.3), but chemical defenses continue to operate even if a specialist gains access, thus reinforcing the specialization of reproductive structure herbivores. For example, *Curculio* performance in different acorn species is determined by the chemical composition of the acorns (Munoz et al. 2014); access by one species does not allow access by others. Similarly, Janzen (1969) found that dry tropical hardwood legumes use a wide range of chemicals for defense against bruchid beetles, including pentose, methylpentose, saponins, endopeptidase inhibitors, alkaloids, and free amino acids. The latter group frequently accounted for host specificity (see Sect. 16.4.3).

16.3.1.3   Physical Defenses

There is considerable evidence that herbivory acted as a selective agent in the evolution of many physical traits of reproductive structures. Fossil records show an increase in seed cone size and compactness (percent of interlocking scales) in Cupressaceae and Pinaceae species during the Early to Middle Jurassic (Leslie 2011). During the same period, insect mouthparts show significant diversification (e.g. the appearance of new piercing and sucking types) and new insect groups such as weevils emerged.

Since the number of seeds per cone did not increase with size, and pollen cones, which are less susceptible to herbivory, remained similar during the same period, the increase in cone size suggests a greater investment in protective tissue in response to increased severity and diversification of seed herbivory.

Seed coat toughness is another form of physical defense, such as is found in the pericarp of the baobab tree. The baobab weevil is one of the few insects capable of breaching this defense system (Lukasic and Johnson 2007). Another physical defense is resin, which frequently kills eggs and larvae, especially in conifers. For example, *Pinus cembra* L. is unique among European conifers in successfully defending against the pine cone weevil, *Pissodes validirostris* Gyll. by copious resin flow during moist conditions (Dormont and Roques 2001).

Other physical traits may have been selected by herbivory such as the number of seeds per structure, seed size and seed coat thickness (Fenner et al. 2002). For example, the large seed mass and early (i.e. autumn) germination of subtropical oak species, provide some tolerance for partial consumption of the seed without totally preventing germination and even the establishment of a viable seedling (Xiao et al. 2007). The number of seeds per reproductive structure may also be under selective pressure from herbivory. Seifert et al. (2000) found indications of an interaction between the seed quantity in spruce cones and seed insect infestation. On the contrary, no correlation was found between the total number of seeds in a cone and their infestation by the seed wasp *Megastigmus suspectus* Borries (Skrzypczynska 1998).

Lignification of the reproductive structure or the presence of a waxy surface are other physical defenses that may make it hard for herbivores to oviposit inside or at the surface of the structure. Lignification will generally occur progressively during the development of the reproductive structure. For herbivores that oviposit inside the reproductive structure (e.g. endophytic seed feeders, see Sect. 16.2.3.2), oviposition success will be primarily linked to their ability to penetrate through the tissues as they harden. Often, hardening is accompanied by intensive structural changes that render oviposition possible for a narrow temporal window so herbivores have to synchronize their development with the phenology of their hosts.

When physical defenses are breached, it is usually by specialist species, which then allows access to reproductive structures by generalists. For example, in many leguminous trees, a single specialist species can breach physical barriers, which allows entry of multiple non-specialists (Meiado et al. 2013). Thus, the failure of physical defenses can result in a complete defense breakdown.

Selective pressure exerted by reproductive structure herbivores may be countered by selection pressures from a variety of other sources. For example, selection for smaller seed in response to seed herbivores might be countered by the advantage larger seeds provide to the seedling. Physical characteristics of reproductive structures have also influenced the evolution of herbivores. For example, seed size appears to be an important selective agent in the evolution of rostrum size in acorn weevils (Hughes and Vogler 2004).

16.3.1.4   Ontogenetic Defenses

Seed ontogeny can be broadly divided in two groups. In most angiosperms and some gymnosperms (e.g. Cupressaceae), the accumulation of nutrients in the megagametophyte (conifers) or endosperm (angiosperms) occurs only if pollination and fertilization were successful. In other gymnosperms (Pinaceae and Cycadales), the accumulation of nutrients is independent of fertilization and sometimes pollination. Recent histological studies suggest that the endophytic seed feeders of the genus *Megastigmus* synchronize their oviposition to the onset of nutrient accumulation during seed ontogeny (Boivin et al. 2015). More specifically, in a Pinaceae/*Megastigmus* system, the wasp oviposits early during megagametogenesis, a stage at which intense cell death results in the production of a mucilage-like matrix rich in polysaccharides and proteins. In addition to targeting an early phase of the seed development, this seed wasp can redirect ovule development to prevent the degeneration and death of unfertilized ovules and induce the accumulation of nutrients used as food resource for the developing larvae (von Aderkas et al. 2005). In comparison, species parasitizing Cupressaceae lay their eggs later, during embryogenesis. At this stage, a corrosion cavity forms by intense cell lysis of the megagametophyte, creating space for the embryo to grow. The cell lysis provides readily available nutrients to the developing larva. Thus, it appears that these endophytic wasps have evolved towards the use of a plant structure that is a natural sink for nutrients. Moreover, in an interesting and possibly new type of insect-plant interaction, post-fertilization wasps can prevent megagametophyte degeneration and induce differentiation of plant storage tissue even in the absence of fertilization. Extending their observations to a larger group of seed parasitoids, Boivin et al. (2015) suggested that these insect-plant interactions evolved from passive host exploitation when oviposition occurs post-fertilization, to active host manipulation when eggs are laid pre-fertilization, and finally to active host manipulation with creation of new host structures for gall-inducing insects.

In some instances, reproductive structures and their faunistic complement may display parallel successional stages through time. Lukasic and Johnson (2007) found that baobab fruit go through distinct successional stages, with specialist curculionids invading first, using adaptations to breach the defensive pericarp, followed by less specialized lepidopteran larvae, then generalist dipteran larvae as the attacked fruit defenses progressively degrade. *Orsillus maculatus* (Lygaeidae) attacking Mediterranean cypress waits until *Megastigmus* spp. have emerged, then uses their emergence holes for oviposition (Ciesla 2011).

Some forest trees can also defend viable seeds using "decoys". Delayed self-incompatibility in the tropical dipterocarps (see Sect. 16.2.4) means that self-pollinated ovules, which would otherwise be non-viable, are maintained for a period of time. Thus, insect herbivores might target ovules destined to die anyway, thus protecting a higher proportion of viable ovules (Ghazoul and Satake 2009). This theory holds up well when modeled on dipterocarps, and generally holds for any herbivore-satiating mass-flowering forest tree species, but no data exist to verify the models.

16.3.2 Host-Finding and Selection

According to the preference/performance hypothesis (Thompson 1988), if the quality of sites in which an insect lays its eggs influences offspring fitness, natural selection should favor females that oviposit in high-quality sites. In obligate cone-feeding insects, the adult female selects the host, while the larvae must live with her selection (Turgeon et al. 1994). In facultative cone-feeders (inflorescence feeding guild), the larvae can also find and select vegetative structures, thus expanding their potential resource to both reproductive and vegetative structures and reducing the risk of poor host acceptability.

Host-finding is influenced by the spatial structure of forests. In temperate forests with only a few dominant species, insects can move freely between closely spaced hosts, ensuring easy host-finding. This is even more the case in seed orchards where non-host trees are eliminated. However, in tropical forests with a high diversity of tree species, the hosts of any given herbivore species tend to be distributed as islands of widely spaced individuals, separated by a sea of non-hosts. This leads to difficult host-finding and high mortality during the host-finding stage (Janzen 1971).

Cues used in host finding are both long- and short-range. Adults of obligate conifer cone-feeders tend to use whole-tree cues (visual or chemical cues) for long-range orientation, and cone cues (primarily chemical) for short-range orientation (Hulme and Benkman 2002), whereas adults of non-obligate cone-feeders use both whole-tree and cone cues for long-range orientation (Turgeon et al. 1994).

Host finding cues have been explored with many species, and include:

- Chemical cues: These are volatile compounds that can function as long- and short-range cues, and surface chemicals for short-range cues. They are usually attractive, such as the spruce cone terpenes α-pinene, β-pinene, and myrcene that attract *Cydia strobilella* (Jakobsson et al, 2016). However, repellent compounds can also guide long-range orientation. Bedard et al (2002) found that certain repellent non-host aldehydes, alcohols, and (+)-conophthorin also mediate host-finding in the *C. strobilella* spruce system, by repelling females from non-host trees (thus effectively steering them towards host trees). Turgeon et al. (1994) implicated terpenes as being important in long-distance host-finding. Short-distance discrimination of foliage from cones could be mediated by differences in chemicals: in *Picea*, terpinolene and aliphatic acids differ between foliage and cones; in *Pseudotsuga* it is primarily monoterpenes; and in *Pinus*, several sesquiterpenes have been identified. Whether these short-distance potential cues are actually used, though, has not been investigated. Very short distance differentiation between cone bracts and scales can be determined by terpenoid differences (Turgeon et al. 1994), but again it is unknown if these cues are used in host selection.
- Visual cues: These are used when there is a difference between cone and foliage colour. Colour cues are important in some species (e.g. *Contarinia oregonensis* in Douglas-fir, Zahradnik et al. 2012), but not in others (e.g. *Dioryctria abietivorella* in Douglas-fir and *Leptoglossus occidentalis* in western white pine, W. Strong unpublished). It has been suggested that the difference in infrared emissions by

cones and foliage is used by *L. occidentalis* in finding its host (Takacs et al. 2008), though this was later challenged by Schneider (2014).

- Shape and size of reproductive structures: Though this has been implicated in host defense and utilization, little has been found regarding host orientation. Zahradnik et al. (2012) found that branch-shaped (long and thin) but not barrel-shaped (short and wide) silhouettes are attractive to *C. oregonensis*, possibly indicating that branches are used in orientation towards the cone-bearing portions of Douglas-fir.

Host-selection following host-finding requires a means of assessing host quality. Bruchid beetles in tropical legume forests probably select hosts based upon the types of alkaloids and free amino acids present (Janzen 1969). These are very toxic compounds that are present in different types and quantities in different tree species; each species of bruchid has evolved to cope with the chemical suite in a restricted host range. Some of these cues can be co-opted for mate finding: *Cydia strobilella* males (but not females) are frequently caught in traps baited with volatiles specific to their spruce hosts (Jakobsson et al. 2016), suggesting that they use the volatiles to find mates or locations where mates are likely to be.

16.3.3 Temporal Transience

16.3.3.1 Within Season

Reproductive phenology (i.e. the seasonal onset and the duration of the different phases of reproduction) of individuals of the same host species can vary broadly among populations depending on local conditions, suggesting that these traits can evolve rapidly in response to bottom-up selective pressures such as climate and photoperiod. Top-down selective pressures such as biotic interactions may also affect the selection of the onset and duration of flowering period depending on the nature of the interaction (mutualist vs antagonist, Elzinga et al. 2007). Two types of flowering phenology can minimize predispersal seed herbivory. One strategy is to desynchronize flowering with herbivore phenology, i.e. producing fruits before or after seed herbivory peak. The opposite strategy is to apply 'herbivore satiation' (Janzen 1971). In this case, fruits are produced in massive quantities over a short period of time so that the herbivore cannot manage to attack all of them. However, escaping herbivory in time can have negative selective consequences. Flowering might occur during suboptimal periods such as times when fewer pollinators or seed dispersers are available, or times when frost or drought can impact seed production.

In some conifer species, the timing of reproductive structure infestation seems to depend on the lifestyle of the herbivore. Obligate cone-feeders prefer the middle stages of cone development, which have high sugar content and low indigestible fiber levels, while facultative cone-feeders prefer early or late stages, which might have nutritional characteristics more similar to the foliage they also feed on (Roques 1991).

Finally, the timing of attack with regard to the host reproductive phase may affect the outcome. For example, *Megastigmus* species that attack Cupressaceae oviposit only after ovule fertilization, while those that attack Pinaceae can exploit the host before or after fertilization (Rouault et al. 2004). Thus, the Pinaceae group can parasitize a higher proportion of ovules. In another example, the synchrony of acorn production and acorn weevil larval growth has implications for the success of regeneration. When phenological synchrony is good, fewer acorns survive to maturity and canopy recruitment suffers, while asynchrony leads to improved oak reproduction (Munoz et al. 2014).

16.3.3.2 Between Seasons (Masting)

The reproductive pattern of many tree species is characterized by high annual variability and spatial synchrony. This phenomenon, referred to as masting or mast seeding (Silvertown 1980), manifests itself in intermittent and synchronous episodes of abundant reproductive structures (mast years) followed by one or more years of low abundance (non-mast years, Kelly 1994). The mechanisms involved in mast seeding have been debated for many years. According to Poncet et al (2009), three conditions are required for masting to be successful: mast year crops must exceed the consumptive capacity of the herbivores; mast years must be separated by sufficient time to reduce herbivore numbers between mast years; and mast crops must be synchronized over a greater spatial scale than the herbivore dispersal distance. Linhart et al. (2014) clarifies Poncet's first condition as a requirement that the herbivore must follow a Type II functional response, in which the proportion of reproductive structure consumed declines with reproductive structure density, due to satiation.

From an evolutionary point of view, several lines of evidence support the theory that herbivory favored the selection of spatial and temporal variation in reproductive structure dynamics, particularly large-scale reproductive synchrony (masting) that leads to herbivore satiation (Kelly 1994). Models that combine masting and herbivore population dynamics have been used to investigate the role of herbivores in selection for mast fruiting. Surprisingly, even if masting is often considered as the most common strategy to escape herbivory, results show herbivory is not required for masting to evolve. The presence of seedling banks with some seedlings surviving more than one year is required for masting to develop (Tachiki and Iwasa 2013). In the absence of this condition, even strong herbivory does not in itself promote masting. When multi-age seedling banks are available for recruitment to fill forest gaps, herbivory can promote the evolution of masting.

Although the theoretical framework for the role of masting in herbivore satiation is now well developed, there has been limited empirical evidence to support it until the recent publication of several long-term studies. In a 17-year study, Poncet et al (2009) found that European larch cone crop size was synchronized across a wide region in the French Alps. Low seed predation by a species complex of the highly mobile cone-tunneling *Strobilomyia* (see Sect. 16.2.2.2.1), in years of low cone crops following mast years, supported the satiation theory (Fig. 16.2). A high rate of reproductive synchrony was also found in a 29-year study of ponderosa pine (Linhart et al. 2014) (Fig. 16.3), as well as a Type II functional response and higher overall predation (but lower proportion of seeds consumed) in mast years (Fig. 16.4). Lower predation rates on individual trees with more cones in mast years suggested the intensified effect on insects with lower dispersal capabilities. Kobro et al. (2003) found that fruit production in the rowan tree, *Sorbus aucuparia* L. (Rosaceae) was synchronized spatially and temporally over the course of a 22-year study. The apple fruit moth *Argyresthia conjugella* Zeller used apple during years of low rowan fruit production, which reduced the success of masting for rowan trees. In the same system, an earlier study (Sperens 1997) found a type II functional response, with more moths in high fruit years but a lower proportion of damage. However, masting was determined to be less effective because tree populations were small with high local synchrony but poor regional synchrony, and moths are highly vagile.

Other examples of the effect of temporally variable production of reproductive structures on herbivore satiation have been observed in Mediterranean oaks (Espelta et al. 2008), European juniper (Mezquida et al. 2016), Antarctic beech (Soler et al. 2017), Japanese beech (Yasaka et al. 2003), European rowan (Żywiec et al. 2013), Japanese oaks (Fukumoto and Kajimura 2011) and American ponderosa pine (Linhart et al. 2014).

Masting is generally most successful against host-specific obligate cone, fruit, and seed feeders (Hulme and Benkman 2002). Herbivores with a broad host range can feed upon reproductive structures of alternative tree species during non-mast years. Non-obligate feeders can survive low crop years by feeding on vegetative plant parts, such as western spruce budworm on Douglas-fir (see Sect. 16.2.1). Less

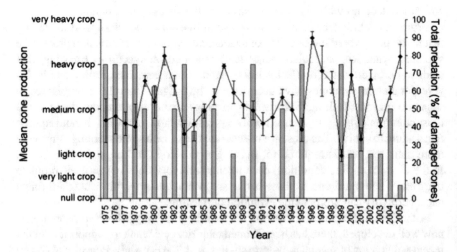

Fig. 16.2 Annual fluctuation of cone production in European larch in the 20 sites studied in the French Alps (*bars*) and of global predation rates by *Strobilomyia* species (*circles*; mean ± SE). In 1987, 1992, 1996, 1997 and 1998, the median larch cone production was nill but enough cones were sampled to calculate a predation rate. From Poncet et al. (2009)

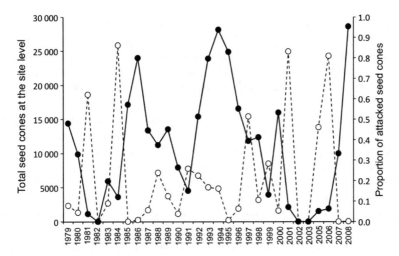

Fig. 16.3 Annual estimates (from 1979 to 2008, except 2004) of the total number of seed cones at the site level (white dots, dashed line and left axis) and proportion of attacked seed cones by specialist insect seed predators (black dots, solid line and right axis). Each point represents the average of 217 ponderosa pine trees. Error bars are omitted for clarity. From Linhart et al. (2014)

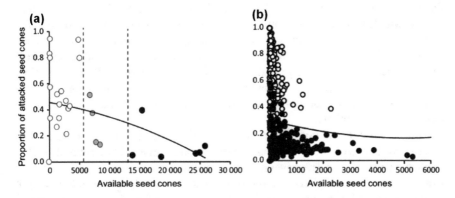

Fig. 16.4 **a** Relationship between the number of available seed cones and the proportion of attacked seed cones by insect seed predators at the population level (white dots for non-mast years, grey dots for intermediate years and black dots for mast years, $r = -0.44$, $P = 0.017$). Each point represents a year ($N = 29$). **b** Relationship between the number of available seed cones and the proportion of attacked seed cones by insect seed predators at the individual level in non-mast years (white dots, $r = -0.11$, $P = 0.117$) and mast years (black dots, $r = -0.20$, $P = 0.003$). Each point represents an individual ponderosa pine tree ($N = 217$). The single lines shows the functional response across all trees in both mast and non-mast years. From Linhart et al. (2014)

mobile insects are more impacted by masting because greater mobility allows insects to find widely dispersed crops, outside of the masting region. This can be particularly relevant in obligate, host-specific seed-feeders with high mobility. The timing of reproductive structure herbivory within a year can also be important in the success of masting. Harris et al. (1996) found that pecan nut casebearer, *Acrobasis nuxvorella* Neunzig, which attacks seeds early in the year, lives at low densities and feeds on the few nuts produced in non-mast years, thus reducing nut density to zero. The pecan weevil (see Sect. 16.2.2.1.2), which feeds in later season, must then deal with intensified masting effects (i.e. very low nut availability in non-mast years).

Forest structure is important in the success of masting as well. Tropical forests with widely separated individuals of a species can use masting successfully against obligate specialists because of the difficulty of locating and moving to another individual of the same species (Janzen 1971). Tropical generalist insects are less affected by masting because of the ability to use other tree species or plant parts. Masting in temperate monoculture forests is often less effective, particularly against insects of high mobility, because of the ease of finding and moving to non-masting hosts, perhaps outside the masting area. In this way, as host density increases, herbivore satiation becomes less effective.

Herbivore satiation in the absence of masting is another defensive mechanism. Dry tropical *Acacia pennatula* (Chamb. and Schltdl.) Berth. produce abundant seed in mid-altitude areas where seed production is greatest and energetically least costly; this satiates seed-feeding bruchids (*Mimosestes* spp.) and ensures the preservation of some seeds (Peguero et al. 2014). On the other hand, in high-altitude areas where seed production is more energetically expensive, *A. pennatula* uses seed abortion: the tree aborts seeds upon bruchid oviposition, thus allowing the diversion of energy to uninfested seeds, and decoying the bruchids into a reproductive dead-end, thereby reducing insect numbers.

Janzen (1969) found that different dry tropical *Acacia* spp. responded to feeding by bruchids in one of two ways. Some species produce more total weight of smaller seeds that have reduced defensive chemicals, resulting in herbivore satiation allowing a few escapes despite high seed mortality. Other species produce a lower total weight of larger seeds, with moderate or high levels of defensive chemicals, resulting in low seed mortality and a higher percentage of escapes. Herbivore satiation is aided by spatial discontinuity of the trees, thus favouring high production of small seeds. Satiation is hindered by a greater number of bruchid species feeding on one tree species, which encourages lower production of well-defended seeds.

16.3.4   Evolutionary Consequences at the Host Species Level

While reproductive structure herbivory has considerable impacts on host seed output (Crawley 1989) it is less clear how strong a selective force it is in shaping host tree reproductive and non-reproductive traits. In general, trees produce far more immature ovules than can possibly mature, the loss of some to herbivory might not affect host fitness. Even if the host's capacity to produce seeds is reduced, other factors may compensate, such as an increase in maternal resources or a decrease in seedling competition. To affect evolutionary trajectories, reproductive structure herbivory has to meet the following conditions: (1) herbivory rates have to consistently relate to certain plant phenotypes (e.g. plant height, shape or color, flowering phenology, number of flowers, seed size), (2) these phenotypes have to be heritable, (3) plant trait-fitness relationships have to change as a result of herbivory, (4) the magnitude of the herbivory has to be large in comparison to the sensitivity of overall plant population performance to changes in seed production. Although the data are still sparse, the overall pattern of reproductive structure herbivory suggests that the main conditions for selection are present, and a growing number of studies confirm the selective forces of reproductive structure herbivores and their role in the evolution of traits (Hulme and Benkman 2002; Kolb et al. 2007).

Assuming that herbivory meets the conditions listed above, it has likely selected for plant traits that tend to minimize the negative effects of seed loss and for plant tolerance mechanisms that do not preclude seed consumption but reduce seed loss among seed crops. In a review of the ecological literature, Kolb et al. (2007) found that the most common traits affected by seed herbivory are the number, morphology and size of reproductive structures, flowering phenology, plant size and flower number. The strength and the direction of the relationship between a plant phenotypic trait and reproductive structure herbivory is generally species dependent. For example, seed herbivory was related to flowering phenology in 80% of the species reviewed by Kolb et al. (2007) but the direction of selection depended on the species. In some cases herbivory selected for early flowering, in other cases for later flowering. Conversely, since the generation time of reproductive structure herbivores is generally much shorter than their host, it is expected that the herbivore will likely be able to track the evolution of host traits.

16.3.5   Evolutionary Consequences at the Community Level

Associational resistance exists in many plant species. This is the close spatial association of hosts with non-hosts, which increases volatile diversity and reduces the host-finding ability of pests (Barbosa et al. 2009). Though this is a common phenomenon in general plant resistance, no papers concerning this mechanism were located, which are specific to insects of reproductive structures.

Reproductive structure herbivory can play an important role in the dynamics of plant populations by limiting seed production. At the community level, inter-specific differences in reproductive structure herbivory may influence the relative recruitment and ultimately the abundance of co-existing species. Specifically, the coexistence of tree species sharing the same habitat may be maintained by differential temporal or spatial variations in reproductive structure herbivory that counteract competitive exclusion (Hulme 1996). For example, Espelta et al. (2009) studied pre-dispersal acorn herbivory of two co-occurring weevil species on the Mediterranean oaks, *Quercus ilex* L. and *Q. humilis* Miller, and compared the relevance of this herbivory to that of other processes involved in recruitment. Herbivory significantly contributed to inter-specific differences in recruitment relative to other factors such as post-dispersal herbivory and germination. Herbivory rate cannot be used as a surrogate for the effects on plant fitness or population dynamics as there might not always be a direct link between seed herbivory, seed abundance, and recruitment (Kolb et al. 2007). In some cases, high seed herbivory rates may even enhance seedling survival by reducing post-dispersal intraspecific competition (Halpern and Underwood 2006).

Among all sources of herbivory, reproductive structure herbivory can be of particular importance because of its direct and obvious impacts on plant fitness and often strong effects on recruitment patterns. The Janzen-Connell hypothesis illustrates this concept: the high diversity of tropical rainforest trees is explained by the spatial variation in species-specific seed and seedling herbivores (Janzen 1970; Connell 1971). Under this hypothesis, insect herbivores of reproductive structures may promote the stable coexistence of different tree species because these insects are often specialized on a single host species and are more prone to depress recruitment of locally abundant species, thus giving advantage to rare species. The Janzen-Connell hypothesis also states that for reproductive structure herbivores to play a role in the maintenance of biodiversity, they must cause positive distance- or density-dependent mortality (i.e. mortality increases with host density). These conditions are in direct opposition to the herbivore satiation hypothesis that requires negative density-dependent mortality. However, both processes may act simultaneously at different scales. In an investigation of East Asian oak, *Quercus serrata* Thunb. ex Murray acorn herbivory, Xiao et al. (2016) observed that overall herbivore satiation limited the occurrence of Janzen–Connell effects but also that the direction and magnitude of density-dependent seed herbivory by host-specific insects differed between individual tree and tree population scales.

Reproductive structure herbivores that are not species-specific may also link the dynamics of host species that are not otherwise competing for resources. This form of herbivore-mediated interaction is known as apparent competition (Holt and Lawton 1993), and occurs when different host species share the same herbivore. Shared reproductive structure herbivores may especially reduce the occurrence of congeneric host species at close proximity as those are more likely to host shared herbivores (Lewis and Gripenberg 2008).

16.4 Diversity in Insect Strategies and Community Structures

With a rich and varied resource comes many strategies for its exploitation by insects. Competition, plant protective strategies, and spatial and temporal heterogeneity have led to distinct structuring of the insect community.

16.4.1 Species Diversity

A wide range of insect species have specialized on reproductive structures of trees. Sam et al. (2017) reared 122 species of Lepidoptera alone from the fruiting structures of 326 woody plant species. In general, host specificity was low: 69% of species attacked hosts from >1 tree families, and only 17% were monophagous. Each kg of fruit contained an average of 0.81 generalists, and only 0.07 specialists (defined as feeding within a single host genus).

Most of the community diversity studies have been conducted in the tropical Dipterocarpaceae (see Sect. 16.2.4). In Southeast Asia, dipterocarps are strictly masting host species, leading to reduced fruit herbivore species diversity (Hosaka et al. 2009). Two nanophyid weevils emerged only from immature fruit; two *Alcidodes* weevils emerged only from mature fruit, and a single *Andrioplecta* (Tortricidae) moth species was found in all stages of fruit. All five insect species were found in all dipterocarp species studied. This helps explain the evolution of synchronous mast events among congeneric host species, because if all dipterocarp species mast in the same year, then food resources are minimal across the landscape in non-mast years, thus effectively limiting populations of seed feeding insects.

Dipterocarps in Borneo, on the other hand, support a richer community of reproductive structure herbivores. Nakagawa et al. (2003) found 51 species of insects feeding on reproductive structures of 24 dipterocarp species. Herbivores were grouped into "smaller moths" and scolytids, including weevils. Feeding was non-specific, with abundant overlap in host ranges, and the dominant herbivores were not consistent among host species or among years (Fig. 16.5).

Lyal and Curran (2000) examined *Alcidoides* weevil associations in 70 species of dipterocarps throughout Asia. Though many weevil species feed on a range of host species, and up to five weevil species were found in a single host species, no weevils fed on sympatric congeneric hosts. Because of this, mast fruiting can be successful, and it is found almost universally among the Dipterocarpaceae.

Looking at the community structure of acorn weevils in oaks, Govindan and Swihart (2015) found that species richness and community similarity were highest when mast production of three host tree species were in phase. Multispecies, multiseason models show that differential suitability of hosts as resources for *Curculio* created a spatial storage effect that, when coupled with a temporal storage effect induced by prolonged diapause common among *Curculio*, facilitated species coexistence.

16.4.2 Host Resource Partitioning

Insects that feed on reproductive structures are often limited by their host resources. Competition in seeds is frequent because seeds are small, finite, and of high nutritional quality (Janzen 1971). With most plant hosts, seeds are less prone to high levels of

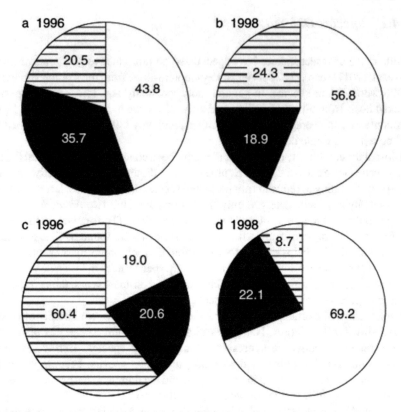

Fig. 16.5 Porportions of species (a, b) and individual numbers (c, d) of three major taxonomic groups of insect seed predators: smaller moths (horizontal bars), scolytids (solid black), and weevils (solid white), during general flowering and seeding events in 1996 and 1998. From Nakagawa et al. (2003)

herbivory after dispersal. When the bulk of herbivory is on pre-dispersal seeds, hosts tend to phenologically advance dispersal (Janzen 1971). For example, Douglas-fir, whose seeds are heavily fed upon by *Leptoglossus occidentalis* Heidemann and *Dioryctria abietivorella* Groté (see Sects. 16.2.2.1.1, 16.2.3.1), sheds seeds as soon as they are mature, while black spruce, which has few pre-dispersal pests, sheds its seeds over a protracted time period after maturity.

Phenological adjustment is not the only plant response to insect feeding pressure. Many plants have evolved phytochemicals to deter herbivory (seed escape through toxicity). These phytochemicals differ among plant structures and change as the season progresses. Thus, reproductive structure herbivores can respond by specializing, essentially partitioning the habitat in time or space, allowing a herbivore species to deal with a subset of the plant's defenses. One consequence is that habitat partitioning helps spread risk of dependence on a limited resource (Janzen 1971). Overexploitation in these situations tends to lead to contest-type competition (Atkinson and Shorrocks 1981), where multiple individuals infest a single resource unit, such as a seed, but only a single individual emerges alive. If on the other hand plants evolve high seed volumes and masting systems (seed escape through satiation), there tends to be an overlap of resource partitioning, and an increase in scramble-type competition. In scramble competition, resources are shared by all individuals, so that if resources are limiting, every individual gets less than optimal resources, leading to undersized pupae and adults, or death of all individuals in extreme cases (Atkinson and Shorrocks 1981).

Spatial division of host resources often results in a uniform, rather than random or aggregated, distribution of insects. Quiring et al (1998) found that larvae of the spruce cone fly *Strobilomyia neanthracina* Michelsen have a non-random (uniform) distribution among cones, thus reducing competition when uninfested cones are still available. This was found to be mediated through a host-marking pheromone applied by the adult female's mouthparts after oviposition. With *Bruchidius dorsalis* (Fahraeus) feeding on seeds of the Japanese honey-locust *Gleditsia japonica* Miquel, if the number of eggs is less than the number of seeds, larvae search for seeds not previously infested, giving a relatively uniform larval distribution (Shimada et al 2001). If the number of eggs exceeds the number of seeds, up to 10 bruchid larvae can feed on a single seed, but usually only one adult emerges, with no decrease in body size. This cannibalistic contest competition ensures some survival despite limited seed resources.

Studies on the bruchid beetle *Callosobruchus maculatus* (F.) have shown that two evolutionarily distinct strains exist. "S" strain adults lay only one egg per seed, and larvae resort to contest competition, whereas the "I" strain adults lay >1 egg per seed, and larvae display scramble competition (Messina 1991). These strains are genetically determined, having evolved as a consequence of the long association of the bruchid with a small-seeded host. It was found that the S strain outcompetes the I strain unless I strain larvae have a 2-day head start. Neither form can be completely overtaken by the other in competition models, a result that was confirmed and explained in the following subsequent studies. Toquenaga (1993) showed that the type of competition was density dependent, and which strain predominated depended on seed size. Tuda and Iwasa (1998) further studied the system in the laboratory and found that the initial laboratory population engaged in scramble type competition. As the population density increased, individuals with contest-type traits appeared. In a seed with multiple larvae, if one contest type was present, only it emerged as an adult, if no contest-types were present, all the scramble types emerged. After 20 generations the system stabilized: under abundant resources, the scramble type predominated, while under limited resources, the contest type was selected for. These studies show that contest and scramble competition are coexisting genetic traits within a population, and this plasticity can lead to superior overall exploitation of variable host resources.

Resource partitioning occurs not just within an insect species, but also between species. Meiado et al. (2013) found that insects feeding on the fruit and seeds of *Enterolobium* (Leguminosae) trees partitioned the resource spatially and temporally. Some species fed on the fruit, others on the seed; species that fed on only one structure attacked at discrete and separate points of time. In chestnuts, the weevil

Curculio elephas (Gyllenhal) avoids nuts previously attacked by the moth larva *Cydia splendana* (Hübner), while *Cydia* does not avoid nuts previously attacked by *Curculio* (Debouzie et al. 1996). This is very apparent on the scale of individual nuts, less so on the scale of husks (containing two nuts), not at all at a whole-tree scale or larger. In pecan, there exists a temporal division of resources among two seed-feeding species (Harris et al. 1996). The nut casebearer *Acrobosis nuxvorella* Neunzig kills young nuts, while the pecan weevil *Curculio caryae* (Horn) feeds on mature nuts later on in the season. Early feeding by the casebearer can eliminate the entire crop in a light year, which increases masting effects on the later-feeding weevil. Thus, the weevil is faced with a more variable resource environment than the casebearer.

16.4.3   Host Specificity

Insects that feed on tree reproductive structures display a range of host specificity, from generalists that feed on multiple structures and multiple host species to specialists that feed on only a single structure, with a very limited range of host species. In European conifers, about one third of cone-feeding insects can also feed on other plant structures such as foliage; the rest are obligate cone-feeders (Roques 1991). Of the generalists, most feed on cones only occasionally, when other structures are limited. These generalists also have a wider host species range than the obligate cone-feeders. None of the obligate cone-feeders is host-specific to the species level. *Megastigmus* spp., the seed chalcids, are the most host-specific group perhaps because the larva is encased in the seed capsule, in intimate physiological and hormonal contact with the ovule (von Aderkas et al. 2005). Such conditions are ideal for speciation and the development of host-specificity. Even so, different sympatric host species are often occupied by a single *Megastigmus* species, and allopatric hosts brought into contact with European hosts are attacked by European *Megastigmus*. So, there do not appear to be any examples of host specificity to the species level in cone-feeding insects, at least among European conifers.

   Host specificity can arise as a result of the seed escape strategy used by tree species, including escape by satiation (in which the host produces massive crops of seeds (masting, see Sect. 16.3.3.2) that overwhelm herbivore consumption) and seed defense, both physical and chemical (see Sects. 16.3.1.2, 16.3.1.3). In general, escape by herbivore satiation leads to lower host specificity. For example, African acacia tree species with large-seeded indehiscent pods (a large stable resource) harbour more bruchid species than tree species with small-seeded dehiscent pods (a small ephemeral resource) (Miller 1996). In frugivorous lepidoptera larvae of Papua New Guinea, Sam et al. (2017) found that tree species whose fruit had large seeds and a thin mesocarp hosted generalists only, while tree species with small seeds and thick mesocarp hosted both generalists and specialists. Janzen (1969), studying legume/bruchid systems in Central America and Kansas, found that trees fell into 2 groups: (a) small-seeded, with many seeds/tree and few chemical/physical defenses; (b) Large-seeded, with few seeds/tree, but well-defended chemically and physically. The first group has a wide complex of bruchid beetles attacking them, while the second group has none. There is likely a self-reinforcing coevolutionary shift leading to the divergence of these two plant groups, resulting in each tree strategy, with its guild of seed feeders, being evolutionarily stable.

Conifer cone size and masting can also play a role in host specificity. Red pine *Pinus resinosa* cones are too small for most *Conophthorus* cone beetle species, but red pine cone beetle *C. resinosae* lays smaller clutches of eggs on more cones so can use the smaller cones of red pine (Mattson 1980). Red pine is also a masting species, which further limits species diversity, but *C. resinosae* can oviposit, and larvae develop, on shoot tips in years of low cone density. This adaptation allows *C. resinosae* to specialize on red pine, while other *Conophthorus* species cannot (see Sect. 16.2.2.1.1).

The chemical defenses (see Sect. 16.3.1.2) that can lead to specificity include attractants, deterrents, toxicants, and feeding stimulants (Janzen 1971). The evolution of mechanisms in reproductive structure herbivores to overcome these defenses is uncommon and specific to the chemical defense, leading to a limited number of insect species capable of exploiting a narrow range of hosts: in other words, specialists. For example, bruchid beetles have evolved resistance to the alkaloids and free amino acids of many legume tree species that are toxic to generalists. The basis of this host specificity may be bacterial symbionts that break down the toxins.

The toxicity of non-seed structures can also be a source of host selection for seed-feeding insects. Tuda et al. (2014) found that seeds of the Mimosa tree *Leucaena leucocephala* (Lam.) de Wit are eaten by the bruchid beetle *Acanthoscelides macrophthalmus* (Schaeffer). However, the bruchid will not eat seeds of the related pigeon pea (*Cajanus* spp.) even though its chemical composition is very similar to the mimosa. The basis of host selection in this case is the seed pod, which in pigeon pea contains repellents and toxicants that deter *A. macrophthalmus*.

Less commonly, host specificity can be caused by adaptation to factors other than the host tree reproductive strategy. Conifer seed chalcids, *Megastigmus* spp., often specialize on hosts based on the phenology of their reproductive structure development (Rouault et al. 2004) (see Sect. 16.3.3.1). These authors also found that chalcids infesting the Pinaceae can oviposit in unfertilized ovules, whereas those attacking the Cupressaceae cannot. These differential abilities have evolved in response to different ovule physiologies between the plant families, leading to host specialization.

16.4.4 Extended Diapause

Diapause is a means of surviving periods of low resource availability (e.g. winter). When host resources are low for a variable number of years, it can be beneficial for insects to extend their diapause by one or more years. According to Soula and Menu (2003), "[Extended] diapause allows individuals to survive when conditions are unfavorable for development and reproduction, and ensures synchronization of active stages with favorable conditions. However, diapause is associated with both metabolic (consumption of energetic reserves without feeding) and reproductive (missed reproductive occasion) costs." Therefore, the benefits of extended diapause must exceed costs for it to evolve. In other words, overall fitness must increase due to extended diapause or it will be selected against.

The benefits of extended diapause are greatest when crops of reproductive structures are temporally non-uniform. Hanski (1988) found that conifer seed feeders have evolved extended diapause on host species with highly variable cone crops, but not on host species with more temporally uniform cone crops. For example, *Megastigmus spermotrophus* in Douglas-fir (which has highly variable seed production) makes use of extended diapause (Roux et al. 1997), while *M. specularis* on *Abies sibirica* (which has uniformly moderate crops through time, and virtually no crop failures) does not.

Diapause that is highly synchronized with host masting might mitigate the effects of masting on herbivore populations and increase the proportion of the resource consumed in mast years (Janzen 1971). However, synchronization can be difficult for herbivores to achieve, particularly with host species (e.g. some oaks and conifers) that initiate flower development 2–3 years prior to seed maturation. In tropical forests, synchrony tends to be reduced due to the high species diversity of trees and a lack of environmental synchronizing cues, though local or seasonal synchrony might develop. Larval densities typically do not affect diapause length, though inverse correlations between larval density and current and subsequent cone crops can be strong (Turgeon et al. 1994).

Researchers have described several kinds of extended diapause. Soula and Menu (2003) identified three evolutionary models, while Hanski (1988) described four different kinds of extended diapause in insects. All seven models can be classified to two general types: stochastic (in which the insect responds to cues that do not influence tree reproduction) and predictive (in which the insect responds to the environmental cues responsible for initiating tree reproduction) (Menu and Debouzie 1993). Turgeon et al. (1994) found that 55% of obligate conifer cone-feeders in Western Europe are capable of extended diapause. The majority use the stochastic type.

While stochastic extended diapause may not accurately predict future host resources, it can still be an effective means of spreading risk. For example the chestnut weevil *Curculio elephas*, which diapauses in the ground, has been shown to follow the stochastic type of extended diapause (Menu and Debouzie 1993; Soula and Menu 2003). After 2 years, 32% of individuals had emerged, and 56% had emerged after three years. Poor nut years, with near-zero insect recruitment, have almost no effect on subsequent insect populations, illustrating the effectiveness of stochastic extended diapause.

Stochastic diapause is also found in other trophic levels. Parasitoids of cone-feeding insects often undergo extended diapause in response to population variation in their host insects (Turgeon et al. 1994). Kobro et al. (2003) examined two fruit-feeding insects of rowan, *Sorbus aucuparia* L. Both the fruit moth *Agyresthia conjugella* Zeller and the seed chalcid *Megastigmus brevicaudis* Ratzeburg have

Table 16.1 Pooled number of seed predators and their parasitoids hatched from rowanberries in years after harvesting. From Kobro et al. (2003)

Number of years	1	2	3	4	5
Argyresthia conjugella Zeller	4873	135	20		
Microgaster politus Marsh	891	155	1	42	
Megastigmus brevicaudis Ratzeburg	1986	1809	442	106	4
Torymus aucuparia (Rodzinako)	101	196	222	57	2

Approximately 3–5 kg from each of 6 years

extended diapause that most closely fit the stochastic model. Each of their main parasitoids also displays stochastic extended diapause (Table 16.1). Such multi-trophic effects have not been found in predictive diapause.

Predictive extended diapause results in better synchrony between insect abundance and host resources, but well-described examples are less common, and the cues are typically poorly characterized. One example is *Curculio* weevils feeding on acorns in Japanese oak forests. Maeto and Ozaki (2003) report that the oaks mast every second year, and the weevils have a two-year extended diapause that is synchronized with acorn production. Predictive diapause thus allows use of an abundant resource in mast years, and avoids starvation in non-mast years. However, a failed mast event (e.g. if a late frost kills the acorn crop) can result in much lower weevil recruitment and crop damage two years later.

Some species apparently are able to use both predictive and stochastic methods, described as the stable genetic polymorphism of Menu and Debouzie (1993). The seed chalcid *Megastigmus pseudotsugae* displays prolonged diapause up to five years. Emergence after the first year of diapause is strongly correlated with the size of the seed crop that year. Because the Douglas-fir host crop size is highly variable, this component of diapause is best described as predictive. Those individuals that do not emerge after the first year emerge randomly in the following years. The proportion of chalcids that remain in diapause for up to five years are best described as stochastic. Thus, induction of prolonged diapause can be multi-factorial.

16.4.5   Natural Enemies

Due to their cryptic nature, insects of reproductive structures host few predators, but a variety of parasitoids. Parasitoids might find cryptic hosts by using a complex of volatiles, which change when reproductive structures are infested by insects (Turgeon et al. 1994). About two-thirds of Western European conifer cone-feeding species host natural enemies (Roques 1991); this is partly determined by how cryptic their larvae are. For example, when infesting shoots, *Rhyacionia bouliana* (Schiff.) has twenty species of natural enemies, but when infesting cones, only three species were found. On all cone-feeders examined in this study, sixteen families of natural enemies were found, with the parasitoid families Ichneumonidae, Braconidae, and Pteromalidae representing 76% of parasitoid species.

Cone and seed insects of Canadian tamarack, *Larix laricina* (Du Roi) K. Koch, were surveyed by Prévost (2002). One hundred percent of the cones were infested, yielding only a single viable seed per cone on average. This was despite the presence of six predators and parasitoids, indicating that natural enemies did not provide effective control of cone and seed insects. In Norway spruce in Poland, ten cone-feeding insect species were also found, and the complex had an elevational gradient (Koziol 2000). In this system some cone insect densities were reduced by natural enemies: the cone axis moth *Cydia strobilella* L. numbers were reduced by 12.3% by five ichneumonoid species; and the scale gall midge *Kaltenbachiola strobi* (Winn.) were reduced by 22.6% by six chalcidoid species. The influence of parasitoids also varied with elevation (Fig. 16.6).

Refugia play a role in natural enemy biology. Hosts can escape natural enemies by oviposition location. The cypress seed bug, *Orsillus maculatus* (Feiber), oviposits preferentially in exit holes of the cypress seed chalcid *Megastigmus wachtli* Seitner (Rouault et al. 2007). When not available, it lays eggs under scales. Eggs laid under scales are attacked more frequently by the seed bug parasitoid *Telenomus* gr. *floridanus*. Oviposition in exit holes is an adaptive strategy to escape parasitoids. Refugia also provide shelter for natural enemies. A proportion of pods of a Chilean *Acacia* tree persist over the winter; these persistent pods provide a refuge for four bruchid seed beetles (Rojas-Rousse 2006). In turn, these beetles become a refuge for four parasitoids, thus providing a mechanism for parasitoids to survive *Acacia* masting events.

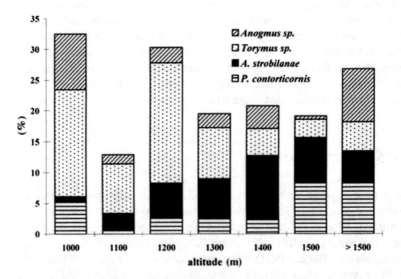

Fig. 16.6 Percentage of species from Chalcidoidea in the parasitization of *Kaltenbachiola strobi* (Winn.) at different altitudes above sea level. From Koziol (2000)

16.5 Implications for Management

16.5.1 Seed Collection in Natural Stands and Seed Orchards

16.5.1.1 Seed Collection in Natural Stands

From the early 1900's until the advent of tree improvement programs in the late 1950's, foresters collected seeds from natural stands with little concern about seed quality (Wakeley 1935). As demand for seeds intensified, seed collection areas—natural stands with undesirable phenotypes removed—provided quality seeds. Seed collection areas are chosen and managed to promote seed crops (Rudolph 1959). Collection from natural stands and seed collection areas is still the primary source for tree seeds in tropical ecosystems (Schmidt 2000). In intensively managed temperate systems, seed orchards are the primary source of seeds.

16.5.1.2 Seed Orchards

A seed orchard is a plantation of clones or progenies from selected trees. It is isolated to minimize pollination from outside sources and managed to provide frequent, abundant and easily harvested seed crops (Rudolph 1959; Zobel and Talbert 1984). Seed orchards are also used for breeding and research by forest geneticists. The intensive management includes management of insect pests and effective pest management is critical to successful orchard production (Zobel and Talbert 1984).

16.5.2 Why Management is Necessary

Management of insects of reproductive structures is seldom done except in seed orchards, seed collection stands and nut-production orchards. These insects can be devastating to seed orchards in particular. Seed orchards typically are small in area and often isolated from general forest stands. Seed orchards are monocultures intensively managed to produce large seed crops each year; this consistent, abundant and nutrient-rich resource is readily taken advantage of by opportunistic seed-feeding insects (Whitehouse et al. 2011). Genetically improved orchard trees represent years of breeding investment and resource managers want the greatest possible production of improved seed from every crop. The threshold for insect damage is low; consequently, insect pest management is necessary (Turgeon et al. 1994; Mangini et al. 2003). The situation is complicated by the fact that, for many species, the damaging stages, usually larvae, are hidden inside the cone or nut and are safe from external manipulation such as foliar pesticide application.

16.5.3 Integrated Pest Management for Insects of Reproductive Structures

Historically, seed and cone insects were controlled with routine applications of synthetic chemical insecticides with complete elimination of all insect pests as the goal. Pesticide applications included chlorinated hydrocarbons, organophosphates and carbamates, often at very high rates. For example, pine seed orchard managers in the southern United States used the organophosphate azinphos-methyl (Guthion®) to control coneworms, *Dioryctria* spp., and the leaffooted pine seed bug, *Leptoglossus corculus* (Say). Carbofuran (Furadan®), a systemic carbamate, was applied to soil under the trees. Both were effective but very toxic to non-target insects, birds and mammals.

16.5.3.1 What is Integrated Pest Management

Fortunately, resource managers have adopted an ecologically compatible approach to pest management for production of conifer and hardwood seeds and nut crops; this approach, Integrated Pest Management, can be concisely defined:

A pest management program employing the optimal combination of control methods to reduce and maintain a pest insect population below an economic threshold, with as few harmful effects as possible on the environment and other non-target organisms, and is based on: (1) the amount of damage that is tolerable, (2) the cost of reducing this damage to the acceptable level, and (3) the impact of the management on the environment (modified from Borror et al. 1989).

Integrated pest management aims to reduce injury to an acceptable level, not to eliminate the pest completely (which is rarely possible). Integrated pest management requires knowledge of the pest species present, the level of damage caused, a defined economic threshold (the number of pests that will result in unacceptable economic loss), and consideration of all possible management options (Coulson and Witter 1984; Turgeon and de Groot 1992). Integrated pest management considers the evolved interactions of host-plant and pest insect (see Sect. 16.5.5).

16.5.3.2 Integrated Pest Management Concepts

Integrated pest management requires a detailed knowledge of how the life cycle of the insect interacts with host-tree reproductive phenology. This allows the manager to focus management efforts on the most vulnerable stage in the life cycle. Small, early-instar larvae are delicate and can be killed by a very small amount of insecticide. Any life stage exposed outside the fruit, cone or seed is more susceptible than those hidden inside these structures. Exposed over-wintering stages, for example pupae or adults in fallen cones, are susceptible to control measures (see Sect. 16.5.3.5).

In managed seed orchards, it is necessary to determine if control efforts are needed and justified. Monitoring is crucial (see Sect. 16.5.3.3). This begins with estimating the impact of pests on seed yield. Historical seed yields, the estimated crop size and the value of the seeds are assessed to estimate damage thresholds based on balancing cost of control with value of the crop. The appropriate management efforts can then be taken. An effective integrated pest management strategy considers all pests present. Sometimes the methods prioritized to control the most damaging species will, by default, also control other minor pests; for example, insecticide treatments for *Dioryctria* species also impact seed bugs, *Leptoglossus* species (Hanula et al. 2002). All methods of control are considered including cultural, sanitation, biological and chemical (Turgeon and de Groot 1992).

Insecticides are still used when necessary but with proper planning and timing of applications (Turgeon and de Groot 1992). Integrated pest management also takes advantage of modern insecticides. New biopesticides, with low avian and mammalian toxicities, can be applied at relatively low rates. Growth regulators mimic insect hormones and impact only certain groups, for example, Lepidoptera. Biological agents, such as the bacterium *Bacillus thuringiensis*, are effective against Lepidoptera (Rosenberg and Weslien 2005).

When insecticide applications must, on occasion, be applied, the type of herbivory and the mode of action of the insecticide together determine efficacy. Contact insecticides, such as the synthetic pyrethroids, need only be touched by the insect to be effective; the material is absorbed through the cuticle. The growth regulators and biological agents typically must be ingested by the feeding insect; they are effective against lepidopteran larvae with chewing mouthparts that readily consume treated material. Since many insects of reproductive structures feed inside the cone or fruit, systemic insecticides, those transported through plant tissue, are potentially effective (Grosman et al. 2002; Cook et al. 2013). For example, emamectin benzoate, injected into pine trees, is effective for several years against *Dioryctria* spp. However, it is less effective against seed bugs, *Leptoglossus* spp., presumably because the material does not pass through the seed coat into the megagametophyte. Bugs, using their piercing-sucking mouthparts, are not exposed to an effective dose (Grosman et al. 2002).

16.5.3.3 The Importance of Monitoring

Monitoring of the crop and insect population is essential for effective integrated pest management in seed orchards and seed collection stands; it is an inventory control process that allows one to know the size and health of the cone crop as well as the impact of insect pests (Turgeon et al. 2005). Other benefits include identification of good and poor crop trees, identification of insect pests present, and estimation of efficacy of control measures if implemented (Turgeon and de Groot 1992; Turgeon et al. 2005). Monitoring protocols for insect damage, using traditional life-table methods, have been used for years (Fatzinger et al. 1980; Bramlett and Godbee 1982). Cones are chosen, tagged and examined at intervals as they develop. Cone condition is recorded, and crop health is determined. These protocols have been formalized into a computer-based system by the Canadian Forest Service; this includes a decision support system for managers (Turgeon et al. 2005). Pheromone-baited traps allow estimation of some insects, especially Lepidoptera (De Barr et al. 1982). Monitoring efforts must consider the extended diapause that occurs in many species (see Sect. 16.4.4).

16.5.3.4 Pesticide Application Timing

Control measures must be timed to coincide with the most susceptible host stage of the insect. This is critical for efficacy and economy, but timing is also very difficult. Insect physiology is temperature dependent; life cycle events vary with the weather from year to year (see Sect. 16.3.3.1). In the past, managers made insecticide applications at routine intervals, thus ensuring exposure to any potential pest throughout the season. Now, applications can be timed to target the susceptible stage and reduce pesticide use. For example, insecticides are often aerially applied four to six times per season to control the insect complex of loblolly pine seed orchards in the southern United States, particularly *Dioryctria amatella* (Nord et al. 1985; Lowe et al. 1994). Hanula et al. (2002) developed a degree-day model that predicts hatch times of proportions of the *D. amatella* spring-generation egg population, based on previous studies of development (Hanula et al. 1984, 1987). Two model-timed applications of the insecticide fenvalerate were as efficacious as four monthly applications for not only the coneworm but also for the seed bug, *Leptoglossus corculus*.

16.5.3.5 Environmental Manipulation

The cone beetles *Conophthorus coniperda* (Schwartz) and *C. resinosae* Hopkins are major pests of white pine, *Pinus strobus* L., and red pine, *P. resinosa* Ait., respectively. Both cause significant loss in seed orchards (see Sect. 16.2.2.1.1). *Conophthorus coniperda* adults overwinter in infested cones on the ground; *C. resinosae* overwinters in pine twigs that have fallen to the ground (Drooz 1985). Low-intensity prescribed burns in early spring kill these overwintering beetles before they emerge, resulting in reduced cone beetle attacks (DeBarr et al. 1989; Miller 1978).

16.5.3.6 Continuous Improvement of Methods

Research and new technology continually refine integrated pest management methods. Traditional management for *Contarinia oregonensis*, the Douglas-fir cone gall midge, requires insecticide application during the brief interval from conelet scale closure to conelets becoming horizontal (Hedlin 1961; Morewood et al. 2002). To assess the need for treatment, a sampling protocol using the number of egg-infested scales per conelet as an estimator of damage at harvest was developed (Miller 1986).

The method is labor-intensive and must be timed precisely and done quickly for optimal timing of treatments (Morewood et al. 2002). Gries et al. (2002) identified the sex pheromone of female *C. oregonensis*. Catches of male midges in pheromone-baited traps proved as effective as egg counts. Traps are species-specific, inexpensive and easily deployed (Morewood et al. 2002). However, pheromone-baited traps are indirect estimators of female abundance. Zahradnik et al. (2012) found that adult midges were attracted to specific infrared radiation and traps constructed to emit this radiation attract both males and females, providing a better estimate of midge numbers.

16.5.4 Seed Loss Versus Extractability

In conifers, insect feeding results in aborted conelets, cones killed or partially killed before maturity, and empty seeds. Healthy seeds that cannot be extracted from insect-damaged cones can also be a source of loss. When kiln-dried and tumbled, the scales of deformed or resin-encrusted infested cones remain closed and the seeds are not shaken out. In other instances, gall tissue or resin may fuse seeds to the scales such that they cannot be removed even if the scales open. Pre-harvest damage assessments can be done to avoid harvest of such cones (Kolotelo et al. 2001; Turgeon and de Groot 1992).

Many coarse internal feeders deform and partially kill cones. The ponderosa pine coneworm, *Dioryctria auranticella* (Groté), often causes twisted cone scales that do not open to release seed (Hedlin et al. 1981). The seedworms, *Cydia* spp., cone tunnel makers, typically do not consume every seed in a cone; however, they lay down silken threads along their feeding tunnels. This silk fuses seeds to the scale and holds them in the opened cone (Turgeon and de Groot 1992). The cone resin midge, *Asynapta hopkinsi* Felt (Cecidomyiidae), a cone scale feeder, causes distorted scales and flakes of resin that may impede extraction of host pine cones (Turgeon and de Groot 1992). Among the cone and fruit gallers, *Contarinia oregonensis*, the Douglas-fir cone gall midge, can fuse healthy seeds to scales when galls are abundant in a cone. Another cecidomyiid gall-former, *Kaltenbachiola canadensis* (Felt), forms galls in spruce cones; extraction of seeds from infested cones can be difficult (Turgeon and de Groot 1992).

16.5.5 Evolutionary Implications for Management

Stand management techniques to improve seed recruitment and forest regeneration have rarely been studied because managed conifer seed orchards have been the primary source for operational forestry in temperate areas. Lombardo and McCarthy

(2008) found that neither prescribed burns nor canopy thinning influenced seed predation or acorn recruitment in a mixed oak forest. Natural masting cycles were more influential than stand level management in oak regeneration.

Estimation of forest seed crops must take reproductive structure herbivory into account. Seed herbivory in crop estimation is particularly important in masting species. Hulme and Benkman (2002) found that specialist herbivores in masting trees cause lower percentage damage in high crop years, and higher damage in the following year. So, the proportion of seeds damaged is correlated to the size of the previous year's crop. In non-masting trees such as lodgepole pine, populations of reproductive structure herbivores can build up over time and result in an overall higher proportion of seed loss. Also, non-obligatory herbivores survive low seed years better, so they can be a greater problem in high seed years (see Sect. 16.3.3.2).

16.6 Future Perspectives

Reproductive structure herbivores and their hosts are ideal model systems for the study of many evolutionary aspects of plant–insect interactions. Because they are often highly specialized and have to rely on an unpredictable resource, reproductive structure insects exhibit a wide range of evolutionary adaptations to host detection and spatio-temporal dispersal (Boivin et al. 2017). A little-explored issue is how reproductive structure herbivory affects host populations through its effects on genetic variation in reproductive traits and the possible co-selection of non-reproductive traits as opposed to its effects on fitness. An interesting opportunity to study this particular type of selective pressure is in seed orchards where host trees are often represented by several clones, each clone having an often-well-known set of reproductive and non-reproductive traits and reproductive structure herbivory is monitored at the individual level. It is surprising that, although seed orchards constitute experimental designs well adapted to studying the impacts of reproductive structure herbivory on host genetic variation, they have yet to be fully exploited for this purpose.

Emerging anthropogenic disturbances such as climate change, land-use change and biological invasions have placed a new impetus on clarifying the ecological and evolutionary consequences of reproductive structure herbivory (Ramsfield et al. 2016). Predicted changes in temperature regimes will likely affect many insect life-history traits including diapause, developmental phenology and oviposition timing. Climate change may also affect the phenology and the interannual patterns of tree reproductive dynamics. The consequences for reproductive structure insects that rely generally on a close synchrony between their developmental phenology and the reproductive phenology of their hosts is the likelihood of a partial or total desynchrony between the herbivore and its host (Parmesan and Yohe 2003; Voigt et al. 2003). As a result, insects may exert a selective pressure on new host traits such as early reproductive phenology. The ecological and evolutionary consequences of such changes for the host populations are largely unknown. Climate change also has considerable potential to modify the masting patterns, particularly when the cue for masting is climatic. Changes in climatic drivers of inter-annual reproductive synchrony might therefore lead to a more regular production of reproductive structures. It would likely have strong effects on specialized herbivores and ultimately on the host's natural regeneration. This potential impact is of particular concern for endangered tree species with poor natural regeneration success (Guido and Roques 1996). Reproductive structure herbivory may also hamper adaptation strategies such as assisted migration by limiting seed orchard production and natural regeneration when tree hosts are at or beyond the limit of their natural range (Zocca et al. 2008; Jameson et al. 2015).

The rapid change in climatic conditions that is expected to occur in the coming decades provides a unique opportunity to observe ecological and evolutionary changes in plant–insect interactions. Although reproductive structure herbivores are often less conspicuous than other forest pests and, as such, have historically attracted less attention in climate change studies, they will constitute a key driver in the future evolution of natural and managed forest ecosystems.

References

Abe Y, Melica G, Stone GN (2007) The diversity and phylogeography of cynipid gallwasps (Hymenoptera: Cynipidae) of the Oriental and eastern Palearctic regions, and their associated communities. Orient Insects 41(1):169–212

Acheampong S, Strong WB (2016) First Canadian records for two invasive seed-feeding bugs, *Arocatus melanocephalus* (Fabricius, 1798) and *Raglius alboacuminatus* (Goeze, 1778), and a range extension for a third species, *Rhyparochromus vulgaris* (Schilling, 1829)(Hemiptera: Heteroptera). J Entomol Soc British Columbia 113:74–78

Andersen AN, New TR (1987) Insect inhabitants of fruits of *Leptospermum, Eucalyptus* and *Casuarina* in south-east Australia. Aust J Zool 35:327–336

AliNiazee MT (1983) Pest status of filbert (hazelnut) insects: a 10-year study. Can Entomol 115:1155–1162

AliNiazee MT (1997) Integrated pest management of hazelnut pests: a worldwide perspective. In: Köskal AÏ, Okay Y, Günes NT (eds) International Society for Horticultural Science. Fourth International Symposium Hazelnut. Acta Horticulturae 445:469–476

Allen RC (1969) A revision of the genus *Leptoglossus* Guerin (Hemiptera: Coreidae). Entomologica Americanca 45:35–140

Asano I, Nakagawa M, Takeuchi Y, Sakai S, Kishimoto-Yamada K, Shimizu-kaya U, Mohammed F, Yazid Hossman M, Bunyok A, abd Raman MY, Meleng P, Itioka T (2016) The population dynamics and biodiversity of insect seed predators in tropical rainforests of Sarawak. In: Proceedings of the symposium "Frontier in tropical forest research: progress in joint projects between the Forest Department Sarawak and the Japan Research Consortium for Tropical Forests in Sarawak" (2016), 179–186, 2016-06

Atkinson WD, Shorrocks B (1981) Competition on a divided and ephemeral resource: a simulation model. J Anim Ecol 50:461–471

Auger-Rozenberg M-A, Kerdelhué C, Magnoux E, Turgeon J, Rasplus J-Y, Roques A (2006) Molecular phylogeny and evolution of host-plant use in conifer seed chalcids in the genus *Megastigmus* (Hymenoptera: Torymidae). Syst Entomol 31:47–64

Bakke A (1970) Effect of temperature on termination of diapause in larvae of *Laspeyresia strobilella* (L.) (Lepidoptera: Tortricidae). Ent Scand 1:209–214

Barbosa P, Hines J, Kaplan I, Martinson H, Szczepaniec A, Szendrei Z (2009) Associational resistance and associational susceptibility: having right or wrong neighbors. Annu Rev Ecol Evol Syst 40:1–20

Barger JH, Davidson RH (1967) A life history study of the ash seed weevils, *Thysanocnemis bischoffi* Blatchley and *T. helvola* Leconte (Coleoptera: Curculionidae). Ohio J Sci 67(2):123–127

Basset Y, Ctvrtecka R, Dahl C, Miller SE, Quicke DLJ, Segar ST, Barrios H, Beaver RA, Brown JW, Bunyavejchewin S, Gripenberg S, Knizek M, Kongnoo P, Lewis OT, Pongpattanurak N, Pramual P, Sakchoowong W, Schutze M (2019) Insect assemblages attacking seeds and fruits in Thailand. Entomol Sci 22:137–150

Bates SL (2005) Damage to common plumbing materials caused by overwintering *Leptoglossus occidentalis* (Hemiptera: Coreidae). Can Entomol 137:492–496

Bates SL, Borden JH, Kermode AR, Bennett RG (2000) Impact of *Leptoglossus occidentalis* (Hemiptera: Coreidae) on Douglas-fir seed. J Econ Entomol 93(5):1444–1451

Bates SL, Strong WB, Borden JH (2002) Abortion and seed set in lodgepole and western white pine conelets following feeding by *Leptoglossus occidentalis* (Heteroptera: Coreidae). Environ Entomol 31(6):1023–1029

Bedard C, Gries R, Gries G, Bennett R (2002) *Cydia strobilella* (Lepidoptera: Tortricidae): antennal and behavioral responses to host and nonhost volatiles. Can Entomol 134:803–804

Bell EA (1978) Toxins in seeds. In: Harborne JB (ed) Biochemical aspects of plant and animal coevolution. Academic Press, New York, pp 143–161

Boivin T, Doublet V, Candau J-N (2017) The ecology of predispersal insect herbivory on tree reproductive structures in natural forest ecosystems. Insect Sci 00:1–17. https://doi.org/10.1111/1744-7917.12549

Boivin T, Auger-Rozenberg MA (2016) Native fruit, cone and seed insects in the Mediterranean Basin. In: Paine TD, Lieutier F (eds.) Insects and diseases of Mediterranean forest systems. Springer International Publishing. https://doi.org/10.1007/978-3-319-24744-1_4

Boivin T, Gidoin C, von Aderkas P, Safrana J, Candau J-N, Chalon A, Sondo M, El Maataoui M (2015) Host-parasite interactions from the inside: plant reproductive ontogeny drives specialization in parasitic insects. PLoS ONE 10(10):e0139634. https://doi.org/10.1371/journal.pone.0139634

Boivin T, Henri H, Vavre F, Gidoin C, Veber P, Candau J-N, Magnou E, Roques A, Auger-Rozenberg M-A (2014) Epidemiology of asexuality induced by the endosymbiotic *Wolbachia* across phytophagous wasp species: host plant specialization matters. Mol Ecol 23:2362–2375

Bonner FT, Karrfalt RP (2008) The woody plant seed manual. United States Department of Agriculture, Forest Service, Agriculture Handbook 727, 1228p

Borror DJ, Triplehorn CA, Johnson NF (1989) An introduction to the study of insects. Saunders College Publishing, Harcourt Brace College Publishers, Orlando, FL, 875p

Boucher DH, Sork VL (1979) Early drop of nuts in response to insect infestation. Oikos 33(3):440–443

Bracalini M, Benedettelli S, Croci F, Terreni P, Tiberi R, Panzavolta T (2013) Cone and seed pests of *Pinus pinea*: assessment and characterization of damage. J Econ Entomol 106(1):229–234

Bramlett DL, Godbee JF, Jr (1982) Inventory-monitoring system for southern pine seed orchards. Georgia Forestry Commission, Research Division. Forest Research Paper No. 28. 19p

Branco M, Branco C, Merouani H, Alemida MH (2002) Germination success, survival and seedling vigour of *Quercus suber* acorns in relation to insect damage. For Ecol Manage 166:159–164

Brockerhoff EG, Kenis M (1996) Oviposition, life cycle, and parasitoids of the spruce cone maggot, *Strobilomyia anthracina* (Diptera: Anthomyiidae), in the Alps. Bull Entomol Res 87:555–562

Brown JW, Komai F (2008) Key to larvae of Castanea-feeding Olethreutinae frequently intercepted at U.S. ports-of-entry. Trop Lepid 18(1):2–4

Burdick DJ (1961) A taxonomic and biological study of the genus *Xyela* Dalman in North America. University of California Publications in Entomology. Volume XVII, 1960–61:285–354

Cameron RS (1981) Toward insect pest management in southern pine seed orchards with emphasis on the biology of *Tetyra bipunctata* (Hemiptera: Pentatomidae) and the pheromone of *Dioryctria clarioralis* (Lepidoptera: Pyralidae). Texas Forest Service, Texas A&M University, Publication 126, 149p

Cañellas I, Roig S, Poblaciones MJ, Gea-Izquierdo G, Olea L (2007) An approach to acorn production in Iberian dehesas. Agroforest Syst 70:3–9. https://doi.org/10.1007/s10457-007-9034-0

Cibrián-Tovar D, Ebel BH, Yates HO III, Méndez-Montiel JT (1986) Insectos de conos y semillas de las coníferas de México / Cone and seed insects of the Mexican conifers. Universidad Autónoma Chapingo, Secretaria de Agricultura y Recursos Hidráulicos, México / U.S. Department of Agriculture, Forest Service. Gen. Tech. Rep. SE-40. Asheville NC: U.S. Department of Agriculture, Forest Service, Southeastern Forest Experiment Station, 110p

Ciesla WM (2011) Forest entomology: a global perspective. Wiley-Blackwell, John Wiley & Sons Ltd, West Sussex, UK, 400p

Cognato AI, Gillette NE, Campos Bolaños C, Sperling FAH (2005) Mitochondrial phylogeny of pine cone beetles (Scolytinae, *Conophthorus*) and their affiliation with geographic area and host. Mol Phylogenet Evol 36(3):494–508

Connell JH (1971) On the role of natural enemies in preventing competitive exclusion in some marine animals and in rain forest trees. In: Den Boer PJ, Gradwell GR (eds) Dynamics of populations. Pudoc, Wageningen, 1970

Connelly AE, Schowalter TD (1991) Seed losses to feeding by *Leptoglossus occidentalis* (Heteroptera: Coreidae) during two periods of second-year cone development in western white pine. J Econ Entomol 84(1):215–217

Cook SP, Sloniker BD, Rust ML (2013) Using systematically applied insecticides for management of ponderosa pine cone beetle and Dioryctria coneworms in seed orchards. West J Appl for 28(2):66–70

Coulson RN, Franklin RT (1970) The biology of *Dioryctria amatella* (Lepidoptera: Phycitidae). Can Entomol 102:679–684

Coulson RN, Witter JA (1984) Forest entomology. Ecology and management. Wiley-Interscience. John Wiley & Sons, New York, 669p

Coyne JF (1968) *Laspeyresia ingens*, a seedworm infesting cones of longleaf pine. Ann Entomol Soc Am 61(5):1116–1122

Crawley MJ (1989) Insect herbivores and plant population dynamics. Annu Rev Entomol 34:531–564

Crawley MJ (2000) Seed predators and plant population dynamics (Chapter 7: 167–182). In: Fenner M (ed) Seeds: The ecology of regeneration in plant communities, 2nd edn. CABI Publishing, Oxford and New York.

Csóka G, Hirka A (2006) Direct effects of carpophagous insects on the germination ability and early abscission of oak acorns. Acta Silv Lign Hung 2:57–68

de Groot P (1998) Life history and habits of the white pine cone borer, *Eucosma tocullionana* (Lepidoptera: Tortricidae). Can Entomol 130:79–90

DeBarr GL (1967) Two new sucking insect pests of seed in southern pine seed orchards. U.S. Department of Agriculture, Forest Service. Research Note SE-78. Asheville NC: U.S. Department of Agriculture, Forest Service, Southeastern Forest Experiment Station, 3p.

DeBarr GL (1970) Characteristics and radiographic detection of seed bug damage to slash pine seed. Fla Entomol 53(2):109–121

DeBarr GL (1974) Harvest counts underestimate the impact of Dioryctria on second-year slash pine cone crops. U.S. Department of Agriculture, Forest Service. Research Note SE-203. Asheville, NC: U.S. Department of Agriculture, Forest Service, Southeastern Forest Experiment Station, 3p

DeBarr GL, Ebel BE (1973) How seedbugs reduce the quantity and quality of pine seed yields. In: 12th Southern Forest Tree Improvement Conference. Baton Rouge, Louisiana June 12–13, 1973. Proceedings: 233–242. https://rngr.net/publications/tree-improvement-proceedings/sftic/1973

DeBarr GL, Ebel BE (1974) Conelet abortion and seed damage of shortleaf and loblolly pines by a seedbug, *Leptoglossus corculus*. For Sci 20(2):165–170

DeBarr GL, Kormanik PP (1975) Anatomical basis for conelet abortion on *Pinus echinata* following feeding by *Leptoglossus corculus* (Hemiptera: Coreidae). Can Entomol 107:81–86

DeBarr GL, Barber LR, Manchester E (1989) Use of prescribed fire to control the white pine cone beetle in an eastern white pine seed orchard. 20th Biennial Southern Forest Tree Improvement Conference. Charleston, South Carolina June 26–30, 1989. Proceedings: 430. https://rngr.net/pub lications/tree-improvement-proceedings/sftic/1989

DeBarr GL, Barber LR, Berisford CW, Weatherby JC (1982) Pheromone traps detect webbing coneworms in loblolly pine seed orchards. South J Appl for 6:122–127

Debouzie D, Heizmann A, Desouhant E, Menu F (1996) Interference at several temporal and spatial scales between two chestnut insects. Oecologia 108:151–158

Derbel S, Noumi Z, Anton KW, Chaieb M (2007) Life cycle of the coleopter *Bruchidius raddianae* and the seed predation of the *Acacia tortilis* Subsp. *raddiana* in Tunisia. C R Biologies 330:49–54

Dioli P (1991) Presenza di *Orsillus depressus* Dallas, 1852 nella zona alpine e osservazioni sulle specie italiane del genere (Insecta, Heteroptera, Lygaeidae). Atti Mus Civ Stor Nat Morbegno 2:47–51

Dormont L, Roques A (2001) Why are seed cones of Swiss stone pine (*Pinus cembra*) not attacked by the specialized pine cone weevil, *Pissodes validirostris*? A case of host selection vs. host suitability. Entomol Exp Appl 99:157–163

Drooz AT (1985) Insects of eastern forests. U.S. Department of Agriculture, Forest Service. Miscellaneous Publication 1426. Washington, DC: U.S. Department of Agriculture, Forest Service. 654p

Dunning CE, Paine TD, Redak RA (2002) Insect-oak interactions with coast live oak (*Quercus agrifolia*) and Engelmann oak (*Q. engelmannii*) at the acorn and seedling stage. In: Standiford RB, McCreary D, Purcell KL, technical coordinators. Proceedings of the fifth symposium on oak woodlands: oaks in California's changing landscape. 2001 October 22–25; San Diego, CA. Gen. Tech. Rep. PSW-GTR-184. Albany, CA: U.S. Department of Agriculture, Forest Service, Pacific Southwest Research Station; 846p

Ebel BH, Flavell TH, Drake LE, Yates III HO, DeBarr GL (1980) Seed and cone insects of southern pines. U.S. Department of Agriculture, Forest Service, Gen. Tech. Rep. SE-8, rev. Southeastern Forest Experimental Station, Asheville, N.C., and Southeastern Area, State and Private Forestry, Atlanta, GA, 43p

Elzinga JA, Atlan A, Biere A, Gigord L, Weis AE, Bernasconi G (2007) Time after time: flowering phenology and biotic interactions. Trends Ecol Evol 22:432–439

Espelta JM, Cortés P, Molowny-Horas R, Retana J (2009) Acorn crop size and pre-dispersal herbivory determine inter-specific differences in the recruitment of co-occurring oaks. Oecologia 161(3):559–568

Espelta JM, Cortés P, Molowny-Horas R, Sánchez-Humanes B, Retana J (2008) Masting mediated by summer drought reduces acorn herbivory in Mediterranean oak forests. Ecology 89(3):805–817

Fatzinger CW, Hertel GW, Merkel EP, Pepper, WD, Cameron RS (1980) Identification and sequential occurrence of mortality factors affecting seed yields of southern pine seed orchards. U.S. Department of Agriculture, Forest Service. Res. Pap. SE-216. Asheville, NC: Southeastern Forest Experiment Station, 43p

Fenner M, Cresswell J, Hurley R, Baldwin T (2002) Relationship between capitulum size and pre-dispersal seed predation by insect larvae in common Asteraceae. Oecologia 130:72–77

Ferracini C, Alma A (2008) *Arocatus melanocephalus*, a hemipteran pest on elm in the urban environment. Bull Insectol 61(1):193–194

Fidgen LL, Quiring DT, Sweeney JD (1998) Effect of cone size on adult and larval foraging behavior of *Strobilomyia neanthracina* and *Strobilomyia appalachensis* (Diptera: Anthomyiidae). Environ Entomol 27(4):877–884

Fukumoto H, Kajimura H (2011) Effects of asynchronous acorn production by co-occurring Quercus trees on resource utilization by acorn-feeding insects. J For Res 16(1):62–67

Fukumoto H, Kajimura H (2001) Guild structures of seed insects in relation to acorn development in two oak species. Ecol Res 16:145–155

Gagné RJ, Jaschhof M (2014) A catalog of the Cecidomyiidae (Diptera) of the world. 3rd Edition. Digital version 2. U.S. Department of Agriculture, Agricultural Research Service. Washington, DC: Systematic Entomology Laboratory, 493p

Gao C, Kondorosy E, Bu W (2013) A review of the genus *Arocatus* from palearctic and oriental regions (Hemiptera: Heteroptera: Lygaeidae). Raffles Bull Zool 61(2):687–704

Gessé F, Ribes J, Goula M (2009) *Belonochilus numenius*, the sycamore seed bug, new record for the Iberian fauna. Bull Insectol 62(1):121–123

Ghazoul J, Satake A (2009) Nonviable seed set enhances plant fitness: the sacrificial sibling hypothesis. Ecology 90(2):369–377

Gibson LP (1964) Biology and life history of acorn-infesting weevils of the genus *Conotrachelus* (Coleoptera: Curculionidae). Ann Entomol Soc Am 57:521–526

Gibson LP (1971) Insects of bur oak acorns. Ann Entomol Soc Am 64(1):232–234

Gibson LP (1982) Insects that damage northern red oak acorns. U.S. Department of Agriculture, Forest Service. Res. Pap. NE-492. Broomall, PA: Northeast. For. Exp. Stn., 6p

Gosik R, Łętowski J, Mokrzycki T, Wanat M (2001) *Lignyodes bischoffi* (Blatchley, 1916) (Coleoptera: Curculionidae) – new to the fauna of Poland (in Polish with English summary). Wiad Entomol 20(1–2):43–48

Govindan BN, Swihart RK (2015) Community structure of acorn weevils (Curculio): inferences from multispecies occupancy models. Can J Zool 93:31–39

Graber RE (1964) Impact of the white-pine cone beetle on a pine seed crop. J For, July 1964:499–500

Greene E (1989) A diet-induced developmental polymorphism in a caterpillar. Science 243:643–646

Gries R, Khaskin G, Gries G, Bennett RG, Skip King GG, Morewood P, Slessor KN, Morewood WD (2002) (Z, Z)-4,7-tridecadien-(S)-2-yl acetate: sex pheromone of Douglas-fir cone gall midge, *Contarinia oregonensis*. J Chem Ecol 28(11):2283–2297

Grissell EE (1999) An annotated catalog of world Megastigminae (Hymenoptera: Chalcidoidea: Torymidae). Contributions of the American Entomological Institute 31(4):1–92

Grosman DM, Upton WW, McCook FA, Billings RF (2002) Systemic insecticide injections for control of cone and seed insects in loblolly pine seed orchards – 2 year results. South J Appl For 26(3):146–152

Guido M, Roques A (1996) Impact of the phytophagous insect and mite complex associated with cones of Junipers (*Juniperus phoenicea* L. and *J. cedrus* Webb and Berth.) in the Canary Islands. Ecologica Mediteraranea XXII 112:1–10

Halpern SL, Underwood N (2006) Approaches for testing herbivore effects on plant population dynamics. J Appl Ecol 43:922–929

Hanski I (1988) Four kinds of extra long diapause in insects: a review of theory and observations. Ann Zool Fennici 25:37–53

Hanula JL, DeBarr GL, Berisford CW (1984) Oviposition behavior and temperature effects on egg development of the southern pine coneworm, *Dioryctria amatella* (Lepidoptera: Pyralidae). Environ Entomol 13:1624–1626

Hanula JL, DeBarr GL, Berisford CW (1987) Threshold temperature and degree-day estimates for development of immature southern pine coneworms (Lepidoptera: Pyralidae) at constant and fluctuating temperatures. J Econ Entomol 80:62–64

Hanula JL, DeBarr GL, Weatherby JC, Barber LR, Berisford CW (2002) Degree-day model for timing insecticide applications to control *Dioryctria amatella* (Lepidoptera: Pyralidae) in loblolly pine seed orchards. Can Entomol 134:255–268

Harborne JB (1993) Introduction to ecological biochemistry. Academic Press

Harris MK, Chung CS, Jackman JA (1996) Masting and pecan interaction with insectan predehiscent nut feeders. Environ Entomol 25(5):1068–1076

Hawkins CP, MacMahon JA (1989) Guilds: The multiple meanings of a concept. Annu Rev Entomol 34:423–451

Hedlin AF (1956) Studies on the balsam-fir seed chalcid, *Megastigmus specularis* Walley (Hymenoptera: Chalcididae). Can Entomol 88:691–697

Hedlin AF (1961) The life history and habits of a midge, *Contarinia oregonensis* Foote (Diptera: Cecidomyiidae) in Douglas-fir cones. Can Entomol 93:952–967

Hedlin AF (1967) The pine seedworm, *Laspeyresia piperana* (Lepidoptera: Olethreutidae), in cones of ponderosa pine. Can Entomol 99:264–267

Hedlin AF, Yates III HO, Tovar DC, Ebel BH, Koerber TW, Merkel EP (1981) Cone and seed insects of North American conifers. Canadian Forestry Service/USDA Forest Service/Secretaría de Agricultura y Recursos Hidráulicos, México. Victoria B.C., 122p

Herrera CM, Jordano P, Guitián J, Traveset A (1998) Annual variability in seed production by woody plants and the masting concept: reassessment of principles and relationships to pollination and seed dispersal. Am Nat 152:576–594

Holt RD, Lawton JH (1993) Apparent competition and enemy-free space in insect host-parasitoid communities. Am Nat 142(4):623–645

Hosaka T, Yumoto T, Kojima H, Komai F, Noor MSM (2009) Community structure of pre-dispersal seed predatory insects on eleven *Shorea* (Dipterocarpaceae) species. J Trop Ecol 25:625–636

Hosaka T, Yumoto T, Chen Y-Y, Sun I-F, Wright SJ, Numata S, Supardi NMN (2017) Responses of pre-dispersal seed predators to sequential flowering of Dipterocarps in Malaysia. Biotropica 49:177–179

Hughes J, Vogler AP (2004) The phylogeny of acorn weevils (genus *Curculio*) from mitochondrial and nuclear DNA sequences: the problem of incomplete data. Mol Phylogenet Evol 32:601–615

Hulme PE (1996) Herbivory, plant regeneration, and species coexistence. J Ecol 84(4):609–615

Hulme PE (1998) Post-dispersal seed herbivory: consequences for plant demography and evolution. Perspect Plant Ecol, Evol Syst 1:32–46

Hulme PE, Benkman CW (2002) Granivory. In: Herrera CM, Pellmyr O (eds) Plant-animal interactions: an evolutionary approach. Blackwell, Oxford, pp 132–154

Hussey NW (1955) The life-histories of *Megastigmus spermotrophus* Wachtl (Hymenoptera: Chalcidoidea) and its principle parasite, with descriptions of the developmental stages. Trans R Entomol Soc Lond 106(2):133–151

Idaho State Department of Agriculture (2013) Elm seed bug, *Arocatus melanocephalus*: an exotic invasive pest new to the U.S. http://extension.oregonstate.edu/malheur/sites/default/files/spring_2013_esb_fact_sheet.pdf

Jakobsson J, Svensson GP, Lofstedt C, Anderbrant O (2016) Antennal and behavioural responses of the spruce seed moth, *Cydia strobilella*, to floral volatiles of Norway spruce, *Picea abies*, and temporal variation in emission of active compounds. Entomol Exp Appl 160:209–218

Jameson RG, Trant AJ, Hermanutz L (2015) Insects can limit seed productivity at the treeline. Can J For Res 45(3):286–296

Janzen DH (1969) Seed-eaters versus seed size, number, toxicity and dispersal. Evolution 23(1):1–27

Janzen DH (1971) Seed predation by animals. Annu Rev Ecol Syst 2:465–492

Janzen DH (1970) Herbivores and the number of tree species in tropical forests. Am Nat 104(940):501–528

Jarry M, Candau J-N, Roques A, Ycart B (1997) Impact of emigrating seed chalcid, *Megastigmus spermotrophus* Wachtl (Hymenoptera: Torymidae), on seed production in a Douglas-fir seed orchard in France and modelling of orchard invasion. Can Entomol 129:7–19

Jimenez-Pino A, Maistrello L, Lopez-Martinez MA, Ocete-Rubio ME, Soria-Iglesias FJ (2011) Spatial distribution of *Cydia fagiglandana* (Zeller) in an exploited holm oak (*Quercus ilex* L.) forest. Span J Agric Res 9(2):570–579

Johnson NE (1963) *Contarinia washingtonensis* (Diptera: Cecidomyiidae), new species infesting the cones of Douglas-fir. Ann Entomol Soc Am 56:94–103

Jose S (2009) Agroforestry for ecosystem services and environmental benefits: an overview. Agrofor Syst 76:1–10

Keen FP (1958) Cone and seed insects of western forest trees. U.S. Department of Agriculture, Forest Service. Tech. Bull. 1169. Berkeley, CA: California Forest and Range Experimental Station, 168p

Kelly D (1994) The evolutionary ecology of mast seeding. Trends Ecol Evol 9:465–470

Kestring D, Menezes LC, Tomaz CA, Lima GP, Rossi MN (2009) Relationship among phenolic contents, seed herbivory, and physical seed traits in Mimosa bimucronata plants. J Plant Biol 52(6):569

Kinzer HG, Ridgill BJ, Watts JG (1972) Seed and cone insects of ponderosa pine. Las Cruces, NM: New Mexico State University. Agricultural Experiment Station Bulletin 594, 36p

Knölke S (2007) A revision of the European representatives of the microlepidopteran genus *Dioryctria* Zeller, 1846. Dissertation zur Erlangung des Doktorgrades der Fakultät für Biologie der Ludwig-Maximilians-Universität München. München: 113p. + Appendix

Kobro S, Søreide L, Djønne E, Rafoss T, Jaastad G, Witzgal P (2003) Masting of rowan *Sorbus aucuparia* L. and consequences for the apple fruit moth *Argyresthia conjugella* Zeller. Popul Ecol 45:25–30

Koerber TW (1963) *Leptoglossus occidentalis* (Hemiptera: Coreidae), a newly discovered pest of coniferous seed. Ann Entomol Soc Am 56:229–234

Kolb A, Ehrlén J, Eriksson O (2007) Ecological and evolutionary consequences of spatial and temporal variation in pre-dispersal seed herbivory. Perspect Plant Ecol, Evol Syst 9:79–100

Kolotelo D, Van Steenis E, Peterson M, Bennett R, Trotter D, Dennis J (2001) Seed handling guidebook. Surrey, BC: British Columbia Ministry of Forests, Tree Improvement Branch, 106p

Koziol M (2000) Cono- and seminiphagous insects of Norway spruce *Picea abies* (L.) Karst. and their parasitoids in lower and upper montane zone of the Tatra National Park in Poland. J Appl Ent 124:259–266

Kraft KJ (1968) Ecology of the cone moth *Laspeyresia toreuta* in *Pinus banksiana* stands. Ann Entomol Soc Am 61(6):1462–1465

Krugman SL, Koerber TW (1969) Effect of cone feeding by *Leptoglossus occidentalis* on ponderosa pine seed development. For Sci 15(1):104–111

Labandeira C (2006) Silurian to Triassic plant and hexapod clades and their associations: new data, a review and interpretations. Arthropod Syst Biol 64(1):53–94

Lander TA, Klein EK, Oddou-Muratorio S, Candau J-N, Gidoin C, Chalon A, Roig A, Fallour D, Auger-Rozenberg M-A, Boivin T (2014) Reconstruction of a windborne insect invasion using a particle dispersal model, historical wind data, and Bayesian analysis of genetic data. Ecol Evol 4(24):4609–4625

Leidy NA, Neunzig HH (1989) Taxonomic study of the larvae of six eastern North American *Dioryctria* (Lepidoptera: Pyralidae: Phycitinae). Proc Entomol Soc Wash 91:325–341

Leiva MJ, Fernández-Alés R (2005) Holm-oak (*Quercus ilex* subsp. *Ballota*) acorns infestation by insects in Mediterranean dehesas and shrublands Its effect on acorn germination and seedling emergence. For Ecol Manage 212:221–229

Lesieur V, Lombaert E, Guillemaud T, Courtial B, Strong W, Roques A, Auger-Rozenberg M-A (2019) The rapid spread of *Leptoglossus occidentalis* in Europe: a bridgehead invasion. J Pest Sci 92(1):189–200

Lesieur V, Yart A, Guilbon S, Lorme P, Auger-Rozenberg M-A, Roques A (2014) The invasive *Leptoglossus* seed bug, a threat for commercial seed crops, but for conifer diversity? Biol Invasions 16:1833–1849

Leslie AB (2011) Herbivory and protection in the macroevolutionary history of conifer cones. Proc R Soc B 278:3003–3008

Lewis OT, Gripenberg S (2008) Insect seed herbivores and environmental change. J Appl Ecol 45(6):1593–1599

Lewis V (1992) Within-tree distribution of acorns Infested by *Curculio occidentalis* (Coleoptera: Curculionidae) and *Cydia latiferreana* (Lepidoptera: Tortricidae) on the Coast Live Oak. Environ Entomol 21(5):975–982

Linhart YB, Moreira X, Snyder MA, Mooney K (2014) Variability in seed cone production and functional response of seed herbivores to seed cone availability: support for the herbivore satiation hypothesis. J Ecol 102:576–583

Lombardo JA, McCarthy BC (2008) Silvicultural treatment effects on oak seed production and predation by acorn weevils in southeastern Ohio. For Ecol Manage 255:2566–2576

Lowe WJ, Barber LR, Cameron RS, DeBarr GL, Hodge GR, Jett JB, McConnell JL, Mangini AC, Nord JC, Taylor JC (1994) A southwide test of bifenthrin (Capture®) for cone and seed insect control in seed orchards. South J Appl For 18:72–75

Luchi N, Mancini V, Feducci M, Santini A, Capretti P (2012) *Leptoglossus occidentalis* and *Diplodia pinea*: a new insect-fungus association in Mediterranean forests. Forest Pathol 42:246–251

Lukasic P, Johnson T (2007) Arthropod communities and succession in baobab, *Adansonia rubrostipa*, fruits in a dry deciduous forest in Kirindy Forest Reserve, Madagascar. African Entomol 15:214–220

Lyal CHC, Curran LM (2000) Seed-feeding beetles of the weevil tribe Mecysolobini (Insecta: Coleoptera: Curculionidae) developing in seeds of trees in the Dipterocarpaceae. J Nat Hist 34(9):1743–1847

Maeto K, Ozaki K (2003) Prolonged diapause of specialist seed-feeders makes predator satiation unstable in masting of *Quercus crispula*. Oecologia 137:392–398

Mailleux A-C, Roques A, Molenberg J-M, Grégoire J-C (2008) A North American invasive seed pest, *Megastigmus spermotrophus* (Wachtl) (Hymenoptera: Torymidae): Its populations and parasitoids in a European introduction zone. Biol Control 44:137–141

Mangini AC, Bruce WW, Hanula JW (2004) Radiographic analysis of shortleaf pine seeds from the Ouachita and Ozark National Forests. In: Guldin JM, tech. comp. 2004. Ouachita and Ozark Mountains symposium: ecosystem management research. Gen. Tech. Rep. SRS-74. Asheville, NC: U.S. Department of Agriculture, Forest Service, Southern Research Station: 89–91

Mangini AC, Duerr DA, Taylor JW (2003) Seed and cone insect pest management: challenges and solutions. In: Proceedings, Society of American Foresters 2002 National Convention, 2002, October 5–9, Winston-Salem, NC. SAF Publication 03-01, Bethesda, MD: Society of American Foresters: 170–175

Mattson WJ Jr (1980) Cone resources and the ecology of the red pine cone beetle, *Conophthorus resinosae* (Coleoptera: Scolytidae). Ann Entomol Soc Am 73:390–396

Mattson WJ (1971) Relationship between cone crop size and cone damage by insects in red pine seed-production areas. Can Entomol 103:617–621

McAlpine JF (1956) Cone-infesting lonchaeids of the genus *Earomyia* Zett., with descriptions of five new species from western North America (Diptera: Lonchaeidae). Can Entomol 88:178–196

McPherson JE, Packauskas RJ, Taylor SJ, O'Brien MF (1990) Eastern range extension of *Leptoglossus occidentalis* with a key to *Leptoglossus* species of America north of Mexico. Great Lakes Entomol 23(2):99–104

Meiado MV, Simabukuro EA, Iannuzzi L (2013) Entomofauna associated to fruits and seeds of two species of *Enterolobium* Mart. (Leguminosae): Harm or benefit? Revista Brasileira de Entomologia 57(1): 100–104

Melika G, Abrahamson WG (2002) Review of the world genera of oak cynipid wasps (Hymenoptera: Cynipidae: Cynipini). In: Melika, G., and Thuróczy, C. (eds.). 2002. Parasitic Wasps: Evolution, Systematics, Biodiversity and Biological Control: International Symposium: "Parasitic hymenoptera: taxonomy and biological control", 14–17 May 2001, Köszeg, Hungary. Budapest: Agroinform: 150–190

Menu F, Debouzie D (1993) Coin-flipping plasticity and prolonged diapause in insects: example of the chestnut weevil *Curculio elephas* (Coleoptera: Curculionidae). Oecologia 93:367–373

Merkel EP (1963) Distribution of the pine seedworm, *Laspeyresia anaranjada*, with notes on the occurrence of *Laspeyresia ingens*. Ann Entomol Soc Am 56:667–669

Messina FJ (1991) Life-history variation in a seed beetle: adult egg-laying vs. larval competitive ability. Oecologia 85:447–455

Mezquida ET, Rodríguez-García E, Olano JM (2016) Efficiency of pollination and satiation of herbivores determine reproductive output in Iberian *Juniperus thurifera* woodlands. Plant Biol 18:1438–8677

Michelsen V (1988) A world revision of *Strobilomyia* gen.n.: the anthomyiid seed pests of conifers (Diptera: Anthomyiidae). Syst Entomol 13:271–314

Miller GE (1986) Damage prediction for *Contarinia oregonensis* Foote (Diptera: Cecidomyiidae) in Douglas-fir seed orchards. Can Entomol 118:1297–1306

Miller M (1996) *Acacia* seed predation by bruchids in an African savanna ecosystem. J Appl Ecol 33(1137):1144

Miller WE (1978) Use of prescribed burning in seed production areas to control red pine cone beetle. Environ Entomol 7(5):698–702

Morewood P, Morewood WD, Bennett RG, Gries G (2002) Potential for pheromone-baited traps to predict seed loss caused by *Contarinia oregonensis* (Diptera: Cecidomyiidae). Can Entomol 134:689–697

Munoz A, Bonal R, Espelta JM (2014) Acorn-weevil interactions in a mixed-oak forest: outcomes for larval growth and plant recruitment. For Ecol Manage 322:98–105

Nakagawa M, Itioka T, Momose K, Yumoto T, Komai F, Morimoto K, Jordal BH, Kato M, Kaliang H, Hamid AA, Inoue T, Nakashizuka T (2003) Bull Entomol Res 93:455–466

Nord JC, DeBarr GL, Barber LR, Weatherby JC, Overgaard NA (1985) Low-volume applications of azinphosmethyl, fenvalerate, and permethrin for control of coneworms (Lepidoptera: Pyralidae) and seed bugs (Hemiptera: Coreidae and Pentatomidae) in southern pine seed orchards. J Econ Entomol 78:445–450

Ollieu MM, Schenk JA (1966) The biology of *Eucosma rescissoriana* Heinrich in western white pine in Idaho (Lepidoptera: Olethreutidae). Can Entomol 98:268–274

Paparatti B, Speranza S (2004) Management of chestnut weevil (*Curculio* spp.), insect key-pest in central Italy. In: Abreu CG, Rosa E, Monteiro AA (eds) Proceedings of the Third International Chestnut Congress: International Society for Horticultural Science. Working Group on Chestnuts: 551–556

Parmesan C, Yohe G (2003) A globally coherent fingerprint of climate change impacts across natural systems. Nature 421(6918):37–42

Paulson AR, von Aderkas P, Perlman SJ (2014) Bacterial associates of seed-parasitic wasps (Hymenoptera: Megastigmus). BMC Microbiol 14:224–239

Pedrazzoli F, Salvadori C, De Christofaro A, Di Santo P, Endrizzi E, Peverieri GS, Roversi PF, Ziccardi A, Angeli G (2012) A new strategy of environmentally safe control of chestnut tortricid moths. In: De Cristofaro A, Di Palma A, Escudero-Colomar LA, Ioriatti C, Molinari F (eds) Proceedings of the workshop on "Sustainable protection of fruit crops in the Mediterranean area" at Vico del Gargano (Italy), 12–17 September, 2010. Integrated Protection of Fruit Crops Subgroups, Pome fruit arthropods and Stone fruits. International Organisation for Biological Control/West Palearctic Regional Section. Bulletin 74:117–123

Peguero G, Bonal R, Espelta JM (2014) Variation of predator satiation and seed abortion as seed defense mechanisms across an altitudinal range. Basic Appl Ecol 15:269–276

Poncet BN, Garat P, Manel S, Bru N, Sachet J-M, Roques A, Despres L (2009) The effect of climate on masting in the European larch and on its specific seed predators. Oecologia 159:527–537

Prévost YH (2002) Seasonal feeding patterns of insects in cones of tamarack, *Larix laricina* (Du Roi) K. Koch (Pinaceae). For Ecol Man 168:101–109

Price PW, Fernandes GW, Waring GL (1987) Adaptive nature of insect galls. Environ Entomol 16(1):15–24

Quiring DT, Sweeney JW, Bennett RG (1998) Evidence for a host-marking pheromone in white spruce cone fly, *Strobilomyia neanthracina*. J Chem Ecol 24(4):709–721

Ramsfield TD, Bentz BJ, Faccoli M, Jactel H, Brockerhoff EG (2016) Forest health in a changing world: effects of globalization and climate change on forest insect and pathogen impacts. Forestry 89(3):245–252

Rauf A, Cecich RA, Benjamin DM (1984) Conelet abortion in jack pine caused by *Platylygus luridus* (Hemiptera: Miridae). Can Entomol 116:1213–1218

Redfern M, Shirley P (2002) British plant galls: identification of galls on plants and fungi. Field Stud 10:207–531

Roe AD, Stein JD, Gillette NE, Sperling FAH (2006) Identification of *Dioryctria* (Lepidoptera: Pyralidae) in a seed orchard at Chico, California. Ann Entomol Soc Am 99:433–448. https://doi.org/10.1603/0013-8746(2006)99[433:IODLPI]2.0.CO;2

Rohlfs DA (1999) A study of acorn feeding insects: Filbert weevil (*Curculio occidentis* (Casey)) and filbertworm (*Cydia latiferreana* (Walsingham)) on Garry oak (*Quercus garryana* (Dougl.) in the southeastern Vancouver Island area. Victoria, BC: The University of British Columbia, M.Sc. Thesis, 157p. https://open.library.ubc.ca/cIRcle/collections/ubctheses/831/items/1.0099321

Rojas-Rousse DD (2006) Persistent pods of the tree *Acacia caven*: a natural refuge for diverse insects including Bruchid beetles and the parasitoids Trichogrammatidae, Pteromalidae and Eulophidae. J Insect Sci 6, article 08

Root RB (1967) The niche exploitation pattern of the blue-gray gnatcatcher. Ecol Monogr 37(4):317–350

Roques A (1991) Structure, specificity, and evolution of insect guilds related to cones of conifers in western Europe. In: Baranchikov YN, Mattson WJ, Hain FP, Payne TL (eds) Forest insect guilds: Patterns of Interaction with host trees. U.S. Department of Agriculture, Forest Service. Gen. Tech. Rep. NE-153. Radnor, PA: U.S. Department of Agriculture, Forest Service, Northeastern Forest Experiment Station: 300–315

Roques A, Skrzypczyńska M (2003) Seed-infesting chalcids of the genus *Megastigmus* Dalman, 1820 (Hymenoptera: Torymidae) native and introduced to the West Palearctic region: taxonomy, host specificity and distribution. J Nat Hist 37(2):127–238. https://doi.org/10.1080/713834669

Roques A, Raimbault JP, Delplanque A (1984) Les diptères Anthomyiidae du genre *Lasiomma* Stein. Ravageurs des cônes et grains de mélèze d'Europe (*Larix decidua* Mill.) en France (In French). Zeitschrift Für Angewandte Entomologie 98:350–367

Roques A, Sun J-H, Zhang, X-D (1996) Cone flies, *Strobilomyia* spp. (Diptera: Anthomyiidae), attacking larch cones in China, with description of a new species. J Swiss Entomol Soc 69(3–4):417–429

Roques A, Sun J-H, Zhang X-D, Turgeon JJ, Xu S-B (1995) Visual trapping of the *Strobilomyia* spp. (Dipt., Anthomyiidae) flies damaging Siberian larch cones in north-eastern China. J Appl Entomol 119:659–665

Rosenberg O, Weslien J (2005) Assessment of cone-damaging insects in a Swedish spruce seed orchard and the efficacy of large-scale application of *Bacillus thuringiensis* variety *aizawai* x *kurstaki* against Lepidoptera. J Econ Entomol 98(2):402–408

Rosenberg O, Nordlander G, Weslien J (2015) Effects of different insect species on seed quantity and quality in Norway spruce. Agric for Entomol 17:158–163

Roskam JC (1977) Biosystematics of insects living in female birch catkins. I. Gall midges of the genus *Semudobia* Keiffer (Diptera, Cecidomyiidae). Tijdschrift Voor Entomologie 120:153–197

Roskam JC (2013) Biosystematics of insects living in female birch catkins. V. Chalcidoid ectoparasitoids of the genera *Torymus* Dalman, *Aprostocetus* Westwood, *Psilonotus* Walker and *Eupelmus* Dalman (Hymenoptera, Chalcidoidea). Tijdschrift Voor Entomologie 156:21–34

Rouault G, Cantini R, Battisti A, Roques A (2005) Geographic distribution and ecology of two species of *Orsillus* (Hemiptera: Lygaeidae) associated with cones of native and introduced Cupressaceae in Europe and the Mediterranean Basin. Can Entomol 137:450–470

Rouault G, Battisti A, Roques A (2007) Oviposition sites of the cypress seed bug *Orsillus maculatus* and response of the egg parasitoid *Telenomus* gr. *floridanus*. Biocontrol 52:9–24

Rouault G, Turgeon J, Candau J-N, Roques A, von Aderkas P (2004) Oviposition strategies of conifer seed chalcids in relation to host phenology. Naturwissenschaften 91:472–480

Roux G, Roques A, Menu F (1997) Effect of temperature and photoperiod on diapause development in a Douglas fir seed chalcid, *Megastigmus spermotrophus*. Oecologia 111:172–177

Roux-Morabito G, Gillette NE, Roques A, Dormont L, Stein J, Sperling FAH (2008) Systematics of the *Dioryctria abietella* species group (Lepidoptera: Pyralidae) based on mitochondrial DNA. Ann Entomol Soc Am 101(5):845–859

Roversi PF, Strong WB, Caleca V, Maltese M, Sabbatini Peverieri G, Marianelli L, Marziali L, Strangi A (2011) Introduction into Italy of *Gryon pennsylvanicum* (Ashmead), an egg parasitoid of the alien invasive bug *Leptoglossus occidentalis* Heidemann. Européenne Et Méditerranéenne Pour La Protection Des Plantes/european and Mediterranean Plant Protection Organization Bulletin 41:72–75

Rudolf PO (1959) Seed production areas in the Lake States. Guidelines for their establishment and management. Station Paper 73. U.S. Department of Agriculture, Forest Service, Lake States Forest Experimental Station, 17p

Sachet J-M, Poncet B, Roques A, Després L (2009) Adaptive radiation through phenological shift: the importance of the temporal niche in species diversification. Ecol Entomol 34:81–89

Sachet J-M, Roques A, Després L (2006) Linking patterns and processes of species diversification in the cone flies, *Strobilomyia* (Diptera: Anthomyiidae). Mol Phylogenet Evol 41:606–621

Sallabanks R, Courtney SP (1992) Frugivory, seed predation, and insect-vertebrate interactions. Annu Rev Entomol 37:377–400

Sam K, Ctvrtecka R, Miller SE, Rosati ME, Molem K, Damas K, Gewa B, Novotny V (2017) Low host specificity and abundance of frugivorous lepidoptera in the lowland rain forests of Papua New Guinea. PLoS ONE 12(2):e0171843. https://doi.org/10.1371/journal.pone.0171843

Schmidt LH (2000) Guide to handling of tropical and subtropical forest seed. Danida Forest Seed Centre, Humlebaek, Denmark, 511p

Schneider Erik S (2014) Funktionsmorphologische Untersuchungen abdominaler Infrarot-Rezeptoren von Insekten. Ph.D. Thesis, Rheinischen Friedrich-Wilhelms-Universität, Bonn, Germany

Seifert M, Wermelinger B, Schneider D (2000) The effect of spruce cone insects on seed production in Switzerland. J Appl Entomol 124(7–8):269–278

Shea PJ (1989) Interactions among phytophagous insect species colonizing cones of white fir (Abies concolor). Oecologia 81:104–110

Shimada M, Kurota H, Toquenaga Y (2001) Regular distribution of larvae and resource monopolization in the seed beetle Bruchidius dorsalis infesting seeds of the Japanese honey locust Gleditsia japonica. Popul Ecol 43:245–252

Shin Y-M, Nam J-W, Kim D-K, Byun B-K, Kim I-K (2018) Two lepidopteran pests and damage on the cones of *Abies koreana* (Pinaceae) in Jeju Island, Korea. J Asia-Pacific Biodivers 11:80–86

Silvertown JW (1980) The evolutionary ecology of mast seeding in trees. Biol J Lin Soc 14:235–250

Silvius KM, Fragoso JMV (2002) Pulp handling by vertebrate seed dispersers increases palm seed predation by bruchid beetles in the northern Amazon. J Ecol 90:1024–1032

Skrzypczyńska M (1978) *Megastigmus suspectus* Borries, 1895 (Hymenoptera, Torymidae), its morphology, biology and economic significance. Zeitschrift Für Angewandte Entomologie 85:204–215

Skrzypczyńska M (1985) Gall-midge (Cecidomyiidae, Diptera) pests in seeds and cones of coniferous trees in Poland. Zeitschrift Für Angewandte Entomologie 100:448–450

Skrzypczyńska M (1998) Insect pests and their parasitoids inhabiting cones of fir *Abies alba* Mill. in Poland. Anzeiger Für Schädlingskunde Pflanzenschutz Umweltschutz 71:50–52

Slater JA (1972) Lygaeid bugs (Hemiptera: Lygaeidae) as seed predators of figs. Biotropica 4:145–151

Smith DR (1978) Family Xyelidae. In: van der Vecht J, Shenefelt RD (eds) Hymenopterorum Catalogus, pars 14. Dr. W. Junk B.V., The Hague, pp 1–27

Smith DR (1979) Symphyta. In: Krombein KV, et al (eds) Catalog of Hymenoptera in America north of Mexico, Vol. 1. Smithsonian Institution Press, Washington DC, pp 3–137

Soler R, Espelta JM, Lencinas MV, Peri PL, Pastur GM (2017) Masting has different effects on seed herbivory by insects and birds in antarctic beech forests with no influence of forest management. For Ecol Manage 400:173–180

Soula B, Menu F (2003) Variability in diapause duration in the chestnut weevil: mixed ESS, genetic polymorphism or bet-hedging? Oikos 100:574–580

Soula B, Menu F (2005) Extended life cycle in the chestnut weevil prolonged or repeated diapause? Entomol Exp Appl 115:333–340

Southgate BJ (1979) Biology of the Bruchidae. Annu Rev Entomol 24:449–473

Speranza S (1999) Chestnut pests in central Italy. In: Saleses G (ed) Proceedings of the second international symposium on Chestnut: International society for horticultural science. Working Group on Chestnuts. Acta Horticulturae 494: 417–423

Sperens U (1997) Fruit production in *Sorbus aucuparia* L. (Rosaceae) and pre-dispersal seed predation by the apple fruit moth (*Argyresthia conjugella* Zell.). Oecologia 110:368–373

Strong WB (2006) Seasonal changes in seed reduction in lodgepole pine cones caused by feeding of *Leptoglossus occidentalis* (Hemiptera: Coreidae). Can Entomol 138:888–896

Strong WB (2015) Lodgepole pine seedset increase by mesh bagging is due to exclusion of Leptoglossus occidentalis (Hemiptera: Coreidae). J Entomol Soc British Columbia 112:3–18

Suez M, Gidoin C, Lefèvre F, Candau J-N, Chalon A, Boivin T (2013) Temporal population genetics of time travelling insects: a long term study in a seed-specialized wasp. PLoS ONE 8(8):e70818. https://doi.org/10.1371/journal.pone.0070818

Svensson GP, Wang H-L, Lassance J-M, Anderbrant O, Chen G-F, Gregorsson B, Guertin C, Harala E, Jirle EV, Liblikas I, Petko V, Roques A, Rosenberg O, Strong W, Voolma K, Ylioja T, Want Y-J, Zhou X-M, Löfstedt C (2012) Assessment of genetic and pheromonal diversity of the *Cydia strobilella* species complex (Lepidoptera: Tortricidae). Syst Entomol 38:305–315

Sweeney JD, Turgeon JJ (1994) Life cycle and phenology of a cone maggot, *Strobilomyia appalachensis* Michelsen (Diptera: Anthomyiidae), on black spruce, *Picea marinana* (Mill.) B.S.P., in eastern Canada. Can Entomol 126:49–59

Tachiki Y, Iwasa Y (2013) Coevolution of mast seeding in trees and extended diapause of seed herbivores. J Theor Biol 339:129–139

Takács S, Bottomley H, Andreller I, Zaradnik T, Schwarz J, Bennett R, Strong W, Gries G (2009) Infrared radiation from hot cones on cool conifers attracts seed-feeding insects. Proc R Soc B 276:649–655. https://doi.org/10.1098/rspb.2008.0742

Takács S, Hardin K, Gries G (2008) Vibratory communication signal produced by male western conifer seed bugs (Hemiptera: Coreidae). Can Entomol 140:174–183

Tamburini M, Maresi G, Salvadori C, Battisti A, Zottele F, Pedrazzoli F (2012) Adaptation of the invasive western conifer seed bug *Leptoglossus occidentalis* to Trentino, an alpine region (Italy). Bull Insectol 65(2):161–170

Thompson JN (1988) Evolutionary ecology of the relationship between oviposition preference and performance of offspring in phytophagous insects. Entomol Exp Appl 47(1):3–14

Tinker ME (1952) The seasonal behavior and ecology of the boxelder but *Leptocoris trivittatus* in Minnesota. Ecology 33(3):407–414

Toquenaga Y (1993) Contest and scramble competitions in *Callosobruchus maculatus* (Coleoptera: Bruchidae) II. Larval competition and interference mechanisms. Res Popul Ecol 35:57–68

Triplehorn CA, Johnson NF (2005) Borror and DeLong's introduction to the study of insects. Thomson Brooks/Cole, Belmont, CA, 864p

Tripp HA (1954) Description and habits of the spruce seedworm (*Laspeyresia youngana* (Kft.) (Lepidoptera: Olethreutidae). Can Entomol 86:385–402

Trudel R, Bauce E, Guertin C, Cabana J (1999) Performance of the fir coneworm *Dioryctria abietivorella* (Grote) as affected by host species and presence or absence of seed cones. Agric For Entomol 1:189–194

Tuda M, Iwasa Y (1998) Evolution of contest competition and its effect on host-parasitoid dynamics. Evol Ecol 12:855–870

Tuda M, Wu L-H, Yamada N, Wang CP, Wu W-J, Buranapanichpan S, Kagoshima K, Chen Z-Q, Teramoto KK, Kumashiro BR, Heu RR (2014) Host shift capability of a specialist seed predator of an invasive plant: roles of competition, population genetics and plant chemistry. Biol Invasions 16:303–313

Turgeon JJ, Roques A, de Groot P (1994) Insect fauna of coniferous seed cones: diversity, host plant interactions, and management. Ann Rev Entomol 39:179–212

Turgeon JJ, de Groot P (1992) Management of insect pests of cones in seed orchards in Eastern Canada. Ontario Forest Research Institute and Forest Pest Management Institute. Queen's Printer for Ontario, Sault Ste. Marie, Ontario, 98p

Turgeon JJ, de Groot P, Sweeney JD (2005) Insects of seed cones in Eastern Canada: Field Guide. Ontario Forest Research Institute and Forest Pest Management Institute. Sault Ste. Marie, Ontario: Queen's Printer for Ontario, 127p

Voigt W, Perner J, Davis AJ, Eggers T, Schumacher J, Bährmann R, Fabian B, Heinrich W, Köhler G, Lichter D, Marstaller R (2003) Trophic levels are differentially sensitive to climate. Ecology 84(9):2444–2453

von Aderkas P, Rouault G, Wagner R, Rohr R, Roques A (2005) Seed parasitism redirects ovule development in Douglas fir. Proceedings of the Royal Society of London b: Biological Sciences 272(1571):1491–1496

Wakeley PC (1935) Collecting, extracting, and marketing southern pine seed. Occasional Paper No. 51. U.S. Department of Agriculture, Forest Service, Southern Forest Experimental Station, 10p

Wang H-L, Svensson GP, Rosenberg O, Bengtsson M, Erling VJ, Löfstedt C (2010) Identification of the sex pheromone of the spruce seed moth, *Cydia strobilella* L. J Chem Ecol 36:305–313

Wheeler AG Jr (1976) Life history of *Kleidocerys resedae* on European white birch and ericaceous shrubs. Ann Entomol Soc Am 69(3):459–463

Whitehouse CM, Roe AD, Strong WB, Evenden ML, Sperling FAH (2011) Biology and management of North American cone-feeding *Dioryctria* species. Can Entomol 143(1):1–34

Williams CE (1989) Checklist of North American nut-infesting insects and host plants. J Entomol Sci 24(4):550–562

Xiao Z, Harris MK, Zhang Z (2007) Acorn defenses to herbivory from insects: implications for the joint evolution of resistance, tolerance and escape. For Ecol Manage 238:302–308

Xiao Z, Mi X, Holyoak M, Xie W, Cao K, Yang X, Huang X, Krebs CJ (2016) Seed–herbivore satiation and Janzen-Connell effects vary with spatial scales for seed-feeding insects. Ann Bot 119(1):109–116

Yasaka M, Terazawa K, Koyama H, Kon H (2003) Masting behavior of *Fagus crenata* in northern Japan: spatial synchrony and pre-dispersal seed herbivory. For Ecol Manage 184(1):277–284

Yates HO III (1986) Checklist of insect and mite species attacking cones and seeds of world conifers. J Entomol Sci 21(2):142–168

Zahradnik T, Takács S, Strong W, Bennett R, Kuzmin A, Gries G (2012) Douglas-fir cone gall midges respond to shape and infrared wavelength attributes of host tree branches. Can Entomol 144:658–666

Zobel B, Talbert J (1984) Applied forest tree improvement. The Blackburn Press, Caldwell, NJ, p 505p

Zocca A, Zanini C, Aimi A, Frigimelica G, La Porta N, Battisti A (2008) Spread of plant pathogens and insect vectors at the northern range margin of cypress in Italy. Acta Oecologica 33(3):307–313

Żywiec M, Holeksa J, Ledwoń M, Seget P (2013) Reproductive success of individuals with different fruit production patterns. What does it mean for the herbivore satiation hypothesis? Oecologia 172:461–467

Chapter 17
IPM: The Forest Context

Jon Sweeney, Kevin J. Dodds, Christopher J. Fettig, and Angus J. Carnegie

Integrated pest management (IPM) is perhaps best described as "…the maintenance of destructive agents, including insects, at tolerable levels by the planned use of a variety of preventative, suppressive or regulatory tactics that are ecologically and economically efficient and socially acceptable. It is implicit that the actions taken are fully integrated into the total resource management process in both planning and operation" (Waters 1974). Another useful definition of IPM is "an ecosystem-based strategy that focuses on long-term prevention of pests or their damage through a combination of techniques such as biological control, habitat manipulation, modification of cultural practices, and use of resistant varieties. Pesticides are used only after monitoring indicates they are needed according to established guidelines, and treatments are made with the goal of removing only the target organism. Pest control materials are selected and applied in a manner that minimizes risks to human health, beneficial and non-target organisms, and the environment" (University of California, Davis 2015). The spatial and temporal scale of forests demands landscape-level and long-term planning with an emphasis on preventive measures, e.g. silviculture. IPM programs in forests have historically been concerned mainly with pests that have large

J. Sweeney (✉)
Natural Resources Canada, Canadian Forest Service – Atlantic Forestry Centre, Fredericton, NB, Canada
e-mail: jon.sweeney@nrcan-rncan.gc.ca

K. J. Dodds
U.S. Forest Service, Eastern Region, State, Private, and Tribal Forestry, Durham, NH, USA
e-mail: kevin.j.dodds@usda.gov

C. J. Fettig
U.S. Forest Service, Pacific Southwest Research Station, Davis, CA, USA

A. J. Carnegie
Forest Science, Department of Primary Industries, Sydney, NSW, Australia

© The Author(s) 2023
J. D. Allison et al. (eds.), *Forest Entomology and Pathology*,
https://doi.org/10.1007/978-3-031-11553-0_17

impacts on fibre and wood supply and the livelihood of resource-dependent communities, e.g. the spruce budworm, *Choristoneura fumiferana* (Clemens), and mountain pine beetle, *Dendroctonus ponderosae* Hopkins, in North America. However, in Europe and more recently in North America, there has been a shift away from a focus on individual pests and towards IPM as part of ecosystem management (Häusler and Scherer-Lorenzen 2001; Alfaro and Langor 2016). Ideally, IPM should be an integral part of sustainable forest management, which in addition to sustained forest productivity, includes principles such as maintenance of biodiversity and ecological processes, carbon sequestration, and protection of soil and water quality (Holvoet and Muys 2004).

17.1   Components of IPM

17.1.1   *Biology and Ecology of the Pest-Tree-Forest System*

A central component of IPM in forests is knowledge and understanding of the biology and ecology of the pests, their host trees, and the forest system in which they interact. Effective strategies for reducing the negative impact of an insect pest requires sufficient knowledge of the pest's life history and the factors that affect its population dynamics, such as host susceptibility and natural enemies. A key aim is to reduce pest impacts while minimizing negative effects on ecosystem services and function.

17.1.1.1   Systematics and Taxonomy

The first step in IPM is accurate identification of the causative pest(s) and that requires some knowledge of taxonomy, and not infrequently, the assistance of taxonomic specialists. The next step would be to determine what is known about the pest species' biology, and suitable methods for its survey and control. If the pest can be identified to genus only (e.g. it may be a non-native species accidentally introduced to a region) it may still be possible to determine some of its biology based on what is known of other species in the same genus (Huber and Langor 2004). Accurate identification can be difficult for species with only subtle morphological differences from other species. Misidentifications can be costly, as illustrated in Box 17.1.

> **Box 17.1 Importance of taxonomy and accurate species identification**
> Specimens of the brown spruce longhorn beetle, *Tetropium fuscum* (Fabr.), a European native, were collected in Point Pleasant Park, Halifax, Nova Scotia, Canada in 1990 during a trapping survey for spruce beetle, *Dendroctonus rufipennis* (Kirby), but were misidentified as the native species, *Tetropium*

cinnamopterum Kirby. It was not until 1999 that the causal organism was correctly identified as *T. fuscum* (Smith and Hurley 2000). A quarantine and eradication program was initiated in 2000 at an estimated cost of CAN$4–6 million per year (Huber and Langor 2004). The goal of eradicating *T. fuscum* was abandoned in 2007 when it was clear that the beetle had established itself over a large area. Although it is quite possible that *T. fuscum* was already established in Nova Scotia several years before specimens were first collected in 1990, the delay of almost a decade in the accurate identification of *T. fuscum* likely made effective containment and eradication more difficult (Huber and Langor 2004) (see Chapter 19).

17.1.1.2 Pest Life History and Factors Affecting Pest Populations

Knowing a pest's life history is fundamental to developing effective survey and control methods. Furthermore, understanding the key factors that affect pest population biology makes it possible to develop tools and tactics that have less interference with natural mortality factors. For example, the discovery that nucleopolyhedrosis viruses cause the collapse of outbreaks of defoliators such as Douglas-fir tussock moth, *Orgyia pseudotsugata* (McDunnough), and balsam fir sawfly, *Neodiprion abietis* (Harris), has led to the mass production and application of species-specific viruses to suppress defoliator populations (Shepherd et al. 1984; Otvos et al. 1987; Lucarotti et al. 2007). Below, we briefly highlight some of the natural factors affecting pest distribution and abundance that must be considered when developing IPM programs. The myriad of interacting abiotic and biotic factors and their effects on insect populations is beyond the scope of this chapter. For more information, see Price et al. (2011) and Schowalter (2016).

Climate

Climate, especially temperature and precipitation patterns, has a substantial influence on the distribution of plants and the animals that feed on them (Merriam 1894). All insect pests have upper and lower temperature limits beyond which they do not survive, and these limits are useful for predicting their potential geographic distribution, e.g. an exotic species introduced to a new continent or a native pest expanding its range in a changing climate. Knowing a pest's distribution in the landscape and how it may vary in response to climate is a prerequisite for efficient targeting of IPM tactics. Warming temperatures in the last couple of decades have enabled range expansions of some species like the pine processionary moth, *Thaumetopoea pityocampa* Schiff. (Battisti et al. 2005), mountain pine beetle (Logan and Powell 2001; Carroll et al.

2003; Weed et al. 2013), and southern pine beetle, *Dendroctonus frontalis* Zimmermann (Dodds et al. 2018), and generated the need for temperature-based models to predict where range expansions may occur (e.g. Buffo et al. 2007; Lesk et al. 2017). Exposure to unseasonal cold temperature is often the largest single source of mortality in mountain pine beetle populations (Safranyik 1978). In Alberta, Canada, ground surveys are conducted every spring to estimate mountain pine beetle overwintering survival and forecast population trends, which are in turn used to focus management activities where they are most effective in slowing the beetle's spread (Anon. 2007b). In addition to overwintering survival, temperature affects the rate of development, voltinism (number of generations per year), dispersal, reproduction, and degree of phenological synchrony with their hosts (Hansen and Bentz 2003). Favourable temperatures during larval development can shift spruce beetle populations from a 2-year life cycle to a 1-year life cycle and contribute to large-scale outbreaks, whereas cold temperatures that occur before spruce beetles have acclimatized can contribute to outbreak collapse (Aukema et al. 2016). Knowledge of temperature-phenology relationships is useful in models for predicting the impact of pests (Powell and Bentz 2014) and the need for management actions.

Natural Enemies

Most insect herbivores serve as food or brood hosts for a large assortment of natural enemies (i.e. predators, parasitoids, and pathogens) which have been implicated as major mortality factors of about half of the pest species for which long-term population studies have been conducted (Price et al. 2011). For example, pupal mortality from parasitism (Fitzgerald 1995) and bird predation (Parry et al. 1997) contribute to population regulation of the forest tent caterpillar, *Malacosoma disstria* Hübner, and density-dependent pupal predation regulates low-density populations of the winter moth, *Operophtera brumata* L. (Varley and Gradwell 1968; Roland 1994). Baculoviruses infect many species of forest Lepidoptera and sawflies (Cory et al. 1997) and have been used to control defoliators like the balsam fir sawfly, *Neodiprion abietis* (Harris) (Moreau and Lucarotti 2007). Knowledge of a pest's natural enemies and their impacts on pest populations is beneficial when developing an IPM program, e.g. to reduce negative impacts when using insecticides (Williams et al. 2003) or for classical biological control of exotic, invasive forest pests (Bauer et al. 2015).

Host Tree-Insect Interactions and Food Quality

In addition to the top-down effects exerted on herbivore populations by natural enemies, the quality and availability of food (e.g. host trees) exert considerable bottom-up effects, and this is the basis of IPM tactics that affect tree vigour (e.g. thinning) and breeding for genetic resistance. Host resistance is one of the main factors regulating endemic populations of bark beetles (Aukema et al. 2016).

Factors that stress trees and reduce their vigour, such as root rots, overstocked growing conditions, drought, defoliation, or root damage from wind events, reduce host defenses and make trees more susceptible to colonization and mortality by bark beetles (Fettig et al. 2007; Kolb et al. 2016). When this occurs on a large scale it often leads to greater reproductive success and higher bark beetle populations (Werner et al. 2006; Aukema et al. 2016). When bark beetle populations reach the epidemic phase, they can overcome the defenses of healthy trees thanks to mass attack facilitated by aggregation pheromones (Wallin and Raffa 2004; Boone et al. 2011). Lodgepole pines, *Pinus contorta* Dougl. ex Loud. var. *latifolia* Englem., with thicker phloem offer more food and space for larval development and produce more mountain pine beetle brood than trees with thinner phloem (Amman 1972). Suppressed trees with smaller diameters are habitat for *Sirex noctilio* F. and removing these through silvicultural treatments can reduce tree mortality attributed to *S. noctilio* in a stand (Neumann et al. 1987; Dodds et al. 2014).

In contrast to bark beetles and *S. noctilio* that perform better in stressed or weakened hosts, some species, such as the white pine weevil, *Pissodes strobi* (Peck), prefer vigorous hosts (Alfaro et al. 1995). In addition, many defoliators prefer more vigorous hosts with leaves of high protein and water content (Dury et al. 1998). Furthermore, the species, size, and age of trees and foliage may also affect the development rate and survival of insect herbivores. For example, the spruce beetle has larger broods in white spruce, *Picea glauca* Moench (Voss), than in Sitka spruce, *Picea sitchensis* (Bong.) Carr., and Lutz spruce, *Picea × lutzii* Little (i.e. a hybrid of white and Sitka spruce) (Holsten and Werner 1990). However, although vigorous growth (foliage) may be more susceptible to herbivores, the tree as a whole may have other defence mechanisms to tolerate such damage (Stone 2001). For example, the foliage of fast-growing species of *Eucalyptus* is highly susceptible to herbivorous insects, but the trees can tolerate the defoliation because of their fast growth, i.e. they use foliage replacement as their defence mechanism. Compare this to slow-growing *Eucalyptus* species that utilise phytochemical and physical properties within leaves as the main defence against herbivorous insects.

Dispersal

Emigration and immigration are key processes in the life history of many forest insects, allowing some species to move out of unsuitable habitats, and to expand, or contract their range. Factors affecting dispersal include pest density, host density, body size, lipid content, and weather conditions (e.g. temperature, wind) (Smith et al. 2001; Evenden et al. 2014; Jones et al. 2019). Although dispersal is often short range within forest stands, wind-assisted long-range dispersal flights (30–100 km) of the mountain pine beetle are considered to be partly responsible for the beetle's range expansion across the Rocky Mountains into northern Alberta (Jackson et al. 2008; Safranyik et al. 2010).

Pests may also be spread by people via the inadvertent movement of infested material such as firewood (Jacobi et al. 2012) and solid wood packaging of goods in shipping containers arriving at international ports (Haack 2006; Haack et al. 2014). This pathway has led to the establishment of invasive bark and wood boring insects outside their native ranges, such as the polyphagous shot hole borer, *Euwallacea fornicatus* (Eichhoff) (Eskalen et al. 2012; Smith et al. 2019), and emerald ash borer, *Agrilus planipennis* Fairmaire (Herms and McCullough 2014), in North America, and Asian longhorned beetle, *Anoplophora glabripennis* Motschulsky, in North America and Europe (Haack et al. 2010). Importation of unprocessed logs from North America is thought to be the likely pathway that led to establishment of the red turpentine beetle, *Dendroctonus valens* LeConte, in China (Yan et al. 2005). Regulatory controls and phytosanitary treatment of wood packaging has reduced the risk of anthropogenic dispersal of pests (Haack et al. 2014). However, exotic, invasive species like spongy moth, *Lymantria dispar* (L.) (formerly gypsy moth), and emerald ash borer, spread by a process referred to as 'stratified dispersal', involving long distance movement of the insect by people to locations far beyond the area where the pest is established, combined with natural dispersal from outlier populations that establish at these new locations (Sharov and Liebhold 1998a, b; Herms and McCullough 2014). The better we understand factors that influence both natural and human-assisted dispersal of pests, the better we can predict rates and direction of spread and develop effective IPM strategies.

Forest/Stand Structure and Susceptibility to Pests

Tree species composition, age class distribution, stand density and host tree condition affect the susceptibility of forests to insect pests. By knowing stand conditions that favour a particular pest, or vice-versa, it is possible to develop *risk and hazard rating* models (see Sect. 17.1.4) that predict the potential impact of a pest in different stands. For high impact pests such as the mountain pine beetle and spruce budworm, this information can be used in decision-support systems to direct where and when management is implemented.

Monocultures often tend to be more susceptible to insect herbivory than mixed species forests (Jactel and Brockerhoff 2007; Guyot et al. 2015, 2016). One mechanism thought to be responsible for greater herbivory in less diverse plant communities is greater host availability and increased foraging efficiency. For example, vast areas of mature, even aged lodgepole pine, along with warmer than average winter temperatures that increase overwintering survival of mountain pine beetle, are considered important factors inciting mountain pine beetle outbreaks (Bentz et al. 2010; Safranyik et al. 2010). In contrast, host plants are less plentiful and also more patchy and more difficult for herbivores to locate in diverse stands. Diverse forests often have more complementary resources (pollen, nectar) and alternative hosts than monocultures, and this supports a more robust assemblage of natural enemies that can exert greater top-down regulation of herbivores (Lawton and Strong 1981).

17.1.2 Survey and Monitoring

Effective techniques to survey and monitor forest insect populations are critical components of IPM, and when available, provide natural resource managers with important information on how to prioritize management actions (Edmonds et al. 2000; Carnegie et al. 2005b). For example, decisions to harvest an area prior to an anticipated outbreak; to perform sanitation of infested trees to reduce pest populations; or to increase survey efforts in surrounding forests, are best made from predictions of tree damage or tree mortality estimated from survey data. However, sampling insects in forested environments presents unique challenges that may not be encountered in agricultural systems or urban forests where damage is more easily observed and quantified (Fettig et al. 2001, 2005). In addition, the cryptic nature of many insects and their presence in portions of trees that are difficult to sample (e.g. upper tree crowns) can make surveys of forest insects challenging (Ric et al. 2007).

17.1.2.1 Pest Density-Damage Relationships

Estimates of pest density based on regular surveillance provide managers with the opportunity to pre-emptively plan and implement IPM. Pest density-damage estimates are much more common for defoliators than other feeding guilds. For example, systems have been developed for spruce budworm using light traps to predict population trends (Simmons and Elliott 1985) and pheromone traps to predict larval densities (that can then be related to tree damage) (Sanders 1988; Rhainds et al. 2016). Pheromone-baited trapping systems have also been developed for the western spruce budworm, *Choristoneura occidentalis* Freeman (Niwa and Overhulser 2015), European sawfly, *Neodiprion sertifer* (Geoffroy) (Lyytikäinen-Saarenmaa et al. 2006) and spongy moth (Gage et al. 1990), among other species. Pheromone-baited traps are a preferred sampling tool in many cases as they are effective at detecting low-density populations, are often species-specific, especially with moth pests, and are relatively easy to use.

There are very few examples of using bark beetle trap catches to successfully predict tree mortality. However, trap catches of spruce beetle can be used to estimate its population phase (i.e. endemic vs. epidemic), which is linked to tree mortality (Hansen et al. 2006). Damage thresholds predicted from pheromone-baited traps have also been developed for European spruce beetle, *Ips typographus* (L.), in Italy (Faccoli and Stergulc 2004) and Sweden (Weslien 1992b). In the southeastern U.S., Billings (1988) developed a practical method of forecasting population trends and infestation levels of southern pine beetle based on captures of southern pine beetle and the ratio of southern pine beetle to one of its major predators. Attempts to predict tree mortality in western North America from trap catches of western pine beetle, *Dendroctonus brevicomis* LeConte, have been unsuccessful (Hayes et al. 2009).

17.1.2.2 Trap Trees

Trap trees are tools used to survey or monitor bole-infesting insects. Trees selected as trap trees are either artificially stressed through chemical or mechanical means, or pheromones are used to initiate insect colonization. Depending on the life history traits of the target insect, trap trees are either left standing (Neumann et al. 1982) or felled (Hodgkinson 1985). Trap trees provide multiple opportunities to detect target insects, including the capture of insects in traps attached to trap trees, collecting adult insects that emerge from sections of trap trees removed from the field and placed in rearing containers, or through signs such as galleries, resinosis or emergence holes that are reliably diagnostic (McCullough et al. 2009; Zylstra et al. 2010).

Trap trees have been used operationally to detect exotic species. Probably the best example of this is the use of chemically girdled pine trees as detection tools for *S. noctilio* in the Southern Hemisphere. Positive trap trees (those colonized by *S. noctilio*) are then integrated into the biological control program for managing *S. noctilio* using a parasitic nematode (Neumann et al. 1982; see Sect. 17.3.3). Trap trees have also been used operationally to detect emerald ash borer in North America (McCullough et al. 2009). In this context, ash trap trees are girdled using a chainsaw and a section of bark is removed at about breast height. Later in the summer, these trap trees are felled, and the bark is peeled to determine if larvae are present (Fig. 17.1).

A benefit of trap trees over pheromone-baited detection/monitoring traps, especially for an insect that does not utilize long-range sex pheromones, is that they provide a more complete suite of chemical and visual cues to attract the target insect, which often results in a more sensitive survey tool (Mercader et al. 2013). However, using trap trees is logistically more difficult than semiochemical-baited traps and creating dead trees in many areas can often create safety hazards if precautions are not undertaken. Felling, handling, and transporting tree sections into rearing facilities can also be challenging, expensive and time consuming. Where colonizing insects do not make signs of infestation that result in species determination, there can be an extensive lag time between when trap trees are colonized and when adults emerge from the wood allowing for species identification. However, using molecular techniques to identify pest species from larvae or frass can reduce the lag time for some species (Kethidi et al. 2003; Wilson and Schiff 2010; Ide et al. 2016).

17.1.2.3 Semiochemical-Baited Traps

Semiochemicals, including pheromones and kairomones, are used by insects to find mates or to locate suitable habitats and hosts (Roelofs and Cardé 1977; Wood 1982). These chemicals can be strong sources of attraction for insects and provide excellent survey and monitoring tools. Some semiochemicals are attractive to only one or very few species (e.g. *L. dispar* moth sex pheromone) while others may attract a broad range of species (e.g. alpha-pinene, a host plant volatile that is emitted from many tree species). Pheromones, used alone or with host volatiles, have been used extensively to detect and monitor Lepidoptera (Elkinton and Cardé 1981; Grant 1991; Jactel

Fig. 17.1 Ash trap trees, girdled in spring to increase their stress levels and attraction to the emerald ash borer, *Agrilus planipennis*, have been used in surveys to detect and delimit infestations of emerald ash borer (Photo credit: Pennsylvania Department of Conservation and Natural Resources)

et al. 2006; Jones et al. 2009), Coleoptera (Weslien 1992b; Brockerhoff et al. 2006; Sweeney et al. 2006; Billings and Upton 2010), and Hymenoptera (Lyytikäinen-Saarenmaa et al. 1999; Dodds and de Groot 2012) (Fig. 17.2). Combining more than one host volatile (e.g. alpha-pinene and ethanol) or combining pheromones with host volatiles can synergize attraction and increase trap captures of some Coleoptera (Chenier and Philogene 1989; Silk et al. 2007; Allison et al. 2013).

A wide array of traps are available for use with semiochemicals (Fig. 17.3). Multiple-funnel traps and panel intercept traps were designed specifically for bark beetles (Lindgren 1983) or bark beetles and woodborers (Czokajlo et al. 2001), respectively (Fig. 17.3). Modifications of these traps, including enlarging funnel holes (Miller et al. 2013), applying lubricants (de Groot and Nott 2003; Graham et al. 2010; Allison et al. 2011), and extending a collar above the bottom funnel (Allison et al. 2014) can improve trap captures. Canopy malaise traps have also shown promise for sampling bark beetles and woodborers (Vance et al. 2003; Dodds et al. 2015). Traps commonly used to monitor lepidopteran pests include pheromone-baited delta sticky traps or non-sticky traps that use a dry collecting cup and pesticide strip to kill captured insects (e.g. the Unitrap [Fig. 17.3d]). Sticky traps are efficient at catching moths but the sticky surface becomes saturated with moths (and moth

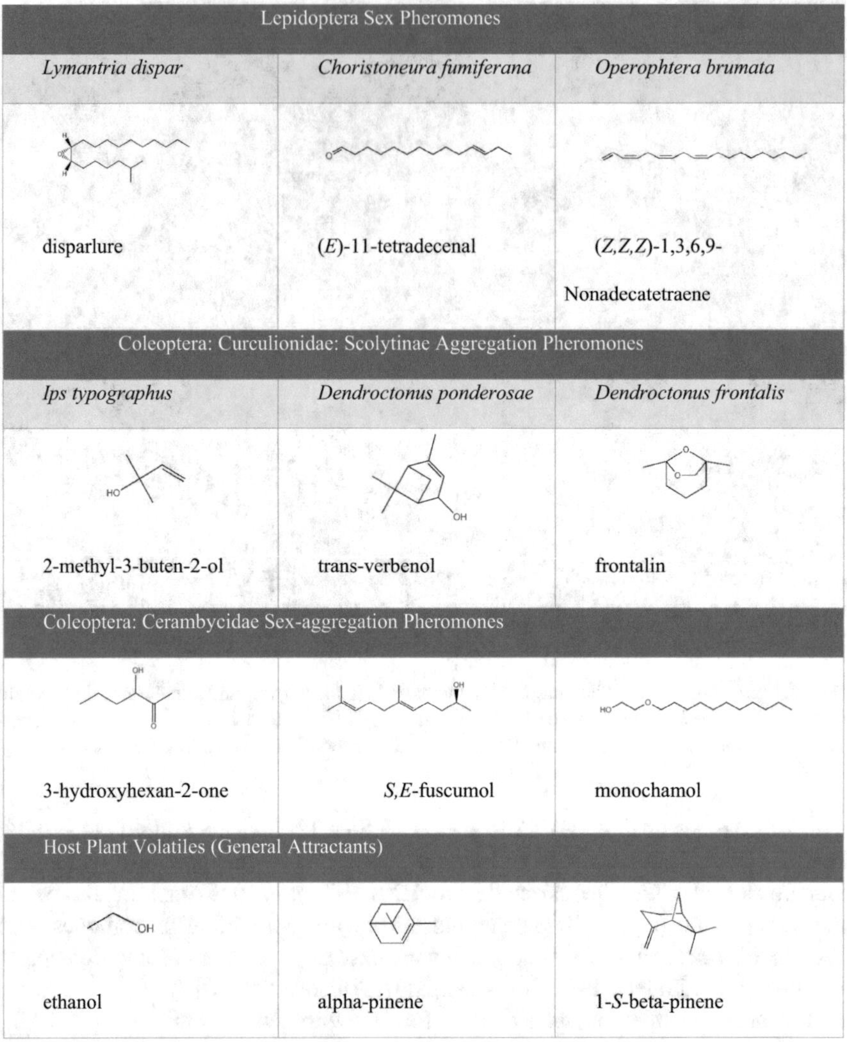

Lepidoptera Sex Pheromones		
Lymantria dispar	*Choristoneura fumiferana*	*Operophtera brumata*
disparlure	(*E*)-11-tetradecenal	(*Z,Z,Z*)-1,3,6,9- Nonadecatetraene
Coleoptera: Curculionidae: Scolytinae Aggregation Pheromones		
Ips typographus	*Dendroctonus ponderosae*	*Dendroctonus frontalis*
2-methyl-3-buten-2-ol	trans-verbenol	frontalin
Coleoptera: Cerambycidae Sex-aggregation Pheromones		
3-hydroxyhexan-2-one	*S,E*-fuscumol	monochamol
Host Plant Volatiles (General Attractants)		
ethanol	alpha-pinene	1-*S*-beta-pinene

Fig. 17.2 Examples of chemical structures (El-Sayed 2022) of some insect pheromones and plant volatiles used in forest insect pest surveys. Lepidoptera and Scolytinae pheromones are usually more specific than Cerambycidae pheromones and all of these pheromones are more specific than host volatiles

scales) at relatively low population densities, so they are not as suitable as non-sticky traps for monitoring large changes in population densities (Sanders 1986).

Various factors influence trap captures and can broadly be categorized as intrinsic and extrinsic. Intrinsic factors include the type of trap (Flechtmann et al. 2000; Sweeney et al. 2006; Dodds et al. 2015), trap color (Campbell and Borden 2009; Francese et al. 2010; Rassati et al. 2019), trap surface treatments (de Groot and Nott

Fig. 17.3 Examples of traps used to survey for forest insects: **a** multiple funnel and **b** intercept panel traps are commonly employed to survey for bark beetles and woodborers; **c** canopy malaise traps are used for bark beetles and woodborers as well as other taxa (Photo credits: K. J. Dodds); and **d** Unitraps are used to collect Lepidoptera (Photo credit: M. MacDonnell, University of New Brunswick, Fredericton, NB)

2003; Graham et al. 2010; Allison et al. 2011), type of collection cup or adhesive (Miller and Duerr 2008), placement of traps along environmental gradients (Dodds 2014; Schmeelk et al. 2016; Allison et al. 2019; Sweeney et al. 2020), and other trap modifications (Allison et al. 2014). Extrinsic factors include variables such as local forest disturbance history (wildfire and silvicultural treatments) (Sullivan et al. 2003; Dodds 2011), forest stand composition and structure (Ohsawa 2004), volume and decay class of downed wood (Lee et al. 2014), and local insect population levels where traps are deployed. Meteorological variables including temperature, relative humidity, precipitation, and wind may also influence trapping results (Salom and Mclean 1991; Peng et al. 1992; Jönsson and Anderbrant 1993).

The potential economic and ecological impacts of exotic, invasive forest insects have been the impetus for nationwide detection surveys using semiochemical-baited traps. Many countries, including Canada (Canadian Food Inspection Agency 2016), New Zealand (Brockerhoff et al. 2006), Australia (Wylie et al. 2008; Carnegie et al. 2018), and the U.S. (Rabaglia et al. 2008; Jackson et al. 2014), among others, have well-developed annual surveys that target bark beetles, woodborers, moths, and other damaging insects. These surveys may focus on individual target species (Wylie et al. 2008; Jackson et al. 2014) or on broader target taxa, e.g. species of bark and wood boring insects in the families Cerambycidae, Buprestidae, Curculionidae (Scolytinae), and Siricidae, at risk of transcontinental movement in wood packaging. In both cases, the goal is to detect introduced and established pests as early as possible, when populations and infested areas are small and the chances of eradication are good (Tobin et al. 2014). If a newly detected exotic species is considered to pose a threat, surveys with semiochemical-baited traps are implemented to delimit the population in the invaded region (Liebhold et al. 2016). Examples of large-scale delimitation efforts include *S. noctilio* (Dodds and de Groot 2012) and emerald ash borer (USDA APHIS PPQ 2017) in North America and *Uraba lugens* Walker in New Zealand (Suckling et al. 2005).

Semiochemical-based detection and monitoring traps are easy to deploy and consequently frequently used in IPM programs. Lures and traps are relatively inexpensive and most traps can be used for many years. However, understanding and interpreting what trap captures mean, and do not mean, is critical. Presence of an insect in a trap may or may not indicate the existence of a local population, as many insects can disperse long distances. Conversely, for several reasons a trap that is negative for a target species cannot be interpreted as evidence that the area is free of that species. Traps have a defined active sampling space, capture efficiencies are often low, and for some species, retention of individuals captured in traps is low (Elkinton and Childs 1983; Byers 2008; Allison and Redak 2017). For example, mark-release-recapture studies have shown that only 1–29% of bark beetles that are marked and released are recaptured in traps (Birch et al. 1982; Weslien and Lindelöw 1989; Zolubas and Byers 1995).

17.1.2.4   Ground-Based Surveys

When signs and/or symptoms of infestation are obvious, ground surveys can be very effective, especially in small stands or other areas where trees are easily accessible. Signs and symptoms of infestation, such as resin or staining (Coleman and Seybold 2008; Ryan et al. 2013), oviposition sites on tree boles (Ric et al. 2007), defoliation in crowns (MacLean and Lidstone 1982), tree crown fade (Billings and Pase 1979), bark flaking by woodpeckers (de Groot et al. 2006) or egg masses (Shepherd and Brown 1971; Liebhold et al. 1994) can all provide evidence of insect presence and in some cases, population levels. (Fig. 17.4). An advantage of ground surveys over pheromone-baited trap surveys is that signs or symptoms that are strongly correlated with a specific insect provide direct evidence that the species is established in the area. Further sampling of trees detected during ground surveys often yields more information, such as pest population estimates and identification of mortality factors. Although they are laborious and time consuming, ground surveys of infested trees are conducted on a systematic basis throughout the year in Europe during European spruce beetle outbreaks (Fettig and Hilszczański 2015). Once identified, infested trees are marked, numbered, and mapped for sanitation (see Sect. 17.1.4.2.4).

17.1.2.5   Remote Sensing and Aerial Detection Surveys

Remote detection methods are also useful tools for assessing the effects of insects on forest resources (Ciesla 2000). Surveys may be conducted using manned or unmanned aircraft (McConnell et al. 2000; Lehmann et al. 2015) or may integrate detailed information related to plant growth and stress from satellite systems or aircraft-based sensors. Remote sensing provides opportunities to rapidly gather information on changes in forest condition over large spatial scales (Stone and Mohammed 2017).

Aerial detection surveys using aircraft have been used in parts of the U.S. since the 1940s (Wear and Buckhorn 1955) and have occurred annually over much of the forested lands in the U.S. since the 1970s. Aerial surveys are also an important component of insect and disease monitoring in Canada (British Columbia Ministry of Forests 2000), New Zealand (Kershaw 1989), and Australia (Carnegie et al. 2008). Originally used to track tree mortality and damage, these surveys can also detect new pests or damage in new areas, estimate levels of damage, and provide guidance for further survey or management (McConnell et al. 2000; Johnson and Wittwer 2008). Aerial detection surveys are occasionally followed with *ground-based surveys* to more precisely delimit damage observed from the air.

Satellite technologies, including multispectral and hyperspectral sensors (airborne and spaceborne) and LiDAR (Light Detection and Ranging—a laser-based method of mapping landscape features) have been useful for survey and monitoring of forest insects. These approaches have also been used to identify tree species and to map tree distributions (Somers and Asner 2012), determine areas where tree stress is occurring (Hanavan et al. 2015), as well as mapping locations where insects have caused damage (White et al. 2006; Fassnacht et al. 2014; Stone and Mohammed 2017).

Fig. 17.4 Examples of signs used to survey for insects: **a** resin at the sites of southern pine beetle, *Dendroctonus frontalis*, attack (Photo credit: K. J. Dodds); **b** resin beading associated with *Sirex noctilio* oviposition sites (Photo credit: A. J. Carnegie); **c** bark staining resulting from goldspotted oak borer, *Agrilus auroguttatus*, attacks (Photo credit: T. W. Coleman, Forest Health Protection, USDA Forest Service); **d** woodpecker flaking on a ponderosa pine, *Pinus ponderosa*, colonized by western pine beetle, *Dendroctonus brevicomis* (Photo credit: C. J. Fettig); **e** Asian longhorned beetle, *Anoplophora glabripennis*, oviposition sites (Photo credit: K. J. Dodds); and **f** spongy moth, *Lymantria dispar*, egg mass (Photo credit: K. J. Dodds)

A drawback of remote sensing is that information collected at such large scales is often incomplete, e.g. it may be possible to detect tree stress or tree mortality but not necessarily determine a specific causal agent of tree mortality. Another challenge is the time lag between when the information is collected and when it is processed into a format that is useful to managers, although this is rapidly improving. Integration of a number of survey and monitoring tools provides the best chances to provide reliable information within the context of an IPM program.

17.1.3 Pest Impact Assessment and Cost–Benefit Analysis

Pest impact assessment considers the ecological and economic impacts of a pest. Defining the ecological impacts or changes to an ecosystem associated with particular pests is difficult (Swank et al. 1981; Reynolds et al. 2000; Lovett et al. 2002, 2006; Lewis and Liken 2007). Assessment of cumulative impacts must consider effects on individuals (e.g. reproductive success, growth), populations (e.g. genetics, population dynamics), communities (e.g. species diversity, species composition), ecosystems (e.g. nutrient cycling), and regions (Parker et al. 1999; Ricciardi et al. 2013) and is often context dependent. For example, bark beetle outbreaks are often detrimental to many ecological goods and services, while at the same time benefiting other ecological goods and services (Morris et al. 2018). By opening forest canopies and creating large gaps, grazing habitat may be enhanced. In rural areas, real-estate values may also increase due to better scenic views and transition to tree species more appealing to landowners.

Depending on how widespread tree damage or tree mortality is, different approaches to quantify ecological impacts are used. Among stands of similar type and conditions, comparisons can be made between infested and uninfested sites. The first step is often focused on plot-level vegetation assessment and establishment of permanent plots. Through this type of data collection, information can be gained on changes in forest condition, often with an emphasis on tree structure and composition. These types of assessments have been conducted for both native (Donato et al. 2013; Zeppenfeld et al. 2015) and invasive insect species (Dodds et al. 2010; Dodds and Orwig 2011; Simmons et al. 2014; Haavik et al. 2015). While standard vegetation plots provide strong information on impacts occurring at the stand level, knowledge of factors acting at larger spatial scales can also be investigated by dispersing vegetation plots or increasing their size throughout an impacted landscape (Orwig et al. 2008). Coarser landscape-scale assessments may be made through aerial surveys or remote sensing data.

Quantifying the costs associated with ecological impacts of forest pests is difficult due to challenges valuating and monetizing ecological goods and services (Boyd and Banzhaf 2007; Holmes et al. 2009; Stenger et al. 2009). However, the market value of forest trees for timber and fiber can easily be estimated and compared to management costs to determine the net cost or benefit of actions to reduce tree losses. Considering treatment options for insects that do not kill trees but cause growth reduction is more challenging, as economic projections over the life of trees must be considered. Models may be used to predict the value of a stand that can then be used to guide pest management decision-making; these can range from simple stand growth models to more elaborate models that incorporate non-traditional forest products (Fox et al. 1997). Comparing growth over the rotation of treated and untreated stands can provide cost-benefits of control programs (Cameron et al. 2018b; Wardlaw et al. 2018), but such analyses are rare.

Although costs and benefits are underlying principles of IPM, full cost–benefit analysis rarely occurs in forest pest management (MacQuarrie et al. 2016; Niquidet et al. 2016; Cameron et al. 2018b). More frequently, cost–benefit analysis is done to determine the lowest cost option to achieve a specific management objective, for example, protecting foliage from defoliation by spruce budworm by using the lowest effective dose of *Bacillus thuringiensis kurstaki* (*Btk*, a soil-dwelling bacterium commonly used as a biological pesticide) (Morris 1984). Alternatively, cost–benefit analysis may be used to detect pest threshold levels that justify treatment (Niquidet et al. 2016).

Estimating the benefits of management interventions is often difficult in forestry due to long delays (25–60 years) between management actions and harvests combined with volatility in forest product prices and economic parameters (Niquidet et al. 2016). However, there have been some thorough cost–benefit analyses of pest management programs against exotic, invasive forest pests (e.g. Sharov and Liebhold 1998a; Cameron et al. 2018b). For example, the costs of protecting urban ash trees from the emerald ash borer have been demonstrated to be substantially lower than costs of tree removal and replacement following mortality due to emerald ash borer infestation (McCullough and Mercader 2012). Tobin (2008) showed that slowing the spread of spongy moth in North America is a cost-effective strategy in spite of the large infested area because it delays the costs associated with maintaining expanded quarantine zones and managing spongy moth outbreaks.

17.1.4   Management Strategies

There are two basic strategies to reduce the negative impacts of insects on forests. *Prevention* is designed to reduce the probability and severity of future infestations by manipulating stand, forest and/or landscape conditions. *Suppression* is designed to reduce current infestations by manipulating pest populations using *remedial tactics*. In some cases, *risk and hazard rating systems* or *decision support systems* are available to identify stands that should be prioritized for management (see Sect. 17.1.4.2). For example, the Spruce Budworm Decision Support System (SBWDSS) is used to quantify returns in marginal timber supply from protecting stands against spruce budworm infestations in Canada (MacLean et al. 2001).

When implementing prevention or suppression, managers should be cognisant of opportunities to address additional objectives with little or no additional cost. For example, in pine-dominated forests of the southern and western U.S., fuel reduction treatments, such as mechanical thinning and prescribed fire, are frequently used to reduce forest fuels (Stephens et al. 2012). While prescriptions differ between thinning treatments implemented for fuels reduction and those for managing pest infestations, there are opportunities to alter fuel reduction treatments without reducing their efficacy while increasing the effectiveness of these treatments to reduce the susceptibility of forests to certain pests. In the latter case, *crown* or *selection thinning* (removal of larger trees in the dominant and codominant crown classes) may be necessary to achieve suitable reductions in the abundance of preferred hosts of certain pests. In other situations, some resource objectives may be negatively impacted by preventive and suppressive tactics, and it is prudent to identify as many of these impacts as possible and to adjust management strategies accordingly (e.g. changing the timing, scale, frequency and/or intensity of treatments) (e.g. Fettig et al. 2008, 2014).

As indicated earlier, in many cases, management strategies may not be justified due to ecological or social constraints. In other cases, the benefits of intervention may not justify the costs. Furthermore, metrics used to assess impacts caused by forest insects have traditionally been based on timber values yet, increasingly, emphasis is placed on the full range of ecological goods and services derived from forests (Morris et al. 2018; Fettig 2019). Unfortunately, empirical estimation of potential market and nonmarket values for most ecological goods and services is in its infancy (Stenger et al. 2009; McCollum and Lundquist 2019), and as such is a major obstacle in establishing credible linkages between management interventions and changes in economic valuations in forests.

17.1.4.1 Prevention

Regulatory Controls

Regulatory controls are designed to prevent the introduction of exotic, invasive species and/or to reduce their spread once established. These are usually informed by *pest risk assessments*, which quantify risks associated with the introduction and/or spread of exotic, invasive species based on assessments of relevant factors, such as invasion pathways, host distributions, and impacts. Data on the ecology and life history of many insect species are limited even in their native environments. Consequently, many pest risk assessments rely heavily on expert judgment and assessment. Although pest risk assessments can be useful in explaining the general causes and consequences of an invasion, more formalized and quantitative estimates of risk based on spatially explicit, multi-scale decision support systems are becoming more common (e.g. due to uncertainty associated with the impacts of climate change). Risk assessments for exotic species are now standard procedure prescribed by the World Trade Organization Agreement on Sanitary and Phytosanitary Procedures (Yemshanov et al. 2009).

Quarantines are used to reduce the spread of exotic, invasive species once established in a new environment, and include information on the regulated species and articles (e.g. host materials), the geographic scope of the quarantine, and penalties

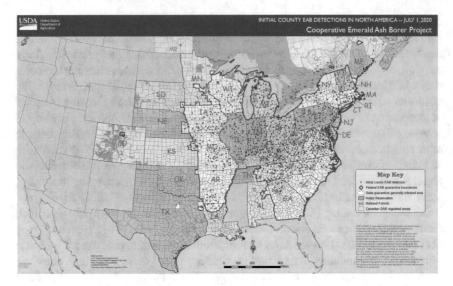

Fig. 17.5 Quarantines are regulatory measures designed to reduce the spread of exotic, invasive species once established in a new environment. The emerald ash borer, *Agrilus planipennis,* was first detected in Michigan in 2002, and by January 2021 had spread to 36 U.S. states and 5 Canadian provinces despite federal, provincial and state quarantines (Stone 2021)

for noncompliance. As an example, the state of Minnesota established a quarantine of pine wood with bark (Minnesota Statute 18G.06, subd. 4, 2013), exclusive of pine mulch or chips, pine Christmas trees and pine nursery stock, from areas of the U.S. determined to have established mountain pine beetle populations. Any person violating the quarantine is subject to civil and criminal penalties. The U.S. government, through the Animal and Plant Health Inspection Service (APHIS), imposed a quarantine on emerald ash borer in the eastern U.S. from 2003 to 2020 (Fig. 17.5). Internationally, standards to prevent the introduction and spread of exotic species are established by the International Plant Protection Convention (IPPC, www.ippc. int/en/). The International Standards for Phytosanitary Measures No. 15 (ISPM 15), first established in 2002, provides standards for wood packing materials (e.g. dunnage, crates, and pallets) used in international trade, and requires that they be heat treated or fumigated and branded with a seal of compliance (Haack et al. 2014).

Tree Breeding

Forest geneticists and tree breeders have traditionally focused on polygenic traits, because very few important traits in trees are controlled by single genes. An important exception is that of some disease resistance genes. As an example, populations of sugar pine, *Pinus lambertiana* Dougl., have been heavily impacted for decades by *Cronartium ribicola* J.C. Fisch., the exotic pathogen that causes white pine blister rust. Although white pine blister rust can be fatal to all species of white pine, a gene occurs at low frequency in sugar pine that confers resistance to *C. ribicola* (Kinloch et al. 1970). Restoring populations of sugar pine involves, among other factors, identifying white pine blister rust-resistant trees in the field followed by selective breeding of these individuals, and eventual outplanting of white pine blister rust-resistant seedlings.

In general, tree breeders have largely ignored opportunities to increase insect resistance. For example, for decades it has been recognized that oleoresin flow characteristics in some pines are predictable and heritable (Smith 1975; Hodges et al. 1979) and potentially could be selected for in tree breeding programs to increase resistance to bark beetles (Strom et al. 2002). However, little progress has been made. Perhaps the most successful example of harnessing natural genetic variation for forest insect resistance involves the white pine weevil in North America. Several seed orchards have been established to grow white pine weevil-resistant spruce seedlings for use in reforestation (Alfaro et al. 2013).

There have been rapid advances in the application of plant biotechnology in the last two decades (Harfouche et al. 2011). While applications in forestry are experimental, there have been achievements in poplars, pines, and eucalypts, for example, involving insertion of *Bt* genes to increase resistance to insect defoliators. For genetically modified trees, significant study is required to evaluate the stable expression of genes after insertion into the tree, as biosafety concerns involving potential drift of genes into the environment must be considered (Vettori et al. 2016). As such, future uses of genetically modified trees in IPM programs will likely be limited to short-rotation woody cropping systems (Fig. 17.6).

Silvicultural Tactics

Silviculture is the backbone of IPM in forests (see Chapter 20), and in some cases begins with the proper selection of planting stock that is pest-free and appropriate for site conditions. When selecting planting stock managers should not only consider the climate of today, but that likely to be experienced in the future. Some experts suggest that *assisted migration*, the practice of planting tree species outside of their current distribution due to anticipated changes in the climatic niche, is important and should be applied more widely than has occurred (Gray et al. 2011). Planting may also provide an opportunity to increase tree species diversity, which as indicated earlier, is often associated with reductions in insect herbivory (Jactel and Brockerhoff 2007).

Managing stand density through thinning is an important silvicultural tactic in several systems (Figs. 17.7 and 17.8). Thinning operations vary in their prescription (e.g. some remove many trees of a particular species or size class whereas others may remove few trees) resulting in different stand structures and tree species compositions that influence susceptibility in different ways to different forest pests. While it is widely accepted that thinning is effective for reducing future levels of tree mortality attributed to some bark beetles (Fettig et al. 2007), there is no clear evidence that

Fig. 17.6 A five-year old poplar, *Populus* spp., research plantation at the Savannah River Site, South Carolina, U.S. This site received irrigation and fertilization throughout the growing season. Future applications of genetically modified trees to increase insect resistance will likely be limited to short-rotation woody cropping systems (Photo credit: D. R. Coyle, Clemson University)

thinning reduces losses from forest defoliators (Muzika and Liebhold 2000). In addition, there are examples where forest insects have greater impacts in thinned than unthinned stands, including the balsam fir sawfly in eastern Canada (Ostaff et al. 2006) and white pine weevil in western Canada (Alfaro and Omule 1990), but these tend to be the exception. In some cases, proper management of logging residues is important to reduce risks of future infestations by species that may breed in this material (Fettig et al. 2006). One unique variation to thinning is pre-emptive removal of certain host species in an attempt to limit the spread of a particular pest (Vannatta et al. 2012).

17.1.4.2 Suppression

Pesticides

Pesticides are an integral part of IPM, but social concerns and environmental considerations restrict their use in many forests, particularly in Europe. They may be applied manually, with ground-based equipment (e.g. soil and tree injection systems,

Fig. 17.7 Thinning ponderosa pine, *Pinus ponderosa,* in California to increase resistance to bark beetles, primarily western pine beetle *Dendroctonus brevicomis,* and mountain pine beetle, *D. ponderosae.* Among other factors, thinning reduces host availability; reduces competition among trees for nutrients, water, and other resources thereby increasing vigor; and affects microclimate decreasing the effectiveness of chemical cues used in host finding, selection and colonization by bark beetles (Photo credit: C. J. Fettig)

sprayers, blowers, and related equipment) or aerially with fixed-wing aircraft or helicopters. In forests, pesticides most commonly used for management of forest insects include contact and systemic insecticides, microbials (bacteria, viruses, pathogens and nematodes), insect growth regulators, soaps and horticultural oils. Most applications are confined to intensively managed areas, such as nurseries, seed orchards, short-rotation woody cropping systems and recreation sites.

Insecticides used to protect individual trees from colonization by bark beetles, and to a lesser extent woodborers, usually consist of ground-based sprays applied to the tree bole (Fig. 17.9). Residual activity varies by active ingredient, bark beetle species, tree species and associated climatic conditions (Fettig et al. 2013). In the western U.S., ten of thousands of trees may be treated annually to protect them from bark beetles during large-scale outbreaks, such as observed with mountain pine beetle in the mid-2000s (Fettig et al. 2021).

In recent years, researchers attempting to find safer, more portable, and longer lasting alternatives to bole sprays have evaluated the effectiveness of injecting small quantities of systemic insecticides directly into the tree bole with pressurized systems. These systems push low volumes of product, generally less than several hundred

Fig. 17.8 A loblolly pine, *Pinus taeda*, plantation thinned in Virginia as part of the Southern Pine Beetle Prevention Program. Since 2003, the Southern Pine Beetle Prevention Program, a joint effort of the USDA Forest Service and Southern Group of State Foresters, has encouraged and provided cost-share assistance for silvicultural treatments to reduce stand and forest susceptibility to southern pine beetle, *Dendroctonus frontalis* (Photo credit: J. T. Nowak, Forest Health Protection, USDA Forest Service)

Fig. 17.9 Protection of individual trees from mortality attributed to bark beetles may involve applications of liquid formulations of contact insecticides to the tree bole (Photo credit: C. J. Fettig)

milliliters for even large trees, into the small vesicles of the sapwood. Following injection, the product is transported throughout the tree to the target tissue (i.e. the phloem where bark beetle feeding occurs) (Fettig et al. 2013). In North America, bole injections have been demonstrated effective for mountain pine beetle, spruce beetle, and western pine beetle (Fettig et al. 2020) but are used most commonly in urban forests for control of exotic, invasive species such as emerald ash borer (Herms and McCullough 2014).

Synthetic formulations of entomopathogenic microorganisms may also be useful for managing bark beetles and wood borers. Research efforts have focused on the fungus *Beauveria bassiana* (Bals.-Criv.) Vuill. Tactics under development include contaminating beetles collected in traps and then releasing these individuals back into field populations to contaminate the pest population (Kreutz et al. 2000; Lyons et al. 2012) and applying various suspensions of spores to the surfaces of felled and standing trees (Davis et al. 2018).

Most large-scale insecticide applications for defoliators involve the use of fixed-wing aircraft (Fig. 17.10). Among the first were applications in eastern Canada in the late 1920s when >85,000 kg of calcium arsenate dust was applied in attempts to control outbreaks of spruce budworm and hemlock looper, *Lambdina fiscellaria* (Guen.) (Holmes and MacQuarrie 2016). The development of synthetic organic insecticides in the early 1940s led to use of dichlorodiphenyltrichloroethane (DDT) for control of spruce budworm and other forest defoliators in North America. DDT remained the preferred control option for spruce budworm throughout the 1950s and 1960s (Nigam 1975) but by the mid-1950s the negative impacts of DDT were recognized (Turusov et al. 2002) and by the mid-1970s several countries banned most uses of DDT (Fig. 17.11). Today, microbial agents such as *B. thuringiensis* and insect growth regulators, such as diflubenzuron, have replaced the use of most synthetic insecticides for management of forest defoliators.

The Slow the Spread Program for management of spongy moth in the U.S. is a great example of the incorporation of insecticides into an IPM program. This combined federal and state effort involves detecting isolated populations of spongy moth with pheromone-baited traps (see Sect. 17.1.4.2.2) placed along the expanding population front from North Carolina to Wisconsin. In most cases, detected colonies are treated with *Btk*, diflubenzuron or mating-disruption pheromone. It has been estimated that this project has reduced the spread of *L. dispar* from infested areas to adjacent uninfested areas by >50% (Sharov et al. 2002).

Semiochemical Tactics

Semiochemicals are used to disrupt mating behaviors, mass trap pest insects, attract and kill insects, and to inhibit colonization of individual trees and forest stands. Semiochemicals have the benefit of being environmentally benign compared to insecticides, and many are species- or genera-specific. It is common practice to combine several semiochemical treatments, such as aggregation and anti-aggregation pheromones, into one IPM program targeting an insect species.

Fig. 17.10 Applications of insecticides for management of forest defoliators usually involve the use of fixed-wing aircraft. This photo shows an application of DDT to control the Douglas-fir tussock moth, *Orgyia pseudotsugata,* in Idaho in 1947. Today, microbial agents such as *Bacillus thuringiensis* and insect growth regulators such as diflubenzuron have replaced DDT and other broad-spectrum insecticides (Photo credit: Furniss [2007])

Mating disruption has been used successfully to manage some lepidopteran pests, primarily in agricultural and orchard settings (Cardé and Minks 1995). Fundamental to mating disruption success is release of a highly attractive sex pheromone from multiple points throughout a treatment area that makes mate location difficult (Cardé and Minks 1995; Miller et al. 2006). The Slow the Spread Program for spongy moth management represents the most extensive example of mating disruption for a forest pest. Synthetic spongy moth pheromone, disparlure, is spread over targeted landscapes through flakes applied by airplane (Tobin and Blackburn 2007). Since 1993, >5.5 million ha have been treated with a spongy moth mating disruption product (USDA 2016). Aerial applications of sex pheromone in large-scale field trials have suppressed spruce budworm mating (Rhainds et al. 2012) but not budworm egg or larval densities, likely due to immigration of mated female moths (Régnière et al. 2019). Other large-scale mating disruption programs for forest Lepidoptera are rare (Rhainds et al. 2012; Svensson et al. 2018). Examples of mating disruption are less common in other insect orders, however promising results have been demonstrated for pinhole borers (Coleoptera: Platypodinae) (Funes et al. 2011), cerambycids (Maki et al. 2011; Sweeney et al. 2017), and sawflies (Anderbrant et al. 1995; Martini et al. 2002).

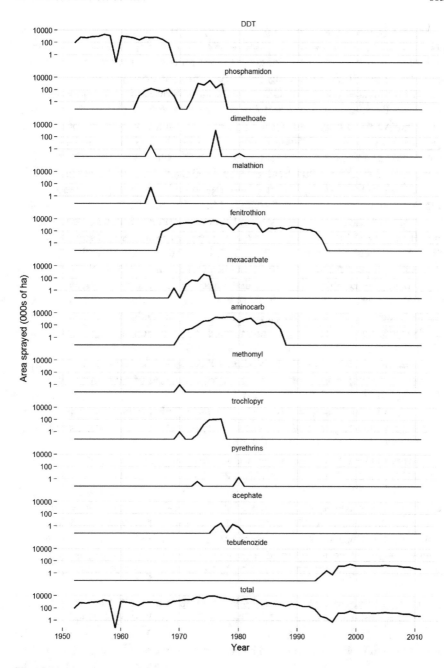

Fig. 17.11 Insecticides used for management of spruce budworm, *Choristoneura fumiferana*, in Canada (Figure credit: C.J.K. MacQuarrie, Natural Resources Canada; reproduced from Holmes and MacQuarrie [2016])

Mass trapping of insects is a population reduction technique in which pest insects are lured into pheromone-baited traps or stressed trees for purposes of collection and removal. Most attempts to mass trap forest pests have focused on bark beetles (Seybold et al. 2018). Evaluating the effectiveness of these programs is difficult, and often hindered by the inability to accurately estimate insect populations and determine the impact of mass-trapping on local pest populations or their damage. An extensive mass trapping program was used to reduce populations of European spruce beetle, but it was difficult to determine whether declining populations were due to mass trapping or to coincidental natural abiotic and biotic factors (Bakke 1991; Weslien 1992a). Despite this, Schlyter et al. (2001) provided evidence that northern spruce bark beetle, *Ips duplicatus* (Sahlberg), populations were reduced by mass trapping using pheromone-baited traps in Mongolia. Similarly, traps and trap logs were used to reduce tree mortality from European spruce beetle in spruce forests of Italy (Faccoli and Stergulc 2008). Another case where success was documented, in terms of total beetle captures and reduction in log degrade, was with ambrosia beetles in western North America. Ambrosia beetles create small holes in the sapwood of logs where they also inoculate fungi that stain and degrade the wood. Pheromone-baited traps were deployed in log yards to reduce populations of ambrosia beetles and to reduce log damage (McLean and Borden 1979; Lindgren and Borden 1983), and later developed into a larger annual mass trapping effort (Lindgren and Fraser 1994; Borden and Stokkink 2021). In other systems, successful reduction in damage has been less clear (Bakke 1991; Weslien 1992a; Ross and Daterman 1997). In an attempt to develop a tool that may aid in the reduction of pine wood nematode, *Bursaphelenchus xylophilus* (Steiner & Buhrer) Nickle, in Europe, mass trapping of the cerambycid, *Monochamus galloprovincialis* (Olivier), was tested and showed promise for reducing *M. galloprovincialis* populations at local scales as part of an IPM program (Sanchez-Husillos et al. 2015).

The use of pheromone-baited traps for mass trapping of bark beetles has disadvantages. First, many of these semiochemicals elicit strong responses from insects and can result in spillover attacks on adjacent healthy trees. Consequently, managers should carefully consider placement of pheromone-baited traps and select areas where tree loss is acceptable and where attacked trees can be removed to further reduce local bark beetle populations (Ross and Daterman 1997). Another issue related to mass trapping is the large numbers of beneficial insects, especially natural enemies, which are also captured, but this can be reduced by trap modifications (Ross and Daterman 1998).

The use of trap trees can serve a similar function to mass trapping. Trap trees are baited with semiochemicals or induced (e.g. herbicide treatment) to release attractive semiochemicals in order to stimulate attack by target taxa. When implemented as part of a suppression plan, trap trees can be chemically treated with contact insecticides or injected with systemic insecticides to kill arriving insects (i.e. attract and kill) (Lister et al. 1976; Lanier and Jones 1985; Gray et al. 1990; Drumont et al. 1992; Hansen et al. 2016; McCullough et al. 2016) or left untreated to allow successful colonization by insects. Untreated trap trees are then removed and destroyed before the next generation of insects emerges (Bakke 1989). Trap trees have been implemented in several countries for management of European spruce beetle (Bakke 1989) and have been effective (Raty et al. 1995), but they were less effective than pheromone-baited traps for suppressing Douglas-fir beetle, *Dendroctonus pseudotsugae* Hopkins, populations in western North America (Dodds et al. 2000; Laidlaw et al. 2003).

Push–pull systems use a combination of tools in an attempt to manipulate insect populations to colonize one area and ignore another. Common components of push–pull systems include attractants, such as aggregation pheromones, with simultaneous use of repellants, such as antiaggregation pheromones (Cook et al. 2007). For this method to be practical, the insect must have a strong response to semiochemicals used as attractants and repellants. Because of this, successful push–pull systems are rare but promising results have been demonstrated for a few bark beetle species (Lindgren and Borden 1993; Ross and Daterman 1994; Borden et al. 2006; Gillette et al. 2012; Seybold et al. 2018). However, because of concerns about spillover attacks induced by the use of synthetic pheromone, push only (repellent) treatments have received more attention. Of note, a push–pull strategy was ineffective in protecting trap trees (used for application of the biocontrol nematode for *S. noctilio*) from colonization by the eastern fivespined ips, *Ips grandicollis* (Eichoff), in Australia (Carnegie and Loch 2010).

Push–pull systems have also been tested on tree pests in settings outside of forests. In a nursery setting, a push–pull system using ethanol-baited traps and verbenone was unsuccessful at protecting trees from attack by non-native ambrosia beetles (Werle et al. 2019). The use of attractive UV lights to pull burnt pine longicorn, *Arhopalus ferus* (Mulsant), away from log storage and processing facilities has been tested in New Zealand as a means of reducing infestation of logs (Pawson and Watt 2009).

Biological Controls

Biological control is the reduction of pests through the activity of one or more biological control agents. We use the terminology of Eilenberg et al. (2001) who describe four different strategies: classical, inoculation, inundation, and conservation. Classical biological control involves *"the introduction of an exotic, usually co-evolved, biological control agent for permanent establishment and long-term pest management"* (Eilenberg et al. 2001). The main goal of classical biological control is the permanent establishment of biological control agent(s) to provide long-term pest control. It usually involves the importation and release of insect parasitoids (or occasionally predators) to control non-native insect pests. Classical biological control has by far been the most common method of biocontrol used in forest pest management, and has had reasonable success, providing long-term control of tree pests in 34% of cases (Kenis et al. 2017). In Canada, more than 150 species of biocontrol agents have been released against 41 different forest insects, resulting in long-term control of nine target species, all defoliators, such as the winter moth, *Opherophtera brumata* (L.), the larch casebearer, *Coleophora laricella* (Hübner), and the European spruce sawfly, *Gilpinia hercyniae* (Hartig) (MacQuarrie et al. 2016). Parasitoid wasps released and established in the Southern Hemisphere for control of *S. noctilio*, primarily *Ibalia leucospoides* Hochenw. and *Megarhyssa nortoni* (Cresson) (Taylor 1976; Cameron 2012), can provide up to 50% parasitism (Carnegie et al. 2005a; Collett and Elms 2009). Some other examples include *Diaeretus essigellae* Starý and Zuparko for control of the Monterey pine aphid, *Essigella californica* (Essig.), in Australia (Kimber et al. 2010) (see Box 17.2), the predatory beetle *Rhisophagus grandis* Gyll. for control of the great spruce bark beetle, *Dendroctonus micans* Kug., in Great Britain (Evans and Fielding 1994), and the egg parasitoid, *Avetianella longoi* Siscaro, for control of the woodborer, *Phoracantha semipunctata* Fabr., in California, U.S. (Hanks et al. 1996).

In contrast to these successes, the introduction of more than 700,000 individuals of about 33 different predator species provided no measurable control of the balsam woolly adelgid, *Adelges piceae* (Ratzeburg), in North America (Kenis et al. 2017). A large-scale classical biological control program underway to control the highly destructive and invasive emerald ash borer in North America has successfully established egg and larval parasitoids and measured some impact on populations (Duan et al. 2014), but its long-term success remains uncertain (Bauer et al. 2015; Jennings et al. 2016). Non-target effects can also be significant, as the case of *Compsilura concinnata* (Meigen) illustrates. A highly generalist tachinid parasitoid introduced into North America in 1906 to control *L. dispar* and browntail moth, *Euproctis chrysorrhoea* (L.), *C. concinnata* had little impact on *L. dispar*, effectively controlled browntail moth, but likely caused the decline of several species of silk moths (Elkinton and Boettner 2012). Analyses of cost: benefit ratios of classical biological control of forest pests are rare but have been estimated at 1:15 for the winter moth and 1:19 for the European spruce sawfly, compared to about 1:2.5 for most chemical control programmes (Tisdell 1990).

Inoculation biological control is "*the intentional release of a living organism as a biological control agent with the expectation that it will multiply and control the pest for an extended period, but not permanently*" (Eilenberg et al. 2001). The distinguishing feature of this strategy is that control is not permanent, and additional releases of the biological control agent are necessary. An example of this strategy is the annual release of the nematode *Deladenus siricidicola* Bedding for *S. noctilio* in the Southern Hemisphere (Bedding and Akhurst 1974). The nematode is mass cultured and inoculated into trap trees weakened by herbicide treatment or girdling to increase attraction and susceptibility to colonization by the woodwasp. The nematodes infect the woodwasp larvae and render adult females sterile, effectively filling the woodwasp eggs with juvenile nematodes. Infected females then spread the infection when they lay their sterile nematode-filled eggs into other host trees (Bedding and Akhurst 1974) (see Sect. 17.3.3).

Inundation biological control is "*the use of living organisms to control pests when the control is achieved exclusively by the released organisms themselves*" (Eilenberg et al. 2001). In this strategy, the released biological control agents must control a sufficiently high proportion of the pest population, or reduce damage significantly, before dispersing or dying. Control relies solely on the released biological control agent(s), not on their progeny. Examples of the use of this strategy in forestry are rare, but field trials in Canada showed that inundative releases of the native egg parasitoid, *Trichogramma minutum* Riley, suppressed spruce budworm populations and reduced defoliation (Smith et al. 1990a, b). Unfortunately, populations collapsed before commercial production could be made viable and the method was never used operationally (MacQuarrie et al. 2016). In the southeastern U.S., inundative releases of encapsulated *Trichogramma exiguum* Pinto and Platner increased rates of egg parasitism in the Nantucket pine tip moth, *Rhyacionia frustrana* (Comstock), but was considered impractical as a control strategy due to high predation of encapsulated *T. exiguum* by ants (Asaro et al. 2003).

Conservation biological control is the "*modification of the environment or existing practices to protect and enhance specific natural enemies or other organisms to reduce the effect of pests*" (Eilenberg et al. 2001). This strategy includes activities that protect or enhance populations of biological control agents, such as reduced or more targeted use of pesticides (e.g. Cadogan et al. 1995) or providing alternate hosts and food sources for natural enemies. For example, supplemental feedings of southern pine beetle parasitoids in the laboratory and field with Eliminade™, an artificial diet consisting largely of sucrose, was shown to increase their longevity and fecundity (Stephen and Browne 2000).

Box 17.2 Case history of classical biological control: Monterey pine aphid, *Essigella californica*

The Monterey pine aphid is native to western North America where it feeds on pines (Sorensen 1994) but is not considered of economic importance (Ohmart 1981). It was detected in Australia in 1998 (Carver and Kent 2000) and once established, spread quickly throughout the major pine growing regions (Anon. 2000; Carver and Kent 2000). The Monterey pine aphid has been associated with severe chlorosis and defoliation across much of the Monterey pine, *Pinus radiata* D. Don, plantation estate in mainland Australia and is considered a significant pest, especially following years of below-average rainfall (Eyles et al. 2011; Stone et al. 2013a, b) (Fig. 17.12a, b, c). Defoliation tends to be more severe in mid-rotation (16–20-year-old) to mature (30–35-year-old) stands, more often in the upper crown, and can cause up to 95% crown loss. In some years, 30–45% of the plantation estate is impacted. Defoliation by the aphid has been estimated to cause losses of AU$21 million in annual wood production (May 2004). Investigations into management options in Australia determined that biological control would be the most cost-effective option, with an estimated net present value of around AU$15 million over 30 years and providing a benefit in perpetuity (May 2004). The aphid's only known parasitoid, *D. essigellae*, was described from museum specimens but live specimens had not been observed in the field (Kimber et al. 2010) (Fig. 17.12d, e). *Diaeretus essigellae* were subsequently located in California and imported to

Fig. 17.12 The Monterey pine aphid, *Essigella californica*, on *Pinus radiata* in Australia—(a) defoliation of upper crown, compared to unaffected tree, (b) chlorosis in mature stand, (c) aphids on branch and needles (note needle chlorosis); (d) female *Diaeretus essigellae*; and (e) *D. essigellae*-parasitized *E. californica* (mummies) on pine needles (Photo credits: A. Carnegie [a, b, c, e], Forests and Wood Products Australia [d])

an approved quarantine facility where host-specificity testing was performed (Kimber et al. 2010). The first releases occurred in late 2009, with subsequent releases during the next three years in all Australian States where Monterey pine aphid was present. Annual monitoring for *D. essigellae* has occurred since 2010. There was initial concern when no established populations were detected by 2012 and no further releases were planned after 2012. However, in 2013, a few established populations were detected, some up to 50 km from release points. By 2017, surveys found *D. essigellae* established in most pine growing regions in New South Wales and Victoria, and more recently in South Australia, but not in Tasmania, Queensland, or Western Australia. Established populations have been found over 100 km from release sites. In terms of impact, the area of New South Wales affected by the aphid has decreased following release of *D. essigellae*. However, in some years (e.g. 2017) there were localized areas with high populations of both the Monterey pine aphid and *D. essigellae*.

Cultural Tactics

Sanitation is a cultural tactic that involves the identification of currently infested trees, and subsequent felling and removal or treatment to destroy pests within the tree in order to reduce pest populations. At the smallest scale, this may include *pruning* of affected portions of the tree (e.g. twigs or branches). Sanitation is commonly employed for management of bark beetles in Europe and North America (Fettig and Hilszczański 2015). Where it is economically feasible, trees may be harvested and transported to mills where broods are killed during processing and milling of lumber and some economic return may be realized (Fig. 17.13). Otherwise, felled trees are burned, chipped, debarked, or treated by solarization (placement of infested material in the direct sun, which is often sufficient to kill brood beneath the bark in warmer climates). In some cases, an emphasis is placed on sanitation of newly infested trees during the early stages of tree colonization in order to reduce the quantity of attractive semiochemicals (e.g. aggregation pheromones) released into the stand. Synthetic attractants may be used to concentrate existing infestations within small groups of trees prior to sanitation (see Sect. 17.1.4.2.2).

Sanitation is used to disrupt the unique attack behavior of southern pine beetle in the southeastern U.S., which relies on the release of aggregation pheromones by pioneering beetles for initial (*spot*) infestations to expand. By harvesting and processing southern pine beetle-infested trees, plus a buffer strip of uninfested trees, spot growth can be halted and some economic return realized. However, timely sanitation is often not possible due to limitations in labor, processing, milling and other factors. In these cases, the best alternative consists of felling all freshly attacked and brood-bearing trees toward the center of the spot. In addition, a horseshoe-shaped buffer of uninfested trees at the spot's expanding front is felled to help disrupt pheromone plumes and recruitment of other southern pine beetle (Fettig et al. 2007). Typically, the width of the buffer is equivalent to the height of the average tree in the stand, although actual buffer width (3–90 m) varies depending on spot size and the rate of spot growth (i.e. numbers of recently attacked trees).

Salvage involves harvesting and processing dead trees usually to recover some economic value that would otherwise be lost or for safety concerns as dead trees

Fig. 17.13 Sanitation is considered the most effective tactic for reducing levels of tree mortality attributed to European spruce beetle, *Ips typographus*, in Germany and many other countries. During outbreaks, it is common for large numbers of currently infested trees to be harvested, decked, and transported to local mills for processing (Photo credit: C. J. Fettig)

pose hazards to forest visitors and workers. Technically, salvage is not a suppressive tactic as its implementation has no immediate effect on insect populations in most cases. However, the term commonly appears in the literature, and in certain cases salvage of damaged, broken or windthrown trees may limit future increases in insect populations. Salvage is commonly used for management of European spruce beetle (Wermelinger 2004; Fettig and Hilszczański 2015) as outbreaks are often incited by windstorms, which provide an abundance of downed and weakened host material that fosters rapid increases in European spruce beetle populations (Marini et al. 2017). In some situations, timing of salvage operations is critical to reduce economic losses, e.g. pine plantations in Australia weakened by fire or windstorms are susceptible to infestation by eastern fivespined ips and its associated blue stain fungus—the sooner timber is salvaged the lower the loss in value due to blue stain (Wylie et al. 1999).

17.1.5 Integrating IPM Within Overall Forest Ecosystem Management

Ideally, pest managers should work directly with foresters and other natural resource managers to develop forest management plans. This ensures that tactics to prevent or reduce pest impacts are considered from the outset (see Sect. 17.1.4.1). Clearly, IPM cannot focus on solving individual pest problems to the exclusion of other natural processes and management actions. For example, use of fertilizers to increase tree growth rates must be balanced against potentially greater feeding damage by pests such as the cottonwood leaf beetle, *Chrysomela scripta* F., and the cottonwood leafcurl mite, *Tetra lobulifera* (Keifer) (Coyle et al. 2005). Society's demand for multiple and sometimes conflicting values from forests, such as wildlife habitat, recreation, biodiversity, and forest products, requires that IPM be practiced within a framework of ecosystem management (Alfaro and Langor 2016). Integration of pest management within overall forest management planning is a start. Tools that can facilitate this process are models, risk analysis frameworks and decision support systems.

17.1.5.1 Modeling as an IPM Tool

Models are useful tools to synthesize what is known about processes affecting pest populations and pest density-damage relationships so that predictions can be made under different management scenarios. They are also useful for revealing gaps in understanding and directing future research. Many models have been developed to describe pest population dynamics and improve IPM of eruptive defoliators like the spruce budworm (Régnière and You 1991; Sturtevant et al. 2015), spongy moth (Liebhold et al. 1998; Sharov and Liebhold 1998a, b; Tobin et al. 2004), and forest tent caterpillar (Cooke et al. 2012), as well as bark beetles (Logan et al. 1998; Perez and Dragicevic 2010; Duncan et al. 2015). Statistical regression models quantify relationships between pest density and damage to determine action thresholds (Johns et al. 2006; Fry et al. 2008). Temperature-driven phenology models like BioSIM (Régnière 1996; Régnière et al. 2014) and the spongy moth life stage model (Gray 2010) forecast the timing of events in a pest's life cycle. Models have been used to improve the efficacy of sampling and control methods (Tobin et al. 2004), to predict changes in pest distributions under climate change scenarios (Carroll et al. 2006; Régnière et al. 2009), and to assess the risk of invasive species establishment (Pitt et al. 2007; Yemshanov et al. 2009; Gray 2010, 2016, 2017). Models have also explored factors affecting dispersal patterns and rates of spread of native and exotic, invasive forest pests (Sharov and Liebhold 1998a, b; Prasad et al. 2010; Křivan et al. 2016). Finally, outputs from models of pest population dynamics and impacts can be integrated with models of stand dynamics and forest inventory to develop decision support systems that can forecast damage and guide management actions.

17.1.5.2   Risk Analysis Frameworks and Decision Support Systems

Risk analysis frameworks can help determine the risks of economic losses from major insect pests and identify optimal management responses (Fuentealba et al. 2013; Nealis 2015). They are particularly useful for facilitating cooperation among multiple landowners and jurisdictions with a common pest problem but with different priorities and policies (Nealis 2015) (see Sect. 17.2.1). The basic elements of risk analysis frameworks are risk assessment, risk response, and risk communication (Nealis 2015). Risk assessment and response address the questions of "what do we know?", "what does it mean?" and "what should we do?" and works in an iterative fashion. Risk communication establishes a back-and-forth dialogue between managers and multiple stakeholders to establish community-based estimates of risk tolerance and to make the process as transparent as possible.

Using a risk assessment framework, Alfaro and Fuentealba (2016) outlined four steps when planning forest stand regeneration. First, the probability of pest damage is evaluated based on site variables such as climate, exposure, soil type, proximity to sources of potential pests, e.g. an early risk rating model for white pine weevil indicated low risk of damage to Sitka spruce in areas with low temperatures and high humidity (McMullen 1976). Secondly, models are used to forecast economic impacts at different infestation levels and integrated with cost and efficacy of control measures to determine the infestation threshold(s) where costs of pest damage exceed those of control actions. Third, the vulnerability of planting stock is determined and the tree species and provenance with least risk of pest damage are selected. For many years, Sitka spruce was not considered for regenerating many areas of British Columbia due to the impact of white pine weevil, but this has changed since the development of weevil-resistant Sitka spruce (Alfaro et al. 2013). The final step is to determine the appropriate management response, based on the previous three steps, and to communicate the level of risk to managers (Alfaro and Fuentealba 2016; Alfaro and Langor 2016).

Decision support systems incorporate models of stand-level and forest-level dynamics with models of pest population dynamics and pest impact to forecast damage and prioritize areas for management. They are particularly useful in regions that experience large-scale cyclical outbreaks of pests with significant impacts on tree mortality or tree growth and yield, e.g. the spruce beetle in Alaska (Reynolds et al. 1994), mountain pine beetle in western North America (Shore and Safranyik 2003) (see Sect. 17.3.2), spruce budworm in eastern North America (MacLean et al. 2001; Hennigar et al. 2007) (see Sect. 17.3.1), and spruce weevil in British Columbia (Alfaro et al. 1997). The SBWDSS forecasts the impact of budworm outbreaks on tree volume growth and tree mortality and integrates stand-level impacts across the forest to set priorities for foliage protection and/or harvest. Management of mountain pine beetle aims to keep populations in the endemic phase using direct controls (e.g. removal and burning of currently infested trees) to suppress beetle numbers, and preventive management to reduce the susceptibility of the forest to the beetle (Berryman 1978; Shore and Safranyik 2003) (Fig. 17.14). Parts of the mountain pine

Fig. 17.14 Factors contributing to a shift from endemic to epidemic populations in mountain pine beetle, *Dendroctonus ponderosae*. Management of mountain pine beetle is aimed at keeping populations in the endemic phase using direct controls to suppress beetle populations on a localized basis and preventive controls to increase resistance of trees and stands [after Shore and Safranyik 2003]

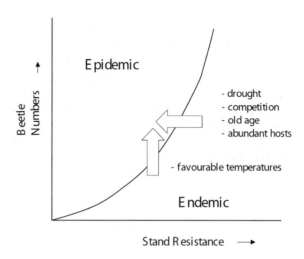

beetle decision support system have been adopted as part of the overall management strategy to contain the spread of mountain pine beetle in Alberta and provide a good example of efforts to integrate IPM within overall forest management planning (Anon. 2007b).

17.2   Constraints to Implementing IPM

17.2.1   Multiple Jurisdictions and Conflicting Priorities

Implementation of IPM within a large and diverse landscape is limited by available knowledge, tools, and budgets, and complicated by multiple landowners with different priorities and responsibilities (Nealis 2015). There are also legal and organizational constraints as well as differences in perspective among stakeholders that affects adoption of IPM in forests (Stark et al. 1985). Pest outbreaks do not recognize jurisdictional boundaries. In some cases, the actions of multiple agencies and jurisdictions are coordinated through pest-specific strategic planning committees. For example, a "Strategic Direction Council" was formed in response to the potential (and eventual) spread of mountain pine beetle into central Alberta, with members representing the Canadian Forest Service, Alberta Sustainable Resource Development, Alberta Community Development, Parks Canada, and forest industry. The council provides broad policy direction and priorities and ensures effective communication among the various agencies, while technical sub-committees develop and recommend specific management actions (Dalman 2003).

Risk analysis frameworks have been used to address multi-jurisdictional impediments to proactive IPM of major forest insect pests (Nealis 2015). A risk analysis framework was used to address the question of whether range expansion of mountain pine beetle to central Alberta represented a significant threat to the boreal forest. It was successful in providing a timely science-based response and led to increased communication and cooperation among jurisdictions, e.g. Saskatchewan partially funds mountain pine beetle monitoring and control actions in Alberta (Nealis 2015). One of the most important benefits of a risk analysis framework is its emphasis on open communication and engagement with the public on the potential risks and uncertainties of pest impacts and management responses. When a risk analysis framework was applied to the spongy moth eradication program in British Columbia, public opposition to the program was reduced and some communities initiated programs to survey and monitor spongy moth populations (Nealis 2009).

17.2.2   Legal, Policy, and Economic Constraints

Most countries have laws and policies designed to protect the environment and to reduce the risk of negative impacts on environmental quality from human activities. For example, the National Environmental Policy Act (NEPA) in the U.S. and the Environmental Assessment Act in Canada require the completion of environmental impact assessments before undertaking any major management actions. Other levels of government have similar laws designed to protect people and the environment from potential harm from pest management activities, e.g. pesticide applications. While these laws are necessary to protect the environment and human health, the requirements can be complex and may constrain the implementation of IPM programs. For example, 60–120 m buffer zones are required around streams and other water bodies to protect aquatic ecosystems from aerially-applied pesticides, which can present a significant operational challenge. Fortunately, the efficacy and efficiency of aerial applications of pest control products has increased significantly in the last couple of decades with ultra-low volume sprays, use of process-oriented models (Cooke and Régnière 1996; Régnière and Cooke 1998), and on-board electronic guidance systems that optimize aircraft flight lines to compensate for changes in altitude and wind direction, and buffer zones (McLeod et al. 2012). These advances have greatly minimized drift of pesticides onto non-target areas and buffer zones (Thompson et al. 2010).

The Environmental Protection Agency (EPA) in the U.S. and the Pest Management Regulatory Agency (PMRA) in Canada have strict requirements for the registration and use of pest control products. Before approval of a new control product, much research and development are necessary to determine its efficacy against the target pest(s) and its impact on human health and the environment, and this requires substantial investment. For example, it costs an estimated $286 million and 11 years to develop and register a new pesticide; not surprisingly, <0.001% of newly discovered active ingredients become registered pest control products (Anon. 2017). For private companies, this largely makes it cost prohibitive to develop anything other than broad-spectrum pesticides with large potential markets to ensure a return on investment. Development and registration of more environmentally benign, species-specific products like baculoviruses or semiochemicals have usually been done with government assistance due to limited markets and return on investment. This is particularly evident in the forestry sector (Gillette and Fettig 2021). Fortunately, since around 2001, regulatory agencies have reduced the requirements for registration of pest control products like pheromones and microbial insecticides that, due to their less toxic nature and greater species-specificity, are considered lower risk (Anon. 2001, 2002). However, even for environmentally benign products, strict data requirements must be met, and costs can be substantial (Lucarotti et al. 2007).

17.2.3 Attitudinal Constraints and Social License

Public opposition to the application of pesticides has severely limited their use in most forests. To that end, it is difficult to imagine present day public acceptance of DDT applications in Stanley Park, Vancouver, Canada, but such applications were made in the 1950s for control of hemlock looper and the greenstriped forest looper, *Melanolophia imitata* (Walker) (Holmes and MacQuarrie 2016). Between 1985 and 1990, *Btk* gradually replaced synthetic insecticides in spruce budworm control programmes in Canada (Van Frankenhuyzen et al. 2000). However, aerial applications of *Btk* in urban settings for control of exotic, invasive insects like the Asian spongy moth, *Lymantria dispar asiatica* Vnukovskij, remains contentious (Ginsburg 2006) in spite of studies that indicate no effects on public health (Green et al. 1990). Interestingly, a survey of public attitudes toward control of forest insects in Ontario, Canada found that acceptance of insecticide use by the public was higher in people that had experienced outbreaks of pest insects (MacDonald et al. 1998).

The Forest Stewardship Council (FSC) and similar organizations were established to promote sustainable forest management and environmental integrity and they influence many pest management practices in forestry. Many consumers and some companies will not purchase forest products from industries that do not follow FSC standards. In 2015, an estimated 181 million ha of forests in 80 countries were FSC certified (Zanuncio et al. 2016). Pesticides designated as highly hazardous are severely restricted and cannot be used in FSC-certified estates "...without specific derogation, regardless of prevailing national approvals system" (Anon. 2007a). FSC prohibition of certain insecticides has made it difficult to manage some forest insect pests, e.g. chrysomelid beetles in eucalyptus plantations (Carnegie et al. 2005b) and leaf-cutting ants, *Atta* and *Acromyrmex* spp., in Brazil (Zanuncio et al. 2016).

17.3 IPM Programs for Major Forest Insect Pests

17.3.1 Spruce Budworm in Eastern North America

The spruce budworm is a major defoliating pest of balsam fir, *Abies balsamea* (L.) Mill., and spruce forests in North America. Populations erupt every 30–40 years and outbreaks last 1–20 years (Gray et al. 2000), causing extensive tree growth loss and tree mortality (MacLean 2016). Between 1975 and 2000, budworm caused moderate to severe defoliation on >450,000,000 ha of forests in Canada, reducing radial growth rates by as much as 75% and killing an average of 85% of trees in mature balsam fir stands and 36% of trees in mature spruce stands (MacLean 2016; Miller 1977). In the past few decades, spruce budworm management has consisted mainly of aerial applications of insecticides to protect foliage and reduce volume loss and tree mortality. Surveys are done by aerial sketch mapping to locate areas of moderate to severe defoliation, and by branch sampling of the overwintering second instar larvae to estimate budworm abundance and to predict subsequent levels of defoliation. Phenology models like BioSIM aid managers in the most effective timing of *Btk* applications (Régnière and Sharov 1998; Régnière et al. 2014).

The SBWDSS (Erdle 1989; Hennigar et al. 2007; MacLean et al. 2001), which projects the effects of budworm outbreaks on stand volume growth and tree mortality, is built on four relationships that quantify budworm impacts: (1) forecasts of budworm population levels over time, based on population dynamics models and previous outbreak cycles; (2) the relationship between pest population level and defoliation level; (3) the relationship between damage level (defoliation) and tree/stand growth loss and mortality; and (4) effects on the forest landscape as a function of accumulated stand-level impacts (Erdle and MacLean 1999). For defoliators such as spruce budworm, the main factor influencing the impact on tree growth loss and mortality is cumulative defoliation (MacLean 2016). The model allows managers to compare the effects of various scenarios of foliage protection, harvest scheduling, and salvage on future wood supply. Additional software, called the *Accuair Forest Protection Optimization System* (ForPRO), integrates the SBWDSS with the Woodstock timber supply model and allows users to simulate the impacts of spruce budworm on stand- and forest-level growth and yield under different protection scenarios (Hennigar et al. 2007) (Fig. 17.15). The SBWDSS and ForPRO have been tested in Maine, U.S., and Quebec, Alberta, Saskatchewan, New Brunswick, and Newfoundland, Canada, and were used to plan insecticide spray operations for several years during the latest budworm outbreak in Saskatchewan.

Faced with the likelihood of another spruce budworm outbreak in New Brunswick, the 'Healthy Forest Partnership' (a consortium of federal and provincial governments, forest industry, and universities) was formed in 2014 with the goal of testing an 'early intervention strategy' against spruce budworm (Anon. 2022). The strategy involves intensive monitoring and early detection of populations in supposed "hotspots" and suppression of these populations with applications of *Btk* or tebufenozide (Mimic®) to prevent the outbreak from spreading via moth dispersal (MacLean 2016). The

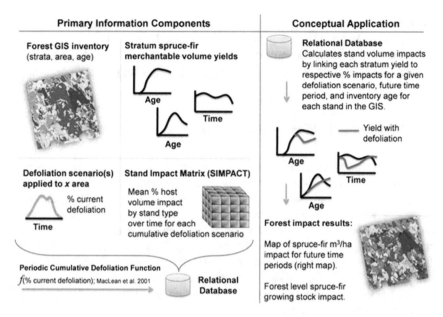

Fig. 17.15 Information sources required and conceptual application of the SBWDSS to calculate spruce-fir stand volume impacts for alternative spruce budworm, *Choristoneura fumiferana*, defoliation scenarios (From: C. H. Hennigar 2009)

most significant change in management strategy is the switch from foliage protection to that of population suppression. This has been controversial and is based on the notion that spruce budworm outbreaks spread from localized foci or hotspots, as suggested by early models of spruce budworm population dynamics (Morris 1963; Clark et al. 1979). That early notion had been rejected in subsequent analyses of budworm population dynamics (Royama 1984, 2012), which instead hypothesized that the oscillation in budworm populations was driven by density-dependent mortality from a complex of natural enemies (Eveleigh et al. 2007), and that dispersal of egg-carrying moths contributed only to secondary fluctuations or "noise" about the basic cycle. Subsequent studies supported Royama's hypothesis that mortality from natural enemies is the main driver of budworm population cycles, but also suggested that food availability and moth dispersal may play larger roles in population dynamics, e.g. an influx of moths might counteract mate-finding Allee effects observed in low-density populations (Régnière et al. 2013; Pureswaran et al. 2016).

As part of the early intervention strategy, SBWDSS and ForPRO have been modified from the original objective of foliage protection to that of population suppression based on overwintering budworm larval populations and were used to plan budworm suppression treatments in New Brunswick and Newfoundland from 2015–2020 (Johns et al. 2019). It is too early to determine the success of the budworm early intervention strategy, but results from the first five years are promising (MacLean et al. 2019). The program has generated data on dynamics of low-density budworm populations, fostered communication and collaboration among government, industry, and academia, and is stimulating public involvement through citizen science initiatives to help monitor budworm dispersal events through an extensive network of pheromone-baited traps and smart phone applications (Carleton et al. 2020).

17.3.2 Mountain Pine Beetle in Western North America

In western North America, about 15 species of bark beetles are capable of causing large amounts of tree mortality (Bentz et al. 2020). Most notable is the mountain pine beetle, which colonizes several tree species, including lodgepole pine, ponderosa pine, *Pinus ponderosa* Dougl. ex Laws., sugar pine, limber pine, *Pinus flexilis* James, western white pine, *Pinus monticola* Dougl., and whitebark pine, *Pinus albicaulis* Engelm., among others (Negrón and Fettig 2014). Recent outbreaks of mountain pine beetle have been severe, long lasting, and well documented (Audley et al. 2020; Fettig et al. 2021). While a formal IPM program is not universally recognized for mountain pine beetle, many of the associated components have been developed and are being implemented at different scales.

Information on the intensity and extent of mountain pine beetle infestations is most often accomplished by aerial detection surveys using fixed-wing aircraft and/or helicopters (Wulder et al. 2006) followed by ground-based surveys of areas with noticeable levels of tree mortality. During surveys, a common method of estimating when trees were colonized and killed by mountain pine beetle uses needle color and retention with three stages: *green stage* (within one year of attack; green foliage or foliage just beginning to fade); *red stage* (1–3 years since death; red foliage); and *grey stage* (>3 years since death; grey, limited or no foliage). However, it is important to emphasize that these are crude estimates that may vary by several years from the actual time since tree death.

Several risk and hazard rating systems have also been developed to describe the susceptibility of a stand to infestation by mountain pine beetle. The most frequently used was developed by Shore and Safranyik (1992) for lodgepole pine. Stand susceptibility is calculated based on four factors: (1) percentage of susceptible basal area (trees ≥15 cm dbh), (2) average stand age of dominant and co-dominant trees, (3) stand density of all trees ≥7.5 cm dbh, and (4) the geographic location of the stand in terms of latitude, longitude, and elevation. Insect population data, referred to as a "beetle pressure index", incorporates the proximity and size of the mountain pine beetle population. The stand susceptibility index and the beetle pressure index are then used to compute an overall stand risk index (Shore and Safranyik 1992).

The first documented use of suppressive tactics for mountain pine beetle occurred in the early 1900s in the Black Hills of South Dakota and Wyoming, U.S. (Hopkins 1905). Today, strategies often aim to reduce localized populations, slow the rate of spread of infestations, and to provide protection of individual trees or stands. These focus on the use of insecticides, semiochemicals and sanitation harvests. Coggins et al. (2011) found that mitigation rates of >50% (sanitation harvests) coupled with ongoing detection and monitoring of infested trees within treated sites in British Columbia was sufficient to control mountain pine beetle infestations, especially with persistent implementation. Alternatively, other researchers have stressed that many large-scale, well-funded and well-coordinated sanitation efforts were largely ineffective, and that resources would be better allocated to prevention (e.g. Wickman 1987). Sanitation is likely to be most effective if the following IPM principles are followed: (1) early detection, (2) rapid response, (3) continued monitoring, and (4) persistent application of suppressive treatments until populations return to endemic levels.

Age-class structure and tree species composition are dominant factors influencing the extent and severity of mountain pine beetle infestations (Taylor and Carroll 2003). Preventive tactics, such as thinning, that address these factors will influence the susceptibility of forests to mountain pine beetle infestations. Among other factors, thinning reduces host availability; reduces competition among trees for nutrients, water, and other resources thereby increasing vigor; and affects microclimate, decreasing the effectiveness of chemical stimuli used by mountain pine beetle in host finding, selection and colonization (Progar et al. 2014). Thinning implemented for mountain pine beetle in lodgepole pine, where the species has its greatest impacts, include thinning from above or diameter-limit thinning, and thinning from below (i.e. focusing on removal of trees in the suppressed and intermediate crown classes) applied to reduce basal area, remove trees with thick phloem, and/or increase residual tree spacing (Fettig et al. 2014). Thinning from below may optimize the effects of microclimate, inter-tree spacing, and tree vigor even though residual trees are of diameter classes considered more susceptible to mountain pine beetle (Mitchell et al. 1983). Bollenbacher and Gibson (1986) provide a list of attributes useful for assessing the potential effectiveness of thinning for reducing the probability of mountain pine beetle infestation and extent of tree mortality in lodgepole pine forests (Table 17.1).

Cottrell et al. (2020) examined the current state of knowledge regarding institutional, social, and environmental factors that influence the ability to manage

Table 17.1 Favourable conditions for reducing the probability of mountain pine beetle, *Dendroctonus ponderosae,* infestation and extent of tree mortality by thinning of lodgepole pine, *Pinus contorta,* forests in the Intermountain West, U.S. (adapted from Bollenbacher and Gibson 1986)

Parameter	Value
Stand composition	>80% *Pinus contorta*
Stand age	60–110 years
Basal area	>29.8 m^2/ha
Stand density	750–1500 trees/ha (>7.5 cm dbh[1])
Average diameter	>20 cm dbh
Elevation	<1800 m
Percentage of trees currently infested	<10%

[1] Diameter at breast height

mountain pine beetle (i.e. "adaptive capacity"). Three main categories were identified: (1) environment including stressor (i.e. mountain pine beetle), exposure (i.e. system connectivity) and sensitivity (i.e. forest health) factors; (2) society including impacts (i.e. metrics), public opinion (i.e. communication, perceptions, and attitudes), and management (i.e. proactive and reactive); and (3) ecosystem services including aesthetics, air quality, carbon sink/source, timber resources and water quality/quantity. Their research provides a framework for managers and policy-makers that is useful in identifying limitations in adaptive capacity in hopes of addressing them more effectively in the future. Public opinion, and the availability of human and financial capital, were identified as significant constraints.

17.3.3   Sirex noctilio *in Australia*

Sirex noctilio is native to Eurasia and northern Africa and is now a significant pest in exotic pine plantations in the Southern Hemisphere (Slippers et al. 2011) (Fig. 17.16a, b). Females lay eggs into trees along with a white rot fungus, *Amylostereum areolatum* (Chaillet ex Fries) Boidin, and a phytotoxic mucus (or toxin); the combination of which eventually kills the tree (Ryan and Hurley 2012). First reported in New Zealand in 1900, it was not until severe droughts occurred in unthinned, over-stocked stands that *S. noctilio* became a serious pest, with 33% of trees killed on 120,000 ha between 1946 and 1951 (Morgan and Stewart 1966; Bain et al. 2012). *Sirex noctilio* was first detected in Australia in Tasmania in 1952 (Gilbert and Miller 1952), on the mainland in 1961 (Irvine 1962), and is now established in major pine growing regions in Victoria, South Australia, New South Wales, and Queensland (Neumann et al. 1987; Carnegie and Bashford 2012). Between 1987 and 1989 >5 million trees were killed in a single area in southern Australia (Haugen 1990), and a contemporary analysis of this outbreak calculated the value of lost wood production at AU$22.3 million (Cameron et al. 2018a). *Sirex noctilio* is also a significant pest in major pine growing regions in South America (Iede et al. 2012; Klasmer and Botto 2012) and South Africa (Hurley et al. 2012).

In response to severe losses in Australia, a *Sirex Management Strategy* was developed (Haugen et al. 1990), which includes biological control (Bedding et al. 1993), forest surveillance, quarantine and silvicultural methods (National Sirex Control Committee 2022). The first attempts at biological control occurred in New Zealand in the late 1920s and 1930s, with releases of *Rhyssa persuasoria* (L.) (Cameron 2012). *Ibalia leucospoides* was later introduced and established in the 1950s, followed by *M. nortoni* in the 1960s (Fig. 17.16c). In response to the establishment of *S. noctilio* on the mainland of Australia, the Sirex Biological Control Unit was established in the United Kingdom by the Commonwealth Scientific and Industrial Research Institute (CSIRO) with the aim to collect and identify parasitoids of potential use in Australia (Carnegie and Bashford 2012). Over 20 parasitoids were collected and sent to Australia and New Zealand, including *Schlettererius cinctipes* (Cresson), *I. leucospoides, M. nortoni,* and *R. persuasoria* (Taylor 1976; Nuttall 1989; Hurley

Fig. 17.16 *Sirex noctilio*: **a** tree mortality from an outbreak in a 13-year-old Monterey pine, *Pinus radiata*, plantation in Australia; **b** *S. noctilio* female; **c** *Megarhyssa nortoni* female; **d** *Deladenus siricidicola* being injected into trap tree (Photo credits: A. Carnegie [a, b, d], New South Wales Department of Primary Industries [c])

et al. 2007). Releases of parasitoids in South America and South Africa mostly originated from Australia and New Zealand (Hurley et al. 2007; Cameron 2012). The most successful parasitoids in Australia and elsewhere have been *I. leucospoides* and *M. nortoni* (Carnegie et al. 2005a; Hurley et al. 2007; Collett and Elms 2009). *Ibalia leucospoides* is the most abundant and effective parasitoid in Australia, with parasitism generally from 30–50% (Carnegie et al. 2005a; Collett and Elms 2009). It is possible that *M. nortoni* abundance is quite high in some areas in Australia, but current sampling techniques are not optimal for monitoring this parasitoid, due to asynchronous emergence times of the pest and parasitoid (Carnegie et al. 2005a; Collett and Elms 2009).

The nematode *D. siricidicola* was first found parasitizing *S. noctilio* in New Zealand (Zondag 1969), likely introduced with *S. noctilio* (Hurley et al. 2007). Subsequent surveys revealed up to 90% parasitism of *S. noctilio* in some forests (Zondag 1979). Concurrent with the CSIRO survey for *S. noctilio* parasitoids was a search for parasitic nematodes, with several hundred strains from seven species screened and tested for parasitism (Bedding and Akhurst 1978; Bedding and Iede 2005). A single strain, Sopron, grew well on *A. areolatum* cultures, did not parasitize the siricid-attacking parasitoids, and emerging wasps were larger than those parasitized by other strains (Bedding and Akhurst 1978; Bedding 2009). This strain resulted in almost 100% parasitism of *S. noctilio* in Australia (Bedding and Akhurst 1974) and was released operationally into areas where *S. noctilio* spread (Fig. 17.16d). In addition to the use of biological controls, silvicultural treatments such as thinning from below, to remove suppressed trees and to increase vigor of the remaining trees, are important IPM components, and are commonly used to reduce the impact of *S. noctilio* in commercial forests throughout the Southern Hemisphere (Dodds et al. 2014).

17.3.4 Eucalyptus Leaf Beetles in Australia

The leaf beetle, *Paropsisterna bimaculata* (Olivier), is endemic to Tasmania, and has long been a significant defoliator of eucalypts in native forests (Greaves 1966; de Little 1979) (Fig. 17.17a, b). Early attempts to establish eucalypt plantations in Tasmania resulted in severe damage to trees from *P. bimaculata* (de Little 1989; Candy et al. 1992), and it was realized that an effective management strategy was needed (Elliott et al. 1992). A large research program ensued from the 1970s to the 1990s that included study of the biology of *P. bimaculata* and its natural enemies (Elliott and de Little 1980; de Little 1983; de Little et al. 1990) and the impacts of defoliation on tree growth (Candy et al. 1992; Elliott et al. 1992; Elek 1997; Candy 2000). From this research, an IPM program for *P. bimaculata* in eucalypt plantations was developed (Elliott et al. 1992) and later refined (Candy 2000, 2003). This program includes: (1) monitoring egg and larval populations in young plantations; (2) predicting damage based on population estimates; and (3) determining economic injury thresholds to guide control decisions, (i.e. aerial application of insecticides) (Fig. 17.17c, d).

Forestry Tasmania (now Sustainable Timber Tasmania) implemented the IPM program operationally in 1998 coinciding with a rapid expansion of the eucalypt plantings in Tasmania (Wardlaw et al. 2018). Research into non-insecticidal control options was initiated during the early stages of the IPM program to avoid disrupting the high levels of natural predation (Elliott et al. 1992) and threats to non-target aquatic animals (de Little 1989), but more recently the focus has turned toward gaining FSC certification (Wardlaw et al. 2018). The use of *Bacillus thuringiensis* var. *tenebrionis* (*Btt*) was investigated (Beveridge and Elek 1999; Elek and Beveridge 1999), but its efficacy was limited to first instar larvae, and it was insufficiently

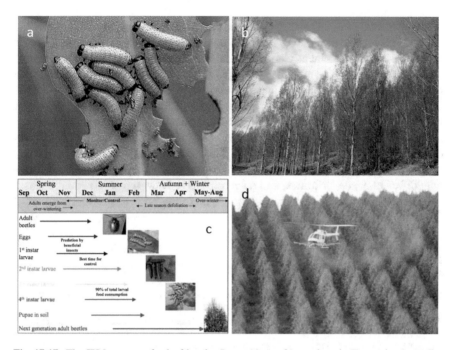

Fig. 17.17 The IPM program for leaf beetle, *Paropsisterna bimaculata*, in Tasmania, Australia: **a** *P. bimaculata* larvae feeding on a young eucalypt leaf; **b** severe defoliation of an 11-year old *Eucalyptus nitens* plantation; **c** life cycle of *P. bimaculata,* used to monitor population levels to optimize control program; and **d** aerial application of insecticides (Photo credits: T. Wardlaw and J. Elek, Forestry Tasmania)

effective when applied operationally (Wardlaw et al. 2018). Success™, one of the spinosyn group of biological insecticides, was shown to be effective against young *P. bimaculata* with no impact on non-target insects (Elek et al. 2004). Although used operationally from 2003–2011, its use progressively declined due to the higher cost and operational complexity compared to α-cypermethrin (Wardlaw et al. 2018). The IPM program for leaf beetles in Tasmania has proven to be cost beneficial over the long term (Cameron et al. 2018b; Wardlaw et al. 2018).

17.4   Summary

Insects have important roles in forest ecosystems as disturbance agents, decomposers, nutrient cyclers, and natural enemies, but a small fraction are considered pests because they compete directly with people for ecological goods and services. The goal of IPM in forests is to keep pest populations and their damage at acceptable levels using a variety of tactics that are ecologically based, cost-effective, and socially and environmentally acceptable. This requires effective tools and methods for survey and monitoring, and an understanding of the relationships between pest populations and their impact on forest resources to determine when and where management actions are necessary. Management tactics may be preventive, e.g. regulatory controls to reduce the risk that wood used to pack shipping containers contains live wood boring beetles, or remedial, e.g. the aerial application of *Btk* or sex pheromone to slow the spread of spongy moth in North America. Ideally, IPM programs should be part of overall ecosystem management and integrated within forest management plans from the outset. Though many IPM strategies and decision support tools have been developed, their operational implementation in forests is often limited by legal, attitudinal, organizational, and financial constraints, and complicated by multiple landowners with different needs, priorities, and responsibilities. However, IPM is continuously evolving through more research, application, and adaptive management.

References

Alfaro RI, Fuentealba A (2016) Insects affecting regenerating conifers in Canada: natural history and management. Can Entomol 148(S1):S111–S137

Alfaro RI, Langor D (2016) Changing paradigms in the management of forest insect disturbances. Can Entomol 148(S1):S7–S18

Alfaro RI, Omule SAY (1990) The effect of spacing on Sitka spruce weevil damage to Sitka spruce. Can J For Res 20(2):179–184

Alfaro RI, Borden JH, Fraser RG, Yanchuk A (1995) The white pine weevil in British Columbia: basis for an integrated pest management system. Forestry Chronicle 71(1):66–73

Alfaro RI, Brown R, Mitchell K, Polsson K, McDonald R (1997). SWAT: a decision support system for spruce weevil management. In: Shore TL, MacLean DA (eds) Decision support systems for forest pest management. Natural Resources Canada Canadian Forest Service Forest Resource Development Agreement Report 260, pp 31–41

Alfaro RI, King JN, vanAkker L (2013) Delivering Sitka spruce with resistance against white pine weevil in British Columbia, Canada. For Chron 89(2):235–245

Allison JD, Redak RA (2017) The impact of trap type and design features on survey and detection of bark and woodboring beetles and their associates: a review and meta-analysis. Annu Rev Entomol 62:127–146

Allison JD, Johnson CW, Meeker JR, Strom BL, Butler SM (2011) Effect of aerosol surface lubricants on the abundance and richness of selected forest insects captured in multiple-funnel and panel traps. J Econ Entomol 104(4):1258–1264

Allison JD, McKenney JL, Miller DR, Gimmel ML (2013) Kairomonal responses of natural enemies and associates of the southern *Ips* (Coleoptera: Curculionidae: Scolytinae) to ipsdienol, ipsenol and cis-verbenol. J Insect Behav 26:321–335

Allison JD, Bhandari BD, Mc Kenney JL, Millar JG (2014) Design factors that influence the performance of flight intercept traps for the capture of longhorned beetles (Coleoptera: Cerambycidae) from the subfamilies Lamiinae and Cerambycinae. PLoS One 9(3):e93203

Allison JD, Strom B, Sweeney J, Mayo P (2019) Trap deployment along linear transects perpendicular to forest edges: impact on capture of longhorned beetles (Coleoptera: Cerambycidae). J Pest Sci 92:299–308

Amman GD (1972) Mountain pine beetle brood production in relation to thickness of lodgepole pine phloem. J Econ Entomol 65(1):138–140

Anderbrant O, Lofqvist J, Hogberg HE, Hedenstrom E (1995) Development of mating disruption for control of pine sawfly populations. Entomol Exp Appl 74(1):83–90

Anon (2000) A new introduction to Tasmania. In: Doyle R (ed) Forest health bulletin. Hobart, Tasmania, pp 1–2

Anon (2001) Regulatory Directive DIR2001-02. Guidelines for the registration of microbial pest control agents and products. Retrieved from https://www.canada.ca/en/health-canada/services/consumer-product-safety/pesticides-pest-management.html

Anon (2002) Regulatory Directive: the PMRA initiative for reduced-risk pesticides. Retrieved from https://www.canada.ca/en/health-canada/services/consumer-product-safety/reports-public ations/pesticides-pest-management/policies-guidelines/regulatory-directive/2002/initiative-red uced-risk-pesticides-dir2002-02.html?=undefined&wbdisable=true#microbial

Anon (2007a) Forest Stewardship Council pesticide policy: guidance on implementation. Retrieved from https://us.fsc.org/preview.fsc-pesticide-policy-guidance-on-implementation.a-193.pdf

Anon (2007b) Mountain pine beetle management strategy. Alberta Sustainable Resource Development. Retrieved from http://www1.agric.gov.ab.ca/$department/deptdocs.nsf/all/formain15803/$file/MPB-ManagementStrategy-Dec2007.pdf

Anon (2017) Discovery to launch—The slow road to bring pesticides to market. Retrieved from http://www.westernfarmpress.com/tree-nuts/discovery-launch-slow-road-bring-pesticides-market

Anon (2022) Healthy Forest Partnership Early Intervention Strategy. Retrieved from https://health yforestpartnership.ca/who-we-are/our-story/

Asaro C, Fettig CJ, McCravy KW, Nowak JT, Berisford CW (2003) The Nantucket pine tip moth: a literature review with management implications. J Entomol Sci 38:1–40

Audley JP, Fettig CJ, Munson AS, Runyon JB, Mortenson LA, Steed BE, Gibson KE, Jørgensen CL, McKelvey SR, McMillin JD, Negrón JF (2020) Impacts of mountain pine beetle outbreaks on lodgepole pine forests in the Intermountain West, U.S., 2004–2019. Forest Ecology and Management. https://doi.org/10.1016/j.foreco.2020.118403

Aukema BH, McKee FR, Wytrykush DL, Carroll AL (2016) Population dynamics and epidemiology of four species of *Dendroctonus* (Coleoptera: Curculionidae): 100 years since J.M. Swaine. Can Entomol 148(S1):S82-S110

Bain J, Sopow SL, Bulman LS (2012) The *Sirex* woodwasp in New Zealand: history and current status. In: Slippers B, de Groot P, Wingfield MJ (eds) The sirex woodwasp and its fungal symbiont: research and management of a worldwide invasive pest. Springer, Dordrecht, pp 167–173

Bakke A (1989) The recent *Ips typographus* outbreak in Norway—Experiences from a control program. Holarct Ecol 12:515–519

Bakke A (1991) Using pheromones in the management of bark beetle outbreaks. In: Baranchikov YN, Mattson WJ, Hain FP, Payne TL (eds) Forest insect guilds: patterns of interactions with host trees. USDA Forest Service General Technical Report NE-GTR-153, Radnor, PA, pp 371–377

Battisti A, Stastny M, Netherer S, Robinet C, Schopf A, Roques A, Larsson S (2005) Expansion of geographic range in the pine processionary moth caused by increased winter temperatures. Ecol Appl 15:2084–2096

Bauer LS, Duan JJ, Gould JR, Van Driesche R (2015) Progress in the classical biological control of *Agrilus planipennis* Fairmaire (Coleoptera: Buprestidae) in North America. Can Entomol 147(3):300–317

Bedding R (2009) Controlling the pine-killing woodwasp, *Sirex noctilio*, with nematodes. In: Use of microbes for control and eradication of invasive arthropods, 213–235

Bedding RA, Akhurst RJ (1974) Use of the nematode *Deladenus siricidicola* in the biological control of *Sirex noctilio* in Australia. Aust J Entomol 13(2):129–135

Bedding R, Akhurst R (1978) Geographical distribution and host preferences of *Deladenus* species (Nematoda: Neotylenchidae) parasitic in siricid woodwasps and associated hymenopterous parasitoids. Nematologica 24(3):286–294

Bedding R, Iede E (2005) Application of *Beddingia siricidicola* for Sirex woodwasp control. In: Grewal PS, Ehlers R-U, Shapiro-Ilan DI (eds) Nematodes as biocontrol agents. CABI Publishing, Cambridge, MA, pp 385–400

Bedding R, Akhurst R, Kaya H (eds) (1993) Nematodes and the biological control of insect pests. CSIRO Publications, Melbourne, Australia

Bentz BJ, Régnière J, Fettig CJ, Hansen EM, Hayes JL, Hicke JA, Kelsey RG, Lundquist J, Negrón JF, Seybold SJ (2010) Climate change and bark beetles of the western United States and Canada: direct and indirect effects. Bioscience 60:602–613

Bentz BJ, Bonello E, Delb H, Fettig CJ, Poland T, Pureswaran D, Seybold SJ (2020) Advances in understanding and managing insect pests of forest trees. In: Stanturf J (ed) Achieving sustainable management of boreal and temperate forests. Burleigh Dodds Science Publishing Ltd, Cambridge, pp 515–585

Berryman A (1978) A synoptic model of the lodgepole pine/mountain pine beetle interaction and its potential application in forest management. In: Kibbee DL, Berryman AA, Amman GD, Stark RW (eds) Theory and practice of mountain pine beetle management in lodgepole pine forests. Proceedings of a Meeting, 25–27 Apr. 1978, Pullman, Wash. University of Idaho, Moscow, ID, pp 95–103

Beveridge N, Elek JA (1999) *Bacillus thuringiensis* var. *tenebrionis* shows no toxicity to the predator *Chauliognathus lugubris* (F.) (Coleoptera: Cantharidae). Aust J Entomol 38:34–39

Billings RF (1988) Forecasting southern pine beetle infestation trends with pheromone traps. In: Payne TL, Saarenmaa H (eds) Proceedings of the symposium: integrated control of scolytid bark beetles. IUFRO Working Party and XVII International Congress of Entomology, 4 July 1988, Vancouver, BC, Canada. Virginia Polytechnical Institute and State University, Blacksburg, VA, pp 295–306

Billings RF, Pase HA (1979) A field guide for ground checking southern pine beetle spots. USDA Agriculture Handbook No. 558, Washington, DC

Billings RF, Upton WW (2010) A methodology for assessing annual risk of southern pine beetle outbreaks across the southern region using pheromone traps. In: Pye JM, Rauscher HM, Sands Y, Lee DC, Beatty JS (eds) Advances in threat assessment and their application to forest and range-land management. USDA Forest Service General Technical Report PNW-GTR-802, Portland, OR, pp 73–86

Birch MC, Miller JC, Paine TD (1982) Evaluation of two attempts to trap defined populations of *Scolytus multistriatus*. J Chem Ecol 8:25–136

Bollenbacher B, Gibson K (1986) Mountain pine beetle: a land manager's perspective. USDA Forest Service Forest Pest Management Report 86-15, Missoula, MT

Boone CK, Aukema BH, Bohlmann J, Carroll AL, Raffa KF (2011) Efficacy of tree defense physiology varies with bark beetle population density: a basis for positive feedback in eruptive species. Can J For Res 41(6):1174–1188

Borden JH, Birmingham AL, Burleigh JS (2006) Evaluation of the push-pull tactic against the mountain pine beetle using verbenone and non-host volatiles in combination with pheromone-baited trees. Forestry Chronicle 82(4):579–590

Borden JH, Stokkink E (2021) Semiochemical-based integrated pest management of ambrosia beetles (Coleoptera: Curculionidae: Scolytinae) in British Columbia's forest industry: implemented in 1982 and still running. Can Entomol 153:79–90

Boyd J, Banzhaf S (2007) What are ecosystem services? The need for standardized environmental accounting units. Ecol Econ 63(2–3):616–626

British Columbia Ministry of Forests (2000) Forest health aerial overview survey standards for British Columbia, Version 2.0. Victoria, BC

Brockerhoff EG, Jones DC, Kimberley MO, Suckling DM, Donaldson T (2006) Nationwide survey for invasive wood-boring and bark beetles (Coleoptera) using traps baited with pheromones and kairomones. For Ecol Manage 228:234–240

Buffo E, Battisti A, Stastny M, Larsson S (2007) Temperature as a predictor of survival of the pine processionary moth in the Italian Alps. Agric For Entomol 9:65–72

Byers JA (2008) Active space of pheromone plume and its relationship to effective attraction radius in applied models. J Chem Ecol 34:1134–1145

Cadogan BL, Nealis VG, van Frankenhuyzen K (1995) Control of spruce budworm (Lepidoptera: Tortricidae) with *Bacillus thuringiensis* applications timed to conserve a larval parasitoid. Crop Prot 14(1):31–36

Cameron EA (2012) Parasitoids in the management of *Sirex noctilio*: looking back and looking ahead. In: Slippers B, de Groot P, Wingfield MJ (eds) The sirex woodwasp and its fungal symbiont: research and management of a worldwide invasive pest. Springer, Dordrecht, pp 103–117

Cameron NL, Carnegie AJ, Wardlaw T, Lawson SA, Venn T (2018a) An economic evaluation of Sirex wood wasp (*Sirex noctilio*) control in Australian pine plantations. Aust For 81:37–45

Cameron N, Wardlaw T, Venn T, Carnegie A, Lawson S (2018b) Costs and benefits of a leaf beetle Integrated Pest Management program II. Cost-Benefit Analysis. Aust For 81:53–59

Campbell SA, Borden JH (2009) Additive and synergistic integration of multimodal cues of both hosts and non-hosts during host selection by woodboring insects. Oikos 118:553–563

Canadian Food Inspection Agency (CFIA) (2016) Plant protection survey report 2015–2016. CFIA, Ottawa, ON. Retrieved from http://publications.gc.ca/site/eng/9.831610/publication.html. Accessed 15 Aug 2017

Candy SG (2000) Predictive models for Integrated Pest Management of the leaf beetle *Chrysophtharta bimaculata* in *Eucalyptus nitens* plantations in Tasmania. PhD Thesis, University of Tasmania, Hobart

Candy SG (2003) Predicting time to peak occurrence of insect life-stages using regression models calibrated from stage-frequency data and ancillary stage-mortality data. Agric For Entomol 5:43–49

Candy SG, Elliott HJ, Bashford R, Greener A (1992) Modelling the impact of defoliation by the leaf beetle, *Chrysophtharta bimaculata* (Coleoptera: Chrysomelidae), on height growth of *Eucalyptus regnans*. For Ecol Manage 54:69–87

Cardé RT, Minks AK (1995) Control of moth pests by mating disruption - successes and constraints. Annu Rev Entomol 40:559–585

Carleton RD, Owens E, Blaquière H, Bourassa S, Bowden JJ, Candau J-N, DeMerchant I, Edwards S, Heustis A, James PMA, Kanoti AM, MacQuarrie CJK, Martel V, Moise ERD, Pureswaran DS, Shanks E, Johns RC (2020) Tracking insect outbreaks: a case study of community-assisted moth monitoring using sex pheromone traps. Facets 5:91–104

Carnegie A, Bashford R (2012) Sirex woodwasp in Australia: current management strategies, research and emerging issues. In: Slippers B, de Groot P, Wingfield MJ (eds) The sirex woodwasp and its fungal symbiont: research and management of a worldwide invasive pest. Springer, Dordrecht, pp 175–201

Carnegie AJ, Loch AD (2010) Is *Ips grandicollis* disrupting the biological control of *Sirex noctilio* in Australia? In: McManus KA, Gottschalk KW (eds) Proceedings, 21st U.S. Department of Agriculture interagency research forum on invasive species 12–15 January 2010; Annapolis, MD. USDA Forest Service Proceedings NRS-P-75, Newton Square, PA, pp 8–11

Carnegie AJ, Eldridge RH, Waterson DG (2005a) History and management of sirex wood wasp in pine plantations in New South Wales, Australia. NZ J Forest Sci 35(1):3–24

Carnegie A, Stone C, Lawson S, Matsuki M (2005b) Can we grow certified eucalypt plantations in subtropical Australia? An insect pest management perspective. NZ J Forest Sci 35(2/3):223–245

Carnegie AJ, Cant RG, Eldridge RH (2008) Forest health surveillance in New South Wales Australia. Aust For 71(3):164–176

Carnegie AJ, Lawson S, Wardlaw T, Cameron N, Venn T (2018) Benchmarking forest health surveillance and biosecurity activities for managing Australia's exotic forest pest and pathogen risks. Aust For 81:14–23

Carroll AL, Taylor SW, Régnière J, Safranyik L (2003) Effect of climate change on range expansion by the mountain pine beetle in British Columbia. In: Shore TL, Brooks JE, Stone JE (eds) Mountain pine beetle symposium: challenges and solutions, Oct 30–31, 2003. Kelowna BC. Natural Resources Canada Information Report BC-X-399, Victoria, BC, pp 223–232

Carroll A, Régnière J, Logan J, Taylor S, Bentz B, Powell J (2006) Impacts of climate change on range expansion by the mountain pine beetle. Mountain Pine Beetle Initiative Working Paper 2006-14. Natural Resources Canada Canadian Forest Service, Victoria, BC

Carver M, Kent DS (2000) *Essigella californica* (Essig) and *Eulachnus thunbergii* Wilson (Hemiptera: Aphididae: Lachninae) on *Pinus* in south-eastern Australia. Austral Entomol 39(2):62–69

Chenier JVR, Philogene BJR (1989) Field responses of certain forest Coleoptera to conifer monoterpenes and ethanol. J Chem Ecol 15:1729–1746

Ciesla WM (2000) Remote sensing in forest health protection. USDA Forest Service Forest Health Technology Enterprise Team Report FHTET-00-03, Fort Collins, CO

Clark WC, Jones DD, Holling CS (1979) Lessons for ecological policy design: a case study of ecosystem management. Ecol Model 7(1):1–53

Coggins SB, Coops NC, Wulder MA, Bater CW, Ortlepp SM (2011) Comparing the impacts of mitigation and non-mitigation on mountain pine beetle populations. J Environ Manage 92:112–120

Coleman TW, Seybold SJ (2008) Previously unrecorded damage to oak, *Quercus* spp., in southern California by the goldspotted oak borer, *Agrilus coxalis* Waterhouse (Coleoptera: Buprestidae). The Pan-Pacific Entomol 84:288–300

Collett N, Elms S (2009) The control of sirex wood wasp using biological control agents in Victoria Australia. Agric For Entomol 11(3):283–294

Cook SM, Khan ZR, Pickett JA (2007) The use of push-pull strategies in integrated pest management. Annu Rev Entomol 52(1):375–400

Cooke BJ, Régnière J (1996) An object-oriented, process-based stochastic simulation model of *Bacillus thuringiensis* efficacy against spruce budworm, *Choristoneura fumiferana* (Lepidoptera: Tortricidae). Int J Pest Manag 42(4):291–306

Cooke BJ, MacQuarrie CJK, Lorenzetti F (2012) The dynamics of forest tent caterpillar outbreaks across east-central Canada. Ecography 35(5):422–435

Cory J, Hails R, Sait S (1997) Baculovirus ecology. In: Miller LK (ed) The Baculoviruses. Plenum Press, New York, pp 301–339

Cottrell S, Mattor KM, Morris JL, Fettig CJ, McGrady P, Maguire D, James PMA, Clear J, Wurtzebach Z, Wei Y, Brunelle A, Western J, Maxwell R, Rotar M, Gallagher L, Roberts R (2020) Adaptive capacity in social-ecological systems: a framework for addressing bark beetle disturbances in natural resource management. Sustain Sci 15:555–567

Coyle DR, Nebeker TE, Hart ER, Mattson WJ (2005) Biology and management of insect pests in North American intensively managed hardwood forest systems. Annu Rev Entomol 50:1–29

Czokajlo D, Ross D, Kirsch P (2001) Intercept panel trap, a novel trap for monitoring forest Coleoptera. J For Sci 47(Special Issue No. 2):63–65

Dalman D (2003) Mountain pine beetle management in Canada's mountain national parks. In: Shore TL, Brooks JE, Stone JE (eds) Mountain pine beetle symposium: challenges and solutions, Oct 30–31, 2003. Kelowna BC. Natural Resources Canada Information Report BC-X-399, Victoria, BC, pp 87–96

Davis TS, Mann AJ, Malesky D, Jankowski E, Bradley C (2018) Laboratory and field evaluation of the entomopathogenic fungus *Beauveria bassiana* (Deuteromycotina: Hyphomycetes) for population management of spruce beetle, *Dendroctonus rufipennis* (Coleoptera: Scolytinae), in felled trees and factors limiting pathogen success. Environ Entomol 47:594–602

de Groot P, Nott RW (2003) Response of *Monochamus* (Col., Cerambycidae) and some Buprestidae to flight intercept traps. J Appl Entomol 127:548–552

de Groot P, Biggs WD, Lyons DB, Scarr T, Czerwinski E, Evans HJ, Ingram W, Marchant K (2006). A visual guide to detecting emerald ash borer damage. Natural Resources Canada Canadian Forest Service, Sault Ste. Marie, ON

de Little D (1979) A preliminary review of the genus *Paropsis* Olivier (Coleoptera: Chrysomelidae) in Tasmania. J Aust Entomol Soc 18:91–107 de Little DW (1983) Life-cycle and aspects of the biology of Tasmanian *Eucalyptus* leaf beetle, *Chrysophtharta bimaculata* (Olivier) (Coleoptera: Chrysomelidae). J Aust Entomol Soc 22:15–18

de Little DW (1989) Paropsine chrysomelid attack on plantations of *Eucalyptus nitens* in Tasmania. NZ J Forest Sci 19:223–227

de Little DW, Elliott HJ, Madden JL, Bashford R (1990) Stage-specific mortality in two field populations of immature *Chrysophtharta bimaculata* (Olivier) (Coleoptera: Chrysomelidae). J Aust Entomol Soc 29:51–55

Dodds KJ (2011) Effects of habitat type and trap placement on captures of bark (Coleoptera: Scolytidae) and longhorned (Coleoptera: Cerambycidae) beetles in semiochemical-baited traps. J Econ Entomol 104:879–888

Dodds KJ (2014) Effects of trap height on captures of arboreal insects in pine stands of northeastern United States of America. Can Entomol 146:80–89

Dodds KJ, de Groot P (2012) Sirex, surveys, and management: challenges of having *Sirex noctilio* in North America. In: Slippers B, de Groot P, Wingfield MJ (eds) The sirex woodwasp and its fungal symbiont: research and management of a worldwide invasive pest. Springer, Dordrecht, pp 265–286

Dodds KJ, Orwig DA (2011) An invasive urban forest pest invades natural environments—Asian longhorned beetle in northeastern US hardwood forests. Can J For Res 41:1729–1742

Dodds KJ, Ross DW, Daterman GE (2000) A comparison of traps and trap trees for capturing Douglas-fir beetle, *Dendroctonus pseudotsugae* (Coleoptera: Scolytidae). J Entomol Soc British Columbia 97:33–38

Dodds KJ, de Groot P, Orwig DA (2010) The impact of *Sirex noctilio* in *Pinus resinosa* and *Pinus sylvestris* stands in New York and Ontario. Can J For Res 40:212–223

Dodds KJ, Cooke RR, Hanavan RP (2014) The effects of silvicultural treatment on *Sirex noctilio* attacks and tree health in northeastern United States. Forests 5(11):2810–2824

Dodds KJ, Allison JD, Miller DR, Hanavan RP, Sweeney J (2015) Considering species richness and rarity when selecting optimal survey traps: comparisons of semiochemical baited flight intercept traps for Cerambycidae in eastern North America. Agric For Entomol 17(1):36–47

Dodds KJ, Aoki CF, Arango-Velez A, Cancelliere J, D'Amato AW, DiGirolomo MF, Rabaglia RJ (2018) Expansion of southern pine beetle into northeastern forests: management and impact of a primary bark beetle in a new region. J Forest 116:178–191

Donato DC, Harvey BJ, Romme WH, Simard M, Turner MG (2013) Bark beetle effects on fuel profiles across a range of stand structures in Douglas-fir forests of Greater Yellowstone. Ecol Appl 23(1):3–20

Drumont A, Gonzalez R, de Windt N, Grégoire JC, Proft MD, Seutin E (1992) Semiochemicals and the integrated management of *Ips typographus* (L.) (Col., Scolytidae) in Belgium. J Appl Entomol 114(1–5):333–337

Duan JJ, Abell KJ, Bauer LS, Gould J, Van Driesche R (2014) Natural enemies implicated in the regulation of an invasive pest: a life table analysis of the population dynamics of the emerald ash borer. Agric For Entomol 16(4):406–416

Duncan JP, Powell JA, Gordillo LF, Eason J (2015) A model for mountain pine beetle outbreaks in an age-structured forest: predicting severity and outbreak-recovery cycle period. Bull Math Biol 77(7):1256–1284

Dury SJ, Good JEG, Perrins CM, Buse A, Kaye T (1998) The effects of increasing CO_2 and temperature on oak leaf palatability and the implications for herbivorous insects. Glob Change Biol 4:55–61

Edmonds RL, Agree JK, Gara RI (2000) Forest health and protection. Waveland Press, Long Grove, IL

Eilenberg J, Hajek A, Lomer C (2001) Suggestions for unifying the terminology in biological control. Biocontrol 46(4):387–400

Elek JA (1997) Assessing the impact of leaf beetles in eucalypt plantations and exploring options for their management. Tasforests 9:139–154

Elek J, Allen GR, Matsuki M (2004) Effects of spraying with Dominex and Success on target and non-target species and rate of recolonisation after spraying in *Eucalyptus nitens* plantations in Tasmania. Technical Report No TR133. Cooperative Research Centre for Sustainable Production Forestry, Australia

Elek J, Beveridge N (1999) Effect of a *Bacillus thuringiensis subsp. tenebrionis insecticidal spray* on the mortality, feeding, and development rates of larval Tasmanian eucalyptus leaf beetles (Coleoptera: Chrysomelidae). J Econ Entomol 92:1062–1071

Elkinton JS, Boettner GH (2012) Benefits and harm caused by the introduced generalist tachinid, *Compsilura concinnata*, in North America. Biocontrol 57:277–288

Elkinton JS, Cardé RT (1981) The use of pheromone traps to monitor distribution and population trends of the gypsy moth. In: Mitchell ER (ed) Management of insect pests with semiochemicals: concepts and practice. Springer, Boston, MA, pp 41–55

Elkinton JS, Childs RD (1983) Efficiency of two gypsy moth (Lepidoptera: Lymantriidae) pheromone-baited traps. Environ Entomol 12(5):1519–1525

Elliott HJ, de Little DW (1980) Laboratory studies on predation of *Chrysophtharta bimaculata* (Olivier) (Coleoptera: Chrysomelidae) eggs by coccinellids *Cleobora mellyi* Mulsant and *Harmonia conformis* (Biosduval). Gen Appl Entomol 12:33–36

Elliott HJ, Bashford R, Greener A, Candy SG (1992) Integrated pest management of the Tasmanian eucalyptus leaf beetle, *Chrysophtharta bimaculata* (Olivier) (Coleoptera: Chrysomelidae). For Ecol Manage 53:29–38

El-Sayed A (2022) The Pherobase: database of pheromones and semiochemicals [online] Pherobase.com. Retrieved from https://www.pherobase.com/

Erdle T (1989) Concept and practise of integrated harvest and protection design in the management of eastern spruce-fir forests. PhD Thesis, University of New Brunswick, Fredericton, New Brunswick

Erdle TA, MacLean DA (1999) Stand growth model calibration for use in forest pest impact assessment. For Chron 75(1):141–152

Eskalen A, Gonzalez A, Wang DH, Twizeyimana M, Mayorquin JS, Lynch SC (2012) First report of a *Fusarium* sp. and its vector tea shot hole borer (*Euwallacea fornicatus*) causing Fusarium dieback on avocado in California. Plant Dis 96:1070

Evans HF, Fielding NJ (1994) Integrated management of *Dendroctonus micans* in the UK. For Ecol Manage 65(1):17–30

Eveleigh ES, McCann KS, McCarthy PC, Pollock SJ, Lucarotti CJ, Morin, McDougall GA, Strongman DB, Huber JT, Umbanhowar J, Faria LDB (2007) Fluctuations in density of an outbreak species drive diversity cascades in food webs. Proc Natl Acad Sci U S A 104(43):16976–16981

Evenden ML, Whitehouse CM, Sykes J (2014) Factors influencing flight capacity of the mountain pine beetle (Coleoptera: Curculionidae: Scolytinae). Environ Entomol 43:187–196

Eyles A, Robinson AP, Smith D, Carnegie A, Smith I, Stone C, Mohammed C (2011) Quantifying stem growth loss at the tree-level in a *Pinus radiata* plantation to repeated attack by the aphid, *Essigella Californica*. For Ecol Manage 261(1):120–127

Faccoli M, Stergulc F (2004) *Ips typographus* (L.) pheromone trapping in south Alps: spring catches determine damage thresholds. J Appl Entomol 128(4):307–311

Faccoli M, Stergulc F (2008) Damage reduction and performance of mass trapping devices for forest protection against the spruce bark beetle, *Ips typographus* (Coleoptera Curculionidae Scolytinae). Ann For Sci 65:309

Fassnacht FE, Latifi H, Ghosh A, Joshi PK, Koch B (2014) Assessing the potential of hyperspectral imagery to map bark beetle-induced tree mortality. Remote Sens Environ 140:533–548

Fettig CJ (2019) Socioecological impacts of the western pine beetle outbreak in southern California: lessons for the future. J Forest 117:138–143

Fettig C, Hilszczański J (2015) Management strategies for bark beetles in conifer forests. In: Vega FE, Hofstetter RW (eds) Bark beetles: biology and ecology of native and invasive species. Elsevier, London, pp 555–584

Fettig CJ, Fidgen JG, McClellan QC, Salom SM (2001) Sampling methods for forest and shade tree insects of North America. USDA Forest Service Forest Health Technology Enterprise Team Report FHTET-2001-02, Morgantown, WV

Fettig CJ, Fidgen JG, Salom SM (2005) A review of sampling procedures available for IPM decision-making of forest and shade tree insects in North America. J Arboric 31:38–47

Fettig CJ, McMillin JD, Anhold JA, Hamud SM, Borys RR, Dabney CP, Seybold SJ (2006) The effects of mechanical fuel reduction treatments on the activity of bark beetles (Coleoptera: Scolytidae) infesting ponderosa pine. For Ecol Manage 230:55–68

Fettig CJ, Klepzig KD, Billings RF, Munson AS, Nebeker TE, Negrón JF, Nowak JT (2007) The effectiveness of vegetation management practices for prevention and control of bark beetle infestations in coniferous forests of the western and southern United States. For Ecol Manage 238(1–3):24–53

Fettig CJ, Munson AS, McKelvey SR, Bush PB, Borys RR (2008) Spray deposition from ground-based applications of carbaryl to protect individual trees from bark beetle attack. J Environ Qual 37:1170–1179

Fettig CJ, Grosman DM, Munson AS (2013) Advances in insecticide tools and tactics for protecting conifers from bark beetle attack in the western United States. In: Trdan S (ed) Insecticides-development of safer and more effective technologies. InTech, Rijeka, Croatia, pp 472–492

Fettig CJ, Gibson KE, Munson AS, Negrón JF (2014) Cultural practices for prevention and mitigation of mountain pine beetle infestations. For Sci 60(3):450–463

Fettig CJ, Munson AS, Grosman DM, Blackford DC (2020) Evaluations of emamectin benzoate and propiconazole for protecting Engelmann spruce from mortality attributed to colonization by spruce beetle (Coleoptera: Curculionidae) in the Intermountain West, U.S. J Entomol Sci 55:301–309

Fettig CJ, Progar RA, Paschke J, Sapio FJ (2021) Forest insects. In: Robertson G, Barrett T (eds) Disturbance and sustainability in forests of the western United States. USDA Forest Service General Technical Report PNW-GTR-992, Portland, OR, pp 81–121

Fitzgerald TD (1995) The tent caterpillars. Cornell University Press, Ithaca, NY

Flechtmann CAH, Ottati ALT, Berisford CW (2000) Comparison of four trap types for ambrosia beetles (Coleoptera, Scolytidae) in Brazilian eucalyptus stands. J Econ Entomol 93:1701–1707

Fox G, Beke J, Hopkin T, McKenney D (1997) A framework for the use of economic thresholds in forest pest management. For Chron 73(3):331–339

Francese JA, Crook DJ, Fraser I, Lance DR, Sawyer AJ, Mastro VC (2010) Optimization of trap color for emerald ash borer (Coleoptera: Buprestidae). J Econ Entomol 103:1235–1241

Fry HRC, Quiring DT, Ryall KL, Dixon PL (2008) Relationships between elm spanworm, *Ennomos subsignaria*, juvenile density and defoliation on mature sycamore maple in an urban environment. For Ecol Manage 255(7):2726–2732

Fuentealba A, Alfaro R, Bauce T (2013) Theoretical framework for assessment of risks posed to Canadian forests by invasive insect species. For Ecol Manage 302:97–106

Funes H, Griffo R, Zerba E, Gonzalez-Audino P (2011) Mating disruption of the ambrosia beetle *Megaplatypus mutatus* in poplar and hazelnut plantations using reservoir systems for pheromones. Entomol Exp Appl 139(3):226–234

Furniss MM (2007) A history of forest entomology in the Intermountain and Rocky Mountain areas, 1901 to 1982. USDA Forest Service General Technical Report RMRS-GTR-195, Fort Collins, CO

Gage SH, Wirth TM, Simmons GA (1990) Predicting regional gypsy moth (Lymantriidae) population trends in an expanding population using pheromone trap catch and spatial analysis. Environ Entomol 19(2):370–377

Gilbert J, Miller L (1952) An outbreak of *Sirex noctilio* F. in Tasmania. Aust For 16(2):63–69

Gillette NE, Fettig CJ (2021) Semiochemicals for bark beetle (Coleoptera: Curculionidae) management in western North America: where do we go from here? Can Entomol 153(1):121–135

Gillette NE, Mehmel CJ, Mori SR, Webster JN, Wood DL, Erbilgin N, Owen DR (2012) The push-pull tactic for mitigation of mountain pine beetle (Coleoptera: Curculionidae) damage in lodgepole and whitebark pines. Environ Entomol 41(6):1575–1586

Ginsburg C (2006) Aerial spraying of *Bacillus thuringiensis* kurstaki (Btk). J Pestic Reform 20:13–16

Graham EE, Mitchell RF, Reagel PF, Barbour JD, Millar JG, Hanks LM (2010) Treating panel traps with a fluoropolymer enhances their efficiency in capturing cerambycid beetles. J Econ Entomol 103:641–647

Grant GG (1991) Development and use of pheromones for monitoring lepidopteran forest defoliators in North America. For Ecol Manage 39(1–4):153–162

Gray DR (2010) Hitchhikers on trade routes: a phenology model estimates the probabilities of gypsy moth introduction and establishment. Ecol Appl 20(8):2300–2309

Gray DR (2016) Risk reduction of an invasive insect by targeting surveillance efforts with the assistance of a phenology model and international maritime shipping routes and schedules. Risk Anal 36(5):914–925

Gray DR (2017) Risk analysis of the invasion pathway of the Asian gypsy moth: a known forest invader. Biol Invasions 19(11):3259–3272

Gray DR, Holsten EH, Pascuzzo M (1990) Effects of semiochemical baiting on the attractiveness of felled and unfelled lethal trap trees for spruce beetle, *Dendroctonus rufipennis* (Kirby) (Coleoptera: Scolytidae), management in areas of high and low beetle populations. Can Entomol 122:373–379

Gray DR, Régnière J, Boulet B (2000) Analysis and use of historical patterns of spruce budworm defoliation to forecast outbreak patterns in Quebec. For Ecol Manage 127(1):217–231

Gray LK, Gylander T, Mbogga MS, Chen P-Y, Hamann A (2011) Assisted migration to address climate change: recommendations for aspen reforestation in western Canada. Ecol Appl 21:1591–1603

Greaves R (1966) Insect defoliation in the eucalypt regrowth in the Florentine Valley, Tasmania. Appita 19:119–126

Green M, Heumann M, Sokolow R, Foster LR, Bryant R, Skeels M (1990) Public health implications of the microbial pesticide *Bacillus thuringiensis*: an epidemiological study, Oregon, 1985–86. Am J Public Health 80(7):848–852

Guyot V, Castagneyrol B, Vialatte A, Deconchat M, Selvi F, Bussotti F, Jactel H (2015) Tree diversity limits the impact of an invasive forest pest. PLoS One 10(9)

Guyot V, Castagneyrol B, Vialatte A, Deconchat M, Jactel H (2016) Tree diversity reduces pest damage in mature forests across Europe. Biol Let 12:20151037

Haack RA (2006) Exotic bark and wood-boring Coleoptera in the United States: recent establishments and interceptions. Can J For Res 36:269–288

Haack RA, Hérard F, Sun J, Turgeon JJ (2010) Managing invasive populations of Asian longhorned beetle and citrus longhorned beetle: a worldwide perspective. Annu Rev Entomol 55:521–546

Haack RA, Britton KO, Brockerhoff EG, Cavey JF, Garrett LJ, Kimberley M, Lowenstein F, Nuding A, Olson LJ, Turner J, Vasilaky KN (2014) Effectiveness of the international phytosanitary standard ISPM No. 15 on reducing wood borer infestation rates in wood packaging material entering the United States. PLoS One 9(5)

Haavik LJ, Flint ML, Coleman TW, Venette RC, Seybold SJ (2015) Goldspotted oak borer effects on tree health and colonization patterns at six newly-established sites. Agric for Entomol 17(2):146–157

Hanavan RP, Pontius J, Hallett R (2015) A 10-year assessment of hemlock decline in the Catskill Mountain region of New York State using hyperspectral remote sensing techniques. J Econ Entomol 108(1):339–349

Hanks LM, Paine TD, Millar JG (1996) Tiny wasp helps protect eucalypts from eucalyptus longhorned borer. Calif Agric 50:14–16

Hansen EM, Bentz BJ (2003) Comparison of reproductive capacity among univoltine, semivoltine, and re-emerged parent spruce beetles (Coleoptera: Scolytidae). Can Entomol 135(5):697–712

Hansen EM, Bentz BJ, Munson AS, Vandygriff JC, Turner DL (2006) Evaluation of funnel traps for estimating tree mortality and associated population phase of spruce beetle in Utah. Can J For Res 36(10):2574–2584

Hansen EM, Munson AS, Blackford DC, Wakarchuk D, Baggett LS (2016) Lethal trap trees and semiochemical repellents as area host protection strategies for spruce beetle (Coleoptera: Curculionidae, Scolytinae) in Utah. J Econ Entomol 109(5):2137–2144

Harfouche A, Meilan R, Altman A (2011) Tree genetic engineering and applications to sustainable forestry and biomass production. Trends Biotechnol 29(1):9–17

Haugen D (1990) Control procedures for *Sirex noctilio* in the Green Triangle: review from detection to severe outbreak (1977–1987). Aust For 53(1):24–32

Haugen D, Bedding R, Underdown M, Neumann F (1990) National strategy for control of *Sirex noctilio* in Australia. Aust For Grow 13(2):8

Häusler A, Scherer-Lorenzen M (2001) Sustainable forest management in Germany: the ecosystem approach of the biodiversity convention reconsidered. Results of the R+D–Project 800 83 001. German Federal Agency for Nature Conservation, Bonn, Germany, 65 p

Hayes CJ, Fettig CJ, Merrill LD (2009) Evaluation of multiple funnel traps and stand characteristics for estimating western pine beetle-caused tree mortality. J Econ Entomol 102(6):2170–2182

Hennigar CR (2009) Modeling the influence of forest management and spruce budworm epidemics on timber supply and carbon sequestration. PhD thesis, Univ. of New Brunswick. Fredericton, NB, Canada, 225 p

Hennigar CR, MacLean DA, Porter KB, Quiring DT (2007) Optimized harvest planning under alternative foliage-protection scenarios to reduce volume losses to spruce budworm. Can J For Res 37(9):1755–1769

Herms DA, McCullough DG (2014) Emerald ash borer invasion of North America: history, biology, ecology, impacts, and management. Annu Rev Entomol 59:13–30

Hodges JD, Elam WW, Watson WF, Nebeker TE (1979) Oleoresin characteristics and susceptibility of four southern pines to southern pine beetle (Coleoptera: Scolytidae) attacks. Can Entomol 111(8):889–896

Hodgkinson RS (1985) Use of trap trees for spruce beetle management in British Columbia, 1979–1984. British Columbia Ministry of Forests Pest Management Report No. 5, Victoria, BC

Holmes SB, MacQuarrie CJK (2016) Chemical control in forest pest management. Can Entomol 148(S1):S270–S295

Holmes TP, Aukema JE, Von Holle B, Liebhold A, Sills E (2009) Economic impacts of invasive species in forests. Ann N Y Acad Sci 1162(1):18–38

Holsten EH, Werner RA (1990) Comparison of white, Sitka, and Lutz spruce as hosts of the spruce beetle in Alaska. Can J For Res 20(3):292–297

Holvoet B, Muys B (2004) Sustainable forest management worldwide: a comparative assessment of standards. Int For Rev 6:99–122

Hopkins AD (1905) The Black Hills beetle: with further notes on its distribution, life history, and methods of control. USDA Bureau of Entomology Bulletin 56, Washington, DC

Huber J, Langor D (2004) Systematics: its role in supporting sustainable forest management. For Chron 80(4):451–457

Hurley BP, Croft P, Verleur M, Wingfield MJ, Slippers B (2012) The control of the Sirex woodwasp in diverse environments: the South African experience. In: Slippers B, de Groot P, Wingfield MJ (eds) The sirex woodwasp and its fungal symbiont: research and management of a worldwide invasive pest. Springer, Dordrecht, pp 247–264

Hurley BP, Slippers B, Wingfield MJ (2007) A comparison of control results for the alien invasive woodwasp, *Sirex noctilio*, in the southern hemisphere. Agric For Entomol 9(3):159–171

Ide T, Kanzaki N, Ohmura W, Okabe K (2016) Molecular identification of an invasive wood-boring insect *Lyctus brunneus* (Coleoptera: Bostrichidae: Lyctinae) using frass by loop-mediated isothermal amplification and nested PCR assays. J Econ Entomol 109:1410–1414

Iede ET, Penteado SR, Wilson Filho R (2012) The woodwasp *Sirex noctilio* in Brazil: monitoring and control. In: Slippers B, de Groot P, Wingfield MJ (eds) The sirex woodwasp and its fungal symbiont: research and management of a worldwide invasive pest. Springer, Dordrecht, pp 217–228

Irvine C (1962) Forest and timber insects in Victoria. Victoria's Resources 4:40–43

Jackson L, Molet T, Smith G, Campbell N, Stiers E (2014) Exotic wood borer/bark beetle survey reference, Raleigh, NC

Jackson PL, Straussfogel D, Lindgren BS, Mitchell S, Murphy B (2008) Radar observation and aerial capture of mountain pine beetle, *Dendroctonus ponderosae* Hopk. (Coleoptera: Scolytidae) in flight above the forest canopy. Can J For Res 38:2313–2327

Jacobi WR, Hardin JG, Goodrich BA, Cleaver CM (2012) Retail firewood can transport live tree pests. J Econ Entomol 105(5):1645–1658

Jactel H, Brockerhoff EG (2007) Tree diversity reduces herbivory by forest insects. Ecol Lett 10(9):835–848

Jactel H, Menassieu P, Vetillard F, Barthelemy B, Piou D, Ferot B, Rousselet J, Goussard F, Branco M, Battisti A (2006) Population monitoring of the pine processionary moth (Lepidoptera: Thaumetopoeidae) with pheromone-baited traps. For Ecol Manage 235(1–3):96–106

Jennings DE, Duan JJ, Bean D, Gould JR, Rice KA, Shrewsbury PM (2016) Monitoring the establishment and abundance of introduced parasitoids of emerald ash borer larvae in Maryland, U.S.A. Biol Control 101:138–144

Johns R, Ostaff D, Quiring D (2006) Relationships between yellowheaded spruce sawfly, *Pikonema alaskensis*, density and defoliation on juvenile black spruce. For Ecol Manage 228(1–3):51–60

Johns RC, Bowden JJ, Carleton RD, Cooke BJ, Edwards S, Emilson EJS, James PMA, Kneeshaw D, MacLean DA, Martel V, Moise ERD, Mott DJ, Norfolk CJ, Owens E, Pureswaran DS, Quiring DT, Régnière J, Richard B, Stastny M (2019) A conceptual framework for the spruce budworm Early Intervention Strategy: can outbreaks be stopped? Forests 10:910

Johnson EW, Wittwer D (2008) Aerial detection surveys in the United States. Aust For 71:212–215

Jones BC, Roland J, Evenden ML (2009) Development of a combined sex pheromone-based monitoring system for *Malacosoma disstria* (Lepidoptera: Lasiocampidae) and *Choristoneura conflictana* (Lepidoptera: Tortricidae). Environ Entomol 38(2):459–471

Jones KL, Shegelski VA, Marculis NG, Wijerathna AN, Evenden ML (2019) Factors influencing dispersal by flight in bark beetles (Coleoptera: Curculionidae: Scolytinae): from genes to landscapes. Can J For Res 49:1024–1041

Jönsson P, Anderbrant O (1993) Weather factors influencing catch of *Neodiprion sertifer* (Hymenoptera: Diprionidae) in pheromone traps. Environ Entomol 22(2):445–452

Kenis M, Hurley BP, Hajek AE, Cock MJW (2017) Classical biological control of insect pests of trees: facts and figures. Biol Invasions 19(11):3401–3417

Kershaw DJ (1989) History of forest health surveillance in New Zealand. NZ J Forest Sci 19:357–377

Kethidi DR, Roden DB, Ladd TR, Krell PJ, Retnakaran A, Feng QL (2003) Development of SCAR markers for the DNA-based detection of the Asian long-horned beetle, *Anoplophora glabripennis* (Motschulsky). Arch Insect Biochem Physiol 52(4):193–204

Kimber W, Glatz R, Caon G, Roocke D (2010) *Diaeretus essigellae* Starỳ and Zuparko (Hymenoptera: Braconidae: Aphidiini), a biological control for Monterey pine aphid, *Essigella californica* (Essig) (Hemiptera: Aphididae: Cinarini): host-specificity testing and historical context. Aust J Entomol 49(4):377–387

Kinloch BB Jr, Parks GK, Fowler CW (1970) White pine blister rust: simply inherited resistance in sugar pine. Science 167:193–195

Klasmer P, Botto E (2012) The ecology and biological control of the woodwasp *Sirex noctilio* in Patagonia, Argentina. In: Slippers B, de Groot P, Wingfield MJ (eds) The sirex woodwasp and its fungal symbiont: research and management of a worldwide invasive pest. Springer, Dordrecht, pp 203–215

Kolb TE, Fettig CJ, Ayres MP, Bentz BJ, Hicke JA, Mathiasen R, Stewart JE, Weed AS (2016) Observed and anticipated impacts of drought on forests insects and diseases in the United States. For Ecol Manage 380:321–334

Kreutz J, Zimmermann G, Marohn H, Vaupel O, Mosbacher G (2000) Preliminary investigations on the use of *Beauveria bassiana* (Bals.) Vuill. and other control methods against the bark beetle *Ips typographus* (Col., Scolytidae) in the field. Bulletin of the International Organization for Biological Control of Noxious Animals and Plants, West Palearctic Regional Section, 23, 167–173

Křivan V, Lewis M, Bentz BJ, Bewick S, Lenhart SM, Liebhold A (2016) A dynamical model for bark beetle outbreaks. J Theor Biol 407:25–37

Laidlaw WG, Prenzel BG, Reid ML, Fabris S, Wieser H (2003) Comparison of the efficacy of pheromone-baited traps, pheromone-baited trees, and felled trees for the control of *Dendroctonus pseudotsugae* (Coleoptera: Scolytidae). Environ Entomol 32:477–483

Lanier GN, Jones AH (1985) Trap trees for elm bark beetles. J Chem Ecol 11:11–20

Lawton JH, Strong D Jr (1981) Community patterns and competition in folivorous insects. Am Nat 118(3):317–338

Lee S-I, Spence JR, Langor DW (2014) Succession of saproxylic beetles associated with decomposition of boreal white spruce logs. Agric For Entomol 16(4):391–405

Lehmann J, Nieberding F, Prinz T, Knoth C (2015) Analysis of unmanned aerial system-based CIR images in forestry—A new perspective to monitor pest infestation levels. Forests 6(3):594–612

Lesk C, Coffel E, D'Amato AW, Dodds K, Horton R (2017) Threats to North American forests from southern pine beetle with warming winters. Nat Clim Chang 7:713–717

Lewis GP, Likens GE (2007) Changes in stream chemistry associated with insect defoliation in a Pennsylvania hemlock-hardwoods forest. For Ecol Manage 238(1–3):199–211

Liebhold A, Thorpe K, Ghent J, Lyons DB (1994) Gypsy moth egg mass sampling for decision-making: A users' guide. USDA Forest Service Technical Paper NA-TP-04-94, Morgantown, WV

Liebhold A, Luzader E, Reardon R, Roberts A, Ravlin FW, Sharov A, Zhou GF (1998) Forecasting gypsy moth (Lepidoptera: Lymantriidae) defoliation with a geographical information system. J Econ Entomol 91(2):464–472

Liebhold AM, Berec L, Brockerhoff EG, Epanchin-Niell RS, Hastings A, Herms DA, Kean JM, McCullough DG, Suckling DM, Tobin PC, Yamanaka T (2016) Eradication of invading insect populations: from concepts to applications. Annu Rev Entomol 61:335–352

Lindgren BS (1983) A multiple funnel trap for scolytid beetles (Coleoptera). Can Entomol 115:299–302

Lindgren BS, Borden JH (1983) Survey and mass trapping of ambrosia beetles (Coleoptera: Scolytidae) in timber processing areas on Vancouver Island. Can J For Res 13:481–493

Lindgren BS, Borden JH (1993) Displacement and aggregation of mountain pine beetles, *Dendroctonus ponderosae* (Coleoptera: Scolytidae), in response to their antiaggregation and aggregation pheromones. Can J For Res 23(2):286–290

Lindgren BS, Fraser RG (1994) Control of ambrosia beetle damage by mass trapping at a dryland log sorting area in British Columbia. Forestry Chronicle 70:159–163

Lister CK, Schmid JM, Minnemeyer CD, Frye RH (1976) Refinement of lethal trap tree method for spruce beetle control. J Econ Entomol 69(3):415–418

Logan JA, Powell JA (2001) Ghost forests, global warming, and the mountain pine beetle (Coleoptera: Scolytidae). Am Entomol 47(3):160–173

Logan JA, White P, Bentz BJ, Powell JA (1998) Model analysis of spatial patterns in mountain pine beetle outbreaks. Theor Popul Biol 53(3):236–255

Lovett GM, Christenson LM, Groffman PM, Jones CG, Hart JE, Mitchell MJ (2002) Insect defoliation and nitrogen cycling in forests. Bioscience 52(4):335–341

Lovett GM, Canham CD, Arthur MA, Weathers KC, Fitzhugh RD (2006) Forest ecosystem responses to exotic pests and pathogens in eastern North America. Bioscience 56:395–405

Lucarotti CJ, Moreau G, Kettela EG (2007) Abietiv™, a viral biopesticide for control of the balsam fir sawfly. In: Vincent C, Goettel MS, Lazarovits G (eds) Biological control: a global perspective. CABI Publishing, Cambridge, MA, pp 353–361

Lyons DB, Lavallée R, Kyei-Poku G, Van Frankenhuyzen K, Johny S, Guertin C, Francese JA, Jones GC, Blais M (2012) Towards the development of an autocontamination trap system to manage populations of emerald ash borer (Coleoptera: Buprestidae) with the native entomopathogenic fungus, *Beauveria bassiana*. J Econ Entomol 105(6):1929–1939

Lyytikäinen-Saarenmaa P, Anderbrant O, Löfqvist J, Hedenström E, Högberg HE (1999) Monitoring European pine sawfly population densities with pheromone traps in young pine plantations. For Ecol Manage 124:113–121

Lyytikäinen-Saarenmaa P, Varama M, Anderbrant O, Kukkola M, Kokkonen AM, Hedenström E, Högberg HE (2006) Monitoring the European pine sawfly with pheromone traps in maturing Scots pine stands. Agric For Entomol 8(1):7–15

MacDonald H, McKenney D, Nealis V (1998) A survey on attitudes toward control of forest insects. Forestry Chronicle 74(4):554–560

MacLean DA (2016) Impacts of insect outbreaks on tree mortality, productivity, and stand development. Can Entomol 148(S1):S138–S159

MacLean DA, Lidstone RG (1982) Defoliation by spruce budworm: estimation by ocular and shoot-count methods and variability among branches, trees, and stands. Can J For Res 12(3):582–594

MacLean DA, Erdle T, MacKinnon W, Porter K, Beaton K, Cormier G, Morehouse S, Budd M (2001) The spruce budworm decision support system: forest protection planning to sustain long-term wood supply. Can J For Res 31(10):1742–1757

MacLean DA, Amirault P, Amos-Binks L, Carleton D, Hennigar C, Johns R, Régnière J (2019) Positive results of an early intervention strategy to suppress a spruce budworm outbreak after five years of trials. Forests 10(5):448

MacQuarrie CJ, Lyons D, Seehausen ML, Smith SM (2016) A history of biological control in Canadian forests, 1882–2014. Can Entomol 148(S1):S239–S269

Maki EC, Millar JG, Rodstein J, Hanks LM, Barbour JD (2011) Evaluation of mass trapping and mating disruption for managing *Prionus californicus* (Coleoptera: Cerambycidae) in hop production yards. J Econ Entomol 104(3):933–938

Marini L, Økland B, Jönsson AM, Bentz B, Carroll A, Forster B, Grégoire JC, Hurling R, Nageleisen LM, Netherer S (2017) Climate drivers of bark beetle outbreak dynamics in Norway spruce forests. Ecography 40:1426–1435

Martini A, Baldassari N, Baronio P, Anderbrant O, Hedenstrom E, Hogberg HE, Rocchetta G (2002) Mating disruption of the pine sawfly *Neodiprion sertifer* (Hymenoptera: Diprionidae) in isolated pine stands. Agric For Entomol 4(3):195–201

May B (2004) Assessment of the causality of *Essigella*-ascribed defoliation of mid-rotation radiata pine and its national impact in terms of cost of lost wood production. Forest and Wood Products Research and Development Corporation, Victoria, Australia

McCollum DW, Lundquist JE (2019) Bark beetle infestation of western US forests: a context for assessing and evaluating impacts. J Forest 117:171–177

McConnell TJ, Johnson EW, Burns B (2000) A guide to conducting aerial sketchmapping surveys. USDA Forest Service Forest Health Technology Enterprise Team Report FHTET 00-01, Fort Collins, CO

McCullough DG, Mercader RJ (2012) Evaluation of potential strategies to Slow Ash Mortality (SLAM) caused by emerald ash borer (*Agrilus planipennis*): SLAM in an urban forest. Int J Pest Manag 58:9–23

McCullough DG, Poland TM, Cappaert D (2009) Attraction of the emerald ash borer to ash trees stressed by girdling, herbicide treatment, or wounding. Can J For Res 39(7):1331–1345

McCullough DG, Poland TM, Lewis PA (2016) Lethal trap trees: a potential option for emerald ash borer (*Agrilus planipennis* Fairmaire) management. Pest Manag Sci 72(5):1023–1030

McLean JA, Borden JH (1979) An operational pheromone-based suppression program for an ambrosia beetle, *Gnathotrichus sulcatus*, in a commercial sawmill. J Econ Entomol 72:165–172

McLeod IM, Lucarotti CJ, Hennigar CR, MacLean DA, Holloway AGL, Cormier GA, Davies DC (2012) Advances in aerial application technologies and decision support for integrated pest management. In: Soloneski S (ed) Integrated pest management and pest control-current and future tactics. InTech, Rijeka, Croatia, pp 651–668

McMullen L (1976) Spruce weevil damage: ecological basis and hazard rating for Vancouver Island. Natural Resources Canada Canadian Forest Service Information Report BC-X-141, Victoria, BC

Mercader RJ, McCullough DG, Bedford JM (2013) A comparison of girdled ash detection trees and baited artificial traps for *Agrilus planipennis* (Coleoptera: Buprestidae) detection. Environ Entomol 42(5):1027–1039

Merriam CH (1894) Laws of temperature control of the geographic distribution of terrestrial animals and plants. National Geographic Magazine 6:229–238

Miller C (1977) The feeding impact of spruce budworm on balsam fir. Can J For Res 7(1):76–84

Miller DR, Duerr DA (2008) Comparison of arboreal beetle catches in wet and dry collection cups with Lindgren multiple funnel traps. J Econ Entomol 101:107–113

Miller DR, Crowe CM, Barnes BF, Gandhi KJK, Duerr DA (2013) Attaching lures to multiple-funnel traps targeting saproxylic beetles (Coleoptera) in pine stands: inside or outside funnels? J Econ Entomol 106:206–214

Miller JR, Gut LJ, de Lame FM, Stelinski LL (2006) Differentiation of competitive vs. non-competitive mechanisms mediating disruption of moth sexual communication by point sources of sex pheromone (part I): theory. J Chem Ecol 32:2089–2114

Minnesota Statute 18G.06, subd. 4 (2013) Retrieved from https://www.revisor.mn.gov/statutes/cite/18G.06

Mitchell R, Waring RH, Pitman G (1983) Thinning lodgepole pine increases tree vigor and resistance to mountain pine beetle. For Sci 29(1):204–211

Moreau G, Lucarotti CJ (2007) A brief review of the past use of baculoviruses for the management of eruptive forest defoliators and recent developments on a sawfly virus in Canada. For Chron 83(1):105–112

Morgan FD, Stewart NC (1966) The biology and behaviour of the wood-wasp *Sirex noctilio* F. in New Zealand. Trans Proc R Soc New Zealand 7(14):195–204

Morris ON (1984) Field response of the spruce budworm, *Choristoneura fumiferana* (Lepidoptera: Tortricidae), to dosage and volume rates of commercial *Bacillus thuringiensis*. Can Entomol 116(7):983–990

Morris R (1963) The dynamics of epidemic spruce budworm populations. Memoirs of the Entomological Society of Canada, Vol. 31

Morris JL, Cottrell S, Fettig CJ, Hansen WD, Sherriff RL, Carter VA, Clear J, Clement J, DeRose RJ, Hicke JA, Higuera PE, Mattor KM, Seddon AWR, Seppä H, Stednick JD, Seybold SJ (2018) Bark beetles as agents of change in social-ecological systems. Front Ecol Environ 16(S1):S34–S43

Muzika RM, Liebhold AM (2000) A critique of silvicultural approaches to managing defoliating insects in North America. Agric For Entomol 2:97–105

National Sirex Control Committee (2022) Biological control of Sirex woodwasp in Australia. Retrieved from http://australiansirex.com.au/

Nealis V (2009) Still invasive after all these years: keeping gypsy moth out of British Columbia. For Chron 85(4):593–603

Nealis VG (2015) A risk analysis framework for forest pest management. Forestry Chronicle 91(1):32–39

Negrón JF, Fettig CJ (2014) Mountain pine beetle, a major disturbance agent in US western coniferous forests: a synthesis of the state of knowledge. For Sci 60:409–413

Neumann F, Harris J, Kassaby F, Minko G (1982) An improved technique for early detection and control of the Sirex wood wasp in radiata pine plantations. Aust For 45(2):117–124

Neumann F, Morey J, McKimm R (1987) The sirex wasp in Victoria. Department of Conservation Forest and Lands Bulletin No. 29, Victoria, BC

Nigam P (1975) Chemical insecticides. In: Prebble ML (ed) Aerial control of forest insects in Canada. Canadian Forestry Service Information Canada Cat(F023/19), Ottawa, ON, pp 8–24

Niquidet K, Tang J, Peter B (2016) Economic analysis of forest insect pests in Canada. Can Entomol 148:S357–S366

Niwa CG, Overhulser DL (2015) Monitoring western spruce budworm with pheromone-baited sticky traps to predict subsequent defoliation. USDA Forest Service Research Note PNW-RN-571, Portland, OR.

Nuttall M (1989) *Sirex noctilio* F., Sirex wood wasp (Hymenoptera: Siricidae). In: Cameron PJ, Hill RL, Bain J, Thomas WP (eds) Review of biological control of invertebrate pests and weeds in New Zealand, 1874–1986. CAB International Institute of Biological Control, Ascot, Technical communication No. 10

Ohmart CP (1981) An annotated list of insects associated with *Pinus radiata* D. Don in California. CSIRO Division of Forest Research Divisional Report No. 8, Australia

Ohsawa M (2004) Species richness of Cerambycidae in larch plantations and natural broad-leaved forests of the central mountainous region of Japan. For Ecol Manage 189:375–385

Orwig DA, Cobb RC, D'Amato AW, Kizlinski ML, Foster DR (2008) Multi-year ecosystem response to hemlock woolly adelgid infestation in southern New England forests. Can J For Res 38(4):834–843

Ostaff DP, Piene H, Quiring DT, Moreau G, Farrell JCG, Scarr T (2006) Influence of pre-commercial thinning of balsam fir on defoliation by the balsam fir sawfly. For Ecol Manage 223:342–348

Otvos I, Cunningham J, Alfaro R (1987) Aerial application of nuclear polyhedrosis virus against Douglas-fir tussock moth, *Orgyia pseudotsugata* (McDunnough) (Lepidoptera: Lymantriidae): II. Impact 1 and 2 years after application. Can Entomol 119(7–8):707–715

Parker IM, Simberloff D, Lonsdale WM, Goodell K, Wonham M, Kareiva PM, Williamson MH, Von Holle B, Moyle PB, Byers JE, Goldwasser L (1999) Impact: toward a framework for understanding the ecological effects of invaders. Biol Invasions 1:3–19

Parry D, Spence J, Volney W (1997) Responses of natural enemies to experimentally increased populations of the forest tent caterpillar, *Malacosoma Disstria*. Ecol Entomol 22(1):97–108

Pawson SM, Watt MS (2009) An experimental test of a visual-based push-pull strategy for control of wood boring phytosanitary pests. Agric For Entomol 11(3):239–245

Peng RK, Fletcher CR, Sutton SL (1992) The effect of microclimate on flying dipterans. Int J Biometeorol 36:69–76

Perez L, Dragicevic S (2010) Modeling mountain pine beetle infestation with an agent-based approach at two spatial scales. Environ Model Softw 25(2):223–236

Pitt J, Régnière J, Worner S (2007) Risk assessment of the gypsy moth, *Lymantria dispar* (L.), in New Zealand based on phenology modelling. Int J Biometeorol 51(4):295–305

Powell JA, Bentz BJ (2014) Phenology and density-dependent dispersal predict patterns of mountain pine beetle (*Dendroctonus ponderosae*) impact. Ecol Model 273:173–185

Prasad AM, Iverson LR, Peters MP, Bossenbroek JM, Matthews SN, Sydnor TD, Schwartz MW (2010) Modeling the invasive emerald ash borer risk of spread using a spatially explicit cellular model. Landscape Ecol 25(3):353–369

Price PW, Denno RF, Eubanks MD, Finke DL, Kaplan I (2011) Insect ecology: behavior, populations and communities. Cambridge University Press, Cambridge

Progar RA, Gillette N, Fettig CJ, Hrinkevich K (2014) Applied chemical ecology of the mountain pine beetle. For Sci 60(3):414–433

Pureswaran DS, Johns R, Heard SB, Quiring D (2016) Paradigms in eastern spruce budworm (Lepidoptera: Tortricidae) population ecology: a century of debate. Environ Entomol 45(6):1333–1342

Rabaglia R, Duerr D, Acciavatti RE, Ragenovich I (2008) Early detection and rapid response for non-native bark and ambrosia beetles. USDA Forest Service Forest Health Protection Report, Washington, DC

Rassati D, Marini L, Marchioro M, Rapuzzi P, Magnani G, Poloni R, Di Giovanni F, Mayo P, Sweeney J (2019) Developing trapping protocols for wood-boring beetles associated with broadleaf trees. J Pest Sci 92:267–279

Raty L, Drumont A, De Windt N, Gregoire J-C (1995) Mass trapping of the spruce bark beetle *Ips typographus* L.: traps or trap trees. For Ecol Manage 78:191–205

Régnière J (1996) Generalized approach to landscape-wide seasonal forecasting with temperature-driven simulation models. Environ Entomol 25(5):869–881

Régnière J, Cooke B (1998) Validation of a process-oriented model of *Bacillus thuringiensis* variety *kurstaki* efficacy against spruce budworm (Lepidoptera: Tortricidae). Environ Entomol 27(4):801–811

Régnière J, Sharov A (1998) Phenology of *Lymantria dispar* (Lepidoptera: Lymantriidae), male flight and the effect of moth dispersal in heterogeneous landscapes. Int J Biometeorol 41(4):161–168

Régnière J, You M (1991) A simulation model of spruce budworm (Lepidoptera: Tortricidae) feeding on balsam fir and white spruce. Ecol Model 54(3–4):277–297

Régnière J, Nealis V, Porter K (2009) Climate suitability and management of the gypsy moth invasion into Canada. Biol Invasions 11(1):135–148

Régnière J, Delisle J, Pureswaran D, Trudel R (2013) Mate-finding Allee effect in spruce budworm population dynamics. Entomol Exp Appl 146(1):112–122

Régnière J, St-Amant R, Béchard A (2014) BioSIM 10–User's manual

Régnière J, Delisle J, Dupont A, Trudel R (2019) The impact of moth migration on apparent fecundity overwhelms mating disruption as a method to manage spruce budworm populations. Forests 10:775

Reynolds BC, Hunter MD, Crossley DAJ (2000) Effects of canopy herbivory on nutrient cycling in a northern hardwood forest in western North Carolina. Selbyana 21:74–78

Reynolds KM, Holsten EH, Werner RA (1994) SBexpert users guide (version 1.0): a knowledge-based decision-support system for spruce beetle management. USDA Forest Service General Technical Report PNW-GTR-345, Portland, OR

Rhainds M, Kettela EG, Silk PJ (2012) Thirty-five years of pheromone-based mating disruption studies with *Choristoneura fumiferana* (Clemens) (Lepidoptera: Tortricidae). Can Entomol 144(3):379–395

Rhainds M, Therrien P, Morneau L (2016) Pheromone-based monitoring of spruce budworm (Lepidoptera: Tortricidae) larvae in relation to trap position. J Econ Entomol 109(2):717–723

Ric J, Groot P, Gasman B, Orr M, Doyle J, Smith MT, Dumouchel L, Scarr T, Turgeon JJ (2007) Detecting signs and symptoms of Asian longhorned beetle injury: a training guide. Agriculture and Agri-Food Canada, Ottawa, ON

Ricciardi A, Hoopes MF, Marchetti MP, Lockwood JL (2013) Progress toward understanding the ecological impacts of nonnative species. Ecol Monogr 83:263–282

Roelofs WL, Cardé RT (1977) Responses of Lepidoptera to synthetic sex pheromone chemicals and their analogues. Annu Rev Entomol 22(1):377–405

Roland J (1994) After the decline: what maintains low winter moth density after successful biological control? J Anim Ecol 63(2):392–398

Ross DW, Daterman GE (1994) Reduction of Douglas-fir beetle infestations of high-risk stands by antiaggregation and aggregation pheromones. Can J For Res 24:2184–2190

Ross DW, Daterman GE (1997) Using pheromone-baited traps to control the amount and distribution of tree mortality during outbreaks of the Douglas-fir beetle. For Sci 43:65–70

Ross DW, Daterman GE (1998) Pheromone-baited traps for *Dendroctonus pseudotsugae* (Coleoptera: Scolytidae): influence of selected release rates and trap designs. J Econ Entomol 91:500–506

Royama T (1984) Population dynamics of the spruce budworm *Choristoneura fumiferana*. Ecol Monogr 54(4):429–462

Royama T (2012) Analytical population dynamics, vol 10. Springer, Dordrecht

Ryan K, Hurley B (2012) Life history and biology of *Sirex noctilio*. In: Slippers B, de Groot P, Wingfield MJ (eds) The sirex woodwasp and its fungal symbiont: research and management of a worldwide invasive pest. Springer, Dordrecht, pp 15–30

Ryan K, De Groot P, Smith SM, Turgeon JJ (2013) Seasonal occurrence and spatial distribution of resinosis, a symptom of *Sirex noctilio* (Hymenoptera: Siricidae) injury, on boles of *Pinus sylvestris* (Pinaceae). Can Entomol 145(1):117–122

Safranyik L (1978) Effects of climate and weather on mountain pine beetle populations. In: Kibbee DL, Berryman AA, Amman GD, Stark RW (eds) Theory and practice of mountain pine beetle management in lodgepole pine forests. Proceedings of a Meeting, 25–27 Apr. 1978, Pullman, Wash. University of Idaho, Moscow, ID, pp. 77–84

Safranyik L, Carroll AL, Régnière J, Langor DW, Riel WG, Shore TL, Peter B, Cooke BJ, Nealis VG, Taylor SW (2010) Potential for range expansion of mountain pine beetle into the boreal forest of North America. Can Entomol 142(5):415–442

Salom SM, Mclean JA (1991) Environmental influences on dispersal of *Trypodendron lineatum* (Coleoptera: Scolytidae). Environ Entomol 20(2):565–576

Sanchez-Husillos E., Etxebeste, I., & Pajares, J. (2015). Effectiveness of mass trapping in the reduction of *Monochamus galloprovincialis* Olivier (Col.: Cerambycidae) populations. J Appl Entomol 139(10):747–758

Sanders CJ (1986) Evaluation of high-capacity, nonsaturating sex pheromone traps for monitoring population densities of spruce budworm (Lepidoptera: Tortricidae). Can Entomol 118(7):611–619

Sanders CJ (1988) Monitoring spruce budworm population density with sex pheromone traps. Can Entomol 120(2):175–183

Schlyter F, Zhang Q-H, Liu G-T, Ji L-Z (2001) A successful case of pheromone mass trapping of the bark beetle *Ips duplicatus* in a forest island, analysed by 20-year time-series data. Integr Pest Manag Rev 6:185–196

Schowalter TD (2016) Insect ecology: an ecosystem approach, 4th edn. Elsevier, London

Schmeelk TC, Millar JG, Hanks LM (2016) Influence of trap height and bait type on abundance and species diversity of cerambycid beetles captured in forests of east-central Illinois. J Econ Entomol 109:1750–1757

Seybold SJ, Bentz BJ, Fettig CJ, Lundquist JE, Progar RA, Gillette NE (2018) Management of western North American bark beetles with semiochemicals. Annu Rev Entomol 63:407–432

Sharov A, Liebhold A (1998a) Bioeconomics of managing the spread of exotic pest species with barrier zones. Ecol Appl 8(3):833–845

Sharov AA, Liebhold AM (1998b) Model of slowing the spread of gypsy moth (Lepidoptera: Lymantriidae) with a barrier zone. Ecol Appl 8(4):1170–1179

Sharov A, Leonard D, Liebhold A, Roberts E, Dickerson W (2002) "Slow the spread": a national program to contain the gypsy moth. J Forest 100:30–36

Shepherd RF, Brown CE (1971) Sequential egg-band sampling and probability methods of predicting defoliation by *Malacosoma disstria* (Lasiocampidae: Lepidoptera). Can Entomol 103(10):1371–1379

Shepherd RF, Otvos IS, Chorney RJ, Cunningham JC (1984) Pest management of Douglas-fir tussock moth (Lepidoptera: Lymantriidae): prevention of an outbreak through early treatment with a nuclear polyhedrosis virus by ground and aerial applications. Can Entomol 116:1533–1542

Shore TL, Safranyik L (1992) Susceptibility and risk rating systems for the mountain pine beetle in lodgepole pine stands. Forestry Canada, Pacific and Yukon Region, Pacific Forestry Centre, Information Report BC-X-336. 12 p

Shore T, Safranyik L (2003) Mountain pine beetle management and decision support. In: Shore TL, Brooks JE, Stone JE (eds) *Mountain pine beetle symposium: challenges and solutions*, Oct 30–31, 2003. Kelowna BC. Natural Resources Canada Information Report BC-X-399, Victoria, BC, pp 97–105

Silk PJ, Sweeney J, Wu JP, Price J, Gutowski JM, Kettela EG (2007) Evidence for a male-produced pheromone in *Tetropium fuscum* (F.) and *Tetropium cinnamopterum* (Kirby) (Coleoptera: Cerambycidae). Naturwissenschaften 94(8):697–701

Simmons GA, Elliott NC (1985) Use of moths caught in light traps for predicting outbreaks of the spruce budworm (Lepidoptera, Tortricidae) in Maine. J Econ Entomol 78(2):362–365

Simmons MJ, Lee TD, Ducey MJ, Dodds KJ (2014) Invasion of winter moth in New England: effects of defoliation and site quality on tree mortality. Forests 5(10):2440–2463

Slippers B, De Groot P, Wingfield MJ (eds) (2011) The sirex woodwasp and its fungal symbiont: research and management of a worldwide invasive pest. Springer, Dordrecht

Smith G, Hurley JE (2000) First North American record of the palearctic species *Tetropium fuscum* (Fabricius) (Coleoptera: Cerambycidae). Coleopt Bull 54(4):540–540

Smith MT, Bancroft J, Li G, Gao R, Teale S (2001) Dispersal of *Anoplophora glabripennis* (Cerambycidae). Environ Entomol 30:1036–1040

Smith RH (1975) Formula for describing effect of insect and host tree factors on resistance to western pine beetle attack. J Econ Entomol 68:841–844

Smith SM, Carrow JR, Laing JE (eds) (1990a) Inundative release of the egg parasitoid, *Trichogramma minutum* (Hymenoptera: Trichogrammatidae), against forest insect pests such as the spruce budworm, *Choristoneura fumiferana* (Lepidoptera: Tortricidae): the Ontario Project 1982–1986. Memoirs of the Entomological Society of Canada 153:1–87

Smith SM, Gomez DF, Beaver RA, Hulcr J, Cognato AI (2019) Reassessment of the species in the *Euwallacea fornicatus* (Coleoptera: Curculionidae: Scolytinae) complex after the rediscovery of the "lost" type specimen. Insects 10:261

Smith S, Wallace D, Howse G, Meating J (1990b) Suppression of spruce budworm populations by *Trichogramma minutum* Riley, 1882-1986. Memoirs of the Entomological Society of Canada, 153, 56–81

Somers B, Asner GP (2012) Hyperspectral time series analysis of native and invasive species in Hawaiian rainforests. Remote Sens 4(9):2510

Sorensen J (1994) A revision of the aphid genus *Essigella* (Homoptera: Aphididae: Lachninae): its ecological associations with, and evolution on, Pinaceae Hosts. Pan Pac Entomol 70(1):1–102

Stark RW, Waters WE, Wood DL (1985) Summary. In: Waters WE, Stark RW, Wood DL (eds) Integrated pest management in pine-bark beetle ecosystems. John Wiley & Sons, New York, pp 191–202

Stenger A, Harou P, Navrud S (2009) Valuing environmental goods and services derived from the forests. J For Econ 15(1):1–14

Stephen FP, Browne LE (2000) Application of Eliminade^TM parasitoid food to boles and crowns of pines (Pinaceae) infested with *Dendroctonus frontalis* (Coleoptera: Scolytidae). Can Entomol 132:983–985

Stephens SL, McIver JD, Boerner RE, Fettig CJ, Fontaine JB, Hartsough BR, Schwilk DW (2012) The effects of forest fuel-reduction treatments in the United States. Bioscience 62(6):549–560

Stone A (2021) Emerald ash borer update for the New Year. Buckeye Yard and Garden Online. Ohio State University. Retrieved from https://bygl.osu.edu/node/1739

Stone C (2001) Reducing the impact of insect herbivory in eucalypt plantations through management of intrinsic influences on tree vigour. Austral Ecol 26:482–488

Stone C, Mohammed C (2017) Application of remote sensing technologies for assessing planted forests damaged by insect pests and fungal pathogens: a review. Current Forestry Reports 3:75–92

Stone C, Carnegie A, Melville G, Smith D, Nagel M (2013a) Aerial mapping canopy damage by the aphid *Essigella californica* in a *Pinus radiata* plantation in southern New South Wales: what are the challenges? Aust For 76(2):101–109

Stone C, Melville G, Carnegie A, Smith D, Eyles A, Nagel M (2013b) Crown damage by the aphid *Essigella californica* in a *Pinus radiata* plantation in southern New South Wales: causality and related management issues. Aust For 76(1):16–24

Strom B, Goyer R, Ingram L, Boyd G, Lott L (2002) Oleoresin characteristics of progeny of loblolly pines that escaped attack by the southern pine beetle. For Ecol Manage 158(1):169–178

Sturtevant BR, Cooke BJ, Kneeshaw DD, Mac Lean DA (2015) Modeling insect disturbance across forested landscapes: insights from the spruce budworm. In: Perera AH, Sturtevant BR, Buse LJ (eds) Simulation modeling of forest landscape disturbances. Springer International Publishing, Switzerland, pp 93–134

Suckling DM, Gibb AR, Dentener PR, Seldon DS, Clare GK, Jamieson L, Baird D, Kriticos DJ, El-Sayed AM (2005) *Uraba lugens* (Lepidoptera: Nolidae) in New Zealand: pheromone trapping for delimitation and phenology. J Econ Entomol 98(4):1187–1192

Sullivan BT, Fettig CJ, Otrosina WJ, Dalusky MJ, Berisford CW (2003) Association between severity of prescribed burns and subsequent activity of conifer-infesting beetles in stands of longleaf pine. For Ecol Manage 185:327–340

Svensson GP, Wang H, Jirle EV, Rosenberg O, Liblikas I, Chong JM, Löfstedt C, Anderbrant O (2018) Challenges of pheromone-based mating disruption of *Cydia strobilella* and *Dioryctria abietella* in spruce seed orchards. J Pest Sci 91:639–650

Swank WT, Waide JB, Crossley DA, Todd RL (1981) Insect defoliation enhances nitrate export from forest ecosystems. Oecologia 51(3):297–299

Sweeney J, Gutowski JM, Price J, de Groot P (2006) Effect of semiochemical release rate, killing agent, and trap design on detection of *Tetropium fuscum* (F.) and other longhorn beetles (Coleoptera: Cerambycidae). Environ Entomol 35:645–654

Sweeney J, Silk PJ, Rhainds M, MacKay W, Hughes C, Van Rooyen K, MacKinnon W, LeClair G, Kettela EG (2017) First report of mating disruption with an aggregation pheromone: a case study with *Tetropium fuscum* (Coleoptera: Cerambycidae). J Econ Entomol 110(3):1078–1086

Sweeney J, Hughes C, Webster V, Kostanowicz C, Webster R, Mayo P, Allison JD (2020) Impact of horizontal edge-interior and vertical canopy-understory gradients on the abundance and diversity of bark and woodboring beetles in survey traps. Insects 573. https://doi.org/10.3390/insects11 090573

Taylor K (1976) The introduction and establishment of insect parasitoids to control *Sirex noctilio* in Australia. Entomophaga 21(4):429–440

Taylor S, Carroll A (2003) Disturbance, forest age, and mountain pine beetle outbreak dynamics in BC: a historical perspective. In: Shore TL, Books, JE, Stone, JE (eds) Mountain pine beetle symposium: challenges and solutions. Natural Resources Canada Canadian Forest Service Information Report BC-X-399, Victoria, BC, pp 41–51

Thompson D, Chartrand D, Staznik B, Leach J, Hodgins P (2010) Integrating advanced technologies for optimization of aerial herbicide applications. New For 40(1):45–66

Tisdell C (1990) Economic impact of biological control of weeds and insects. In: Mackauer, M, Ehler LE, Roland J (eds) Critical issues in biological control. Intercept, Andover, UK, pp 301–316

Tobin PC (2008) Cost analysis and biological ramifications for implementing the gypsy moth slow the spread program. USDA Forest Service, General Technical Report NRS-GTR-37, Newton Square, PA.

Tobin PC, Blackburn LM (2007) Slow the Spread: a national program to manage the gypsy moth. USDA Forest Service General Technical Report NRS-GTR-6, Newtown Square, PA

Tobin P, Sharov A, Liebhold A, Leonard D, Roberts E, Learn M (2004) Management of the gypsy moth through a decision algorithm under the STS project. Am Entomol 50(4):200–209

Tobin PC, Kean JM, Suckling DM, McCullough DG, Herms DA, Stringer LD (2014) Determinants of successful arthropod eradication programs. Biol Invasions 16(2):401–414

Turusov V, Rakitsky V, Tomatis L (2002) Dichlorodiphenyltrichloroethane (DDT): ubiquity, persistence, and risks. Environ Health Perspect 110(2):125–128

University of California, Davis (2015) Definition of integrated pest management. Retrieved from https://www2.ipm.ucanr.edu/What-is-IPM/

USDA (2016) Gypsy moth digest. Retrieved from https://www.na.fs.fed.us/fhp/gm/index.shtml

USDA APHIS PPQ (2017) 2017 emerald ash Borer survey guidelines. Retrieved from https://www. aphis.usda.gov/plant_health/plant_pest_info/emerald_ash_b/downloads/survey_guidelines.pdf

Van Frankenhuyzen K, Nystrom C, Dedes J, Seligy V (2000) Mortality, feeding inhibition, and recovery of spruce budworm (Lepidoptera: Tortricidae) larvae following aerial application of a high-potency formulation of *Bacillus thuringiensis* subsp. *kurstaki*. Can Entomol 132(4):505–518

Vance CC, Kirby KR, Malcolm JR, Smith SM (2003) Community composition of longhorned beetles (Coleoptera: Cerambycidae) in the canopy and understorey of sugar maple and white pine stands in south-central Ontario. Environ Entomol 32:1066–1074

Vannatta AR, Hauer RH, Schuettpelz NM (2012) Economic analysis of emerald ash borer (Coleoptera: Buprestidae) management options. J Econ Entomol 105:196–206

Varley G, Gradwell G (1968) Population models for the winter moth. Blackwell Scientific Publications, Oxford

Vettori C, Gallardo F, Häggman H, Kazana V, Migliacci F, Pilate G, Fladung M (2016) Biosafety of forest transgenic trees: improving the scientific basis for safe tree development and implementation of EU policy directives (Vol. 82). Springer, Netherlands

Wallin KF, Raffa KF (2004) Feedback between individual host selection behavior and population dynamics in an eruptive herbivore. Ecol Monogr 74(1):101–116

Wardlaw T, Cameron N, Carnegie A, Lawson S, Venn T (2018) Costs and benefits of a leaf beetle Integrated Pest Management (IPM) program. I. Modeling changes in wood volume yields from pest management. Aust For 81(1):46–52

Waters W (1974) Systems approach to managing pine bark beetles. In: Coulson R, Thatcher R (eds) Proceedings of the Southern Pine Beetle Symposium, College Station, TX: Texas Agricultural Experiment Station, Texas A&M University, pp 12–74

Wear JF, Buckhorn WJ (1955) Organization and conduct of forest insect aerial surveys in Oregon and Washington. USDA Forest Service Paper, Portland, OR. Retrieved from https://archive.org/details/CAT10507178

Weed AS, Ayres MP, Hicke JA (2013) Consequences of climate change for biotic disturbances in North American forests. Ecol Monogr 83(4):441–470

Werle CT, Ranger CM, Schultz PB, Reding ME, Addesso KM, Oliver JB, Sampson BJ (2019) Integrating repellent and attractant semiochemicals into a push–pull strategy for ambrosia beetles (Coleoptera: Curculionidae). J Appl Entomol 143(4):333–343

Wermelinger B (2004) Ecology and management of the spruce bark beetle *Ips typographus*—A review of recent research. For Ecol Manage 202(1–3):67–82

Werner RA, Holsten EH, Matsuoka SM, Burnside RE (2006) Spruce beetles and forest ecosystems in south-central Alaska: a review of 30 years of research. For Ecol Manage 227(3):195–206

Weslien J (1992a) Effects of mass trapping on *Ips typographus* (L.) populations. J Appl Entomol 114(3):228–232

Weslien J (1992b) Monitoring *Ips typographus* (L.) populations and forecasting damage. J Appl Entomol 114(4):338–340

Weslien J, Lindelöw A (1989) Trapping a local population of spruce bark beetles *Ips typographus* (L.): population size and origin of trapped beetles. Holarct Ecol 12(4):511–514

White JC, Wulder MA, Grills D (2006) Detecting and mapping mountain pine beetle red-attack damage with SPOT-5 10 m multispectral imagery. BC J Ecosyst Manag 7:105–118

Wickman BE (1987) The battle against bark beetles in Crater Lake National Park: 1925–34. USDA Forest Service General Technical Report PNW-GTR-259. Portland, OR

Williams T, Valle J, Viñuela E (2003) Is the naturally derived insecticide Spinosad® compatible with insect natural enemies? Biocontrol Sci Tech 13:459–475

Wilson AD, Schiff NM (2010) Identification of *Sirex noctilio* and native North American woodwasp larvae using DNA barcode. J Entomol 7:60–79

Wood DL (1982) The role of pheromones, kairomones, and allomones in the host selection and colonization behavior of bark beetles. Annu Rev Entomol 27(1):411–446

Wulder M, Dymond C, White J, Erickson B, Safranyik L, Wilson B (2006). Detection, mapping, and monitoring of the mountain pine beetle. In: The mountain pine beetle: a synthesis of biology, management, and impacts on lodgepole pine. Natural Resources Canada Canadian Forest Service, Victoria, BC, pp 123–154

Wylie FR, Griffiths M, King J (2008) Development of hazard site surveillance programs for forest invasive species: a case study from Brisbane, Australia. Aust For 71(3):229–235

Wylie FR, Peters B, DeBarr M, King J, Fitzgerald C (1999) Managing attack by bark and ambrosia beetles (Coleoptera: Scolytidae) in fire-damaged *Pinus* plantations and salvaged logs in Queensland, Australia. Aust For 62(2):148–153

Yan Z, Sun J, Don O, Zhang Z (2005) The red turpentine beetle, *Dendroctonus valens* LeConte (Scolytidae): an exotic invasive pest of pine in China. Biodivers Conserv 14(7):1735–1760

Yemshanov D, Koch FH, McKenney DW, Downing MC, Sapio F (2009) Mapping invasive species risks with stochastic models: a cross-border United States-Canada application for *Sirex noctilio* Fabricius. Risk Anal 29(6):868–884

Zanuncio JC, Lemes PG, Antunes LR, Maia JLS, Mendes JEP, Tanganelli KM, Salvador JF, Serrão JE (2016) The impact of the Forest Stewardship Council (FSC) pesticide policy on the management of leaf-cutting ants and termites in certified forests in Brazil. Ann For Sci 73(2):205–214

Zeppenfeld T, Svoboda M, DeRose RJ, Heurich M, Muller J, Cizkova P, Starý M, Bače R, Donato DC (2015) Response of mountain *Picea abies* forests to stand-replacing bark beetle outbreaks: neighbourhood effects lead to self-replacement. J Appl Ecol 52(5):1402–1411

Zolubas P, Byers J (1995) Recapture of dispersing bark beetle *Ips typographus* L. (Col., Scolytidae) in pheromone-baited traps: regression models. J Appl Entomol 119:285–289

Zondag R (1969) A nematode infection of *Sirex noctilio* (F.) in New Zealand. N Z J Sci 12(4):732–747

Zondag R (1979) PART 11. Introductions and establishments in the South Island-1968–75. N Z J For Sci 9(1):68–76

Zylstra KE, Dodds KJ, Francese JA, Mastro VC (2010) *Sirex noctilio* (Hymenoptera: Siricidae) in North America: the effect of stem-injection timing on the attractiveness and suitability of trap trees. Agric For Entomol 12:243–250

Chapter 18
Spatial Dynamics of Forest Insects

Patrick C. Tobin, Kyle J. Haynes, and Allan L. Carroll

18.1   Introduction

The study of the spatial dynamics of forest insects has a long history, and many forest insect species have served as model systems for studying conceptual processes of population biology and ecology. Some of the earliest works by A.D Hopkins, considered as the founding scholar of forest entomology in North America, focused on forest insects and their interactions with natural enemies (Hopkins 1899a), or the role that forest insects play in patterns of tree mortality (Hopkins 1899b). Not surprisingly, the study of forest insect spatial dynamics long predates computers, geodatabases, and spatial statistical software, as forest insect population data were often collected at georeferenced locations. For example, aerial surveys of forest stands affected by biotic disturbance agents, including insects, date to the late 1940s in both Canada and the United States.

Advances in geostatistics and computer processing power over the past several decades have enabled forest entomologists to consider forest insect dynamics over multiple spatial and temporal scales, and vast spatial and temporal extents. In this chapter, we first introduce the importance of scaling in studies of spatial dynamics, and review spatial pattern formation in forest insect populations. We conclude the chapter by addressing metapopulation dynamics, and the concept of spatial synchrony in outbreaking forest insects.

P. C. Tobin (✉)
School of Environmental and Forest Sciences, University of Washington, Seattle, WA, USA
e-mail: pctobin@uw.edu

K. J. Haynes
Department of Environmental Sciences, University of Virginia, Boyce, VA, USA

A. L. Carroll
Faculty of Forestry, University of British Columbia, Vancouver, BC, Canada

© The Author(s) 2023   647
J. D. Allison et al. (eds.), *Forest Entomology and Pathology*,
https://doi.org/10.1007/978-3-031-11553-0_18

18.2 Spatial Scales

The concept of forest insect spatial dynamics is ultimately dependent on the scale at which spatial dynamics are considered. On the level of an individual woody plant host, herbivorous forest insect species are generally restricted to certain plant parts, such as the roots, subcortical regions, leaves or needles, or plant reproductive parts, and consequently many forest entomology courses focus on the groups (i.e. guilds) of insects that feed on each plant part (Berryman 1986). Several species that exploit the same plant concurrently may exploit different parts of the plant due to interspecific competition. This is a concept known as niche partitioning (Schoener 1974), and has been observed in competing bark beetle species, some of which attack the lower bole whereas others attack the middle or upper bole (Paine et al. 1981; Ayres et al. 2001). Moreover, species that attack the same host plant may also exhibit temporal niche partitioning and thus avoid competition by feeding on the same host plant at different times. For example, lepidopteran folivores of Eurasian pines, primarily *Pinus sylvestris*, display dramatic differences in the seasonal occurrence of the larval feeding stage; *Panolis flammea* (Denis & Schiffermüller) feeds from March to July, *Lymantria monacha* (L.) from April to June, *Dendrolimus pini* (L.) from June to July, and *Bupalus piniarius* (L.) from July to November (Altenkirch et al. 2002). Lastly, different insects will feed on woody plants over the life and death of the host. For example, many bark beetle species, most notably *Dendroctonus* spp., are primary species that only attack live host trees, and are followed by secondary species that attack dying or dead trees, which are followed by saproxylic and detritivorous species that play important roles in nutrient cycling (Paine et al. 1997; Grove 2002; Jonsson et al. 2005).

The level of a forest stand presents another scale, and is often the one most commonly addressed in studies of the spatial dynamics of forest insects. A stand is defined as a contiguous community of trees sufficiently uniform in composition, structure, age and size class distribution, spatial arrangement, site quality, condition or location to distinguish it from adjacent communities (e.g. Nyland 2007). Within a stand, forest insects interact with a number of mutualists, competitors, and natural enemies (Janzen 1987; Komonen 2003). For example, Safranyik et al. (2000) collected 30 different species of Scolytinae over two years in one mature stand of *Pinus contorta* (lodgepole pine) following an outbreak of *Dendroctonus ponderosae* Hopkins.

Stands can be aggregated into landscapes, and landscapes into biomes in studies of processes that affect forest insect spatial dynamics. Depending on scale, different patterns of spatial structuring might be revealed. Indeed, fundamental processes operating at one scale may be entirely obscured when the system is considered at a different scale (Raffa et al. 2008). Thus, it is critically important to recognize the spatial scale of a study and how it can influence and limit inference with regard to spatial dynamics.

18.3 Spatial Pattern Formation

Insect populations are distributed in space. The spatial structure of insect populations is of paramount importance in sampling and management plans, as well as in efforts to quantify the underlying factors that affect insect population dynamics (Rossi et al. 1992; Liebhold et al. 1993; Tobin 2004). Spatial patterns occur at multiple spatial scales. For example, the spatial arrangement of a species on a single host plant will have a structure, as will its arrangement within a single forest stand, or across a landscape consisting of a number of forest stands.

There are three basic types of spatial distributions common to not only insect populations but also to life in general, regardless of taxonomic Kingdom: random, uniform, and clustered or aggregated (Fig. 18.1). Randomly distributed populations are rare in nature, and perhaps it is best to think of a random spatial arrangement as a null hypothesis of insect spatial structure. Uniform patterns are also rare, but are present in nature under certain conditions, within specific spatial scales, and at specific population densities. For example, sessile feeders, such as *Adelges tsugae* (Annand), might be expected to be uniformly distributed on a single hemlock shoot in the absence of overcrowding conditions given their feeding behavior. Each *A. tsugae* individual occupies a certain amount of space and inserts their stylet into the petiole of a hemlock needle, which furthermore tend to be uniformly arranged on a shoot. The vast majority of insect species, especially as spatial scales increase, are undoubtedly aggregated (Taylor 1961).

One basic explanation for spatial aggregation by most herbivorous insects is that they have life histories characteristic of *r*-strategists in which females oviposit several to many eggs (or other immature life stages) in one area at once. Even though neonates may be capable of dispersing, such dispersal is normally limited to short distances. Thus, each new cohort is initiated with a high degree of aggregation. Insects, regardless of feeding guild, are also often dependent upon resources that are spatially structured. For example, plants generally follow elevational and latitudinal gradients due to variation in a number of factors, such as temperature, precipitation, solar energy, and soil characteristics. The spatial pattern of plants spatially structures the insect herbivores that rely on those plants, which in turn spatially structures natural enemies of those herbivores, and so forth (Taylor 1984; McCoy 1990; Hodkinson 2005). Some forest insects may also be engaged in gregarious behaviors; for example, semiochemicals such as aggregation pheromones in tree-killing bark beetles facilitate mass-attacks on host trees (Borden 1989; Raffa 2001; Gitau et al. 2013). Other species may use sex pheromones or engage in lekking behaviors that could result in the aggregation of adults for mating (Landolt 1997; Wickman and Rutowski 1999).

Historical methods of spatial pattern analyses relied on frequency distribution models and mean-to-variance relationships (e.g. Southwood 1978). These approaches involved examining the ratio of the sample variance-to-the-sample mean of a collection of samples from a sampling quadrat or area (Taylor 1961; Southwood 1978). If the sample variance was less than the sample mean, the population was considered uniformly distributed. In contrast, if the sample variance was greater

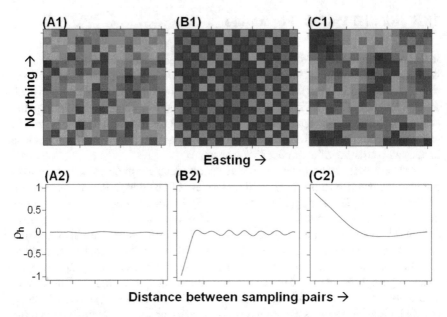

Fig. 18.1 Spatial representation of a random (**A1**), uniform (**B1**), and clustered (**C1**) spatial pattern, and the corresponding spatial correlogram (ρ_h) shown in **A2**, **B2**, and **C2**, respectively. In random patterns, the correlation between values from pairs of sampling locations is ~0 regardless of the distance that separates the sampling locations (**A2**). This is in contrast to clustered populations (**C2**) in which there is high correlation between pairs of sampling locations as the distances that separates these locations → 0, with the range of spatial dependency extending to the distance at which ρ_h ~0. In uniformly-distributed populations, high values are generally located next to low values, which results in a negative correlation as the distance that separates sampling locations → 0 (**B2**). © Patrick Tobin

than the sample mean, the population was considered to be aggregated. If the sample variance was approximately the same as the sample mean, then the population was considered to be randomly distributed. This simple approach was certainly useful in the days before computers, and did shed light onto the basic spatial patterns of insects, but was not necessarily spatially explicit or amenable to statistical hypothesis testing.

More sophisticated spatial statistical techniques have been available for some time (Legendre and Fortin 1989; Rossi et al. 1992; Bjørnstad and Falck 2001). These modern techniques rely on the estimation of the spatial correlogram, which considers the spatial correlation between values of pairs of samples as a function of the distance separating the two samples (Rossi et al. 1992; Fig. 18.1). An underlying premise is that the values of a given variable collected from two locations that are close in space are more likely to be similar in value than data collected from two locations that are farther away in space. The correlation of a variable with itself across space is known as spatial autocorrelation (Getis 2008).

The behavior of the spatial correlogram provides information as to the degree of local spatial autocorrelation, which is the correlation of a variable between sample pairs as the distance between sample pairs approaches 0 (i.e. the y-intercept). As the distance between sample pairs approaches 0, the theoretical expectation of the spatial autocorrelation is 1, or perfect positive autocorrelation. However, in field-collected data, the spatial autocorrelation is often <1 as the distance approaches 0, in part due to random variation and measurement error. In the geological and mining literature, upon which the foundation of spatial statistics was developed, the difference between estimates of the local spatial autocorrelation and its theoretical value of 1 is known as the "nugget effect"; a term motivated by the occurrence of a large mineral deposit, such as a gold nugget, in a theoretically unexpected location in space based on nearby samples (Krige 1999). The spatial correlogram also provides an estimate of the spatial range, which is the distance over which sample pairs are correlated; thus, at this distance, the estimated spatial autocorrelation approaches 0 (i.e. the x-intercept). The spatial range can be used to estimate the distance that samples need to be apart to acquire spatially independent data, and the spatial extent of aggregation in an insect population.

Quantification of spatial pattern formation, and the approach used to do so, has a number of important ramifications for the management of forest insect populations. For example, there are benefits to using prior knowledge of population structure, such as the degree and range of spatial correlation, to design sampling protocols with the goal of obtaining spatially independent data. By collecting spatially independent data, sampling efforts can be reduced yet still allow georeferenced data to be used in interpolation efforts, such as through kriging (Liebhold et al. 1993; Fleischer et al. 1999). However, it should be noted that in cases where estimates of population density are readily available at scales finer than the range of spatial autocorrelation, such as in studies where the proportion of forest defoliated in a given area of forested land was used as a proxy of the local population density of *Lymantria dispar* (L.) (Haynes et al. 2018), statistical methods have been developed to account for the non-independence of data values from nearby sample areas. An application of the spatial autocorrelation based upon field-collected data of *L. dispar* is presented in Box 18.1.

Box 18.1: The *Lymantria dispar* Invasion of North America

Life stages of *L. dispar* were introduced to Medford, Massachusetts, USA, by an amateur entomologist, Étienne Léopold Trouvelot, in 1869 (Riley and Vasey 1870). It is believed that following a storm, life stages escaped from the rearing conditions maintained by Trouvelot (Forbush and Fernald 1896). It has subsequently spread in North America such that it now occupies an area from Minnesota to North Carolina to Maine in the U.S., and southern Ontario to Nova Scotia in Canada. Current management efforts include outbreak suppression in its established area, slowing it spread along its expanding population front, and eradication in areas outside of the established area (Tobin et al. 2012).

Fig. 1 *Lymantria dispar* larvae on *Betula papyrifera* (paper birch), Stockton Island, Wisconsin, USA (Photo credit: P. Tobin)

Lymantria dispar undergoes one generation per year. Overwintering eggs hatch in spring, and larval and pupal development occurs over ~8 and 2 weeks, respectively. Female adults are not capable of sustained flight, and produce a sex pheromone to attract male mates. Adults are short-lived (~2–3 days). In summer, females oviposit 200–500 eggs in an egg mass, which will not hatch until the following year.

Larvae (Fig. 1) are highly polyphagous and are capable of consuming >300 species of host plants, including ~80 species that are highly preferred. Highly preferred hosts include species within *Betula, Crataegus, Larix, Populus, Quercus*, and *Salix* (Liebhold et al. 1995).

Along its expanding population front, *L. dispar* generally spreads through stratified dispersal in which short-range dispersal is coupled with long distance 'jumps' in areas ahead of the leading edge. Spatial analyses of *L. dispar* spread using the spatial autocorrelation are indicative of a spatial trend as it invades across a region (Fig. 2).

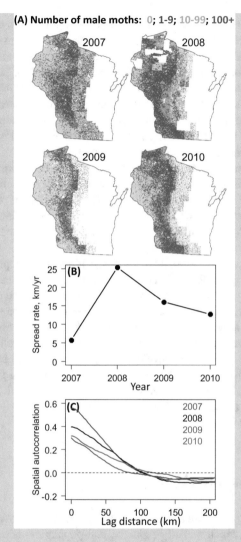

Fig. 2 Spread and spatial dynamics of *L. dispar* in Wisconsin, 2007–2010. (**A**) Counts of male moths from deployed pheromone-baited traps. (**B**) Mean rates of spread (km/yr) from year-to-year; for example, the spread rate in 2007 reflects the change from 2006 to 2007 (Tobin et al. 2007). (**C**) Estimates of the spatial autocorrelation (Bjørnstad and Falck 2001) in trap catch for each year. In each year, spatial autocorrelation was detected at distances up to ~100 km (i.e. the *x*-intercept), and the linear pattern of spatial autocorrelation is indicative of a spatial trend as *L. dispar* invades Wisconsin from the east to the west

Accurate delimitation of the spatial extent of a population has important implications for both forest insect pest management and conservation management. For example, understanding the spatial extent of a pest population allows for the deployment of site-specific control interventions, and by extension, reduced non-target effects of control tactics (Sharov et al. 2002; Tobin et al. 2004; Blackburn et al. 2011). In forest insects that are threatened or endangered, or in areas of conservation concerns, understanding their spatial dynamics helps to develop better conservation plans (Didham et al. 1996; Gering et al. 2003). Many forest insect species have important ecosystem roles, and some provide important ecosystem services (Noriega et al. 2018). Knowledge of their spatial structure can provide insight as to the spatial extent of these ecosystem services. Lastly, long-term and baseline knowledge of forest insect spatial dynamics permits the study of how species respond to climate change, habitat fragmentation and changes in land use, and the introduction of invasive species (Harrington et al. 2001; Knops et al. 2002; Walther et al. 2002; Logan et al. 2003; Opdam and Wascher 2004; Turner 2010).

It is important to recognize that the spatial patterns of insect populations are not static; rather, they vary both within and among generations. Consider the phenology (i.e. the seasonal timing of specific events in an organism's life cycle) and spatial arrangement of an insect population that inhabits a stand on both a south-facing slope and a north-facing slope. Reproductive asynchrony, which occurs when the adults within a population are present at different times, owing to, for example, temperature variation leading to variation in developmental rate, could lead to spatial variation in mating success rates and hence spatial variation in population growth through time (Calabrese and Fagan 2004; Robinet et al. 2008; Walter et al. 2015). Thus, both space and time are fundamental for understanding the processes influencing insect population dynamics.

18.4 Metapopulation Dynamics

Many forest insect populations exist, especially at endemic population densities, as metapopulations in which spatially-separated sub-populations of a species exist over a large landscape (Levins 1969; Hanski 1998). Often in forest ecosystems, these spatially-separated subpopulations exist due to fragmented host plant resources. The fragmentation of host plant resources could be the result of human activities, such a logging, or environmental conditions, such as host trees adapted to mid-elevations or valleys and are thus separated by mountain peaks. Hanski (1997) proposed a set of conditions that define metapopulations, and one condition is that subpopulations are close enough to be connected by dispersal. Thus, depending on the dispersal ability of the insect, a metapopulation can exist over a range of distances between subpopulations. Another important condition of a metapopulation is that patches of host resources are fragmented over a larger landscape, and some of these patches must be of sufficient host quality and abundance to allow for population persistence. Nevertheless, the subpopulations within all patches are theoretically prone to extinction, although extinction rates can differ from patch-to-patch. Even insects inhabiting a forest stand with a high abundance of high-quality host plant resources have some rate of extinction due to, for example, stochastic mortality factors such as winter conditions during which temperatures drop below supercooling points. A final condition of metapopulations is that local population dynamics are independent of each other and thus are not necessarily synchronous; consequently, densities in one patch could be high, which theoretically allow it to serve as a source, while other patches could be going extinct. These conditions comprise the classic model of metapopulations (Fig. 18.2).

One key aspect of metapopulation dynamics that is applicable to the study of forest insect ecology is the underlying spatial heterogeneity that fragments an insect population (Hunter 2002). Past work has highlighted how this spatial heterogeneity affects natural enemy-victim interactions (Hastings 1990; Taylor 1990), which can play a large role in the population dynamics of forest insect species. For example, Roland (1993) examined outbreak duration in the forest tent caterpillar, *Malacosoma disstria* Hübner, a defoliator native to North America, and observed that an increase in forest fragmentation due to logging spatially decoupled *M. disstria* from parasitoids and pathogens to the benefit of the defoliator. The result was longer and more intense outbreaks in areas with high spatial heterogeneity. Although outbreak dynamics are inherently spatially synchronized (see Sect. 18.5), the underlying fragmentation of local populations, which are likely independent at endemic population levels, can provide sufficient escape from natural enemies that would otherwise provide population control.

Fig. 18.2 Classic metapopulation model. Subpopulations are fragmented in space, all metapopulations are connected by dispersal, all metapopulations have the probability of going extinct (i.e. empty patches), and metapopulations are asynchronous. © Patrick Tobin

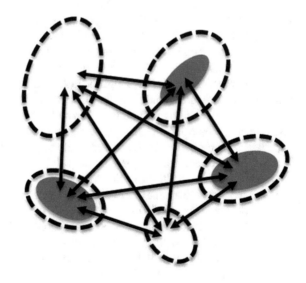

18.5 Spatial Synchrony and Outbreak Dynamics

In contrast to one of the core conditions of metapopulations, specifically the independence of dynamics among subpopulations, spatial synchrony refers to the congruence in temporal variation of abundance across geographically disjunct populations (Bjørnstad et al. 1999; Liebhold et al. 2004). In other words, spatial synchrony is the phenomenon in which the densities of populations distributed across a region tend to rise and fall synchronously. Spatial synchrony has been found in populations of a wide variety of taxa including many forest insect species (Peltonen et al. 2002; Liebhold et al. 2004). Spatial synchrony in forest insect populations has been observed over distances of hundreds (Peltonen et al. 2002) to thousands of kilometers (Royama 1984). At times, these forest insect populations can be irruptive and increase dramatically across a large region over short periods of time, which is often the case in the development of insect outbreaks (Aukema et al. 2006).

Several mechanisms have been proposed to give rise to spatial synchrony in insect populations. One biotic mechanism is dispersal between and among populations of a species (Peltonen et al. 2002), and especially density-dependent dispersal in which individuals from areas with high population densities disperse to lower density populations to reduce intraspecific competition. Another biotic mechanism arises from trophic interactions with populations of other species that are spatially synchronous, thus inducing spatial synchrony in the forest insect under consideration (Ims and Steen 1990).

Perhaps the most important factors affecting the spatial synchrony of poikilothermic species, such as insects, are the abiotic effects of weather. Excessively harsh or mild winter temperatures, for example, can have dramatic region-wide effects on insect populations. Generally, exogenous weather factors, such as temperature or precipitation, are highly spatially autocorrelated in a given year and thereby affect, concurrently, insect populations over large spatial extents; a phenomenon known as the Moran effect (Moran 1953; Royama 1992; Myers 1998; Hudson and Cattadori 1999). Moran's theorem states that the correlation through time (spatial synchrony) between two populations will be approximately the same as the synchrony of the environment (Moran 1953). Thus, when insect populations are strongly affected by a spatially synchronous weather factor or factors, the Moran theorem predicts the densities of the affected insect populations will be strongly synchronous.

On its simplest level, the quantification of synchrony involves the estimation of the correlation between two characteristics of a population measured through time (i.e. a time series) such as population growth rate or population density for a collection of spatially disjunct subpopulations. Spatial synchrony is then a measurement of the extent to which this synchrony exists over spatial scales. Past work has reviewed the basis of quantifying synchrony and spatial synchrony (Bjørnstad et al. 1999; Buonaccorsi et al. 2001; Liebhold et al. 2004). Briefly, the statistical techniques used to quantify synchrony are an extension of the tools used in estimating spatial autocorrelation in which the estimate of the range (i.e. the x-intercept) provides an estimate of the spatial extent over which synchronous fluctuations in populations are similar. A conceptual figure of a spatially synchronous insect outbreak and the resulting estimates of synchrony is presented in Fig. 18.3.

Fig. 18.3 Hypothetical spatial and temporal progression in the severity of, or the area affected by, an insect outbreak through time steps t (**A**), $t + 1$ (**B**), and $t + 2$ (**C**). Typically, the strength of spatial synchrony, measured as the correlation in the severity of outbreak severity through time, declines with increasing distance between locations (**D**). © Patrick Tobin

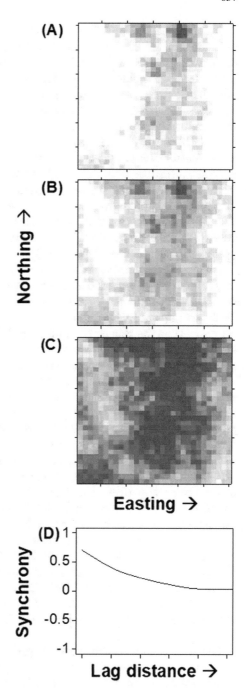

An outbreak, which in entomological terms is defined as an explosive increase in the abundance of an insect population over a relatively short time period (Barbosa and Schultz 1987), is inherently spatially synchronized in that high densities are present over a large geographic area at roughly the same time. Forest insect outbreaks, much like forest fires, can be an extremely important component of forest ecosystem dynamics, but also like forest fires, they can have profound ecological and economic ramifications (Barbosa and Schultz 1987; Mattson and Haack 1987; McCullough et al. 1998; Raffa et al. 2008). This is especially the case in outbreaks of non-native forest insects, or when outbreaks of native species are occurring at different intervals, intensities or in different habitats than the historical norm. Although all insect outbreaks are spatially synchronized at some spatial scale, the extent at which the outbreak occurs often defines a forest insect as a pest or not. Small scale outbreaks that affect a locally distributed forest resource can certainly have measurable impacts; however, it is the large and spatially synchronous outbreaks that are most damaging (Raffa et al. 2008; Liebhold et al. 2012).

An important economic consequence of large-scale forest insect outbreaks is that they can exacerbate the economic burden on individual stake-holders and land owners due to the fact that a large portion of their forested area is often affected. From a management perspective, outbreaks that are spatially synchronized over large areas can overwhelm the budgetary and logistical abilities of federal, state/provincial or industrial agencies to implement control tactics intended to mitigate impacts and potentially suppress populations. The spatially synchronous behavior of outbreaks can also reduce, or in extreme cases eliminate, undisturbed areas that would otherwise serve as refuge against the effects of an outbreak. Lastly, spatially synchronous outbreaks can dilute the regulating effects of any natural enemy that could otherwise provide local control, which in itself could be a contributing factor to the development of high-density forest insect populations.

Forest insect outbreaks, especially in defoliators, may be cyclical (i.e. periodic) and at times, populations exist at endemic levels despite the widespread availability of susceptible host trees. The time between outbreak peaks is referred to as an outbreak interval or period length. Fascinatingly, many outbreaking forest insects exhibit cycles at relatively fixed period lengths. For example, prior work has high-lighted a 8–12-year or a 4–5-year cycle in *L. dispar* outbreaks depending on forest stand composition (Johnson et al. 2005, 2006a), a 7–11 year cycle in *L. monacha* outbreaks (Haynes et al. 2014), and a 35–40-year cycle in *Choristoneura fumiferana* (Clemens) outbreaks (Royama 1984; Royama et al. 2005). The periodicity of cyclic species can persist for a very long time, with evidence for an 8–9 year cycle in *Zeiraphera diniana* (Guenée) extending back approximately 1200 years (Esper et al. 2007).

Statistical techniques to quantify the periodicity of forest insect outbreaks include the estimation of periodograms through, for example, spectral analysis. A wavelet-based spectral analysis (Torrence and Compo 1998; Cazelles et al. 2014) is one technique used in the quantification of time series that describe insect population dynamics including time series of insect outbreaks (Johnson et al. 2006b). An advantage of this technique relative to others, such as those using Fourier transformations, is that the wavelet transform can be applied to non-stationary time series, where characteristics such as period length and the amplitude of fluctuations vary through time (Torrence and Compo 1998); this is often the case in time series of the abundance of outbreaking insect species (Aukema et al. 2006; Liebhold et al. 2012). The process of a wavelet analysis is essentially akin to taking wavelet functions of known period lengths and sliding them across a time series, in this case a time series consisting of insect abundances surveyed at regular intervals. Then, at each point in time, the degree of overlap between the wavelet functions and the population abundance data is measured. In doing so, one can determine the degree to which fluctuations in abundance are cyclical, the period lengths of any such cycles, and changes over time in the presence or period lengths of cycles (Torrence and Compo 1998). An example of the application of wavelet analysis to a hypothetical time series of insect outbreaks in presented in Fig. 18.4.

Applications of the study of the periodicity and intensity of forest insect outbreaks include providing background knowledge to forest health managers, who might use these findings to anticipate the next forest outbreak and preemptively apply management practices such as silvicultural strategies (Sartwell and Stevens 1975; Bergeron et al. 1999; Muzika and Liebhold 2000; Coyle et al. 2005). The study of the spatial synchrony of insect outbreaks can also shed light on the extent of the affected area. For example, Aukema et al. (2006) measured spatial synchrony in a *D. ponderosae* outbreak in British Columbia, Canada, that was significant beyond 900 km, which not only refuted popular perception that the outbreak began in a protected area but also provided evidence that *D. ponderosae* populations were erupting throughout its range. A case study of the *D. ponderosae* outbreak in western Canada is presented in Box 18.2.

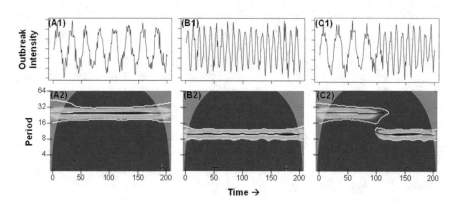

Fig. 18.4 Hypothetical time series of an insect outbreak showing different periods of time (periodicity) between outbreak peaks including a long period (**A1**), short period (**B1**), and one in which there is a transition from a long period to a short period (**C1**). The corresponding wavelet analyses are shown in **A2**, **B2**, and **C2**, with the solid black line representing the expected periodicity in time, while the colored region bounded by white lines represents the confidence intervals. For **A1**, the measured periodicity is ~25 units in time (**A2**), and for **B1**, the measured periodicity is ~10 units of time (**B2**). The periodicity for the time series with the transition (shown in **C1**) is presented in **C2**. This approach can be useful in statistically quantifying changes in the periodicity of insect outbreaks, or any other measured demographic trait, through time. © Patrick Tobin

**Box 18.2: The *Dendroctonus ponderosae* (Mountain Pine Beetle)
Outbreak in Western Canada**

The mountain pine beetle is native to western North America. It feeds and reproduces within the phloem tissues of most species of pine trees. During mid to late summer, beetles select host trees and initiate attacks by boring through the bark. Trees respond by producing sticky, toxic resin (Fig. 1). Beetles ingest the defensive resin and chemically convert some of its constituents into aggregation pheromones that attract more beetles. The result is a mass attack that overwhelms tree defenses and leads to rapid tree mortality.

Fig. 1 Mountain pine beetles attacking a *Pinus contorta* (lodgepole pine) tree. Note the tree's defensive resin (Photo credit: A. Carroll)

Normally, mountain pine beetle populations are innocuous, infesting occasional vigor-impaired trees within a forest; however, they periodically erupt synchronously into large-scale epidemics that cause the mortality of trees over large areas (Fig. 2A). This is a likely consequence of the Moran effect (Moran 1953; Aukema et al. 2006).

Most mountain pine beetles disperse short distances through the forest when seeking new hosts, but a small percentage will fly above the canopy (Safranyik et al. 1992). Thus, sub-outbreak populations are largely independent across landscapes. During outbreaks, large numbers of beetles may be carried above the forest canopy by prevailing winds (Jackson et al. 2008), leading to synchronized dynamics across very large distances (Fig. 2B).

Due to an increase in the number of susceptible trees as a result of fire suppression, and an expansion of climatically suitable habitats as a consequence of global warming, mountain pine beetle populations erupted during the mid-1990s and rapidly increased to unprecedented levels, establishing within historically climatically unsuitable pine forests at higher latitudes and elevations (Carroll et al. 2004; Safranyik et al. 2010).

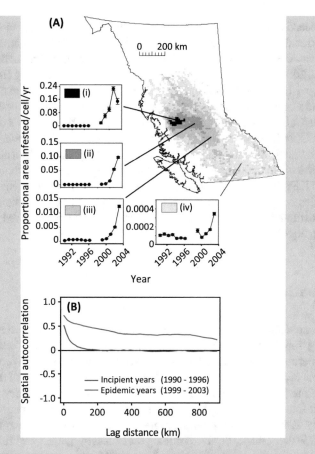

Fig. 2 (**A**) Time series patterns of tree mortality caused by the mountain pine beetle within 12 × 12 km cells in British Columbia, Canada, between 1999 and 2003, based on hierarchical cluster analysis (e.g. Swanson & Johnson 1999). Although the outbreak intensified earliest in the west-central portion of the province [cluster (i)], populations increased concurrently throughout the region [clusters (ii), (iii) and (iv)] indicating that many localized infestations erupted in geographically disjunct areas rather than originating and spreading from an epicenter. (**B**) Estimates of the spatial autocorrelation (Bjørnstad and Falck 2001) in tree mortality caused by the mountain pine beetle during incipient years (1990–1996) and epidemic years (1999–2003). Note that prior to the extensive outbreak, populations were largely independent at scales >200 km; however, during epidemic years populations were synchronous at distances >900 km. Adapted from Aukema et al. (2006)

More recently, quantifying forest insect outbreak dynamics has allowed studies of how outbreak intensity and periodicity might be changing as a consequence of climate change. For example, the ~1200 years of consistent outbreaks by *Z. diniana* (Esper et al. 2007) collapsed in recent decades due to climate warming (Johnson et al. 2010). Haynes et al. (2014) used long-term data on forest defoliators to quantify both positive and negative changes in their respective outbreak intensity and periodicity in response to climate change. Lastly, spatial analyses of the *D. ponderosae* outbreak in western Canada provided evidence and a quantification of *D. ponderosae* range expansion owing to climate change (Aukema et al. 2008; Sambaraju et al. 2012). The use of analytical techniques such as wavelet analyses provides opportunities to better understand the relationship between climate change and insect outbreaks, which for some species could become more intense and frequent while in others, outbreaks could be disrupted with yet unknown ecological consequences (Weed et al. 2013; Tobin et al. 2014).

18.6   Conclusion

Certain phenomena, such as spatial autocorrelation and spatial synchrony, are pervasive among forest insect populations, but for any given species these and other spatial properties are dynamic. In recent years, quantifying how such properties shift through time or across space has opened new avenues for exploration of the processes underlying the population dynamics of forest insect species. One developing area of research focusses on understanding the drivers of population spatial synchrony by studying factors associated with geographic variation in the strength of spatial synchrony (Walter et al. 2017). Determining the causes of spatial synchrony in a given study organism is often difficult, in part, because different mechanisms can lead to similar spatial patterns, such as the tendency for the strength of synchrony to decline as the distance between populations increases. Furthermore, spatial synchrony in forest insect populations often extends over such large distances that field experiments are impractical. By exploiting geographic variation in the strength of spatial synchrony, however, researchers have begun to discover relationships between spatial synchrony and factors considered as potential drivers (Haynes et al. 2013, 2018; Walter et al. 2017).

The dynamic nature of the ranges of forest insect species reveals much about biotic processes underlying patterns of range expansion or contraction, as well as anthropogenic impacts. Temporal patterns and spatial variability in the local rate of spread of native and non-native invasive forest insects, for example, have underscored the importance of factors including forest management practices, accidental human transport of invasive insects, Allee effects (positive density dependent population growth at low densities) operating at the leading edge of invasion fronts, and population cycles in rates of spread (Johnson et al. 2006b; Tobin et al. 2007; Walter et al. 2015; Cooke and Carroll 2017). But global-scale impacts of human-induced climate change are also changing the spatial distributions of forest insect pest species.

The outbreaks of some forest insect pests are occurring at higher latitudes or higher elevations than they did historically, likely because warming temperatures have led to geographic shifts in the occurrence of optimal temperatures for population growth (Carroll et al. 2004; Battisti et al. 2005; Jepsen et al. 2008, 2011; Johnson et al. 2010; Safranyik et al. 2010). Other aspects of climate change, such as milder winter temperatures and increasing summertime drought, have increased the spatial extent and duration of bark beetle outbreaks, leading to dramatic increases in tree mortality (Taylor et al. 2006; Raffa et al. 2008; Bentz et al. 2010; Cooke and Carroll 2017). The effects of climate change on forest insect pests and forest ecosystems also involve feedbacks between relatively short-term events, such as insect outbreaks, and long-term processes including regional diebacks of tree species and the increased release into the atmosphere of CO_2 due to increased tree mortality (Raffa et al. 2008). For example, the implications of the positive feedback involving climate change leading to increased tree mortality due to mountain pine beetle outbreaks, increasing flux of CO_2 into the atmosphere, resulting in increased climate change and thus greater likelihood of further outbreaks seem relatively clear cut (Kurz et al. 2008). However, the ramifications of climate-pest-ecosystem feedbacks are generally difficult to predict. Given the importance of understanding such interactions, shifts in the spatial dynamics of forest insect species and their impacts will likely represent a major research area for decades to come.

References

Altenkirch W, Majunke C, Ohnesorge B (2002) Waldschutz auf ökologischer Grundlage. Ulmer Verlag, Stuttgart, Germany

Aukema BH, Carroll AL, Zhu J, Raffa KF, Sickley TA, Taylor SW (2006) Landscape level analysis of mountain pine beetle in British Columbia, Canada: spatiotemporal development and spatial synchrony within the present outbreak. Ecography 29:427–441

Aukema BH, Carroll AL, Zheng Y, Zhu J, Raffa KF, Moore RD, Stahl K, Taylor SW (2008) Movement of outbreak populations of mountain pine beetle: influences of spatiotemporal patterns and climate. Ecography 31:348–358

Ayres BD, Ayres MP, Abrahamson MD, Teale SA (2001) Resource partitioning and overlap in three sympatric species of *Ips* bark beetles (Coleoptera: Scolytidae). Oecologia 128:443–453

Barbosa P, Schultz JC (eds) (1987) Insect outbreaks. Academic Press Inc, San Diego, CA

Battisti A, Stastny M, Netherer S, Robinet C, Schopf A, Roques A, Larsson S (2005) Expansion of geographic range in the pine processionary moth caused by increased winter temperatures. Ecol Appl 15:2084–2096

Bentz BJ, Régnière J, Fettig CJ, Hansen EM, Hayes JL, Hicke JA, Kelsey RG, Negrón JF, Seybold SJ (2010) Climate change and bark beetles of the western United States and Canada: direct and indirect effects. Bioscience 60:602–613

Bergeron Y, Harvey B, Leduc A, Gauthier S (1999) Forest management guidelines based on natural disturbance dynamics: stand- and forest-level considerations. For Chron 75:49–54

Berryman AA (1986) Forest insects. Principles and practice of population management. Plenum Press, New York

Bjørnstad O, Falck W (2001) Nonparametric spatial covariance functions: estimating and testing. Environ Ecol Stat 8:53–70

Bjørnstad ON, Ims RA, Lambin X (1999) Spatial population dynamics: analysing patterns and processes of population synchrony. Trends Ecol Evol 14:427–431

Blackburn LM, Leonard DS, Tobin PC (2011) The use of *Bacillus thuringiensis kurstaki* for managing gypsy moth populations under the Slow-the-Spread program, 1996–2010, relative to the distributional range of threatened and endangered species. USDA Forest Service, Research Paper NRS-18

Borden JH (1989) Semiochemicals and bark beetle populations: exploitation of natural phenomena by pest management strategists. Ecography 12:501–510

Buonaccorsi JP, Elkinton JS, Evans SR, Liebhold AM (2001) Measuring and testing for spatial synchrony. Ecology 82:1668–1679

Calabrese JM, Fagan WF (2004) Lost in time, lonely, and single: reproductive asynchrony and the Allee effect. Am Nat 164:24–37

Carroll AL, Taylor SW, Régnière J, Safranyik L (2004) Effects of climate change on range expansion by the mountain pine beetle in British Columbia. In: Shore TL, Brooks JE, Stone JE (ed) Mountain Pine Beetle symposium: challenges and solutions. Vol. Pacific Forestry Centre Information Report BC-X-399. Natural Resources Canada, Canadian Forest Service, Victoria, pp 221–230

Cazelles B, Cazelles K, Chavez M (2014) Wavelet analysis in ecology and epidemiology: impact of statistical tests. J R Soc Interface 11:20130585

Cooke BJ, Carroll AL (2017) Predicting the risk of mountain pine beetle spread to eastern pine forests: considering uncertainty in uncertain times. For Ecol Manage 396:11–25

Coyle DR, Nebeker TE, Hart ER, Mattson WJ (2005) Biology and management of insect pests in north american intensively managed hardwood forest systems. Annu Rev Entomol 50:1–29

Didham RK, Ghazoul J, Stork NE, Davis AJ (1996) Insects in fragmented forests: a functional approach. Trends Ecol Evol 11:255–260

Esper J, Büntgen U, Frank DC, Nievergelt D, Liebhold AM (2007) 1200 years of regular outbreaks in alpine insects. Proc R Soc B: Biol Sci 274:671–679

Fleischer SJ, Blom PE, Weisz R (1999) Sampling in precision IPM: when the objective is a map. Phytopathology 89:1112–1118

Forbush EH, Fernald CH (1896) The gypsy moth. Wright and Potter, Boston, MA

Gering JC, Crist TO, Veech JA (2003) Additive partitioning of species diversity across multiple spatial scales: implications for regional conservation of biodiversity. Conserv Biol 17:488–499

Getis A (2008) A history of the concept of spatial autocorrelation: a geographer's perspective. Geogr Anal 40:297–309

Gitau CW, Bashford R, Carnegie AJ, Gurr GM (2013) A review of semiochemicals associated with bark beetle (Coleoptera: Curculionidae: Scolytinae) pests of coniferous trees: a focus on beetle interactions with other pests and their associates. For Ecol Manage 297:1–14

Grove SJ (2002) Saproxylic insect ecology and the sustainable management of forests. Annu Rev Ecol Syst 33:1–23

Hanski I (1997) Metapopulation dynamics: from concepts and observations to predictive models. In: Hanski I, Gilpin ME (eds) Metapopulation biology. Academic Press, San Diego, pp 69–91

Hanski I (1998) Metapopulation dynamics. Nature (London) 396:41–49

Harrington R, Fleming RA, Woiwod IP (2001) Climate change impacts on insect management and conservation in temperate regions: can they be predicted? Agric For Entomol 3:233–240

Hastings A (1990) Spatial heterogeneity and ecological models. Ecology 71:426–428

Haynes KJ, Bjørnstad ON, Allstadt AJ, Liebhold AM (2013) Geographical variation in the spatial synchrony of a forest-defoliating insect: isolation of spatial and environmental drivers. Proc R Soc B: Biol Sci 280:20130112

Haynes KJ, Allstadt AJ, Klimetzek D (2014) Forest defoliator outbreaks under climate change: effects on the frequency and severity of outbreaks of five pine insect pests. Glob Change Biol 20:2004–2018

Haynes KJ, Liebhold AM, Bjørnstad ON, Allstadt AJ, Morin RS (2018) Geographic variation in forest composition and precipitation predict the synchrony of forest insect outbreaks. Oikos 127:634–642

Hodkinson ID (2005) Terrestrial insects along elevation gradients: species and community responses to altitude. Biol Rev 80:489–513

Hopkins AD (1899a) Preliminary report on the insect enemies of forests in the Northwest. U.S. Department of Agriculture Bulletin No. 21

Hopkins AD (1899b) Report on investigations to determine the cause of unhealthy conditions of the spruce and pine from 1880–1893. West Virginia Agricultural Experiment Station Bulletin #56

Hudson P, Cattadori I (1999) The Moran effect: a cause of population synchrony. Trends Ecol Evol 14:1–2

Hunter MD (2002) Landscape structure, habitat fragmentation, and the ecology of insects. Agric For Entomol 4:159–166

Ims RA, Steen H (1990) Geographical synchrony in microtine population cycles: a theoretical evaluation of the role of nomadic avian predators. Oikos 57:381–387

Jackson PL, Straussfogel D, Lindgren BS, Mitchell S, Murphy B (2008) Radar observation and aerial capture of mountain pine beetle, *Dendroctonus ponderosae* Hopk (Coleoptera: Scolytidae) in flight above the forest canopy. Can J For Res 38:2313–2327

Janzen DH (1987) Insect diversity of a Costa Rican dry forest: why keep it, and how? Biol J Lin Soc 30:343–356

Jepsen JU, Hagen SB, Ims RA, Yoccoz NG (2008) Climate change and outbreaks of the geometrids *Operophtera brumata* and *Epirrita autumnata* in subarctic birch forest: evidence of a recent outbreak range expansion. J Anim Ecol 77:257–264

Jepsen JU, Kapari L, Hagen SB, Schott T, Vindstad OPL, Nilssen AC, Ims RA (2011) Rapid northwards expansion of a forest insect pest attributed to spring phenology matching with sub-Arctic birch. Glob Change Biol 17:2071–2083

Johnson DM, Liebhold AM, Bjørnstad ON, McManus ML (2005) Circumpolar variation in periodicity and synchrony among gypsy moth populations. J Anim Ecol 74:882–892

Johnson DM, Liebhold AM, Bjørnstad ON (2006a) Geographical variation in the periodicity of gypsy moth outbreaks. Ecography 29:367–374

Johnson DM, Liebhold AM, Tobin PC, Bjørnstad ON (2006b) Allee effects and pulsed invasion of the gypsy moth. Nature 444:361–363

Johnson DM, Büntgen U, Frank DC, Kausrud K, Haynes KJ, Liebhold AM, Esper J, Stenseth NC (2010) Climatic warming disrupts recurrent Alpine insect outbreaks. Proc Natl Acad Sci 107:20576–20581

Jonsson BG, Kruys N, Ranius T (2005) Ecology of species living on dead wood—Lessons for dead wood management. Silva Fennica 39:289–309

Knops JMH, Tilman D, Haddad NM, Naeem S, Mitchell CE, Haarstad J, Ritchie ME, Howe KM, Reich PB, Siemann E, Groth J (2002) Effects of plant species richness on invasion dynamics, disease outbreaks, insect abundances and diversity. Ecol Lett 2:286–293

Komonen A (2003) Hotspots of insect diversity in boreal forests. Conserv Biol 17:976–981

Krige DE (1999) Essential basic concepts in mining geostatitics and their links with geology and classical statistics. S Afr J Geol 102:147–151

Kurz WA, Dymond CC, Stinson G, Rampley GJ, Neilson ET, Carroll AL, Ebata T, Safranyik L (2008) Mountain pine beetle and forest carbon feedback to climate change. Nature 452:987–990

Landolt PJ (1997) Sex attractant and aggregation pheromones of male phytophagous insects. Am Entomol 43:12–22

Legendre P, Fortin M-J (1989) Spatial pattern and ecological analysis. Vegetatio 80:107–138

Levins R (1969) Some demographic and genetic consequences of environmental heterogeneity for biological control. Bull Entomol Soc Am 15:237–240

Liebhold AM, Rossi RE, Kemp WP (1993) Geostatistics and geographical information systems in applied insect ecology. Annu Rev Entomol 38:303–327

Liebhold AM, Koenig WD, Bjørnstad ON (2004) Spatial synchrony in population dynamics. Annu Rev Ecol Evol Syst 35:467–490

Liebhold AM, Haynes KJ, Bjørnstad ON (2012) Spatial synchrony of insect outbreaks. In: Barbosa P, Letourneau DK, Agrawal AA (eds) Insect outbreaks revisited. Wiley-Blackwell, Oxford, UK, pp 113–125

Liebhold AM, Gottschalk KW, Muzika RM, Montgomery ME, Young R, O'Day K, Kelley B (1995) Suitability of North American tree species to the gypsy moth: a summary of field and laboratory tests. USDA Forest Service, General Technical Report NE-211

Logan JA, Régnière J, Powell JA (2003) Assessing the impacts of global warming on forest pest dynamics. Front Ecol Environ 1:130–137

Mattson WJ, Haack RA (1987) The role of drought in outbreaks of plant-eating insects. Bioscience 37:110–118

McCoy ED (1990) The distribution of insects along elevational gradients. Oikos 58:313–322

McCullough DG, Werner RA, Neumann D (1998) Fire and insects in Northern and boreal forest ecosystems of North America. Annu Rev Entomol 43:107–127

Moran PAP (1953) The statistical analysis of the Canadian lynx cycle. II. Synchronization and meteorology. Aust J Zool 1:291–298

Muzika RM, Liebhold AM (2000) A critique of silvicultural approaches to managing defoliating insects in North America. Agric For Entomol 2:97–105

Myers JH (1998) Synchrony in outbreaks of forest Lepidoptera: a possible example of the Moran effect. Ecology 79:1111–1117

Noriega JA, Hortal J, Azcárate FM, Berg MP, Bonada N, Briones MJI, Del Toro I, Goulson D, Ibanez S, Landis DA, Moretti M, Potts SG, Slade EM, Stout JC, Ulyshen MD, Wackers FL, Woodcock BA, Santos AMC (2018) Research trends in ecosystem services provided by insects. Basic Appl Ecol 26:8–23

Nyland RD (2007) Silviculture: concepts and applications, 2nd edn. Waveland Press, Long Grove, IL

Opdam P, Wascher D (2004) Climate change meets habitat fragmentation: linking landscape and biogeographical scale levels in research and conservation. Biol Cons 117:285–297

Paine TD, Birch MC, Švihra P (1981) Niche breadth and resource partitioning by four sympatric species of bark beetles (Coleoptera: Scolytidae). Oecologia 48:1–6

Paine TD, Raffa KF, Harrington TC (1997) Interactions among scolytid bark beetles, their associated fungi, and live host conifers. Annu Rev Entomol 42:179–206

Peltonen M, Liebhold AM, Bjørnstad ON, Williams DW (2002) Spatial synchrony in forest insect outbreaks: roles of regional stochasticity and dispersal. Ecology 83:3120–3129

Raffa KF (2001) Mixed messages across multiple trophic levels: the ecology of bark beetle chemical communication systems. Chemoecology 11:49–65

Raffa KF, Aukema BH, Bentz BJ, Carroll AL, Hicke JA, Turner MG, Romme WH (2008) Cross-scale drivers of natural disturbances prone to anthropogenic amplification: the dynamics of bark beetle eruptions. Bioscience 58:501–517

Riley CV, Vasey G (1870) Imported insects and native American insects. Am Entomol 2:110–112

Robinet C, Lance DR, Thorpe KW, Tcheslavskaia KS, Tobin PC, Liebhold AM (2008) Dispersion in time and space affect mating success and Allee effects in invading gypsy moth populations. J Anim Ecol 77:966–973

Roland J (1993) Large-scale forest fragmentation increases the duration of tent caterpillar outbreak. Oecologia 93:25–30

Rossi RE, Mulla DJ, Journel AG, Franz EH (1992) Geostatistical tools for modeling and interpreting ecological spatial dependence. Ecol Monogr 62:277–314

Royama T (1984) Population dynamics of the spruce budworm, *Choristoneura fumiferana*. Ecol Monogr 54:429–492

Royama T, MacKinnon WE, Kettela EG, Carter NE, Hartling LK (2005) Analysis of spruce budworm outbreak cycles in New Brunswick, Canada, since 1952. Ecology 86:1212–1224

Royama T (1992) Analytical population dynamics. Chapman & Hall, London, UK

Safranyik L, Linton DA, Silversides R, McMullen LH (1992) Dispersal of released mountain pine beetles under the canopy of a mature lodgepole pine stand. J Appl Entomol 113:441–450

Safranyik L, Linton DA, Shore TL (2000) Temporal and vertical distribution of bark beetles (Coleoptera: Scolytidae) captured in barrier traps at baited and unbaited lodgepole pines the year following attack by the mountain pine beetle. Can Entomol 132:799–810

Safranyik L, Carroll AL, Régnière J, Langor DW, Riel WG, Shore TL, Peter B, Cooke BJ, Nealis VG, Taylor SW (2010) Potential for range expansion of mountain pine beetle into the boreal forest of North America. Can Entomol 142:415–442

Sambaraju KR, Carroll AL, Zhu J, Stahl K, Moore RD, Aukema BH (2012) Climate change could alter the distribution of mountain pine beetle outbreaks in western Canada. Ecography 35:211–223

Sartwell C, Stevens RE (1975) Mountain pine beetle in ponderosa pine–Prospects for silvicultural control in second-growth stands. J Forest 73:136–140

Schoener TW (1974) Resource partitioning in ecological communities. Science 185:27–39

Sharov AA, Leonard D, Liebhold AM, Clemens NS (2002) Evaluation of preventive treatments in low-density gypsy moth populations. J Econ Entomol 95:1205–1215

Southwood TRE (1978) Ecological methods with particular reference to the study of insect populations. Chapman and Hall, London, UK, 524 pp

Swanson BJ, Johnson DR (1999) Distinguishing causes of intraspecific synchrony in population dynamics. Oikos 86:265–274

Taylor LR (1961) Aggregation, variance, and the mean. Nature 189:732–735

Taylor LR (1984) Assessing and interpreting the spatial distributions of insect populations. Annu Rev Entomol 29:321–357

Taylor AD (1990) Metapopulations, dispersal, and predator-prey dynamics: an overview. Ecology 71:429–433

Taylor SW, Carroll AL, Alfaro RI, Safranyik L (2006) Forest, climate and mountain pine beetle outbreak dynamics in western Canada. In: Safranyik L, Wilson B (eds) The mountain pine beetle a synthesis of biology, management, and impacts on lodgepole pine. Natural Resources Canada, Canadian Forest Service, Victoria, BC, pp 67–94

Tobin PC (2004) Estimation of the spatial autocorrelation function: consequences of sampling dynamic populations in space and time. Ecography 27:767–775

Tobin PC, Sharov AA, Liebhold AM, Leonard DS, Roberts EA, Learn MR (2004) Management of the gypsy moth through a decision algorithm under the Slow-the-Spread project. Am Entomol 50:200–209

Tobin PC, Whitmire SL, Johnson DM, Bjørnstad ON, Liebhold AM (2007) Invasion speed is affected by geographic variation in the strength of Allee effects. Ecol Lett 10:36–43

Tobin PC, Bai BB, Eggen DA, Leonard DS (2012) The ecology, geopolitics, and economics of managing Lymantria dispar in the United States. International Journal of Pest Management 53:195–210

Tobin PC, Parry D, Aukema BH (2014) The influence of climate change on insect invasions in temperate forest ecosystems. In: Fenning T (ed) Challenges and opportunities for the world's forests in the 21st century. Springer, pp 267–296

Torrence C, Compo GP (1998) A practical guide to wavelet analysis. Bull Am Meteor Soc 79:61–78

Turner MG (2010) Disturbance and landscape dynamics in a changing world. Ecology 91:2833–2849

Walter JA, Meixler MS, Mueller T, Fagan WF, Tobin PC, Haynes KJ (2015) How topography induces reproductive asynchrony and alters gypsy moth invasion dynamics. J Anim Ecol 84:188–198

Walter JA, Sheppard LW, Anderson TL, Kastens JH, Bjørnstad ON, Liebhold AM, Reuman DC (2017) The geography of spatial synchrony. Ecol Lett 20:801–814

Walther G-R, Post E, Convey P, Menzel A, Parmesan C, Beebee TJC, Fromentin J-M, Hoegh-Guldberg O, Bairlein F (2002) Ecological responses to recent climate change. Nature 416:389

Weed AS, Ayres MP, Hicke JA (2013) Consequences of climate change for biotic disturbances in North American forests. Ecol Monogr 83:441–470

Wickman P-O, Rutowski RL (1999) The evolution of mating dispersion in insects. Oikos 84:463–472

Chapter 19
Monitoring and Surveillance of Forest Insects

Eckehard G. Brockerhoff, Juan C. Corley, Hervé Jactel, Daniel R. Miller, Robert J. Rabaglia, and Jon Sweeney

19.1 Introduction and Overview

Monitoring of insect populations is widely used in entomology in the context of biodiversity studies, as an aspect of pest management, and for the detection of non-native invasive species (e.g. Prasad and Prabhakar 2012; Rabaglia et al. 2019; Seibold et al. 2019). Here we focus on monitoring and surveillance of forest insect 'pests' as well as the detection of non-native invasive species. In general, monitoring is undertaken to (i) obtain information on the presence or abundance of particular species; (ii) study their phenology (e.g. oviposition or flight periods); (iii) predict pest population size, spread and damage; or (iv) to determine if pest management activities such as insecticide treatments or mating disruption are required. These activities are critical aspects of integrated pest management (IPM) programs (Ravlin 1991; Ehler 2006; Chapter 17, this volume).

E. G. Brockerhoff (✉)
Swiss Federal Research Institute WSL, Birmensdorf, Switzerland
e-mail: eckehard.brockerhoff@wsl.ch

J. C. Corley
IFAB (INTA EEA Bariloche-CONICET) and Departamento de Ecología, Universidad Nacional del Comahue, Bariloche, Argentina

H. Jactel
INRAE, University of Bordeaux, BIOGECO, Cestas, Bordeaux, France

D. R. Miller
USDA Forest Service, Southern Research Station, Athens, GA, USA

R. J. Rabaglia
USDA Forest Service, State and Private Forestry, Forest Health Protection, Washington, DC, USA

J. Sweeney
Natural Resources Canada-Canadian Forest Service, Atlantic Forestry Centre, Fredericton, NB, Canada
e-mail: jon.sweeney@nrcan-rncan.gc.ca

© The Author(s) 2023
J. D. Allison et al. (eds.), *Forest Entomology and Pathology*,
https://doi.org/10.1007/978-3-031-11553-0_19

Insect monitoring and surveillance can be done with a variety of methods including physical surveys, the use of insect traps, molecular methods, as well as aerial surveys and remote sensing (Prasad and Prabhakar 2012; Poland and Rassati 2019). Physical field surveys (i.e. by direct observation) focus on insect life stages, characteristic damage symptoms on host plants (e.g. defoliation) or other noticeable signs. Such surveys usually involve a combination of observations in the field, collecting and counting specimens, and recording and analyzing these data. Tools that have long been used to facilitate and standardize insect 'sampling' include sweep-nets and tree-beating sheets (e.g. Morris 1960; Harris et al. 1972). However, these methods are labor-intensive, time-consuming, and can only sample species and life stages that are present at the time when the activity is undertaken by a person in the forest.

An alternative method that is widely used and often more efficient involves the use of insect traps that are based on a variety of mechanisms that draw insects to traps and/or intercept their flights. There is a wide range of trap types such as passive interception traps, light traps, colored sticky traps, and traps baited with certain chemical attractants (e.g. Muirhead-Thompson 1991). In recent years, molecular methods have become increasingly important not only for diagnostic purposes (i.e. species identification) but also for insect monitoring. For example, analyzing eDNA collected from plant surfaces can be a very effective method to detect the presence of target species in an area (Valentin et al. 2018). Remote sensing and aerial surveys are useful for monitoring insect damage across larger geographic areas and where forest access on the ground is limited (Hall et al. 2016; Stone and Mohammed 2017).

Monitoring insects is a very broad and complex subject. This chapter focusses on some of the more important methods to provide an overview of the objectives and applications of monitoring and surveillance of forest insects. These are illustrated with several case studies on monitoring and surveillance of prominent forest insects.

19.2   Monitoring Insect Populations and Damage

There is no single monitoring method that is suitable for all species and purposes. If and how monitoring is done ultimately depends on one's objectives and the availability and suitability of monitoring tools for the target species. Some species can be easily observed because their damage or other signs are highly visible by a trained observer and sufficiently specific. Other insects are rather cryptic and difficult to observe, for example because they are feeding under the bark or in the wood of trees. In such cases, alternative methods such as attractant-baited traps can be very helpful if effective attractants and traps for the target species are available. In this section we introduce the most common conventional monitoring methods.

19.2.1 Ground-Based Monitoring Methods for Insect Life Stages, Damage Symptoms and Other Signs

19.2.1.1 Visual Surveys for Insect Life Stages

Field surveys for eggs, larvae, pupae or adults of target species are a common practice for many species. For example, in the United States, egg masses of spongy moth (*Lymantria dispar*, Erebidae) are counted to determine whether infestation levels are so high that treatments may be necessary to prevent defoliation (Liebhold et al. 1994) (Fig. 19.1). Counting egg masses on tree trunks and branches can be done from the ground, ideally during winter when there is no foliage to obscure egg masses and to provide sufficient lead time for planning management actions. Several procedures have been developed to obtain reliable estimates of spongy moth population density, such as the "fixed-radius" plot method where all trees within several 100 m^2 plots are counted and the average density of egg masses is calculated (Liebhold et al. 1994). Leaf miners and gall makers are also easily identified based on their characteristic symptoms and surveys looking for these symptoms are feasible. Other insects and life stages are commonly sampled with specific tools developed for this purpose.

Fig. 19.1 Egg masses of spongy moth on an oak tree trunk. Credit: Milan Zubrik, Forest Research Institute—Slovakia, Bugwood.org

19.2.1.2   Tools for Sampling Insects

Surveys for foliage-feeding insects are often done using 'beat sheets' in which a pole is used to beat branches and dislodge specimens onto a drop sheet where they can be collected and counted. The number of replicates depends on the size of the area of interest and the sampling accuracy required, but at least three trees should be sampled (Harris et al. 1972). This method has been used, for example, to sample and study the host range of conifer aphids in New Zealand (Redlich et al. 2019) and to sample predators of hemlock woolly adelgid (*Adelges tsugae*, Adelgidae), a severe pest of Eastern hemlock (*Tsuga canadensis*) in eastern North America (Mayfield et al. 2020) (Fig. 19.2). Suction traps using air suction are often used for sampling insects dispersing in large numbers such as aphids and thrips (e.g. Allison and Pike 1988) but they are used less with forest insects. Insects that are concealed inside wood or other plant tissues (e.g. bark beetles, wood borers) may be sampled by enclosing sections of tree stems, branches and twigs in emergence cages or by collecting tree parts and incubating them in chambers to collect the emerging adults (Ferro and Carlton 2011; Chapter 3, this volume).

Fig. 19.2   Using a beat sheet to sample *Laricobius* beetles, predators of the hemlock woolly adelgid. Credit: A. Mayfield, USDA Forest Service

19.2.1.3   Surveys for Symptoms and Signs

The extensive mortality of pines caused by the southern pine beetle (*Dendroctonus frontalis*, Scolytinae) in the southern United States is highly visible. To monitor earlier signs of attack, before trees have succumbed to the beetles and when management interventions are still feasible to avert damage, surveys of boring dust and 'pitch tubes' created by the resin response of attacked trees are an effective method (Billings 2011) (Fig. 19.3).

Monitoring for the presence and relative abundance of the pine processionary moth (*Thaumetopoea pityocampa*, Thaumetopoeidae), a serious defoliator of pines and a public health risk in southern Europe, is done by counting the easily visible silken winter nests made by larvae in the crowns of pine trees (Gery and Miller 1985) (Fig. 19.4) (see also the case study on the pine processionary moth below).

19.2.2   Insect Monitoring Using Traps

Ground-based visual surveys for insect life stages or symptoms of attack may be labour-intensive and time-consuming. Trapping can be more effective, especially if an effective attractant is available that increases the catch rate and specificity of traps. Trapping is widely used for insect monitoring and there is a variety of trap types and mechanisms that may be generic or optimised for particular target species (e.g. Muirhead-Thompson 1991; Häuser and Riede 2015).

Fig. 19.3 'Pitch tubes' on a loblolly pine trunk caused by southern pine beetle attack. Credit: James R. Meeker, USDA Forest Service, Bugwood.org

Fig. 19.4 Nests of the pine processionary moth on Scots pine in Switzerland. Credit: Beat Forster, Swiss Federal Institute for Forest, Snow and Landscape Research, Bugwood.org

19.2.2.1   Passive Traps

Passive traps do not use any particular mode of attraction but simply intercept and trap insects as they are moving about. Examples include pitfall traps (cups buried at ground level that are filled with a liquid preservative that trap walking insects), Malaise traps (tent-like structures that intercept flying insects and trap them in a jar filled with a liquid preservative), window traps and other types of flight intercept traps (see Häuser and Riede (2015) and Knuff et al. (2019) for further references and Fig. 19.5). These trap types are commonly used for biodiversity studies but less so to sample forest pests, partly because they are non-specific and collect large numbers of insects from many species, which results in considerable sorting effort. Such passive traps are typically less sensitive than traps that involve some means of attraction.

19.2.2.2   Traps Involving Attraction of Insects by Light or Color

There are many trap types that attract insects with light, specific colors or silhouettes, chemical attractants (odorants such as insect pheromones and host plant volatiles), or a combination of two or more of these (Muirhead-Thompson 1991). Historically, light trapping was used for monitoring populations of insect pests that fly at night (such as moths and certain beetles). An advantage of light traps is that they capture both males and females (whereas traps baited with sex pheromones typically capture only males). Light traps used to require access to the electricity grid (i.e. mains power) which prohibited their use at most field sites but this is less of a problem now

Fig. 19.5 A malaise trap for capturing flying insects. Credit: D. Miller, USDA Forest Service

with the wide availability of portable power sources. Still, today light trapping is used mainly in biodiversity studies because other methods are more species-specific and more effective.

Trap color on its own is exploited, for example, in yellow traps which are used mainly for monitoring agricultural and greenhouse pests. However, trap color can also affect captures of certain forest insects by synergizing attraction of bark beetles to chemical attractants (e.g. Kerr et al. 2017). Several species of longhorned wood boring beetles (Cerambycidae) respond more to black traps than clear or white traps (Campbell and Borden 2009; Allison and Redak 2017) while other cerambycids and jewel beetles (Buprestidae) are attracted to bright green traps or purple traps (Rassati et al. 2019). Bright green or yellow sticky traps mimic the color of foliage and can be used to monitor defoliators such as the beech leaf-mining weevil (Goodwin et al. 2020). Certain trap colors may also reduce catches of non-target species (e.g. Sukovata et al. 2020).

19.2.2.3   Traps Baited with Pheromones and Host Plant Volatiles

The most widely used traps for forest insects are those baited with odorant lures such as pheromones and host plant volatiles. Pheromones are chemicals that insects release for communication with conspecifics (Howse et al. 1998). The best-known pheromones are moth 'sex pheromones' that are released by females to attract males. Many bark beetles (Scolytinae) release 'aggregation pheromones' that facilitate aggregation on host trees (Byers 1989), and many wood boring longhorned beetles (Cerambycidae) emit 'sex-aggregation pheromones' that attract both sexes, primarily for mating (Hanks and Millar 2016). There are several other types of pheromones (Howse et al. 1998) but they are less important in the context of monitoring.

The chemical structures of pheromones have been identified for many forest insects, especially those of economic importance, and synthetic lures may be commercially available (El-Sayed 2020). Pheromones are often composed of several components and are more or less specific to their species or genus, especially in moths (Lepidoptera) (Löfstedt et al. 2016). For example, traps baited with the main pheromone component of spongy moth (7,8-epoxy-2-methyloctadecane, a 19-carbon epoxide), also known as 'disparlure', catch mainly spongy moth and several congenerics and are widely used for monitoring and detection purposes. The complete blend of the pheromone of spongy moth contains minor components which increase its species specificity (Gries et al. 1996). On the other hand, longhorned wood boring beetles share many of the same sex-aggregation pheromone components. For example, traps baited with racemic 3-hydroxy-2-hexanone can attract several species of Cerambycidae (Millar and Hanks 2017).

Not all insect species use pheromones, and those of many other species remain to be identified. However, host plant volatiles may be used as an alternative attractant for plant-feeding insects because many species use these cues when searching for their hosts. For example, many conifer-feeding bark beetles and woodborers are attracted to alpha-pinene and ethanol, two components that are commonly associated with conifers. Hence, alpha-pinene and ethanol are used to monitor beetles associated with conifers including species of *Arhopalus* (Cerambycidae), *Hylastes* and *Ips* (Scolytinae) (Brockerhoff et al. 2006; Miller and Rabaglia 2009). Likewise, many ambrosia beetles are attracted to ethanol which is an effective lure for species such as *Xyleborus* spp. and *Xylosandrus crassiusculus* (Scolytinae) (Miller and Rabaglia 2009; Reding et al. 2011). Plant volatiles that assist insects with finding their host plants are often referred to as 'kairomones'. While pheromones are 'information chemicals' that are involved in intraspecific communication, kairomones are used as cues for interspecific interactions.

Traps used with pheromones and host plant attractants come in a variety of shapes, sizes, and colours. They use different mechanisms for trapping insects either on a sticky surface or in a collection jar that is easy to enter for an insect but very difficult to exit (effectively a one-way entry). Multiple-funnel traps (also called Lindgren funnel traps after their inventor) are used mainly for bark beetles (Lindgren 1983). They consist of a stack of several funnels and a collection cup at the base (Fig. 19.6a). Panel traps are an alternative design that involves intersecting panels with a single funnel and a collection jar at the base (Fig. 19.6b). These panel traps are typically used for longhorned beetles, weevils and bark beetles. A fluoropolymer may be applied to traps to make them more 'slippery' so that beetles can't hold on to the panel surface (Graham et al. 2010). Funnel and panel traps are mainly colored black so that they resemble the silhouette of a tree trunk, but they are available in other colors. For example, for monitoring emerald ash borer (*Agrilus planipennis*, Buprestidae), green funnel traps (with an attractant) are preferable (Poland et al. 2019). The most common trap design used for bark beetle monitoring in Europe is the so-called Theysohn slot-trap which is based on an alternative flight interception design (Fig. 19.6c).

Neither of these traps work well for Lepidoptera, Hymenoptera and other less 'robust' taxa with a comparatively soft cuticle. For these species, trap types with sticky surfaces are commonly chosen. Perhaps the most widely used of these is the Delta trap which has a roof-shaped design with a sticky substance either on the entire internal surface or on a removable sheet in the trap. A lure is placed inside the trap and insects attracted by this lure are trapped when they land on the sticky internal surface (Fig. 19.6d). An advantage of this design is that the captured insects are spread out on the sticky area which makes examining the catches easy, unless they need to be removed for closer inspection, which may be difficult. A potential disadvantage of delta traps is their propensity to become saturated with the target species. When that is a problem, bucket traps with a larger holding capacity can be used. Unwanted by-catch can be reduced by choosing traps colored green which attract fewer flower-visiting insects than yellow or white traps, for example (Sukovata et al. 2020).

19.2.3 Important Considerations for Trap-Based Monitoring Programs Targeting Bark and Wood Boring Beetles

There are many successful monitoring programs for bark and woodboring beetles in Europe, North America and elsewhere. For example, in Europe, trapping is widely used to monitor populations of the European spruce bark beetle (*Ips typographus*, Scolytinae), the most serious insect pest of spruce forests in Europe. The main purpose is to follow population trends, as described, for example, by Faccoli and Stergulc (2005). Typically, Theysohn slot-traps baited with pheromone (ipsdienol and methyl-butenol) dispensers are used to attract *I. typographus*, and the ratio of trap captures of the summer generation and the spring generation can be calculated to determine whether populations are growing or declining. However, there is some controversy about the extent to which trap captures reflect *I. typographus* population sizes and trends (see Sect. 19.4).

In the southern USA, forest managers use a trap-based monitoring system as part of an IPM program to manage the southern pine beetle (SPB), a major pest of southern pines (Clarke 2012). In the spring of every year, funnel traps baited with pheromone (frontalin) and kairomones (alpha-pinene and beta-pinene) are deployed

Fig. 19.6 Various traps used for insect monitoring and surveillance: **a** Lindgren-funnel trap. Credit: D. Miller, USDA Forest Service; **b** Panel trap with alpha-pinene and ethanol lures attached. Credit: J. Kerr, Scion, New Zealand; **c** Theysohn bark beetle trap. Credit: Gernot Hoch, BFW Institut für Waldschutz, Vienna, Austria; **d** Delta trap. Credit: Karla Salp, Washington State Department of Agriculture, Bugwood.org; **e** Sticky plate trap with pheromone lures in the center and a trapped pine processionary moth. Credit: Hervé Jactel, INRAe, France

at key locations in and around pine stands. Managers consider the number of SPB captured as well as the ratio of predators (the checkered beetle *Thanasimus dubius*, Cleridae) to SPB to determine if local epidemics are increasing, stable or collapsing. This information is used to determine the need for management efforts against SPB.

Operationally, the choice of trap type, lure type and trap position is a major concern for managers planning a trapping program, and these parameters depend on the target species. The efficacy of a trapping program for a single species or broad diversity can be affected by numerous factors such as trap location (canopy vs ground, forest edge vs forest interior), trap type and color, and trapping period and duration (e.g. Brockerhoff et al. 2012; Dodds 2014; Flaherty et al. 2019; Sweeney et al. 2020). Managers need to be clear about their objectives for a trapping program as there is no single scheme that can target all species equally.

Relative species-specificity of lures can be achieved for some species such as the engraver bark beetle *Ips paraconfusus* (Scolytinae) that uses a combination of (–)-ipsenol, (+)-ipsdienol and *cis*-verbenol as its pheromone blend, while frontalin is a common pheromone for various species of *Dendroctonus* (Scolytinae) (Byers 1989). Traps baited with genus-specific monochamol lures are attractive specifically to sawyer beetles (*Monochamus* spp., Cerambycidae) in North America, Europe and Asia, although traps baited with the bark beetle pheromone ipsenol may be equally attractive for *Monochamus* species (Ryall et al. 2015; Miller et al. 2016).

To capture multiple species, blends of multiple attractants can be used. For example, blends of certain hexanediols and hydroxyketones are broadly attractive to numerous woodborers in the longhorn beetle subfamily Cerambycinae (Hanks and Millar 2016). Traps baited with the host plant volatiles alpha-pinene and ethanol are broadly attractive to many bark and ambrosia beetles (Miller and Rabaglia 2009). A combination of alpha-pinene and ethanol and bark beetle pheromones attracts numerous species of woodborers including *Monochamus* species as well as numerous species of bark and ambrosia beetles, and associated predators (e.g. Miller et al. 2013, 2015; Alvarez et al. 2016; Chase et al. 2018).

19.2.4 Monitoring the Population Dynamics of Pine Processionary Moth with Pheromone Trapping

The pine processionary moth (PPM) is the main insect defoliator of pine forests in southern Europe and North Africa (Roques 2015). Severe defoliations by PPM caterpillars feeding on needles result in reduced tree growth (Jacquet et al. 2012) and increase the risk of mortality (Jacquet et al. 2014). The larvae have urticating hairs which can cause serious health problems in people and domestic animals (Vega et al. 2011). PPM populations exhibit cyclic outbreaks (Li et al. 2015) and even though the year of the next peak infestation can be forecasted, the amplitude of defoliation remains unpredictable (Toïgo et al. 2017). It was therefore important to develop a reliable method for monitoring and predicting PPM infestation levels in order to warn forest users and implement necessary control measures such as applications of the toxin of *Bacillus thuringiensis* (*Bt*) when populations get too large.

The conventional population monitoring of PPM is based on counts of winter nests made by larvae in the tree crown (Gery and Miller 1985), but this is tedious and inaccurate in mature or dense pine stands. Pheromone trapping has been considered an alternative method and has proven highly effective in the field (Einhorn et al. 1983). To develop pheromone trapping as a reliable sampling technique, a suitable trap design and trap position had to be identified and it needed to be shown that trap captures were indicative of actual population levels. Sticky plate traps hung at user-friendly heights of about 1.5 m above ground (Fig. 19.6e) appeared to be the most efficient (Jactel et al. 2006). It was also necessary to optimise the pheromone dose and the density of traps to improve the statistical correlations between mean trap capture and other measures of population density. Four sticky plate traps baited with 0.2 mg of the commercial pheromone ("pityolure") provide an accurate and cost-effective estimate of the total number of PPM per hectare (Jactel et al. 2006). This method was tested and further refined in a large operational trial in France (see Sect. 19.4).

19.2.5 Monitoring Populations of the Invasive Woodwasp Sirex Noctilio

Among the non-native invasive forest insects observed in commercial plantation forests in many southern hemisphere countries, the woodwasp *Sirex noctilio* F. (Siricidae) is probably the best known. The species is capable of widespread damage on cultivated pines within the invaded range, especially during population outbreaks (Lantschner and Corley 2015). *Sirex noctilio* is a woodboring species with a solitary lifestyle that infests pine trees. Following mating, females lay eggs by drilling holes in pine stems which they locate by following volatile cues associated with tree stress. During oviposition, the female introduces a symbiotic fungus (*Amylostereum areolatum*) and a phytotoxic venom which together can kill attacked trees (Slippers et al. 2015).

Population monitoring is an important aspect of *S. noctilio* pest management and is often carried out within the invaded range by looking for trees with signs of attack, rather than the insect itself. Attacked pines typically show crown chlorosis, and resin droplets on their stems resulting from oviposition by *S. noctilio*. Sequential sampling protocols and/or aerial surveys support estimations of tree damage and the application of control measures. However, sequential sampling is somewhat flawed as attacks are typically highly aggregated. This approach may underestimate attack levels, especially when populations are low such as in recently invaded sites (Carnegie et al. 2005; Lantschner and Corley 2015).

Alternatively, the trap-tree technique is used to detect early-stage populations. This consists of treating 4–10 trees with low doses of herbicide or careful girdling prior to the wasp flight season (Fig. 19.7). Foraging females are attracted to these artificially stressed trees which can then reveal the presence of *S. noctilio*. Felling of any attacked trees after the flight season may be necessary to avoid the build-up of local populations (Lantschner and Corley 2015). When billets (stem sections) from these trees are caged, the presence and abundance of natural enemies (especially parasitoids attacking the wood wasps) and their potential impact on the *S. noctilio* population can be estimated.

Fig. 19.7 Trap trees to attract Sirex wood wasps in a Pinus contorta plantation in Patagonia, Argentina. Credit: Juan Corley

Flight intercept traps (panel traps or funnel traps) baited with combinations of alpha-pinene and beta-pinene, which are also emitted by stressed trees, can be used to sample *S. noctilio* populations. However, trapping with these lures is usually not as effective as it is for many other insects (Batista et al. 2018). The development of new pheromone and kairomone lures which are based on attractive volatiles from conspecifics or from the wasp's fungal symbiont, may prove important as this type of lure can be highly specific and works well also at low population densities (Fernández Ajó et al. 2015).

The development of effective sampling methods to monitor *S. noctilio* populations within the invaded range is especially important since detecting small populations as early as possible during the invasion phase and understanding when and why *S. noctilio* populations increase is key to preventing regional spread and major economic impact in invaded areas. These should not only include effective trap and lure designs but also statistically valid sampling efforts, to provide quantitative data in diverse environmental conditions. This information is also needed to interpret the success of the control practices deployed.

19.3 Surveillance to Detect Invaders

Preventing the introduction of non-native species is the most effective and first line of defense, although some species may inevitably escape detection and become established. The greatest opportunity for eradication and cost-effective management is immediately after their introduction when their populations are still small and limited to a small area. Early detection followed by rapid assessment and response increases the likelihood of successful eradication or containment (Brockerhoff et al. 2010; Liebhold et al. 2016). There are a number of other purposes of surveillance including to demonstrate freedom from certain pests within an area (a potential requirement for international trade) and to verify the effectiveness of biosecurity measures (Kalaris et al. 2014).

Numerous methods and tools can be applied for surveillance and detection of non-native insects (e.g. Augustin et al. 2012; Kalaris et al. 2014; Poland and Rassati 2019). Many are similar to those used for monitoring native insects (see Sect. 19.2). But there are several key differences: (i) the main initial goal is to detect the *presence* of a non-native species, whereas determining its population size and spatial extent (i.e. delimitation) is a subsequent step; (ii) there is a rather large number of potential invaders, and surveillance often aims at detecting any of multiple species, although some programs are aimed at just one specific unwanted species; and (iii) one is virtually looking for a needle in a hay stack as the aim is to find a small population that could be anywhere. Consequently, methods that are highly sensitive and can cover large areas are preferable. However, if the identity of the target is unknown, methods suitable for a wide range of species are needed. For both cases, trapping with suitable trap type and lure combinations is a preferred option (e.g. Quilici et al. 2012). Below we describe two trapping programs to detect invaders (for spongy moth and non-native bark- and woodboring beetles). But as trapping can only target a limited number of species, more generic surveillance methods that can detect a wider range of species are also needed. Physical searching by trained biosecurity specialists to detect new non-native species is being carried out in several countries, often with a focus on high-risk sites. Engagement of the wider public in surveillance activities can also be highly effective. Examples of these approaches are given below.

19.3.1 High-Risk Site Surveillance

Early detection of non-native species is very important for successful responses to detections. Because the resources for surveillance are limited, efforts need to be focused on locations where non-native species are most likely to arrive and become established. By definition, such locations can be characterized by the likelihood of arrival of non-native pests and by the likelihood of establishment at those sites.

Insights about the likelihood of arrival can be gained from information about trade patterns, particularly regarding the volume and destinations of those types of imports that are known to be associated with species of concern (Colunga-Garcia et al. 2013; Kalaris et al. 2014). These sites tend to be concentrated around commercial and industrial areas, rather than in the forests that are at risk. The surroundings of air and sea ports are also considered high-risk sites although with today's fast and often containerized trade, there is more opportunity for organisms to escape at the eventual destinations of shipments, rather than at ports where shipments pass through. Larger metropolitan areas that are the destination of a large proportion of imported goods and insects transported with these (Branco et al. 2019) are focus areas for surveillance. Therefore, human population size and density can be used as simple proxies if more detailed information about trade flows is not available.

Sites that warrant particular attention are those where imported high-risk commodities arrive such as live plants (e.g. nurseries and garden centers) and wood packaging materials (e.g. recipients of large volumes of paving stones and tiles that are typically packed with pallets and case wood) (Liebhold et al. 2012; Haack 2006). Such information has been used to identify potential hotspots for invasion pressure in the United States where surveillance efforts should be particularly intensive (Colunga-Garcia et al. 2013). Similar concepts have been developed and implemented in other countries. For example, the New Zealand government operates a high-risk site surveillance system in the main urban areas with thousands of transect inspections every year, focusing on urban trees and parks near commercial and industrial areas as well as campsites in natural areas where overseas tourists may introduce pests inadvertently (Bulman 2008; Stevens 2008).

19.3.2 Engaging the Public in Surveillance Activities

Although most members of the public are not experienced in insect identification and detection of non-native species, they are far more numerous than trained professional surveillance staff. It is not uncommon for citizens to notice unusual tree damage and unusual insects in their neighborhood. Consequently, the public should be considered in a surveillance framework as contributing to 'passive surveillance' (e.g. Froud et al. 2008; Hester and Cacho 2017). In New Zealand, there is an established system by which the general public can contribute and report suspicious finds of insects and other species via a toll-free phone number, with about 4,000 notifications per year (Froud et al. 2008). Approximately 8% of all detections of new incursions were reported by the general public, slightly more than those reported by industry. The public is especially encouraged to assist with reporting particular high-risk species and New Zealand's biosecurity authority runs campaigns with newspaper advertisements, tv commercials and social media posts such as the "Catch it - call us" campaign (Fig. 19.8).

The development of a biosecurity board game targeted at both children and adults has proven useful as another way to increase the awareness of the public about biosecurity issues, including the purpose of surveillance. To enhance the ability of the public to identify and report potential biosecurity threats, mobile phone-based

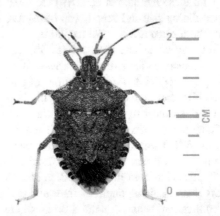

IF YOU FIND ONE OF THESE IN YOUR GARDEN:

CATCH IT.
CALL US.

EXOTIC PEST & DISEASE HOTLINE **0800 80 99 66**

The brown marmorated stink bug can ruin gardens and infest your home.
They're also a major threat to our primary industries and environment.
If you find one: Catch it. Call us.

Fig. 19.8 Advertising used for the "Catch it - call us" campaign by New Zealand's national biosecurity agency MPI to encourage reporting finds of an invasive insect. *Source* https://twitter.com/MPI_NZ/status/662489108065812480

apps have been developed including 'Wild Spotter' in the United States (www.wil dspotter.org, Wild Spotter 2020), 'Observatree' in the UK (https://www.observatree. org.uk) and 'Find-a-Pest' in New Zealand (www.findapest.nz, Pawson et al. 2020). The Find-a-Pest app is effective in reducing the number of false positives (i.e. reports that were of no concern). False positives can be a problem because they occupy the attention and time of biosecurity officials.

19.3.3 Spongy Moth Detection Trapping

The program to detect new infestations of spongy moth along its invasion front and in uninfested regions of the United States is perhaps the largest trap-based pest detection and surveillance program in the world. Approximately 250,000 pheromone-baited spongy moth traps are placed annually by the Animal and Plant Health Inspection Service of the United States Department of Agriculture (USDA APHIS) to detect new populations (USDA 2019). In addition, the USDA Forest Service deploys more than 100,000 traps as part of the spongy moth 'slow the spread' program (Sharov et al. 2002; Bloem et al. 2014). The goal of this program is to minimize the rate of spongy moth spread into uninfested areas in central and southern US forests. Traps along the expanding population front are used to identify newly established populations. Any such populations are then treated to prevent them from growing and coalescing into larger infestations. This approach has successfully reduced the spread rate of spongy moth by > 70% from the historical spread rate of approximately 21 km per year to an average of approximately 6 km per year between 1990 and 2005 (Fig. 5.11a in Tobin and Blackburn 2007), and has a projected benefit-to-cost ratio of approximately 3:1 by delaying the onset of impacts and management expenditures that occur as spongy moth invades new areas (Tobin and Blackburn 2007). This intensive targeted surveillance has enabled the very high success rate of eradications of spongy moth populations, close to 100%, that were detected (Kean et al. 2020).

A similar but smaller detection program is carried out in New Zealand and in Australia. But intensive surveillance is costly and it would be difficult to fund similar programs multiple times for a large number of potential pests. However, it is possible to add lures for other species to spongy moth traps, and this was examined for pairs of spongy moth and 18 other well-known pest moths (Brockerhoff et al. 2013). Lures for more than half of the species could be combined without a substantial reduction in trap sensitivity for either species, and most of the other pairs still caught moths in numbers sufficient for detection purposes. Therefore, combining compatible lures for multiple target species could increase the number of targeted species without greatly increasing the cost of such surveillance programs.

19.3.4 Trapping Programs to Detect Non-Native Bark Beetles and Wood Borers

Bark beetles (Scolytinae) have long been a focus of surveillance programs for non-native forest insects. For example, following the detection of the European pine shoot beetle (*Tomicus piniperda*, Scolytinae) in 1992 in Ohio, a surveillance trapping program was initiated in 1993 in the northeastern United States to enable early detection of other non-native bark beetles (Bridges 1995). Trapping with attractant-baited traps focused on high risk sites including areas near ports, importer warehouses and lumberyards. In 1996, when the first established population of the Asian longhorned beetle (*Anoplophora glabripennis*, Cerambycidae) outside its native range was discovered in New York City (Haack et al. 2010), the threat posed by longhorned beetles became more obvious. There was a growing realization that the large-scale use of solid wood packaging material (WPM) in international trade was a dangerous pathway that made invasions of both wood borers and bark beetles more likely. Between 1985 and 2005, established populations of 25 exotic species of bark beetles and wood borers (Scolytinae, Cerambycidae, Buprestidae) were detected in the United States (Haack 2006) and most of these probably arrived with WPM. Subsequently, a nationwide surveillance trapping program for bark beetles and ambrosia beetles was initiated in the United States (see Sect. 19.3.5).

Several other countries have developed surveillance programs for bark and wood-boring insects, albeit on a smaller scale. For example, such a program was trialed in New Zealand from 2002–2005 using funnel traps baited with host plant attractants and/or bark beetle pheromones, targeting a range of conifer-infesting wood borers and bark beetles (Brockerhoff et al. 2006). Although that particular surveillance program did not lead to the detection of any species not already known to be present, it did confirm the suitability of the approach as numerous non-native Scolytinae and Cerambycidae were trapped near seaports, airports, cargo unloading sites, and in forests near such locations. The surveillance trapping program for bark beetles and wood borers in New Zealand was discontinued mainly because there was uncertainty whether expenditures for the program were justified. However, a benefit–cost analysis carried out later indicated that such a surveillance program is expected to provide economic net benefits even at a high trap density because the economic benefits of early detection, a greater likelihood of successful eradication and less pest damage, likely exceeded the costs of the surveillance program (Epanchin-Niell et al. 2014).

Intercept panel traps or multiple-funnel traps (described above) are used in most detection programs. However, Malaise traps may be more effective in the detection of numerous species of bark and wood boring beetles (Dodds et al. 2015) but there is a trade-off because Malaise-type traps are about five times more expensive than intercept or funnel traps. In addition, Malaise traps tend to capture many more non-target species and consequently require more labor for sorting samples. Given the apparent variability in trapping efficiency even at short distances, detection programs might be more cost effective by using a larger number of panel or funnel traps than Malaise-type traps. Another method that may be suitable for increasing the efficiency of detection trapping is to use a combination of lure blends so that each trap targets multiple species (rather than using separate traps each baited only for a particular species) (Chase et al. 2018; Fan et al. 2019; Rassati et al. 2019). This would either reduce the number of traps needed, or it would lead to an increased number of traps available for detecting particular species. There is a potential disadvantage of using lure blends in that it may reduce the number of insects caught of some species (Miller et al. 2017). However, for the purpose of detection, it is only necessary to trap at least one individual of a target species, so this disadvantage may be tolerable.

19.3.5 Early Detection of Bark and Ambrosia Beetles in the US

Bark and ambrosia beetles (Scolytinae) are some of the most important insects affecting forests in North America, and are the most commonly intercepted group of beetles at US ports of entry (Haack 2006). From 1984–2008, more than 8,000 interceptions of bark and ambrosia beetles, from 85 different countries, were reported at US ports (Haack and Rabaglia 2013). To increase the likelihood of early detection of such beetles, the USDA Forest Service began a pilot project in 2001 (Rabaglia et al. 2008) and then implemented in 2007 a national project for the early detection and rapid response (EDRR) of non-native bark and ambrosia beetles across the US (Rabaglia et al. 2019). The target species were selected based on their frequency of interception, the potential damage a species may cause in the US, and the availability of effective traps and lures for the species. The Scolytinae species selected were *Hylurgops palliatus, Hylurgus ligniperda, Ips sexdentatus, Ips typographus, Orthotomicus erosus, Pityogenes chalcographus, Tomicus minor, Tomicus piniperda, Trypodendron domesticum,* and *Xyleborus* species.

Three Lindgren funnel traps were used at each survey location, and each trap was baited with one of the following three lures or lure combinations: (i) ultra-high release (UHR) ethanol lure only, a general attractant for woodboring insects in hardwood and some coniferous hosts, (ii) UHR alpha-pinene and UHR ethanol lures together, which are general attractants for woodboring insects in coniferous hosts (Miller and Rabaglia 2009), and (iii) a three-component exotic *Ips* lure of ipsdienol, cis-verbenol and methyl-butenol, a specific combination for *I. typographus* and several other conifer-feeding exotic bark beetles (Bakke et al. 1977). Trapping began based on local phenology of bud break and knowledge of early emergence of bark and ambrosia beetles, from late February to early May, depending on the State, and lasted typically for 12 weeks.

Since 2010, the project focused on five high- risk states (California, Florida, Georgia, New York, and Texas), based on interceptions at ports-of-entry, the number of established non-native species, the amount of forest land, and transportation corridors. Other states were surveyed only every 3–7 years, depending on their risk and available funding. Within each state, trapping was carried out in wooded areas or parks near high-risk sites where potentially infested solid wood packing material (e.g. wooden crates and pallets) were imported, stored, or recycled. Taxonomists identified all of the bark and ambrosia beetles and the data were shared at www.bar kbeetles.info.

More than 840,000 specimens of bark and ambrosia beetles had been collected and identified in forty-eight states (including Alaska and Hawaii), Puerto Rico, and Guam from 2007–2016 (Rabaglia et al. 2019). Within the continental U.S., the survey captured specimens of approximately 300 species out of the approximately 550 that occur in the U.S. Forty-three of the species collected were non-native species established in the U.S. The three most common species in traps were *Xyleborinus saxesenii*, *Xylosandrus crassiusculus*, and *Xylosandrus germanus*, three well-established non-native species with strong responses to ethanol-baited traps.

The primary goal of EDRR is the early detection of species new to North America. In the first few years of the pilot phase of EDRR, five species of scolytines new to North America were found in traps, and since 2007, three additional species new to North America were found: *Xyleborinus octiesdentatus*, *Xylosandrus amputatus*, and *Xyleborinus artestriatus* (Rabaglia et al. 2019). Assessments and follow up surveys to delimit the distribution of the new species were conducted soon after but these beetles were established over large areas and eradication was not feasible. Eradication of xyleborine ambrosia beetles, such as these three species, can be particularly challenging. Their cryptic nature, wide host range (these species breed in most hardwood trees), and their inbred, polygynous biology, allows them to go undetected and to quickly spread from just a few individuals.

It is likely that some, if not most, of the species newly detected during the beginning years of EDRR were present in the U.S. for decades. These legacy species were soon detected with the start of surveys such as EDRR. Since 2010, there have been no detections of species new to North America in EDRR traps. It is possible that all non-native species established in the states surveyed before 2010 have been detected and any new detections will be of recent introductions allowing for a more effective rapid response. It is also possible that the implementation of international protocols, such as ISPM 15, and awareness of the risk of moving wood products has reduced the number of wood boring insects introduced into the U.S.

19.3.6   Development of Survey Tools for an Invasive Longhorn Beetle in Canada

The brown spruce longhorn beetle (BSLB), *Tetropium fuscum* (Cerambycidae), native to Europe, was discovered in Halifax, Nova Scotia, Canada in 1999, infesting mature red spruce (Smith and Hurley 2000). About one third of trees displaying signs of resin flow on the trunk and spheroidal exit holes were dead but most were alive and appeared healthy, suggesting BSLB was successfully colonizing and killing trees (O'Leary et al. 2003). The Canadian Food Inspection Agency (CFIA) declared

BSLB a regulated quarantine pest in spring of 2000 and led a multiagency "BSLB task Force" in a survey and eradication program. The regulated area was delimited using intensive ground surveys and the visual signs of infestation, examining > 52,000 conifers on > 47,000 residential properties in greater Halifax in 2000.

Lindgren funnel traps (Lindgren 1983) baited with the same three lure combinations used by the EDRR program in the US (i.e. ethanol and alpha-pinene, ethanol alone, or a three-component exotic *Ips* lure) had been deployed in Halifax by CFIA since 1995 for exotic woodborer surveillance, but had failed to detect BSLB. Thus, members of the Task Force collaborated to develop survey tools to detect spread of BSLB and monitor the progress of the eradication program. Decks of freshly cut spruce logs (Post and Werner 1988) were deployed along major highways from Halifax in 2000–2002. Log decks detected BSLB in two new locations outside of the regulated area but were labor-intensive and slow. In 2003, log decks were replaced by intercept panel traps (Czokajlo et al. 2001; de Groot and Nott 2001) baited with a synthetic "spruce blend" lure, consisting of five major monoterpenes emitted from infested spruce (Sweeney et al. 2004). Adding an ethanol lure increased detection rates (Sweeney et al. 2004, 2006) and from 2004–2006, these baited traps detected BSLB in 25 sites outside of the regulated area, prompting CFIA to expand the regulated area in spring of 2007.

In 2006, Silk et al. (2007) identified a male-produced sex-aggregation pheromone, (*E*)-6,10-dimethyl-5,9,-undecadien-2-ol ("fuscumol"), that synergized attraction of both sexes of BSLB when combined with spruce blend and ethanol. In 2007, operational surveys with the more sensitive pheromone-baited traps detected BSLB in 16 sites outside of the newly expanded regulated area, and CFIA switched the goal from eradication to slowing the spread (CFIA 2017). By spring of 2015, BSLB had been detected in more than 100 sites outside of the 2007 regulated area and CFIA declared the entire province of Nova Scotia infested (CFIA 2017).

This case study highlights the importance of inter-agency collaboration and rapid technology transfer in the development of operational survey tools. It also highlights the critical need for effective survey tools for early detection when containment or eradication of an invasive species is still feasible (Brockerhoff et al. 2010; Tobin et al. 2014; Liebhold and Keen 2018).

19.4 Making Sense of Trap Catch Data, and Statistical Considerations

19.4.1 Relationships Between Trap Catch and Local Population Size

The relationship between trap catch and local population density of forest insects, tree damage or tree mortality is not always strong. For example, while pheromone-baited traps can be useful for determining whether *I. typographus* populations are growing or declining (Faccoli and Stergulc 2006), and low catches were indicative of low levels of damage occurring, high catches were not well correlated with infestation levels near traps (Lindelöw and Schroeder 2001). In another study, no relationship at all was found between trap catch of *I. typographus* and attacks of trees nearby (Wichmann and Ravn 2001). Likewise, in North America, a study of western pine beetle (*Dendroctonus brevicomis*) suggested that pheromone-baited funnel traps were not useful for predicting mortality of pines nearby (Hayes et al. 2009). Conversely, pheromone trap catch of spruce beetle (*Dendroctonus rufipennis*) provided reliable estimates of Engelmann spruce mortality around the trap, albeit with large variance (Hansen et al. 2006; Negrón and Popp 2018).

Relationships between pheromone trap catch and indicators of population size were found to be more reliable for several Lepidoptera species. For example, catches of eastern spruce budworm (*Choristoneura fumiferana*, Tortricidae) by traps baited with sex pheromone showed a strong relationship with densities of spruce budworm larvae in the following year, which allowed prediction of outbreaks in eastern Canada up to six years in advance (Sanders 1988). However, at high population densities, trap catch was less indicative of population trends. Nevertheless, pheromone traps have been used for decades to monitor spruce budworm populations. Pheromone trap catch of the Nantucket pine tip moth (*Rhyacionia frustrana*, Tortricidae) in Georgia was moderately to highly correlated with population density and damage for the first adult generation but less so for subsequent generations within a year (Asaro and Berisford 2001). In France, pheromone trapping of the pine processionary moth was developed for population monitoring (Jactel et al. 2006) and tested from 2010 to 2016 across 50 pine plantations. This showed that trap catch is highly correlated with the annual number of attacked trees and can be used to predict infestations in the following year. Pheromone trap catch of a close relative, the oak processionary moth (*Thaumetopoea processionea*, Thaumetopoeidae), was less well correlated with local population densities in the U.K. (Straw et al. 2019). Nevertheless, the presence of nests within 250 m from a trap was successfully determined in 91% of cases.

Several important points need to be taken into consideration when evaluating relationships between trap catch and other indicators of insect presence, abundance, and damage: (i) traps can capture insects that have flown tens or hundreds of meters from where they had been feeding on a tree so that trap catch is not necessarily related to populations in the immediate neighborhood of a trap; (ii) insect populations can be highly patchy in space (Safranyik et al. 2004) and small numbers of traps may not provide an accurate indication of larger-scale abundance or damage, but a larger number of traps deployed at a site may do so (Schroeder 2013); (iii) when local populations are large, pheromone traps "compete" with many natural pheromone sources, and the same applies to traps baited with host plant volatiles when these are located in areas with an abundance of naturally occurring host plant volatiles (Wermelinger 2004; Schroeder 2013); (iv) the relationship between trap catch and population size may or may not be relevant depending on whether the purpose of trapping is for prediction of damage or just for detection of the presence of a species (as in pest detection and delimitation surveys) (Brockerhoff et al. 2013). Consequently, the choice of trapping or an alternative method depends on the purpose of the activity. If prediction of population size is important, then a larger number of traps across a forest may be needed to obtain a better estimate and other factors such as the amount of host trees and the condition of sites need to be considered (Schroeder 2013). Furthermore, conclusions or inferences from trap catch data strongly depend on context such as catches of the same insect species in previous years or in traps at other locations in the same year.

19.4.2 Pheromone Trap Attraction Range

Beyond the intrinsic capture efficiency of an attractant-baited trap, it is important to know its attraction range, the area around a trap over which the target species is drawn towards the trap. The attraction range is relevant for validating correlations between trap catch and local population level at the same spatial scale. It is important for making inferences about the effective sampling area, i.e. the portion of the landscape where the target species can be detected, especially for surveillance of alien invasive pests (Kriticos et al. 2007). Additionally, knowledge of when the interception zones of adjacent traps overlap assists with designing pheromone trap networks (Manoukis et al. 2014) to optimize trap density, save time and reduce costs of trapping programs.

A common and convenient method of estimating the attraction range is based on analyzing interference between adjacent attractant-baited traps, considering that competition for insect capture would occur if two neighboring traps are sufficiently close to have overlapping attraction ranges (i.e. are at a distance shorter than twice their attraction range) (Schlyter 1992). To evaluate the distance between adjacent pheromone traps that would minimize competition and thus approximate the attraction range (or radius), a number of studies have been conducted with more or less complex grids, circles or groups of traps (Wall and Perry 1987; Schlyter et al. 1987; Elkinton and Cardé 1988; Oehlschlager et al. 2003; Jactel et al. 2019). Although the attraction range of pheromone traps for forest insects can vary greatly depending on trap design and the rate of release of pheromone lures, it is typically in the order of a few tens to hundreds of meters.

19.5 Other Detection Techniques Including Detector Dogs, E-Noses, Acoustic Detection and Molecular Techniques

19.5.1 Detection of Volatiles Emitted by Target Species

Most insects have a particular smell that may be related to pheromone production, some other biochemical process or other organisms associated with them. This can be exploited for surveillance purposes either by using chemical detection devices or with trained dogs. In several countries, trained detector dogs (or 'sniffer dogs') are used at airports to detect imports infested with insects or to find smuggled or prohibited goods (USDA 2012). However, detector dogs can also be used in urban areas and in plant nurseries to detect trees or plants for planting that are infested by an unwanted insect. In Austria and other countries in Europe, dogs have been trained to detect *Anoplophora* beetles in wood packaging material, live plant imports, and in urban or rural areas (Hoyer-Tomiczek and Sauseng 2013). Such dogs can be very effective; for example, 15,000 plants imported from Asia were screened over a period of three days, and the dogs detected five plants that were infested by citrus longhorned beetle (*Anoplophora chinensis*, Cerambycidae) (Hoyer-Tomiczek and Sauseng 2013). In the US, an *Anoplophora* dog detection program was found to be 80–90% successful in detecting infested trees (Errico 2013). However, detector dogs are mainly suitable for particular target species; their use for generic detection of insects and fungi is limited due to the ubiquitous presence of these organisms.

Conventional analytical identification of volatile organic compounds (VOCs) can also be used for surveillance purposes. Typically, this involves headspace analysis by gas chromatography (GC) and mass spectrometry to characterize the volatiles associated with a target species. Once identified, the environment can be screened for these volatiles using a similar procedure. For example, volatiles emitted by the brown marmorated stinkbug (*Halyomorpha halys*, Pentatomidae) in a confined space were identified in this way, and it was then tested whether detectable concentrations of these volatiles could be isolated in a larger environment (Nixon et al. 2018). However, the highly diluted volatiles proved difficult to detect, and the sensitivity of this technique may rarely be sufficient for practical application in the field.

Another potentially suitable approach for detecting volatiles of target species is the use of electronic noses (e-noses). Proof-of concept studies have demonstrated the potential suitability of bio-electronic noses for detection purposes, but no such devices are ready for application on an operational basis, although considerable progress has been achieved (e.g. Oh et al. 2011; Du et al. 2013). It is expected that such devices will be available for practical use sometime in the 2020s (Glatz and Bailey-Hill 2011).

19.5.2 Acoustic Detection

Many insects produce sounds or vibrations for communication or in conjunction with movement or feeding (e.g. Hill 2008; Mankin et al. 2011). These acoustic and vibrational signals can be detected with a variety of sensors and devices, most of which are portable (Mankin et al. 2011). A key advantage of this technique is that it allows the detection of species that are hidden from sight such as wood borers and bark beetles inside wood, and it is non-destructive. As many species produce characteristic sounds, it may be possible to identify the type of organism or even the species by acoustic analysis (Bedoya et al. 2021). This technique has its limitations, though, as these signals are often very quiet and sensors need to be very close to the source, and background noise can be a problem (Mankin et al. 2011). For example, the detection of bark beetle chirps under the bark of trees or logs is only possible within a distance of less than one meter and preferably much closer (Bedoya et al. 2022). Although operational application has been limited so far, acoustic detection of the red palm weevil (*Rhynchophorus ferrugineus*, Curculionidae), an invasive pest of palms that feeds inside palm trees is possible with a mobile acoustic detection system with > 80% accuracy (Herrick and Mankin 2012). Acoustic and low-frequency vibrational signals can also be detected with laser vibrometers. A portable laser vibrometer can be used to detect Asian longhorned beetles infesting trees or logs (Zorović and Čokl 2015).

19.5.3 Molecular Techniques and eDNA

Molecular techniques are increasingly used in a monitoring and surveillance context to identify insects. Eggs, larvae and pupae, which are difficult to identify using morphological characters, can often be identified with DNA barcoding using the mitochondrial COI gene (Frewin et al. 2013; Madden et al. 2019). There are also a wide range of molecular tools that are suitable for the detection and diagnosis of potentially invasive organisms on infested imports. These commonly use polymerase chain reaction (PCR) amplification in the laboratory but mobile PCR-based or loop-mediated isothermal amplification (LAMP) devices that can be used in the field are now available (Arif et al. 2013; Baldi and La Porta 2020), although these are used much more for pathogens than for insects. However, the use of environmental DNA (eDNA) has been shown to be effective in revealing the presence of small populations of invasive insects that may be difficult to detect with other methods (Valentin et al. 2018). Analysis by eDNA techniques of samples of plant material or rain water run-off on tree trunks could be a useful approach for surveillance and early detection of known target species.

19.6 Aerial Surveys and Remote Sensing

19.6.1 Aerial Surveys

When surveys are required for very large areas and ground-based surveillance and trapping programs are not practical, aerial surveys are often used. In North America, for example, aerial overview surveys of forest lands have been one of the foundations of forest pest management for decades (Hall et al. 2016). Aerial surveys are critical for assessing pest impacts in remote areas as well as for insects that impact forests at the landscape level. Yearly identification and mapping of numerous forest insect pests such as eastern spruce budworm, southern pine beetle, Douglas-fir tussock moth (*Orgyia pseudotsugata*, Erebidae) and mountain pine beetle (*Dendroctonus ponderosae*, Scolytinae), provide assessments of infestations on forest lands (Aukema et al. 2006; Bouchard et al. 2006; Taylor and MacLean 2008; Hall et al. 2016). Aerial surveys can be affected by weather conditions and navigation but they are relatively accurate. For example, a comparison of aerial sketch mapping of annual defoliation by eastern spruce budworm and defoliation assessments on the ground showed that 85% of aerial mapping correctly classified defoliation as either nil to light (0–30%) or moderate to severe (31–100%) (Taylor and MacLean 2008). Apart from assessing current impacts, these data can be analyzed together with data on historical outbreak patterns to predict spatiotemporal patterns of future epidemics (see Aukema et al. 2006 for an example on mountain pine beetle).

Considerable effort goes into aerial forest health surveys. For example, in British Columbia, aerial overview surveys in 2019 were conducted for 80% of the province with 658 flight hours logged over 129 flight days (Westfall et al. 2019a). These revealed that a total of 5.9 million ha of forested lands were damaged by ≥ 46 agents (biotic and abiotic). Combined with directed ground inspections, these identified major infestations of 15 insect species and 10 diseases in coniferous forests while deciduous forests recorded impacts from 6 insect species and 2 diseases. Areas damaged by insects were greatest for the western balsam bark beetle in coniferous stands (3.2 million ha) and the aspen leaf miner in hardwood stands (1.3 million ha). Linking the incidences and expansions of tree mortality and defoliation with inventory databases permits accurate determinations of tree mortality and potential losses from such infestations, thereby broadly guiding management efforts such as stand thinning, sanitation and salvage logging, and insecticide applications.

Typically, aerial surveys are conducted by trained professionals per specific guidelines (see Westfall et al. 2019b, for example, for British Columbia). Surveyors identify tree species and damage agents from small planes or helicopters, sketch mapping types of damage and boundaries of disturbances directly on forest cover maps. The use of GIS and GPS has greatly improved the accuracy of aerial surveys. The use of aerial photography and remote sensing (see below) adds additional overlays to maps. Ground truthing of infestations is an important step to verify the accuracy of aerial surveys. In addition to species identification of causal agents, ground truthing can provide important information on the stage of infestations. In pine stands attacked by the mountain pine beetle, forest health professionals can assess attack densities on trees and the ratio of trees currently under attack to those that were attacked the previous year, providing a measure of risk for further attacks the following year. Integrating such data with inventory data on susceptible volumes of trees in the area helps determine the likelihood of further expansion of infestations.

Ground truthing can also help prevent over-reactions to apparent insect damage by forest managers. For example, sawflies can cause extensive defoliation on hemlocks in coastal forests of British Columbia (Nealis and Turnquist 2010). The visibility of red foliage over thousands of hectares can cause concern with forest managers resulting in initial impulses to log the area before timber is degraded by disease or checking. Ground truthing provides the opportunity to document that damage occurs on old foliage while new, current year foliage is untouched by sawflies. Moreover, sawfly infestations are generally short-lived due to the effects of natural enemies. Examinations of branches in the field can readily verify high rates of parasitism of sawfly pupae. The use of drones or unmanned aerial vehicles (UAVs) with cameras can add significant benefits to ground truthing efforts, enabling surveyors the chance to examine crowns of tall trees and survey expansive regeneration stands that are difficult to traverse in person. Potential UAV applications are covered in the following section.

19.6.2   Remote Sensing of Forest Insect Damage

The use of remote sensing for forest health monitoring has increased substantially in recent years as research progress has made this an increasingly accessible and potentially powerful tool. Remote sensing involves high-resolution multi-spectral imagery acquired by satellites, aircraft or UAVs, which is processed (e.g. corrected for topography and atmospheric conditions) and analyzed (Hall et al. 2016; Stone and Mohammed 2017; Torresan et al. 2017). Satellite imagery can be of sufficient spatial resolution to enable identification of individual tree crowns or even individual branches, although there is a trade-off between resolution and the area displayed (i.e. the high-resolution 1.2-m pixel size of the Worldview-3 satellite sensor has an image width of only 13 km whereas the Landsat-8 satellite sensor has an image width of 185 km but a pixel size of 30 m, too coarse to display individual tree crowns) (Hall et al. 2016). Optical remote sensing captures the reflection of sunlight from trees and other structures, and the more separate spectral bands are recorded by a sensor, the better the spectral resolution and visualization of symptoms. The detection of insect damage is typically done by identifying damage-specific changes in spectral reflectance between images recorded from the same location in successive years, although a single image may sometimes suffice. The detection of change can be automated and there are many different approaches for doing this (Hall et al. 2016).

A review of uses of satellite imagery for detection of forest insect damage in North America has been compiled by Hall et al. (2016), including some 50 examples for mountain pine beetle, spruce beetle, eastern spruce budworm, western spruce budworm (*Choristoneura occidentalis*, Tortricidae), jack pine budworm (*Choristoneura pinus*, Tortricidae), spongy moth, and others. However, the uptake for operational use of satellite-based remote sensing data for forest health surveys has been limited so far. This has been attributed to several complicating factors including the requirement for species-specific spectral identification of insect damage, the limited time window when damage can be detected and atmospheric conditions/cloud cover need to be suitable, and difficulty with damage classification which typically occurs on a continuum rather than in specific classes (such as light, moderate, and severe) (Hall et al. 2016).

Despite some challenges, there is rapid progress with image resolution and analysis, and it can be expected that this technology will be adopted increasingly for operational use. When insect damage is sufficiently severe and detectable in satellite images, then this methodology is already relatively powerful. For example, a study in Sweden investigated the onset of infestations of Norway spruce by the invading Hungarian spruce scale insect (*Physokermes inopinatus*, Coccidae) which causes characteristic black 'sooty mold' on the foliage (Olsson et al. 2012). Using SPOT satellite data, 78% of damage was detected successfully, and retrospective data analysis was able to identify the year when this characteristic damage first occurred (Olsson et al. 2012). One way in which damage symptoms can be identified with greater reliability is by combining data from passive light sensors with data from active systems like LiDAR (Light Detection and Ranging) and Radar sensors (Stone and Mohammed 2017).

Multispectral analysis of aerial imagery taken by aircraft uses the same principles as that of satellite imagery but it has the advantage of user-controlled timing of image acquisition when symptoms and atmospheric conditions are ideal. However, taking images by manned aircraft can become expensive when larger areas need to be surveyed. Using UAVs for this purpose is increasingly feasible and may be more cost-effective than using larger manned aircraft, especially when surveys involve smaller areas. Torresan et al. (2017) reviewed several studies that tested UAVs equipped with visible and near-infrared or hyperspectral cameras to detect and classify forest insect damage. The use of UAVs for this purpose was promising with a detection reliability of ca. 75–90%. A UAV remote sensing application for detecting bark beetle damage on individual urban trees was developed by Näsi et al. (2018) with similar levels of accuracy of identification of healthy, infested, and dead trees.

19.7 Outlook

The need for monitoring and surveillance of forest insects is likely to grow in importance. Insect outbreaks appear to become more frequent and more severe as multiple disturbance factors including climate change and other anthropogenic impacts disturb forest ecosystems. Likewise, international trade is expected to increase and involve ever more trading partners around the world, which will facilitate more arrivals and establishments of non-native species, despite our efforts to curb these. To keep up with these trends, early detection of both incursions of non-native species and outbreaks of native species will be critical to enable effective responses.

There is a large pool of methods for monitoring and surveillance and more are becoming available with the rapid progress of science and technology. Conventional methods such as surveillance of forests and high-risk sites by trained experts as well as trapping using targeted and broad-spectrum attractants will remain important. Trapping programs are likely to become more effective for a wider range of species as new attractants are being developed. Nevertheless, many species will remain for which trapping is not an option. A disadvantage of these conventional methods is their limited spatial coverage.

Several new technologies are being developed or refined that enable monitoring and surveillance over larger areas including enhanced aerial surveillance and remote sensing using a variety of platforms. Progress with big data analysis and modelling also plays a role here. New developments in acoustic, chemical, and molecular detection methods and tools are also playing an increasingly important role. For example, the use of eDNA is promising for a range of surveillance applications. However, many of these methods are costly, and large-scale implementation would require large budgets. Conversely, better education and raised awareness among the wider public would be valuable without necessarily being costly. Citizen science projects are emerging in many countries and this is a promising development.

Acknowledgements This publication was supported in part by the HOMED project (http://homed-project.eu/), which received funding from the European Union's Horizon 2020 research and innovation program under grant agreement No. 771271. Contributions by E.G.B. were also supported by the New Zealand government via MBIE core funding to Scion under contract C04X1104 and the Better Border Biosecurity Collaboration (www.b3nz.org).

References

Allison D, Pike KS (1988) An inexpensive suction trap and its use in an aphid monitoring network. J Agr Entomol 5(2):103–107

Allison JD, Redak RA (2017) The impact of trap type and design features on survey and detection of bark and woodboring beetles and their associates: a review and meta-analysis. Annu Rev Entomol 62:127–146

Alvarez G, Gallego D, Hall DR, Jactel H, Pajares JA (2016) Combining pheromone and kairomones for effective trapping of the pine sawyer beetle *Monochamus galloprovincialis*. J Appl Entomol 140(1–2):58–71

Arif M, Fletcher J, Marek SM, Melcher U, Ochoa-Corona FM (2013) Development of a rapid, sensitive, and field-deployable razor ex biodetection system and quantitative PCR assay for detection of *Phymatotrichopsis omnivora* using multiple gene targets. Appl Environ Microbiol 79(7):2312–2320

Asaro C, Berisford CW (2001) Predicting infestation levels of the Nantucket pine tip moth (Lepidoptera: Tortricidae) using pheromone traps. Environ Entomol 30(4):776–784

Augustin S, Boonham N, De Kogel WJ, Donner P, Faccoli M, Lees DC, ... Roques A (2012) A review of pest surveillance techniques for detecting quarantine pests in Europe. EPPO Bull 42(3):515–551

Aukema BH, Carroll AL, Zhu J, Raffa KF, Sickley TA, Taylor SW (2006) Landscape level analysis of T pine beetle in British Columbia, Canada: spatiotemporal development and spatial synchrony within the present outbreak. Ecography 29(3):427–441

Bakke A, Froyen P, Skattebol L (1977) Field response to a new pheromonal compound isolated from *Ips typographus*. Naturwissenschaften 64:98–99

Baldi P, La Porta N (2020) Molecular approaches for low-cost point-of-care pathogen detection in agriculture and forestry. Front Plant Sci 11:570862

Batista ES, Redak RA, Busoli AC, Camargo MB, Allison JD (2018) Trapping for *Sirex* woodwasp in Brazilian pine plantations: lure, trap type and height of deployment. J Insect Behav 31(2):210–221

Bedoya CL, Hofstetter RW, Nelson XJ, Hayes M, Miller DR, Brockerhoff EG (2021) Sound production in bark and ambrosia beetles. Bioacoustics 30(1):58–73

Bedoya CL, Nelson XJ, Brockerhoff EG, Pawson S, Hayes M (2022) Experimental characterization and automatic identification of stridulatory sounds inside wood. Roy Soc Open Sci 9: 220217. https://doi.org/10.1098/rsos.220217

Billings RF (2011) Aerial detection, ground evaluation, and monitoring of the southern pine beetle: state perspectives. In: Coulson RN, Klepzig KD (eds) Southern Pine Beetle II. Gen. Tech. Rep. SRS-140. US Department of Agriculture Forest Service, Southern Research Station, Asheville, NC, pp 245–246

Bloem K, Brockerhoff EG, Mastro V, Simmons GS, Sivinski J, Suckling DM (2014) Insect eradication and containment of invasive alien species. In: Gordh G, McKirdy S (eds) The handbook of plant biosecurity. Springer, Dordrecht, pp 417–446

Bouchard M, Kneeshaw D, Bergeron Y (2006) Forest dynamics after successive spruce budworm outbreaks in mixedwood forests. Ecology 87(9):2319–2329

Branco M, Nunes P, Roques A, Fernandes MR, Orazio C, Jactel H (2019) Urban trees facilitate the establishment of non-native forest insects. NeoBiota 52:25–46

Bridges JR (1995) Exotic pests: major threats to forest health. In Eskew LG, comp. Forest Health Through Silviculture: Proceedings of the 1995 National Silviculture Workshop, Mescalero, New Mexico, May 8–11, 1995. Gen. Tech. Rep. RM-GTR-267. US Department of Agriculture, Forest Service, Rocky Mountain Forest and Range Experiment Station, Fort Collins, CO, pp 105–113

Brockerhoff EG, Jones DC, Kimberley MO, Suckling DM, Donaldson T (2006) Nationwide survey for invasive wood-boring and bark beetles (Coleoptera) using traps baited with pheromones and kairomones. For Ecol Manag 228:234–240

Brockerhoff EG, Liebhold AM, Richardson B, Suckling DM (2010) Eradication of invasive forest insects: concepts, methods, costs and benefits. New Zealand J For Sci 40(Supplement):S117–S135

Brockerhoff EG, Suckling, DM, Kimberley M, Richardson B, Coker G, Gous S, ... Zhang A (2012) Aerial application of pheromones for mating disruption of an invasive moth as a potential eradication tool. PLoS ONE 7(8):e43767

Brockerhoff EG, Suckling DM, Roques A, Jactel H, Branco M, Twidle AM, ... and Kimberley MO (2013) Improving the efficiency of lepidopteran pest detection and surveillance: constraints and opportunities for multiple-species trapping. J Chem Ecol 39(1):50–58

Bulman LS (2008) Pest detection surveys on high-risk sites in New Zealand. Aust For 71(3):242–244

Byers JA (1989) Chemical ecology of bark beetles. Experientia 45:271–283

Campbell SA, Borden JH (2009) Additive and synergistic integration of multimodal cues of both hosts and non-hosts during host selection by woodboring insects. Oikos 118:553–563

Carnegie AJ, Eldridge RH, Waterson DG (2005) History and management of *Sirex* wood wasp in pine plantations in New South Wales Australia. New Zealand J For Sci 35(1):3–24

CFIA [Canadian Food Inspection Agency] (2017) Brown spruce longhorn beetle—*Tetropium fuscum*. Retrieved July 2017, http://www.inspection.gc.ca/plants/plant-pests-invasive-species/insects/brown-spruce-longhorn-beetle/eng/1330656129493/1330656721978

Chase KD, Stringer LD, Butler RC, Liebhold AM, Miller DR, Shearer PW, Brockerhoff EG (2018) Multiple-lure surveillance trapping for *Ips* bark beetles, *Monochamus* longhorn beetles, and *Halyomorpha halys* (Hemiptera: Pentatomidae). J Econ Entomol 111(5):2255–2263

Clarke S (2012) Implications of population phases on the integrated pest management of the Southern pine beetle, *Dendroctonus frontalis*. J Integ Pest Manag 3:F1–F7

Colunga-Garcia M, Haack RA, Magarey RD, Borchert DM (2013) Understanding trade pathways to target biosecurity surveillance. NeoBiota 18:103–118

Czokajlo D, Ross D, Kirsch P (2001) Intercept panel trap, a novel trap for monitoring forest Coleoptera. J For Sci 47:63–65

de Groot P, Nott RW (2001) Evaluation of traps of six different designs to capture pine sawyer beetles (Coleoptera: Cerambycidae). Agr For Entomol 3:107–111

Dodds KJ (2014) Effects of trap height on captures of arboreal insects in pine stands of Northeastern United States of America. Can Entomol 146(1):80–89

Dodds KJ, Allison JD, Miller DR, Hanavan RP, Sweeney J (2015) Considering species richness and rarity when selecting optimal survey traps: comparisons of semiochemical baited flight intercept traps for Cerambycidae in Eastern North America. Agr For Entomol 17(1):36–47

Du L, Wu C, Liu Q, Huang L, Wang P (2013) Recent advances in olfactory receptor-based biosensors. Biosens Bioelectron 42:570–580

Ehler LE (2006) Integrated pest management (IPM): definition, historical development and implementation, and the other IPM. Pest Manag Sci 62(9):787–789

Einhorn J, Menassieu P, Michelot D, Riom J (1983) The use of sex-traps baited with synthetic attractants against the pine processionary, *Thaumetopoea pityocampa* Schiff. (Lep., Notodontidae) First Experiments in South-Western France. Agronomie 3:499–505

Elkinton JS, Cardé RT (1988) Effects of intertrap distance and wind direction on the interaction of gypsy moth (Lepidoptera: Lymantriidae) pheromone-baited traps. Environ Entomol 17(5):764–769

El-Sayed AM (2020) The Pherobase: database of pheromones and semiochemicals. https://www.pherobase.com

Epanchin-Niell RS, Brockerhoff EG, Kean JM, Turner JA (2014) Designing cost-efficient surveillance for early detection and control of multiple biological invaders. Ecol Appl 24(6):1258–1274

Errico M (2013) Asian longhorned beetle detector dog pilot project. In: McManus KA, Gottschalk KW (eds) Proceedings of the 23rd US Department of Agriculture Interagency Research Forum on Invasive Species 2012. General Technical Report NRS-P-114. United States Department of Agriculture Forest Service Northern Research Station, p 18

Faccoli M, Stergulc F (2006) A practical method for predicting the short-time trend of bivoltine populations of *Ips typographus* (L.) (Col., Scolytidae). J Appl Entomol 130(1):61–66

Fan JT, Denux O, Courtin C, Bernard A, Javal M, Millar JG, ... Roques A (2019) Multi-component blends for trapping native and exotic longhorn beetles at potential points-of-entry and in forests. J Pest Sci 92(1):281–297

Fernández Ajó AA, Martínez AS, Villacide JM, Corley JC (2015) Behavioural response of the woodwasp *Sirex noctilio* to volatile emissions of its fungal symbiont. J Appl Entomol 139(9):654–659

Ferro ML, Carlton CE (2011) A practical emergence chamber for collecting Coleoptera from rotting wood, with a review of emergence chamber designs to collect saproxylic insects. Coleopts Bull 65:115–124

Flaherty L, Gutowski JMG, Hughes C, Mayo P, Mokrzycki T, Pohl G, ... Sweeney J (2019) Pheromone-enhanced lure blends and multiple trap heights improve detection of bark and wood-boring beetles potentially moved in solid wood packaging. J Pest Sci 92(1):309–325

Frewin A, Scott-Dupree C, Hanner R (2013) DNA barcoding for plant protection: applications and summary of available data for arthropod pests. CAB Rev 8:18

Froud PM, Oliver TM, Bingham PC, Flynn AR, Rowswell NJ (2008) Passive surveillance of new exotic pests and diseases in New Zealand. In: Froud KJ, Popay AI, Zydenbos SM (eds) Surveillance for biosecurity: pre-border to pest management. New Zealand Plant Protection Society. https://nzpps.org/_oldsite/books/2008_Surveillance/Surveillance.pdf

Gery C, Miller C (1985) Evaluation of the populations of pine processionary caterpillar (*Thaumetopoea pityocampa* Schiff. Lepidopterae-Thaumetopoeidae) in Mont Ventoux, France. Annu For Sci 42:143–183

Glatz R, Bailey-Hill K (2011) Mimicking nature's noses: From receptor deorphaning to olfactory biosensing. Prog Neurobiol 93(2):270–296

Goodwin JT, Pawlowski SP, Mayo PD, Silk PJ, Sweeney JD, Hillier NK (2020) Influence of trap colour, type, deployment height, and a host volatile on monitoring *Orchestes fagi* (Coleoptera: Curculionidae) in Nova Scotia Canada. Can Entomol 152(1):98–109

Graham EE, Mitchell RF, Reagel PF, Barbour JD, Millar JG, Hanks LM (2010) Treating panel traps with a fluoropolymer enhances their efficiency in capturing cerambycid beetles. J Econ Entomol 103(3):641–647

Gries G, Gries R, Khaskin G, Slessor KN, Grant GG, Liška J, Kapitola P (1996) Specificity of nun and gypsy moth sexual communication through multiple-component pheromone blends. Naturwissenschaften 83(8):382–385

Haack RA (2006) Exotic bark-and wood-boring Coleoptera in the United States: recent establishments and interceptions. Can J For Res 36(2):269–288

Haack RA, Rabaglia RJ (2013) Exotic bark and ambrosia beetles in the USA: potential and current invaders. In: Pena J (ed) Potential invasive pests of agricultural crop species. Wallingford, UK, CAB International, pp 48–74

Haack RA, Hérard F, Sun J, Turgeon JJ (2010) Managing invasive populations of Asian longhorned beetle and citrus longhorned beetle: a worldwide perspective. Annu Rev Entomol 55:521–546

Häuser CL, Riede K (2015) Field methods for inventorying insects. In: Watson MF, Lyal C, Pendry C (eds) Descriptive taxonomy: the foundation of biodiversity research. Cambridge University Press, Cambridge, pp 190–213

Hall RJ, Castilla G, White JC, Cooke BJ, Skakun RS (2016) Remote sensing of forest pest damage: a review and lessons learned from a Canadian perspective. Can Entomol 148(S1):S296–S356

Hanks LM, Millar JG (2016) Sex and aggregation-sex pheromones of cerambycid beetles: basic science and practical applications. J Chem Ecol 42(7):631–654

Hansen EM, Bentz BJ, Munson AS, Vandygriff JC, Turner DL (2006) Evaluation of funnel traps for estimating tree mortality and associated population phase of spruce beetle in Utah. Can J For Res 36(10):2574–2584

Harris JWE, Collis DG, Magar KM (1972) Evaluation of the tree-beating method for sampling defoliating forest insects. Can Entomol 104(5):723–729

Hayes CJ, Fettig CJ, Merrill LD (2009) Evaluation of multiple funnel traps and stand characteristics for estimating western pine beetle-caused tree mortality. J Econ Entomol 102(6):2170–2182

Herrick NJ, Mankin RW (2012) Acoustical detection of early instar *Rhynchophorus ferrugineus* (Coleoptera: Curculionidae) in Canary Island date palm, *Phoenix canariensis* (Arecales: Arecaceae). Fla Entomol 95(4):983–990

Hester SM, Cacho OJ (2017) The contribution of passive surveillance to invasive species management. Biol Invas 19(3):737–748

Hill PS (2008) Vibrational communication in animals. Harvard University Press, Boston

Howse P, Stevens JM, Jones OT (1998) Insect pheromones and their use in pest management. Chapman and Hall, London

Hoyer-Tomiczek U, Sauseng G (2013) Sniffer dogs to find *Anoplophora* spp. infested plants. J Entomol Acarol Res 45:10–12

Jacquet JS, Orazio C, Jactel H (2012) Defoliation by processionary moth significantly reduces tree growth: a quantitative review. Annu For Sci 69(8):857–866

Jacquet JS, Bosc A, O'Grady A, Jactel H (2014) Combined effects of defoliation and water stress on pine growth and non-structural carbohydrates. Tree Physiol 34(4):367–376

Jactel H, Menassieu P, Vétillard F, Barthélémy B, Piou D, Frérot B, ... Battisti A (2006) Population monitoring of the pine processionary moth (Lepidoptera: Thaumetopoeidae) with pheromone-baited traps. For Ecol Manag 235:96–106

Jactel H, Bonifacio L, Van Halder I, Vétillard F, Robinet C, David G (2019) A novel, easy method for estimating pheromone trap attraction range: application to the pine sawyer beetle *Monochamus galloprovincialis*. Agr For Entomol 21(1):8–14

Kalaris T, Fieselmann D, Magarey R, Colunga-Garcia M, Roda A, Hardie D, ... Whittle P (2014) The role of surveillance methods and technologies in plant biosecurity. In: Gordh G, McKirdy S (eds), The handbook of plant biosecurity. Dordrecht, Springer, pp 309–337

Kean JM, Suckling DM, Sullivan NJ, Tobin PC, Stringer LD, Smith GR, Kimber B, Lee DC, Flores Vargas R, Fletcher J, Macbeth F, McCullough DG, Herms DA, et al (2020) Global eradication and response database. http://b3.net.nz/gerda (accessed 29 July 2020)

Kerr JL, Kelly D, Bader MKF, Brockerhoff EG (2017) Olfactory cues, visual cues, and semiochemical diversity interact during host location by invasive forest beetles. J Chem Ecol 43(1):17–25

Knuff AK, Winiger N, Klein AM, Segelbacher G, Staab M (2019) Optimizing sampling of flying insects using a modified window trap. Methods Ecol Evol 10(10):1820–1825

Kriticos DJ, Potter KJ, Alexander NS, Gibb AR, Suckling DM (2007) Using a pheromone lure survey to establish the native and potential distribution of an invasive Lepidopteran *Uraba Lugens*. J Appl Ecol 44(4):853–863

Lantschner MV, Corley JC (2015) Spatial pattern of attacks of the invasive woodwasp *Sirex noctilio*, at landscape and stand scales. PLoS ONE 10(5):e0127099

Li S, Daudin JJ, Piou D, Robinet C, Jactel H (2015) Periodicity and synchrony of pine processionary moth outbreaks in France. For Ecol Manag 354:309–317

Liebhold A, Thorpe K, Ghent J, Lyons DB (1994) Gypsy moth egg mass sampling for decision-making: a users' guide. USDA Forest Service Northeastern Area, Forest Health Protection, Report NA-TP-04-94

Liebhold AM, Brockerhoff EG, Garrett LJ, Parke JL, Britton KO (2012) Live plant imports: the major pathway for forest insect and pathogen invasions of the US. Front Ecol Environ 10(3):135–143

Liebhold AM, Berec L, Brockerhoff EG, Epanchin-Niell RS, Hastings A, Herms DA, ... Yamanaka T (2016) Eradication of invading insect populations: from concepts to applications. Annu Rev Entomol 61:335–352

Liebhold AM, Kean JM (2018) Eradication and containment of non-native forest insects: successes and failures. J Pest Sci 92:83–91

Lindelöw Å, Schroeder M (2001) Spruce bark beetle, *Ips typographus* (L.), in Sweden: monitoring and risk assessment. J For Sci 47:40–42

Lindgren BS (1983) A multiple funnel trap for scolytid beetles (Coleoptera). Can Entomol 115(3):299–302

Löfstedt C, Wahlberg N, Millar JG (2016) Evolutionary patterns of pheromone diversity in Lepidoptera. In: Allison JD, Cardé RT (eds) Pheromone communication in moths: evolution, behavior and application. University of California Press, pp 43–82

Madden MJ, Young RG, Brown JW, Miller SE, Frewin AJ, Hanner RH (2019) Using DNA barcoding to improve invasive pest identification at US ports-of-entry. PLoS ONE 14(9):e0222291

Mankin RW, Hagstrum DW, Smith MT, Roda AL, Kairo MT (2011) Perspective and promise: a century of insect acoustic detection and monitoring. American Entomol 57(1):30–44

Manoukis NC, Hall B, Geib SM (2014) A computer model of insect traps in a landscape. Sci Rep UK 4:7015

Mayfield AE, III, Salom SM, Sumpter K, McAvoy T, Schneeberger NF, Rhea R (2020) Integrating chemical and biological control of the hemlock Woolly Adelgid: a resource manager's guide. FHAAST-2018-04. USDA Forest Service, Forest Health Assessment and Applied Sciences Team, Morgantown, West Virginia

Miller DR, Rabaglia RJ (2009) Ethanol and (−)-α-pinene: attractant kairomones for bark and ambrosia beetles in the Southeastern US. J Chem Ecol 35(4):435–448

Miller DR, Dodds KJ, Eglitis A, Fettig CJ, Hofstetter RW, Langor DW, Mayfield AE III, Munson AS, Poland TM, Raffa KF (2013) Trap lure blend of pine volatiles and bark beetle pheromones for Monochamus spp. (Coleoptera: Cerambycidae) in pine forests of Canada and the United States. J Econ Entomol 106:1684–1692

Miller DR, Crowe CM, Dodds KJ, Galligan LD, de Groot P, Hoebeke ER, Mayfield AE III, Poland TM, Raffa KF, Sweeney JD (2015) Ipsenol, ipsdienol, ethanol and α-pinene: Trap lure blend for Cerambycidae and Buprestidae (Coleoptera) in pine forests of Eastern North America. J Econ Entomol 108:1837–1851

Miller DR, Allison JD, Crowe CM, Dickinson DM, Eglitis A, Hofstetter RW, Munson AS, Poland TM, Reid LS, Steed BE, Sweeney JD (2016) Pine sawyers (Coleoptera: Cerambycidae) attracted to α-pinene, monochamol, and ipsenol in North America. J Econ Entomol 109:1205–1214

Miller DR, Crowe CM, Mayo P, Silk PJ, Sweeney JD (2017) Interactions between ethanol, syn-2,3-hexanediol, 3-hydroxyhexan-2-one, and 3-hydroxyoctan-2-one lures on trap catches of hardwood longhorn beetles in Southeastern United States. J Econ Entomol 110:2119–2128

Millar JG, Hanks LM (2017) Chemical ecology of cerambycids. In: Wang Q (ed) Cerambycidae of the world: biology and pest management. CRC Press, Boca Raton, London, New York, pp 161–208

Morris RF (1960) Sampling insect populations. Annu Rev Entomol 5(1):243–264

Muirhead-Thompson RC (1991) Trap responses of flying insects: the influence of trap design on capture efficiency. UK, Academic Press, London

Näsi R, Honkavaara E, Blomqvist M, Lyytikäinen-Saarenmaa P, Hakala T, Viljanen N, ... Holopainen M (2018) Remote sensing of bark beetle damage in urban forests at individual tree level using a novel hyperspectral camera from UAV and aircraft. Urban For Urban Green 30:72–83

Nealis VG, Turnquist R (2010) Impact and recovery of Western hemlock following disturbances by forestry and insect defoliation. For Ecol Manag 260(5):699–706

Negrón JF, Popp JB (2018) Can spruce beetle (*Dendroctonus rufipennis* Kirky) pheromone trap catches or stand conditions predict engelmann spruce (*Picea engelmannii* Parry ex Engelm.) tree mortality in Colorado? Agr For Entomol 20(2):162–169

Nixon LJ, Morrison WR, Rice KB, Brockerhoff EG, Leskey TC, Guzman F, ... Rostás M (2018) Identification of volatiles released by diapausing brown marmorated stink bug, *Halyomorpha halys* (Hemiptera: Pentatomidae). PLoS ONE 13(1):e0191223

Oehlschlager AC, Leal WS, Gonzalez L, Chacon M, Andrade R (2003) Trapping of *Phyllophaga elenans* with a female-produced pheromone. J Chem Ecol 29(1):27–36

Oh EH, Song HS, Park TH (2011) Recent advances in electronic and bioelectronic noses and their biomedical applications. Enzyme Microb Technol 48(6):427–437

O'Leary K, Hurley JE, MacKay W, Sweeney J (2003) Radial growth rate and susceptibility of *Picea rubens* Sarg. to *Tetropium fuscum* (Fabr.). In: McManus ML, Liebhold AM (eds) Proceedings: ecology, survey, and management of forest insects; 2002 September 1–5; Krakow, Poland. Gen.Tech. Rep. NE-311, U.S. Department of Agriculture, Forest Service, Northeastern Research Station, Newtown Square, PA, pp 107–114. https://www.fs.usda.gov/treesearch/pubs/download/6071.pdf

Olsson PO, Jönsson AM, Eklundh L (2012) A new invasive insect in Sweden-*Physokermes inopinatus*: tracing forest damage with satellite based remote sensing. For Ecol Manag 285:29–37

Pawson SM, Sullivan JJ, Grant A (2020) Expanding general surveillance of invasive species by integrating citizens as both observers and identifiers. J Pest Sci 93:1155–1166. www.findapest.nz

Poland TM, Petrice TR, Ciaramitaro TM (2019) Trap designs, colors, and lures for emerald ash borer detection. Front Forests Global Change 2:80

Poland TM, Rassati D (2019) Improved biosecurity surveillance of non-native forest insects: a review of current methods. J Pest Sci 92(1):37–49

Post KE, Werner RA (1988) Wood borer distribution and damage in decked white spruce logs. Northern J Appl Forest 5:49–51

Prasad Y, Prabhakar M (2012) Pest monitoring and forecasting. In: Abrol D, Shankar U (eds) Integrated pest management: principles and practice. CABI, Oxfordshire, UK, pp 41–57

Quilici S, Donner P, Battisti A (2012) Surveillance techniques for non-native insect pest detection. EPPO Bull 42(1):95–101

Rabaglia R, Duerr D, Acciavatti R, Ragenovich I (2008) Early detection and rapid response for non-native bark and ambrosia beetles: summary of the 2001–2005 pilot project. USDA Forest Service, Forest Health Protection

Rabaglia RJ, Cognato AI, Hoebeke ER, Johnson CW, LaBonte JR, Carter ME, Vlach JJ (2019) Early detection and rapid response: a 10-year summary of the USDA forest service program of surveillance for non-native bark and ambrosia beetles. American Entomol 65(1):29–42

Rassati D, Marini L, Marchioro M, Rapuzzi P, Magnani RG, Poloni R, Di Giovanni F, Mayo P, Sweeney J (2019) Developing trapping protocols for wood-boring beetles associated with broadleaf trees. J Pest Sci 92:267–279

Ravlin FW (1991) Development of monitoring and decision-support systems for integrated pest management of forest defoliators in North America. Forest Ecol Manag 39:3–13

Reding ME, Schultz PB, Ranger CM, Oliver JB (2011) Optimizing ethanol-baited traps for monitoring damaging ambrosia beetles (Coleoptera: Curculionidae, Scolytinae) in ornamental nurseries. J Econ Entomol 104(6):2017–2024

Redlich S, Clemens J, Bader MKF, Pendrigh D, Perret-Gentil A, Godsoe W, ... Brockerhoff EG (2019) Identifying new associations between invasive aphids and pinaceae trees using plant sentinels in botanic gardens. Biol Invas 21(1):217–228

Roques A (ed) (2015) Processionary moths and climate change: an update. Springer, Dordrecht

Ryall K, Silk P, Webster RP, Gutowski JM, Meng Q, Li Y, Gao W, Fidgen J, Kimoto T, Scarr T, Mastro V, Sweeney JD (2015) Further evidence that monochamol is attractive to *Monochamus* (Coleoptera: Cerambycidae) species, with attraction synergized by host plant volatiles and bark beetle (Coleoptera: Curculionidae) pheromones. Can Entomol 147:564–579

Safranyik L, Shore TL, Linton DA (2004) Measuring trap efficiency for bark beetles (Col., Scolytidae). J Appl Entomol 128(5):337–341

Sanders CJ (1988) Monitoring spruce budworm population density with sex pheromone traps. Can Entomol 120(2):175–183

Schlyter F (1992) Sampling range, attraction range, and effective attraction radius: estimates of trap efficiency and communication distance in coleopteran pheromone and host attractant systems 1. J Appl Entomol 114(1–5):439–454

Schlyter F, Byers JA, Löfqvist J (1987) Attraction to pheromone sources of different quantity, quality, and spacing: density-regulation mechanisms in bark beetle *Ips typographus*. J Chem Ecol 13(6):1503–1523

Schroeder LM (2013) Monitoring of *Ips typographus* and *Pityogenes chalcographus*: influence of trapping site and surrounding landscape on catches. Agr Forest Entomol 15(2):113–119

Seibold S, Gossner MM, Simons NK, Blüthgen N, Müller J, Ambarlı D, ... and Linsenmair KE (2019) Arthropod decline in grasslands and forests is associated with landscape-level drivers. Nature 574:671–674

Sharov AA, Leonard D, Liebhold AM, Roberts EA, Dickerson W (2002) "Slow the spread" a national program to contain the gypsy moth. J Forest 100:30–36

Silk P, Sweeney J, Wu J, Price J, Gutowski J, Kettela E (2007) Evidence for a male produced pheromone in *Tetropium fuscum* (F.) and *Tetropium cinnamopterum* (Kirby) (Coleoptera: Cerambycidae). Naturwissenschaften 94:697–701

Slippers B, Hurley BP, Wingfield MJ (2015) *Sirex* woodwasp: a model for evolving management paradigms of invasive forest pests. Annu Rev Entomol 60:601–619

Smith G, Hurley JE (2000) First North American record of the Palearctic species *Tetropium fuscum* (Fabricius) (Coleoptera: Cerambcyidae). Coleopts Bull 54:540

Stevens PM (2008) High risk site surveillance (HRSS): an example of best practice plant pest surveillance. In: Froud KJ, Popay AI, Zydenbos SM (eds) Surveillance for biosecurity: pre-border to pest management. New Zealand Plant Protection Society, pp 127–134. https://nzpps.org/_oldsite/books/2008_Surveillance/Surveillance.pdf

Stone C, Mohammed C (2017) Application of remote sensing technologies for assessing planted forests damaged by insect pests and fungal pathogens: a review. Curr For Rep 3(2):75–92

Straw NA, Hoppit A, Branson J (2019) The relationship between pheromone trap catch and local population density of the oak processionary moth *Thaumetopoea processionea* (Lepidoptera: Thaumetopoeidae). Agr Forest Entomol 21(4):424–430

Sukovata L, Dziuk A, Parratt M, Bystrowski C, Dainton K, Polaszek A, Moore R (2020) The importance of trap type, trap colour and capture liquid for catching *Dendrolimus pini* and their impact on by-catch of beneficial insects. Agr For Entomol 22(4):319–327

Sweeney J, de Groot P, MacDonald L, Smith S, Cocquempot C, Kenis M, Gutowski J (2004) Host volatile attractants and traps for detection of *Tetropium fuscum* (F.), *Tetropium castaneum* (L.), and other longhorned beetles (Coleoptera: Cerambycidae). Environ Entomol 33:844–854

Sweeney J, Gutowski J, Price J, de Groot P (2006) Effect of semiochemical release rate, killing agent, and trap design on capture of *Tetropium fuscum* (F.), and other longhorn beetles (Coleoptera: Cerambycidae). Environ Entomol 35:645–654

Sweeney J, Hughes C, Webster V, Kostanowicz C, Webster R, Mayo P, Allison JD (2020) Impact of horizontal edge–interior and vertical canopy–understory gradients on the abundance and diversity of bark and woodboring beetles in survey traps. Insects 11(9):573

Taylor SL, MacLean DA (2008) Validation of spruce budworm outbreak history developed from aerial sketch mapping of defoliation in New Brunswick. Northern J Appl Forest 25(3):139–145

Tobin PC, Blackburn LM (eds) (2007) Slow the spread: a national program to manage the gypsy moth. Gen. Tech. Rep. NRS-6. U.S. Department of Agriculture, Forest Service, Northern Research Station, Newtown Square, PA

Tobin PC, Kean JM, Suckling DM, McCullough DG, Herms DA, Stringer LD (2014) Determinants of successful eradication programs. Biol Invas 16:401–414

Toïgo M, Barraquand F, Barnagaud JY, Piou D, Jactel H (2017) Geographical variation in climatic drivers of the pine processionary moth population dynamics. Forest Ecol Manag 404:141–155

Torresan C, Berton A, Carotenuto F, Di Gennaro SF, Gioli B, Matese A, ... Wallace L (2017) Forestry applications of UAVs in Europe: a review. Int J Remote Sens 38(8-10):2427–2447

USDA [United States Department of Agriculture] (2012) National detector dog manual. United States Department of Agriculture APHIS PPQ report, p 262. http://www.aphis.usda.gov/import_export/plants/manuals/ports/downloads/detector_dog.pdf

USDA [United States Department of Agriculture] (2019) Gypsy moth program manual. United States Department of Agriculture, Washington DC. https://www.aphis.usda.gov/import_export/plants/manuals/domestic/downloads/gypsy_moth.pdf

Valentin RE, Fonseca DM, Nielsen AL, Leskey TC, Lockwood JL (2018) Early detection of invasive exotic insect infestations using eDNA from crop surfaces. Front Ecol Environ 16(5):265–270

Vega JM, Moneo I, Ortiz JCG, Palla PS, Sanchís ME, Vega J, ... Roques A (2011) Prevalence of cutaneous reactions to the pine processionary moth (Thaumetopoea pityocampa) in an adult population. Contact Dermat 64(4):220–228

Wall C, Perry JN (1987) Range of action of moth sex-attractant sources. Entomol Exp Appl 44(1):5–14

Wermelinger B (2004) Ecology and management of the spruce bark beetle *Ips typographus*—a review of recent research. Forest Ecol Manag 202(1–3):67–82

Westfall J, Ebata T, Bains B (2019a) Summary of forest health conditions in British Columbia. Pest Management Report Number 15. Ministry of Forests, Lands, Natural Resource Operations and Rural Development, Victoria, British Columbia. https://www2.gov.bc.ca/assets/gov/environment/research-monitoring-and-reporting/monitoring/aerial-overview-survey-documents/aos_2019a.pdf

Westfall J, Ebata T, HR GISolutions Inc (2019b) Forest health aerial overview survey standards for British Columbia. Ministry of Forests, Lands, Natural Resource Operations and Rural Development, Victoria, British Columbia. https://www.for.gov.bc.ca/ftp/HFP/external/!publish/Aerial_Overview/Data_stds/AOS%20Standards%202019b.pdf

Wichmann L, Ravn HP (2001) The spread of *Ips typographus* (L.) (Coleoptera, Scolytidae) attacks following heavy windthrow in Denmark, analysed using GIS. Forest Ecol Manag 148(1–3):31–39

Wild Spotter (2020) *Wild spotter—mapping invasives in America's wild places*. University of Georgia, USA, www.wildspotter.org

Zorović M, Čokl A (2015) Laser vibrometry as a diagnostic tool for detecting wood-boring beetle larvae. J Pest Sci 88(1):107–112

Chapter 20
Silviculture

Kristen M. Waring and Ethan Bucholz

20.1 Introduction

Silviculture is the art and science of managing forest stands to meet landowner goals
and objectives (see Box 20.1); traditional examples of goals and objectives include
managing for timber production, improved wildlife habitat, fuels reduction, and
maintenance or improvement of forest health. Within forest health, objectives often
involve mitigating negative impacts of forest insects while recognizing the beneficial
role of insects in provision of ecosystem services. Goals tend to be broad, encompassing perspective on desired conditions at large scales, such as the forest or landscape. Objectives are more specific, and often target specific outcomes (e.g. reduced
levels of insect-caused mortality following treatment) and are typically focused at
the stand-scale. In this chapter, we have focused on the stand-scale unless explicitly
noted otherwise. Silviculture, through prescriptions and treatment implementation
(see Box 20.1 for definitions) can be used to manipulate the species composition,
vertical and horizontal structure of the stand, individual and stand-level tree vigor, and
numerous other stand characteristics that might influence susceptibility to insects.
Numerous silvicultural treatments exist (e.g. prescribed fire); however, mechanical
removal of trees is perhaps the most common association people make with silvicultural treatments to meet management objectives. The outcomes targeted by silvicultural prescriptions will depend upon the site, existing stand characteristics, specific
insect(s) of concern, and any other management objectives.

K. M. Waring (✉) · E. Bucholz
Northern Arizona University, Flagstaff, AZ, USA
e-mail: Kristen.Waring@nau.edu

E. Bucholz
Colorado State Forest Service, Fort Collins, CO, USA

J. D. Allison et al. (eds.), *Forest Entomology and Pathology*,
https://doi.org/10.1007/978-3-031-11553-0_20

Box 20.1: Silviculture definitions used in this chapter. From *The Dictionary of Forestry* (Helms 1998) unless otherwise indicated

Term	Definition
Silviculture	The art and science of controlling the establishment, growth, composition, health, and quality of forests and woodlands to meet the diverse needs and values of landowners and society on a sustainable basis
Silviculture Prescription	A planned series of treatments designed to change current stand structure to one that meets management goals
Silvicultural Treatment	A management intervention conducted to achieve desired goals (definition by authors)
Stand	A contiguous group of trees sufficiently uniform in age-class distribution, composition and structure and growing on a site of sufficiently uniform quality to be a distinguishable unit
Even-aged stand	A stand of trees composed of a single age class
Uneven-aged stand	A stand of trees of three or more distinct age classes, either intimately mixed or in small groupings
Multi-aged stand	A stand of trees with two or more distinct age classes
Regeneration	Seedlings or saplings existing in a stand
Residual tree(s)	A tree or snag remaining after an intermediate or partial cutting of a stand
Stand density	A quantitative measure of stocking expressed either absolutely in terms of number of trees, basal area, or volume per unit area or relative to some standard condition
Stand development	Changes in forest stand structure over time
Stand structure	The horizontal and vertical distribution of components of a forest stand including the height, diameter, crown layers, and stems of trees, shrubs, herbaceous understory, snags, and down woody debris
Intermediate treatment	Any treatment or tending designed to enhance growth, quality, vigor and composition of the stand after establishment or regeneration and prior to final harvest

(continued)

(continued)

Term	Definition
Thinning	An intermediate treatment made to reduce stand density of trees primarily to improve growth, enhance forest health, or recover potential mortality. Variations on the most common types of thinning (defined below) are common • Low thinning: removal of trees in the suppressed/overtopped crown class in order to favor those in the upper crown classes. *Syn thin from below* • Mechanical thinning: thinning of trees involving removal of trees in rows, strips or by using fixed spacing intervals. *Syn geometric thinning* • Crown thinning: removal of trees from the dominant or co-dominant crown classes in order to favor the best trees of those same crown classes • Dominant thinning: removal of trees in the dominant crown class in order to favor the lower crown classes. *Syn selection thinning; thin from above*
Sanitation cutting	The removal of trees to improve stand health by stopping or reducing the actual or anticipated spread of insects and disease
Salvage cutting	The removal of dead trees or trees damaged or dying because of injurious agents other than competition, to recover economic value that would otherwise be lost
Regeneration method	A cutting procedure by which a new age class is created. Traditional methods are: • Clearcut: the cutting of essentially all trees, producing a fully exposed microclimate for the development of a new age class • Coppice: All trees from the previous stand are cut and the majority of regeneration is from sprouts or root suckers. *Syn. clearfell* • Seed tree: the cutting of all trees except for a small number of widely dispersed trees retained for seed production and to produce a new age class in fully exposed microenvironments • Shelterwood: the cutting of most trees, leaving those needed to produce sufficient shade to produce a new age class in a moderated environment. Modifications are numerous, and include group shelterwood with non-uniform spacing of residual trees post-harvest and shelterwoods with reserves, in which the residual trees are not removed, creating a two-aged stand • Group Selection: trees are removed and new age classes are established in small groups • Single tree selection: individual trees of all size classes are removed more or less uniformly throughout the stand, to promote growth of remaining trees and to provide space for regeneration

The goals of this chapter are:

1. To identify broad approaches and specific silvicultural strategies and tools managers can use to alleviate or prevent forest insect problems such as mortality or reduced growth and vigor; and
2. To discuss the impact of silvicultural strategies and tools on forest structure, stand development, and other biotic and abiotic factors as well as forest insect population dynamics.

20.2   Silvicultural Strategies for Management of Forest Insects

From a silvicultural perspective, managing forest insects can be considered in two broad approaches: (1) those that increase resistance, and/or (2) those that increase resilience (DeRose and Long 2014; Table 20.1). Resistance is the ability of a system to withstand change; that is, a resistant forest stand will have the same condition, structure, and species composition before and after a disturbance (Walker et al. 2004). Resilience is the ability of a system to change but maintain its basic attributes; a resilient forest stand subjected to disturbance will return to conditions similar to those present prior to the disturbance but may have changes in structure (Walker et al. 2004). A more entomological perspective would place silvicultural strategies into the categories of reducing susceptibility or vulnerability along with increasing regeneration potential (Muzika and Liebhold 2000). This chapter takes the silvicultural perspective in terminology, but the underlying theoretical basis for treatments between the two perspectives is highly compatible.

Strategies designed to increase stand resistance focus on the influence of structure and species composition on the potential severity of a given insect disturbance (DeRose and Long 2014). Severity is principally determined by how much mortality or die-back is associated with an insect outbreak. Strategies to increase stand resilience are longer-term and focus on how the disturbance influences stand structure and species composition (DeRose and Long 2014). Silviculture can be used in both approaches to mitigate anticipated negative impacts, with prescriptions based on characteristics of, and predictions for, individual stands.

Resistance and resilience strategies can be applied separately or as complementary short- and long-term treatments to ensure that live trees remain in a stand over longer time horizons. For example, the spruce beetle (*Dendroctonus rufipennis*) in the western United States tends to target mature overstory spruce (*Picea* spp.) and may cause extensive mortality in stands dominated by large spruce (>90%) (DeRose and Long 2007, and references therein). In the short-term, reducing overstory density may increase resistance of existing trees to spruce beetle attack, thus maintaining similar stand conditions by preventing extensive overstory mortality. Over a longer time period, resilience is necessary to maintain healthy stand conditions. Windmuller-Campione and Long (2015) defined resilience of spruce-fir stands to spruce beetle

Table 20.1 Approaches to increasing resistance and resilience using silvicultural strategies to adjust vertical and horizontal stand structure, their impact on residual trees and potential impact on bark beetle and defoliator damage. Assumes species composition is not altered during treatment. See Box 20.1 for definitions

Broad approach	Silvicultural strategy	Residual vertical structure	Residual horizontal structure	Impact for bark beetle hazard	Impact for defoliator damage
Intermediate silvicultural operations					
Resistance	Thinning from below/geometric thinning/Low intensity prescribed fire	Simple: one canopy layer	Regular	Increases tree size, vigor and phloem thickness/may increase susceptibility if trees were low vigor prior to treatment	Reduces complexity of vertical structure, decreasing dispersal through canopy; faster tree recovery at lower tree densities
Resistance	Thinning from above/crown thinning	Simple: one canopy layer	Regular, assuming even distribution of lower canopy classes in stand	Decreases tree size; tree vigor and phloem thickness decrease or remain unchanged	May leave small suppressed trees susceptible to dispersal from residual overstory trees (if present); faster tree recovery at lower tree densities
Resistance or resilience	Variable density thinning/free thinning	Dependent upon pre-treatment structure; generally won't change vertical structure	Patchy, with openings and dense areas across stand	Increases tree size, vigor and phloem thickness in patches; denser areas have no change in susceptibility	No change due to vertical canopy structure remaining the same; patches not large enough to prevent dispersal onto regeneration; faster tree recovery at lower tree densities

(continued)

Table 20.1 (continued)

Broad approach	Silvicultural strategy	Residual vertical structure	Residual horizontal structure	Impact for bark beetle hazard	Impact for defoliator damage
	Regeneration methods				
Resistance	Clearcut/traditional shelterwood/seed tree	Simple: one canopy layer	Dependent upon regeneration methods and patterns	Immediate decrease; will increase with time since treatment as trees age and grow larger	Immediate decrease if shelterwood or seed trees are removed within 10 years; small clearcuts will be susceptible to re-infestation from neighboring stands
Resilience	Group shelterwood or shelterwood with reserves[a]	2 canopy layers (shelterwood with reserves)	Patchy	Overstory trees remain susceptible if clumped (lower vigor) and if left in the stand	Increases susceptibility of new cohort to dispersal from overstory
Resilience	Single tree and Group selection	Multiple canopy layers	Patchy but can be regular within groups if group selection	Varies by canopy layer, tree size, and whether increased tree vigor is greater than insect pressure on large trees with thick phloem	High susceptibility of dispersal to all canopy layers which may be lessened at low residual densities; groups not large enough to prevent dispersal to regeneration; faster tree recovery at lower tree densities

[a] also applies to clearcut or seed tree with reserves

outbreaks as adequate stocking of Engelmann spruce (*Picea engelmannii*) regeneration following an outbreak. Resilience is provided through the use of young spruce to replace overstory spruce trees lost during the outbreak, providing for live trees in the stand over a long time period. Silviculture can be used proactively (prior to an outbreak) to create conditions conducive to Engelmann spruce regeneration, thus increasing long-term resilience.

Silvicultural treatments are also commonly categorized according to whether they target direct or indirect control of forest insects, primarily in bark beetle management (Fettig et al. 2014). Direct control strategies are meant to immediately reduce current insect populations. Indirect strategies focus on proactive management meant to reduce the potential for future tree damage. Most silvicultural strategies are indirect and consequently the primary focus of this chapter. However, a few common direct control tools are identified where appropriate.

20.2.1 Structural Strategies

Silvicultural strategies that adjust the vertical or horizontal stand structure can target both increased resistance or resilience at the stand-scale. Such strategies attempt to reduce the potential for large-scale insect infestations and can include a number of silvicultural treatments that result in a wide range of vertical and horizontal stand structures. Adjustments to vertical and horizontal stand structure can be effective because some stand structures are more susceptible to damage from forest insect pests. Silviculture can be used to shift stand structures from more susceptible to less susceptible states. Susceptible stand structures vary depending upon the insect pest species, corresponding tree host species characteristics and underlying site conditions.

Two common guilds of forest insect pests are bark beetles (see Chapter 10, Bark Beetles) and foliage feeders (defoliators; see Chapter 9, Foliage Feeders). Susceptible forest structures associated with damage by some of the most damaging agents within these guilds can be quite different, leading to trade-offs between structures: a structure that creates resistance or resilience to a bark beetle may lessen these attributes when considering a defoliator, for example. It is important to understand the mechanisms driving these relationships and why shifting structures can be an effective management strategy.

Bark beetles need to successfully find host trees and overcome tree defenses; they also require bark with thick enough phloem to complete their development and ensure reproductive success. Some bark beetle species require relatively large trees as hosts (e.g. mountain pine beetle (*Dendroctonus ponderosae*) in lodgepole pine (*Pinus contorta*)) while other bark beetle species need smaller diameters to successfully complete their life cycle (e.g. pine engraver (*Ips pini*) in ponderosa pine (*Pinus ponderosa*)). Defoliators need to find appropriate host trees, but some species are limited to relatively short distance dispersal, often by wind and gravity from upper to lower crowns or trees, or by crawling between individual trees. Hence, complex, multilayered vertical structures are more conducive to defoliator success than simple, single canopy layers. Conversely, bark beetle populations are favored by simple structures of even-sized trees.

Silvicultural treatments that remove trees alter stand structure immediately, and the indirect control of insect damage is based on changes to the microclimate within the stand and the ability for insects to find appropriate host trees. Microclimatic changes include disruption of the chemical signals used by insects to find host trees and mates (Progar et al. 2014) and changing individual tree microclimates enough to reduce the suitability of host trees (e.g. by creating warmer conditions along the tree bole). Microclimates within the stand may also be altered enough to affect insect success. For example, increased temperatures or insolation may result in increased mortality during the dispersal phase and/or the early larval stage. In order to reduce the ability of insects to find appropriate host trees, managers can reduce the number of host trees available, change the average tree size, and/or create a vertical or horizontal structure not conducive to successful host location by the insect (Fettig et al. 2014).

Tree vigor in general refers to the overall health of trees, and can be assessed qualitatively, by visually rating tree crowns (Miller and Keen 1960) or quantitatively, by comparing growth rates of trees to each other and their potential to succumb to insect attack. Quantitative assessments of tree vigor require additional field measurements, and may be assessed along with qualitative crown ratings, typically through the use of tree cores to measure annual or periodic basal area growth, sapwood area (water conducting tissue) and density or size of defensive structures (resin ducts) (Kane and Kolb 2010). While early research often related sapwood area to leaf area (photosynthetic capacity of the tree) to define vigor (Waring and Pitman 1980), other researchers have found a simple measure of basal area increment adequately captures individual tree vigor (defined by increased resin flow) (McDowell et al. 2007). Trees that produce less sapwood per unit leaf area typically require fewer bark beetle attacks for successful colonization (Waring and Pittman 1980) and Mitchell et al. (1983) related this to stand density, finding that reducing tree density was an effective method for increasing relative resistance to bark beetle attack by increasing tree growth per unit of leaf area. Ultimately, silviculture can shift stand structure to increase resistance and/or resilience, with the underlying cause of the increase likely a combination of multiple factors working together (Fig. 20.1).

Silvicultural treatments to reduce structural complexity include thinning from below and traditional even-aged regeneration methods (Table 20.1). Silvicultural strategies to reduce defoliation and its impacts have not been researched as thoroughly as strategies for bark beetle management and damage mitigation. The lack of empirical studies documenting post-treatment reductions in defoliation and/or defoliation damage means treatment effects are largely hypothetical, based on expected stand responses (Muzika and Liebhold 2000). Additionally, increasing tree vigor through density reduction may not alleviate defoliation severity, but may enable trees to recover more quickly following defoliation (Fajvan and Gottschalk 2012). A wide variety of traditional and modified silvicultural treatments are used to alter vertical and horizontal stand structure, many of which are identified, along with the

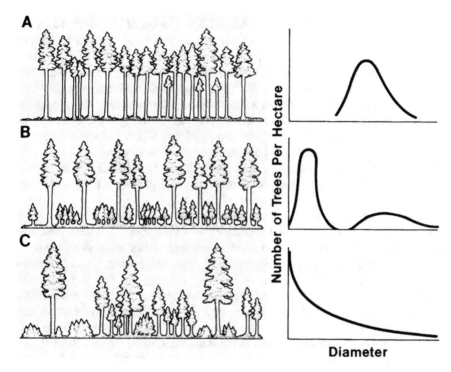

Fig. 20.1 Susceptibility to defoliation damage increases as structural complexity increases, from A through **C**. From Brookes et al. (1987)

anticipated impact of treatment on bark beetle and defoliator damage (Table 20.1). Additionally, each strategy is placed into either the resistance or resilience approach.

Traditional thinning results in a regular spatial pattern, creating similar spacing between residual trees. This pattern may be more resilient to bark beetle outbreaks from a tree vigor perspective, than leaving trees irregularly spaced where inter-tree competition remains high within groups of trees. However, inter-tree distance can also influence microclimate and negatively affect dispersal, and mate- and host-finding ability; a factor to consider when designing thinning regimes and spatial patterns of residual (leave) trees.

Much research has focused on the use of thinning to prevent bark beetle outbreaks in the United States, and the majority of research indicates that thinning can be effective at reducing tree mortality during outbreaks (i.e. thinned stands have less mortality than denser, unthinned stands) (Fettig et al. 2007). Dense, unthinned stands are generally considered to be at high hazard of bark beetle infestation and subsequent tree mortality, and hazard rating systems include metrics such as stand basal area or trees per unit area as an indicator variable. While thinning may reduce the probability of future mortality from bark beetles in most conifer species, some tree mortality should be expected when bark beetle populations rise to very high levels and pressure on the stand is high, except at low to moderate stand densities (15–20 m^2 ha^{-1})

(McGregor et al. 1987; Schmid and Mata 2005; Hansen et al. 2010). However, different bark beetle species, sites, and host species may have different thresholds. For example, stand susceptibility to southern pine beetle (*Dendroctonus frontalis*) decreases when stands are thinned to under 7.5 m^2 ha^{-1} basal area (Nowak et al. 2008 and references within, Nowak et al. 2015). Additionally, bark beetle mortality may create conditions more resilient to future outbreaks by increasing the proportion of unfavorable size classes or host species (Kashian et al. 2011). Ultimately, the reduction of stand density to a critical threshold that is site and species specific is more important than whether silviculture, bark beetles, or some other damaging agent causes the density reduction. In stands impacted by defoliators, thinning can improve the ability of defoliated trees to recover to previous rates of growth (Fajvan and Gottschalk 2012).

Regeneration methods fall into both the resistance and resilience categories given their effects on the overstory and understory over both short- and long-term time frames (Table 20.1). Most even- and uneven-aged regeneration methods reduce overstory density and stand susceptibility while providing for regeneration, which is not an objective of intermediate treatments, including thinning (see Box 20.1 for definitions). The exceptions are clearcuts, which reduce density to zero, do not increase vigor because no overstory trees remain, alter the microclimate dramatically, and provide for regeneration when implemented correctly. Traditional seed tree and shelterwood regeneration methods result in the same stand structure as a clearcut, and all three eliminate the potential for bark beetle-caused mortality until the newly regenerated trees reach a susceptible size.

Even-aged regeneration methods can be modified (e.g. group shelterwood or any even-aged system with reserves; Table 20.1) to provide additional structure by leaving residual overstory trees. These trees would have increased vigor and experience an altered microclimate, both factors which can influence bark beetle attacks. These methods result in two-aged or multi-aged stands (Table 20.1) and can also be resistant and resilient to bark beetle outbreaks. The large overstory trees will be at a low density and, depending on spatial pattern, spaced at a distance far enough from each other to reduce inter-tree competition and create conditions less conducive to successful insect mating, dispersal, and host-finding. Until the youngest age class reaches a susceptible size and density, extensive mortality from bark beetles is unlikely. Regeneration methods can also be used to enhance development of a new age class of trees, creating long-term resilience by providing for young trees if bark beetles kill the overstory (Windmuller-Campione and Long 2015). Group shelterwood methods may be useful in promoting such resilience in spruce stands dominated by large diameter, even-aged trees. These stands are highly susceptible to spruce beetle, which is a particularly aggressive bark beetle that may kill the entire overstory during an outbreak. Prior to an outbreak, implementing a group shelterwood to create conditions for a new spruce age class in the understory results in a stand that will have live trees, albeit young and small, following the outbreak (Windmuller-Campione and Long 2015).

Other insects less common than bark beetles and defoliators can also cause stand-scale damage. White pine weevils (*Pissodes stobi*) infest the leaders of seedlings, resulting in multiple forks and stem deformities. White pine weevils are most abundant in open areas that promote higher temperatures in the understory and thicker leader diameters in seedlings (Ostry et al. 2010; Pitt et al. 2016). Group shelterwood or shelterwood with reserves methods (Table 20.1) that leave the residual trees intact can be used to successfully regenerate eastern white pine (*Pinus strobus*) while mitigating white pine weevil damage. The overstory cover provided through these systems (50–75% full sunlight or up to 26 m^2 ha^{-1}) provides enough cover to moderate the microclimate and reduce eastern white pine regeneration leader diameters, thus reducing damage from the white pine weevil (Stiell and Berry 1985; Pitt et al. 2016).

Multi-aged regeneration methods can result in structures that are both resistant and resilient to bark beetle outbreaks due to the vertical complexity that results (O'Hara 2014). However, resistance may vary across the stand, as a complex horizontal structure can also result in dense groups of trees that are competing heavily under a similar microclimate as pre-treatment. Such pockets of trees may remain susceptible to bark beetle attack. However, Kollenberg and O'Hara (1999) found multiaged stands tended to have higher leaf area indices and basal area increment compared to even-aged stands.

The benefits of structural complexity and the overall increased resistance and resilience are likely to outweigh the consequences of small-scale pockets of lower vigor trees. In uneven-aged, single-species stands, treatments that reduce density only marginally are not likely to alter the microclimate or tree vigor enough to reduce bark beetle hazard and may have the opposite effect. For example, a low thinning that removes only suppressed/overtopped trees increases average tree size—a factor that could increase bark beetle hazard. However, if the stand is being converted from a simple structure to a more complex structure, resistance and resilience will increase to bark beetles while decreasing to defoliators following harvest. The opposite would be expected if a stand is shifted from a more complex structure to a simplified vertical and / or horizontal structure. It therefore requires a careful balancing of objectives to arrive at a vertical and horizontal structure that is both resistant and resilient to bark beetles and defoliators while also meeting other objectives, such as timber production or fire hazard reduction. In the western United States, timber production is becoming less of a societal value and healthy forested landscapes resilient to large-scale mortality events that provide biodiversity and wildlife habitat are taking precedence. In these forests, reducing overall stand density to a low basal area (~35% of carrying capacity) has the potential to meet these new objectives without creating increased insect susceptibility or wildfire hazard.

Sanitation is an intermediate treatment and direct control approach used to reduce insect population levels in a stand (Box 20.1). The objective of sanitation treatment is to improve stand health by removing trees infested or likely to be infested by insects. Controlling a bark beetle population using sanitation is not considered a viable option, with the exception of the southern pine beetle. Spot infestations (Fig. 20.2) of southern pine beetle can be controlled, thus avoiding a landscape-scale outbreak, using either cut-and-remove or cut-and-leave strategies. If trees can be removed and handled appropriately following removal from the site, cut-and-remove strategies are preferred (Fig. 20.3; Fettig et al. 2007). However, cut-and-leave strategies, in which cut trees are left onsite, can also be effective and do not appear to increase the hazard of attacks in nearby trees (Fettig et al. 2007 and references therein).

Southern pine beetle outbreaks have decreased in frequency since the 1950's despite a concomitant increase in the acreage of pine plantations. One hypothesis related to the decline in outbreaks is that intensive silviculture practices have resulted in less susceptible stands (lower density, higher average tree vigor) than were present in earlier decades (Asaro et al. 2017). Widespread use of sanitation strategies may also

Fig. 20.2 Spot infestation of southern pine beetle, from above (left) and below (right). Modified from Asaro et al. (2017)

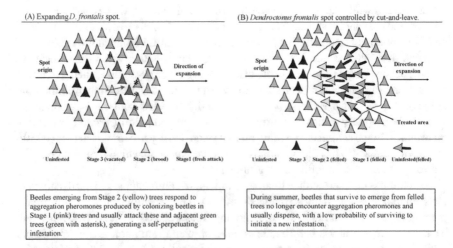

Fig. 20.3 Illustration of an expanding spot infestation (A) and the cut-and-leave sanitation treatment implemented to control southern pine beetle (B). From Fettig et al. (2007)

have a role in outbreak frequency reduction, as cut-and-remove and cut-and-leave strategies are implemented quickly in new spot infestations (Asaro et al. 2017).

Coniferous forests composed of a single species, size and age class will be highly susceptible to bark beetle outbreaks when these factors align with insect pest host preferences. When bark beetle populations are high, even trees in lower density stands may be attacked and overcome by beetles. In these situations, sanitation may be the best option if attacked and dead trees need to be removed. For example, dead trees in recreation areas are hazard trees and pose a safety threat to visitors and should be removed using a sanitation treatment. Numerous dead trees in more remote locations may not warrant removal if they do not pose a safety issue and recovering economic value from these trees is not an objective.

20.2.2 Strategies to Adjust Species Composition

Many forest insect pests are considered specialists, preferring specific tree host species over others. In some insects, this host preference is quite strong and attacks on non-preferred species are rare (e.g. spruce beetle). Other insects have a range of tree hosts, with one generally preferred over others but finding several species attacked in a stand would not be considered unusual. Bark beetles tend to have narrower host ranges than defoliators. Defoliators frequently infest a range of host species, with an order of preference. For example, western spruce budworm (*Choristoneura freemanii*) is an unfortunately named species, as it preferentially attacks true fir (*Abies* spp.). Infestation of spruce (*Picea* spp.) occurs, but damage and mortality may be less severe or occur after the true fir have been fully infested and are dead or declining from multiple, successive defoliation events (Polinko 2014; Vane et al. 2017). Western spruce budworm inhabits a wide geographic range across western North America, and preferentially feeds on tree hosts in order of tree shade tolerance patterns (Brookes et al. 1978). Other defoliators vary more in their host preferences; Douglas-fir tussock moth (*Orgyia pseudotsugata*) primarily feeds on white fir (*Abies concolor*) in the southwestern US, switching to a preference for either grand fir (*Abies grandis*) or Douglas-fir (*Pseudotsuga mensiezii*) in the northern Rocky Mountains, depending on site conditions, and even further north, in Canada's Rocky Mountains, feeds primarily on Douglas-fir (Brookes et al. 1985). However, even with changing geographic host tree preference (i.e. when a species' host tree preference differs throughout its range), preferred host tree species still exhibit higher shade tolerance than less preferred species in the same stand (e.g. pine species (*Pinus* spp.). Under certain circumstances, such as at high larval population levels or when non-preferred tree hosts are surrounded by more preferred tree hosts, feeding will occur on all tree species in the area. Defoliator damage to host trees ranges from short- and long-term growth reductions to widespread mortality following multiple, recurring defoliation events (Naidoo and Lechowicz 2001; Vane et al. 2017; Rapp 2017).

Silvicultural strategies designed to adjust species composition are primarily used to mitigate defoliator damage and mortality but could also be used to prevent or mitigate other insect infestations, particularly if tree host preference is known. Defoliators disperse from upper to lower tree canopies; the most susceptible stands are dense with a species composition composed primarily of the most preferred host species in multiple vertical canopy layers. Abiotic site factors, including warmer, drier sites that are more prone to drought (e.g. upper ridges), can also play a pre-disposing role in defoliator hazard. If one or more, less preferred host tree species are present or planned for after treatment, silviculture can be an effective indirect control method of reducing the potential for future insect damage. Intermediate treatments or regeneration methods can be used (Box 20.1, Table 20.1); the prescription should remove dead and dying infested trees and live trees of the most preferred tree host species. Such a prescription should also adjust the vertical and horizontal stand structure in a complementary manner to increase both resistance and resilience. Additionally, other stand objectives are typically accounted for in the prescription, including fire hazard reduction, timber production, and wildlife habitat.

Eastern spruce budworm (*Choristoneura fumiferana*) prefers balsam fir (*Abies balsmaea*) over white spruce (*Picea glauca*) and black spruce (*Picea mariana*) in eastern Canada, and also tends to cause the highest levels of mortality in dense mature balsam fir stands. A silvicultural prescription that both reduces density and preferentially removes mature balsam fir will result in a stand with a lower probability of future damage (DeGroot et al. 2005). Similar strategies are being implemented to reduce western spruce budworm damage in the southwestern US; the prescription reduces density to increase overall tree vigor and shifts species composition towards less preferred host trees such as quaking aspen (*Populus tremuloides*) and ponderosa pine. White fir and defoliated Douglas-fir are preferentially removed (Fig. 20.4). When developing silvicultural prescriptions, it is important to understand differences in the ecology of insect species. For example, while eastern spruce budworm causes mortality in mature balsam fir first, western spruce budworm mortality tends to occur first in the smaller size classes (Brookes et al. 1978; DeGroot et al. 2005).

Another opportunity to shift species composition occurs during the regeneration phase. Silviculture can be used to encourage certain species to naturally regenerate over others or artificial regeneration can be used to select a specific species composition and density for the new age class. Ensuring adequate natural regeneration can be challenging following widespread overstory mortality if live trees are not available to provide a seed source. In the case of defoliators, heavily defoliated live trees will often have limited capacity for seed production following defoliation (Brookes et al. 1978). In these stands, natural regeneration of less preferred host tree species is more likely than regeneration of the most susceptible host species. A shift toward less preferred host tree species can be encouraged even more by removing preferred host trees from the overstory and leaving only less preferred host trees to regenerate the stand. Such a composition shift may or may not be desirable, depending upon the objectives of the silvicultural treatment.

Planting is the best way to ensure regeneration by less preferred host trees. In most situations, complete replacement of preferred host tree species with less preferred

Fig. 20.4 Silviculture used to reduce western spruce budworm damage and mortality on the Kaibab National Forest, Arizona, USA. The treatment reduced stand density, created openings to promote regeneration, and favored less preferred host species as residual trees. Photo by K. Waring

host tree species will not be desirable, as this represents a stand conversion. Single-species plantations may also be vulnerable to a different suite of insect and/or disease problems but may be warranted to meet landowner goals and objectives, such as timber production. Generally, planting will entail a subtle shift from dominance by preferred host tree species to dominance by less preferred host tree species by planting a reduced density of the preferred tree host species.

20.2.3 Potential Drawbacks to the Use of Silviculture

It is possible to create conditions more conducive to insect damage and mortality through silviculture. For example, regenerating eastern white pine under full sun will lead to white pine weevil problems in certain regions (Ostry et al. 2010). It is the responsibility of the silviculturist to know and understand the silvics and ecology of the trees and their pests in a given stand to avoid creating these problems. Silviculturists frequently rely on forest health experts to provide information about specific, stand-level insect or disease issues that may be a concern before or after treatment. Pruning large live branches during bark beetle flight periods can result in attacks leading to mortality, thus pruning treatments should be timed to occur outside of these flight periods whenever possible. Generally, the objectives of pruning for wood quality will not create conditions conducive to bark beetle attack as the stands targeted for pruning treatments are young, and small live branch removal from conifers has not been found to increase bark beetle susceptibility. Hadfield and Flanagan (2000) found pruning increased susceptibility to Douglas-fir beetle attack in campgrounds where large live branches were pruned to meet a hazard tree objective (removal of dwarf mistletoe (*Arceuthobium douglasii*)-infected branches with large brooms).

Prescribed burning, even at low intensity and severity that does not outright kill the overstory trees, can increase susceptibility to bark beetle attacks through crown scorch and injuries to the cambium (McHugh et al. 2003; Billings et al. 2004). Post-fire tree mortality due to bark beetle attack tends to be short-term (Kane et al. 2017) but as we increase the use of prescribed fire as a management tool, caution is warranted (Bentz et al. 2009). Frequent use of prescribed fire also reduces stand resilience by removing tree regeneration. Central American forests were subject to management practices that reduced both resistance and resilience, resulting in a large, landscape-scale southern pine beetle outbreak (Billings et al. 2004).

The interactions between tree physiology (including tree defenses), herbivory, and abiotic stresses are complex and a review of these is beyond the scope of this chapter (see Massad and Dyer (2010) and Ryan et al. (2015) and literature cited within, for a review and overview of these concepts).

From a silvicultural perspective, thinning has the potential to not just increase tree vigor, but also increases residual tree growth, leading to thicker phloem. Very dense stands have small individual trees with thin phloem that limits bark beetle development and reproduction. Such stands may have reduced susceptibility to bark beetle attacks; thinning may increase susceptibility by increasing average tree size and phloem thickness (Anhold et al. 1996). Very low stand densities have historically been resistant to bark beetle attacks (as described previously in this chapter). Recent research indicates that individual trees in such stands may be less resilient to drought, possibly due to an inability to maintain large crowns when water is limiting (D'Amato et al. 2013).

Drought stress has been linked to increased insect activity in multiple tree species (Savage 1994; Gaylord et al. 2013; Anderegg et al. 2015; Kolb et al. 2016;). Very low and very high stand densities may not be conducive to long-term resistance or resilience given this interaction. A recent study suggests that drought lowers tree resistance to infection by some bark beetle fungal symbionts (Klutsch et al. 2017). During drought conditions, stress is often manifested within individual trees as reduced growth (Fischer et al. 2010; Thomas and Waring 2015; Sohn et al. 2016).

The ability of individual trees within a stand to recover to pre-drought growth rates can be an indicator of susceptibility to bark beetles. Fischer et al. (2010) found that at high stand densities (~14 m^2 ha^{-1}) ponderosa pine trees that failed to return to pre-drought growth rates were preferentially attacked by the rounded pine beetle (*Dendroctonus adjunctus*). Douglas-fir tussock moth and western spruce budworm damage tends to be higher on sites more prone to drought conditions (Brookes et al. 1978, 1985). This effect is likely linked to the preferred host species being among the least drought tolerant at these sites. Thinning may also change the chemical composition of residual tree foliage, leading to increased susceptibility. In spruce-fir stands of northeastern North America, thinning altered the foliar monoterpene concentrations of both spruce and fir, making them more susceptible to defoliation from eastern spruce budworm (Fuentealba and Bauce 2011). Due to the complex interactions described above, the response to thinning is not always predictable, nor does it always lead to reductions in herbivory.

Implementing silvicultural treatments can result in logging damage to residual trees and increases slash on the forest floor. To avoid increasing residual tree susceptibility to bark beetle attack, logging operations should be timed to occur when bark beetle flights are low or not occurring, and care should be taken to avoid damaging live trees. Slash piles can serve as suitable host material for many *Ips* species, which may then 'spill-over' into the tops of neighboring trees (Kegley et al. 1997). Slash piles should be removed, chipped or burned in a timely manner to avoid this problem. Freshly cut logs and log decks of large trees can result in fast build-up of certain bark beetles as well (such as the spruce beetle), which then move on to attack live trees nearby (Reynolds and Holsten 1994). Logging activities may damage the soil, increasing compaction, erosion, and/or rutting. Soil damage can lead to increased tree stress, and susceptibility to insect damage, such as the Douglas-fir tussock moth (Brookes et al. 1978).

20.2.4 Linkages with Integrated Pest Management

As discussed in Chapter 17, Integrated Pest Management (IPM) is an integrated approach, which considers multiple strategies and tactics to manage pests efficiently while incorporating economic, social and ecological components. In forest entomology, IPM has primarily focused on efforts to reduce or describe more targeted approaches for land managers using insecticides, and silviculture adds another tool to help reduce potentially environmentally damaging chemical agents on the landscape (McIntire 1988).

It should be noted that silviculture is an IPM tactic. Generally considered cultural strategies, these tactics are generally defined as any treatment that involves a modification of established practices to make a host less favorable for pests or minimize the loss of a particular commodity. Concepts of preventative management are readily applicable to silvicultural strategies. In stands where pest outbreaks are a concern, using management tactics to foster resistance and/or resilience in the resulting stand is crucial (Table 20.1). Silvicultural tactics can be used in tandem with other management activities to increase resistance and/or resilience, while also providing opportunities for other, more immediate tactics to be implemented should pest populations increase. In this section, we cover the use of silviculture in combination with monitoring, chemical control, biological control, and genetic selection.

20.2.5 Silviculture and Monitoring

As discussed previously in Chapter 19, effective monitoring of insect activity is the critical first step of developing an appropriate IPM response. Monitoring should be conducted in a way that is both regular and economically feasible, in order to continually update information on insect population sizes and activity. Management actions should be based on regular assessments of both the insect pest population size and their potential to inflict damage. Conducting regular stand assessments for insect activity, in addition to more stationary and passive approaches, i.e. insect traps, should be both conducted annually, and monitored frequently, to best identify areas where insect activity is increasing. Land managers use this information to prioritize stands for management and abate potential large-scale insect damage or mortality.

Proactive management entails preparing unaffected areas such that if the problem occurs (i.e. non-native invasive expands its non-native range) stands are better able to cope with these changes (e.g. Schoettle and Sniezko 2007). Monitoring pest spread is a key component of proactive management facilitating the identification of high-risk areas (i.e. as characterized by stand conditions, species compositions, vertical/horizontal structure, edaphic and abiotic features of the landscape). Silvicultural actions triggered through monitoring demonstrate the potential of the combination of these two strategies to better prepare forested stands for potential or imminent pest expansion and movement.

In long-term forestry projects, regular monitoring is crucial to determine if silvicultural approaches are warranted (i.e. the identification of emerging threats). Posttreatment, they can be used to evaluate treatment impact on target pest populations. Favorable environmental conditions, or certain disturbances (wind-throw events, storm damage, etc.) can lead to rapid insect population growth. Regular monitoring facilitates the identification of both changes in insect populations and above-threshold population levels [levels above which severe economic damage occurs (see Chapter 19)], both of which are critical to maintaining the health and vigor of forest stands.

Monitoring is critical for effective management of non-native, invasive insects. For example, the sirex woodwasp (*Sirex noctilio*), an invasive insect of pines that recently established in northeastern North America (Hoebeke et al. 2005), the combined approach of proper silvicultural management and monitoring population expansion, whether through trapping or categorizing infestations aerially, helps land managers determine a proper course of action. Stand resistance to sirex woodwasp can be increased through thinning prior to insect invasion. Maintenance of both host tree vigor through basal area reductions (for eastern white and red (*Pinus resinosa*) pines these are reported between 9.3 and 14 m^2 ha^{-1}), creates stands that are optimal for tree growth and therefore production of defensive compounds (Gilmore and Palik 2006; Dodds et al. 2007). Monitoring allows managers to prioritize treatment of pre-invasion stands while considering location of those stands across the landscape.

Monitoring is also an important consideration for native insect pests. Bark beetles are especially damaging during epidemic population cycles. Due to their ubiquity in the Northern Hemisphere, methods such as aerial detection, trapping, ground surveys and remote sensing have been developed and implemented widely for monitoring, and newer technologies, such as unmanned aerial vehicles, are being considered (Wulder et al. 2005; Fettig and Hilszczanski 2015; Morris et al. 2017). Ultimately, proactive monitoring in combination with silvicultural strategies, such as direct control of potential infestations, can be effective preventative measures to make stands and landscapes less susceptible to widespread mortality from the activity of both non-native and native pests.

20.2.6 Silviculture and Chemical Control (Insecticides)

As discussed previously the impetus for the development of IPM was largely generated by an over-reliance on insecticides and the subsequent development of insecticide resistance. However, chemical control is still a large part of any IPM strategy, and proper timing of applications and insecticide selection can yield multiple benefits. For example, the North Carolina Department of Agriculture recommends spraying Fraser fir (*Abies fraseri*) plantations with a number of pyrethroid insecticides during specific times of the year to control for multiple pests such as balsam woolly adelgid (*Adelges piceae*), balsam twig aphid (*Mindarus abietinus*) and hemlock rust mite (*Nalepella tsugifoliae*) (Sidebottom 2009). The timing of the applications, coupled with adequate tree spacing in these plantation settings, highlights an effective IPM strategy combining silviculture (spacing, tree growth) with insecticide use. Pest populations are reduced when problematic, while minimizing the number of insecticide applications required to reach the management goal.

Effective and economical use of chemicals cannot always be achieved in forest settings. Chemical control is expensive and difficult to apply at landscape-scales or in remote areas, highlighting the necessity of having multiple management strategies to manage pests. Imidacloprid, a neonicotinoid systemic pesticide, has been used by the National Park Service to protect eastern hemlock (*Tsuga canadensis*) from damage caused by the invasive insect, hemlock woolly adelgid (*Adelges tsugae*) in both trunk and soil applications (NPS Environmental Assessment 2007). Current research is showing that hemlock woolly adelgid responds negatively to increased light and that releasing these shade-tolerant species using silviculture (e.g. crown thinning, where eastern hemlock are the favored residual trees, with the objective of sustaining the species) may be a strategy to reduce pest populations through stand manipulations. This strategy may be particularly useful for releasing understory hemlock, especially in riparian areas and other areas not feasibly sprayed with insecticides (Brantley et al. 2017).

Carlson et al. (1983) suggest simplifying stand vertical structure (i.e. single-canopy or two-aged), and varying species composition are viable silvicultural strategies to mitigate damage and potential population increase of western spruce budworm in spruce-fir forests. By simplifying canopy strata/altering composition, land managers build natural barriers to population expansion on longer time scales, while using insecticides in untreated and susceptible stands. These examples highlight how insecticide use can be minimized by the creation of less susceptible stand conditions through active IPM management strategies.

Targeted insecticide use can reduce impacts on non-target species and can effectively reduce pest populations during outbreaks. When coupled with regeneration methods (Table 20.1), chemical control can be utilized to protect the future stand. For example, Gottschalk (1993) recommended shelterwood regeneration methods in stands vulnerable to spongy moth (*Lymantria dispar*), followed by aerial application of insecticides. This strategy reduces insect population numbers while building resilience through the regenerating trees. While chemical control may still be an effective management tool to reduce pest numbers during outbreaks, using silviculture to maintain tree vigor and maintain or enhance understory species diversity and abundance [as habitat for potential biological control agents (e.g. natural predators and parasitoids)], can provide useful components of IPM programs that help to alleviate the need for chemical control (Elek and Wardlaw 2013).

20.2.7 Semiochemicals

Chemical control also includes the use of semiochemicals, organic molecules produced by plants or animals that mediate behavioral interactions between organisms. Semiochemicals involved with intraspecific (within species) communication are pheromones, and those involved with interspecific (between species) communication are allelochemicals. Synthetic copies of these signals and cues can be used in monitoring and management programs for forest insects. For example, verbenone, an anti-aggregation pheromone released by both mountain pine beetle and western pine beetle (*Dendroctonus brevicomis*), has been utilized to directly protect many different species of western North American conifers (e.g. Gillette et al. 2012; Borden et al. 2006; Fettig and Munson 2020). Site factors such as lower stand densities and higher temperatures diminish its efficacy on a stand-scale when deployed as individual slow-release packets (Fettig et al. 2009), while area-wide deployment on the forest floor in flake releasing formulations effectively reduce beetle mass-attacks on individuals (Gillette et al. 2014).

These strategies, referred to as push/pull strategies, exploit bark beetle behavior to repel pests from the desired resource (e.g. a stand or individual tree) and pull them towards a resource that can then be managed to explicitly eradicate attracted individuals (Cook et al. 2007). Push strategies use numerous tactics including but not limited to semiochemicals (both host- and pest-derived) such as anti-feedants (host-derived chemicals that deter insect feeding activities), anti-aggregants (such as verbenone) and alarm pheromones (pest-derived pheromones that elicit fight-or-flight responses) (Cook et al. 2007).

Push strategies emphasize keeping the pest away from resources (e.g. host trees), while pull strategies tend to use attractants to concentrate individuals in an area. Trap trees represent a common tactic used as a pull strategy in controlling endemic and epidemic bark beetle populations (e.g. Fettig et al. 2007). Felled trees, which mimic windthrown trees, are targeted by some species of bark beetle, therefore felling and baiting trap trees with an aggregation pheromone can be an effective pull strategy (e.g. Schmid and Frye 1977). Trap trees then need to be removed from the stand in a sanitation operation to limit population build-up in stands. Combined with silvicultural strategies such as harvesting infested individuals (as in sanitation treatments; Table 20.1), trap trees (both baited and non-baited) are effective at controlling endemic populations of beetles (e.g. Bentz and Munson 2000).

Generally, large diameter trees tend to be more attractive to infestation by bark beetles, indicating the usefulness of selecting trap trees that are most likely to become infested (Mezei et al. 2014) and effectively timing treatments for greatest impact. Use of felled or standing trap trees is a common sanitation tactic, but their effective use is dependent upon the environment (e.g. Fettig and Hilszczanski 2015). For example, during warm, dry winters with low snowpack, Holusa et al. (2017) recommend land managers fell trap trees just before bark beetle emergence in the spring to maximize efficacy, but during cooler, wetter winters with more snowpack, trap trees can be felled earlier in the winter, as these conditions maintain characteristics of the trap trees attractive to emerging beetles. Coupling push–pull strategies with silvicultural strategies designed to maintain vigorous trees and favoring less susceptible host trees for retention can aid in reducing pest population growth.

20.2.8 Silviculture and Biological Control

Biological control involves utilizing natural enemies (parasites, parasitoids, pathogens etc.) to achieve a reduction or control of pest populations. Increasing the size of established natural enemy populations (parasitoids, predators etc.) by releasing large numbers of individuals as defense against pests is referred to as augmentative biological control (Hoy 2004a). In contrast, classical biological control

(Hoy 2004b) involves introducing non-native natural enemies to establish populations to reduce non-native, invasive pest populations. A third option, conservation biological control, involves altering the vertical or horizontal structure, including species composition, of a given land unit to provide more habitat for natural enemies and thus maintain a reserve of beneficial insects within your forested stand.

Silviculture can actively promote conservation biological control, by manipulating the overstory composition or structure to increase understory growth or shift species composition to increase habitat reservoirs of beneficial natural enemy species, illustrating the direct link between silvicultural strategies and biological control in pest management. Classical and augmentative biological control can be used in concert with silvicultural treatments designed to promote individual tree vigor or increase or maintain horizontal and vertical stand structural complexity, including the use of species mixtures. For example, Perez-Alvarez et al. (2019) found classical and augmentative biological control to be more effective in complex than in simple landscapes. This highlights the potential for creation of complex forest structures, and landscape heterogeneity, to potentially increase the impact of biological control programs.

Traditional silviculture practice to meet timber production objectives has primarily utilized monocultures and even-aged regeneration methods (clearcut, seed tree and shelterwood methods) and thus result in reduced stand complexity. Even-aged monocultures can be more susceptible to insect outbreaks and large-scale damage and mortality. Increasing stand structural and compositional complexity increases natural enemy populations and relatively low pest populations (Klapwijk et al. 2016) while also enhancing resilience. For example, single-tree selection in uneven-aged stands increases shading of cut stumps, lowering the temperature of the stump surface and increasing development times for the large pine weevil larvae (*Hylobius abietis*), making them more vulnerable to predation (Inward et al. 2012). Predator population increases help to prevent the buildup of pest populations and thus can aid in preventing epidemic outbreaks (Klapwijk et al. 2016).

Warzée et al. (2006) calculated predator/prey ratios for the native European spruce bark beetle, *Ips typographus,* which primarily attacks spruce species, and the predator, the ant beetle (*Thanasimus formicarius*), in stands of different species compositions. They found that these ratios were significantly greater in mixed species stands, especially those stands with a substantial pine component, as the ant beetle finds more favorable pupation sites on thick barked pines compared to thinner barked spruce (Warzée et al. 2006). In this study, pine species were present on two sites, one composed of 26% pine, the other 80% pine, suggesting that pine as either a minor or major component can positively influence predator/prey ratios for this species (Warzée et al. 2006). Similarly, promotion of certain flowering species in agricultural settings can increase longevity of parasitoids, showing promise for similar use in forested stands (Russell 2015). Mixed species management can influence the life cycle and population levels of natural enemies thus additionally impacting pest species populations (Klapwijk et al. 2016). Incorporating native biodiversity into the silvicultural prescription allows for multiple objectives to be met in a single treatment.

When considering biological control of invasive species, natural enemies from their native habitat are often used as biological control agents in their introduced ranges (e.g. Cheah 2011; Bauer et al. 2015; Kenis et al. 2017). The abundance of invasive species is often greater in their invaded ranges, potentially due to their release from predation, and as such, invasive species often do not have natural enemies in their new habitats (the Enemy Release Hypothesis; Williamson 1996). This generally means that within their native ranges, populations are controlled by natural enemies and tree host defenses, however, when freed from natural predation and host defenses, they become much more damaging as populations rise. Many recent insect invasions around the world exhibit population growth supportive of this hypothesis, including the recent invasion of the emerald ash borer (*Agrilus planipennis*) in the eastern United States. This invasive beetle kills overstory ash species (*Fraxinus* spp.), significantly altering forest succession, and causing economic losses. Researchers and managers, as part of a classical biological control program, released a parasitoid, *Tetrastichus planipennisi,* of the emerald ash borer, which effectively reduced sapling mortality in emerald ash borer-infested stands in Michigan (Duan et al. 2017). While most ash species show little to no resilience to emerald ash borer, green ash (*Fraxinus pennsylvanica*) regenerates quickly after disturbance and reaches reproductive maturity relatively quickly (Kashian 2016). There is potential to sustain green ash by combining classical biological control with silviculture, creating stand conditions conducive to maintaining or increasing populations of the biological control agent and regeneration of green ash.

The sirex woodwasp has been an established non-native invasive pest for decades, recently arriving in the northeastern United States. Current silvicultural strategies involve thinning stands and removing smaller and suppressed size classes (Dodds et al. 2014). Establishment of the parasitic nematode, *Deladenus siricidicola,* for biological control has been successfully utilized in Australia (and elsewhere in the southern hemisphere) and shows promise, albeit with serious reservations, for expansion to North America (Haugen 1990; Bedding and Iede 2005; Bittner et al. 2019). Pines are introduced to Australia, meaning the risk to non-target organisms is minimal as insects in Australia did not co-evolve with pines and are rarely associated with the trees. In North America, there are communities of native insects associated with pines and, consequently, there are potential negative impacts for non-target organisms that warrant pause in applying this strategy. In a recent study, Bittner et al. (2019) evaluated strains of these nematodes within North America, and observed that native nematodes may both positively and negatively influence the sterilization success of sirex woodwasp. Other invasive insects, such as the balsam and hemlock woolly adelgids, have both been successfully preyed upon by a single species of beetle, *Sasajiscymnus tsugae,* in a laboratory setting, showing potential for this agent to be released as a classic biological control agent and further advance IPM strategies for both invasive species (Jetton et al. 2011).

Elkinton et al. (1996) found evidence that increased white-footed mouse (*Peromyscus leucopus*) density resulted in reduced spongy moth population density. Further, they also found a strong positive association between acorn density and the white-footed mouse population, indicating the importance of acorns for over-wintering populations of white-footed mice. Strategies aimed at maximizing acorn production (e.g. low thinning, crop-tree release), as well as species composition manipulation, especially in low-risk stands, may help to maintain conditions less conducive to high spongy moth populations. This example illustrates how silviculture can promote conditions conducive to native predators (conservation biocontrol) for the control of non-native invasive insects, thereby aiding in reduction of pest populations and maintaining forest health.

20.2.9 Silviculture and Genetic Selection

Genetic selection, or selecting host trees that show promise of resistance to insect attack, and the establishment of breeding programs to propagate these "plus" trees, is a widely researched topic (Kinloch and Stonecypher 1969; McKeand et al. 2003; Roberds et al. 2003). Outside of traditional tree breeding programs, selection of trees in natural settings requires managers to select trees based on their phenotype; the underlying genotype is usually unknown. Exploiting these pre-adapted traits through the utilization of existing genetic variation in breeding programs is an effective method of characterizing resistance mechanisms within species, and then propagating progeny that show increased defensive capabilities. For example, Zas et al. (2017) characterized existing genetic variation in Norway spruce (*Picea abies*) traits related to increased resistance to the large pine weevil.

Land mangers currently use silvicultural strategies to minimize damage from this pest on artificial regeneration, including soil preparation and shelterwood treatments (Nordlander et al. 2011), however these may be difficult to apply or expensive. Therefore, the decision-making process land managers use to select treatment options is important. Consider Fig. 20.5, which highlights a general decision model including both silvicultural strategies and genetic selection can be utilized to manage plantations, as well as the research requirements for IPM (Alfaro et al. 1995). This demonstrates how genetic selection of putatively resistant and susceptible individuals, and subsequent silvicultural interventions along with other IPM strategies (biological control, etc.) create a framework to help guide land managers in establishing productive plantations that demonstrate the core principles of IPM. By including both resistant and susceptible individuals, one can assess how alternative management strategies (pruning, spacing, biological control) can reduce infestation levels.

Genetic host tree resistance can be categorized as constitutive or inducible. Constitutive defenses are those defenses that are always expressed, whereas induced defenses are those defenses a plant expresses in response to herbivory (Larsson 2002). Antibiosis indicates that some aspect of the host plant (chemical composition of tissues, defenses) has a negative impact on the pest biology (i.e. survival, development) (Painter 1958). For example, Bucholz et al. (2017) found that without direct contact, volatile organic compounds associated with the resistant Veitch fir (*Abies veitchii*) compared to a susceptible species, Fraser fir, resulted in significantly

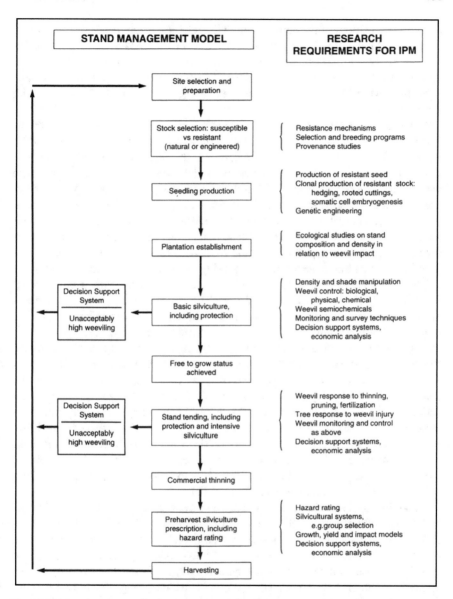

Fig. 20.5 Decision key for integrated pest management of Sitka spruce plantations to mitigate damage from the white pine weevil. Modified from Alfaro et al. (1995)

reduced eclosion success of balsam woolly adelgid eggs. This suggests an antibiotic effect of the constitutive chemicals released by Veitch fir on balsam woolly adelgid eggs.

Antixenosis or non-preference (e.g. Painter 1958), occurs when some aspect of the host, either chemical or morphological, results in reduced interaction (e.g.

feeding, oviposition) of the herbivore with the host. An example of this can be seen with the codling moth (*Cydia pomonella*) and leaf tissue metabolites from resistant and susceptible cultivars of apple trees (*Malus* spp). Significantly more oviposition occurred on cloth containing metabolites from susceptible than resistant cultivars, indicating oviposition preference based on chemical cues (Lombarkia and Derridj 2008).

Tolerance describes hosts that can sustain insect feeding activity without serious loss in productivity and is therefore subtly different from resistance (Painter 1958). Strauss and Agrawal (1999) defined tolerance as the degree to which plant fitness is impacted by herbivore damage relative to an undamaged state, whereas resistance was defined as any plant trait which reduced herbivore preference or performance. Tolerance mechanisms are therefore related to increased net photosynthetic rate after damage or compensatory action, high relative growth rates, root carbon storage for above-ground reproduction, and increased branching and resource allocations after damage (Strauss and Agrawal 1999 and references therein). An example of tolerance involves tannin concentration in quaking aspen leaves that does not serve as a defensive compound (i.e. resistance) but one that facilitates nutrient uptake post-defoliation (Madritch and Lindroth 2015). This is viewed as a tolerance mechanism, as the production of greater amounts of these types of secondary metabolites influences nutrient recoveries that may be hindered by defoliation damage.

Other resistance mechanisms, such as oleoresin production, are quantitative genetic traits that can be selected for during tree breeding programs. These traits are complex in that they are composed of many different "small effect" loci that contribute to tree phenotype (e.g. Mundt 2014). Ultimately, the goal is to breed host varieties resistant to certain pest species. Oleoresin flow, along with number of canals or preformed defensive (resin) ducts, has been shown to be positively correlated with survival following bark beetle attack. Bark beetle feeding activity slices through these canals or ducts, releasing their resin, which may envelop or remove the beetles; the more resin ducts a tree has, the more likely it is to successfully eject the attacking beetles (Strom et al. 2002; Kane and Kolb 2010).

In addition, seasonality, and its impact on physiological processes is an important consideration. Lorio (1986) used the framework of Loomis' (1932) growth-differentiation balance hypothesis to examine conditions ideal for southern pine beetle population expansion. He concluded that this hypothesis was useful in explaining seasonal demand for photosynthate, ultimately driving seasonal vulnerability to southern pine beetle. The trade-off between spring growth and defense suggests that fast-growing trees can be susceptible during these periods when growth processes use more available photosynthate, leaving less allocated towards defense production. The variance in this trait among populations of loblolly pine (*Pinus taeda*) is heritable.

Westbrook et al. (2013) developed a genomic prediction model across the range of loblolly pine, identifying specific genetic regions associated with increased oleoresin production. This work yielded a guide for making genetic selections to provide increased resistance to southern pine beetle. Trees with increased resistance can be incorporated into silviculture when regenerating stands. Planting all or part of the stand with more resistant trees in anticipation of future insect herbivory, coupled with ongoing silvicultural strategies that promote tree vigor (thinning, adequate spacing etc.) would be an approach to increase resilience (Table 20.1) and may be an important step forward for bark beetle IPM.

Resistance within stands can be promoted by maintenance of stand density index [SDI; a measure of relative density using the relationship between average tree size and stand density (Reineke 1933)] below certain thresholds. For example, Long and Shaw (2005) review strategies associated with size/density relationships surrounding mountain pine beetle, and found that maintaining a SDI below 250 minimizes susceptibility to mountain pine beetle attack. The difficult aspect of management is prioritizing stands for treatment and connecting treatments across landscapes to decrease susceptibility. Reforestation, including planting with genetically improved genotypes where available and economically feasible, aids in contributing to decreased landscape-level susceptibility.

Earlier in the chapter, we discussed using silviculture to shift vertical and horizontal stand structure, with one outcome being increased vigor of residual trees. Individual tree response to reduced competition and increased resource availability is related to the genetic profile of the tree and the surrounding abiotic site conditions (the environment). Growth response to treatment can be optimized through appropriate silviculture in combination with genetic selection by retaining high vigor trees (those that allocate more stem wood per square meter leaf area), with the assumption that this trait is partially determined by genetics. Selecting trees with high growth rates prior to treatment can be challenging (Fischer et al. 2010) but may be possible through additional measurements of tree cores and crown area. Remote sensing applications can detect thinned stands with increased growth rates and therefore resistance to bark beetle attack, establishing a relatively easy method of monitoring overall stand resistance at large scales and across multiple land ownerships (Coops et al. 2009). Silviculturists need to consider the evolutionary adaptation occurring in the stand between bark beetles, host trees, and climate; for example bark beetles may be able to select for host trees least adapted to the changing climate (e.g. Six et al. 2018), resulting in a more resilient stand following bark beetle mortality.

Abiotic site conditions also play a key role in determining phenotypes. Abiotic site conditions (e.g. slope, aspect, topography, soil conditions) tend to change slowly through time or not at all. The abiotic capacity of a site to produce vegetation is often used as a proxy for site quality; high (good) sites produce more vegetation than low (poor) sites. Vegetation production is less on low sites due to limiting resources for plant growth, often related to poor soil resources. The relationship between site conditions and resistance to insects is highly dependent upon the tree host species and corresponding pest species. Slow-growing individuals may be more susceptible to attack by certain insects (i.e. subalpine fir (*Abies lasiocarpa*) by the western balsam bark beetle (*Dryocoetes confusus*), therefore considerations of site quality and genetics of growing stock (whether natural or artificial) are important for decreasing susceptibility to insect attack (Bleiker et al. 2005). For example, slower-growing Eucalypts (*Eucalyptus* spp.) are recommended for low quality sites, since slower growing tree foliage may be better defended against defoliation (Stone

2001). Managers face a complexity of decisions related to the interactions between silviculture, genetic selection, and underlying site factors. After carefully considering these interactions, silviculture and genetic selection can be important components of successful IPM programs.

20.3 Silviculture Over Long Temporal and Large Spatial Scales

The impacts of silviculture extend over long temporal and large spatial scales. Understanding the role of individual stands at these large scales is an important consideration when selecting the appropriate silvicultural strategy because individual stands are connected to form a landscape. Silvicultural treatments necessitate understanding and predicting patterns of tree and stand growth at large spatial and long temporal scales, including interactions with various disturbances and incorporating uncertainty into predictive models. However, understanding these predictions and then designing appropriate strategies that meet multiple goals and objectives is a necessary component of building resistant and resilient landscapes.

Building resistance to insect pests at smaller spatial scales highlights the difficulties faced by silviculturists by both long time scales and scaling up to a landscape. For example, stands at risk to spruce bark beetle have common structural characteristics that can be manipulated through management thereby reducing risk. At the stand-scale, this would entail reducing the relative proportion of overstory basal area in host spruce, reducing the average size of spruce in the stand, or reducing stand basal area (Schmid and Frye 1976). However, while an individual stand may be treated, building this resistance at landscape-scales in practice has proved unrealistic due to economic and political constraints (DeRose and Long 2014). Having adjacent stands that are left untreated provides environments capable of allowing pest insect populations to grow. Once populations have reached epidemic levels, resistant stands become susceptible. Strategically placed area treatments (SPLATs, Finney 2001) are useful in reducing fire severity while only treating ~ 20% of the landscape, however, the efficacy of this practice for insect outbreak remains untested. Resistance is a temporally defined window that changes continually as stands grow and develop after treatment. In the case of spruce bark beetle, the maintenance of resistance at a stand-scale would require multiple treatments to maintain vigor of residual spruce, eventually resulting in structures susceptible to spruce beetle outbreak (Schmid and Frye 1976; DeRose and Long 2014). Therefore, focusing solely on building resistance to a pest may be unproductive. Instead, land managers should focus on a dual approach of targeted treatments in high-risk stands, as well as building resilience through maintaining diversity in both age class structure and species composition across the landscape. In many instances where public and private lands are interspersed, training and shared stewardship programs can help bring private landowners and other stakeholders into the decision-making process alongside silviculturists and other land managers (e.g. Neely et al. 2011).

Building resilience across larger temporal and spatial scales adds complexity to assessing silvicultural strategies at smaller scales. For example, quaking aspen across the western and southwestern United States has experienced large-scale droughts over the past decade. As a result, Sudden Aspen Decline, which is a complex progression of physiological stress, insect infestation and disease, has degraded and caused widespread mortality in many stands (Worrall et al. 2010). This decline is complex because it involves multiple agents of mortality, starting with abiotic stress (drought) creating conditions conducive to attack by mostly secondary pests. Increased numbers of susceptible, stressed host trees have allowed an increase in secondary mortality-causing pest populations, and therefore their increased ability to be a major driver of mortality within these stands. Insects such as bronze poplar borer (*Agrilus liragus*) and aspen bark beetles (e.g. *Trypophloeus populi*) are viewed as contributing agents in this decline complex, where abiotic factors are considered inciting events (Worrall et al. 2010). Therefore, assessing site characteristics that predispose stands to drought may help managers prioritize stands for silvicultural strategies designed to increase the ability to recover following abiotic disturbance. Landscape-scale resilience can be increased by reducing the proportion of drought-susceptible stands in the landscape. Examples of strategies to increase drought resistance and resilience include thinning to increase individual tree vigor, clearfelling the overstory to regenerate the stand, or shifting species composition toward more drought tolerant species.

The rate at which climate change is occurring highlights the challenge in adapting management. Understanding how abiotic conditions can both cause mortality and stress, therefore creating conditions conducive for attack by biotic agents, is an important concept in promoting resilience at a landscape scale. Although trees have the ability to cope with climate stressors (e.g. stomatal regulation, migration to new areas), rapid climatic change and the concomitant alteration of insect pest populations creates uncertainty in tree host species acclimation potential (Rehfeldt 2006). Evolutionary adaptation and migration work on much slower scales in perennial woody species than in annual species. Generation times are slower in forested ecosystems, and therefore large-scale abiotic changes, along with accompanying biotic changes (e.g. native/invasive species ranges, increased reproductive generations) may inhibit their natural abilities to adapt to altered conditions. Concepts like assisted migration (Sensu Aitken et al. 2008) and assisted gene flow (Sensu Aitken and Whitlock 2013) exist to represent this human-aided transition of species to new areas currently outside their range, but require adequate forethought and forecasting to help determine where to move species and how to genetically bolster species in situ.

20.3.1 Adaptive Silviculture for Climate Change

Studies aimed at developing ecologically-based silvicultural treatments for the future in different ecosystems are needed to understand the complex interactions between

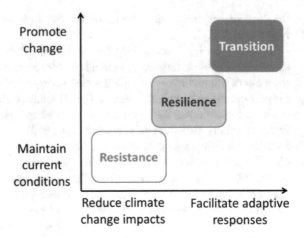

Fig. 20.6 Silvicultural strategies being investigated in the adaptive silviculture for climate change program. From Nagel et al. (2017)

ecological components under rapid climatic change. An ongoing effort in the United States, referred to as the Adaptive Silviculture for Climate Change (ASCC) program, is one such study (Nagel et al. 2017). As a result of the continuing impact of climate change and the primarily unknown effects of interactions between climate change and both native and non-native insect pests (Weed et al. 2013), there is a need to develop silvicultural strategies now that can benefit forests in the future. The overarching goal of the ASCC is to understand different silvicultural strategies focused around three central approaches: Resistance, Resilience and Transition (Nagel et al. 2017). Figure 20.6 details how each of the above categories fits into management, and the level of change associated with each (Nagel et al. 2017).

Given the uncertainty of climate change predictions, as well as the heterogeneous impact of various abiotic and biotic stressors at different locations, ASCC attempts to address how different silvicultural strategies can be used to meet land management goals at varying time scales and across regions and ecotypes. The three approaches represent an increasing scale of change. The resistance approach maintains the "status quo", the resilience approach maintains overstory tree vigor while opening growing space for natural regeneration and the transition approach focuses on shifting composition toward trees considered better suited for an uncertain climatic future. The resistance approach increases the ability of current stands to withstand change, while the resilience and transition approaches attempt to accommodate a moderate-to-large amount of change and a shift away from the current structure and/or species composition. This large-scale research project will yield valuable information for silviculturists attempting to sustain healthy stands and forests under an increasingly uncertain and complex future.

20.4 Synthesis and Conclusion

Use of silviculture to manipulate either vertical and horizontal structure or species composition will also impact the trajectory of stand development and the timing of changes within the stand (stand dynamics) (Oliver and Larson 1996). Silviculture results in a disturbance, and depending upon the number and pattern of trees removed, can effectively shift stands in different directions along a stand development continuum. For example, a dense, even-aged stand under high competition that has the overstory density reduced to below full site occupancy will shift from stem exclusion into understory re-initiation as a new age class develops in the understory. While this transition would occur naturally without silvicultural intervention, with silviculture, a stand can shift overnight from one stage to another, greatly increasing the rate of change and altering the process of stand development.

Silviculturists must be able to predict changes to stand development patterns following treatment. This is most frequently achieved using models (e.g. the Forest Vegetation Simulator; Dixon 2002) and before-after monitoring data. Analysis of before-after data allows the silviculturist to adapt the treatment plan as necessary through time. Silviculturists must also understand and watch for the interactions between silviculture, forest insects and diseases, and other disturbances and provide for appropriate mitigation strategies where necessary.

The approach of managing forest insects through increased resistance and/or resilience can be effectively met using silvicultural strategies. These include strategies developed in conjunction with other management tools in an IPM program. Specific silvicultural prescriptions will vary depending upon stand conditions, site factors, and host tree and pest ecology. However, research and experience indicate that similar results can be expected under specific stand vertical and horizontal structures (Table 20.1) and species composition. From simple to quite complex, using silviculture to manage forest insects can be challenging. Only those (e.g. forest health specialists, forest managers) with advanced training should attempt to resolve forest insect problems in multiaged, mixed species stands without aid from a more experienced silviculturist. Silviculture continues to be an important addition to most forest insect management strategies, and approaching it from a resistance and resilience framework is likely to be successful under rapidly changing environmental and social conditions.

References

Aitken SN, Yeaman S, Holliday JA, Wang T, Curtis-McLane S (2008) Adaptation, migration or extirpation: climate change outcomes for tree populations. Evol Appl 95–111. https://doi.org/10.1111/j.1752-4571.2007.00013.x

Aitken SN, Whitlock MC (2013) Assisted gene flow to facilitate local adaptation to climate change. Annu Rev Ecol Evol Syst 44:367–388. https://doi.org/10.1146/annurev-ecolsys-110512-135747

Alfaro RI, Borden JH, Fraser RG, Yanchuck AD (1995) The white pine weevil in British Columbia: basis for an integrated pest management system. For Chron 7(1):66–73

Anderegg WRL, Hicke JA, Fisher RA, Allen CD, Aukema J, Bentz B, … Zeppel M (2015) Tree mortality from drought, insects, and their interactions in a changing climate. New Phytol 208(3):674–683. https://doi.org/10.1111/nph.13477

Anhold JA, Jenkins MJ, Long JN (1996) Technical commentary: management of lodgepole pine stand density to reduce susceptibility to mountain pine beetle attack. West J Appl for 11(2):50–53. https://doi.org/10.1093/wjaf/11.2.50

Asaro C, Nowak JT, Elledge A (2017) Why have southern pine beetle outbreaks declined in the Southeastern U.S. with the expansion of intensive pine silviculture? a brief review of hypotheses. For Ecol Manage 391:338–348. https://doi.org/10.1016/J.FORECO.2017.01.035

Bauer LS, Duan JJ, Gould JR, Driesche RV (2015) Progress in the classical biological control of *Agrilus planipennis* Fairmaire (Coleoptera:Buprestidae) in North America. Can Entomol 317:300–317. https://doi.org/10.4039/tce.2015.18

Bedding RA, Iede ET (2005) Ch. 21: Application of *Beddingia siricidicola* for sirex woodwasp control. In: Grewal PS, Ehlers RU, Shapiro-Ilan DI (eds) Nematodes as biocontrol agents. CAB International.

Bentz B, Logan J, MacMahon J, Allen CD, Ayres M, Berg E, … Wood D (2009) Bark beetle outbreaks in western North America: causes and consequences. Bark Beetle Symposium; Snowbird, Utah; November, 2005. University of Utah Press, Salt Lake City, UT, p 42.

Bentz BJ, Munson AS (2000) Spruce beetle population suppression in Northern Utah. West J Appl For 15(3):122–128. https://doi.org/10.1093/wjaf/15.3.122

Billings RF, Clarke SR, Mendoza VE, Cabrera PC, Figueroa BM, Campos JR, Baeza G (2004) Bark beetle outbreaks and fire: a devastating combination for Central America's pine forests. Forest Chem Rev 124(6):10–15

Bittner TD, Havill N, Caetano IAL, Hajek AE (2019) Efficacy of Kamona strain *Deladenus siricidicola* nematodes for biological control of *Sirex noctilio* in North America and hybridisation with invasive conspecifics. NeoBiota 44:39–55

Bleiker KP, Lindgren BS, Maclauchlan LE (2005) Resistance of fast- and slow-growing subalpine fir to pheromone- induced attack by Western balsam bark beetle (Coleoptera: Scolytinae). Agric For Entomol 7:237–244

Borden JH, Birmingham AL, Burleigh JS (2006) Evaluation of the push-pull tactic against the mountain pine beetle using verbenone and non-host volatiles in combination with pheromone-baited trees. For Chron 82(4):579–590

Brantley ST, Mayfield AE, Jetton RM, Miniat CF, Zietlow DR, Brown CL, Rhea JR (2017) Elevated light levels reduce hemlock woolly adelgid infestation and improve carbon balance of infested Eastern hemlock seedlings. For Ecol Manage 385:150–160. https://doi.org/10.1016/j.foreco.2016.11.028

Brookes M, Colbert J, Mitchell R, Stark R (eds) (1985) Managing trees and stands susceptible to western spruce budworm. USDA Forest Service Cooperative State Research Service Technical Bulletin No. 1695, Washington, DC

Brookes M, Campbell RW, Colbert JJ, Mitchell RG, Stark RW (eds) (1987) Western spruce budworm. USDA Forest Service Cooperative State Research Service Technical Bulletin No. 1694, Washington, DC

Brookes M, Stark R, Campbell R (eds) (1978) The Douglas-fir tussock moth: a synthesis. USDA Forest Service Science and Education Agency Technical Bulletin 1585, Washington, DC

Bucholz E, Frampton J, Jetton RM, Tilotta D, Lucia L (2017) Effect of different headspace concentrations of bornyl acetate on fecundity of green peach aphid and balsam woolly adelgid. Scand J For Res. https://doi.org/10.1080/02827581.2016.1275769

Carlson CE, Fellin DG, Schmidt WC (1983) The Western spruce budworm in Northern Rocky mountain forests: a review of ecology, insecticidal treatments and silvicultural practices. In: OLaughlin J, Pfister RD (eds) Management of second growth forests, the state of knowledge and research needs. Montana Forest Conservation Experiment Station, University of Montana, Missoula, Missoula, MT, pp 76–103

Cheah C (2011) *Sasjiscymnus* (= *Pseudoscymnus*) *tsugae*, a ladybeetle from Japan. In: Onken B, Reardon R (eds) Implementation and status of biological control of the hemlock woolly adelgid. FHTET-2011–04. U.S. Department of Agriculture, Forest Service, Forest Health Technology Enterprise Team: 43–52. Chapter 4, Morgantown, WV

Cook SM, Khan ZR, Pickett JA (2007) The use of push-pull strategies in integrated pest management. Ann Rev Entomol 52:375–400. https://doi.org/10.1146/annurev.ento.52.110405.091407

Coops NC, Waring RH, Wulder MA, White JC (2009) Remote sensing of environment prediction and assessment of bark beetle-induced mortality of lodgepole pine using estimates of stand vigor derived from remotely sensed data. Remote Sens Environ 113(5):1058–1066. https://doi.org/10.1016/j.rse.2009.01.013

D'Amato AW, Bradford JB, Fraver S, Palik BJ (2013) Effects of thinning on drought vulnerability and climate response in North temperate forest ecosystems. Ecol Appl 23(8):1735–1742

De Groot P, Hopkin AA, Sajan RJ (2005) Silvicultural techniques and guidelines for the management of major insects and diseases of spruce, pine and aspen in Eastern Canada. Natural Resources Canada, Canadian Forest Service, Great Lakes Forestry Centre, Sault Ste. Marie, Ont, p 65

DeRose RJ, Long JN (2007) Disturbance, structure, and composition: spruce beetle and engelmann spruce forests on the Markagunt Plateau, Utah. For Ecol Manag 244(1–3):16–23. https://doi.org/10.1016/j.foreco.2007.03.065

DeRose RJ, Long JN (2014) Resistance and resilience: a conceptual framework for silviculture. Forest Sci 60(6):1205–1212. https://doi.org/10.5849/forsci.13-507

Dixon GE, C. (2002). Essential FVS: A user's guide to the forest vegetation simulator. U.S. Department of Agriculture Forest Service, Forest Management Center, Internal Rep. Fort Collins, CO

Dodds KJ, Cooke RR, Gilmore DW (2007) Silvicultural options to reduce pine susceptibility to attack by a newly detected invasive species, *Sirex noctilio*. North J Appl For 24(3):165–167

Dodds K, Cooke R, Hanavan R (2014) The effects of silvicultural treatment on sirex noctilio attacks and tree health in Northeastern United States. Forests 5(11):2810–2824. https://doi.org/10.3390/f5112810

Duan JJ, Bauer LS, Van Driesche RG (2017) Emerald ash borer biocontrol in ash saplings: The potential for early stage recovery of North American ash trees. For Ecol Manage 394:64–72. https://doi.org/10.1016/j.foreco.2017.03.024

Elek J, Wardlaw T (2013) Options for managing Chrysomelid leaf beetles in Australian eucalypt plantations: reducing the chemical footprint. Agric For Entomol 15:351–365. https://doi.org/10.1111/afe.12021

Elkinton JS, Healy WM, Buonaccorsi JP, Boettner GH, Hazzard AM, Smith HR (1996) Interactions among gypsy moths, white-footed mice, and acorns. Ecology 77(8):2332–2342

Fajvan MA, Gottschalk KW (2012) The effects of silvicultural thinning and *Lymantria dispar* L. defoliation on wood volume growth of *Quercus* spp. American J Plant Sci 03(02):276–282. https://doi.org/10.4236/ajps.2012.32033

Fettig C, Gibson K, Munson A, Negrón J (2014) A comment on "management for mountain pine beetle outbreak suppression: does relevant science support current policy?" Forests 5(4):822–826. https://doi.org/10.3390/f5040822

Fettig CJ, Hilszczanski J (2015) Management strategies for bark beetles in conifer forests. In: Vega F, Hofstetter R (eds) Bark beetles: biology and ecology of native and invasive species. Academic Press

Fettig CJ, Klepzig KD, Billings RF, Munson AS, Nebeker TE, Negrón JF, Nowak JT (2007) The effectiveness of vegetation management practices for prevention and control of bark beetle infestations in coniferous forests of the Western and Southern United States. For Ecol Manag 238(1–3):24–53. https://doi.org/10.1016/J.FORECO.2006.10.011

Fettig CJ, McKelvey SR, Borys RR, Dabney P, Hamud SM, Nelson LJ, Seybold SJ (2009) Efficacy of verbenone for protecting ponderosa pine stands from western pine beetle (Coleoptera: Curculionidae: Scolytinae) attack in California. J Econ Entomol 102(5):1

Fettig CJ, Munson AS (2020) Efficacy of verbenone and a blend of verbenone and nonhost volatiles for protecting lodgepole pine from mountain pine beetle (Coleoptera: Curculionidae). Agric For Entomol. https://doi.org/10.1111/afe.12392

Finney MA (2001) Design of regular landscape fuel treatment patterns for modifying fire growth and behavior. Forest Sci 47(2):219–228. https://doi.org/10.1093/forestscience/47.2.219

Fischer MJ, Waring KM, Hofstetter RW, Kolb TE (2010) Ponderosa pine characteristics associated with attack by the roundheaded pine beetle. Forest Sci 56(5):473–483

Fuentealba A, Bauce E (2011) Site factors and management influence short-term host resistance to spruce budworm, Choristoneura fumiferana (Clem.), in a species-specific manner. Pest Manag Sci, 245–253. https://doi.org/10.1002/ps.2253

Gaylord ML, Kolb TE, Pockman WT, Plaut JA, Yepez EA, Macalady AK, Pangle RE, McDowell NG (2013) Drought predisposes piñon–juniper woodlands to insect attacks and mortality. New Phytol 198:567–578. https://doi.org/10.1111/nph.12174

Gillette NE, Kegley SJ, Costello SL, Mori SR, Webster JN, Mehmel CJ, Wood DL (2014) Efficacy of verbenone and green leaf volatiles for protecting whitebark and limber pines from attack by mountain pine beetle (Coleoptera: Curculionidae: Scolytinae). Environ Entomol 43(4):1019–1026. https://doi.org/10.1603/EN12330

Gillette NE, Mehmel CJ, Mori SR, Webster JN, Wood DL, Erbilgin N, Owen DR (2012) The push—pull tactic for mitigation of mountain pine beetle (Coleoptera: Curculionidae) damage in lodgepole and whitebark pines. Environ Entomol 41(6):1575–1586

Gilmore DW, Palik BJ (2006) A revised managers handbook for red pine in the North Central Region. U.S. Department of Agriculture, Forest Service, North Central Research Station, Saint Paul, MN. General Technical Report NC-264, 1–55

Gottschalk KW (1993) Silvicultural guidelines for forest stands threatened by the gypsy moth. US Department of Agriculture, Forest Service, Northeastern Forest Experiment Station. General Technical Report NE-171

Hadfield JS, Flanagan PT (2000) Dwarf mistletoe pruning may induce Douglas-fir beetle attacks. West J Appl for 15(1):34–36. https://doi.org/10.1093/wjaf/15.1.34

Hansen EM, Negron JF, Munson AS, Anhold JA (2010) A retrospective assessment of partial cutting to reduce spruce beetle-caused mortality in the Southern Rocky Mountains. West J Appl for 25(2):81–87

Haugen D (1990) Control procedures for Sirex noctilio in the green triangle: review from detection to severe outbreak. Aust For 53:24–32

Helms J (ed) (1998) The dictionary of forestry. The Society of American Foresters, Bethesda, MD

Hoebeke ER, Haugen DA, Haack RA (2005) Sirex noctilio: discovery of a Palearctic wood wasp in New York. Newsletter Michigan Entomol Soc 50(1 and 2):24–25

Holusa J, Hlasny T, Modlinger R, Lukásova K, Kula E (2017) Felled trap trees as the traditional method for bark beetle control: can the trapping performance be increased? For Ecol Manage 404:165–173. https://doi.org/10.1016/j.foreco.2017.08.019

Hoy M.A. (2004a). Augmentative biological control. In: Encyclopedia of entomology. Springer, Dordrecht

Hoy M.A. (2004b). Classical biological control. In: Encyclopedia of entomology. Springer, Dordrecht

Inward DJG, Wainhouse D, Peace A (2012) The effect of temperature on the development and life cycle regulation of the pine weevil Hylobius abietis and the potential impacts of climate change. Agric For Entomol 348–357. https://doi.org/10.1111/j.1461-9563.2012.00575

Jetton RM, Monahan JF, Hain FP (2011) Laboratory studies of feeding and oviposition preference, developmental performance, and survival of the predatory beetle, Sasajiscymnus tsugae on diets of the woolly adelgids, Adelges tsugae and Adelges piceae. J Insect Sci 11(1):14. https://doi.org/10.1673/031.011.6801

Kane JM, Kolb TE (2010) Importance of resin ducts in reducing ponderosa pine mortality from bark beetle attack. Oecologia 601–609. https://doi.org/10.1007/s00442-010-1683-4

Kane JM, Varner JM, Metz MR, van Mantgem PJ (2017) Characterizing interactions between fire and other disturbances and their impacts on tree mortality in Western U.S forests. For Ecol Manage 405:188–199. https://doi.org/10.1016/J.FORECO.2017.09.037

Kashian DM (2016) Sprouting and seed production may promote persistence of green ash in the presence of the emerald ash borer. Ecosphere 7:1–15

Kashian DM, Jackson RM, Lyons HD (2011) Forest structure altered by mountain pine beetle outbreaks affects subsequent attack in a Wyoming lodgepole pine forest, USA. Can J for Res 41(12):2403–2412. https://doi.org/10.1139/x11-142

Kegley SJ, Livingston RL, Gibson KE (1997) Pine engraver, Ips pini (Say), in the Western United States. U.S. Department of Agriculture Forest Service, Forest Insect and Disease Leaflet 122

Kenis M, Hurley BP, Hajek AE, Cock MJW (2017) Classical biological control of insect pests of trees: facts and figures. Biol Invas 19(11):3401–3417. https://doi.org/10.1007/s10530-017-1414-4

Kinloch BB, Stonecypher RW (1969) Genetic variation in susceptibility to fusiform rust in seedlings from a wild population of loblolly pine. Phytopathology 59:1246–1255

Klapwijk MJ, Bylund H, Schroeder M, Bjorkman C (2016) Forest management and natural biocontrol of insect pests. Forestry 89:253–262. https://doi.org/10.1093/forestry/cpw019

Klutsch JG, Shamoun SF, Erbilgin N (2017) Drought stress leads to systemic induced susceptibility to a nectrotrophic fungus associated with mountain pine beetle in *Pinus banksiana* seedlings. PLoS ONE 12(12):e0189203. https://doi.org/10.1371/journal.pone.0189203

Kolb TE, Fettig CJ, Ayres MP, Bentz BJ, Hicke JA, Mathiasen R, … Weed AS (2016) Observed and anticipated impacts of drought on forest insects and diseases in the United States. For Ecol Manage 380:321–334. https://doi.org/10.1016/J.FORECO.2016.04.051

Kollenberg CL, O'Hara KL (1999) Leaf area and tree increment dynamics of even-aged and multiaged lodgepole pine stands in Montana. Can J For Res 29:687–695

Larsson S (2002) Chapter 1: Resistance in trees to insects—an overview of mechanisms and interactions. In: Wagner MR, Clancy KM, Lieutier F, Paine TD (eds) Mechanisms and deployment of resistance in trees to insects. Kluwer Academic Publishers, Norwell, MA, pp 1–31

Lombarkia N, Derridj S (2008) Resistance of apple trees to *Cydia pomonella* egg-laying due to leaf surface metabolites. Entomol Exp Appl 128:57–65. https://doi.org/10.1111/j.1570-7458.2008.00741.x

Long JN, Shaw JD (2005) A density management diagram for even-aged ponderosa pine stands. West J Appl For 20(4):205–215

Loomis WE (1932) Growth-differentiation balance vs carbohydrate-nitrogen ratio. Proc American Horticult Soc 29:240–245

Lorio PL (1986) Growth-differentiation balance: A basis for understanding Southern pine beetle-tree interactions. For Ecol Manage 14(4):259–273. https://doi.org/10.1016/0378-1127(86)90172-6

Madritch MD, Lindroth RL (2015) Condensed tannins increase nitrogen recovery by trees following insect defoliation. New Phytol 208:410–420

Massad TJ, Dyer LA (2010) A meta-analysis of the effects of global environmental change on plant-herbivore interactions. Arthropod-Plant Interac 4:181–188. https://doi.org/10.1007/s11829-010-9102-7

McDowell NG, Adams HD, Bailey JD, Kolb TE (2007) The role of stand density on growth efficiency, leaf area index, and resin flow in Southwestern ponderosa pine forests. Can J For Res 37(2):343–355. https://doi.org/10.1139/X06-233

McGregor MD, Amman GD, Schmitz RF, Oakes RD (1987) Partial cutting lodgepole pine stands to reduce losses to the mountain pine beetle. Can J for Res 17(10):1234–1239. https://doi.org/10.1139/x87-191

McHugh CW, Kolb TE, Wilson JL (2003) Bark beetle attacks on ponderosa pine following fire in Northern Arizona. Environ Entomol 32(3):510–522. https://doi.org/10.1603/0046-225X-32. 3.510

McIntire T (1988) Forest health through silviculture and integrated pest management: a strategic plan. U.S. Department of Agriculture, Forest Service

McKeand SE, Amerson HV, Li B, Mullin TJ (2003) Families of loblolly pine that are the most stable for resistance to fusiform rust are the least predictable. Can J For Res 33:1335–1339

Mezei P, Grodzki W, Bazenec M, Škvarenina J, Brandysova V, Jakus R (2014) Host and site factors affecting tree mortality caused by the spruce bark beetle (*Ips typographus*) in mountainous conditions. For Ecol Manage 331:196–207. https://doi.org/10.1016/j.foreco.2014.07.031

Miller JM, Keen FP (1960) Biology and control of the western pine beetle. USDA Forest Service Misc. Pub. 800

Mitchell R, Waring RH, Pitman G (1983) Thinning lodgepole pine increases tree vigor and resistance to mountain pine beetle. For Sci 29(1):204–211

Morris JL, Cottrell S, Fettig CJ, Hansen WD, Sherriff L, Carter VA, … Sepp HT (2017) Managing bark beetle impacts on ecosystems and society: priority questions to motivate future research. J Appl Ecol 54:750–760. https://doi.org/10.1111/1365-2664.12782

Mundt CC (2014) Durable resistance: a key to sustainable management of pathogens and pests. Infect Genet Evol 27:446–455. https://doi.org/10.1016/j.meegid.2014.01.011

Muzika RM, Liebhold AM (2000) A critique of silvicultural approaches to managing defoliating insects in North America. Agric For Entomol 2:97–105. https://doi.org/10.1046/j.1461-9563. 2000.00063.x

Nagel LM, Palik BJ, Battaglia MA, D'Amato AW, Guldin JM, Swanston CW, … Peterson DL (2017) Adaptive silviculture for climate change: a national experiment in manager-scientist partnerships to apply an adaptation framework. J Forest 115(3):167–178

Naidoo R, Lechowicz MJ (2001) Effects of gypsy moth on radial growth of deciduous trees. For Sci 47(3):338–348. https://doi.org/10.1093/forestscience/47.3.338

Neely B, Rondeau R, Sanderson J, Pague C, Kuhn B, Siemers J, Grunau L, Robertson J, McCarthy P, Barsugli J, Schulz T, Knapp C (2011) Gunnison basin: climate change vulnerability assessment for the Gunnison Climate Working Group, The Nature Conservancy, Colorado Natural Heritage Program, Western Water Assessment, University of Colorado, Boulder, and University of Alaska, Fairbanks. Project of the Southwest Climate Change Initiative, Boulder, CO, USA

Nordlander G, Hellqvist C, Johansson K, Nordenhem H (2011) Regeneration of European boreal forests: effectiveness of measures against seedling mortality caused by the pine weevil *Hylobius abietis*. For Ecol Manage 262:2354–2363

Nowak J, Asaro C, Klepzig K, Billings R (2008) The southern pine beetle prevention initiative: working for healthier forests. J Forest 106(5):261–267

Nowak JT, Meeker JR, Coyle DR, Steiner CA, Brownie C (2015) Southern pine beetle infestations in relation to forest stand conditions, previous thinning, and prescribed burning: evaluation of the Southern pine beetle prevention program. J Forest 113(5):454–462. https://doi.org/10.5849/jof.15-002

NPS Environmental Assessment (2007) Hemlock woolly adelgid control strategies along the Blue Ridge Parkway. US National Park Service: Blue Ridge Parkway

O'Hara KL (2014) Multiaged silviculture: managing for complex forest stand structure. Oxford University Press, Oxford, United Kingdom

Oliver CD, Larson BC (1996) Forest stand dynamics, Update. John Wiley and Sons Inc, New York

Ostry ME, Laflamme G, Katovich SA (2010) Silvicultural approaches for management of Eastern white pine to minimize impacts of damaging agents. For Pathol 40(3–4):332–346. https://doi.org/10.1111/j.1439-0329.2010.00661.x

Painter RH (1958) Resistance of plants to insects. Ann Rev Entomol 3:267–290

Perez-Alvarez R, Nault BA, Poveda K (2019) Effectiveness of augmentative biological control depends on landscape context. Sci Rep 9(1):1–15. https://doi.org/10.1038/s41598-019-45041-1

Pitt D, Hoepting M, Parker W, Morneault A, Lanteigne L, Stinson A, ... Farrell JCG (2016) Optimum vegetation conditions for successful establishment of planted Eastern white pine (*Pinus strobus* L.). Forests 7(12):175. https://doi.org/10.3390/f7080175

Polinko A (2014) Stand response to western spruce budworm defoliation and mortality in New Mexico. Master of Science Thesis, School of Forestry, Northern Arizona University

Progar RA, Gillette N, Fettig CJ, Hrinkevich K (2014) Applied chemical ecology of the mountain pine beetle. For Sci 60(3):414–433. https://doi.org/10.5849/forsci.13-010

Rapp M (2017) Effects of western spruce budworm in mixed conifer forests of New Mexico. Master of Science Thesis, School of Forestry, Northern Arizona University

Rehfeldt GE (2006) A spline model of climate for the Western United States. USDA Forest Service General Technical Report RMRS-GTR-165, Rocky Mountain Research Station, Fort Collins, Colorado, USA

Reineke LH (1933) Perfecting a stand-density index for even-aged forests. J Agric Res 46:627–638

Reynolds KM, Holsten EH (1994) Relative importance of risk factors for spruce beetle outbreaks. Can J For Res 24(10):2089–2095. https://doi.org/10.1139/x94-268

Roberds JH, Strom BL, Hain FP, Gwaze DP, Mckeand SE, Lott LH (2003) Estimates of genetic parameters for oleoresin and growth traits in juvenile loblolly pine. Can J For Res 33:2469–2476. https://doi.org/10.1139/X03-186

Russell M (2015) A meta-analysis of physiological and behavioral responses of parasitoid wasps to flowers of individual plant species. Biol Control 82:96–103. https://doi.org/10.1016/j.biocontrol.2014.11.014

Ryan MG, Sapes G, Sala A, Hood SM (2015) Tree physiology and bark beetles. New Phytol 205:955–957. https://doi.org/10.1111/nph.13256

Savage M (1994) Anthropogenic and natural disturbance and patterns of mortality in a mixed conifer forest in California. Can J For Res 24(6):1149–1159. https://doi.org/10.1139/x94-152

Schmid JM, Frye RH (1976) Stand ratings for spruce beetles. Research Note RM-309, US Department of Agriculture, Forest Service, Rocky Mountain Forest and Range Experiment Station

Schmid JM, Frye RH (1977) Spruce beetle in the rockies. USDA Forest Service General Technical Report RM-49, Rocky Mountain Forest and Range Experiment Station

Schmid JM, Mata SA (2005) Mountain pine beetle-caused tree mortality in partially cut plots surrounded by unmanaged stands. Research Paper RMRS-RP-54. US Department of Agriculture, Forest Service, Rocky Mountain Research Station. https://doi.org/10.2737/RMRS-RP-54

Schoettle AW, Sniezko RA (2007) Proactive intervention to sustain high-elevation pine ecosystems threatened by white pine blister rust. J for Res 12:327–336. https://doi.org/10.1007/s10310-007-0024-x

Sidebottom J (2009) Balsam woolly adelgid: Christmas tree notes. North Carolina State University Extension Publications

Six DL, Vergobbi C, Cutter M (2018) Are survivors different? genetic-based selection of trees by mountain pine beetle during a climate change-driven outbreak in a high-elevation pine forest. Front Plant Sci 9:993

Sohn JA, Saha S, Bauhus J (2016) Potential of forest thinning to mitigate drought stress: a meta-analysis. For Ecol Manage 380:261–273. https://doi.org/10.1016/J.FORECO.2016.07.046

Stiell WM, Berry AB (1985) Limiting white pine weevil attacks by side shade. For Chron 61(1):5–9. https://doi.org/10.5558/tfc61005-1

Stone C (2001) Reducing the impact of insect herbivory in eucalypt plantations through management of extrinsic influences on tree vigour. Austral Ecol 482–488

Strauss SY, Agrawal AA (1999) The ecology and evolution of plant tolerance to herbivory. Trends Ecol Evol 14(5):179–185. https://doi.org/10.1016/S0169-5347(98)01576-6

Strom BL, Goyer RA, Ingram LL Jr, Boyd GDL, Lott LH (2002) Oleoresin characteristics of progeny of loblolly pines that escaped attack by the Southern pine beetle. For Ecol Manage 158(1–3):169–178

Thomas Z, Waring KM (2015) Enhancing resiliency and restoring ecological attributes in second-growth ponderosa pine stands in Northern New Mexico, USA. Forest Sci 61(1). https://doi.org/10.5849/forsci.13-085

Vane E, Waring KM, Polinko A (2017) The influence of Western spruce budworm on fire in spruce-fir forests. Fire Ecol 13(1):16–33. https://doi.org/10.4996/fireecology.1301016

Walker B, Holling CS, Carpenter SR, Kinzig A (2004) Resilience, adaptability and transformability in social–ecological systems. Ecol Soc 9(2):5. https://doi.org/10.2307/26267673

Waring RH, Pitman GB (1980) A simple model of host resistance to bark beetles. Oregon State University Forest Research Laboratory Research Note 65

Warzée N, Gilbert M, Grégoire J (2006) Predator/prey ratios: a measure of bark-beetle population status influenced by stand composition in different French stands after the 1999 storms. Ann For Sci 63:301–308. https://doi.org/10.1051/forest

Weed AS, Ayres MP, Hicke JA (2013) Consequences of climate change for biotic disturbances in North American forests. Ecol Monogr 83(4):441–470. https://doi.org/10.1890/13-0160.1

Westbrook JW, Resende MF Jr, Munoz P, Walker AR, Wegrzyn JL, Nelson CD, … Peter GF (2013) Association genetics of oleoresin flow in loblolly pine: discovering genes and predicting phenotype for improved resistance to bark beetles and bioenergy potential. New Phytol 199(1):89–100

Williamson M (1996) Biological invasions. Springer, Netherlands

Windmuller-Campione MA, Long JN (2015) If long-term resistance to a spruce beetle epidemic is futile, can silvicultural treatments increase resilience in spruce-fir forests in the Central Rocky Mountains? Forests 6:1157–1178. https://doi.org/10.3390/f6041157

Worrall JJ, Marchetti SB, Egeland L, Mask RA, Eager T, Howell B (2010) Forest ecology and management effects and etiology of sudden aspen decline in Southwestern Colorado, USA. For Ecol Manage 260(5):638–648. https://doi.org/10.1016/j.foreco.2010.05.020

Wulder MA, Dymond CC, White JC, Leckie DG, Carroll AL (2005) Surveying mountain pine beetle damage of forests: a review of remote sensing opportunities. For Ecol Manage 221:27–41

Zas R, Björklund N, Sampedro L, Hellqvist C, Karlsson B, Jansson S, Nordlander G (2017) Genetic variation in resistance of Norway spruce seedlings to damage by the pine weevil *Hylobius abietis*. Tree Genet Genomes 13:12. https://doi.org/10.1007/s11295-017-1193-1

Chapter 21
Forest Health in the Anthropocene

Allan L. Carroll

21.1 Introduction

Forests cover approximately one third of Earth's terrestrial surface (FAO and UNEP 2020). They provide a wide range of vital environmental and socioeconomic benefits to all people in the form of ecosystem services. These services include fibre, fuel, non-timber forest products, biodiversity, carbon sequestration, soil and water protection and socio-cultural values (Shvidenko et al. 2005; Brandt et al. 2013; Sing et al. 2017). As the global population rises, the demand for ecosystem services has increased while the capacity of forests to deliver them has declined due to high rates of deforestation (Carpenter et al. 2009; Seidl et al.2016; FAO and UNEP 2020), and increased rates of disturbance (Johnstone et al. 2016; Seidl et al. 2017). The capacity to quantify the health of forests and assess their ability to sustain ecosystem services into the future has become a fundamental challenge to resource managers in a rapidly changing world.

All forests are adapted in some way to disturbance events that alter ecosystem processes [(White and Pickett 1985; Turner 2010;) see Box 21.1 for definitions]. Following disturbance, forest ecosystems will either regenerate or reorganize. If an ecosystem is resistant to disturbance and returns to a similar pre-disturbance state, it is considered resilient (Holling 1973; Gunderson 2000; Folke et al. 2004; Scheffer 2009). If instead the disturbed ecosystem is sufficiently changed that it regenerates to a different state (e.g. a forest becomes a grassland; Fig. 21.1), then it has undergone a regime shift (Folke et al. 2004; Scheffer 2009; Allen et al. 2016; Johnstone et al. 2016). Relationships between forms of disturbance and the probability of a regime shift are highly non-linear and characterized by thresholds where a relatively small

A. L. Carroll (✉)
Department of Forest & Conservation Sciences, University of British Columbia, Vancouver, BC, Canada
e-mail: allan.carroll@ubc.ca

© The Author(s) 2023
J. D. Allison et al. (eds.), *Forest Entomology and Pathology*,
https://doi.org/10.1007/978-3-031-11553-0_21

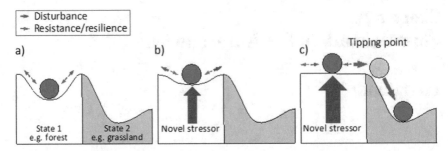

Fig. 21.1 Conventional cup-and-ball model of ecosystem resilience (Lamothe et al. 2019). The ball represents the current ecosystem, the valleys indicate the possible ecosystem states [e.g. forests (white region) and grasslands (grey region)] and the weight of the arrows indicates the relative strength of interactions. **a)** Forests are resistant and resilient to disturbance where ecosystem processes (blue arrows) maintain them in or return them to their original state following pertur-bation (pink arrows). **b)** Forests are less resistant and resilient to disturbance due to alteration of ecosystem processes by a novel stressor (red arrow) such as climate change and are therefore less likely to return to their original state following perturbation. **c)** Forests have lost resistance and resilience due to a novel stressor and disturbance has perturbed them beyond their original state to a tipping point where they undergo a regime shift and rapidly reorganize into a new ecosystem

change may lead to a large shift in the state of an ecosystem (Scheffer et al. 2001)—a process known as a tipping point (Brook et al. 2013; Reyer et al. 2015). Over large spatial scales and long time spans, and without significant human intervention, disturbances tend to recur within a natural range of variability (Landres et al.1999). At these scales the characteristics of disturbances together with their return inter-vals make up a disturbance regime (Turner 2010). Whereas disturbance instigates processes of ecosystem renewal (White and Pickett 1985; Thom et al. 2016), distur-bance regimes generate diverse landscapes (Turner 2010; Turner and Gardner 2015; Thom and Seidl 2016).

Box 21.1 Terms and definitions associated with forest ecosystem health

Term	Definition
Disturbance	Any relatively discrete event that disrupts the structure of an ecosystem, community, or population, and changes resource availability or the physical environment (White and Pickett 1985)
Natural range of variability	The ecological conditions, and the spatial and temporal variation in these conditions, that are relatively unaffected by people, within a period of time and geographical area appropriate to an expressed goal (Landres et al. 1999)

(continued)

(continued)

Term	Definition
Disturbance regime	The spatial and temporal dynamics of disturbances that include spatial distribution, frequency, return interval, rotation period, size, intensity, and severity (Turner 2010)
Resistance	The influence of structure and composition on the severity of disturbance (DeRose and Long 2014)
Resilience	The ability of an ecosystem to absorb disturbances and re-organize under change to maintain similar functioning and structure (Scheffer 2009)
Tipping point	A threshold at which a small change in conditions leads to a strong change in the state of a system (Brook et al. 2013)
Regime shift	A rapid modification of ecosystem organization and dynamics with prolonged consequences (Scheffer and Carpenter 2003)

Sustainable extraction of services from forests is contingent upon ecosystems that are resistant and resilient to disturbance (Seidl et al. 2016; Grimm et al. 2016). However, forests around the world are increasingly forced to contend with anthropogenic stressors that influence disturbances both directly via fragmentation, pollution and introduced alien invasive species (Vilà et al. 2010; Paoletti et al. 2010; FAO and UNEP 2020) and indirectly through climate change-mediated alterations to ecosystem processes (Raffa et al. 2009; Seidl et al. 2017; Williams et al. 2019). These novel stressors may reduce the resiliency of forest ecosystems (Fig. 21.1b), increase the probability of abrupt tipping points and regime shifts (Fig. 21.1c), and ultimately threaten the sustainability of ecosystem services. Quantification of the resilience of forest ecosystems and detection of critical changes in condition that may compromise ecosystem service sustainability grows more essential with ongoing global change. In this chapter I will review the concept of forest health, its utility as an indicator of forest ecosystem resistance and resilience to disturbance, and its relevance in an era of extensive global change known as the Anthropocene.

21.2 A Working Definition of Forest Health

The concept of "forest health" as an indicator of ecosystem sustainability is widely accepted; however, its broad adoption has been associated with applications that do not correspond with the term's intent to describe the health of forest ecosystems (Raffa et al. 2009). Thus, a clear and concise definition of forest health is required before it is possible to fully consider its utility and relevance in a changing world. Edmonds et al. (2011) provide a list of eight definitions of forest health. Several refer to management objectives and human needs, and are considered "utilitarian" (Kolb et al. 1994; Edmonds et al. 2011; Trumbore et al. 2015), while the remainder are based on aspects of ecosystem function and processes. Utilitarian concepts of health are appropriate in agriculture or agroforestry systems that have well-defined management objectives such as the plantation shown in Fig. 21.2a established for the production of fibre. These systems provide valuable services, but they are limited in most aspects of ecological function and are unlikely to be very resistant or resilient to disturbance. Moreover, allowing such systems to behave naturally, for example permitting the growth of competing vegetation, would likely lead to their failure because their goals are to provide socioeconomic benefits often at the expense of ecological processes (Raffa et al. 2009). Based on a utilitarian definition of forest health, the success or failure of a plantation to meet the objective of fibre production would cause it to be deemed a healthy or unhealthy forest, respectively, regardless of ecological condition.

The pitfalls of utilitarian definitions of forest health become more obvious when applied to natural forests. If a disturbance like the native bark beetle outbreak in Fig. 21.2b were to occur in a working forest, the beetle would be considered a pest and the forest unhealthy; however, if the forest was part of a park or protected area,

Fig. 21.2 a) A red pine (Pinus resinosa) plantation in central Wisconsin established to produce fibre (*Source* Steven Katovich, Bugwood.org). **b)** A lodgepole pine (*P. contorta* var. *latifolia*) forest in southern British Columbia affected by an outbreak of the mountain pine beetle (*Dendroctonus ponderosae*)

then the beetle would be considered a natural disturbance agent and part of the normal healthy functioning of such an ecosystem (Raffa et al. 2009). Layering of human expectations onto natural forest ecosystems leads to conflicts that preclude the general use of the term "forest health" as an indicator of forest vitality. Processes that make up a functioning forest ecosystem do so independent of human expectations. They include not only the inherent biological, geochemical and physical elements that form the basis of the ecosystem, but also natural disturbances such as windstorms, insect and disease outbreaks and wildfire that arise from interactions among them. All of these processes are essential to resilient ecosystems (Folke et al. 2004; Turner 2010; Johnstone et al. 2016) and should therefore be the basis of a healthy forest. Indeed, several recent studies have emphasized that processes associated with ecosystem resilience must be emphasized when considering forest health, and that health should be measured against ecosystem responses to external drivers and perturbations arising from global change (Raffa et al. 2009; Millar and Stephenson 2015; Trumbore et al. 2015).

Based on the preceding argument, I propose the following definition of forest health that is free from human values and expectations:

> Forests are healthy when their underlying ecological processes operate within a natural range of variability so that on any temporal or spatial scale they are resistant and resilient to disturbance.

It is important to note that this definition is not intended to imply that management of forests toward objectives associated with human values should be abandoned in favour of natural ecological processes. Indeed, careful management of both natural and planted forests can deliver products and services while maintaining ecosystem function (Brandt et al. 2013; Gauthier et al. 2015; Trumbore et al. 2015; Wingfield et al. 2015; Pohjanmies et al. 2017). Instead, restricting the definition of forest health to ecosystem processes allows assessments of the potential of forests (natural, planted or combinations) to remain resilient and provide services in an era of global change.

21.3 Forest Health:From Stands to Landscapes

Since forest health has been defined in terms of resistance and resilience to disturbance, the processes of disturbance and how they interact with ecosystems must be considered in detail. Forest disturbances comprise discrete events that can be manmade (e.g. harvesting or land clearing) or natural. Natural disturbances are either biotic, such as insect or pathogen outbreaks, or abiotic such as wildfires, windstorms, floods, avalanches and volcanic eruptions. By definition, disturbances can operate at spatial scales ranging from individual trees to entire landscapes. However, from the perspective of forest health, a stand[1] is the finest scale at which disturbance will

---

[1] Defined as an area of forest or woodland whose structure or composition is different from adjacent areas (Lindenmayer and Franklin 2002).

be considered because the stand (i) is the fundamental unit of forest management programs, and (ii) it captures key processes associated with ecosystem resilience (McElhinny et al. 2005). The broadest scale of consideration will be the forest landscape which is simply defined as multiple sets of stands that cover an area ranging from hundreds to tens of thousands of hectares (Lindenmayer and Franklin 2002).

21.3.1 Health of Forest Stands

At the scale of a forest stand, the outcome of a disturbance event, and the potential for an ecosystem to either regenerate or reorganize, is a result of complex interactions among disturbance type, severity, structure and composition, and topography of the stand in question (White and Jentsch 2001). Abiotic disturbances such as fire can cause the direct mortality of the majority of plants and animals in an individual stand depending on its severity (Turner et al. 1998). If a fire results in destruction of propagules from the original stand (e.g. a seed bank), then the reduced likelihood of regeneration to an equivalent pre-disturbance state means that the stand was neither resistant to disturbance nor resilient, and therefore unhealthy prior to being disturbed. In contrast, biotic disturbance by an insect defoliator may not directly cause the mortality of any component of a stand, but simply alter the competitive advantage of dominant trees within the overstory leading to a change in canopy composition (Cooke et al. 2007). In this case the stand was largely resistant and resilient to the disturbance, and therefore healthy. Between these extremes, disturbance by both abiotic and biotic agents can be less or more severe, respectively. The severity continuum is further influenced by stand structure and composition. A young stand, or one with a low density of trees, may comprise insufficient fuels to support a high-severity fire (Turner et al. 1994) allowing the stand to regenerate and remain resilient. Similarly, stands without suitable and susceptible host-tree species would be completely resistant to an outbreak of a specialist pathogen or insect disturbance agent (Jactel et al. 2017). Lastly, topographical features of a stand, such as slope and aspect, may influence the severity of both abiotic and biotic disturbances (White and Jentsch 2001) thereby affecting the health of a given stand.

The resilience and health of stands is also potentially influenced by biological legacies that persist through the disturbance event such as surviving trees, seedbanks and/or other below ground organs (Seidl et al. 2014; Johnstone et al. 2016). Given that forest ecosystems have evolved with disturbance, species within them may also display long-term biological legacies in the form of adaptive traits that improve their resistance and/or resilience (Keeley et al. 2011). For example, cone serotiny (the release of seeds in response to an environmental trigger) in some *Pinus* species facilitates the dissemination of seeds immediately following a stand-replacing fire, thus ensuring regeneration of a similar pre-disturbance ecosystem (Turner et al. 1998). Alternatively, many tree species resist disturbance by insect herbivores through adaptations that allow them tolerate tissue loss such as increased photosynthetic and growth rates, and reallocation of stored resources (Strauss and Agrawal 1999). Similarly, following high-severity fires *Eucalyptus* species resprout epicormically from suppressed, dormant buds along their boles and replace stand canopies within a year of disturbance (Keeley et al. 2011).

21.3.2 Health of Forest Landscapes

While the same disturbances that affect stands will affect landscapes, their relevance to resiliency and forest health may change as spatial and temporal scales increase. For example, disturbance that results in the local destruction of propagules, as with our example of fire above, may lead to the conclusion that a stand was unhealthy prior to fire. But if the stand is situated among other stands (i.e. in a landscape) capable of dispersing seeds into the disturbed area, then regeneration is possible and resiliency is likely. In contrast, local eruption of an aggressive bark beetle population may cause the mortality of a relatively small proportion of mature trees in a mixed species stand, leaving it largely intact. But if surrounding stands contain susceptible host trees the eruption may propagate over the landscape causing extensive tree mortality and threatening ecological processes such as biodiversity and carbon sequestration (Kurz et al. 2008; Raffa et al. 2008).

Disturbance creates gaps in vegetation and alters available light and nutrients, initiating secondary succession within the openings (White and Pickett 1985; White and Jentsch 2001; Turner and Gardner 2015). Variation in these processes will, over time, produce a mosaic of stands across a landscape in different states of regeneration or reorganization (Fig. 21.3). Although the impacts of disturbance may be scale dependent, some forms of disturbance to stands such as that caused by fire or insects can have long-term, persistent impacts on species, communities and ecosystems (White and Jentsch 2001) as a consequence of the biological legacies described above. These forms of disturbance have been referred to as key structuring processes that dominate the formation of patterns over spatial scales of hundreds of metres to hundreds of kilometers (Holling 1992), leading to heterogeneous landscapes. The resultant heterogeneity will influence interactions and exchanges among stands, and ultimately the biotic and abiotic processes associated with forest health at the landscape scale (Turner 1989; Krawchuk et al. 2020).

Heterogeneity influences the resistance of forest landscapes to disturbance through impacts on the susceptibility of stands and the capacity for disturbances to spread within landscapes (Turner and Gardner 2015; Krawchuk et al. 2020). Tree species composition, physiological condition, age and climatic conditions are well known factors that influence the susceptibility of forest stands within a landscape to biotic disturbances by insects and pathogens (Cooke et al. 2007; Raffa et al. 2008; Jactel et al. 2017). The susceptibility of stands to abiotic disturbances will also vary across forested landscapes. For example, areas that are more exposed (edges, gaps, ridgelines) will suffer more windthrow, and drier regions (south-facing slopes, valley

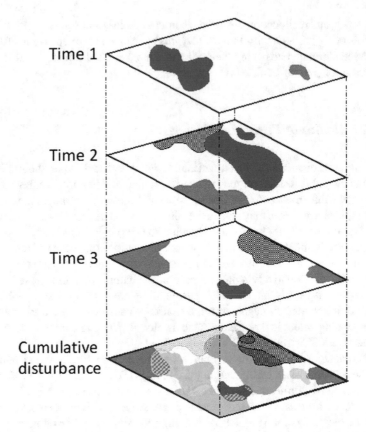

Fig. 21.3 Representation of multiple disturbances acting on the same landscape through time and cumulatively. Colour and pattern indicate different processes, darker shading in the cumulative landscape indicates more recent events. Modified from Parker and Pickett (1998)

bottoms) are more conducive to fire (Turner and Gardner 2015). The spread of disturbances through landscapes may also be impeded by heterogeneity. For example, the distribution of susceptible stands in a landscape will affect the ability of bark beetles to traverse it (Barclay et al. 2005; Raffa et al. 2008). Similarly, low- and moderate-severity wildfires in coniferous forests may be constrained by natural fire breaks and young stands (Turner et al. 1994; Turner and Gardner 2015). Due to the influence of landscape heterogeneity on disturbance susceptibility and spread, even extensive, potentially homogenizing disturbances such as large wildfires will perpetuate further heterogeneity (Turner et al. 1994; Turner 2010; Turner and Gardner 2015). Consider the landscape in Fig. 21.4. The disturbed area within the fire boundary contains areas of varying size with fire severities ranging from none to severe. Such a landscape may be considered resilient and healthy due to the increased probability that areas of severe disturbance can recruit key species to maintain ecological processes from nearby intact areas (Loreau et al. 2001; Krawchuk et al. 2020). By contrast, the

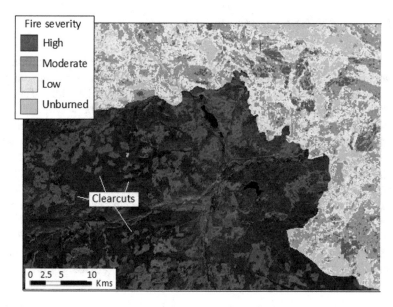

Fig. 21.4 Satellite image of a working forest in central British Columbia, Canada, and adjacent area that burned in a wildfire in 2017. Note the size, distribution and varying severity of the patches disturbed by fire as compared to the clearcuts in the unaffected forest. *Source* ESRI, DigitalGlobe, GeoEye, Earthstar Geographics, CNES/Airbus DS, USDA, USGS, AeroGrid, IGN, and the GIS User Community

unburned portion of the landscape in Fig. 21.4 has been disturbed by more regular clear-cut harvesting and is relatively less heterogeneous. Maintenance of hetero-geneity over landscapes provides "spatial insurance" for healthy ecosystem function by facilitating spatial exchanges among local systems (Loreau et al. 2003).

Over long time-spans patterns of forest disturbance (i.e. disturbance regimes) become apparent with distinct distributions of type, severity, frequency and size (White and Jentsch 2001; Turner 2010; Turner and Gardner 2015). In the absence of anthropogenic alterations, disturbance regimes function within an historic or natural range of variability that can be used to represent the envelope of possible ecosystem conditions over a landscape (Landres et al. 1999). Implicit within the concept of a natural range of variability are the assumptions that ecosystems are dynamic and their responses to change are represented by past variability, and that they have a range of conditions within which they are self-sustaining, beyond which they are not (Keane et al. 2009). Thus, historical conditions can serve as a proxy for forest health (Swetnam et al. 1999) where the resilience of ecosystems is considered in the context of the type, severity, frequency, size, spatial distribution, and return intervals of disturbance. Deviation of disturbance processes within a forest landscape beyond the natural range of variability would threaten its resilience and health.

21.4 Forest Health and Global Change

Global change refers to the independent and interacting effects of anthropogenic stressors on ecosystems at a planetary scale. The primary drivers of global change affecting forests are climate change, land-use change and biotic invasions (Tylianakis et al. 2008). Impacts by these broad stressors on forests can be very complex and sometimes difficult to distinguish from natural variability. Given that the definition of a healthy forest defined above is derived from the processes of disturbance, resistance and resilience, I will examine the interactions of global change drivers with each process in turn using a range of recent examples. My emphasis will be on impacts by novel stressors that perturb forests beyond their natural range of variability (see Fig. 21.1). It is important to note that interactions seldom operate in isolation, and so I will also consider interdependencies for which there is documented evidence and acknowledge that many more interactions are likely at work. Finally, while most examples originate from the northern hemisphere, this is simply a byproduct of available data. The concepts they illustrate are relevant around the world.

21.4.1 Climate Change

Climate change refers to both global warming caused by human emissions of greenhouse gases and the resultant large-scale shifts in weather patterns and extremes. Warming of the climate system is unequivocal, and since the 1950s, many of the observed changes are unprecedented over decades to millennia (IPCC 2014). Increasingly, impacts to natural and human systems have been documented on all continents and across the oceans.

21.4.1.1 Disturbance Versus Climate Change

Changing the tempo, intensity, or spatial attributes of disturbance can alter disturbance regimes (Turner 2010; Johnstone et al. 2016; Seidl et al. 2017). As discussed above, when a regime has been modified beyond its natural range of variability then forest landscapes may no longer be resistant and/or resilient, and their health will have been compromised. Perhaps the greatest impact that climate change will have on forest ecosystems in the coming decades will arise from altered disturbance regimes (Lindner et al. 2010). Indeed, many forms of disturbance have already been influenced by climate change (Seidl et al. 2011, 2017). Among the most significant forms of disturbance affected to date are insect outbreaks and wildfire.

Biotic disturbances, primarily caused by insects, affect almost 44 million ha of forests in the northern hemisphere each year (Kautz et al. 2017). Insects are ectothermic, and therefore highly sensitive to changing climate. Not surprisingly, climate change has been implicated in alterations to many aspects of the spatial and temporal dynamics of forest insects and their potential to cause disturbance. These alterations include shortened life cycle durations (Berg et al. 2006; Choi et al. 2011), increased thermally benign habitats (Carroll et al. 2004; Battisti et al. 2005; Jepsen et al. 2008), enhanced seasonal synchrony among trophic levels and/or the environment (Logan and Powell 2001; Jepsen et al. 2011) and reduced mortality from natural enemies (Stireman et al. 2005; Menéndez et al. 2008). The predominant outcome of these altered dynamics has been a general increase in the rate of biotic disturbance (Kautz et al. 2017; Seidl et al. 2017) potentially leading to modified disturbance regimes.

Despite the general perception that wildfires are increasing in severity around the world, evidence suggests that there is actually less fire in the global landscape today than centuries ago (Doerr and Santín 2016). That said, there are regions where disturbance by wildfire has increased, particularly in western North America. These increases have been attributed to warming-induced changes in atmospheric aridity leading to elevated evaporative demand and reduced fuel moisture, snowpack, and summer precipitation frequency (Abatzoglou and Williams 2016; Williams et al. 2019). Between 1972 and 2018, the area burned in California increased by over 400% (Williams et al. 2019), and virtually all projections based on climate scenarios suggest wildfire potential will continue to rise across western North America (Liu et al. 2010; Jolly et al. 2015). Increased rates of disturbance by fire beyond historic levels raise uncertainties regarding the capacity for forest ecosystem to remain healthy (Turner 2010; Kelly et al. 2013; Millar and Stephenson 2015; Coop et al. 2020).

Interactions among disturbance agents are a major component of disturbance regimes that create heterogeneous, resistant and resilient landscapes (see Fig. 21.3). However, increasing disturbance activity under climate change also means an increasing propensity for disturbance interactions, potentially exacerbating their severity (Buma 2015). In a review and synthesis of climate change effects on important abiotic and biotic disturbances, Seidl et al. (2017) found that links between an initial abiotic agent and subsequent biotic disturbances, especially by bark beetles in conifer forests, were particularly strong and led to amplification of disturbance in the majority of interactions. Bark beetle outbreaks generally arise following an acute pulse of defensively impaired trees that facilitate rapid population increases (Raffa et al. 2008, 2015). This resource pulse is often a result of an initial abiotic disturbance such as a wind storm (Kausrud et al. 2012), wildfire (Hood and Bentz 2007), or drought (Seidl et al. 2016a, b). These interactions can lead to the mortality of trees over many millions of hectares (Raffa et al. 2008). Disturbances at these scales are of particular concern since they are very likely to exceed natural ranges of variation.

21.4.1.2 Resistance Versus Climate Change

Rising temperatures have amplified drought-induced stress in forests around the world (Young et al. 2017; Stephens et al. 2018) and have affected the capacity of ecosystems to resist disturbance. This aspect of climate change is most evident in interactions of forest ecosystems with phloem-feeding insects such as the bark beetles whose attack and colonization success are constrained by tree defenses that are sensitive to water availability (Raffa et al. 2015; Marini et al. 2017). Many conifers close stomata to protect xylem cells from cavitation during drought, reducing photosynthesis to near zero (Koepke and Kolb 2013). However, production and deployment of defensive resin is reduced under conditions of limited photosynthesis, thus lowering tree resistance to bark beetle attacks during droughts (Raffa et al. 2015).

Climate change-exacerbated droughts have also affected forest resistance to abiotic disturbances such as wildfire. Drought not only causes increased amounts of fuels in forests in the form of dead wood, it also reduces the moisture content within those fuels and alters the ratio of dead to live fuels within the canopy of living trees, thus reducing the resistance of some forests to fire and facilitating larger, more severe fires (Stephens et al. 2018; Nolan et al. 2020).

21.4.1.3   Resilience Versus Climate Change

The structural and functional changes in forests in response to disturbance may compromise their capacity to recover in a warming environment. Evidence is accumulating that forest ecosystem resilience may be affected by climate change-exacerbated wildfires. In the western region of the North American boreal forest, drier and warmer weather associated with climate change has decreased the resilience of ecosystems by reducing the interval between wildfires leading to altered patterns of regeneration (Whitman et al. 2019; Coop et al. 2020).

Similarly, in the western US, increasingly unfavorable post-fire growing conditions due to a changing climate have compromised ecosystem resilience by reducing seedling establishment and increasing regeneration failures (Harvey et al. 2016; Stevens-Rumann et al. 2018; Davis et al. 2019; Coop et al. 2020). These impacts comprise an abrupt tipping point given that fire has killed the adult trees that could have persisted in the warmer conditions, but since those conditions are no longer suitable for seedling establishment and survival, ecosystems cannot return to similar pre-disturbance conditions (Davis et al. 2019).

21.4.2   Land-Use Change

Land-use change typically refers to the permanent conversion of forests as opposed to temporary losses from wildfires or harvesting. Where land-use change leads to loss of forest, it results in disturbance well beyond the natural range of variability and complete negation of forest health. This form of global change is a significant impact to forested landscapes. Deforestation through land-use change is responsible for over one-quarter of forest loss around the world (Curtis et al. 2018). By contrast, partial land-use changes, also known as forest degradation, may be less severe and involve retention of some ecological processes (Ghazoul et al. 2015; Ghazoul and Chazdon

2017), allowing consideration of forest health. Since partial land-use changes are associated with diminished or constrained ecological function within forests, examples are broad and include extraction of non-timber forest products, collection of fuel wood, free-range livestock grazing, shifting cultivation, selective logging, urban encroachment and wildfire suppression (Thompson et al. 2013). These activities have the potential to alter all aspects of forest health.

21.4.2.1 Disturbance Versus Land-Use Change

Partial land-use changes can significantly alter the behaviour and characteristics of disturbances, especially abiotic disturbance. Wildfires depend on the coincidence of dry weather, available fuel and ignition sources (Jolly et al. 2015). As outlined above, weather conditions conducive to fire have increased due to climate change-related drought in many regions. In western North America, land-use changes have also affected the remaining two requirements for severe wildfires. The legacy of human settlement and fire suppression has contributed to increased fuel loads in forests (Higuera et al. 2015; Parks et al. 2015). Moreover, growing populations and urban encroachment have resulted in increased frequency and type of human-caused ignitions (Balch et al. 2017; Radeloff et al. 2018). Thus, land-use changes have further exacerbated the impacts of climate change in terms of fire severity, particularly in the western US as evidenced by recent record-breaking fire seasons.

21.4.2.2 Resistance Versus Land-Use Change

In many cases, partial land-use changes have reduced the resistance of forests to disturbances by constraining or removing critical ecosystem functions. For example, widespread fire suppression in biomes adapted to frequent wildfires can severely compromise resistance to both abiotic and biotic disturbances. In western Canada, aggressive fire suppression over the past century allowed large areas of pine-dominated forests to age to the point of becoming highly susceptible (i.e. less resistant) to the mountain pine beetle (Taylor and Carroll 2004), leading to a "hyper-epidemic" that reached an order of magnitude greater extent and severity than any previously recorded (Sambaraju et al. 2019).

Wind is one of the most important abiotic forest disturbances in many parts of the world (Seidl et al. 2017). In tropical forests prone to cyclones and hurricanes, altered forest structure (increased gaps, edges) and shifts in plant species composition as a result of forest fragmentation reduce the resistance of forests to storm damage (Laurance and Curran 2008). Similarly, in the Norway spruce forests of Europe, resistance to wind disturbances is compromised by fragmentation (Zeng et al. 2009). In these forests wind disturbance is further amplified by outbreaks of the European spruce beetle that erupt from freshly broken or uprooted trees and spread into intact forests (Stadelmann et al. 2014).

21.4.2.3   Resilience Versus Land-Use Change

Reduced resilience of forest ecosystems associated with partial land-use change is common when the change interrupts biological legacies and impairs regeneration. For example, repeated burning of forested areas to promote livestock grazing in the Amazon has led to reduced seed availability and seedling recruitment and subsequent reorganization to shrub-dominated landscapes (Mesquita et al. 2015). Similarly, the resilience of some old-growth eucalypt forests in Australia have been diminished by clear cut logging that increases both fine fuels and the prevalence of young densely stocked stands that together support elevated fire severity compromising the capacity for systems to regenerate to equivalent pre-logging conditions (Lindenmayer et al. 2011).

21.4.3   Biotic Invasions

Biological invasions have become a defining feature of the Anthropocene (Lewis and Maslin 2015). Dramatic increases in human transport and commerce have increased the rate of introductions of non-native species into virtually all habitats around the world. Although most species introduced into new habitats will not survive, some will establish and persist. A small percentage of those that persist can become invasive where they proliferate and spread to the detriment of the environment (Mack et al. 2000; Aukema et al. 2010). Invasive species can affect all ecological processes within forests causing altered diversity, nutrient cycling, succession, and frequency and intensity of wildfires (Kenis et al, 2009; Liebhold et al. 2017). Non-native organisms from nearly every taxon have been introduced into forests; however, insects represent the most diverse group of invaders (Brockerhoff and Liebhold 2017; Liebhold et al. 2017). Thus, the examples discussed below will be mostly derived from invasive forest insects.

21.4.3.1   Disturbance Versus Biotic Invasions

The most apparent impact of biotic invasions within forest ecosystems involves altered disturbance rates as a result of direct tree mortality caused by the invasive organisms. There are many examples of these types of disturbances (Gandhi and Herms 2009; Kenis et al. 2009; Brockerhoff and Liebhold 2017). Among the most severe are the hemlock woolly adelgid and the emerald ash borer. The hemlock woolly adelgid, a sap feeder, was accidentally introduced from Japan to the eastern US during the early decades of the last century (Brockerhoff and Liebhold 2017). It has caused extensive mortality of eastern hemlock, causing its decline as a dominant forest species throughout eastern North America (Morin and Liebhold 2015). More recently, the emerald ash borer was introduced from north-eastern Asia to both North America and western Russia (Herms and McCullough 2014). Since its arrival it has caused extensive mortality and eliminated the majority of ash trees (*Fraxinus* spp.) within the areas it has invaded (Straw et al. 2013; Herms and McCullough 2014; Morin et al. 2017).

Disturbances associated with biotic invasions will potentially worsen under climate change. Hellman et al. (2008) discuss the potential impacts of a warming environment on invasive species. Three impacts in particular are relevant to forest disturbance; (i) altered climatic constraints on invasive species, (ii) altered distribution of existing invasive species, and (iii) altered impact of existing invasive species. The sum of these impacts implies a general increase in thermally benign habitats available to invasive species that may lead to higher rates of disturbances in forests as the climate continues to warm. Indeed, the number of established alien species is projected to continue increasing through the current century (Seebens et al. 2021).

21.4.3.2 Resistance Versus Biotic Invasions

The term biotic resistance is used to describe the ability of communities to resist invasive species. In general, forests tend to be more resistant to invasions than other terrestrial systems due to their inherently high diversity and the resultant interactions of introduced organisms with native competitors, predators, etc. (Iannone et al. 2016; Nunez-Mir et al. 2017). However, when invasive species cause extensive forest disturbances, it is most often a consequence of an insufficient or inadequate response on the part of trees to defend themselves from herbivores (Brockerhoff and Liebhold 2017). This is referred to as the defense-free space hypothesis in which population growth and spread of an invader is facilitated by low resistance of evolutionarily naïve host plants (Gandhi and Herms 2009). Defense-free space has been implicated in the exacerbated impacts of many invasive forest insects and pathogens including hemlock woolly adelgid and emerald ash borer mentioned above (Showalter et al. 2018).

The concept of defense-free space is not confined to interactions of non-native organisms with forest ecosystems. Native herbivorous insects are often constrained by climate to a portion of the range of their host trees. As discussed above, a warming environment has been associated with increases in the availability of thermally benign habitats for several insect species, facilitating an expansion of ranges into evolutionarily naïve populations and species of host trees (Burke et al. 2017). Due to an insufficiently evolved defensive response, the resistance of naïve host tree populations and species to native climate migrants is inadequate to prevent severe disturbance (Cudmore et al. 2010; Raffa et al. 2013; Clark et al. 2014). This phenomenon is best exemplified by the recent expansion of the mountain pine beetle across the Rocky Mountains of North American and invasion of the transcontinental boreal forest (Cooke and Carroll 2017).

21.4.3.3 Resilience Versus Biotic Invasions

Biotic invasions that result in extensive disturbances to evolutionarily naïve forests are by definition beyond the range of historic variability. Hence, impacts to the processes associated with ecosystem resilience are often extreme and forests are forced to reorganize. Many examples exist of drastically altered ecosystems as a consequence of biotic invasion, but perhaps the best known is that associated with a fungal pathogen, the chestnut blight, accidentally introduced from Asia into North America in the early 1900s (Griffin 1986). The resultant devastation of the American chestnut by the fungus represents one of the greatest recorded changes to a forest biome caused by an introduced organism (Liebhold et al. 1995). Within a relatively short period of its introduction, the pathogen spread and functionally eliminated the American chestnut through most of its range. The loss of chestnut trees throughout eastern North America has had spectacular and long-term effects on forest ecosystems including reorganization to oak-dominated overstories, altered disturbance regimes and loss of wildlife habitat.

21.5 Forest Health in Practice

This chapter has defined forest health, outlined its constituent components across spatial and temporal scales, and reviewed the impacts of global change on each. How then are changes in forest health detected and how can forests be managed to allow sustainable extraction of ecosystem services? The foundation of the definition of a healthy forest is that its ecological processes operate within an envelope of possible ecosystem conditions. This concept of a natural range of variability (Landres et al. 1999) provides a framework for understanding the ecological context of a forest and in evaluating changes in its health.

Quantifying natural variability in forests requires information on the ecological processes and conditions of interest and their variation through time and space. This information is obtained from studies in the fields of dendroecology, dendroclimatology, palynology, landscape ecology and remote sensing that provide measurements over a sufficiently long time period and spatial extent so that meaningful information can be gained about changes in populations, ecosystem structures, disturbance frequencies, process rates, trends, periodicities, and other dynamical behaviors (Swetnam et al. 1999). Application of the concept of natural range of variability to ecosystem management is based on the following premises as reviewed by Landres et al. (1999):

- contemporary anthropogenic change may diminish the viability of many species that are adapted to past or historical conditions and processes;
- approximating historical conditions will sustain the viability of diverse species, even for those for which we have limited information;

- natural variability is a reference for evaluating the influence of anthropogenic change in ecological systems at local and shorter time scales;
- natural variability encompasses the dynamic ecological processes that drive both spatial and temporal variation in ecological systems, as well as the influence of this variation on evolution and biological diversity;
- disturbances have a strong and lasting influence on species, communities, and ecosystems;
- spatial heterogeneity is an integral component of ecological systems that is positively related to biodiversity, and resistance and resilience to disturbance.

Although difficult to generate, considerable information regarding the natural range of variability of ecological processes within many forest types has been amassed in recent decades (Keane et al. 2009). Indeed, it is now widely recognized that forest management should seek to emulate the natural range of variability of forests to maintain biodiversity and ecological function (Drever et al. 2006; Keane et al. 2009; Čada et al. 2020; Donato et al. 2020). This recognition has stimulated efforts to minimize differences between managed and natural forests by, for example, modifying harvesting practices to generate spatial and temporal patterns consistent with historical disturbance regimes (Bergeron et al. 2002; Harvey et al. 2002; Kuuluvainen and Grenfell 2012; Leclerc et al. 2021).

Consideration of disturbance, resistance and resilience within the context of natural range of variability may at first seem overly simplistic since it assumes that the record of historical conditions must reflect the range of possible conditions for future landscapes, thus ignoring the potential impacts of global change. However, determination of the natural range of variability of forest ecosystems necessarily captures large variations in the conditions of past centuries (Swetnam et al. 1999), and therefore it remains relevant even when faced with anthropogenic change. Moreover, the potential impacts of global change may be buffered by aspects of forest health. Landscape heterogeneity is directly related to species diversity (Tews et al. 2004; Fahrig et al. 2011) and diversity improves resistance and resilience to disturbance by virtue of spatial exchanges among local systems in heterogeneous landscapes (Loreau et al. 2003; Brockerhoff et al. 2017; Krawchuk et al. 2020). Thus, a forest type with higher species diversity will be healthier than an otherwise equivalent, but depauperate one, and better able to withstand novel stressors. Species diversity can be quantified at each spatial scale relevant to forest health using the concept of α-, β- and γ-diversity (Whittaker 1972; Veech et al. 2002), where α-diversity refers to species diversity within stands, β-diversity refers to species diversity among stands in a landscape, and γ-diversity is the total species diversity of the biome (i.e. sets of landscapes comprising distinct biological communities that have formed in response to a shared physical climate).

Indicators of forest health vary from stands to biomes and can be expressed in terms of each of our components of forest health—disturbance, resistance and resilience (Fig. 21.5). At the finest scale, a healthy stand is one where the type, severity and frequency of any disturbance falls within the range of natural variability. The capacity for a stand to respond to disturbance and remain within the historic range of ecosystem conditions (i.e. avoid tipping points and regime shifts) is contingent upon its inherent resistance and resilience. Since resistance and resilience increase with increasing diversity, then at any point in time, a stand will be healthier with greater α-diversity relative to equivalent stands at similar successional stages. And as discussed above, stands are further considered healthy if following disturbance, they successfully regenerate (naturally or by planting) along a successional trajectory that will return them to a functionally equivalent pre-disturbance state. Similarly, indicators of forest health at the scale of landscapes comprise disturbance regimes (type, severity, frequency, size and return interval) that remain within the natural range of variability, high relative β-diversity and heterogeneous structures derived from diverse seral stages with high connectivity. And finally, forest biomes will be healthy when disturbance regimes within constituent landscapes remain within the natural range of variability, there is high absolute γ -diversity and all constituent landscapes persist through time (Trumbore et al. 2015).

21.5.1 Forest Health Monitoring

Given that the processes of forest health vary across scales (Fig. 21.5), forest health monitoring programs must collect and synthesize data within and among scales to support managers, decision makers, and politicians in their decisions regarding forest management. Within stands, health conditions are often measured directly from individual forest inventory plots where species diversity, and the status of trees, vegetation, soils and other ecosystem properties are quantified. These data may be augmented with high-resolution remote-sensing techniques such RADAR or LiDAR which have the potential to reconstruct forest structures within and below the canopy (Lausch et al. 2017). Data at broader scales can be derived from networks of forest inventory plots (Woodall et al. 2011) and from a wide variety of broad-scale remote sensing techniques (Lausch et al. 2016).

Despite considerable efforts by many countries to develop comprehensive forest health monitoring programs, there still remains some discrepancy between the information required by forest managers and the data that are available for understanding and assessing the complexity of forest health processes (Lausch et al. 2018). Long-term monitoring based on forest inventory plot networks provides valuable information regarding trends in forest health processes (Tkacz et al. 2008; Woodall et al. 2011); however, short-term perturbations that may trigger abrupt nonlinear declines in health are not sufficiently assessed since measurement intervals are often multiple years (Lausch et al. 2017). More recently, integration of forest inventory plot networks with remote sensing tools has facilitated generalization of intensive and expensive ground-based measurements to temporal and spatial scales required by forest managers (McDowell et al. 2015).

Forest health indicators

- *Disturbance:*
  Type, severity, frequency
- *Resistance:*
  Relative α diversity
- *Resilience:*
  Regeneration success,
  Successional trajectory

- *Disturbance:*
  Type, severity, frequency,
  size, return interval
- *Resistance:*
  Relative β diversity
- *Resilience:*
  Seral diversity, connectivity
  heterogeneity

- *Disturbance:*
  Type, severity, frequency,
  size, return interval
- *Resistance:*
  γ diversity
- *Resilience:*
  Landscape persistence

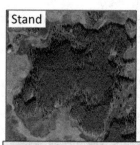  

Forest health monitoring tools

- Forest inventory plot
- High-resolution remote
  sensing

- Forest inventory plot
  network
- Remote sensing

- Remote sensing

Fig. 21.5 Indicators and monitoring tools for forest health at the stand-, landscape- and biome-scale. Indicators are scale dependent, and refer to (i) disturbance (type, severity, frequency, size, return interval) within a natural range of variability, (ii) resistance defined by stand- (α) and landscape- (β) diversity relative to equivalent stands and landscapes in the biome, or total species diversity (γ) in the biome, and (iii) resilience to disturbance defined as the capacity for forests to return to equivalent pre-disturbance states at each scale. Forest health monitoring tools are also scale dependent and range from individual plots in stands to broad-scale remote sensing

21.6 Forest Health Versus Nonconventional Forests

Although planted forests comprise approximately 7% of forests around the world, they account for roughly 70% of industrial wood products (Carle and Homgren 2008). The demand for wood products from plantations has been growing, and so has the area devoted to plantations such that the area of planted forest is likely to double by the end of the century (Brockerhoff et al. 2013). The majority of plantation forests comprise non-native *Pinus*, *Eucalyptus* and *Acacia* species in the southern hemisphere and mostly native species in several northern hemisphere countries (Payn et al. 2015). As the emphasis on plantation forestry has grown, so has the need for assessments of ecosystem health.

As discussed above, the definition of forest health derived from processes of disturbance, resistance and resilience does not necessarily exclude intensively managed plantations, so how is it applied? Since plantations are largely artificial constructs the concepts of natural range of variability and disturbance regimes are not applicable. Indeed, disturbances are mainly restricted to biotic agents such as insects and pathogens (Wingfield et al. 2015). Further, issues of resilience to disturbance (and associated tipping points and regime shifts) are rendered largely irrelevant since intensive management will lead to immediate investments toward regeneration of any disturbed areas. In contrast, resistance processes that influence the severity of disturbance are vital to forest plantations (Brockerhoff et al. 2013; Wingfield et al. 2015) and provide the basis for the assessment of their health.

Planted forests are typically of a single, non-native species grown primarily for efficient wood production. They tend to be characterised by lower levels of biodiversity than natural and semi-natural forests (Brockerhoff et al. 2008) and they achieve high productivity mainly through intensive pest control efforts or the outright exclusion of pests (Liebhold et al. 2017; Wingfield et al. 2015). When pest impacts do occur, large amounts of damage can result.

Mixed-species forests are more resistant than monocultures to biotic disturbance due to the greater abundance of trophic interactions that regulate biotic disturbance agents (Jactel et al. 2017). Although plantations are normally established as monocultures, they afford an ideal opportunity to create diversity and increase resistance to disturbance given that replanting after harvesting is a frequent and recurrent process (Brockerhoff et al. 2017; Paquette and Messier 2010). Based on the expectation of increasing disturbance in a warmer world (see discussion above), greater resistance to disturbance through creation of mixed-species plantations should offset any short-term costs associated with their establishment.

In light of the capacity for innovative management of intensive plantations to promote diversity and resistance, Brockerhoff et al. (2013) recommend that clearing natural vegetation should be avoided prior to planting, native tree species should be preferred, and where possible mixed-species plantations should be established. Furthermore, in keeping with the expectations of a healthy forest landscape, they recommend the protection and enhancement of remnants of natural vegetation, the creation of mosaics of stand ages and tree species and the establishment of corridors linking habitat patches.

21.7   Conclusions

Increasingly forests are threatened by anthropogenic stressors arising from global change that compromise provisioning of vital ecosystem services. Therefore, the need to promote forest ecosystems that are resistant and resilient to stressors has never been greater. Forests are highly complex and their response to natural and anthropogenic stressors is difficult to quantify. Given an increasing global population, forests cannot and should not be viewed as independent of human influences; however, the capacity to identify the source and impact of natural and anthropogenic stressors is essential for effective forest management intended to foster forest health. A definition of forest health derived from ecological processes, and independent of human values, facilitates baseline assessments of forest function at all temporal and spatial scales and allows partitioning of the relative impacts of natural and anthropogenic stressors and their interactions. Defining forest health in terms of the processes of disturbance, resistance and resilience within natural ranges of variability allows quantification of the vitality of any forest type in any conceivable state and determine its probability of persistence.

The ability of forests to be resistant and resilient to disturbance is dependent upon species diversity and landscape heterogeneity. Resistant and resilient forests will retain ecological processes and the capacity to deliver ecosystem services. Therefore, management toward healthy forests should seek to maximize relative biodiversity at all scales as much as is practicable. In so doing, we can provide spatial insurance for ecosystem functioning (Loreau et al. 2003) by increasing the probability of robust resistance and resilience in the face of global change.

References

Abatzoglou JT, Williams AP (2016) Impact of anthropogenic climate change on wildfire across western US forests. Proc Natl Acad Sci 113:11770–11775

Allen CR, Angeler DG, Cumming GS, Folke C, Twidwell D, Uden DR (2016) Quantifying spatial resilience. J Appl Ecol 53:625–635

Aukema JE, McCullough DG, Holle BV, Liebhold AM, Britton K, Frankel SJ (2010) Historical accumulation of nonindigenous forest pests in the Continental United States. Bioscience 60:886–897

Balch JK, Bradley BA, Abatzoglou JT, Nagy RC, Fusco EJ, Mahood AL (2017) Human-started wildfires expand the fire niche across the United States. Proc Natl Acad Sci 114:2946–2951

Barclay HJ, Li C, Benson L, Taylor S, Shore T (2005) Effects of fire return rates on traversability of lodgepole pine forests for mountain pine beetle (Coleoptera: Scolytidae) and the use of patch metrics to estimate traversability. Can Entomol 137:566–583

Battisti A, Stastny M, Netherer S, Robinet C, Schopf A, Roques A, Larsson S (2005) Expansion of geographic range in the pine processionary moth caused by increased winter temperatures. Ecol Appl 15:2084–2096

Berg EE, Henry JD, Fastie CL, Volder ADD, Matsuoka SM (2006) Spruce beetle outbreaks on the Kenai Peninsula, Alaska, and Kluane National Park and Reserve, Yukon Territory: Relationship to summer temperatures and regional differences in disturbance regimes. For Ecol Manage 227:219–232

Bergeron Y, Leduc A, Harvey B, Gauthier S (2002). Natural fire regime: a guide for sustainable management of the Canadian boreal forest. Silva Fenn 36. https://doi.org/10.14214/sf.553

Brandt JP, Flannigan MD, Maynard DG, Thompson ID, Volney WJA (2013) An introduction to Canada's boreal zone: ecosystem processes, health, sustainability, and environmental issues1. Environ Rev 21:207–226

Brockerhoff EG, Barbaro L, Castagneyrol B, Forrester DI, Gardiner B, González-Olabarria JR et al (2017) Forest biodiversity, ecosystem functioning and the provision of ecosystem services. Biodivers Conserv 26:3005–3035

Brockerhoff EG, Jactel H, Parrotta JA, Ferraz SFB (2013) Role of eucalypt and other planted forests in biodiversity conservation and the provision of biodiversity-related ecosystem services. For Ecol Manage 301:43–50

Brockerhoff EG, Jactel H, Parrotta JA, Quine CP, Sayer J (2008) Plantation forests and biodiversity: oxymoron or opportunity? Biodivers Conserv 17:925–951

Brockerhoff EG, Liebhold AM (2017) Ecology of forest insect invasions. Biol Invasions 19:3141–3159

Brook BW, Ellis EC, Perring MP, Mackay AW, Blomqvist L (2013) Does the terrestrial biosphere have planetary tipping points? Trends Ecol Evol 28:396–401

Buma B (2015) Disturbance interactions: characterization, prediction, and the potential for cascading effects. Ecosphere 6:1–15

Burke JL, Bohlmann J, Carroll AL (2017) Consequences of distributional asymmetry in a warming environment: invasion of novel forests by the mountain pine beetle. Ecosphere 8:e01778

Čada V, Trotsiuk V, Janda P, Mikoláš M, Bače R, Nagel TA et al (2020) Quantifying natural disturbances using a large-scale dendrochronological reconstruction to guide forest management. Ecol Appl 30. https://doi.org/10.1002/eap.2189

Carle J, Homgren P (2008) Wood from planted forests: a global outlook 2005–2030. For Prod J 58:6–18

Carpenter SR, Mooney HA, Agard J, Capistrano D, DeFries RS, Díaz S et al (2009) Science for managing ecosystem services: beyond the Millennium Ecosystem Assessment. Proc Natl Acad Sci 106:1305–1312

Carroll AL, Taylor SW, Régnière J, Safranyik L (2004). Effects of climate change on range expansion by the mountain Pine Beetle in British Columbia. In: Shore TL, Stone JE Stone (eds) Mountain Pine Beetle symposium: challenges and solutions. Natural Resources Canada, Canadian Forest Service, Victoria, BC, Canada, pp 223–232

Choi WI, Park Y-K, Park Y-S, Ryoo MI, Lee H-P (2011) Changes in voltinism in a pine moth Dendrolimus spectabilis (Lepidoptera: Lasiocampidae) population: implications of climate change. Appl Entomol Zool 46:319–325

Clark EL, Pitt C, Carroll AL, Lindgren BS, Huber DPW (2014) Comparison of lodgepole and jack pine resin chemistry: implications for range expansion by the mountain pine beetle, Dendroctonus ponderosae (Coleoptera: Curculionidae). PeerJ 2:e240

Cooke BJ, Carroll AL (2017) Predicting the risk of mountain pine beetle spread to eastern pine forests: Considering uncertainty in uncertain times. For Ecol Manage 396:11–25

Cooke BJ, Nealis VG, Régnière J (2007) Insect defoliators as periodic disturbances in northern forest ecosystems. In: Johnson EA, Miyanishi K (eds) Plant disturbance ecology: the process and the response. Elsevier Academic Press, Burlington, Mass, USA, pp 487–525

Coop JD, Parks SA, Stevens-Rumann CS, Crausbay SD, Higuera PE, Hurteau MD et al (2020) Wildfire-Driven forest conversion in Western North American landscapes. Bioscience 70:659–673

Cudmore TJ, Björklund N, Carroll AL, Lindgren BS (2010) Climate change and range expansion of an aggressive bark beetle: evidence of higher beetle reproduction in naïve host tree populations. J Appl Ecol 47:1036–1043

Curtis PG, Slay CM, Harris NL, Tyukavina A, Hansen MC (2018) Classifying drivers of global forest loss. Science 361:1108–1111

Davis KT, Dobrowski SZ, Higuera PE, Holden ZA, Veblen TT, Rother MT et al (2019) Wildfires and climate change push low-elevation forests across a critical climate threshold for tree regeneration. Proc Natl Acad Sci 116:201815107

DeRose RJ, Long JN (2014) Resistance and resilience: a conceptual framework for silviculture. For Sci 60:1205–1212

de Mesquita R, C. G., Massoca, P. E. dos S., Jakovac, C. C., Bentos, T. V., & Williamson, G. B. (2015) Amazon rain forest succession: stochasticity or land-use legacy? Bioscience 65:849–861

Doerr SH, Santín C (2016) Global trends in wildfire and its impacts: perceptions versus realities in a changing world. Philosophical transactions of the royal society : Biol Sci 371:20150345

Donato DC, Halofsky JS, Reilly MJ (2020) Corralling a black swan: natural range of variation in a forest landscape driven by rare, extreme events. Ecol Appl 30. https://doi.org/10.1002/eap.2013

Drever CR, Peterson G, Messier C, Bergeron Y, Flannigan M (2006) Can forest management based on natural disturbances maintain ecological resilience? Can J for Res 36:2285–2299

Edmonds RL, Agee JK, Gara RI (2011) Forest health and protection. Waveland Press, Long Grove, IL, USA

Fahrig L, Baudry J, Brotons L, Burel FG, Crist TO, Fuller RJ et al (2011) Functional landscape heterogeneity and animal biodiversity in agricultural landscapes. Ecol Lett 14:101–112

FAO, and UNEP (2020) The state of the world's forests 2020. forests, biodiversity and people. Rome: food and agriculture organization of the United Nations and UN Environment Program.

Folke C, Carpenter S, Walker B, Scheffer M, Elmqvist T, Gunderson L, Holling CS (2004) Regime shifts, resilience, and biodiversity in ecosystem management. Annu Rev Ecol Evol Syst 35:557–581

Gandhi KJK, Herms DA (2009) Direct and indirect effects of alien insect herbivores on ecological processes and interactions in forests of eastern North America. Biol Invasions. https://doi.org/10.1007/s10530-009-9627-9

Gauthier S, Bernier P, Kuuluvainen T, Shvidenko AZ, Schepaschenko DG (2015) Boreal forest health and global change. Science 349:819–822

Ghazoul J, Burivalova Z, Garcia-Ulloa J, King LA (2015) Conceptualizing forest degradation. Trends Ecol Evol 30:622–632

Ghazoul J, Chazdon R (2017) Degradation and recovery in changing forest landscapes: a multiscale conceptual framework. Annu Rev Environ Resour 42:161–188

Griffin GJ (1986) Chestnut blight and its control. Horticultural Review 8:291–335

Grimm NB, Groffman P, Staudinger M, Tallis H (2016) Climate change impacts on ecosystems and ecosystem services in the United States: process and prospects for sustained assessment. Clim Change 135:97–109

Gunderson LH (2000) Ecological resilience—in theory and application. Annu Rev Ecol Syst 31:425–439

Harvey BJ, Donato DC, Turner MG (2016) High and dry: post-fire tree seedling establishment in subalpine forests decreases with post-fire drought and large stand-replacing burn patches. Glob Ecol Biogeogr 25:655–669

Harvey BJ, Leduc A, Gauthier S, Bergeron Y (2002) Stand-landscape integration in natural disturbance-based management of the southern boreal forest. For Ecol Manage 155:369–385

Hellman JJ, Byers JE, Biergarden BG, Dukes JS (2008) Five potential consequences of climate change for invasive species. Conserv Biol 22:534–543

Herms DA, McCullough DG (2014) Emerald ash borer invasion of North America: history, biology, ecology, impacts, and management. Annu Rev Entomol 59:13–30

Higuera PE, Abatzoglou JT, Littell JS, Morgan P (2015) The changing strength and nature of fire-climate relationships in the Northern rocky mountains, U.S.A., 1902–2008. PLOS One, 10:e0127563.

Holling CS (1973) Resilience and stability of ecological systems. Annu Rev Ecol Syst 4:1–23

Holling CS (1992) Cross-scale morphology, geometry, and dynamics of ecosystems. Ecol Monogr 62:447–502

Hood S, Bentz B (2007) Predicting postfire Douglas-fir beetle attacks and tree mortality in the northern Rocky Mountains. Can J for Res 37:1058–1069

Iannone BV, Potter KM, Hamil K-AD, Huang W, Zhang H, Guo Q et al (2016) Evidence of biotic resistance to invasions in forests of the Eastern USA. Landscape Ecol 31:85–99

IPCC. (2014). Climate change 2014: synthesis report. Contribution of working groups I, II and III to the fifth assessment report of the intergovernmental panel on climate change, Geneva, Switzerland: IPCC, p 151

Jactel H, Bauhus J, Boberg J, Bonal D, Castagneyrol B, Gardiner B et al (2017) Tree diversity drives forest stand resistance to natural disturbances. Curr For Rep 3:223–243

Jepsen JU, Hagen SB, Ims RA, Yoccoz NG (2008) Climate change and outbreaks of the geometrids Operophtera brumata and Epirrita autumnata in subarctic birch forest: evidence of a recent outbreak range expansion. J Anim Ecol 77:257–264

Jepsen JU, Kapari L, Hagen SB, Schott T, Vindstad OPL, Nilssen AC, Ims RA (2011) Rapid northwards expansion of a forest insect pest attributed to spring phenology matching with sub-Arctic birch. Glob Change Biol 17:2071–2083

Johnstone JF, Allen CD, Franklin JF, Frelich LE, Harvey BJ, Higuera PE et al (2016) Changing disturbance regimes, ecological memory, and forest resilience. Front Ecol Environ 14:369–378

Jolly WM, Cochrane MA, Freeborn PH, Holden ZA, Brown TJ, Williamson GJ, Bowman DMJS (2015) Climate-induced variations in global wildfire danger from 1979 to 2013. Nat Commun 6:7537

Kausrud K, Okland B, Skarpaas O, Grégoire J-C, Erbilgin N, Stenseth NC (2012) Population dynamics in changing environments: the case of an eruptive forest pest species. Biol Rev Camb Philos Soc 87:34–51

Kautz M, Meddens AJH, Hall RJ, Arneth A (2017) Biotic disturbances in Northern Hemisphere forests—a synthesis of recent data, uncertainties and implications for forest monitoring and modelling. Glob Ecol Biogeogr 26:533–552

Keane RE, Hessburg PF, Landres PB, Swanson FJ (2009) The use of historical range and variability (HRV) in landscape management. For Ecol Manage 258:1025–1037

Keeley JE, Pausas JG, Rundel PW, Bond WJ, Bradstock RA (2011) Fire as an evolutionary pressure shaping plant traits. Trends Plant Sci 16:406–411

Kelly R, Chipman ML, Higuera PE, Stefanova I, Brubaker LB, Hu FS (2013) Recent burning of boreal forests exceeds fire regime limits of the past 10,000 years. Proc Natl Acad Sci 110:13055–13060

Kenis M, Auger-Rozenberg M-A, Roques A, Timms L, Péré C, Cock MJW et al (2009) Ecological effects of invasive alien insects. Biol Invasions 11:21–45

Koepke DF, Kolb TE (2013) Species variation in water relations and xylem vulnerability to cavitation at a forest-woodland ecotone. For Sci 59:524–535

Kolb TE, Wagner MR, Covington WW (1994) Utilitarian and ecosystem perspectives: concepts of forest health. J Forest 92:10–15

Krawchuk MA, Meigs GW, Cartwright JM, Coop JD, Davis R, Holz A et al (2020) Disturbance refugia within mosaics of forest fire, drought, and insect outbreaks. Front Ecol Environ 18:235–244

Kurz WA, Dymond CC, Stinson G, Rampley GJ, Neilson ET, Carroll AL et al (2008) Mountain pine beetle and forest carbon feedback to climate change. Nature 452:987–990

Kuuluvainen T, Grenfell R (2012) Natural disturbance emulation in boreal forest ecosystem management—theories, strategies, and a comparison with conventional even-aged management. Can J for Res 42:1185–1203

Lamothe KA, Somers KM, Jackson DA (2019) Linking the ball-and-cup analogy and ordination trajectories to describe ecosystem stability, resistance, and resilience. Ecosphere 10:e02629

Landres PB, Morgan P, Swanson FJ (1999) Overview of the use of natural variability concepts in managing ecological systems. Ecol Appl 9:1179–1188

Laurance WF, Curran TJ (2008) Impacts of wind disturbance on fragmented tropical forests: A review and synthesis. Austral Ecol 33:399–408

Lausch A, Borg E, Bumberger J, Dietrich P, Heurich M, Huth A et al (2018) Understanding forest health with remote sensing, part III: requirements for a scalable multi-source forest health monitoring network based on data science approaches. Remote Sens 10:1120

Lausch A, Erasmi S, King DJ, Magdon P, Heurich M (2017) Understanding forest health with Rrmote sensing-part II—A review of approaches and data models. Remote Sens 9:129

Lausch A, Erasmi S, King D, Magdon P, Heurich M (2016) Understanding forest health with remote sensing–part I—A review of spectral traits. processes and remote-sensing characteristics. Remote Sens 8:1029

Leclerc M-AF, Daniels LD, Carroll AL (2021) Managing wildlife habitat: complex interactions with biotic and abiotic disturbances. Front Ecol Evol 9:613371

Lewis SL, Maslin MA (2015) Defining the anthropocene. Nature 519:171–180

Liebhold AM, Brockerhoff EG, Kalisz S, Nuñez MA, Wardle DA, Wingfield MJ (2017) Biological invasions in forest ecosystems. Biol Invasions 19:3437–3458

Liebhold AM, MacDonald WL, Bergdahl D, Mastro VC (1995) Invasion by exotic forest pests: a threat to forest ecosystems. Forest Science Monographs 30:1–49

Lindenmayer DB, Hobbs RJ, Likens GE, Krebs CJ, Banks SC (2011) Newly discovered landscape traps produce regime shifts in wet forests. Proc Natl Acad Sci 108:15887–15891

Lindenmayer DB, Franklin JF (2002) Conserving forest biodiversity: a comprehensive multiscaled approach. Island Press, Washington

Lindner M, Maroschek M, Netherer S, Kremer A, Barbati A, Garcia-Gonzalo J et al (2010) Climate change impacts, adaptive capacity, and vulnerability of European forest ecosystems. For Ecol Manage 259:698–709

Liu Y, Stanturf J, Goodrick S (2010) Trends in global wildfire potential in a changing climate. For Ecol Manage 259:685–697

Logan JA, Powell JA (2001) Ghost forests, global warming, and the mountain pine beetle (Coleoptera: Scolytidae). Am Entomol 47:160–172

Loreau M, Mouquet N, Gonzalez A (2003) Biodiversity as spatial insurance in heterogeneous landscapes. Proc Natl Acad Sci 100:12765–12770

Loreau M, Naeem S, Inchausti P, Bengtsson J, Grime JP, Hector A et al (2001) Biodiversity and ecosystem functioning: current knowledge and future challenges. Science 294:804–808

Mack RN, Simberloff D, Lonsdale WM, Evans H, Clout M, Bazzaz FA (2000) Biotic invasions: causes, epidemiology, global consequences, and control. Ecol Appl 10:689–710

Marini L, Okland B, Jönsson AM, Bentz B, Carroll AL, Forster B et al (2017) Climate drivers of bark beetle outbreak dynamics in Norway spruce forests. Ecography. https://doi.org/10.1111/ecog.02769

McDowell NG, Coops NC, Beck PSA, Chambers JQ, Gangodagamage C, Hicke JA et al (2015) Global satellite monitoring of climate-induced vegetation disturbances. Trends Plant Sci 20:114–123

McElhinny C, Gibbons P, Brack C, Bauhus J (2005) Forest and woodland stand structural complexity: Its definition and measurement. For Ecol Manage 218:1–24

Menéndez R, González-Megías A, Lewis OT, Shaw MR, Thomas CD (2008) Escape from natural enemies during climate-driven range expansion: a case study. Ecological Entomology 33:413–421

Millar CI, Stephenson NL (2015) Temperate forest health in an era of emerging megadisturbance. Science 349:823–826

Morin RS, Liebhold AM (2015) Invasions by two non-native insects alter regional forest species composition and successional trajectories. For Ecol Manage 341:67–74

Morin RS, Liebhold AM, Pugh SA, Crocker SJ (2017) Regional assessment of emerald ash borer, Agrilus planipennis, impacts in forests of the Eastern United States. Biol Invasions 19:703–711

Nolan RH, Blackman CJ, de Dios VR, Choat B, Medlyn BE, Li X et al (2020) Linking forest flammability and plant vulnerability to drought. For 11:779

Nunez-Mir GC, Liebhold AM, Guo Q, Brockerhoff EG, Jo I, Ordonez K, Fei S (2017) Biotic resistance to exotic invasions: its role in forest ecosystems, confounding artifacts, and future directions. Biol Invasions 19:3287–3299

Paoletti E, Schaub M, Matyssek R, Wieser G, Augustaitis A, Bastrup-Birk AM et al (2010) Advances of air pollution research: From forest decline to multiple-stress effects on forest ecosystem services. Environ Pollut 158:1986–1989

Paquette A, Messier C (2010) The role of plantations in managing the world's forests in the Anthropocene. Front Ecol Environ 8:27–34

Parker, V. T., & Pickett, S. T. A. (1998). Historical contingency and multiple scales of dynamics within plant communities. In Ecological scale: theory and applications, Columbia University Press, pp. 171–192

Parks SA, Miller C, Parisien M-A, Holsinger LM, Dobrowski SZ, Abatzoglou J (2015) Wildland fire deficit and surplus in the western United States, 1984–2012. Ecosphere 6:1–13

Payn T, Carnus J-M, Freer-Smith P, Kimberley M, Kollert W, Liu S et al (2015) Changes in planted forests and future global implications. For Ecol Manage 352:57–67

Pohjanmies T, Triviño M, Tortorec EL, Mazziotta A, Snäll T, Mönkkönen M (2017) Impacts of forestry on boreal forests: An ecosystem services perspective. Ambio 46:743–755

Radeloff VC, Helmers DP, Kramer HA, Mockrin MH, Alexandre PM, Bar-Massada A et al (2018) Rapid growth of the US wildland-urban interface raises wildfire risk. Proc Natl Acad Sci 115:201718850

Raffa KF, Aukema B, Bentz BJ, Carroll A, Erbilgin N, Herms DA et al (2009) A Literal Use of 'Forest Health' Safeguards against Misuse and Misapplication. J Forest 107:276–277

Raffa KF, Aukema B, Bentz BJ, Carroll AL, Hicke JA., Kolb TE (2015) Responses of tree-killing bark beetles to a changing climate. In Climate change and insect pests. CABI, pp 173–201

Raffa KF, Aukema BH, Bentz BJ, Carroll AL, Hicke JA, Turner MG, Romme WH (2008) Cross-scale drivers of natural disturbances prone to anthropogenic amplification: the dynamics of bark beetle eruptions. Bioscience 58:501–517

Raffa KF, Powell EN, Townsend PA (2013) Temperature-driven range expansion of an irruptive insect heightened by weakly coevolved plant defenses. Proc Natl Acad Sci USA 110:2193–2198

Reyer CPO, Brouwers N, Rammig A, Brook BW, Epila J, Grant RF et al (2015) Forest resilience and tipping points at different spatio-temporal scales: approaches and challenges. J Ecol 103:5–15

Sambaraju KR, Carroll AL, Aukema BH (2019) Multiyear weather anomalies associated with range shifts by the mountain pine beetle preceding large epidemics. For Ecol Manage 438:86–95

Scheffer M (2009) Critical transitions in nature and society. Princeton University Press, Princeton, USA

Scheffer M, Carpenter SR (2003) Catastrophic regime shifts in ecosystems: linking theory to observation. Trends Ecol Evol 18:648–656

Scheffer M, Carpenter S, Foley JA, Folke C, Walker B (2001) Catastrophic shifts in ecosystems. Nature 413:591–596

Seebens H, Bacher S, Blackburn TM, Capinha C, Dawson W, Dullinger S et al (2021) Projecting the continental accumulation of alien species through to 2050. Glob Change Biol 27:970–982

Seidl R, Müller J, Hothorn T, Bässler C, Heurich M, Kautz M (2016) Small beetle, large-scale drivers: how regional and landscape factors affect outbreaks of the European spruce bark beetle. J Appl Ecol 53:530–540

Seidl R, Rammer W, Spies TA (2014) Disturbance legacies increase the resilience of forest ecosystem structure, composition, and functioning. Ecol Appl 24:2063–2077

Seidl R, Schelhaas M-J, Lexer MJ (2011) Unraveling the drivers of intensifying forest disturbance regimes in Europe. Glob Change Biol 17:2842–2852

Seidl R, Spies TA, Peterson DL, Stephens SL, Hicke JA (2016b) Searching for resilience: addressing the impacts of changing disturbance regimes on forest ecosystem services. J Appl Ecol 53:120–129

Seidl R, Thom D, Kautz M, Martin-Benito D, Peltoniemi M, Vacchiano G et al (2017) Forest disturbances under climate change. Nat Clim Chang 7:395–402

Showalter DN, Raffa KF, Sniezko RA, Herms DA, Liebhold AM, Smith JA, Bonello P (2018) Strategic development of tree resistance against forest pathogen and insect invasions in defense-free space. Front Ecol Evol 6:124

Shvidenko A, Barber CV, Persson R, Gonzales P, Hassan R, Lakyda P et al (2005). Forest and woodland systems. In M. de los Angeles & C. Sastry (eds) Millennium ecosystem assessment: current state & trends Aasessment. Island Press, Washington, pp 587–614

Sing L, Metzger MJ, Paterson JS, Ray D (2017) A review of the effects of forest management intensity on ecosystem services for northern European temperate forests with a focus on the UK. For: Int J For Res, 91:151–164.

Stadelmann G, Bugmann H, Wermelinger B, Bigler C (2014) Spatial interactions between storm damage and subsequent infestations by the European spruce bark beetle. For Ecol Manage 318:167–174

Stephens SL, Collins BM, Fettig CJ, Finney MA, Hoffman CM, Knapp EE et al (2018) Drought, tree mortality, and wildfire in forests adapted to frequent fire. Bioscience 68:77–88

Stevens-Rumann CS, Kemp KB, Higuera PE, Harvey BJ, Rother MT, Donato DC et al (2018) Evidence for declining forest resilience to wildfires under climate change. Ecol Lett 21:243–252

Stireman JO, Dyer LA, Janzen DH, Singer MS, Lill JT, Marquis RJ et al (2005) Climatic unpredictability and parasitism of caterpillars: implications of global warming. Proc Natl Acad Sci USA 102:17384–17387

Strauss SY, Agrawal AA (1999) The ecology and evolution of plant tolerance to herbivory. Trends Ecol Evol 14:179–185

Straw NA, Williams DT, Kulinich O, Gninenko YI (2013) Distribution, impact and rate of spread of emerald ash borer Agrilus planipennis (Coleoptera: Buprestidae) in the Moscow region of Russia. For: Int J For Res, 86, 515–522.

Swetnam TW, Allen CD, Betancourt JL (1999) Applied historical ecology: using the past to manage for the future. Ecol Appl 9:1189–1206

Taylor, S. W., & Carroll, A. L. (2004). Disturbance, forest age, and mountain pine beetle outbreak dynamics in BC: a historical perspective. In: Shore TL , Brooks JE, Stone JE (eds) Mountain Pine Beetle Symposium: Challenges and Solutions . Natural Resources Canada, Canadian Forest Service, Victoria, BC, Canada, pp 41–51

Tews J, Brose U, Grimm V, Tielbörger K, Wichmann MC, Schwager M, Jeltsch F (2004) Animal species diversity driven by habitat heterogeneity/diversity: the importance of keystone structures. J Biogeogr 31:79–92

Thom D, Rammer W, Seidl R (2016) Disturbances catalyze the adaptation of forest ecosystems to changing climate conditions. Glob Change Biol 23:269–282

Thom D, Seidl R (2016) Natural disturbance impacts on ecosystem services and biodiversity in temperate and boreal forests. Biol Rev 91:760–781

Thompson ID, Guariguata MR, Okabe K, Bahamondez C, Nasi R, Heymell V, Sabogal C (2013) An operational framework for defining and monitoring forest degradation. Ecol Soc 18:1–23

Tkacz B, Moody B, Castillo JV, Fenn ME (2008) Forest health conditions in North America. Environ Pollut 155:409–425

Trumbore S, Brando P, Hartmann H (2015) Forest health and global change. Science 349:814–818

Turner MG (1989) Landscape Ecology: the effect of pattern on process. Annu Rev Ecol Syst 20:171–197

Turner MG (2010) Disturbance and landscape dynamics in a changing world. Ecology 91:2833–2849

Turner MG, Baker WL, Peterson CJ, Peet RK (1998) Factors Influencing Succession: lessons from large, infrequent natural disturbances. Ecosystems 1:511–523

Turner MG, Gardner RH (2015) Landscape ecology in theory and practice, pattern and process. 175–228.

Turner MG, Hargrove WW, Gardner RH, Romme WH (1994) Effects of fire on landscape heterogeneity in Yellowstone National Park, Wyoming. J Veg Sci 5:731–742

Tylianakis JM, Didham RK, Bascompte J, Wardle DA (2008) Global change and species interactions in terrestrial ecosystems. Ecol Lett 11:1351–1363

Veech JA, Summerville KS, Crist TO, Gering JC (2002) The additive partitioning of species diversity: recent revival of an old idea. Oikos 99:3–9

Vilà M, Basnou C, Pyšek P, Josefsson M, Genovesi P, Gollasch S et al (2010) How well do we understand the impacts of alien species on ecosystem services? A pan-European, cross-taxa assessment. Front Ecol Environ 8:135–144

White PS, Jentsch A (2001) The search for generality in studies of disturbance and ecosystem dynamics. Progress in Botany 62:399–450

White PS, Pickett STA (1985) Natural disturbance and patch dynamics: an introduction. In: Pickett STA, White PS (eds) The ecology of natural disturbance and patch dynamics. Academic Press, New York, NY, USA, pp 3–13

Whitman E, Parisien M-A, Thompson DK, Flannigan MD (2019) Short-interval wildfire and drought overwhelm boreal forest resilience. Sci Rep 9:18796

Whittaker RH (1972) Evolution and measurement of species diversity. Taxon 21:213–251

Williams AP, Abatzoglou JT, Gershunov A, Guzman-Morales J, Bishop DA, Balch JK, Lettenmaier DP (2019) Observed impacts of anthropogenic climate change on wildfire in California. Earth's Futur 7:892–910

Wingfield MJ, Brockerhoff EG, Wingfield BD, Slippers B (2015) Planted forest health: the need for a global strategy. Science 349:832–836

Woodall CW, Amacher MC, Bechtold WA, Coulston JW, Jovan S, Perry CH et al (2011) Status and future of the forest health indicators program of the USA. Environ Monit Assess 177:419–436

Young DJN, Stevens JT, Earles JM, Moore J, Ellis A, Jirka AL, Latimer AM (2017) Long-term climate and competition explain forest mortality patterns under extreme drought. Ecol Lett 20:78–86

Zeng H, Peltola H, Väisänen H, Kellomäki S (2009) The effects of fragmentation on the susceptibility of a boreal forest ecosystem to wind damage. For Ecol Manage 257:1165–1173

Chapter 22
Climate Change and Forest Insect Pests

Andrea Battisti and Stig Larsson

22.1   Introduction

Climate change and the underlying causal factors have been thoroughly described (Field et al. 2014). Climate change, particularly increased temperature, has several consequences for the functioning of ecosystems. For instance, we know that the distribution range of some organisms has changed (Parmesan et al. 1999), tree phenology altered (Walther et al. 2002), and phenological asynchrony developed, e.g. between tree and associated insects (Visser and Both 2006). Although these effects are well understood and documented, we are only beginning to understand the effects of climate change on insect communities. This is in large part because of the complexity of their interactions with the abiotic and biotic environment.

It seems obvious that insect pest problems will be more important in a warmer climate because of the strong positive effect that temperature has on insect physiology and demography (Ayres and Lombardero 2000). However, temperature increases above optimal ranges may also be detrimental to insect fitness (Lehmann et al. 2020). In addition, it must be remembered that insect distribution and abundance are controlled by many factors other than temperature.

Klapwijk et al. (2012) reviewed climate-change associated factors affecting the outbreak potential of forest insects. They identified direct and indirect factors and provided a theoretical framework for assessing how changes in climate can be incorporated into predictive models of insect population dynamics. Similarly, Battisti and Larsson (2015) and Jactel et al. (2019) reviewed how climate change can affect the distribution range of insect pests, and provided examples of forest insect species whose ranges have been changed in a manner consistent with changes in climate.

A. Battisti (✉)
University of Padova, DAFNAE, Legnaro, Padova, Italy
e-mail: andrea.battisti@unipd.it

S. Larsson
Swedish University of Agricultural Sciences, Uppsala, Sweden

© The Author(s) 2023   773
J. D. Allison et al. (eds.), *Forest Entomology and Pathology*,
https://doi.org/10.1007/978-3-031-11553-0_22

This chapter summarizes empirical evidence for climate-change induced insect pest problems, i.e. changed distribution range and frequency of insect outbreaks. Climate change can interact with non-native insect species accidently introduced into novel areas (Brockerhoff and Liebhold 2017). The issue of invasions is discussed in Chapter 23 of this volume. In this chapter we briefly discuss, in general terms, if and how climate change can be a factor that contributes to non-native insect species being established and becoming invasive, i.e. acting as novel pests in the forest.

Throughout the chapter the focus is on how climate change affects the distribution and abundance of forest pests (directly and indirectly through biotic interactions). We acknowledge that climate change will also influence host tree vulnerability and tolerance, and thus potential future damage (Toïgo et al. 2020; Forzieri et al. 2021) (discussed in Chapter 20).

22.2 Climatic Drivers

There is general consensus among scientists that the global climate is changing at an unprecedented rate, with many regions experience warming trends, shifts in precipitation patterns, and more frequent extreme weather events (Field et al. 2014). Factors potentially affecting forest insects include temperature, precipitation, rare weather events such as wind storms and heat waves, and atmospheric carbon dioxide concentration. All these factors can act both directly and indirectly (through host plant or natural enemies) on insect pests.

Temperature is the most important driver because it has steadily increased since the beginning of the twentieth century (0.61 °C in global mean temperature from 1850–1900 to 1986–2005; Field et al. 2014), and is predicted to increase further. Forests experience different levels of climate change depending on geographic position. Upper latitudes of northern and southern hemispheres, where most of the world's temperate and boreal forests grow, are expected to experience a higher warming. Insects, being poikilothermic organisms, respond directly to temperature as described by their specific reaction norms. Temperature also affects insects indirectly through effects on the host plant (bottom-up, see Chapter 7) and natural enemies (top-down, see Chapter 6).

Patterns in precipitation are due to a complex interaction between air circulation and temperature. Thus, an effect of temperature increase on precipitation patterns is expected. The result, however, is not as clear as the one depicted for temperature alone. Predictions on the total amount of annual precipitation vary according to the geographic area, with upper latitudes of both hemispheres experiencing more precipitation than mid latitudes, while the tropical and subtropical regions show a patchy effect (Field et al. 2014). Precipitation is also characterized by two more aspects, i.e. its distribution in the year and the intensity of the precipitation events. At upper latitudes of the northern hemisphere, precipitation increase will mainly occur in winter, while intense precipitation events will be more likely everywhere. Forests will thus experience different precipitation regimes according to geographic region. The interactions of these changes with those of temperature and solar radiation (through modified cloudiness) will likely modify the microclimatic niche experienced by forest insects. Although precipitation is known to directly affect forest insects, most of its action is indirect because water availability is crucial for tree growth, and consequently, host quality for insect herbivores.

Extreme rare weather events, such as high/low temperature and rainfall, strong wind, and their combinations, will probably occur in higher frequency, and this is considered a potentially important component of climate change (Field et al. 2014). Two factors characterize the nature of these events, timing and intensity. For example, a heat wave may suppress all the insects active in that moment because the upper thermal threshold is achieved (see Chapter 4), or a wind storm may simultaneously fell a large number of trees that may facilitate a bark beetle outbreak (see Chapter 10). The periods when such events may happen are roughly predictable, because they are associated with the yearly variation of both temperature and precipitation, although it is impossible to define exactly when and where they will occur.

Carbon dioxide, together with other greenhouse gases, is a major determinant of temperature increase (Field et al. 2014). Being a fundamental molecule for photosynthesis, the increase in carbon dioxide in the atmosphere may affect the metabolism of forest trees, including molecules of importance to tree-feeding insects (Lindroth et al. 1993), although the general effects on herbivorous insects are weak and idiosyncratic (Hillstrom et al. 2014).

The climate in the future will most likely differ in all the above-mentioned aspects. However, with respect to effects on forest insects temperature has by far been the most discussed in the literature, and thus is the factor for which there exists a reasonable amount of data. Therefore, it will be the focus of discussion in the following sections.

22.3 Insect Response to Increased Temperature

In this section we deal with temperature effects at the level of the individual insect. Temperature has a direct effect on insect development rate and survival. Development rate generally increases with increasing temperature to some maximum, above which development slows down and mortality increases (see Chapter 4). Increased development rate could lead to increased voltinism in facultative multivoltine species. Increased development rate in insect larvae could result in reduced temporal exposure to enemies or other mortality agents, with resulting higher survival.

Winter mortality is likely to decrease under increasing temperatures (e.g. Ayres and Lombardero 2000), although decreased snow cover (and therefore decreased insulation of overwintering sites) can reverse that pattern (Petrucco-Toffolo and Battisti 2008). Warmer winters may permit some non–diapausing species to continue feeding and development during months that were previously too cold (Schneider et al. 2021). For example, larvae of the pine processionary moth *Thaumetopoea pityocampa* have a higher probability of survival if winter temperatures do not often fall below specific feeding thresholds (Battisti et al. 2005, Fig. 22.1A, B).

A

B

Fig. 22.1 **A** The same tree can be colonized in subsequent years as shown in the photo where the remains of an old nest are visible close to two new nests with white silk. The tree is a Scots pine (*Pinus sylvestris*) growing at high elevation (>1,400 m) in the Southern Alps (Venosta/Vinschgau valley) where the insect has expanded its distribution in recent decades and reached the upper limit of host plant range. At even higher elevations in the Alps, the pine processionary moth is massively colonizing the dwarf mountain pine (*Pinus mugo*), historically not a suitable host because of being covered by snow. **B** The photo shows a colony that survived the winter because the limited snow cover, and higher temperatures, permitted suitable conditions for larval feeding across the winter.

Several indirect effects mediated through the host tree exist. Many insect species match their feeding activity with certain developmental stages in the host plant; for example, species associated with deciduous trees, such as the autumnal moth *Epirrita autumnata*, match their feeding with nutritious immature foliage during spring and early summer (Haukioja 2003).

If host trees are reasonably well matched to historically favorable climatic conditions, then it is inevitable that changing climate will lead to situations where trees are poorly matched to the new conditions, i.e. trees can be stressed. Stress-induced changes in plant tissue quality and their effects on insect survival and reproduction are well documented in experimental studies (Koricheva et al. 1998).

That plant stress can trigger insect outbreaks is a long-standing hypothesis in forest entomology. Insect outbreaks have been commonly correlated with conditions that induce stress in their host plants (e.g. Mattson and Haack 1987). This has led to speculation that there is a causal link between stress-induced changes in plant quality, and thus insect performance, and the start of outbreaks (e.g. White 1974). Experimental tests of the plant stress hypothesis, most often at the level of individual insects, have produced mixed results; species from some feeding guilds respond to experimentally stressed trees with increased performance, some are unaffected, and some respond negatively (Larsson 1989). Bark beetles constitute a globally important group of insects for which plant stress seems relevant; a long-standing paradigm is that healthy trees are resistant to most bark beetle species (and other boring insects), but that periods of stress make trees susceptible, although at high beetle density even non-stressed trees can be attacked and killed (Raffa et al. 2008).

Arthropod natural enemies can exert powerful forces on the performance of herbivorous insects (see Chapter 6), and climate change may affect their activity as much as that of their prey. In addition, the phenological synchrony between natural enemies and their hosts/prey can also be affected. However, specialist enemies should be under strong selection to track phenological changes in their prey, which might make them less likely to become temporally uncoupled from their prey (Klapwijk et al. 2010). Higher temperatures can influence parasitism and predation rates by increasing searching activity of individual parasitoids and predators. When the prey are relatively immobile (e.g. many immature insect herbivores), this should generally increase rates of detection and attack.

Insect pathogens, e.g. fungi, bacteria, and viruses, can also limit the performance of herbivorous insects (Hajek 1997). Temperature can be important for both infection rate and defense responses within the host. Different thermal optima for host and pathogen might lead to a situation where high temperatures favor the host by both optimizing defense responses and directly limiting pathogen growth (Blanford and Thomas 1999).

Insects rarely, if ever, act independently from other organisms. Therefore, it is necessary to consider the position of the forest insect in the trophic web. In other words, not only should we consider direct effects of climate change on the target insect, but we need to recognize likely interactions of climatic variables with host tree, natural enemies, and insect diseases (indirect effects). Such an approach is necessary to fully understand the potential consequences that climate change can have for pests. For climate-change driven effects to have an impact on forest ecosystems, and thus be of economic concern for forest managers, effects at the level of individual insects need to be confirmed at the level of the population. This is not a trivial step as a multitude of biotic interactions, each with its inherent uncertainty, can modify the effects when it comes to populations, as discussed in the following section.

22.4 Insect Population Response to Increased Temperature

Climate warming can influence two important aspects of insect population ecology: distribution and abundance. Many insect species have been documented to change their distribution range in response to increased temperature (Battisti and Larsson 2015). It is important to realize, however, that the dynamics of range expansion are rarely known in any detail. This is simply because populations in expansion areas are initially at very low density and thus can remain undetected for a long time. If the expanding population is a forest pest, then the expansion is likely to be discovered if the population reaches outbreak numbers.

In general, the majority of insect populations are controlled by a number of different agents and thus occur at low density. Under some conditions an insect species may escape from the controlling agents and reach outbreak densities. Outbreaks are easily observed because they are generally defined by managers as population densities so high that they are of economic concern in forestry. Outbreaks can thus be seen as a proxy for high-density events, and of course as a warning signal of potential forest health problems. Low-density populations will also vary in size, but their dynamics will most of the time be unnoticed because their densities will not result in damage to the forest.

In the following section, we present case studies to illustrate effects of climate change on insect populations. The focus is on outbreaks, which we assign to three main groups. The first group of case studies refer to *Outbreaks at the core of historical range of distribution,* thus evidence of climate change effects (or lack of effects) on populations in the historic range of the distribution. The second group deals with *New areas of outbreaks within historic species distribution.* This refers to species where no outbreaks were recorded for a portion of their historical distribution, typically in the colder areas, but where outbreaks in recent years have been observed. The third group includes *Outbreaks in recently invaded geographic areas,* in other words, species that have expanded their distribution range and occurred at outbreak densities that clearly can be related to warming.

22.4.1   Outbreaks at the Core of Historical Range of Distribution

Long-term surveillance data of European insect populations report a large variability in the responses of key forest pests to climate change: positive, negative, and no response to increased temperatures (Haynes et al. 2014; Lehmann et al. 2020).

The European spruce beetle *Ips typographus* is the most aggressive bark beetle in Europe. Analysis of 17 time-series spanning from 1980 to 2010 shows density-dependent factors to be the main drivers of population dynamics, although high temperature and summer precipitation deficit also play a role (Marini et al. 2017). In addition, temperature appears to be important for the voltinism of *I. typographus,* as populations may become bi- or multivoltine under favorable conditions (Wermelinger 2004). Results suggest that greater efforts should be made to integrate temperature increase, drought, and storm effects into future scenarios of outbreaks under climate change (Marini et al. 2017).

The larch bud moth *Zeiraphera griseana* is an example of a pest where climate warming has had negative effects on population growth. Dendrochronological analyses of host trees associated with *Z. griseana* outbreaks over 500 years reveal periodicities of 4, 8, and 16 years throughout the time series, except during the period 1690–1790, and since 1980. The data suggest a disruption of periodicity probably related to changes in climate; temperature decreased in the period 1690–1790 (Little Ice Age) wheras it increased since 1980 (Saulnier et al. 2017).

Responses at the species level appear idiosyncratic and no general patterns were observed in several species of defoliating insects associated with coniferous trees in southern Germany for more than 200 years (Haynes et al. 2014). A similar study in Hungary involving five species of defoliating insects associated with broadleaved tree species for a period of about 60 years also observed no clear pattern in responses (Klapwijk et al. 2013). It should be noted that for both these studies data refer to large scale events, and that changes at local scale could have gone undetected. More precise data are available for the pine processionary moth *Thaumetopoea pityocampa* from eight geographic zones in France (but for a shorter period, 1981–2014). Although in general, populations were controlled mainly by density-dependent agents, population growth was negatively related to precipitation in five regions and positively related to winter temperature in four regions; thus, these data suggest that the effects of weather-related factors need to be considered at a local scale using appropriate measures of population density (Toïgo et al. 2017).

22.4.2   New Areas of Outbreaks Within Historic Species Distribution

Both the autumnal moth *Epirrita autumnata* and the winter moth *Operophtera brumata* have expanded their outbreak range in recent years, presumably as a result of improved winter survival of eggs, and maintenance of synchrony (through adaptive phenological plasticity) with bud burst of their main host, the mountain birch *Betula pubescens* ssp. *tortuosa* (Jepsen et al. 2008). Winter moth populations show a pronounced north-eastern expansion of outbreaks into areas previously dominated by the autumnal moth, which in turn has expanded historically into colder areas (Tenow 1996). This has been possible because eggs of the autumnal moth are more cold tolerant than those of the winter moth. This important direct effect of increased temperature can be affected by indirect effects in the trophic interactions and in the synchronization with the bud break of the host plants. In subarctic mountain birch forests, predation rates on *E. autumnata* and *O. brumata* larvae were almost twice as high in low versus high elevation sites, indicating that release from predation pressure at high elevations can favor outbreaks in these cooler habitats (Pepi et al. 2017).

Records of spruce budworm (*Choristoneura fumiferana*) defoliation and tree-ring analysis indicate that the outbreak range of this insect has expanded to the north. A regional tree-ring chronology performed by Boulanger et al. (2012) represents the longest and most replicated reconstruction of outbreak dynamics in North America (1551–1995). The authors identified nine potential outbreaks and three uncertain outbreaks in a 400-year period and concluded that outbreak frequency varied with temperature, being less frequent during the 1660–1850 period (every ~ 50 years,

Little Ice Age) and more frequent in warmer periods like prior to 1660 (every ~ 28 years) and during the twentieth century (every ~ 30 years). The simultaneous occurrence of a general increase in temperature in northern latitudes at the start of the last outbreak indicates a relation with climate change (Candau and Fleming 2011). An interesting indirect effect involving the host plant has been suggested. The main host of the spruce budworm is balsam fir *Abies balsamea*, whereas black spruce *Picea mariana* is a secondary host. Climate change is predicted to advance the phenology of the secondary host that is more abundant at the upper latitudinal edge, making it more susceptible to defoliation, and thus facilitating expansion of the outbreak area into higher latitudes (Pureswaran et al. 2015). This factor has been hypothesized to explain the occurrence of the new outbreak that started in 2006 about four degrees (445 km) of latitude north of the previous one (1966–1992), with a prediction for a more northern expansion in 2041–2070 (Régnière et al. 2012).

In the southern hemisphere, the defoliation of *Nothofagus* forests by the saturnid moths of the genus *Ormiscodes* have been associated with drier and warmer seasons. The outbreaks have been more frequent in southern than in northern Patagonia. Results are consistent with recent warming in southern Patagonia and suggest that outbreak frequency may continue to increase with further warming (Paritsis and Veblen 2011).

22.4.3 Outbreaks in Recently Invaded Geographic Areas

In recent decades, the pine processionary moth *T. pityocampa* has expanded its latitudinal and elevational distribution range (Battisti et al. 2005). Improved survival during the feeding period in winter has contributed to outbreaks in pine forests previously unoccupied in France, Italy, Spain, and Turkey. Rapid range expansion is facilitated by warm summer nights that contribute to long-distance (more than 2 km) dispersal of female moths (Battisti et al. 2006). In the newly occupied areas, however, population dynamics are driven more by density-dependent agents than by climatic drivers (Tamburini et al. 2013). Thus, once the expansion area is occupied population dynamics seem to be determined by the same factors as in the historical range, provided that specialist enemies have tracked the host in the new areas. Interestingly, *T. pityocampa* shows prolonged diapause facilitating persistence in the newly colonized areas even if the weather turns unfavorable for one or more years; diapause can last up to eight years with some individuals emerging every year (Salman et al. 2016).

In western Canada recent outbreaks of the mountain pine beetle *Dendroctonus ponderosae* have led to extensive tree mortality within at least 14 million hectares of lodgepole pine *Pinus contorta* forests. The start of the outbreak was facilitated by fire suppression during the last century, which created large tracts of over-mature pine stands, in combination with recent climatic patterns, viz. mild winters and warm dry summers (Raffa et al. 2008; Bentz and Jönsson 2015). However, the relative importance of large areas of susceptible pine forests and suitable climatic conditions for beetle population growth is not entirely clear (Cooke and Carroll 2017).

In general, bark beetle species associated with weakened trees are difficult to detect at low-densities, whereas damage and tree mortality become obvious during outbreaks. Therefore, the range edge generally considered is that of the epidemic range, whereas the margins of the endemic range remain largely unknown. In southeast USA, the distribution of the southern pine beetle *Dendroctonus frontalis* has been moved northwards due to milder winters that enhance beetle performance (Ungerer et al. 1999). Similarly, spruce bark beetle (*Dendroctonus rufipennis*) outbreaks may occur throughout the range of spruce in North America in the future. In its coldest locations, *D. rufipennis* is semivoltine, having a generation every two years and outbreaks are rare in these populations (Schebeck et al., 2017).

22.5 Invasive Species and Climate Change

The increasing problem with invasive species during the last decades may be linked to climate change although the evidence for this remain limited. Global trade and travel are the major drivers of the invasion process (Ramsfield et al. 2016; Brockerhoff and Liebhold 2017). The process of invasion is often divided into several phases (pre-transport, transport, arrival, establishment, and spread; see Chapter 23). Here we briefly discuss how changes in climate can interact with trade and travel in each of the invasion phases.

Very little can be said about the pre-establishment phases (pre-transport, transport, arrival). It is obvious that propagule pressure in the area of origin is important in order to assess the probability of transportation, but it is unclear to what extent changes in climate affect propagule pressure. The next two phases (establishment and spread) are clearly linked to climate change as they depend on the matching between the area of origin and the area of arrival. The impact of climate change on climate matching between areas of origin and destination on the establishment and spread of non-native species is difficult to assess as data about failure to establish are rarely available for forest insects. The increasing number of the incursions of ambrosia beetles from tropical and subtropical regions in temperate forests could be an example of how this category of organisms is favored by climate change (Rassati et al. 2016). The inclusion of climatic responses of pests in the risk assessment of invasive species may help to predict which ones are the most likely to get established and threat newly colonized habitats (Grousset et al. 2020).

The spread of invasive species in a newly colonized area depend on niche availability and dispersal traits of the insect, and in principle does not differ from that of native species (Pureswaran et al. 2018). The hemlock woolly adelgid *Adelges tsugae* was introduced into Virgina, eastern USA in the mid 1900s. Increase in mean minimum winter temperature resulted in higher survival in overwintering life stages and facilitated the expansion northwards in the eastern USA (Paradis et al. 2008). The build up of high density populations in the already colonized areas contributed greatly to hemlock (*Tsuga canadensis, T. caroliniana*) dieback (Fitzpatrick et al. 2012).

Once established, the response of invasive species to climatic factors may be similar to that of native species, as illustrated by the spongy moth *Lymantria dispar*. The population dynamics of *L. dispar* have been thoroughly documented in the USA, showing periods of cyclic outbreaks intermingled by periods with no cycles. The dynamics seem to be driven by trophic interactions while the role of climate appears to be negligible (Allstadt et al. 2013). It is not clear if a changing climate would cause a net increase in suitable habitat for invading insects such as spongy moth in North America, as there should be some areas that become more favorable and others that become less favorable (Tobin et al. 2014).

22.6 Conclusions

Climate change, in particular increased temperature, is certain to have qualitative and quantitative effects on insect populations, primarily because temperature ultimately sets the limit for most insect distribution ranges (Battisti and Larsson 2015). For insects on trees, however, the availability of the host tree(s) will be a critical factor because most insect species are associated with one or a few host tree species only. The expected slower range expansion by trees compared with that of the insects, because of the much longer generation time of trees, will likely slow down the successful expansion of the insects. Overall, this probably means that at a certain point in time host tree availability, rather than temperature, may set the limit for future insect range expansion. This scenario would only apply to insect species not able to switch to novel host tree species in the expansion area.

Climate-change attention is mostly on insect species expanding their range into geographic areas that have become climatically more favorable. We have to assume, however, that an equally large area may become unfavorable. In contrast to expansion, such retraction of the range at the lower edge of the distribution will not be as immediate as the expansion, mainly because plants do not react as quickly to the warming as insects do. This is why a net increase in areal distribution is expected in the short-term while in a medium-long term a general shift of the range is predicted.

A difficult task is to assess whether or not damage to forests will be more, or less, severe under climate change (Jactel et al. 2019; Lehmann et al. 2020). The degree of damage is usually positively related to the density of the insect pest population. Thus, we can reformulate the issue using outbreak as a proxy for damage and ask: are outbreaks likely to be more common under climate change?

Ideally, in order to scientifically analyze this issue we should be able to refer to the frequency of outbreaks for a scenario of no climate warming. Obviously, this is not a straightforward matter, but the literature provides important information about insect populations that can be used as a simple null model of outbreak frequency (Barbosa and Schultz 1987). Most insect populations in forest ecosystems thrive around low mean densities, far below outbreak densities (Landsberg and Ohmart 1989), meaning that they are efficiently controlled by several, mostly unknown factors. It also indicates that many insect populations often remain unnoticed for a long time (endemic) and are considered pests only when they build up epidemic populations (outbreak) (Barbosa et al. 2012).

The categorization of forest insect populations outlined above is simplistic, but still useful as a basis for the following discussion of forest damage and insect pests under future climate change. We envisage four situations:

1. The extent of range expansion of non-outbreak insects is virtually unknown; this should come as no surprise because, by definition, these insect species occur at low density. It is quite likely, however, that such expansion has occurred but should be of minor importance from a management point of view, given that the population ecology of these putative species in the new area is similar to that in their core area.
2. A bias exists in the literature with almost all evidence of climate-change effects coming from outbreak species, for obvious reasons (easy to observe). In the event that outbreaks occur in the expanded range, an important question will then be whether the outbreak dynamics are similar to those in the original distribution range or show new characteristics (the mountain pine beetle outbreak may be an example of this as it has invaded new host tree species, such as *Pinus banksiana* in Alberta, creating the potential for massive range expansion into north central and eastern north America).
3. Outbreaks in the historical area can be more, or less, frequent under climate change depending on the life history of the insect and how climate affects biotic interactions (with host tree, natural enemies, insect pathogens). Forest management is changing in many parts of the world, e.g. with stands being overall more intensively managed than in the past. So far, there are no data to suggest that pest dynamics are significantly different under intensive forestry, such as nitrogen fertilization of natural stands (Kytö et al. 1996). If novel management practices, e.g. for maximizing carbon sequestration, will be introduced on a large scale, then there is certainly a risk that pest problems will follow.
4. Forest health problems due to non-native insect species will most likely continue to increase in the future. Some non-natives will establish but with dynamics of the low-density type. The distribution of other non-natives will expand, perhaps as a consequence of climate change, and establish in natural forests where populations increase to outbreak level (thus becoming an invasive). An especially serious threat is the situation where non-native insects establish in plantations of non-native tree species. Here managers may be faced with a situation of intensive control practice most often not necessary in traditional forestry, such as the application of biological control with a parasitic nematode against the *Sirex* wood wasp in pine plantations (Slippers et al. 2015).

The science of outbreak dynamics includes data from economically important insect populations whose dynamics appear to be driven by factors that differ from those of non-outbreak species (e.g. Larsson et al. 1993). Thus, there is no overarching hypothesis based on logic (or data) that allows for specific predictions at the species (or population) level. Our approach has been to use information from the past in order to understand the future. This approach allows us to take advantage of existing scientific knowledge. Although we advocate this approach we emphasize that we also need to appreciate that the available data, and thus predictions based on these data, have a substantial degree of uncertainty. Very rarely, if ever, can outbreak data be considered replicated, due to different boundary conditions. This is a situation that is true for many ecological data sets meant to be used in policy, but is especially troublesome here because we are interested in changes over long periods of time, hundreds of years.

Acknowledgements Many thanks to the Editors Jeremy D. Allison and Tim Paine, and to Barbara Bentz, Sandy Liebhold, and Deepa Pureswaran who kindly commented an earlier version of the chapter. The work benefited of a grant of Swedish University of Agricultural Sciences (SLU) and the European Union's Horizon 2020 Program for Research and Innovation 'HOMED' (grant no. 771271).

References

Allstadt AJ, Haynes KJ, Liebhold AM, Johnson DM (2013) Long-term shifts in the cyclicity of outbreaks of a forest-defoliating insect. Oecologia 172:141–151

Ayres MP, Lombardero MJ (2000) Assessing the consequences of global change for forest disturbance from herbivores and pathogens. Sci Total Environ 262:263–286

Barbosa P, Schultz JC (1987) Insect outbreaks. Academic Press, New York

Barbosa P, Letourneau DK, Agrawal AA (2012) Insect outbreaks revisited. Academic Press, New York

Battisti A, Stastny M, Netherer S, Robinet C, Schopf A, Roques A, Larsson S (2005) Expansion of geographic range in the pine processionary moth caused by increased winter temperatures. Ecol Appl 15:2084–2096

Battisti A, & Larsson S (2015) Climate change and insect pest distribution range. In C. Björkman & P. Niemelä, editors. Climate change and insect pests. CABI International, pp 1–15

Battisti A, Stastny M, Buffo E, Larsson S (2006) A rapid altitudinal range expansion in the pine processionary moth produced by the 2003 climatic anomaly. Glob Change Biol 12:662–671

Bentz BJ, Jönsson AM (2015) Modeling bark beetle responses to climate change. In: Vega FE, Hofstetter RW (eds) Bark Beetles. Academic Press, San Diego, pp 533–553

Blanford S, Thomas MB (1999) Host thermal biology: the key to understanding host–pathogen interactions and microbial pest control? Agric for Entomol 1:195–202

Boulanger Y, Arseneault D, Morin H, Jardon Y, Bertrand P, Dagneau C (2012) Dendrochronological reconstruction of spruce budworm (*Choristoneura fumiferana*) outbreaks in southern Quebec for the last 400 years. Can J For Res 42:1264–1276

Brockerhoff EG, Liebhold AM (2017) Ecology of forest insect invasions. Biol Invasions 19:3141–3159

Candau J-N, Fleming R (2011) Forecasting the response of spruce budworm defoliation to climate change in Ontario. Can J For Res 41:1948–1960

Cooke BJ, Carroll AL (2017) Predicting the risk of mountain pine beetle spread to eastern pine forests: considering uncertainty in uncertain times. For Ecol Manage 396:11–25

Field CB, Barros VR, Mach KJ et al. (2014). Climate change 2014: impacts, adaptation, and vulnerability. part A: global and sectoral aspects. Contribution of Working Group II to the fifth assessment report of the intergovernmental panel on climate change. Cambridge University Press, Cambridge, United Kingdom and New York, NY, USA, pp 35–94

Fitzpatrick MC, Preisser EL, Porter A, Elkinton J, Ellison AM (2012) Modeling range dynamics in heterogeneous landscapes: invasion of the hemlock woolly adelgid in eastern North America. Ecol Appl 22:472–486

Forzieri G, Girardello M, Ceccherini G, Spinoni J, Feyen L, Hartmann H, Beck PSA, Camps-Valls G, Chirici G, Mauri A, Cescatti A (2021) Emergent vulnerability to climate-driven disturbances in European forests. Nat Commun 12:1081

Grousset F, Grégoire J-C, Jactel H, Battisti A, Benko Beloglavec A, Hrašovec B, Hulcr J, Inward D, Orlinski A, Petter F (2020) The risk of bark and ambrosia beetles associated with imported non-coniferous wood and potential horizontal phytosanitary measures. Forests 11:342

Hajek AE (1997) Ecology of terrestrial fungal entomopathogens. Adv Microb Ecol 15:193–249

Haukioja E (2003) Putting the insect into the birch–insect interaction. Oecologia 136:161–168

Haynes KJ, Allstadt AJ, Klimetzek D (2014) Forest defoliator outbreaks under climate change: Effects on the frequency and severity of outbreaks of five pine insect pests. Glob Change Biol 20:2004–2018

Hillstrom ML, Couture JJ, Lindroth RL (2014) Elevated carbon dioxide and ozone have weak, idiosyncratic effects on herbivorous forest insect abundance, species richness, and community composition. Insect Conservation and Diversity 7:553–562

Jactel H, Koricheva J, Castagneyrol B (2019) Responses of forest insect pests to climate change: not so simple. Current opinion in Insect Science 35:103–108

Jepsen JU, Hagen SB, Ims RA, Yoccoz NG (2008) Climate change and outbreaks of the geometrids *Operophtera brumata* and *Epirrita autumnata* in subarctic birch forest: evidence of a recent outbreak range expansion. J Anim Ecol 77:257–264

Klapwijk MJ, Grobler BC, Ward K, Wheeler D, Lewis OT (2010) Influence of experimental warming and shading on host–parasitoid synchrony. Glob Change Biol 16:102–112

Klapwijk MJ, Ayres MP, Battisti A, Larsson S (2012) Assessing the impact of climate change on outbreak potential. In: Barbosa P, Letourneau DK, Agrawal AA (eds) Insect outbreaks revisited. Academic Press, New York, pp 429–450

Klapwijk MJ, Csóka G, Hirka A, Björkman C (2013) Forest insects and climate change: long-term trends in herbivore damage. Ecol Evol 3:4183–4196

Koricheva J, Larsson S, Haukioja E (1998) Insect performance on experimentally stressed woody plants: a meta-analysis. Annu Rev Entomol 43:195–216

Kytö M, Niemelä P, Larsson S (1996) Insects on trees: population and individual response to fertilization. Oikos 75:148–159

Landsberg J, Ohmart C (1989) Levels of insect defoliation in forests: patterns and concepts. Trends Ecol Evol 4:96–100

Larsson S (1989) Stressful times for the plant stress—insect performance hypothesis. Oikos 56:277–283

Larsson S, Björkman C, Kidd NAC (1993) Outbreaks in diprionid sawflies: why some species and not others? In: Wagner MR, Raffa KF (eds) Sawfly life history adaptations to woody plants. Academic Press, San Diego, pp 453–483

Lehmann P, Ammunét T, Barton M, Battisti A, Eigenbrode SD, Jepsen JU, Kalinkat G, Neuvonen S, Niemelä P, Terblanche JS, Økland B, Björkman C (2020) Complex responses of global insect pests to climate warming. Front Ecol Environ 18:141–150

Lindroth RL, Kinney KK, Platz CL (1993) Responses of deciduous trees to elevated atmospheric CO_2: productivity, phytochemistry, and insect performance. Ecology 74:763–777

Marini L, Økland B, Jönsson AM, Bentz B, Carroll A, Forster B, Grégoire J-C, Hurling R, Nageleisen LM, Netherer S, Ravn HP, Weed A, Schroeder M (2017) Climate drivers of bark beetle outbreak dynamics in Norway spruce forests. Ecography 40:1426–1435

Mattson WJ, Haack RA (1987) The role of drought stress in provoking outbreaks of phytophagous insects. In: Barbosa P, Schultz JC (eds) Insect Outbreaks. Academic Press, San Diego, pp 365–394

Paradis A, Elkinton J, Hayhoe K, Buonaccorsi J (2008) Role of winter temperature and climate change on the survival and future range expansion of the hemlock woolly adelgid (*Adelges tsugae*) in eastern North America. Mitig Adapt Strat Glob Change 13:541–554

Paritsis J, Veblen TT (2011) Dendroecological analysis of defoliator outbreaks on *Nothofagus pumilio* and their relation to climate variability in the Patagonian Andes. Glob Change Biol 17:239–253

Parmesan C, Ryrholm N, Stefanescu C, Hill JK, Thomas CD, Descimon H, Huntley B, Kaila L, Kullberg J, Tammaru T, Tennent WJ, Thomas JA, Warren M (1999) Poleward shifts in geographical ranges of butterfly species associated with regional warming. Nature 399:579–583

Pepi AA, Vinstad OPL, Ek M, Jepsen JU (2017) Elevationally biased avian predation as a contributor to the spatial distribution of geometrid moth outbreaks in sub-arctic mountain birch forest. Ecological Entomology 42:430–438

Petrucco-Toffolo E, Battisti A (2008) Performances of an expanding insect under elevated CO_2 and snow cover in the Alps. Iforest Biogeosciences For 1:126–131

Pureswaran DS, De Grandpré LD, Paré D, Taylor A, Barrette M, Morin H, Régnière J, Kneeshaw DD (2015) Climate-induced changes in host tree-insect phenology may drive ecological state-shift in boreal forests. Ecology 96:1480–1491

Pureswaran DS, Roques A, Battisti A (2018) Forest insects and climate change. Curr For Rep 4:35–50

Raffa KF, Aukema BH, Bentz BJ, Carroll AL, Hicke JA, Turner MG, Romme WH (2008) Cross-scale drivers of natural disturbances prone to anthropogenic amplification: the dynamics of bark beetle eruptions. Bioscience 58:501–517

Ramsfield TD, Bentz BJ, Faccoli M, Jactel H, Brockerhoff EG (2016) Forest health in a changing world: effects of globalization and climate change on forest insect and pathogen impacts. Forestry 89:245–252

Rassati D, Faccoli M, Haack RA, Battisti A, Marini L (2016) Bark and ambrosia beetles show different invasion patterns in the USA. PLoS One 11(7):e0158519

Régnière J, St-Amant R, Duval P (2012) Predicting insect distributions under climate change from physiological responses: spruce budworm as an example. Biol Invasions 14:1571–1586

Salman HR, Hellrigl K, Minerbi S, Battisti A (2016) Prolonged pupal diapause drives population dynamics of the pine processionary moth (*Thaumetopoea pityocampa*) in an outbreak expansion area. For Ecol Manage 361:375–381

Saulnier M, Roques A, Guibal F, Rozenberg P, Saracco G, Corona C, Edouard J-L (2017) Spatiotemporal heterogeneity of larch budmoth outbreaks in the French Alps over the last 500 years. Can J For Res 47:667–680

Schebeck M, Hansen E, Schopf A, Gregory R, Stauffer C, Bentz B (2017) Diapause and overwintering of two spruce bark beetle species. Physiol Entomol 42:200–210

Schneider L, Comte V, Rebetez M (2021) Increasingly favourable winter temperature conditions for major crop and forest insect pest species in Switzerland. Agric for Meteorol 298–299:108315

Slippers B, Hurley BP, Wingfield MJ (2015) Sirex woodwasp: A model for evolving management paradigms of invasive forest pests. Annu Rev Entomol 60:601–619

Tamburini G, Marini L, Hellrigl K, Salvadori C, Battisti A (2013) Effects of climate and density-dependent factors on population dynamics of the pine processionary moth in the Southern Alps. Clim Change 121:701–712

Tenow O (1996) Hazards to a mountain birch forest—Abisko in perspective. Ecol Bull 45:104–114

Tobin PC, Gray DR, Liebhold AM (2014) Supraoptimal temperatures influence the range dynamics of a non-native insect. Divers Distrib 20:813–823

Toïgo M, Barraquand F, Barnagaud J-Y, Piou D, Jactel H (2017) Geographical variation in climatic drivers of the pine processionary moth population dynamics. For Ecol Manage 404:141–155

Toïgo M, Nicolas M, Jonard M, Croisé L, Nageleisen LM, Jactel H (2020) Temporal trends in tree defoliation and response to multiple biotic and abiotic stresses. For Ecol Manage 477:118476

Ungerer MJ, Ayres MP, Lombardero MJ (1999) Climate and the northern distribution limits of *Dendroctonus frontalis* Zimmermann (Coleoptera: Scolytidae). J Biogeogr 26:1133–1145

Visser ME, Both C (2006) Shifts in phenology due to global climate change: the need for a yardstick. Proceedings of the royal society of London series B—Biol Sci 272:2561–2569

Walther GR, Post E, Convey P, Menzel A, Parmesan C, Beebee TJC, Fromentin JM, Hoegh-Guldberg O, Bairlein F (2002) Ecological responses to recent climate change. Nature 416:389–395

Wermelinger B (2004) Ecology and management of the spruce bark beetle, *Ips typographus*—a review of recent research. For Ecol Manage 202:67–82

White TCR (1974) Hypothesis to explain outbreaks of looper caterpillars, with special reference to population of *Selidosema suavis* in a plantation of *Pinus radiata* in New-Zealand. Oecologia 16:279–301

Chapter 23
Forest Insect Invasions and Their Management

Andrew M. Liebhold, Eckehard G. Brockerhoff, and Deborah G. McCullough

23.1 Introduction

The problem of biological invasions is largely an inadvertent result of globalization. Global trade and human travel have resulted in the accidental movement of organisms across geographic barriers such as oceans and major mountain ranges that previously compartmentalized the world's flora and fauna through millions of years of evolution. Most non-native organisms established outside their range are inconsequential, with little noticeable impact on invaded ecosystems. However, a fraction of non-native species become extremely abundant and/or greatly alter ecosystem processes and properties (Lovett et al. 2006, 2016). Ever-increasing rates of international trade and travel are likely to provide further opportunities for transport of non-native organisms into new regions.

Given that insects are the most diverse group of organisms in the world, it comes as no surprise that they comprise a large portion of all invading species worldwide (Seebens et al. 2017). Insects exhibit remarkable variation in life histories, and many species require plants for feeding, habitat or both. Among non-native insects that feed on forest trees, there are four major groups of that are particularly damaging: insects that bore through the outer bark of trees to feed on phloem (inner bark) and/or wood, defoliating insects that feed on foliage or within shoots, sap-feeding insects, and seed-eaters. These types of insects are common among non-native forest insects

A. M. Liebhold (✉)
US Forest Service Northern Research Station, Morgantown, WV, USA
e-mail: Andrew.Liebhold@usda.gov

E. G. Brockerhoff
Swiss Federal Research Institute WSL, Birmensdorf, Switzerland

D. G. McCullough
Department of Entomology and Department of Forestry, Michigan State University, East Lansing, MI, USA

© The Author(s) 2023
J. D. Allison et al. (eds.), *Forest Entomology and Pathology*,
https://doi.org/10.1007/978-3-031-11553-0_23

in all regions of the world, with many affecting native forests, tree plantations and urban forests.

Fortunately, most species of non-native forest insects have little impact on trees in their new habitat (Aukema et al. 2010). A fraction of these insects, however, affect tree appearance, growth or vigor, and a small number of species have had catastrophic impacts on invaded forests (Table 23.1). In some cases, such as the invasion of the emerald ash borer, *Agrilus planipennis* (Buprestidea), in North America (Herms and McCullough 2014), invasion can result in local extirpation of their hosts. Several invasive forest species have greatly altered silvicultural practices. For example, damage caused by the green spruce aphid, *Elatiobium abietinum* (Aphididae), was so severe in Iceland that planting of spruce was largely abandoned in southern regions (Halldórsson et al. 2003).

Less destructive insect species don't necessarily kill their host trees and their impacts may be more difficult to quantify. Defoliators, for example, may reduce growth rates, affect form of young trees, and increase vulnerability of severely affected trees to other, secondary pests. A few species facilitate infection by tree pathogens, which can result in considerable damage. For example, the beech scale, *Cryptococcus fagisuga* (Eriococcidae), which was accidentally introduced to North America and Europe, creates punctures in the outer bark where the tiny beech scales feed. These punctures, allow entry of pathogenic fungi, *Nectria* spp., which cause beech bark disease and ultimately tree death (Houston 1994). Feeding by the European elm bark beetle, *Scolytus multistriatus*, (Curculionidae) rarely damages trees but the beetles vector Dutch elm disease, which is typically fatal to American elms. Numerous invasive forest insects, either directly or indirectly, alter ecosystem processes such as nutrient cycling or competitive interactions among plant species (Lovett et al. 2016).

Here we provide a general overview of the causes, ecology and impacts of forest insect invasions, including strategies for managing invasions. We limit our coverage to plant-feeding species, though other feeding guilds (e.g. predators, pollinators) can also have ecological impacts. Other reviews covering forest insect invasions with different areas of focus may be found elsewhere (Niemelä and Mattson 1996; Aukema et al. 2010, 2011; Brockerhoff and Liebhold 2017). Insect invasions are part of a larger problem of biological invasions in forests, and other reviews cover that subject (e.g. Liebhold et al. 1995, 2017a; NRC 2002; Ghelardini et al. 2017; Seebens et al. 2017). Our treatment of this subject is structured using the three universal phases of invasions: arrival, establishment and spread of invading populations.

23.2 Arrival

The problem of biological invasions is largely caused by people moving organisms from their native range into new regions. Mechanisms by which organisms are inadvertently moved around the world are varied and referred to as "invasion pathways". There are many different invasion pathways responsible for insect introductions

Table 23.1 Examples of damaging non-native forest insects

Species	Family	Invaded region	Native range	Type of Damage	Reference
Emerald ash borer, *Agrilus planipennis*	Coleoptera: Buprestidae	N. America, European Russia	East Asia	Phloem-feeder, tree mortality	Herms and McCullough 2014
Redneck longhorned beetle, *Aromia bungii*	Coleoptera: Cerambycidae	Japan	China	Phloem-feeder, tree mortality	Xu et al. 2017
Sirex woodwasp, Sirex noctilio	Hymenoptera: Siricidae	New Zealand, Australia, Africa, S. America, N. America	Europe	Xylem-feeder, tree mortality	Slippers et al. 2015
Beech scale, *Cryptococcus fagisuga*	Hemiptera: Eriococcidae	Western Europe, N. America	Caucus Mtn. regions	Sap-feeding (phloem), facilitates infection by pathogenic fungi	Houston 1994
Eucalyptus snout beetle, *Gonipterus scutellatus*	Coleoptera: Curculionidae	New Zealand, Europe, N. America, S. America	Australia	Foliage-feeder causing defoliation	Paine et al. 2011
Horse chestnut leaf-miner, *Cameraria ohridella*	Lepidoptera: Gracillariidae	Central and northern Europe	Southern Europe	Leaf-miner, defoliation	Straw and Bellett-Travers 2004

(McCullough et al. 2006; Meurisse et al. 2019). Forest insects, however, are most often introduced through one of four invasion pathways: international movement of (i) wood, (ii) plants and plant parts, (iii) hitchhiking (i.e. movement on inanimate or non-host objects) and (iv) intentional introductions (including biological control agents) (Table 23.2). Relatively few forest insect species are thought to have dispersed naturally (e.g. by flight or on wind or water) to new world regions.

An analysis of 62 species of invasive forest insect pests (excluding seed-feeding insects) established in the USA indicated that historically, imports of live plants was responsible for more invasions than any other pathway (Liebhold et al. 2012).

Table 23.2 Principal pathways by which forest insects are transported outside of their native ranges

Pathway	Insect groups	Mitigation methods
Wood including round wood, wood packaging material	Bark and wood-boring insects	Quarantine bans, fumigation, heat treatment
Plants including cuttings, bare-root plants, cut flowers, seeds	Sap-feeding insects, foliage-feeding insects, sap-feeding insects, seed & cone insects	Quarantine bans, fumigation
Hitchhiking—i.e. transport on non-host material such as sea containers, machinery, automobiles	Foliage-feeding insects, ants, wasps	Steam-cleaning of cargo
Intentional introductions	Insect predators and parasitoids, weed biological control agents	Regulation of biological control, risk analysis

Plants imported for propagation are a particularly dominant invasion pathway for sap-feeding and foliage-feeding insects. Live plants represent the "perfect" pathway for many herbivorous insects since most can live and feed on their host plant throughout their journey and upon arrival, the insects already have a suitable host plant that is likely to be nurtured and tended. The importance of live plant imports as a pathway for insect invasions has been confirmed in many world regions (Kiritani and Yamamura 2003; Roques et al. 2009). There have been substantial advances in developing biosecurity measures designed to limit accidental insect invasions with commercially imported plants (Liebhold and Griffin 2016). These policies vary among world regions, however, and regulation of plant imports in some regions is weak (Eschen et al. 2015).

Not surprisingly, most non-native insects that feed beneath bark on phloem or wood are introduced with imported logs or wood. Some invasions of bark- and wood-borers, such as the introduction of the European elm bark beetle to North America, are attributed to international shipments of unprocessed logs (i.e. logs with bark) (May 1934). More recently, however, solid wood packaging material is considered the dominant invasion pathway for insects that feed beneath bark. Solid wood packaging material refers to crating, pallets, spools and dunnage (e.g. timbers used to prop up maritime cargo or containers). Increases in global movement of solid wood packaging material corresponds to the surge in the use of large container ships for cargo transport beginning in the 1980's. Increased use of solid wood packaging material has resulted in a notable jump in the number of bark and wood-boring insects introduced to North America and elsewhere over the last few decades (Brockerhoff et al. 2006a; Aukema et al. 2010). Wood packaging material is typically made from relatively poor quality trees and low cost wood. Such wood is often infested with wood-boring insects. Many of these insects can survive for months, especially if some amount of bark remains on the wood to retain moisture. Insects in wood may complete development then emerge in a new habitat or region.

A substantial number of non-native forest insect species have been introduced intentionally, mostly for biological control of damaging invasive pests. In classical biological control, natural enemies from the native range of an invasive pest are imported, reared and released into the new habitat, ideally to provide long-term control of the invasive pest. If successful, such efforts can provide long-term control of a pest across a broad area. Insect parasitoids and predators are most often used in classical biocontrol programs, although in a few instances, a specialized pathogen may be considered.

Kenis et al. (2017) reported that worldwide, 6158 species of parasitoids or predators have been introduced for control of 588 forest insect pests and of those, 172 pest species were controlled with some success.

Historical rates of establishment of non-native insect species in various world regions (Fig. 23.1) indicate that the accumulation of non-native forest insects has not slowed during the last century (Brockerhoff and Liebhold 2017; Seebens et al. 2017). In fact, the rate of establishment may even be increasing in some regions, such as Europe. However, in other areas, numbers of new establishments per year of certain insect groups have declined. Such declines are sometimes a result of improved biosecurity practices (Liebhold and Griffin 2016). In other cases, declines may reflect the depletion of the supply of species capable of invading a specific new range (Levine and D'Antonio 2003). For example, numbers of bark beetle (Scolytinae) species invading N. America from Europe have declined, at least in part because centuries of trade between these continents have depleted the pool of European species capable of arriving and establishing in North America. In contrast, invasions of Asian bark beetle species have continued to increase. Substantial imports of commodities from Asia into N. America are comparatively recent and species pools have not yet been depleted (Liebhold et al. 2017b).

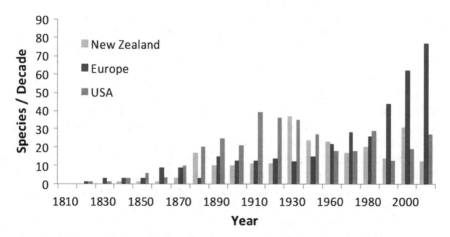

Fig. 23.1 Numbers of new non-native forest insects discovered by decade in New Zealand, Europe and USA. Redrawn from Brockerhoff and Liebhold (2017)

Economic analyses (e.g. Leung et al. 2002, 2014; NRC 2002) consistently suggest that the most effective strategy for mitigating the biological invasion problem is prevention—i.e. taking measures to prevent the transport and arrival of non-native species. This is generally true for forest insect pests and there are several approaches for preventing their arrival. Most prevention strategies focus on managing the two dominant invasion pathways for forest insects: live plants and wood.

Importation of plants has historically played a crucial role in their domestication and genetic improvement for agricultural, forestry and ornamental purposes. Because imported plants obviously represent a high-risk invasion pathway for insects and plant pathogens, several measures have been identified to limit this risk. First, import of high-risk plant species can be simply banned. Some countries implement "black list" systems (e.g. imports of certain plant taxa are banned), while other countries use "white list" systems (plant taxa are banned unless they are known to be of relatively low risk). Although this is a somewhat simplistic representation, the latter system is considered more effective at preventing introductions of unknown organisms (Eschen et al. 2015). Second, phytosanitary treatments, such as fumigation or treatment of plants with pesticides, can reduce the likelihood that pests will be introduced on high-risk plants. Post-entry quarantines may also be applied. In this practice, imported plants are initially cultivated in a quarantine facility or secure location and monitored to ensure they are free of insects or pathogens before they are released for sale or cultivation.

"Integrated measures" can also be used to reduce risks of introducing new insect pests with imported plants. This involves applying multiple measures, often before the plants are shipped and again when the plants arrive. This approach can include phytosanitary treatments and inspections of plants before and after shipping. This may also include efforts to suppress pests at overseas plant production facilities. Insecticides or other pest management tactics may be used to ensure plant material is pest-free when it is exported (International Plant Protection Convention 2012).

Considerable variation exists among nations with regard to their regulation of plant imports (Eschen et al. 2015). New Zealand and Australia apply strict regulations based on a white list system, which limits the plant taxa that can be imported without phytosanitary treatments and/or post-entry quarantine. In contrast, the European Union implements relatively relaxed regulations. Some plants can be imported to Europe without any permit and in soil, which could harbor nematodes, plant pathogens and other pests. Regulations in other countries such as the USA, Canada and Japan, fall somewhere in between these extremes (Eschen et al. 2015). Such biosecurity measures have successfully reduced risks of pest introductions with legally imported plants (Liebhold and Griffin 2016). Illegal importation of plants, however, either by members of the public unaware of regulations or by importers deliberately avoiding oversight, represents a relatively uncontrolled invasion pathway. Many countries with developing economies lack the resources to implement biosecurity practices. These areas potentially can serve as "bridgeheads" enabling alien species to become established and eventually invade other regions (Hurley et al. 2017).

Effectively regulating imports of wood to prevent insects and diseases from accidentally being transported remains challenging. Several countries ban imports of logs with bark, since many insects are associated with bark and phloem. Other countries allow logs to be imported but require fumigation either before or during international shipment (Allen et al. 2017). In 2002, the International Plant Protection Convention implemented a harmonized phytosanitary standard, ISPM 15 to reduce the risk of introducing live pests in solid wood packing material with imported cargo. The ISPM 15 standard requires heat treatment, fumigation or other measures be applied to solid wood packaging material used with cargo moving between countries. All countries that implemented ISPM 15 agreed to abide by these same regulations; hence the standard is "harmonized." Data from cargo inspections at ports conducted by regulatory officials from 2003–2009 showed the implementation of ISPM 15 decreased rates of insect contamination in solid wood packaging material by 36–52% (Haack et al. 2014). Although 100% effectiveness of these treatments would be highly desirable, a cost–benefit analysis showed the economic costs of ISPM 15 were substantially lower than the economic benefits resulting from reducing rates of insect invasions (Leung 2014).

Although vast amounts of cargo arrive at many ports and border crossings, only a small fraction of any shipment can be inspected. Additionally, many insects and plant pathogens are small, cryptic and difficult to observe. Consequently, inspection may not be highly effective as a method of directly preventing arrival and introduction of unwanted organisms. It does, however, serve an important purpose as an incentive for producers and importers to reduce pest contamination of shipments and as a source of information about the species of pests associated with particular imports (Whattam et al. 2014). Inspections can also provide information about organisms that are relatively abundant in specific pathways, which may play a crucial role in identifying the need for new quarantine or phytosanitary measures (McCullough et al. 2006).

Inspection of air passengers to detect biosecurity threats occurs to varying degrees in different countries. Quarantine officers inspecting the baggage of air passengers arriving in the US have reported 10,000—20,000 interceptions of insects each year. These interceptions include insects from all orders and numerous species of concern to biosecurity (Liebhold et al. 2006). In New Zealand, quarantine officers inspect baggage and question all arriving passengers about items that could be infested with insects or other biosecurity threats. For example, several camping tents carried by passengers were found to contain live insects (Gadgil and Flint 1983).

23.3 Establishment

Although many different species of non-native forest insects are transported across borders or through ports every year, only a small fraction of those species actually become established in the new region (NRC 2002; Blackburn et al. 2011). Generally, only a few "colonists" are transported to a new region on imported plants, wood or other materials. Upon arrival, this very small population must be able to survive the local climate, locate and successfully feed on a suitable host, and reproduce. Very low-density populations of a non-native insect face a high probability of extinction, much like native endangered species. When populations are at very low densities, they are especially vulnerable to unpredictable events that can lead to extinction. Such events, termed stochastic effects, can include unfavorable weather. Unusually cold temperatures in spring, a wildfire or a bad storm, for example, can wipe out a small, low-density population.

Allee effects (see Chapter 5) can also cause a newly or recently established population to go extinct (Lande 1998). Named after the University of Chicago professor Warder Allee who first described the phenomenon in 1949, the Allee effect refers to the phenomenon of decreasing population growth with decreasing density of an organism. In other words, very small populations of some species tend to become even smaller over time. For example, when spongy moth (*Lymantria dispar*) populations are at very low population densities, males may be unable to locate femalesfor mating. Low reproduction success causes the density to drop even further (Tobin et al. 2009). If the density of a population drops below a critical threshold level, the population will decline to extinction (Liebhold and Tobin 2008). Most populations of non-native insects arrive at densities below these thresholds, which helps to explain why so many fail to establish.

Several mechanisms can cause Allee effects in forest insect populations. Most insect species reproduce sexually and if densities are very low, may be unable to find a mate, leading to Allee dynamics (Gascoigne et al. 2009). Some species, including certain bark beetles, must mass attack their host tree to successfully overcome host defenses and reproduce within the tree. Such a phenomenon, a form of "group feeding," is also capable of producing an Allee effect (Chase 2016). Additionally, attack by predators may create a weak Allee effect; predation levels are typically higher in small populations but with large populations there may be "safety in numbers" and therefore greater survival.

Given the assorted mechanisms that may cause Allee effects, plus the variation in their strength in affecting different species, it is not surprising that the probability of establishment varies considerably among insect species. Many species of Hemiptera, such as scales and adelgids, for example, reproduce asexually and population growth is not limited by the need to find mates. This life history trait may be one reason why Hemiptera are generally over-represented in non-native insect assemblages and relatively more successful invaders (Liebhold et al. 2016a). Aggressive tree-killing bark beetles such as the North American species, *Dendroctonus ponderosae,* and the European species, *Ips typographus* (Curculionidae), reach very high densities during outbreaks. Although both species have repeatedly been intercepted at overseas ports of entry, neither has become established outside of their native range. This likely reflects a strong Allee effect; density of the introduced populations is not high enough to enable the beetles to overcome resistance of their host trees and successfully reproduce (Brockerhoff et al. 2006a).

Management to prevent establishment of non-native populations plays a key role in biosecurity strategies. The general approach shared among all such efforts is a combination of surveillance, to find newly arrived reproducing populations, and eradication, the forced extinction of a population (Liebhold and Tobin 2008; Liebhold et al. 2016b). Techniques for surveillance of non-native insect populations may take a variety of forms, depending upon the biology and behavior of the target species (see Chapter 4) and the potential impacts of the species.

Detection surveys commonly rely on traps baited with lures containing pheromones or compounds produced and emitted into the air by host plants. If lures are highly attractive to the target pest, baited traps can be very sensitive tools and effectively detect low-density populations. For example, traps baited with synthetic sex pheromones are used in many countries for detecting newly arrived and very low-density populations of moths, such as the spongy moth. In contrast, traps baited with compounds produced by host trees, such as alpha-pinene, are widely used for detecting an array of conifer bark beetles, as well as phloem- and wood-boring insects (Brockerhoff et al. 2006b; Rabaglia et al. 2019). Traps baited with host volatiles are typically less sensitive than pheromone-baited traps. Often the lures with host volatiles do not strongly attract the target pest or the lures may be overwhelmed by complex blends of compounds produced by nearby live trees.

Some insects are not attracted to any type of chemical lure and other surveillance options, such as light traps or visual searches for evidence of infestation, may be the only option available (Chapter 19). Analysis of historical insect eradication programs indicates that the availability of a sensitive tool, such as attractant-baited traps, greatly increases the likelihood of early detection and successful eradication of invading populations (Tobin et al. 2014). Sensitive detection tools also provide an effective means to delimit (i.e. delineate spatial boundaries) invading populations and evaluate the success or failure of eradication programs.

Although eradication may be difficult or even impossible when there are no effective options to detect low-density populations or when a non-native population has already spread across a large area, there are many examples of forest insect species that have been successfully eradicated (Brockerhoff et al. 2010). Painted apple moth, *Orgyia anartoides* (Erebidae), an Australian Lymantriinae, was eradicated from New Zealand between 2001 and 2003 using a combination of tactics including host plant removal, aerial application of a microbial insecticide (*Bacillus thuringiensis*) and sterile male releases (Suckling et al. 2007) (Fig. 23.2). A remarkable aspect of this program is that pheromone-baited traps were not used in the eradication. Synthetic pheromone was found to be unstable and could not be used in lures. Instead, traps used for delimitation were baited with live female moths that were reared in the laboratory then placed in small cages attached to traps.

In another example, an extensive population of Asian longhorned beetle (ALB), *Anoplophora glabripennis* (Cerambycidae), was successfully eradicated from Chicago (1998–2008) without the use of any traps or attractants for delimitation. Instead, delimitation was accomplished via visual surveys for characteristic holes on tree boles and branches left by emerging adult beetles. Eradication was accomplished by a combination of host removal and injections of systemic insecticides into all potential host trees within 400 m of positive finds (Haack et al. 2010). Other recent ALB eradication projects, such as those in Ontario and New Jersey,

Fig. 23.2 Scenes from a spongy moth eradication program in Washington, USA. **A**. Arial application of *Bacillus thuriengensis*; **B**. Fifth instar spongy moth larva; **C**. Public notification of aerial spraying; **D**. Checking traps to confirm eradication success [Photo credits: (A and C) James Marra, Washington State Department of Agriculture; (B) Jon Yuschock, Bugwood.org; (D) USDA APHIS PPQ, USDA APHIS PPQ, Bugwood.org]

have relied strictly on removal of potential host trees within either a 400 m or 800 m radius of every infested tree (Turgeon et al. 2007). The success of these and other ALB eradication programs can be attributed, in part, to the limited dispersal behavior of beetles which constrained spread of the invading populations.

One of the challenges managers of surveillance and eradication strategies may face is the negative reaction to control activities that is sometimes expressed by residents of the treatment area. Invasions of forest insects characteristically occur in urban / suburban habitats (Poland and McCullough 2006) and residents may strongly oppose activities such as removal of apparently healthy (but possibly infested) trees or widespread pesticide applications (Liebhold et al. 2016b). Successful eradication programs, such as the painted apple moth eradication from Auckland, New Zealand and the ALB eradications in Chicago and Toronto, typically involve considerable effort in public engagement to explain the need for and value of such efforts. Other cases, such as the failed eradication of the light brown apple moth from California

(Lindeman 2013; Suckling et al. 2014) demonstrate that public engagement and outreach efforts are essential to build support for the program. Organized opponents may otherwise step in and disseminate misinformation, eroding support for these programs.

23.4 Spread

Once a non-native population of a forest insect becomes established in a region, populations can build and typically begin expanding further into suitable habitats (Blackburn et al. 2011). This spread often occurs via two mechanisms; natural dispersal of the insects and accidental transport of insects by people. Natural dispersal can occur when insects fly or are transported by wind, birds or other animals. Long distance spread occurs when people move insect life stages or infested plant material into uninfested areas. Domestic invasion pathways refer to the means by which non-native forest insects are accidentally introduced into new states, provinces or currently uninfested regions. These domestic pathways often resemble those for intercontinental invasions. Human movement of live plants, and infested firewood, logs or solid wood packaging material may inadvertently transport non-native insects. Life stages of certain insect species can also hitchhike on non-host goods shipped from an infested area.

Rates of spread vary considerably among non-native forest insect species (Fig. 23.3). Spread of a non-native species represents the combination of population growth and dispersal; factors that affect either of these components will likely affect rates of spread (Liebhold and Tobin 2008). Rates of historical spread of invasive forest insects and diseases in the USA are positively related to human population density, host tree density and voltinism (Hudgins et al. 2017; Fahrner and Aukema 2018). Similarly, human population density is positively related to historical spread of the horse chestnut leafminer, *Cameraria ohridella* (Gracillariidae), in Europe (Gilbert et al. 2004).

One approach to managing spread involves regulating the pest and often the tree species or commodity that is likely to introduce the pest. For example, transport of ash trees from nurseries, ash logs and firewood were regulated in North America by federal and parallel state quarantines imposed to limit spread of the emerald ash borer (Herms and McCullough 2014). States, provinces or other regional governments may also impose their own quarantines to prevent the introduction of an invasive pest established in other regions. Quarantines typically prohibit transport of potentially infested host material unless specific phytosanitary treatments are applied to ensure the trees or wood is not infested.

Another approach to controlling spread involves using barrier zones to slow or stop the spread of an invading species. Perhaps the most extensive of such programs is the spongy moth 'slow the spread' program (Sharov et al. 2002) in the US. Each year, a grid of ca. 100,000 pheromone traps is deployed across a 100 km wide band along the leading edge of the spongy moth invasion in the US. When a new, isolated

a

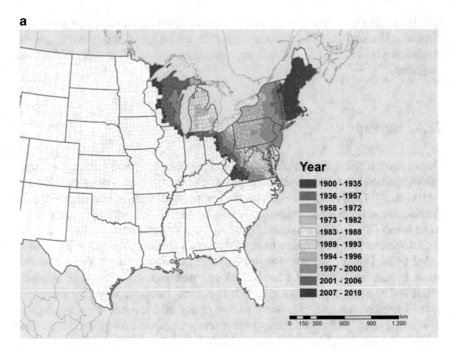

b

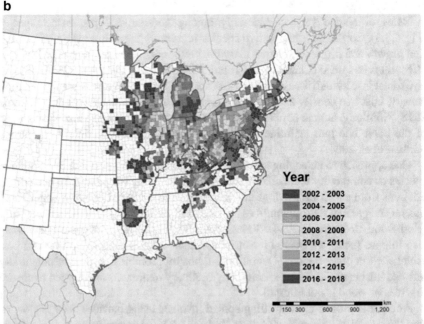

Fig. 23.3 Historical spread in the USA of **a**) the spongy moth, *Lymantria dispar* (dates are year of first quarantine by the US Dept. Agriculture), **b**) the emerald ash borer, *Agrilus planipennis* (dates are year of first detection of reproducing populations by the US Dept. Agriculture)

colony of spongy moth is detected, it is delimited, then treated to either eradicate the population or to dramatically reduce the density of the population. This approach has substantially slowed spongy moth spread in the US, yielding economic benefits by delaying establishment and thus impacts of this pest species (Epanchin-Niell and Liebhold 2015).

23.5 Established Populations

Ecological and economic impacts of non-native forest insect pests vary considerably ranging from minor defoliation to widespread mortality of host trees. Invasive forest insects that can kill their hosts are obviously of great concern. Emerald ash borer, first discovered in N. America in 2002, has become the most destructive and costly forest insect to invade that continent (Herms and McCullough 2014). This beetle has already killed hundreds of millions of ash (*Fraxinus* spp) in forests and landscapes in the US and Canada and continues to spread (Morin et al. 2017). Impacts of other invasive insects, such as hemlock woolly adelgid, *Adelges tsugae* (Adelgidae) and beech scale, *Cryptococcus fagisuga* (Eriococcidae), have resulted in large-scale shifts in forest composition, altering trajectories of regional forest and tree species succession (Morin and Liebhold 2015; Lovett et al. 2016).

In many parts of the world non-native tree species are planted for fiber production; much of the exceptional growth of such tree species can be attributed to their escape from insects and diseases present in their native ranges. However, these plantations can be particularly susceptible to invasions of insects and diseases from the native range of the tree species (Wingfield et al. 2015). In some cases, insect invasions have caused forest practices involving certain tree species to be totally abandoned (Hurley et al. 2016). For example, planting of *Eucalyptus* spp. in New Zealand was largely discontinued following the establishment of the eucalyptus tortoise beetle, *Paropsis charybdis* (Chrysomelidae), and other insects and pathogens from Australia (Withers 2001).

While invasive forest insects can have substantial impacts on market resources such as timber, the primary economic impacts of many of these species are largely in non-market economic sectors. Aukema et al. (2011) compiled a comprehensive analysis of economic costs associated with invasive phloem- and wood-boring, sap-feeding and foliage feeding insects in the US. The greatest economic impacts for all three feeding guilds were sustained by local governments, i.e. municipalities, and private property owners. These costs reflect the high value of trees growing in public and private landscapes, along roadways, and in parks or recreation areas. Property owners incur costs when trees must be protected from a pest with insecticides or when dead, dying or severely declining trees require removal (e.g. Kovacs et al. 2010). Dead and declining trees also reduce property values (Holmes et al. 2006). Additional costs are sustained because of the loss of ecological services provided by urban trees, such as storm water uptake and pollutant capture (Nowak et al. 2001; Jones 2017). Impacts of invasive forest insects on ecosystem services provided by forests can be complex and are not well understood (Boyd et al. 2013). Information to date indicates considerable variation in the impact of invasive insect pests on forest ecosystem processes such as nutrient cycling, and food web structure (Lovett et al. 2006).

What explains the often unusually high population growth rates that many introduced non-native forest insects exhibit following initial invasion? The answer may lie with evolutionary history. Most species evolved during millions of years, with natural selection shaping their interactions with other insect species and with their host trees. But when insects establish in alien habitats, the species with which they co-existed in their native range are typically absent. This suggests two mechanisms likely contribute to the often explosive population growth. First, the complex of natural enemies that regulate populations in their native range are typically absent in invaded regions. When a non-native forest insect species becomes established in the absence of predators, parasitoids and pathogens, populations may quickly grow to high levels (Colautti et al. 2004). Several foliage-feeding Lepidoptera represent examples of this 'enemy release' phenomenon. For example, the brown tail moth, *Euproctis chrysorrhoea* (Erebidae), which was introduced to North America in the late 1800's, caused major defoliation of forests in the northeastern US until a generalist parasitoid, *Compsilura concinnata* (Tachinidae), was introduced in 1906. This Dipteran parasitoid is credited with causing the collapse of brown tail moth populations, including their virtual extinction over most of the invaded range (Elkinton et al. 2006). Unfortunately, this parasitoid, which is now well established across much of the US, also attacks many moth species native to North America and is credited with dramatic reductions in populations of native saturniid moths in the northeast (Elkinton and Boettner 2012).

Severe impacts by invasive forest insects in their new habitat often reflect a lack of host resistance. Most forest insects have co-evolved with the host trees they colonize in their native range for millions of years. Over time, trees usually have evolved at least some amount of resistance to insect herbivores, which acts to constrain population growth of these insect populations. When insects invade a new region, however, they are often able to feed and develop on tree species in the new habitat that are similar to their original hosts. However, without any previous evolutionary exposure, those tree species may lack resistance to the non-native insect, especially if there are no similar insect species in the invaded region. Insects that encounter such a "defense free space" may thus thrive at the expense of their novel host (Gandhi and Herms 2010). For example, emerald ash borer populations are native to China and other regions of Asia, where they act as secondary pests, colonizing Asian ash trees that are only stressed or dying. North American ash species however have no co-evolutionary history with emerald ash borer and healthy, as well as stressed, ash trees can be colonized and killed by this invader (Herms and McCullough 2014).

Novel associations between non-native forest insects and other non-host organisms, such as symbiotic fungi, can also lead to severe impacts, particularly by non-native bark beetles (Wingfield et al. 2017). Many bark beetles (Scolytinae) have important associations with fungi that they introduce into trees. In some tree species, certain mutualistic fungi improve nutrient levels or other conditions for the bark beetles (Six and Wingfield 2011). These mutualistic fungi may play a key role in determining whether a non-native bark beetle species can colonize healthy trees in a new region. As an example, the red turpentine beetle, *Dendroctonus valens* (Curculionidae), is native to North America where it acts as a secondary pest, colonizing stressed pine trees. Where it has invaded China, however, associations with novel Ophiostomatoid fungi enable the beetles to colonize healthy pines and thus act as a primary pest (Sun et al. 2013).

Understanding the mechanisms that contribute to severe impacts of invasive forest insects can help to identify possible strategies for managing a given pest. In the case of enemy release, classical biological control, which involves importing, rearing and releasing natural enemies from the pest's native range, may be appropriate. Many of these efforts are not successful, but there are cases where an imported biological control agent has virtually eliminated outbreaks and damage caused by an invasive species (Kenis et al. 2017). For example, an egg parasitoid *Anaphes nitens* (Mymaridae), was imported and has provided effective control of the Eucalyptus snout beetle, *Gonipterus* spp. (Curculionidae), which was previously a serious defoliating pest of *Eucalyptus* in Africa, New Zealand, South America and California (Tribe 2005; Schröder et al. 2020). While classical biological control can reduce impacts of invasive forest insects over the long term, potential effects of imported natural enemies on native insects can create unintended problems (Hajek et al. 2016). Imported natural enemies may begin to prey on native non-target insects that cause little or no damage. They may also displace or outcompete native natural enemies that control native pests. Currently, biological control introductions are closely regulated in most countries. Ideally, potential biological control agents should be carefully screened to verify that they will attack the invasive target pest without adversely affecting native species.

When damage by an invasive forest insect is driven primarily by a lack of host resistance in the new region, selection for and breeding resistant trees may help reduce impacts. Different strategies, such as backcrossing susceptible native species with resistant species or genetic editing may be used to increase resistance to a particular pest (Sniezko and Koch 2017; Showalter et al. 2018). While selecting for or breeding host resistance holds promise, there are few examples where this has been successfully applied to overcome impacts of a damaging invasive forest insect. A major challenge with this strategy is deployment—i.e. establishing resistant trees in forests over large regions. While resistant genotypes can be cultivated and planted in ornamental settings or in forest plantations relatively easily, deployment of resistant strains is much more difficult in forests where natural regeneration dominates and competition with other species may be intense.

Today, chemical insecticides are rarely used to control invasive insect pests in forested settings because of an array of environmental and economic concerns. In urban forests, however, insecticides are frequently used to protect valuable landscape trees from an array of insect pests, including invasive species. Systemic products, which are applied by injecting the insecticide into the base of the trunk, pouring the insecticide around the base of the trunk, or spraying the lower portion of the trunk and allowing the insecticide to penetrate the outer bark, have largely replaced insecticide cover sprays. Systemic products greatly reduce insecticide impacts on beneficial insects and other non-target organisms, environmental residues and applicator exposure, and can effectively control insects feeding in phloem or in tree canopies (McCullough 2019). While these insecticides can obviously protect individual trees, in a few cases, systemic insecticides are applied to reduce the impacts of an invasive forest insect over large areas. In one large-scale project, a relatively small proportion of ash trees were treated with a systemic insecticide to successfully slow the rate of emerald ash borer population growth and the rate at which ash trees declined and died (Mercader et al. 2015, 2016). Many states in the eastern U.S. apply systemic insecticides to suppress hemlock woolly adelgid populations, protecting watersheds and riparian areas where hemlock trees are abundant (e.g. Benton et al. 2015). The bacterial pesticide *Bacillus thuringiensis* is sometimes applied aerially to suppress outbreak populations of foliage-feeding invasive pests such as the spongy moth, and thereby reduce impacts caused by defoliation (van Frankenhuyzen 2000).

Silviculture represents another strategy for reducing the impacts of damaging invasive forest insects (Muzika 2017). Silvicultural practices can increase stand level resistance to invasive, as well as native, forest insect pests. Increasing diversity at genetic, species and landscape scales, for example, will generally decrease susceptibility of forests to invasive, as well as native, insect pests. Diverse forests also provide habitat for an array of insect predators and parasitoids, which can often help control pest populations. Thinning and other practices to decrease competition and maintain healthy stands can contribute to reducing invasive pest impacts. For example, the invasion of *Sirex noctilio* (Siricidae) in New Zealand around 1900 resulted in high mortality in dense plantations of non-native pines. Over time, outbreaks of this invasive woodwasp generally subsided, probably as a result of the introduction of biological control agents (a nematode and two parasitoid species) plus an emphasis on thinning overstocked pine stands. Damage was much less severe in thinned forests than in dense stands, where competition reduced tree vigor (Hurley et al. 2007). While silvicultural practices may help to reduce damage caused by invasive forest insects that exploit low-vigor trees, there are often few options for pests that can colonize healthy trees aside from conversion of stands to favor non-host tree species.

23.6 Conclusions

Only a few decades ago, most of the important insect pests that forest entomologists focused on were native species. As the world has globalized, however, an ever-increasing proportion of the significant forest pests are invasive. Invasive forest insects have largely transformed the field of forest entomology and have changed our overall approach to forest pest problems. They have also greatly affected silvicultural management of stands dominated by affected tree species. Additionally, plantations of non-native trees are likely to play an increasing role in fulfilling the world's demand for wood products in the future. Excluding invading pest species will play a critical role in maintaining the high productivity of these stands; hence forest biosecurity is likely to increase in importance (Wingfield et al. 2015).

There is little question that with current trends of globalization, insect species will continue to be introduced to new regions and some of these species will become established. Although the pool of species that could eventually become invasive is decreasing in some regions (Liebhold et al. 2017b), more invasions are inevitable, reflecting a combination of increased rates of imports, new trading partners and creation of new invasion pathways. These mechanisms, alone or collectively, can increase the exposure of one region to new, previously untapped species pools. Furthermore, there is often a considerable delay of 10–50 years between the establishment of a non-native species and its "discovery" when damage becomes extensive and readily apparent (Epanchin-Niell and Liebhold 2015). This means that many new but currently unknown species have probably already established. A portion of these species will inevitably emerge sometime in the future as serious problems.

Climate change, which is covered in Chapter 22, will affect future impacts of forest insects, including those resulting from invasive species. Range expansion or shifts, altered development rates, and changes in how forest insects interact with their hosts and natural enemies will undoubtedly be influenced by changing temperature and precipitation patterns in the future (e.g. Battisti et al. 2005).

Acknowledgements We thank E. Luzader for assistance with figures. A. Liebhold acknowledges support from the USDA Forest Service, the Swedish Agricultural University and grant "EVA4.0", No. CZ.02.1.01/0.0/0.0/16_019/0000803 financed by OP RDE. This publication was supported in part by the HOMED project (http://homed-project.eu/), which received funding from the European Union's Horizon 2020 research and innovation program under grant agreement No. 771271. Contributions by E.G.B. were also supported by the New Zealand government via MBIE core funding to Scion under contract C04X1104 and the Better Border Biosecurity Collaboration (www.b3nz.org).

References

Allen E, Noseworthy M, Ormsby M (2017) Phytosanitary measures to reduce the movement of forest pests with the international trade of wood products. Biol Invasions 19:3365–3376

Aukema JE, McCullough DG, Von Holle B, Liebhold AM, Britton K, Frankel SJ (2010) Historical accumulation of nonindigenous forest pests in the continental United States. Bioscience 60:886–897

Aukema JE, Leung B, Kovacs K, Chivers C, Britton KO, Englin J, Frankel SJ, Haight RG, Holmes TP, Liebhold AM, McCullough DG (2011) Economic impacts of non-native forest insects in the continental United States. PLoS One 6:e24587

Battisti A, Stastny M, Netherer S, Robinet C, Schopf A, Roques A, Larsson S (2005) Expansion of geographic range in the pine processionary moth caused by increased winter temperatures. Ecol Appl 15:2084–2096

Benton EP, Grant JF, Webster RJ, Nichols RJ, Cowles RS, Lagalante AF, Coots CI (2015) Assessment of imidacloprid and its metabolites in foliage of eastern hemlock multiple years following treatment for hemlock woolly adelgid, *Adelges tsugae* (Hemiptera: Adelgidae), in forested conditions. J Econ Entomol 108:2672–2682

Blackburn TM, Pyšek P, Bacher S, Carlton JT, Duncan RP, Jarošík V, Wilson JRU, Richardson DM (2011) A proposed unified framework for biological invasions. Trends Ecol Evol 26:333–339

Boyd IL, Freer-Smith PH, Gilligan CA, Godfray HCJ (2013) The consequence of tree pests and diseases for ecosystem services. Science 342:1235773

Brockerhoff EG, Liebhold AM (2017) Ecology of forest insect invasions. Biol Invasions 19:3141–3159

Brockerhoff EG, Bain J, Kimberley M, Knížek M (2006a) Interception frequency of exotic bark and ambrosia beetles (Coleoptera: Scolytinae) and relationship with establishment in New Zealand and worldwide. Can J for Res 36:289–298

Brockerhoff EG, Jones DC, Kimberley MO, Suckling DM, Donaldson T (2006b) Nationwide survey for invasive wood-boring and bark beetles (Coleoptera) using traps baited with pheromones and kairomones. For Ecol Manage 228:234–240

Brockerhoff EG, Liebhold AM, Richardson B, Suckling DM (2010) Eradication of invasive forest insects: concepts, methods, costs and benefits. NZ J Forest Sci 40(Suppl):S117–S135

Chase KD (2016) Allee effects, host tree density and the establishment of invasive bark beetles. Doctoral Dissertation, Dept. of Biology, University of Canterbury. http://hdl.handle.net/10092/12581

Colautti RI, Ricciardi A, Grigorovich IA, MacIsaac HJ (2004) Is invasion success explained by the enemy release hypothesis? Ecol Lett 7:721–733

Elkinton JS, Parry D, Boettner GH, G.H., (2006) Implicating an introduced generalist parasitoid in the invasive browntail moth's enigmatic demise. Ecology 87:2664–2672

Elkinton JS, Boettner GH (2012) Benefits and harm caused by the introduced generalist tachinid, Compsilura concinnata, in North America. Biocontrol 57:277–288

Epanchin-Niell RS, Liebhold AM (2015) Benefits of invasion prevention: effect of time lags, spread rates, and damage persistence. Ecol Econ 116:146–153

Eschen R, Britton K, Brockerhoff E, Burgess T, Dalley V, Epanchin-Niell RS, Gupta K, Hardy G, Huang Y, Kenis M, Kimani E (2015) International variation in phytosanitary legislation and regulations governing importation of plants for planting. Environ Sci Policy 51:228–237

Fahrner S, Aukema BH (2018) Correlates of spread rates for introduced insects. Glob Ecol Biogeogr. https://doi.org/10.1111/geb.12737

Gadgil PD, Flint TN (1983) Assessment of the risk of introduction of exotic forest insects and diseases with imported plants. N Z J for 28:58–67

Gandhi KJ, Herms DA (2010) Direct and indirect effects of alien insect herbivores on ecological processes and interactions in forests of eastern North America. Biol Invasions 12:389–405

Gascoigne J, Berec L, Gregory S, Courchamp F (2009) Dangerously few liaisons: a review of mate-finding Allee effects. Popul Ecol 51:355–372

Ghelardini LN, Luchi F, Pecori AL, Pepori R, Danti GD, Rocca P, Capretti PT, Ssantini A (2017) Ecology of invasive forest pathogens. Biol Invasions 19:3183–3200

Gilbert M, Grégoire JC, Freise JF, Heitland W (2004) Long-distance dispersal and human population density allow the prediction of invasive patterns in the horse chestnut leafminer *Cameraria ohridella*. J Anim Ecol 73:459–468

Hajek AE, Hurley BP, Kenis M, Garnas JR, Bush SJ, Wingfield MJ, van Lenteren JC, Cock MJW (2016) Exotic biological control agents: a solution or contribution to arthropod invasions? Biol Invasions 18:953–969

Haack RA, Hérard F, Sun J, Turgeon JJ (2010) Managing invasive populations of Asian longhorned beetle and citrus longhorned beetle: a worldwide perspective. Annu Rev Entomol 55:521–546

Haack RA, Britton KO, Brockerhoff EG, Cavey JF, Garrett LJ, Kimberley M, Lowenstein F, Nuding A, Olson LJ, Turner J, Vasilaky KN (2014) Effectiveness of the International Phytosanitary Standard ISPM No. 15 on reducing wood borer infestation rates in wood packaging material entering the United States. PLoS One 9:5, p e96611

Halldórsson G, Th Benedikz O, Eggertsson ES, Oddsdóttir, and Óskarsson H (2003) The impact of the green spruce aphid Elatobium abietinum (Walker) on long-term growth of Sitka spruce in Iceland. For Ecol Manage 181:281–287

Herms DA, McCullough DG (2014) The emerald ash borer invasion of North America: history, biology, ecology, impacts and management. Annu Rev Entomol 59:13–30

Holmes TP, Murphy EA, Bell KP (2006) Exotic forest insects and residential property values. Agric Resour Econ Rev 35:155–166

Houston DR (1994) Major new tree disease epidemics: beech bark disease. Annu Rev Phytopathol 32:75–87

Hudgins EJ, Liebhold AM, Leung B (2017) Predicting the spread of all invasive forest pests in the United States. Ecol Lett 20:426–435

Hurley BP, Slippers B, Wingfield MJ (2007) A comparison of control results for the alien invasive woodwasp, Sirex noctilio, in the southern hemisphere. Agric for Entomol 9:159–171

Hurley BP, Garnas J, Wingfield MJ, Branco M, Richardson DM, Slippers B (2016) Increasing numbers and intercontinental spread of invasive insects on eucalypts. Biol Invasions 18:921–933

Hurley BP, Slippers B, Sathyapala S, Wingfield MJ (2017) Challenges to planted forest health in developing economies. Biol Invasions 19:3273–3285

International Plant Protection Convention. (2012). ISPM-36, Integrated measures for plants for planting. United Nations food and agriculture organization http://www.fao.org/3/a-k8114e.pdf

Jones BA (2017) Invasive species impacts on human well-being using the life satisfaction index. Ecol Econ 134:250–257

Kenis M, Hurley BP, Hajek AE, Cock MJ (2017) Classical biological control of insect pests of trees: facts and figures. Biol Invasions 19:3401–3417

Kiritani K, Yamamura K (2003) Exotic insects and their pathways for invasion. In: Ruiz GM, Carlton JT (eds) Invasive species—vectors and management strategies. Island Press, Washington, pp 44–67

Kovacs KF, Haight RG, McCullough DG, Mercader RJ, Siegert NW, Liebhold AM (2010) Cost of potential emerald ash borer damage in US communities, 2009–2019. Ecol Econ 69:569–578

Lande R (1998) Demographic stochasticity and Alle effect on a scale with isotropic noise. Oikos 83:353–358

Leung B, Lodge DM, Finnoff D, Shogren JF, Lewis MA, Lamberti G (2002) An ounce of prevention or a pound of cure: bioeconomic risk analysis of invasive species. Proceedings of the Royal Society of London: Biological Sciences 269:2407–2413

Leung B, Springborn MR, Turner JA, Brockerhoff EG (2014) Pathway-level risk analysis: the net present value of an invasive species policy in the US. Front Ecol Environ 12:273–279

Levine JM, D'Antonio CM, CM (2003) Forecasting biological invasions with increasing international trade. Conserv Biol 17:322–326

Liebhold AM, Griffin RL (2016) The legacy of Charles Marlatt and efforts to limit plant pest invasions. Bull Entomol Soc Am 62:218–227

Liebhold AM, Tobin PC (2008) Population ecology of insect invasions and their management. Annu Rev Entomol 53:387–408

Liebhold AM, Macdonald WL, Bergdahl D, Mastro VC (1995) Invasion by exotic forest pests: a threat to forest ecosystems. For Sci Monogr 30:1–49

Liebhold AM, Work TT, McCullough DG, Cavey JF (2006) Airline baggage as a pathway for alien insect species invading the United States. Am Entomol 52:48–54

Liebhold AM, Brockerhoff EG, Garrett LJ, Parke JL, Britton KO (2012) Live plant imports: the major pathway for forest insect and pathogen invasions of the United States. Front Ecol Environ 10:135–143

Liebhold AM, Yamanaka T, Roques A, Augustin S, Chown SL, Brockerhoff EG, Pyšek P (2016a) Global compositional variation among native and non-native regional insect assemblages emphasizes the importance of pathways. Biol Invasions 18:893–905

Liebhold AM, Berec L, Brockerhoff EG, Epanchin-Niell RS, Hastings A, Herms DA, Kean JM, McCullough DG, Suckling DM, Tobin PC, Yamanaka T (2016b) Eradication of invading insect populations: from concepts to applications. Annu Rev Entomol 61:335–352

Liebhold AM, Brockerhoff EG, Kalisz S, Nuñez MA, Wardle DA, Wingfield MJ (2017a) Biological invasions in forest ecosystems. Biol Invasions 19:3437–3458

Liebhold AM, Brockerhoff EG, Kimberley M (2017b) Depletion of heterogeneous source species pools predicts future invasion rates. J Appl Ecol 54:1968–1977

Lindeman N (2013) Subjectivized knowledge and grassroots advocacy: an analysis of an environmental controversy in Northern California. J Bus Tech Commun 27:62–90

Lovett GM, Canham CD, Arthur MA, Weathers KC, Fitzhugh RD (2006) Forest ecosystem responses to exotic pests and pathogens in eastern North America. Bioscience 56:395–405

Lovett GM, Weiss M, Liebhold AM, Holmes TP, Leung B, Lambert KF, Orwig DA, Campbell FT, Rosenthal J, McCullough DG, Wildova R (2016) Nonnative forest insects and pathogens in the United States: impacts and policy options. Ecol Appl 26:1437–1455

May C (1934) Outbreaks of the Dutch elm disease in the United States. Circular 322, US Department of Agriculture, Washington, D.C., USA

McCullough DG, Work TT, Cavey JF, Liebhol AM, Marshall D (2006) Interceptions of nonindigenous plant pests at US ports of entry and border crossings over a 17-year period. Biol Invasions 8:611–630

McCullough DG (2019) Challenges, tactics and integrated management of emerald ash borer. Forestry: an International Journal of Forest Research. Forestry 93:197–211

Mercader RJ, McCullough DG, Storer AJ, Bedford J, Poland TM, Katovich S (2015) Evaluation of the potential use of a systemic insecticide and girdled trees in area wide management of the emerald ash borer. For Ecol Manage 350:70–80

Mercader RJ, McCullough DG, Storer AJ, Bedford JM, Heyd R, Siegert NW, Katovich S, Poland TM (2016) Estimating local spread of recently established emerald ash borer, Agrilus planipennis, infestations and the potential to influence it with a systemic insecticide and girdled ash trees. For Ecol Manage 366:87–97

Meurisse N, Rassati D, Hurley BP, Brockerhoff EG, Haack RA (2019) Common pathways by which non-native forest insects move internationally and domestically. J Pest Sci 92:13–27

Morin RS, Liebhold AM (2015) Invasions by two non-native insects alter regional forest species composition and successional trajectories. For Ecol Manage 341:67–74

Morin RS, Liebhold AM, Pugh SA, Crocker SJ (2017) Regional assessment of emerald ash borer, Agrilus planipennis, impacts in forests of the Eastern United States. Biol Invasions 19:703–711

Muzika RM (2017) Opportunities for silviculture in management and restoration of forests affected by invasive species. Biol Invasions 19:1–17

National Research Council, Board of Agriculture and Natural Resources, National Academy of Science (2002) Predicting invasions of nonindigenous plants and plant pests. National Academy Press, Washington, DC, p 198

Niemelä P, Mattson WJ (1996) Invasion of North American forests by European phytophagous insects. Bioscience 46:741–753

Nowak DJ, Pasek JE, Sequeira RA, Crane DE, Mastro VC (2001) Potential effect of Afzoplophoru glabripennis (Coleoptera: Cerambycidae) on urban tree in the United States. J Econ Entomol 94:116–122

Paine TD, Steinbauer MJ, Lawson SA (2011) Native and exotic pests of Eucalyptus: a worldwide perspective. Annu Rev Entomol 56:181–201

Poland TM, McCullough DG (2006) Emerald ash borer: invasion of the urban forest and the threat to North America's ash resource. J Forest 104:118–124

Rabaglia RJ, Cognato AI, Hoebeke ER, Johnson CW, LaBonte JR, Carter ME, Vlach JJ (2019) Early detection and rapid response: a 10-Year summary of the USDA forest service program of surveillance for non-native bark and ambrosia beetles. Am Entomol 65:29–42

Roques A, Rabitsch W, Rasplus JY, Lopez-Vaamonde C, Nentwig W, Kenis M (2009) In: Nentwig W, Hulme P, Pysek P, Vila M (eds) Handbook of alien species in Europe. Springer, Dordrecht, pp 63–79

Showalter DN, Raffa KF, Sniezko RA, Herms DA, Liebhold AM, Smith JA, Bonello P (2018) Strategic development of tree resistance against forest pathogen and insect invasions in defense-free space. Front Ecol Evol 6:124

Schröder ML, Slippers B, Wingfield MJ, Hurley BP (2020) Invasion history and management of Eucalyptus snout beetles in the Gonipterus scutellatus species complex. J Pest Sci 93:11–25

Seebens H, Blackburn TM, Dyer EE, Genovesi P, Hulme PE, Jeschke JM, Pagad S et al (2017) No saturation in the accumulation of alien species worldwide. Nat Commun 8:14435

Sharov AAD, Leonard AM, Liebhold EA, Roberts WD (2002) "Slow The Spread": A National Program to Contain the Gypsy Moth. J Forest 100:30–36

Six DL, Wingfield MJ (2011) The role of phytopathogenicity in bark beetle–fungus symbioses: a challenge to the classic paradigm. Annu Rev Entomol 56:255–272

Slippers B, Hurley BP, Wingfield MJ (2015) Sirex woodwasp: a model for evolving management paradigms of invasive forest pests. Annu Rev Entomol 60:601–619

Sniezko RA, Koch J (2017) Breeding trees resistant to insects and diseases: putting theory into application. Biol Invasions 19:3377–3400

Straw NA, Bellett-Travers M (2004) Impact and management of the horse chestnut leaf-miner (*Cameraria ohridella*). Arboric J 28:67–83

Suckling DM, Barrington AM, Chhagan A, Stephens AEA, Burnip GM, Charles JG, Wee SL (2007) Eradication of the Australian painted apple moth *Teia anartoides* in New Zealand: trapping, inherited sterility, and male competitiveness. In: Vreysen MJB, Robinson AS, Hendrichs J (eds) Area-wide control of insect pests. Springer, Dordrecht, pp 603–615

Suckling DM, Stringer LD, Baird DB, Butler RC, Sullivan TES, Lance DR, Simmons GS (2014) Light brown apple moth (*Epiphyas postvittana*) (Lepidoptera: Tortricidae) colonization of California. Biol Invasions 16:1851–1863

Sun J, Lu M, Gillette NE, Wingfield MJ (2013) Red turpentine beetle: innocuous native becomes invasive tree killer in China. Annu Rev Entomol 58:293–311

Tobin PC, Robinet C, Johnson DM, Whitmire SL, Bjørnstad ON, Liebhold AM (2009) The role of Allee effects in gypsy moth, *Lymantria dispar* (L.), invasions. Popul Ecol 51:373–384

Tobin PC, Kean JM, Suckling DM, McCullough DG, Herms DA, Stringer LD (2014) Determinants of successful arthropod eradication programs. Biol Invasions 16:401–414

Tribe GD (2005) The present status of Anaphes nitens (Hymenoptera: Mymaridae), an egg parasitoid of the Eucalyptus snout beetle *Gonipterus scutellatus*, in the Western Cape Province of South Africa. South Afr for J 203:49–54

Turgeon, J.J., J. Ric, P. de Groot, B. Gasman, M. Orr, J. Doyle, M. T. Smith, L. Dumouchel and T. Scarr. 2007. Détection des signes et des symptômes d'attaque par le longicorne étoilé: guide de formation. Service canadien des forêts, Ressources naturelles Canada, Ottawa, Ont.

van Frankenhuyzen K (2000) Application of Bacillus thuringiensis in forestry. In: Charles JF, Delécluse A, Roux CNL (eds) Entomopathogenic bacteria: from laboratory to field application. Springer, Dordrecht, pp 371–382

Whattam M, Clover G, Firko M, Kalaris T (2014) The biosecurity continuum and trade: border operations. In: Gordh G, McKirdy S (eds) The Handbook of Plant Biosecurity. Springer, Dordrecht, pp 149–188

Withers TM (2001) Colonization of eucalypts in New Zealand by Australian insects. Austral Ecol 26:467–476

Wingfield MJ, Brockerhoff EG, Wingfield BD, Slippers B (2015) Planted forest health: the need for a global strategy. Science 349:832–836

Wingfield MJ, Barnes I, de Beer ZW, Roux J, Wingfield BD, Taerum SJ, S.J. (2017) Novel associations between ophiostomatoid fungi, insects and tree hosts: current status—future prospects. Biol Invasions 19:3215–3228

Xu T, Hiroe Y, Teale SA, Fujiwara-Tsujii N, Wickham JD, Fukaya M, Hansen L et al (2017) Identification of a male-produced sex-aggregation pheromone for a highly invasive cerambycid beetle. Aromia Bungii. Sci Rep 7:7330

Printed in the United States
by Baker & Taylor Publisher Services